Single Variable

THOMAS'
CALCULUS
Early Transcendentals

FOURTEENTH EDITION

Single Variable

THOMAS'
CALCULUS
Early Transcendentals

FOURTEENTH EDITION

Based on the original work by
GEORGE B. THOMAS, JR.
Massachusetts Institute of Technology

as revised by

JOEL HASS
University of California, Davis

CHRISTOPHER HEIL
Georgia Institute of Technology

MAURICE D. WEIR
Naval Postgraduate School

Director, Portfolio Management: Deirdre Lynch
Executive Editor: Jeff Weidenaar
Editorial Assistant: Jennifer Snyder
Content Producer: Rachel S. Reeve
Managing Producer: Scott Disanno
Producer: Stephanie Green
TestGen Content Manager: Mary Durnwald
Manager: Content Development, Math: Kristina Evans
Product Marketing Manager: Claire Kozar
Field Marketing Manager: Evan St. Cyr
Marketing Assistants: Jennifer Myers, Erin Rush
Senior Author Support/Technology Specialist: Joe Vetere
Rights and Permissions Project Manager: Gina M. Cheselka
Manufacturing Buyer: Carol Melville, LSC Communications
Program Design Lead: Barbara T. Atkinson
Associate Director of Design: Blair Brown
Text and Cover Design, Production Coordination, Composition: Cenveo® Publisher Services
Illustrations: Network Graphics, Cenveo Publisher Services

Cover Image: Te Rewa Rewa Bridge, Getty Images/Kanwal Sandhu

Library of Congress Cataloging-in-Publication Data

Names: Hass, Joel. | Heil, Christopher, 1960- | Weir, Maurice D.
Title: Thomas' calculus : early transcendentals : single variable : based on the original work by George B. Thomas, Jr., Massachusetts Institute of Technology.
Other titles: Early transcendentals : single variable
Description: Fourteenth edition / as revised by Joel Hass, University of California, Davis, Christopher Heil, Georgia Institute of Technology, Maurice D. Weir, Naval Postgraduate School. | New York, New York : Pearson, [2018] | Includes index.
Identifiers: LCCN 2016036095| ISBN 9780134443419 | ISBN 0134443414
Subjects: LCSH: Calculus--Textbooks. | Geometry, Analytic--Textbooks. | Transcendental numbers--Textbooks.
Classification: LCC QA303.2 .F565 2018 | DDC 515--dc23 LC record available at https://lccn.loc.gov/2016036095

4 18

Student Edition
ISBN-13: 978-0-13-443941-9
ISBN-10: 0-13-443941-4

Contents

8 Techniques of Integration 461

9 First-Order Differential Equations 540

10 Infinite Sequences and Series 577

Preface

Thomas' Calculus: Early Transcendentals, Fourteenth Edition, provides a modern introduction to calculus that focuses on developing conceptual understanding of the underlying mathematical ideas. This text supports a calculus sequence typically taken by students in STEM fields over several semesters. Intuitive and precise explanations, thoughtfully chosen examples, superior figures, and time-tested exercise sets are the foundation of this text. We continue to improve this text in keeping with shifts in both the preparation and the goals of today's students, and in the applications of calculus to a changing world.

Many of today's students have been exposed to calculus in high school. For some, this translates into a successful experience with calculus in college. For others, however, the result is an overconfidence in their computational abilities coupled with underlying gaps in algebra and trigonometry mastery, as well as poor conceptual understanding. In this text, we seek to meet the needs of the increasingly varied population in the calculus sequence. We have taken care to provide enough review material (in the text and appendices), detailed solutions, and a variety of examples and exercises, to support a complete understanding of calculus for students at varying levels. Additionally, the MyMathLab course that accompanies the text provides adaptive support to meet the needs of all students. Within the text, we present the material in a way that supports the development of mathematical maturity, going beyond memorizing formulas and routine procedures, and we show students how to generalize key concepts once they are introduced. References are made throughout, tying new concepts to related ones that were studied earlier. After studying calculus from *Thomas*, students will have developed problem-solving and reasoning abilities that will serve them well in many important aspects of their lives. Mastering this beautiful and creative subject, with its many practical applications across so many fields, is its own reward. But the real gifts of studying calculus are acquiring the ability to think logically and precisely; understanding what is defined, what is assumed, and what is deduced; and learning how to generalize conceptually. We intend this book to encourage and support those goals.

New to This Edition

We welcome to this edition a new coauthor, Christopher Heil from the Georgia Institute of Technology. He has been involved in teaching calculus, linear algebra, analysis, and abstract algebra at Georgia Tech since 1993. He is an experienced author and served as a consultant on the previous edition of this text. His research is in harmonic analysis, including time-frequency analysis, wavelets, and operator theory.

This is a substantial revision. Every word, symbol, and figure was revisited to ensure clarity, consistency, and conciseness. Additionally, we made the following text-wide updates:

- Updated graphics to bring out clear visualization and mathematical correctness.

- Added examples (in response to user feedback) to overcome conceptual obstacles. See Example 3 in Section 9.1.

- Added new types of homework exercises throughout, including many with a geometric nature. The new exercises are not just more of the same, but rather give different perspectives on and approaches to each topic. We also analyzed aggregated student usage and performance data from MyMathLab for the previous edition of this text. The results of this analysis helped improve the quality and quantity of the exercises.

- Added short URLs to historical links that allow students to navigate directly to online information.

- Added new marginal notes throughout to guide the reader through the process of problem solution and to emphasize that each step in a mathematical argument is rigorously justified.

New to MyMathLab

Many improvements have been made to the overall functionality of MyMathLab (MML) since the previous edition. Beyond that, we have also increased and improved the content specific to this text.

- Instructors now have more exercises than ever to choose from in assigning homework. There are approximately 8080 assignable exercises in MML.

- The MML exercise-scoring engine has been updated to allow for more robust coverage of certain topics, including differential equations.

- A full suite of Interactive Figures have been added to support teaching and learning. The figures are designed to be used in lecture, as well as by students independently. The figures are editable using the freely available GeoGebra software. The figures were created by Marc Renault (Shippensburg University), Kevin Hopkins (Southwest Baptist University), Steve Phelps (University of Cincinnati), and Tim Brzezinski (Berlin High School, CT).

- Enhanced Sample Assignments include just-in-time prerequisite review, help keep skills fresh with distributed practice of key concepts (based on research by Jeff Hieb, Keith Lyle, and Pat Ralston of University of Louisville), and provide opportunities to work exercises without learning aids (to help students develop confidence in their ability to solve problems independently).

- Additional Conceptual Questions augment text exercises to focus on deeper, theoretical understanding of the key concepts in calculus. These questions were written by faculty at Cornell University under an NSF grant. They are also assignable through Learning Catalytics.

- An Integrated Review version of the MML course contains pre-made quizzes to assess the prerequisite skills needed for each chapter, plus personalized remediation for any gaps in skills that are identified.

- Setup & Solve exercises now appear in many sections. These exercises require students to show how they set up a problem as well as the solution, better mirroring what is required of students on tests.

- Over 200 new instructional videos by Greg Wisloski and Dan Radelet (both of Indiana University of PA) augment the already robust collection within the course. These videos support the overall approach of the text—specifically, they go beyond routine procedures to show students how to generalize and connect key concepts.

Content Enhancements

Chapter 1

- Shortened 1.4 to focus on issues arising in use of mathematical software and potential pitfalls. Removed peripheral material on regression, along with associated exercises.

- Clarified explanation of definition of exponential function in 1.5.

- Replaced $\sin^{-1}$ notation for the inverse sine function with arcsin as default notation in 1.6, and similarly for other trig functions.

- Added new Exercises: **1.1:** 59–62, **1.2:** 21–22; **1.3:** 64–65, **1.6:** 61–64, 79cd; **PE:** 29–32.

Chapter 2

- Added definition of average speed in 2.1.

- Clarified definition of limits to allow for arbitrary domains. The definition of limits is now consistent with the definition in multivariable domains later in the text and with more general mathematical usage.

- Reworded limit and continuity definitions to remove implication symbols and improve comprehension.

- Added new Example 7 in 2.4 to illustrate limits of ratios of trig functions.

- Rewrote 2.5 Example 11 to solve the equation by finding a zero, consistent with previous discussion.

- Added new Exercises: **2.1:** 15–18; **2.2:** 3h–k, 4f–I; **2.4:** 19–20, 45–46; **2.5:** 31–32; **2.6:** 69–74; **PE:** 57–58; **AAE:** 35–38.

Chapter 3

- Clarified relation of slope and rate of change.

- Added new Figure 3.9 using the square root function to illustrate vertical tangent lines.

- Added figure of $x \sin(1/x)$ in 3.2 to illustrate how oscillation can lead to nonexistence of a derivative of a continuous function.

- Revised product rule to make order of factors consistent throughout text, including later dot product and cross product formulas.

- Added new Exercises: **3.2:** 36, 43–44; **3.3:** 65–66; **3.5:** 43–44, 61bc; **3.6:** 79–80, 111–113; **3.7:** 27–28; **3.8:** 97–100; **3.9:** 43–46; **3.10:** 47; **AAE:** 14–15, 26–27.

Chapter 4

- Added summary to 4.1.

- Added new Example 3 with new Figure 4.27 and Example 12 with new Figure 4.35 to give basic and advanced examples of concavity.

- Added new Exercises: **4.1:** 53–56, 67–70; **4.3:** 45–46, 67–68; **4.4:** 107–112; **4.6:** 37–42; **4.7:** 7–10; **4.8:** 115–118; **PE:** 1–16, 101–102; **AAE:** 19–20, 38–39. Moved Exercises 4.1: 53–68 to PE.

Chapter 5

- Improved discussion in 5.4 and added new Figure 5.18 to illustrate the Mean Value Theorem.

- Added new Exercises: **5.2:** 33–36; **5.4:** 71–72; **5.6:** 47–48; **PE:** 43–44, 75–76.

Chapter 6

- Clarified cylindrical shell method.

- Converted 6.5 Example 4 to metric units.

- Added introductory discussion of mass distribution along a line, with figure, in 6.6.

- Added new Exercises: **6.1:** 15–16; **6.2:** 49–50; **6.3:** 13–14; **6.5:** 1–2; **6.6:** 1–6, 21–22; **PE:** 17–18, 23–24, 37–38.

Chapter 7

- Clarified discussion of separable differential equations in 7.2.

- Added new Exercises: **7.1:** 61–62, 73; **PE:** 41–42.

Chapter 8

- Updated 8.2 Integration by Parts discussion to emphasize $u(x)\,v'(x)\,dx$ form rather than $u\,dv$. Rewrote Examples 1–3 accordingly.

- Removed discussion of tabular integration and associated exercises.

- Updated discussion in 8.5 on how to find constants in the method of partial fractions.

- Updated notation in 8.8 to align with standard usage in statistics.

- Added new Exercises: **8.1:** 41–44; **8.2:** 53–56, 72–73; **8.3:** 75–76; **8.4:** 49–52; **8.5:** 51–66, 73–74; **8.8:** 35–38, 77–78; **PE:** 69–88.

Chapter 9

- Added new Example 3 with Figure 9.3 to illustrate how to construct a slope field.

- Added new Exercises: **9.1:** 11–14; **PE:** 17–22, 43–44.

Chapter 10

- Clarified the differences between a sequence and a series.

- Added new Figure 10.9 to illustrate sum of a series as area of a histogram.

- Added to 10.3 a discussion on the importance of bounding errors in approximations.

- Added new Figure 10.13 illustrating how to use integrals to bound remainder terms of partial sums.

- Rewrote Theorem 10 in 10.4 to bring out similarity to the integral comparison test.

- Added new Figure 10.16 to illustrate the differing behaviors of the harmonic and alternating harmonic series.

- Renamed the nth-Term Test the "nth-Term Test for Divergence" to emphasize that it says nothing about convergence.

- Added new Figure 10.19 to illustrate polynomials converging to $\ln (1 + x)$, which illustrates convergence on the half-open interval $(-1, 1]$.

- Used red dots and intervals to indicate intervals and points where divergence occurs, and blue to indicate convergence, throughout Chapter 10.

- Added new Figure 10.21 to show the six different possibilities for an interval of convergence.

- Added new Exercises: **10.1:** 27–30, 72–77; **10.2:** 19–22, 73–76, 105; **10.3:** 11–12, 39–42; **10.4:** 55–56; **10.5:** 45–46, 65–66; **10.6:** 57–82; **10.7:** 61–65; **10.8:** 23–24, 39–40; **10.9:** 11–12, 37–38; **PE:** 41–44, 97–102.

Chapter 11

- Added new Example 1 and Figure 11.2 in 11.1 to give a straightforward first example of a parametrized curve.

- Updated area formulas for polar coordinates to include conditions for positive r and nonoverlapping θ.

- Added new Example 3 and Figure 11.37 in 11.4 to illustrate intersections of polar curves.

- Added new Exercises: **11.1:** 19–28; **11.2:** 49–50; **11.4:** 21–24.

Chapter 12

- Added new Figure 12.13(b) to show the effect of scaling a vector.

- Added new Example 7 and Figure 12.26 in 12.3 to illustrate projection of a vector.

- Added discussion on general quadric surfaces in 12.6, with new Example 4 and new Figure 12.48 illustrating the description of an ellipsoid not centered at the origin via completing the square.

- Added new Exercises: **12.1:** 31–34, 59–60, 73–76; **12.2:** 43–44; **12.3:** 17–18; **12.4:** 51–57; **12.5:** 49–52.

Chapter 13

- Added sidebars on how to pronounce Greek letters such as kappa, tau, etc.

- Added new Exercises: **13.1:** 1–4, 27–36; **13.2:** 15–16, 19–20; **13.4:** 27–28; **13.6:** 1–2.

Chapter 14

- Elaborated on discussion of open and closed regions in 14.1.

- Standardized notation for evaluating partial derivatives, gradients, and directional derivatives at a point, throughout the chapter.

- Renamed "branch diagrams" as "dependency diagrams," which clarifies that they capture dependence of variables.

- Added new Exercises: **14.2:** 51–54; **14.3:** 51–54, 59–60, 71–74, 103–104; **14.4:** 20–30, 43–46, 57–58; **14.5:** 41–44; **14.6:** 9–10, 61; **14.7:** 61–62.

Chapter 15

- Added new Figure 15.21b to illustrate setting up limits of a double integral.

- Added new 15.5 Example 1, modified Examples 2 and 3, and added new Figures 15.31, 15.32, and 15.33 to give basic examples of setting up limits of integration for a triple integral.

- Added new material on joint probability distributions as an application of multivariable integration.

- Added new Examples 5, 6 and 7 to Section 15.6.

- Added new Exercises: **15.1:** 15–16, 27–28; **15.6:** 39–44; **15.7:** 1–22.

Chapter 16

- Added new Figure 16.4 to illustrate a line integral of a function.

- Added new Figure 16.17 to illustrate a gradient field.

- Added new Figure 16.19 to illustrate a line integral of a vector field.

- Clarified notation for line integrals in 16.2.

- Added discussion of the sign of potential energy in 16.3.

- Rewrote solution of Example 3 in 16.4 to clarify connection to Green's Theorem.

- Updated discussion of surface orientation in 16.6 along with Figure 16.52.

- Added new Exercises: **16.2:** 37–38, 41–46; **16.4:** 1–6; **16.6:** 49–50; **16.7:** 1–6; **16.8:** 1–4.

Appendices: Rewrote Appendix A7 on complex numbers.

Continuing Features

Rigor The level of rigor is consistent with that of earlier editions. We continue to distinguish between formal and informal discussions and to point out their differences. Starting with a more intuitive, less formal approach helps students understand a new or difficult concept so they can then appreciate its full mathematical precision and outcomes. We pay attention to defining ideas carefully and to proving theorems appropriate for calculus students, while mentioning deeper or subtler issues they would study in a more advanced course. Our organization and distinctions between informal and formal discussions give the instructor a degree of flexibility in the amount and depth of coverage of the various topics. For example, while we do not prove the Intermediate Value Theorem or the Extreme Value Theorem for continuous functions on a closed finite interval, we do state these theorems precisely, illustrate their meanings in numerous examples, and use them to prove other important results. Furthermore, for those instructors who desire greater depth of coverage, in Appendix 6 we discuss the reliance of these theorems on the completeness of the real numbers.

Writing Exercises Writing exercises placed throughout the text ask students to explore and explain a variety of calculus concepts and applications. In addition, the end of each chapter contains a list of questions for students to review and summarize what they have learned. Many of these exercises make good writing assignments.

End-of-Chapter Reviews and Projects In addition to problems appearing after each section, each chapter culminates with review questions, practice exercises covering the entire chapter, and a series of Additional and Advanced Exercises with more challenging or synthesizing problems. Most chapters also include descriptions of several **Technology Application Projects** that can be worked by individual students or groups of students over a longer period of time. These projects require the use of *Mathematica* or *Maple*, along with pre-made files that are available for download within MyMathLab.

Writing and Applications This text continues to be easy to read, conversational, and mathematically rich. Each new topic is motivated by clear, easy-to-understand examples and is then reinforced by its application to real-world problems of immediate interest to students. A hallmark of this book has been the application of calculus to science and engineering. These applied problems have been updated, improved, and extended continually over the last several editions.

Technology In a course using the text, technology can be incorporated according to the taste of the instructor. Each section contains exercises requiring the use of technology; these are marked with a T if suitable for calculator or computer use, or they are labeled **Computer Explorations** if a computer algebra system (CAS, such as *Maple* or *Mathematica*) is required.

Additional Resources

MyMathLab® Online Course (access code required)

Built around Pearson's best-selling content, MyMathLab is an online homework, tutorial, and assessment program designed to work with this text to engage students and improve results. MyMathLab can be successfully implemented in any classroom environment—lab-based, hybrid, fully online, or traditional.

Used by more than 37 million students worldwide, MyMathLab delivers consistent, measurable gains in student learning outcomes, retention, and subsequent course success. Visit **www.mymathlab.com/results** to learn more.

Preparedness   One of the biggest challenges in calculus courses is making sure students are adequately prepared with the prerequisite skills needed to successfully complete their course work. MyMathLab supports students with just-in-time remediation and key-concept review.

- **Integrated Review Course** can be used for just-in-time prerequisite review. These courses contain pre-made quizzes to assess the prerequisite skills needed for each chapter, plus personalized remediation for any gaps in skills that are identified.

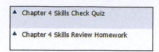

Motivation   Students are motivated to succeed when they're engaged in the learning experience and understand the relevance and power of mathematics. MyMathLab's online homework offers students immediate feedback and tutorial assistance that motivates them to do more, which means they retain more knowledge and improve their test scores.

- **Exercises with immediate feedback**—the over 8080 assignable exercises for this text regenerate algorithmically to give students unlimited opportunity for practice and mastery. MyMathLab provides helpful feedback when students enter incorrect answers and includes optional learning aids such as Help Me Solve This, View an Example, videos, and an eText.

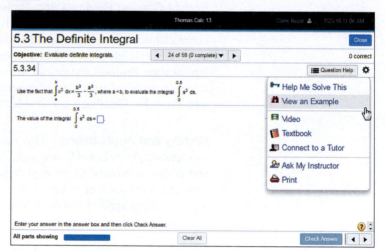

- **Setup and Solve Exercises** ask students to first describe how they will set up and approach the problem. This reinforces students' conceptual understanding of the process they are applying and promotes long-term retention of the skill.

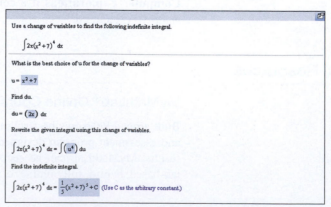

- **Additional Conceptual Questions** focus on deeper, theoretical understanding of the key concepts in calculus. These questions were written by faculty at Cornell University under an NSF grant and are also assignable through Learning Catalytics.

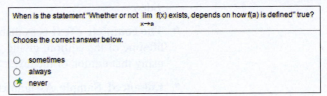

- **Learning Catalytics™** is a student response tool that uses students' smartphones, tablets, or laptops to engage them in more interactive tasks and thinking during lecture. Learning Catalytics fosters student engagement and peer-to-peer learning with real-time analytics. Learning Catalytics is available to all MyMathLab users.

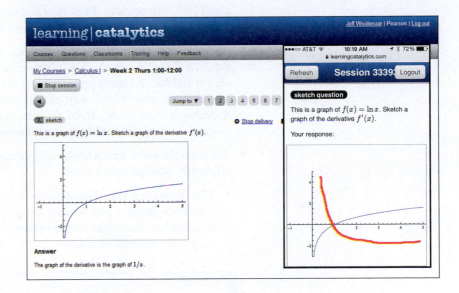

Learning and Teaching Tools

- **Interactive Figures** illustrate key concepts and allow manipulation for use as teaching and learning tools. We also include videos that use the Interactive Figures to explain key concepts.

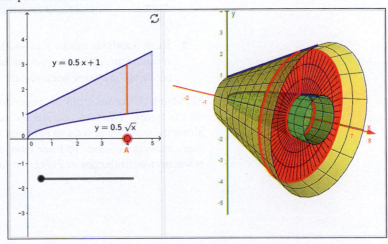

- **Instructional videos**—hundreds of videos are available as learning aids within exercises and for self-study. The Guide to Video-Based Assignments makes it easy to assign videos for homework by showing which MyMathLab exercises correspond to each video.

- **The complete eText** is available to students through their MyMathLab courses for the lifetime of the edition, giving students unlimited access to the eText within any course using that edition of the text.

- **Enhanced Sample Assignments** These assignments include just-in-time prerequisite review, help keep skills fresh with distributed practice of key concepts, and provide opportunities to work exercises without learning aids so students can check their understanding.

- **PowerPoint Presentations** that cover each section of the book are available for download.

- **Mathematica manual and projects, Maple manual and projects, TI Graphing Calculator manual**—These manuals cover *Maple* 17, *Mathematica 8,* and the TI-84 Plus and TI-89, respectively. Each provides detailed guidance for integrating the software package or graphing calculator throughout the course, including syntax and commands.

- **Accessibility** and achievement go hand in hand. MyMathLab is compatible with the JAWS screen reader, and it enables students to read and interact with multiple-choice and free-response problem types via keyboard controls and math notation input. MyMathLab also works with screen enlargers, including ZoomText, MAGic, and SuperNova. And, all MyMathLab videos have closed-captioning. More information is available at **http://mymathlab.com/accessibility**.

- **A comprehensive gradebook** with enhanced reporting functionality allows you to efficiently manage your course.

  - **The Reporting Dashboard** offers insight as you view, analyze, and report learning outcomes. Student performance data is presented at the class, section, and program levels in an accessible, visual manner so you'll have the information you need to keep your students on track.

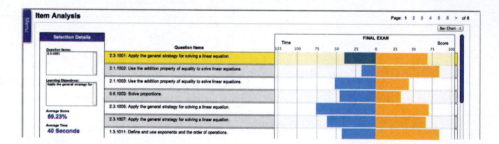

  - **Item Analysis** tracks class-wide understanding of particular exercises so you can refine your class lectures or adjust the course/department syllabus. Just-in-time teaching has never been easier!

MyMathLab comes from an experienced partner with educational expertise and an eye on the future. Whether you are just getting started with MyMathLab, or have a question along the way, we're here to help you learn about our technologies and how to incorporate them into your course. To learn more about how MyMathLab helps students succeed, visit **www.mymathlab.com** or contact your Pearson rep.

Instructor's Solutions Manual (downloadable)
ISBN: 0-13-443932-5 | 978-0-13-443932-7

The Instructor's Solutions Manual contains complete worked-out solutions to all the exercises in *Thomas' Calculus: Early Transcendentals*. It can be downloaded from within MyMathLab or the Pearson Instructor Resource Center, **www.pearsonhighered.com/irc**.

Student's Solutions Manual
Single Variable Calculus (Chapters 1–11), ISBN: 0-13-443933-3 | 978-0-13-443933-4
Multivariable Calculus (Chapters 10–16), ISBN: 0-13-443916-3 | 978-0-13-443916-7

The Student's Solutions Manual contains worked-out solutions to all the odd-numbered exercises in *Thomas' Calculus: Early Transcendentals*. These manuals are available in print and can be downloaded from within MyMathLab.

Just-In-Time Algebra and Trigonometry for Early Transcendentals Calculus, Fourth Edition
ISBN 0-321-67103-1 | 978-0-321-67103-5

Sharp algebra and trigonometry skills are critical to mastering calculus, and *Just-in-Time Algebra and Trigonometry for Early Transcendentals Calculus* by Guntram Mueller and Ronald I. Brent is designed to bolster these skills while students study calculus. As students make their way through calculus, this brief supplementary text is with them every step of the way, showing them the necessary algebra or trigonometry topics and pointing out potential problem spots. The easy-to-use table of contents has topics arranged in the order in which students will need them as they study calculus. This supplement is available in printed form only (note that MyMathLab contains a separate diagnostic and remediation system for gaps in algebra and trigonometry skills).

Technology Manuals and Projects (downloadable)
Maple Manual and Projects by Marie Vanisko, Carroll College
Mathematica Manual and Projects by Marie Vanisko, Carroll College
TI-Graphing Calculator Manual by Elaine McDonald-Newman, Sonoma State University

These manuals and projects cover Maple 17, Mathematica 9, and the TI-84 Plus and TI-89. Each manual provides detailed guidance for integrating a specific software package or graphing calculator throughout the course, including syntax and commands. The projects include instructions and ready-made application files for Maple and Mathematica. These materials are available to download within MyMathLab.

TestGen®
ISBN: 0-13-443936-8 | 978-0-13-443936-5

TestGen® (**www.pearsoned.com/testgen**) enables instructors to build, edit, print, and administer tests using a computerized bank of questions developed to cover all the objectives of the text. TestGen is algorithmically based, allowing instructors to create multiple but equivalent versions of the same question or test with the click of a button. Instructors can also modify test bank questions or add new questions. The software and test bank are available for download from Pearson Education's online catalog, **www.pearsonhighered.com**.

PowerPoint® Lecture Slides
ISBN: 0-13-443943-0 | 978-0-13-443943-3

These classroom presentation slides were created for the *Thomas' Calculus* series. Key graphics from the book are included to help bring the concepts alive in the classroom. These files are available to qualified instructors through the Pearson Instructor Resource Center, **www.pearsonhighered.com/irc**, and within MyMathLab.

Acknowledgments

We are grateful to Duane Kouba, who created many of the new exercises. We would also like to express our thanks to the people who made many valuable contributions to this edition as it developed through its various stages:

Accuracy Checkers
Thomas Wegleitner
Jennifer Blue
Lisa Collette

Reviewers for the Fourteenth Edition
Alessandro Arsie, *University of Toledo*
Doug Baldwin, *SUNY Geneseo*
Steven Heilman, *UCLA*
David Horntrop, *New Jersey Institute of Technology*
Eric B. Kahn, *Bloomsburg University*
Colleen Kirk, *California Polytechnic State University*
Mark McConnell, *Princeton University*

Niels Martin Møller, *Princeton University*
James G. O'Brien, *Wentworth Institute of Technology*
Alan Saleski, *Loyola University Chicago*
Alan Von Hermann, *Santa Clara University*
Don Gayan Wilathgamuwa, *Montana State University*
James Wilson, *Iowa State University*

The following faculty members provided direction on the development of the MyMathLab course for this edition.

Charles Obare, *Texas State Technical College, Harlingen*
Elmira Yakutova-Lorentz, *Eastern Florida State College*
C. Sohn, *SUNY Geneseo*
Ksenia Owens, *Napa Valley College*
Ruth Mortha, *Malcolm X College*
George Reuter, *SUNY Geneseo*
Daniel E. Osborne, *Florida A&M University*
Luis Rodriguez, *Miami Dade College*
Abbas Meigooni, *Lincoln Land Community College*
Nader Yassin, *Del Mar College*
Arthur J. Rosenthal, *Salem State University*
Valerie Bouagnon, *DePaul University*
Brooke P. Quinlan, *Hillsborough Community College*
Shuvra Gupta, *Iowa State University*
Alexander Casti, *Farleigh Dickinson University*
Sharda K. Gudehithlu, *Wilbur Wright College*
Deanna Robinson, *McLennan Community College*

Kai Chuang, *Central Arizona College*
Vandana Srivastava, *Pitt Community College*
Brian Albright, *Concordia University*
Brian Hayes, *Triton College*
Gabriel Cuarenta, *Merced College*
John Beyers, *University of Maryland University College*
Daniel Pellegrini, *Triton College*
Debra Johnsen, *Orangeburg Calhoun Technical College*
Olga Tsukernik, *Rochester Institute of Technology*
Jorge Sarmiento, *County College of Morris*
Val Mohanakumar, *Hillsborough Community College*
MK Panahi, *El Centro College*
Sabrina Ripp, *Tulsa Community College*
Mona Panchal, *East Los Angeles College*
Gail Illich, *McLennan Community College*
Mark Farag, *Farleigh Dickinson University*
Selena Mohan, *Cumberland County College*

Dedication

We regret that prior to the writing of this edition our coauthor Maurice Weir passed away. Maury was dedicated to achieving the highest possible standards in the presentation of mathematics. He insisted on clarity, rigor, and readability. Maury was a role model to his students, his colleagues, and his coauthors. He was very proud of his daughters, Maia Coyle and Renee Waina, and of his grandsons, Matthew Ryan and Andrew Dean Waina. He will be greatly missed.

1

Functions

OVERVIEW Functions are fundamental to the study of calculus. In this chapter we review what functions are and how they are visualized as graphs, how they are combined and transformed, and ways they can be classified.

1.1 Functions and Their Graphs

Functions are a tool for describing the real world in mathematical terms. A function can be represented by an equation, a graph, a numerical table, or a verbal description; we will use all four representations throughout this book. This section reviews these ideas.

Functions; Domain and Range

The temperature at which water boils depends on the elevation above sea level. The interest paid on a cash investment depends on the length of time the investment is held. The area of a circle depends on the radius of the circle. The distance an object travels depends on the elapsed time.

In each case, the value of one variable quantity, say y, depends on the value of another variable quantity, which we often call x. We say that "y is a function of x" and write this symbolically as

$$y = f(x) \qquad \text{("y equals f of x")}.$$

The symbol f represents the function, the letter x is the **independent variable** representing the input value to f, and y is the **dependent variable** or output value of f at x.

DEFINITION A **function** f from a set D to a set Y is a rule that assigns a *unique* value $f(x)$ in Y to each x in D.

The set D of all possible input values is called the **domain** of the function. The set of all output values of $f(x)$ as x varies throughout D is called the **range** of the function. The range might not include every element in the set Y. The domain and range of a function can be any sets of objects, but often in calculus they are sets of real numbers interpreted as points of a coordinate line. (In Chapters 13–16, we will encounter functions for which the elements of the sets are points in the plane, or in space.)

Often a function is given by a formula that describes how to calculate the output value from the input variable. For instance, the equation $A = \pi r^2$ is a rule that calculates the area A of a circle from its radius r. When we define a function $y = f(x)$ with a formula and the domain is not stated explicitly or restricted by context, the domain is assumed to

1

be the largest set of real x-values for which the formula gives real y-values. This is called the **natural domain** of f. If we want to restrict the domain in some way, we must say so. The domain of $y = x^2$ is the entire set of real numbers. To restrict the domain of the function to, say, positive values of x, we would write "$y = x^2, x > 0$."

Changing the domain to which we apply a formula usually changes the range as well. The range of $y = x^2$ is $[0, \infty)$. The range of $y = x^2, x \geq 2$, is the set of all numbers obtained by squaring numbers greater than or equal to 2. In set notation (see Appendix 1), the range is $\{x^2 \mid x \geq 2\}$ or $\{y \mid y \geq 4\}$ or $[4, \infty)$.

When the range of a function is a set of real numbers, the function is said to be **real-valued**. The domains and ranges of most real-valued functions we consider are intervals or combinations of intervals. Sometimes the range of a function is not easy to find.

A function f is like a machine that produces an output value $f(x)$ in its range whenever we feed it an input value x from its domain (Figure 1.1). The function keys on a calculator give an example of a function as a machine. For instance, the $\sqrt{x}$ key on a calculator gives an output value (the square root) whenever you enter a nonnegative number x and press the $\sqrt{x}$ key.

A function can also be pictured as an **arrow diagram** (Figure 1.2). Each arrow associates to an element of the domain D a single element in the set Y. In Figure 1.2, the arrows indicate that $f(a)$ is associated with a, $f(x)$ is associated with x, and so on. Notice that a function can have the same *output value* for two different input elements in the domain (as occurs with $f(a)$ in Figure 1.2), but each input element x is assigned a *single* output value $f(x)$.

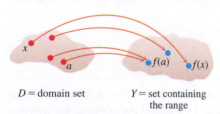

FIGURE 1.1 A diagram showing a function as a kind of machine.

D = domain set Y = set containing the range

FIGURE 1.2 A function from a set D to a set Y assigns a unique element of Y to each element in D.

EXAMPLE 1   Verify the natural domains and associated ranges of some simple functions. The domains in each case are the values of x for which the formula makes sense.

Function	Domain (x)	Range (y)
$y = x^2$	$(-\infty, \infty)$	$[0, \infty)$
$y = 1/x$	$(-\infty, 0) \cup (0, \infty)$	$(-\infty, 0) \cup (0, \infty)$
$y = \sqrt{x}$	$[0, \infty)$	$[0, \infty)$
$y = \sqrt{4 - x}$	$(-\infty, 4]$	$[0, \infty)$
$y = \sqrt{1 - x^2}$	$[-1, 1]$	$[0, 1]$

Solution   The formula $y = x^2$ gives a real y-value for any real number x, so the domain is $(-\infty, \infty)$. The range of $y = x^2$ is $[0, \infty)$ because the square of any real number is nonnegative and every nonnegative number y is the square of its own square root: $y = (\sqrt{y})^2$ for $y \geq 0$.

The formula $y = 1/x$ gives a real y-value for every x except $x = 0$. For consistency in the rules of arithmetic, *we cannot divide any number by zero*. The range of $y = 1/x$, the set of reciprocals of all nonzero real numbers, is the set of all nonzero real numbers, since $y = 1/(1/y)$. That is, for $y \neq 0$ the number $x = 1/y$ is the input that is assigned to the output value y.

The formula $y = \sqrt{x}$ gives a real y-value only if $x \geq 0$. The range of $y = \sqrt{x}$ is $[0, \infty)$ because every nonnegative number is some number's square root (namely, it is the square root of its own square).

In $y = \sqrt{4 - x}$, the quantity $4 - x$ cannot be negative. That is, $4 - x \geq 0$, or $x \leq 4$. The formula gives nonnegative real y-values for all $x \leq 4$. The range of $\sqrt{4 - x}$ is $[0, \infty)$, the set of all nonnegative numbers.

The formula $y = \sqrt{1 - x^2}$ gives a real y-value for every x in the closed interval from -1 to 1. Outside this domain, $1 - x^2$ is negative and its square root is not a real number. The values of $1 - x^2$ vary from 0 to 1 on the given domain, and the square roots of these values do the same. The range of $\sqrt{1 - x^2}$ is $[0, 1]$.   ∎

Graphs of Functions

If f is a function with domain D, its **graph** consists of the points in the Cartesian plane whose coordinates are the input-output pairs for f. In set notation, the graph is

$$\{(x, f(x)) \mid x \in D\}.$$

The graph of the function $f(x) = x + 2$ is the set of points with coordinates (x, y) for which $y = x + 2$. Its graph is the straight line sketched in Figure 1.3.

The graph of a function f is a useful picture of its behavior. If (x, y) is a point on the graph, then $y = f(x)$ is the height of the graph above (or below) the point x. The height may be positive or negative, depending on the sign of $f(x)$ (Figure 1.4).

x	$y = x^2$
-2	4
-1	1
0	0
1	1
$\dfrac{3}{2}$	$\dfrac{9}{4}$
2	4

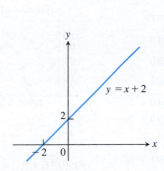

FIGURE 1.3 The graph of $f(x) = x + 2$ is the set of points (x, y) for which y has the value $x + 2$.

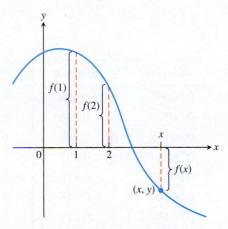

FIGURE 1.4 If (x, y) lies on the graph of f, then the value $y = f(x)$ is the height of the graph above the point x (or below x if $f(x)$ is negative).

EXAMPLE 2   Graph the function $y = x^2$ over the interval $[-2, 2]$.

Solution   Make a table of xy-pairs that satisfy the equation $y = x^2$. Plot the points (x, y) whose coordinates appear in the table, and draw a *smooth* curve (labeled with its equation) through the plotted points (see Figure 1.5). ∎

How do we know that the graph of $y = x^2$ doesn't look like one of these curves?

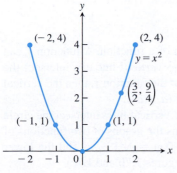

FIGURE 1.5 Graph of the function in Example 2.

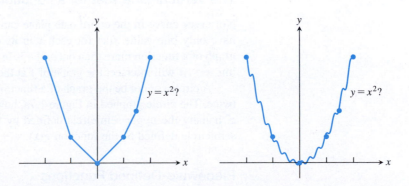

To find out, we could plot more points. But how would we then connect *them*? The basic question still remains: How do we know for sure what the graph looks like between the points we plot? Calculus answers this question, as we will see in Chapter 4. Meanwhile, we will have to settle for plotting points and connecting them as best we can.

Representing a Function Numerically

We have seen how a function may be represented algebraically by a formula and visually by a graph (Example 2). Another way to represent a function is **numerically**, through a table of values. Numerical representations are often used by engineers and experimental scientists. From an appropriate table of values, a graph of the function can be obtained using the method illustrated in Example 2, possibly with the aid of a computer. The graph consisting of only the points in the table is called a **scatterplot**.

EXAMPLE 3   Musical notes are pressure waves in the air. The data associated with Figure 1.6 give recorded pressure displacement versus time in seconds of a musical note produced by a tuning fork. The table provides a representation of the pressure function (in micropascals) over time. If we first make a scatterplot and then connect the data points (t, p) from the table, we obtain the graph shown in the figure.

Time	Pressure	Time	Pressure
0.00091	−0.080	0.00362	0.217
0.00108	0.200	0.00379	0.480
0.00125	0.480	0.00398	0.681
0.00144	0.693	0.00416	0.810
0.00162	0.816	0.00435	0.827
0.00180	0.844	0.00453	0.749
0.00198	0.771	0.00471	0.581
0.00216	0.603	0.00489	0.346
0.00234	0.368	0.00507	0.077
0.00253	0.099	0.00525	−0.164
0.00271	−0.141	0.00543	−0.320
0.00289	−0.309	0.00562	−0.354
0.00307	−0.348	0.00579	−0.248
0.00325	−0.248	0.00598	−0.035
0.00344	−0.041		

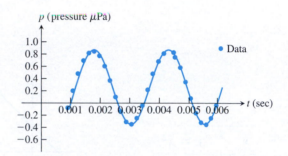

FIGURE 1.6   A smooth curve through the plotted points gives a graph of the pressure function represented by the accompanying tabled data (Example 3).

The Vertical Line Test for a Function

Not every curve in the coordinate plane can be the graph of a function. A function f can have only one value $f(x)$ for each x in its domain, so *no vertical* line can intersect the graph of a function more than once. If a is in the domain of the function f, then the vertical line $x = a$ will intersect the graph of f at the single point $(a, f(a))$.

A circle cannot be the graph of a function, since some vertical lines intersect the circle twice. The circle graphed in Figure 1.7a, however, contains the graphs of two functions of x, namely the upper semicircle defined by the function $f(x) = \sqrt{1 - x^2}$ and the lower semicircle defined by the function $g(x) = -\sqrt{1 - x^2}$ (Figures 1.7b and 1.7c).

Piecewise-Defined Functions

Sometimes a function is described in pieces by using different formulas on different parts of its domain. One example is the **absolute value function**

$$|x| = \begin{cases} x, & x \geq 0 \quad \text{First formula} \\ -x, & x < 0 \quad \text{Second formula} \end{cases}$$

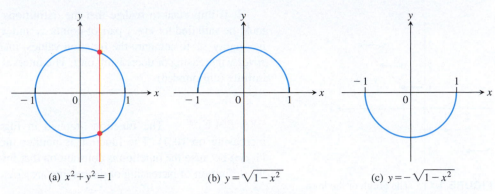

(a) $x^2 + y^2 = 1$ (b) $y = \sqrt{1 - x^2}$ (c) $y = -\sqrt{1 - x^2}$

FIGURE 1.7 (a) The circle is not the graph of a function; it fails the vertical line test. (b) The upper semicircle is the graph of the function $f(x) = \sqrt{1 - x^2}$. (c) The lower semicircle is the graph of the function $g(x) = -\sqrt{1 - x^2}$.

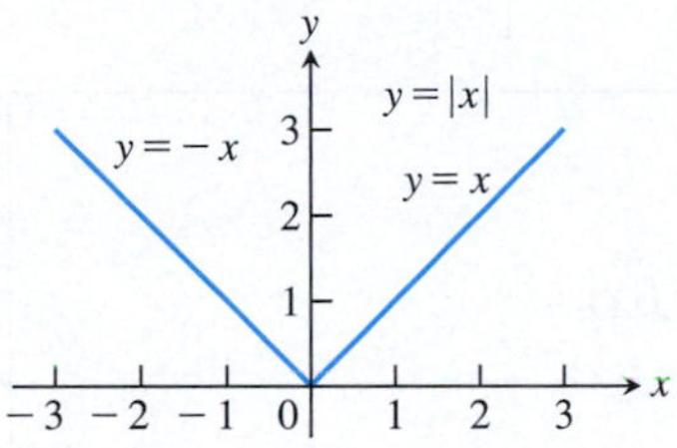

whose graph is given in Figure 1.8. The right-hand side of the equation means that the function equals x if $x \geq 0$, and equals $-x$ if $x < 0$. Piecewise-defined functions often arise when real-world data are modeled. Here are some other examples.

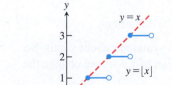

FIGURE 1.8 The absolute value function has domain $(-\infty, \infty)$ and range $[0, \infty)$.

EXAMPLE 4 The function

$$f(x) = \begin{cases} -x, & x < 0 & \text{First formula} \\ x^2, & 0 \leq x \leq 1 & \text{Second formula} \\ 1, & x > 1 & \text{Third formula} \end{cases}$$

is defined on the entire real line but has values given by different formulas, depending on the position of x. The values of f are given by $y = -x$ when $x < 0$, $y = x^2$ when $0 \leq x \leq 1$, and $y = 1$ when $x > 1$. The function, however, is *just one function* whose domain is the entire set of real numbers (Figure 1.9). ■

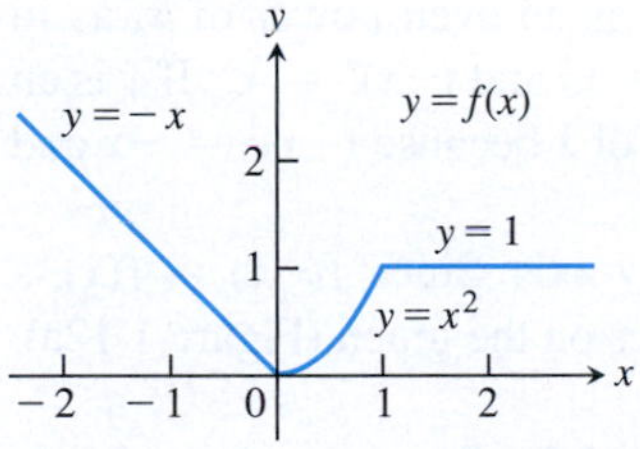

FIGURE 1.9 To graph the function $y = f(x)$ shown here, we apply different formulas to different parts of its domain (Example 4).

EXAMPLE 5 The function whose value at any number x is the *greatest integer less than or equal to x* is called the **greatest integer function** or the **integer floor function**. It is denoted $\lfloor x \rfloor$. Figure 1.10 shows the graph. Observe that

$$\lfloor 2.4 \rfloor = 2, \quad \lfloor 1.9 \rfloor = 1, \quad \lfloor 0 \rfloor = 0, \quad \lfloor -1.2 \rfloor = -2,$$
$$\lfloor 2 \rfloor = 2, \quad \lfloor 0.2 \rfloor = 0, \quad \lfloor -0.3 \rfloor = -1, \quad \lfloor -2 \rfloor = -2. \quad ■$$

EXAMPLE 6 The function whose value at any number x is the *smallest integer greater than or equal to x* is called the **least integer function** or the **integer ceiling function**. It is denoted $\lceil x \rceil$. Figure 1.11 shows the graph. For positive values of x, this function might represent, for example, the cost of parking x hours in a parking lot that charges $1 for each hour or part of an hour. ■

Increasing and Decreasing Functions

If the graph of a function climbs or rises as you move from left to right, we say that the function is *increasing*. If the graph descends or falls as you move from left to right, the function is *decreasing*.

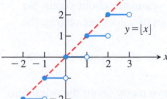

FIGURE 1.10 The graph of the greatest integer function $y = \lfloor x \rfloor$ lies on or below the line $y = x$, so it provides an integer floor for x (Example 5).

DEFINITIONS Let f be a function defined on an interval I and let x_1 and x_2 be two distinct points in I.

1. If $f(x_2) > f(x_1)$ whenever $x_1 < x_2$, then f is said to be **increasing** on I.

2. If $f(x_2) < f(x_1)$ whenever $x_1 < x_2$, then f is said to be **decreasing** on I.

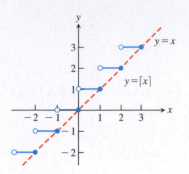

FIGURE 1.11   The graph of the least integer function $y = \lceil x \rceil$ lies on or above the line $y = x$, so it provides an integer ceiling for x (Example 6).

It is important to realize that the definitions of increasing and decreasing functions must be satisfied for *every* pair of points x_1 and x_2 in I with $x_1 < x_2$. Because we use the inequality $<$ to compare the function values, instead of $\leq$, it is sometimes said that f is *strictly* increasing or decreasing on I. The interval I may be finite (also called bounded) or infinite (unbounded).

EXAMPLE 7   The function graphed in Figure 1.9 is decreasing on $(-\infty, 0)$ and increasing on $(0, 1)$. The function is neither increasing nor decreasing on the interval $(1, \infty)$ because the function is constant on that interval, and hence the strict inequalities in the definition of increasing or decreasing are not satisfied on $(1, \infty)$.   ∎

Even Functions and Odd Functions: Symmetry

The graphs of *even* and *odd* functions have special symmetry properties.

DEFINITIONS   A function $y = f(x)$ is an

> **even function of x**   if $f(-x) = f(x)$,
> **odd function of x**   if $f(-x) = -f(x)$,

for every x in the function's domain.

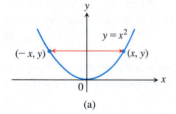

(a)

The names *even* and *odd* come from powers of x. If y is an even power of x, as in $y = x^2$ or $y = x^4$, it is an even function of x because $(-x)^2 = x^2$ and $(-x)^4 = x^4$. If y is an odd power of x, as in $y = x$ or $y = x^3$, it is an odd function of x because $(-x)^1 = -x$ and $(-x)^3 = -x^3$.

The graph of an even function is **symmetric about the y-axis**. Since $f(-x) = f(x)$, a point (x, y) lies on the graph if and only if the point $(-x, y)$ lies on the graph (Figure 1.12a). A reflection across the y-axis leaves the graph unchanged.

The graph of an odd function is **symmetric about the origin**. Since $f(-x) = -f(x)$, a point (x, y) lies on the graph if and only if the point $(-x, -y)$ lies on the graph (Figure 1.12b). Equivalently, a graph is symmetric about the origin if a rotation of $180°$ about the origin leaves the graph unchanged. Notice that the definitions imply that both x and $-x$ must be in the domain of f.

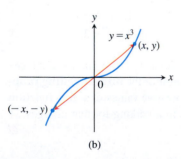

(b)

FIGURE 1.12   (a) The graph of $y = x^2$ (an even function) is symmetric about the y-axis. (b) The graph of $y = x^3$ (an odd function) is symmetric about the origin.

EXAMPLE 8   Here are several functions illustrating the definitions.

$f(x) = x^2$ — Even function: $(-x)^2 = x^2$ for all x; symmetry about y-axis. So $f(-3) = 9 = f(3)$. Changing the sign of x does not change the value of an even function.

$f(x) = x^2 + 1$ — Even function: $(-x)^2 + 1 = x^2 + 1$ for all x; symmetry about y-axis (Figure 1.13a).

$f(x) = x$ — Odd function: $(-x) = -x$ for all x; symmetry about the origin. So $f(-3) = -3$ while $f(3) = 3$. Changing the sign of x changes the sign of an odd function.

$f(x) = x + 1$ — Not odd: $f(-x) = -x + 1$, but $-f(x) = -x - 1$. The two are not equal.

Not even: $(-x) + 1 \neq x + 1$ for all $x \neq 0$ (Figure 1.13b).   ∎

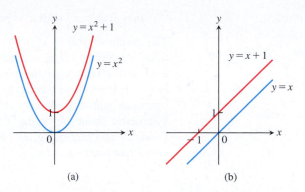

FIGURE 1.13 (a) When we add the constant term 1 to the function $y = x^2$, the resulting function $y = x^2 + 1$ is still even and its graph is still symmetric about the y-axis. (b) When we add the constant term 1 to the function $y = x$, the resulting function $y = x + 1$ is no longer odd, since the symmetry about the origin is lost. The function $y = x + 1$ is also not even (Example 8).

Common Functions

A variety of important types of functions are frequently encountered in calculus.

Linear Functions A function of the form $f(x) = mx + b$, where m and b are fixed constants, is called a **linear function**. Figure 1.14a shows an array of lines $f(x) = mx$. Each of these has $b = 0$, so these lines pass through the origin. The function $f(x) = x$ where $m = 1$ and $b = 0$ is called the **identity function**. Constant functions result when the slope is $m = 0$ (Figure 1.14b).

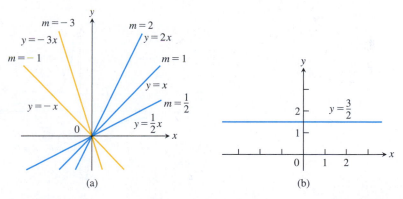

FIGURE 1.14 (a) Lines through the origin with slope m. (b) A constant function with slope $m = 0$.

> **DEFINITION** Two variables y and x are **proportional** (to one another) if one is always a constant multiple of the other—that is, if $y = kx$ for some nonzero constant k.

If the variable y is proportional to the reciprocal $1/x$, then sometimes it is said that y is **inversely proportional** to x (because $1/x$ is the multiplicative inverse of x).

Power Functions A function $f(x) = x^a$, where a is a constant, is called a **power function**. There are several important cases to consider.

(a) $f(x) = x^a$ with $a = n$, a positive integer.

The graphs of $f(x) = x^n$, for $n = 1, 2, 3, 4, 5$, are displayed in Figure 1.15. These functions are defined for all real values of x. Notice that as the power n gets larger, the curves tend to flatten toward the x-axis on the interval $(-1, 1)$ and to rise more steeply for $|x| > 1$. Each curve passes through the point $(1, 1)$ and through the origin. The graphs of functions with even powers are symmetric about the y-axis; those with odd powers are symmetric about the origin. The even-powered functions are decreasing on the interval $(-\infty, 0]$ and increasing on $[0, \infty)$; the odd-powered functions are increasing over the entire real line $(-\infty, \infty)$.

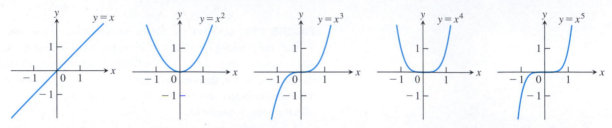

FIGURE 1.15   Graphs of $f(x) = x^n$, $n = 1, 2, 3, 4, 5$, defined for $-\infty < x < \infty$.

(b) $f(x) = x^a$ with $a = -1$ or $a = -2$.

The graphs of the functions $f(x) = x^{-1} = 1/x$ and $g(x) = x^{-2} = 1/x^2$ are shown in Figure 1.16. Both functions are defined for all $x \neq 0$ (you can never divide by zero). The graph of $y = 1/x$ is the hyperbola $xy = 1$, which approaches the coordinate axes far from the origin. The graph of $y = 1/x^2$ also approaches the coordinate axes. The graph of the function f is symmetric about the origin; f is decreasing on the intervals $(-\infty, 0)$ and $(0, \infty)$. The graph of the function g is symmetric about the y-axis; g is increasing on $(-\infty, 0)$ and decreasing on $(0, \infty)$.

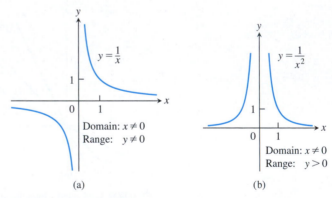

FIGURE 1.16   Graphs of the power functions $f(x) = x^a$. (a) $a = -1$, (b) $a = -2$.

(c) $a = \dfrac{1}{2}, \dfrac{1}{3}, \dfrac{3}{2},$ and $\dfrac{2}{3}$.

The functions $f(x) = x^{1/2} = \sqrt{x}$ and $g(x) = x^{1/3} = \sqrt[3]{x}$ are the **square root** and **cube root** functions, respectively. The domain of the square root function is $[0, \infty)$, but the cube root function is defined for all real x. Their graphs are displayed in Figure 1.17, along with the graphs of $y = x^{3/2}$ and $y = x^{2/3}$. (Recall that $x^{3/2} = (x^{1/2})^3$ and $x^{2/3} = (x^{1/3})^2$.)

Polynomials   A function p is a **polynomial** if

$$p(x) = a_n x^n + a_{n-1} x^{n-1} + \cdots + a_1 x + a_0$$

where n is a nonnegative integer and the numbers $a_0, a_1, a_2, \ldots, a_n$ are real constants (called the **coefficients** of the polynomial). All polynomials have domain $(-\infty, \infty)$. If the

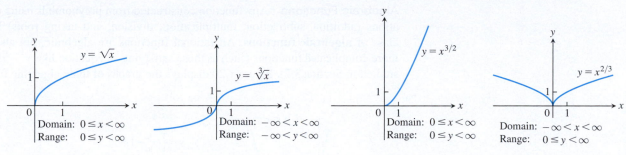

FIGURE 1.17 Graphs of the power functions $f(x) = x^a$ for $a = \frac{1}{2}, \frac{1}{3}, \frac{3}{2},$ and $\frac{2}{3}$.

leading coefficient $a_n \neq 0$, then n is called the **degree** of the polynomial. Linear functions with $m \neq 0$ are polynomials of degree 1. Polynomials of degree 2, usually written as $p(x) = ax^2 + bx + c$, are called **quadratic functions**. Likewise, **cubic functions** are polynomials $p(x) = ax^3 + bx^2 + cx + d$ of degree 3. Figure 1.18 shows the graphs of three polynomials. Techniques to graph polynomials are studied in Chapter 4.

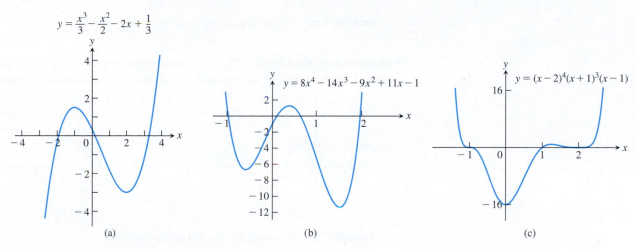

FIGURE 1.18 Graphs of three polynomial functions.

Rational Functions   A **rational function** is a quotient or ratio $f(x) = p(x)/q(x)$, where p and q are polynomials. The domain of a rational function is the set of all real x for which $q(x) \neq 0$. The graphs of several rational functions are shown in Figure 1.19.

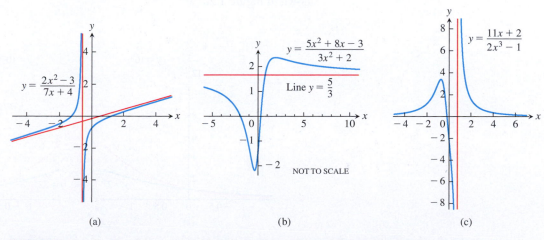

FIGURE 1.19 Graphs of three rational functions. The straight red lines approached by the graphs are called *asymptotes* and are not part of the graphs. We discuss asymptotes in Section 2.6.

Algebraic Functions Any function constructed from polynomials using algebraic operations (addition, subtraction, multiplication, division, and taking roots) lies within the class of **algebraic functions**. All rational functions are algebraic, but also included are more complicated functions (such as those satisfying an equation like $y^3 - 9xy + x^3 = 0$, studied in Section 3.7). Figure 1.20 displays the graphs of three algebraic functions.

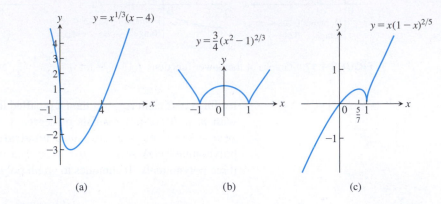

(a) (b) (c)

FIGURE 1.20 Graphs of three algebraic functions.

Trigonometric Functions The six basic trigonometric functions are reviewed in Section 1.3. The graphs of the sine and cosine functions are shown in Figure 1.21.

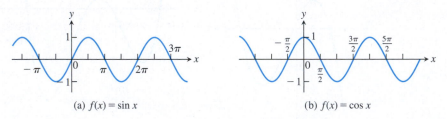

(a) $f(x) = \sin x$ (b) $f(x) = \cos x$

FIGURE 1.21 Graphs of the sine and cosine functions.

Exponential Functions A function of the form $f(x) = a^x$, where $a > 0$ and $a \neq 1$, is called an **exponential function** (with base a). All exponential functions have domain $(-\infty, \infty)$ and range $(0, \infty)$, so an exponential function never assumes the value 0. We discuss exponential functions in Section 1.5. The graphs of some exponential functions are shown in Figure 1.22.

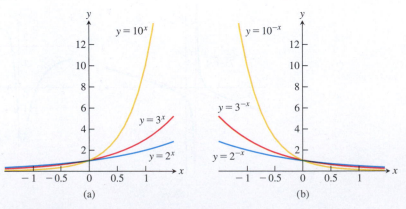

(a) (b)

FIGURE 1.22 Graphs of exponential functions.

Logarithmic Functions These are the functions $f(x) = \log_a x$, where the base $a \neq 1$ is a positive constant. They are the *inverse functions* of the exponential functions, and we discuss these functions in Section 1.6. Figure 1.23 shows the graphs of four logarithmic functions with various bases. In each case the domain is $(0, \infty)$ and the range is $(-\infty, \infty)$.

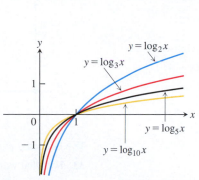

FIGURE 1.23 Graphs of four logarithmic functions.

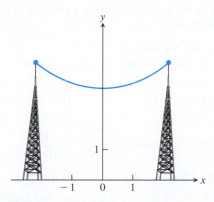

FIGURE 1.24 Graph of a catenary or hanging cable. (The Latin word *catena* means "chain.")

Transcendental Functions These are functions that are not algebraic. They include the trigonometric, inverse trigonometric, exponential, and logarithmic functions, and many other functions as well. The **catenary** is one example of a transcendental function. Its graph has the shape of a cable, like a telephone line or electric cable, strung from one support to another and hanging freely under its own weight (Figure 1.24). The function defining the graph is discussed in Section 7.3.

EXERCISES 1.1

Functions

In Exercises 1–6, find the domain and range of each function.

1. $f(x) = 1 + x^2$ **2.** $f(x) = 1 - \sqrt{x}$

3. $F(x) = \sqrt{5x + 10}$ **4.** $g(x) = \sqrt{x^2 - 3x}$

5. $f(t) = \dfrac{4}{3 - t}$ **6.** $G(t) = \dfrac{2}{t^2 - 16}$

In Exercises 7 and 8, which of the graphs are graphs of functions of x, and which are not? Give reasons for your answers.

7. a.

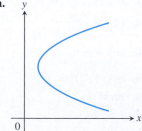

b.

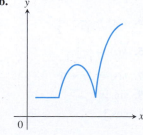

8. a.

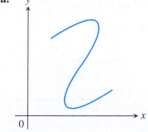

b.

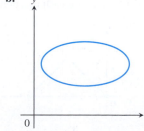

Finding Formulas for Functions

9. Express the area and perimeter of an equilateral triangle as a function of the triangle's side length x.

10. Express the side length of a square as a function of the length d of the square's diagonal. Then express the area as a function of the diagonal length.

11. Express the edge length of a cube as a function of the cube's diagonal length d. Then express the surface area and volume of the cube as a function of the diagonal length.

12. A point P in the first quadrant lies on the graph of the function $f(x) = \sqrt{x}$. Express the coordinates of P as functions of the slope of the line joining P to the origin.

13. Consider the point (x, y) lying on the graph of the line $2x + 4y = 5$. Let L be the distance from the point (x, y) to the origin $(0, 0)$. Write L as a function of x.

14. Consider the point (x, y) lying on the graph of $y = \sqrt{x - 3}$. Let L be the distance between the points (x, y) and $(4, 0)$. Write L as a function of y.

Functions and Graphs

Find the natural domain and graph the functions in Exercises 15–20.

15. $f(x) = 5 - 2x$

16. $f(x) = 1 - 2x - x^2$

17. $g(x) = \sqrt{|x|}$

18. $g(x) = \sqrt{-x}$

19. $F(t) = t/|t|$

20. $G(t) = 1/|t|$

21. Find the domain of $y = \dfrac{x + 3}{4 - \sqrt{x^2 - 9}}$.

22. Find the range of $y = 2 + \sqrt{9 + x^2}$.

23. Graph the following equations and explain why they are not graphs of functions of x.

   a. $|y| = x$ **b.** $y^2 = x^2$

24. Graph the following equations and explain why they are not graphs of functions of x.

   a. $|x| + |y| = 1$ **b.** $|x + y| = 1$

Piecewise-Defined Functions

Graph the functions in Exercises 25–28.

25. $f(x) = \begin{cases} x, & 0 \le x \le 1 \\ 2 - x, & 1 < x \le 2 \end{cases}$

26. $g(x) = \begin{cases} 1 - x, & 0 \le x \le 1 \\ 2 - x, & 1 < x \le 2 \end{cases}$

27. $F(x) = \begin{cases} 4 - x^2, & x \le 1 \\ x^2 + 2x, & x > 1 \end{cases}$

28. $G(x) = \begin{cases} 1/x, & x < 0 \\ x, & 0 \le x \end{cases}$

Find a formula for each function graphed in Exercises 29–32.

29. a.
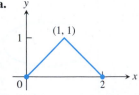
  b.
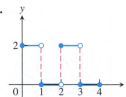

30. a.
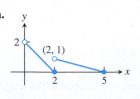
  b.
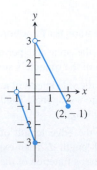

31. a.
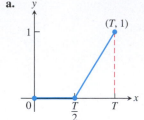
  b.
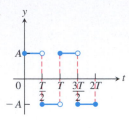

32. a.
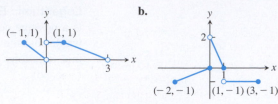
  b.

The Greatest and Least Integer Functions

33. For what values of x is

   a. $\lfloor x \rfloor = 0$?   **b.** $\lceil x \rceil = 0$?

34. What real numbers x satisfy the equation $\lfloor x \rfloor = \lceil x \rceil$?

35. Does $\lceil -x \rceil = -\lfloor x \rfloor$ for all real x? Give reasons for your answer.

36. Graph the function

$$f(x) = \begin{cases} \lfloor x \rfloor, & x \ge 0 \\ \lceil x \rceil, & x < 0. \end{cases}$$

Why is $f(x)$ called the *integer part* of x?

Increasing and Decreasing Functions

Graph the functions in Exercises 37–46. What symmetries, if any, do the graphs have? Specify the intervals over which the function is increasing and the intervals where it is decreasing.

37. $y = -x^3$

38. $y = -\dfrac{1}{x^2}$

39. $y = -\dfrac{1}{x}$

40. $y = \dfrac{1}{|x|}$

41. $y = \sqrt{|x|}$

42. $y = \sqrt{-x}$

43. $y = x^3/8$

44. $y = -4\sqrt{x}$

45. $y = -x^{3/2}$

46. $y = (-x)^{2/3}$

Even and Odd Functions

In Exercises 47–58, say whether the function is even, odd, or neither. Give reasons for your answer.

47. $f(x) = 3$

48. $f(x) = x^{-5}$

49. $f(x) = x^2 + 1$

50. $f(x) = x^2 + x$

51. $g(x) = x^3 + x$

52. $g(x) = x^4 + 3x^2 - 1$

53. $g(x) = \dfrac{1}{x^2 - 1}$

54. $g(x) = \dfrac{x}{x^2 - 1}$

55. $h(t) = \dfrac{1}{t - 1}$

56. $h(t) = |t^3|$

57. $h(t) = 2t + 1$

58. $h(t) = 2|t| + 1$

59. $\sin 2x$

60. $\sin x^2$

61. $\cos 3x$

62. $1 + \cos x$

Theory and Examples

63. The variable s is proportional to t, and $s = 25$ when $t = 75$. Determine t when $s = 60$.

64. Kinetic energy The kinetic energy K of a mass is proportional to the square of its velocity v. If $K = 12,960$ joules when $v = 18$ m/sec, what is K when $v = 10$ m/sec?

65. The variables r and s are inversely proportional, and $r = 6$ when $s = 4$. Determine s when $r = 10$.

66. Boyle's Law Boyle's Law says that the volume V of a gas at constant temperature increases whenever the pressure P decreases, so that V and P are inversely proportional. If $P = 14.7 \text{ lb/in}^2$ when $V = 1000 \text{ in}^3$, then what is V when $P = 23.4 \text{ lb/in}^2$?

67. A box with an open top is to be constructed from a rectangular piece of cardboard with dimensions 14 in. by 22 in. by cutting out equal squares of side x at each corner and then folding up the sides as in the figure. Express the volume V of the box as a function of x.

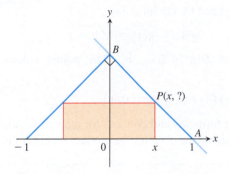

68. The accompanying figure shows a rectangle inscribed in an isosceles right triangle whose hypotenuse is 2 units long.

   a. Express the y-coordinate of P in terms of x. (You might start by writing an equation for the line AB.)

   b. Express the area of the rectangle in terms of x.

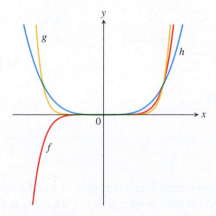

In Exercises 69 and 70, match each equation with its graph. Do not use a graphing device, and give reasons for your answer.

69. a. $y = x^4$ **b.** $y = x^7$ **c.** $y = x^{10}$

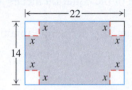

70. a. $y = 5x$ **b.** $y = 5^x$ **c.** $y = x^5$

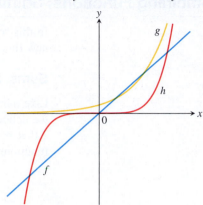

T 71. a. Graph the functions $f(x) = x/2$ and $g(x) = 1 + (4/x)$ together to identify the values of x for which

$$\frac{x}{2} > 1 + \frac{4}{x}.$$

   b. Confirm your findings in part (a) algebraically.

T 72. a. Graph the functions $f(x) = 3/(x - 1)$ and $g(x) = 2/(x + 1)$ together to identify the values of x for which

$$\frac{3}{x - 1} < \frac{2}{x + 1}.$$

   b. Confirm your findings in part (a) algebraically.

73. For a curve to be *symmetric about the x-axis*, the point (x, y) must lie on the curve if and only if the point $(x, -y)$ lies on the curve. Explain why a curve that is symmetric about the x-axis is not the graph of a function, unless the function is $y = 0$.

74. Three hundred books sell for $40 each, resulting in a revenue of $(300)(\$40) = \$12,000$. For each $5 increase in the price, 25 fewer books are sold. Write the revenue R as a function of the number x of $5 increases.

75. A pen in the shape of an isosceles right triangle with legs of length x ft and hypotenuse of length h ft is to be built. If fencing costs $5/ft for the legs and $10/ft for the hypotenuse, write the total cost C of construction as a function of h.

76. Industrial costs A power plant sits next to a river where the river is 800 ft wide. To lay a new cable from the plant to a location in the city 2 mi downstream on the opposite side costs $180 per foot across the river and $100 per foot along the land.

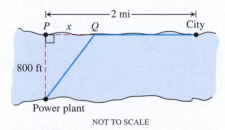

NOT TO SCALE

   a. Suppose that the cable goes from the plant to a point Q on the opposite side that is x ft from the point P directly opposite the plant. Write a function $C(x)$ that gives the cost of laying the cable in terms of the distance x.

   b. Generate a table of values to determine if the least expensive location for point Q is less than 2000 ft or greater than 2000 ft from point P.

1.2   Combining Functions; Shifting and Scaling Graphs

In this section we look at the main ways functions are combined or transformed to form new functions.

Sums, Differences, Products, and Quotients

Like numbers, functions can be added, subtracted, multiplied, and divided (except where the denominator is zero) to produce new functions. If f and g are functions, then for every x that belongs to the domains of both f and g (that is, for $x \in D(f) \cap D(g)$), we define functions $f + g$, $f - g$, and fg by the formulas

$$(f + g)(x) = f(x) + g(x)$$
$$(f - g)(x) = f(x) - g(x)$$
$$(fg)(x) = f(x)g(x).$$

Notice that the $+$ sign on the left-hand side of the first equation represents the operation of addition of *functions*, whereas the $+$ on the right-hand side of the equation means addition of the real numbers $f(x)$ and $g(x)$.

At any point of $D(f) \cap D(g)$ at which $g(x) \neq 0$, we can also define the function f/g by the formula

$$\left(\frac{f}{g}\right)(x) = \frac{f(x)}{g(x)} \qquad \text{(where } g(x) \neq 0\text{)}.$$

Functions can also be multiplied by constants: If c is a real number, then the function cf is defined for all x in the domain of f by

$$(cf)(x) = cf(x).$$

EXAMPLE 1   The functions defined by the formulas

$$f(x) = \sqrt{x} \qquad \text{and} \qquad g(x) = \sqrt{1 - x}$$

have domains $D(f) = [0, \infty)$ and $D(g) = (-\infty, 1]$. The points common to these domains are the points in

$$[0, \infty) \cap (-\infty, 1] = [0, 1].$$

The following table summarizes the formulas and domains for the various algebraic combinations of the two functions. We also write $f \cdot g$ for the product function fg.

Function	Formula	Domain
$f + g$	$(f + g)(x) = \sqrt{x} + \sqrt{1 - x}$	$[0, 1] = D(f) \cap D(g)$
$f - g$	$(f - g)(x) = \sqrt{x} - \sqrt{1 - x}$	$[0, 1]$
$g - f$	$(g - f)(x) = \sqrt{1 - x} - \sqrt{x}$	$[0, 1]$
$f \cdot g$	$(f \cdot g)(x) = f(x)g(x) = \sqrt{x(1 - x)}$	$[0, 1]$
f/g	$\dfrac{f}{g}(x) = \dfrac{f(x)}{g(x)} = \sqrt{\dfrac{x}{1 - x}}$	$[0, 1)$ $(x = 1$ excluded$)$
g/f	$\dfrac{g}{f}(x) = \dfrac{g(x)}{f(x)} = \sqrt{\dfrac{1 - x}{x}}$	$(0, 1]$ $(x = 0$ excluded$)$

The graph of the function $f + g$ is obtained from the graphs of f and g by adding the corresponding y-coordinates $f(x)$ and $g(x)$ at each point $x \in D(f) \cap D(g)$, as in Figure 1.25. The graphs of $f + g$ and $f \cdot g$ from Example 1 are shown in Figure 1.26.

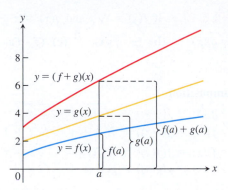

FIGURE 1.25 Graphical addition of two functions.

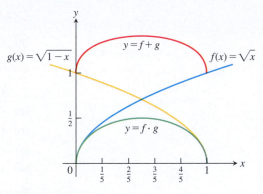

FIGURE 1.26 The domain of the function $f + g$ is the intersection of the domains of f and g, the interval $[0, 1]$ on the x-axis where these domains overlap. This interval is also the domain of the function $f \cdot g$ (Example 1).

Composite Functions

Composition is another method for combining functions. In this operation the output from one function becomes the input to a second function.

DEFINITION If f and g are functions, the **composite** function $f \circ g$ ("f composed with g") is defined by

$$(f \circ g)(x) = f(g(x)).$$

The domain of $f \circ g$ consists of the numbers x in the domain of g for which $g(x)$ lies in the domain of f.

The definition implies that $f \circ g$ can be formed when the range of g lies in the domain of f. To find $(f \circ g)(x)$, *first* find $g(x)$ and *second* find $f(g(x))$. Figure 1.27 pictures $f \circ g$ as a machine diagram, and Figure 1.28 shows the composition as an arrow diagram.

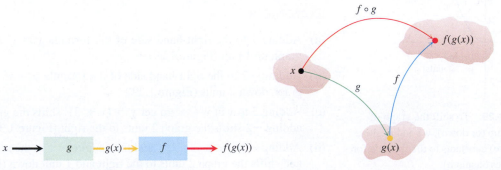

FIGURE 1.27 A composite function $f \circ g$ uses the output $g(x)$ of the first function g as the input for the second function f.

FIGURE 1.28 Arrow diagram for $f \circ g$. If x lies in the domain of g and $g(x)$ lies in the domain of f, then the functions f and g can be composed to form $(f \circ g)(x)$.

To evaluate the composite function $g \circ f$ (when defined), we find $f(x)$ first and then find $g(f(x))$. The domain of $g \circ f$ is the set of numbers x in the domain of f such that $f(x)$ lies in the domain of g.

The functions $f \circ g$ and $g \circ f$ are usually quite different.

EXAMPLE 2   If $f(x) = \sqrt{x}$ and $g(x) = x + 1$, find

(a) $(f \circ g)(x)$ **(b)** $(g \circ f)(x)$ **(c)** $(f \circ f)(x)$ **(d)** $(g \circ g)(x)$.

Solution

Composition	Domain
(a) $(f \circ g)(x) = f(g(x)) = \sqrt{g(x)} = \sqrt{x + 1}$	$[-1, \infty)$
(b) $(g \circ f)(x) = g(f(x)) = f(x) + 1 = \sqrt{x} + 1$	$[0, \infty)$
(c) $(f \circ f)(x) = f(f(x)) = \sqrt{f(x)} = \sqrt{\sqrt{x}} = x^{1/4}$	$[0, \infty)$
(d) $(g \circ g)(x) = g(g(x)) = g(x) + 1 = (x + 1) + 1 = x + 2$	$(-\infty, \infty)$

To see why the domain of $f \circ g$ is $[-1, \infty)$, notice that $g(x) = x + 1$ is defined for all real x but $g(x)$ belongs to the domain of f only if $x + 1 \ge 0$, that is to say, when $x \ge -1$. ∎

Notice that if $f(x) = x^2$ and $g(x) = \sqrt{x}$, then $(f \circ g)(x) = \left(\sqrt{x}\right)^2 = x$. However, the domain of $f \circ g$ is $[0, \infty)$, not $(-\infty, \infty)$, since $\sqrt{x}$ requires $x \ge 0$.

Shifting a Graph of a Function

A common way to obtain a new function from an existing one is by adding a constant to each output of the existing function, or to its input variable. The graph of the new function is the graph of the original function shifted vertically or horizontally, as follows.

Shift Formulas

Vertical Shifts

$y = f(x) + k$ Shifts the graph of f *up* k units if $k > 0$
Shifts it *down* $|k|$ units if $k < 0$

Horizontal Shifts

$y = f(x + h)$ Shifts the graph of f *left* h units if $h > 0$
Shifts it *right* $|h|$ units if $h < 0$

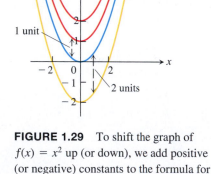

FIGURE 1.29 To shift the graph of $f(x) = x^2$ up (or down), we add positive (or negative) constants to the formula for f (Examples 3a and b).

EXAMPLE 3

(a) Adding 1 to the right-hand side of the formula $y = x^2$ to get $y = x^2 + 1$ shifts the graph up 1 unit (Figure 1.29).

(b) Adding -2 to the right-hand side of the formula $y = x^2$ to get $y = x^2 - 2$ shifts the graph down 2 units (Figure 1.29).

(c) Adding 3 to x in $y = x^2$ to get $y = (x + 3)^2$ shifts the graph 3 units to the left, while adding -2 shifts the graph 2 units to the right (Figure 1.30).

(d) Adding -2 to x in $y = |x|$, and then adding -1 to the result, gives $y = |x - 2| - 1$ and shifts the graph 2 units to the right and 1 unit down (Figure 1.31). ∎

Scaling and Reflecting a Graph of a Function

To scale the graph of a function $y = f(x)$ is to stretch or compress it, vertically or horizontally. This is accomplished by multiplying the function f, or the independent variable x, by an appropriate constant c. Reflections across the coordinate axes are special cases where $c = -1$.

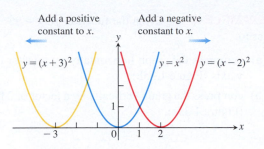

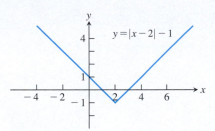

FIGURE 1.30 To shift the graph of $y = x^2$ to the left, we add a positive constant to x (Example 3c). To shift the graph to the right, we add a negative constant to x.

FIGURE 1.31 The graph of $y = |x|$ shifted 2 units to the right and 1 unit down (Example 3d).

Vertical and Horizontal Scaling and Reflecting Formulas

For $c > 1$, the graph is scaled:

$y = cf(x)$ Stretches the graph of f vertically by a factor of c.

$y = \dfrac{1}{c}f(x)$ Compresses the graph of f vertically by a factor of c.

$y = f(cx)$ Compresses the graph of f horizontally by a factor of c.

$y = f(x/c)$ Stretches the graph of f horizontally by a factor of c.

For $c = -1$, the graph is reflected:

$y = -f(x)$ Reflects the graph of f across the x-axis.

$y = f(-x)$ Reflects the graph of f across the y-axis.

EXAMPLE 4 Here we scale and reflect the graph of $y = \sqrt{x}$.

(a) Vertical: Multiplying the right-hand side of $y = \sqrt{x}$ by 3 to get $y = 3\sqrt{x}$ stretches the graph vertically by a factor of 3, whereas multiplying by $1/3$ compresses the graph vertically by a factor of 3 (Figure 1.32).

(b) Horizontal: The graph of $y = \sqrt{3x}$ is a horizontal compression of the graph of $y = \sqrt{x}$ by a factor of 3, and $y = \sqrt{x/3}$ is a horizontal stretching by a factor of 3 (Figure 1.33). Note that $y = \sqrt{3x} = \sqrt{3}\sqrt{x}$ so a horizontal compression *may* correspond to a vertical stretching by a different scaling factor. Likewise, a horizontal stretching may correspond to a vertical compression by a different scaling factor.

(c) Reflection: The graph of $y = -\sqrt{x}$ is a reflection of $y = \sqrt{x}$ across the x-axis, and $y = \sqrt{-x}$ is a reflection across the y-axis (Figure 1.34). ◼

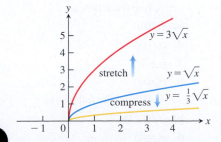

FIGURE 1.32 Vertically stretching and compressing the graph $y = \sqrt{x}$ by a factor of 3 (Example 4a).

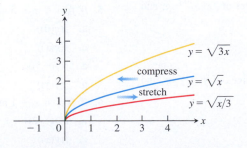

FIGURE 1.33 Horizontally stretching and compressing the graph $y = \sqrt{x}$ by a factor of 3 (Example 4b).

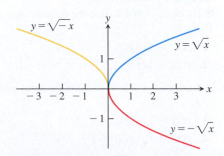

FIGURE 1.34 Reflections of the graph $y = \sqrt{x}$ across the coordinate axes (Example 4c).

EXAMPLE 5 Given the function $f(x) = x^4 - 4x^3 + 10$ (Figure 1.35a), find formulas to

(a) compress the graph horizontally by a factor of 2 followed by a reflection across the y-axis (Figure 1.35b).

(b) compress the graph vertically by a factor of 2 followed by a reflection across the x-axis (Figure 1.35c).

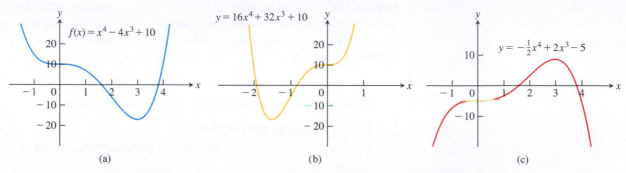

(a)   (b)   (c)

FIGURE 1.35 (a) The original graph of f. (b) The horizontal compression of $y = f(x)$ in part (a) by a factor of 2, followed by a reflection across the y-axis. (c) The vertical compression of $y = f(x)$ in part (a) by a factor of 2, followed by a reflection across the x-axis (Example 5).

Solution

(a) We multiply x by 2 to get the horizontal compression, and by -1 to give reflection across the y-axis. The formula is obtained by substituting $-2x$ for x in the right-hand side of the equation for f:

$$y = f(-2x) = (-2x)^4 - 4(-2x)^3 + 10$$
$$= 16x^4 + 32x^3 + 10.$$

(b) The formula is

$$y = -\frac{1}{2}f(x) = -\frac{1}{2}x^4 + 2x^3 - 5.$$

EXERCISES 1.2

Algebraic Combinations

In Exercises 1 and 2, find the domains and ranges of f, g, $f + g$, and $f \cdot g$.

1. $f(x) = x$, $g(x) = \sqrt{x - 1}$

2. $f(x) = \sqrt{x + 1}$, $g(x) = \sqrt{x - 1}$

In Exercises 3 and 4, find the domains and ranges of f, g, f/g, and g/f.

3. $f(x) = 2$, $g(x) = x^2 + 1$

4. $f(x) = 1$, $g(x) = 1 + \sqrt{x}$

Compositions of Functions

5. If $f(x) = x + 5$ and $g(x) = x^2 - 3$, find the following.

  a. $f(g(0))$ **b.** $g(f(0))$

  c. $f(g(x))$ **d.** $g(f(x))$

  e. $f(f(-5))$ **f.** $g(g(2))$

  g. $f(f(x))$ **h.** $g(g(x))$

6. If $f(x) = x - 1$ and $g(x) = 1/(x + 1)$, find the following.

  a. $f(g(1/2))$ **b.** $g(f(1/2))$

  c. $f(g(x))$ **d.** $g(f(x))$

  e. $f(f(2))$ **f.** $g(g(2))$

  g. $f(f(x))$ **h.** $g(g(x))$

In Exercises 7–10, write a formula for $f \circ g \circ h$.

7. $f(x) = x + 1$, $g(x) = 3x$, $h(x) = 4 - x$

8. $f(x) = 3x + 4$, $g(x) = 2x - 1$, $h(x) = x^2$

9. $f(x) = \sqrt{x + 1}$, $g(x) = \dfrac{1}{x + 4}$, $h(x) = \dfrac{1}{x}$

10. $f(x) = \dfrac{x + 2}{3 - x}$, $g(x) = \dfrac{x^2}{x^2 + 1}$, $h(x) = \sqrt{2 - x}$

Let $f(x) = x - 3$, $g(x) = \sqrt{x}$, $h(x) = x^3$, and $j(x) = 2x$. Express each of the functions in Exercises 11 and 12 as a composition involving one or more of f, g, h, and j.

11. a. $y = \sqrt{x} - 3$ **b.** $y = 2\sqrt{x}$

     c. $y = x^{1/4}$ **d.** $y = 4x$

     e. $y = \sqrt{(x - 3)^3}$ **f.** $y = (2x - 6)^3$

12. a. $y = 2x - 3$ **b.** $y = x^{3/2}$

     c. $y = x^9$ **d.** $y = x - 6$

     e. $y = 2\sqrt{x - 3}$ **f.** $y = \sqrt{x^3 - 3}$

13. Copy and complete the following table.

	$g(x)$	$f(x)$	$(f \circ g)(x)$
a.	$x - 7$	$\sqrt{x}$	?
b.	$x + 2$	$3x$	?
c.	?	$\sqrt{x - 5}$	$\sqrt{x^2 - 5}$
d.	$\dfrac{x}{x - 1}$	$\dfrac{x}{x - 1}$	?
e.	?	$1 + \dfrac{1}{x}$	x
f.	$\dfrac{1}{x}$	?	x

14. Copy and complete the following table.

	$g(x)$	$f(x)$	$(f \circ g)(x)$
a.	$\dfrac{1}{x - 1}$	$\lvert x \rvert$	?
b.	?	$\dfrac{x - 1}{x}$	$\dfrac{x}{x + 1}$
c.	?	$\sqrt{x}$	$\lvert x \rvert$
d.	$\sqrt{x}$	?	$\lvert x \rvert$

15. Evaluate each expression using the given table of values:

x	-2	-1	0	1	2
$f(x)$	1	0	-2	1	2
$g(x)$	2	1	0	-1	0

     a. $f(g(-1))$ **b.** $g(f(0))$ **c.** $f(f(-1))$

     d. $g(g(2))$ **e.** $g(f(-2))$ **f.** $f(g(1))$

16. Evaluate each expression using the functions

$$f(x) = 2 - x, \quad g(x) = \begin{cases} -x, & -2 \le x < 0 \\ x - 1, & 0 \le x \le 2. \end{cases}$$

     a. $f(g(0))$ **b.** $g(f(3))$ **c.** $g(g(-1))$

     d. $f(f(2))$ **e.** $g(f(0))$ **f.** $f(g(1/2))$

In Exercises 17 and 18, **(a)** write formulas for $f \circ g$ and $g \circ f$ and find the **(b)** domain and **(c)** range of each.

17. $f(x) = \sqrt{x + 1}$, $g(x) = \dfrac{1}{x}$

18. $f(x) = x^2$, $g(x) = 1 - \sqrt{x}$

19. Let $f(x) = \dfrac{x}{x - 2}$. Find a function $y = g(x)$ so that $(f \circ g)(x) = x$.

20. Let $f(x) = 2x^3 - 4$. Find a function $y = g(x)$ so that $(f \circ g)(x) = x + 2$.

21. A balloon's volume V is given by $V = s^2 + 2s + 3 \text{ cm}^3$, where s is the ambient temperature in °C. The ambient temperature s at time t minutes is given by $s = 2t - 3\,°C$. Write the balloon's volume V as a function of time t.

22. Use the graphs of f and g to sketch the graph of $y = f(g(x))$.

   a.   **b.**

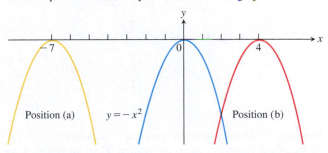

Shifting Graphs

23. The accompanying figure shows the graph of $y = -x^2$ shifted to two new positions. Write equations for the new graphs.

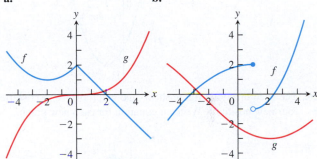

24. The accompanying figure shows the graph of $y = x^2$ shifted to two new positions. Write equations for the new graphs.

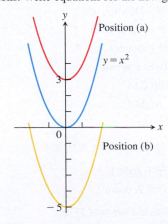

25. Match the equations listed in parts (a)–(d) to the graphs in the accompanying figure.

a. $y = (x - 1)^2 - 4$ **b.** $y = (x - 2)^2 + 2$

c. $y = (x + 2)^2 + 2$ **d.** $y = (x + 3)^2 - 2$

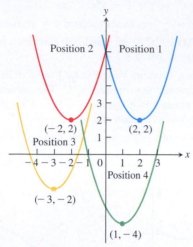

26. The accompanying figure shows the graph of $y = -x^2$ shifted to four new positions. Write an equation for each new graph.

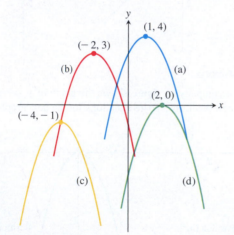

Exercises 27–36 tell how many units and in what directions the graphs of the given equations are to be shifted. Give an equation for the shifted graph. Then sketch the original and shifted graphs together, labeling each graph with its equation.

27. $x^2 + y^2 = 49$ Down 3, left 2

28. $x^2 + y^2 = 25$ Up 3, left 4

29. $y = x^3$ Left 1, down 1

30. $y = x^{2/3}$ Right 1, down 1

31. $y = \sqrt{x}$ Left 0.81

32. $y = -\sqrt{x}$ Right 3

33. $y = 2x - 7$ Up 7

34. $y = \frac{1}{2}(x + 1) + 5$ Down 5, right 1

35. $y = 1/x$ Up 1, right 1

36. $y = 1/x^2$ Left 2, down 1

Graph the functions in Exercises 37–56.

37. $y = \sqrt{x + 4}$

38. $y = \sqrt{9 - x}$

39. $y = |x - 2|$

40. $y = |1 - x| - 1$

41. $y = 1 + \sqrt{x - 1}$

42. $y = 1 - \sqrt{x}$

43. $y = (x + 1)^{2/3}$

44. $y = (x - 8)^{2/3}$

45. $y = 1 - x^{2/3}$

46. $y + 4 = x^{2/3}$

47. $y = \sqrt[3]{x - 1} - 1$

48. $y = (x + 2)^{3/2} + 1$

49. $y = \frac{1}{x - 2}$

50. $y = \frac{1}{x} - 2$

51. $y = \frac{1}{x} + 2$

52. $y = \frac{1}{x + 2}$

53. $y = \frac{1}{(x - 1)^2}$

54. $y = \frac{1}{x^2} - 1$

55. $y = \frac{1}{x^2} + 1$

56. $y = \frac{1}{(x + 1)^2}$

57. The accompanying figure shows the graph of a function $f(x)$ with domain $[0, 2]$ and range $[0, 1]$. Find the domains and ranges of the following functions, and sketch their graphs.

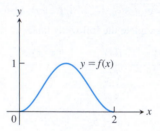

a. $f(x) + 2$ **b.** $f(x) - 1$

c. $2f(x)$ **d.** $-f(x)$

e. $f(x + 2)$ **f.** $f(x - 1)$

g. $f(-x)$ **h.** $-f(x + 1) + 1$

58. The accompanying figure shows the graph of a function $g(t)$ with domain $[-4, 0]$ and range $[-3, 0]$. Find the domains and ranges of the following functions, and sketch their graphs.

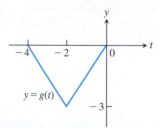

a. $g(-t)$ **b.** $-g(t)$

c. $g(t) + 3$ **d.** $1 - g(t)$

e. $g(-t + 2)$ **f.** $g(t - 2)$

g. $g(1 - t)$ **h.** $-g(t - 4)$

Vertical and Horizontal Scaling

Exercises 59–68 tell by what factor and direction the graphs of the given functions are to be stretched or compressed. Give an equation for the stretched or compressed graph.

59. $y = x^2 - 1$, stretched vertically by a factor of 3

60. $y = x^2 - 1$, compressed horizontally by a factor of 2

61. $y = 1 + \frac{1}{x^2}$, compressed vertically by a factor of 2

62. $y = 1 + \dfrac{1}{x^2}$, stretched horizontally by a factor of 3

63. $y = \sqrt{x + 1}$, compressed horizontally by a factor of 4

64. $y = \sqrt{x + 1}$, stretched vertically by a factor of 3

65. $y = \sqrt{4 - x^2}$, stretched horizontally by a factor of 2

66. $y = \sqrt{4 - x^2}$, compressed vertically by a factor of 3

67. $y = 1 - x^3$, compressed horizontally by a factor of 3

68. $y = 1 - x^3$, stretched horizontally by a factor of 2

Graphing

In Exercises 69–76, graph each function, not by plotting points, but by starting with the graph of one of the standard functions presented in Figures 1.14–1.17 and applying an appropriate transformation.

69. $y = -\sqrt{2x + 1}$

70. $y = \sqrt{1 - \dfrac{x}{2}}$

71. $y = (x - 1)^3 + 2$

72. $y = (1 - x)^3 + 2$

73. $y = \dfrac{1}{2x} - 1$

74. $y = \dfrac{2}{x^2} + 1$

75. $y = -\sqrt[3]{x}$

76. $y = (-2x)^{2/3}$

77. Graph the function $y = |x^2 - 1|$.

78. Graph the function $y = \sqrt{|x|}$.

Combining Functions

79. Assume that f is an even function, g is an odd function, and both f and g are defined on the entire real line $(-\infty, \infty)$. Which of the following (where defined) are even? odd?

a. fg	**b.** f/g	**c.** g/f
d. $f^2 = ff$	**e.** $g^2 = gg$	**f.** $f \circ g$
g. $g \circ f$	**h.** $f \circ f$	**i.** $g \circ g$

80. Can a function be both even and odd? Give reasons for your answer.

T 81. (*Continuation of* Example 1.) Graph the functions $f(x) = \sqrt{x}$ and $g(x) = \sqrt{1 - x}$ together with their (a) sum, (b) product, (c) two differences, (d) two quotients.

T 82. Let $f(x) = x - 7$ and $g(x) = x^2$. Graph f and g together with $f \circ g$ and $g \circ f$.

1.3 Trigonometric Functions

This section reviews radian measure and the basic trigonometric functions.

Angles

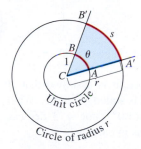

FIGURE 1.36 The radian measure of the central angle $A'CB'$ is the number $\theta = s/r$. For a unit circle of radius $r = 1$, θ is the length of arc AB that central angle ACB cuts from the unit circle.

Angles are measured in degrees or radians. The number of **radians** in the central angle $A'CB'$ within a circle of radius r is defined as the number of "radius units" contained in the arc s subtended by that central angle. If we denote this central angle by θ when measured in radians, this means that $\theta = s/r$ (Figure 1.36), or

$$s = r\theta \qquad (\theta \text{ in radians}). \tag{1}$$

If the circle is a unit circle having radius $r = 1$, then from Figure 1.36 and Equation (1), we see that the central angle θ measured in radians is just the length of the arc that the angle cuts from the unit circle. Since one complete revolution of the unit circle is $360°$ or 2π radians, we have

$$\pi \text{ radians} = 180° \tag{2}$$

and

$$1 \text{ radian} = \frac{180}{\pi} (\approx 57.3) \text{ degrees} \quad \text{or} \quad 1 \text{ degree} = \frac{\pi}{180} (\approx 0.017) \text{ radians}.$$

Table 1.1 shows the equivalence between degree and radian measures for some basic angles.

TABLE 1.1 Angles measured in degrees and radians

Degrees	−180	−135	−90	−45	0	30	45	60	90	120	135	150	180	270	360
θ (radians)	$-\pi$	$\dfrac{-3\pi}{4}$	$\dfrac{-\pi}{2}$	$\dfrac{-\pi}{4}$	0	$\dfrac{\pi}{6}$	$\dfrac{\pi}{4}$	$\dfrac{\pi}{3}$	$\dfrac{\pi}{2}$	$\dfrac{2\pi}{3}$	$\dfrac{3\pi}{4}$	$\dfrac{5\pi}{6}$	π	$\dfrac{3\pi}{2}$	2π

An angle in the *xy*-plane is said to be in **standard position** if its vertex lies at the origin and its initial ray lies along the positive *x*-axis (Figure 1.37). Angles measured counterclockwise from the positive *x*-axis are assigned positive measures; angles measured clockwise are assigned negative measures.

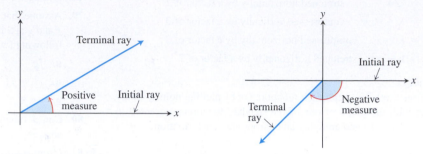

FIGURE 1.37   Angles in standard position in the *xy*-plane.

Angles describing counterclockwise rotations can go arbitrarily far beyond 2π radians or $360°$. Similarly, angles describing clockwise rotations can have negative measures of all sizes (Figure 1.38).

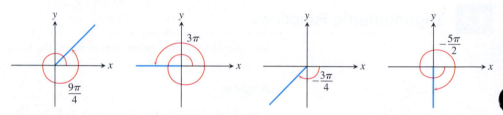

FIGURE 1.38   Nonzero radian measures can be positive or negative and can go beyond 2π.

Angle Convention: Use Radians   From now on, in this book it is assumed that all angles are measured in radians unless degrees or some other unit is stated explicitly. When we talk about the angle $\pi/3$, we mean $\pi/3$ radians (which is $60°$), not $\pi/3$ degrees. Using radians simplifies many of the operations and computations in calculus.

The Six Basic Trigonometric Functions

The trigonometric functions of an acute angle are given in terms of the sides of a right triangle (Figure 1.39). We extend this definition to obtuse and negative angles by first placing the angle in standard position in a circle of radius *r*. We then define the trigonometric functions in terms of the coordinates of the point $P(x, y)$ where the angle's terminal ray intersects the circle (Figure 1.40).

sine: $\sin\theta = \dfrac{y}{r}$ **cosecant:** $\csc\theta = \dfrac{r}{y}$

cosine: $\cos\theta = \dfrac{x}{r}$ **secant:** $\sec\theta = \dfrac{r}{x}$

tangent: $\tan\theta = \dfrac{y}{x}$ **cotangent:** $\cot\theta = \dfrac{x}{y}$

These extended definitions agree with the right-triangle definitions when the angle is acute.

Notice also that whenever the quotients are defined,

$$\tan\theta = \frac{\sin\theta}{\cos\theta} \qquad \cot\theta = \frac{1}{\tan\theta}$$

$$\sec\theta = \frac{1}{\cos\theta} \qquad \csc\theta = \frac{1}{\sin\theta}$$

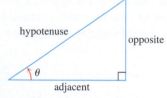

$$\sin\theta = \frac{\text{opp}}{\text{hyp}} \qquad \csc\theta = \frac{\text{hyp}}{\text{opp}}$$

$$\cos\theta = \frac{\text{adj}}{\text{hyp}} \qquad \sec\theta = \frac{\text{hyp}}{\text{adj}}$$

$$\tan\theta = \frac{\text{opp}}{\text{adj}} \qquad \cot\theta = \frac{\text{adj}}{\text{opp}}$$

FIGURE 1.39   Trigonometric ratios of an acute angle.

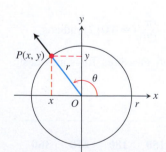

FIGURE 1.40   The trigonometric functions of a general angle θ are defined in terms of *x*, *y*, and *r*.

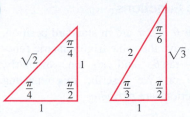
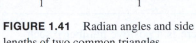

FIGURE 1.41 Radian angles and side lengths of two common triangles.

As you can see, $\tan \theta$ and $\sec \theta$ are not defined if $x = \cos \theta = 0$. This means they are not defined if θ is $\pm \pi/2$, $\pm 3\pi/2$, ... Similarly, $\cot \theta$ and $\csc \theta$ are not defined for values of θ for which $y = 0$, namely $\theta = 0$, $\pm \pi$, $\pm 2\pi$, ...

The exact values of these trigonometric ratios for some angles can be read from the triangles in Figure 1.41. For instance,

$$\sin \frac{\pi}{4} = \frac{1}{\sqrt{2}} \qquad \sin \frac{\pi}{6} = \frac{1}{2} \qquad \sin \frac{\pi}{3} = \frac{\sqrt{3}}{2}$$

$$\cos \frac{\pi}{4} = \frac{1}{\sqrt{2}} \qquad \cos \frac{\pi}{6} = \frac{\sqrt{3}}{2} \qquad \cos \frac{\pi}{3} = \frac{1}{2}$$

$$\tan \frac{\pi}{4} = 1 \qquad \tan \frac{\pi}{6} = \frac{1}{\sqrt{3}} \qquad \tan \frac{\pi}{3} = \sqrt{3}$$

The ASTC rule (Figure 1.42) is useful for remembering when the basic trigonometric functions are positive or negative. For instance, from the triangle in Figure 1.43, we see that

$$\sin \frac{2\pi}{3} = \frac{\sqrt{3}}{2}, \qquad \cos \frac{2\pi}{3} = -\frac{1}{2}, \qquad \tan \frac{2\pi}{3} = -\sqrt{3}.$$

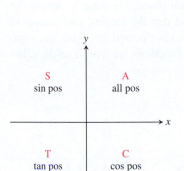

FIGURE 1.42 The ASTC rule, remembered by the statement "All Students Take Calculus," tells which trigonometric functions are positive in each quadrant.

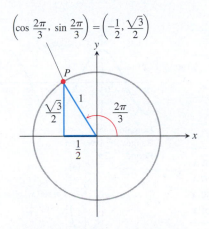

FIGURE 1.43 The triangle for calculating the sine and cosine of $2\pi/3$ radians. The side lengths come from the geometry of right triangles.

Using a similar method we obtain the values of $\sin \theta$, $\cos \theta$, and $\tan \theta$ shown in Table 1.2.

TABLE 1.2 Values of $\sin \theta$, $\cos \theta$, and $\tan \theta$ for selected values of θ

Degrees	-180	-135	-90	-45	0	30	45	60	90	120	135	150	180	270	360
θ (radians)	$-\pi$	$\dfrac{-3\pi}{4}$	$\dfrac{-\pi}{2}$	$\dfrac{-\pi}{4}$	0	$\dfrac{\pi}{6}$	$\dfrac{\pi}{4}$	$\dfrac{\pi}{3}$	$\dfrac{\pi}{2}$	$\dfrac{2\pi}{3}$	$\dfrac{3\pi}{4}$	$\dfrac{5\pi}{6}$	π	$\dfrac{3\pi}{2}$	2π
$\sin \theta$	0	$\dfrac{-\sqrt{2}}{2}$	-1	$\dfrac{-\sqrt{2}}{2}$	0	$\dfrac{1}{2}$	$\dfrac{\sqrt{2}}{2}$	$\dfrac{\sqrt{3}}{2}$	1	$\dfrac{\sqrt{3}}{2}$	$\dfrac{\sqrt{2}}{2}$	$\dfrac{1}{2}$	0	-1	0
$\cos \theta$	-1	$\dfrac{-\sqrt{2}}{2}$	0	$\dfrac{\sqrt{2}}{2}$	1	$\dfrac{\sqrt{3}}{2}$	$\dfrac{\sqrt{2}}{2}$	$\dfrac{1}{2}$	0	$-\dfrac{1}{2}$	$\dfrac{-\sqrt{2}}{2}$	$\dfrac{-\sqrt{3}}{2}$	-1	0	1
$\tan \theta$	0	1		-1	0	$\dfrac{\sqrt{3}}{3}$	1	$\sqrt{3}$		$-\sqrt{3}$	-1	$\dfrac{-\sqrt{3}}{3}$	0		0

Periodicity and Graphs of the Trigonometric Functions

When an angle of measure θ and an angle of measure $\theta + 2\pi$ are in standard position, their terminal rays coincide. The two angles therefore have the same trigonometric function values: $\sin(\theta + 2\pi) = \sin\theta$, $\tan(\theta + 2\pi) = \tan\theta$, and so on. Similarly, $\cos(\theta - 2\pi) = \cos\theta$, $\sin(\theta - 2\pi) = \sin\theta$, and so on. We describe this repeating behavior by saying that the six basic trigonometric functions are *periodic*.

Periods of Trigonometric Functions

Period π:
$\tan(x + \pi) = \tan x$
$\cot(x + \pi) = \cot x$

Period 2π:
$\sin(x + 2\pi) = \sin x$
$\cos(x + 2\pi) = \cos x$
$\sec(x + 2\pi) = \sec x$
$\csc(x + 2\pi) = \csc x$

> **DEFINITION**   A function $f(x)$ is **periodic** if there is a positive number p such that $f(x + p) = f(x)$ for every value of x. The smallest such value of p is the **period** of f.

When we graph trigonometric functions in the coordinate plane, we usually denote the independent variable by x instead of θ. Figure 1.44 shows that the tangent and cotangent functions have period $p = \pi$, and the other four functions have period 2π. Also, the symmetries in these graphs reveal that the cosine and secant functions are even and the other four functions are odd (although this does not prove those results).

Even

$\cos(-x) = \cos x$
$\sec(-x) = \sec x$

Odd

$\sin(-x) = -\sin x$
$\tan(-x) = -\tan x$
$\csc(-x) = -\csc x$
$\cot(-x) = -\cot x$

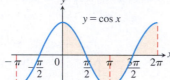

Domain: $-\infty < x < \infty$
Range: $-1 \le y \le 1$
Period: 2π
(a)

Domain: $-\infty < x < \infty$
Range: $-1 \le y \le 1$
Period: 2π
(b)

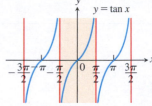

Domain: $x \ne \pm\dfrac{\pi}{2},\ \pm\dfrac{3\pi}{2}, \ldots$
Range: $-\infty < y < \infty$
Period: π (c)

Domain: $x \ne \pm\dfrac{\pi}{2},\ \pm\dfrac{3\pi}{2}, \ldots$
Range: $y \le -1$ or $y \ge 1$
Period: 2π

(d)

Domain: $x \ne 0,\ \pm\pi,\ \pm 2\pi, \ldots$
Range: $y \le -1$ or $y \ge 1$
Period: 2π

(e)

Domain: $x \ne 0,\ \pm\pi,\ \pm 2\pi, \ldots$
Range: $-\infty < y < \infty$
Period: π

(f)

FIGURE 1.44   Graphs of the six basic trigonometric functions using radian measure. The shading for each trigonometric function indicates its periodicity.

Trigonometric Identities

The coordinates of any point $P(x, y)$ in the plane can be expressed in terms of the point's distance r from the origin and the angle θ that ray OP makes with the positive x-axis (Figure 1.40). Since $x/r = \cos\theta$ and $y/r = \sin\theta$, we have

$$x = r\cos\theta, \qquad y = r\sin\theta.$$

When $r = 1$ we can apply the Pythagorean theorem to the reference right triangle in Figure 1.45 and obtain the equation

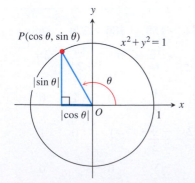

FIGURE 1.45   The reference triangle for a general angle θ.

$$\cos^2\theta + \sin^2\theta = 1. \tag{3}$$

This equation, true for all values of θ, is the most frequently used identity in trigonometry. Dividing this identity in turn by $\cos^2 \theta$ and $\sin^2 \theta$ gives

$$1 + \tan^2 \theta = \sec^2 \theta$$
$$1 + \cot^2 \theta = \csc^2 \theta$$

The following formulas hold for all angles A and B (Exercise 58).

Addition Formulas

$$\cos(A + B) = \cos A \cos B - \sin A \sin B$$
$$\sin(A + B) = \sin A \cos B + \cos A \sin B$$

(4)

There are similar formulas for $\cos(A - B)$ and $\sin(A - B)$ (Exercises 35 and 36). All the trigonometric identities needed in this book derive from Equations (3) and (4). For example, substituting θ for both A and B in the addition formulas gives

Double-Angle Formulas

$$\cos 2\theta = \cos^2 \theta - \sin^2 \theta$$
$$\sin 2\theta = 2 \sin \theta \cos \theta$$

(5)

Additional formulas come from combining the equations

$$\cos^2 \theta + \sin^2 \theta = 1, \qquad \cos^2 \theta - \sin^2 \theta = \cos 2\theta.$$

We add the two equations to get $2\cos^2 \theta = 1 + \cos 2\theta$ and subtract the second from the first to get $2\sin^2 \theta = 1 - \cos 2\theta$. This results in the following identities, which are useful in integral calculus.

Half-Angle Formulas

$$\cos^2 \theta = \frac{1 + \cos 2\theta}{2}$$

(6)

$$\sin^2 \theta = \frac{1 - \cos 2\theta}{2}$$

(7)

The Law of Cosines

If a, b, and c are sides of a triangle ABC and if θ is the angle opposite c, then

$$c^2 = a^2 + b^2 - 2ab \cos \theta.$$

(8)

This equation is called the **law of cosines**.

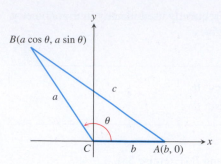

FIGURE 1.46 The square of the distance between A and B gives the law of cosines.

To see why the law holds, we position the triangle in the xy-plane with the origin at C and the positive x-axis along one side of the triangle, as in Figure 1.46. The coordinates of A are $(b, 0)$; the coordinates of B are $(a\cos\theta, a\sin\theta)$. The square of the distance between A and B is therefore

$$c^2 = (a\cos\theta - b)^2 + (a\sin\theta)^2$$
$$= a^2\underbrace{(\cos^2\theta + \sin^2\theta)}_{1} + b^2 - 2ab\cos\theta$$
$$= a^2 + b^2 - 2ab\cos\theta.$$

The law of cosines generalizes the Pythagorean theorem. If $\theta = \pi/2$, then $\cos\theta = 0$ and $c^2 = a^2 + b^2$.

Two Special Inequalities

For any angle θ measured in radians, the sine and cosine functions satisfy

$$-|\theta| \le \sin\theta \le |\theta| \qquad \text{and} \qquad -|\theta| \le 1 - \cos\theta \le |\theta|.$$

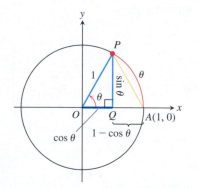

FIGURE 1.47 From the geometry of this figure, drawn for $\theta > 0$, we get the inequality $\sin^2\theta + (1 - \cos\theta)^2 \le \theta^2$.

To establish these inequalities, we picture θ as a nonzero angle in standard position (Figure 1.47). The circle in the figure is a unit circle, so $|\theta|$ equals the length of the circular arc AP. The length of line segment AP is therefore less than $|\theta|$.

Triangle APQ is a right triangle with sides of length

$$QP = |\sin\theta|, \qquad AQ = 1 - \cos\theta.$$

From the Pythagorean theorem and the fact that $AP < |\theta|$, we get

$$\sin^2\theta + (1 - \cos\theta)^2 = (AP)^2 \le \theta^2. \tag{9}$$

The terms on the left-hand side of Equation (9) are both positive, so each is smaller than their sum and hence is less than or equal to θ^2:

$$\sin^2\theta \le \theta^2 \qquad \text{and} \qquad (1 - \cos\theta)^2 \le \theta^2.$$

By taking square roots, this is equivalent to saying that

$$|\sin\theta| \le |\theta| \qquad \text{and} \qquad |1 - \cos\theta| \le |\theta|,$$

so

$$-|\theta| \le \sin\theta \le |\theta| \qquad \text{and} \qquad -|\theta| \le 1 - \cos\theta \le |\theta|.$$

These inequalities will be useful in the next chapter.

Transformations of Trigonometric Graphs

The rules for shifting, stretching, compressing, and reflecting the graph of a function summarized in the following diagram apply to the trigonometric functions we have discussed in this section.

Vertical stretch or compression;
reflection about $y = d$ if negative

Vertical shift

$$y = af(b(x + c)) + d$$

Horizontal stretch or compression;
reflection about $x = -c$ if negative

Horizontal shift

The transformation rules applied to the sine function give the **general sine function** or **sinusoid** formula

$$f(x) = A \sin\left(\frac{2\pi}{B}(x - C)\right) + D,$$

where $|A|$ is the *amplitude*, $|B|$ is the *period*, C is the *horizontal shift*, and D is the *vertical shift*. A graphical interpretation of the various terms is given below.

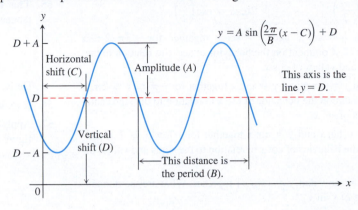

EXERCISES   1.3

Radians and Degrees

1. On a circle of radius 10 m, how long is an arc that subtends a central angle of (**a**) $4\pi/5$ radians? (**b**) $110°$?

2. A central angle in a circle of radius 8 is subtended by an arc of length 10π. Find the angle's radian and degree measures.

3. You want to make an $80°$ angle by marking an arc on the perimeter of a 12-in.-diameter disk and drawing lines from the ends of the arc to the disk's center. To the nearest tenth of an inch, how long should the arc be?

4. If you roll a 1-m-diameter wheel forward 30 cm over level ground, through what angle will the wheel turn? Answer in radians (to the nearest tenth) and degrees (to the nearest degree).

Evaluating Trigonometric Functions

5. Copy and complete the following table of function values. If the function is undefined at a given angle, enter "UND." Do not use a calculator or tables.

θ	$-\pi$	$-2\pi/3$	0	$\pi/2$	$3\pi/4$
$\sin \theta$					
$\cos \theta$					
$\tan \theta$					
$\cot \theta$					
$\sec \theta$					
$\csc \theta$					

6. Copy and complete the following table of function values. If the function is undefined at a given angle, enter "UND." Do not use a calculator or tables.

θ	$-3\pi/2$	$-\pi/3$	$-\pi/6$	$\pi/4$	$5\pi/6$
$\sin \theta$					
$\cos \theta$					
$\tan \theta$					
$\cot \theta$					
$\sec \theta$					
$\csc \theta$					

In Exercises 7–12, one of $\sin x$, $\cos x$, and $\tan x$ is given. Find the other two if x lies in the specified interval.

7. $\sin x = \dfrac{3}{5}$, $x \in \left[\dfrac{\pi}{2}, \pi\right]$ 8. $\tan x = 2$, $x \in \left[0, \dfrac{\pi}{2}\right]$

9. $\cos x = \dfrac{1}{3}$, $x \in \left[-\dfrac{\pi}{2}, 0\right]$ 10. $\cos x = -\dfrac{5}{13}$, $x \in \left[\dfrac{\pi}{2}, \pi\right]$

11. $\tan x = \dfrac{1}{2}$, $x \in \left[\pi, \dfrac{3\pi}{2}\right]$ 12. $\sin x = -\dfrac{1}{2}$, $x \in \left[\pi, \dfrac{3\pi}{2}\right]$

Graphing Trigonometric Functions

Graph the functions in Exercises 13–22. What is the period of each function?

13. $\sin 2x$ 14. $\sin(x/2)$

15. $\cos \pi x$ 16. $\cos \dfrac{\pi x}{2}$

17. $-\sin \dfrac{\pi x}{3}$ 18. $-\cos 2\pi x$

19. $\cos\left(x - \dfrac{\pi}{2}\right)$ 20. $\sin\left(x + \dfrac{\pi}{6}\right)$

21. $\sin\left(x - \dfrac{\pi}{4}\right) + 1$ **22.** $\cos\left(x + \dfrac{2\pi}{3}\right) - 2$

Graph the functions in Exercises 23–26 in the *ts*-plane (*t*-axis horizontal, *s*-axis vertical). What is the period of each function? What symmetries do the graphs have?

23. $s = \cot 2t$ **24.** $s = -\tan \pi t$

25. $s = \sec\left(\dfrac{\pi t}{2}\right)$ **26.** $s = \csc\left(\dfrac{t}{2}\right)$

T **27. a.** Graph $y = \cos x$ and $y = \sec x$ together for $-3\pi/2 \le x \le 3\pi/2$. Comment on the behavior of $\sec x$ in relation to the signs and values of $\cos x$.

     b. Graph $y = \sin x$ and $y = \csc x$ together for $-\pi \le x \le 2\pi$. Comment on the behavior of $\csc x$ in relation to the signs and values of $\sin x$.

T **28.** Graph $y = \tan x$ and $y = \cot x$ together for $-7 \le x \le 7$. Comment on the behavior of $\cot x$ in relation to the signs and values of $\tan x$.

29. Graph $y = \sin x$ and $y = \lfloor \sin x \rfloor$ together. What are the domain and range of $\lfloor \sin x \rfloor$?

30. Graph $y = \sin x$ and $y = \lceil \sin x \rceil$ together. What are the domain and range of $\lceil \sin x \rceil$?

Using the Addition Formulas

Use the addition formulas to derive the identities in Exercises 31–36.

31. $\cos\left(x - \dfrac{\pi}{2}\right) = \sin x$ **32.** $\cos\left(x + \dfrac{\pi}{2}\right) = -\sin x$

33. $\sin\left(x + \dfrac{\pi}{2}\right) = \cos x$ **34.** $\sin\left(x - \dfrac{\pi}{2}\right) = -\cos x$

35. $\cos(A - B) = \cos A \cos B + \sin A \sin B$ (Exercise 57 provides a different derivation.)

36. $\sin(A - B) = \sin A \cos B - \cos A \sin B$

37. What happens if you take $B = A$ in the trigonometric identity $\cos(A - B) = \cos A \cos B + \sin A \sin B$? Does the result agree with something you already know?

38. What happens if you take $B = 2\pi$ in the addition formulas? Do the results agree with something you already know?

In Exercises 39–42, express the given quantity in terms of $\sin x$ and $\cos x$.

39. $\cos(\pi + x)$ **40.** $\sin(2\pi - x)$

41. $\sin\left(\dfrac{3\pi}{2} - x\right)$ **42.** $\cos\left(\dfrac{3\pi}{2} + x\right)$

43. Evaluate $\sin \dfrac{7\pi}{12}$ as $\sin\left(\dfrac{\pi}{4} + \dfrac{\pi}{3}\right)$.

44. Evaluate $\cos \dfrac{11\pi}{12}$ as $\cos\left(\dfrac{\pi}{4} + \dfrac{2\pi}{3}\right)$.

45. Evaluate $\cos \dfrac{\pi}{12}$.   **46.** Evaluate $\sin \dfrac{5\pi}{12}$.

Using the Half-Angle Formulas

Find the function values in Exercises 47–50.

47. $\cos^2 \dfrac{\pi}{8}$ **48.** $\cos^2 \dfrac{5\pi}{12}$

49. $\sin^2 \dfrac{\pi}{12}$ **50.** $\sin^2 \dfrac{3\pi}{8}$

Solving Trigonometric Equations

For Exercises 51–54, solve for the angle θ, where $0 \le \theta \le 2\pi$.

51. $\sin^2 \theta = \dfrac{3}{4}$ **52.** $\sin^2 \theta = \cos^2 \theta$

53. $\sin 2\theta - \cos \theta = 0$ **54.** $\cos 2\theta + \cos \theta = 0$

Theory and Examples

55. The tangent sum formula The standard formula for the tangent of the sum of two angles is

$$\tan(A + B) = \frac{\tan A + \tan B}{1 - \tan A \tan B}.$$

Derive the formula.

56. (*Continuation of Exercise 55.*) Derive a formula for $\tan(A - B)$.

57. Apply the law of cosines to the triangle in the accompanying figure to derive the formula for $\cos(A - B)$.

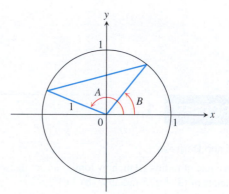

58. a. Apply the formula for $\cos(A - B)$ to the identity $\sin \theta = \cos\left(\dfrac{\pi}{2} - \theta\right)$ to obtain the addition formula for $\sin(A + B)$.

     b. Derive the formula for $\cos(A + B)$ by substituting $-B$ for B in the formula for $\cos(A - B)$ from Exercise 35.

59. A triangle has sides $a = 2$ and $b = 3$ and angle $C = 60°$. Find the length of side c.

60. A triangle has sides $a = 2$ and $b = 3$ and angle $C = 40°$. Find the length of side c.

61. The law of sines *The law of sines* says that if a, b, and c are the sides opposite the angles A, B, and C in a triangle, then

$$\frac{\sin A}{a} = \frac{\sin B}{b} = \frac{\sin C}{c}.$$

Use the accompanying figures and the identity $\sin(\pi - \theta) = \sin \theta$, if required, to derive the law.

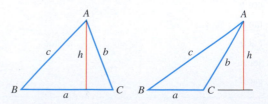

62. A triangle has sides $a = 2$ and $b = 3$ and angle $C = 60°$ (as in Exercise 59). Find the sine of angle B using the law of sines.

63. A triangle has side $c = 2$ and angles $A = \pi/4$ and $B = \pi/3$. Find the length a of the side opposite A.

64. Consider the length h of the perpendicular from point B to side b in the given triangle. Show that

$$h = \frac{b \tan \alpha \tan \gamma}{\tan \alpha + \tan \gamma}$$

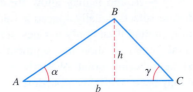

65. Refer to the given figure. Write the radius r of the circle in terms of α and θ.

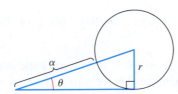

T **66. The approximation $\sin x \approx x$** It is often useful to know that, when x is measured in radians, $\sin x \approx x$ for numerically small values of x. In Section 3.11, we will see why the approximation holds. The approximation error is less than 1 in 5000 if $|x| < 0.1$.

   a. With your grapher in radian mode, graph $y = \sin x$ and $y = x$ together in a viewing window about the origin. What do you see happening as x nears the origin?

   b. With your grapher in degree mode, graph $y = \sin x$ and $y = x$ together about the origin again. How is the picture different from the one obtained with radian mode?

General Sine Curves

For

$$f(x) = A \sin\left(\frac{2\pi}{B}(x - C)\right) + D,$$

identify A, B, C, and D for the sine functions in Exercises 67–70 and sketch their graphs.

67. $y = 2 \sin(x + \pi) - 1$

68. $y = \frac{1}{2} \sin(\pi x - \pi) + \frac{1}{2}$

69. $y = -\frac{2}{\pi} \sin\left(\frac{\pi}{2} t\right) + \frac{1}{\pi}$

70. $y = \frac{L}{2\pi} \sin \frac{2\pi t}{L}, \quad L > 0$

COMPUTER EXPLORATIONS

In Exercises 71–74, you will explore graphically the general sine function

$$f(x) = A \sin\left(\frac{2\pi}{B}(x - C)\right) + D$$

as you change the values of the constants A, B, C, and D. Use a CAS or computer grapher to perform the steps in the exercises.

71. The period B Set the constants $A = 3$, $C = D = 0$.

   a. Plot $f(x)$ for the values $B = 1, 3, 2\pi, 5\pi$ over the interval $-4\pi \le x \le 4\pi$. Describe what happens to the graph of the general sine function as the period increases.

   b. What happens to the graph for negative values of B? Try it with $B = -3$ and $B = -2\pi$.

72. The horizontal shift C Set the constants $A = 3$, $B = 6$, $D = 0$.

   a. Plot $f(x)$ for the values $C = 0, 1$, and 2 over the interval $-4\pi \le x \le 4\pi$. Describe what happens to the graph of the general sine function as C increases through positive values.

   b. What happens to the graph for negative values of C?

   c. What smallest positive value should be assigned to C so the graph exhibits no horizontal shift? Confirm your answer with a plot.

73. The vertical shift D Set the constants $A = 3$, $B = 6$, $C = 0$.

   a. Plot $f(x)$ for the values $D = 0, 1$, and 3 over the interval $-4\pi \le x \le 4\pi$. Describe what happens to the graph of the general sine function as D increases through positive values.

   b. What happens to the graph for negative values of D?

74. The amplitude A Set the constants $B = 6$, $C = D = 0$.

   a. Describe what happens to the graph of the general sine function as A increases through positive values. Confirm your answer by plotting $f(x)$ for the values $A = 1, 5$, and 9.

   b. What happens to the graph for negative values of A?

1.4 Graphing with Software

Many computers, calculators, and smartphones have graphing applications that enable us to graph very complicated functions with high precision. Many of these functions could not otherwise be easily graphed. However, some care must be taken when using such graphing software, and in this section we address some of the issues that can arise. In Chapter 4 we will see how calculus helps us determine that we are accurately viewing the important features of a function's graph.

Graphing Windows

When software is used for graphing, a portion of the graph is visible in a **display** or **viewing window**. Depending on the software, the default window may give an incomplete or misleading picture of the graph. We use the term *square window* when the units or scales used on both axes are the same. This term does not mean that the display window itself is square (usually it is rectangular), but instead it means that the *x*-unit is the same length as the *y*-unit.

When a graph is displayed in the default mode, the *x*-unit may differ from the *y*-unit of scaling in order to capture essential features of the graph. This difference in scaling can cause visual distortions that may lead to erroneous interpretations of the function's behavior. Some graphing software allows us to set the viewing window by specifying one or both of the intervals, $a \leq x \leq b$ and $c \leq y \leq d$, and it may allow for equalizing the scales used for the axes as well. The software selects equally spaced *x*-values in $[a, b]$ and then plots the points $(x, f(x))$. A point is plotted if and only if *x* lies in the domain of the function and $f(x)$ lies within the interval $[c, d]$. A short line segment is then drawn between each plotted point and its next neighboring point. We now give illustrative examples of some common problems that may occur with this procedure.

EXAMPLE 1   Graph the function $f(x) = x^3 - 7x^2 + 28$ in each of the following display or viewing windows:

(a) $[-10, 10]$ by $[-10, 10]$ **(b)** $[-4, 4]$ by $[-50, 10]$ **(c)** $[-4, 10]$ by $[-60, 60]$

Solution

(a) We select $a = -10$, $b = 10$, $c = -10$, and $d = 10$ to specify the interval of *x*-values and the range of *y*-values for the window. The resulting graph is shown in Figure 1.48a. It appears that the window is cutting off the bottom part of the graph and that the interval of *x*-values is too large. Let's try the next window.

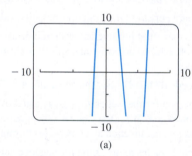
(a)

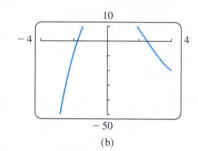
(b)

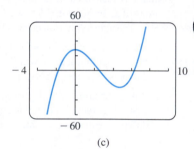
(c)

FIGURE 1.48   The graph of $f(x) = x^3 - 7x^2 + 28$ in different viewing windows. Selecting a window that gives a clear picture of a graph is often a trial-and-error process (Example 1). The default window used by the software may automatically display the graph in (c).

(b) We see some new features of the graph (Figure 1.48b), but the top is missing and we need to view more to the right of $x = 4$ as well. The next window should help.

(c) Figure 1.48c shows the graph in this new viewing window. Observe that we get a more complete picture of the graph in this window, and it is a reasonable graph of a third-degree polynomial.   ■

EXAMPLE 2   When a graph is displayed, the *x*-unit may differ from the *y*-unit, as in the graphs shown in Figures 1.48b and 1.48c. The result is distortion in the picture, which may be misleading. The display window can be made square by compressing or stretching the units on one axis to match the scale on the other, giving the true graph. Many software systems have built-in options to make the window "square." If yours does not, you may have to bring to your viewing some foreknowledge of the true picture.

Figure 1.49a shows the graphs of the perpendicular lines $y = x$ and $y = -x + 3\sqrt{2}$, together with the semicircle $y = \sqrt{9 - x^2}$, in a nonsquare $[-4, 4]$ by $[-6, 8]$ display window. Notice the distortion. The lines do not appear to be perpendicular, and the semicircle appears to be elliptical in shape.

Figure 1.49b shows the graphs of the same functions in a square window in which the x-units are scaled to be the same as the y-units. Notice that the scaling on the x-axis for Figure 1.49a has been compressed in Figure 1.49b to make the window square. Figure 1.49c gives an enlarged view of Figure 1.49b with a square $[-3, 3]$ by $[0, 4]$ window.   ■

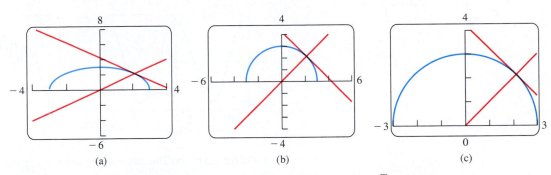

FIGURE 1.49 Graphs of the perpendicular lines $y = x$ and $y = -x + 3\sqrt{2}$ and of the semicircle $y = \sqrt{9 - x^2}$ appear distorted (a) in a nonsquare window, but clear (b) and (c) in square windows (Example 2). Some software may not provide options for the views in (b) or (c).

If the denominator of a rational function is zero at some x-value within the viewing window, graphing software may produce a steep near-vertical line segment from the top to the bottom of the window. Example 3 illustrates steep line segments.

Sometimes the graph of a trigonometric function oscillates very rapidly. When graphing software plots the points of the graph and connects them, many of the maximum and minimum points are actually missed. The resulting graph is then very misleading.

EXAMPLE 3   Graph the function $f(x) = \sin 100x$.

Solution   Figure 1.50a shows the graph of f in the viewing window $[-12, 12]$ by $[-1, 1]$. We see that the graph looks very strange because the sine curve should oscillate periodically between -1 and 1. This behavior is not exhibited in Figure 1.50a. We might experiment with a smaller viewing window, say $[-6, 6]$ by $[-1, 1]$, but the graph is not better (Figure 1.50b). The difficulty is that the period of the trigonometric function $y = \sin 100x$ is very small ($2\pi/100 \approx 0.063$). If we choose the much smaller viewing window $[-0.1, 0.1]$ by $[-1, 1]$ we get the graph shown in Figure 1.50c. This graph reveals the expected oscillations of a sine curve.   ■

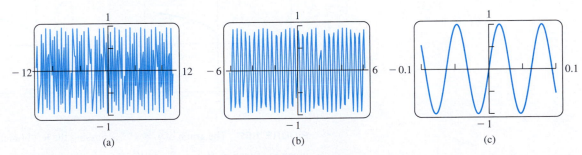

FIGURE 1.50 Graphs of the function $y = \sin 100x$ in three viewing windows. Because the period is $2\pi/100 \approx 0.063$, the smaller window in (c) best displays the true aspects of this rapidly oscillating function (Example 3).

EXAMPLE 4   Graph the function $y = \cos x + \dfrac{1}{200} \sin 200x$.

Solution   In the viewing window $[-6, 6]$ by $[-1, 1]$ the graph appears much like the cosine function with some very small sharp wiggles on it (Figure 1.51a). We get a better look when we significantly reduce the window to $[-0.2, 0.2]$ by $[0.97, 1.01]$, obtaining the graph in Figure 1.51b. We now see the small but rapid oscillations of the second term, $(1/200)\sin 200x$, added to the comparatively larger values of the cosine curve.

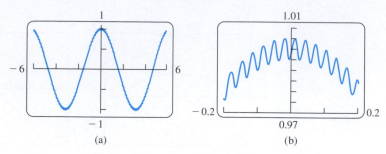

(a)   (b)

FIGURE 1.51   (a) The function $y = \cos x + \dfrac{1}{200}\sin 200x$. (b) A close up view, blown up near the y-axis. The term $\cos x$ clearly dominates the second term, $\dfrac{1}{200}\sin 200x$, which produces the rapid oscillations along the cosine curve. Both views are needed for a clear idea of the graph (Example 4).

Obtaining a Complete Graph

Some graphing software will not display the portion of a graph for $f(x)$ when $x < 0$. Usually that happens because of the algorithm the software is using to calculate the function values. Sometimes we can obtain the complete graph by defining the formula for the function in a different way, as illustrated in the next example.

EXAMPLE 5   Graph the function $y = x^{1/3}$.

Solution   Some graphing software displays the graph shown in Figure 1.52a. When we compare it with the graph of $y = x^{1/3} = \sqrt[3]{x}$ in Figure 1.17, we see that the left branch for $x < 0$ is missing. The reason the graphs differ is that the software algorithm calculates $x^{1/3}$ as $e^{(1/3)\ln x}$. Since the logarithmic function is not defined for negative values of x, the software can produce only the right branch, where $x > 0$. (Logarithmic and exponential functions are introduced in the next two sections.)

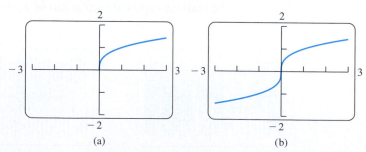

(a)   (b)

FIGURE 1.52   The graph of $y = x^{1/3}$ is missing the left branch in (a). In (b) we graph the function $f(x) = \dfrac{x}{|x|} \cdot |x|^{1/3}$, obtaining both branches. (See Example 5.)

To obtain the full picture showing both branches, we can graph the function

$$f(x) = \frac{x}{|x|} \cdot |x|^{1/3}.$$

This function equals $x^{1/3}$ except at $x = 0$ (where f is undefined, although $0^{1/3} = 0$). A graph of f is displayed in Figure 1.52b.

EXERCISES 1.4

Choosing a Viewing Window

T In Exercises 1–4, use graphing software to determine which of the given viewing windows displays the most appropriate graph of the specified function.

1. $f(x) = x^4 - 7x^2 + 6x$

  a. $[-1, 1]$ by $[-1, 1]$
  b. $[-2, 2]$ by $[-5, 5]$
  c. $[-10, 10]$ by $[-10, 10]$
  d. $[-5, 5]$ by $[-25, 15]$

2. $f(x) = x^3 - 4x^2 - 4x + 16$

  a. $[-1, 1]$ by $[-5, 5]$
  b. $[-3, 3]$ by $[-10, 10]$
  c. $[-5, 5]$ by $[-10, 20]$
  d. $[-20, 20]$ by $[-100, 100]$

3. $f(x) = 5 + 12x - x^3$

  a. $[-1, 1]$ by $[-1, 1]$
  b. $[-5, 5]$ by $[-10, 10]$
  c. $[-4, 4]$ by $[-20, 20]$
  d. $[-4, 5]$ by $[-15, 25]$

4. $f(x) = \sqrt{5 + 4x - x^2}$

  a. $[-2, 2]$ by $[-2, 2]$
  b. $[-2, 6]$ by $[-1, 4]$
  c. $[-3, 7]$ by $[0, 10]$
  d. $[-10, 10]$ by $[-10, 10]$

Finding a Viewing Window

T In Exercises 5–30, find an appropriate graphing software viewing window for the given function and use it to display its graph. The window should give a picture of the overall behavior of the function. There is more than one choice, but incorrect choices can miss important aspects of the function.

5. $f(x) = x^4 - 4x^3 + 15$

6. $f(x) = \dfrac{x^3}{3} - \dfrac{x^2}{2} - 2x + 1$

7. $f(x) = x^5 - 5x^4 + 10$

8. $f(x) = 4x^3 - x^4$

9. $f(x) = x\sqrt{9 - x^2}$

10. $f(x) = x^2(6 - x^3)$

11. $y = 2x - 3x^{2/3}$

12. $y = x^{1/3}(x^2 - 8)$

13. $y = 5x^{2/5} - 2x$

14. $y = x^{2/3}(5 - x)$

15. $y = |x^2 - 1|$

16. $y = |x^2 - x|$

17. $y = \dfrac{x + 3}{x + 2}$

18. $y = 1 - \dfrac{1}{x + 3}$

19. $f(x) = \dfrac{x^2 + 2}{x^2 + 1}$

20. $f(x) = \dfrac{x^2 - 1}{x^2 + 1}$

21. $f(x) = \dfrac{x - 1}{x^2 - x - 6}$

22. $f(x) = \dfrac{8}{x^2 - 9}$

23. $f(x) = \dfrac{6x^2 - 15x + 6}{4x^2 - 10x}$

24. $f(x) = \dfrac{x^2 - 3}{x - 2}$

25. $y = \sin 250x$

26. $y = 3 \cos 60x$

27. $y = \cos\left(\dfrac{x}{50}\right)$

28. $y = \dfrac{1}{10}\sin\left(\dfrac{x}{10}\right)$

29. $y = x + \dfrac{1}{10}\sin 30x$

30. $y = x^2 + \dfrac{1}{50}\cos 100x$

Use graphing software to graph the functions specified in Exercises 31–36. Select a viewing window that reveals the key features of the function.

31. Graph the lower half of the circle defined by the equation $x^2 + 2x = 4 + 4y - y^2$.

32. Graph the upper branch of the hyperbola $y^2 - 16x^2 = 1$.

33. Graph four periods of the function $f(x) = -\tan 2x$.

34. Graph two periods of the function $f(x) = 3\cot\dfrac{x}{2} + 1$.

35. Graph the function $f(x) = \sin 2x + \cos 3x$.

36. Graph the function $f(x) = \sin^3 x$.

1.5 Exponential Functions

Exponential functions occur in a wide variety of applications, including interest rates, radioactive decay, population growth, the spread of a disease, consumption of natural resources, the earth's atmospheric pressure, temperature change of a heated object placed in a cooler environment, and the dating of fossils. In this section we introduce these functions informally, using an intuitive approach. We give a rigorous development of them in Chapter 7, based on the ideas of integral calculus.

Exponential Behavior

When a positive quantity P doubles, it increases by a factor of 2 and the quantity becomes $2P$. If it doubles again, it becomes $2(2P) = 2^2P$, and a third doubling gives $2(2^2P) = 2^3P$. Continuing to double in this fashion leads us to consider the function $f(x) = 2^x$. We call this an *exponential* function because the variable x appears in the exponent of 2^x. Functions such as $g(x) = 10^x$ and $h(x) = (1/2)^x$ are other examples of exponential functions. In general, if $a \neq 1$ is a positive constant, the function

$$f(x) = a^x, \quad a > 0$$

is the **exponential function with base** a.

Don't confuse the exponential 2^x with the power function x^2. In the exponential, the variable x is in the exponent, whereas the variable x is the base in the power function.

EXAMPLE 1   In 2014, \$100 is invested in a savings account, where it grows by accruing interest that is compounded annually (once a year) at an interest rate of 5.5%. Assuming no additional funds are deposited to the account and no money is withdrawn, give a formula for a function describing the amount A in the account after x years have elapsed.

Solution   If $P = 100$, at the end of the first year the amount in the account is the original amount plus the interest accrued, or

$$P + \left(\frac{5.5}{100}\right)P = (1 + 0.055)P = (1.055)P.$$

At the end of the second year the account earns interest again and grows to

$$(1 + 0.055) \cdot (1.055P) = (1.055)^2 P = 100 \cdot (1.055)^2. \qquad {\color{blue}P = 100}$$

Continuing this process, after x years the value of the account is

$$A = 100 \cdot (1.055)^x.$$

This is a multiple of the exponential function $f(x) = (1.055)^x$ with base 1.055. Table 1.3 shows the amounts accrued over the first four years. Notice that the amount in the account each year is always 1.055 times its value in the previous year.

TABLE 1.3 Savings account growth

Year	Amount (dollars)	Yearly increase
2014	100	
2015	$100(1.055) = 105.50$	5.50
2016	$100(1.055)^2 = 111.30$	5.80
2017	$100(1.055)^3 = 117.42$	6.12
2018	$100(1.055)^4 = 123.88$	6.46

In general, the amount after x years is given by $P(1 + r)^x$, where P is the starting amount and r is the interest rate (expressed as a decimal).   ■

For integer and rational exponents, the value of an exponential function $f(x) = a^x$ is obtained arithmetically by taking an appropriate number of products, quotients, or roots. If $x = n$ is a positive integer, the number a^n is given by multiplying a by itself n times:

$$a^n = \underbrace{a \cdot a \cdot \cdots \cdot a}_{n \text{ factors}}.$$

If $x = 0$, then we set $a^0 = 1$, and if $x = -n$ for some positive integer n, then

$$a^{-n} = \frac{1}{a^n} = \left(\frac{1}{a}\right)^n.$$

If $x = 1/n$ for some positive integer n, then

$$a^{1/n} = \sqrt[n]{a},$$

which is the positive number that when multiplied by itself n times gives a. If $x = p/q$ is any rational number, then

$$a^{p/q} = \sqrt[q]{a^p} = \left(\sqrt[q]{a}\right)^p.$$

When x is *irrational*, the meaning of a^x is not immediately apparent. The value of a^x can be approximated by raising a to rational numbers that get closer and closer to the irrational

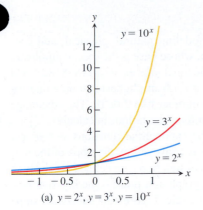

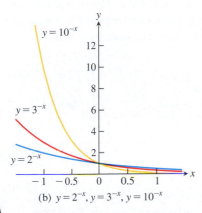

FIGURE 1.53 Graphs of exponential functions.

TABLE 1.4 Values of $2^{\sqrt{3}}$ for rational r closer and closer to $\sqrt{3}$

r	2^r
1.0	2.000000000
1.7	3.249009585
1.73	3.317278183
1.732	3.321880096
1.7320	3.321880096
1.73205	3.321995226
1.732050	3.321995226
1.7320508	3.321997068
1.73205080	3.321997068
1.732050808	3.321997086

number x. We will describe this informally now and will give a rigorous definition in Chapter 7.

The graphs of several exponential functions are shown in Figure 1.53. These graphs show the values of the exponential functions for real inputs x. We choose the value of a^x when x is irrational so that there are no "holes" or "jumps" in the graph of a^x (these words are not rigorous mathematical terms, but they informally convey the underlying idea). The value of a^x when x is irrational is chosen so that the function $f(x) = a^x$ is *continuous*, a notion that will be carefully developed in Chapter 2. This choice ensures that the graph is increasing when $a > 1$ and is decreasing when $0 < a < 1$ (see Figure 1.53).

We illustrate how to define the value of an exponential function at an irrational power using the exponential function $f(x) = 2^x$. How do we make sense of the expression $2^{\sqrt{3}}$? Any particular irrational number, say $x = \sqrt{3}$, has a decimal expansion

$$\sqrt{3} = 1.732050808\ldots.$$

We then consider the list of powers of 2 with more and more digits in the decimal expansion,

$$2^1, 2^{1.7}, 2^{1.73}, 2^{1.732}, 2^{1.7320}, 2^{1.73205}, \ldots. \tag{1}$$

We know the meaning of each number in list (1) because the successive decimal approximations to $\sqrt{3}$ given by 1, 1.7, 1.73, 1.732, and so on are all *rational* numbers. As these decimal approximations get closer and closer to $\sqrt{3}$, it seems reasonable that the list of numbers in (1) gets closer and closer to some fixed number, which we specify to be $2^{\sqrt{3}}$.

Table 1.4 illustrates how taking better approximations to $\sqrt{3}$ gives better approximations to the number $2^{\sqrt{3}} \approx 3.321997086$. It is the *completeness* property of the real numbers (discussed in Appendix 6) which guarantees that this procedure gives a single number we define to be $2^{\sqrt{3}}$ (although it is beyond the scope of this text to give a proof). In a similar way, we can identify the number 2^x (or $a^x, a > 0$) for any irrational x. By identifying the number a^x for both rational and irrational x, we eliminate any "holes" or "gaps" in the graph of a^x.

Exponential functions obey the rules of exponents listed below. It is easy to check these rules using algebra when the exponents are integers or rational numbers. We prove them for all real exponents in Chapter 7.

Rules for Exponents

If $a > 0$ and $b > 0$, the following rules hold for all real numbers x and y.

1. $a^x \cdot a^y = a^{x+y}$ **2.** $\dfrac{a^x}{a^y} = a^{x-y}$

3. $(a^x)^y = (a^y)^x = a^{xy}$ **4.** $a^x \cdot b^x = (ab)^x$

5. $\dfrac{a^x}{b^x} = \left(\dfrac{a}{b}\right)^x$

EXAMPLE 2   We use the rules for exponents to simplify some numerical expressions.

1. $3^{1.1} \cdot 3^{0.7} = 3^{1.1+0.7} = 3^{1.8}$ Rule 1

2. $\dfrac{\left(\sqrt{10}\right)^3}{\sqrt{10}} = \left(\sqrt{10}\right)^{3-1} = \left(\sqrt{10}\right)^2 = 10$ Rule 2

3. $\left(5^{\sqrt{2}}\right)^{\sqrt{2}} = 5^{\sqrt{2}\cdot\sqrt{2}} = 5^2 = 25$ Rule 3

4. $7^{\pi} \cdot 8^{\pi} = (56)^{\pi}$ Rule 4

5. $\left(\dfrac{4}{9}\right)^{1/2} = \dfrac{4^{1/2}}{9^{1/2}} = \dfrac{2}{3}$ Rule 5

The Natural Exponential Function e^x

The most important exponential function used for modeling natural, physical, and economic phenomena is the **natural exponential function**, whose base is the special number e. The number e is irrational, and its value to nine decimal places is 2.718281828. (In Section 3.8 we will see a way to calculate the value of e.) It might seem strange that we would use this number for a base rather than a simple number like 2 or 10. The advantage in using e as a base is that it greatly simplifies many of the calculations in calculus.

In Figure 1.53a you can see that the graphs of the exponential functions $y = a^x$ get steeper as the base a gets larger. This idea of steepness is conveyed by the slope of the tangent line to the graph at a point. Tangent lines to graphs of functions are defined precisely in the next chapter, but intuitively the tangent line to the graph at a point is the line that best approximates the graph at the point, like a tangent to a circle. Figure 1.54 shows the slope of the graph of $y = a^x$ as it crosses the y-axis for several values of a. Notice that the slope is exactly equal to 1 when a equals the number e. The slope is smaller than 1 if $a < e$, and larger than 1 if $a > e$. The graph of $y = e^x$ has slope 1 when it crosses the y-axis.

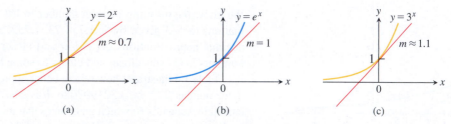

FIGURE 1.54   Among the exponential functions, the graph of $y = e^x$ has the property that the slope m of the tangent line to the graph is exactly 1 when it crosses the y-axis. The slope is smaller for a base less than e, such as 2^x, and larger for a base greater than e, such as 3^x.

Exponential Growth and Decay

The function $y = y_0 e^{kx}$, where k is a nonzero constant, is a model for **exponential growth** if $k > 0$ and a model for **exponential decay** if $k < 0$. Here y_0 is a constant that represents the value of the function when $x = 0$. An example of exponential growth occurs when computing interest **compounded continuously.** This is modeled by the formula $y = Pe^{rt}$, where P is the initial monetary investment, r is the interest rate as a decimal, and t is time in units consistent with r. An example of exponential decay is the model $y = Ae^{-1.2 \times 10^{-4} t}$, which represents how the radioactive isotope carbon-14 decays over time. Here A is the original amount of carbon-14 and t is the time in years. Carbon-14 decay is used to date the remains of dead organisms such as shells, seeds, and wooden artifacts. Figure 1.55 shows graphs of exponential growth and exponential decay.

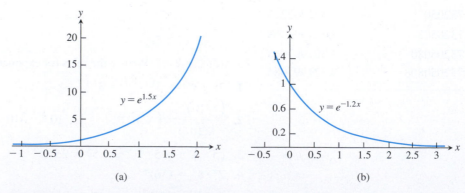

FIGURE 1.55   Graphs of (a) exponential growth, $k = 1.5 > 0$, and (b) exponential decay, $k = -1.2 < 0$.

EXAMPLE 3 Investment companies often use the model $y = Pe^{rt}$ in calculating the growth of an investment. Use this model to track the growth of $100 invested in 2014 at an annual interest rate of 5.5%.

Solution Let $t = 0$ represent 2014, $t = 1$ represent 2015, and so on. Then the exponential growth model is $y(t) = Pe^{rt}$, where $P = 100$ (the initial investment), $r = 0.055$ (the annual interest rate expressed as a decimal), and t is time in years. To predict the amount in the account in 2018, after four years have elapsed, we take $t = 4$ and calculate

$$y(4) = 100e^{0.055(4)}$$
$$= 100e^{0.22}$$
$$= 124.61. \qquad \text{Nearest cent using calculator}$$

This compares with $123.88 in the account when the interest is compounded annually, as was done in Example 1.

EXAMPLE 4 Laboratory experiments indicate that some atoms emit a part of their mass as radiation, with the remainder of the atom re-forming to make an atom of some new element. For example, radioactive carbon-14 decays into nitrogen; radium eventually decays into lead. If y_0 is the number of radioactive nuclei present at time zero, the number still present at any later time t will be

$$y = y_0 e^{-rt}, \qquad r > 0.$$

The number r is called the **decay rate** of the radioactive substance. (We will see how this formula is obtained in Section 7.2.) For carbon-14, the decay rate has been determined experimentally to be about $r = 1.2 \times 10^{-4}$ when t is measured in years. Predict the percent of carbon-14 present after 866 years have elapsed.

Solution If we start with an amount y_0 of carbon-14 nuclei, after 866 years we are left with the amount

$$y(866) = y_0 e^{(-1.2 \times 10^{-4})(866)}$$
$$\approx (0.901)y_0. \qquad \text{Calculator evaluation}$$

That is, after 866 years, we are left with about 90% of the original amount of carbon-14, so about 10% of the original nuclei have decayed.

You may wonder why we use the family of functions $y = e^{kx}$ for different values of the constant k instead of the general exponential functions $y = a^x$. In the next section, we show that the exponential function a^x is equal to e^{kx} for an appropriate value of k. So the formula $y = e^{kx}$ covers the entire range of possibilities, and it is generally easier to use.

EXERCISES 1.5

Sketching Exponential Curves
In Exercises 1–6, sketch the given curves together in the appropriate coordinate plane and label each curve with its equation.

1. $y = 2^x$, $y = 4^x$, $y = 3^{-x}$, $y = (1/5)^x$
2. $y = 3^x$, $y = 8^x$, $y = 2^{-x}$, $y = (1/4)^x$
3. $y = 2^{-t}$ and $y = -2^t$ 4. $y = 3^{-t}$ and $y = -3^t$
5. $y = e^x$ and $y = 1/e^x$ 6. $y = -e^x$ and $y = -e^{-x}$

In each of Exercises 7–10, sketch the shifted exponential curves.

7. $y = 2^x - 1$ and $y = 2^{-x} - 1$
8. $y = 3^x + 2$ and $y = 3^{-x} + 2$
9. $y = 1 - e^x$ and $y = 1 - e^{-x}$
10. $y = -1 - e^x$ and $y = -1 - e^{-x}$

Applying the Laws of Exponents
Use the laws of exponents to simplify the expressions in Exercises 11–20.

11. $16^2 \cdot 16^{-1.75}$ 12. $9^{1/3} \cdot 9^{1/6}$
13. $\dfrac{4^{4.2}}{4^{3.7}}$ 14. $\dfrac{3^{5/3}}{3^{2/3}}$
15. $\left(25^{1/8}\right)^4$ 16. $\left(13^{\sqrt{2}}\right)^{\sqrt{2}/2}$

17. $2^{\sqrt{3}} \cdot 7^{\sqrt{3}}$

18. $\left(\sqrt{3}\right)^{1/2} \cdot \left(\sqrt{12}\right)^{1/2}$

19. $\left(\dfrac{2}{\sqrt{2}}\right)^{4}$

20. $\left(\dfrac{\sqrt{6}}{3}\right)^{2}$

Compositions Involving Exponential Functions

Find the domain and range for each of the functions in Exercises 21–24.

21. $f(x) = \dfrac{1}{2 + e^x}$

22. $g(t) = \cos\left(e^{-t}\right)$

23. $g(t) = \sqrt{1 + 3^{-t}}$

24. $f(x) = \dfrac{3}{1 - e^{2x}}$

Applications

[T] In Exercises 25–28, use graphs to find approximate solutions.

25. $2^x = 5$

26. $e^x = 4$

27. $3^x - 0.5 = 0$

28. $3 - 2^{-x} = 0$

[T] In Exercises 29–36, use an exponential model and a graphing calculator to estimate the answer in each problem.

29. Population growth   The population of Knoxville is 500,000 and is increasing at the rate of 3.75% each year. Approximately when will the population reach 1 million?

30. Population growth   The population of Silver Run in the year 1890 was 6250. Assume the population increased at a rate of 2.75% per year.

a. Estimate the population in 1915 and 1940.

b. Approximately when did the population reach 50,000?

31. Radioactive decay   The half-life of phosphorus-32 is about 14 days. There are 6.6 grams present initially.

a. Express the amount of phosphorus-32 remaining as a function of time t.

b. When will there be 1 gram remaining?

32. If Jean invests $2300 in a retirement account with a 6% interest rate compounded annually, how long will it take until Jean's account has a balance of $4150?

33. Doubling your money   Determine how much time is required for an investment to double in value if interest is earned at the rate of 6.25% compounded annually.

34. Tripling your money   Determine how much time is required for an investment to triple in value if interest is earned at the rate of 5.75% compounded continuously.

35. Cholera bacteria   Suppose that a colony of bacteria starts with 1 bacterium and doubles in number every half hour. How many bacteria will the colony contain at the end of 24 hr?

36. Eliminating a disease   Suppose that in any given year the number of cases of a disease is reduced by 20%. If there are 10,000 cases today, how many years will it take

a. to reduce the number of cases to 1000?

b. to eliminate the disease; that is, to reduce the number of cases to less than 1?

1.6   Inverse Functions and Logarithms

A function that undoes, or inverts, the effect of a function f is called the *inverse* of f. Many common functions, though not all, are paired with an inverse. In this section we present the natural logarithmic function $y = \ln x$ as the inverse of the exponential function $y = e^x$, and we also give examples of several inverse trigonometric functions.

One-to-One Functions

A function is a rule that assigns a value from its range to each element in its domain. Some functions assign the same range value to more than one element in the domain. The function $f(x) = x^2$ assigns the same value, 1, to both of the numbers -1 and $+1$. Similarly the sines of $\pi/3$ and $2\pi/3$ are both $\sqrt{3}/2$. Other functions assume each value in their range no more than once. The square roots and cubes of different numbers are always different. A function that has distinct values at distinct elements in its domain is called one-to-one.

> **DEFINITION**   A function $f(x)$ is **one-to-one** on a domain D if $f(x_1) \neq f(x_2)$ whenever $x_1 \neq x_2$ in D.

EXAMPLE 1   Some functions are one-to-one on their entire natural domain. Other functions are not one-to-one on their entire domain, but by restricting the function to a smaller domain we can create a function that is one-to-one. The original and restricted functions are not the same functions, because they have different domains. However, the two functions have the same values on the smaller domain.

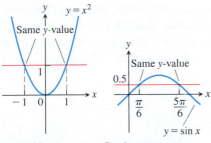

(a) One-to-one: Graph meets each horizontal line at most once.

(b) Not one-to-one: Graph meets one or more horizontal lines more than once.

FIGURE 1.56 (a) $y = x^3$ and $y = \sqrt{x}$ are one-to-one on their domains $(-\infty, \infty)$ and $[0, \infty)$. (b) $y = x^2$ and $y = \sin x$ are not one-to-one on their domains $(-\infty, \infty)$.

Caution

Do not confuse the inverse function f^{-1} with the reciprocal function $1/f$.

(a) $f(x) = \sqrt{x}$ is one-to-one on any domain of nonnegative numbers because $\sqrt{x_1} \neq \sqrt{x_2}$ whenever $x_1 \neq x_2$.

(b) $g(x) = \sin x$ is *not* one-to-one on the interval $[0, \pi]$ because $\sin(\pi/6) = \sin(5\pi/6)$. In fact, for each element x_1 in the subinterval $[0, \pi/2)$ there is a corresponding element x_2 in the subinterval $(\pi/2, \pi]$ satisfying $\sin x_1 = \sin x_2$. The sine function *is* one-to-one on $[0, \pi/2]$, however, because it is an increasing function on $[0, \pi/2]$ and hence gives distinct outputs for distinct inputs in that interval.   ∎

The graph of a one-to-one function $y = f(x)$ can intersect a given horizontal line at most once. If the function intersects the line more than once, then it assumes the same y-value for at least two different x-values and is therefore not one-to-one (Figure 1.56).

> **The Horizontal Line Test for One-to-One Functions**
>
> A function $y = f(x)$ is one-to-one if and only if its graph intersects each horizontal line at most once.

Inverse Functions

Since each output of a one-to-one function comes from just one input, the effect of the function can be inverted to send each output back to the input from which it came.

> **DEFINITION** Suppose that f is a one-to-one function on a domain D with range R. The **inverse function** f^{-1} is defined by
> $$f^{-1}(b) = a \quad \text{if} \quad f(a) = b.$$
> The domain of f^{-1} is R and the range of f^{-1} is D.

The symbol f^{-1} for the inverse of f is read "f inverse." The "-1" in f^{-1} is *not* an exponent; $f^{-1}(x)$ does not mean $1/f(x)$. Notice that the domains and ranges of f and f^{-1} are interchanged.

EXAMPLE 2 Suppose a one-to-one function $y = f(x)$ is given by a table of values

x	1	2	3	4	5	6	7	8
$f(x)$	3	4.5	7	10.5	15	20.5	27	34.5

A table for the values of $x = f^{-1}(y)$ can then be obtained by simply interchanging the values in each column of the table for f:

y	3	4.5	7	10.5	15	20.5	27	34.5
$f^{-1}(y)$	1	2	3	4	5	6	7	8

∎

If we apply f to send an input x to the output $f(x)$ and follow by applying f^{-1} to $f(x)$, we get right back to x, just where we started. Similarly, if we take some number y in the range of f, apply f^{-1} to it, and then apply f to the resulting value $f^{-1}(y)$, we get back the value y from which we began. Composing a function and its inverse has the same effect as doing nothing.

$$(f^{-1} \circ f)(x) = x, \quad \text{for all } x \text{ in the domain of } f$$
$$(f \circ f^{-1})(y) = y, \quad \text{for all } y \text{ in the domain of } f^{-1} \text{ (or range of } f)$$

Only a one-to-one function can have an inverse. The reason is that if $f(x_1) = y$ and $f(x_2) = y$ for two distinct inputs x_1 and x_2, then there is no way to assign a value to $f^{-1}(y)$ that satisfies both $f^{-1}(f(x_1)) = x_1$ and $f^{-1}(f(x_2)) = x_2$.

A function that is increasing on an interval satisfies the inequality $f(x_2) > f(x_1)$ when $x_2 > x_1$, so it is one-to-one and has an inverse. A function that is decreasing on an interval also has an inverse. Functions that are neither increasing nor decreasing may still be one-to-one and have an inverse, as with the function $f(x) = 1/x$ for $x \neq 0$ and $f(0) = 0$, defined on $(-\infty, \infty)$ and passing the horizontal line test.

Finding Inverses

The graphs of a function and its inverse are closely related. To read the value of a function from its graph, we start at a point x on the x-axis, go vertically to the graph, and then move horizontally to the y-axis to read the value of y. The inverse function can be read from the graph by reversing this process. Start with a point y on the y-axis, go horizontally to the graph of $y = f(x)$, and then move vertically to the x-axis to read the value of $x = f^{-1}(y)$ (Figure 1.57).

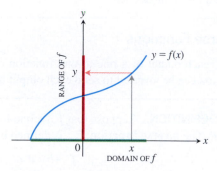

(a) To find the value of f at x, we start at x, go up to the curve, and then over to the y-axis.

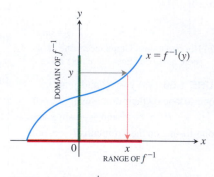

(b) The graph of f^{-1} is the graph of f, but with x and y interchanged. To find the x that gave y, we start at y and go over to the curve and down to the x-axis. The domain of f^{-1} is the range of f. The range of f^{-1} is the domain of f.

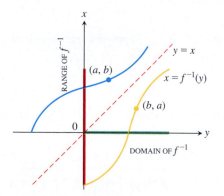

(c) To draw the graph of f^{-1} in the more usual way, we reflect the system across the line $y = x$.

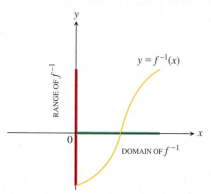

(d) Then we interchange the letters x and y. We now have a normal-looking graph of f^{-1} as a function of x.

FIGURE 1.57 The graph of $y = f^{-1}(x)$ is obtained by reflecting the graph of $y = f(x)$ about the line $y = x$.

We want to set up the graph of f^{-1} so that its input values lie along the x-axis, as is usually done for functions, rather than on the y-axis. To achieve this we interchange the x- and y-axes by reflecting across the 45° line $y = x$. After this reflection we have a new graph that represents f^{-1}. The value of $f^{-1}(x)$ can now be read from the graph in the usual way, by starting with a point x on the x-axis, going vertically to the graph, and then horizontally to the y-axis to get the value of $f^{-1}(x)$. Figure 1.57 indicates the relationship between the graphs of f and f^{-1}. The graphs are interchanged by reflection through the line $y = x$.

The process of passing from f to f^{-1} can be summarized as a two-step procedure.

1. Solve the equation $y = f(x)$ for x. This gives a formula $x = f^{-1}(y)$ where x is expressed as a function of y.

2. Interchange x and y, obtaining a formula $y = f^{-1}(x)$ where f^{-1} is expressed in the conventional format with x as the independent variable and y as the dependent variable.

EXAMPLE 3   Find the inverse of $y = \dfrac{1}{2}x + 1$, expressed as a function of x.

Solution

1. *Solve for x in terms of y:* $y = \dfrac{1}{2}x + 1$ *The graph satisfies the horizontal line test, so it is one-to-one (Fig. 1.58).*

$$2y = x + 2$$
$$x = 2y - 2.$$

2. *Interchange x and y:* $y = 2x - 2.$ *Expresses the function in the usual form where y is the dependent variable.*

The inverse of the function $f(x) = (1/2)x + 1$ is the function $f^{-1}(x) = 2x - 2$. (See Figure 1.58.) To check, we verify that both compositions give the identity function:

$$f^{-1}(f(x)) = 2\left(\frac{1}{2}x + 1\right) - 2 = x + 2 - 2 = x$$

$$f(f^{-1}(x)) = \frac{1}{2}(2x - 2) + 1 = x - 1 + 1 = x.$$ ∎

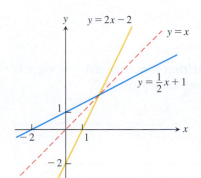

FIGURE 1.58 Graphing $f(x) = (1/2)x + 1$ and $f^{-1}(x) = 2x - 2$ together shows the graphs' symmetry with respect to the line $y = x$ (Example 3).

EXAMPLE 4   Find the inverse of the function $y = x^2$, $x \geq 0$, expressed as a function of x.

Solution   For $x \geq 0$, the graph satisfies the horizontal line test, so the function is one-to-one and has an inverse. To find the inverse, we first solve for x in terms of y:

$$y = x^2$$
$$\sqrt{y} = \sqrt{x^2} = |x| = x \qquad |x| = x \text{ because } x \geq 0$$

We then interchange x and y, obtaining

$$y = \sqrt{x}.$$

The inverse of the function $y = x^2$, $x \geq 0$, is the function $y = \sqrt{x}$ (Figure 1.59). ∎

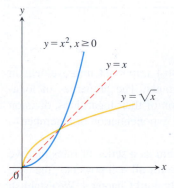

FIGURE 1.59 The functions $y = \sqrt{x}$ and $y = x^2$, $x \geq 0$, are inverses of one another (Example 4).

Notice that the function $y = x^2$, $x \geq 0$, with domain *restricted* to the nonnegative real numbers, *is* one-to-one (Figure 1.59) and has an inverse. On the other hand, the function $y = x^2$, with no domain restrictions, *is not* one-to-one (Figure 1.56b) and therefore has no inverse.

Logarithmic Functions

If a is any positive real number other than 1, then the base a exponential function $f(x) = a^x$ is one-to-one. It therefore has an inverse. Its inverse is called the *logarithm function with base a.*

DEFINITION   The **logarithm function with base** a, written $y = \log_a x$, is the inverse of the base a exponential function $y = a^x$ $(a > 0, a \neq 1)$.

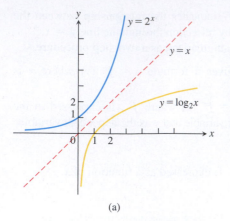

(a)

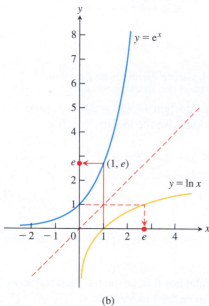

(b)

FIGURE 1.60 (a) The graph of 2^x and its inverse, $\log_2 x$. (b) The graph of e^x and its inverse, $\ln x$.

HISTORICAL BIOGRAPHY

John Napier
(1550–1617)
www.goo.gl/BvGOua

The domain of $\log_a x$ is $(0, \infty)$, the same as the range of a^x. The range of $\log_a x$ is $(-\infty, \infty)$, the same as the domain of a^x.

Figure 1.60a shows the graph of $y = \log_2 x$. The graph of $y = a^x$, $a > 1$, increases rapidly for $x > 0$, so its inverse, $y = \log_a x$, increases slowly for $x > 1$.

Because we have no technique yet for solving the equation $y = a^x$ for x in terms of y, we do not have an explicit formula for computing the logarithm at a given value of x. Nevertheless, we can obtain the graph of $y = \log_a x$ by reflecting the graph of the exponential $y = a^x$ across the line $y = x$. Figure 1.60a shows the graphs for $a = 2$ and $a = e$.

Logarithms with base 2 are often used when working with binary numbers, as is common in computer science. Logarithms with base e and base 10 are so important in applications that many calculators have special keys for them. They also have their own special notation and names:

$$\log_e x \quad \text{is written as} \quad \ln x.$$
$$\log_{10} x \quad \text{is written as} \quad \log x.$$

The function $y = \ln x$ is called the **natural logarithm function**, and $y = \log x$ is often called the **common logarithm function**. For the natural logarithm,

$$\ln x = y \iff e^y = x.$$

In particular, because $e^1 = e$, we obtain

$$\ln e = 1.$$

Properties of Logarithms

Logarithms, invented by John Napier, were the single most important improvement in arithmetic calculation before the modern electronic computer. The properties of logarithms reduce multiplication of positive numbers to addition of their logarithms, division of positive numbers to subtraction of their logarithms, and exponentiation of a number to multiplying its logarithm by the exponent.

We summarize these properties for the natural logarithm as a series of rules that we prove in Chapter 3. Although here we state the Power Rule for all real powers r, the case when r is an irrational number cannot be dealt with properly until Chapter 4. We establish the validity of the rules for logarithmic functions with any base a in Chapter 7.

THEOREM 1—Algebraic Properties of the Natural Logarithm

For any numbers $b > 0$ and $x > 0$, the natural logarithm satisfies the following rules:

1. *Product Rule*: $\quad\quad\quad\quad \ln bx = \ln b + \ln x$

2. *Quotient Rule*: $\quad\quad\quad \ln \dfrac{b}{x} = \ln b - \ln x$

3. *Reciprocal Rule*: $\quad\quad \ln \dfrac{1}{x} = -\ln x \quad$ Rule 2 with $b = 1$

4. *Power Rule*: $\quad\quad\quad\quad \ln x^r = r \ln x$

EXAMPLE 5   We use the properties in Theorem 1 to simplify three expressions.

(a) $\ln 4 + \ln \sin x = \ln (4 \sin x)$ Product Rule

(b) $\ln \dfrac{x + 1}{2x - 3} = \ln (x + 1) - \ln (2x - 3)$ Quotient Rule

(c) $\ln \dfrac{1}{8} = -\ln 8$ Reciprocal Rule

$\qquad = -\ln 2^3 = -3 \ln 2$ Power Rule   ■

Because a^x and $\log_a x$ are inverses, composing them in either order gives the identity function.

Inverse Properties for a^x and $\log_a x$

1. Base a: $\quad a^{\log_a x} = x, \qquad \log_a a^x = x, \qquad a > 0, a \neq 1, x > 0$

2. Base e: $\quad e^{\ln x} = x, \qquad \ln e^x = x, \qquad x > 0$

Substituting a^x for x in the equation $x = e^{\ln x}$ enables us to rewrite a^x as a power of e:

$$a^x = e^{\ln (a^x)} \qquad \text{Substitute } a^x \text{ for } x \text{ in } x = e^{\ln x}.$$
$$ = e^{x \ln a} \qquad \text{Power Rule for logs}$$
$$ = e^{(\ln a) x}. \qquad \text{Exponent rearranged}$$

Thus, the exponential function a^x is the same as e^{kx} with $k = \ln a$.

Every exponential function is a power of the natural exponential function.

$$a^x = e^{x \ln a}$$

That is, a^x is the same as e^x raised to the power $\ln a$: $\quad a^x = e^{kx}$ for $k = \ln a$.

For example,

$$2^x = e^{(\ln 2) x} = e^{x \ln 2}, \qquad \text{and} \qquad 5^{-3x} = e^{(\ln 5)(-3x)} = e^{-3x \ln 5}.$$

Returning once more to the properties of a^x and $\log_a x$, we have

$$\ln x = \ln (a^{\log_a x}) \qquad \text{Inverse Property for } a^x \text{ and } \log_a x$$
$$ = (\log_a x)(\ln a). \qquad \text{Power Rule for logarithms, with } r = \log_a x$$

Rewriting this equation as $\log_a x = (\ln x)/(\ln a)$ shows that every logarithmic function is a constant multiple of the natural logarithm $\ln x$. This allows us to extend the algebraic properties for $\ln x$ to $\log_a x$. For instance, $\log_a bx = \log_a b + \log_a x$.

Change of Base Formula

Every logarithmic function is a constant multiple of the natural logarithm.

$$\log_a x = \frac{\ln x}{\ln a} \qquad (a > 0, a \neq 1)$$

Applications

In Section 1.5 we looked at examples of exponential growth and decay problems. Here we use properties of logarithms to answer more questions concerning such problems.

EXAMPLE 6   If $1000 is invested in an account that earns 5.25% interest compounded annually, how long will it take the account to reach $2500?

Solution   From Example 1, Section 1.5, with $P = 1000$ and $r = 0.0525$, the amount in the account at any time t in years is $1000(1.0525)^t$, so to find the time t when the account reaches \$2500 we need to solve the equation

$$1000(1.0525)^t = 2500.$$

Thus we have

$$(1.0525)^t = 2.5 \qquad \text{Divide by 1000.}$$
$$\ln(1.0525)^t = \ln 2.5 \qquad \text{Take logarithms of both sides.}$$
$$t \ln 1.0525 = \ln 2.5 \qquad \text{Power Rule}$$
$$t = \frac{\ln 2.5}{\ln 1.0525} \approx 17.9 \qquad \text{Values obtained by calculator}$$

The amount in the account will reach \$2500 in 18 years, when the annual interest payment is deposited for that year. ∎

EXAMPLE 7   The **half-life** of a radioactive element is the time expected to pass until half of the radioactive nuclei present in a sample decay. The half-life is a constant that does not depend on the number of radioactive nuclei initially present in the sample, but only on the radioactive substance.

To compute the half-life, let y_0 be the number of radioactive nuclei initially present in the sample. Then the number y present at any later time t will be $y = y_0 e^{-kt}$. We seek the value of t at which the number of radioactive nuclei present equals half the original number:

$$y_0 e^{-kt} = \frac{1}{2}y_0$$

$$e^{-kt} = \frac{1}{2}$$

$$-kt = \ln\frac{1}{2} = -\ln 2 \qquad \text{Reciprocal Rule for logarithms}$$

$$t = \frac{\ln 2}{k}. \tag{1}$$

This value of t is the half-life of the element. It depends only on the value of k; the number y_0 does not have any effect.

The effective radioactive lifetime of polonium-210 is so short that we measure it in days rather than years. The number of radioactive atoms remaining after t days in a sample that starts with y_0 radioactive atoms is

$$y = y_0 e^{-5 \times 10^{-3}\, t}.$$

The element's half-life is

$$\text{Half-life} = \frac{\ln 2}{k} \qquad \text{Eq. (1)}$$

$$= \frac{\ln 2}{5 \times 10^{-3}} \qquad \text{The } k \text{ from polonium's decay equation}$$

$$\approx 139 \text{ days.}$$

This means that after 139 days, $1/2$ of y_0 radioactive atoms remain; after another 139 days (278 days altogether) half of those remain, or $1/4$ of y_0 radioactive atoms remain, and so on (see Figure 1.61). ∎

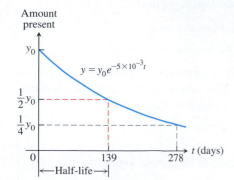

FIGURE 1.61   Amount of polonium-210 present at time t, where y_0 represents the number of radioactive atoms initially present (Example 7).

Inverse Trigonometric Functions

The six basic trigonometric functions are not one-to-one (since their values repeat periodically). However, we can restrict their domains to intervals on which they are one-to-one.

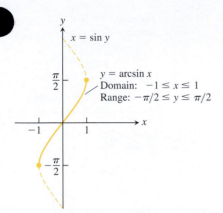

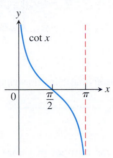

(a)

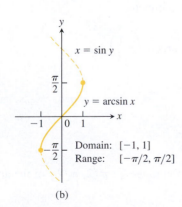

(b)

FIGURE 1.62 The graph of $y = \arcsin x$.

FIGURE 1.63 The graphs of (a) $y = \sin x,\ -\pi/2 \le x \le \pi/2$, and (b) its inverse, $y = \arcsin x$. The graph of $\arcsin x$, obtained by reflection across the line $y = x$, is a portion of the curve $x = \sin y$.

The sine function increases from -1 at $x = -\pi/2$ to $+1$ at $x = \pi/2$. By restricting its domain to the interval $[-\pi/2, \pi/2]$ we make it one-to-one, so that it has an inverse which is called arcsin x (Figure 1.62). Similar domain restrictions can be applied to all six trigonometric functions.

Domain restrictions that make the trigonometric functions one-to-one

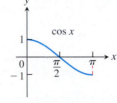

$y = \sin x$
Domain: $[-\pi/2, \pi/2]$
Range: $[-1, 1]$

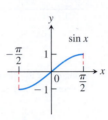

$y = \cos x$
Domain: $[0, \pi]$
Range: $[-1, 1]$

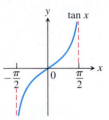

$y = \tan x$
Domain: $(-\pi/2, \pi/2)$
Range: $(-\infty, \infty)$

$y = \cot x$
Domain: $(0, \pi)$
Range: $(-\infty, \infty)$

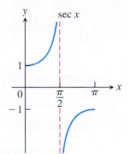

$y = \sec x$
Domain: $[0, \pi/2) \cup (\pi/2, \pi]$
Range: $(-\infty, -1] \cup [1, \infty)$

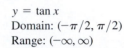

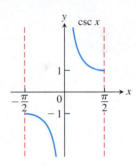

$y = \csc x$
Domain: $[-\pi/2, 0) \cup (0, \pi/2]$
Range: $(-\infty, -1] \cup [1, \infty)$

Since these restricted functions are now one-to-one, they have inverses, which we denote by

$$y = \sin^{-1} x \quad \text{or} \quad y = \arcsin x, \qquad y = \cos^{-1} x \quad \text{or} \quad y = \arccos x$$
$$y = \tan^{-1} x \quad \text{or} \quad y = \arctan x, \qquad y = \cot^{-1} x \quad \text{or} \quad y = \text{arccot } x$$
$$y = \sec^{-1} x \quad \text{or} \quad y = \text{arcsec } x, \qquad y = \csc^{-1} x \quad \text{or} \quad y = \text{arccsc } x$$

These equations are read "y equals the arcsine of x" or "y equals arcsin x" and so on.

Caution The -1 in the expressions for the inverse means "inverse." It does *not* mean reciprocal. For example, the *reciprocal* of $\sin x$ is $(\sin x)^{-1} = 1/\sin x = \csc x$. ●

The graphs of the six inverse trigonometric functions are obtained by reflecting the graphs of the restricted trigonometric functions through the line $y = x$. Figure 1.63b shows the graph of $y = \arcsin x$ and Figure 1.64 shows the graphs of all six functions. We now take a closer look at two of these functions.

The Arcsine and Arccosine Functions

We define the arcsine and arccosine as functions whose values are angles (measured in radians) that belong to restricted domains of the sine and cosine functions.

Domain: $-1 \le x \le 1$
Range: $-\dfrac{\pi}{2} \le y \le \dfrac{\pi}{2}$

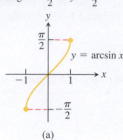

(a)

Domain: $-1 \le x \le 1$
Range: $0 \le y \le \pi$

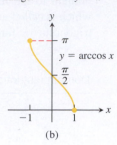

(b)

Domain: $-\infty < x < \infty$
Range: $-\dfrac{\pi}{2} < y < \dfrac{\pi}{2}$

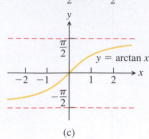

(c)

Domain: $x \le -1$ or $x \ge 1$
Range: $0 \le y \le \pi, y \ne \dfrac{\pi}{2}$

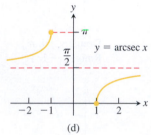

(d)

Domain: $x \le -1$ or $x \ge 1$
Range: $-\dfrac{\pi}{2} \le y \le \dfrac{\pi}{2}, y \ne 0$

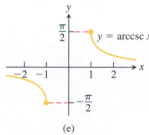

(e)

Domain: $-\infty < x < \infty$
Range: $0 < y < \pi$

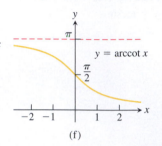

(f)

FIGURE 1.64   Graphs of the six basic inverse trigonometric functions.

DEFINITION

$y = \mathbf{arcsin}\, x$ is the number in $[-\pi/2, \pi/2]$ for which $\sin y = x$.

$y = \mathbf{arccos}\, x$ is the number in $[0, \pi]$ for which $\cos y = x$.

The "Arc" in Arcsine and Arccosine
For a unit circle and radian angles, the arc length equation $s = r\theta$ becomes $s = \theta$, so central angles and the arcs they subtend have the same measure. If $x = \sin y$, then, in addition to being the angle whose sine is x, y is also the length of arc on the unit circle that subtends an angle whose sine is x. So we call y "the arc whose sine is x."

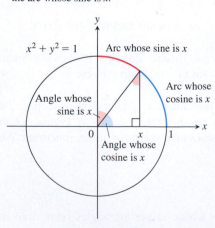

The graph of $y = \arcsin x$ (Figure 1.63b) is symmetric about the origin (it lies along the graph of $x = \sin y$). The arcsine is therefore an odd function:

$$\arcsin(-x) = -\arcsin x. \tag{2}$$

The graph of $y = \arccos x$ (Figure 1.65b) has no such symmetry.

EXAMPLE 8   Evaluate **(a)** $\arcsin\left(\dfrac{\sqrt{3}}{2}\right)$ and **(b)** $\arccos\left(-\dfrac{1}{2}\right)$.

Solution

(a) We see that

$$\arcsin\left(\frac{\sqrt{3}}{2}\right) = \frac{\pi}{3}$$

because $\sin(\pi/3) = \sqrt{3}/2$ and $\pi/3$ belongs to the range $[-\pi/2, \pi/2]$ of the arcsine function. See Figure 1.66a.

(b) We have

$$\arccos\left(-\frac{1}{2}\right) = \frac{2\pi}{3}$$

because $\cos(2\pi/3) = -1/2$ and $2\pi/3$ belongs to the range $[0, \pi]$ of the arccosine function. See Figure 1.66b.   ■

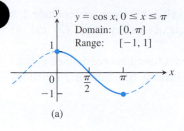

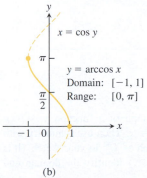

FIGURE 1.65 The graphs of
(a) $y = \cos x$, $0 \le x \le \pi$, and
(b) its inverse, $y = \arccos x$. The graph
of $\arccos x$, obtained by reflection across
the line $y = x$, is a portion of the curve
$x = \cos y$.

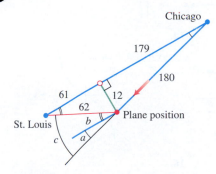

FIGURE 1.67 Diagram for drift
correction (Example 9), with distances
surrounded to the nearest mile (drawing
not to scale).

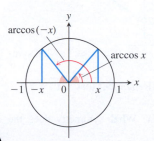

FIGURE 1.68 $\arccos x$ and $\arccos(-x)$
are supplementary angles (so their sum
is π).

Using the same procedure illustrated in Example 8, we can create the following table of
common values for the arcsine and arccosine functions.

x	$\arcsin x$	$\arccos x$
$\sqrt{3}/2$	$\pi/3$	$\pi/6$
$\sqrt{2}/2$	$\pi/4$	$\pi/4$
$1/2$	$\pi/6$	$\pi/3$
$-1/2$	$-\pi/6$	$2\pi/3$
$-\sqrt{2}/2$	$-\pi/4$	$3\pi/4$
$-\sqrt{3}/2$	$-\pi/3$	$5\pi/6$

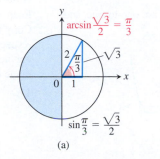

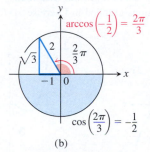

FIGURE 1.66 Values of the arcsine and arccosine functions
(Example 8).

EXAMPLE 9 During a 240 mi airplane flight from Chicago to St. Louis, after flying
180 mi the navigator determines that the plane is 12 mi off course, as shown in Figure
1.67. Find the angle a for a course parallel to the original correct course, the angle b, and
the drift correction angle $c = a + b$.

Solution From the Pythagorean theorem and given information, we compute an approximate hypothetical flight distance of 179 mi, had the plane been flying along the original
correct course (see Figure 1.67). Knowing the flight distance from Chicago to St. Louis, we
next calculate the remaining leg of the original course to be 61 mi. Applying the Pythagorean
theorem again then gives an approximate distance of 62 mi from the position of the plane to
St. Louis. Finally, from Figure 1.67, we see that $180 \sin a = 12$ and $62 \sin b = 12$, so

$$a = \arcsin\frac{12}{180} \approx 0.067 \text{ radian} \approx 3.8°$$

$$b = \arcsin\frac{12}{62} \approx 0.195 \text{ radian} \approx 11.2°$$

$$c = a + b \approx 15°. \qquad \blacksquare$$

Identities Involving Arcsine and Arccosine

As we can see from Figure 1.68, the arccosine of x satisfies the identity

$$\arccos x + \arccos(-x) = \pi, \qquad (3)$$

or

$$\arccos(-x) = \pi - \arccos x. \qquad (4)$$

Also, we can see from the triangle in Figure 1.69 that for $x > 0$,

$$\arcsin x + \arccos x = \pi/2. \qquad (5)$$

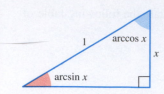

FIGURE 1.69 $\arcsin x$ and $\arccos x$ are complementary angles (so their sum is $\pi/2$).

Equation (5) holds for the other values of x in $[-1, 1]$ as well, but we cannot conclude this from the triangle in Figure 1.69. It is, however, a consequence of Equations (2) and (4) (Exercise 80).

The arctangent, arccotangent, arcsecant, and arccosecant functions are defined in Section 3.9. There we develop additional properties of the inverse trigonometric functions using the identities discussed here.

EXERCISES 1.6

Identifying One-to-One Functions Graphically

Which of the functions graphed in Exercises 1–6 are one-to-one, and which are not?

1.

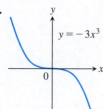

$y = -3x^3$

2.

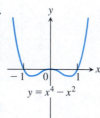

$y = x^4 - x^2$

3.

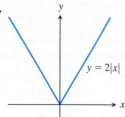

$y = 2|x|$

4.

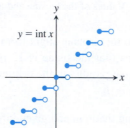

$y = \text{int } x$

5.

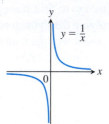

$y = \dfrac{1}{x}$

6.

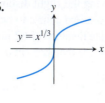

$y = x^{1/3}$

In Exercises 7–10, determine from its graph if the function is one-to-one.

7. $f(x) = \begin{cases} 3 - x, & x < 0 \\ 3, & x \geq 0 \end{cases}$

8. $f(x) = \begin{cases} 2x + 6, & x \leq -3 \\ x + 4, & x > -3 \end{cases}$

9. $f(x) = \begin{cases} 1 - \dfrac{x}{2}, & x \leq 0 \\ \dfrac{x}{x + 2}, & x > 0 \end{cases}$

10. $f(x) = \begin{cases} 2 - x^2, & x \leq 1 \\ x^2, & x > 1 \end{cases}$

Graphing Inverse Functions

Each of Exercises 11–16 shows the graph of a function $y = f(x)$. Copy the graph and draw in the line $y = x$. Then use symmetry with respect to the line $y = x$ to add the graph of f^{-1} to your sketch. (It is not necessary to find a formula for f^{-1}.) Identify the domain and range of f^{-1}.

11.

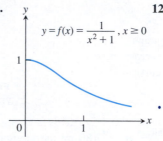

$y = f(x) = \dfrac{1}{x^2 + 1}, x \geq 0$

12.

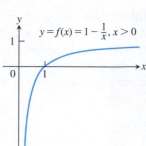

$y = f(x) = 1 - \dfrac{1}{x}, x > 0$

13.

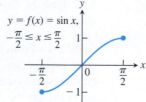

$y = f(x) = \sin x,$ $-\dfrac{\pi}{2} \leq x \leq \dfrac{\pi}{2}$

14.

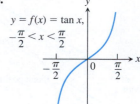

$y = f(x) = \tan x,$ $-\dfrac{\pi}{2} < x < \dfrac{\pi}{2}$

15.

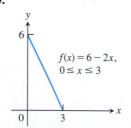

$f(x) = 6 - 2x,$ $0 \leq x \leq 3$

16.

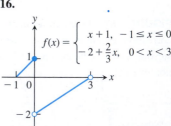

$f(x) = \begin{cases} x + 1, & -1 \leq x \leq 0 \\ -2 + \dfrac{2}{3}x, & 0 < x < 3 \end{cases}$

17. a. Graph the function $f(x) = \sqrt{1 - x^2}, 0 \leq x \leq 1$. What symmetry does the graph have?

   b. Show that f is its own inverse. (Remember that $\sqrt{x^2} = x$ if $x \geq 0$.)

18. a. Graph the function $f(x) = 1/x$. What symmetry does the graph have?

   b. Show that f is its own inverse.

Formulas for Inverse Functions

Each of Exercises 19–24 gives a formula for a function $y = f(x)$ and shows the graphs of f and f^{-1}. Find a formula for f^{-1} in each case.

19. $f(x) = x^2 + 1, \quad x \geq 0$ **20.** $f(x) = x^2, \quad x \leq 0$

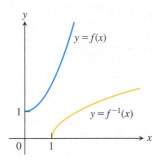 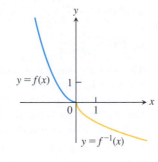

21. $f(x) = x^3 - 1$ **22.** $f(x) = x^2 - 2x + 1, \quad x \geq 1$

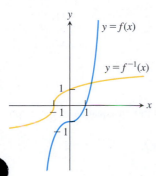 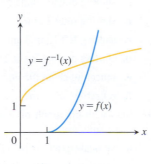

23. $f(x) = (x + 1)^2, \quad x \geq -1$ **24.** $f(x) = x^{2/3}, \quad x \geq 0$

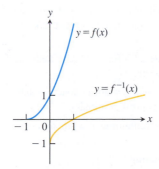 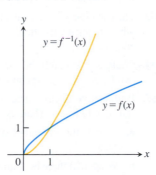

Each of Exercises 25–36 gives a formula for a function $y = f(x)$. In each case, find $f^{-1}(x)$ and identify the domain and range of f^{-1}. As a check, show that $f(f^{-1}(x)) = f^{-1}(f(x)) = x$.

25. $f(x) = x^5$ **26.** $f(x) = x^4, \quad x \geq 0$

27. $f(x) = x^3 + 1$ **28.** $f(x) = (1/2)x - 7/2$

29. $f(x) = 1/x^2, \quad x > 0$ **30.** $f(x) = 1/x^3, \quad x \neq 0$

31. $f(x) = \dfrac{x + 3}{x - 2}$ **32.** $f(x) = \dfrac{\sqrt{x}}{\sqrt{x} - 3}$

33. $f(x) = x^2 - 2x, \quad x \leq 1$ **34.** $f(x) = (2x^3 + 1)^{1/5}$

    (*Hint:* Complete the square.)

35. $f(x) = \dfrac{x + b}{x - 2}, \quad b > -2$ and constant

36. $f(x) = x^2 - 2bx, \quad b > 0$ and constant, $x \leq b$

Inverses of Lines

37. a. Find the inverse of the function $f(x) = mx$, where m is a constant different from zero.

    b. What can you conclude about the inverse of a function $y = f(x)$ whose graph is a line through the origin with a nonzero slope m?

38. Show that the graph of the inverse of $f(x) = mx + b$, where m and b are constants and $m \neq 0$, is a line with slope $1/m$ and y-intercept $-b/m$.

39. a. Find the inverse of $f(x) = x + 1$. Graph f and its inverse together. Add the line $y = x$ to your sketch, drawing it with dashes or dots for contrast.

    b. Find the inverse of $f(x) = x + b$ (*b* constant). How is the graph of f^{-1} related to the graph of f?

    c. What can you conclude about the inverses of functions whose graphs are lines parallel to the line $y = x$?

40. a. Find the inverse of $f(x) = -x + 1$. Graph the line $y = -x + 1$ together with the line $y = x$. At what angle do the lines intersect?

    b. Find the inverse of $f(x) = -x + b$ (*b* constant). What angle does the line $y = -x + b$ make with the line $y = x$?

    c. What can you conclude about the inverses of functions whose graphs are lines perpendicular to the line $y = x$?

Logarithms and Exponentials

41. Express the following logarithms in terms of $\ln 2$ and $\ln 3$.

    a. $\ln 0.75$ **b.** $\ln (4/9)$

    c. $\ln (1/2)$ **d.** $\ln \sqrt[3]{9}$

    e. $\ln 3\sqrt{2}$ **f.** $\ln \sqrt{13.5}$

42. Express the following logarithms in terms of $\ln 5$ and $\ln 7$.

    a. $\ln (1/125)$ **b.** $\ln 9.8$

    c. $\ln 7\sqrt{7}$ **d.** $\ln 1225$

    e. $\ln 0.056$ **f.** $(\ln 35 + \ln (1/7))/(\ln 25)$

Use the properties of logarithms to write the expressions in Exercises 43 and 44 as a single term.

43. a. $\ln \sin \theta - \ln \left(\dfrac{\sin \theta}{5} \right)$ **b.** $\ln (3x^2 - 9x) + \ln \left(\dfrac{1}{3x} \right)$

    c. $\dfrac{1}{2} \ln (4t^4) - \ln b$

44. a. $\ln \sec \theta + \ln \cos \theta$ **b.** $\ln (8x + 4) - 2 \ln c$

    c. $3 \ln \sqrt[3]{t^2 - 1} - \ln (t + 1)$

Find simpler expressions for the quantities in Exercises 45–48.

45. a. $e^{\ln 7.2}$ **b.** $e^{-\ln x^2}$ **c.** $e^{\ln x - \ln y}$

46. a. $e^{\ln (x^2 + y^2)}$ **b.** $e^{-\ln 0.3}$ **c.** $e^{\ln \pi x - \ln 2}$

47. a. $2 \ln \sqrt{e}$ **b.** $\ln (\ln e^e)$ **c.** $\ln (e^{-x^2 - y^2})$

48. a. $\ln (e^{\sec \theta})$ **b.** $\ln (e^{(e^x)})$ **c.** $\ln (e^{2 \ln x})$

In Exercises 49–54, solve for y in terms of t or x, as appropriate.

49. $\ln y = 2t + 4$ **50.** $\ln y = -t + 5$

51. $\ln (y - b) = 5t$ **52.** $\ln (c - 2y) = t$

53. $\ln (y - 1) - \ln 2 = x + \ln x$

54. $\ln (y^2 - 1) - \ln (y + 1) = \ln (\sin x)$

In Exercises 55 and 56, solve for k.

55. a. $e^{2k} = 4$ **b.** $100e^{10k} = 200$ **c.** $e^{k/1000} = a$

56. a. $e^{5k} = \dfrac{1}{4}$ **b.** $80e^k = 1$ **c.** $e^{(\ln 0.8)k} = 0.8$

In Exercises 57–64, solve for t.

57. a. $e^{-0.3t} = 27$ **b.** $e^{kt} = \dfrac{1}{2}$ **c.** $e^{(\ln 0.2)t} = 0.4$

58. a. $e^{-0.01t} = 1000$ **b.** $e^{kt} = \dfrac{1}{10}$ **c.** $e^{(\ln 2)t} = \dfrac{1}{2}$

59. $e^{\sqrt{t}} = x^2$ **60.** $e^{(x^2)}e^{(2x+1)} = e^t$

61. $e^{2t} - 3e^t = 0$ **62.** $e^{-2t} + 6 = 5e^{-t}$

63. $\ln\left(\dfrac{t}{t-1}\right) = 2$ **64.** $\ln(t-2) = \ln 8 - \ln t$

Simplify the expressions in Exercises 65–68.

65. a. $5^{\log_5 7}$ **b.** $8^{\log_8 \sqrt{2}}$ **c.** $1.3^{\log_{1.3} 75}$

     d. $\log_4 16$ **e.** $\log_3 \sqrt{3}$ **f.** $\log_4\left(\dfrac{1}{4}\right)$

66. a. $2^{\log_2 3}$ **b.** $10^{\log_{10}(1/2)}$ **c.** $\pi^{\log_\pi 7}$

     d. $\log_{11} 121$ **e.** $\log_{121} 11$ **f.** $\log_3\left(\dfrac{1}{9}\right)$

67. a. $2^{\log_4 x}$ **b.** $9^{\log_3 x}$ **c.** $\log_2(e^{(\ln 2)(\sin x)})$

68. a. $25^{\log_5(3x^2)}$ **b.** $\log_e(e^x)$ **c.** $\log_4(2^{e^x \sin x})$

Express the ratios in Exercises 69 and 70 as ratios of natural logarithms and simplify.

69. a. $\dfrac{\log_2 x}{\log_3 x}$ **b.** $\dfrac{\log_2 x}{\log_8 x}$ **c.** $\dfrac{\log_x a}{\log_{x^2} a}$

70. a. $\dfrac{\log_9 x}{\log_3 x}$ **b.** $\dfrac{\log_{\sqrt{10}} x}{\log_{\sqrt{2}} x}$ **c.** $\dfrac{\log_a b}{\log_b a}$

Arcsine and Arccosine

In Exercises 71–74, find the exact value of each expression.

71. a. $\sin^{-1}\left(\dfrac{-1}{2}\right)$ **b.** $\sin^{-1}\left(\dfrac{1}{\sqrt{2}}\right)$ **c.** $\sin^{-1}\left(\dfrac{-\sqrt{3}}{2}\right)$

72. a. $\cos^{-1}\left(\dfrac{1}{2}\right)$ **b.** $\cos^{-1}\left(\dfrac{-1}{\sqrt{2}}\right)$ **c.** $\cos^{-1}\left(\dfrac{\sqrt{3}}{2}\right)$

73. a. $\arccos(-1)$ **b.** $\arccos(0)$

74. a. $\arcsin(-1)$ **b.** $\arcsin\left(-\dfrac{1}{\sqrt{2}}\right)$

Theory and Examples

75. If $f(x)$ is one-to-one, can anything be said about $g(x) = -f(x)$? Is it also one-to-one? Give reasons for your answer.

76. If $f(x)$ is one-to-one and $f(x)$ is never zero, can anything be said about $h(x) = 1/f(x)$? Is it also one-to-one? Give reasons for your answer.

77. Suppose that the range of g lies in the domain of f so that the composition $f \circ g$ is defined. If f and g are one-to-one, can anything be said about $f \circ g$? Give reasons for your answer.

78. If a composition $f \circ g$ is one-to-one, must g be one-to-one? Give reasons for your answer.

79. Find a formula for the inverse function f^{-1} and verify that $(f \circ f^{-1})(x) = (f^{-1} \circ f)(x) = x$.

     a. $f(x) = \dfrac{100}{1 + 2^{-x}}$ **b.** $f(x) = \dfrac{50}{1 + 1.1^{-x}}$

     c. $f(x) = \dfrac{e^x - 1}{e^x + 1}$ **d.** $f(x) = \dfrac{\ln x}{2 - \ln x}$

80. The identity $\sin^{-1} x + \cos^{-1} x = \pi/2$ Figure 1.69 establishes the identity for $0 < x < 1$. To establish it for the rest of $[-1, 1]$, verify by direct calculation that it holds for $x = 1$, 0, and -1. Then, for values of x in $(-1, 0)$, let $x = -a$, $a > 0$, and apply Eqs. (3) and (5) to the sum $\sin^{-1}(-a) + \cos^{-1}(-a)$.

81. Start with the graph of $y = \ln x$. Find an equation of the graph that results from

     a. shifting down 3 units.

     b. shifting right 1 unit.

     c. shifting left 1, up 3 units.

     d. shifting down 4, right 2 units.

     e. reflecting about the y-axis.

     f. reflecting about the line $y = x$.

82. Start with the graph of $y = \ln x$. Find an equation of the graph that results from

     a. vertical stretching by a factor of 2.

     b. horizontal stretching by a factor of 3.

     c. vertical compression by a factor of 4.

     d. horizontal compression by a factor of 2.

83. The equation $x^2 = 2^x$ has three solutions: $x = 2$, $x = 4$, and one other. Estimate the third solution as accurately as you can by graphing.

84. Could $x^{\ln 2}$ possibly be the same as $2^{\ln x}$ for $x > 0$? Graph the two functions and explain what you see.

85. Radioactive decay   The half-life of a certain radioactive substance is 12 hours. There are 8 grams present initially.

     a. Express the amount of substance remaining as a function of time t.

     b. When will there be 1 gram remaining?

86. Doubling your money   Determine how much time is required for a \$500 investment to double in value if interest is earned at the rate of 4.75% compounded annually.

87. Population growth   The population of Glenbrook is 375,000 and is increasing at the rate of 2.25% per year. Predict when the population will be 1 million.

88. Radon-222   The decay equation for radon-222 gas is known to be $y = y_0 e^{-0.18t}$, with t in days. About how long will it take the radon in a sealed sample of air to fall to 90% of its original value?

CHAPTER 1   Questions to Guide Your Review

1. What is a function? What is its domain? Its range? What is an arrow diagram for a function? Give examples.

2. What is the graph of a real-valued function of a real variable? What is the vertical line test?

3. What is a piecewise-defined function? Give examples.

4. What are the important types of functions frequently encountered in calculus? Give an example of each type.

5. What is meant by an increasing function? A decreasing function? Give an example of each.

6. What is an even function? An odd function? What symmetry properties do the graphs of such functions have? What advantage can we take of this? Give an example of a function that is neither even nor odd.

7. If f and g are real-valued functions, how are the domains of $f + g, f - g, fg$, and f/g related to the domains of f and g? Give examples.

8. When is it possible to compose one function with another? Give examples of compositions and their values at various points. Does the order in which functions are composed ever matter?

9. How do you change the equation $y = f(x)$ to shift its graph vertically up or down by $|k|$ units? Horizontally to the left or right? Give examples.

10. How do you change the equation $y = f(x)$ to compress or stretch the graph by a factor $c > 1$? Reflect the graph across a coordinate axis? Give examples.

11. What is radian measure? How do you convert from radians to degrees? Degrees to radians?

12. Graph the six basic trigonometric functions. What symmetries do the graphs have?

13. What is a periodic function? Give examples. What are the periods of the six basic trigonometric functions?

14. Starting with the identity $\sin^2 \theta + \cos^2 \theta = 1$ and the formulas for $\cos (A + B)$ and $\sin (A + B)$, show how a variety of other trigonometric identities may be derived.

15. How does the formula for the general sine function $f(x) = A \sin ((2\pi/B)(x - C)) + D$ relate to the shifting, stretching, compressing, and reflection of its graph? Give examples. Graph the general sine curve and identify the constants A, B, C, and D.

16. Name three issues that arise when functions are graphed using a calculator or computer with graphing software. Give examples.

17. What is an exponential function? Give examples. What laws of exponents does it obey? How does it differ from a simple power function like $f(x) = x^n$? What kind of real-world phenomena are modeled by exponential functions?

18. What is the number e, and how is it defined? What are the domain and range of $f(x) = e^x$? What does its graph look like? How do the values of e^x relate to x^2, x^3, and so on?

19. What functions have inverses? How do you know if two functions f and g are inverses of one another? Give examples of functions that are (are not) inverses of one another.

20. How are the domains, ranges, and graphs of functions and their inverses related? Give an example.

21. What procedure can you sometimes use to express the inverse of a function of x as a function of x?

22. What is a logarithmic function? What properties does it satisfy? What is the natural logarithm function? What are the domain and range of $y = \ln x$? What does its graph look like?

23. How is the graph of $\log_a x$ related to the graph of $\ln x$? What truth is in the statement that there is really only one exponential function and one logarithmic function?

24. How are the inverse trigonometric functions defined? How can you sometimes use right triangles to find values of these functions? Give examples.

CHAPTER 1   Practice Exercises

Functions and Graphs

1. Express the area and circumference of a circle as functions of the circle's radius. Then express the area as a function of the circumference.

2. Express the radius of a sphere as a function of the sphere's surface area. Then express the surface area as a function of the volume.

3. A point P in the first quadrant lies on the parabola $y = x^2$. Express the coordinates of P as functions of the angle of inclination of the line joining P to the origin.

4. A hot-air balloon rising straight up from a level field is tracked by a range finder located 500 ft from the point of liftoff. Express the balloon's height as a function of the angle the line from the range finder to the balloon makes with the ground.

In Exercises 5–8, determine whether the graph of the function is symmetric about the y-axis, the origin, or neither.

5. $y = x^{1/5}$

6. $y = x^{2/5}$

7. $y = x^2 - 2x - 1$

8. $y = e^{-x^2}$

In Exercises 9–16, determine whether the function is even, odd, or neither.

9. $y = x^2 + 1$

10. $y = x^5 - x^3 - x$

11. $y = 1 - \cos x$

12. $y = \sec x \tan x$

13. $y = \dfrac{x^4 + 1}{x^3 - 2x}$

14. $y = x - \sin x$

15. $y = x + \cos x$

16. $y = x \cos x$

17. Suppose that f and g are both odd functions defined on the entire real line. Which of the following (where defined) are even? odd?

    a. fg **b.** f^3 **c.** $f(\sin x)$ **d.** $g(\sec x)$ **e.** $|g|$

18. If $f(a - x) = f(a + x)$, show that $g(x) = f(x + a)$ is an even function.

In Exercises 19–32, find the **(a)** domain and **(b)** range.

19. $y = |x| - 2$ **20.** $y = -2 + \sqrt{1 - x}$

21. $y = \sqrt{16 - x^2}$ **22.** $y = 3^{2-x} + 1$

23. $y = 2e^{-x} - 3$ **24.** $y = \tan(2x - \pi)$

25. $y = 2\sin(3x + \pi) - 1$ **26.** $y = x^{2/5}$

27. $y = \ln(x - 3) + 1$ **28.** $y = -1 + \sqrt[3]{2 - x}$

29. $y = 5 - \sqrt{x^2 - 2x - 3}$ **30.** $y = 2 + \dfrac{3x^2}{x^2 + 4}$

31. $y = 4\sin\left(\dfrac{1}{x}\right)$ **32.** $y = 3\cos x + 4\sin x$
                                       (*Hint:* A trig identity is required.)

33. State whether each function is increasing, decreasing, or neither.

    a. Volume of a sphere as a function of its radius

    b. Greatest integer function

    c. Height above Earth's sea level as a function of atmospheric pressure (assumed nonzero)

    d. Kinetic energy as a function of a particle's velocity

34. Find the largest interval on which the given function is increasing.

    a. $f(x) = |x - 2| + 1$ **b.** $f(x) = (x + 1)^4$

    c. $g(x) = (3x - 1)^{1/3}$ **d.** $R(x) = \sqrt{2x - 1}$

Piecewise-Defined Functions

In Exercises 35 and 36, find the **(a)** domain and **(b)** range.

35. $y = \begin{cases} \sqrt{-x}, & -4 \le x \le 0 \\ \sqrt{x}, & 0 < x \le 4 \end{cases}$

36. $y = \begin{cases} -x - 2, & -2 \le x \le -1 \\ x, & -1 < x \le 1 \\ -x + 2, & 1 < x \le 2 \end{cases}$

In Exercises 37 and 38, write a piecewise formula for the function.

37.
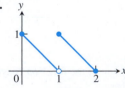

38.
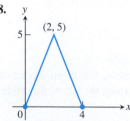

Composition of Functions

In Exercises 39 and 40, find

    a. $(f \circ g)(-1)$.   **b.** $(g \circ f)(2)$.

    c. $(f \circ f)(x)$.   **d.** $(g \circ g)(x)$.

39. $f(x) = \dfrac{1}{x}$, $g(x) = \dfrac{1}{\sqrt{x + 2}}$

40. $f(x) = 2 - x$, $g(x) = \sqrt[3]{x + 1}$

In Exercises 41 and 42, **(a)** write formulas for $f \circ g$ and $g \circ f$ and find the **(b)** domain and **(c)** range of each.

41. $f(x) = 2 - x^2$, $g(x) = \sqrt{x + 2}$

42. $f(x) = \sqrt{x}$, $g(x) = \sqrt{1 - x}$

For Exercises 43 and 44, sketch the graphs of f and $f \circ f$.

43. $f(x) = \begin{cases} -x - 2, & -4 \le x \le -1 \\ -1, & -1 < x \le 1 \\ x - 2, & 1 < x \le 2 \end{cases}$

44. $f(x) = \begin{cases} x + 1, & -2 \le x < 0 \\ x - 1, & 0 \le x \le 2 \end{cases}$

Composition with absolute values In Exercises 45–52, graph f_1 and f_2 together. Then describe how applying the absolute value function in f_2 affects the graph of f_1.

	$f_1(x)$	$f_2(x)$
45.	x	$|x|$
46.	x^2	$|x|^2$
47.	x^3	$|x^3|$
48.	$x^2 + x$	$|x^2 + x|$
49.	$4 - x^2$	$|4 - x^2|$
50.	$\dfrac{1}{x}$	$\dfrac{1}{|x|}$
51.	$\sqrt{x}$	$\sqrt{|x|}$
52.	$\sin x$	$\sin |x|$

Shifting and Scaling Graphs

53. Suppose the graph of g is given. Write equations for the graphs that are obtained from the graph of g by shifting, scaling, or reflecting, as indicated.

    a. Up $\dfrac{1}{2}$ unit, right 3

    b. Down 2 units, left $\dfrac{2}{3}$

    c. Reflect about the y-axis

    d. Reflect about the x-axis

    e. Stretch vertically by a factor of 5

    f. Compress horizontally by a factor of 5

54. Describe how each graph is obtained from the graph of $y = f(x)$.

    a. $y = f(x - 5)$ **b.** $y = f(4x)$

    c. $y = f(-3x)$ **d.** $y = f(2x + 1)$

    e. $y = f\left(\dfrac{x}{3}\right) - 4$ **f.** $y = -3f(x) + \dfrac{1}{4}$

In Exercises 55–58, graph each function, not by plotting points, but by starting with the graph of one of the standard functions presented in Figures 1.15–1.17, and applying an appropriate transformation.

55. $y = -\sqrt{1 + \dfrac{x}{2}}$ **56.** $y = 1 - \dfrac{x}{3}$

57. $y = \dfrac{1}{2x^2} + 1$ **58.** $y = (-5x)^{1/3}$

Trigonometry

In Exercises 59–62, sketch the graph of the given function. What is the period of the function?

59. $y = \cos 2x$

60. $y = \sin \dfrac{x}{2}$

61. $y = \sin \pi x$

62. $y = \cos \dfrac{\pi x}{2}$

63. Sketch the graph $y = 2\cos\left(x - \dfrac{\pi}{3}\right)$.

64. Sketch the graph $y = 1 + \sin\left(x + \dfrac{\pi}{4}\right)$.

In Exercises 65–68, ABC is a right triangle with the right angle at C. The sides opposite angles A, B, and C are a, b, and c, respectively.

65. a. Find a and b if $c = 2$, $B = \pi/3$.

   b. Find a and c if $b = 2$, $B = \pi/3$.

66. a. Express a in terms of A and c.

   b. Express a in terms of A and b.

67. a. Express a in terms of B and b.

   b. Express c in terms of A and a.

68. a. Express $\sin A$ in terms of a and c.

   b. Express $\sin A$ in terms of b and c.

69. Height of a pole Two wires stretch from the top T of a vertical pole to points B and C on the ground, where C is 10 m closer to the base of the pole than is B. If wire BT makes an angle of $35°$ with the horizontal and wire CT makes an angle of $50°$ with the horizontal, how high is the pole?

70. Height of a weather balloon Observers at positions A and B 2 km apart simultaneously measure the angle of elevation of a weather balloon to be $40°$ and $70°$, respectively. If the balloon is directly above a point on the line segment between A and B, find the height of the balloon.

T 71. a. Graph the function $f(x) = \sin x + \cos(x/2)$.

   b. What appears to be the period of this function?

   c. Confirm your finding in part (b) algebraically.

T 72. a. Graph $f(x) = \sin(1/x)$.

   b. What are the domain and range of f?

   c. Is f periodic? Give reasons for your answer.

Transcendental Functions

In Exercises 73–76, find the domain of each function.

73. a. $f(x) = 1 + e^{-\sin x}$ **b.** $g(x) = e^x + \ln \sqrt{x}$

74. a. $f(x) = e^{1/x^2}$ **b.** $g(x) = \ln|4 - x^2|$

75. a. $h(x) = \sin^{-1}\left(\dfrac{x}{3}\right)$ **b.** $f(x) = \cos^{-1}(\sqrt{x} - 1)$

76. a. $h(x) = \ln(\cos^{-1} x)$ **b.** $f(x) = \sqrt{\pi - \sin^{-1} x}$

77. If $f(x) = \ln x$ and $g(x) = 4 - x^2$, find the functions $f \circ g$, $g \circ f$, $f \circ f$, $g \circ g$, and their domains.

78. Determine whether f is even, odd, or neither.

   a. $f(x) = e^{-x^2}$ **b.** $f(x) = 1 + \sin^{-1}(-x)$

   c. $f(x) = |e^x|$ **d.** $f(x) = e^{\ln|x|+1}$

T 79. Graph $\ln x$, $\ln 2x$, $\ln 4x$, $\ln 8x$, and $\ln 16x$ (as many as you can) together for $0 < x \le 10$. What is going on? Explain.

T 80. Graph $y = \ln(x^2 + c)$ for $c = -4, -2, 0, 3$, and 5. How does the graph change when c changes?

T 81. Graph $y = \ln|\sin x|$ in the window $0 \le x \le 22$, $-2 \le y \le 0$. Explain what you see. How could you change the formula to turn the arches upside down?

T 82. Graph the three functions $y = x^a$, $y = a^x$, and $y = \log_a x$ together on the same screen for $a = 2, 10$, and 20. For large values of x, which of these functions has the largest values and which has the smallest values?

Theory and Examples

In Exercises 83 and 84, find the domain and range of each composite function. Then graph the compositions on separate screens. Do the graphs make sense in each case? Give reasons for your answers and comment on any differences you see.

83. a. $y = \sin^{-1}(\sin x)$ **b.** $y = \sin(\sin^{-1} x)$

84. a. $y = \cos^{-1}(\cos x)$ **b.** $y = \cos(\cos^{-1} x)$

85. Use a graph to decide whether f is one-to-one.

   a. $f(x) = x^3 - \dfrac{x}{2}$ **b.** $f(x) = x^3 + \dfrac{x}{2}$

T 86. Use a graph to find to 3 decimal places the values of x for which $e^x > 10{,}000{,}000$.

87. a. Show that $f(x) = x^3$ and $g(x) = \sqrt[3]{x}$ are inverses of one another.

T   b. Graph f and g over an x-interval large enough to show the graphs intersecting at $(1, 1)$ and $(-1, -1)$. Be sure the picture shows the required symmetry in the line $y = x$.

88. a. Show that $h(x) = x^3/4$ and $k(x) = (4x)^{1/3}$ are inverses of one another.

T   b. Graph h and k over an x-interval large enough to show the graphs intersecting at $(2, 2)$ and $(-2, -2)$. Be sure the picture shows the required symmetry in the line $y = x$.

CHAPTER 1   Additional and Advanced Exercises

Functions and Graphs

1. Are there two functions f and g such that $f \circ g = g \circ f$? Give reasons for your answer.

2. Are there two functions f and g with the following property? The graphs of f and g are not straight lines but the graph of $f \circ g$ is a straight line. Give reasons for your answer.

3. If $f(x)$ is odd, can anything be said of $g(x) = f(x) - 2$? What if f is even instead? Give reasons for your answer.

4. If $g(x)$ is an odd function defined for all values of x, can anything be said about $g(0)$? Give reasons for your answer.

5. Graph the equation $|x| + |y| = 1 + x$.

6. Graph the equation $y + |y| = x + |x|$.

Derivations and Proofs

7. Prove the following identities.

a. $\dfrac{1 - \cos x}{\sin x} = \dfrac{\sin x}{1 + \cos x}$ **b.** $\dfrac{1 - \cos x}{1 + \cos x} = \tan^2 \dfrac{x}{2}$

8. Explain the following "proof without words" of the law of cosines. (*Source:* Kung, Sidney H., "Proof Without Words: The Law of Cosines," *Mathematics Magazine*, Vol. 63, no. 5, Dec. 1990, p. 342.)

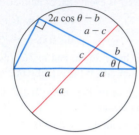

9. Show that the area of triangle *ABC* is given by $(1/2)ab \sin C = (1/2)bc \sin A = (1/2)ca \sin B$.

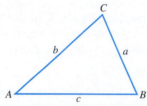

10. Show that the area of triangle *ABC* is given by $\sqrt{s(s-a)(s-b)(s-c)}$ where $s = (a+b+c)/2$ is the semiperimeter of the triangle.

11. Show that if f is both even and odd, then $f(x) = 0$ for every x in the domain of f.

12. a. Even-odd decompositions Let f be a function whose domain is symmetric about the origin, that is, $-x$ belongs to the domain whenever x does. Show that f is the sum of an even function and an odd function:

$$f(x) = E(x) + O(x),$$

where E is an even function and O is an odd function. (*Hint:* Let $E(x) = (f(x) + f(-x))/2$. Show that $E(-x) = E(x)$, so that E is even. Then show that $O(x) = f(x) - E(x)$ is odd.)

b. Uniqueness Show that there is only one way to write f as the sum of an even and an odd function. (*Hint:* One way is given in part (a). If also $f(x) = E_1(x) + O_1(x)$ where E_1 is even and O_1 is odd, show that $E - E_1 = O_1 - O$. Then use Exercise 11 to show that $E = E_1$ and $O = O_1$.)

Effects of Parameters on Graphs

[T] **13.** What happens to the graph of $y = ax^2 + bx + c$ as

   a. a changes while b and c remain fixed?

   b. b changes (a and c fixed, $a \neq 0$)?

   c. c changes (a and b fixed, $a \neq 0$)?

[T] **14.** What happens to the graph of $y = a(x + b)^3 + c$ as

   a. a changes while b and c remain fixed?

   b. b changes (a and c fixed, $a \neq 0$)?

   c. c changes (a and b fixed, $a \neq 0$)?

Geometry

15. An object's center of mass moves at a constant velocity v along a straight line past the origin. The accompanying figure shows the coordinate system and the line of motion. The dots show positions that are 1 sec apart. Why are the areas $A_1, A_2, \ldots, A_5$ in the figure all equal? As in Kepler's equal area law (see Section 13.6), the line that joins the object's center of mass to the origin sweeps out equal areas in equal times.

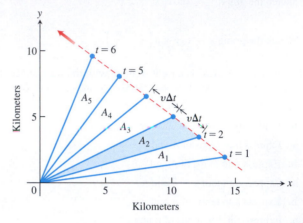

16. a. Find the slope of the line from the origin to the midpoint P of side AB in the triangle in the accompanying figure ($a, b > 0$).

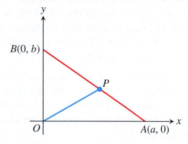

   b. When is OP perpendicular to AB?

17. Consider the quarter-circle of radius 1 and right triangles ABE and ACD given in the accompanying figure. Use standard area formulas to conclude that

$$\frac{1}{2}\sin\theta\cos\theta < \frac{\theta}{2} < \frac{1}{2}\frac{\sin\theta}{\cos\theta}.$$

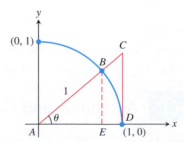

18. Let $f(x) = ax + b$ and $g(x) = cx + d$. What condition must be satisfied by the constants a, b, c, d in order that $(f \circ g)(x) = (g \circ f)(x)$ for every value of x?

Theory and Examples

19. Domain and range Suppose that $a \neq 0, b \neq 1$, and $b > 0$. Determine the domain and range of the function.

a. $y = a(b^{c-x}) + d$ **b.** $y = a \log_b(x - c) + d$

20. Inverse functions Let

$$f(x) = \frac{ax + b}{cx + d}, \qquad c \neq 0, \qquad ad - bc \neq 0.$$

a. Give a convincing argument that f is one-to-one.

b. Find a formula for the inverse of f.

21. Depreciation Smith Hauling purchased an 18-wheel truck for $100,000. The truck depreciates at the constant rate of $10,000 per year for 10 years.

a. Write an expression that gives the value y after x years.

b. When is the value of the truck $55,000?

22. Drug absorption A drug is administered intravenously for pain. The function

$$f(t) = 90 - 52 \ln(1 + t), \qquad 0 \leq t \leq 4$$

gives the number of units of the drug remaining in the body after t hours.

a. What was the initial number of units of the drug administered?

b. How much is present after 2 hours?

c. Draw the graph of f.

23. Finding investment time If Juanita invests $1500 in a retirement account that earns 8% compounded annually, how long will it take this single payment to grow to $5000?

24. The rule of 70 If you use the approximation $\ln 2 \approx 0.70$ (in place of $0.69314\ldots$), you can derive a rule of thumb that says, "To estimate how many years it will take an amount of money to double when invested at r percent compounded continuously, divide r into 70." For instance, an amount of money invested at 5% will double in about $70/5 = 14$ years. If you want it to double in 10 years instead, you have to invest it at $70/10 = 7\%$. Show how the rule of 70 is derived. (A similar "rule of 72" uses 72 instead of 70, because 72 has more integer factors.)

25. For what $x > 0$ does $x^{(x^x)} = (x^x)^x$? Give reasons for your answer.

T **26. a.** If $(\ln x)/x = (\ln 2)/2$, must $x = 2$?

b. If $(\ln x)/x = -2 \ln 2$, must $x = 1/2$?

Give reasons for your answers.

27. The quotient $(\log_4 x)/(\log_2 x)$ has a constant value. What value? Give reasons for your answer.

T **28.** $\log_x (2)$ **vs.** $\log_2 (x)$ How does $f(x) = \log_x (2)$ compare with $g(x) = \log_2 (x)$? Here is one way to find out.

a. Use the equation $\log_a b = (\ln b)/(\ln a)$ to express $f(x)$ and $g(x)$ in terms of natural logarithms.

b. Graph f and g together. Comment on the behavior of f in relation to the signs and values of g.

| CHAPTER 1 | **Technology Application Projects** |

Mathematica/Maple Projects

Projects can be found within MyMathLab.

- **An Overview of Mathematica**
  An overview of *Mathematica* sufficient to complete the *Mathematica* modules appearing on the Web site.

- **Modeling Change: Springs, Driving Safety, Radioactivity, Trees, Fish, and Mammals**
  Construct and interpret mathematical models, analyze and improve them, and make predictions using them.

2

Limits and Continuity

OVERVIEW In this chapter we develop the concept of a limit, first intuitively and then formally. We use limits to describe the way a function varies. Some functions vary *continuously*; small changes in x produce only small changes in $f(x)$. Other functions can have values that jump, vary erratically, or tend to increase or decrease without bound. The notion of limit gives a precise way to distinguish among these behaviors.

2.1 Rates of Change and Tangent Lines to Curves

Average and Instantaneous Speed

In the late sixteenth century, Galileo discovered that a solid object dropped from rest (initially not moving) near the surface of the earth and allowed to fall freely will fall a distance proportional to the square of the time it has been falling. This type of motion is called **free fall**. It assumes negligible air resistance to slow the object down, and that gravity is the only force acting on the falling object. If y denotes the distance fallen in feet after t seconds, then Galileo's law is

$$y = 16t^2 \text{ ft,}$$

where 16 is the (approximate) constant of proportionality. (If y is measured in meters instead, then the constant is close to 4.9.)

More generally, suppose that a moving object has traveled distance $f(t)$ at time t. The object's **average speed** during an interval of time $[t_1, t_2]$ is found by dividing the distance traveled $f(t_2) - f(t_1)$ by the time elapsed $t_2 - t_1$. The unit of measure is length per unit time: kilometers per hour, feet (or meters) per second, or whatever is appropriate to the problem at hand.

Average Speed

When $f(t)$ measures the distance traveled at time t,

$$\text{Average speed over } [t_1, t_2] = \frac{\text{distance traveled}}{\text{elapsed time}} = \frac{f(t_2) - f(t_1)}{t_2 - t_1}$$

EXAMPLE 1 A rock breaks loose from the top of a tall cliff. What is its average speed

(a) during the first 2 sec of fall?

(b) during the 1-sec interval between second 1 and second 2?

Δ is the capital Greek letter Delta

Solution The average speed of the rock during a given time interval is the change in distance, Δy, divided by the length of the time interval, Δt. (The capital Greek letter Delta, written Δ, is traditionally used to indicate the increment, or change, in a variable. Increments like Δy and Δt are reviewed in Appendix 3, and pronounced "delta y" and "delta t.") Measuring distance in feet and time in seconds, we have the following calculations:

(a) For the first 2 sec: $\dfrac{\Delta y}{\Delta t} = \dfrac{16(2)^2 - 16(0)^2}{2 - 0} = 32\,\dfrac{\text{ft}}{\text{sec}}$

(b) From sec 1 to sec 2: $\dfrac{\Delta y}{\Delta t} = \dfrac{16(2)^2 - 16(1)^2}{2 - 1} = 48\,\dfrac{\text{ft}}{\text{sec}}$ ■

We want a way to determine the speed of a falling object at a single instant t_0, instead of using its average speed over an interval of time. To do this, we examine what happens when we calculate the average speed over shorter and shorter time intervals starting at t_0. The next example illustrates this process. Our discussion is informal here but will be made precise in Chapter 3.

EXAMPLE 2 Find the speed of the falling rock in Example 1 at $t = 1$ and $t = 2$ sec.

Solution We can calculate the average speed of the rock over a time interval $[t_0, t_0 + h]$, having length $\Delta t = (t_0 + h) - (t_0) = h$, as

$$\frac{\Delta y}{\Delta t} = \frac{16(t_0 + h)^2 - 16t_0^2}{h}\,\frac{\text{ft}}{\text{sec}}. \tag{1}$$

We cannot use this formula to calculate the "instantaneous" speed at the exact moment t_0 by simply substituting $h = 0$, because we cannot divide by zero. But we *can* use it to calculate average speeds over shorter and shorter time intervals starting at either $t_0 = 1$ or $t_0 = 2$. When we do so, by taking smaller and smaller values of h, we see a pattern (Table 2.1).

TABLE 2.1 **Average speeds over short time intervals** $[t_0, t_0 + h]$

$$\text{Average speed:} \quad \frac{\Delta y}{\Delta t} = \frac{16(t_0 + h)^2 - 16t_0^2}{h}$$

Length of time interval h	Average speed over interval of length h starting at $t_0 = 1$	Average speed over interval of length h starting at $t_0 = 2$
1	48	80
0.1	33.6	65.6
0.01	32.16	64.16
0.001	32.016	64.016
0.0001	32.0016	64.0016

The average speed on intervals starting at $t_0 = 1$ seems to approach a limiting value of 32 as the length of the interval decreases. This suggests that the rock is falling at a speed of 32 ft/sec at $t_0 = 1$ sec. Let's confirm this algebraically.

If we set $t_0 = 1$ and then expand the numerator in Equation (1) and simplify, we find that

$$\frac{\Delta y}{\Delta t} = \frac{16(1 + h)^2 - 16(1)^2}{h} = \frac{16(1 + 2h + h^2) - 16}{h}$$

$$= \frac{32h + 16h^2}{h} = 32 + 16h. \qquad \text{Can cancel } h \text{ when } h \neq 0$$

For values of h different from 0, the expressions on the right and left are equivalent and the average speed is $32 + 16h$ ft/sec. We can now see why the average speed has the limiting value $32 + 16(0) = 32$ ft/sec as h approaches 0.

Similarly, setting $t_0 = 2$ in Equation (1), for values of h different from 0 the procedure yields

$$\frac{\Delta y}{\Delta t} = 64 + 16h.$$

As h gets closer and closer to 0, the average speed has the limiting value 64 ft/sec when $t_0 = 2$ sec, as suggested by Table 2.1. ∎

The average speed of a falling object is an example of a more general idea, an average rate of change.

Average Rates of Change and Secant Lines

Given any function $y = f(x)$, we calculate the average rate of change of y with respect to x over the interval $[x_1, x_2]$ by dividing the change in the value of y, $\Delta y = f(x_2) - f(x_1)$, by the length $\Delta x = x_2 - x_1 = h$ of the interval over which the change occurs. (We use the symbol h for Δx to simplify the notation here and later on.)

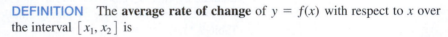

> **DEFINITION** The **average rate of change** of $y = f(x)$ with respect to x over the interval $[x_1, x_2]$ is
>
> $$\frac{\Delta y}{\Delta x} = \frac{f(x_2) - f(x_1)}{x_2 - x_1} = \frac{f(x_1 + h) - f(x_1)}{h}, \qquad h \neq 0.$$

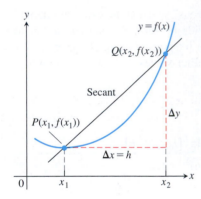

FIGURE 2.1 A secant to the graph $y = f(x)$. Its slope is $\Delta y / \Delta x$, the average rate of change of f over the interval $[x_1, x_2]$.

Geometrically, the rate of change of f over $[x_1, x_2]$ is the slope of the line through the points $P(x_1, f(x_1))$ and $Q(x_2, f(x_2))$ (Figure 2.1). In geometry, a line joining two points of a curve is called a **secant line**. Thus, the average rate of change of f from x_1 to x_2 is identical with the slope of secant line PQ. As the point Q approaches the point P along the curve, the length h of the interval over which the change occurs approaches zero. We will see that this procedure leads to the definition of the slope of a curve at a point.

Defining the Slope of a Curve

We know what is meant by the slope of a straight line, which tells us the rate at which it rises or falls—its rate of change as a linear function. But what is meant by the *slope of a curve* at a point P on the curve? If there is a *tangent line* to the curve at P—a line that grazes the curve like the tangent line to a circle—it would be reasonable to identify the slope of the tangent line as the slope of the curve at P. We will see that, among all the lines that pass through the point P, the tangent line is the one that gives the best approximation to the curve at P. We need a precise way to specify the tangent line at a point on a curve.

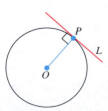

FIGURE 2.2 L is tangent to the circle at P if it passes through P perpendicular to radius OP.

Specifying a tangent line to a circle is straightforward. A line L is tangent to a circle at a point P if L passes through P and is perpendicular to the radius at P (Figure 2.2). But what does it mean to say that a line L is tangent to a more general curve at a point P?

To define tangency for general curves, we use an approach that analyzes the behavior of the secant lines that pass through P and nearby points Q as Q moves toward P along the curve (Figure 2.3). We start with what we *can* calculate, namely the slope of the secant line PQ. We then compute the limiting value of the secant line's slope as Q approaches P along the curve. (We clarify the limit idea in the next section.) If the limit exists, we take it to be the slope of the curve at P and *define* the tangent line to the curve at P to be the line through P with this slope.

The next example illustrates the geometric idea for finding the tangent line to a curve.

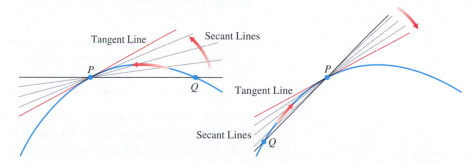

FIGURE 2.3 The tangent line to the curve at P is the line through P whose slope is the limit of the secant line slopes as $Q \to P$ from either side.

EXAMPLE 3 Find the slope of the tangent line to the parabola $y = x^2$ at the point $(2, 4)$ by analyzing the slopes of secant lines through $(2, 4)$. Write an equation for the tangent line to the parabola at this point.

Solution We begin with a secant line through $P(2, 4)$ and a nearby point $Q(2 + h, (2 + h)^2)$. We then write an expression for the slope of the secant line PQ and investigate what happens to the slope as Q approaches P along the curve:

$$\text{Secant line slope} = \frac{\Delta y}{\Delta x} = \frac{(2 + h)^2 - 2^2}{h} = \frac{h^2 + 4h + 4 - 4}{h}$$

$$= \frac{h^2 + 4h}{h} = h + 4.$$

If $h > 0$, then Q lies above and to the right of P, as in Figure 2.4. If $h < 0$, then Q lies to the left of P (not shown). In either case, as Q approaches P along the curve, h approaches zero and the secant line slope $h + 4$ approaches 4. We take 4 to be the parabola's slope at P.

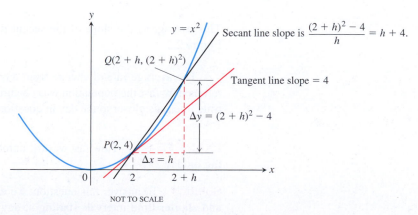

NOT TO SCALE

FIGURE 2.4 Finding the slope of the parabola $y = x^2$ at the point $P(2, 4)$ as the limit of secant line slopes (Example 3).

The tangent line to the parabola at P is the line through P with slope 4:

$$y = 4 + 4(x - 2) \qquad \text{Point-slope equation}$$
$$y = 4x - 4.$$

Rates of Change and Tangent Lines

The rates at which the rock in Example 2 was falling at the instants $t = 1$ and $t = 2$ are called *instantaneous rates of change*. Instantaneous rates of change and slopes of tangent lines are closely connected, as we see in the following examples.

EXAMPLE 4 Figure 2.5 shows how a population p of fruit flies (*Drosophila*) grew in a 50-day experiment. The number of flies was counted at regular intervals, the counted values plotted with respect to the number of elapsed days t, and the points joined by a smooth curve (colored blue in Figure 2.5). Find the average growth rate from day 23 to day 45.

Solution There were 150 flies on day 23 and 340 flies on day 45. Thus the number of flies increased by $340 - 150 = 190$ in $45 - 23 = 22$ days. The average rate of change of the population from day 23 to day 45 was

$$\text{Average rate of change:} \quad \frac{\Delta p}{\Delta t} = \frac{340 - 150}{45 - 23} = \frac{190}{22} \approx 8.6 \text{ flies/day.}$$

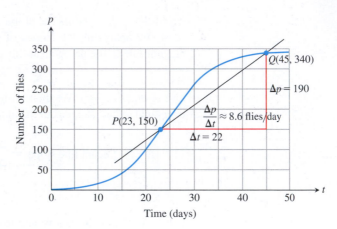

FIGURE 2.5 Growth of a fruit fly population in a controlled experiment. The average rate of change over 22 days is the slope $\Delta p / \Delta t$ of the secant line (Example 4).

This average is the slope of the secant line through the points P and Q on the graph in Figure 2.5. ■

The average rate of change from day 23 to day 45 calculated in Example 4 does not tell us how fast the population was changing on day 23 itself. For that we need to examine time intervals closer to the day in question.

EXAMPLE 5 How fast was the number of flies in the population of Example 4 growing on day 23?

Solution To answer this question, we examine the average rates of change over shorter and shorter time intervals starting at day 23. In geometric terms, we find these rates by calculating the slopes of secant lines from P to Q, for a sequence of points Q approaching P along the curve (Figure 2.6).

Q	Slope of $PQ = \Delta p / \Delta t$ (flies / day)
(45, 340)	$\dfrac{340 - 150}{45 - 23} \approx 8.6$
(40, 330)	$\dfrac{330 - 150}{40 - 23} \approx 10.6$
(35, 310)	$\dfrac{310 - 150}{35 - 23} \approx 13.3$
(30, 265)	$\dfrac{265 - 150}{30 - 23} \approx 16.4$

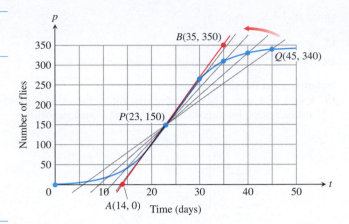

FIGURE 2.6 The positions and slopes of four secant lines through the point P on the fruit fly graph (Example 5).

The values in the table show that the secant line slopes rise from 8.6 to 16.4 as the t-coordinate of Q decreases from 45 to 30, and we would expect the slopes to rise slightly higher as t continued decreasing toward 23. Geometrically, the secant lines rotate counterclockwise about P and seem to approach the red tangent line in the figure. Since the line appears to pass through the points (14, 0) and (35, 350), its slope is approximately

$$\frac{350 - 0}{35 - 14} = 16.7 \text{ flies/day.}$$

On day 23 the population was increasing at a rate of about 16.7 flies / day. ∎

The instantaneous rate of change is the value the average rate of change approaches as the length h of the interval over which the change occurs approaches zero. The average rate of change corresponds to the slope of a secant line; the instantaneous rate corresponds to the slope of the tangent line at a fixed value. So instantaneous rates and slopes of tangent lines are closely connected. We give a precise definition for these terms in the next chapter, but to do so we first need to develop the concept of a *limit*.

EXERCISES 2.1

Average Rates of Change

In Exercises 1–6, find the average rate of change of the function over the given interval or intervals.

1. $f(x) = x^3 + 1$

   a. $[2, 3]$ **b.** $[-1, 1]$

2. $g(x) = x^2 - 2x$

   a. $[1, 3]$ **b.** $[-2, 4]$

3. $h(t) = \cot t$

   a. $[\pi/4, 3\pi/4]$ **b.** $[\pi/6, \pi/2]$

4. $g(t) = 2 + \cos t$

   a. $[0, \pi]$ **b.** $[-\pi, \pi]$

5. $R(\theta) = \sqrt{4\theta + 1}; \quad [0, 2]$

6. $P(\theta) = \theta^3 - 4\theta^2 + 5\theta; \quad [1, 2]$

Slope of a Curve at a Point

In Exercises 7–18, use the method in Example 3 to find **(a)** the slope of the curve at the given point P, and **(b)** an equation of the tangent line at P.

7. $y = x^2 - 5, \quad P(2, -1)$

8. $y = 7 - x^2, \quad P(2, 3)$

9. $y = x^2 - 2x - 3, \quad P(2, -3)$

10. $y = x^2 - 4x, \quad P(1, -3)$

11. $y = x^3, \quad P(2, 8)$

12. $y = 2 - x^3, \quad P(1, 1)$

13. $y = x^3 - 12x, \quad P(1, -11)$

14. $y = x^3 - 3x^2 + 4, \quad P(2, 0)$

15. $y = \dfrac{1}{x}, \quad P(-2, -1/2)$

16. $y = \dfrac{x}{2 - x}$, $P(4, -2)$

17. $y = \sqrt{x}$, $P(4, 2)$

18. $y = \sqrt{7 - x}$, $P(-2, 3)$

Instantaneous Rates of Change

19. Speed of a car The accompanying figure shows the time-to-distance graph for a sports car accelerating from a standstill.

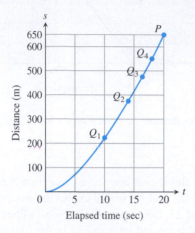

a. Estimate the slopes of secant lines PQ_1, PQ_2, PQ_3, and PQ_4, arranging them in order in a table like the one in Figure 2.6. What are the appropriate units for these slopes?

b. Then estimate the car's speed at time $t = 20\,\text{sec}$.

20. The accompanying figure shows the plot of distance fallen versus time for an object that fell from the lunar landing module a distance 80 m to the surface of the moon.

a. Estimate the slopes of the secant lines PQ_1, PQ_2, PQ_3, and PQ_4, arranging them in a table like the one in Figure 2.6.

b. About how fast was the object going when it hit the surface?

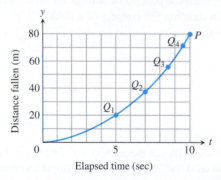

T 21. The profits of a small company for each of the first five years of its operation are given in the following table:

Year	Profit in $1000s
2010	6
2011	27
2012	62
2013	111
2014	174

a. Plot points representing the profit as a function of year, and join them by as smooth a curve as you can.

b. What is the average rate of increase of the profits between 2012 and 2014?

c. Use your graph to estimate the rate at which the profits were changing in 2012.

T 22. Make a table of values for the function $F(x) = (x + 2)/(x - 2)$ at the points $x = 1.2$, $x = 11/10$, $x = 101/100$, $x = 1001/1000$, $x = 10001/10000$, and $x = 1$.

a. Find the average rate of change of $F(x)$ over the intervals $[1, x]$ for each $x \neq 1$ in your table.

b. Extending the table if necessary, try to determine the rate of change of $F(x)$ at $x = 1$.

T 23. Let $g(x) = \sqrt{x}$ for $x \geq 0$.

a. Find the average rate of change of $g(x)$ with respect to x over the intervals $[1, 2]$, $[1, 1.5]$ and $[1, 1 + h]$.

b. Make a table of values of the average rate of change of g with respect to x over the interval $[1, 1 + h]$ for some values of h approaching zero, say $h = 0.1, 0.01, 0.001, 0.0001, 0.00001$, and 0.000001.

c. What does your table indicate is the rate of change of $g(x)$ with respect to x at $x = 1$?

d. Calculate the limit as h approaches zero of the average rate of change of $g(x)$ with respect to x over the interval $[1, 1 + h]$.

T 24. Let $f(t) = 1/t$ for $t \neq 0$.

a. Find the average rate of change of f with respect to t over the intervals (i) from $t = 2$ to $t = 3$, and (ii) from $t = 2$ to $t = T$.

b. Make a table of values of the average rate of change of f with respect to t over the interval $[2, T]$, for some values of T approaching 2, say $T = 2.1, 2.01, 2.001, 2.0001, 2.00001$, and 2.000001.

c. What does your table indicate is the rate of change of f with respect to t at $t = 2$?

d. Calculate the limit as T approaches 2 of the average rate of change of f with respect to t over the interval from 2 to T. You will have to do some algebra before you can substitute $T = 2$.

25. The accompanying graph shows the total distance s traveled by a bicyclist after t hours.

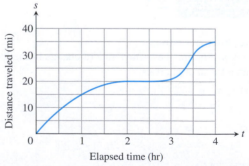

a. Estimate the bicyclist's average speed over the time intervals $[0, 1]$, $[1, 2.5]$, and $[2.5, 3.5]$.

b. Estimate the bicyclist's instantaneous speed at the times $t = \frac{1}{2}$, $t = 2$, and $t = 3$.

c. Estimate the bicyclist's maximum speed and the specific time at which it occurs.

26. The accompanying graph shows the total amount of gasoline A in the gas tank of an automobile after being driven for t days.

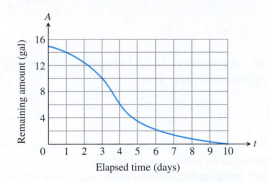

Elapsed time (days)

a. Estimate the average rate of gasoline consumption over the time intervals $[0, 3]$, $[0, 5]$, and $[7, 10]$.

b. Estimate the instantaneous rate of gasoline consumption at the times $t = 1$, $t = 4$, and $t = 8$.

c. Estimate the maximum rate of gasoline consumption and the specific time at which it occurs.

2.2 Limit of a Function and Limit Laws

In Section 2.1 we saw how limits arise when finding the instantaneous rate of change of a function or the tangent line to a curve. We begin this section by presenting an informal definition of the limit of a function. We then describe laws that capture the behavior of limits. These laws enable us to quickly compute limits for a variety of functions, including polynomials and rational functions. We present the precise definition of a limit in the next section.

HISTORICAL ESSAY

Limits

www.goo.gl/5V45iK

Limits of Function Values

Frequently when studying a function $y = f(x)$, we find ourselves interested in the function's behavior *near* a particular point c, but not *at* c itself. An important example occurs when the process of trying to evaluate a function at c leads to division by zero, which is undefined. We encountered this when seeking the instantaneous rate of change in y by considering the quotient function $\Delta y / h$ for h closer and closer to zero. In the next example we explore numerically how a function behaves near a particular point at which we cannot directly evaluate the function.

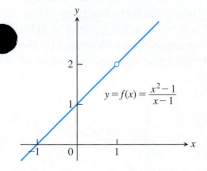

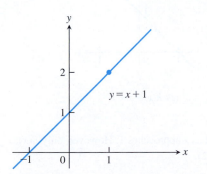

FIGURE 2.7 The graph of f is identical with the line $y = x + 1$ except at $x = 1$, where f is not defined (Example 1).

EXAMPLE 1   How does the function

$$f(x) = \frac{x^2 - 1}{x - 1}$$

behave near $x = 1$?

Solution   The given formula defines f for all real numbers x except $x = 1$ (since we cannot divide by zero). For any $x \neq 1$, we can simplify the formula by factoring the numerator and canceling common factors:

$$f(x) = \frac{(x - 1)(x + 1)}{x - 1} = x + 1 \qquad \text{for} \qquad x \neq 1.$$

The graph of f is the line $y = x + 1$ with the point $(1, 2)$ *removed*. This removed point is shown as a "hole" in Figure 2.7. Even though $f(1)$ is not defined, it is clear that we can make the value of $f(x)$ as close as we want to 2 by choosing x close enough to 1 (Table 2.2). ∎

An Informal Description of the Limit of a Function

We now give an informal definition of the limit of a function f at an interior point of the domain of f. Suppose that $f(x)$ is defined on an open interval about c, *except possibly at c*

TABLE 2.2 As x gets closer to 1, $f(x)$ gets closer to 2.

x	$f(x) = \dfrac{x^2-1}{x-1}$
0.9	1.9
1.1	2.1
0.99	1.99
1.01	2.01
0.999	1.999
1.001	2.001
0.999999	1.999999
1.000001	2.000001

itself. If $f(x)$ is arbitrarily close to the number L (as close to L as we like) for all x sufficiently close to c, other than c itself, then we say that f approaches the **limit** L as x approaches c, and write

$$\lim_{x \to c} f(x) = L,$$

which is read "the limit of $f(x)$ as x approaches c is L." In Example 1 we would say that $f(x)$ approaches the *limit* 2 as x approaches 1, and write

$$\lim_{x \to 1} f(x) = 2, \quad \text{or} \quad \lim_{x \to 1} \frac{x^2 - 1}{x - 1} = 2.$$

Essentially, the definition says that the values of $f(x)$ are close to the number L whenever x is close to c. The value of the function at c itself is not considered.

Our definition here is informal, because phrases like *arbitrarily close* and *sufficiently close* are imprecise; their meaning depends on the context. (To a machinist manufacturing a piston, *close* may mean *within a few thousandths of an inch*. To an astronomer studying distant galaxies, *close* may mean *within a few thousand light-years*.) Nevertheless, the definition is clear enough to enable us to recognize and evaluate limits of many specific functions. We will need the precise definition given in Section 2.3, when we set out to prove theorems about limits or study complicated functions. Here are several more examples exploring the idea of limits.

EXAMPLE 2 The limit of a function does not depend on how the function is defined at the point being approached. Consider the three functions in Figure 2.8. The function f has limit 2 as $x \to 1$ even though f is not defined at $x = 1$. The function g has limit 2 as $x \to 1$ even though $2 \neq g(1)$. The function h is the only one of the three functions in Figure 2.8 whose limit as $x \to 1$ equals its value at $x = 1$. For h, we have $\lim_{x \to 1} h(x) = h(1)$. This equality of limit and function value has an important meaning. As illustrated by the three examples in Figure 2.8, equality of limit and function value captures the notion of "continuity." We study this in detail in Section 2.5. ∎

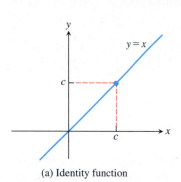

(a) Identity function

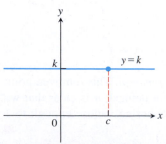

(b) Constant function

FIGURE 2.9 The functions in Example 3 have limits at all points c.

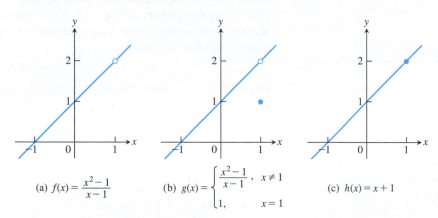

(a) $f(x) = \dfrac{x^2-1}{x-1}$

(b) $g(x) = \begin{cases} \dfrac{x^2-1}{x-1}, & x \neq 1 \\ 1, & x = 1 \end{cases}$

(c) $h(x) = x + 1$

FIGURE 2.8 The limits of $f(x)$, $g(x)$, and $h(x)$ all equal 2 as x approaches 1. However, only $h(x)$ has the same function value as its limit at $x = 1$ (Example 2).

The process of finding a limit can be broken up into a series of steps involving limits of basic functions, which are combined using a sequence of simple operations that we will develop. We start with two basic functions.

EXAMPLE 3 We find the limits of the identity function and of a constant function as x approaches $x = c$.

(a) If f is the **identity function** $f(x) = x$, then for any value of c (Figure 2.9a),

$$\lim_{x \to c} f(x) = \lim_{x \to c} x = c.$$

(b) If f is the **constant function** $f(x) = k$ (function with the constant value k), then for any value of c (Figure 2.9b),

$$\lim_{x \to c} f(x) = \lim_{x \to c} k = k.$$

For instances of each of these rules we have

$$\lim_{x \to 3} x = 3 \qquad \text{Limit of identity function at } x = 3$$

and

$$\lim_{x \to -7} (4) = \lim_{x \to 2} (4) = 4. \qquad \begin{array}{l} \text{Limit of constant function} \\ f(x) = 4 \text{ at } x = -7 \text{ or at } x = 2 \end{array}$$

We prove these rules in Example 3 in Section 2.3. ◼

A function may not have a limit at a particular point. Some ways that limits can fail to exist are illustrated in Figure 2.10 and described in the next example.

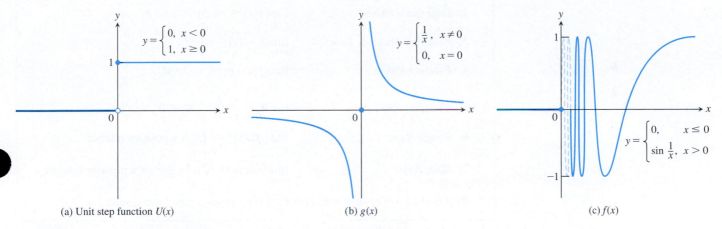

(a) Unit step function $U(x)$ (b) $g(x)$ (c) $f(x)$

FIGURE 2.10   None of these functions has a limit as x approaches 0 (Example 4).

EXAMPLE 4   Discuss the behavior of the following functions, explaining why they have no limit as $x \to 0$.

(a) $U(x) = \begin{cases} 0, & x < 0 \\ 1, & x \geq 0 \end{cases}$

(b) $g(x) = \begin{cases} \dfrac{1}{x}, & x \neq 0 \\ 0, & x = 0 \end{cases}$

(c) $f(x) = \begin{cases} 0, & x \leq 0 \\ \sin \dfrac{1}{x}, & x > 0 \end{cases}$

Solution

(a) The function *jumps*: The **unit step function** $U(x)$ has no limit as $x \to 0$ because its values jump at $x = 0$. For negative values of x arbitrarily close to zero, $U(x) = 0$. For positive values of x arbitrarily close to zero, $U(x) = 1$. There is no *single* value L approached by $U(x)$ as $x \to 0$ (Figure 2.10a).

(b) The function *grows too "large"* to have a limit: $g(x)$ has no limit as $x \to 0$ because the values of g grow arbitrarily large in absolute value as $x \to 0$ and therefore do not stay close to *any* fixed real number (Figure 2.10b). We say the function is *not bounded*.

(c) The function *oscillates too much to have a limit*: $f(x)$ has no limit as $x \to 0$ because the function's values oscillate between $+1$ and -1 in every open interval containing 0. The values do not stay close to any single number as $x \to 0$ (Figure 2.10c). ∎

The Limit Laws

A few basic rules allow us to break down complicated functions into simple ones when calculating limits. By using these laws, we can greatly simplify many limit computations.

THEOREM 1—Limit Laws

If L, M, c, and k are real numbers and

$$\lim_{x \to c} f(x) = L \quad \text{and} \quad \lim_{x \to c} g(x) = M, \quad \text{then}$$

1. *Sum Rule:* $\qquad\qquad\qquad \lim_{x \to c}(f(x) + g(x)) = L + M$

2. *Difference Rule:* $\qquad\qquad \lim_{x \to c}(f(x) - g(x)) = L - M$

3. *Constant Multiple Rule:* $\qquad \lim_{x \to c}(k \cdot f(x)) = k \cdot L$

4. *Product Rule:* $\qquad\qquad\quad \lim_{x \to c}(f(x) \cdot g(x)) = L \cdot M$

5. *Quotient Rule:* $\qquad\qquad\quad \lim_{x \to c}\dfrac{f(x)}{g(x)} = \dfrac{L}{M}, \quad M \neq 0$

6. *Power Rule:* $\qquad\qquad\quad\ \lim_{x \to c}[\,f(x)\,]^n = L^n$, n a positive integer

7. *Root Rule:* $\qquad\qquad\qquad \lim_{x \to c}\sqrt[n]{f(x)} = \sqrt[n]{L} = L^{1/n}$, n a positive integer

(If n is even, we assume that $f(x) \geq 0$ for x in an interval containing c.)

The Sum Rule says that the limit of a sum is the sum of the limits. Similarly, the next rules say that the limit of a difference is the difference of the limits; the limit of a constant times a function is the constant times the limit of the function; the limit of a product is the product of the limits; the limit of a quotient is the quotient of the limits (provided that the limit of the denominator is not 0); the limit of a positive integer power (or root) of a function is the integer power (or root) of the limit (provided that the root of the limit is a real number).

There are simple intuitive arguments for why the properties in Theorem 1 are true (although these do not constitute proofs). If x is sufficiently close to c, then $f(x)$ is close to L and $g(x)$ is close to M, from our informal definition of a limit. It is then reasonable that $f(x) + g(x)$ is close to $L + M$; $f(x) - g(x)$ is close to $L - M$; $kf(x)$ is close to kL; $f(x)g(x)$ is close to LM; and $f(x)/g(x)$ is close to L/M if M is not zero. We prove the Sum Rule in Section 2.3, based on a rigorous definition of the limit. Rules 2–5 are proved in Appendix 4. Rule 6 is obtained by applying Rule 4 repeatedly. Rule 7 is proved in more advanced texts. The Sum, Difference, and Product Rules can be extended to any number of functions, not just two.

EXAMPLE 5 Use the observations $\lim_{x \to c} k = k$ and $\lim_{x \to c} x = c$ (Example 3) and the limit laws in Theorem 1 to find the following limits.

(a) $\lim_{x \to c}(x^3 + 4x^2 - 3)$

(b) $\lim_{x \to c}\dfrac{x^4 + x^2 - 1}{x^2 + 5}$

(c) $\lim_{x \to -2}\sqrt{4x^2 - 3}$

Solution

(a) $\lim_{x \to c} (x^3 + 4x^2 - 3) = \lim_{x \to c} x^3 + \lim_{x \to c} 4x^2 - \lim_{x \to c} 3$ Sum and Difference Rules

$= c^3 + 4c^2 - 3$ Power and Multiple Rules

(b) $\lim_{x \to c} \dfrac{x^4 + x^2 - 1}{x^2 + 5} = \dfrac{\lim_{x \to c} (x^4 + x^2 - 1)}{\lim_{x \to c} (x^2 + 5)}$ Quotient Rule

$= \dfrac{\lim_{x \to c} x^4 + \lim_{x \to c} x^2 - \lim_{x \to c} 1}{\lim_{x \to c} x^2 + \lim_{x \to c} 5}$ Sum and Difference Rules

$= \dfrac{c^4 + c^2 - 1}{c^2 + 5}$ Power or Product Rule

(c) $\lim_{x \to -2} \sqrt{4x^2 - 3} = \sqrt{\lim_{x \to -2} (4x^2 - 3)}$ Root Rule with $n = 2$

$= \sqrt{\lim_{x \to -2} 4x^2 - \lim_{x \to -2} 3}$ Difference Rule

$= \sqrt{4(-2)^2 - 3}$ Product and Multiple Rules and limit of a constant function

$= \sqrt{16 - 3}$

$= \sqrt{13}$ ∎

Evaluating Limits of Polynomials and Rational Functions

Theorem 1 simplifies the task of calculating limits of polynomials and rational functions. To evaluate the limit of a polynomial function as x approaches c, just substitute c for x in the formula for the function. To evaluate the limit of a rational function as x approaches a point c *at which the denominator is not zero*, substitute c for x in the formula for the function. (See Examples 5a and 5b.) We state these results formally as theorems.

THEOREM 2—Limits of Polynomials

If $P(x) = a_n x^n + a_{n-1} x^{n-1} + \cdots + a_0$, then

$$\lim_{x \to c} P(x) = P(c) = a_n c^n + a_{n-1} c^{n-1} + \cdots + a_0.$$

THEOREM 3—Limits of Rational Functions

If $P(x)$ and $Q(x)$ are polynomials and $Q(c) \neq 0$, then

$$\lim_{x \to c} \frac{P(x)}{Q(x)} = \frac{P(c)}{Q(c)}.$$

EXAMPLE 6   The following calculation illustrates Theorems 2 and 3:

$$\lim_{x \to -1} \frac{x^3 + 4x^2 - 3}{x^2 + 5} = \frac{(-1)^3 + 4(-1)^2 - 3}{(-1)^2 + 5} = \frac{0}{6} = 0$$

Since the denominator of this rational expression does not equal 0 when we substitute -1 for x, we can just compute the value of the expression at $x = -1$ to evaluate the limit.   ∎

Eliminating Common Factors from Zero Denominators

Theorem 3 applies only if the denominator of the rational function is not zero at the limit point c. If the denominator is zero, canceling common factors in the numerator and

Identifying Common Factors

If $Q(x)$ is a polynomial and $Q(c) = 0$, then $(x - c)$ is a factor of $Q(x)$. Thus, if the numerator and denominator of a rational function of x are both zero at $x = c$, they have $(x - c)$ as a common factor.

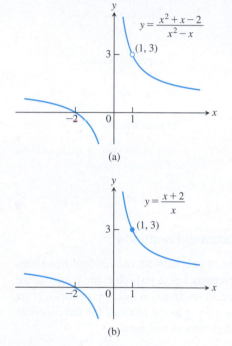

(a)

(b)

FIGURE 2.11 The graph of $f(x) = (x^2 + x - 2)/(x^2 - x)$ in part (a) is the same as the graph of $g(x) = (x + 2)/x$ in part (b) except at $x = 1$, where f is undefined. The functions have the same limit as $x \to 1$ (Example 7).

denominator may reduce the fraction to one whose denominator is no longer zero at c. If this happens, we can find the limit by substitution in the simplified fraction.

EXAMPLE 7   Evaluate

$$\lim_{x \to 1} \frac{x^2 + x - 2}{x^2 - x}.$$

Solution   We cannot substitute $x = 1$ because it makes the denominator zero. We test the numerator to see if it, too, is zero at $x = 1$. It is, so it has a factor of $(x - 1)$ in common with the denominator. Canceling this common factor gives a simpler fraction with the same values as the original for $x \neq 1$:

$$\frac{x^2 + x - 2}{x^2 - x} = \frac{(x - 1)(x + 2)}{x(x - 1)} = \frac{x + 2}{x}, \qquad \text{if } x \neq 1.$$

Using the simpler fraction, we find the limit of these values as $x \to 1$ by evaluating the function at $x = 1$, as in Theorem 3:

$$\lim_{x \to 1} \frac{x^2 + x - 2}{x^2 - x} = \lim_{x \to 1} \frac{x + 2}{x} = \frac{1 + 2}{1} = 3.$$

See Figure 2.11.

Using Calculators and Computers to Estimate Limits

We can try using a calculator or computer to guess a limit numerically. However, calculators and computers can sometimes give false values and misleading evidence about limits. Usually the problem is associated with rounding errors, as we now illustrate.

EXAMPLE 8   Estimate the value of $\lim_{x \to 0} \dfrac{\sqrt{x^2 + 100} - 10}{x^2}$.

Solution   Table 2.3 lists values of the function obtained on a calculator for several points approaching $x = 0$. As x approaches 0 through the points ± 1, ± 0.5, ± 0.10, and ± 0.01, the function seems to approach the number 0.05.

As we take even smaller values of x, ± 0.0005, ± 0.0001, ± 0.00001, and ± 0.000001, the function appears to approach the number 0.

Is the answer 0.05 or 0, or some other value? We resolve this question in the next example.

TABLE 2.3 Computed values of $f(x) = \dfrac{\sqrt{x^2 + 100} - 10}{x^2}$ near $x = 0$

x	$f(x)$	
± 1	0.049876	
± 0.5	0.049969	approaches 0.05?
± 0.1	0.049999	
± 0.01	0.050000	
± 0.0005	0.050000	
± 0.0001	0.000000	approaches 0?
± 0.00001	0.000000	
± 0.000001	0.000000	

Using a computer or calculator may give ambiguous results, as in Example 8. A computer cannot always keep track of enough digits to avoid rounding errors in computing the values of $f(x)$ when x is very small. We cannot substitute $x = 0$ in the problem, and the numerator and denominator have no obvious common factors (as they did in Example 7). Sometimes, however, we can create a common factor algebraically.

EXAMPLE 9   Evaluate

$$\lim_{x \to 0} \frac{\sqrt{x^2 + 100} - 10}{x^2}.$$

Solution   This is the limit we considered in Example 8. We can create a common factor by multiplying both numerator and denominator by the conjugate radical expression $\sqrt{x^2 + 100} + 10$ (obtained by changing the sign after the square root). The preliminary algebra rationalizes the numerator:

$$\frac{\sqrt{x^2 + 100} - 10}{x^2} = \frac{\sqrt{x^2 + 100} - 10}{x^2} \cdot \frac{\sqrt{x^2 + 100} + 10}{\sqrt{x^2 + 100} + 10} \qquad \text{Multiply and divide by the conjugate.}$$

$$= \frac{x^2 + 100 - 100}{x^2(\sqrt{x^2 + 100} + 10)} \qquad \text{Simplify}$$

$$= \frac{x^2}{x^2(\sqrt{x^2 + 100} + 10)} \qquad \text{Common factor } x^2$$

$$= \frac{1}{\sqrt{x^2 + 100} + 10}. \qquad \text{Cancel } x^2 \text{ for } x \neq 0.$$

Therefore,

$$\lim_{x \to 0} \frac{\sqrt{x^2 + 100} - 10}{x^2} = \lim_{x \to 0} \frac{1}{\sqrt{x^2 + 100} + 10}$$

$$= \frac{1}{\sqrt{0^2 + 100} + 10} \qquad \text{Limit Quotient Rule: Denominator not 0 at } x = 0 \text{ so can substitute.}$$

$$= \frac{1}{20} = 0.05.$$

This calculation provides the correct answer, resolving the ambiguous computer results in Example 8. ■

We cannot always manipulate the terms in an expression to find the limit of a quotient where the denominator becomes zero. In some cases the limit might then be found with geometric arguments (see the proof of Theorem 7 in Section 2.4), or through methods of calculus (developed in Section 4.5). The next theorem shows how to evaluate difficult limits by comparing them with functions having known limits.

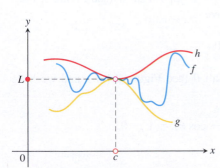

FIGURE 2.12   The graph of f is sandwiched between the graphs of g and h.

The Sandwich Theorem

The following theorem enables us to calculate a variety of limits. It is called the Sandwich Theorem because it refers to a function f whose values are sandwiched between the values of two other functions g and h that have the same limit L at a point c. Being trapped between the values of two functions that approach L, the values of f must also approach L (Figure 2.12). A proof is given in Appendix 4.

THEOREM 4—The Sandwich Theorem

Suppose that $g(x) \leq f(x) \leq h(x)$ for all x in some open interval containing c, except possibly at $x = c$ itself. Suppose also that

$$\lim_{x \to c} g(x) = \lim_{x \to c} h(x) = L.$$

Then $\lim_{x \to c} f(x) = L$.

The Sandwich Theorem is also called the Squeeze Theorem or the Pinching Theorem.

EXAMPLE 10 Given a function u that satisfies

$$1 - \frac{x^2}{4} \leq u(x) \leq 1 + \frac{x^2}{2} \qquad \text{for all } x \neq 0,$$

find $\lim_{x \to 0} u(x)$, no matter how complicated u is.

Solution Since

$$\lim_{x \to 0} (1 - (x^2/4)) = 1 \qquad \text{and} \qquad \lim_{x \to 0} (1 + (x^2/2)) = 1,$$

the Sandwich Theorem implies that $\lim_{x \to 0} u(x) = 1$ (Figure 2.13).

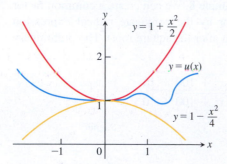

FIGURE 2.13 Any function $u(x)$ whose graph lies in the region between $y = 1 + (x^2/2)$ and $y = 1 - (x^2/4)$ has limit 1 as $x \to 0$ (Example 10).

EXAMPLE 11 The Sandwich Theorem helps us establish several important limit rules:

(a) $\lim_{\theta \to 0} \sin \theta = 0$

(b) $\lim_{\theta \to 0} \cos \theta = 1$

(c) For any function f, $\lim_{x \to c} |f(x)| = 0$ implies $\lim_{x \to c} f(x) = 0$.

Solution

(a) In Section 1.3 we established that $-|\theta| \leq \sin \theta \leq |\theta|$ for all θ (see Figure 2.14a). Since $\lim_{\theta \to 0}(-|\theta|) = \lim_{\theta \to 0} |\theta| = 0$, we have

$$\lim_{\theta \to 0} \sin \theta = 0.$$

(b) From Section 1.3, $0 \leq 1 - \cos \theta \leq |\theta|$ for all θ (see Figure 2.14b), and we have $\lim_{\theta \to 0} (1 - \cos \theta) = 0$ so

$$\lim_{\theta \to 0} 1 - (1 - \cos \theta) = 1 - \lim_{\theta \to 0} (1 - \cos \theta) = 1 - 0,$$

$$\lim_{\theta \to 0} \cos \theta = 1. \qquad\qquad \text{Simplify}$$

(c) Since $-|f(x)| \leq f(x) \leq |f(x)|$ and $-|f(x)|$ and $|f(x)|$ have limit 0 as $x \to c$, it follows that $\lim_{x \to c} f(x) = 0$.

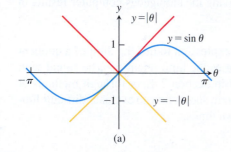

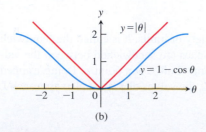

FIGURE 2.14 The Sandwich Theorem confirms the limits in Example 11.

Example 11 shows that the sine and cosine functions are equal to their limits at $\theta = 0$. We have not yet established that for any c, $\lim_{\theta \to c} \sin \theta = \sin c$, and $\lim_{\theta \to c} \cos \theta = \cos c$. These limit formulas do hold, as will be shown in Section 2.5.

EXERCISES 2.2

Limits from Graphs

1. For the function $g(x)$ graphed here, find the following limits or explain why they do not exist.

a. $\lim\limits_{x \to 1} g(x)$ **b.** $\lim\limits_{x \to 2} g(x)$ **c.** $\lim\limits_{x \to 3} g(x)$ **d.** $\lim\limits_{x \to 2.5} g(x)$

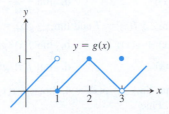

2. For the function $f(t)$ graphed here, find the following limits or explain why they do not exist.

a. $\lim\limits_{t \to -2} f(t)$ **b.** $\lim\limits_{t \to -1} f(t)$ **c.** $\lim\limits_{t \to 0} f(t)$ **d.** $\lim\limits_{t \to -0.5} f(t)$

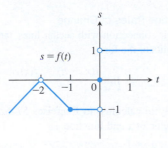

3. Which of the following statements about the function $y = f(x)$ graphed here are true, and which are false?

a. $\lim\limits_{x \to 0} f(x)$ exists.

b. $\lim\limits_{x \to 0} f(x) = 0$

c. $\lim\limits_{x \to 0} f(x) = 1$

d. $\lim\limits_{x \to 1} f(x) = 1$

e. $\lim\limits_{x \to 1} f(x) = 0$

f. $\lim\limits_{x \to c} f(x)$ exists at every point c in $(-1, 1)$.

g. $\lim\limits_{x \to 1} f(x)$ does not exist.

h. $f(0) = 0$

i. $f(0) = 1$

j. $f(1) = 0$

k. $f(1) = -1$

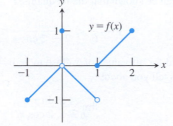

4. Which of the following statements about the function $y = f(x)$ graphed here are true, and which are false?

a. $\lim\limits_{x \to 2} f(x)$ does not exist.

b. $\lim\limits_{x \to 2} f(x) = 2$

c. $\lim\limits_{x \to 1} f(x)$ does not exist.

d. $\lim\limits_{x \to c} f(x)$ exists at every point c in $(-1, 1)$.

e. $\lim\limits_{x \to c} f(x)$ exists at every point c in $(1, 3)$.

f. $f(1) = 0$

g. $f(1) = -2$

h. $f(2) = 0$

i. $f(2) = 1$

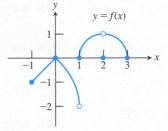

Existence of Limits

In Exercises 5 and 6, explain why the limits do not exist.

5. $\lim\limits_{x \to 0} \dfrac{x}{|x|}$ **6.** $\lim\limits_{x \to 1} \dfrac{1}{x - 1}$

7. Suppose that a function $f(x)$ is defined for all real values of x except $x = c$. Can anything be said about the existence of $\lim_{x \to c} f(x)$? Give reasons for your answer.

8. Suppose that a function $f(x)$ is defined for all x in $[-1, 1]$. Can anything be said about the existence of $\lim_{x \to 0} f(x)$? Give reasons for your answer.

9. If $\lim_{x \to 1} f(x) = 5$, must f be defined at $x = 1$? If it is, must $f(1) = 5$? Can we conclude *anything* about the values of f at $x = 1$? Explain.

10. If $f(1) = 5$, must $\lim_{x \to 1} f(x)$ exist? If it does, then must $\lim_{x \to 1} f(x) = 5$? Can we conclude *anything* about $\lim_{x \to 1} f(x)$? Explain.

Calculating Limits

Find the limits in Exercises 11–22.

11. $\lim\limits_{x \to -3} (x^2 - 13)$ **12.** $\lim\limits_{x \to 2} (-x^2 + 5x - 2)$

13. $\lim\limits_{t \to 6} 8(t - 5)(t - 7)$ **14.** $\lim\limits_{x \to -2} (x^3 - 2x^2 + 4x + 8)$

15. $\lim\limits_{x \to 2} \dfrac{2x + 5}{11 - x^3}$ **16.** $\lim\limits_{s \to 2/3} (8 - 3s)(2s - 1)$

17. $\lim\limits_{x \to -1/2} 4x(3x + 4)^2$ **18.** $\lim\limits_{y \to 2} \dfrac{y + 2}{y^2 + 5y + 6}$

19. $\lim\limits_{y \to -3} (5 - y)^{4/3}$ **20.** $\lim\limits_{z \to 4} \sqrt{z^2 - 10}$

21. $\lim\limits_{h \to 0} \dfrac{3}{\sqrt{3h + 1} + 1}$ **22.** $\lim\limits_{h \to 0} \dfrac{\sqrt{5h + 4} - 2}{h}$

Limits of quotients   Find the limits in Exercises 23–42.

23. $\lim\limits_{x \to 5} \dfrac{x - 5}{x^2 - 25}$ **24.** $\lim\limits_{x \to -3} \dfrac{x + 3}{x^2 + 4x + 3}$

25. $\lim\limits_{x \to -5} \dfrac{x^2 + 3x - 10}{x + 5}$ **26.** $\lim\limits_{x \to 2} \dfrac{x^2 - 7x + 10}{x - 2}$

27. $\lim\limits_{t \to 1} \dfrac{t^2 + t - 2}{t^2 - 1}$ **28.** $\lim\limits_{t \to -1} \dfrac{t^2 + 3t + 2}{t^2 - t - 2}$

29. $\lim\limits_{x \to -2} \dfrac{-2x - 4}{x^3 + 2x^2}$ **30.** $\lim\limits_{y \to 0} \dfrac{5y^3 + 8y^2}{3y^4 - 16y^2}$

31. $\lim\limits_{x \to 1} \dfrac{x^{-1} - 1}{x - 1}$

32. $\lim\limits_{x \to 0} \dfrac{\frac{1}{x-1} + \frac{1}{x+1}}{x}$

33. $\lim\limits_{u \to 1} \dfrac{u^4 - 1}{u^3 - 1}$

34. $\lim\limits_{v \to 2} \dfrac{v^3 - 8}{v^4 - 16}$

35. $\lim\limits_{x \to 9} \dfrac{\sqrt{x} - 3}{x - 9}$

36. $\lim\limits_{x \to 4} \dfrac{4x - x^2}{2 - \sqrt{x}}$

37. $\lim\limits_{x \to 1} \dfrac{x - 1}{\sqrt{x + 3} - 2}$

38. $\lim\limits_{x \to -1} \dfrac{\sqrt{x^2 + 8} - 3}{x + 1}$

39. $\lim\limits_{x \to 2} \dfrac{\sqrt{x^2 + 12} - 4}{x - 2}$

40. $\lim\limits_{x \to -2} \dfrac{x + 2}{\sqrt{x^2 + 5} - 3}$

41. $\lim\limits_{x \to -3} \dfrac{2 - \sqrt{x^2 - 5}}{x + 3}$

42. $\lim\limits_{x \to 4} \dfrac{4 - x}{5 - \sqrt{x^2 + 9}}$

Limits with trigonometric functions Find the limits in Exercises 43–50.

43. $\lim\limits_{x \to 0} (2 \sin x - 1)$

44. $\lim\limits_{x \to 0} \sin^2 x$

45. $\lim\limits_{x \to 0} \sec x$

46. $\lim\limits_{x \to 0} \tan x$

47. $\lim\limits_{x \to 0} \dfrac{1 + x + \sin x}{3 \cos x}$

48. $\lim\limits_{x \to 0} (x^2 - 1)(2 - \cos x)$

49. $\lim\limits_{x \to -\pi} \sqrt{x + 4} \cos (x + \pi)$

50. $\lim\limits_{x \to 0} \sqrt{7 + \sec^2 x}$

Using Limit Rules

51. Suppose $\lim\limits_{x \to 0} f(x) = 1$ and $\lim\limits_{x \to 0} g(x) = -5$. Name the rules in Theorem 1 that are used to accomplish steps (a), (b), and (c) of the following calculation.

$$\lim\limits_{x \to 0} \dfrac{2f(x) - g(x)}{(f(x) + 7)^{2/3}} = \dfrac{\lim\limits_{x \to 0} (2f(x) - g(x))}{\lim\limits_{x \to 0} (f(x) + 7)^{2/3}} \quad \text{(a)}$$

$$= \dfrac{\lim\limits_{x \to 0} 2f(x) - \lim\limits_{x \to 0} g(x)}{\left(\lim\limits_{x \to 0} (f(x) + 7)\right)^{2/3}} \quad \text{(b)}$$

$$= \dfrac{2 \lim\limits_{x \to 0} f(x) - \lim\limits_{x \to 0} g(x)}{\left(\lim\limits_{x \to 0} f(x) + \lim\limits_{x \to 0} 7\right)^{2/3}} \quad \text{(c)}$$

$$= \dfrac{(2)(1) - (-5)}{(1 + 7)^{2/3}} = \dfrac{7}{4}$$

52. Let $\lim\limits_{x \to 1} h(x) = 5$, $\lim\limits_{x \to 1} p(x) = 1$, and $\lim\limits_{x \to 1} r(x) = 2$. Name the rules in Theorem 1 that are used to accomplish steps (a), (b), and (c) of the following calculation.

$$\lim\limits_{x \to 1} \dfrac{\sqrt{5h(x)}}{p(x)(4 - r(x))} = \dfrac{\lim\limits_{x \to 1} \sqrt{5h(x)}}{\lim\limits_{x \to 1} (p(x)(4 - r(x)))} \quad \text{(a)}$$

$$= \dfrac{\sqrt{\lim\limits_{x \to 1} 5h(x)}}{\left(\lim\limits_{x \to 1} p(x)\right)\left(\lim\limits_{x \to 1} (4 - r(x))\right)} \quad \text{(b)}$$

$$= \dfrac{\sqrt{5 \lim\limits_{x \to 1} h(x)}}{\left(\lim\limits_{x \to 1} p(x)\right)\left(\lim\limits_{x \to 1} 4 - \lim\limits_{x \to 1} r(x)\right)} \quad \text{(c)}$$

$$= \dfrac{\sqrt{(5)(5)}}{(1)(4 - 2)} = \dfrac{5}{2}$$

53. Suppose $\lim\limits_{x \to c} f(x) = 5$ and $\lim\limits_{x \to c} g(x) = -2$. Find

a. $\lim\limits_{x \to c} f(x)g(x)$

b. $\lim\limits_{x \to c} 2f(x)g(x)$

c. $\lim\limits_{x \to c} (f(x) + 3g(x))$

d. $\lim\limits_{x \to c} \dfrac{f(x)}{f(x) - g(x)}$

54. Suppose $\lim\limits_{x \to 4} f(x) = 0$ and $\lim\limits_{x \to 4} g(x) = -3$. Find

a. $\lim\limits_{x \to 4} (g(x) + 3)$

b. $\lim\limits_{x \to 4} xf(x)$

c. $\lim\limits_{x \to 4} (g(x))^2$

d. $\lim\limits_{x \to 4} \dfrac{g(x)}{f(x) - 1}$

55. Suppose $\lim\limits_{x \to b} f(x) = 7$ and $\lim\limits_{x \to b} g(x) = -3$. Find

a. $\lim\limits_{x \to b} (f(x) + g(x))$

b. $\lim\limits_{x \to b} f(x) \cdot g(x)$

c. $\lim\limits_{x \to b} 4g(x)$

d. $\lim\limits_{x \to b} f(x)/g(x)$

56. Suppose that $\lim\limits_{x \to -2} p(x) = 4$, $\lim\limits_{x \to -2} r(x) = 0$, and $\lim\limits_{x \to -2} s(x) = -3$. Find

a. $\lim\limits_{x \to -2} (p(x) + r(x) + s(x))$

b. $\lim\limits_{x \to -2} p(x) \cdot r(x) \cdot s(x)$

c. $\lim\limits_{x \to -2} (-4p(x) + 5r(x))/s(x)$

Limits of Average Rates of Change

Because of their connection with secant lines, tangents, and instantaneous rates, limits of the form

$$\lim\limits_{h \to 0} \dfrac{f(x + h) - f(x)}{h}$$

occur frequently in calculus. In Exercises 57–62, evaluate this limit for the given value of x and function f.

57. $f(x) = x^2, \quad x = 1$

58. $f(x) = x^2, \quad x = -2$

59. $f(x) = 3x - 4, \quad x = 2$

60. $f(x) = 1/x, \quad x = -2$

61. $f(x) = \sqrt{x}, \quad x = 7$

62. $f(x) = \sqrt{3x + 1}, \quad x = 0$

Using the Sandwich Theorem

63. If $\sqrt{5 - 2x^2} \le f(x) \le \sqrt{5 - x^2}$ for $-1 \le x \le 1$, find $\lim\limits_{x \to 0} f(x)$.

64. If $2 - x^2 \le g(x) \le 2 \cos x$ for all x, find $\lim\limits_{x \to 0} g(x)$.

65. a. It can be shown that the inequalities

$$1 - \dfrac{x^2}{6} < \dfrac{x \sin x}{2 - 2 \cos x} < 1$$

hold for all values of x close to zero. What, if anything, does this tell you about

$$\lim\limits_{x \to 0} \dfrac{x \sin x}{2 - 2 \cos x}?$$

Give reasons for your answer.

T **b.** Graph $y = 1 - (x^2/6)$, $y = (x \sin x)/(2 - 2 \cos x)$, and $y = 1$ together for $-2 \le x \le 2$. Comment on the behavior of the graphs as $x \to 0$.

66. a. Suppose that the inequalities

$$\frac{1}{2} - \frac{x^2}{24} < \frac{1 - \cos x}{x^2} < \frac{1}{2}$$

hold for values of x close to zero. (They do, as you will see in Section 9.9.) What, if anything, does this tell you about

$$\lim_{x \to 0} \frac{1 - \cos x}{x^2}?$$

Give reasons for your answer.

T **b.** Graph the equations $y = (1/2) - (x^2/24)$, $y = (1 - \cos x)/x^2$, and $y = 1/2$ together for $-2 \le x \le 2$. Comment on the behavior of the graphs as $x \to 0$.

Estimating Limits

T You will find a graphing calculator useful for Exercises 67–76.

67. Let $f(x) = (x^2 - 9)/(x + 3)$.

a. Make a table of the values of f at the points $x = -3.1$, $-3.01, -3.001$, and so on as far as your calculator can go. Then estimate $\lim_{x \to -3} f(x)$. What estimate do you arrive at if you evaluate f at $x = -2.9, -2.99, -2.999, \ldots$ instead?

b. Support your conclusions in part (a) by graphing f near $c = -3$ and using Zoom and Trace to estimate y-values on the graph as $x \to -3$.

c. Find $\lim_{x \to -3} f(x)$ algebraically, as in Example 7.

68. Let $g(x) = (x^2 - 2)/(x - \sqrt{2})$.

a. Make a table of the values of g at the points $x = 1.4, 1.41$, 1.414, and so on through successive decimal approximations of $\sqrt{2}$. Estimate $\lim_{x \to \sqrt{2}} g(x)$.

b. Support your conclusion in part (a) by graphing g near $c = \sqrt{2}$ and using Zoom and Trace to estimate y-values on the graph as $x \to \sqrt{2}$.

c. Find $\lim_{x \to \sqrt{2}} g(x)$ algebraically.

69. Let $G(x) = (x + 6)/(x^2 + 4x - 12)$.

a. Make a table of the values of G at $x = -5.9, -5.99, -5.999$, and so on. Then estimate $\lim_{x \to -6} G(x)$. What estimate do you arrive at if you evaluate G at $x = -6.1, -6.01, -6.001, \ldots$ instead?

b. Support your conclusions in part (a) by graphing G and using Zoom and Trace to estimate y-values on the graph as $x \to -6$.

c. Find $\lim_{x \to -6} G(x)$ algebraically.

70. Let $h(x) = (x^2 - 2x - 3)/(x^2 - 4x + 3)$.

a. Make a table of the values of h at $x = 2.9, 2.99, 2.999$, and so on. Then estimate $\lim_{x \to 3} h(x)$. What estimate do you arrive at if you evaluate h at $x = 3.1, 3.01, 3.001, \ldots$ instead?

b. Support your conclusions in part (a) by graphing h near $c = 3$ and using Zoom and Trace to estimate y-values on the graph as $x \to 3$.

c. Find $\lim_{x \to 3} h(x)$ algebraically.

71. Let $f(x) = (x^2 - 1)/(|x| - 1)$.

a. Make tables of the values of f at values of x that approach $c = -1$ from above and below. Then estimate $\lim_{x \to -1} f(x)$.

b. Support your conclusion in part (a) by graphing f near $c = -1$ and using Zoom and Trace to estimate y-values on the graph as $x \to -1$.

c. Find $\lim_{x \to -1} f(x)$ algebraically.

72. Let $F(x) = (x^2 + 3x + 2)/(2 - |x|)$.

a. Make tables of values of F at values of x that approach $c = -2$ from above and below. Then estimate $\lim_{x \to -2} F(x)$.

b. Support your conclusion in part (a) by graphing F near $c = -2$ and using Zoom and Trace to estimate y-values on the graph as $x \to -2$.

c. Find $\lim_{x \to -2} F(x)$ algebraically.

73. Let $g(\theta) = (\sin \theta)/\theta$.

a. Make a table of the values of g at values of θ that approach $\theta_0 = 0$ from above and below. Then estimate $\lim_{\theta \to 0} g(\theta)$.

b. Support your conclusion in part (a) by graphing g near $\theta_0 = 0$.

74. Let $G(t) = (1 - \cos t)/t^2$.

a. Make tables of values of G at values of t that approach $t_0 = 0$ from above and below. Then estimate $\lim_{t \to 0} G(t)$.

b. Support your conclusion in part (a) by graphing G near $t_0 = 0$.

75. Let $f(x) = x^{1/(1-x)}$.

a. Make tables of values of f at values of x that approach $c = 1$ from above and below. Does f appear to have a limit as $x \to 1$? If so, what is it? If not, why not?

b. Support your conclusions in part (a) by graphing f near $c = 1$.

76. Let $f(x) = (3^x - 1)/x$.

a. Make tables of values of f at values of x that approach $c = 0$ from above and below. Does f appear to have a limit as $x \to 0$? If so, what is it? If not, why not?

b. Support your conclusions in part (a) by graphing f near $c = 0$.

Theory and Examples

77. If $x^4 \le f(x) \le x^2$ for x in $[-1, 1]$ and $x^2 \le f(x) \le x^4$ for $x < -1$ and $x > 1$, at what points c do you automatically know $\lim_{x \to c} f(x)$? What can you say about the value of the limit at these points?

78. Suppose that $g(x) \le f(x) \le h(x)$ for all $x \ne 2$ and suppose that

$$\lim_{x \to 2} g(x) = \lim_{x \to 2} h(x) = -5.$$

Can we conclude anything about the values of f, g, and h at $x = 2$? Could $f(2) = 0$? Could $\lim_{x \to 2} f(x) = 0$? Give reasons for your answers.

79. If $\lim_{x \to 4} \dfrac{f(x) - 5}{x - 2} = 1$, find $\lim_{x \to 4} f(x)$.

80. If $\lim_{x \to -2} \dfrac{f(x)}{x^2} = 1$, find

a. $\lim_{x \to -2} f(x)$ **b.** $\lim_{x \to -2} \dfrac{f(x)}{x}$

81. a. If $\lim_{x \to 2} \dfrac{f(x) - 5}{x - 2} = 3$, find $\lim_{x \to 2} f(x)$.

b. If $\lim_{x \to 2} \dfrac{f(x) - 5}{x - 2} = 4$, find $\lim_{x \to 2} f(x)$.

82. If $\lim\limits_{x\to 0} \dfrac{f(x)}{x^2} = 1$, find

  a. $\lim\limits_{x\to 0} f(x)$ **b.** $\lim\limits_{x\to 0} \dfrac{f(x)}{x}$

T **83. a.** Graph $g(x) = x\sin(1/x)$ to estimate $\lim_{x\to 0} g(x)$, zooming in on the origin as necessary.

  b. Confirm your estimate in part (a) with a proof.

T **84. a.** Graph $h(x) = x^2\cos(1/x^3)$ to estimate $\lim_{x\to 0} h(x)$, zooming in on the origin as necessary.

  b. Confirm your estimate in part (a) with a proof.

COMPUTER EXPLORATIONS

Graphical Estimates of Limits

In Exercises 85–90, use a CAS to perform the following steps:

  a. Plot the function near the point c being approached.

  b. From your plot guess the value of the limit.

85. $\lim\limits_{x\to 2} \dfrac{x^4 - 16}{x - 2}$

86. $\lim\limits_{x\to -1} \dfrac{x^3 - x^2 - 5x - 3}{(x + 1)^2}$

87. $\lim\limits_{x\to 0} \dfrac{\sqrt[3]{1 + x} - 1}{x}$

88. $\lim\limits_{x\to 3} \dfrac{x^2 - 9}{\sqrt{x^2 + 7} - 4}$

89. $\lim\limits_{x\to 0} \dfrac{1 - \cos x}{x \sin x}$

90. $\lim\limits_{x\to 0} \dfrac{2x^2}{3 - 3\cos x}$

2.3 The Precise Definition of a Limit

We now turn our attention to the precise definition of a limit. The early history of calculus saw controversy about the validity of the basic concepts underlying the theory. Apparent contradictions were argued over by both mathematicians and philosophers. These controversies were resolved by the precise definition, which allows us to replace vague phrases like "gets arbitrarily close to" in the informal definition with specific conditions that can be applied to any particular example. With a rigorous definition, we can avoid misunderstandings, prove the limit properties given in the preceding section, and establish many important limits.

To show that the limit of $f(x)$ as $x \to c$ equals the number L, we need to show that the gap between $f(x)$ and L can be made "as small as we choose" if x is kept "close enough" to c. Let us see what this requires if we specify the size of the gap between $f(x)$ and L.

EXAMPLE 1 Consider the function $y = 2x - 1$ near $x = 4$. Intuitively it seems clear that y is close to 7 when x is close to 4, so $\lim_{x\to 4}(2x - 1) = 7$. However, how close to $x = 4$ does x have to be so that $y = 2x - 1$ differs from 7 by, say, less than 2 units?

Solution We are asked: For what values of x is $|y - 7| < 2$? To find the answer we first express $|y - 7|$ in terms of x:

$$|y - 7| = |(2x - 1) - 7| = |2x - 8|.$$

The question then becomes: what values of x satisfy the inequality $|2x - 8| < 2$? To find out, we solve the inequality:

$$|2x - 8| < 2$$
$$-2 < 2x - 8 < 2 \qquad \text{\color{blue}Removing absolute value gives two inequalities.}$$
$$6 < 2x < 10 \qquad \text{\color{blue}Add 8 to each term.}$$
$$3 < x < 5 \qquad \text{\color{blue}Solve for } x.$$
$$-1 < x - 4 < 1. \qquad \text{\color{blue}Solve for } x - 4.$$

Keeping x within 1 unit of $x = 4$ will keep y within 2 units of $y = 7$ (Figure 2.15). ∎

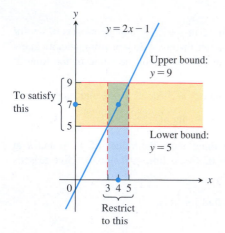

FIGURE 2.15 Keeping x within 1 unit of $x = 4$ will keep y within 2 units of $y = 7$ (Example 1).

In the previous example we determined how close x must be to a particular value c to ensure that the outputs $f(x)$ of some function lie within a prescribed interval about a limit value L. To show that the limit of $f(x)$ as $x \to c$ actually equals L, we must be able to show that the gap between $f(x)$ and L can be made less than *any prescribed error*, no matter how

δ is the Greek letter delta
ε is the Greek letter epsilon

small, by holding x close enough to c. To describe arbitrary prescribed errors, we introduce two constants, δ (delta) and ε (epsilon). These Greek letters are traditionally used to represent small changes in a variable or a function.

Definition of Limit

Suppose we are watching the values of a function $f(x)$ as x approaches c (without taking on the value c itself). Certainly we want to be able to say that $f(x)$ stays within one-tenth of a unit from L as soon as x stays within some distance δ of c (Figure 2.16). But that in itself is not enough, because as x continues on its course toward c, what is to prevent $f(x)$ from jumping around within the interval from $L - (1/10)$ to $L + (1/10)$ without tending toward L? We can be told that the error can be no more than $1/100$ or $1/1000$ or $1/100,000$. Each time, we find a new δ-interval about c so that keeping x within that interval satisfies the new error tolerance. And each time the possibility exists that $f(x)$ might jump away from L at some later stage.

The figures on the next page illustrate the problem. You can think of this as a quarrel between a skeptic and a scholar. The skeptic presents ε-challenges to show there is room for doubt that the limit exists. The scholar counters every challenge with a δ-interval around c which ensures that the function takes values within ε of L.

How do we stop this seemingly endless series of challenges and responses? We can do so by proving that for *every* error tolerance ε that the challenger can produce, we can present a matching distance δ that keeps x "close enough" to c to keep $f(x)$ within that ε-tolerance of L (Figure 2.17). This leads us to the precise definition of a limit.

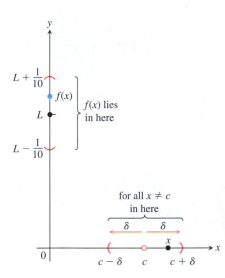

FIGURE 2.16 How should we define $\delta > 0$ so that keeping x within the interval $(c - \delta, c + \delta)$ will keep $f(x)$ within the interval $\left(L - \dfrac{1}{10}, L + \dfrac{1}{10} \right)$?

DEFINITION   Let $f(x)$ be defined on an open interval about c, except possibly at c itself. We say that the **limit of $f(x)$ as x approaches c is the number L**, and write
$$\lim_{x \to c} f(x) = L,$$
if, for every number $\varepsilon > 0$, there exists a corresponding number $\delta > 0$ such that
$$|f(x) - L| < \varepsilon \quad \text{whenever} \quad 0 < |x - c| < \delta.$$

To visualize the definition, imagine machining a cylindrical shaft to a close tolerance. The diameter of the shaft is determined by turning a dial to a setting measured by a variable x. We try for diameter L, but since nothing is perfect we must be satisfied with a diameter $f(x)$ somewhere between $L - \varepsilon$ and $L + \varepsilon$. The number δ is our control tolerance for the dial; it tells us how close our dial setting must be to the setting $x = c$ in order to guarantee that the diameter $f(x)$ of the shaft will be accurate to within ε of L. As the tolerance for error becomes stricter, we may have to adjust δ. The value of δ, how tight our control setting must be, depends on the value of ε, the error tolerance.

The definition of limit extends to functions on more general domains. It is only required that each open interval around c contains points in the domain of the function other than c. See Additional and Advanced Exercises 39–43 for examples of limits for functions with complicated domains. In the next section we will see how the definition of limit applies at points lying on the boundary of an interval.

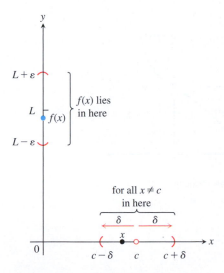

FIGURE 2.17   The relation of δ and ε in the definition of limit.

Examples: Testing the Definition

The formal definition of limit does not tell how to find the limit of a function, but it does enable us to verify that a conjectured limit value is correct. The following examples show how the definition can be used to verify limit statements for specific functions. However, the real purpose of the definition is not to do calculations like this, but rather to prove general theorems so that the calculation of specific limits can be simplified, such as the theorems stated in the previous section.

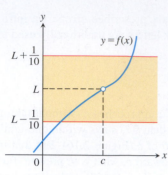

The challenge:
Make $|f(x) - L| < \varepsilon = \frac{1}{10}$

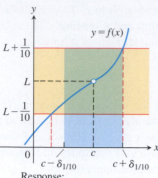

Response:
$|x - c| < \delta_{1/10}$ (a number)

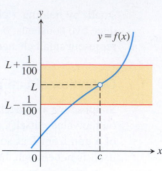

New challenge:
Make $|f(x) - L| < \varepsilon = \frac{1}{100}$

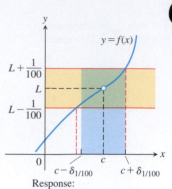

Response:
$|x - c| < \delta_{1/100}$

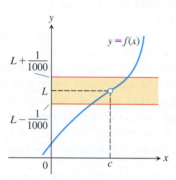

New challenge:
$\varepsilon = \frac{1}{1000}$

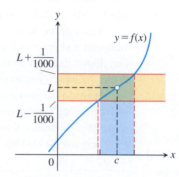

Response:
$|x - c| < \delta_{1/1000}$

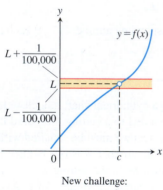

New challenge:
$\varepsilon = \frac{1}{100,000}$

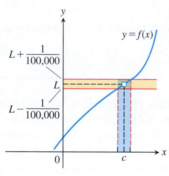

Response:
$|x - c| < \delta_{1/100,000}$

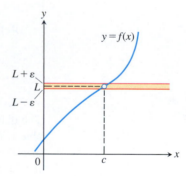

New challenge:
$\varepsilon = \cdots$

EXAMPLE 2   Show that

$$\lim_{x \to 1} (5x - 3) = 2.$$

Solution   Set $c = 1$, $f(x) = 5x - 3$, and $L = 2$ in the definition of limit. For any given $\varepsilon > 0$, we have to find a suitable $\delta > 0$ so that if $x \neq 1$ and x is within distance δ of $c = 1$, that is, whenever

$$0 < |x - 1| < \delta,$$

it is true that $f(x)$ is within distance ε of $L = 2$, so

$$|f(x) - 2| < \varepsilon.$$

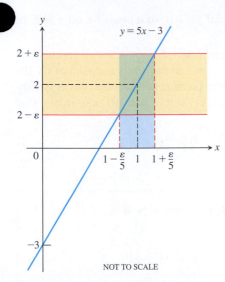

FIGURE 2.18 If $f(x) = 5x - 3$, then $0 < |x - 1| < \varepsilon/5$ guarantees that $|f(x) - 2| < \varepsilon$ (Example 2).

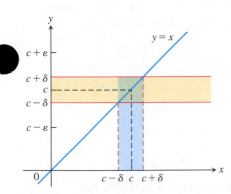

FIGURE 2.19 For the function $f(x) = x$, we find that $0 < |x - c| < \delta$ will guarantee $|f(x) - c| < \varepsilon$ whenever $\delta \le \varepsilon$ (Example 3a).

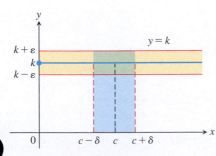

FIGURE 2.20 For the function $f(x) = k$, we find that $|f(x) - k| < \varepsilon$ for any positive δ (Example 3b).

We find δ by working backward from the ε-inequality:

$$|(5x - 3) - 2| = |5x - 5| < \varepsilon$$
$$5|x - 1| < \varepsilon$$
$$|x - 1| < \varepsilon/5.$$

Thus, we can take $\delta = \varepsilon/5$ (Figure 2.18). If $0 < |x - 1| < \delta = \varepsilon/5$, then

$$|(5x - 3) - 2| = |5x - 5| = 5|x - 1| < 5(\varepsilon/5) = \varepsilon,$$

which proves that $\lim_{x \to 1}(5x - 3) = 2$.

The value of $\delta = \varepsilon/5$ is not the only value that will make $0 < |x - 1| < \delta$ imply $|5x - 5| < \varepsilon$. Any smaller positive δ will do as well. The definition does not ask for the "best" positive δ, just one that will work. ∎

EXAMPLE 3   Prove the following results presented graphically in Section 2.2.

(a) $\lim_{x \to c} x = c$

(b) $\lim_{x \to c} k = k$ (k constant)

Solution

(a) Let $\varepsilon > 0$ be given. We must find $\delta > 0$ such that

$$|x - c| < \varepsilon \qquad \text{whenever} \qquad 0 < |x - c| < \delta.$$

The implication will hold if δ equals ε or any smaller positive number (Figure 2.19). This proves that $\lim_{x \to c} x = c$.

(b) Let $\varepsilon > 0$ be given. We must find $\delta > 0$ such that

$$|k - k| < \varepsilon \qquad \text{whenever} \qquad 0 < |x - c| < \delta.$$

Since $k - k = 0$, we can use any positive number for δ and the implication will hold (Figure 2.20). This proves that $\lim_{x \to c} k = k$. ∎

Finding Deltas Algebraically for Given Epsilons

In Examples 2 and 3, the interval of values about c for which $|f(x) - L|$ was less than ε was symmetric about c and we could take δ to be half the length of that interval. When the interval around c on which we have $|f(x) - L| < \varepsilon$ is not symmetric about c, we can take δ to be the distance from c to the interval's *nearer* endpoint.

EXAMPLE 4   For the limit $\lim_{x \to 5}\sqrt{x - 1} = 2$, find a $\delta > 0$ that works for $\varepsilon = 1$. That is, find a $\delta > 0$ such that

$$|\sqrt{x - 1} - 2| < 1 \qquad \text{whenever} \qquad 0 < |x - 5| < \delta.$$

Solution   We organize the search into two steps.

1. *Solve the inequality* $|\sqrt{x - 1} - 2| < 1$ *to find an interval containing* $x = 5$ *on which the inequality holds for all* $x \ne 5$.

$$|\sqrt{x - 1} - 2| < 1$$
$$-1 < \sqrt{x - 1} - 2 < 1$$
$$1 < \sqrt{x - 1} < 3$$
$$1 < x - 1 < 9$$
$$2 < x < 10$$

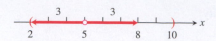

FIGURE 2.21 An open interval of radius 3 about $x = 5$ will lie inside the open interval (2, 10).

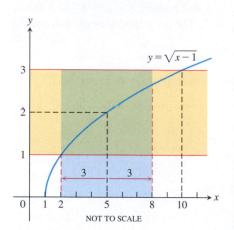

FIGURE 2.22 The function and intervals in Example 4.

The inequality holds for all x in the open interval (2, 10), so it holds for all $x \neq 5$ in this interval as well.

2. *Find a value of $\delta > 0$ to place the centered interval $5 - \delta < x < 5 + \delta$ (centered at $x = 5$) inside the interval (2, 10).* The distance from 5 to the nearer endpoint of (2, 10) is 3 (Figure 2.21). If we take $\delta = 3$ or any smaller positive number, then the inequality $0 < |x - 5| < \delta$ will automatically place x between 2 and 10 and imply that $|\sqrt{x - 1} - 2| < 1$ (Figure 2.22):

$$|\sqrt{x - 1} - 2| < 1 \qquad \text{whenever} \qquad 0 < |x - 5| < 3. \qquad \blacksquare$$

How to Find Algebraically a δ for a Given f, L, c, and $\varepsilon > 0$

The process of finding a $\delta > 0$ such that

$$|f(x) - L| < \varepsilon \qquad \text{whenever} \qquad 0 < |x - c| < \delta$$

can be accomplished in two steps.

1. *Solve the inequality $|f(x) - L| < \varepsilon$ to find an open interval (a, b) containing c on which the inequality holds for all $x \neq c$.* Note that we do not require the inequality to hold at $x = c$. It may hold there or it may not, but the value of f at $x = c$ does not influence the existence of a limit.

2. *Find a value of $\delta > 0$ that places the open interval $(c - \delta, c + \delta)$ centered at c inside the interval (a, b).* The inequality $|f(x) - L| < \varepsilon$ will hold for all $x \neq c$ in this δ-interval.

EXAMPLE 5 Prove that $\lim_{x \to 2} f(x) = 4$ if

$$f(x) = \begin{cases} x^2, & x \neq 2 \\ 1, & x = 2. \end{cases}$$

Solution Our task is to show that given $\varepsilon > 0$ there exists a $\delta > 0$ such that

$$|f(x) - 4| < \varepsilon \qquad \text{whenever} \qquad 0 < |x - 2| < \delta.$$

1. *Solve the inequality $|f(x) - 4| < \varepsilon$ to find an open interval containing $x = 2$ on which the inequality holds for all $x \neq 2$.*

For $x \neq c = 2$, we have $f(x) = x^2$, and the inequality to solve is $|x^2 - 4| < \varepsilon$:

$$|x^2 - 4| < \varepsilon$$
$$-\varepsilon < x^2 - 4 < \varepsilon$$
$$4 - \varepsilon < x^2 < 4 + \varepsilon$$
$$\sqrt{4 - \varepsilon} < |x| < \sqrt{4 + \varepsilon} \qquad \text{Assumes } \varepsilon < 4; \text{ see below.}$$
$$\sqrt{4 - \varepsilon} < x < \sqrt{4 + \varepsilon}. \qquad \text{An open interval about } x = 2 \text{ that solves the inequality.}$$

The inequality $|f(x) - 4| < \varepsilon$ holds for all $x \neq 2$ in the open interval $\left(\sqrt{4 - \varepsilon}, \sqrt{4 + \varepsilon}\right)$ (Figure 2.23).

2. *Find a value of $\delta > 0$ that places the centered interval $(2 - \delta, 2 + \delta)$ inside the interval $\left(\sqrt{4 - \varepsilon}, \sqrt{4 + \varepsilon}\right)$.*

Take δ to be the distance from $x = 2$ to the nearer endpoint of $\left(\sqrt{4 - \varepsilon}, \sqrt{4 + \varepsilon}\right)$. In other words, take $\delta = \min\{2 - \sqrt{4 - \varepsilon}, \sqrt{4 + \varepsilon} - 2\}$, the *minimum* (the

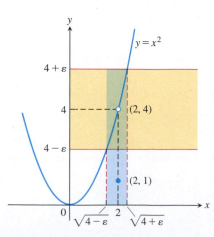

FIGURE 2.23 An interval containing $x = 2$ so that the function in Example 5 satisfies $|f(x) - 4| < \varepsilon$.

smaller) of the two numbers $2 - \sqrt{4 - \varepsilon}$ and $\sqrt{4 + \varepsilon} - 2$. If δ has this or any smaller positive value, the inequality $0 < |x - 2| < \delta$ will automatically place x between $\sqrt{4 - \varepsilon}$ and $\sqrt{4 + \varepsilon}$ to make $|f(x) - 4| < \varepsilon$. For all x,

$$|f(x) - 4| < \varepsilon \qquad \text{whenever} \qquad 0 < |x - 2| < \delta.$$

This completes the proof for $\varepsilon < 4$.

If $\varepsilon \geq 4$, then we take δ to be the distance from $x = 2$ to the nearer endpoint of the interval $\left(0, \sqrt{4 + \varepsilon}\right)$. In other words, take $\delta = \min\left\{2, \sqrt{4 + \varepsilon} - 2\right\}$. (See Figure 2.23.)   ■

Using the Definition to Prove Theorems

We do not usually rely on the formal definition of limit to verify specific limits such as those in the preceding examples. Rather, we appeal to general theorems about limits, in particular the theorems of Section 2.2. The definition is used to prove these theorems (Appendix 5). As an example, we prove part 1 of Theorem 1, the Sum Rule.

EXAMPLE 6   Given that $\lim_{x \to c} f(x) = L$ and $\lim_{x \to c} g(x) = M$, prove that

$$\lim_{x \to c} (f(x) + g(x)) = L + M.$$

Solution   Let $\varepsilon > 0$ be given. We want to find a positive number δ such that

$$|f(x) + g(x) - (L + M)| < \varepsilon \qquad \text{whenever} \qquad 0 < |x - c| < \delta.$$

Regrouping terms, we get

$$|f(x) + g(x) - (L + M)| = |(f(x) - L) + (g(x) - M)| \qquad \text{Triangle Inequality:}$$
$$\leq |f(x) - L| + |g(x) - M|. \qquad |a + b| \leq |a| + |b|$$

Since $\lim_{x \to c} f(x) = L$, there exists a number $\delta_1 > 0$ such that

$$|f(x) - L| < \varepsilon/2 \qquad \text{whenever} \qquad 0 < |x - c| < \delta_1. \qquad \begin{array}{l} \text{Can find } \delta_1 \text{ since} \\ \lim_{x \to c} f(x) = L \end{array}$$

Similarly, since $\lim_{x \to c} g(x) = M$, there exists a number $\delta_2 > 0$ such that

$$|g(x) - M| < \varepsilon/2 \qquad \text{whenever} \qquad 0 < |x - c| < \delta_2. \qquad \begin{array}{l} \text{Can find } \delta_2 \text{ since} \\ \lim_{x \to c} g(x) = M \end{array}$$

Let $\delta = \min\{\delta_1, \delta_2\}$, the smaller of δ_1 and δ_2. If $0 < |x - c| < \delta$ then $|x - c| < \delta_1$, so $|f(x) - L| < \varepsilon/2$, and $|x - c| < \delta_2$, so $|g(x) - M| < \varepsilon/2$. Therefore

$$|f(x) + g(x) - (L + M)| < \frac{\varepsilon}{2} + \frac{\varepsilon}{2} = \varepsilon.$$

This shows that $\lim_{x \to c} (f(x) + g(x)) = L + M.$ ■

EXERCISES 2.3

Centering Intervals About a Point

In Exercises 1–6, sketch the interval (a, b) on the x-axis with the point c inside. Then find a value of $\delta > 0$ such that $a < x < b$ whenever $0 < |x - c| < \delta$.

1. $a = 1, \quad b = 7, \quad c = 5$

2. $a = 1, \quad b = 7, \quad c = 2$

3. $a = -7/2, \quad b = -1/2, \quad c = -3$

4. $a = -7/2, \quad b = -1/2, \quad c = -3/2$

5. $a = 4/9, \quad b = 4/7, \quad c = 1/2$

6. $a = 2.7591, \quad b = 3.2391, \quad c = 3$

Finding Deltas Graphically

In Exercises 7–14, use the graphs to find a $\delta > 0$ such that

$$|f(x) - L| < \varepsilon \quad \text{whenever} \quad 0 < |x - c| < \delta.$$

7.

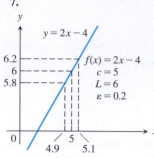

$y = 2x - 4$

$f(x) = 2x - 4$
$c = 5$
$L = 6$
$\varepsilon = 0.2$

6.2
6
5.8

0   5
4.9   5.1

NOT TO SCALE

8.

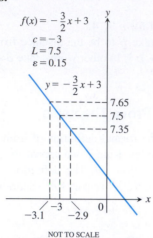

$f(x) = -\dfrac{3}{2}x + 3$

$c = -3$
$L = 7.5$
$\varepsilon = 0.15$

$y = -\dfrac{3}{2}x + 3$

7.65
7.5
7.35

-3
-3.1 -2.9 0

NOT TO SCALE

9.

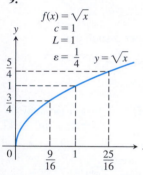

$f(x) = \sqrt{x}$
$c = 1$
$L = 1$
$\varepsilon = \dfrac{1}{4}$ $y = \sqrt{x}$

$\dfrac{5}{4}$
1
$\dfrac{3}{4}$

0   $\dfrac{9}{16}$ 1 $\dfrac{25}{16}$

10.

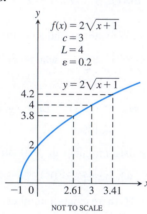

$f(x) = 2\sqrt{x + 1}$
$c = 3$
$L = 4$
$\varepsilon = 0.2$

$y = 2\sqrt{x + 1}$

4.2
4
3.8

2

-1 0   2.61   3   3.41

NOT TO SCALE

11.

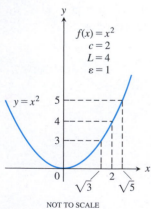

$f(x) = x^2$
$c = 2$
$L = 4$
$\varepsilon = 1$

$y = x^2$ 5

4

3

0   $\sqrt{3}$ 2 $\sqrt{5}$

NOT TO SCALE

12.

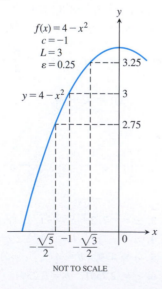

$f(x) = 4 - x^2$
$c = -1$
$L = 3$
$\varepsilon = 0.25$

$y = 4 - x^2$

3.25
3
2.75

$-\dfrac{\sqrt{5}}{2}$ -1 $-\dfrac{\sqrt{3}}{2}$ 0

NOT TO SCALE

13.

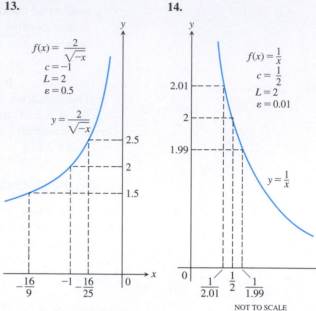

$f(x) = \dfrac{2}{\sqrt{-x}}$
$c = -1$
$L = 2$
$\varepsilon = 0.5$

$y = \dfrac{2}{\sqrt{-x}}$

2.5
2
1.5

$-\dfrac{16}{9}$ -1 $-\dfrac{16}{25}$ 0

NOT TO SCALE

14.

$f(x) = \dfrac{1}{x}$
$c = \dfrac{1}{2}$
$L = 2$
$\varepsilon = 0.01$

2.01
2
1.99

$y = \dfrac{1}{x}$

0   $\dfrac{1}{2.01}$ $\dfrac{1}{2}$ $\dfrac{1}{1.99}$

NOT TO SCALE

Finding Deltas Algebraically

Each of Exercises 15–30 gives a function $f(x)$ and numbers L, c, and $\varepsilon > 0$. In each case, find an open interval about c on which the inequality $|f(x) - L| < \varepsilon$ holds. Then give a value for $\delta > 0$ such that for all x satisfying $0 < |x - c| < \delta$ the inequality $|f(x) - L| < \varepsilon$ holds.

15. $f(x) = x + 1, \quad L = 5, \quad c = 4, \quad \varepsilon = 0.01$

16. $f(x) = 2x - 2, \quad L = -6, \quad c = -2, \quad \varepsilon = 0.02$

17. $f(x) = \sqrt{x + 1}, \quad L = 1, \quad c = 0, \quad \varepsilon = 0.1$

18. $f(x) = \sqrt{x}, \quad L = 1/2, \quad c = 1/4, \quad \varepsilon = 0.1$

19. $f(x) = \sqrt{19 - x}, \quad L = 3, \quad c = 10, \quad \varepsilon = 1$

20. $f(x) = \sqrt{x - 7}, \quad L = 4, \quad c = 23, \quad \varepsilon = 1$

21. $f(x) = 1/x, \quad L = 1/4, \quad c = 4, \quad \varepsilon = 0.05$

22. $f(x) = x^2, \quad L = 3, \quad c = \sqrt{3}, \quad \varepsilon = 0.1$

23. $f(x) = x^2, \quad L = 4, \quad c = -2, \quad \varepsilon = 0.5$

24. $f(x) = 1/x, \quad L = -1, \quad c = -1, \quad \varepsilon = 0.1$

25. $f(x) = x^2 - 5, \quad L = 11, \quad c = 4, \quad \varepsilon = 1$

26. $f(x) = 120/x, \quad L = 5, \quad c = 24, \quad \varepsilon = 1$

27. $f(x) = mx, \quad m > 0, \quad L = 2m, \quad c = 2, \quad \varepsilon = 0.03$

28. $f(x) = mx, \quad m > 0, \quad L = 3m, \quad c = 3, \quad \varepsilon = c > 0$

29. $f(x) = mx + b, \quad m > 0, \quad L = (m/2) + b,$
$c = 1/2, \quad \varepsilon = c > 0$

30. $f(x) = mx + b, \quad m > 0, \quad L = m + b, \quad c = 1, \quad \varepsilon = 0.05$

Using the Formal Definition

Each of Exercises 31–36 gives a function $f(x)$, a point c, and a positive number ε. Find $L = \lim\limits_{x \to c} f(x)$. Then find a number $\delta > 0$ such that

$$|f(x) - L| < \varepsilon \quad \text{whenever} \quad 0 < |x - c| < \delta.$$

31. $f(x) = 3 - 2x, \quad c = 3, \quad \varepsilon = 0.02$

32. $f(x) = -3x - 2, \quad c = -1, \quad \varepsilon = 0.03$

33. $f(x) = \dfrac{x^2 - 4}{x - 2}, \quad c = 2, \quad \varepsilon = 0.05$

34. $f(x) = \dfrac{x^2 + 6x + 5}{x + 5}, \qquad c = -5, \qquad \varepsilon = 0.05$

35. $f(x) = \sqrt{1 - 5x}, \qquad c = -3, \qquad \varepsilon = 0.5$

36. $f(x) = 4/x, \qquad c = 2, \qquad \varepsilon = 0.4$

Prove the limit statements in Exercises 37–50.

37. $\lim\limits_{x \to 4} (9 - x) = 5$

38. $\lim\limits_{x \to 3} (3x - 7) = 2$

39. $\lim\limits_{x \to 9} \sqrt{x - 5} = 2$

40. $\lim\limits_{x \to 0} \sqrt{4 - x} = 2$

41. $\lim\limits_{x \to 1} f(x) = 1 \quad \text{if} \quad f(x) = \begin{cases} x^2, & x \neq 1 \\ 2, & x = 1 \end{cases}$

42. $\lim\limits_{x \to -2} f(x) = 4 \quad \text{if} \quad f(x) = \begin{cases} x^2, & x \neq -2 \\ 1, & x = -2 \end{cases}$

43. $\lim\limits_{x \to 1} \dfrac{1}{x} = 1$

44. $\lim\limits_{x \to \sqrt{3}} \dfrac{1}{x^2} = \dfrac{1}{3}$

45. $\lim\limits_{x \to -3} \dfrac{x^2 - 9}{x + 3} = -6$

46. $\lim\limits_{x \to 1} \dfrac{x^2 - 1}{x - 1} = 2$

47. $\lim\limits_{x \to 1} f(x) = 2 \quad \text{if} \quad f(x) = \begin{cases} 4 - 2x, & x < 1 \\ 6x - 4, & x \geq 1 \end{cases}$

48. $\lim\limits_{x \to 0} f(x) = 0 \quad \text{if} \quad f(x) = \begin{cases} 2x, & x < 0 \\ x/2, & x \geq 0 \end{cases}$

49. $\lim\limits_{x \to 0} x \sin \dfrac{1}{x} = 0$

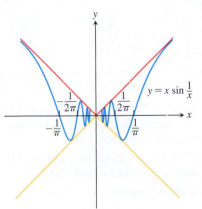

$y = x \sin \dfrac{1}{x}$

50. $\lim\limits_{x \to 0} x^2 \sin \dfrac{1}{x} = 0$

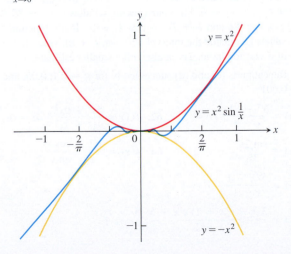

$y = x^2$

$y = x^2 \sin \dfrac{1}{x}$

$y = -x^2$

Theory and Examples

51. Define what it means to say that $\lim\limits_{x \to 0} g(x) = k$.

52. Prove that $\lim\limits_{x \to c} f(x) = L$ if and only if $\lim\limits_{h \to 0} f(h + c) = L$.

53. A wrong statement about limits Show by example that the following statement is wrong.

> The number L is the limit of $f(x)$ as x approaches c if $f(x)$ gets closer to L as x approaches c.

Explain why the function in your example does not have the given value of L as a limit as $x \to c$.

54. Another wrong statement about limits Show by example that the following statement is wrong.

> The number L is the limit of $f(x)$ as x approaches c if, given any $\varepsilon > 0$, there exists a value of x for which $|f(x) - L| < \varepsilon$.

Explain why the function in your example does not have the given value of L as a limit as $x \to c$.

T **55. Grinding engine cylinders** Before contracting to grind engine cylinders to a cross-sectional area of 9 in², you need to know how much deviation from the ideal cylinder diameter of $c = 3.385$ in. you can allow and still have the area come within 0.01 in² of the required 9 in². To find out, you let $A = \pi(x/2)^2$ and look for the interval in which you must hold x to make $|A - 9| \leq 0.01$. What interval do you find?

56. Manufacturing electrical resistors Ohm's law for electrical circuits like the one shown in the accompanying figure states that $V = RI$. In this equation, V is a constant voltage, I is

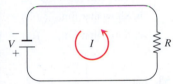

the current in amperes, and R is the resistance in ohms. Your firm has been asked to supply the resistors for a circuit in which V will be 120 volts and I is to be 5 ± 0.1 amp. In what interval does R have to lie for I to be within 0.1 amp of the value $I_0 = 5$?

When Is a Number L Not the Limit of $f(x)$ as $x \to c$?

Showing L is not a limit We can prove that $\lim\limits_{x \to c} f(x) \neq L$ by providing an $\varepsilon > 0$ such that no possible $\delta > 0$ satisfies the condition

$$|f(x) - L| < \varepsilon \quad \text{whenever} \quad 0 < |x - c| < \delta.$$

We accomplish this for our candidate ε by showing that for each $\delta > 0$ there exists a value of x such that

$$0 < |x - c| < \delta \quad \text{and} \quad |f(x) - L| \geq \varepsilon.$$

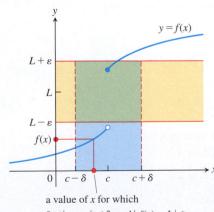

a value of x for which

$0 < |x - c| < \delta$ and $|f(x) - L| \geq \varepsilon$

57. Let $f(x) = \begin{cases} x, & x < 1 \\ x + 1, & x > 1. \end{cases}$

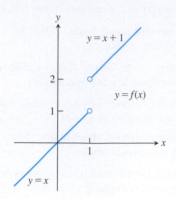

a. Let $\varepsilon = 1/2$. Show that no possible $\delta > 0$ satisfies the following condition:

$$|f(x) - 2| < 1/2 \quad \text{whenever} \quad 0 < |x - 1| < \delta.$$

That is, for each $\delta > 0$ show that there is a value of x such that

$$0 < |x - 1| < \delta \quad \text{and} \quad |f(x) - 2| \geq 1/2.$$

This will show that $\lim_{x \to 1} f(x) \neq 2$.

b. Show that $\lim_{x \to 1} f(x) \neq 1$.

c. Show that $\lim_{x \to 1} f(x) \neq 1.5$.

58. Let $h(x) = \begin{cases} x^2, & x < 2 \\ 3, & x = 2 \\ 2, & x > 2. \end{cases}$

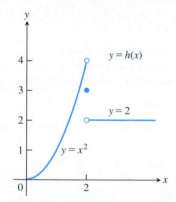

Show that

a. $\lim_{x \to 2} h(x) \neq 4$

b. $\lim_{x \to 2} h(x) \neq 3$

c. $\lim_{x \to 2} h(x) \neq 2$

59. For the function graphed here, explain why

a. $\lim_{x \to 3} f(x) \neq 4$

b. $\lim_{x \to 3} f(x) \neq 4.8$

c. $\lim_{x \to 3} f(x) \neq 3$

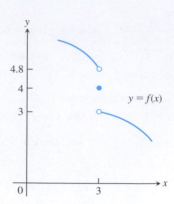

60. a. For the function graphed here, show that $\lim_{x \to -1} g(x) \neq 2$.

b. Does $\lim_{x \to -1} g(x)$ appear to exist? If so, what is the value of the limit? If not, why not?

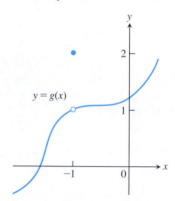

COMPUTER EXPLORATIONS

In Exercises 61–66, you will further explore finding deltas graphically. Use a CAS to perform the following steps:

a. Plot the function $y = f(x)$ near the point c being approached.

b. Guess the value of the limit L and then evaluate the limit symbolically to see if you guessed correctly.

c. Using the value $\varepsilon = 0.2$, graph the banding lines $y_1 = L - \varepsilon$ and $y_2 = L + \varepsilon$ together with the function f near c.

d. From your graph in part (c), estimate a $\delta > 0$ such that

$$|f(x) - L| < \varepsilon \quad \text{whenever} \quad 0 < |x - c| < \delta.$$

Test your estimate by plotting f, y_1, and y_2 over the interval $0 < |x - c| < \delta$. For your viewing window use $c - 2\delta \leq x \leq c + 2\delta$ and $L - 2\varepsilon \leq y \leq L + 2\varepsilon$. If any function values lie outside the interval $[L - \varepsilon, L + \varepsilon]$, your choice of δ was too large. Try again with a smaller estimate.

e. Repeat parts (c) and (d) successively for $\varepsilon = 0.1, 0.05,$ and 0.001.

61. $f(x) = \dfrac{x^4 - 81}{x - 3}, \quad c = 3$ **62.** $f(x) = \dfrac{5x^3 + 9x^2}{2x^5 + 3x^2}, \quad c = 0$

63. $f(x) = \dfrac{\sin 2x}{3x}, \quad c = 0$ **64.** $f(x) = \dfrac{x(1 - \cos x)}{x - \sin x}, \quad c = 0$

65. $f(x) = \dfrac{\sqrt[3]{x} - 1}{x - 1}, \quad c = 1$

66. $f(x) = \dfrac{3x^2 - (7x + 1)\sqrt{x} + 5}{x - 1}, \quad c = 1$

2.4 One-Sided Limits

In this section we extend the limit concept to *one-sided limits*, which are limits as x approaches the number c from the left-hand side (where $x < c$) or the right-hand side ($x > c$) only. These allow us to describe functions that have different limits at a point, depending on whether we approach the point from the left or from the right. One-sided limits also allow us to say what it means for a function to have a limit at an endpoint of an interval.

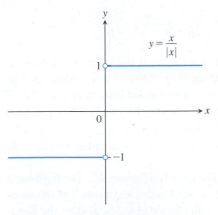

FIGURE 2.24 Different right-hand and left-hand limits at the origin.

Approaching a Limit from One Side

Suppose a function f is defined on an interval that extends to both sides of a number c. In order for f to have a limit L as x approaches c, the values of $f(x)$ must approach the value L as x approaches c from either side. Because of this, we sometimes say that the limit is **two-sided**.

If f fails to have a two-sided limit at c, it may still have a one-sided limit, that is, a limit if the approach is only from one side. If the approach is from the right, the limit is a **right-hand limit** or **limit from the right**. From the left, it is a **left-hand limit** or **limit from the left**.

The function $f(x) = x/|x|$ (Figure 2.24) has limit 1 as x approaches 0 from the right, and limit -1 as x approaches 0 from the left. Since these one-sided limit values are not the same, there is no single number that $f(x)$ approaches as x approaches 0. So $f(x)$ does not have a (two-sided) limit at 0.

Intuitively, if we only consider the values of $f(x)$ on an interval (c, b), where $c < b$, and the values of $f(x)$ become arbitrarily close to L as x approaches c from within that interval, then f has **right-hand limit** L at c. In this case we write

$$\lim_{x \to c^+} f(x) = L.$$

The notation "$x \to c^+$" means that we consider only values of $f(x)$ for x greater than c. We don't consider values of $f(x)$ for $x \leq c$.

Similarly, if $f(x)$ is defined on an interval (a, c), where $a < c$ and $f(x)$ approaches arbitrarily close to M as x approaches c from within that interval, then f has **left-hand limit** M at c. We write

$$\lim_{x \to c^-} f(x) = M.$$

The symbol "$x \to c^-$" means that we consider the values of f only at x-values less than c.

These informal definitions of one-sided limits are illustrated in Figure 2.25. For the function $f(x) = x/|x|$ in Figure 2.24 we have

$$\lim_{x \to 0^+} f(x) = 1 \qquad \text{and} \qquad \lim_{x \to 0^-} f(x) = -1.$$

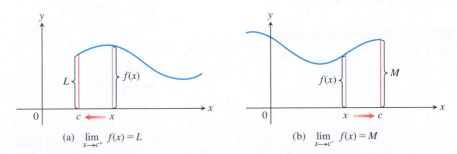

FIGURE 2.25 (a) Right-hand limit as x approaches c. (b) Left-hand limit as x approaches c.

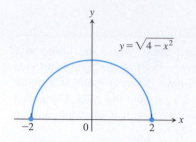

FIGURE 2.26 The function $f(x) = \sqrt{4 - x^2}$ has a right-hand limit 0 at $x = -2$ and a left-hand limit 0 at $x = 2$ (Example 1).

We now give the definition of the limit of a function at a boundary point of its domain. This definition is consistent with limits at boundary points of regions in the plane and in space, as we will see in Chapter 14. When the domain of f is an interval lying to the left of c, such as $(a, c]$ or (a, c), then we say that f has a limit at c if it has a left-hand limit at c. Similarly, if the domain of f is an interval lying to the right of c, such as $[c, b)$ or (c, b), then we say that f has a limit at c if it has a right-hand limit at c.

EXAMPLE 1   The domain of $f(x) = \sqrt{4 - x^2}$ is $[-2, 2]$; its graph is the semicircle in Figure 2.26. We have

$$\lim_{x \to -2^+} \sqrt{4 - x^2} = 0 \quad \text{and} \quad \lim_{x \to 2^-} \sqrt{4 - x^2} = 0.$$

This function has a two-sided limit at each point in $(-2, 2)$. It has a left-hand limit at $x = 2$ and a right-hand limit at $x = -2$. The function does not have a left-hand limit at $x = -2$ or a right-hand limit at $x = 2$. It does not have a two-sided limit at either -2 or 2 because f is not defined on both sides of these points. At the domain boundary points, where the domain is an interval on one side of the point, we have $\lim_{x \to -2} \sqrt{4 - x^2} = 0$ and $\lim_{x \to 2} \sqrt{4 - x^2} = 0$. The function f does have a limit at $x = -2$ and at $x = 2$.   ■

One-sided limits have all the properties listed in Theorem 1 in Section 2.2. The right-hand limit of the sum of two functions is the sum of their right-hand limits, and so on. The theorems for limits of polynomials and rational functions hold with one-sided limits, as does the Sandwich Theorem. One-sided limits are related to limits at interior points in the following way.

THEOREM 6   Suppose that a function f is defined on an open interval containing c, except perhaps at c itself. Then $f(x)$ has a limit as x approaches c if and only if it has left-hand and right-hand limits there and these one-sided limits are equal:

$$\lim_{x \to c} f(x) = L \quad \Longleftrightarrow \quad \lim_{x \to c^-} f(x) = L \quad \text{and} \quad \lim_{x \to c^+} f(x) = L.$$

Theorem 6 applies at interior points of a function's domain. At a boundary point of its domain, a function has a limit when it has an appropriate one-sided limit.

EXAMPLE 2   For the function graphed in Figure 2.27,

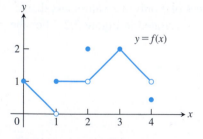

FIGURE 2.27 Graph of the function in Example 2.

At $x = 0$: $\lim_{x \to 0^-} f(x)$ does not exist,   *f is not defined to the left of $x = 0$.*
$\lim_{x \to 0^+} f(x) = 1,$ *f has a right-hand limit at $x = 0$.*
$\lim_{x \to 0} f(x) = 1.$ *f has a limit at domain endpoint $x = 0$.*

At $x = 1$: $\lim_{x \to 1^-} f(x) = 0,$ *Even though $f(1) = 1$.*
$\lim_{x \to 1^+} f(x) = 1,$
$\lim_{x \to 1} f(x)$ does not exist.   *Right- and left-hand limits are not equal.*

At $x = 2$: $\lim_{x \to 2^-} f(x) = 1,$
$\lim_{x \to 2^+} f(x) = 1,$
$\lim_{x \to 2} f(x) = 1.$ *Even though $f(2) = 2$.*

At $x = 3$: $\lim_{x \to 3^-} f(x) = \lim_{x \to 3^+} f(x) = \lim_{x \to 3} f(x) = f(3) = 2.$

At $x = 4$: $\lim_{x \to 4^-} f(x) = 1,$ *Even though $f(4) \neq 1$.*
$\lim_{x \to 4^+} f(x)$ does not exist,   *f is not defined to the right of $x = 4$.*
$\lim_{x \to 4} f(x) = 1.$ *f has a limit at domain endpoint $x = 4$.*

At every other point c in $[0, 4]$, $f(x)$ has limit $f(c)$.   ■

Precise Definitions of One-Sided Limits

The formal definition of the limit in Section 2.3 is readily modified for one-sided limits.

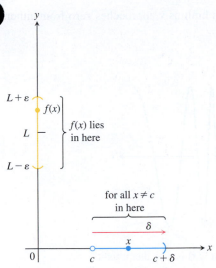

FIGURE 2.28 Intervals associated with the definition of right-hand limit.

> **DEFINITIONS** (a) Assume the domain of f contains an interval (c, d) to the right of c. We say that $f(x)$ has **right-hand limit L at c**, and write
>
> $$\lim_{x \to c^+} f(x) = L$$
>
> if for every number $\varepsilon > 0$ there exists a corresponding number $\delta > 0$ such that
>
> $$|f(x) - L| < \varepsilon \quad \text{whenever} \quad c < x < c + \delta.$$
>
> (b) Assume the domain of f contains an interval (b, c) to the left of c. We say that f has **left-hand limit L at c**, and write
>
> $$\lim_{x \to c^-} f(x) = L$$
>
> if for every number $\varepsilon > 0$ there exists a corresponding number $\delta > 0$ such that
>
> $$|f(x) - L| < \varepsilon \quad \text{whenever} \quad c - \delta < x < c.$$

The definitions are illustrated in Figures 2.28 and 2.29.

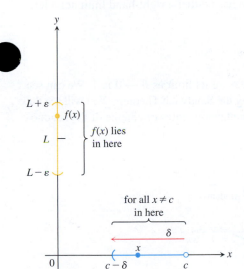

FIGURE 2.29 Intervals associated with the definition of left-hand limit.

EXAMPLE 3 Prove that

$$\lim_{x \to 0^+} \sqrt{x} = 0.$$

Solution Let $\varepsilon > 0$ be given. Here $c = 0$ and $L = 0$, so we want to find a $\delta > 0$ such that

$$|\sqrt{x} - 0| < \varepsilon \quad \text{whenever} \quad 0 < x < \delta,$$

or

$$\sqrt{x} < \varepsilon \quad \text{whenever} \quad 0 < x < \delta. \qquad \sqrt{x} \geq 0 \text{ so } |\sqrt{x}| = \sqrt{x}$$

Squaring both sides of this last inequality gives

$$x < \varepsilon^2 \quad \text{if} \quad 0 < x < \delta.$$

If we choose $\delta = \varepsilon^2$ we have

$$\sqrt{x} < \varepsilon \quad \text{whenever} \quad 0 < x < \delta = \varepsilon^2,$$

or

$$|\sqrt{x} - 0| < \varepsilon \quad \text{whenever} \quad 0 < x < \varepsilon^2.$$

According to the definition, this shows that $\lim_{x \to 0^+} \sqrt{x} = 0$ (Figure 2.30). ∎

The functions examined so far have had some kind of limit at each point of interest. In general, that need not be the case.

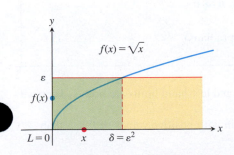

FIGURE 2.30 $\lim_{x \to 0^+} \sqrt{x} = 0$ in Example 3.

EXAMPLE 4   Show that $y = \sin(1/x)$ has no limit as x approaches zero from either side (Figure 2.31).

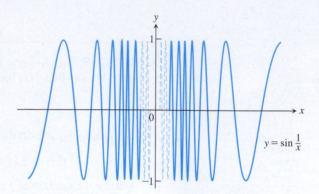

FIGURE 2.31   The function $y = \sin(1/x)$ has neither a right-hand nor a left-hand limit as x approaches zero (Example 4). The graph here omits values very near the y-axis.

Solution   As x approaches zero, its reciprocal, $1/x$, grows without bound and the values of $\sin(1/x)$ cycle repeatedly from -1 to 1. There is no single number L that the function's values stay increasingly close to as x approaches zero. This is true even if we restrict x to positive values or to negative values. The function has neither a right-hand limit nor a left-hand limit at $x = 0$.   ■

Limits Involving $(\sin \theta)/\theta$

A central fact about $(\sin \theta)/\theta$ is that in radian measure its limit as $\theta \to 0$ is 1. We can see this in Figure 2.32 and confirm it algebraically using the Sandwich Theorem. You will see the importance of this limit in Section 3.5, where instantaneous rates of change of the trigonometric functions are studied.

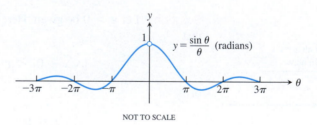

NOT TO SCALE

FIGURE 2.32   The graph of $f(\theta) = (\sin \theta)/\theta$ suggests that the right- and left-hand limits as θ approaches 0 are both 1.

THEOREM 7—Limit of the Ratio $\sin \theta/\theta$ as $\theta \to 0$

$$\lim_{\theta \to 0} \frac{\sin \theta}{\theta} = 1 \qquad (\theta \text{ in radians}) \tag{1}$$

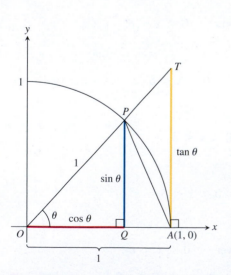

FIGURE 2.33   The ratio $TA/OA = \tan \theta$, and $OA = 1$, so $TA = \tan \theta$.

Proof   The plan is to show that the right-hand and left-hand limits are both 1. Then we will know that the two-sided limit is 1 as well.

To show that the right-hand limit is 1, we begin with positive values of θ less than $\pi/2$ (Figure 2.33). Notice that

$$\text{Area } \triangle OAP < \text{ area sector } OAP < \text{ area } \triangle OAT.$$

We can express these areas in terms of θ as follows:

$$\text{Area } \Delta OAP = \frac{1}{2}\text{base} \times \text{height} = \frac{1}{2}(1)(\sin\theta) = \frac{1}{2}\sin\theta$$

$$\text{Area sector } OAP = \frac{1}{2}r^2\theta = \frac{1}{2}(1)^2\theta = \frac{\theta}{2} \qquad (2)$$

$$\text{Area } \Delta OAT = \frac{1}{2}\text{base} \times \text{height} = \frac{1}{2}(1)(\tan\theta) = \frac{1}{2}\tan\theta.$$

The use of radians to measure angles is essential in Equation (2): The area of sector OAP is $\theta/2$ only if θ is measured in radians.

Thus,

$$\frac{1}{2}\sin\theta < \frac{1}{2}\theta < \frac{1}{2}\tan\theta.$$

This last inequality goes the same way if we divide all three terms by the number $(1/2)\sin\theta$, which is positive, since $0 < \theta < \pi/2$:

$$1 < \frac{\theta}{\sin\theta} < \frac{1}{\cos\theta}.$$

Taking reciprocals reverses the inequalities:

$$1 > \frac{\sin\theta}{\theta} > \cos\theta.$$

Since $\lim_{\theta\to0^+}\cos\theta = 1$ (Example 11b, Section 2.2), the Sandwich Theorem gives

$$\lim_{\theta\to0^+}\frac{\sin\theta}{\theta} = 1.$$

To consider the left-hand limit, we recall that $\sin\theta$ and θ are both *odd functions* (Section 1.1). Therefore, $f(\theta) = (\sin\theta)/\theta$ is an *even function*, with a graph symmetric about the y-axis (see Figure 2.32). This symmetry implies that the left-hand limit at 0 exists and has the same value as the right-hand limit:

$$\lim_{\theta\to0^-}\frac{\sin\theta}{\theta} = 1 = \lim_{\theta\to0^+}\frac{\sin\theta}{\theta},$$

so $\lim_{\theta\to0}(\sin\theta)/\theta = 1$ by Theorem 6. ∎

EXAMPLE 5   Show that **(a)** $\displaystyle\lim_{y\to0}\frac{\cos y - 1}{y} = 0$ and   **(b)** $\displaystyle\lim_{x\to0}\frac{\sin 2x}{5x} = \frac{2}{5}$.

Solution

(a) Using the half-angle formula $\cos y = 1 - 2\sin^2(y/2)$, we calculate

$$\lim_{y\to0}\frac{\cos y - 1}{y} = \lim_{y\to0} -\frac{2\sin^2(y/2)}{y}$$

$$= -\lim_{\theta\to0}\frac{\sin\theta}{\theta}\sin\theta \qquad \text{Let } \theta = y/2.$$

$$= -(1)(0) = 0. \qquad \text{Eq. (1) and Example 11a in Section 2.2}$$

(b) Equation (1) does not apply to the original fraction. We need a $2x$ in the denominator, not a $5x$. We produce it by multiplying numerator and denominator by $2/5$:

$$\lim_{x\to0}\frac{\sin 2x}{5x} = \lim_{x\to0}\frac{(2/5)\cdot\sin 2x}{(2/5)\cdot 5x}$$

$$= \frac{2}{5}\lim_{x\to0}\frac{\sin 2x}{2x} \qquad \text{Eq. (1) applies with } \theta = 2x.$$

$$= \frac{2}{5}(1) = \frac{2}{5}.$$

EXAMPLE 6   Find $\displaystyle\lim_{t\to 0}\frac{\tan t \sec 2t}{3t}$.

Solution   From the definition of $\tan t$ and $\sec 2t$, we have

$$\lim_{t\to 0}\frac{\tan t \sec 2t}{3t} = \lim_{t\to 0}\frac{1}{3}\cdot\frac{1}{t}\cdot\frac{\sin t}{\cos t}\cdot\frac{1}{\cos 2t}$$

$$= \frac{1}{3}\lim_{t\to 0}\frac{\sin t}{t}\cdot\frac{1}{\cos t}\cdot\frac{1}{\cos 2t}$$

$$= \frac{1}{3}(1)(1)(1) = \frac{1}{3}. \qquad \text{Eq. (1) and Example 11b in Section 2.2}$$

EXAMPLE 7   Show that for nonzero constants A and B.

$$\lim_{\theta\to 0}\frac{\sin A\theta}{\sin B\theta} = \frac{A}{B}.$$

Solution

$$\lim_{\theta\to 0}\frac{\sin A\theta}{\sin B\theta} = \lim_{\theta\to 0}\frac{\sin A\theta}{A\theta}A\theta\frac{B\theta}{\sin B\theta}\frac{1}{B\theta} \qquad \text{Multiply and divide by } A\theta \text{ and } B\theta.$$

$$= \lim_{\theta\to 0}\frac{\sin A\theta}{A\theta}\frac{B\theta}{\sin B\theta}\frac{A}{B} \qquad \lim_{u\to 0}\frac{\sin u}{u} = 1, \text{ with } u = A\theta$$

$$= \lim_{\theta\to 0}(1)(1)\frac{A}{B} \qquad \lim_{v\to 0}\frac{v}{\sin v} = 1, \text{ with } v = B\theta$$

$$= \frac{A}{B}.$$

EXERCISES 2.4

Finding Limits Graphically

1. Which of the following statements about the function $y = f(x)$ graphed here are true, and which are false?

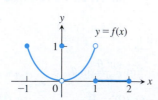

$y = f(x)$

a. $\displaystyle\lim_{x\to -1^+} f(x) = 1$ b. $\displaystyle\lim_{x\to 0^-} f(x) = 0$

c. $\displaystyle\lim_{x\to 0^-} f(x) = 1$ d. $\displaystyle\lim_{x\to 0^-} f(x) = \lim_{x\to 0^+} f(x)$

e. $\displaystyle\lim_{x\to 0} f(x)$ exists.   f. $\displaystyle\lim_{x\to 0} f(x) = 0$

g. $\displaystyle\lim_{x\to 0} f(x) = 1$ h. $\displaystyle\lim_{x\to 1} f(x) = 1$

i. $\displaystyle\lim_{x\to 1} f(x) = 0$ j. $\displaystyle\lim_{x\to 2^-} f(x) = 2$

k. $\displaystyle\lim_{x\to -1^-} f(x)$ does not exist.   l. $\displaystyle\lim_{x\to 2^+} f(x) = 0$

2. Which of the following statements about the function $y = f(x)$ graphed here are true, and which are false?

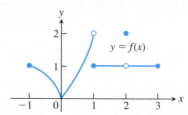

$y = f(x)$

a. $\displaystyle\lim_{x\to -1^+} f(x) = 1$ b. $\displaystyle\lim_{x\to 2} f(x)$ does not exist.

c. $\displaystyle\lim_{x\to 2} f(x) = 2$ d. $\displaystyle\lim_{x\to 1^-} f(x) = 2$

e. $\displaystyle\lim_{x\to 1^+} f(x) = 1$ f. $\displaystyle\lim_{x\to 1} f(x)$ does not exist.

g. $\displaystyle\lim_{x\to 0^+} f(x) = \lim_{x\to 0^-} f(x)$

h. $\displaystyle\lim_{x\to c} f(x)$ exists at every c in the open interval $(-1, 1)$.

i. $\displaystyle\lim_{x\to c} f(x)$ exists at every c in the open interval $(1, 3)$.

j. $\displaystyle\lim_{x\to -1^-} f(x) = 0$ k. $\displaystyle\lim_{x\to 3^+} f(x)$ does not exist.

3. Let $f(x) = \begin{cases} 3 - x, & x < 2 \\ \dfrac{x}{2} + 1, & x > 2. \end{cases}$

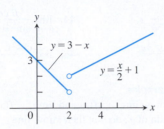

a. Find $\lim_{x \to 2^+} f(x)$ and $\lim_{x \to 2^-} f(x)$.

b. Does $\lim_{x \to 2} f(x)$ exist? If so, what is it? If not, why not?

c. Find $\lim_{x \to 4^-} f(x)$ and $\lim_{x \to 4^+} f(x)$.

d. Does $\lim_{x \to 4} f(x)$ exist? If so, what is it? If not, why not?

4. Let $f(x) = \begin{cases} 3 - x, & x < 2 \\ 2, & x = 2 \\ \dfrac{x}{2}, & x > 2. \end{cases}$

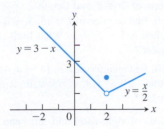

a. Find $\lim_{x \to 2^+} f(x)$, $\lim_{x \to 2^-} f(x)$, and $f(2)$.

b. Does $\lim_{x \to 2} f(x)$ exist? If so, what is it? If not, why not?

c. Find $\lim_{x \to -1^-} f(x)$ and $\lim_{x \to -1^+} f(x)$.

d. Does $\lim_{x \to -1} f(x)$ exist? If so, what is it? If not, why not?

5. Let $f(x) = \begin{cases} 0, & x \le 0 \\ \sin \dfrac{1}{x}, & x > 0. \end{cases}$

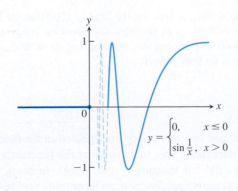

$$y = \begin{cases} 0, & x \le 0 \\ \sin \dfrac{1}{x}, & x > 0 \end{cases}$$

a. Does $\lim_{x \to 0^+} f(x)$ exist? If so, what is it? If not, why not?

b. Does $\lim_{x \to 0^-} f(x)$ exist? If so, what is it? If not, why not?

c. Does $\lim_{x \to 0} f(x)$ exist? If so, what is it? If not, why not?

6. Let $g(x) = \sqrt{x} \sin(1/x)$.

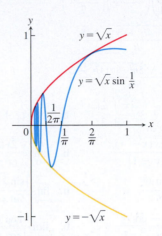

a. Does $\lim_{x \to 0^+} g(x)$ exist? If so, what is it? If not, why not?

b. Does $\lim_{x \to 0^-} g(x)$ exist? If so, what is it? If not, why not?

c. Does $\lim_{x \to 0} g(x)$ exist? If so, what is it? If not, why not?

7. a. Graph $f(x) = \begin{cases} x^3, & x \ne 1 \\ 0, & x = 1. \end{cases}$

b. Find $\lim_{x \to 1^-} f(x)$ and $\lim_{x \to 1^+} f(x)$.

c. Does $\lim_{x \to 1} f(x)$ exist? If so, what is it? If not, why not?

8. a. Graph $f(x) = \begin{cases} 1 - x^2, & x \ne 1 \\ 2, & x = 1. \end{cases}$

b. Find $\lim_{x \to 1^+} f(x)$ and $\lim_{x \to 1^-} f(x)$.

c. Does $\lim_{x \to 1} f(x)$ exist? If so, what is it? If not, why not?

Graph the functions in Exercises 9 and 10. Then answer these questions.

a. What are the domain and range of f?

b. At what points c, if any, does $\lim_{x \to c} f(x)$ exist?

c. At what points does the left-hand limit exist but not the right-hand limit?

d. At what points does the right-hand limit exist but not the left-hand limit?

9. $f(x) = \begin{cases} \sqrt{1 - x^2}, & 0 \le x < 1 \\ 1, & 1 \le x < 2 \\ 2, & x = 2 \end{cases}$

10. $f(x) = \begin{cases} x, & -1 \le x < 0, \text{ or } 0 < x \le 1 \\ 1, & x = 0 \\ 0, & x < -1 \text{ or } x > 1 \end{cases}$

Finding One-Sided Limits Algebraically

Find the limits in Exercises 11–20.

11. $\lim_{x \to -0.5^-} \sqrt{\dfrac{x + 2}{x + 1}}$

12. $\lim_{x \to 1^+} \sqrt{\dfrac{x - 1}{x + 2}}$

13. $\lim_{x \to -2^+} \left(\dfrac{x}{x + 1} \right) \left(\dfrac{2x + 5}{x^2 + x} \right)$

14. $\lim\limits_{x \to 1^-} \left(\dfrac{1}{x+1} \right) \left(\dfrac{x+6}{x} \right) \left(\dfrac{3-x}{7} \right)$

15. $\lim\limits_{h \to 0^+} \dfrac{\sqrt{h^2 + 4h + 5} - \sqrt{5}}{h}$

16. $\lim\limits_{h \to 0^-} \dfrac{\sqrt{6} - \sqrt{5h^2 + 11h + 6}}{h}$

17. a. $\lim\limits_{x \to -2^+} (x+3) \dfrac{|x+2|}{x+2}$ **b.** $\lim\limits_{x \to -2^-} (x+3) \dfrac{|x+2|}{x+2}$

18. a. $\lim\limits_{x \to 1^+} \dfrac{\sqrt{2x}\,(x-1)}{|x-1|}$ **b.** $\lim\limits_{x \to 1^-} \dfrac{\sqrt{2x}\,(x-1)}{|x-1|}$

19. a. $\lim\limits_{x \to 0^+} \dfrac{|\sin x|}{\sin x}$ **b.** $\lim\limits_{x \to 0^-} \dfrac{|\sin x|}{\sin x}$

20. a. $\lim\limits_{x \to 0^+} \dfrac{1 - \cos x}{|\cos x - 1|}$ **b.** $\lim\limits_{x \to 0^-} \dfrac{\cos x - 1}{|\cos x - 1|}$

Use the graph of the greatest integer function $y = \lfloor x \rfloor$, Figure 1.10 in Section 1.1, to help you find the limits in Exercises 21 and 22.

21. a. $\lim\limits_{\theta \to 3^+} \dfrac{\lfloor \theta \rfloor}{\theta}$ **b.** $\lim\limits_{\theta \to 3^-} \dfrac{\lfloor \theta \rfloor}{\theta}$

22. a. $\lim\limits_{t \to 4^+} (t - \lfloor t \rfloor)$ **b.** $\lim\limits_{t \to 4^-} (t - \lfloor t \rfloor)$

Using $\lim\limits_{\theta \to 0} \dfrac{\sin \theta}{\theta} = 1$

Find the limits in Exercises 23–46.

23. $\lim\limits_{\theta \to 0} \dfrac{\sin \sqrt{2}\theta}{\sqrt{2}\theta}$

24. $\lim\limits_{t \to 0} \dfrac{\sin kt}{t}$ (k constant)

25. $\lim\limits_{y \to 0} \dfrac{\sin 3y}{4y}$

26. $\lim\limits_{h \to 0^-} \dfrac{h}{\sin 3h}$

27. $\lim\limits_{x \to 0} \dfrac{\tan 2x}{x}$

28. $\lim\limits_{t \to 0} \dfrac{2t}{\tan t}$

29. $\lim\limits_{x \to 0} \dfrac{x \csc 2x}{\cos 5x}$

30. $\lim\limits_{x \to 0} 6x^2 (\cot x)(\csc 2x)$

31. $\lim\limits_{x \to 0} \dfrac{x + x \cos x}{\sin x \cos x}$

32. $\lim\limits_{x \to 0} \dfrac{x^2 - x + \sin x}{2x}$

33. $\lim\limits_{\theta \to 0} \dfrac{1 - \cos \theta}{\sin 2\theta}$

34. $\lim\limits_{x \to 0} \dfrac{x - x \cos x}{\sin^2 3x}$

35. $\lim\limits_{t \to 0} \dfrac{\sin(1 - \cos t)}{1 - \cos t}$

36. $\lim\limits_{h \to 0} \dfrac{\sin(\sin h)}{\sin h}$

37. $\lim\limits_{\theta \to 0} \dfrac{\sin \theta}{\sin 2\theta}$

38. $\lim\limits_{x \to 0} \dfrac{\sin 5x}{\sin 4x}$

39. $\lim\limits_{\theta \to 0} \theta \cos \theta$

40. $\lim\limits_{\theta \to 0} \sin \theta \cot 2\theta$

41. $\lim\limits_{x \to 0} \dfrac{\tan 3x}{\sin 8x}$

42. $\lim\limits_{y \to 0} \dfrac{\sin 3y \cot 5y}{y \cot 4y}$

43. $\lim\limits_{\theta \to 0} \dfrac{\tan \theta}{\theta^2 \cot 3\theta}$

44. $\lim\limits_{\theta \to 0} \dfrac{\theta \cot 4\theta}{\sin^2 \theta \cot^2 2\theta}$

45. $\lim\limits_{x \to 0} \dfrac{1 - \cos 3x}{2x}$

46. $\lim\limits_{x \to 0} \dfrac{\cos^2 x - \cos x}{x^2}$

Theory and Examples

47. Once you know $\lim_{x \to a^+} f(x)$ and $\lim_{x \to a^-} f(x)$ at an interior point of the domain of f, do you then know $\lim_{x \to a} f(x)$? Give reasons for your answer.

48. If you know that $\lim_{x \to c} f(x)$ exists, can you find its value by calculating $\lim_{x \to c^+} f(x)$? Give reasons for your answer.

49. Suppose that f is an odd function of x. Does knowing that $\lim_{x \to 0^+} f(x) = 3$ tell you anything about $\lim_{x \to 0^-} f(x)$? Give reasons for your answer.

50. Suppose that f is an even function of x. Does knowing that $\lim_{x \to 2^-} f(x) = 7$ tell you anything about either $\lim_{x \to -2^-} f(x)$ or $\lim_{x \to -2^+} f(x)$? Give reasons for your answer.

Formal Definitions of One-Sided Limits

51. Given $\varepsilon > 0$, find an interval $I = (5, 5 + \delta)$, $\delta > 0$, such that if x lies in I, then $\sqrt{x - 5} < \varepsilon$. What limit is being verified and what is its value?

52. Given $\varepsilon > 0$, find an interval $I = (4 - \delta, 4)$, $\delta > 0$, such that if x lies in I, then $\sqrt{4 - x} < \varepsilon$. What limit is being verified and what is its value?

Use the definitions of right-hand and left-hand limits to prove the limit statements in Exercises 53 and 54.

53. $\lim\limits_{x \to 0^-} \dfrac{x}{|x|} = -1$ **54.** $\lim\limits_{x \to 2^+} \dfrac{x - 2}{|x - 2|} = 1$

55. Greatest integer function Find **(a)** $\lim_{x \to 400^+} \lfloor x \rfloor$ and **(b)** $\lim_{x \to 400^-} \lfloor x \rfloor$; then use limit definitions to verify your findings. **(c)** Based on your conclusions in parts (a) and (b), can you say anything about $\lim_{x \to 400} \lfloor x \rfloor$? Give reasons for your answer.

56. One-sided limits   Let $f(x) = \begin{cases} x^2 \sin(1/x), & x < 0 \\ \sqrt{x}, & x > 0. \end{cases}$

Find **(a)** $\lim_{x \to 0^+} f(x)$ and **(b)** $\lim_{x \to 0^-} f(x)$; then use limit definitions to verify your findings. **(c)** Based on your conclusions in parts (a) and (b), can you say anything about $\lim_{x \to 0} f(x)$? Give reasons for your answer.

2.5   Continuity

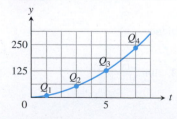

FIGURE 2.34   Connecting plotted points.

When we plot function values generated in a laboratory or collected in the field, we often connect the plotted points with an unbroken curve to show what the function's values are likely to have been at the points we did not measure (Figure 2.34). In doing so, we are assuming that we are working with a *continuous function*, so its outputs vary regularly and consistently with the inputs, and do not jump abruptly from one value to another without taking on the values in between. Intuitively, any function $y = f(x)$ whose graph can be sketched over its domain in one unbroken motion is an example of a continuous function. Such functions play an important role in the study of calculus and its applications.

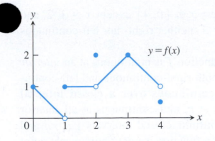

FIGURE 2.35 The function is not continuous at $x = 1$, $x = 2$, and $x = 4$ (Example 1).

Continuity at a Point

To understand continuity, it helps to consider a function like that in Figure 2.35, whose limits we investigated in Example 2 in the last section.

EXAMPLE 1   At which numbers does the function f in Figure 2.35 appear to be not continuous? Explain why. What occurs at other numbers in the domain?

Solution   First we observe that the domain of the function is the closed interval $[0, 4]$, so we will be considering the numbers x within that interval. From the figure, we notice right away that there are breaks in the graph at the numbers $x = 1$, $x = 2$, and $x = 4$. The break at $x = 1$ appears as a jump, which we identify later as a "jump discontinuity." The break at $x = 2$ is called a "removable discontinuity" since by changing the function definition at that one point, we can create a new function that is continuous at $x = 2$. Similarly $x = 4$ is a removable discontinuity.

Numbers at which the graph of f has breaks:

At the interior point $x = 1$, the function fails to have a limit. It does have both a left-hand limit, $\lim_{x \to 1^-} f(x) = 0$, as well as a right-hand limit, $\lim_{x \to 1^+} f(x) = 1$, but the limit values are different, resulting in a jump in the graph. The function is not continuous at $x = 1$. However the function value $f(1) = 1$ is equal to the limit from the right, so the function *is* continuous from the right at $x = 1$.

At $x = 2$, the function does have a limit, $\lim_{x \to 2} f(x) = 1$, but the value of the function is $f(2) = 2$. The limit and function values are not the same, so there is a break in the graph and f is not continuous at $x = 2$.

At $x = 4$, the function does have a left-hand limit at this right endpoint, $\lim_{x \to 4^-} f(x) = 1$, but again the value of the function $f(4) = \frac{1}{2}$ differs from the value of the limit. We see again a break in the graph of the function at this endpoint and the function is not continuous from the left.

Numbers at which the graph of f has no breaks:

At $x = 3$, the function has a limit, $\lim_{x \to 3} f(x) = 2$. Moreover, the limit is the same value as the function there, $f(3) = 2$. The function is continuous at $x = 3$.

At $x = 0$, the function has a right-hand limit at this left endpoint, $\lim_{x \to 0^+} f(x) = 1$, and the value of the function is the same, $f(0) = 1$. The function is continuous from the right at $x = 0$. Because $x = 0$ is a left endpoint of the function's domain, we have that $\lim_{x \to 0} f(x) = 1$ and so f is continuous at $x = 0$.

At all other numbers $x = c$ in the domain, the function has a limit equal to the value of the function, so $\lim_{x \to c} f(x) = f(c)$. For example, $\lim_{x \to 5/2} f(x) = f\left(\frac{5}{2}\right) = \frac{3}{2}$. No breaks appear in the graph of the function at any of these numbers and the function is continuous at each of them. ∎

The following definitions capture the continuity ideas we observed in Example 1.

DEFINITIONS   Let c be a real number that is either an interior point or an endpoint of an interval in the domain of f.

The function f is **continuous at c** if

$$\lim_{x \to c} f(x) = f(c).$$

The function f is **right-continuous at c (or continuous from the right)** if

$$\lim_{x \to c^+} f(x) = f(c).$$

The function f is **left-continuous at c (or continuous from the left)** if

$$\lim_{x \to c^-} f(x) = f(c).$$

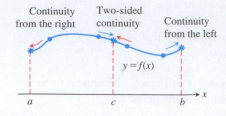

FIGURE 2.36 Continuity at points a, b, and c.

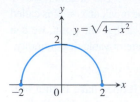

FIGURE 2.37 A function that is continuous over its domain (Example 2).

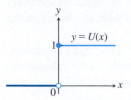

FIGURE 2.38 A function that has a jump discontinuity at the origin (Example 3).

The function f in Example 1 is continuous at every x in $[0, 4]$ except $x = 1, 2$, and 4. It is right-continuous but not left-continuous at $x = 1$, neither right- nor left-continuous at $x = 2$, and not left-continuous at $x = 4$.

From Theorem 6, it follows immediately that a function f is continuous at an *interior* point c of an interval in its domain if and only if it is both right-continuous and left-continuous at c (Figure 2.36). We say that a function is **continuous over a closed interval** $[a, b]$ if it is right-continuous at a, left-continuous at b, and continuous at all interior points of the interval. This definition applies to the infinite closed intervals $[a, \infty)$ and $(-\infty, b]$ as well, but only one endpoint is involved. If a function is not continuous at point c of its domain, we say that f is **discontinuous at c**, and that f has a discontinuity at c. Note that a function f can be continuous, right-continuous, or left-continuous only at a point c for which $f(c)$ is defined.

EXAMPLE 2   The function $f(x) = \sqrt{4 - x^2}$ is continuous over its domain $[-2, 2]$ (Figure 2.37). It is right-continuous at $x = -2$, and left-continuous at $x = 2$. ∎

EXAMPLE 3   The unit step function $U(x)$, graphed in Figure 2.38, is right-continuous at $x = 0$, but is neither left-continuous nor continuous there. It has a jump discontinuity at $x = 0$. ∎

At an interior point or an endpoint of an interval in its domain, a function is continuous at points where it passes the following test.

Continuity Test

A function $f(x)$ is continuous at a point $x = c$ if and only if it meets the following three conditions.

1. $f(c)$ exists   (c lies in the domain of f).
2. $\lim_{x \to c} f(x)$ exists   (f has a limit as $x \to c$).
3. $\lim_{x \to c} f(x) = f(c)$ (the limit equals the function value).

For one-sided continuity, the limits in parts 2 and 3 of the test should be replaced by the appropriate one-sided limits.

EXAMPLE 4   The function $y = \lfloor x \rfloor$ introduced in Section 1.1 is graphed in Figure 2.39. It is discontinuous at every integer n, because the left-hand and right-hand limits are not equal as $x \to n$:

$$\lim_{x \to n^-} \lfloor x \rfloor = n - 1 \quad \text{and} \quad \lim_{x \to n^+} \lfloor x \rfloor = n.$$

Since $\lfloor n \rfloor = n$, the greatest integer function is right-continuous at every integer n (but not left-continuous).

The greatest integer function is continuous at every real number other than the integers. For example,

$$\lim_{x \to 1.5} \lfloor x \rfloor = 1 = \lfloor 1.5 \rfloor.$$

In general, if $n - 1 < c < n$, n an integer, then

$$\lim_{x \to c} \lfloor x \rfloor = n - 1 = \lfloor c \rfloor.$$ ∎

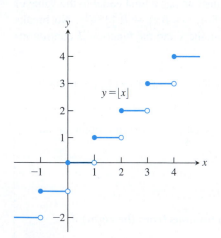

FIGURE 2.39 The greatest integer function is continuous at every noninteger point. It is right-continuous, but not left-continuous, at every integer point (Example 4).

Figure 2.40 displays several common ways in which a function can fail to be continuous. The function in Figure 2.40a is continuous at $x = 0$. The function in Figure 2.40b does not contain $x = 0$ in its domain. It would be continuous if its domain extended so that $f(0) = 1$. The function in Figure 2.40c would be continuous if $f(0)$ were 1 instead of 2. The discontinuity in Figure 2.40c is **removable**. The function has a limit as $x \to 0$, and we can remove the discontinuity by setting $f(0)$ equal to this limit.

The discontinuities in Figure 2.40d through f are more serious: $\lim_{x \to 0} f(x)$ does not exist, and there is no way to improve the situation by appropriately defining f at 0. The step function in Figure 2.40d has a **jump discontinuity**: The one-sided limits exist but have different values. The function $f(x) = 1/x^2$ in Figure 2.40e has an **infinite discontinuity**. The function in Figure 2.40f has an **oscillating discontinuity**: It oscillates so much that its values approach each number in $[-1, 1]$ as $x \to 0$. Since it does not approach a single number, it does not have a limit as x approaches 0.

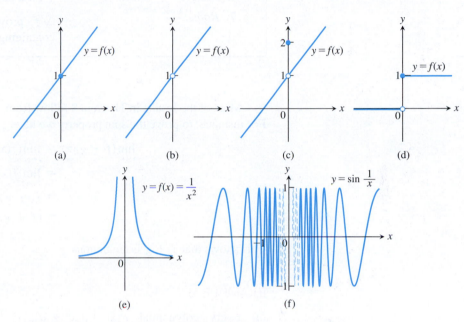

FIGURE 2.40 The function in (a) is continuous at $x = 0$; the functions in (b) through (f) are not.

Continuous Functions

We now describe the continuity behavior of a function throughout its entire domain, not only at a single point. We define a **continuous function** to be one that is continuous at every point in its domain. This is a property of the *function*. A function always has a specified domain, so if we change the domain then we change the function, and this may change its continuity property as well. If a function is discontinuous at one or more points of its domain, we say it is a **discontinuous** function.

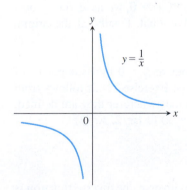

FIGURE 2.41 The function $f(x) = 1/x$ is continuous over its natural domain. It is not defined at the origin, so it is not continuous on any interval containing $x = 0$ (Example 5).

EXAMPLE 5

(a) The function $f(x) = 1/x$ (Figure 2.41) is a continuous function because it is continuous at every point of its domain. The point $x = 0$ is not in the domain of the function f, so f is not continuous on any interval containing $x = 0$. Moreover, there is no way to extend f to a new function that is defined and continuous at $x = 0$. The function f does not have a removable discontinuity at $x = 0$.

(b) The identity function $f(x) = x$ and constant functions are continuous everywhere by Example 3, Section 2.3.   ■

Algebraic combinations of continuous functions are continuous wherever they are defined.

THEOREM 8—Properties of Continuous Functions

If the functions f and g are continuous at $x = c$, then the following algebraic combinations are continuous at $x = c$.

1. *Sums:* $\qquad\qquad\qquad$ $f + g$
2. *Differences:* $\qquad\qquad$ $f - g$
3. *Constant multiples:* $\qquad$ $k \cdot f$, for any number k
4. *Products:* $\qquad\qquad$ $f \cdot g$
5. *Quotients:* $\qquad\qquad$ f/g, provided $g(c) \neq 0$
6. *Powers:* $\qquad\qquad\quad$ f^n, n a positive integer
7. *Roots:* $\qquad\qquad\qquad$ $\sqrt[n]{f}$, provided it is defined on an interval containing c, where n is a positive integer

---

Most of the results in Theorem 8 follow from the limit rules in Theorem 1, Section 2.2. For instance, to prove the sum property we have

$$
\begin{aligned}
\lim_{x \to c}(f + g)(x) &= \lim_{x \to c}(f(x) + g(x)) \\
&= \lim_{x \to c} f(x) + \lim_{x \to c} g(x) \qquad \text{Sum Rule, Theorem 1} \\
&= f(c) + g(c) \qquad\qquad\quad \text{Continuity of } f, g \text{ at } c \\
&= (f + g)(c).
\end{aligned}
$$

This shows that $f + g$ is continuous.

EXAMPLE 6

(a) Every polynomial $P(x) = a_n x^n + a_{n-1} x^{n-1} + \cdots + a_0$ is continuous because $\lim_{x \to c} P(x) = P(c)$ by Theorem 2, Section 2.2.

(b) If $P(x)$ and $Q(x)$ are polynomials, then the rational function $P(x)/Q(x)$ is continuous wherever it is defined ($Q(c) \neq 0$) by Theorem 3, Section 2.2. ∎

EXAMPLE 7 The function $f(x) = |x|$ is continuous. If $x > 0$, we have $f(x) = x$, a polynomial. If $x < 0$, we have $f(x) = -x$, another polynomial. Finally, at the origin, $\lim_{x \to 0} |x| = 0 = |0|$. ∎

The functions $y = \sin x$ and $y = \cos x$ are continuous at $x = 0$ by Example 11 of Section 2.2. Both functions are continuous everywhere (see Exercise 72). It follows from Theorem 8 that all six trigonometric functions are continuous wherever they are defined. For example, $y = \tan x$ is continuous on $\cdots \cup (-\pi/2, \pi/2) \cup (\pi/2, 3\pi/2) \cup \cdots$.

Inverse Functions and Continuity

When a continuous function defined on an interval has an inverse, the inverse function is itself a continuous function over its own domain. This result is suggested by the observation that the graph of f^{-1}, being the reflection of the graph of f across the line $y = x$, cannot have any breaks in it when the graph of f has no breaks. A rigorous proof that f^{-1} is continuous whenever f is continuous on an interval is given in more advanced texts. As an example, the inverse trigonometric functions are all continuous over their domains.

We defined the exponential function $y = a^x$ in Section 1.5 informally. The graph was obtained from the graph of $y = a^x$ for x a rational number by "filling in the holes" at the irrational points x, so as to make the function $y = a^x$ continuous over the entire real line. The inverse function $y = \log_a x$ is also continuous. In particular, the natural exponential function $y = e^x$ and the natural logarithm function $y = \ln x$ are both continuous over their domains. Proofs of continuity for these functions will be given in Chapter 7.

Continuity of Compositions of Functions

Functions obtained by composing continuous functions are continuous. If $f(x)$ is continuous at $x = c$ and $g(x)$ is continuous at $x = f(c)$, then $g \circ f$ is also continuous at $x = c$ (Figure 2.42). In this case, the limit of $g \circ f$ as $x \rightarrow c$ is $g(f(c))$.

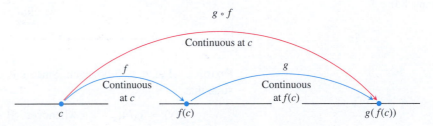

FIGURE 2.42 Compositions of continuous functions are continuous.

THEOREM 9—Compositions of Continuous Functions

If f is continuous at c and g is continuous at $f(c)$, then the composition $g \circ f$ is continuous at c.

Intuitively, Theorem 9 is reasonable because if x is close to c, then $f(x)$ is close to $f(c)$, and since g is continuous at $f(c)$, it follows that $g(f(x))$ is close to $g(f(c))$.

The continuity of compositions holds for any finite number of compositions of functions. The only requirement is that each function be continuous where it is applied. An outline of a proof of Theorem 9 is given in Exercise 6 in Appendix 4.

EXAMPLE 8   Show that the following functions are continuous on their natural domains.

(a) $y = \sqrt{x^2 - 2x - 5}$

(b) $y = \dfrac{x^{2/3}}{1 + x^4}$

(c) $y = \left| \dfrac{x - 2}{x^2 - 2} \right|$

(d) $y = \left| \dfrac{x \sin x}{x^2 + 2} \right|$

Solution

(a) The square root function is continuous on $[0, \infty)$ because it is a root of the continuous identity function $f(x) = x$ (Part 7, Theorem 8). The given function is then the composition of the polynomial $f(x) = x^2 - 2x - 5$ with the square root function $g(t) = \sqrt{t}$, and is continuous on its natural domain.

(b) The numerator is the cube root of the identity function squared; the denominator is an everywhere-positive polynomial. Therefore, the quotient is continuous.

(c) The quotient $(x - 2)/(x^2 - 2)$ is continuous for all $x \neq \pm\sqrt{2}$, and the function is the composition of this quotient with the continuous absolute value function (Example 7).

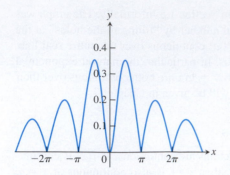

FIGURE 2.43 The graph suggests that $y = |(x \sin x)/(x^2 + 2)|$ is continuous (Example 8d).

(d) Because the sine function is everywhere-continuous (Exercise 72), the numerator term $x \sin x$ is the product of continuous functions, and the denominator term $x^2 + 2$ is an everywhere-positive polynomial. The given function is the composite of a quotient of continuous functions with the continuous absolute value function (Figure 2.43). ■

Theorem 9 is actually a consequence of a more general result, which we now prove. It states that if the limit of $f(x)$ as x approaches c is equal to b, then the limit of the composition function $g \circ f$ as x approaches c is equal to $g(b)$.

THEOREM 10—Limits of Continuous Functions

If $\lim_{x \to c} f(x) = b$ and g is continuous at the point b, then

$$\lim_{x \to c} g(f(x)) = g(b).$$

Proof   Let $\varepsilon > 0$ be given. Since g is continuous at b, there exists a number $\delta_1 > 0$ such that

$$|g(y) - g(b)| < \varepsilon \quad \text{whenever} \quad 0 < |y - b| < \delta_1. \qquad \text{\color{blue}$\lim_{y \to b} g(y) = g(b)$ since g is continuous at $y = b$.}$$

Since $\lim_{x \to c} f(x) = b$, there exists a $\delta > 0$ such that

$$|f(x) - b| < \delta_1 \quad \text{whenever} \quad 0 < |x - c| < \delta. \qquad \text{\color{blue}Definition of $\lim_{x \to c} f(x) = b$}$$

If we let $y = f(x)$, we then have that

$$|y - b| < \delta_1 \quad \text{whenever} \quad 0 < |x - c| < \delta,$$

which implies from the first statement that $|g(y) - g(b)| = |g(f(x)) - g(b)| < \varepsilon$ whenever $0 < |x - c| < \delta$. From the definition of limit, it follows that $\lim_{x \to c} g(f(x)) = g(b)$. This gives the proof for the case where c is an interior point of the domain of f. The case where c is an endpoint of the domain is entirely similar, using an appropriate one-sided limit in place of a two-sided limit. ■

EXAMPLE 9   As an application of Theorem 10, we have the following calculations.

(a) $\displaystyle \lim_{x \to \pi/2} \cos\left(2x + \sin\left(\frac{3\pi}{2} + x\right)\right) = \cos\left(\lim_{x \to \pi/2} 2x + \lim_{x \to \pi/2} \sin\left(\frac{3\pi}{2} + x\right)\right)$

$$= \cos(\pi + \sin 2\pi) = \cos \pi = -1.$$

(b) $\displaystyle \lim_{x \to 1} \sin^{-1}\left(\frac{1 - x}{1 - x^2}\right) = \sin^{-1}\left(\lim_{x \to 1} \frac{1 - x}{1 - x^2}\right)$ {\color{blue}Arcsine is continuous.}

$$= \sin^{-1}\left(\lim_{x \to 1} \frac{1}{1 + x}\right) \qquad \text{\color{blue}Cancel common factor $(1 - x)$.}$$

$$= \sin^{-1}\frac{1}{2} = \frac{\pi}{6}$$

We sometimes denote e^u by $\exp(u)$ when u is a complicated mathematical expression.

(c) $\displaystyle \lim_{x \to 0} \sqrt{x + 1}\, e^{\tan x} = \lim_{x \to 0} \sqrt{x + 1} \cdot \exp\left(\lim_{x \to 0} \tan x\right)$ {\color{blue}exp is continuous.}

$$= 1 \cdot e^0 = 1 \qquad ■$$

Intermediate Value Theorem for Continuous Functions

A function is said to have the **Intermediate Value Property** if whenever it takes on two values, it also takes on all the values in between.

THEOREM 11—The Intermediate Value Theorem for Continuous Functions

If f is a continuous function on a closed interval $[a, b]$, and if y_0 is any value between $f(a)$ and $f(b)$, then $y_0 = f(c)$ for some c in $[a, b]$.

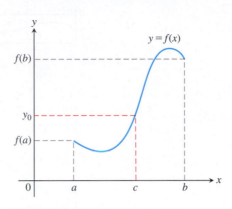

Theorem 11 says that continuous functions over *finite closed* intervals have the Intermediate Value Property. Geometrically, the Intermediate Value Theorem says that any horizontal line $y = y_0$ crossing the y-axis between the numbers $f(a)$ and $f(b)$ will cross the curve $y = f(x)$ at least once over the interval $[a, b]$.

The proof of the Intermediate Value Theorem depends on the completeness property of the real number system. The completeness property implies that the real numbers have no holes or gaps. In contrast, the rational numbers do not satisfy the completeness property, and a function defined only on the rationals would not satisfy the Intermediate Value Theorem. See Appendix 7 for a discussion and examples.

The continuity of f on the interval is essential to Theorem 11. If f fails to be continuous at even one point of the interval, the theorem's conclusion may fail, as it does for the function graphed in Figure 2.44 (choose y_0 as any number between 2 and 3).

A Consequence for Graphing: Connectedness   Theorem 11 implies that the graph of a function that is continuous on an interval cannot have any breaks over the interval. It will be **connected**—a single, unbroken curve. It will not have jumps such as the ones found in the graph of the greatest integer function (Figure 2.39), or separate branches as found in the graph of $1/x$ (Figure 2.41).

A Consequence for Root Finding   We call a solution of the equation $f(x) = 0$ a **root** of the equation or **zero** of the function f. The Intermediate Value Theorem tells us that if f is continuous, then any interval on which f changes sign contains a zero of the function. Somewhere between a point where a continuous function is positive and a second point where it is negative, the function must be equal to zero.

In practical terms, when we see the graph of a continuous function cross the horizontal axis on a computer screen, we know it is not stepping across. There really is a point where the function's value is zero.

EXAMPLE 10   Show that there is a root of the equation $x^3 - x - 1 = 0$ between 1 and 2.

Solution   Let $f(x) = x^3 - x - 1$. Since $f(1) = 1 - 1 - 1 = -1 < 0$ and $f(2) = 2^3 - 2 - 1 = 5 > 0$, we see that $y_0 = 0$ is a value between $f(1)$ and $f(2)$. Since f is a polynomial, it is continuous, and the Intermediate Value Theorem says there is a zero of f between 1 and 2. Figure 2.45 shows the result of zooming in to locate the root near $x = 1.32$. ∎

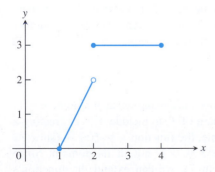

FIGURE 2.44   The function

$$f(x) = \begin{cases} 2x - 2, & 1 \le x < 2 \\ 3, & 2 \le x \le 4 \end{cases}$$

does not take on all values between $f(1) = 0$ and $f(4) = 3$; it misses all the values between 2 and 3.

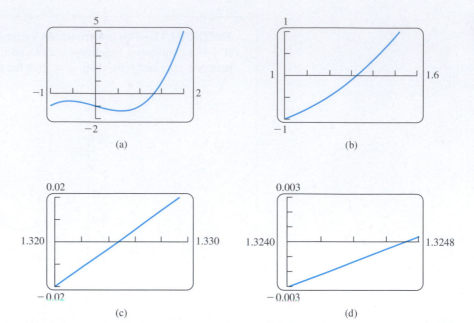

FIGURE 2.45 Zooming in on a zero of the function $f(x) = x^3 - x - 1$. The zero is near $x = 1.3247$ (Example 10).

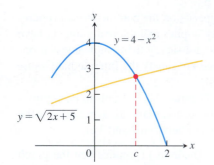

FIGURE 2.46 The curves $y = \sqrt{2x + 5}$ and $y = 4 - x^2$ have the same value at $x = c$ where $\sqrt{2x + 5} + x^2 - 4 = 0$ (Example 11).

EXAMPLE 11   Use the Intermediate Value Theorem to prove that the equation

$$\sqrt{2x + 5} = 4 - x^2$$

has a solution (Figure 2.46).

Solution   We rewrite the equation as

$$\sqrt{2x + 5} + x^2 - 4 = 0,$$

and set $f(x) = \sqrt{2x + 5} + x^2 - 4$. Now $g(x) = \sqrt{2x + 5}$ is continuous on the interval $[-5/2, \infty)$ since it is formed as the composition of two continuous functions, the square root function with the nonnegative linear function $y = 2x + 5$. Then f is the sum of the function g and the quadratic function $y = x^2 - 4$, and the quadratic function is continuous for all values of x. It follows that $f(x) = \sqrt{2x + 5} + x^2 - 4$ is continuous on the interval $[-5/2, \infty)$. By trial and error, we find the function values $f(0) = \sqrt{5} - 4 \approx -1.76$ and $f(2) = \sqrt{9} = 3$. Note that f is continuous on the finite closed interval $[0, 2] \subset [-5/2, \infty)$. Since the value $y_0 = 0$ is between the numbers $f(0) = -1.76$ and $f(2) = 3$, by the Intermediate Value Theorem there is a number $c \in [0, 2]$ such that $f(c) = 0$. The number c solves the original equation.   ∎

Continuous Extension to a Point

Sometimes the formula that describes a function f does not make sense at a point $x = c$. It might nevertheless be possible to extend the domain of f, to include $x = c$, creating a new function that is continuous at $x = c$. For example, the function $y = f(x) = (\sin x)/x$ is continuous at every point except $x = 0$, since $x = 0$ is not in its domain. Since $y = (\sin x)/x$ has a finite limit as $x \to 0$ (Theorem 7), we can extend the function's domain to include the point $x = 0$ in such a way that the extended function is continuous at $x = 0$. We define the new function

$$F(x) = \begin{cases} \dfrac{\sin x}{x}, & x \neq 0 \qquad \text{\color{blue}Same as original function for } x \neq 0 \\ 1, & x = 0. \qquad \text{\color{blue}Value at domain point } x = 0 \end{cases}$$

The new function $F(x)$ is continuous at $x = 0$ because

$$\lim_{x \to 0} \frac{\sin x}{x} = F(0),$$

so it meets the requirements for continuity (Figure 2.47).

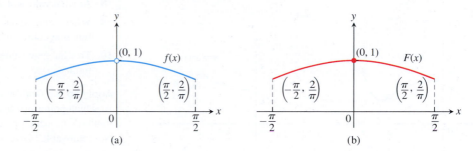

FIGURE 2.47 (a) The graph of $f(x) = (\sin x)/x$ for $-\pi/2 \le x \le \pi/2$ does not include the point $(0, 1)$ because the function is not defined at $x = 0$. (b) We can extend the domain to include $x = 0$ by defining the new function $F(x)$ with $F(0) = 1$ and $F(x) = f(x)$ everywhere else. Note that $F(0) = \lim_{x \to 0} f(x)$ and $F(x)$ is a continuous function at $x = 0$.

More generally, a function (such as a rational function) may have a limit at a point where it is not defined. If $f(c)$ is not defined, but $\lim_{x \to c} f(x) = L$ exists, we can define a new function $F(x)$ by the rule

$$F(x) = \begin{cases} f(x), & \text{if } x \text{ is in the domain of } f \\ L, & \text{if } x = c. \end{cases}$$

The function F is continuous at $x = c$. It is called the **continuous extension of f** to $x = c$. For rational functions f, continuous extensions are often found by canceling common factors in the numerator and denominator.

EXAMPLE 12   Show that

$$f(x) = \frac{x^2 + x - 6}{x^2 - 4}, \quad x \ne 2$$

has a continuous extension to $x = 2$, and find that extension.

Solution   Although $f(2)$ is not defined, if $x \ne 2$ we have

$$f(x) = \frac{x^2 + x - 6}{x^2 - 4} = \frac{(x - 2)(x + 3)}{(x - 2)(x + 2)} = \frac{x + 3}{x + 2}.$$

The new function

$$F(x) = \frac{x + 3}{x + 2}$$

is equal to $f(x)$ for $x \ne 2$, but is continuous at $x = 2$, having there the value of $5/4$. Thus F is the continuous extension of f to $x = 2$, and

$$\lim_{x \to 2} \frac{x^2 + x - 6}{x^2 - 4} = \lim_{x \to 2} f(x) = \frac{5}{4}.$$

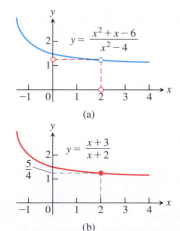

FIGURE 2.48 (a) The graph of $f(x)$ and (b) the graph of its continuous extension $F(x)$ (Example 12).

The graph of f is shown in Figure 2.48. The continuous extension F has the same graph except with no hole at $(2, 5/4)$. Effectively, F is the function f extended across the missing domain point at $x = 2$ so as to give a continuous function over the larger domain. ∎

EXERCISES 2.5

Continuity from Graphs

In Exercises 1–4, say whether the function graphed is continuous on $[-1, 3]$. If not, where does it fail to be continuous and why?

1.

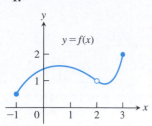

2.

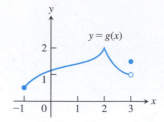

3.

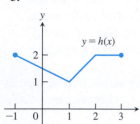

4.

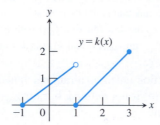

Exercises 5–10 refer to the function

$$f(x) = \begin{cases} x^2 - 1, & -1 \le x < 0 \\ 2x, & 0 < x < 1 \\ 1, & x = 1 \\ -2x + 4, & 1 < x < 2 \\ 0, & 2 < x < 3 \end{cases}$$

graphed in the accompanying figure.

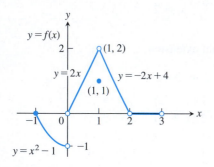

The graph for Exercises 5–10.

5. a. Does $f(-1)$ exist?

   b. Does $\lim_{x \to -1^+} f(x)$ exist?

   c. Does $\lim_{x \to -1^+} f(x) = f(-1)$?

   d. Is f continuous at $x = -1$?

6. a. Does $f(1)$ exist?

   b. Does $\lim_{x \to 1} f(x)$ exist?

   c. Does $\lim_{x \to 1} f(x) = f(1)$?

   d. Is f continuous at $x = 1$?

7. a. Is f defined at $x = 2$? (Look at the definition of f.)

   b. Is f continuous at $x = 2$?

8. At what values of x is f continuous?

9. What value should be assigned to $f(2)$ to make the extended function continuous at $x = 2$?

10. To what new value should $f(1)$ be changed to remove the discontinuity?

Applying the Continuity Test

At which points do the functions in Exercises 11 and 12 fail to be continuous? At which points, if any, are the discontinuities removable? Not removable? Give reasons for your answers.

11. Exercise 1, Section 2.4

12. Exercise 2, Section 2.4

At what points are the functions in Exercises 13–32 continuous?

13. $y = \dfrac{1}{x - 2} - 3x$

14. $y = \dfrac{1}{(x + 2)^2} + 4$

15. $y = \dfrac{x + 1}{x^2 - 4x + 3}$

16. $y = \dfrac{x + 3}{x^2 - 3x - 10}$

17. $y = |x - 1| + \sin x$

18. $y = \dfrac{1}{|x| + 1} - \dfrac{x^2}{2}$

19. $y = \dfrac{\cos x}{x}$

20. $y = \dfrac{x + 2}{\cos x}$

21. $y = \csc 2x$

22. $y = \tan \dfrac{\pi x}{2}$

23. $y = \dfrac{x \tan x}{x^2 + 1}$

24. $y = \dfrac{\sqrt{x^4 + 1}}{1 + \sin^2 x}$

25. $y = \sqrt{2x + 3}$

26. $y = \sqrt[4]{3x - 1}$

27. $y = (2x - 1)^{1/3}$

28. $y = (2 - x)^{1/5}$

29. $g(x) = \begin{cases} \dfrac{x^2 - x - 6}{x - 3}, & x \ne 3 \\ 5, & x = 3 \end{cases}$

30. $f(x) = \begin{cases} \dfrac{x^3 - 8}{x^2 - 4}, & x \ne 2, x \ne -2 \\ 3, & x = 2 \\ 4, & x = -2 \end{cases}$

31. $f(x) = \begin{cases} 1 - x, & x < 0 \\ e^x, & 0 \le x \le 1 \\ x^2 + 2, & x > 1 \end{cases}$

32. $f(x) = \dfrac{x + 3}{2 - e^x}$

Limits Involving Trigonometric Functions

Find the limits in Exercises 33–40. Are the functions continuous at the point being approached?

33. $\lim_{x \to \pi} \sin(x - \sin x)$

34. $\lim_{t \to 0} \sin\left(\dfrac{\pi}{2} \cos(\tan t)\right)$

35. $\lim_{y \to 1} \sec(y \sec^2 y - \tan^2 y - 1)$

36. $\lim\limits_{x \to 0} \tan\left(\dfrac{\pi}{4}\cos\left(\sin x^{1/3}\right)\right)$

37. $\lim\limits_{t \to 0} \cos\left(\dfrac{\pi}{\sqrt{19 - 3\sec 2t}}\right)$ **38.** $\lim\limits_{x \to \pi/6} \sqrt{\csc^2 x + 5\sqrt{3}\tan x}$

39. $\lim\limits_{x \to 0^+} \sin\left(\dfrac{\pi}{2}e^{\sqrt{x}}\right)$ **40.** $\lim\limits_{x \to 1} \cos^{-1}\left(\ln\sqrt{x}\right)$

Continuous Extensions

41. Define $g(3)$ in a way that extends $g(x) = (x^2 - 9)/(x - 3)$ to be continuous at $x = 3$.

42. Define $h(2)$ in a way that extends $h(t) = (t^2 + 3t - 10)/(t - 2)$ to be continuous at $t = 2$.

43. Define $f(1)$ in a way that extends $f(s) = (s^3 - 1)/(s^2 - 1)$ to be continuous at $s = 1$.

44. Define $g(4)$ in a way that extends

$$g(x) = (x^2 - 16)/(x^2 - 3x - 4)$$

to be continuous at $x = 4$.

45. For what value of a is

$$f(x) = \begin{cases} x^2 - 1, & x < 3 \\ 2ax, & x \geq 3 \end{cases}$$

continuous at every x?

46. For what value of b is

$$g(x) = \begin{cases} x, & x < -2 \\ bx^2, & x \geq -2 \end{cases}$$

continuous at every x?

47. For what values of a is

$$f(x) = \begin{cases} a^2 x - 2a, & x \geq 2 \\ 12, & x < 2 \end{cases}$$

continuous at every x?

48. For what values of b is

$$g(x) = \begin{cases} \dfrac{x - b}{b + 1}, & x \leq 0 \\ x^2 + b, & x > 0 \end{cases}$$

continuous at every x?

49. For what values of a and b is

$$f(x) = \begin{cases} -2, & x \leq -1 \\ ax - b, & -1 < x < 1 \\ 3, & x \geq 1 \end{cases}$$

continuous at every x?

50. For what values of a and b is

$$g(x) = \begin{cases} ax + 2b, & x \leq 0 \\ x^2 + 3a - b, & 0 < x \leq 2 \\ 3x - 5, & x > 2 \end{cases}$$

continuous at every x?

In Exercises 51–54, graph the function f to see whether it appears to have a continuous extension to the origin. If it does, use Trace and Zoom to find a good candidate for the extended function's value at

$x = 0$. If the function does not appear to have a continuous extension, can it be extended to be continuous at the origin from the right or from the left? If so, what do you think the extended function's value(s) should be?

51. $f(x) = \dfrac{10^x - 1}{x}$ **52.** $f(x) = \dfrac{10^{|x|} - 1}{x}$

53. $f(x) = \dfrac{\sin x}{|x|}$ **54.** $f(x) = (1 + 2x)^{1/x}$

Theory and Examples

55. A continuous function $y = f(x)$ is known to be negative at $x = 0$ and positive at $x = 1$. Why does the equation $f(x) = 0$ have at least one solution between $x = 0$ and $x = 1$? Illustrate with a sketch.

56. Explain why the equation $\cos x = x$ has at least one solution.

57. Roots of a cubic Show that the equation $x^3 - 15x + 1 = 0$ has three solutions in the interval $[-4, 4]$.

58. A function value Show that the function $F(x) = (x - a)^2 \cdot (x - b)^2 + x$ takes on the value $(a + b)/2$ for some value of x.

59. Solving an equation If $f(x) = x^3 - 8x + 10$, show that there are values c for which $f(c)$ equals **(a)** π; **(b)** $-\sqrt{3}$; **(c)** 5,000,000.

60. Explain why the following five statements ask for the same information.

  a. Find the roots of $f(x) = x^3 - 3x - 1$.

  b. Find the x-coordinates of the points where the curve $y = x^3$ crosses the line $y = 3x + 1$.

  c. Find all the values of x for which $x^3 - 3x = 1$.

  d. Find the x-coordinates of the points where the cubic curve $y = x^3 - 3x$ crosses the line $y = 1$.

  e. Solve the equation $x^3 - 3x - 1 = 0$.

61. Removable discontinuity Give an example of a function $f(x)$ that is continuous for all values of x except $x = 2$, where it has a removable discontinuity. Explain how you know that f is discontinuous at $x = 2$, and how you know the discontinuity is removable.

62. Nonremovable discontinuity Give an example of a function $g(x)$ that is continuous for all values of x except $x = -1$, where it has a nonremovable discontinuity. Explain how you know that g is discontinuous there and why the discontinuity is not removable.

63. A function discontinuous at every point

  a. Use the fact that every nonempty interval of real numbers contains both rational and irrational numbers to show that the function

$$f(x) = \begin{cases} 1, & \text{if } x \text{ is rational} \\ 0, & \text{if } x \text{ is irrational} \end{cases}$$

  is discontinuous at every point.

  b. Is f right-continuous or left-continuous at any point?

64. If functions $f(x)$ and $g(x)$ are continuous for $0 \leq x \leq 1$, could $f(x)/g(x)$ possibly be discontinuous at a point of $[0, 1]$? Give reasons for your answer.

65. If the product function $h(x) = f(x) \cdot g(x)$ is continuous at $x = 0$, must $f(x)$ and $g(x)$ be continuous at $x = 0$? Give reasons for your answer.

66. Discontinuous composite of continuous functions Give an example of functions f and g, both continuous at $x = 0$, for which the composite $f \circ g$ is discontinuous at $x = 0$. Does this contradict Theorem 9? Give reasons for your answer.

67. Never-zero continuous functions Is it true that a continuous function that is never zero on an interval never changes sign on that interval? Give reasons for your answer.

68. Stretching a rubber band Is it true that if you stretch a rubber band by moving one end to the right and the other to the left, some point of the band will end up in its original position? Give reasons for your answer.

69. A fixed point theorem Suppose that a function f is continuous on the closed interval $[0, 1]$ and that $0 \le f(x) \le 1$ for every x in $[0, 1]$. Show that there must exist a number c in $[0, 1]$ such that $f(c) = c$ (c is called a **fixed point** of f).

70. The sign-preserving property of continuous functions Let f be defined on an interval (a, b) and suppose that $f(c) \ne 0$ at some c where f is continuous. Show that there is an interval $(c - \delta, c + \delta)$ about c where f has the same sign as $f(c)$.

71. Prove that f is continuous at c if and only if

$$\lim_{h \to 0} f(c + h) = f(c).$$

72. Use Exercise 71 together with the identities

$$\sin(h + c) = \sin h \cos c + \cos h \sin c,$$

$$\cos(h + c) = \cos h \cos c - \sin h \sin c$$

to prove that both $f(x) = \sin x$ and $g(x) = \cos x$ are continuous at every point $x = c$.

Solving Equations Graphically

T Use the Intermediate Value Theorem in Exercises 73–80 to prove that each equation has a solution. Then use a graphing calculator or computer grapher to solve the equations.

73. $x^3 - 3x - 1 = 0$

74. $2x^3 - 2x^2 - 2x + 1 = 0$

75. $x(x - 1)^2 = 1$ (one root)

76. $x^x = 2$

77. $\sqrt{x} + \sqrt{1 + x} = 4$

78. $x^3 - 15x + 1 = 0$ (three roots)

79. $\cos x = x$ (one root). Make sure you are using radian mode.

80. $2 \sin x = x$ (three roots). Make sure you are using radian mode.

2.6 Limits Involving Infinity; Asymptotes of Graphs

In this section we investigate the behavior of a function when the magnitude of the independent variable x becomes increasingly large, or $x \to \pm\infty$. We further extend the concept of limit to *infinite limits*. Infinite limits provide useful symbols and language for describing the behavior of functions whose values become arbitrarily large in magnitude. We use these ideas to analyze the graphs of functions having *horizontal* or *vertical asymptotes*.

Finite Limits as $x \to \pm\infty$

The symbol for infinity (∞) does not represent a real number. We use ∞ to describe the behavior of a function when the values in its domain or range outgrow all finite bounds. For example, the function $f(x) = 1/x$ is defined for all $x \ne 0$ (Figure 2.49). When x is positive and becomes increasingly large, $1/x$ becomes increasingly small. When x is negative and its magnitude becomes increasingly large, $1/x$ again becomes small. We summarize these observations by saying that $f(x) = 1/x$ has limit 0 as $x \to \infty$ or $x \to -\infty$, or that 0 is a *limit of $f(x) = 1/x$ at infinity and at negative infinity*. Here are precise definitions for the limit of a function whose domain contains positive or negative numbers of unbounded magnitude.

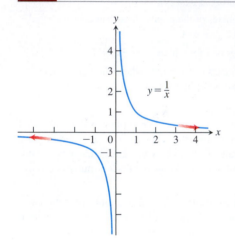

FIGURE 2.49 The graph of $y = 1/x$ approaches 0 as $x \to \infty$ or $x \to -\infty$.

DEFINITIONS

1. We say that $f(x)$ has the **limit L as x approaches infinity** and write

$$\lim_{x \to \infty} f(x) = L$$

if, for every number $\varepsilon > 0$, there exists a corresponding number M such that for all x in the domain of f

$$|f(x) - L| < \varepsilon \quad \text{whenever} \quad x > M.$$

2. We say that $f(x)$ has the **limit L as x approaches negative infinity** and write

$$\lim_{x \to -\infty} f(x) = L$$

if, for every number $\varepsilon > 0$, there exists a corresponding number N such that for all x in the domain of f

$$|f(x) - L| < \varepsilon \quad \text{whenever} \quad x < N.$$

Intuitively, $\lim_{x \to \infty} f(x) = L$ if, as x moves increasingly far from the origin in the positive direction, $f(x)$ gets arbitrarily close to L. Similarly, $\lim_{x \to -\infty} f(x) = L$ if, as x moves increasingly far from the origin in the negative direction, $f(x)$ gets arbitrarily close to L.

The strategy for calculating limits of functions as $x \to +\infty$ or as $x \to -\infty$ is similar to the one for finite limits in Section 2.2. There we first found the limits of the constant and identity functions $y = k$ and $y = x$. We then extended these results to other functions by applying Theorem 1 on limits of algebraic combinations. Here we do the same thing, except that the starting functions are $y = k$ and $y = 1/x$ instead of $y = k$ and $y = x$.

The basic facts to be verified by applying the formal definition are

$$\lim_{x \to \pm\infty} k = k \qquad \text{and} \qquad \lim_{x \to \pm\infty} \frac{1}{x} = 0. \qquad (1)$$

We prove the second result in Example 1, and leave the first to Exercises 93 and 94.

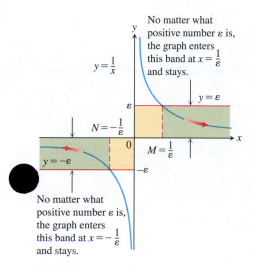

No matter what positive number ε is, the graph enters this band at $x = \frac{1}{\varepsilon}$ and stays.

$y = \frac{1}{x}$

$y = \varepsilon$

$N = -\frac{1}{\varepsilon}$

$M = \frac{1}{\varepsilon}$

$y = -\varepsilon$

No matter what positive number ε is, the graph enters this band at $x = -\frac{1}{\varepsilon}$ and stays.

FIGURE 2.50 The geometry behind the argument in Example 1.

EXAMPLE 1 Show that

(a) $\lim_{x \to \infty} \frac{1}{x} = 0$ 　　　　**(b)** $\lim_{x \to -\infty} \frac{1}{x} = 0.$

Solution

(a) Let $\varepsilon > 0$ be given. We must find a number M such that

$$\left| \frac{1}{x} - 0 \right| = \left| \frac{1}{x} \right| < \varepsilon \qquad \text{whenever} \qquad x > M.$$

The implication will hold if $M = 1/\varepsilon$ or any larger positive number (Figure 2.50). This proves $\lim_{x \to \infty}(1/x) = 0$.

(b) Let $\varepsilon > 0$ be given. We must find a number N such that

$$\left| \frac{1}{x} - 0 \right| = \left| \frac{1}{x} \right| < \varepsilon \qquad \text{whenever} \qquad x < N.$$

The implication will hold if $N = -1/\varepsilon$ or any number less than $-1/\varepsilon$ (Figure 2.50). This proves $\lim_{x \to -\infty}(1/x) = 0$. ∎

Limits at infinity have properties similar to those of finite limits.

> **THEOREM 12** All the Limit Laws in Theorem 1 are true when we replace $\lim_{x \to c}$ by $\lim_{x \to \infty}$ or $\lim_{x \to -\infty}$. That is, the variable x may approach a finite number c or $\pm\infty$.

EXAMPLE 2 The properties in Theorem 12 are used to calculate limits in the same way as when x approaches a finite number c.

(a) $\lim_{x \to \infty} \left(5 + \frac{1}{x} \right) = \lim_{x \to \infty} 5 + \lim_{x \to \infty} \frac{1}{x}$ 　　　Sum Rule

$$= 5 + 0 = 5$$ 　　　Known limits

(b) $\lim_{x \to -\infty} \frac{\pi\sqrt{3}}{x^2} = \lim_{x \to -\infty} \pi\sqrt{3} \cdot \frac{1}{x} \cdot \frac{1}{x}$

$$= \lim_{x \to -\infty} \pi\sqrt{3} \cdot \lim_{x \to -\infty} \frac{1}{x} \cdot \lim_{x \to -\infty} \frac{1}{x}$$ 　　　Product Rule

$$= \pi\sqrt{3} \cdot 0 \cdot 0 = 0$$ 　　　Known limits 　■

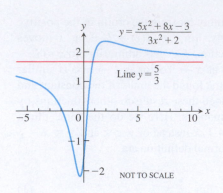

FIGURE 2.51 The graph of the function in Example 3a. The graph approaches the line $y = 5/3$ as $|x|$ increases.

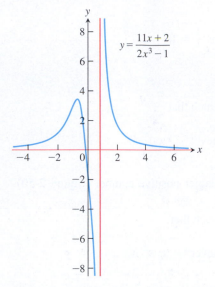

FIGURE 2.52 The graph of the function in Example 3b. The graph approaches the x-axis as $|x|$ increases.

Limits at Infinity of Rational Functions

To determine the limit of a rational function as $x \to \pm\infty$, we first divide the numerator and denominator by the highest power of x in the denominator. The result then depends on the degrees of the polynomials involved.

EXAMPLE 3   These examples illustrate what happens when the degree of the numerator is less than or equal to the degree of the denominator.

(a) $\displaystyle \lim_{x \to \infty} \frac{5x^2 + 8x - 3}{3x^2 + 2} = \lim_{x \to \infty} \frac{5 + (8/x) - (3/x^2)}{3 + (2/x^2)}$ Divide numerator and denominator by x^2.

$\displaystyle = \frac{5 + 0 - 0}{3 + 0} = \frac{5}{3}$ See Fig. 2.51.

(b) $\displaystyle \lim_{x \to -\infty} \frac{11x + 2}{2x^3 - 1} = \lim_{x \to -\infty} \frac{(11/x^2) + (2/x^3)}{2 - (1/x^3)}$ Divide numerator and denominator by x^3.

$\displaystyle = \frac{0 + 0}{2 - 0} = 0$ See Fig. 2.52.   ■

Cases for which the degree of the numerator is greater than the degree of the denominator are illustrated in Examples 10 and 14.

Horizontal Asymptotes

If the distance between the graph of a function and some fixed line approaches zero as a point on the graph moves increasingly far from the origin, we say that the graph approaches the line *asymptotically* and that the line is an *asymptote* of the graph.

Looking at $f(x) = 1/x$ (see Figure 2.49), we observe that the x-axis is an asymptote of the curve on the right because

$$\lim_{x \to \infty} \frac{1}{x} = 0$$

and on the left because

$$\lim_{x \to -\infty} \frac{1}{x} = 0.$$

We say that the x-axis is a *horizontal asymptote* of the graph of $f(x) = 1/x$.

DEFINITION   A line $y = b$ is a **horizontal asymptote** of the graph of a function $y = f(x)$ if either

$$\lim_{x \to \infty} f(x) = b \qquad \text{or} \qquad \lim_{x \to -\infty} f(x) = b.$$

The graph of a function can have zero, one, or two horizontal asymptotes, depending on whether the function has limits as $x \to \infty$ and as $x \to -\infty$.

The graph of the function

$$f(x) = \frac{5x^2 + 8x - 3}{3x^2 + 2}$$

sketched in Figure 2.51 (Example 3a) has the line $y = 5/3$ as a horizontal asymptote on both the right and the left because

$$\lim_{x \to \infty} f(x) = \frac{5}{3} \qquad \text{and} \qquad \lim_{x \to -\infty} f(x) = \frac{5}{3}.$$

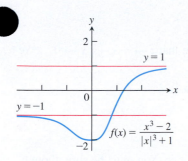

FIGURE 2.53 The graph of the function in Example 4 has two horizontal asymptotes.

EXAMPLE 4   Find the horizontal asymptotes of the graph of

$$f(x) = \frac{x^3 - 2}{|x|^3 + 1}.$$

Solution   We calculate the limits as $x \to \pm\infty$.

For $x \geq 0$: $\lim\limits_{x \to \infty} \dfrac{x^3 - 2}{|x|^3 + 1} = \lim\limits_{x \to \infty} \dfrac{x^3 - 2}{x^3 + 1} = \lim\limits_{x \to \infty} \dfrac{1 - (2/x^3)}{1 + (1/x^3)} = 1.$

For $x < 0$: $\lim\limits_{x \to -\infty} \dfrac{x^3 - 2}{|x|^3 + 1} = \lim\limits_{x \to -\infty} \dfrac{x^3 - 2}{(-x)^3 + 1} = \lim\limits_{x \to -\infty} \dfrac{1 - (2/x^3)}{-1 + (1/x^3)} = -1.$

The horizontal asymptotes are $y = -1$ and $y = 1$. The graph is displayed in Figure 2.53. Notice that the graph crosses the horizontal asymptote $y = -1$ for a positive value of x.   ■

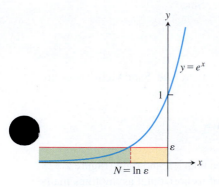

FIGURE 2.54 The graph of $y = e^x$ approaches the x-axis as $x \to -\infty$ (Example 5).

EXAMPLE 5   The x-axis (the line $y = 0$) is a horizontal asymptote of the graph of $y = e^x$ because

$$\lim_{x \to -\infty} e^x = 0.$$

To see this, we use the definition of a limit as x approaches $-\infty$. So let $\varepsilon > 0$ be given, but arbitrary. We must find a constant N such that

$$\left| e^x - 0 \right| < \varepsilon \quad \text{whenever} \quad x < N.$$

Now $\left| e^x - 0 \right| = e^x$, so the condition that needs to be satisfied whenever $x < N$ is

$$e^x < \varepsilon.$$

Let $x = N$ be the number where $e^x = \varepsilon$. Since e^x is an increasing function, if $x < N$, then $e^x < \varepsilon$. We find N by taking the natural logarithm of both sides of the equation $e^N = \varepsilon$, so $N = \ln \varepsilon$ (see Figure 2.54). With this value of N the condition is satisfied, and we conclude that $\lim_{x \to -\infty} e^x = 0$.   ■

EXAMPLE 6   Find **(a)** $\lim\limits_{x \to \infty} \sin(1/x)$ and **(b)** $\lim\limits_{x \to \pm\infty} x \sin(1/x)$.

Solution

(a)   We introduce the new variable $t = 1/x$. From Example 1, we know that $t \to 0^+$ as $x \to \infty$ (see Figure 2.49). Therefore,

$$\lim_{x \to \infty} \sin\frac{1}{x} = \lim_{t \to 0^+} \sin t = 0.$$

(b)   We calculate the limits as $x \to \infty$ and $x \to -\infty$:

$$\lim_{x \to \infty} x \sin\frac{1}{x} = \lim_{t \to 0^+} \frac{\sin t}{t} = 1 \quad \text{and} \quad \lim_{x \to -\infty} x \sin\frac{1}{x} = \lim_{t \to 0^-} \frac{\sin t}{t} = 1.$$

The graph is shown in Figure 2.55, and we see that the line $y = 1$ is a horizontal asymptote.   ■

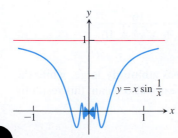

FIGURE 2.55 The line $y = 1$ is a horizontal asymptote of the function graphed here (Example 6b).

Similarly, we can investigate the behavior of $y = f(1/x)$ as $x \to 0$ by investigating $y = f(t)$ as $t \to \pm\infty$, where $t = 1/x$.

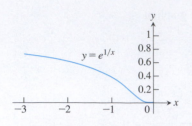

FIGURE 2.56 The graph of $y = e^{1/x}$ for $x < 0$ shows $\lim_{x \to 0^-} e^{1/x} = 0$ (Example 7).

EXAMPLE 7   Find $\lim_{x \to 0^-} e^{1/x}$.

Solution   We let $t = 1/x$. From Figure 2.49, we can see that $t \to -\infty$ as $x \to 0^-$. (We make this idea more precise further on.) Therefore,

$$\lim_{x \to 0^-} e^{1/x} = \lim_{t \to -\infty} e^t = 0 \qquad \text{Example 5}$$

(Figure 2.56).

The Sandwich Theorem also holds for limits as $x \to \pm\infty$. You must be sure, though, that the function whose limit you are trying to find stays between the bounding functions at very large values of x in magnitude consistent with whether $x \to \infty$ or $x \to -\infty$.

EXAMPLE 8   Using the Sandwich Theorem, find the horizontal asymptote of the curve

$$y = 2 + \frac{\sin x}{x}.$$

Solution   We are interested in the behavior as $x \to \pm\infty$. Since

$$0 \le \left| \frac{\sin x}{x} \right| \le \left| \frac{1}{x} \right|$$

and $\lim_{x \to \pm\infty} |1/x| = 0$, we have $\lim_{x \to \pm\infty} (\sin x)/x = 0$ by the Sandwich Theorem. Hence,

$$\lim_{x \to \pm\infty} \left(2 + \frac{\sin x}{x} \right) = 2 + 0 = 2,$$

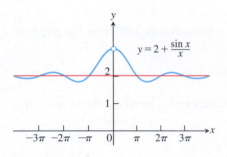

FIGURE 2.57 A curve may cross one of its asymptotes infinitely often (Example 8).

and the line $y = 2$ is a horizontal asymptote of the curve on both left and right (Figure 2.57). This example illustrates that a curve may cross one of its horizontal asymptotes many times.

EXAMPLE 9   Find $\lim_{x \to \infty} \left(x - \sqrt{x^2 + 16} \right)$.

Solution   Both of the terms x and $\sqrt{x^2 + 16}$ approach infinity as $x \to \infty$, so what happens to the difference in the limit is unclear (we cannot subtract ∞ from ∞ because the symbol does not represent a real number). In this situation we can multiply the numerator and the denominator by the conjugate radical expression to obtain an equivalent algebraic expression:

$$\lim_{x \to \infty} \left(x - \sqrt{x^2 + 16} \right) = \lim_{x \to \infty} \left(x - \sqrt{x^2 + 16} \right) \frac{x + \sqrt{x^2 + 16}}{x + \sqrt{x^2 + 16}} \qquad \begin{array}{l}\text{Multiply and}\\\text{divide by the}\\\text{conjugate.}\end{array}$$

$$= \lim_{x \to \infty} \frac{x^2 - (x^2 + 16)}{x + \sqrt{x^2 + 16}} = \lim_{x \to \infty} \frac{-16}{x + \sqrt{x^2 + 16}}.$$

As $x \to \infty$, the denominator in this last expression becomes arbitrarily large, while the numerator remains constant, so we see that the limit is 0. We can also obtain this result by a direct calculation using the Limit Laws:

$$\lim_{x \to \infty} \frac{-16}{x + \sqrt{x^2 + 16}} = \lim_{x \to \infty} \frac{-\dfrac{16}{x}}{1 + \sqrt{\dfrac{x^2}{x^2} + \dfrac{16}{x^2}}} = \frac{0}{1 + \sqrt{1 + 0}} = 0.$$

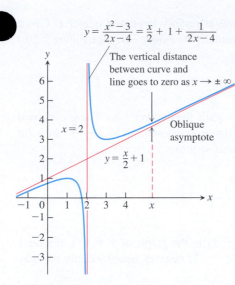

$$y = \frac{x^2 - 3}{2x - 4} = \frac{x}{2} + 1 + \frac{1}{2x - 4}$$

The vertical distance between curve and line goes to zero as $x \to \pm\infty$

Oblique asymptote

$y = \frac{x}{2} + 1$

$x = 2$

FIGURE 2.58 The graph of the function in Example 10 has an oblique asymptote.

Oblique Asymptotes

If the degree of the numerator of a rational function is 1 greater than the degree of the denominator, the graph has an **oblique** or **slant line asymptote**. We find an equation for the asymptote by dividing numerator by denominator to express f as a linear function plus a remainder that goes to zero as $x \to \pm\infty$.

EXAMPLE 10   Find the oblique asymptote of the graph of

$$f(x) = \frac{x^2 - 3}{2x - 4}$$

in Figure 2.58.

Solution   We are interested in the behavior as $x \to \pm\infty$. We divide $(2x - 4)$ into $(x^2 - 3)$:

$$
\begin{array}{r}
\frac{x}{2} + 1 \\
2x - 4 \overline{) x^2 + 0x - 3} \\
\underline{x^2 - 2x} \\
2x - 3 \\
\underline{2x - 4} \\
1
\end{array}
$$

This tells us that

$$f(x) = \frac{x^2 - 3}{2x - 4} = \underbrace{\left(\frac{x}{2} + 1 \right)}_{\text{linear } g(x)} + \underbrace{\left(\frac{1}{2x - 4} \right)}_{\text{remainder}}.$$

As $x \to \pm\infty$, the remainder, whose magnitude gives the vertical distance between the graphs of f and g, goes to zero, making the slanted line

$$g(x) = \frac{x}{2} + 1$$

an asymptote of the graph of f (Figure 2.58). The line $y = g(x)$ is an asymptote both to the right and to the left. ∎

Notice in Example 10 that if the degree of the numerator in a rational function is greater than the degree of the denominator, then the limit as $|x|$ becomes large is $+\infty$ or $-\infty$, depending on the signs assumed by the numerator and denominator.

Infinite Limits

Let us look again at the function $f(x) = 1/x$. As $x \to 0^+$, the values of f grow without bound, eventually reaching and surpassing every positive real number. That is, given any positive real number B, however large, the values of f become larger still (Figure 2.59).

Thus, f has no limit as $x \to 0^+$. It is nevertheless convenient to describe the behavior of f by saying that $f(x)$ approaches ∞ as $x \to 0^+$. We write

$$\lim_{x \to 0^+} f(x) = \lim_{x \to 0^+} \frac{1}{x} = \infty.$$

In writing this equation, we are *not* saying that the limit exists. Nor are we saying that there is a real number ∞, for there is no such number. Rather, this expression is just a concise way of saying that $\lim_{x \to 0^+} (1/x)$ *does not exist because* $1/x$ *becomes arbitrarily large and positive as* $x \to 0^+$.

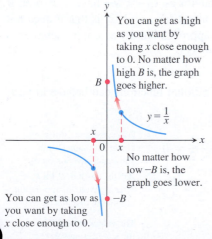

You can get as high as you want by taking x close enough to 0. No matter how high B is, the graph goes higher.

B

$y = \frac{1}{x}$

No matter how low $-B$ is, the graph goes lower.

You can get as low as you want by taking x close enough to 0.

$-B$

FIGURE 2.59 One-sided infinite limits:

$$\lim_{x \to 0^+} \frac{1}{x} = \infty \quad \text{and} \quad \lim_{x \to 0^-} \frac{1}{x} = -\infty.$$

As $x \to 0^-$, the values of $f(x) = 1/x$ become arbitrarily large and negative. Given any negative real number $-B$, the values of f eventually lie below $-B$. (See Figure 2.59.) We write

$$\lim_{x \to 0^-} f(x) = \lim_{x \to 0^-} \frac{1}{x} = -\infty.$$

Again, we are not saying that the limit exists and equals the number $-\infty$. There *is* no real number $-\infty$. We are describing the behavior of a function whose limit as $x \to 0^-$ *does not exist because its values become arbitrarily large and negative.*

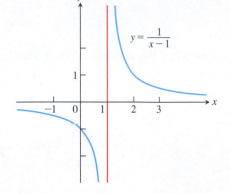

EXAMPLE 11   Find $\lim_{x \to 1^+} \dfrac{1}{x-1}$ and $\lim_{x \to 1^-} \dfrac{1}{x-1}$.

Geometric Solution   The graph of $y = 1/(x-1)$ is the graph of $y = 1/x$ shifted 1 unit to the right (Figure 2.60). Therefore, $y = 1/(x-1)$ behaves near 1 exactly the way $y = 1/x$ behaves near 0:

$$\lim_{x \to 1^+} \frac{1}{x-1} = \infty \qquad \text{and} \qquad \lim_{x \to 1^-} \frac{1}{x-1} = -\infty.$$

FIGURE 2.60   Near $x = 1$, the function $y = 1/(x-1)$ behaves the way the function $y = 1/x$ behaves near $x = 0$. Its graph is the graph of $y = 1/x$ shifted 1 unit to the right (Example 11).

Analytic Solution   Think about the number $x - 1$ and its reciprocal. As $x \to 1^+$, we have $(x-1) \to 0^+$ and $1/(x-1) \to \infty$. As $x \to 1^-$, we have $(x-1) \to 0^-$ and $1/(x-1) \to -\infty$. ∎

EXAMPLE 12   Discuss the behavior of

$$f(x) = \frac{1}{x^2} \qquad \text{as} \qquad x \to 0.$$

Solution   As x approaches zero from either side, the values of $1/x^2$ are positive and become arbitrarily large (Figure 2.61). This means that

$$\lim_{x \to 0} f(x) = \lim_{x \to 0} \frac{1}{x^2} = \infty.$$

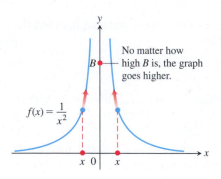

FIGURE 2.61   The graph of $f(x)$ in Example 12 approaches infinity as $x \to 0$.

The function $y = 1/x$ shows no consistent behavior as $x \to 0$. We have $1/x \to \infty$ if $x \to 0^+$, but $1/x \to -\infty$ if $x \to 0^-$. All we can say about $\lim_{x \to 0} (1/x)$ is that it does not exist. The function $y = 1/x^2$ is different. Its values approach infinity as x approaches zero from either side, so we can say that $\lim_{x \to 0} (1/x^2) = \infty$. ∎

EXAMPLE 13   These examples illustrate that rational functions can behave in various ways near zeros of the denominator.

(a) $\displaystyle \lim_{x \to 2} \frac{(x-2)^2}{x^2 - 4} = \lim_{x \to 2} \frac{(x-2)^2}{(x-2)(x+2)} = \lim_{x \to 2} \frac{x-2}{x+2} = 0$ *Can substitute 2 for x after algebraic manipulation eliminates division by 0.*

(b) $\displaystyle \lim_{x \to 2} \frac{x-2}{x^2 - 4} = \lim_{x \to 2} \frac{x-2}{(x-2)(x+2)} = \lim_{x \to 2} \frac{1}{x+2} = \frac{1}{4}$ *Again substitute 2 for x after algebraic manipulation eliminates division by 0.*

(c) $\displaystyle \lim_{x \to 2^+} \frac{x-3}{x^2 - 4} = \lim_{x \to 2^+} \frac{x-3}{(x-2)(x+2)} = -\infty$ *The values are negative for $x > 2$, x near 2.*

(d) $\displaystyle \lim_{x \to 2^-} \frac{x-3}{x^2 - 4} = \lim_{x \to 2^-} \frac{x-3}{(x-2)(x+2)} = \infty$ *The values are positive for $x < 2$, x near 2.*

(e) $\lim_{x \to 2} \dfrac{x-3}{x^2-4} = \lim_{x \to 2} \dfrac{x-3}{(x-2)(x+2)}$ does not exist.   *Limits from left and from right differ.*

(f) $\lim_{x \to 2} \dfrac{2-x}{(x-2)^3} = \lim_{x \to 2} \dfrac{-(x-2)}{(x-2)^3} = \lim_{x \to 2} \dfrac{-1}{(x-2)^2} = -\infty$ *Denominator is positive, so values are negative near $x = 2$.*

In parts (a) and (b) the effect of the zero in the denominator at $x = 2$ is canceled because the numerator is zero there also. Thus a finite limit exists. This is not true in part (f), where cancellation still leaves a zero factor in the denominator.   ■

EXAMPLE 14   Find $\displaystyle\lim_{x \to -\infty} \dfrac{2x^5 - 6x^4 + 1}{3x^2 + x - 7}$.

Solution   We are asked to find the limit of a rational function as $x \to -\infty$, so we divide the numerator and denominator by x^2, the highest power of x in the denominator:

$$\lim_{x \to -\infty} \dfrac{2x^5 - 6x^4 + 1}{3x^2 + x - 7} = \lim_{x \to -\infty} \dfrac{2x^3 - 6x^2 + x^{-2}}{3 + x^{-1} - 7x^{-2}}$$

$$= \lim_{x \to -\infty} \dfrac{2x^2(x - 3) + x^{-2}}{3 + x^{-1} - 7x^{-2}}$$

$$= -\infty, \qquad x^{-n} \to 0,\ x - 3 \to -\infty$$

because the numerator tends to $-\infty$ while the denominator approaches 3 as $x \to -\infty$.   ■

Precise Definitions of Infinite Limits

Instead of requiring $f(x)$ to lie arbitrarily close to a finite number L for all x sufficiently close to c, the definitions of infinite limits require $f(x)$ to lie arbitrarily far from zero. Except for this change, the language is very similar to what we have seen before. Figures 2.62 and 2.63 accompany these definitions.

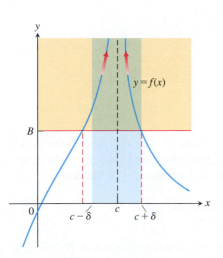

FIGURE 2.62   For $c - \delta < x < c + \delta$, the graph of $f(x)$ lies above the line $y = B$.

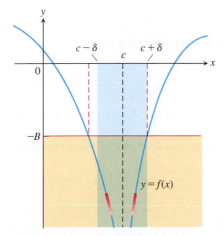

FIGURE 2.63   For $c - \delta < x < c + \delta$, the graph of $f(x)$ lies below the line $y = -B$.

> **DEFINITIONS**
>
> **1.** We say that $f(x)$ **approaches infinity** as x approaches c, and write
>
> $$\lim_{x \to c} f(x) = \infty,$$
>
> if for every positive real number B there exists a corresponding $\delta > 0$ such that
>
> $$f(x) > B \quad \text{whenever} \quad 0 < |x - c| < \delta.$$
>
> **2.** We say that $f(x)$ **approaches negative infinity** as x approaches c, and write
>
> $$\lim_{x \to c} f(x) = -\infty,$$
>
> if for every negative real number $-B$ there exists a corresponding $\delta > 0$ such that
>
> $$f(x) < -B \quad \text{whenever} \quad 0 < |x - c| < \delta.$$

The precise definitions of one-sided infinite limits at c are similar and are stated in the exercises.

EXAMPLE 15   Prove that $\displaystyle\lim_{x \to 0} \dfrac{1}{x^2} = \infty$.

Solution   Given $B > 0$, we want to find $\delta > 0$ such that

$$0 < |x - 0| < \delta \quad \text{implies} \quad \dfrac{1}{x^2} > B.$$

Now,

$$\frac{1}{x^2} > B \qquad \text{if and only if} \qquad x^2 < \frac{1}{B}$$

or, equivalently,

$$|x| < \frac{1}{\sqrt{B}}.$$

Thus, choosing $\delta = 1/\sqrt{B}$ (or any smaller positive number), we see that

$$|x| < \delta \quad \text{implies} \quad \frac{1}{x^2} > \frac{1}{\delta^2} \geq B.$$

Therefore, by definition,

$$\lim_{x \to 0} \frac{1}{x^2} = \infty.$$

∎

Vertical Asymptotes

Notice that the distance between a point on the graph of $f(x) = 1/x$ and the y-axis approaches zero as the point moves vertically along the graph and away from the origin (Figure 2.64). The function $f(x) = 1/x$ is unbounded as x approaches 0 because

$$\lim_{x \to 0^+} \frac{1}{x} = \infty \qquad \text{and} \qquad \lim_{x \to 0^-} \frac{1}{x} = -\infty.$$

We say that the line $x = 0$ (the y-axis) is a *vertical asymptote* of the graph of $f(x) = 1/x$. Observe that the denominator is zero at $x = 0$ and the function is undefined there.

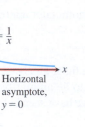

> **DEFINITION**   A line $x = a$ is a **vertical asymptote** of the graph of a function $y = f(x)$ if either
>
> $$\lim_{x \to a^+} f(x) = \pm\infty \qquad \text{or} \qquad \lim_{x \to a^-} f(x) = \pm\infty.$$

FIGURE 2.64   The coordinate axes are asymptotes of both branches of the hyperbola $y = 1/x$.

EXAMPLE 16   Find the horizontal and vertical asymptotes of the curve

$$y = \frac{x + 3}{x + 2}.$$

Solution   We are interested in the behavior as $x \to \pm\infty$ and the behavior as $x \to -2$, where the denominator is zero.

The asymptotes are revealed if we recast the rational function as a polynomial with a remainder, by dividing $(x + 2)$ into $(x + 3)$:

$$\begin{array}{r} 1 \\ x + 2 \overline{)x + 3} \\ \underline{x + 2} \\ 1 \end{array}$$

This result enables us to rewrite y as:

$$y = 1 + \frac{1}{x + 2}.$$

As $x \to \pm\infty$, the curve approaches the horizontal asymptote $y = 1$; as $x \to -2$, the curve approaches the vertical asymptote $x = -2$. We see that the curve in question is the graph of $f(x) = 1/x$ shifted 1 unit up and 2 units left (Figure 2.65). The asymptotes, instead of being the coordinate axes, are now the lines $y = 1$ and $x = -2$.

∎

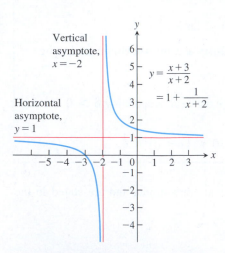

FIGURE 2.65   The lines $y = 1$ and $x = -2$ are asymptotes of the curve in Example 16.

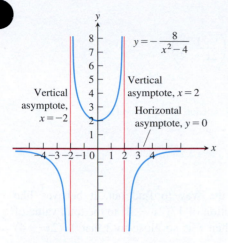

FIGURE 2.66 Graph of the function in Example 17. Notice that the curve approaches the x-axis from only one side. Asymptotes do not have to be two-sided.

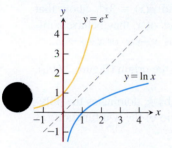

FIGURE 2.67 The line $x = 0$ is a vertical asymptote of the natural logarithm function (Example 18).

EXAMPLE 17 Find the horizontal and vertical asymptotes of the graph of

$$f(x) = -\frac{8}{x^2 - 4}.$$

Solution We are interested in the behavior as $x \to \pm\infty$ and as $x \to \pm 2$, where the denominator is zero. Notice that f is an even function of x, so its graph is symmetric with respect to the y-axis.

(a) *The behavior as $x \to \pm\infty$.* Since $\lim_{x\to\infty} f(x) = 0$, the line $y = 0$ is a horizontal asymptote of the graph to the right. By symmetry it is an asymptote to the left as well (Figure 2.66). Notice that the curve approaches the x-axis from only the negative side (or from below). Also, $f(0) = 2$.

(b) *The behavior as $x \to \pm 2$.* Since

$$\lim_{x\to 2^+} f(x) = -\infty \qquad \text{and} \qquad \lim_{x\to 2^-} f(x) = \infty,$$

the line $x = 2$ is a vertical asymptote both from the right and from the left. By symmetry, the line $x = -2$ is also a vertical asymptote.

There are no other asymptotes because f has a finite limit at all other points. ∎

EXAMPLE 18 The graph of the natural logarithm function has the y-axis (the line $x = 0$) as a vertical asymptote. We see this from the graph sketched in Figure 2.67 (which is the reflection of the graph of the natural exponential function across the line $y = x$) and the fact that the x-axis is a horizontal asymptote of $y = e^x$ (Example 5). Thus,

$$\lim_{x\to 0^+} \ln x = -\infty.$$

The same result is true for $y = \log_a x$ whenever $a > 1$. ∎

EXAMPLE 19 The curves

$$y = \sec x = \frac{1}{\cos x} \qquad \text{and} \qquad y = \tan x = \frac{\sin x}{\cos x}$$

both have vertical asymptotes at odd-integer multiples of $\pi/2$, where $\cos x = 0$ (Figure 2.68).

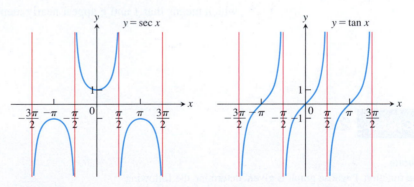

FIGURE 2.68 The graphs of $\sec x$ and $\tan x$ have infinitely many vertical asymptotes (Example 19). ∎

Dominant Terms

In Example 10 we saw that by using long division, we can rewrite the function

$$f(x) = \frac{x^2 - 3}{2x - 4}$$

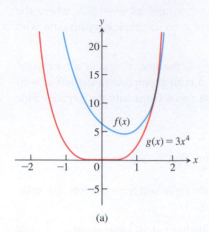

(a)

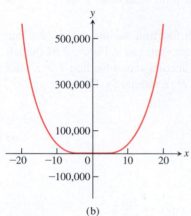

(b)

FIGURE 2.69 The graphs of f and g are (a) distinct for $|x|$ small, and (b) nearly identical for $|x|$ large (Example 20).

as a linear function plus a remainder term:

$$f(x) = \left(\frac{x}{2} + 1\right) + \left(\frac{1}{2x - 4}\right).$$

This tells us immediately that

$$f(x) \approx \frac{x}{2} + 1 \qquad \text{For } |x| \text{ large, } \frac{1}{2x - 4} \text{ is near 0.}$$

$$f(x) \approx \frac{1}{2x - 4} \qquad \text{For } x \text{ near 2, this term is very large in absolute value.}$$

If we want to know how f behaves, this is the way to find out. It behaves like $y = (x/2) + 1$ when $|x|$ is large and the contribution of $1/(2x - 4)$ to the total value of f is insignificant. It behaves like $1/(2x - 4)$ when x is so close to 2 that $1/(2x - 4)$ makes the dominant contribution.

We say that $(x/2) + 1$ **dominates** when x approaches ∞ or $-\infty$, and we say that $1/(2x - 4)$ dominates when x approaches 2. **Dominant terms** like these help us predict a function's behavior.

EXAMPLE 20   Let $f(x) = 3x^4 - 2x^3 + 3x^2 - 5x + 6$ and $g(x) = 3x^4$. Show that although f and g are quite different for numerically small values of x, they behave similarly for $|x|$ very large, in the sense that their ratios approach 1 as $x \to \infty$ or $x \to -\infty$.

Solution   The graphs of f and g behave quite differently near the origin (Figure 2.69a), but appear as virtually identical on a larger scale (Figure 2.69b).

We can test that the term $3x^4$ in f, represented graphically by g, dominates the polynomial f for numerically large values of x by examining the ratio of the two functions as $x \to \pm\infty$. We find that

$$\lim_{x \to \pm\infty} \frac{f(x)}{g(x)} = \lim_{x \to \pm\infty} \frac{3x^4 - 2x^3 + 3x^2 - 5x + 6}{3x^4}$$

$$= \lim_{x \to \pm\infty} \left(1 - \frac{2}{3x} + \frac{1}{x^2} - \frac{5}{3x^3} + \frac{2}{x^4}\right)$$

$$= 1,$$

which means that f and g appear nearly identical when $|x|$ is large. ∎

EXERCISES 2.6

Finding Limits

1. For the function f whose graph is given, determine the following limits.

a. $\lim_{x \to 2} f(x)$ **b.** $\lim_{x \to -3^+} f(x)$

c. $\lim_{x \to -3^-} f(x)$ **d.** $\lim_{x \to -3} f(x)$

e. $\lim_{x \to 0^+} f(x)$ **f.** $\lim_{x \to 0^-} f(x)$

g. $\lim_{x \to 0} f(x)$ **h.** $\lim_{x \to \infty} f(x)$

i. $\lim_{x \to -\infty} f(x)$

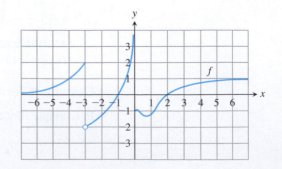

2. For the function f whose graph is given, determine the following limits.

a. $\lim\limits_{x\to 4} f(x)$ b. $\lim\limits_{x\to 2^+} f(x)$ c. $\lim\limits_{x\to 2^-} f(x)$

d. $\lim\limits_{x\to 2} f(x)$ e. $\lim\limits_{x\to -3^+} f(x)$ f. $\lim\limits_{x\to -3^-} f(x)$

g. $\lim\limits_{x\to -3} f(x)$ h. $\lim\limits_{x\to 0^+} f(x)$ i. $\lim\limits_{x\to 0^-} f(x)$

j. $\lim\limits_{x\to 0} f(x)$ k. $\lim\limits_{x\to\infty} f(x)$ l. $\lim\limits_{x\to -\infty} f(x)$

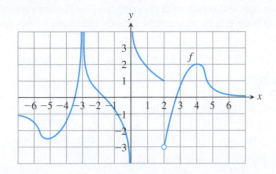

In Exercises 3–8, find the limit of each function (a) as $x\to\infty$ and (b) as $x\to -\infty$. (You may wish to visualize your answer with a graphing calculator or computer.)

3. $f(x) = \dfrac{2}{x} - 3$ **4.** $f(x) = \pi - \dfrac{2}{x^2}$

5. $g(x) = \dfrac{1}{2 + (1/x)}$ **6.** $g(x) = \dfrac{1}{8 - (5/x^2)}$

7. $h(x) = \dfrac{-5 + (7/x)}{3 - (1/x^2)}$ **8.** $h(x) = \dfrac{3 - (2/x)}{4 + (\sqrt{2}/x^2)}$

Find the limits in Exercises 9–12.

9. $\lim\limits_{x\to\infty} \dfrac{\sin 2x}{x}$ **10.** $\lim\limits_{\theta\to -\infty} \dfrac{\cos\theta}{3\theta}$

11. $\lim\limits_{t\to -\infty} \dfrac{2 - t + \sin t}{t + \cos t}$ **12.** $\lim\limits_{r\to\infty} \dfrac{r + \sin r}{2r + 7 - 5\sin r}$

Limits of Rational Functions

In Exercises 13–22, find the limit of each rational function (a) as $x\to\infty$ and (b) as $x\to -\infty$.

13. $f(x) = \dfrac{2x + 3}{5x + 7}$ **14.** $f(x) = \dfrac{2x^3 + 7}{x^3 - x^2 + x + 7}$

15. $f(x) = \dfrac{x + 1}{x^2 + 3}$ **16.** $f(x) = \dfrac{3x + 7}{x^2 - 2}$

17. $h(x) = \dfrac{7x^3}{x^3 - 3x^2 + 6x}$ **18.** $h(x) = \dfrac{9x^4 + x}{2x^4 + 5x^2 - x + 6}$

19. $g(x) = \dfrac{10x^5 + x^4 + 31}{x^6}$ **20.** $g(x) = \dfrac{x^3 + 7x^2 - 2}{x^2 - x + 1}$

21. $f(x) = \dfrac{3x^7 + 5x^2 - 1}{6x^3 - 7x + 3}$ **22.** $h(x) = \dfrac{5x^8 - 2x^3 + 9}{3 + x - 4x^5}$

Limits as $x\to\infty$ or $x\to -\infty$

The process by which we determine limits of rational functions applies equally well to ratios containing noninteger or negative powers of x:

Divide numerator and denominator by the highest power of x in the denominator and proceed from there. Find the limits in Exercises 23–36.

23. $\lim\limits_{x\to\infty} \sqrt{\dfrac{8x^2 - 3}{2x^2 + x}}$ **24.** $\lim\limits_{x\to -\infty} \left(\dfrac{x^2 + x - 1}{8x^2 - 3}\right)^{1/3}$

25. $\lim\limits_{x\to -\infty} \left(\dfrac{1 - x^3}{x^2 + 7x}\right)^5$ **26.** $\lim\limits_{x\to\infty} \sqrt{\dfrac{x^2 - 5x}{x^3 + x - 2}}$

27. $\lim\limits_{x\to\infty} \dfrac{2\sqrt{x} + x^{-1}}{3x - 7}$ **28.** $\lim\limits_{x\to\infty} \dfrac{2 + \sqrt{x}}{2 - \sqrt{x}}$

29. $\lim\limits_{x\to -\infty} \dfrac{\sqrt[3]{x} - \sqrt[5]{x}}{\sqrt[3]{x} + \sqrt[5]{x}}$ **30.** $\lim\limits_{x\to\infty} \dfrac{x^{-1} + x^{-4}}{x^{-2} - x^{-3}}$

31. $\lim\limits_{x\to\infty} \dfrac{2x^{5/3} - x^{1/3} + 7}{x^{8/5} + 3x + \sqrt{x}}$ **32.** $\lim\limits_{x\to -\infty} \dfrac{\sqrt[3]{x} - 5x + 3}{2x + x^{2/3} - 4}$

33. $\lim\limits_{x\to\infty} \dfrac{\sqrt{x^2 + 1}}{x + 1}$ **34.** $\lim\limits_{x\to -\infty} \dfrac{\sqrt{x^2 + 1}}{x + 1}$

35. $\lim\limits_{x\to\infty} \dfrac{x - 3}{\sqrt{4x^2 + 25}}$ **36.** $\lim\limits_{x\to -\infty} \dfrac{4 - 3x^3}{\sqrt{x^6 + 9}}$

Infinite Limits

Find the limits in Exercises 37–48.

37. $\lim\limits_{x\to 0^+} \dfrac{1}{3x}$ **38.** $\lim\limits_{x\to 0^-} \dfrac{5}{2x}$

39. $\lim\limits_{x\to 2^-} \dfrac{3}{x - 2}$ **40.** $\lim\limits_{x\to 3^+} \dfrac{1}{x - 3}$

41. $\lim\limits_{x\to -8^+} \dfrac{2x}{x + 8}$ **42.** $\lim\limits_{x\to -5^-} \dfrac{3x}{2x + 10}$

43. $\lim\limits_{x\to 7} \dfrac{4}{(x - 7)^2}$ **44.** $\lim\limits_{x\to 0} \dfrac{-1}{x^2(x + 1)}$

45. a. $\lim\limits_{x\to 0^+} \dfrac{2}{3x^{1/3}}$ **b.** $\lim\limits_{x\to 0^-} \dfrac{2}{3x^{1/3}}$

46. a. $\lim\limits_{x\to 0^+} \dfrac{2}{x^{1/5}}$ **b.** $\lim\limits_{x\to 0^-} \dfrac{2}{x^{1/5}}$

47. $\lim\limits_{x\to 0} \dfrac{4}{x^{2/5}}$ **48.** $\lim\limits_{x\to 0} \dfrac{1}{x^{2/3}}$

Find the limits in Exercises 49–52.

49. $\lim\limits_{x\to(\pi/2)^-} \tan x$ **50.** $\lim\limits_{x\to(-\pi/2)^+} \sec x$

51. $\lim\limits_{\theta\to 0^-} (1 + \csc\theta)$ **52.** $\lim\limits_{\theta\to 0} (2 - \cot\theta)$

Find the limits in Exercises 53–58.

53. $\lim \dfrac{1}{x^2 - 4}$ as

a. $x\to 2^+$ b. $x\to 2^-$

c. $x\to -2^+$ d. $x\to -2^-$

54. $\lim \dfrac{x}{x^2 - 1}$ as

a. $x\to 1^+$ b. $x\to 1^-$

c. $x\to -1^+$ d. $x\to -1^-$

55. $\lim \left(\dfrac{x^2}{2} - \dfrac{1}{x} \right)$ as

   a. $x \to 0^+$ **b.** $x \to 0^-$

   c. $x \to \sqrt[3]{2}$ **d.** $x \to -1$

56. $\lim \dfrac{x^2 - 1}{2x + 4}$ as

   a. $x \to -2^+$ **b.** $x \to -2^-$

   c. $x \to 1^+$ **d.** $x \to 0^-$

57. $\lim \dfrac{x^2 - 3x + 2}{x^3 - 2x^2}$ as

   a. $x \to 0^+$ **b.** $x \to 2^+$

   c. $x \to 2^-$ **d.** $x \to 2$

   e. What, if anything, can be said about the limit as $x \to 0$?

58. $\lim \dfrac{x^2 - 3x + 2}{x^3 - 4x}$ as

   a. $x \to 2^+$ **b.** $x \to -2^+$

   c. $x \to 0^-$ **d.** $x \to 1^+$

   e. What, if anything, can be said about the limit as $x \to 0$?

Find the limits in Exercises 59–62.

59. $\lim \left(2 - \dfrac{3}{t^{1/3}} \right)$ as

   a. $t \to 0^+$ **b.** $t \to 0^-$

60. $\lim \left(\dfrac{1}{t^{3/5}} + 7 \right)$ as

   a. $t \to 0^+$ **b.** $t \to 0^-$

61. $\lim \left(\dfrac{1}{x^{2/3}} + \dfrac{2}{(x - 1)^{2/3}} \right)$ as

   a. $x \to 0^+$ **b.** $x \to 0^-$

   c. $x \to 1^+$ **d.** $x \to 1^-$

62. $\lim \left(\dfrac{1}{x^{1/3}} - \dfrac{1}{(x - 1)^{4/3}} \right)$ as

   a. $x \to 0^+$ **b.** $x \to 0^-$

   c. $x \to 1^+$ **d.** $x \to 1^-$

Graphing Simple Rational Functions

Graph the rational functions in Exercises 63–68. Include the graphs and equations of the asymptotes and dominant terms.

63. $y = \dfrac{1}{x - 1}$ **64.** $y = \dfrac{1}{x + 1}$

65. $y = \dfrac{1}{2x + 4}$ **66.** $y = \dfrac{-3}{x - 3}$

67. $y = \dfrac{x + 3}{x + 2}$ **68.** $y = \dfrac{2x}{x + 1}$

Domains, Ranges, and Asymptotes

Determine the domain and range of each function. Use various limits to find the asymptotes and the ranges.

69. $y = 4 + \dfrac{3x^2}{x^2 + 1}$ **70.** $y = \dfrac{2x}{x^2 - 1}$

71. $y = \dfrac{8 - e^x}{2 + e^x}$ **72.** $y = \dfrac{4e^x + e^{2x}}{e^x + e^{2x}}$

73. $y = \dfrac{\sqrt{x^2 + 4}}{x}$ **74.** $y = \dfrac{x^3}{x^3 - 8}$

Inventing Graphs and Functions

In Exercises 75–78, sketch the graph of a function $y = f(x)$ that satisfies the given conditions. No formulas are required—just label the coordinate axes and sketch an appropriate graph. (The answers are not unique, so your graphs may not be exactly like those in the answer section.)

75. $f(0) = 0$, $f(1) = 2$, $f(-1) = -2$, $\lim\limits_{x \to -\infty} f(x) = -1$, and
   $$\lim\limits_{x \to \infty} f(x) = 1$$

76. $f(0) = 0$, $\lim\limits_{x \to \pm\infty} f(x) = 0$, $\lim\limits_{x \to 0^+} f(x) = 2$, and $\lim\limits_{x \to 0^-} f(x) = -2$

77. $f(0) = 0$, $\lim\limits_{x \to \pm\infty} f(x) = 0$, $\lim\limits_{x \to 1^-} f(x) = \lim\limits_{x \to -1^+} f(x) = \infty$,
   $\lim\limits_{x \to 1^+} f(x) = -\infty$, and $\lim\limits_{x \to -1^-} f(x) = -\infty$

78. $f(2) = 1$, $f(-1) = 0$, $\lim\limits_{x \to \infty} f(x) = 0$, $\lim\limits_{x \to 0^+} f(x) = \infty$,
   $\lim\limits_{x \to 0^-} f(x) = -\infty$, and $\lim\limits_{x \to -\infty} f(x) = 1$

In Exercises 79–82, find a function that satisfies the given conditions and sketch its graph. (The answers here are not unique. Any function that satisfies the conditions is acceptable. Feel free to use formulas defined in pieces if that will help.)

79. $\lim\limits_{x \to \pm\infty} f(x) = 0$, $\lim\limits_{x \to 2^-} f(x) = \infty$, and $\lim\limits_{x \to 2^+} f(x) = \infty$

80. $\lim\limits_{x \to \pm\infty} g(x) = 0$, $\lim\limits_{x \to 3^-} g(x) = -\infty$, and $\lim\limits_{x \to 3^+} g(x) = \infty$

81. $\lim\limits_{x \to -\infty} h(x) = -1$, $\lim\limits_{x \to \infty} h(x) = 1$, $\lim\limits_{x \to 0^-} h(x) = -1$, and
   $\lim\limits_{x \to 0^+} h(x) = 1$

82. $\lim\limits_{x \to \pm\infty} k(x) = 1$, $\lim\limits_{x \to 1^-} k(x) = \infty$, and $\lim\limits_{x \to 1^+} k(x) = -\infty$

83. Suppose that $f(x)$ and $g(x)$ are polynomials in x and that $\lim_{x \to \infty} (f(x)/g(x)) = 2$. Can you conclude anything about $\lim_{x \to -\infty} (f(x)/g(x))$? Give reasons for your answer.

84. Suppose that $f(x)$ and $g(x)$ are polynomials in x. Can the graph of $f(x)/g(x)$ have an asymptote if $g(x)$ is never zero? Give reasons for your answer.

85. How many horizontal asymptotes can the graph of a given rational function have? Give reasons for your answer.

Finding Limits of Differences When $x \to \pm\infty$

Find the limits in Exercises 86–92. (*Hint*: Try multiplying and dividing by the conjugate.)

86. $\lim\limits_{x \to \infty} (\sqrt{x + 9} - \sqrt{x + 4})$

87. $\lim\limits_{x \to \infty} (\sqrt{x^2 + 25} - \sqrt{x^2 - 1})$

88. $\lim\limits_{x \to -\infty} (\sqrt{x^2 + 3} + x)$

89. $\lim\limits_{x \to -\infty} (2x + \sqrt{4x^2 + 3x - 2})$

90. $\lim\limits_{x \to \infty} (\sqrt{9x^2 - x} - 3x)$

91. $\lim\limits_{x \to \infty} (\sqrt{x^2 + 3x} - \sqrt{x^2 - 2x})$

92. $\lim\limits_{x \to \infty} (\sqrt{x^2 + x} - \sqrt{x^2 - x})$

Using the Formal Definitions

Use the formal definitions of limits as $x \to \pm\infty$ to establish the limits in Exercises 93 and 94.

93. If f has the constant value $f(x) = k$, then $\lim\limits_{x \to \infty} f(x) = k$.

94. If f has the constant value $f(x) = k$, then $\lim\limits_{x \to -\infty} f(x) = k$.

Use formal definitions to prove the limit statements in Exercises 95–98.

95. $\lim\limits_{x \to 0} \dfrac{-1}{x^2} = -\infty$

96. $\lim\limits_{x \to 0} \dfrac{1}{|x|} = \infty$

97. $\lim\limits_{x \to 3} \dfrac{-2}{(x-3)^2} = -\infty$

98. $\lim\limits_{x \to -5} \dfrac{1}{(x+5)^2} = \infty$

99. Here is the definition of **infinite right-hand limit**.

> Suppose that an interval (c, d) lies in the domain of f. We say that $f(x)$ approaches infinity as x approaches c from the right, and write
> $$\lim\limits_{x \to c^+} f(x) = \infty,$$
> if, for every positive real number B, there exists a corresponding number $\delta > 0$ such that
> $$f(x) > B \quad \text{whenever} \quad c < x < c + \delta.$$

Modify the definition to cover the following cases.

   a. $\lim\limits_{x \to c^-} f(x) = \infty$

   b. $\lim\limits_{x \to c^+} f(x) = -\infty$

   c. $\lim\limits_{x \to c^-} f(x) = -\infty$

Use the formal definitions from Exercise 99 to prove the limit statements in Exercises 100–104.

100. $\lim\limits_{x \to 0^+} \dfrac{1}{x} = \infty$

101. $\lim\limits_{x \to 0^-} \dfrac{1}{x} = -\infty$

102. $\lim\limits_{x \to 2^-} \dfrac{1}{x - 2} = -\infty$

103. $\lim\limits_{x \to 2^+} \dfrac{1}{x - 2} = \infty$

104. $\lim\limits_{x \to 1^-} \dfrac{1}{1 - x^2} = \infty$

Oblique Asymptotes

Graph the rational functions in Exercises 105–110. Include the graphs and equations of the asymptotes.

105. $y = \dfrac{x^2}{x - 1}$

106. $y = \dfrac{x^2 + 1}{x - 1}$

107. $y = \dfrac{x^2 - 4}{x - 1}$

108. $y = \dfrac{x^2 - 1}{2x + 4}$

109. $y = \dfrac{x^2 - 1}{x}$

110. $y = \dfrac{x^3 + 1}{x^2}$

Additional Graphing Exercises

T Graph the curves in Exercises 111–114. Explain the relationship between the curve's formula and what you see.

111. $y = \dfrac{x}{\sqrt{4 - x^2}}$

112. $y = \dfrac{-1}{\sqrt{4 - x^2}}$

113. $y = x^{2/3} + \dfrac{1}{x^{1/3}}$

114. $y = \sin\left(\dfrac{\pi}{x^2 + 1}\right)$

T Graph the functions in Exercises 115 and 116. Then answer the following questions.

   a. How does the graph behave as $x \to 0^+$?

   b. How does the graph behave as $x \to \pm\infty$?

   c. How does the graph behave near $x = 1$ and $x = -1$?

Give reasons for your answers.

115. $y = \dfrac{3}{2}\left(x - \dfrac{1}{x}\right)^{2/3}$

116. $y = \dfrac{3}{2}\left(\dfrac{x}{x - 1}\right)^{2/3}$

CHAPTER 2   **Questions to Guide Your Review**

1. What is the average rate of change of the function $g(t)$ over the interval from $t = a$ to $t = b$? How is it related to a secant line?

2. What limit must be calculated to find the rate of change of a function $g(t)$ at $t = t_0$?

3. Give an informal or intuitive definition of the limit
$$\lim\limits_{x \to c} f(x) = L.$$
Why is the definition "informal"? Give examples.

4. Does the existence and value of the limit of a function $f(x)$ as x approaches c ever depend on what happens at $x = c$? Explain and give examples.

5. What function behaviors might occur for which the limit may fail to exist? Give examples.

6. What theorems are available for calculating limits? Give examples of how the theorems are used.

7. How are one-sided limits related to limits? How can this relationship sometimes be used to calculate a limit or prove it does not exist? Give examples.

8. What is the value of $\lim\limits_{\theta \to 0} ((\sin\theta)/\theta)$? Does it matter whether θ is measured in degrees or radians? Explain.

9. What exactly does $\lim_{x\to c} f(x) = L$ mean? Give an example in which you find a $\delta > 0$ for a given $f, L, c,$ and $\varepsilon > 0$ in the precise definition of limit.

10. Give precise definitions of the following statements.

 a. $\lim_{x\to 2^-} f(x) = 5$ b. $\lim_{x\to 2^+} f(x) = 5$

 c. $\lim_{x\to 2} f(x) = \infty$ d. $\lim_{x\to 2} f(x) = -\infty$

11. What conditions must be satisfied by a function if it is to be continuous at an interior point of its domain? At an endpoint?

12. How can looking at the graph of a function help you tell where the function is continuous?

13. What does it mean for a function to be right-continuous at a point? Left-continuous? How are continuity and one-sided continuity related?

14. What does it mean for a function to be continuous on an interval? Give examples to illustrate the fact that a function that is not continuous on its entire domain may still be continuous on selected intervals within the domain.

15. What are the basic types of discontinuity? Give an example of each. What is a removable discontinuity? Give an example.

16. What does it mean for a function to have the Intermediate Value Property? What conditions guarantee that a function has this property over an interval? What are the consequences for graphing and solving the equation $f(x) = 0$?

17. Under what circumstances can you extend a function $f(x)$ to be continuous at a point $x = c$? Give an example.

18. What exactly do $\lim_{x\to\infty} f(x) = L$ and $\lim_{x\to-\infty} f(x) = L$ mean? Give examples.

19. What are $\lim_{x\to\pm\infty} k$ (k a constant) and $\lim_{x\to\pm\infty} (1/x)$? How do you extend these results to other functions? Give examples.

20. How do you find the limit of a rational function as $x \to \pm\infty$? Give examples.

21. What are horizontal and vertical asymptotes? Give examples.

CHAPTER 2 Practice Exercises

Limits and Continuity

1. Graph the function

$$f(x) = \begin{cases} 1, & x \le -1 \\ -x, & -1 < x < 0 \\ 1, & x = 0 \\ -x, & 0 < x < 1 \\ 1, & x \ge 1. \end{cases}$$

 Then discuss, in detail, limits, one-sided limits, continuity, and one-sided continuity of f at $x = -1, 0,$ and 1. Are any of the discontinuities removable? Explain.

2. Repeat the instructions of Exercise 1 for

$$f(x) = \begin{cases} 0, & x \le -1 \\ 1/x, & 0 < |x| < 1 \\ 0, & x = 1 \\ 1, & x > 1. \end{cases}$$

3. Suppose that $f(t)$ and $f(t)$ are defined for all t and that $\lim_{t\to t_0} f(t) = -7$ and $\lim_{t\to t_0} g(t) = 0$. Find the limit as $t \to t_0$ of the following functions.

 a. $3f(t)$ b. $(f(t))^2$

 c. $f(t) \cdot g(t)$ d. $\dfrac{f(t)}{g(t) - 7}$

 e. $\cos(g(t))$ f. $|f(t)|$

 g. $f(t) + g(t)$ h. $1/f(t)$

4. Suppose the functions $f(x)$ and $g(x)$ are defined for all x and that $\lim_{x\to 0} f(x) = 1/2$ and $\lim_{x\to 0} g(x) = \sqrt{2}$. Find the limits as $x \to 0$ of the following functions.

 a. $-g(x)$ b. $g(x) \cdot f(x)$

 c. $f(x) + g(x)$ d. $1/f(x)$

 e. $x + f(x)$ f. $\dfrac{f(x) \cdot \cos x}{x - 1}$

In Exercises 5 and 6, find the value that $\lim_{x\to 0} g(x)$ must have if the given limit statements hold.

5. $\lim_{x\to 0}\left(\dfrac{4 - g(x)}{x}\right) = 1$ 6. $\lim_{x\to -4}\left(x \lim_{x\to 0} g(x)\right) = 2$

7. On what intervals are the following functions continuous?

 a. $f(x) = x^{1/3}$ b. $g(x) = x^{3/4}$

 c. $h(x) = x^{-2/3}$ d. $k(x) = x^{-1/6}$

8. On what intervals are the following functions continuous?

 a. $f(x) = \tan x$

 b. $g(x) = \csc x$

 c. $h(x) = \dfrac{\cos x}{x - \pi}$

 d. $k(x) = \dfrac{\sin x}{x}$

Finding Limits

In Exercises 9–28, find the limit or explain why it does not exist.

9. $\lim \dfrac{x^2 - 4x + 4}{x^3 + 5x^2 - 14x}$

 a. as $x \to 0$ b. as $x \to 2$

10. $\lim \dfrac{x^2 + x}{x^5 + 2x^4 + x^3}$

 a. as $x \to 0$ b. as $x \to -1$

11. $\lim_{x\to 1} \dfrac{1 - \sqrt{x}}{1 - x}$ 12. $\lim_{x\to a} \dfrac{x^2 - a^2}{x^4 - a^4}$

13. $\lim_{h\to 0} \dfrac{(x + h)^2 - x^2}{h}$ 14. $\lim_{x\to 0} \dfrac{(x + h)^2 - x^2}{h}$

15. $\lim\limits_{x\to 0} \dfrac{\dfrac{1}{2+x} - \dfrac{1}{2}}{x}$

16. $\lim\limits_{x\to 0} \dfrac{(2+x)^3 - 8}{x}$

17. $\lim\limits_{x\to 1} \dfrac{x^{1/3} - 1}{\sqrt{x} - 1}$

18. $\lim\limits_{x\to 64} \dfrac{x^{2/3} - 16}{\sqrt{x} - 8}$

19. $\lim\limits_{x\to 0} \dfrac{\tan(2x)}{\tan(\pi x)}$

20. $\lim\limits_{x\to \pi^-} \csc x$

21. $\lim\limits_{x\to \pi} \sin\left(\dfrac{x}{2} + \sin x\right)$

22. $\lim\limits_{x\to \pi} \cos^2(x - \tan x)$

23. $\lim\limits_{x\to 0} \dfrac{8x}{3\sin x - x}$

24. $\lim\limits_{x\to 0} \dfrac{\cos 2x - 1}{\sin x}$

25. $\lim\limits_{t\to 3^+} \ln(t - 3)$

26. $\lim\limits_{t\to 1} t^2 \ln(2 - \sqrt{t})$

27. $\lim\limits_{\theta\to 0^+} \sqrt{\theta}\, e^{\cos(\pi/\theta)}$

28. $\lim\limits_{z\to 0^+} \dfrac{2e^{1/z}}{e^{1/z} + 1}$

In Exercises 29–32, find the limit of $g(x)$ as x approaches the indicated value.

29. $\lim\limits_{x\to 0^+} (4g(x))^{1/3} = 2$

30. $\lim\limits_{x\to \sqrt{5}} \dfrac{1}{x + g(x)} = 2$

31. $\lim\limits_{x\to 1} \dfrac{3x^2 + 1}{g(x)} = \infty$

32. $\lim\limits_{x\to -2} \dfrac{5 - x^2}{\sqrt{g(x)}} = 0$

⊤ Roots

33. Let $f(x) = x^3 - x - 1$.

   a. Use the Intermediate Value Theorem to show that f has a zero between -1 and 2.

   b. Solve the equation $f(x) = 0$ graphically with an error of magnitude at most 10^{-8}.

   c. It can be shown that the exact value of the solution in part **(b)** is

   $$\left(\dfrac{1}{2} + \dfrac{\sqrt{69}}{18}\right)^{1/3} + \left(\dfrac{1}{2} - \dfrac{\sqrt{69}}{18}\right)^{1/3}.$$

   Evaluate this exact answer and compare it with the value you found in part (b).

⊤ **34.** Let $f(\theta) = \theta^3 - 2\theta + 2$.

   a. Use the Intermediate Value Theorem to show that f has a zero between -2 and 0.

   b. Solve the equation $f(\theta) = 0$ graphically with an error of magnitude at most 10^{-4}.

   c. It can be shown that the exact value of the solution in part (b) is

   $$\left(\sqrt{\dfrac{19}{27}} - 1\right)^{1/3} - \left(\sqrt{\dfrac{19}{27}} + 1\right)^{1/3}.$$

   Evaluate this exact answer and compare it with the value you found in part (b).

Continuous Extension

35. Can $f(x) = x(x^2 - 1)/|x^2 - 1|$ be extended to be continuous at $x = 1$ or -1? Give reasons for your answers. (Graph the function—you will find the graph interesting.)

36. Explain why the function $f(x) = \sin(1/x)$ has no continuous extension to $x = 0$.

⊤ In Exercises 37–40, graph the function to see whether it appears to have a continuous extension to the given point a. If it does, use Trace and Zoom to find a good candidate for the extended function's value at a. If the function does not appear to have a continuous extension, can it be extended to be continuous from the right or left? If so, what do you think the extended function's value should be?

37. $f(x) = \dfrac{x - 1}{x - \sqrt[4]{x}}, \quad a = 1$

38. $g(\theta) = \dfrac{5\cos\theta}{4\theta - 2\pi}, \quad a = \pi/2$

39. $h(t) = (1 + |t|)^{1/t}, \quad a = 0$

40. $k(x) = \dfrac{x}{1 - 2^{|x|}}, \quad a = 0$

Limits at Infinity

Find the limits in Exercises 41–54.

41. $\lim\limits_{x\to\infty} \dfrac{2x + 3}{5x + 7}$

42. $\lim\limits_{x\to-\infty} \dfrac{2x^2 + 3}{5x^2 + 7}$

43. $\lim\limits_{x\to-\infty} \dfrac{x^2 - 4x + 8}{3x^3}$

44. $\lim\limits_{x\to\infty} \dfrac{1}{x^2 - 7x + 1}$

45. $\lim\limits_{x\to-\infty} \dfrac{x^2 - 7x}{x + 1}$

46. $\lim\limits_{x\to\infty} \dfrac{x^4 + x^3}{12x^3 + 128}$

47. $\lim\limits_{x\to\infty} \dfrac{\sin x}{\lfloor x \rfloor}$ (If you have a grapher, try graphing the function for $-5 \le x \le 5$.)

48. $\lim\limits_{\theta\to\infty} \dfrac{\cos\theta - 1}{\theta}$ (If you have a grapher, try graphing $f(x) = x(\cos(1/x) - 1)$ near the origin to "see" the limit at infinity.)

49. $\lim\limits_{x\to\infty} \dfrac{x + \sin x + 2\sqrt{x}}{x + \sin x}$

50. $\lim\limits_{x\to\infty} \dfrac{x^{2/3} + x^{-1}}{x^{2/3} + \cos^2 x}$

51. $\lim\limits_{x\to\infty} e^{1/x} \cos\dfrac{1}{x}$

52. $\lim\limits_{t\to\infty} \ln\left(1 + \dfrac{1}{t}\right)$

53. $\lim\limits_{x\to-\infty} \tan^{-1} x$

54. $\lim\limits_{t\to-\infty} e^{3t} \sin^{-1}\dfrac{1}{t}$

Horizontal and Vertical Asymptotes

55. Use limits to determine the equations for all vertical asymptotes.

   a. $y = \dfrac{x^2 + 4}{x - 3}$

   b. $f(x) = \dfrac{x^2 - x - 2}{x^2 - 2x + 1}$

   c. $y = \dfrac{x^2 + x - 6}{x^2 + 2x - 8}$

56. Use limits to determine the equations for all horizontal asymptotes.

   a. $y = \dfrac{1 - x^2}{x^2 + 1}$

   b. $f(x) = \dfrac{\sqrt{x} + 4}{\sqrt{x} + 4}$

   c. $g(x) = \dfrac{\sqrt{x^2 + 4}}{x}$

   d. $y = \sqrt{\dfrac{x^2 + 9}{9x^2 + 1}}$

57. Determine the domain and range of $y = \dfrac{\sqrt{16 - x^2}}{x - 2}$.

58. Assume that constants a and b are positive. Find equations for all horizontal and vertical asymptotes for the graph of

$$y = \dfrac{\sqrt{ax^2 + 4}}{x - b}.$$

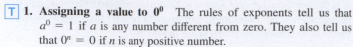

CHAPTER 2   Additional and Advanced Exercises

T **1. Assigning a value to 0^0** The rules of exponents tell us that $a^0 = 1$ if a is any number different from zero. They also tell us that $0^n = 0$ if n is any positive number.

    If we tried to extend these rules to include the case 0^0, we would get conflicting results. The first rule would say $0^0 = 1$, whereas the second would say $0^0 = 0$.

    We are not dealing with a question of right or wrong here. Neither rule applies as it stands, so there is no contradiction. We could, in fact, define 0^0 to have any value we wanted as long as we could persuade others to agree.

    What value would you like 0^0 to have? Here is an example that might help you to decide. (See Exercise 2 below for another example.)

  a. Calculate x^x for $x = 0.1, 0.01, 0.001$, and so on as far as your calculator can go. Record the values you get. What pattern do you see?

  b. Graph the function $y = x^x$ for $0 < x \leq 1$. Even though the function is not defined for $x \leq 0$, the graph will approach the y-axis from the right. Toward what y-value does it seem to be headed? Zoom in to further support your idea.

T **2. A reason you might want 0^0 to be something other than 0 or 1**
As the number x increases through positive values, the numbers $1/x$ and $1/(\ln x)$ both approach zero. What happens to the number

$$f(x) = \left(\frac{1}{x}\right)^{1/(\ln x)}$$

as x increases? Here are two ways to find out.

  a. Evaluate f for $x = 10, 100, 1000$, and so on as far as your calculator can reasonably go. What pattern do you see?

  b. Graph f in a variety of graphing windows, including windows that contain the origin. What do you see? Trace the y-values along the graph. What do you find?

3. Lorentz contraction In relativity theory, the length of an object, say a rocket, appears to an observer to depend on the speed at which the object is traveling with respect to the observer. If the observer measures the rocket's length as L_0 at rest, then at speed v the length will appear to be

$$L = L_0 \sqrt{1 - \frac{v^2}{c^2}}.$$

This equation is the Lorentz contraction formula. Here, c is the speed of light in a vacuum, about 3×10^8 m/sec. What happens to L as v increases? Find $\lim_{v \to c^-} L$. Why was the left-hand limit needed?

4. Controlling the flow from a draining tank Torricelli's law says that if you drain a tank like the one in the figure shown, the rate y at which water runs out is a constant times the square root of the water's depth x. The constant depends on the size and shape of the exit valve.

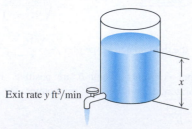

Exit rate y ft^3/min

Suppose that $y = \sqrt{x}/2$ for a certain tank. You are trying to maintain a fairly constant exit rate by adding water to the tank with a hose from time to time. How deep must you keep the water if you want to maintain the exit rate

  a. within 0.2 ft^3/min of the rate $y_0 = 1$ ft^3/min?

  b. within 0.1 ft^3/min of the rate $y_0 = 1$ ft^3/min?

5. Thermal expansion in precise equipment As you may know, most metals expand when heated and contract when cooled. The dimensions of a piece of laboratory equipment are sometimes so critical that the shop where the equipment is made must be held at the same temperature as the laboratory where the equipment is to be used. A typical aluminum bar that is 10 cm wide at 70°F will be

$$y = 10 + (t - 70) \times 10^{-4}$$

centimeters wide at a nearby temperature t. Suppose that you are using a bar like this in a gravity wave detector, where its width must stay within 0.0005 cm of the ideal 10 cm. How close to $t_0 = 70$°F must you maintain the temperature to ensure that this tolerance is not exceeded?

6. Stripes on a measuring cup The interior of a typical 1-L measuring cup is a right circular cylinder of radius 6 cm (see accompanying figure). The volume of water we put in the cup is therefore a function of the level h to which the cup is filled, the formula being

$$V = \pi 6^2 h = 36\pi h.$$

How closely must we measure h to measure out 1 L of water (1000 cm^3) with an error of no more than 1% (10 cm^3)?

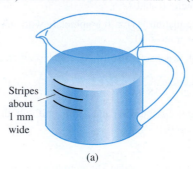

Stripes about 1 mm wide

(a)

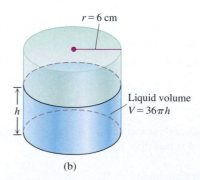

$r = 6$ cm

h

Liquid volume $V = 36\pi h$

(b)

A 1-L measuring cup (a), modeled as a right circular cylinder (b) of radius $r = 6$ cm

Precise Definition of Limit

In Exercises 7–10, use the formal definition of limit to prove that the function is continuous at c.

7. $f(x) = x^2 - 7, \quad c = 1$ **8.** $g(x) = 1/(2x), \quad c = 1/4$

9. $h(x) = \sqrt{2x - 3}, \quad c = 2$ **10.** $F(x) = \sqrt{9 - x}, \quad c = 5$

11. Uniqueness of limits Show that a function cannot have two different limits at the same point. That is, if $\lim_{x \to c} f(x) = L_1$ and $\lim_{x \to c} f(x) = L_2$, then $L_1 = L_2$.

12. Prove the limit Constant Multiple Rule:

$$\lim_{x \to c} kf(x) = k \lim_{x \to c} f(x) \quad \text{for any constant } k.$$

13. One-sided limits If $\lim_{x \to 0^+} f(x) = A$ and $\lim_{x \to 0^-} f(x) = B$, find

a. $\lim_{x \to 0^+} f(x^3 - x)$ **b.** $\lim_{x \to 0^-} f(x^3 - x)$

c. $\lim_{x \to 0^+} f(x^2 - x^4)$ **d.** $\lim_{x \to 0^-} f(x^2 - x^4)$

14. Limits and continuity Which of the following statements are true, and which are false? If true, say why; if false, give a counterexample (that is, an example confirming the falsehood).

a. If $\lim_{x \to c} f(x)$ exists but $\lim_{x \to c} g(x)$ does not exist, then $\lim_{x \to c}(f(x) + g(x))$ does not exist.

b. If neither $\lim_{x \to c} f(x)$ nor $\lim_{x \to c} g(x)$ exists, then $\lim_{x \to c}(f(x) + g(x))$ does not exist.

c. If f is continuous at x, then so is $|f|$.

d. If $|f|$ is continuous at c, then so is f.

In Exercises 15 and 16, use the formal definition of limit to prove that the function has a continuous extension to the given value of x.

15. $f(x) = \dfrac{x^2 - 1}{x + 1}, \quad x = -1$ **16.** $g(x) = \dfrac{x^2 - 2x - 3}{2x - 6}, \quad x = 3$

17. A function continuous at only one point Let

$$f(x) = \begin{cases} x, & \text{if } x \text{ is rational} \\ 0, & \text{if } x \text{ is irrational.} \end{cases}$$

a. Show that f is continuous at $x = 0$.

b. Use the fact that every nonempty open interval of real numbers contains both rational and irrational numbers to show that f is not continuous at any nonzero value of x.

18. The Dirichlet ruler function If x is a rational number, then x can be written in a unique way as a quotient of integers m/n where $n > 0$ and m and n have no common factors greater than 1. (We say that such a fraction is in *lowest terms*. For example, $6/4$ written in lowest terms is $3/2$.) Let $f(x)$ be defined for all x in the interval $[0, 1]$ by

$$f(x) = \begin{cases} 1/n, & \text{if } x = m/n \text{ is a rational number in lowest terms} \\ 0, & \text{if } x \text{ is irrational.} \end{cases}$$

For instance, $f(0) = f(1) = 1$, $f(1/2) = 1/2$, $f(1/3) = f(2/3) = 1/3$, $f(1/4) = f(3/4) = 1/4$, and so on.

a. Show that f is discontinuous at every rational number in $[0, 1]$.

b. Show that f is continuous at every irrational number in $[0, 1]$. (*Hint*: If ε is a given positive number, show that there are only finitely many rational numbers r in $[0, 1]$ such that $f(r) \geq \varepsilon$.)

c. Sketch the graph of f. Why do you think f is called the "ruler function"?

19. Antipodal points Is there any reason to believe that there is always a pair of antipodal (diametrically opposite) points on Earth's equator where the temperatures are the same? Explain.

20. If $\lim_{x \to c}(f(x) + g(x)) = 3$ and $\lim_{x \to c}(f(x) - g(x)) = -1$, find $\lim_{x \to c} f(x)g(x)$.

21. Roots of a quadratic equation that is almost linear The equation $ax^2 + 2x - 1 = 0$, where a is a constant, has two roots if $a > -1$ and $a \neq 0$, one positive and one negative:

$$r_+(a) = \frac{-1 + \sqrt{1 + a}}{a}, \qquad r_-(a) = \frac{-1 - \sqrt{1 + a}}{a},$$

a. What happens to $r_+(a)$ as $a \to 0$? As $a \to -1^+$?

b. What happens to $r_-(a)$ as $a \to 0$? As $a \to -1^+$?

c. Support your conclusions by graphing $r_+(a)$ and $r_-(a)$ as functions of a. Describe what you see.

d. For added support, graph $f(x) = ax^2 + 2x - 1$ simultaneously for $a = 1, 0.5, 0.2, 0.1$, and 0.05.

22. Root of an equation Show that the equation $x + 2 \cos x = 0$ has at least one solution.

23. Bounded functions A real-valued function f is **bounded from above** on a set D if there exists a number N such that $f(x) \leq N$ for all x in D. We call N, when it exists, an **upper bound** for f on D and say that f is bounded from above by N. In a similar manner, we say that f is **bounded from below** on D if there exists a number M such that $f(x) \geq M$ for all x in D. We call M, when it exists, a **lower bound** for f on D and say that f is bounded from below by M. We say that f is **bounded** on D if it is bounded from both above and below.

a. Show that f is bounded on D if and only if there exists a number B such that $|f(x)| \leq B$ for all x in D.

b. Suppose that f is bounded from above by N. Show that if $\lim_{x \to c} f(x) = L$, then $L \leq N$.

c. Suppose that f is bounded from below by M. Show that if $\lim_{x \to c} f(x) = L$, then $L \geq M$.

24. Max $\{a, b\}$ and min $\{a, b\}$

a. Show that the expression

$$\max \{a, b\} = \frac{a + b}{2} + \frac{|a - b|}{2}$$

equals a if $a \geq b$ and equals b if $b \geq a$. In other words, $\max \{a, b\}$ gives the larger of the two numbers a and b.

b. Find a similar expression for $\min \{a, b\}$, the smaller of a and b.

Generalized Limits Involving $\dfrac{\sin \theta}{\theta}$

The formula $\lim_{\theta \to 0}(\sin \theta)/\theta = 1$ can be generalized. If $\lim_{x \to c} f(x) = 0$ and $f(x)$ is never zero in an open interval containing the point $x = c$, except possibly at c itself, then

$$\lim_{x \to c} \frac{\sin f(x)}{f(x)} = 1.$$

Here are several examples.

a. $\lim_{x \to 0} \dfrac{\sin x^2}{x^2} = 1$

b. $\lim_{x \to 0} \dfrac{\sin x^2}{x} = \lim_{x \to 0} \dfrac{\sin x^2}{x^2} \lim_{x \to 0} \dfrac{x^2}{x} = 1 \cdot 0 = 0$

c. $\lim\limits_{x \to -1} \dfrac{\sin(x^2 - x - 2)}{x + 1} = \lim\limits_{x \to -1} \dfrac{\sin(x^2 - x - 2)}{(x^2 - x - 2)}.$

$\lim\limits_{x \to -1} \dfrac{(x^2 - x - 2)}{x + 1} = 1 \cdot \lim\limits_{x \to -1} \dfrac{(x + 1)(x - 2)}{x + 1} = -3$

d. $\lim\limits_{x \to 1} \dfrac{\sin\left(1 - \sqrt{x}\right)}{x - 1} = \lim\limits_{x \to 1} \dfrac{\sin\left(1 - \sqrt{x}\right)}{1 - \sqrt{x}} \dfrac{1 - \sqrt{x}}{x - 1}$

$= \lim\limits_{x \to 1} \dfrac{\left(1 - \sqrt{x}\right)\left(1 + \sqrt{x}\right)}{(x - 1)\left(1 + \sqrt{x}\right)}$

$= \lim\limits_{x \to 1} \dfrac{1 - x}{(x - 1)\left(1 + \sqrt{x}\right)} = -\dfrac{1}{2}$

Find the limits in Exercises 25–30.

25. $\lim\limits_{x \to 0} \dfrac{\sin(1 - \cos x)}{x}$

26. $\lim\limits_{x \to 0^+} \dfrac{\sin x}{\sin\sqrt{x}}$

27. $\lim\limits_{x \to 0} \dfrac{\sin(\sin x)}{x}$

28. $\lim\limits_{x \to 0} \dfrac{\sin(x^2 + x)}{x}$

29. $\lim\limits_{x \to 2} \dfrac{\sin(x^2 - 4)}{x - 2}$

30. $\lim\limits_{x \to 9} \dfrac{\sin\left(\sqrt{x} - 3\right)}{x - 9}$

Oblique Asymptotes

Find all possible oblique asymptotes in Exercises 31–34.

31. $y = \dfrac{2x^{3/2} + 2x - 3}{\sqrt{x} + 1}$

32. $y = x + x \sin\dfrac{1}{x}$

33. $y = \sqrt{x^2 + 1}$

34. $y = \sqrt{x^2 + 2x}$

Showing an Equation Is Solvable

35. Assume that $1 < a < b$ and $\dfrac{a}{x} + x = \dfrac{1}{x - b}$. Show that this equation is solvable.

More Limits

36. Find constants a and b so that each of the following limits is true.

a. $\lim\limits_{x \to 0} \dfrac{\sqrt{a + bx} - 1}{x} = 2$ **b.** $\lim\limits_{x \to 1} \dfrac{\tan(ax - a) + b - 2}{x - 1} = 3$

37. Evaluate $\lim\limits_{x \to 1} \dfrac{x^{2/3} - 1}{1 - \sqrt{x}}.$ **38.** Evaluate $\lim\limits_{x \to 0} \dfrac{|3x + 4| - |x| - 4}{x}.$

Limits on Arbitrary Domains

The definition of the limit of a function at $x = c$ extends to functions whose domains near c are more complicated than intervals.

> **General Definition of Limit**
>
> Suppose every open interval containing c contains a point other than c in the domain of f. We say that $\lim\limits_{x \to c} f(x) = L$ if for every number $\varepsilon > 0$ there exists a corresponding number $\delta > 0$ such that for all x in the domain of f, $|f(x) - L| < \varepsilon$ whenever $0 < |x - c| < \delta$.

For the functions in Exercises 39–42,

    a. Find the domain.

    b. Show that at $c = 0$ the domain has the property described above.

    c. Evaluate $\lim\limits_{x \to 0} f(x)$.

39. The function f is defined as follows: $f(x) = x$ if $x = 1/n$ where n is a positive integer, and $f(0) = 1$.

40. The function f is defined as follows: $f(x) = 1 - x$ if $x = 1/n$ where n is a positive integer, and $f(0) = 1$.

41. $f(x) = \sqrt{x} \sin(1/x)$

42. $f(x) = \sqrt{\ln(\sin(1/x))}$

43. Let g be a function with domain the rational numbers, defined by

$$g(x) = \dfrac{2}{x - \sqrt{2}} \text{ for rational } x.$$

    a. Sketch the graph of g as well as you can, keeping in mind that g is only defined at rational points.

    b. Use the general definition of a limit to prove that $\lim\limits_{x \to 0} g(x) = -\sqrt{2}$.

    c. Prove that g is continuous at the point $x = 0$ by showing that the limit in part (b) equals $g(0)$.

    d. Is g continuous at other points of its domain?

CHAPTER 2 Technology Application Projects

Mathematica/Maple Projects

Projects can be found within MyMathLab.

Take It to the Limit
Part I
Part II (Zero Raised to the Power Zero: What Does It Mean?)
Part III (One-Sided Limits)
Visualize and interpret the limit concept through graphical and numerical explorations.
Part IV (What a Difference a Power Makes)
See how sensitive limits can be with various powers of x.

Going to Infinity
Part I (Exploring Function Behavior as $x \to \infty$ or $x \to -\infty$)
This module provides four examples to explore the behavior of a function as $x \to \infty$ or $x \to -\infty$.
Part II (Rates of Growth)
Observe graphs that *appear* to be continuous, yet the function is not continuous. Several issues of continuity are explored to obtain results that you may find surprising.

3

Derivatives

OVERVIEW In Chapter 2 we discussed how to determine the slope of a curve at a point and how to measure the rate at which a function changes. Now that we have studied limits, we can make these notions precise and see that both are interpretations of the *derivative* of a function at a point. We then extend this concept from a single point to the *derivative function*, and we develop rules for finding this derivative function easily, without having to calculate limits directly. These rules are used to find derivatives of most of the common functions reviewed in Chapter 1, as well as combinations of them.

The derivative is used to study a wide range of problems in mathematics, science, economics, and medicine. These problems include finding solutions to very general equations, calculating the velocity and acceleration of a moving object, describing the path followed by a light ray going from a point in air to a point in water, finding the number of items a manufacturing company should produce in order to maximize its profits, studying the spread of an infectious disease within a given population, and calculating the amount of blood the heart pumps per minute based on how well the lungs are functioning.

3.1 Tangent Lines and the Derivative at a Point

In this section we define the slope and tangent to a curve at a point, and the derivative of a function at a point. The derivative gives a way to find both the slope of a graph and the instantaneous rate of change of a function.

Finding a Tangent Line to the Graph of a Function

To find a tangent line to an arbitrary curve $y = f(x)$ at a point $P(x_0, f(x_0))$, we use the procedure introduced in Section 2.1. We calculate the slope of the secant line through P and a nearby point $Q(x_0 + h, f(x_0 + h))$. We then investigate the limit of the slope as $h \to 0$ (Figure 3.1). If the limit exists, we call it the slope of the curve at P and define the tangent line at P to be the line through P having this slope.

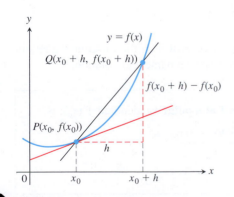

FIGURE 3.1 The slope of the tangent line at P is $\lim\limits_{h \to 0} \dfrac{f(x_0 + h) - f(x_0)}{h}$.

DEFINITIONS The **slope of the curve** $y = f(x)$ at the point $P(x_0, f(x_0))$ is the number

$$\lim_{h \to 0} \frac{f(x_0 + h) - f(x_0)}{h} \qquad \text{(provided the limit exists)}.$$

The **tangent line** to the curve at P is the line through P with this slope.

In Section 2.1, Example 3, we applied these definitions to find the slope of the parabola $f(x) = x^2$ at the point $P(2, 4)$ and the tangent line to the parabola at P. Let's look at another example.

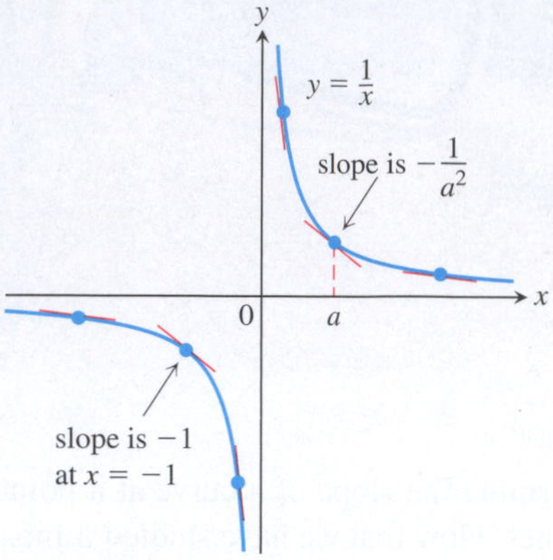

EXAMPLE 1

(a) Find the slope of the curve $y = 1/x$ at any point $x = a \neq 0$. What is the slope at the point $x = -1$?

(b) Where does the slope equal $-1/4$?

(c) What happens to the tangent line to the curve at the point $(a, 1/a)$ as a changes?

Solution

(a) Here $f(x) = 1/x$. The slope at $(a, 1/a)$ is

$$\lim_{h \to 0} \frac{f(a + h) - f(a)}{h} = \lim_{h \to 0} \frac{\dfrac{1}{a + h} - \dfrac{1}{a}}{h} = \lim_{h \to 0} \frac{1}{h} \frac{a - (a + h)}{a(a + h)}$$

$$= \lim_{h \to 0} \frac{-h}{ha(a + h)} = \lim_{h \to 0} \frac{-1}{a(a + h)} = -\frac{1}{a^2}.$$

Notice how we had to keep writing "$\lim_{h \to 0}$" before each fraction until the stage at which we could evaluate the limit by substituting $h = 0$. The number a may be positive or negative, but not 0. When $a = -1$, the slope is $-1/(-1)^2 = -1$ (Figure 3.2).

(b) The slope of $y = 1/x$ at the point where $x = a$ is $-1/a^2$. It will be $-1/4$ provided that

$$-\frac{1}{a^2} = -\frac{1}{4}.$$

This equation is equivalent to $a^2 = 4$, so $a = 2$ or $a = -2$. The curve has slope $-1/4$ at the two points $(2, 1/2)$ and $(-2, -1/2)$ (Figure 3.3).

(c) The slope $-1/a^2$ is always negative if $a \neq 0$. As $a \to 0^+$, the slope approaches $-\infty$ and the tangent line becomes increasingly steep (Figure 3.2). We see this situation again as $a \to 0^-$. As a moves away from the origin in either direction, the slope approaches 0 and the tangent line levels off, becoming more and more horizontal. ∎

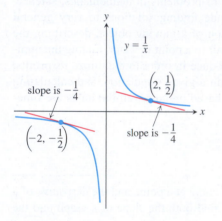

FIGURE 3.2 The tangent line slopes, steep near the origin, become more gradual as the point of tangency moves away (Example 1).

FIGURE 3.3 The two tangent lines to $y = 1/x$ having slope $-1/4$ (Example 1).

Rates of Change: Derivative at a Point

The expression

$$\frac{f(x_0 + h) - f(x_0)}{h}, \quad h \neq 0$$

is called the **difference quotient of f at x_0 with increment h**. If the difference quotient has a limit as h approaches zero, that limit is given a special name and notation.

> **DEFINITION** The **derivative of a function f at a point x_0**, denoted $f'(x_0)$, is
>
> $$f'(x_0) = \lim_{h \to 0} \frac{f(x_0 + h) - f(x_0)}{h}$$
>
> provided this limit exists.

The notation $f'(x_0)$ is read "f prime of x_0."

The derivative has more than one meaning, depending on what problem we are considering. The formula for the derivative is the same as the formula for the slope of the curve $y = f(x)$ at a point. If we interpret the difference quotient as the slope of a secant line, then the derivative gives the slope of the curve $y = f(x)$ at the point $P(x_0, f(x_0))$. If we interpret the difference quotient as an average rate of change (Section 2.1), then the derivative gives the function's instantaneous rate of change with respect to x at the point $x = x_0$. We study this interpretation in Section 3.4.

EXAMPLE 2   In Examples 1 and 2 in Section 2.1, we studied the speed of a rock falling freely from rest near the surface of the earth. The rock fell $y = 16t^2$ feet during the first t sec, and we used a sequence of average rates over increasingly short intervals to estimate the rock's speed at the instant $t = 1$. What was the rock's *exact* speed at this time?

Solution   We let $f(t) = 16t^2$. The average speed of the rock over the interval between $t = 1$ and $t = 1 + h$ seconds, for $h > 0$, was found to be

$$\frac{f(1 + h) - f(1)}{h} = \frac{16(1 + h)^2 - 16(1)^2}{h} = \frac{16(h^2 + 2h)}{h} = 16(h + 2).$$

The rock's speed at the instant $t = 1$ is then

$$f'(1) = \lim_{h \to 0} \frac{f(1 + h) - f(1)}{h} = \lim_{h \to 0} 16(h + 2) = 16(0 + 2) = 32 \text{ ft/sec.} \quad \blacksquare$$

Summary

We have been discussing slopes of curves, lines tangent to a curve, the rate of change of a function, and the derivative of a function at a point. All of these ideas are based on the same limit.

The following are all interpretations for the limit of the difference quotient

$$\lim_{h \to 0} \frac{f(x_0 + h) - f(x_0)}{h}.$$

1. The slope of the graph of $y = f(x)$ at $x = x_0$
2. The slope of the tangent line to the curve $y = f(x)$ at $x = x_0$
3. Rate of change of $f(x)$ with respect to x at the $x = x_0$
4. The derivative $f'(x_0)$ at $x = x_0$

In the next sections, we allow the point x_0 to vary across the domain of the function f.

EXERCISES 3.1

Slopes and Tangent Lines
In Exercises 1–4, use the grid and a straight edge to make a rough estimate of the slope of the curve (in y-units per x-unit) at the points P_1 and P_2.

1.
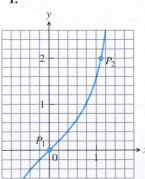

2.
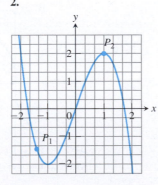

3.
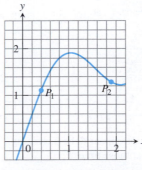

4.
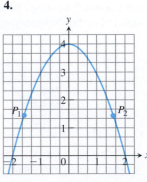

In Exercises 5–10, find an equation for the tangent line to the curve at the given point. Then sketch the curve and tangent line together.

5. $y = 4 - x^2$, $(-1, 3)$ **6.** $y = (x - 1)^2 + 1$, $(1, 1)$

7. $y = 2\sqrt{x}$, $(1, 2)$ **8.** $y = \dfrac{1}{x^2}$, $(-1, 1)$

9. $y = x^3$, $(-2, -8)$ **10.** $y = \dfrac{1}{x^3}$, $\left(-2, -\dfrac{1}{8}\right)$

In Exercises 11–18, find the slope of the function's graph at the given point. Then find an equation for the line tangent to the graph there.

11. $f(x) = x^2 + 1$, $(2, 5)$ **12.** $f(x) = x - 2x^2$, $(1, -1)$

13. $g(x) = \dfrac{x}{x - 2}$, $(3, 3)$ **14.** $g(x) = \dfrac{8}{x^2}$, $(2, 2)$

15. $h(t) = t^3$, $(2, 8)$ **16.** $h(t) = t^3 + 3t$, $(1, 4)$

17. $f(x) = \sqrt{x}$, $(4, 2)$ **18.** $f(x) = \sqrt{x + 1}$, $(8, 3)$

In Exercises 19–22, find the slope of the curve at the point indicated.

19. $y = 5x - 3x^2$, $x = 1$ **20.** $y = x^3 - 2x + 7$, $x = -2$

21. $y = \dfrac{1}{x - 1}$, $x = 3$ **22.** $y = \dfrac{x - 1}{x + 1}$, $x = 0$

Interpreting Derivative Values

23. Growth of yeast cells   In a controlled laboratory experiment, yeast cells are grown in an automated cell culture system that counts the number P of cells present at hourly intervals. The number after t hours is shown in the accompanying figure.

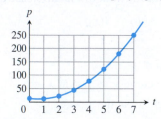

a. Explain what is meant by the derivative $P'(5)$. What are its units?

b. Which is larger, $P'(2)$ or $P'(3)$? Give a reason for your answer.

c. The quadratic curve capturing the trend of the data points (see Section 1.4) is given by $P(t) = 6.10t^2 - 9.28t + 16.43$. Find the instantaneous rate of growth when $t = 5$ hours.

24. Effectiveness of a drug   On a scale from 0 to 1, the effectiveness E of a pain-killing drug t hours after entering the bloodstream is displayed in the accompanying figure.

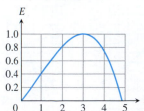

a. At what times does the effectiveness appear to be increasing? What is true about the derivative at those times?

b. At what time would you estimate that the drug reaches its maximum effectiveness? What is true about the derivative at that time? What is true about the derivative as time increases in the 1 hour *before* your estimated time?

At what points do the graphs of the functions in Exercises 25 and 26 have horizontal tangent lines?

25. $f(x) = x^2 + 4x - 1$ **26.** $g(x) = x^3 - 3x$

27. Find equations of all lines having slope -1 that are tangent to the curve $y = 1/(x - 1)$.

28. Find an equation of the straight line having slope $1/4$ that is tangent to the curve $y = \sqrt{x}$.

Rates of Change

29. Object dropped from a tower   An object is dropped from the top of a 100-m-high tower. Its height above ground after t sec is $100 - 4.9t^2$ m. How fast is it falling 2 sec after it is dropped?

30. Speed of a rocket   At t sec after liftoff, the height of a rocket is $3t^2$ ft. How fast is the rocket climbing 10 sec after liftoff?

31. Circle's changing area   What is the rate of change of the area of a circle $(A = \pi r^2)$ with respect to the radius when the radius is $r = 3$?

32. Ball's changing volume   What is the rate of change of the volume of a ball $(V = (4/3)\pi r^3)$ with respect to the radius when the radius is $r = 2$?

33. Show that the line $y = mx + b$ is its own tangent line at any point $(x_0, mx_0 + b)$.

34. Find the slope of the tangent line to the curve $y = 1/\sqrt{x}$ at the point where $x = 4$.

Testing for Tangent Lines

35. Does the graph of

$$f(x) = \begin{cases} x^2 \sin(1/x), & x \neq 0 \\ 0, & x = 0 \end{cases}$$

have a tangent line at the origin? Give reasons for your answer.

36. Does the graph of

$$g(x) = \begin{cases} x \sin(1/x), & x \neq 0 \\ 0, & x = 0 \end{cases}$$

have a tangent line at the origin? Give reasons for your answer.

Vertical Tangent Lines

We say that a continuous curve $y = f(x)$ has a **vertical tangent line** at the point where $x = x_0$ if the limit of the difference quotient is ∞ or $-\infty$. For example, $y = x^{1/3}$ has a vertical tangent line at $x = 0$ (see accompanying figure):

$$\lim_{h \to 0} \frac{f(0 + h) - f(0)}{h} = \lim_{h \to 0} \frac{h^{1/3} - 0}{h}$$

$$= \lim_{h \to 0} \frac{1}{h^{2/3}} = \infty.$$

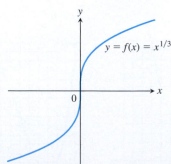

VERTICAL TANGENT LINE AT ORIGIN

However, $y = x^{2/3}$ has *no* vertical tangent line at $x = 0$ (see next figure):

$$\lim_{h \to 0} \frac{g(0 + h) - g(0)}{h} = \lim_{h \to 0} \frac{h^{2/3} - 0}{h}$$

$$= \lim_{h \to 0} \frac{1}{h^{1/3}}$$

does not exist, because the limit is ∞ from the right and $-\infty$ from the left.

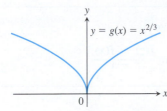

$y = g(x) = x^{2/3}$

NO VERTICAL TANGENT LINE AT ORIGIN

37. Does the graph of

$$f(x) = \begin{cases} -1, & x < 0 \\ 0, & x = 0 \\ 1, & x > 0 \end{cases}$$

have a vertical tangent line at the origin? Give reasons for your answer.

38. Does the graph of

$$U(x) = \begin{cases} 0, & x < 0 \\ 1, & x \geq 0 \end{cases}$$

have a vertical tangent line at the point $(0, 1)$? Give reasons for your answer.

T Graph the curves in Exercises 39–48.

a. Where do the graphs appear to have vertical tangent lines?

b. Confirm your findings in part (a) with limit calculations. But before you do, read the introduction to Exercises 37 and 38.

39. $y = x^{2/5}$

40. $y = x^{4/5}$

41. $y = x^{1/5}$

42. $y = x^{3/5}$

43. $y = 4x^{2/5} - 2x$

44. $y = x^{5/3} - 5x^{2/3}$

45. $y = x^{2/3} - (x - 1)^{1/3}$

46. $y = x^{1/3} + (x - 1)^{1/3}$

47. $y = \begin{cases} -\sqrt{|x|}, & x \leq 0 \\ \sqrt{x}, & x > 0 \end{cases}$

48. $y = \sqrt{|4 - x|}$

COMPUTER EXPLORATIONS

Use a CAS to perform the following steps for the functions in Exercises 49–52:

a. Plot $y = f(x)$ over the interval $(x_0 - 1/2) \leq x \leq (x_0 + 3)$.

b. Holding x_0 fixed, the difference quotient

$$q(h) = \frac{f(x_0 + h) - f(x_0)}{h}$$

at x_0 becomes a function of the step size h. Enter this function into your CAS workspace.

c. Find the limit of q as $h \to 0$.

d. Define the secant lines $y = f(x_0) + q \cdot (x - x_0)$ for $h = 3, 2,$ and 1. Graph them together with f and the tangent line over the interval in part (a).

49. $f(x) = x^3 + 2x$, $x_0 = 0$

50. $f(x) = x + \dfrac{5}{x}$, $x_0 = 1$

51. $f(x) = x + \sin(2x)$, $x_0 = \pi/2$

52. $f(x) = \cos x + 4\sin(2x)$, $x_0 = \pi$

3.2 The Derivative as a Function

In the last section we defined the derivative of $y = f(x)$ at the point $x = x_0$ to be the limit

$$f'(x_0) = \lim_{h \to 0} \frac{f(x_0 + h) - f(x_0)}{h}.$$

We now investigate the derivative as a *function* derived from f by considering the limit at each point x in the domain of f.

DEFINITION The **derivative** of the function $f(x)$ with respect to the variable x is the function f' whose value at x is

$$f'(x) = \lim_{h \to 0} \frac{f(x + h) - f(x)}{h},$$

provided the limit exists.

We use the notation $f(x)$ in the definition, rather than $f(x_0)$ as before, to emphasize that f' is a function of the independent variable x with respect to which the derivative function $f'(x)$ is being defined. The domain of f' is the set of points in the domain of f for

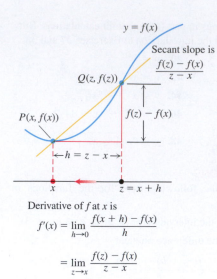

$y = f(x)$

Secant slope is
$$\frac{f(z) - f(x)}{z - x}$$

$Q(z, f(z))$

$P(x, f(x))$

$f(z) - f(x)$

$\leftarrow h = z - x \rightarrow$

x $z = x + h$

Derivative of f at x is
$$f'(x) = \lim_{h \to 0} \frac{f(x + h) - f(x)}{h}$$

$$= \lim_{z \to x} \frac{f(z) - f(x)}{z - x}$$

FIGURE 3.4 Two forms for the difference quotient.

Derivative of the Reciprocal Function

$$\frac{d}{dx}\left(\frac{1}{x}\right) = -\frac{1}{x^2}, \quad x \neq 0$$

---

which the limit exists, which means that the domain may be the same as or smaller than the domain of f. If f' exists at a particular x, we say that f is **differentiable (has a derivative) at x**. If f' exists at every point in the domain of f, we call f **differentiable**.

If we write $z = x + h$, then $h = z - x$ and h approaches 0 if and only if z approaches x. Therefore, an equivalent definition of the derivative is as follows (see Figure 3.4). This formula is sometimes more convenient to use when finding a derivative function, and focuses on the point z that approaches x.

Alternative Formula for the Derivative

$$f'(x) = \lim_{z \to x} \frac{f(z) - f(x)}{z - x}$$

Calculating Derivatives from the Definition

The process of calculating a derivative is called **differentiation**. To emphasize the idea that differentiation is an operation performed on a function $y = f(x)$, we use the notation

$$\frac{d}{dx} f(x)$$

as another way to denote the derivative $f'(x)$. Example 1 of Section 3.1 illustrated the differentiation process for the function $y = 1/x$ when $x = a$. For x representing any point in the domain, we get the formula

$$\frac{d}{dx}\left(\frac{1}{x}\right) = -\frac{1}{x^2}.$$

Here are two more examples in which we allow x to be any point in the domain of f.

EXAMPLE 1   Differentiate $f(x) = \dfrac{x}{x - 1}$.

Solution   We use the definition of derivative, which requires us to calculate $f(x + h)$ and then subtract $f(x)$ to obtain the numerator in the difference quotient. We have

$$f(x) = \frac{x}{x - 1} \quad \text{and} \quad f(x + h) = \frac{(x + h)}{(x + h) - 1}, \text{ so}$$

$$f'(x) = \lim_{h \to 0} \frac{f(x + h) - f(x)}{h} \qquad \text{Definition}$$

$$= \lim_{h \to 0} \frac{\dfrac{x + h}{x + h - 1} - \dfrac{x}{x - 1}}{h} \qquad \text{Substitute.}$$

$$= \lim_{h \to 0} \frac{1}{h} \cdot \frac{(x + h)(x - 1) - x(x + h - 1)}{(x + h - 1)(x - 1)} \qquad \frac{a}{b} - \frac{c}{d} = \frac{ad - cb}{bd}$$

$$= \lim_{h \to 0} \frac{1}{h} \cdot \frac{-h}{(x + h - 1)(x - 1)} \qquad \text{Simplify.}$$

$$= \lim_{h \to 0} \frac{-1}{(x + h - 1)(x - 1)} = \frac{-1}{(x - 1)^2}. \qquad \text{Cancel } h \neq 0 \text{ and evaluate.} \blacksquare$$

EXAMPLE 2

(a) Find the derivative of $f(x) = \sqrt{x}$ for $x > 0$.

(b) Find the tangent line to the curve $y = \sqrt{x}$ at $x = 4$.

Derivative of the Square Root Function

$$\frac{d}{dx}\sqrt{x} = \frac{1}{2\sqrt{x}}, \quad x > 0$$

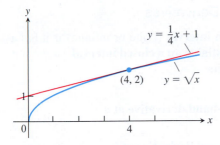

FIGURE 3.5 The curve $y = \sqrt{x}$ and its tangent line at $(4, 2)$. The tangent line's slope is found by evaluating the derivative at $x = 4$ (Example 2).

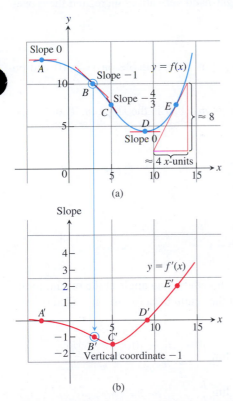

(a)

(b)

FIGURE 3.6 We made the graph of $y = f'(x)$ in (b) by plotting slopes from the graph of $y = f(x)$ in (a). The vertical coordinate of B' is the slope at B and so on. The slope at E is approximately $8/4 = 2$. In (b) we see that the rate of change of f is negative for x between A' and D'; the rate of change is positive for x to the right of D'.

Solution

(a) We use the alternative formula to calculate f':

$$f'(x) = \lim_{z \to x} \frac{f(z) - f(x)}{z - x}$$

$$= \lim_{z \to x} \frac{\sqrt{z} - \sqrt{x}}{z - x}$$

$$= \lim_{z \to x} \frac{\sqrt{z} - \sqrt{x}}{\left(\sqrt{z} - \sqrt{x}\right)\left(\sqrt{z} + \sqrt{x}\right)} \qquad \frac{1}{a^2 - b^2} = \frac{1}{(a-b)(a+b)}$$

$$= \lim_{z \to x} \frac{1}{\sqrt{z} + \sqrt{x}} = \frac{1}{2\sqrt{x}}. \qquad \text{Cancel and evaluate.}$$

(b) The slope of the curve at $x = 4$ is

$$f'(4) = \frac{1}{2\sqrt{4}} = \frac{1}{4}.$$

The tangent is the line through the point $(4, 2)$ with slope $1/4$ (Figure 3.5):

$$y = 2 + \frac{1}{4}(x - 4)$$

$$y = \frac{1}{4}x + 1.$$

Notation

There are many ways to denote the derivative of a function $y = f(x)$, where the independent variable is x and the dependent variable is y. Some common alternative notations for the derivative include

$$f'(x) = y' = \frac{dy}{dx} = \frac{df}{dx} = \frac{d}{dx}f(x) = D(f)(x) = D_x f(x).$$

The symbols d/dx and D indicate the operation of differentiation. We read dy/dx as "the derivative of y with respect to x," and df/dx and $(d/dx)f(x)$ as "the derivative of f with respect to x." The "prime" notations y' and f' originate with Newton. The d/dx notations are similar to those used by Leibniz. The symbol dy/dx should not be regarded as a ratio; it is simply a notation that denotes a derivative.

To indicate the value of a derivative at a specified number $x = a$, we use the notation

$$f'(a) = \frac{dy}{dx}\bigg|_{x=a} = \frac{df}{dx}\bigg|_{x=a} = \frac{d}{dx}f(x)\bigg|_{x=a}.$$

For instance, in Example 2

$$f'(4) = \frac{d}{dx}\sqrt{x}\bigg|_{x=4} = \frac{1}{2\sqrt{x}}\bigg|_{x=4} = \frac{1}{2\sqrt{4}} = \frac{1}{4}.$$

Graphing the Derivative

We can often make an approximate plot of the derivative of $y = f(x)$ by estimating the slopes on the graph of f. That is, we plot the points $(x, f'(x))$ in the xy-plane and connect them with a smooth curve, which represents $y = f'(x)$.

EXAMPLE 3   Graph the derivative of the function $y = f(x)$ in Figure 3.6a.

Solution   We sketch the tangent lines to the graph of f at frequent intervals and use their slopes to estimate the values of $f'(x)$ at these points. We plot the corresponding $(x, f'(x))$ pairs and connect them with a smooth curve as sketched in Figure 3.6b.

What can we learn from the graph of $y = f'(x)$? At a glance we can see

1. where the rate of change of f is positive, negative, or zero;

2. the rough size of the growth rate at any x and its size in relation to the size of $f(x)$;

3. where the rate of change itself is increasing or decreasing.

Differentiable on an Interval; One-Sided Derivatives

A function $y = f(x)$ is **differentiable on an open interval** (finite or infinite) if it has a derivative at each point of the interval. It is **differentiable on a closed interval** $[a, b]$ if it is differentiable on the interior (a, b) and if the limits

$$\lim_{h \to 0^+} \frac{f(a + h) - f(a)}{h} \qquad \textbf{Right-hand derivative at } a$$

$$\lim_{h \to 0^-} \frac{f(b + h) - f(b)}{h} \qquad \textbf{Left-hand derivative at } b$$

exist at the endpoints (Figure 3.7).

Right-hand and left-hand derivatives may or may not be defined at any point of a function's domain. Because of Theorem 6, Section 2.4, a function has a derivative at an interior point if and only if it has left-hand and right-hand derivatives there, and these one-sided derivatives are equal.

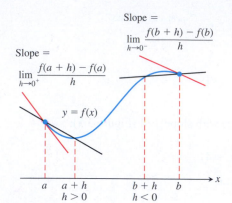

FIGURE 3.7 Derivatives at endpoints of a closed interval are one-sided limits.

EXAMPLE 4   Show that the function $y = |x|$ is differentiable on $(-\infty, 0)$ and on $(0, \infty)$ but has no derivative at $x = 0$.

Solution   From Section 3.1, the derivative of $y = mx + b$ is the slope m. Thus, to the right of the origin, when $x > 0$,

$$\frac{d}{dx}(|x|) = \frac{d}{dx}(x) = \frac{d}{dx}(1 \cdot x) = 1. \qquad \frac{d}{dx}(mx + b) = m, |x| = x \text{ since } x > 0$$

To the left, when $x < 0$,

$$\frac{d}{dx}(|x|) = \frac{d}{dx}(-x) = \frac{d}{dx}(-1 \cdot x) = -1 \qquad |x| = -x \text{ since } x < 0$$

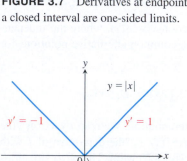

FIGURE 3.8   The function $y = |x|$ is not differentiable at the origin where the graph has a "corner" (Example 4).

(Figure 3.8). The two branches of the graph come together at an angle at the origin, forming a non-smooth corner. There is no derivative at the origin because the one-sided derivatives differ there:

$$\text{Right-hand derivative of } |x| \text{ at zero} = \lim_{h \to 0^+} \frac{|0 + h| - |0|}{h} = \lim_{h \to 0^+} \frac{|h|}{h}$$

$$= \lim_{h \to 0^+} \frac{h}{h} \qquad |h| = h \text{ when } h > 0$$

$$= \lim_{h \to 0^+} 1 = 1$$

$$\text{Left-hand derivative of } |x| \text{ at zero} = \lim_{h \to 0^-} \frac{|0 + h| - |0|}{h} = \lim_{h \to 0^-} \frac{|h|}{h}$$

$$= \lim_{h \to 0^-} \frac{-h}{h} \qquad |h| = -h \text{ when } h < 0$$

$$= \lim_{h \to 0^-} -1 = -1.$$

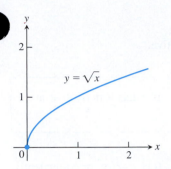

FIGURE 3.9 The square root function is not differentiable at $x = 0$, where the graph of the function has a vertical tangent line.

EXAMPLE 5   In Example 2 we found that for $x > 0$,

$$\frac{d}{dx}\sqrt{x} = \frac{1}{2\sqrt{x}}.$$

We apply the definition to examine if the derivative exists at $x = 0$:

$$\lim_{h \to 0^+} \frac{\sqrt{0 + h} - \sqrt{0}}{h} = \lim_{h \to 0^+} \frac{1}{\sqrt{h}} = \infty.$$

Since the (right-hand) limit is not finite, there is no derivative at $x = 0$. Since the slopes of the secant lines joining the origin to the points $(h, \sqrt{h})$ on a graph of $y = \sqrt{x}$ approach ∞, the graph has a *vertical tangent line* at the origin. (See Figure 3.9.)   ■

When Does a Function *Not* Have a Derivative at a Point?

A function has a derivative at a point x_0 if the slopes of the secant lines through $P(x_0, f(x_0))$ and a nearby point Q on the graph approach a finite limit as Q approaches P. Thus differentiability is a "smoothness" condition on the graph of f. A function can fail to have a derivative at a point for many reasons, including the existence of points where the graph has

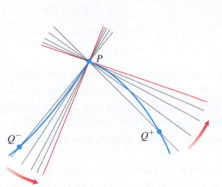

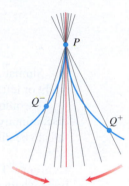

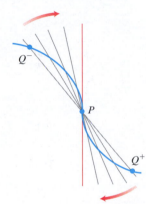

1. a *corner*, where the one-sided derivatives differ

2. a *cusp*, where the slope of PQ approaches ∞ from one side and $-\infty$ from the other

3. a *vertical tangent line*, where the slope of PQ approaches ∞ from both sides or approaches $-\infty$ from both sides (here, $-\infty$)

$$y = \begin{cases} x \sin \frac{1}{x}, & x \neq 0 \\ 0, & x = 0 \end{cases}$$

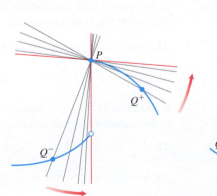

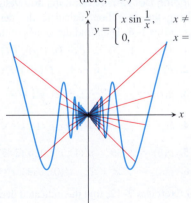

4. a *discontinuity* (two examples shown)

5. wild oscillation

The last example shows a function that is continuous at $x = 0$, but whose graph oscillates wildly up and down as it approaches $x = 0$. The slopes of the secant lines through 0 oscillate between -1 and 1 as x approaches 0, and do not have a limit at $x = 0$.

Differentiable Functions Are Continuous

A function is continuous at every point where it has a derivative.

THEOREM 1—Differentiability Implies Continuity   If f has a derivative at $x = c$, then f is continuous at $x = c$.

Proof   Given that $f'(c)$ exists, we must show that $\lim_{x \to c} f(x) = f(c)$, or equivalently, that $\lim_{h \to 0} f(c + h) = f(c)$. If $h \neq 0$, then

$$f(c + h) = f(c) + (f(c + h) - f(c)) \qquad \text{Add and subtract } f(c).$$

$$= f(c) + \frac{f(c + h) - f(c)}{h} \cdot h. \qquad \text{Divide and multiply by } h.$$

Now take limits as $h \to 0$. By Theorem 1 of Section 2.2,

$$\lim_{h \to 0} f(c + h) = \lim_{h \to 0} f(c) + \lim_{h \to 0} \frac{f(c + h) - f(c)}{h} \cdot \lim_{h \to 0} h$$

$$= f(c) + f'(c) \cdot 0$$

$$= f(c) + 0$$

$$= f(c). \qquad\blacksquare$$

Similar arguments with one-sided limits show that if f has a derivative from one side (right or left) at $x = c$, then f is continuous from that side at $x = c$.

Theorem 1 says that if a function has a discontinuity at a point (for instance, a jump discontinuity), then it cannot be differentiable there. The greatest integer function $y = \lfloor x \rfloor$ fails to be differentiable at every integer $x = n$ (Example 4, Section 2.5).

Caution   The converse of Theorem 1 is false. A function need not have a derivative at a point where it is continuous, as we saw with the absolute value function in Example 4. ●

EXERCISES   3.2

Finding Derivative Functions and Values

Using the definition, calculate the derivatives of the functions in Exercises 1–6. Then find the values of the derivatives as specified.

1. $f(x) = 4 - x^2$; $f'(-3), f'(0), f'(1)$

2. $F(x) = (x - 1)^2 + 1$; $F'(-1), F'(0), F'(2)$

3. $g(t) = \dfrac{1}{t^2}$; $g'(-1), g'(2), g'(\sqrt{3})$

4. $k(z) = \dfrac{1 - z}{2z}$; $k'(-1), k'(1), k'(\sqrt{2})$

5. $p(\theta) = \sqrt{3\theta}$; $p'(1), p'(3), p'(2/3)$

6. $r(s) = \sqrt{2s + 1}$; $r'(0), r'(1), r'(1/2)$

In Exercises 7–12, find the indicated derivatives.

7. $\dfrac{dy}{dx}$ if $y = 2x^3$

8. $\dfrac{dr}{ds}$ if $r = s^3 - 2s^2 + 3$

9. $\dfrac{ds}{dt}$ if $s = \dfrac{t}{2t + 1}$

10. $\dfrac{dv}{dt}$ if $v = t - \dfrac{1}{t}$

11. $\dfrac{dp}{dq}$ if $p = q^{3/2}$

12. $\dfrac{dz}{dw}$ if $z = \dfrac{1}{\sqrt{w^2 - 1}}$

Slopes and Tangent Lines

In Exercises 13–16, differentiate the functions and find the slope of the tangent line at the given value of the independent variable.

13. $f(x) = x + \dfrac{9}{x}$, $x = -3$

14. $k(x) = \dfrac{1}{2 + x}$, $x = 2$

15. $s = t^3 - t^2$, $t = -1$

16. $y = \dfrac{x + 3}{1 - x}$, $x = -2$

In Exercises 17–18, differentiate the functions. Then find an equation of the tangent line at the indicated point on the graph of the function.

17. $y = f(x) = \dfrac{8}{\sqrt{x - 2}}$, $(x, y) = (6, 4)$

18. $w = g(z) = 1 + \sqrt{4 - z}$, $(z, w) = (3, 2)$

In Exercises 19–22, find the values of the derivatives.

19. $\dfrac{ds}{dt}\bigg|_{t=-1}$ if $s = 1 - 3t^2$

20. $\dfrac{dy}{dx}\bigg|_{x=\sqrt{3}}$ if $y = 1 - \dfrac{1}{x}$

21. $\dfrac{dr}{d\theta}\bigg|_{\theta=0}$ if $r = \dfrac{2}{\sqrt{4 - \theta}}$

22. $\dfrac{dw}{dz}\bigg|_{z=4}$ if $w = z + \sqrt{z}$

Using the Alternative Formula for Derivatives

Use the formula

$$f'(x) = \lim_{z \to x} \frac{f(z) - f(x)}{z - x}$$

to find the derivative of the functions in Exercises 23–26.

23. $f(x) = \dfrac{1}{x + 2}$ **24.** $f(x) = x^2 - 3x + 4$

25. $g(x) = \dfrac{x}{x - 1}$ **26.** $g(x) = 1 + \sqrt{x}$

Graphs

Match the functions graphed in Exercises 27–30 with the derivatives graphed in the accompanying figures (a)–(d).

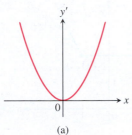

(a)

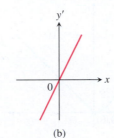

(b)

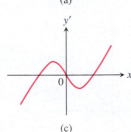

(c)

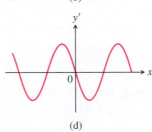

(d)

27.   **28.**

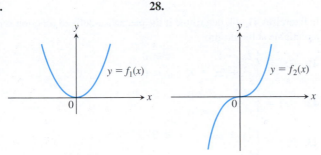

29.   **30.**

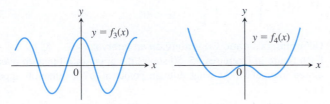

31. a. The graph in the accompanying figure is made of line segments joined end to end. At which points of the interval $[-4, 6]$ is f' not defined? Give reasons for your answer.

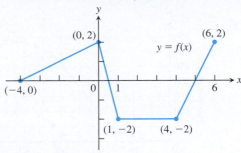

b. Graph the derivative of f.

The graph should show a step function.

32. Recovering a function from its derivative

a. Use the following information to graph the function f over the closed interval $[-2, 5]$.

   i) The graph of f is made of closed line segments joined end to end.

   ii) The graph starts at the point $(-2, 3)$.

   iii) The derivative of f is the step function in the figure shown here.

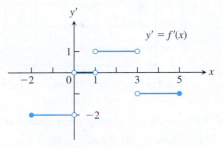

b. Repeat part (a), assuming that the graph starts at $(-2, 0)$ instead of $(-2, 3)$.

33. Growth in the economy   The graph in the accompanying figure shows the average annual percentage change $y = f(t)$ in the U.S. gross national product (GNP) for the years 2005–2011. Graph dy/dt (where defined).

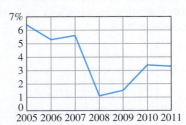

34. Fruit flies   (*Continuation of Example 4, Section 2.1.*) Populations starting out in closed environments grow slowly at first, when there are relatively few members, then more rapidly as the number of reproducing individuals increases and resources are still abundant, then slowly again as the population reaches the carrying capacity of the environment.

a. Use the graphical technique of Example 3 to graph the derivative of the fruit fly population. The graph of the population is reproduced here.

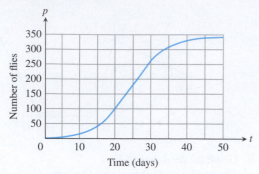

b. During what days does the population seem to be increasing fastest? Slowest?

35. Temperature   The given graph shows the outside temperature T in °F, between 6 A.M. and 6 P.M.

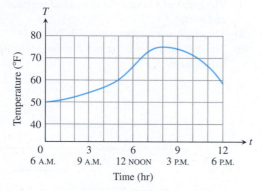

a. Estimate the rate of temperature change at the times

i) 7 A.M.   **ii)** 9 A.M.   **iii)** 2 P.M.   **iv)** 4 P.M.

b. At what time does the temperature increase most rapidly? Decrease most rapidly? What is the rate for each of those times?

c. Use the graphical technique of Example 3 to graph the derivative of temperature T versus time t.

36. Average single-family home prices P (in thousands of dollars) in Sacramento, California, are shown in the accompanying figure from 2006 through 2015.

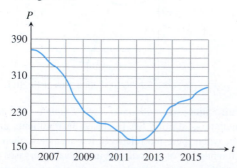

a. During what years did home prices decrease? increase?

b. Estimate home prices at the end of

i. 2007   **ii.** 2012   **iii.** 2015

c. Estimate the rate of change of home prices at the beginning of

i. 2007   **ii.** 2010   **iii.** 2014

d. During what year did home prices drop most rapidly and what is an estimate of this rate?

e. During what year did home prices rise most rapidly and what is an estimate of this rate?

f. Use the graphical technique of Example 3 to graph the derivative of home price P versus time t.

One-Sided Derivatives

Compute the right-hand and left-hand derivatives as limits to show that the functions in Exercises 37–40 are not differentiable at the point P.

37.

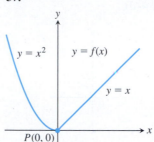

38.

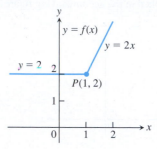

39.

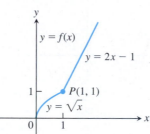

40.

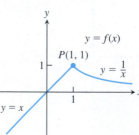

In Exercises 41–44, determine if the piecewise-defined function is differentiable at the origin.

41. $f(x) = \begin{cases} 2x - 1, & x \geq 0 \\ x^2 + 2x + 7, & x < 0 \end{cases}$

42. $g(x) = \begin{cases} x^{2/3}, & x \geq 0 \\ x^{1/3}, & x < 0 \end{cases}$

43. $f(x) = \begin{cases} 2x + \tan x, & x \geq 0 \\ x^2, & x < 0 \end{cases}$

44. $g(x) = \begin{cases} 2x - x^3 - 1, & x \geq 0 \\ x - \dfrac{1}{x + 1}, & x < 0 \end{cases}$

Differentiability and Continuity on an Interval

Each figure in Exercises 45–50 shows the graph of a function over a closed interval D. At what domain points does the function appear to be

a. differentiable?

b. continuous but not differentiable?

c. neither continuous nor differentiable?

Give reasons for your answers.

45.

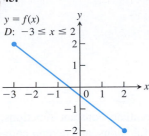

$y = f(x)$
D: $-3 \le x \le 2$

46.

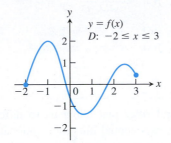

$y = f(x)$
D: $-2 \le x \le 3$

47.

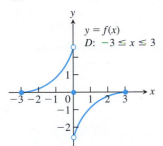

$y = f(x)$
D: $-3 \le x \le 3$

48.

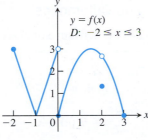

$y = f(x)$
D: $-2 \le x \le 3$

49.

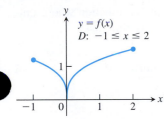

$y = f(x)$
D: $-1 \le x \le 2$

50.

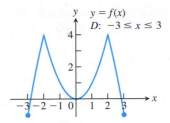

$y = f(x)$
D: $-3 \le x \le 3$

Theory and Examples
In Exercises 51–54,

a. Find the derivative $f'(x)$ of the given function $y = f(x)$.

b. Graph $y = f(x)$ and $y = f'(x)$ side by side using separate sets of coordinate axes, and answer the following questions.

c. For what values of x, if any, is f' positive? Zero? Negative?

d. Over what intervals of x-values, if any, does the function $y = f(x)$ increase as x increases? Decrease as x increases? How is this related to what you found in part (c)? (We will say more about this relationship in Section 4.3.)

51. $y = -x^2$

52. $y = -1/x$

53. $y = x^3/3$

54. $y = x^4/4$

55. Tangent to a parabola Does the parabola $y = 2x^2 - 13x + 5$ have a tangent line whose slope is -1? If so, find an equation for the line and the point of tangency. If not, why not?

56. Tangent to $y = \sqrt{x}$ Does any tangent line to the curve $y = \sqrt{x}$ cross the x-axis at $x = -1$? If so, find an equation for the line and the point of tangency. If not, why not?

57. Derivative of $-f$ Does knowing that a function $f(x)$ is differentiable at $x = x_0$ tell you anything about the differentiability of the function $-f$ at $x = x_0$? Give reasons for your answer.

58. Derivative of multiples Does knowing that a function $g(t)$ is differentiable at $t = 7$ tell you anything about the differentiability of the function $3g$ at $t = 7$? Give reasons for your answer.

59. Limit of a quotient Suppose that functions $g(t)$ and $h(t)$ are defined for all values of t and $g(0) = h(0) = 0$. Can $\lim_{t\to0} (g(t))/(h(t))$ exist? If it does exist, must it equal zero? Give reasons for your answers.

60. a. Let $f(x)$ be a function satisfying $|f(x)| \le x^2$ for $-1 \le x \le 1$. Show that f is differentiable at $x = 0$ and find $f'(0)$.

b. Show that

$$f(x) = \begin{cases} x^2 \sin\dfrac{1}{x}, & x \ne 0 \\ 0, & x = 0 \end{cases}$$

is differentiable at $x = 0$ and find $f'(0)$.

T 61. Graph $y = 1/(2\sqrt{x})$ in a window that has $0 \le x \le 2$. Then, on the same screen, graph

$$y = \frac{\sqrt{x + h} - \sqrt{x}}{h}$$

for $h = 1, 0.5, 0.1$. Then try $h = -1, -0.5, -0.1$. Explain what is going on.

T 62. Graph $y = 3x^2$ in a window that has $-2 \le x \le 2, 0 \le y \le 3$. Then, on the same screen, graph

$$y = \frac{(x + h)^3 - x^3}{h}$$

for $h = 2, 1, 0.2$. Then try $h = -2, -1, -0.2$. Explain what is going on.

63. Derivative of $y = |x|$ Graph the derivative of $f(x) = |x|$. Then graph $y = (|x| - 0)/(x - 0) = |x|/x$. What can you conclude?

T 64. Weierstrass's nowhere differentiable continuous function The sum of the first eight terms of the Weierstrass function $f(x) = \sum_{n-0}^{\infty} (2/3)^n \cos(9^n \pi x)$ is

$$g(x) = \cos(\pi x) + (2/3)^1 \cos(9\pi x) + (2/3)^2 \cos(9^2\pi x)$$
$$+ (2/3)^3 \cos(9^3\pi x) + \cdots + (2/3)^7 \cos(9^7\pi x).$$

Graph this sum. Zoom in several times. How wiggly and bumpy is this graph? Specify a viewing window in which the displayed portion of the graph is smooth.

COMPUTER EXPLORATIONS
Use a CAS to perform the following steps for the functions in Exercises 65–70.

a. Plot $y = f(x)$ to see that function's global behavior.

b. Define the difference quotient q at a general point x, with general step size h.

c. Take the limit as $h \to 0$. What formula does this give?

d. Substitute the value $x = x_0$ and plot the function $y = f(x)$ together with its tangent line at that point.

e. Substitute various values for x larger and smaller than x_0 into the formula obtained in part (c). Do the numbers make sense with your picture?

f. Graph the formula obtained in part (c). What does it mean when its values are negative? Zero? Positive? Does this make sense with your plot from part (a)? Give reasons for your answer.

65. $f(x) = x^3 + x^2 - x, \quad x_0 = 1$

66. $f(x) = x^{1/3} + x^{2/3}, \quad x_0 = 1$

67. $f(x) = \dfrac{4x}{x^2 + 1}, \quad x_0 = 2$

68. $f(x) = \dfrac{x - 1}{3x^2 + 1}, \quad x_0 = -1$

69. $f(x) = \sin 2x, \quad x_0 = \pi/2$

70. $f(x) = x^2 \cos x, \quad x_0 = \pi/4$

3.3 Differentiation Rules

This section introduces several rules that allow us to differentiate constant functions, power functions, polynomials, exponential functions, rational functions, and certain combinations of them, simply and directly, without having to take limits each time.

Powers, Multiples, Sums, and Differences

A basic rule of differentiation is that the derivative of every constant function is zero.

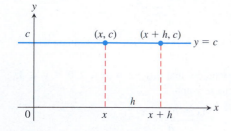

FIGURE 3.10 The rule $(d/dx)(c) = 0$ is another way to say that the values of constant functions never change and that the slope of a horizontal line is zero at every point.

> **Derivative of a Constant Function**
> If f has the constant value $f(x) = c$, then
> $$\frac{df}{dx} = \frac{d}{dx}(c) = 0.$$

Proof We apply the definition of the derivative to $f(x) = c$, the function whose outputs have the constant value c (Figure 3.10). At every value of x, we find that

$$f'(x) = \lim_{h \to 0} \frac{f(x + h) - f(x)}{h} = \lim_{h \to 0} \frac{c - c}{h} = \lim_{h \to 0} 0 = 0.$$

We now consider powers of x. From Section 3.1, we know that

$$\frac{d}{dx}\left(\frac{1}{x}\right) = -\frac{1}{x^2}, \quad \text{or} \quad \frac{d}{dx}(x^{-1}) = -x^{-2}.$$

From Example 2 of the last section we also know that

$$\frac{d}{dx}(\sqrt{x}) = \frac{1}{2\sqrt{x}}, \quad \text{or} \quad \frac{d}{dx}(x^{1/2}) = \frac{1}{2}x^{-1/2}.$$

These two examples illustrate a general rule for differentiating a power x^n. We first prove the rule when n is a positive integer.

> **Derivative of a Positive Integer Power**
> If n is a positive integer, then
> $$\frac{d}{dx}x^n = nx^{n-1}.$$

HISTORICAL BIOGRAPHY

Richard Courant

(1888–1972)

www.goo.gl/PXiQtT

Proof of the Positive Integer Power Rule The formula

$$z^n - x^n = (z - x)(z^{n-1} + z^{n-2}x + \cdots + zx^{n-2} + x^{n-1})$$

can be verified by multiplying out the right-hand side. Then from the alternative formula for the definition of the derivative,

$$f'(x) = \lim_{z \to x} \frac{f(z) - f(x)}{z - x} = \lim_{z \to x} \frac{z^n - x^n}{z - x}$$

$$= \lim_{z \to x} (z^{n-1} + z^{n-2}x + \cdots + zx^{n-2} + x^{n-1}) \qquad n \text{ terms}$$

$$= nx^{n-1}.$$

The Power Rule is actually valid for all real numbers n, not just for positive integers. We have seen examples for a negative integer and fractional power, but n could be an irrational number as well. Here we state the general version of the rule, but postpone its proof until Section 3.8.

Power Rule (General Version)

If n is any real number, then

$$\frac{d}{dx}x^n = nx^{n-1},$$

for all x where the powers x^n and x^{n-1} are defined.

EXAMPLE 1   Differentiate the following powers of x.

(a) x^3 **(b)** $x^{2/3}$ **(c)** $x^{\sqrt{2}}$ **(d)** $\dfrac{1}{x^4}$ **(e)** $x^{-4/3}$ **(f)** $\sqrt{x^{2+\pi}}$

Solution

Applying the Power Rule
Subtract 1 from the exponent and multiply the result by the original exponent.

(a) $\dfrac{d}{dx}(x^3) = 3x^{3-1} = 3x^2$

(b) $\dfrac{d}{dx}(x^{2/3}) = \dfrac{2}{3}x^{(2/3)-1} = \dfrac{2}{3}x^{-1/3}$

(c) $\dfrac{d}{dx}(x^{\sqrt{2}}) = \sqrt{2}x^{\sqrt{2}-1}$

(d) $\dfrac{d}{dx}\left(\dfrac{1}{x^4}\right) = \dfrac{d}{dx}(x^{-4}) = -4x^{-4-1} = -4x^{-5} = -\dfrac{4}{x^5}$

(e) $\dfrac{d}{dx}(x^{-4/3}) = -\dfrac{4}{3}x^{-(4/3)-1} = -\dfrac{4}{3}x^{-7/3}$

(f) $\dfrac{d}{dx}(\sqrt{x^{2+\pi}}) = \dfrac{d}{dx}(x^{1+(\pi/2)}) = \left(1 + \dfrac{\pi}{2}\right)x^{1+(\pi/2)-1} = \dfrac{1}{2}(2 + \pi)\sqrt{x^{\pi}}$ ■

The next rule says that when a differentiable function is multiplied by a constant, its derivative is multiplied by the same constant.

Derivative Constant Multiple Rule

If u is a differentiable function of x, and c is a constant, then

$$\frac{d}{dx}(cu) = c\frac{du}{dx}.$$

Proof

$$\frac{d}{dx}cu = \lim_{h\to 0}\frac{cu(x+h) - cu(x)}{h} \qquad \text{Derivative definition with } f(x) = cu(x)$$

$$= c\lim_{h\to 0}\frac{u(x+h) - u(x)}{h} \qquad \text{Constant Multiple Limit Property}$$

$$= c\frac{du}{dx} \qquad u \text{ is differentiable.}$$

■

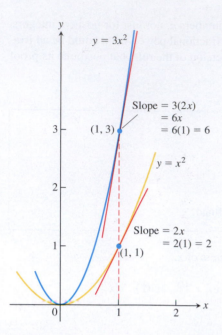

FIGURE 3.11 The graphs of $y = x^2$ and $y = 3x^2$. Tripling the y-coordinate triples the slope (Example 2).

Denoting Functions by u and v
The functions we are working with when we need a differentiation formula are likely to be denoted by letters like f and g. We do not want to use these same letters when stating general differentiation rules, so instead we use letters like u and v that are not likely to be already in use.

EXAMPLE 2

(a) The derivative formula

$$\frac{d}{dx}(3x^2) = 3 \cdot 2x = 6x$$

says that if we rescale the graph of $y = x^2$ by multiplying each y-coordinate by 3, then we multiply the slope at each point by 3 (Figure 3.11).

(b) **Negative of a function**

The derivative of the negative of a differentiable function u is the negative of the function's derivative. The Constant Multiple Rule with $c = -1$ gives

$$\frac{d}{dx}(-u) = \frac{d}{dx}(-1 \cdot u) = -1 \cdot \frac{d}{dx}(u) = -\frac{du}{dx}.$$

∎

The next rule says that the derivative of the sum of two differentiable functions is the sum of their derivatives.

Derivative Sum Rule

If u and v are differentiable functions of x, then their sum $u + v$ is differentiable at every point where u and v are both differentiable. At such points,

$$\frac{d}{dx}(u + v) = \frac{du}{dx} + \frac{dv}{dx}.$$

Proof   We apply the definition of the derivative to $f(x) = u(x) + v(x)$:

$$\frac{d}{dx}[u(x) + v(x)] = \lim_{h \to 0} \frac{[u(x + h) + v(x + h)] - [u(x) + v(x)]}{h}$$

$$= \lim_{h \to 0} \left[\frac{u(x + h) - u(x)}{h} + \frac{v(x + h) - v(x)}{h} \right]$$

$$= \lim_{h \to 0} \frac{u(x + h) - u(x)}{h} + \lim_{h \to 0} \frac{v(x + h) - v(x)}{h} = \frac{du}{dx} + \frac{dv}{dx}. \quad ∎$$

Combining the Sum Rule with the Constant Multiple Rule gives the **Difference Rule**, which says that the derivative of a *difference* of differentiable functions is the difference of their derivatives:

$$\frac{d}{dx}(u - v) = \frac{d}{dx}[u + (-1)v] = \frac{du}{dx} + (-1)\frac{dv}{dx} = \frac{du}{dx} - \frac{dv}{dx}.$$

The Sum Rule also extends to finite sums of more than two functions. If $u_1, u_2, \ldots, u_n$ are differentiable at x, then so is $u_1 + u_2 + \cdots + u_n$, and

$$\frac{d}{dx}(u_1 + u_2 + \cdots + u_n) = \frac{du_1}{dx} + \frac{du_2}{dx} + \cdots + \frac{du_n}{dx}.$$

For instance, to see that the rule holds for three functions we compute

$$\frac{d}{dx}(u_1 + u_2 + u_3) = \frac{d}{dx}((u_1 + u_2) + u_3) = \frac{d}{dx}(u_1 + u_2) + \frac{du_3}{dx} = \frac{du_1}{dx} + \frac{du_2}{dx} + \frac{du_3}{dx}.$$

A proof by mathematical induction for any finite number of terms is given in Appendix 2.

EXAMPLE 3   Find the derivative of the polynomial $y = x^3 + \dfrac{4}{3}x^2 - 5x + 1$.

Solution $\dfrac{dy}{dx} = \dfrac{d}{dx}x^3 + \dfrac{d}{dx}\left(\dfrac{4}{3}x^2\right) - \dfrac{d}{dx}(5x) + \dfrac{d}{dx}(1)$ Sum and Difference Rules

$$= 3x^2 + \dfrac{4}{3}\cdot 2x - 5 + 0 = 3x^2 + \dfrac{8}{3}x - 5 \qquad ■$$

We can differentiate any polynomial term by term, the way we differentiated the polynomial in Example 3. All polynomials are differentiable at all values of x.

EXAMPLE 4   Does the curve $y = x^4 - 2x^2 + 2$ have any horizontal tangent lines? If so, where?

Solution   The horizontal tangent lines, if any, occur where the slope dy/dx is zero. We have

$$\dfrac{dy}{dx} = \dfrac{d}{dx}(x^4 - 2x^2 + 2) = 4x^3 - 4x.$$

Now solve the equation $\dfrac{dy}{dx} = 0$ for x:

$$4x^3 - 4x = 0$$
$$4x(x^2 - 1) = 0$$
$$x = 0, 1, -1.$$

The curve $y = x^4 - 2x^2 + 2$ has horizontal tangents at $x = 0, 1,$ and -1. The corresponding points on the curve are $(0, 2)$, $(1, 1)$, and $(-1, 1)$. See Figure 3.12. $\qquad ■$

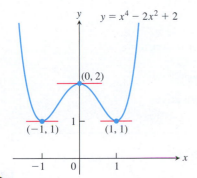

FIGURE 3.12   The curve in Example 4 and its horizontal tangents.

Derivatives of Exponential Functions

We briefly reviewed exponential functions in Section 1.5. When we apply the definition of the derivative to $f(x) = a^x$, we get

$$\dfrac{d}{dx}(a^x) = \lim_{h\to 0}\dfrac{a^{x+h} - a^x}{h} \qquad \text{\color{blue}Derivative definition}$$

$$= \lim_{h\to 0}\dfrac{a^x \cdot a^h - a^x}{h} \qquad \text{\color{blue}}a^{x+h} = a^x \cdot a^h$$

$$= \lim_{h\to 0}a^x \cdot \dfrac{a^h - 1}{h} \qquad \text{\color{blue}Factoring out }a^x$$

$$= a^x \cdot \lim_{h\to 0}\dfrac{a^h - 1}{h} \qquad \text{\color{blue}}a^x\text{ is constant as }h\to 0.$$

$$= \left(\lim_{h\to 0}\dfrac{a^h - 1}{h}\right)\cdot a^x. \qquad (1)$$

$$\underbrace{\phantom{\left(\lim_{h\to 0}\dfrac{a^h - 1}{h}\right)}}_{\text{a fixed number }L}$$

Thus we see that the derivative of a^x is a constant multiple L of a^x. The constant L is a limit we have not encountered before. Note, however, that it equals the derivative of $f(x) = a^x$ at $x = 0$:

$$f'(0) = \lim_{h\to 0}\dfrac{a^h - a^0}{h} = \lim_{h\to 0}\dfrac{a^h - 1}{h} = L.$$

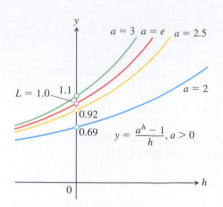

FIGURE 3.13 The position of the curve $y = (a^h - 1)/h, a > 0$, varies continuously with a. The limit L of y as $h \to 0$ changes with different values of a. The number for which $L = 1$ as $h \to 0$ is the number e between $a = 2$ and $a = 3$.

The limit L is therefore the slope of the graph of $f(x) = a^x$ where it crosses the y-axis. In Chapter 7, where we carefully develop the logarithmic and exponential functions, we prove that the limit L exists and has the value $\ln a$. For now we investigate values of L by graphing the function $y = (a^h - 1)/h$ and studying its behavior as h approaches 0.

Figure 3.13 shows the graphs of $y = (a^h - 1)/h$ for four different values of a. The limit L is approximately 0.69 if $a = 2$, about 0.92 if $a = 2.5$, and about 1.1 if $a = 3$. It appears that the value of L is 1 at some number a chosen between 2.5 and 3. That number is given by $a = e \approx 2.718281828$. With this choice of base we obtain the natural exponential function $f(x) = e^x$ as in Section 1.5, and see that it satisfies the property

$$f'(0) = \lim_{h \to 0} \frac{e^h - 1}{h} = 1 \tag{2}$$

because it is the exponential function whose graph has slope 1 when it crosses the y-axis. That the limit is 1 implies an important relationship between the natural exponential function e^x and its derivative:

$$\frac{d}{dx}(e^x) = \lim_{h \to 0}\left(\frac{e^h - 1}{h}\right) \cdot e^x \qquad \text{Eq. (1) with } a = e$$

$$= 1 \cdot e^x = e^x. \qquad \text{Eq. (2)}$$

Therefore the natural exponential function is its own derivative.

Derivative of the Natural Exponential Function

$$\frac{d}{dx}(e^x) = e^x$$

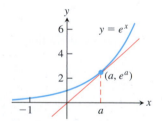

FIGURE 3.14 The line through the origin is tangent to the graph of $y = e^x$ when $a = 1$ (Example 5).

EXAMPLE 5 Find an equation for a line that is tangent to the graph of $y = e^x$ and goes through the origin.

Solution Since the line passes through the origin, its equation is of the form $y = mx$, where m is the slope. If it is tangent to the graph at the point (a, e^a), the slope is $m = (e^a - 0)/(a - 0)$. The slope of the natural exponential at $x = a$ is e^a. Because these slopes are the same, we then have that $e^a = e^a/a$. It follows that $a = 1$ and $m = e$, so the equation of the tangent line is $y = ex$. See Figure 3.14. ∎

We might ask if there are functions *other* than the natural exponential function that are their own derivatives. The answer is that the only functions that satisfy the property that $f'(x) = f(x)$ are functions that are constant multiples of the natural exponential function, $f(x) = c \cdot e^x$, c any constant. We prove this fact in Section 7.2. Note from the Constant Multiple Rule that indeed

$$\frac{d}{dx}(c \cdot e^x) = c \cdot \frac{d}{dx}(e^x) = c \cdot e^x.$$

Products and Quotients

While the derivative of the sum of two functions is the sum of their derivatives, the derivative of the product of two functions is *not* the product of their derivatives. For instance,

$$\frac{d}{dx}(x \cdot x) = \frac{d}{dx}(x^2) = 2x, \qquad \text{while} \qquad \frac{d}{dx}(x) \cdot \frac{d}{dx}(x) = 1 \cdot 1 = 1.$$

The derivative of a product of two functions is the sum of *two* products, as we now explain.

Derivative Product Rule

If u and v are differentiable at x, then so is their product uv, and

$$\frac{d}{dx}(uv) = u\frac{dv}{dx} + \frac{du}{dx}v.$$

The derivative of the product uv is u times the derivative of v plus the derivative of u times v. In prime notation, $(uv)' = uv' + u'v$. In function notation,

$$\frac{d}{dx}[f(x)g(x)] = f(x)g'(x) + f'(x)g(x), \quad \text{or} \quad (fg)' = fg' + f'g. \tag{3}$$

EXAMPLE 6   Find the derivative of **(a)** $y = \frac{1}{x}(x^2 + e^x)$, **(b)** $y = e^{2x}$.

Solution

(a) We apply the Product Rule with $u = 1/x$ and $v = x^2 + e^x$:

$$\frac{d}{dx}\left[\frac{1}{x}(x^2 + e^x)\right] = \frac{1}{x}(2x + e^x) + \left(-\frac{1}{x^2}\right)(x^2 + e^x) \qquad \frac{d}{dx}(uv) = u\frac{dv}{dx} + \frac{du}{dx}v \text{ and}$$

$$\qquad\qquad\qquad\qquad\qquad \frac{d}{dx}\left(\frac{1}{x}\right) = -\frac{1}{x^2}$$

$$= 2 + \frac{e^x}{x} - 1 - \frac{e^x}{x^2}$$

$$= 1 + (x - 1)\frac{e^x}{x^2}.$$

(b) $\dfrac{d}{dx}(e^{2x}) = \dfrac{d}{dx}(e^x \cdot e^x) = e^x \cdot \dfrac{d}{dx}(e^x) + \dfrac{d}{dx}(e^x) \cdot e^x = 2e^x \cdot e^x = 2e^{2x}$ ■

Proof of the Derivative Product Rule

$$\frac{d}{dx}(uv) = \lim_{h \to 0} \frac{u(x + h)v(x + h) - u(x)v(x)}{h}$$

To change this fraction into an equivalent one that contains difference quotients for the derivatives of u and v, we subtract and add $u(x + h)v(x)$ in the numerator:

$$\frac{d}{dx}(uv) = \lim_{h \to 0} \frac{u(x + h)v(x + h) - u(x + h)v(x) + u(x + h)v(x) - u(x)v(x)}{h}$$

$$= \lim_{h \to 0}\left[u(x + h)\frac{v(x + h) - v(x)}{h} + v(x)\frac{u(x + h) - u(x)}{h}\right]$$

$$= \lim_{h \to 0} u(x + h) \cdot \lim_{h \to 0} \frac{v(x + h) - v(x)}{h} + v(x) \cdot \lim_{h \to 0} \frac{u(x + h) - u(x)}{h}.$$

As h approaches zero, $u(x + h)$ approaches $u(x)$ because u, being differentiable at x, is continuous at x. The two fractions approach the values of dv/dx at x and du/dx at x. Therefore,

$$\frac{d}{dx}(uv) = u\frac{dv}{dx} + v\frac{du}{dx}. \qquad ■$$

Picturing the Product Rule

Suppose $u(x)$ and $v(x)$ are positive and increase when x increases, and $h > 0$.

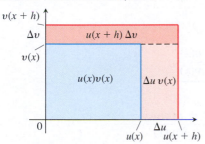

Then the change in the product uv is the difference in areas of the larger and smaller "squares," which is the sum of the upper and right-hand reddish-shaded rectangles. That is,

$$\Delta(uv) = u(x + h)v(x + h) - u(x)v(x)$$
$$= u(x + h)\Delta v + \Delta u v(x).$$

Division by h gives

$$\frac{\Delta(uv)}{h} = u(x + h)\frac{\Delta v}{h} + \frac{\Delta u}{h}v(x).$$

The limit as $h \to 0^+$ gives the Product Rule.

The derivative of the quotient of two functions is given by the Quotient Rule.

Derivative Quotient Rule

If u and v are differentiable at x and if $v(x) \neq 0$, then the quotient u/v is differentiable at x, and

$$\frac{d}{dx}\left(\frac{u}{v}\right) = \frac{v\dfrac{du}{dx} - u\dfrac{dv}{dx}}{v^2}.$$

In function notation,

$$\frac{d}{dx}\left[\frac{f(x)}{g(x)}\right] = \frac{g(x)f'(x) - f(x)g'(x)}{g^2(x)}.$$

EXAMPLE 7   Find the derivative of **(a)** $y = \dfrac{t^2 - 1}{t^3 + 1}$, **(b)** $y = e^{-x}$.

Solution

(a) We apply the Quotient Rule with $u = t^2 - 1$ and $v = t^3 + 1$:

$$\frac{dy}{dt} = \frac{(t^3 + 1) \cdot 2t - (t^2 - 1) \cdot 3t^2}{(t^3 + 1)^2} \qquad \frac{d}{dt}\left(\frac{u}{v}\right) = \frac{v(du/dt) - u(dv/dt)}{v^2}$$

$$= \frac{2t^4 + 2t - 3t^4 + 3t^2}{(t^3 + 1)^2}$$

$$= \frac{-t^4 + 3t^2 + 2t}{(t^3 + 1)^2}.$$

(b) $\dfrac{d}{dx}(e^{-x}) = \dfrac{d}{dx}\left(\dfrac{1}{e^x}\right) = \dfrac{e^x \cdot 0 - 1 \cdot e^x}{(e^x)^2} = \dfrac{-1}{e^x} = -e^{-x}$ ■

Proof of the Derivative Quotient Rule

$$\frac{d}{dx}\left(\frac{u}{v}\right) = \lim_{h \to 0} \frac{\dfrac{u(x + h)}{v(x + h)} - \dfrac{u(x)}{v(x)}}{h}$$

$$= \lim_{h \to 0} \frac{v(x)u(x + h) - u(x)v(x + h)}{hv(x + h)v(x)}$$

To change the last fraction into an equivalent one that contains the difference quotients for the derivatives of u and v, we subtract and add $v(x)u(x)$ in the numerator. We then get

$$\frac{d}{dx}\left(\frac{u}{v}\right) = \lim_{h \to 0} \frac{v(x)u(x + h) - v(x)u(x) + v(x)u(x) - u(x)v(x + h)}{hv(x + h)v(x)}$$

$$= \lim_{h \to 0} \frac{v(x)\dfrac{u(x + h) - u(x)}{h} - u(x)\dfrac{v(x + h) - v(x)}{h}}{v(x + h)v(x)}.$$

Taking the limits in the numerator and denominator now gives the Quotient Rule. Exercise 76 outlines another proof. ■

The choice of which rules to use in solving a differentiation problem can make a difference in how much work you have to do. Here is an example.

EXAMPLE 8   Find the derivative of

$$y = \frac{(x - 1)(x^2 - 2x)}{x^4}.$$

Solution   Using the Quotient Rule here will result in a complicated expression with many terms. Instead, use some algebra to simplify the expression. First expand the numerator and divide by x^4:

$$y = \frac{(x - 1)(x^2 - 2x)}{x^4} = \frac{x^3 - 3x^2 + 2x}{x^4} = x^{-1} - 3x^{-2} + 2x^{-3}.$$

Then use the Sum and Power Rules:

$$\frac{dy}{dx} = -x^{-2} - 3(-2)x^{-3} + 2(-3)x^{-4}$$

$$= -\frac{1}{x^2} + \frac{6}{x^3} - \frac{6}{x^4}.$$ ∎

Second- and Higher-Order Derivatives

If $y = f(x)$ is a differentiable function, then its derivative $f'(x)$ is also a function. If f' is also differentiable, then we can differentiate f' to get a new function of x denoted by f''. So $f'' = (f')'$. The function f'' is called the **second derivative** of f because it is the derivative of the first derivative. It is written in several ways:

$$f''(x) = \frac{d^2y}{dx^2} = \frac{d}{dx}\left(\frac{dy}{dx}\right) = \frac{dy'}{dx} = y'' = D^2(f)(x) = D_x^2 f(x).$$

The symbol D^2 means that the operation of differentiation is performed twice.

If $y = x^6$, then $y' = 6x^5$ and we have

$$y'' = \frac{dy'}{dx} = \frac{d}{dx}(6x^5) = 30x^4.$$

Thus $D^2(x^6) = 30x^4$.

If y'' is differentiable, its derivative, $y''' = dy''/dx = d^3y/dx^3$, is the **third derivative** of y with respect to x. The names continue as you imagine, with

$$y^{(n)} = \frac{d}{dx}y^{(n-1)} = \frac{d^ny}{dx^n} = D^n y$$

denoting the **nth derivative** of y with respect to x for any positive integer n.

We can interpret the second derivative as the rate of change of the slope of the tangent line to the graph of $y = f(x)$ at each point. You will see in the next chapter that the second derivative reveals whether the graph bends upward or downward from the tangent line as we move off the point of tangency. In the next section, we interpret both the second and third derivatives in terms of motion along a straight line.

How to Read the Symbols for Derivatives

y'	"y prime"
y''	"y double prime"
$\dfrac{d^2y}{dx^2}$	"d squared y dx squared"
y'''	"y triple prime"
$y^{(n)}$	"y super n"
$\dfrac{d^ny}{dx^n}$	"d to the n of y by dx to the n"
D^n	"d to the n"

EXAMPLE 9   The first four derivatives of $y = x^3 - 3x^2 + 2$ are

First derivative: $y' = 3x^2 - 6x$

Second derivative: $y'' = 6x - 6$

Third derivative: $y''' = 6$

Fourth derivative: $y^{(4)} = 0.$

All polynomial functions have derivatives of all orders. In this example, the fifth and later derivatives are all zero.   ∎

EXERCISES 3.3

Derivative Calculations

In Exercises 1–12, find the first and second derivatives.

1. $y = -x^2 + 3$

2. $y = x^2 + x + 8$

3. $s = 5t^3 - 3t^5$

4. $w = 3z^7 - 7z^3 + 21z^2$

5. $y = \dfrac{4x^3}{3} - x + 2e^x$

6. $y = \dfrac{x^3}{3} + \dfrac{x^2}{2} + e^{-x}$

7. $w = 3z^{-2} - \dfrac{1}{z}$

8. $s = -2t^{-1} + \dfrac{4}{t^2}$

9. $y = 6x^2 - 10x - 5x^{-2}$

10. $y = 4 - 2x - x^{-3}$

11. $r = \dfrac{1}{3s^2} - \dfrac{5}{2s}$

12. $r = \dfrac{12}{\theta} - \dfrac{4}{\theta^3} + \dfrac{1}{\theta^4}$

In Exercises 13–16, find y' **(a)** by applying the Product Rule and **(b)** by multiplying the factors to produce a sum of simpler terms to differentiate.

13. $y = (3 - x^2)(x^3 - x + 1)$ **14.** $y = (2x + 3)(5x^2 - 4x)$

15. $y = (x^2 + 1)\left(x + 5 + \dfrac{1}{x}\right)$ **16.** $y = (1 + x^2)(x^{3/4} - x^{-3})$

Find the derivatives of the functions in Exercises 17–40.

17. $y = \dfrac{2x + 5}{3x - 2}$

18. $z = \dfrac{4 - 3x}{3x^2 + x}$

19. $g(x) = \dfrac{x^2 - 4}{x + 0.5}$

20. $f(t) = \dfrac{t^2 - 1}{t^2 + t - 2}$

21. $v = (1 - t)(1 + t^2)^{-1}$

22. $w = (2x - 7)^{-1}(x + 5)$

23. $f(s) = \dfrac{\sqrt{s} - 1}{\sqrt{s} + 1}$

24. $u = \dfrac{5x + 1}{2\sqrt{x}}$

25. $v = \dfrac{1 + x - 4\sqrt{x}}{x}$

26. $r = 2\left(\dfrac{1}{\sqrt{\theta}} + \sqrt{\theta}\right)$

27. $y = \dfrac{1}{(x^2 - 1)(x^2 + x + 1)}$ **28.** $y = \dfrac{(x + 1)(x + 2)}{(x - 1)(x - 2)}$

29. $y = 2e^{-x} + e^{3x}$

30. $y = \dfrac{x^2 + 3e^x}{2e^x - x}$

31. $y = x^3 e^x$

32. $w = re^{-r}$

33. $y = x^{9/4} + e^{-2x}$

34. $y = x^{-3/5} + \pi^{3/2}$

35. $s = 2t^{3/2} + 3e^2$

36. $w = \dfrac{1}{z^{1.4}} + \dfrac{\pi}{\sqrt{z}}$

37. $y = \sqrt[7]{x^2} - x^e$

38. $y = \sqrt[3]{x^{9.6}} + 2e^{1.3}$

39. $r = \dfrac{e^s}{s}$

40. $r = e^{\theta}\left(\dfrac{1}{\theta^2} + \theta^{-\pi/2}\right)$

Find the derivatives of all orders of the functions in Exercises 41–44.

41. $y = \dfrac{x^4}{2} - \dfrac{3}{2}x^2 - x$

42. $y = \dfrac{x^5}{120}$

43. $y = (x - 1)(x + 2)(x + 3)$ **44.** $y = (4x^2 + 3)(2 - x)x$

Find the first and second derivatives of the functions in Exercises 45–52.

45. $y = \dfrac{x^3 + 7}{x}$

46. $s = \dfrac{t^2 + 5t - 1}{t^2}$

47. $r = \dfrac{(\theta - 1)(\theta^2 + \theta + 1)}{\theta^3}$ **48.** $u = \dfrac{(x^2 + x)(x^2 - x + 1)}{x^4}$

49. $w = \left(\dfrac{1 + 3z}{3z}\right)(3 - z)$ **50.** $p = \dfrac{q^2 + 3}{(q - 1)^3 + (q + 1)^3}$

51. $w = 3z^2 e^{2z}$ **52.** $w = e^z(z - 1)(z^2 + 1)$

53. Suppose u and v are functions of x that are differentiable at $x = 0$ and that

$$u(0) = 5, \quad u'(0) = -3, \quad v(0) = -1, \quad v'(0) = 2.$$

Find the values of the following derivatives at $x = 0$.

a. $\dfrac{d}{dx}(uv)$ **b.** $\dfrac{d}{dx}\left(\dfrac{u}{v}\right)$ **c.** $\dfrac{d}{dx}\left(\dfrac{v}{u}\right)$ **d.** $\dfrac{d}{dx}(7v - 2u)$

54. Suppose u and v are differentiable functions of x and that

$$u(1) = 2, \quad u'(1) = 0, \quad v(1) = 5, \quad v'(1) = -1.$$

Find the values of the following derivatives at $x = 1$.

a. $\dfrac{d}{dx}(uv)$ **b.** $\dfrac{d}{dx}\left(\dfrac{u}{v}\right)$ **c.** $\dfrac{d}{dx}\left(\dfrac{v}{u}\right)$ **d.** $\dfrac{d}{dx}(7v - 2u)$

Slopes and Tangent Lines

55. a. Normal line to a curve   Find an equation for the line perpendicular to the tangent line to the curve $y = x^3 - 4x + 1$ at the point $(2, 1)$.

b. Smallest slope   What is the smallest slope on the curve? At what point on the curve does the curve have this slope?

c. Tangent lines having specified slope   Find equations for the tangent lines to the curve at the points where the slope of the curve is 8.

56. a. Horizontal tangent lines   Find equations for the horizontal tangent lines to the curve $y = x^3 - 3x - 2$. Also find equations for the lines that are perpendicular to these tangent lines at the points of tangency.

b. Smallest slope   What is the smallest slope on the curve? At what point on the curve does the curve have this slope? Find an equation for the line that is perpendicular to the curve's tangent line at this point.

57. Find the tangent lines to *Newton's serpentine* (graphed here) at the origin and the point $(1, 2)$.

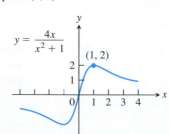

58. Find the tangent line to the *Witch of Agnesi* (graphed here) at the point $(2, 1)$.

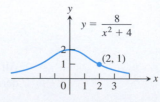

59. Quadratic tangent to identity function The curve $y = ax^2 + bx + c$ passes through the point $(1, 2)$ and is tangent to the line $y = x$ at the origin. Find a, b, and c.

60. Quadratics having a common tangent The curves $y = x^2 + ax + b$ and $y = cx - x^2$ have a common tangent line at the point $(1, 0)$. Find a, b, and c.

61. Find all points (x, y) on the graph of $f(x) = 3x^2 - 4x$ with tangent lines parallel to the line $y = 8x + 5$.

62. Find all points (x, y) on the graph of $g(x) = \frac{1}{3}x^3 - \frac{3}{2}x^2 + 1$ with tangent lines parallel to the line $8x - 2y = 1$.

63. Find all points (x, y) on the graph of $y = x/(x - 2)$ with tangent lines perpendicular to the line $y = 2x + 3$.

64. Find all points (x, y) on the graph of $f(x) = x^2$ with tangent lines passing through the point $(3, 8)$.

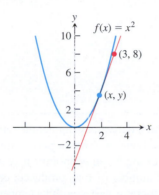

65. Assume that functions f and g are differentiable with $f(1) = 2$, $f'(1) = -3$, $g(1) = 4$, and $g'(1) = -2$. Find the equation of the line tangent to the graph of $F(x) = f(x)g(x)$ at $x = 1$.

66. Assume that functions f and g are differentiable with $f(2) = 3$, $f'(2) = -1$, $g(2) = -4$, and $g'(2) = 1$. Find an equation of the line perpendicular to the graph of $F(x) = \dfrac{f(x) + 3}{x - g(x)}$ at $x = 2$.

67. a. Find an equation for the line that is tangent to the curve $y = x^3 - x$ at the point $(-1, 0)$.

T **b.** Graph the curve and tangent line together. The tangent intersects the curve at another point. Use Zoom and Trace to estimate the point's coordinates.

T **c.** Confirm your estimates of the coordinates of the second intersection point by solving the equations for the curve and tangent line simultaneously.

68. a. Find an equation for the line that is tangent to the curve $y = x^3 - 6x^2 + 5x$ at the origin.

T **b.** Graph the curve and tangent line together. The tangent intersects the curve at another point. Use Zoom and Trace to estimate the point's coordinates.

T **c.** Confirm your estimates of the coordinates of the second intersection point by solving the equations for the curve and tangent line simultaneously.

Theory and Examples

For Exercises 69 and 70 evaluate each limit by first converting each to a derivative at a particular x-value.

69. $\displaystyle\lim_{x \to 1} \frac{x^{50} - 1}{x - 1}$

70. $\displaystyle\lim_{x \to -1} \frac{x^{2/9} - 1}{x + 1}$

71. Find the value of a that makes the following function differentiable for all x-values.

$$g(x) = \begin{cases} ax, & \text{if } x < 0 \\ x^2 - 3x, & \text{if } x \geq 0 \end{cases}$$

72. Find the values of a and b that make the following function differentiable for all x-values.

$$f(x) = \begin{cases} ax + b, & x > -1 \\ bx^2 - 3, & x \leq -1 \end{cases}$$

73. The general polynomial of degree n has the form

$$P(x) = a_n x^n + a_{n-1}x^{n-1} + \cdots + a_2 x^2 + a_1 x + a_0$$

where $a_n \neq 0$. Find $P'(x)$.

74. The body's reaction to medicine The reaction of the body to a dose of medicine can sometimes be represented by an equation of the form

$$R = M^2\left(\frac{C}{2} - \frac{M}{3}\right),$$

where C is a positive constant and M is the amount of medicine absorbed in the blood. If the reaction is a change in blood pressure, R is measured in millimeters of mercury. If the reaction is a change in temperature, R is measured in degrees, and so on.

Find dR/dM. This derivative, as a function of M, is called the sensitivity of the body to the medicine. In Section 4.5, we will see how to find the amount of medicine to which the body is most sensitive.

75. Suppose that the function v in the Derivative Product Rule has a constant value c. What does the Derivative Product Rule then say? What does this say about the Derivative Constant Multiple Rule?

76. The Reciprocal Rule

a. The *Reciprocal Rule* says that at any point where the function $v(x)$ is differentiable and different from zero,

$$\frac{d}{dx}\left(\frac{1}{v}\right) = -\frac{1}{v^2}\frac{dv}{dx}.$$

Show that the Reciprocal Rule is a special case of the Derivative Quotient Rule.

b. Show that the Reciprocal Rule and the Derivative Product Rule together imply the Derivative Quotient Rule.

77. Generalizing the Product Rule The Derivative Product Rule gives the formula

$$\frac{d}{dx}(uv) = u\frac{dv}{dx} + \frac{du}{dx}v$$

for the derivative of the product uv of two differentiable functions of x.

a. What is the analogous formula for the derivative of the product uvw of *three* differentiable functions of x?

b. What is the formula for the derivative of the product $u_1 u_2 u_3 u_4$ of *four* differentiable functions of x?

c. What is the formula for the derivative of a product $u_1 u_2 u_3 \cdots u_n$ of a finite number n of differentiable functions of x?

78. Power Rule for negative integers Use the Derivative Quotient Rule to prove the Power Rule for negative integers, that is,

$$\frac{d}{dx}(x^{-m}) = -mx^{-m-1}$$

where m is a positive integer.

79. Cylinder pressure If gas in a cylinder is maintained at a constant temperature T, the pressure P is related to the volume V by a formula of the form

$$P = \frac{nRT}{V - nb} - \frac{an^2}{V^2},$$

in which a, b, n, and R are constants. Find dP/dV. (See accompanying figure.)

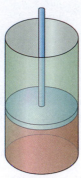

80. The best quantity to order One of the formulas for inventory management says that the average weekly cost of ordering, paying for, and holding merchandise is

$$A(q) = \frac{km}{q} + cm + \frac{hq}{2},$$

where q is the quantity you order when things run low (shoes, TVs, brooms, or whatever the item might be); k is the cost of placing an order (the same, no matter how often you order); c is the cost of one item (a constant); m is the number of items sold each week (a constant); and h is the weekly holding cost per item (a constant that takes into account things such as space, utilities, insurance, and security). Find dA/dq and d^2A/dq^2.

3.4 The Derivative as a Rate of Change

In this section we study applications in which derivatives model the rates at which things change. It is natural to think of a quantity changing with respect to time, but other variables can be treated in the same way. For example, an economist may want to study how the cost of producing steel varies with the number of tons produced, or an engineer may want to know how the power output of a generator varies with its temperature.

Instantaneous Rates of Change

If we interpret the difference quotient $(f(x + h) - f(x))/h$ as the average rate of change in f over the interval from x to $x + h$, we can interpret its limit as $h \to 0$ as the instantaneous rate at which f is changing at the point x. This gives an important interpretation of the derivative.

DEFINITION The **instantaneous rate of change** of f with respect to x at x_0 is the derivative

$$f'(x_0) = \lim_{h \to 0} \frac{f(x_0 + h) - f(x_0)}{h},$$

provided the limit exists.

Thus, instantaneous rates are limits of average rates.

It is conventional to use the word *instantaneous* even when x does not represent time. The word is, however, frequently omitted. When we say *rate of change*, we mean *instantaneous rate of change*.

EXAMPLE 1 The area A of a circle is related to its diameter by the equation

$$A = \frac{\pi}{4}D^2.$$

How fast does the area change with respect to the diameter when the diameter is 10 m?

Solution The rate of change of the area with respect to the diameter is

$$\frac{dA}{dD} = \frac{\pi}{4} \cdot 2D = \frac{\pi D}{2}.$$

When $D = 10$ m, the area is changing with respect to the diameter at the rate of $(\pi/2)10 = 5\pi \text{ m}^2/\text{m} \approx 15.71 \text{ m}^2/\text{m}$. ∎

Motion Along a Line: Displacement, Velocity, Speed, Acceleration, and Jerk

Suppose that an object (or body, considered as a whole mass) is moving along a coordinate line (an s-axis), usually horizontal or vertical, so that we know its position s on that line as a function of time t:

$$s = f(t).$$

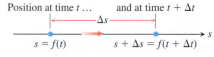

Position at time t ... and at time $t + \Delta t$

$s = f(t)$ $s + \Delta s = f(t + \Delta t)$

FIGURE 3.15 The positions of a body moving along a coordinate line at time t and shortly later at time $t + \Delta t$. Here the coordinate line is horizontal.

The **displacement** of the object over the time interval from t to $t + \Delta t$ (Figure 3.15) is

$$\Delta s = f(t + \Delta t) - f(t),$$

and the **average velocity** of the object over that time interval is

$$v_{av} = \frac{\text{displacement}}{\text{travel time}} = \frac{\Delta s}{\Delta t} = \frac{f(t + \Delta t) - f(t)}{\Delta t}.$$

To find the body's velocity at the exact instant t, we take the limit of the average velocity over the interval from t to $t + \Delta t$ as Δt shrinks to zero. This limit is the derivative of f with respect to t.

DEFINITION **Velocity (instantaneous velocity)** is the derivative of position with respect to time. If a body's position at time t is $s = f(t)$, then the body's velocity at time t is

$$v(t) = \frac{ds}{dt} = \lim_{\Delta t \to 0} \frac{f(t + \Delta t) - f(t)}{\Delta t}.$$

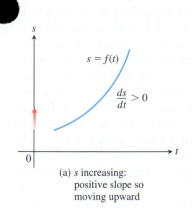

(a) s increasing: positive slope so moving upward

$s = f(t)$

$\frac{ds}{dt} > 0$

Besides telling how fast an object is moving along the horizontal line in Figure 3.15, its velocity tells the direction of motion. When the object is moving forward (s increasing), the velocity is positive; when the object is moving backward (s decreasing), the velocity is negative. If the coordinate line is vertical, the object moves upward for positive velocity and downward for negative velocity. The blue curves in Figure 3.16 represent position along the line over time; they do not portray the path of motion, which lies along the vertical s-axis.

If we drive to a friend's house and back at 30 mph, say, the speedometer will show 30 on the way over but it will not show -30 on the way back, even though our distance from home is decreasing. The speedometer always shows *speed*, which is the absolute value of velocity. Speed measures the rate of progress regardless of direction.

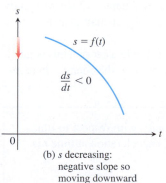

(b) s decreasing: negative slope so moving downward

$s = f(t)$

$\frac{ds}{dt} < 0$

DEFINITION **Speed** is the absolute value of velocity.

$$\text{Speed} = |v(t)| = \left| \frac{ds}{dt} \right|$$

FIGURE 3.16 For motion $s = f(t)$ along a straight line (the vertical axis), $v = ds/dt$ is (a) positive when s increases and (b) negative when s decreases.

EXAMPLE 2 Figure 3.17 shows the graph of the velocity $v = f'(t)$ of a particle moving along a horizontal line (as opposed to showing a position function $s = f(t)$ such as in Figure 3.16). In the graph of the velocity function, it's not the slope of the curve that

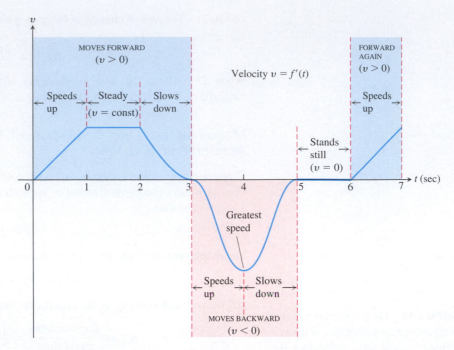

FIGURE 3.17 The velocity graph of a particle moving along a horizontal line, discussed in Example 2.

tells us if the particle is moving forward or backward along the line (which is not shown in the figure), but rather the sign of the velocity. Looking at Figure 3.17, we see that the particle moves forward for the first 3 sec (when the velocity is positive), moves backward for the next 2 sec (the velocity is negative), stands motionless for a full second, and then moves forward again. The particle is speeding up when its positive velocity increases during the first second, moves at a steady speed during the next second, and then slows down as the velocity decreases to zero during the third second. It stops for an instant at $t = 3$ sec (when the velocity is zero) and reverses direction as the velocity starts to become negative. The particle is now moving backward and gaining in speed until $t = 4$ sec, at which time it achieves its greatest speed during its backward motion. Continuing its backward motion at time $t = 4$, the particle starts to slow down again until it finally stops at time $t = 5$ (when the velocity is once again zero). The particle now remains motionless for one full second, and then moves forward again at $t = 6$ sec, speeding up during the final second of the forward motion indicated in the velocity graph. ∎

The rate at which a body's velocity changes is the body's *acceleration*. The acceleration measures how quickly the body picks up or loses speed. In Chapter 13 we will study motion in the plane and in space, where acceleration of an object may also lead to a change in direction.

A sudden change in acceleration is called a *jerk*. When a ride in a car or a bus is jerky, it is not that the accelerations involved are necessarily large but that the changes in acceleration are abrupt.

DEFINITIONS **Acceleration** is the derivative of velocity with respect to time. If a body's position at time t is $s = f(t)$, then the body's acceleration at time t is

$$a(t) = \frac{dv}{dt} = \frac{d^2s}{dt^2}.$$

Jerk is the derivative of acceleration with respect to time:

$$j(t) = \frac{da}{dt} = \frac{d^3s}{dt^3}.$$

Near the surface of Earth all bodies fall with the same constant acceleration. Galileo's experiments with free fall (see Section 2.1) lead to the equation

$$s = \frac{1}{2}gt^2,$$

where s is the distance fallen and g is the acceleration due to Earth's gravity. This equation holds in a vacuum, where there is no air resistance, and closely models the fall of dense, heavy objects, such as rocks or steel tools, for the first few seconds of their fall, before the effects of air resistance are significant.

The value of g in the equation $s = (1/2)gt^2$ depends on the units used to measure t and s. With t in seconds (the usual unit), the value of g determined by measurement at sea level is approximately 32 ft/sec^2 (feet per second squared) in English units, and $g = 9.8 \text{ m/sec}^2$ (meters per second squared) in metric units. (These gravitational constants depend on the distance from Earth's center of mass, and are slightly lower on top of Mt. Everest, for example.)

The jerk associated with the constant acceleration of gravity ($g = 32 \text{ ft/sec}^2$) is zero:

$$j = \frac{d}{dt}(g) = 0.$$

An object does not exhibit jerkiness during free fall.

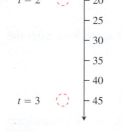

t (seconds)		s (meters)
$t = 0$	●	0
$t = 1$	○	5
		10
		15
$t = 2$	○	20
		25
		30
		35
		40
$t = 3$	○	45

FIGURE 3.18 A ball bearing falling from rest (Example 3).

EXAMPLE 3   Figure 3.18 shows the free fall of a heavy ball bearing released from rest at time $t = 0$ sec.

(a) How many meters does the ball fall in the first 3 sec?

(b) What is its velocity, speed, and acceleration when $t = 3$?

Solution

(a) The metric free-fall equation is $s = 4.9t^2$. During the first 3 sec, the ball falls

$$s(3) = 4.9(3)^2 = 44.1 \text{ m}.$$

(b) At any time t, *velocity* is the derivative of position:

$$v(t) = s'(t) = \frac{d}{dt}(4.9t^2) = 9.8t.$$

At $t = 3$, the velocity is

$$v(3) = 29.4 \text{ m/sec}$$

in the downward (increasing s) direction. The *speed* at $t = 3$ is

$$\text{speed} = |v(3)| = 29.4 \text{ m/sec}.$$

The *acceleration* at any time t is

$$a(t) = v'(t) = s''(t) = 9.8 \text{ m/sec}^2.$$

At $t = 3$, the acceleration is 9.8 m/sec^2.   ■

EXAMPLE 4   A dynamite blast blows a heavy rock straight up with a launch velocity of 160 ft/sec (about 109 mph) (Figure 3.19a). It reaches a height of $s = 160t - 16t^2$ ft after t sec.

(a) How high does the rock go?

(b) What are the velocity and speed of the rock when it is 256 ft above the ground on the way up? On the way down?

(c) What is the acceleration of the rock at any time t during its flight (after the blast)?

(d) When does the rock hit the ground again?

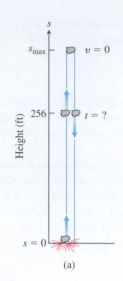

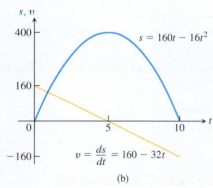

FIGURE 3.19 (a) The rock in Example 4. (b) The graphs of s and v as functions of time; s is largest when $v = ds/dt = 0$. The graph of s is *not* the path of the rock: It is a plot of height versus time. The slope of the plot is the rock's velocity, graphed here as a straight line.

Solution

(a) In the coordinate system we have chosen, s measures height from the ground up, so the velocity is positive on the way up and negative on the way down. The instant the rock is at its highest point is the one instant during the flight when the velocity is 0. To find the maximum height, all we need to do is to find when $v = 0$ and evaluate s at this time.

At any time t during the rock's motion, its velocity is

$$v = \frac{ds}{dt} = \frac{d}{dt}(160t - 16t^2) = 160 - 32t \text{ ft/sec.}$$

The velocity is zero when

$$160 - 32t = 0 \quad \text{or} \quad t = 5 \text{ sec.}$$

The rock's height at $t = 5$ sec is

$$s_{max} = s(5) = 160(5) - 16(5)^2 = 800 - 400 = 400 \text{ ft.}$$

See Figure 3.19b.

(b) To find the rock's velocity at 256 ft on the way up and again on the way down, we first find the two values of t for which

$$s(t) = 160t - 16t^2 = 256.$$

To solve this equation, we write

$$16t^2 - 160t + 256 = 0$$
$$16(t^2 - 10t + 16) = 0$$
$$(t - 2)(t - 8) = 0$$
$$t = 2 \text{ sec}, t = 8 \text{ sec.}$$

The rock is 256 ft above the ground 2 sec after the explosion and again 8 sec after the explosion. The rock's velocities at these times are

$$v(2) = 160 - 32(2) = 160 - 64 = 96 \text{ ft/sec.}$$
$$v(8) = 160 - 32(8) = 160 - 256 = -96 \text{ ft/sec.}$$

At both instants, the rock's speed is 96 ft/sec. Since $v(2) > 0$, the rock is moving upward (s is increasing) at $t = 2$ sec; it is moving downward (s is decreasing) at $t = 8$ because $v(8) < 0$.

(c) At any time during its flight following the explosion, the rock's acceleration is a constant

$$a = \frac{dv}{dt} = \frac{d}{dt}(160 - 32t) = -32 \text{ ft/sec}^2.$$

The acceleration is always downward and is the effect of gravity on the rock. As the rock rises, it slows down; as it falls, it speeds up.

(d) The rock hits the ground at the positive time t for which $s = 0$. The equation $160t - 16t^2 = 0$ factors to give $16t(10 - t) = 0$, so it has solutions $t = 0$ and $t = 10$. At $t = 0$, the blast occurred and the rock was thrown upward. It returns to the ground 10 sec later. ◼

Derivatives in Economics

Economists have a specialized vocabulary for rates of change and derivatives. They call them *marginals*.

In a manufacturing operation, the *cost of production* $c(x)$ is a function of x, the number of units produced. The **marginal cost of production** is the rate of change of cost with respect to level of production, so it is dc/dx.

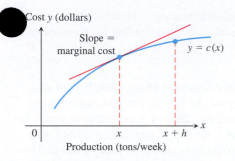

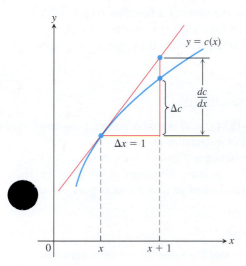

FIGURE 3.20 Weekly steel production: $c(x)$ is the cost of producing x tons per week. The cost of producing an additional h tons is $c(x + h) - c(x)$.

FIGURE 3.21 The marginal cost dc/dx is approximately the extra cost Δc of producing $\Delta x = 1$ more unit.

Suppose that $c(x)$ represents the dollars needed to produce x tons of steel in one week. It costs more to produce $x + h$ tons per week, and the cost difference, divided by h, is the average cost of producing each additional ton:

$$\frac{c(x + h) - c(x)}{h} = \begin{array}{l} \text{average cost of each of the additional} \\ \text{h tons of steel produced.} \end{array}$$

The limit of this ratio as $h \to 0$ is the *marginal cost* of producing more steel per week when the current weekly production is x tons (Figure 3.20):

$$\frac{dc}{dx} = \lim_{h \to 0} \frac{c(x + h) - c(x)}{h} = \text{marginal cost of production.}$$

Sometimes the marginal cost of production is loosely defined to be the extra cost of producing one additional unit:

$$\frac{\Delta c}{\Delta x} = \frac{c(x + 1) - c(x)}{1},$$

which is approximated by the value of dc/dx at x. This approximation is acceptable if the slope of the graph of c does not change quickly near x. Then the difference quotient will be close to its limit dc/dx, which is the rise in the tangent line if $\Delta x = 1$ (Figure 3.21). The approximation often works well for large values of x.

Economists often represent a total cost function by a cubic polynomial

$$c(x) = \alpha x^3 + \beta x^2 + \gamma x + \delta$$

where δ represents *fixed costs*, such as rent, heat, equipment capitalization, and management costs. The other terms represent *variable costs*, such as the costs of raw materials, taxes, and labor. Fixed costs are independent of the number of units produced, whereas variable costs depend on the quantity produced. A cubic polynomial is usually adequate to capture the cost behavior on a realistic quantity interval.

EXAMPLE 5   Suppose that it costs

$$c(x) = x^3 - 6x^2 + 15x$$

dollars to produce x radiators when 8 to 30 radiators are produced and that

$$r(x) = x^3 - 3x^2 + 12x$$

gives the dollar revenue from selling x radiators. Your shop currently produces 10 radiators a day. About how much extra will it cost to produce one more radiator a day, and what is your estimated increase in revenue and increase in profit for selling 11 radiators a day?

Solution   The cost of producing one more radiator a day when 10 are produced is about $c'(10)$:

$$c'(x) = \frac{d}{dx}\left(x^3 - 6x^2 + 15x\right) = 3x^2 - 12x + 15$$

$$c'(10) = 3(100) - 12(10) + 15 = 195.$$

The additional cost will be about \$195. The marginal revenue is

$$r'(x) = \frac{d}{dx}(x^3 - 3x^2 + 12x) = 3x^2 - 6x + 12.$$

The marginal revenue function estimates the increase in revenue that will result from selling one additional unit. If you currently sell 10 radiators a day, you can expect your revenue to increase by about

$$r'(10) = 3(100) - 6(10) + 12 = \$252$$

if you increase sales to 11 radiators a day. The estimated increase in profit is obtained by subtracting the increased cost of $195 from the increased revenue, leading to an estimated profit increase of $252 − $195 = $57.   ■

EXAMPLE 6   Marginal rates frequently arise in discussions of tax rates. If your marginal income tax rate is 28% and your income increases by $1000, you can expect to pay an extra $280 in taxes. This does not mean that you pay 28% of your entire income in taxes. It just means that at your current income level I, the rate of increase of taxes T with respect to income is $dT/dI = 0.28$. You will pay $0.28 in taxes out of every extra dollar you earn. As your income increases, you may land in a higher tax bracket and your marginal rate will increase.   ■

Sensitivity to Change

When a small change in x produces a large change in the value of a function $f(x)$, we say that the function is sensitive to changes in x. The derivative $f'(x)$ is a measure of this sensitivity. The function is more sensitive when $|f'(x)|$ is larger (when the slope of the graph of f is steeper).

EXAMPLE 7   Genetic Data and Sensitivity to Change

The Austrian monk Gregor Johann Mendel (1822–1884), working with garden peas and other plants, provided the first scientific explanation of hybridization.

His careful records showed that if p (a number between 0 and 1) is the frequency of the gene for smooth skin in peas (dominant) and $(1 - p)$ is the frequency of the gene for wrinkled skin in peas, then the proportion of smooth-skinned peas in the next generation will be

$$y = 2p(1 - p) + p^2 = 2p - p^2.$$

The graph of y versus p in Figure 3.22a suggests that the value of y is more sensitive to a change in p when p is small than when p is large. Indeed, this fact is borne out by the derivative graph in Figure 3.22b, which shows that dy/dp is close to 2 when p is near 0 and close to 0 when p is near 1.

The implication for genetics is that introducing a few more smooth skin genes into a population where the frequency of wrinkled skin peas is large will have a more dramatic effect on later generations than will a similar increase when the population has a large proportion of smooth skin peas.   ■

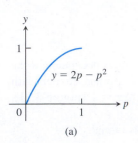

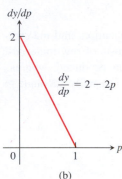

FIGURE 3.22   (a) The graph of $y = 2p - p^2$, describing the proportion of smooth-skinned peas in the next generation. (b) The graph of dy/dp (Example 7).

EXERCISES 3.4

Motion Along a Coordinate Line

Exercises 1–6 give the positions $s = f(t)$ of a body moving on a coordinate line, with s in meters and t in seconds.

   a. Find the body's displacement and average velocity for the given time interval.

   b. Find the body's speed and acceleration at the endpoints of the interval.

   c. When, if ever, during the interval does the body change direction?

1. $s = t^2 - 3t + 2, \quad 0 \le t \le 2$

2. $s = 6t - t^2, \quad 0 \le t \le 6$

3. $s = -t^3 + 3t^2 - 3t, \quad 0 \le t \le 3$

4. $s = (t^4/4) - t^3 + t^2, \quad 0 \le t \le 3$

5. $s = \dfrac{25}{t^2} - \dfrac{5}{t}, \quad 1 \le t \le 5$

6. $s = \dfrac{25}{t + 5}, \quad -4 \le t \le 0$

7. Particle motion   At time t, the position of a body moving along the s-axis is $s = t^3 - 6t^2 + 9t$ m.

   a. Find the body's acceleration each time the velocity is zero.

   b. Find the body's speed each time the acceleration is zero.

   c. Find the total distance traveled by the body from $t = 0$ to $t = 2$.

8. **Particle motion**   At time $t \geq 0$, the velocity of a body moving along the horizontal s-axis is $v = t^2 - 4t + 3$.

   a. Find the body's acceleration each time the velocity is zero.

   b. When is the body moving forward? Backward?

   c. When is the body's velocity increasing? Decreasing?

Free-Fall Applications

9. **Free fall on Mars and Jupiter**   The equations for free fall at the surfaces of Mars and Jupiter (s in meters, t in seconds) are $s = 1.86t^2$ on Mars and $s = 11.44t^2$ on Jupiter. How long does it take a rock falling from rest to reach a velocity of 27.8 m/sec (about 100 km/h) on each planet?

10. **Lunar projectile motion**   A rock thrown vertically upward from the surface of the moon at a velocity of 24 m/sec (about 86 km/h) reaches a height of $s = 24t - 0.8t^2$ m in t sec.

    a. Find the rock's velocity and acceleration at time t. (The acceleration in this case is the acceleration of gravity on the moon.)

    b. How long does it take the rock to reach its highest point?

    c. How high does the rock go?

    d. How long does it take the rock to reach half its maximum height?

    e. How long is the rock aloft?

11. **Finding g on a small airless planet**   Explorers on a small airless planet used a spring gun to launch a ball bearing vertically upward from the surface at a launch velocity of 15 m/sec. Because the acceleration of gravity at the planet's surface was g_s m/sec^2, the explorers expected the ball bearing to reach a height of $s = 15t - (1/2)g_s t^2$ m t sec later. The ball bearing reached its maximum height 20 sec after being launched. What was the value of g_s?

12. **Speeding bullet**   A 45-caliber bullet shot straight up from the surface of the moon would reach a height of $s = 832t - 2.6t^2$ ft after t sec. On Earth, in the absence of air, its height would be $s = 832t - 16t^2$ ft after t sec. How long will the bullet be aloft in each case? How high will the bullet go?

13. **Free fall from the Tower of Pisa**   Had Galileo dropped a cannonball from the Tower of Pisa, 179 ft above the ground, the ball's height above the ground t sec into the fall would have been $s = 179 - 16t^2$.

    a. What would have been the ball's velocity, speed, and acceleration at time t?

    b. About how long would it have taken the ball to hit the ground?

    c. What would have been the ball's velocity at the moment of impact?

14. **Galileo's free-fall formula**   Galileo developed a formula for a body's velocity during free fall by rolling balls from rest down increasingly steep inclined planks and looking for a limiting formula that would predict a ball's behavior when the plank was vertical and the ball fell freely; see part (a) of the accompanying figure. He found that, for any given angle of the plank, the ball's velocity t sec into motion was a constant multiple of t. That is, the velocity was given by a formula of the form $v = kt$. The value of the constant k depended on the inclination of the plank.

In modern notation—part (b) of the figure—with distance in meters and time in seconds, what Galileo determined by experiment was that, for any given angle θ, the ball's velocity t sec into the roll was $v = 9.8(\sin\theta)t$ m/sec.

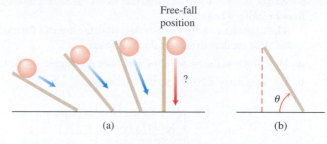

(a)   (b)

a. What is the equation for the ball's velocity during free fall?

b. Building on your work in part (a), what constant acceleration does a freely falling body experience near the surface of Earth?

Understanding Motion from Graphs

15. The accompanying figure shows the velocity $v = ds/dt = f(t)$ (m/sec) of a body moving along a coordinate line.

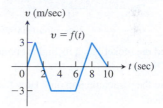

a. When does the body reverse direction?

b. When (approximately) is the body moving at a constant speed?

c. Graph the body's speed for $0 \leq t \leq 10$.

d. Graph the acceleration, where defined.

16. A particle P moves on the number line shown in part (a) of the accompanying figure. Part (b) shows the position of P as a function of time t.

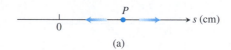

(a)

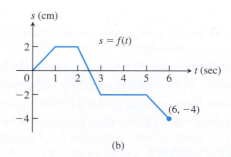

(b)

a. When is P moving to the left? Moving to the right? Standing still?

b. Graph the particle's velocity and speed (where defined).

17. Launching a rocket When a model rocket is launched, the propellant burns for a few seconds, accelerating the rocket upward. After burnout, the rocket coasts upward for a while and then begins to fall. A small explosive charge pops out a parachute shortly after the rocket starts down. The parachute slows the rocket to keep it from breaking when it lands.

The figure here shows velocity data from the flight of the model rocket. Use the data to answer the following.

a. How fast was the rocket climbing when the engine stopped?

b. For how many seconds did the engine burn?

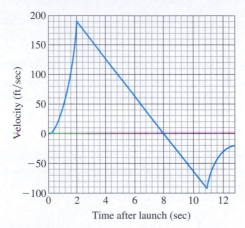
Time after launch (sec)

c. When did the rocket reach its highest point? What was its velocity then?

d. When did the parachute pop out? How fast was the rocket falling then?

e. How long did the rocket fall before the parachute opened?

f. When was the rocket's acceleration greatest?

g. When was the acceleration constant? What was its value then (to the nearest integer)?

18. The accompanying figure shows the velocity $v = f(t)$ of a particle moving on a horizontal coordinate line.

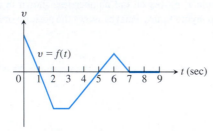

a. When does the particle move forward? Move backward? Speed up? Slow down?

b. When is the particle's acceleration positive? Negative? Zero?

c. When does the particle move at its greatest speed?

d. When does the particle stand still for more than an instant?

19. Two falling balls The multiflash photograph in the accompanying figure shows two balls falling from rest. The vertical rulers are marked in centimeters. Use the equation $s = 490t^2$ (the free-fall equation for s in centimeters and t in seconds) to answer the following questions. (*Source: PSSC Physics*, 2nd ed., Reprinted by permission of Education Development Center, Inc.)

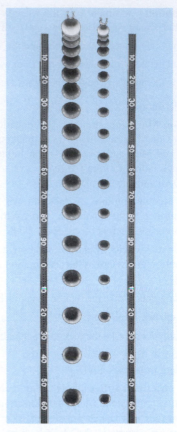

a. How long did it take the balls to fall the first 160 cm? What was their average velocity for the period?

b. How fast were the balls falling when they reached the 160-cm mark? What was their acceleration then?

c. About how fast was the light flashing (flashes per second)?

20. A traveling truck The accompanying graph shows the position s of a truck traveling on a highway. The truck starts at $t = 0$ and returns 15 h later at $t = 15$.

a. Use the technique described in Section 3.2, Example 3, to graph the truck's velocity $v = ds/dt$ for $0 \le t \le 15$. Then repeat the process, with the velocity curve, to graph the truck's acceleration dv/dt.

b. Suppose that $s = 15t^2 - t^3$. Graph ds/dt and d^2s/dt^2 and compare your graphs with those in part (a).

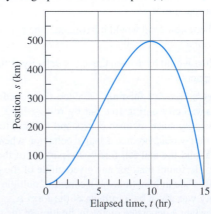
Elapsed time, t (hr)

21. The graphs in the accompanying figure show the position s, velocity $v = ds/dt$, and acceleration $a = d^2s/dt^2$ of a body moving along a coordinate line as functions of time t. Which graph is which? Give reasons for your answers.

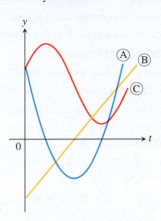

22. The graphs in the accompanying figure show the position s, the velocity $v = ds/dt$, and the acceleration $a = d^2s/dt^2$ of a body moving along a coordinate line as functions of time t. Which graph is which? Give reasons for your answers.

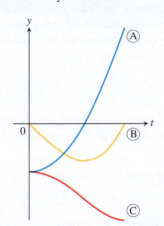

Economics

23. Marginal cost Suppose that the dollar cost of producing x washing machines is $c(x) = 2000 + 100x - 0.1x^2$.

a. Find the average cost per machine of producing the first 100 washing machines.

b. Find the marginal cost when 100 washing machines are produced.

c. Show that the marginal cost when 100 washing machines are produced is approximately the cost of producing one more washing machine after the first 100 have been made, by calculating the latter cost directly.

24. Marginal revenue Suppose that the revenue from selling x washing machines is

$$r(x) = 20{,}000\left(1 - \frac{1}{x}\right)$$

dollars.

a. Find the marginal revenue when 100 machines are produced.

b. Use the function $r'(x)$ to estimate the increase in revenue that will result from increasing production from 100 machines a week to 101 machines a week.

c. Find the limit of $r'(x)$ as $x \to \infty$. How would you interpret this number?

Additional Applications

25. Bacterium population When a bactericide was added to a nutrient broth in which bacteria were growing, the bacterium population continued to grow for a while, but then stopped growing and began to decline. The size of the population at time t (hours) was $b = 10^6 + 10^4t - 10^3t^2$. Find the growth rates at

a. $t = 0$ hours.

b. $t = 5$ hours.

c. $t = 10$ hours.

26. Body surface area A typical male's body surface area S in square meters is often modeled by the formula $S = \frac{1}{60}\sqrt{wh}$, where h is the height in cm, and w the weight in kg, of the person. Find the rate of change of body surface area with respect to weight for males of constant height $h = 180$ cm (roughly 5'9"). Does S increase more rapidly with respect to weight at lower or higher body weights? Explain.

T 27. Draining a tank It takes 12 hours to drain a storage tank by opening the valve at the bottom. The depth y of fluid in the tank t hours after the valve is opened is given by the formula

$$y = 6\left(1 - \frac{t}{12}\right)^2 \text{ m.}$$

a. Find the rate dy/dt (m/h) at which the tank is draining at time t.

b. When is the fluid level in the tank falling fastest? Slowest? What are the values of dy/dt at these times?

c. Graph y and dy/dt together and discuss the behavior of y in relation to the signs and values of dy/dt.

28. Draining a tank The number of gallons of water in a tank t minutes after the tank has started to drain is $Q(t) = 200(30 - t)^2$. How fast is the water running out at the end of 10 min? What is the average rate at which the water flows out during the first 10 min?

29. Vehicular stopping distance Based on data from the U.S. Bureau of Public Roads, a model for the total stopping distance of a moving car in terms of its speed is

$$s = 1.1v + 0.054v^2,$$

where s is measured in ft and v in mph. The linear term $1.1v$ models the distance the car travels during the time the driver perceives a need to stop until the brakes are applied, and the quadratic term $0.054v^2$ models the additional braking distance once they are applied. Find ds/dv at $v = 35$ and $v = 70$ mph, and interpret the meaning of the derivative.

30. Inflating a balloon The volume $V = (4/3)\pi r^3$ of a spherical balloon changes with the radius.

a. At what rate (ft^3/ft) does the volume change with respect to the radius when $r = 2$ ft?

b. By approximately how much does the volume increase when the radius changes from 2 to 2.2 ft?

31. **Airplane takeoff**   Suppose that the distance an aircraft travels along a runway before takeoff is given by $D = (10/9)t^2$, where D is measured in meters from the starting point and t is measured in seconds from the time the brakes are released. The aircraft will become airborne when its speed reaches 200 km/h. How long will it take to become airborne, and what distance will it travel in that time?

32. **Volcanic lava fountains**   Although the November 1959 Kilauea Iki eruption on the island of Hawaii began with a line of fountains along the wall of the crater, activity was later confined to a single vent in the crater's floor, which at one point shot lava 1900 ft straight into the air (a Hawaiian record). What was the lava's exit velocity in feet per second? In miles per hour? (*Hint*: If v_0 is the exit velocity of a particle of lava, its height t sec later will be $s = v_0 t - 16t^2$ ft. Begin by finding the time at which $ds/dt = 0$. Neglect air resistance.)

Analyzing Motion Using Graphs

T   Exercises 33–36 give the position function $s = f(t)$ of an object moving along the s-axis as a function of time t. Graph f together with the velocity function $v(t) = ds/dt = f'(t)$ and the acceleration function $a(t) = d^2s/dt^2 = f''(t)$. Comment on the object's behavior in relation to the signs and values of v and a. Include in your commentary such topics as the following:

   a. When is the object momentarily at rest?

   b. When does it move to the left (down) or to the right (up)?

   c. When does it change direction?

   d. When does it speed up and slow down?

   e. When is it moving fastest (highest speed)? Slowest?

   f. When is it farthest from the axis origin?

33. $s = 200t - 16t^2, \quad 0 \le t \le 12.5$ (a heavy object fired straight up from Earth's surface at 200 ft/sec)

34. $s = t^2 - 3t + 2, \quad 0 \le t \le 5$

35. $s = t^3 - 6t^2 + 7t, \quad 0 \le t \le 4$

36. $s = 4 - 7t + 6t^2 - t^3, \quad 0 \le t \le 4$

3.5   Derivatives of Trigonometric Functions

Many phenomena of nature are approximately periodic (electromagnetic fields, heart rhythms, tides, weather). The derivatives of sines and cosines play a key role in describing periodic changes. This section shows how to differentiate the six basic trigonometric functions.

Derivative of the Sine Function

To calculate the derivative of $f(x) = \sin x$, for x measured in radians, we combine the limits in Example 5a and Theorem 7 in Section 2.4 with the angle sum identity for the sine function:

$$\sin(x + h) = \sin x \cos h + \cos x \sin h.$$

If $f(x) = \sin x$, then

$$f'(x) = \lim_{h \to 0} \frac{f(x + h) - f(x)}{h} = \lim_{h \to 0} \frac{\sin(x + h) - \sin x}{h} \qquad \text{Derivative definition}$$

$$= \lim_{h \to 0} \frac{(\sin x \cos h + \cos x \sin h) - \sin x}{h} \qquad \text{Identity for } \sin(x + h)$$

$$= \lim_{h \to 0} \frac{\sin x (\cos h - 1) + \cos x \sin h}{h}$$

$$= \lim_{h \to 0} \left(\sin x \cdot \frac{\cos h - 1}{h} \right) + \lim_{h \to 0} \left(\cos x \cdot \frac{\sin h}{h} \right)$$

$$= \sin x \cdot \underbrace{\lim_{h \to 0} \frac{\cos h - 1}{h}}_{\text{limit 0}} + \cos x \cdot \underbrace{\lim_{h \to 0} \frac{\sin h}{h}}_{\text{limit 1}} \qquad \begin{array}{l} \text{Example 5a and} \\ \text{Theorem 7, Section 2.4} \end{array}$$

$$= \sin x \cdot 0 + \cos x \cdot 1 = \cos x.$$

> **The derivative of the sine function is the cosine function:**
>
> $$\frac{d}{dx}(\sin x) = \cos x.$$

EXAMPLE 1   We find derivatives of the sine function involving differences, products, and quotients.

(a) $y = x^2 - \sin x$:

$$\frac{dy}{dx} = 2x - \frac{d}{dx}(\sin x) \qquad \text{Difference Rule}$$

$$= 2x - \cos x$$

(b) $y = e^x \sin x$:

$$\frac{dy}{dx} = e^x \frac{d}{dx}(\sin x) + \left(\frac{d}{dx} e^x\right)\sin x \qquad \text{Product Rule}$$

$$= e^x \cos x + e^x \sin x$$

$$= e^x (\cos x + \sin x)$$

(c) $y = \dfrac{\sin x}{x}$:

$$\frac{dy}{dx} = \frac{x \cdot \dfrac{d}{dx}(\sin x) - \sin x \cdot 1}{x^2} \qquad \text{Quotient Rule}$$

$$= \frac{x \cos x - \sin x}{x^2}$$ ■

Derivative of the Cosine Function

With the help of the angle sum formula for the cosine function,

$$\cos(x + h) = \cos x \cos h - \sin x \sin h,$$

we can compute the limit of the difference quotient:

$$\frac{d}{dx}(\cos x) = \lim_{h \to 0} \frac{\cos(x + h) - \cos x}{h} \qquad \text{Derivative definition}$$

$$= \lim_{h \to 0} \frac{(\cos x \cos h - \sin x \sin h) - \cos x}{h} \qquad \begin{matrix}\text{Cosine angle}\\\text{sum identity}\end{matrix}$$

$$= \lim_{h \to 0} \frac{\cos x (\cos h - 1) - \sin x \sin h}{h}$$

$$= \lim_{h \to 0}\left(\cos x \cdot \frac{\cos h - 1}{h}\right) - \lim_{h \to 0}\left(\sin x \cdot \frac{\sin h}{h}\right)$$

$$= \cos x \cdot \lim_{h \to 0}\frac{\cos h - 1}{h} - \sin x \cdot \lim_{h \to 0}\frac{\sin h}{h} \qquad \begin{matrix}\text{Example 5a}\\\text{and Theorem 7,}\\\text{Section 2.4}\end{matrix}$$

$$= \cos x \cdot 0 - \sin x \cdot 1$$

$$= -\sin x.$$

The derivative of the cosine function is the negative of the sine function:

$$\frac{d}{dx}(\cos x) = -\sin x.$$

Figure 3.23 shows a way to visualize this result by graphing the slopes of the tangent lines to the curve $y = \cos x$.

EXAMPLE 2   We find derivatives of the cosine function in combinations with other functions.

(a) $y = 5e^x + \cos x$:

$$\frac{dy}{dx} = \frac{d}{dx}(5e^x) + \frac{d}{dx}(\cos x) \qquad \text{Sum Rule}$$

$$= 5e^x - \sin x$$

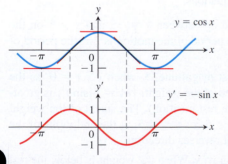

FIGURE 3.23   The curve $y' = -\sin x$ as the graph of the slopes of the tangents to the curve $y = \cos x$.

(b) $y = \sin x \cos x$:

$$\frac{dy}{dx} = \sin x \frac{d}{dx}(\cos x) + \cos x \frac{d}{dx}(\sin x) \qquad \text{Product Rule}$$

$$= \sin x \,(-\sin x) + \cos x \,(\cos x)$$

$$= \cos^2 x - \sin^2 x$$

(c) $y = \dfrac{\cos x}{1 - \sin x}$:

$$\frac{dy}{dx} = \frac{(1 - \sin x)\dfrac{d}{dx}(\cos x) - \cos x \dfrac{d}{dx}(1 - \sin x)}{(1 - \sin x)^2} \qquad \text{Quotient Rule}$$

$$= \frac{(1 - \sin x)(-\sin x) - \cos x\,(0 - \cos x)}{(1 - \sin x)^2}$$

$$= \frac{1 - \sin x}{(1 - \sin x)^2} \qquad \sin^2 x + \cos^2 x = 1$$

$$= \frac{1}{1 - \sin x} \qquad \blacksquare$$

Simple Harmonic Motion

Simple harmonic motion models the motion of an object or weight bobbing freely up and down on the end of a spring, with no resistance. The motion is periodic and repeats indefinitely, so we represent it using trigonometric functions. The next example models motion with no opposing forces (such as friction).

EXAMPLE 3 A weight hanging from a spring (Figure 3.24) is stretched down 5 units beyond its rest position and released at time $t = 0$ to bob up and down. Its position at any later time t is

$$s = 5 \cos t.$$

What are its velocity and acceleration at time t?

Solution We have

Position: $\quad s = 5 \cos t$

Velocity: $\quad v = \dfrac{ds}{dt} = \dfrac{d}{dt}(5 \cos t) = -5 \sin t$

Acceleration: $\quad a = \dfrac{dv}{dt} = \dfrac{d}{dt}(-5 \sin t) = -5 \cos t.$

Notice how much we can learn from these equations:

1. As time passes, the weight moves down and up between $s = -5$ and $s = 5$ on the s-axis. The amplitude of the motion is 5. The period of the motion is 2π, the period of the cosine function.

2. The velocity $v = -5 \sin t$ attains its greatest magnitude, 5, when $\cos t = 0$, as the graphs show in Figure 3.25. Hence, the speed of the weight, $|v| = 5|\sin t|$, is greatest when $\cos t = 0$, that is, when $s = 0$ (the rest position). The speed of the weight is zero when $\sin t = 0$. This occurs when $s = 5 \cos t = \pm 5$, at the endpoints of the interval of motion. At these points the weight reverses direction.

3. The weight is acted on by the spring and by gravity. When the weight is below the rest position, the combined forces pull it up, and when it is above the rest position, they pull it down. The weight's acceleration is always proportional to the negative of its displacement. This property of springs is called *Hooke's Law*, and is studied further in Section 6.5.

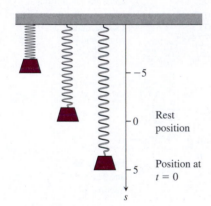

FIGURE 3.24 A weight hanging from a vertical spring and then displaced oscillates above and below its rest position (Example 3).

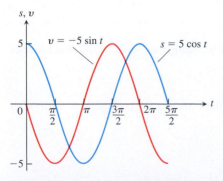

FIGURE 3.25 The graphs of the position and velocity of the weight in Example 3.

4. The acceleration, $a = -5\cos t$, is zero only at the rest position, where $\cos t = 0$ and the force of gravity and the force from the spring balance each other. When the weight is anywhere else, the two forces are unequal and acceleration is nonzero. The acceleration is greatest in magnitude at the points farthest from the rest position, where $\cos t = \pm 1$. ■

EXAMPLE 4   The jerk associated with the simple harmonic motion in Example 3 is

$$j = \frac{da}{dt} = \frac{d}{dt}(-5\cos t) = 5\sin t.$$

It has its greatest magnitude when $\sin t = \pm 1$, not at the extremes of the displacement but at the rest position, where the acceleration changes direction and sign. ■

Derivatives of the Other Basic Trigonometric Functions

Because $\sin x$ and $\cos x$ are differentiable functions of x, the related functions

$$\tan x = \frac{\sin x}{\cos x}, \qquad \cot x = \frac{\cos x}{\sin x}, \qquad \sec x = \frac{1}{\cos x}, \qquad \text{and} \qquad \csc x = \frac{1}{\sin x}$$

are differentiable at every value of x at which they are defined. Their derivatives, calculated from the Quotient Rule, are given by the following formulas. Notice the negative signs in the derivative formulas for the cofunctions.

The derivatives of the other trigonometric functions:

$$\frac{d}{dx}(\tan x) = \sec^2 x \qquad\qquad \frac{d}{dx}(\cot x) = -\csc^2 x$$

$$\frac{d}{dx}(\sec x) = \sec x \tan x \qquad\qquad \frac{d}{dx}(\csc x) = -\csc x \cot x$$

To show a typical calculation, we find the derivative of the tangent function. The other derivations are left to Exercise 62.

EXAMPLE 5   Find $d(\tan x)/dx$.

Solution   We use the Derivative Quotient Rule to calculate the derivative:

$$\frac{d}{dx}(\tan x) = \frac{d}{dx}\left(\frac{\sin x}{\cos x}\right) = \frac{\cos x \dfrac{d}{dx}(\sin x) - \sin x \dfrac{d}{dx}(\cos x)}{\cos^2 x} \qquad \textcolor{blue}{\text{Quotient Rule}}$$

$$= \frac{\cos x \cos x - \sin x (-\sin x)}{\cos^2 x}$$

$$= \frac{\cos^2 x + \sin^2 x}{\cos^2 x}$$

$$= \frac{1}{\cos^2 x} = \sec^2 x. \qquad ■$$

EXAMPLE 6   Find y'' if $y = \sec x$.

Solution   Finding the second derivative involves a combination of trigonometric derivatives.

$$y = \sec x$$

$$y' = \sec x \tan x \qquad\qquad \text{Derivative rule for secant function}$$

$$y'' = \frac{d}{dx}(\sec x \tan x)$$

$$= \sec x \frac{d}{dx}(\tan x) + \tan x \frac{d}{dx}(\sec x) \qquad \text{Derivative Product Rule}$$

$$= \sec x (\sec^2 x) + \tan x (\sec x \tan x) \qquad \text{Derivative rules}$$

$$= \sec^3 x + \sec x \tan^2 x \qquad\qquad\qquad \blacksquare$$

The differentiability of the trigonometric functions throughout their domains implies their continuity at every point in their domains (Theorem 1, Section 3.2). So we can calculate limits of algebraic combinations and compositions of trigonometric functions by direct substitution.

EXAMPLE 7   We can use direct substitution in computing limits involving trigonometric functions. We must be careful to avoid division by zero, which is algebraically undefined.

$$\lim_{x \to 0} \frac{\sqrt{2 + \sec x}}{\cos(\pi - \tan x)} = \frac{\sqrt{2 + \sec 0}}{\cos(\pi - \tan 0)} = \frac{\sqrt{2 + 1}}{\cos(\pi - 0)} = \frac{\sqrt{3}}{-1} = -\sqrt{3} \qquad \blacksquare$$

EXERCISES   3.5

Derivatives

In Exercises 1–18, find dy/dx.

1. $y = -10x + 3 \cos x$

2. $y = \frac{3}{x} + 5 \sin x$

3. $y = x^2 \cos x$

4. $y = \sqrt{x} \sec x + 3$

5. $y = \csc x - 4\sqrt{x} + \frac{7}{e^x}$

6. $y = x^2 \cot x - \frac{1}{x^2}$

7. $f(x) = \sin x \tan x$

8. $g(x) = \frac{\cos x}{\sin^2 x}$

9. $y = xe^{-x} \sec x$

10. $y = (\sin x + \cos x)\sec x$

11. $y = \frac{\cot x}{1 + \cot x}$

12. $y = \frac{\cos x}{1 + \sin x}$

13. $y = \frac{4}{\cos x} + \frac{1}{\tan x}$

14. $y = \frac{\cos x}{x} + \frac{x}{\cos x}$

15. $y = (\sec x + \tan x)(\sec x - \tan x)$

16. $y = x^2 \cos x - 2x \sin x - 2 \cos x$

17. $f(x) = x^3 \sin x \cos x$

18. $g(x) = (2 - x)\tan^2 x$

In Exercises 19–22, find ds/dt.

19. $s = \tan t - e^{-t}$

20. $s = t^2 - \sec t + 5e^t$

21. $s = \frac{1 + \csc t}{1 - \csc t}$

22. $s = \frac{\sin t}{1 - \cos t}$

In Exercises 23–26, find $dr/d\theta$.

23. $r = 4 - \theta^2 \sin \theta$

24. $r = \theta \sin \theta + \cos \theta$

25. $r = \sec \theta \csc \theta$

26. $r = (1 + \sec \theta)\sin \theta$

In Exercises 27–32, find dp/dq.

27. $p = 5 + \frac{1}{\cot q}$

28. $p = (1 + \csc q)\cos q$

29. $p = \frac{\sin q + \cos q}{\cos q}$

30. $p = \frac{\tan q}{1 + \tan q}$

31. $p = \frac{q \sin q}{q^2 - 1}$

32. $p = \frac{3q + \tan q}{q \sec q}$

33. Find y'' if

   a. $y = \csc x$.

   b. $y = \sec x$.

34. Find $y^{(4)} = d^4 y/dx^4$ if

   a. $y = -2 \sin x$.

   b. $y = 9 \cos x$.

Tangent Lines

In Exercises 35–38, graph the curves over the given intervals, together with their tangent lines at the given values of x. Label each curve and tangent line with its equation.

35. $y = \sin x, \quad -3\pi/2 \le x \le 2\pi$

   $x = -\pi, 0, 3\pi/2$

36. $y = \tan x, \quad -\pi/2 < x < \pi/2$

   $x = -\pi/3, 0, \pi/3$

37. $y = \sec x, \quad -\pi/2 < x < \pi/2$

   $x = -\pi/3, \pi/4$

38. $y = 1 + \cos x, \quad -3\pi/2 \le x \le 2\pi$

   $x = -\pi/3, 3\pi/2$

T Do the graphs of the functions in Exercises 39–44 have any horizontal tangent lines in the interval $0 \le x \le 2\pi$? If so, where? If not, why not? Visualize your findings by graphing the functions with a grapher.

39. $y = x + \sin x$

40. $y = 2x + \sin x$

41. $y = x - \cot x$

42. $y = x + 2 \cos x$

43. $y = \frac{\sec x}{3 + \sec x}$

44. $y = \frac{\cos x}{3 - 4 \sin x}$

45. Find all points on the curve $y = \tan x$, $-\pi/2 < x < \pi/2$, where the tangent line is parallel to the line $y = 2x$. Sketch the curve and tangent(s) together, labeling each with its equation.

46. Find all points on the curve $y = \cot x$, $0 < x < \pi$, where the tangent line is parallel to the line $y = -x$. Sketch the curve and tangent(s) together, labeling each with its equation.

In Exercises 47 and 48, find an equation for **(a)** the tangent to the curve at P and **(b)** the horizontal tangent to the curve at Q.

47.

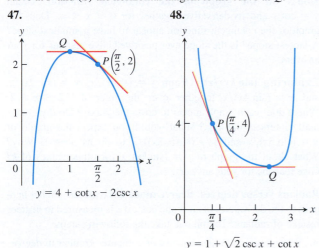

$y = 4 + \cot x - 2\csc x$

48.

$P\left(\frac{\pi}{4}, 4\right)$

$y = 1 + \sqrt{2} \csc x + \cot x$

Trigonometric Limits

Find the limits in Exercises 49–56.

49. $\lim\limits_{x \to 2} \sin\left(\dfrac{1}{x} - \dfrac{1}{2}\right)$

50. $\lim\limits_{x \to -\pi/6} \sqrt{1 + \cos(\pi \csc x)}$

51. $\lim\limits_{\theta \to \pi/6} \dfrac{\sin\theta - \frac{1}{2}}{\theta - \frac{\pi}{6}}$

52. $\lim\limits_{\theta \to \pi/4} \dfrac{\tan\theta - 1}{\theta - \frac{\pi}{4}}$

53. $\lim\limits_{x \to 0} \sec\left[e^x + \pi \tan\left(\dfrac{\pi}{4 \sec x}\right) - 1 \right]$

54. $\lim\limits_{x \to 0} \sin\left(\dfrac{\pi + \tan x}{\tan x - 2\sec x}\right)$

55. $\lim\limits_{t \to 0} \tan\left(1 - \dfrac{\sin t}{t}\right)$

56. $\lim\limits_{\theta \to 0} \cos\left(\dfrac{\pi\theta}{\sin\theta}\right)$

Theory and Examples

The equations in Exercises 57 and 58 give the position $s = f(t)$ of a body moving on a coordinate line (s in meters, t in seconds). Find the body's velocity, speed, acceleration, and jerk at time $t = \pi/4$ sec.

57. $s = 2 - 2\sin t$

58. $s = \sin t + \cos t$

59. Is there a value of c that will make

$$f(x) = \begin{cases} \dfrac{\sin^2 3x}{x^2}, & x \neq 0 \\ c, & x = 0 \end{cases}$$

continuous at $x = 0$? Give reasons for your answer.

60. Is there a value of b that will make

$$g(x) = \begin{cases} x + b, & x < 0 \\ \cos x, & x \geq 0 \end{cases}$$

continuous at $x = 0$? Differentiable at $x = 0$? Give reasons for your answers.

61. By computing the first few derivatives and looking for a pattern, find the following derivatives.

a. $\dfrac{d^{999}}{dx^{999}} (\cos x)$

b. $\dfrac{d^{110}}{dx^{110}} (\sin x - 3\cos x)$

c. $\dfrac{d^{73}}{dx^{73}} (x \sin x)$

62. Derive the formula for the derivative with respect to x of

a. $\sec x$. **b.** $\csc x$. **c.** $\cot x$.

63. A weight is attached to a spring and reaches its equilibrium position ($x = 0$). It is then set in motion resulting in a displacement of

$$x = 10 \cos t,$$

where x is measured in centimeters and t is measured in seconds. See the accompanying figure.

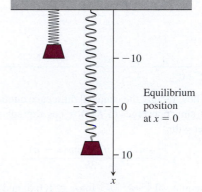

a. Find the spring's displacement when $t = 0$, $t = \pi/3$, and $t = 3\pi/4$.

b. Find the spring's velocity when $t = 0$, $t = \pi/3$, and $t = 3\pi/4$.

64. Assume that a particle's position on the x-axis is given by

$$x = 3\cos t + 4\sin t,$$

where x is measured in feet and t is measured in seconds.

a. Find the particle's position when $t = 0$, $t = \pi/2$, and $t = \pi$.

b. Find the particle's velocity when $t = 0$, $t = \pi/2$, and $t = \pi$.

T **65.** Graph $y = \cos x$ for $-\pi \leq x \leq 2\pi$. On the same screen, graph

$$y = \dfrac{\sin(x + h) - \sin x}{h}$$

for $h = 1, 0.5, 0.3$, and 0.1. Then, in a new window, try $h = -1, -0.5$, and -0.3. What happens as $h \to 0^+$? As $h \to 0^-$? What phenomenon is being illustrated here?

T **66.** Graph $y = -\sin x$ for $-\pi \leq x \leq 2\pi$. On the same screen, graph

$$y = \dfrac{\cos(x + h) - \cos x}{h}$$

for $h = 1, 0.5, 0.3$, and 0.1. Then, in a new window, try $h = -1, -0.5$, and -0.3. What happens as $h \to 0^+$? As $h \to 0^-$? What phenomenon is being illustrated here?

T **67. Centered difference quotients** The *centered difference quotient*

$$\dfrac{f(x + h) - f(x - h)}{2h}$$

is used to approximate $f'(x)$ in numerical work because (1) its limit as $h \to 0$ equals $f'(x)$ when $f'(x)$ exists, and (2) it usually gives a better approximation of $f'(x)$ for a given value of h than the difference quotient

$$\frac{f(x + h) - f(x)}{h}.$$

See the accompanying figure.

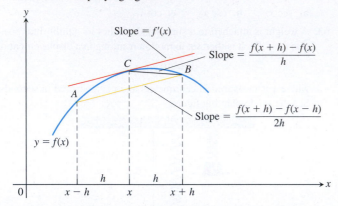

a. To see how rapidly the centered difference quotient for $f(x) = \sin x$ converges to $f'(x) = \cos x$, graph $y = \cos x$ together with

$$y = \frac{\sin(x + h) - \sin(x - h)}{2h}$$

over the interval $[-\pi, 2\pi]$ for $h = 1, 0.5$, and 0.3. Compare the results with those obtained in Exercise 65 for the same values of h.

b. To see how rapidly the centered difference quotient for $f(x) = \cos x$ converges to $f'(x) = -\sin x$, graph $y = -\sin x$ together with

$$y = \frac{\cos(x + h) - \cos(x - h)}{2h}$$

over the interval $[-\pi, 2\pi]$ for $h = 1, 0.5$, and 0.3. Compare the results with those obtained in Exercise 66 for the same values of h.

68. A caution about centered difference quotients (*Continuation of Exercise 67.*) The quotient

$$\frac{f(x + h) - f(x - h)}{2h}$$

may have a limit as $h \to 0$ when f has no derivative at x. As a case in point, take $f(x) = |x|$ and calculate

$$\lim_{h \to 0} \frac{|0 + h| - |0 - h|}{2h}.$$

As you will see, the limit exists even though $f(x) = |x|$ has no derivative at $x = 0$. *Moral:* Before using a centered difference quotient, be sure the derivative exists.

T **69. Slopes on the graph of the tangent function** Graph $y = \tan x$ and its derivative together on $(-\pi/2, \pi/2)$. Does the graph of the tangent function appear to have a smallest slope? A largest slope? Is the slope ever negative? Give reasons for your answers.

T **70. Slopes on the graph of the cotangent function** Graph $y = \cot x$ and its derivative together for $0 < x < \pi$. Does the graph of the cotangent function appear to have a smallest slope? A largest slope? Is the slope ever positive? Give reasons for your answers.

T **71. Exploring $(\sin kx)/x$** Graph $y = (\sin x)/x$, $y = (\sin 2x)/x$, and $y = (\sin 4x)/x$ together over the interval $-2 \le x \le 2$. Where does each graph appear to cross the y-axis? Do the graphs really intersect the axis? What would you expect the graphs of $y = (\sin 5x)/x$ and $y = (\sin(-3x))/x$ to do as $x \to 0$? Why? What about the graph of $y = (\sin kx)/x$ for other values of k? Give reasons for your answers.

T **72. Radians versus degrees: degree mode derivatives** What happens to the derivatives of $\sin x$ and $\cos x$ if x is measured in degrees instead of radians? To find out, take the following steps.

a. With your graphing calculator or computer grapher in *degree mode*, graph

$$f(h) = \frac{\sin h}{h}$$

and estimate $\lim_{h \to 0} f(h)$. Compare your estimate with $\pi/180$. Is there any reason to believe the limit *should* be $\pi/180$?

b. With your grapher still in degree mode, estimate

$$\lim_{h \to 0} \frac{\cos h - 1}{h}.$$

c. Now go back to the derivation of the formula for the derivative of $\sin x$ in the text and carry out the steps of the derivation using degree-mode limits. What formula do you obtain for the derivative?

d. Work through the derivation of the formula for the derivative of $\cos x$ using degree-mode limits. What formula do you obtain for the derivative?

e. The disadvantages of the degree-mode formulas become apparent as you start taking derivatives of higher order. Try it. What are the second and third degree-mode derivatives of $\sin x$ and $\cos x$?

3.6 The Chain Rule

How do we differentiate $F(x) = \sin(x^2 - 4)$? This function is the composition $f \circ g$ of two functions $y = f(u) = \sin u$ and $u = g(x) = x^2 - 4$ that we know how to differentiate. The answer, given by the *Chain Rule*, says that the derivative is the product of the derivatives of f and g. We develop the rule in this section.

Derivative of a Composite Function

The function $y = \dfrac{3}{2}x = \dfrac{1}{2}(3x)$ is the composition of the functions $y = \dfrac{1}{2}u$ and $u = 3x$.

We have

$$\frac{dy}{dx} = \frac{3}{2}, \qquad \frac{dy}{du} = \frac{1}{2}, \qquad \text{and} \qquad \frac{du}{dx} = 3.$$

Since $\dfrac{3}{2} = \dfrac{1}{2} \cdot 3$, we see in this case that

$$\frac{dy}{dx} = \frac{dy}{du} \cdot \frac{du}{dx}.$$

If we think of the derivative as a rate of change, this relationship is intuitively reasonable. If $y = f(u)$ changes half as fast as u and $u = g(x)$ changes three times as fast as x, then we expect y to change $3/2$ times as fast as x. This effect is much like that of a multiple gear train (Figure 3.26). Let's look at another example.

EXAMPLE 1   The function

$$y = (3x^2 + 1)^2$$

is obtained by composing the functions $y = f(u) = u^2$ and $u = g(x) = 3x^2 + 1$. Calculating derivatives, we see that

$$\frac{dy}{du} \cdot \frac{du}{dx} = 2u \cdot 6x$$

$$= 2(3x^2 + 1) \cdot 6x \qquad \text{\color{blue}Substitute for } u$$

$$= 36x^3 + 12x.$$

Calculating the derivative from the expanded formula $(3x^2 + 1)^2 = 9x^4 + 6x^2 + 1$ gives the same result:

$$\frac{dy}{dx} = \frac{d}{dx}(9x^4 + 6x^2 + 1) = 36x^3 + 12x. \qquad \blacksquare$$

The derivative of the composite function $f(g(x))$ at x is the derivative of f at $g(x)$ times the derivative of g at x. This is known as the Chain Rule (Figure 3.27).

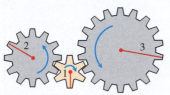

C: y turns   B: u turns   A: x turns

FIGURE 3.26 When gear A makes x turns, gear B makes u turns and gear C makes y turns. By comparing circumferences or counting teeth, we see that $y = u/2$ (C turns one-half turn for each B turn) and $u = 3x$ (B turns three times for A's one), so $y = 3x/2$. Thus, $dy/dx = 3/2 = (1/2)(3) = (dy/du)(du/dx)$.

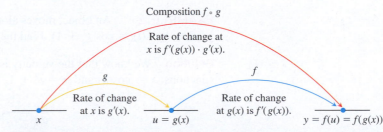

FIGURE 3.27 Rates of change multiply: The derivative of $f \circ g$ at x is the derivative of f at $g(x)$ times the derivative of g at x.

THEOREM 2—The Chain Rule If $f(u)$ is differentiable at the point $u = g(x)$ and $g(x)$ is differentiable at x, then the composite function $(f \circ g)(x) = f(g(x))$ is differentiable at x, and

$$(f \circ g)'(x) = f'(g(x)) \cdot g'(x).$$

In Leibniz's notation, if $y = f(u)$ and $u = g(x)$, then

$$\frac{dy}{dx} = \frac{dy}{du} \cdot \frac{du}{dx},$$

where dy/du is evaluated at $u = g(x)$.

A Proof of One Case of the Chain Rule:

Let Δu be the change in u when x changes by Δx, so that

$$\Delta u = g(x + \Delta x) - g(x).$$

Then the corresponding change in y is

$$\Delta y = f(u + \Delta u) - f(u).$$

If $\Delta u \neq 0$, we can write the fraction $\Delta y/\Delta x$ as the product

$$\frac{\Delta y}{\Delta x} = \frac{\Delta y}{\Delta u} \cdot \frac{\Delta u}{\Delta x} \tag{1}$$

and take the limit as $\Delta x \to 0$:

$$\frac{dy}{dx} = \lim_{\Delta x \to 0} \frac{\Delta y}{\Delta x}$$

$$= \lim_{\Delta x \to 0} \frac{\Delta y}{\Delta u} \cdot \frac{\Delta u}{\Delta x}$$

$$= \lim_{\Delta x \to 0} \frac{\Delta y}{\Delta u} \cdot \lim_{\Delta x \to 0} \frac{\Delta u}{\Delta x}$$

$$= \lim_{\Delta u \to 0} \frac{\Delta y}{\Delta u} \cdot \lim_{\Delta x \to 0} \frac{\Delta u}{\Delta x} \qquad \text{(Note that } \Delta u \to 0 \text{ as } \Delta x \to 0 \text{ since } g \text{ is continuous.)}$$

$$= \frac{dy}{du} \cdot \frac{du}{dx}.$$

The problem with this argument is that if the function $g(x)$ oscillates rapidly near x, then Δu can be zero even when $\Delta x \neq 0$, so the cancelation of Δu in Equation (1) would be invalid. A complete proof requires a different approach that avoids this problem, and we give one such proof in Section 3.11. ∎

EXAMPLE 2 An object moves along the x-axis so that its position at any time $t \geq 0$ is given by $x(t) = \cos(t^2 + 1)$. Find the velocity of the object as a function of t.

Solution We know that the velocity is dx/dt. In this instance, x is a composition of two functions: $x = \cos(u)$ and $u = t^2 + 1$. We have

$$\frac{dx}{du} = -\sin(u) \qquad x = \cos(u)$$

$$\frac{du}{dt} = 2t. \qquad u = t^2 + 1$$

By the Chain Rule,

$$\frac{dx}{dt} = \frac{dx}{du} \cdot \frac{du}{dt}$$
$$= -\sin(u) \cdot 2t$$
$$= -\sin(t^2 + 1) \cdot 2t$$
$$= -2t \sin(t^2 + 1). \qquad \blacksquare$$

Ways to Write the Chain Rule

$(f \circ g)'(x) = f'(g(x)) \cdot g'(x)$

$$\frac{dy}{dx} = \frac{dy}{du} \cdot \frac{du}{dx}$$

$$\frac{dy}{dx} = f'(g(x)) \cdot g'(x)$$

$$\frac{d}{dx} f(u) = f'(u) \frac{du}{dx}$$

"Outside-Inside" Rule

A difficulty with the Leibniz notation is that it doesn't state specifically where the derivatives in the Chain Rule are supposed to be evaluated. So it sometimes helps to write the Chain Rule using functional notation. If $y = f(g(x))$, then

$$\frac{dy}{dx} = f'(g(x)) \cdot g'(x).$$

In words, differentiate the "outside" function f and evaluate it at the "inside" function $g(x)$ left alone; then multiply by the derivative of the "inside function."

EXAMPLE 3   Differentiate $\sin(x^2 + e^x)$ with respect to x.

Solution   We apply the Chain Rule directly and find

$$\frac{d}{dx} \sin \underbrace{(x^2 + e^x)}_{\substack{\text{inside} \\ \text{left alone}}} = \cos \underbrace{(x^2 + e^x)}_{\text{inside}} \cdot \underbrace{(2x + e^x)}_{\substack{\text{derivative of} \\ \text{the inside}}}. \qquad \blacksquare$$

EXAMPLE 4   Differentiate $y = e^{\cos x}$.

Solution   Here the inside function is $u = g(x) = \cos x$ and the outside function is the exponential function $f(x) = e^x$. Applying the Chain Rule, we get

$$\frac{dy}{dx} = \frac{d}{dx}(e^{\cos x}) = e^{\cos x} \frac{d}{dx}(\cos x) = e^{\cos x}(-\sin x) = -e^{\cos x} \sin x. \qquad \blacksquare$$

Generalizing Example 4, we see that the Chain Rule gives the formula

$$\frac{d}{dx} e^u = e^u \frac{du}{dx}.$$

For example,

$$\frac{d}{dx}(e^{kx}) = e^{kx} \cdot \frac{d}{dx}(kx) = ke^{kx}, \qquad \text{for any constant } k$$

and

$$\frac{d}{dx}(e^{x^2}) = e^{x^2} \cdot \frac{d}{dx}(x^2) = 2xe^{x^2}.$$

Repeated Use of the Chain Rule

We sometimes have to use the Chain Rule two or more times to find a derivative.

HISTORICAL BIOGRAPHY

Johann Bernoulli
(1667–1748)
www.goo.gl/Y2KSo4

EXAMPLE 5   Find the derivative of $g(t) = \tan(5 - \sin 2t)$.

Solution   Notice here that the tangent is a function of $5 - \sin 2t$, whereas the sine is a function of $2t$, which is itself a function of t. Therefore, by the Chain Rule,

$$g'(t) = \frac{d}{dt}\tan(5 - \sin 2t)$$

$$= \sec^2(5 - \sin 2t) \cdot \frac{d}{dt}(5 - \sin 2t) \qquad \text{Derivative of } \tan u \text{ with } u = 5 - \sin 2t$$

$$= \sec^2(5 - \sin 2t) \cdot \left(0 - \cos 2t \cdot \frac{d}{dt}(2t)\right) \qquad \text{Derivative of } 5 - \sin u \text{ with } u = 2t$$

$$= \sec^2(5 - \sin 2t) \cdot (-\cos 2t) \cdot 2$$

$$= -2(\cos 2t)\sec^2(5 - \sin 2t).$$

The Chain Rule with Powers of a Function

If n is any real number and f is a power function, $f(u) = u^n$, the Power Rule tells us that $f'(u) = nu^{n-1}$. If u is a differentiable function of x, then we can use the Chain Rule to extend this to the **Power Chain Rule**:

$$\frac{d}{dx}(u^n) = nu^{n-1}\frac{du}{dx}. \qquad\qquad \frac{d}{du}(u^n) = nu^{n-1}$$

EXAMPLE 6   The Power Chain Rule simplifies computing the derivative of a power of an expression.

(a) $\frac{d}{dx}(5x^3 - x^4)^7 = 7(5x^3 - x^4)^6\frac{d}{dx}(5x^3 - x^4)$ $\qquad$ Power Chain Rule with $u = 5x^3 - x^4, n = 7$

$$= 7(5x^3 - x^4)^6(15x^2 - 4x^3)$$

(b) $\frac{d}{dx}\left(\frac{1}{3x - 2}\right) = \frac{d}{dx}(3x - 2)^{-1}$

$$= -1(3x - 2)^{-2}\frac{d}{dx}(3x - 2) \qquad \text{Power Chain Rule with } u = 3x - 2, n = -1$$

$$= -1(3x - 2)^{-2}(3)$$

$$= -\frac{3}{(3x - 2)^2}$$

In part (b) we could also find the derivative with the Quotient Rule.

(c) $\frac{d}{dx}(\sin^5 x) = 5\sin^4 x \cdot \frac{d}{dx}\sin x$ $\qquad$ Power Chain Rule with $u = \sin x, n = 5$, because $\sin^n x$ means $(\sin x)^n, n \neq -1$.

$$= 5\sin^4 x \cos x$$

(d) $\frac{d}{dx}\left(e^{\sqrt{3x+1}}\right) = e^{\sqrt{3x+1}} \cdot \frac{d}{dx}\left(\sqrt{3x + 1}\right)$

$$= e^{\sqrt{3x+1}} \cdot \frac{1}{2}(3x + 1)^{-1/2} \cdot 3 \qquad \text{Power Chain Rule with } u = 3x + 1, n = 1/2$$

$$= \frac{3}{2\sqrt{3x + 1}}e^{\sqrt{3x+1}}$$

EXAMPLE 7   In Example 4 of Section 3.2 we saw that the absolute value function $y = |x|$ is not differentiable at $x = 0$. However, the function is differentiable at all other

Derivative of the Absolute Value Function

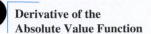

$$\frac{d}{dx}(|x|) = \frac{x}{|x|}, \quad x \neq 0$$

$$= \begin{cases} 1, & x > 0 \\ -1, & x < 0 \end{cases}$$

real numbers, as we now show. Since $|x| = \sqrt{x^2}$, we can derive the following formula, which gives an alternative to the more direct analysis seen before.

$$\frac{d}{dx}(|x|) = \frac{d}{dx}\sqrt{x^2}$$

$$= \frac{1}{2\sqrt{x^2}} \cdot \frac{d}{dx}(x^2) \qquad \text{Power Chain Rule with } u = x^2, n = 1/2, x \neq 0$$

$$= \frac{1}{2|x|} \cdot 2x \qquad \sqrt{x^2} = |x|$$

$$= \frac{x}{|x|}, \quad x \neq 0.$$

EXAMPLE 8   Show that the slope of every line tangent to the curve $y = 1/(1 - 2x)^3$ is positive.

Solution   We find the derivative:

$$\frac{dy}{dx} = \frac{d}{dx}(1 - 2x)^{-3}$$

$$= -3(1 - 2x)^{-4} \cdot \frac{d}{dx}(1 - 2x) \qquad \text{Power Chain Rule with } u = (1 - 2x), n = -3$$

$$= -3(1 - 2x)^{-4} \cdot (-2)$$

$$= \frac{6}{(1 - 2x)^4}.$$

At any point (x, y) on the curve, the denominator is nonzero, and the slope of the tangent line is

$$\frac{dy}{dx} = \frac{6}{(1 - 2x)^4},$$

which is the quotient of two positive numbers.

EXAMPLE 9   The formulas for the derivatives of both $\sin x$ and $\cos x$ were obtained under the assumption that x is measured in radians, *not* degrees. The Chain Rule gives us new insight into the difference between the two. Since $180° = \pi$ radians, $x° = \pi x/180$ radians where $x°$ is the size of the angle measured in degrees.

By the Chain Rule,

$$\frac{d}{dx}\sin(x°) = \frac{d}{dx}\sin\left(\frac{\pi x}{180}\right) = \frac{\pi}{180}\cos\left(\frac{\pi x}{180}\right) = \frac{\pi}{180}\cos(x°).$$

See Figure 3.28. Similarly, the derivative of $\cos(x°)$ is $-(\pi/180)\sin(x°)$.

The factor $\pi/180$ would propagate with repeated differentiation, showing an advantage for the use of radian measure in computations.

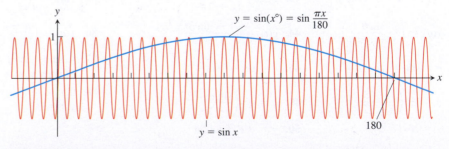

FIGURE 3.28   The function $\sin(x°)$ oscillates only $\pi/180$ times as often as $\sin x$ oscillates. Its maximum slope is $\pi/180$ at $x = 0$ (Example 9).

EXERCISES 3.6

Derivative Calculations

In Exercises 1–8, given $y = f(u)$ and $u = g(x)$, find $dy/dx = dy/dx = f'(g(x))g'(x)$.

1. $y = 6u - 9, \quad u = (1/2)x^4$
2. $y = 2u^3, \quad u = 8x - 1$
3. $y = \sin u, \quad u = 3x + 1$
4. $y = \cos u, \quad u = e^{-x}$
5. $y = \sqrt{u}, \quad u = \sin x$
6. $y = \sin u, \quad u = x - \cos x$
7. $y = \tan u, \quad u = \pi x^2$
8. $y = -\sec u, \quad u = \dfrac{1}{x} + 7x$

In Exercises 9–22, write the function in the form $y = f(u)$ and $u = g(x)$. Then find dy/dx as a function of x.

9. $y = (2x + 1)^5$
10. $y = (4 - 3x)^9$
11. $y = \left(1 - \dfrac{x}{7}\right)^{-7}$
12. $y = \left(\dfrac{\sqrt{x}}{2} - 1\right)^{-10}$
13. $y = \left(\dfrac{x^2}{8} + x - \dfrac{1}{x}\right)^4$
14. $y = \sqrt{3x^2 - 4x + 6}$
15. $y = \sec(\tan x)$
16. $y = \cot\left(\pi - \dfrac{1}{x}\right)$
17. $y = \tan^3 x$
18. $y = 5\cos^{-4} x$
19. $y = e^{-5x}$
20. $y = e^{2x/3}$
21. $y = e^{5-7x}$
22. $y = e^{(4\sqrt{x}+x^2)}$

Find the derivatives of the functions in Exercises 23–50.

23. $p = \sqrt{3 - t}$
24. $q = \sqrt[3]{2r - r^2}$
25. $s = \dfrac{4}{3\pi}\sin 3t + \dfrac{4}{5\pi}\cos 5t$
26. $s = \sin\left(\dfrac{3\pi t}{2}\right) + \cos\left(\dfrac{3\pi t}{2}\right)$
27. $r = (\csc\theta + \cot\theta)^{-1}$
28. $r = 6(\sec\theta - \tan\theta)^{3/2}$
29. $y = x^2 \sin^4 x + x\cos^{-2} x$
30. $y = \dfrac{1}{x}\sin^{-5} x - \dfrac{x}{3}\cos^3 x$
31. $y = \dfrac{1}{18}(3x - 2)^6 + \left(4 - \dfrac{1}{2x^2}\right)^{-1}$
32. $y = (5 - 2x)^{-3} + \dfrac{1}{8}\left(\dfrac{2}{x} + 1\right)^4$
33. $y = (4x + 3)^4(x + 1)^{-3}$
34. $y = (2x - 5)^{-1}(x^2 - 5x)^6$
35. $y = xe^{-x} + e^{x^3}$
36. $y = (1 + 2x)e^{-2x}$
37. $y = (x^2 - 2x + 2)e^{5x/2}$
38. $y = (9x^2 - 6x + 2)e^{x^3}$
39. $h(x) = x\tan(2\sqrt{x}) + 7$
40. $k(x) = x^2\sec\left(\dfrac{1}{x}\right)$
41. $f(x) = \sqrt{7 + x\sec x}$
42. $g(x) = \dfrac{\tan 3x}{(x + 7)^4}$
43. $f(\theta) = \left(\dfrac{\sin\theta}{1 + \cos\theta}\right)^2$
44. $g(t) = \left(\dfrac{1 + \sin 3t}{3 - 2t}\right)^{-1}$
45. $r = \sin(\theta^2)\cos(2\theta)$
46. $r = \sec\sqrt{\theta}\tan\left(\dfrac{1}{\theta}\right)$
47. $q = \sin\left(\dfrac{t}{\sqrt{t + 1}}\right)$
48. $q = \cot\left(\dfrac{\sin t}{t}\right)$
49. $y = \cos\left(e^{-\theta^2}\right)$
50. $y = \theta^3 e^{-2\theta}\cos 5\theta$

In Exercises 51–70, find dy/dt.

51. $y = \sin^2(\pi t - 2)$
52. $y = \sec^2 \pi t$
53. $y = (1 + \cos 2t)^{-4}$
54. $y = (1 + \cot(t/2))^{-2}$
55. $y = (t\tan t)^{10}$
56. $y = (t^{-3/4}\sin t)^{4/3}$
57. $y = e^{\cos^2(\pi t - 1)}$
58. $y = \left(e^{\sin(t/2)}\right)^3$
59. $y = \left(\dfrac{t^2}{t^3 - 4t}\right)^3$
60. $y = \left(\dfrac{3t - 4}{5t + 2}\right)^{-5}$
61. $y = \sin(\cos(2t - 5))$
62. $y = \cos\left(5\sin\left(\dfrac{t}{3}\right)\right)$
63. $y = \left(1 + \tan^4\left(\dfrac{t}{12}\right)\right)^3$
64. $y = \dfrac{1}{6}(1 + \cos^2(7t))^3$
65. $y = \sqrt{1 + \cos(t^2)}$
66. $y = 4\sin\left(\sqrt{1 + \sqrt{t}}\right)$
67. $y = \tan^2(\sin^3 t)$
68. $y = \cos^4(\sec^2 3t)$
69. $y = 3t(2t^2 - 5)^4$
70. $y = \sqrt{3t + \sqrt{2 + \sqrt{1 - t}}}$

Second Derivatives

Find y'' in Exercises 71–78.

71. $y = \left(1 + \dfrac{1}{x}\right)^3$
72. $y = \left(1 - \sqrt{x}\right)^{-1}$
73. $y = \dfrac{1}{9}\cot(3x - 1)$
74. $y = 9\tan\left(\dfrac{x}{3}\right)$
75. $y = x(2x + 1)^4$
76. $y = x^2(x^3 - 1)^5$
77. $y = e^{x^2} + 5x$
78. $y = \sin(x^2 e^x)$

For each of the following functions, solve both $f'(x) = 0$ and $f''(x) = 0$ for x.

79. $f(x) = x(x - 4)^3$
80. $f(x) = \sec^2 x - 2\tan x \quad \text{for} \quad 0 \le x \le 2\pi$

Finding Derivative Values

In Exercises 81–86, find the value of $(f \circ g)'$ at the given value of x.

81. $f(u) = u^5 + 1, \quad u = g(x) = \sqrt{x}, \quad x = 1$
82. $f(u) = 1 - \dfrac{1}{u}, \quad u = g(x) = \dfrac{1}{1 - x}, \quad x = -1$
83. $f(u) = \cot\dfrac{\pi u}{10}, \quad u = g(x) = 5\sqrt{x}, \quad x = 1$
84. $f(u) = u + \dfrac{1}{\cos^2 u}, \quad u = g(x) = \pi x, \quad x = 1/4$
85. $f(u) = \dfrac{2u}{u^2 + 1}, \quad u = g(x) = 10x^2 + x + 1, \quad x = 0$
86. $f(u) = \left(\dfrac{u - 1}{u + 1}\right)^2, \quad u = g(x) = \dfrac{1}{x^2} - 1, \quad x = -1$
87. Assume that $f'(3) = -1, g'(2) = 5, g(2) = 3,$ and $y = f(g(x))$. What is y' at $x = 2$?
88. If $r = \sin(f(t)), f(0) = \pi/3,$ and $f'(0) = 4$, then what is dr/dt at $t = 0$?

89. Suppose that functions f and g and their derivatives with respect to x have the following values at $x = 2$ and $x = 3$.

x	$f(x)$	$g(x)$	$f'(x)$	$g'(x)$
2	8	2	$1/3$	-3
3	3	-4	2π	5

Find the derivatives with respect to x of the following combinations at the given value of x.

a. $2f(x)$, $x = 2$ **b.** $f(x) + g(x)$, $x = 3$

c. $f(x) \cdot g(x)$, $x = 3$ **d.** $f(x)/g(x)$, $x = 2$

e. $f(g(x))$, $x = 2$ **f.** $\sqrt{f(x)}$, $x = 2$

g. $1/g^2(x)$, $x = 3$ **h.** $\sqrt{f^2(x) + g^2(x)}$, $x = 2$

90. Suppose that the functions f and g and their derivatives with respect to x have the following values at $x = 0$ and $x = 1$.

x	$f(x)$	$g(x)$	$f'(x)$	$g'(x)$
0	1	1	5	$1/3$
1	3	-4	$-1/3$	$-8/3$

Find the derivatives with respect to x of the following combinations at the given value of x.

a. $5f(x) - g(x)$, $x = 1$ **b.** $f(x)g^3(x)$, $x = 0$

c. $\dfrac{f(x)}{g(x) + 1}$, $x = 1$ **d.** $f(g(x))$, $x = 0$

e. $g(f(x))$, $x = 0$ **f.** $(x^{11} + f(x))^{-2}$, $x = 1$

g. $f(x + g(x))$, $x = 0$

91. Find ds/dt when $\theta = 3\pi/2$ if $s = \cos\theta$ and $d\theta/dt = 5$.

92. Find dy/dt when $x = 1$ if $y = x^2 + 7x - 5$ and $dx/dt = 1/3$.

Theory and Examples

What happens if you can write a function as a composition in different ways? Do you get the same derivative each time? The Chain Rule says you should. Try it with the functions in Exercises 93 and 94.

93. Find dy/dx if $y = x$ by using the Chain Rule with y as a composition of

a. $y = (u/5) + 7$ and $u = 5x - 35$

b. $y = 1 + (1/u)$ and $u = 1/(x - 1)$.

94. Find dy/dx if $y = x^{3/2}$ by using the Chain Rule with y as a composition of

a. $y = u^3$ and $u = \sqrt{x}$

b. $y = \sqrt{u}$ and $u = x^3$.

95. Find the tangent to $y = ((x - 1)/(x + 1))^2$ at $x = 0$.

96. Find the tangent to $y = \sqrt{x^2 - x + 7}$ at $x = 2$.

97. a. Find the tangent to the curve $y = 2\tan(\pi x/4)$ at $x = 1$.

b. Slopes on a tangent curve What is the smallest value the slope of the curve can ever have on the interval $-2 < x < 2$? Give reasons for your answer.

98. Slopes on sine curves

a. Find equations for the tangents to the curves $y = \sin 2x$ and $y = -\sin(x/2)$ at the origin. Is there anything special about how the tangents are related? Give reasons for your answer.

b. Can anything be said about the tangents to the curves $y = \sin mx$ and $y = -\sin(x/m)$ at the origin (m a constant $\neq 0$)? Give reasons for your answer.

c. For a given m, what are the largest values the slopes of the curves $y = \sin mx$ and $y = -\sin(x/m)$ can ever have? Give reasons for your answer.

d. The function $y = \sin x$ completes one period on the interval $[0, 2\pi]$, the function $y = \sin 2x$ completes two periods, the function $y = \sin(x/2)$ completes half a period, and so on. Is there any relation between the number of periods $y = \sin mx$ completes on $[0, 2\pi]$ and the slope of the curve $y = \sin mx$ at the origin? Give reasons for your answer.

99. Running machinery too fast Suppose that a piston is moving straight up and down and that its position at time t sec is

$$s = A\cos(2\pi bt),$$

with A and b positive. The value of A is the amplitude of the motion, and b is the frequency (number of times the piston moves up and down each second). What effect does doubling the frequency have on the piston's velocity, acceleration, and jerk? (Once you find out, you will know why some machinery breaks when you run it too fast.)

100. Temperatures in Fairbanks, Alaska The graph in the accompanying figure shows the average Fahrenheit temperature in Fairbanks, Alaska, during a typical 365-day year. The equation that approximates the temperature on day x is

$$y = 37\sin\left[\frac{2\pi}{365}(x - 101)\right] + 25$$

and is graphed in the accompanying figure.

a. On what day is the temperature increasing the fastest?

b. About how many degrees per day is the temperature increasing when it is increasing at its fastest?

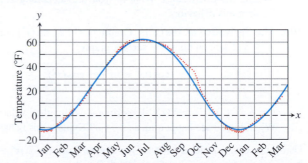

101. Particle motion The position of a particle moving along a coordinate line is $s = \sqrt{1 + 4t}$, with s in meters and t in seconds. Find the particle's velocity and acceleration at $t = 6$ sec.

102. Constant acceleration Suppose that the velocity of a falling body is $v = k\sqrt{s}$ m/sec (k a constant) at the instant the body has fallen s m from its starting point. Show that the body's acceleration is constant.

103. **Falling meteorite**   The velocity of a heavy meteorite entering Earth's atmosphere is inversely proportional to $\sqrt{s}$ when it is s km from Earth's center. Show that the meteorite's acceleration is inversely proportional to s^2.

104. **Particle acceleration**   A particle moves along the x-axis with velocity $dx/dt = f(x)$. Show that the particle's acceleration is $f(x)f'(x)$.

105. **Temperature and the period of a pendulum**   For oscillations of small amplitude (short swings), we may safely model the relationship between the period T and the length L of a simple pendulum with the equation

$$T = 2\pi\sqrt{\frac{L}{g}},$$

where g is the constant acceleration of gravity at the pendulum's location. If we measure g in centimeters per second squared, we measure L in centimeters and T in seconds. If the pendulum is made of metal, its length will vary with temperature, either increasing or decreasing at a rate that is roughly proportional to L. In symbols, with u being temperature and k the proportionality constant,

$$\frac{dL}{du} = kL.$$

Assuming this to be the case, show that the rate at which the period changes with respect to temperature is $kT/2$.

106. **Chain Rule**   Suppose that $f(x) = x^2$ and $g(x) = |x|$. Then the compositions

$$(f \circ g)(x) = |x|^2 = x^2 \quad \text{and} \quad (g \circ f)(x) = |x^2| = x^2$$

are both differentiable at $x = 0$ even though g itself is not differentiable at $x = 0$. Does this contradict the Chain Rule? Explain.

T 107. **The derivative of sin 2x**   Graph the function $y = 2\cos 2x$ for $-2 \le x \le 3.5$. Then, on the same screen, graph

$$y = \frac{\sin 2(x + h) - \sin 2x}{h}$$

for $h = 1.0, 0.5$, and 0.2. Experiment with other values of h, including negative values. What do you see happening as $h \to 0$? Explain this behavior.

108. **The derivative of cos (x²)**   Graph $y = -2x \sin(x^2)$ for $-2 \le x \le 3$. Then, on the same screen, graph

$$y = \frac{\cos((x + h)^2) - \cos(x^2)}{h}$$

for $h = 1.0, 0.7$, and 0.3. Experiment with other values of h. What do you see happening as $h \to 0$? Explain this behavior.

Using the Chain Rule, show that the Power Rule $(d/dx)x^n = nx^{n-1}$ holds for the functions x^n in Exercises 109 and 110.

109. $x^{1/4} = \sqrt{\sqrt{x}}$ 110. $x^{3/4} = \sqrt{x\sqrt{x}}$

111. Consider the function

$$f(x) = \begin{cases} x \sin\left(\dfrac{1}{x}\right), & x > 0 \\ 0, & x \le 0 \end{cases}$$

a. Show that f is continuous at $x = 0$.

b. Determine f' for $x \ne 0$.

c. Show that f is not differentiable at $x = 0$.

112. Consider the function

$$f(x) = \begin{cases} x^2 \cos\left(\dfrac{2}{x}\right), & x \ne 0 \\ 0, & x = 0 \end{cases}$$

a. Show that f is continuous at $x = 0$.

b. Determine f' for $x \ne 0$.

c. Show that f is differentiable at $x = 0$.

d. Show that f' is not continuous at $x = 0$.

113. Verify each of the following statements.

a. If f is even, then f' is odd.

b. If f is odd, then f' is even.

COMPUTER EXPLORATIONS
Trigonometric Polynomials

114. As the accompanying figure shows, the trigonometric "polynomial"

$$s = f(t) = 0.78540 - 0.63662\cos 2t - 0.07074\cos 6t$$
$$- 0.02546\cos 10t - 0.01299\cos 14t$$

gives a good approximation of the sawtooth function $s = g(t)$ on the interval $[-\pi, \pi]$. How well does the derivative of f approximate the derivative of g at the points where dg/dt is defined? To find out, carry out the following steps.

a. Graph dg/dt (where defined) over $[-\pi, \pi]$.

b. Find df/dt.

c. Graph df/dt. Where does the approximation of dg/dt by df/dt seem to be best? Least good? Approximations by trigonometric polynomials are important in the theories of heat and oscillation, but we must not expect too much of them, as we see in the next exercise.

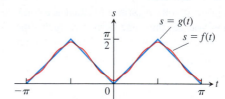

115. (*Continuation of Exercise 114.*) In Exercise 114, the trigonometric polynomial $f(t)$ that approximated the sawtooth function $g(t)$ on $[-\pi, \pi]$ had a derivative that approximated the derivative of the sawtooth function. It is possible, however, for a trigonometric polynomial to approximate a function in a reasonable way without its derivative approximating the function's derivative at all well. As a case in point, the trigonometric "polynomial"

$$s = h(t) = 1.2732\sin 2t + 0.4244\sin 6t + 0.25465\sin 10t$$
$$+ 0.18189\sin 14t + 0.14147\sin 18t$$

graphed in the accompanying figure approximates the step function $s = k(t)$ shown there. Yet the derivative of h is nothing like the derivative of k.

a. Graph dk/dt (where defined) over $[-\pi, \pi]$.

b. Find dh/dt.

c. Graph dh/dt to see how badly the graph fits the graph of dk/dt. Comment on what you see.

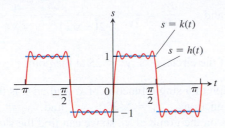

3.7 Implicit Differentiation

Most of the functions we have dealt with so far have been described by an equation of the form $y = f(x)$ that expresses y explicitly in terms of the variable x. We have learned rules for differentiating functions defined in this way. A different situation occurs when we encounter equations like

$$x^3 + y^3 - 9xy = 0, \qquad y^2 - x = 0, \qquad \text{or} \qquad x^2 + y^2 - 25 = 0.$$

(See Figures 3.29, 3.30, and 3.31.) These equations define an *implicit* relation between the variables x and y, meaning that a value of x determines one or more values of y, even though we do not have a simple formula for the y-values. In some cases we may be able to solve such an equation for y as an explicit function (or even several functions) of x. When we cannot put an equation $F(x, y) = 0$ in the form $y = f(x)$ to differentiate it in the usual way, we may still be able to find dy/dx by *implicit differentiation*. This section describes the technique.

Implicitly Defined Functions

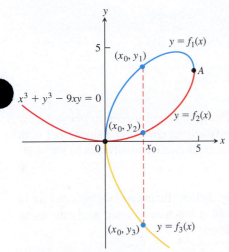

We begin with examples involving familiar equations that we can solve for y as a function of x and then calculate dy/dx in the usual way. Then we differentiate the equations implicitly, and find the derivative. We will see that the two methods give the same answer. Following the examples, we summarize the steps involved in the new method. In the examples and exercises, it is always assumed that the given equation determines y implicitly as a differentiable function of x so that dy/dx exists.

FIGURE 3.29 The curve $x^3 + y^3 - 9xy = 0$ is not the graph of any one function of x. The curve can, however, be divided into separate arcs that *are* the graphs of functions of x. This particular curve, called a *folium*, dates to Descartes in 1638.

EXAMPLE 1 Find dy/dx if $y^2 = x$.

Solution The equation $y^2 = x$ defines two differentiable functions of x that we can actually find, namely $y_1 = \sqrt{x}$ and $y_2 = -\sqrt{x}$ (Figure 3.30). We know how to calculate the derivative of each of these for $x > 0$:

$$\frac{dy_1}{dx} = \frac{1}{2\sqrt{x}} \qquad \text{and} \qquad \frac{dy_2}{dx} = -\frac{1}{2\sqrt{x}}.$$

But suppose that we knew only that the equation $y^2 = x$ defined y as one or more differentiable functions of x for $x > 0$ without knowing exactly what these functions were. Can we still find dy/dx?

The answer is yes. To find dy/dx, we simply differentiate both sides of the equation $y^2 = x$ with respect to x, treating $y = f(x)$ as a differentiable function of x:

$$y^2 = x \qquad \text{The Chain Rule gives}$$

$$2y\frac{dy}{dx} = 1 \qquad \frac{d}{dx}(y^2) = \frac{d}{dx}[f(x)]^2 = 2f(x)f'(x) = 2y\frac{dy}{dx}.$$

$$\frac{dy}{dx} = \frac{1}{2y}.$$

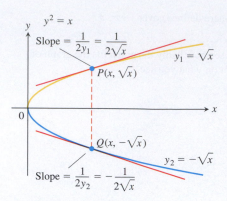

FIGURE 3.30 The equation $y^2 - x = 0$, or $y^2 = x$ as it is usually written, defines two differentiable functions of x on the interval $x > 0$. Example 1 shows how to find the derivatives of these functions without solving the equation $y^2 = x$ for y.

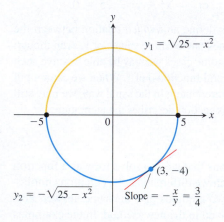

FIGURE 3.31 The circle combines the graphs of two functions. The graph of y_2 is the lower semicircle and passes through $(3, -4)$.

This one formula gives the derivatives we calculated for *both* explicit solutions $y_1 = \sqrt{x}$ and $y_2 = -\sqrt{x}$:

$$\frac{dy_1}{dx} = \frac{1}{2y_1} = \frac{1}{2\sqrt{x}} \qquad \text{and} \qquad \frac{dy_2}{dx} = \frac{1}{2y_2} = \frac{1}{2(-\sqrt{x})} = -\frac{1}{2\sqrt{x}}.$$

EXAMPLE 2 Find the slope of the circle $x^2 + y^2 = 25$ at the point $(3, -4)$.

Solution The circle is not the graph of a single function of x. Rather, it is the combined graphs of two differentiable functions, $y_1 = \sqrt{25 - x^2}$ and $y_2 = -\sqrt{25 - x^2}$ (Figure 3.31). The point $(3, -4)$ lies on the graph of y_2, so we can find the slope by calculating the derivative directly, using the Power Chain Rule:

$$\left.\frac{dy_2}{dx}\right|_{x=3} = -\left.\frac{-2x}{2\sqrt{25 - x^2}}\right|_{x=3} = -\frac{-6}{2\sqrt{25 - 9}} = \frac{3}{4}. \qquad \begin{array}{l} \frac{d}{dx}\left(-(25 - x^2)^{1/2}\right) = \\ -\frac{1}{2}(25 - x^2)^{-1/2}(-2x) \end{array}$$

We can solve this problem more easily by differentiating the given equation of the circle implicitly with respect to x:

$$\frac{d}{dx}(x^2) + \frac{d}{dx}(y^2) = \frac{d}{dx}(25)$$

$$2x + 2y\frac{dy}{dx} = 0 \qquad \text{\color{blue}See Example 1.}$$

$$\frac{dy}{dx} = -\frac{x}{y}.$$

The slope at $(3, -4)$ is $-\left.\frac{x}{y}\right|_{(3,-4)} = -\frac{3}{-4} = \frac{3}{4}.$

Notice that unlike the slope formula for dy_2/dx, which applies only to points below the x-axis, the formula $dy/dx = -x/y$ applies everywhere the circle has a slope; that is, at all circle points (x, y) where $y \neq 0$. Notice also that the derivative involves *both* variables x and y, not just the independent variable x. ■

To calculate the derivatives of other implicitly defined functions, we proceed as in Examples 1 and 2: We treat y as a differentiable implicit function of x and apply the usual rules to differentiate both sides of the defining equation.

Implicit Differentiation

1. Differentiate both sides of the equation with respect to x, treating y as a differentiable function of x.

2. Collect the terms with dy/dx on one side of the equation and solve for dy/dx.

EXAMPLE 3 Find dy/dx if $y^2 = x^2 + \sin xy$ (Figure 3.32).

Solution We differentiate the equation implicitly.

$$y^2 = x^2 + \sin xy$$

$$\frac{d}{dx}(y^2) = \frac{d}{dx}(x^2) + \frac{d}{dx}(\sin xy) \qquad \begin{array}{l}\text{\color{blue}Differentiate both sides with}\\\text{\color{blue}respect to } x \ldots\end{array}$$

$$2y\frac{dy}{dx} = 2x + (\cos xy)\frac{d}{dx}(xy) \qquad \begin{array}{l}\text{\color{blue}\ldots treating } y \text{ as a function of}\\\text{\color{blue}} x \text{ and using the Chain Rule.}\end{array}$$

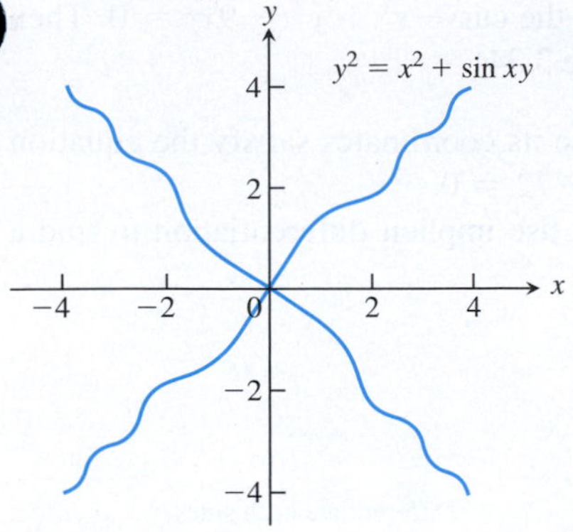

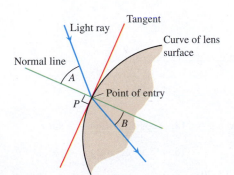

FIGURE 3.32 The graph of the equation in Example 3.

$$2y\frac{dy}{dx} = 2x + (\cos xy)\left(y + x\frac{dy}{dx}\right) \qquad \text{Treat } xy \text{ as a product.}$$

$$2y\frac{dy}{dx} - (\cos xy)\left(x\frac{dy}{dx}\right) = 2x + (\cos xy)y \qquad \text{Collect terms with } dy/dx.$$

$$(2y - x\cos xy)\frac{dy}{dx} = 2x + y\cos xy$$

$$\frac{dy}{dx} = \frac{2x + y\cos xy}{2y - x\cos xy} \qquad \text{Solve for } dy/dx.$$

Notice that the formula for dy/dx applies everywhere that the implicitly defined curve has a slope. Notice again that the derivative involves *both* variables x and y, not just the independent variable x. ∎

Derivatives of Higher Order

Implicit differentiation can also be used to find higher derivatives.

EXAMPLE 4   Find d^2y/dx^2 if $2x^3 - 3y^2 = 8$.

Solution   To start, we differentiate both sides of the equation with respect to x in order to find $y' = dy/dx$.

$$\frac{d}{dx}(2x^3 - 3y^2) = \frac{d}{dx}(8)$$

$$6x^2 - 6yy' = 0 \qquad \text{Treat } y \text{ as a function of } x.$$

$$y' = \frac{x^2}{y}, \qquad \text{when } y \neq 0 \qquad \text{Solve for } y'.$$

We now apply the Quotient Rule to find y''.

$$y'' = \frac{d}{dx}\left(\frac{x^2}{y}\right) = \frac{2xy - x^2y'}{y^2} = \frac{2x}{y} - \frac{x^2}{y^2}\cdot y'$$

Finally, we substitute $y' = x^2/y$ to express y'' in terms of x and y.

$$y'' = \frac{2x}{y} - \frac{x^2}{y^2}\left(\frac{x^2}{y}\right) = \frac{2x}{y} - \frac{x^4}{y^3}, \qquad \text{when } y \neq 0 \qquad ∎$$

Lenses, Tangent Lines, and Normal Lines

In the law that describes how light changes direction as it enters a lens, the important angles are the angles the light makes with the line perpendicular to the surface of the lens at the point of entry (angles A and B in Figure 3.33). This line is called the *normal* to the surface at the point of entry. In a profile view of a lens like the one in Figure 3.33, the **normal** is the line perpendicular (also said to be *orthogonal*) to the tangent of the profile curve at the point of entry.

FIGURE 3.33 The profile of a lens, showing the bending (refraction) of a ray of light as it passes through the lens surface.

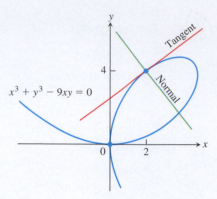

FIGURE 3.34 Example 5 shows how to find equations for the tangent and normal to the folium of Descartes at (2,4).

$x^3 + y^3 - 9xy = 0$

EXAMPLE 5   Show that the point (2, 4) lies on the curve $x^3 + y^3 - 9xy = 0$. Then find the tangent and normal to the curve there (Figure 3.34).

Solution   The point (2, 4) lies on the curve because its coordinates satisfy the equation given for the curve: $2^3 + 4^3 - 9(2)(4) = 8 + 64 - 72 = 0$.

To find the slope of the curve at (2, 4), we first use implicit differentiation to find a formula for dy/dx:

$$x^3 + y^3 - 9xy = 0$$

$$\frac{d}{dx}(x^3) + \frac{d}{dx}(y^3) - \frac{d}{dx}(9xy) = \frac{d}{dx}(0)$$

$$3x^2 + 3y^2\frac{dy}{dx} - 9\left(x\frac{dy}{dx} + y\frac{dx}{dx}\right) = 0 \qquad \text{Differentiate both sides with respect to } x.$$

$$(3y^2 - 9x)\frac{dy}{dx} + 3x^2 - 9y = 0 \qquad \text{Treat } xy \text{ as a product and } y \text{ as a function of } x.$$

$$3(y^2 - 3x)\frac{dy}{dx} = 9y - 3x^2$$

$$\frac{dy}{dx} = \frac{3y - x^2}{y^2 - 3x}. \qquad \text{Solve for } dy/dx.$$

We then evaluate the derivative at $(x, y) = (2, 4)$:

$$\frac{dy}{dx}\bigg|_{(2, 4)} = \frac{3y - x^2}{y^2 - 3x}\bigg|_{(2, 4)} = \frac{3(4) - 2^2}{4^2 - 3(2)} = \frac{8}{10} = \frac{4}{5}.$$

The tangent at (2, 4) is the line through (2, 4) with slope 4/5:

$$y = 4 + \frac{4}{5}(x - 2)$$

$$y = \frac{4}{5}x + \frac{12}{5}.$$

The normal to the curve at (2, 4) is the line perpendicular to the tangent there, the line through (2, 4) with slope $-5/4$:

$$y = 4 - \frac{5}{4}(x - 2)$$

$$y = -\frac{5}{4}x + \frac{13}{2}. \qquad \blacksquare$$

EXERCISES 3.7

Differentiating Implicitly

Use implicit differentiation to find dy/dx in Exercises 1–16.

1. $x^2y + xy^2 = 6$

2. $x^3 + y^3 = 18xy$

3. $2xy + y^2 = x + y$

4. $x^3 - xy + y^3 = 1$

5. $x^2(x - y)^2 = x^2 - y^2$

6. $(3xy + 7)^2 = 6y$

7. $y^2 = \dfrac{x - 1}{x + 1}$

8. $x^3 = \dfrac{2x - y}{x + 3y}$

9. $x = \sec y$

10. $xy = \cot(xy)$

11. $x + \tan(xy) = 0$

12. $x^4 + \sin y = x^3y^2$

13. $y \sin\left(\dfrac{1}{y}\right) = 1 - xy$

14. $x \cos(2x + 3y) = y \sin x$

15. $e^{2x} = \sin(x + 3y)$

16. $e^{x^2y} = 2x + 2y$

Find $dr/d\theta$ in Exercises 17–20.

17. $\theta^{1/2} + r^{1/2} = 1$

18. $r - 2\sqrt{\theta} = \dfrac{3}{2}\theta^{2/3} + \dfrac{4}{3}\theta^{3/4}$

19. $\sin(r\theta) = \dfrac{1}{2}$

20. $\cos r + \cot\theta = e^{r\theta}$

Second Derivatives

In Exercises 21–28, use implicit differentiation to find dy/dx and then d^2y/dx^2. Write the solutions in terms of x and y only.

21. $x^2 + y^2 = 1$

22. $x^{2/3} + y^{2/3} = 1$

23. $y^2 = e^{x^2} + 2x$

24. $y^2 - 2x = 1 - 2y$

25. $2\sqrt{y} = x - y$

26. $xy + y^2 = 1$

27. $3 + \sin y = y - x^3$

28. $\ln y = xe^y - 2$

29. If $x^3 + y^3 = 16$, find the value of d^2y/dx^2 at the point $(2, 2)$.

30. If $xy + y^2 = 1$, find the value of d^2y/dx^2 at the point $(0, -1)$.

In Exercises 31 and 32, find the slope of the curve at the given points.

31. $y^2 + x^2 = y^4 - 2x$ at $(-2, 1)$ and $(-2, -1)$

32. $(x^2 + y^2)^2 = (x - y)^2$ at $(1, 0)$ and $(1, -1)$

Slopes, Tangents, and Normals

In Exercises 33–42, verify that the given point is on the curve and find the lines that are **(a)** tangent and **(b)** normal to the curve at the given point.

33. $x^2 + xy - y^2 = 1$, $(2, 3)$

34. $x^2 + y^2 = 25$, $(3, -4)$

35. $x^2y^2 = 9$, $(-1, 3)$

36. $y^2 - 2x - 4y - 1 = 0$, $(-2, 1)$

37. $6x^2 + 3xy + 2y^2 + 17y - 6 = 0$, $(-1, 0)$

38. $x^2 - \sqrt{3}xy + 2y^2 = 5$, $\left(\sqrt{3}, 2\right)$

39. $2xy + \pi \sin y = 2\pi$, $(1, \pi/2)$

40. $x \sin 2y = y \cos 2x$, $(\pi/4, \pi/2)$

41. $y = 2 \sin(\pi x - y)$, $(1, 0)$

42. $x^2 \cos^2 y - \sin y = 0$, $(0, \pi)$

43. Parallel tangents Find the two points where the curve $x^2 + xy + y^2 = 7$ crosses the x-axis, and show that the tangents to the curve at these points are parallel. What is the common slope of these tangents?

44. Normals parallel to a line Find the normals to the curve $xy + 2x - y = 0$ that are parallel to the line $2x + y = 0$.

45. The eight curve Find the slopes of the curve $y^4 = y^2 - x^2$ at the two points shown here.

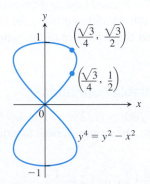

46. The cissoid of Diocles (from about 200 b.c.) Find equations for the tangent and normal to the cissoid of Diocles $y^2(2 - x) = x^3$ at $(1, 1)$.

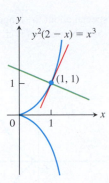

47. The devil's curve (Gabriel Cramer, 1750) Find the slopes of the devil's curve $y^4 - 4y^2 = x^4 - 9x^2$ at the four indicated points.

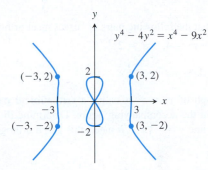

48. The folium of Descartes (See Figure 3.29)

a. Find the slope of the folium of Descartes $x^3 + y^3 - 9xy = 0$ at the points $(4, 2)$ and $(2, 4)$.

b. At what point other than the origin does the folium have a horizontal tangent?

c. Find the coordinates of the point A in Figure 3.29 where the folium has a vertical tangent.

Theory and Examples

49. Intersecting normal The line that is normal to the curve $x^2 + 2xy - 3y^2 = 0$ at $(1, 1)$ intersects the curve at what other point?

50. Power rule for rational exponents Let p and q be integers with $q > 0$. If $y = x^{p/q}$, differentiate the equivalent equation $y^q = x^p$ implicitly and show that, for $y \neq 0$,

$$\frac{d}{dx} x^{p/q} = \frac{p}{q} x^{(p/q)-1}.$$

51. Normals to a parabola Show that if it is possible to draw three normals from the point $(a, 0)$ to the parabola $x = y^2$ shown in the accompanying diagram, then a must be greater than $1/2$. One of the normals is the x-axis. For what value of a are the other two normals perpendicular?

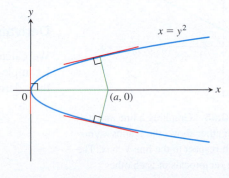

52. Is there anything special about the tangents to the curves $y^2 = x^3$ and $2x^2 + 3y^2 = 5$ at the points $(1, \pm 1)$? Give reasons for your answer.

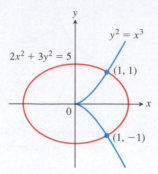

53. Verify that the following pairs of curves meet orthogonally.

   a. $x^2 + y^2 = 4$, $x^2 = 3y^2$

   b. $x = 1 - y^2$, $x = \frac{1}{3}y^2$

54. The graph of $y^2 = x^3$ is called a **semicubical parabola** and is shown in the accompanying figure. Determine the constant b so that the line $y = -\frac{1}{3}x + b$ meets this graph orthogonally.

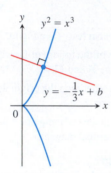

$\boxed{T}$ In Exercises 55 and 56, find both dy/dx (treating y as a differentiable function of x) and dx/dy (treating x as a differentiable function of y). How do dy/dx and dx/dy seem to be related? Explain the relationship geometrically in terms of the graphs.

55. $xy^3 + x^2y = 6$ 56. $x^3 + y^2 = \sin^2 y$

57. **Derivative of arcsine**   Assume that $y = \sin^{-1} x$ is a differentiable function of x. By differentiating the equation $x = \sin y$ implicitly, show that $dy/dx = 1/\sqrt{1 - x^2}$.

58. Use the formula in Exercise 57 to find dy/dx if

   a. $y = (\sin^{-1} x)^2$ b. $y = \sin^{-1}\left(\dfrac{1}{x}\right)$.

COMPUTER EXPLORATIONS

Use a CAS to perform the following steps in Exercises 59–66.

   a. Plot the equation with the implicit plotter of a CAS. Check to see that the given point P satisfies the equation.

   b. Using implicit differentiation, find a formula for the derivative dy/dx and evaluate it at the given point P.

   c. Use the slope found in part (b) to find an equation for the tangent line to the curve at P. Then plot the implicit curve and tangent line together on a single graph.

59. $x^3 - xy + y^3 = 7$, $P(2, 1)$

60. $x^5 + y^3x + yx^2 + y^4 = 4$, $P(1, 1)$

61. $y^2 + y = \dfrac{2 + x}{1 - x}$, $P(0, 1)$ 62. $y^3 + \cos xy = x^2$, $P(1, 0)$

63. $x + \tan\left(\dfrac{y}{x}\right) = 2$, $P\left(1, \dfrac{\pi}{4}\right)$

64. $xy^3 + \tan(x + y) = 1$, $P\left(\dfrac{\pi}{4}, 0\right)$

65. $2y^2 + (xy)^{1/3} = x^2 + 2$, $P(1, 1)$

66. $x\sqrt{1 + 2y} + y = x^2$, $P(1, 0)$

3.8 **Derivatives of Inverse Functions and Logarithms**

In Section 1.6 we saw how the inverse of a function undoes, or inverts, the effect of that function. We defined there the natural logarithm function $f^{-1}(x) = \ln x$ as the inverse of the natural exponential function $f(x) = e^x$. This is one of the most important function-inverse pairs in mathematics and science. We learned how to differentiate the exponential function in Section 3.3. Here we develop a rule for differentiating the inverse of a differentiable function and we apply the rule to find the derivative of the natural logarithm function.

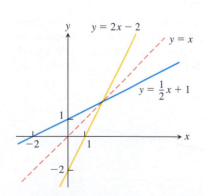

FIGURE 3.35   Graphing a line and its inverse together shows the graphs' symmetry with respect to the line $y = x$. The slopes are reciprocals of each other.

Derivatives of Inverses of Differentiable Functions

We calculated the inverse of the function $f(x) = (1/2)x + 1$ to be $f^{-1}(x) = 2x - 2$ in Example 3 of Section 1.6. Figure 3.35 shows the graphs of both functions. If we calculate their derivatives, we see that

$$\frac{d}{dx}f(x) = \frac{d}{dx}\left(\frac{1}{2}x + 1\right) = \frac{1}{2}$$

$$\frac{d}{dx}f^{-1}(x) = \frac{d}{dx}(2x - 2) = 2.$$

The derivatives are reciprocals of one another, so the slope of one line is the reciprocal of the slope of its inverse line. (See Figure 3.35.)

This is not a special case. Reflecting any nonhorizontal or nonvertical line across the line $y = x$ always inverts the line's slope. If the original line has slope $m \neq 0$, the reflected line has slope $1/m$.

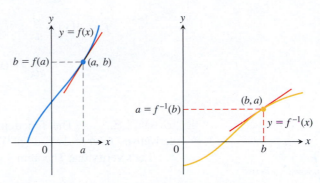

The slopes are reciprocal: $(f^{-1})'(b) = \dfrac{1}{f'(a)}$ or $(f^{-1})'(b) = \dfrac{1}{f'(f^{-1}(b))}$

FIGURE 3.36 The graphs of inverse functions have reciprocal slopes at corresponding points.

The reciprocal relationship between the slopes of f and f^{-1} holds for other functions as well, but we must be careful to compare slopes at corresponding points. If the slope of $y = f(x)$ at the point $(a, f(a))$ is $f'(a)$ and $f'(a) \neq 0$, then the slope of $y = f^{-1}(x)$ at the point $(f(a), a)$ is the reciprocal $1/f'(a)$ (Figure 3.36). If we set $b = f(a)$, then

$$(f^{-1})'(b) = \frac{1}{f'(a)} = \frac{1}{f'(f^{-1}(b))}.$$

If $y = f(x)$ has a horizontal tangent line at $(a, f(a))$, then the inverse function f^{-1} has a vertical tangent line at $(f(a), a)$, and this infinite slope implies that f^{-1} is not differentiable at $f(a)$. Theorem 3 gives the conditions under which f^{-1} is differentiable in its domain (which is the same as the range of f).

THEOREM 3—The Derivative Rule for Inverses

If f has an interval I as domain and $f'(x)$ exists and is never zero on I, then f^{-1} is differentiable at every point in its domain (the range of f). The value of $(f^{-1})'$ at a point b in the domain of f^{-1} is the reciprocal of the value of f' at the point $a = f^{-1}(b)$:

$$(f^{-1})'(b) = \frac{1}{f'(f^{-1}(b))} \tag{1}$$

or

$$\left. \frac{df^{-1}}{dx} \right|_{x=b} = \frac{1}{\left. \dfrac{df}{dx} \right|_{x=f^{-1}(b)}}.$$

Theorem 3 makes two assertions. The first of these has to do with the conditions under which f^{-1} is differentiable; the second assertion is a formula for the derivative of

f^{-1} when it exists. While we omit the proof of the first assertion, the second one is proved in the following way:

$$f(f^{-1}(x)) = x \qquad \text{Inverse function relationship}$$

$$\frac{d}{dx} f(f^{-1}(x)) = 1 \qquad \text{Differentiating both sides}$$

$$f'(f^{-1}(x)) \cdot \frac{d}{dx} f^{-1}(x) = 1 \qquad \text{Chain Rule}$$

$$\frac{d}{dx} f^{-1}(x) = \frac{1}{f'(f^{-1}(x))}. \qquad \text{Solving for the derivative}$$

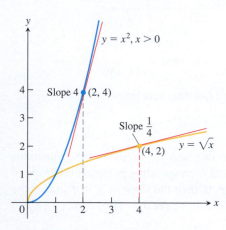

FIGURE 3.37 The derivative of $f^{-1}(x) = \sqrt{x}$ at the point $(4, 2)$ is the reciprocal of the derivative of $f(x) = x^2$ at $(2, 4)$ (Example 1).

EXAMPLE 1   The function $f(x) = x^2, x > 0$ and its inverse $f^{-1}(x) = \sqrt{x}$ have derivatives $f'(x) = 2x$ and $(f^{-1})'(x) = 1/(2\sqrt{x})$.

Let's verify that Theorem 3 gives the same formula for the derivative of $f^{-1}(x)$:

$$(f^{-1})'(x) = \frac{1}{f'(f^{-1}(x))}$$

$$= \frac{1}{2(f^{-1}(x))} \qquad \begin{array}{l} f'(x) = 2x \text{ with } x \text{ replaced} \\ \text{by } f^{-1}(x) \end{array}$$

$$= \frac{1}{2(\sqrt{x})}. \qquad f^{-1}(x) = \sqrt{x}$$

Theorem 3 gives a derivative that agrees with the known derivative of the square root function.

Let's examine Theorem 3 at a specific point. We pick $x = 2$ (the number a) and $f(2) = 4$ (the value b). Theorem 3 says that the derivative of f at 2, which is $f'(2) = 4$, and the derivative of f^{-1} at $f(2)$, which is $(f^{-1})'(4)$, are reciprocals. It states that

$$(f^{-1})'(4) = \frac{1}{f'(f^{-1}(4))} = \frac{1}{f'(2)} = \frac{1}{2x}\Big|_{x=2} = \frac{1}{4}.$$

See Figure 3.37.   ■

We will use the procedure illustrated in Example 1 to calculate formulas for the derivatives of many inverse functions throughout this chapter. Equation (1) sometimes enables us to find specific values of df^{-1}/dx without knowing a formula for f^{-1}.

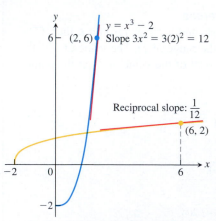

FIGURE 3.38 The derivative of $f(x) = x^3 - 2$ at $x = 2$ tells us the derivative of f^{-1} at $x = 6$ (Example 2).

EXAMPLE 2   Let $f(x) = x^3 - 2, x > 0$. Find the value of df^{-1}/dx at $x = 6 = f(2)$ without finding a formula for $f^{-1}(x)$. See Figure 3.38.

Solution   We apply Theorem 3 to obtain the value of the derivative of f^{-1} at $x = 6$:

$$\frac{df}{dx}\Big|_{x=2} = 3x^2\Big|_{x=2} = 12$$

$$\frac{df^{-1}}{dx}\Big|_{x=f(2)} = \frac{1}{\dfrac{df}{dx}\Big|_{x=2}} = \frac{1}{12}. \qquad \text{Eq. (1)}$$

■

Derivative of the Natural Logarithm Function

Since we know the exponential function $f(x) = e^x$ is differentiable everywhere, we can apply Theorem 3 to find the derivative of its inverse $f^{-1}(x) = \ln x$:

$$(f^{-1})'(x) = \frac{1}{f'(f^{-1}(x))} \qquad \text{Theorem 3}$$

$$= \frac{1}{e^{f^{-1}(x)}} \qquad f'(u) = e^u$$

$$= \frac{1}{e^{\ln x}} \qquad x > 0$$

$$= \frac{1}{x}. \qquad \text{Inverse function relationship}$$

Alternative Derivation Instead of applying Theorem 3 directly, we can find the derivative of $y = \ln x$ using implicit differentiation, as follows:

$$y = \ln x \qquad x > 0$$

$$e^y = x \qquad \text{Inverse function relationship}$$

$$\frac{d}{dx}(e^y) = \frac{d}{dx}(x) \qquad \text{Differentiate implicitly.}$$

$$e^y \frac{dy}{dx} = 1 \qquad \text{Chain Rule}$$

$$\frac{dy}{dx} = \frac{1}{e^y} = \frac{1}{x}. \qquad e^y = x$$

No matter which derivation we use, the derivative of $y = \ln x$ with respect to x is

$$\frac{d}{dx}(\ln x) = \frac{1}{x}, \quad x > 0.$$

The Chain Rule extends this formula to positive functions $u(x)$:

$$\frac{d}{dx}\ln u = \frac{1}{u}\frac{du}{dx}, \qquad u > 0. \tag{2}$$

EXAMPLE 3 We use Equation (2) to find derivatives.

(a) $\dfrac{d}{dx}\ln 2x = \dfrac{1}{2x}\dfrac{d}{dx}(2x) = \dfrac{1}{2x}(2) = \dfrac{1}{x}, \quad x > 0$

(b) Equation (2) with $u = x^2 + 3$ gives

$$\frac{d}{dx}\ln(x^2 + 3) = \frac{1}{x^2 + 3}\cdot\frac{d}{dx}(x^2 + 3) = \frac{1}{x^2 + 3}\cdot 2x = \frac{2x}{x^2 + 3}.$$

(c) Equation (2) with $u = |x|$ gives an important derivative:

$$\frac{d}{dx} \ln|x| = \frac{d}{du} \ln u \cdot \frac{du}{dx} \qquad u = |x|, x \neq 0$$

$$= \frac{1}{u} \cdot \frac{x}{|x|} \qquad \frac{d}{dx}(|x|) = \frac{x}{|x|}$$

$$= \frac{1}{|x|} \cdot \frac{x}{|x|} \qquad \text{Substitute for } u.$$

$$= \frac{x}{x^2}$$

$$= \frac{1}{x}.$$

Derivative of ln $|x|$

$$\frac{d}{dx} \ln|x| = \frac{1}{x}, x \neq 0$$

So $1/x$ is the derivative of $\ln x$ on the domain $x > 0$, and the derivative of $\ln(-x)$ on the domain $x < 0$. ∎

Notice from Example 3a that the function $y = \ln 2x$ has the same derivative as the function $y = \ln x$. This is true of $y = \ln bx$ for any constant b, provided that $bx > 0$:

$$\frac{d}{dx} \ln bx = \frac{1}{x}, \quad bx > 0$$

$$\frac{d}{dx} \ln bx = \frac{1}{bx} \cdot \frac{d}{dx}(bx) = \frac{1}{bx}(b) = \frac{1}{x}. \qquad (3)$$

EXAMPLE 4 A line with slope m passes through the origin and is tangent to the graph of $y = \ln x$. What is the value of m?

Solution Suppose the point of tangency occurs at the unknown point $x = a > 0$. Then we know that the point $(a, \ln a)$ lies on the graph and that the tangent line at that point has slope $m = 1/a$ (Figure 3.39). Since the tangent line passes through the origin, its slope is

$$m = \frac{\ln a - 0}{a - 0} = \frac{\ln a}{a}.$$

Setting these two formulas for m equal to each other, we have

$$\frac{\ln a}{a} = \frac{1}{a}$$

$$\ln a = 1$$

$$e^{\ln a} = e^1$$

$$a = e$$

$$m = \frac{1}{e}. \qquad ∎$$

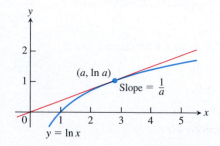

FIGURE 3.39 The tangent line intersects the curve at some point $(a, \ln a)$, where the slope of the curve is $1/a$ (Example 4).

The Derivatives of a^u and $\log_a u$

We start with the equation $a^x = e^{\ln(a^x)} = e^{x \ln a}, a > 0$, which was seen in Section 1.6, where it was used to define the function a^x:

$$\frac{d}{dx} a^x = \frac{d}{dx} e^{x \ln a}$$

$$= e^{x \ln a} \cdot \frac{d}{dx}(x \ln a) \qquad \frac{d}{dx} e^u = e^u \frac{du}{dx}$$

$$= a^x \ln a. \qquad \ln a \text{ is a constant.}$$

That is, if $a > 0$, then a^x is differentiable and

$$\frac{d}{dx} a^x = a^x \ln a. \qquad (4)$$

This equation shows why e^x is the preferred exponential function in calculus. If $a = e$, then $\ln a = 1$ and the derivative of a^x simplifies to

$$\frac{d}{dx}e^x = e^x \ln e = e^x. \qquad \text{ln } e = 1$$

With the Chain Rule, we get a more general form for the derivative of a general exponential function a^u.

If $a > 0$ and u is a differentiable function of x, then a^u is a differentiable function of x and

$$\frac{d}{dx}a^u = a^u \ln a \frac{du}{dx}. \qquad (5)$$

EXAMPLE 5   Here are some derivatives of general exponential functions.

(a) $\dfrac{d}{dx}3^x = 3^x \ln 3$ $\qquad$ Eq. (5) with $a = 3, u = x$

(b) $\dfrac{d}{dx}3^{-x} = 3^{-x}(\ln 3)\dfrac{d}{dx}(-x) = -3^{-x}\ln 3$ $\qquad$ Eq. (5) with $a = 3, u = -x$

(c) $\dfrac{d}{dx}3^{\sin x} = 3^{\sin x}(\ln 3)\dfrac{d}{dx}(\sin x) = 3^{\sin x}(\ln 3)\cos x$ $\qquad$ Eq. (5) with $u = \sin x$ ■

In Section 3.3 we looked at the derivative $f'(0)$ for the exponential functions $f(x) = a^x$ at various values of the base a. The number $f'(0)$ is the limit, $\lim_{h \to 0}(a^h - 1)/h$, and gives the slope of the graph of a^x when it crosses the y-axis at the point $(0, 1)$. We now see from Equation (4) that the value of this slope is

$$\lim_{h \to 0}\frac{a^h - 1}{h} = \ln a. \qquad (6)$$

In particular, when $a = e$ we obtain

$$\lim_{h \to 0}\frac{e^h - 1}{h} = \ln e = 1.$$

However, we have not fully justified that these limits actually exist. While all of the arguments given in deriving the derivatives of the exponential and logarithmic functions are correct, they do assume the existence of these limits. In Chapter 7 we will give another development of the theory of logarithmic and exponential functions which fully justifies that both limits do in fact exist and have the values derived above.

To find the derivative of $\log_a u$ for an arbitrary base ($a > 0, a \neq 1$), we start with the change-of-base formula for logarithms (reviewed in Section 1.6) and express $\log_a u$ in terms of natural logarithms,

$$\log_a x = \frac{\ln x}{\ln a}.$$

Taking derivatives, we have

$$\frac{d}{dx}\log_a x = \frac{d}{dx}\left(\frac{\ln x}{\ln a}\right)$$

$$= \frac{1}{\ln a} \cdot \frac{d}{dx}\ln x \qquad \text{ln } a \text{ is a constant.}$$

$$= \frac{1}{\ln a} \cdot \frac{1}{x}$$

$$= \frac{1}{x \ln a}.$$

If u is a differentiable function of x and $u > 0$, the Chain Rule gives a more general formula.

For $a > 0$ and $a \neq 1$,

$$\frac{d}{dx} \log_a u = \frac{1}{u \ln a} \frac{du}{dx}. \qquad (7)$$

Logarithmic Differentiation

The derivatives of positive functions given by formulas that involve products, quotients, and powers can often be found more quickly if we take the natural logarithm of both sides before differentiating. This enables us to use the laws of logarithms to simplify the formulas before differentiating. The process, called **logarithmic differentiation**, is illustrated in the next example.

EXAMPLE 6   Find dy/dx if

$$y = \frac{(x^2 + 1)(x + 3)^{1/2}}{x - 1}, \qquad x > 1.$$

Solution   We take the natural logarithm of both sides and simplify the result with the algebraic properties of logarithms from Theorem 1 in Section 1.6:

$$\begin{aligned}
\ln y &= \ln \frac{(x^2 + 1)(x + 3)^{1/2}}{x - 1} \\
&= \ln \left((x^2 + 1)(x + 3)^{1/2} \right) - \ln (x - 1) && \text{Rule 2} \\
&= \ln (x^2 + 1) + \ln (x + 3)^{1/2} - \ln (x - 1) && \text{Rule 1} \\
&= \ln (x^2 + 1) + \frac{1}{2} \ln (x + 3) - \ln (x - 1). && \text{Rule 4}
\end{aligned}$$

We then take derivatives of both sides with respect to x, using Equation (2) on the left:

$$\frac{1}{y} \frac{dy}{dx} = \frac{1}{x^2 + 1} \cdot 2x + \frac{1}{2} \cdot \frac{1}{x + 3} - \frac{1}{x - 1}.$$

Next we solve for dy/dx:

$$\frac{dy}{dx} = y \left(\frac{2x}{x^2 + 1} + \frac{1}{2x + 6} - \frac{1}{x - 1} \right).$$

Finally, we substitute for y:

$$\frac{dy}{dx} = \frac{(x^2 + 1)(x + 3)^{1/2}}{x - 1} \left(\frac{2x}{x^2 + 1} + \frac{1}{2x + 6} - \frac{1}{x - 1} \right). \qquad \blacksquare$$

The computation in Example 6 would be much longer if we used the product, quotient, and power rules.

Irrational Exponents and the Power Rule (General Version)

The natural logarithm and the exponential function will be defined precisely in Chapter 7. We can use the exponential function to define the general exponential function, which enables us to raise any positive number to any real power n, rational or irrational. That is, we can define the power function $y = x^n$ for any exponent n.

> **DEFINITION**   For any $x > 0$ and for any real number n,
>
> $$x^n = e^{n \ln x}.$$

Because the logarithm and exponential functions are inverses of each other, the definition gives

$$\ln x^n = n \ln x, \quad \text{for all real numbers } n.$$

That is, the rule for taking the natural logarithm of a power holds for *all* real exponents n, not just for rational exponents.

The definition of the power function also enables us to establish the derivative Power Rule for any real power n, as stated in Section 3.3.

> **General Power Rule for Derivatives**
> For $x > 0$ and any real number n,
>
> $$\frac{d}{dx} x^n = n x^{n-1}.$$
>
> If $x \le 0$, then the formula holds whenever the derivative, x^n, and x^{n-1} all exist.

Proof   Differentiating x^n with respect to x gives

$$
\begin{aligned}
\frac{d}{dx} x^n &= \frac{d}{dx} e^{n \ln x} && \text{Definition of } x^n, x > 0 \\
&= e^{n \ln x} \cdot \frac{d}{dx} (n \ln x) && \text{Chain Rule for } e^u \\
&= x^n \cdot \frac{n}{x} && \text{Definition and derivative of } \ln x \\
&= n x^{n-1}. && x^n \cdot x^{-1} = x^{n-1}
\end{aligned}
$$

In short, whenever $x > 0$,

$$\frac{d}{dx} x^n = n x^{n-1}.$$

For $x < 0$, if $y = x^n$, y', and x^{n-1} all exist, then

$$\ln|y| = \ln|x|^n = n \ln|x|.$$

Using implicit differentiation (which *assumes* the existence of the derivative y') and Example 3(c), we have

$$\frac{y'}{y} = \frac{n}{x}.$$

Solving for the derivative,

$$y' = n \frac{y}{x} = n \frac{x^n}{x} = n x^{n-1}. \qquad y = x^n$$

It can be shown directly from the definition of the derivative that the derivative equals 0 when $x = 0$ and $n \geq 1$ (see Exercise 103). This completes the proof of the general version of the Power Rule for all values of x. ∎

EXAMPLE 7   Differentiate $f(x) = x^x$, $x > 0$.

Solution   We note that $f(x) = x^x = e^{x \ln x}$, so differentiation gives

$$f'(x) = \frac{d}{dx}(e^{x \ln x})$$

$$= e^{x \ln x} \frac{d}{dx}(x \ln x) \qquad \frac{d}{dx}e^u, u = x \ln x$$

$$= e^{x \ln x}\left(\ln x + x \cdot \frac{1}{x}\right) \qquad \text{Product Rule}$$

$$= x^x (\ln x + 1). \qquad x > 0$$

We can also find the derivative of $y = x^x$ using logarithmic differentiation, assuming y' exists. ∎

The Number e Expressed as a Limit

In Section 1.5 we defined the number e as the base value for which the exponential function $y = a^x$ has slope 1 when it crosses the y-axis at $(0, 1)$. Thus e is the constant that satisfies the equation

$$\lim_{h \to 0} \frac{e^h - 1}{h} = \ln e = 1. \qquad \text{Slope equals } \ln e \text{ from Eq. (6).}$$

We now prove that e can be calculated as a certain limit.

THEOREM 4—The Number e as a Limit   The number e can be calculated as the limit

$$e = \lim_{x \to 0}(1 + x)^{1/x}.$$

Proof   If $f(x) = \ln x$, then $f'(x) = 1/x$, so $f'(1) = 1$. But, by the definition of derivative,

$$f'(1) = \lim_{h \to 0} \frac{f(1 + h) - f(1)}{h} = \lim_{x \to 0} \frac{f(1 + x) - f(1)}{x}$$

$$= \lim_{x \to 0} \frac{\ln(1 + x) - \ln 1}{x} = \lim_{x \to 0} \frac{1}{x}\ln(1 + x) \qquad \ln 1 = 0$$

$$= \lim_{x \to 0} \ln(1 + x)^{1/x} = \ln\left[\lim_{x \to 0}(1 + x)^{1/x}\right]. \qquad \begin{array}{l}\text{ln is continuous,} \\ \text{Theorem 10 in} \\ \text{Chapter 2}\end{array}$$

Because $f'(1) = 1$, we have

$$\ln\left[\lim_{x \to 0}(1 + x)^{1/x}\right] = 1.$$

Therefore, exponentiating both sides we get

$$\lim_{x \to 0}(1 + x)^{1/x} = e.$$

See Figure 3.40. ∎

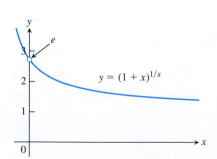

FIGURE 3.40   The number e is the limit of the function graphed here as $x \to 0$.

$y = (1 + x)^{1/x}$

Approximating the limit in Theorem 4 by taking x very small gives approximations to e. Its value is $e \approx 2.718281828459045$ to 15 decimal places.

EXERCISES 3.8

Derivatives of Inverse Functions

In Exercises 1–4:

a. Find $f^{-1}(x)$.

b. Graph f and f^{-1} together.

c. Evaluate df/dx at $x = a$ and df^{-1}/dx at $x = f(a)$ to show that at these points $df^{-1}/dx = 1/(df/dx)$.

1. $f(x) = 2x + 3, \quad a = -1$

2. $f(x) = \dfrac{x + 2}{1 - x}, \quad a = \dfrac{1}{2}$

3. $f(x) = 5 - 4x, \quad a = 1/2$

4. $f(x) = 2x^2, \quad x \geq 0, \quad a = 5$

5. a. Show that $f(x) = x^3$ and $g(x) = \sqrt[3]{x}$ are inverses of one another.

b. Graph f and g over an x-interval large enough to show the graphs intersecting at $(1, 1)$ and $(-1, -1)$. Be sure the picture shows the required symmetry about the line $y = x$.

c. Find the slopes of the tangents to the graphs of f and g at $(1, 1)$ and $(-1, -1)$ (four tangents in all).

d. What lines are tangent to the curves at the origin?

6. a. Show that $h(x) = x^3/4$ and $k(x) = (4x)^{1/3}$ are inverses of one another.

b. Graph h and k over an x-interval large enough to show the graphs intersecting at $(2, 2)$ and $(-2, -2)$. Be sure the picture shows the required symmetry about the line $y = x$.

c. Find the slopes of the tangents to the graphs at h and k at $(2, 2)$ and $(-2, -2)$.

d. What lines are tangent to the curves at the origin?

7. Let $f(x) = x^3 - 3x^2 - 1, x \geq 2$. Find the value of df^{-1}/dx at the point $x = -1 = f(3)$.

8. Let $f(x) = x^2 - 4x - 5, x > 2$. Find the value of df^{-1}/dx at the point $x = 0 = f(5)$.

9. Suppose that the differentiable function $y = f(x)$ has an inverse and that the graph of f passes through the point $(2, 4)$ and has a slope of $1/3$ there. Find the value of df^{-1}/dx at $x = 4$.

10. Suppose that the differentiable function $y = g(x)$ has an inverse and that the graph of g passes through the origin with slope 2. Find the slope of the graph of g^{-1} at the origin.

Derivatives of Logarithms

In Exercises 11–40, find the derivative of y with respect to x, t, or θ, as appropriate.

11. $y = \ln 3x + x$

12. $y = \dfrac{1}{\ln 3x}$

13. $y = \ln(t^2)$

14. $y = \ln(t^{3/2}) + \sqrt{t}$

15. $y = \ln \dfrac{3}{x}$

16. $y = \ln(\sin x)$

17. $y = \ln(\theta + 1) - e^{\theta}$

18. $y = (\cos \theta) \ln(2\theta + 2)$

19. $y = \ln x^3$

20. $y = (\ln x)^3$

21. $y = t(\ln t)^2$

22. $y = t \ln \sqrt{t}$

23. $y = \dfrac{x^4}{4} \ln x - \dfrac{x^4}{16}$

24. $y = (x^2 \ln x)^4$

25. $y = \dfrac{\ln t}{t}$

26. $y = \dfrac{t}{\sqrt{\ln t}}$

27. $y = \dfrac{\ln x}{1 + \ln x}$

28. $y = \dfrac{x \ln x}{1 + \ln x}$

29. $y = \ln(\ln x)$

30. $y = \ln(\ln(\ln x))$

31. $y = \theta(\sin(\ln \theta) + \cos(\ln \theta))$

32. $y = \ln(\sec \theta + \tan \theta)$

33. $y = \ln \dfrac{1}{x\sqrt{x + 1}}$

34. $y = \dfrac{1}{2} \ln \dfrac{1 + x}{1 - x}$

35. $y = \dfrac{1 + \ln t}{1 - \ln t}$

36. $y = \sqrt{\ln \sqrt{t}}$

37. $y = \ln(\sec(\ln \theta))$

38. $y = \ln\left(\dfrac{\sqrt{\sin \theta \cos \theta}}{1 + 2 \ln \theta}\right)$

39. $y = \ln\left(\dfrac{(x^2 + 1)^5}{\sqrt{1 - x}}\right)$

40. $y = \ln \sqrt{\dfrac{(x + 1)^5}{(x + 2)^{20}}}$

Logarithmic Differentiation

In Exercises 41–54, use logarithmic differentiation to find the derivative of y with respect to the given independent variable.

41. $y = \sqrt{x(x + 1)}$

42. $y = \sqrt{(x^2 + 1)(x - 1)^2}$

43. $y = \sqrt{\dfrac{t}{t + 1}}$

44. $y = \sqrt{\dfrac{1}{t(t + 1)}}$

45. $y = (\sin \theta)\sqrt{\theta + 3}$

46. $y = (\tan \theta)\sqrt{2\theta + 1}$

47. $y = t(t + 1)(t + 2)$

48. $y = \dfrac{1}{t(t + 1)(t + 2)}$

49. $y = \dfrac{\theta + 5}{\theta \cos \theta}$

50. $y = \dfrac{\theta \sin \theta}{\sqrt{\sec \theta}}$

51. $y = \dfrac{x\sqrt{x^2 + 1}}{(x + 1)^{2/3}}$

52. $y = \sqrt{\dfrac{(x + 1)^{10}}{(2x + 1)^5}}$

53. $y = \sqrt[3]{\dfrac{x(x - 2)}{x^2 + 1}}$

54. $y = \sqrt[3]{\dfrac{x(x + 1)(x - 2)}{(x^2 + 1)(2x + 3)}}$

Finding Derivatives

In Exercises 55–62, find the derivative of y with respect to x, t, or θ, as appropriate.

55. $y = \ln(\cos^2 \theta)$

56. $y = \ln(3\theta e^{-\theta})$

57. $y = \ln(3te^{-t})$

58. $y = \ln(2e^{-t} \sin t)$

59. $y = \ln\left(\dfrac{e^{\theta}}{1 + e^{\theta}}\right)$

60. $y = \ln\left(\dfrac{\sqrt{\theta}}{1 + \sqrt{\theta}}\right)$

61. $y = e^{(\cos t + \ln t)}$

62. $y = e^{\sin t}(\ln t^2 + 1)$

In Exercises 63–66, find dy/dx.

63. $\ln y = e^y \sin x$

64. $\ln xy = e^{x+y}$

65. $x^y = y^x$

66. $\tan y = e^x + \ln x$

In Exercises 67–88, find the derivative of y with respect to the given independent variable.

67. $y = 2^x$

68. $y = 3^{-x}$

69. $y = 5^{\sqrt{s}}$

70. $y = 2^{(s^2)}$

71. $y = x^\pi$

72. $y = t^{1-e}$

73. $y = \log_2 5\theta$

74. $y = \log_3(1 + \theta \ln 3)$

75. $y = \log_4 x + \log_4 x^2$

76. $y = \log_{25} e^x - \log_5 \sqrt{x}$

77. $y = \log_2 r \cdot \log_4 r$

78. $y = \log_3 r \cdot \log_9 r$

79. $y = \log_3\left(\left(\dfrac{x+1}{x-1}\right)^{\ln 3}\right)$

80. $y = \log_5 \sqrt{\left(\dfrac{7x}{3x+2}\right)^{\ln 5}}$

81. $y = \theta \sin(\log_7 \theta)$

82. $y = \log_7\left(\dfrac{\sin \theta \cos \theta}{e^\theta 2^\theta}\right)$

83. $y = \log_5 e^x$

84. $y = \log_2\left(\dfrac{x^2 e^2}{2\sqrt{x+1}}\right)$

85. $y = 3^{\log_2 t}$

86. $y = 3\log_8(\log_2 t)$

87. $y = \log_2(8t^{\ln 2})$

88. $y = t\log_3\left(e^{(\sin t)(\ln 3)}\right)$

Logarithmic Differentiation with Exponentials

In Exercises 89–100, use logarithmic differentiation to find the derivative of y with respect to the given independent variable.

89. $y = (x + 1)^x$

90. $y = x^{(x+1)}$

91. $y = (\sqrt{t})^t$

92. $y = t^{\sqrt{t}}$

93. $y = (\sin x)^x$

94. $y = x^{\sin x}$

95. $y = x^{\ln x}$

96. $y = (\ln x)^{\ln x}$

97. $y^x = x^3 y$

98. $x^{\sin y} = \ln y$

99. $x = y^{xy}$

100. $e^y = y^{\ln x}$

Theory and Applications

101. If we write $g(x)$ for $f^{-1}(x)$, Equation (1) can be written as

$$g'(f(a)) = \frac{1}{f'(a)}, \quad \text{or} \quad g'(f(a)) \cdot f'(a) = 1.$$

If we then write x for a, we get

$$g'(f(x)) \cdot f'(x) = 1.$$

The latter equation may remind you of the Chain Rule, and indeed there is a connection.

Assume that f and g are differentiable functions that are inverses of one another, so that $(g \circ f)(x) = x$. Differentiate both sides of this equation with respect to x, using the Chain Rule to express $(g \circ f)'(x)$ as a product of derivatives of g and f. What do you find? (This is not a proof of Theorem 3 because we assume here the theorem's conclusion that $g = f^{-1}$ is differentiable.)

102. Show that $\lim_{n \to \infty}\left(1 + \dfrac{x}{n}\right)^n = e^x$ for any $x > 0$.

103. If $f(x) = x^n$, $n \geq 1$, show from the definition of the derivative that $f'(0) = 0$.

104. Using mathematical induction, show that for $n > 1$

$$\frac{d^n}{dx^n} \ln x = (-1)^{n-1}\frac{(n-1)!}{x^n}.$$

COMPUTER EXPLORATIONS

In Exercises 105–112, you will explore some functions and their inverses together with their derivatives and tangent line approximations at specified points. Perform the following steps using your CAS:

a. Plot the function $y = f(x)$ together with its derivative over the given interval. Explain why you know that f is one-to-one over the interval.

b. Solve the equation $y = f(x)$ for x as a function of y, and name the resulting inverse function g.

c. Find the equation for the tangent line to f at the specified point $(x_0, f(x_0))$.

d. Find the equation for the tangent line to g at the point $(f(x_0), x_0)$ located symmetrically across the 45° line $y = x$ (which is the graph of the identity function). Use Theorem 3 to find the slope of this tangent line.

e. Plot the functions f and g, the identity, the two tangent lines, and the line segment joining the points $(x_0, f(x_0))$ and $(f(x_0), x_0)$. Discuss the symmetries you see across the main diagonal.

105. $y = \sqrt{3x - 2}, \quad \dfrac{2}{3} \leq x \leq 4, \quad x_0 = 3$

106. $y = \dfrac{3x + 2}{2x - 11}, \quad -2 \leq x \leq 2, \quad x_0 = 1/2$

107. $y = \dfrac{4x}{x^2 + 1}, \quad -1 \leq x \leq 1, \quad x_0 = 1/2$

108. $y = \dfrac{x^3}{x^2 + 1}, \quad -1 \leq x \leq 1, \quad x_0 = 1/2$

109. $y = x^3 - 3x^2 - 1, \quad 2 \leq x \leq 5, \quad x_0 = \dfrac{27}{10}$

110. $y = 2 - x - x^3, \quad -2 \leq x \leq 2, \quad x_0 = \dfrac{3}{2}$

111. $y = e^x, \quad -3 \leq x \leq 5, \quad x_0 = 1$

112. $y = \sin x, \quad -\dfrac{\pi}{2} \leq x \leq \dfrac{\pi}{2}, \quad x_0 = 1$

In Exercises 113 and 114, repeat the steps above to solve for the functions $y = f(x)$ and $x = f^{-1}(y)$ defined implicitly by the given equations over the interval.

113. $y^{1/3} - 1 = (x + 2)^3, \quad -5 \leq x \leq 5, \quad x_0 = -3/2$

114. $\cos y = x^{1/5}, \quad 0 \leq x \leq 1, \quad x_0 = 1/2$

3.9   Inverse Trigonometric Functions

We introduced the six basic inverse trigonometric functions in Section 1.6, but focused there on the arcsine and arccosine functions. Here we complete the study of how all six inverse trigonometric functions are defined, graphed, and evaluated, and how their derivatives are computed.

Inverses of tan x, cot x, sec x, and csc x

The graphs of these four basic inverse trigonometric functions are shown in Figure 3.41. We obtain these graphs by reflecting the graphs of the restricted trigonometric functions (as discussed in Section 1.6) through the line $y = x$. Let's take a closer look at the arctangent, arccotangent, arcsecant, and arccosecant functions.

Domain: $-\infty < x < \infty$
Range: $-\frac{\pi}{2} < y < \frac{\pi}{2}$

Domain: $-\infty < x < \infty$
Range: $0 < y < \pi$

Domain: $x \le -1$ or $x \ge 1$
Range: $0 \le y \le \pi, y \ne \frac{\pi}{2}$

Domain: $x \le -1$ or $x \ge 1$
Range: $-\frac{\pi}{2} \le y \le \frac{\pi}{2}, y \ne 0$

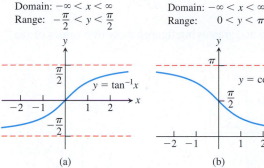
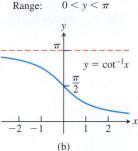
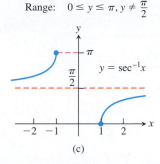
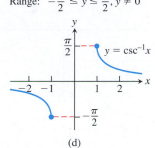

(a)   (b)   (c)   (d)

FIGURE 3.41   Graphs of the arctangent, arccotangent, arcsecant, and arccosecant functions.

The arctangent of x is a radian angle whose tangent is x. The arccotangent of x is an angle whose cotangent is x, and so forth. The angles belong to the restricted domains of the tangent, cotangent, secant, and cosecant functions.

DEFINITIONS

$y = \mathbf{tan^{-1}}x$ is the number in $(-\pi/2, \pi/2)$ for which $\tan y = x$.

$y = \mathbf{cot^{-1}}x$ is the number in $(0, \pi)$ for which $\cot y = x$.

$y = \mathbf{sec^{-1}}x$ is the number in $[0, \pi/2) \cup (\pi/2, \pi]$ for which $\sec y = x$.

$y = \mathbf{csc^{-1}}x$ is the number in $[-\pi/2, 0) \cup (0, \pi/2]$ for which $\csc y = x$.

We use open or half-open intervals to avoid values for which the tangent, cotangent, secant, and cosecant functions are undefined. (See Figure 3.41.)

As we discussed in Section 1.6, the arcsine and arccosine functions are often written as arcsin x and arccos x instead of $\sin^{-1} x$ and $\cos^{-1} x$. Likewise, we sometimes denote the other inverse trigonometric functions by arctan x, arccot x, arcsec x, and arccsc x.

The graph of $y = \tan^{-1}x$ is symmetric about the origin because it is a branch of the graph $x = \tan y$ that is symmetric about the origin (Figure 3.41a). Algebraically this means that

$$\tan^{-1}(-x) = -\tan^{-1}x;$$

the arctangent is an odd function. The graph of $y = \cot^{-1}x$ has no such symmetry (Figure 3.41b). Notice from Figure 3.41a that the graph of the arctangent function has two horizontal asymptotes: one at $y = \pi/2$ and the other at $y = -\pi/2$.

The inverses of the restricted forms of sec x and csc x are chosen to be the functions graphed in Figures 3.41c and 3.41d.

Caution   There is no general agreement about how to define $\sec^{-1}x$ for negative values of x. We chose angles in the second quadrant between $\pi/2$ and π. This choice makes $\sec^{-1}x = \cos^{-1}(1/x)$. It also makes $\sec^{-1}x$ an increasing function on each interval of its domain. Some tables choose $\sec^{-1}x$ to lie in $[-\pi, -\pi/2)$ for $x < 0$ and some texts

Domain: $|x| \geq 1$
Range: $0 \leq y \leq \pi, y \neq \frac{\pi}{2}$

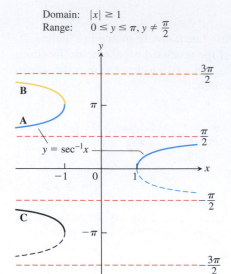

FIGURE 3.42 There are several logical choices for the left-hand branch of $y = \sec^{-1} x$. With choice **A**, $\sec^{-1} x = \cos^{-1}(1/x)$, a useful identity employed by many calculators.

choose it to lie in $[\pi, 3\pi/2)$ (Figure 3.42). These choices simplify the formula for the derivative (our formula needs absolute value signs) but fail to satisfy the computational equation $\sec^{-1} x = \cos^{-1}(1/x)$. From this, we can derive the identity

$$\sec^{-1} x = \cos^{-1}\left(\frac{1}{x}\right) = \frac{\pi}{2} - \sin^{-1}\left(\frac{1}{x}\right) \tag{1}$$

by applying Equation (5) in Section 1.6.

EXAMPLE 1   The accompanying figures show two values of $\tan^{-1} x$.

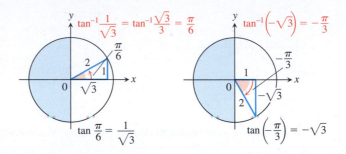

x	$\tan^{-1} x$
$\sqrt{3}$	$\pi/3$
1	$\pi/4$
$\sqrt{3}/3$	$\pi/6$
$-\sqrt{3}/3$	$-\pi/6$
-1	$-\pi/4$
$-\sqrt{3}$	$-\pi/3$

The angles come from the first and fourth quadrants because the range of $\tan^{-1} x$ is $(-\pi/2, \pi/2)$. ∎

The Derivative of $y = \sin^{-1} u$

We know that the function $x = \sin y$ is differentiable in the interval $-\pi/2 < y < \pi/2$ and that its derivative, the cosine, is positive there. Theorem 3 in Section 3.8 therefore assures us that the inverse function $y = \sin^{-1} x$ is differentiable throughout the interval $-1 < x < 1$. We cannot expect it to be differentiable at $x = 1$ or $x = -1$ because the tangents to the graph are vertical at these points (see Figure 3.43).

We find the derivative of $y = \sin^{-1} x$ by applying Theorem 3 with $f(x) = \sin x$ and $f^{-1}(x) = \sin^{-1} x$:

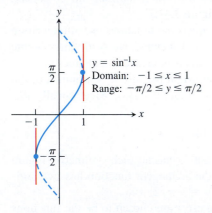

FIGURE 3.43 The graph of $y = \sin^{-1} x$ has vertical tangents at $x = -1$ and $x = 1$.

$$(f^{-1})'(x) = \frac{1}{f'(f^{-1}(x))} \qquad \text{Theorem 3}$$

$$= \frac{1}{\cos(\sin^{-1} x)} \qquad f'(u) = \cos u$$

$$= \frac{1}{\sqrt{1 - \sin^2(\sin^{-1} x)}} \qquad \cos u = \sqrt{1 - \sin^2 u}$$

$$= \frac{1}{\sqrt{1 - x^2}}. \qquad \sin(\sin^{-1} x) = x$$

If u is a differentiable function of x with $|u| < 1$, we apply the Chain Rule to get the general formula

$$\frac{d}{dx}(\sin^{-1} u) = \frac{1}{\sqrt{1 - u^2}}\frac{du}{dx}, \qquad |u| < 1.$$

EXAMPLE 2   Using the Chain Rule, we calculate the derivative

$$\frac{d}{dx}(\sin^{-1} x^2) = \frac{1}{\sqrt{1 - (x^2)^2}} \cdot \frac{d}{dx}(x^2) = \frac{2x}{\sqrt{1 - x^4}}. \qquad \blacksquare$$

The Derivative of $y = \tan^{-1} u$

We find the derivative of $y = \tan^{-1} x$ by applying Theorem 3 with $f(x) = \tan x$ and $f^{-1}(x) = \tan^{-1} x$. Theorem 3 can be applied because the derivative of $\tan x$ is positive for $-\pi/2 < x < \pi/2$:

$$(f^{-1})'(x) = \frac{1}{f'(f^{-1}(x))} \qquad \text{Theorem 3}$$

$$= \frac{1}{\sec^2(\tan^{-1} x)} \qquad f'(u) = \sec^2 u$$

$$= \frac{1}{1 + \tan^2(\tan^{-1} x)} \qquad \sec^2 u = 1 + \tan^2 u$$

$$= \frac{1}{1 + x^2}. \qquad \tan(\tan^{-1} x) = x$$

The derivative is defined for all real numbers. If u is a differentiable function of x, we get the Chain Rule form:

$$\frac{d}{dx}(\tan^{-1} u) = \frac{1}{1 + u^2}\frac{du}{dx}.$$

The Derivative of $y = \sec^{-1} u$

Since the derivative of $\sec x$ is positive for $0 < x < \pi/2$ and $\pi/2 < x < \pi$, Theorem 3 says that the inverse function $y = \sec^{-1} x$ is differentiable. Instead of applying the formula in Theorem 3 directly, we find the derivative of $y = \sec^{-1} x$, $|x| > 1$, using implicit differentiation and the Chain Rule as follows:

$$y = \sec^{-1} x$$

$$\sec y = x \qquad \qquad \text{\color{blue}Inverse function relationship}$$

$$\frac{d}{dx}(\sec y) = \frac{d}{dx}x \qquad \text{\color{blue}Differentiate both sides.}$$

$$\sec y \tan y \frac{dy}{dx} = 1 \qquad \text{\color{blue}Chain Rule}$$

$$\frac{dy}{dx} = \frac{1}{\sec y \tan y}. \qquad \text{\color{blue}Since } |x| > 1, y \text{ lies in } (0, \pi/2) \cup (\pi/2, \pi)$$
$$\text{\color{blue}and } \sec y \tan y \neq 0.$$

To express the result in terms of x, we use the relationships

$$\sec y = x \qquad \text{and} \qquad \tan y = \pm\sqrt{\sec^2 y - 1} = \pm\sqrt{x^2 - 1}$$

to get

$$\frac{dy}{dx} = \pm\frac{1}{x\sqrt{x^2 - 1}}.$$

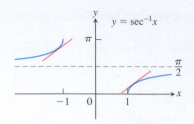

FIGURE 3.44 The slope of the curve $y = \sec^{-1} x$ is positive for both $x < -1$ and $x > 1$.

Can we do anything about the $\pm$ sign? A glance at Figure 3.44 shows that the slope of the graph $y = \sec^{-1} x$ is always positive. Thus,

$$\frac{d}{dx} \sec^{-1} x = \begin{cases} + \dfrac{1}{x\sqrt{x^2 - 1}} & \text{if } x > 1 \\[2ex] - \dfrac{1}{x\sqrt{x^2 - 1}} & \text{if } x < -1. \end{cases}$$

With the absolute value symbol, we can write a single expression that eliminates the "$\pm$" ambiguity:

$$\frac{d}{dx} \sec^{-1} x = \frac{1}{|x|\sqrt{x^2 - 1}}.$$

If u is a differentiable function of x with $|u| > 1$, we have the formula

$$\frac{d}{dx} \left(\sec^{-1} u \right) = \frac{1}{|u|\sqrt{u^2 - 1}} \frac{du}{dx}, \qquad |u| > 1.$$

EXAMPLE 3   Using the Chain Rule and derivative of the arcsecant function, we find

$$\frac{d}{dx} \sec^{-1} (5x^4) = \frac{1}{|5x^4|\sqrt{(5x^4)^2 - 1}} \frac{d}{dx} (5x^4)$$

$$= \frac{1}{5x^4\sqrt{25x^8 - 1}} (20x^3) \qquad \qquad 5x^4 > 1 > 0$$

$$= \frac{4}{x\sqrt{25x^8 - 1}}.$$

Derivatives of the Other Three Inverse Trigonometric Functions

We could use the same techniques to find the derivatives of the other three inverse trigonometric functions—arccosine, arccotangent, and arccosecant—but there is an easier way, thanks to the following identities.

Inverse Function–Inverse Cofunction Identities

$$\cos^{-1} x = \pi/2 - \sin^{-1} x$$
$$\cot^{-1} x = \pi/2 - \tan^{-1} x$$
$$\csc^{-1} x = \pi/2 - \sec^{-1} x$$

We saw the first of these identities in Equation (5) of Section 1.6. The others are derived in a similar way. It follows easily that the derivatives of the inverse cofunctions are the negatives of the derivatives of the corresponding inverse functions. For example, the derivative of $\cos^{-1} x$ is calculated as follows:

$$\frac{d}{dx} (\cos^{-1} x) = \frac{d}{dx} \left(\frac{\pi}{2} - \sin^{-1} x \right) \qquad \text{Identity}$$

$$= -\frac{d}{dx} (\sin^{-1} x)$$

$$= -\frac{1}{\sqrt{1 - x^2}}. \qquad \text{Derivative of arcsine}$$

The derivatives of the inverse trigonometric functions are summarized in Table 3.1.

TABLE 3.1 Derivatives of the inverse trigonometric functions

1. $\dfrac{d(\sin^{-1}u)}{dx} = \dfrac{1}{\sqrt{1-u^2}}\dfrac{du}{dx}, \quad |u| < 1$

2. $\dfrac{d(\cos^{-1}u)}{dx} = -\dfrac{1}{\sqrt{1-u^2}}\dfrac{du}{dx}, \quad |u| < 1$

3. $\dfrac{d(\tan^{-1}u)}{dx} = \dfrac{1}{1+u^2}\dfrac{du}{dx}$

4. $\dfrac{d(\cot^{-1}u)}{dx} = -\dfrac{1}{1+u^2}\dfrac{du}{dx}$

5. $\dfrac{d(\sec^{-1}u)}{dx} = \dfrac{1}{|u|\sqrt{u^2-1}}\dfrac{du}{dx}, \quad |u| > 1$

6. $\dfrac{d(\csc^{-1}u)}{dx} = -\dfrac{1}{|u|\sqrt{u^2-1}}\dfrac{du}{dx}, \quad |u| > 1$

EXERCISES 3.9

Common Values

Use reference triangles in an appropriate quadrant, as in Example 1, to find the angles in Exercises 1–8.

1. a. $\tan^{-1}1$ **b.** $\arctan\left(-\sqrt{3}\right)$ **c.** $\tan^{-1}\left(\dfrac{1}{\sqrt{3}}\right)$

2. a. $\arctan(-1)$ **b.** $\tan^{-1}\sqrt{3}$ **c.** $\tan^{-1}\left(\dfrac{-1}{\sqrt{3}}\right)$

3. a. $\sin^{-1}\left(\dfrac{-1}{2}\right)$ **b.** $\arcsin\left(\dfrac{1}{\sqrt{2}}\right)$ **c.** $\sin^{-1}\left(\dfrac{-\sqrt{3}}{2}\right)$

4. a. $\sin^{-1}\left(\dfrac{1}{2}\right)$ **b.** $\sin^{-1}\left(\dfrac{-1}{\sqrt{2}}\right)$ **c.** $\arcsin\left(\dfrac{\sqrt{3}}{2}\right)$

5. a. $\arccos\left(\dfrac{1}{2}\right)$ **b.** $\cos^{-1}\left(\dfrac{-1}{\sqrt{2}}\right)$ **c.** $\cos^{-1}\left(\dfrac{\sqrt{3}}{2}\right)$

6. a. $\csc^{-1}\sqrt{2}$ **b.** $\csc^{-1}\left(\dfrac{-2}{\sqrt{3}}\right)$ **c.** $\operatorname{arccsc}2$

7. a. $\sec^{-1}\left(-\sqrt{2}\right)$ **b.** $\operatorname{arcsec}\left(\dfrac{2}{\sqrt{3}}\right)$ **c.** $\sec^{-1}(-2)$

8. a. $\operatorname{arccot}(-1)$ **b.** $\cot^{-1}\left(\sqrt{3}\right)$ **c.** $\cot^{-1}\left(\dfrac{-1}{\sqrt{3}}\right)$

Evaluations

Find the values in Exercises 9–12.

9. $\sin\left(\cos^{-1}\left(\dfrac{\sqrt{2}}{2}\right)\right)$ **10.** $\sec\left(\cos^{-1}\dfrac{1}{2}\right)$

11. $\tan\left(\sin^{-1}\left(-\dfrac{1}{2}\right)\right)$ **12.** $\cot\left(\sin^{-1}\left(-\dfrac{\sqrt{3}}{2}\right)\right)$

Limits

Find the limits in Exercises 13–20. (If in doubt, look at the function's graph.)

13. $\lim\limits_{x\to1^-}\sin^{-1}x$ **14.** $\lim\limits_{x\to-1^+}\cos^{-1}x$

15. $\lim\limits_{x\to\infty}\tan^{-1}x$ **16.** $\lim\limits_{x\to-\infty}\tan^{-1}x$

17. $\lim\limits_{x\to\infty}\sec^{-1}x$ **18.** $\lim\limits_{x\to-\infty}\sec^{-1}x$

19. $\lim\limits_{x\to\infty}\csc^{-1}x$ **20.** $\lim\limits_{x\to-\infty}\csc^{-1}x$

Finding Derivatives

In Exercises 21–42, find the derivative of y with respect to the appropriate variable.

21. $y = \cos^{-1}(x^2)$ **22.** $y = \cos^{-1}(1/x)$

23. $y = \sin^{-1}\sqrt{2}\,t$ **24.** $y = \sin^{-1}(1-t)$

25. $y = \sec^{-1}(2s+1)$ **26.** $y = \sec^{-1}5s$

27. $y = \csc^{-1}(x^2+1), \quad x > 0$ **28.** $y = \csc^{-1}\dfrac{x}{2}$

29. $y = \sec^{-1}\dfrac{1}{t}, \quad 0 < t < 1$ **30.** $y = \sin^{-1}\dfrac{3}{t^2}$

31. $y = \cot^{-1}\sqrt{t}$ **32.** $y = \cot^{-1}\sqrt{t-1}$

33. $y = \ln(\tan^{-1}x)$ **34.** $y = \tan^{-1}(\ln x)$

35. $y = \csc^{-1}(e^t)$ **36.** $y = \cos^{-1}(e^{-t})$

37. $y = s\sqrt{1-s^2} + \cos^{-1}s$ **38.** $y = \sqrt{s^2-1} - \sec^{-1}s$

39. $y = \tan^{-1}\sqrt{x^2-1} + \csc^{-1}x, \quad x > 1$

40. $y = \cot^{-1}\dfrac{1}{x} - \tan^{-1}x$ **41.** $y = x\sin^{-1}x + \sqrt{1-x^2}$

42. $y = \ln(x^2+4) - x\tan^{-1}\left(\dfrac{x}{2}\right)$

For problems 43–46 use implicit differentiation to find $\dfrac{dy}{dx}$ at the given point P.

43. $3 \tan^{-1} x + \sin^{-1} y = \dfrac{\pi}{4}$; $P(1, -1)$

44. $\sin^{-1}(x + y) + \cos^{-1}(x - y) = \dfrac{5\pi}{6}$; $P\left(0, \dfrac{1}{2}\right)$

45. $y \cos^{-1}(xy) = \dfrac{-3\sqrt{2}}{4}\pi$; $P\left(\dfrac{1}{2}, -\sqrt{2}\right)$

46. $16(\tan^{-1} 3y)^2 + 9(\tan^{-1} 2x)^2 = 2\pi^2$; $P\left(\dfrac{\sqrt{3}}{2}, \dfrac{1}{3}\right)$

Theory and Examples

47. You are sitting in a classroom next to the wall looking at the blackboard at the front of the room. The blackboard is 12 ft long and starts 3 ft from the wall you are sitting next to. Show that your viewing angle is

$$\alpha = \cot^{-1}\dfrac{x}{15} - \cot^{-1}\dfrac{x}{3}$$

if you are x ft from the front wall.

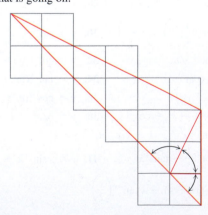

48. Find the angle α.

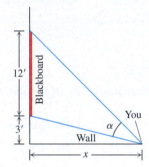

49. Here is an informal proof that $\tan^{-1} 1 + \tan^{-1} 2 + \tan^{-1} 3 = \pi$. Explain what is going on.

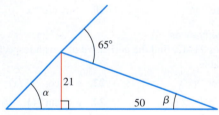

50. **Two derivations of the identity** $\sec^{-1}(-x) = \pi - \sec^{-1}x$

a. (*Geometric*) Here is a pictorial proof that $\sec^{-1}(-x) = \pi - \sec^{-1}x$. See if you can tell what is going on.

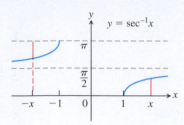

b. (*Algebraic*) Derive the identity $\sec^{-1}(-x) = \pi - \sec^{-1}x$ by combining the following two equations from the text:

$$\cos^{-1}(-x) = \pi - \cos^{-1}x \qquad \text{Eq. (4), Section 1.6}$$
$$\sec^{-1}x = \cos^{-1}(1/x) \qquad \text{Eq. (1)}$$

Which of the expressions in Exercises 51–54 are defined, and which are not? Give reasons for your answers.

51. a. $\tan^{-1} 2$ **b.** $\cos^{-1} 2$

52. a. $\csc^{-1}(1/2)$ **b.** $\csc^{-1} 2$

53. a. $\sec^{-1} 0$ **b.** $\sin^{-1}\sqrt{2}$

54. a. $\cot^{-1}(-1/2)$ **b.** $\cos^{-1}(-5)$

55. Use the identity

$$\csc^{-1} u = \dfrac{\pi}{2} - \sec^{-1} u$$

to derive the formula for the derivative of $\csc^{-1} u$ in Table 3.1 from the formula for the derivative of $\sec^{-1} u$.

56. Derive the formula

$$\dfrac{dy}{dx} = \dfrac{1}{1 + x^2}$$

for the derivative of $y = \tan^{-1} x$ by differentiating both sides of the equivalent equation $\tan y = x$.

57. Use the Derivative Rule in Section 3.8, Theorem 3, to derive

$$\dfrac{d}{dx} \sec^{-1} x = \dfrac{1}{|x|\sqrt{x^2 - 1}}, \quad |x| > 1.$$

58. Use the identity

$$\cot^{-1} u = \dfrac{\pi}{2} - \tan^{-1} u$$

to derive the formula for the derivative of $\cot^{-1} u$ in Table 3.1 from the formula for the derivative of $\tan^{-1} u$.

59. What is special about the functions

$$f(x) = \sin^{-1}\dfrac{x - 1}{x + 1}, \quad x \geq 0, \quad \text{and} \quad g(x) = 2 \tan^{-1}\sqrt{x}?$$

Explain.

60. What is special about the functions

$$f(x) = \sin^{-1}\dfrac{1}{\sqrt{x^2 + 1}} \quad \text{and} \quad g(x) = \tan^{-1}\dfrac{1}{x}?$$

Explain.

61. Find the values of

    a. $\sec^{-1} 1.5$ **b.** $\csc^{-1}(-1.5)$ **c.** $\cot^{-1} 2$

T **62.** Find the values of

    a. $\sec^{-1}(-3)$ **b.** $\csc^{-1} 1.7$ **c.** $\cot^{-1}(-2)$

T In Exercises 63–65, find the domain and range of each composite function. Then graph the composition of the two functions on separate screens. Do the graphs make sense in each case? Give reasons for your answers. Comment on any differences you see.

63. a. $y = \tan^{-1}(\tan x)$ **b.** $y = \tan(\tan^{-1} x)$

64. a. $y = \sin^{-1}(\sin x)$ **b.** $y = \sin(\sin^{-1} x)$

65. a. $y = \cos^{-1}(\cos x)$ **b.** $y = \cos(\cos^{-1} x)$

T Use your graphing utility for Exercises 66–70.

66. Graph $y = \sec(\sec^{-1} x) = \sec(\cos^{-1}(1/x))$. Explain what you see.

67. Newton's serpentine Graph Newton's serpentine, $y = 4x/(x^2 + 1)$. Then graph $y = 2\sin(2\tan^{-1} x)$ in the same graphing window. What do you see? Explain.

68. Graph the rational function $y = (2 - x^2)/x^2$. Then graph $y = \cos(2\sec^{-1} x)$ in the same graphing window. What do you see? Explain.

69. Graph $f(x) = \sin^{-1} x$ together with its first two derivatives. Comment on the behavior of f and the shape of its graph in relation to the signs and values of f' and f''.

70. Graph $f(x) = \tan^{-1} x$ together with its first two derivatives. Comment on the behavior of f and the shape of its graph in relation to the signs and values of f' and f''.

3.10 Related Rates

In this section we look at questions that arise when two or more related quantities are changing. The problem of determining how the rate of change of one of them affects the rate of change of the others is called a *related rates problem*.

Related Rates Equations

Suppose we are pumping air into a spherical balloon. Both the volume and radius of the balloon are increasing over time. If V is the volume and r is the radius of the balloon at an instant of time, then

$$V = \frac{4}{3}\pi r^3.$$

Using the Chain Rule, we differentiate both sides with respect to t to find an equation relating the rates of change of V and r,

$$\frac{dV}{dt} = \frac{dV}{dr}\frac{dr}{dt} = 4\pi r^2 \frac{dr}{dt}.$$

So if we know the radius r of the balloon and the rate dV/dt at which the volume is increasing at a given instant of time, then we can solve this last equation for dr/dt to find how fast the radius is increasing at that instant. Note that it is easier to directly measure the rate of increase of the volume (the rate at which air is being pumped into the balloon) than it is to measure the increase in the radius. The related rates equation allows us to calculate dr/dt from dV/dt.

Very often the key to relating the variables in a related rates problem is drawing a picture that shows the geometric relations between them, as illustrated in the following example.

EXAMPLE 1   Water runs into a conical tank at the rate of 9 ft³/min. The tank stands point down and has a height of 10 ft and a base radius of 5 ft. How fast is the water level rising when the water is 6 ft deep?

Solution   Figure 3.45 shows a partially filled conical tank. The variables in the problem are

$$V = \text{volume (ft}^3\text{) of the water in the tank at time } t \text{ (min)}$$

$$x = \text{radius (ft) of the surface of the water at time } t$$

$$y = \text{depth (ft) of the water in the tank at time } t.$$

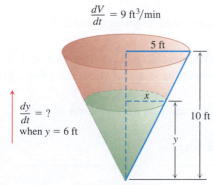

$\frac{dV}{dt} = 9$ ft³/min

$\frac{dy}{dt} = ?$ when $y = 6$ ft

FIGURE 3.45 The geometry of the conical tank and the rate at which water fills the tank determine how fast the water level rises (Example 1).

We assume that V, x, and y are differentiable functions of t. The constants are the dimensions of the tank. We are asked for dy/dt when

$$y = 6 \text{ ft} \quad \text{and} \quad \frac{dV}{dt} = 9 \text{ ft}^3/\text{min}.$$

The water forms a cone with volume

$$V = \frac{1}{3}\pi x^2 y.$$

This equation involves x as well as V and y. Because no information is given about x and dx/dt at the time in question, we need to eliminate x. The similar triangles in Figure 3.45 give us a way to express x in terms of y:

$$\frac{x}{y} = \frac{5}{10} \quad \text{or} \quad x = \frac{y}{2}.$$

Therefore, we find

$$V = \frac{1}{3}\pi \left(\frac{y}{2}\right)^2 y = \frac{\pi}{12}y^3$$

to give the derivative

$$\frac{dV}{dt} = \frac{\pi}{12} \cdot 3y^2 \frac{dy}{dt} = \frac{\pi}{4}y^2 \frac{dy}{dt}.$$

Finally, use $y = 6$ and $dV/dt = 9$ to solve for dy/dt.

$$9 = \frac{\pi}{4}(6)^2 \frac{dy}{dt}$$

$$\frac{dy}{dt} = \frac{1}{\pi} \approx 0.32$$

At the moment in question, the water level is rising at about 0.32 ft/min. ■

Related Rates Problem Strategy

1. *Draw a picture and name the variables and constants.* Use t for time. Assume that all variables are differentiable functions of t.

2. *Write down the numerical information* (in terms of the symbols you have chosen).

3. *Write down what you are asked to find* (usually a rate, expressed as a derivative).

4. *Write an equation that relates the variables.* You may have to combine two or more equations to get a single equation that relates the variable whose rate you want to the variables whose rates you know.

5. *Differentiate with respect to t.* Then express the rate you want in terms of the rates and variables whose values you know.

6. *Evaluate.* Use known values to find the unknown rate.

EXAMPLE 2   A hot air balloon rising straight up from a level field is tracked by a range finder 150 m from the liftoff point. At the moment the range finder's elevation angle is $\pi/4$, the angle is increasing at the rate of 0.14 rad/min. How fast is the balloon rising at that moment?

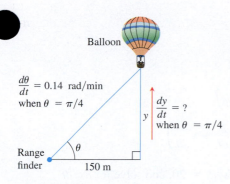

FIGURE 3.46 The rate of change of the balloon's height is related to the rate of change of the angle the range finder makes with the ground (Example 2).

Solution We answer the question in the six strategy steps.

1. *Draw a picture and name the variables and constants* (Figure 3.46). The variables in the picture are

$$\theta = \text{the angle in radians the range finder makes with the ground.}$$
$$y = \text{the height in meters of the balloon above the ground.}$$

We let t represent time in minutes and assume that θ and y are differentiable functions of t.

The one constant in the picture is the distance from the range finder to the liftoff point (150 m). There is no need to give it a special symbol.

2. *Write down the additional numerical information.*

$$\frac{d\theta}{dt} = 0.14 \text{ rad/min} \qquad \text{when} \qquad \theta = \frac{\pi}{4}$$

3. *Write down what we are to find.* We want dy/dt when $\theta = \pi/4$.

4. *Write an equation that relates the variables y and θ.*

$$\frac{y}{150} = \tan \theta \qquad \text{or} \qquad y = 150 \tan \theta$$

5. *Differentiate with respect to t using the Chain Rule.* The result tells how dy/dt (which we want) is related to $d\theta/dt$ (which we know).

$$\frac{dy}{dt} = 150 \,(\sec^2 \theta) \frac{d\theta}{dt}$$

6. *Evaluate with $\theta = \pi/4$ and $d\theta/dt = 0.14$ to find dy/dt.*

$$\frac{dy}{dt} = 150 \left(\sqrt{2} \right)^2 (0.14) = 42 \qquad \sec \frac{\pi}{4} = \sqrt{2}$$

At the moment in question, the balloon is rising at the rate of 42 m/min. ■

EXAMPLE 3 A police cruiser, approaching a right-angled intersection from the north, is chasing a speeding car that has turned the corner and is now moving straight east. When the cruiser is 0.6 mi north of the intersection and the car is 0.8 mi to the east, the police determine with radar that the distance between them and the car is increasing at 20 mph. If the cruiser is moving at 60 mph at the instant of measurement, what is the speed of the car?

Solution We picture the car and cruiser in the coordinate plane, using the positive x-axis as the eastbound highway and the positive y-axis as the southbound highway (Figure 3.47). We let t represent time and set

$$x = \text{position of car at time } t$$
$$y = \text{position of cruiser at time } t$$
$$s = \text{distance between car and cruiser at time } t.$$

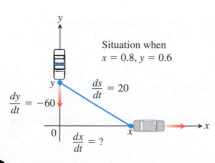

FIGURE 3.47 The speed of the car is related to the speed of the police cruiser and the rate of change of the distance s between them (Example 3).

We assume that x, y, and s are differentiable functions of t.

We want to find dx/dt when

$$x = 0.8 \text{ mi}, \qquad y = 0.6 \text{ mi}, \qquad \frac{dy}{dt} = -60 \text{ mph}, \qquad \frac{ds}{dt} = 20 \text{ mph}.$$

Note that dy/dt is negative because y is decreasing.

We differentiate the distance equation between the car and the cruiser,

$$s^2 = x^2 + y^2$$

(we could also use $s = \sqrt{x^2 + y^2}$), and obtain

$$2s\frac{ds}{dt} = 2x\frac{dx}{dt} + 2y\frac{dy}{dt}$$

$$\frac{ds}{dt} = \frac{1}{s}\left(x\frac{dx}{dt} + y\frac{dy}{dt}\right)$$

$$= \frac{1}{\sqrt{x^2 + y^2}}\left(x\frac{dx}{dt} + y\frac{dy}{dt}\right).$$

Finally, we use $x = 0.8$, $y = 0.6$, $dy/dt = -60$, $ds/dt = 20$, and solve for dx/dt.

$$20 = \frac{1}{\sqrt{(0.8)^2 + (0.6)^2}}\left(0.8\frac{dx}{dt} + (0.6)(-60)\right)$$

$$\frac{dx}{dt} = \frac{20\sqrt{(0.8)^2 + (0.6)^2} + (0.6)(60)}{0.8} = 70$$

At the moment in question, the car's speed is 70 mph. ■

EXAMPLE 4   A particle P moves clockwise at a constant rate along a circle of radius 10 m centered at the origin. The particle's initial position is $(0, 10)$ on the y-axis, and its final destination is the point $(10, 0)$ on the x-axis. Once the particle is in motion, the tangent line at P intersects the x-axis at a point Q (which moves over time). If it takes the particle 30 sec to travel from start to finish, how fast is the point Q moving along the x-axis when it is 20 m from the center of the circle?

Solution   We picture the situation in the coordinate plane with the circle centered at the origin (see Figure 3.48). We let t represent time and let θ denote the angle from the x-axis to the radial line joining the origin to P. Since the particle travels from start to finish in 30 sec, it is traveling along the circle at a constant rate of $\pi/2$ radians in $1/2$ min, or π rad/min. In other words, $d\theta/dt = -\pi$, with t being measured in minutes. The negative sign appears because θ is decreasing over time.

Setting $x(t)$ to be the distance at time t from the point Q to the origin, we want to find dx/dt when

$$x = 20 \text{ m} \qquad \text{and} \qquad \frac{d\theta}{dt} = -\pi \text{ rad/min.}$$

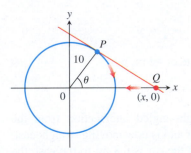

FIGURE 3.48   The particle P travels clockwise along the circle (Example 4).

To relate the variables x and θ, we see from Figure 3.48 that $x \cos \theta = 10$, or $x = 10 \sec \theta$. Differentiation of this last equation gives

$$\frac{dx}{dt} = 10 \sec \theta \tan \theta \frac{d\theta}{dt} = -10\pi \sec \theta \tan \theta.$$

Note that dx/dt is negative because x is decreasing (Q is moving toward the origin). When $x = 20$, $\cos \theta = 1/2$ and $\sec \theta = 2$. Also, $\tan \theta = \sqrt{\sec^2\theta - 1} = \sqrt{3}$. It follows that

$$\frac{dx}{dt} = (-10\pi)(2)\left(\sqrt{3}\right) = -20\sqrt{3}\pi.$$

At the moment in question, the point Q is moving toward the origin at the speed of $20\sqrt{3}\pi \approx 109$ m/min. ■

EXAMPLE 5   A jet airliner is flying at a constant altitude of 12,000 ft above sea level as it approaches a Pacific island. The aircraft comes within the direct line of sight of a radar station located on the island, and the radar indicates the initial angle between sea level and its line of sight to the aircraft is 30°. How fast (in miles per hour) is the aircraft

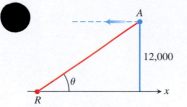

FIGURE 3.49 Jet airliner A traveling at constant altitude toward radar station R (Example 5).

approaching the island when first detected by the radar instrument if it is turning upward (counterclockwise) at the rate of 2/3 deg/sec in order to keep the aircraft within its direct line of sight?

Solution The aircraft A and radar station R are pictured in the coordinate plane, using the positive x-axis as the horizontal distance at sea level from R to A, and the positive y-axis as the vertical altitude above sea level. We let t represent time and observe that $y = 12{,}000$ is a constant. The general situation and line-of-sight angle θ are depicted in Figure 3.49. We want to find dx/dt when $\theta = \pi/6$ rad and $d\theta/dt = 2/3$ deg/sec.

From Figure 3.49, we see that

$$\frac{12{,}000}{x} = \tan \theta \quad \text{or} \quad x = 12{,}000 \cot \theta.$$

Using miles instead of feet for our distance units, the last equation translates to

$$x = \frac{12{,}000}{5280} \cot \theta.$$

Differentiation with respect to t gives

$$\frac{dx}{dt} = -\frac{1200}{528} \csc^2 \theta \, \frac{d\theta}{dt}.$$

When $\theta = \pi/6$, $\sin^2 \theta = 1/4$, so $\csc^2 \theta = 4$. Converting $d\theta/dt = 2/3$ deg/sec to radians per hour, we find

$$\frac{d\theta}{dt} = \frac{2}{3}\left(\frac{\pi}{180}\right)(3600) \text{ rad/hr}. \qquad \text{\textcolor{blue}{1 hr = 3600 sec, 1 deg = $\pi/180$ rad}}$$

Substitution into the equation for dx/dt then gives

$$\frac{dx}{dt} = \left(-\frac{1200}{528}\right)(4)\left(\frac{2}{3}\right)\left(\frac{\pi}{180}\right)(3600) \approx -380.$$

The negative sign appears because the distance x is decreasing, so the aircraft is approaching the island at a speed of approximately 380 mi/hr when first detected by the radar. ■

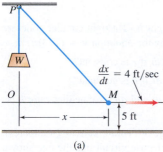

EXAMPLE 6 Figure 3.50a shows a rope running through a pulley at P and bearing a weight W at one end. The other end is held 5 ft above the ground in the hand M of a worker. Suppose the pulley is 25 ft above ground, the rope is 45 ft long, and the worker is walking rapidly away from the vertical line PW at the rate of 4 ft/sec. How fast is the weight being raised when the worker's hand is 21 ft away from PW?

Solution We let OM be the horizontal line of length x ft from a point O directly below the pulley to the worker's hand M at any instant of time (Figure 3.50). Let h be the height of the weight W above O, and let z denote the length of rope from the pulley P to the worker's hand. We want to know dh/dt when $x = 21$ given that $dx/dt = 4$. Note that the height of P above O is 20 ft because O is 5 ft above the ground. We assume the angle at O is a right angle.

At any instant of time t we have the following relationships (see Figure 3.50b):

$$20 - h + z = 45 \qquad \text{\textcolor{blue}{Total length of rope is 45 ft.}}$$
$$20^2 + x^2 = z^2. \qquad \text{\textcolor{blue}{Angle at O is a right angle.}}$$

If we solve for $z = 25 + h$ in the first equation, and substitute into the second equation, we have

$$20^2 + x^2 = (25 + h)^2. \qquad (1)$$

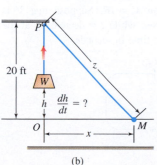

FIGURE 3.50 A worker at M walks to the right, pulling the weight W upward as the rope moves through the pulley P (Example 6).

Differentiating both sides with respect to t gives

$$2x\frac{dx}{dt} = 2(25 + h)\frac{dh}{dt},$$

and solving this last equation for dh/dt we find

$$\frac{dh}{dt} = \frac{x}{25 + h}\frac{dx}{dt}. \tag{2}$$

Since we know dx/dt, it remains only to find $25 + h$ at the instant when $x = 21$. From Equation (1),

$$20^2 + 21^2 = (25 + h)^2$$

so that

$$(25 + h)^2 = 841, \quad \text{or} \quad 25 + h = 29.$$

Equation (2) now gives

$$\frac{dh}{dt} = \frac{21}{29}\cdot 4 = \frac{84}{29} \approx 2.9 \text{ ft/sec}$$

as the rate at which the weight is being raised when $x = 21$ ft. ∎

EXERCISES 3.10

1. **Area**   Suppose that the radius r and area $A = \pi r^2$ of a circle are differentiable functions of t. Write an equation that relates dA/dt to dr/dt.

2. **Surface area**   Suppose that the radius r and surface area $S = 4\pi r^2$ of a sphere are differentiable functions of t. Write an equation that relates dS/dt to dr/dt.

3. Assume that $y = 5x$ and $dx/dt = 2$. Find dy/dt.

4. Assume that $2x + 3y = 12$ and $dy/dt = -2$. Find dx/dt.

5. If $y = x^2$ and $dx/dt = 3$, then what is dy/dt when $x = -1$?

6. If $x = y^3 - y$ and $dy/dt = 5$, then what is dx/dt when $y = 2$?

7. If $x^2 + y^2 = 25$ and $dx/dt = -2$, then what is dy/dt when $x = 3$ and $y = -4$?

8. If $x^2y^3 = 4/27$ and $dy/dt = 1/2$, then what is dx/dt when $x = 2$?

9. If $L = \sqrt{x^2 + y^2}$, $dx/dt = -1$, and $dy/dt = 3$, find dL/dt when $x = 5$ and $y = 12$.

10. If $r + s^2 + v^3 = 12$, $dr/dt = 4$, and $ds/dt = -3$, find dv/dt when $r = 3$ and $s = 1$.

11. If the original 24 m edge length x of a cube decreases at the rate of 5 m/min, when $x = 3$ m at what rate does the cube's
    a. surface area change?
    b. volume change?

12. A cube's surface area increases at the rate of 72 in²/sec. At what rate is the cube's volume changing when the edge length is $x = 3$ in?

13. **Volume**   The radius r and height h of a right circular cylinder are related to the cylinder's volume V by the formula $V = \pi r^2 h$.

    a. How is dV/dt related to dh/dt if r is constant?
    b. How is dV/dt related to dr/dt if h is constant?
    c. How is dV/dt related to dr/dt and dh/dt if neither r nor h is constant?

14. **Volume**   The radius r and height h of a right circular cone are related to the cone's volume V by the equation $V = (1/3)\pi r^2 h$.

    a. How is dV/dt related to dh/dt if r is constant?
    b. How is dV/dt related to dr/dt if h is constant?
    c. How is dV/dt related to dr/dt and dh/dt if neither r nor h is constant?

15. **Changing voltage**   The voltage V (volts), current I (amperes), and resistance R (ohms) of an electric circuit like the one shown here are related by the equation $V = IR$. Suppose that V is increasing at the rate of 1 volt/sec while I is decreasing at the rate of $1/3$ amp/sec. Let t denote time in seconds.

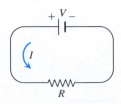

    a. What is the value of dV/dt?
    b. What is the value of dI/dt?
    c. What equation relates dR/dt to dV/dt and dI/dt?
    d. Find the rate at which R is changing when $V = 12$ volts and $I = 2$ amps. Is R increasing, or decreasing?

16. Electrical power   The power P (watts) of an electric circuit is related to the circuit's resistance R (ohms) and current I (amperes) by the equation $P = RI^2$.

　a. How are dP/dt, dR/dt, and dI/dt related if none of P, R, and I are constant?

　b. How is dR/dt related to dI/dt if P is constant?

17. Distance   Let x and y be differentiable functions of t and let $s = \sqrt{x^2 + y^2}$ be the distance between the points $(x, 0)$ and $(0, y)$ in the xy-plane.

　a. How is ds/dt related to dx/dt if y is constant?

　b. How is ds/dt related to dx/dt and dy/dt if neither x nor y is constant?

　c. How is dx/dt related to dy/dt if s is constant?

18. Diagonals   If x, y, and z are lengths of the edges of a rectangular box, the common length of the box's diagonals is $s = \sqrt{x^2 + y^2 + z^2}$.

　a. Assuming that x, y, and z are differentiable functions of t, how is ds/dt related to dx/dt, dy/dt, and dz/dt?

　b. How is ds/dt related to dy/dt and dz/dt if x is constant?

　c. How are dx/dt, dy/dt, and dz/dt related if s is constant?

19. Area   The area A of a triangle with sides of lengths a and b enclosing an angle of measure θ is

$$A = \frac{1}{2}ab \sin \theta.$$

　a. How is dA/dt related to $d\theta/dt$ if a and b are constant?

　b. How is dA/dt related to $d\theta/dt$ and da/dt if only b is constant?

　c. How is dA/dt related to $d\theta/dt$, da/dt, and db/dt if none of a, b, and θ are constant?

20. Heating a plate   When a circular plate of metal is heated in an oven, its radius increases at the rate of 0.01 cm/min. At what rate is the plate's area increasing when the radius is 50 cm?

21. Changing dimensions in a rectangle   The length l of a rectangle is decreasing at the rate of 2 cm/sec while the width w is increasing at the rate of 2 cm/sec. When $l = 12$ cm and $w = 5$ cm, find the rates of change of **(a)** the area, **(b)** the perimeter, and **(c)** the lengths of the diagonals of the rectangle. Which of these quantities are decreasing, and which are increasing?

22. Changing dimensions in a rectangular box   Suppose that the edge lengths x, y, and z of a closed rectangular box are changing at the following rates:

$$\frac{dx}{dt} = 1 \text{ m/sec}, \quad \frac{dy}{dt} = -2 \text{ m/sec}, \quad \frac{dz}{dt} = 1 \text{ m/sec}.$$

Find the rates at which the box's **(a)** volume, **(b)** surface area, and **(c)** diagonal length $s = \sqrt{x^2 + y^2 + z^2}$ are changing at the instant when $x = 4$, $y = 3$, and $z = 2$.

23. A sliding ladder   A 13-ft ladder is leaning against a house when its base starts to slide away (see accompanying figure). By the time the base is 12 ft from the house, the base is moving at the rate of 5 ft/sec.

　a. How fast is the top of the ladder sliding down the wall then?

　b. At what rate is the area of the triangle formed by the ladder, wall, and ground changing then?

　c. At what rate is the angle θ between the ladder and the ground changing then?

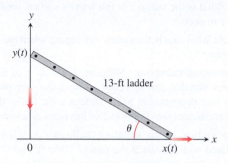

24. Commercial air traffic   Two commercial airplanes are flying at an altitude of 40,000 ft along straight-line courses that intersect at right angles. Plane A is approaching the intersection point at a speed of 442 knots (nautical miles per hour; a nautical mile is 2000 yd). Plane B is approaching the intersection at 481 knots. At what rate is the distance between the planes changing when A is 5 nautical miles from the intersection point and B is 12 nautical miles from the intersection point?

25. Flying a kite   A girl flies a kite at a height of 300 ft, the wind carrying the kite horizontally away from her at a rate of 25 ft/sec. How fast must she let out the string when the kite is 500 ft away from her?

26. Boring a cylinder   The mechanics at Lincoln Automotive are reboring a 6-in.-deep cylinder to fit a new piston. The machine they are using increases the cylinder's radius one-thousandth of an inch every 3 min. How rapidly is the cylinder volume increasing when the bore (diameter) is 3.800 in.?

27. A growing sand pile   Sand falls from a conveyor belt at the rate of 10 m³/min onto the top of a conical pile. The height of the pile is always three-eighths of the base diameter. How fast are the **(a)** height and **(b)** radius changing when the pile is 4 m high? Answer in centimeters per minute.

28. A draining conical reservoir   Water is flowing at the rate of 50 m³/min from a shallow concrete conical reservoir (vertex down) of base radius 45 m and height 6 m.

　a. How fast (centimeters per minute) is the water level falling when the water is 5 m deep?

　b. How fast is the radius of the water's surface changing then? Answer in centimeters per minute.

29. A draining hemispherical reservoir   Water is flowing at the rate of 6 m³/min from a reservoir shaped like a hemispherical bowl of radius 13 m, shown here in profile. Answer the following questions, given that the volume of water in a hemispherical bowl of radius R is $V = (\pi/3)y^2(3R - y)$ when the water is y meters deep.

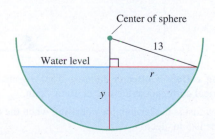

a. At what rate is the water level changing when the water is 8 m deep?

b. What is the radius r of the water's surface when the water is y m deep?

c. At what rate is the radius r changing when the water is 8 m deep?

30. A growing raindrop Suppose that a drop of mist is a perfect sphere and that, through condensation, the drop picks up moisture at a rate proportional to its surface area. Show that under these circumstances the drop's radius increases at a constant rate.

31. The radius of an inflating balloon A spherical balloon is inflated with helium at the rate of 100π ft^3/min. How fast is the balloon's radius increasing at the instant the radius is 5 ft? How fast is the surface area increasing?

32. Hauling in a dinghy A dinghy is pulled toward a dock by a rope from the bow through a ring on the dock 6 ft above the bow. The rope is hauled in at the rate of 2 ft/sec.

a. How fast is the boat approaching the dock when 10 ft of rope are out?

b. At what rate is the angle θ changing at this instant (see the figure)?

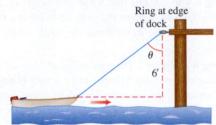
Ring at edge of dock
θ
6'

33. A balloon and a bicycle A balloon is rising vertically above a level, straight road at a constant rate of 1 ft/sec. Just when the balloon is 65 ft above the ground, a bicycle moving at a constant rate of 17 ft/sec passes under it. How fast is the distance $s(t)$ between the bicycle and balloon increasing 3 sec later?

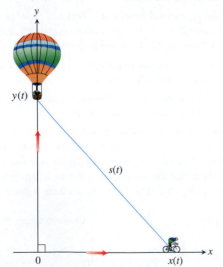
y
$y(t)$
$s(t)$
0
$x(t)$
x

34. Making coffee Coffee is draining from a conical filter into a cylindrical coffeepot at the rate of 10 in^3/min.

a. How fast is the level in the pot rising when the coffee in the cone is 5 in. deep?

b. How fast is the level in the cone falling then?

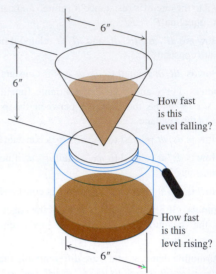
6"
6"
How fast is this level falling?
How fast is this level rising?
6"

35. Cardiac output In the late 1860s, Adolf Fick, a professor of physiology in the Faculty of Medicine in Würzberg, Germany, developed one of the methods we use today for measuring how much blood your heart pumps in a minute. Your cardiac output as you read this sentence is probably about 7 L/min. At rest it is likely to be a bit under 6 L/min. If you are a trained marathon runner running a marathon, your cardiac output can be as high as 30 L/min.

Your cardiac output can be calculated with the formula

$$y = \frac{Q}{D},$$

where Q is the number of milliliters of CO_2 you exhale in a minute and D is the difference between the CO_2 concentration (ml/L) in the blood pumped to the lungs and the CO_2 concentration in the blood returning from the lungs. With $Q = 233$ ml/min and $D = 97 - 56 = 41$ ml/L,

$$y = \frac{233 \text{ ml/min}}{41 \text{ ml/L}} \approx 5.68 \text{ L/min},$$

fairly close to the 6 L/min that most people have at basal (resting) conditions. (Data courtesy of J. Kenneth Herd, M.D., Quillan College of Medicine, East Tennessee State University.)

Suppose that when $Q = 233$ and $D = 41$, we also know that D is decreasing at the rate of 2 units a minute but that Q remains unchanged. What is happening to the cardiac output?

36. Moving along a parabola A particle moves along the parabola $y = x^2$ in the first quadrant in such a way that its x-coordinate (measured in meters) increases at a steady 10 m/sec. How fast is the angle of inclination θ of the line joining the particle to the origin changing when $x = 3$ m?

37. Motion in the plane The coordinates of a particle in the metric xy-plane are differentiable functions of time t with $dx/dt = -1$ m/sec and $dy/dt = -5$ m/sec. How fast is the particle's distance from the origin changing as it passes through the point (5, 12)?

38. Videotaping a moving car You are videotaping a race from a stand 132 ft from the track, following a car that is moving at

180 mi/h (264 ft/sec), as shown in the accompanying figure. How fast will your camera angle θ be changing when the car is right in front of you? A half second later?

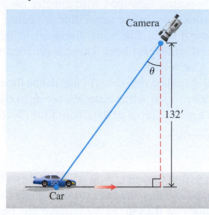

39. A moving shadow A light shines from the top of a pole 50 ft high. A ball is dropped from the same height from a point 30 ft away from the light. (See accompanying figure.) How fast is the shadow of the ball moving along the ground 1/2 sec later? (Assume the ball falls a distance $s = 16t^2$ ft in t sec.)

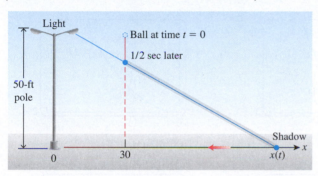

40. A building's shadow On a morning of a day when the sun will pass directly overhead, the shadow of an 80-ft building on level ground is 60 ft long. At the moment in question, the angle θ the sun makes with the ground is increasing at the rate of 0.27°/min. At what rate is the shadow decreasing? (Remember to use radians. Express your answer in inches per minute, to the nearest tenth.)

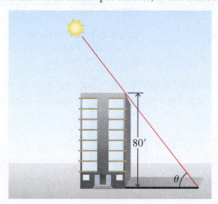

41. A melting ice layer A spherical iron ball 8 in. in diameter is coated with a layer of ice of uniform thickness. If the ice melts at the rate of 10 in³/min, how fast is the thickness of the ice decreasing when it is 2 in. thick? How fast is the outer surface area of ice decreasing?

42. Highway patrol A highway patrol plane flies 3 mi above a level, straight road at a steady 120 mi/h. The pilot sees an oncoming car and with radar determines that at the instant the line-of-sight distance from plane to car is 5 mi, the line-of-sight distance is decreasing at the rate of 160 mi/h. Find the car's speed along the highway.

43. Baseball players A baseball diamond is a square 90 ft on a side. A player runs from first base to second at a rate of 16 ft/sec.

a. At what rate is the player's distance from third base changing when the player is 30 ft from first base?

b. At what rates are angles θ_1 and θ_2 (see the figure) changing at that time?

c. The player slides into second base at the rate of 15 ft/sec. At what rates are angles θ_1 and θ_2 changing as the player touches base?

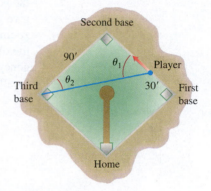

44. Ships Two ships are steaming straight away from a point *O* along routes that make a 120° angle. Ship *A* moves at 14 knots (nautical miles per hour; a nautical mile is 2000 yd). Ship *B* moves at 21 knots. How fast are the ships moving apart when $OA = 5$ and $OB = 3$ nautical miles?

45. Clock's moving hands At what rate is the angle between a clock's minute and hour hands changing at 4 o'clock in the afternoon?

46. Oil spill An explosion at an oil rig located in gulf waters causes an elliptical oil slick to spread on the surface from the rig. The slick is a constant 9 in. thick. After several days, when the major axis of the slick is 2 mi long and the minor axis is 3/4 mi wide, it is determined that its length is increasing at the rate of 30 ft/hr, and its width is increasing at the rate of 10 ft/hr. At what rate (in cubic feet per hour) is oil flowing from the site of the rig at that time?

47. A lighthouse beam A lighthouse sits 1 km offshore, and its beam of light rotates counterclockwise at the constant rate of 3 full circles per minute. At what rate is the image of the beam moving down the shoreline when the image is 1 km from the spot on the shoreline nearest the lighthouse?

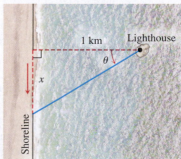

3.11 Linearization and Differentials

It is often useful to approximate complicated functions with simpler ones that give the accuracy we want for specific applications and at the same time are easier to work with than the original functions. The approximating functions discussed in this section are called *linearizations*, and they are based on tangent lines. Other approximating functions, such as polynomials, are discussed in Chapter 10.

We introduce new variables dx and dy, called *differentials*, and define them in a way that makes Leibniz's notation for the derivative dy/dx a true ratio. We use dy to estimate error in measurement, which then provides for a precise proof of the Chain Rule (Section 3.6).

Linearization

As you can see in Figure 3.51, the tangent to the curve $y = x^2$ lies close to the curve near the point of tangency. For a brief interval to either side, the y-values along the tangent line give good approximations to the y-values on the curve. We observe this phenomenon by zooming in on the two graphs at the point of tangency, or by looking at tables of values for the difference between $f(x)$ and its tangent line near the x-coordinate of the point of tangency. The phenomenon is true not just for parabolas; every differentiable curve behaves locally like its tangent line.

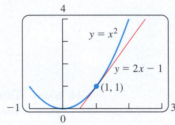

$y = x^2$ and its tangent $y = 2x - 1$ at $(1, 1)$.

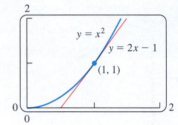

Tangent and curve very close near $(1, 1)$.

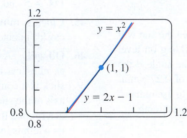

Tangent and curve very close throughout entire x-interval shown.

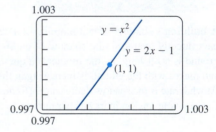

Tangent and curve closer still. Computer screen cannot distinguish tangent from curve on this x-interval.

FIGURE 3.51 The more we magnify the graph of a function near a point where the function is differentiable, the flatter the graph becomes and the more it resembles its tangent.

In general, the tangent to $y = f(x)$ at a point $x = a$, where f is differentiable (Figure 3.52), passes through the point $(a, f(a))$, so its point-slope equation is

$$y = f(a) + f'(a)(x - a).$$

Thus, this tangent line is the graph of the linear function

$$L(x) = f(a) + f'(a)(x - a).$$

As long as this line remains close to the graph of f as we move off the point of tangency, $L(x)$ gives a good approximation to $f(x)$.

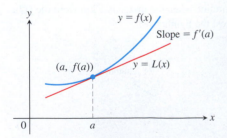

FIGURE 3.52 The tangent to the curve $y = f(x)$ at $x = a$ is the line $L(x) = f(a) + f'(a)(x - a)$.

DEFINITIONS If f is differentiable at $x = a$, then the approximating function

$$L(x) = f(a) + f'(a)(x - a)$$

is the **linearization** of f at a. The approximation

$$f(x) \approx L(x)$$

of f by L is the **standard linear approximation** of f at a. The point $x = a$ is the **center** of the approximation.

EXAMPLE 1   Find the linearization of $f(x) = \sqrt{1 + x}$ at $x = 0$ (Figure 3.53).

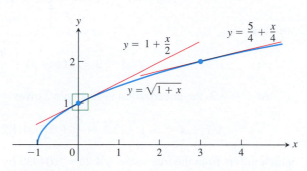
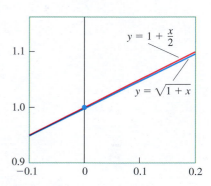

FIGURE 3.53 The graph of $y = \sqrt{1 + x}$ and its linearizations at $x = 0$ and $x = 3$. Figure 3.54 shows a magnified view of the small window about 1 on the y-axis.

FIGURE 3.54 Magnified view of the window in Figure 3.53.

Solution   Since

$$f'(x) = \frac{1}{2}(1 + x)^{-1/2},$$

we have $f(0) = 1$ and $f'(0) = 1/2$, giving the linearization

$$L(x) = f(a) + f'(a)(x - a) = 1 + \frac{1}{2}(x - 0) = 1 + \frac{x}{2}.$$

See Figure 3.54. ■

The following table shows how accurate the approximation $\sqrt{1 + x} \approx 1 + (x/2)$ from Example 1 is for some values of x near 0. As we move away from zero, we lose accuracy. For example, for $x = 2$, the linearization gives 2 as the approximation for $\sqrt{3}$, which is not even accurate to one decimal place.

Approximation	True value	|True value − approximation|
$\sqrt{1.005} \approx 1 + \dfrac{0.005}{2} = 1.00250$	1.002497	$0.000003 < 10^{-5}$
$\sqrt{1.05} \approx 1 + \dfrac{0.05}{2} = 1.025$	1.024695	$0.000305 < 10^{-3}$
$\sqrt{1.2} \approx 1 + \dfrac{0.2}{2} = 1.10$	1.095445	$0.004555 < 10^{-2}$

Do not be misled by the preceding calculations into thinking that whatever we do with a linearization is better done with a calculator. In practice, we would never use a linearization to find a particular square root. The utility of a linearization is its ability to replace a complicated formula by a simpler one over an *entire interval of values*. If we have to work with $\sqrt{1 + x}$ for x in an interval close to 0 and can tolerate the small amount of error involved over that interval, we can work with $1 + (x/2)$ instead. Of course, we then need to know how much error there is. We further examine the estimation of error in Chapter 10.

A linear approximation normally loses accuracy away from its center. As Figure 3.53 suggests, the approximation $\sqrt{1 + x} \approx 1 + (x/2)$ is too crude to be useful near $x = 3$. There, we need the linearization at $x = 3$.

EXAMPLE 2   Find the linearization of $f(x) = \sqrt{1 + x}$ at $x = 3$. (See Figure 3.53.)

Solution   We evaluate the equation defining $L(x)$ at $a = 3$. With

$$f(3) = 2, \qquad f'(3) = \frac{1}{2}(1 + x)^{-1/2}\bigg|_{x=3} = \frac{1}{4},$$

we have

$$L(x) = 2 + \frac{1}{4}(x - 3) = \frac{5}{4} + \frac{x}{4}.$$

At $x = 3.2$, the linearization in Example 2 gives

$$\sqrt{1 + x} = \sqrt{1 + 3.2} \approx \frac{5}{4} + \frac{3.2}{4} = 1.250 + 0.800 = 2.050,$$

which differs from the true value $\sqrt{4.2} \approx 2.04939$ by less than one one-thousandth. The linearization in Example 1 gives

$$\sqrt{1 + x} = \sqrt{1 + 3.2} \approx 1 + \frac{3.2}{2} = 1 + 1.6 = 2.6,$$

a result that is off by more than 25%.   ∎

EXAMPLE 3   Find the linearization of $f(x) = \cos x$ at $x = \pi/2$ (Figure 3.55).

Solution   Since $f(\pi/2) = \cos(\pi/2) = 0$, $f'(x) = -\sin x$, and $f'(\pi/2) = -\sin(\pi/2) = -1$, we find the linearization at $a = \pi/2$ to be

$$L(x) = f(a) + f'(a)(x - a)$$

$$= 0 + (-1)\left(x - \frac{\pi}{2}\right)$$

$$= -x + \frac{\pi}{2}.$$ ∎

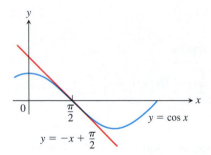

FIGURE 3.55   The graph of $f(x) = \cos x$ and its linearization at $x = \pi/2$. Near $x = \pi/2$, $\cos x \approx -x + (\pi/2)$ (Example 3).

An important linear approximation for roots and powers is

$$(1 + x)^k \approx 1 + kx \qquad (x \text{ near } 0; \text{ any number } k)$$

(Exercise 15). This approximation, good for values of x sufficiently close to zero, has broad application. For example, when x is small,

Approximations Near $x = 0$
$\sqrt{1 + x} \approx 1 + \dfrac{x}{2}$
$\dfrac{1}{1 - x} \approx 1 + x$
$\dfrac{1}{\sqrt{1 - x^2}} \approx 1 + \dfrac{x^2}{2}$

$$\sqrt{1 + x} \approx 1 + \frac{1}{2}x \qquad\qquad k = 1/2$$

$$\frac{1}{1 - x} = (1 - x)^{-1} \approx 1 + (-1)(-x) = 1 + x \qquad\qquad k = -1; \text{ replace } x \text{ by } -x.$$

$$\sqrt[3]{1 + 5x^4} = (1 + 5x^4)^{1/3} \approx 1 + \frac{1}{3}(5x^4) = 1 + \frac{5}{3}x^4 \qquad\qquad k = 1/3; \text{ replace } x \text{ by } 5x^4.$$

$$\frac{1}{\sqrt{1 - x^2}} = (1 - x^2)^{-1/2} \approx 1 + \left(-\frac{1}{2}\right)(-x^2) = 1 + \frac{1}{2}x^2 \qquad\qquad k = -1/2; \text{ replace } x \text{ by } -x^2.$$

Differentials

We sometimes use the Leibniz notation dy/dx to represent the derivative of y with respect to x. Contrary to its appearance, it is not a ratio. We now introduce two new variables dx and dy with the property that when their ratio exists, it is equal to the derivative.

DEFINITION   Let $y = f(x)$ be a differentiable function. The **differential dx** is an independent variable. The **differential dy** is

$$dy = f'(x)\, dx.$$

Unlike the independent variable dx, the variable dy is always a dependent variable. It depends on both x and dx. If dx is given a specific value and x is a particular number in the domain of the function f, then these values determine the numerical value of dy. Often the variable dx is chosen to be Δx, the change in x.

EXAMPLE 4

(a) Find dy if $y = x^5 + 37x$.

(b) Find the value of dy when $x = 1$ and $dx = 0.2$.

Solution

(a) $dy = (5x^4 + 37)\, dx$

(b) Substituting $x = 1$ and $dx = 0.2$ in the expression for dy, we have

$$dy = (5 \cdot 1^4 + 37)0.2 = 8.4. \qquad \blacksquare$$

The geometric meaning of differentials is shown in Figure 3.56. Let $x = a$ and set $dx = \Delta x$. The corresponding change in $y = f(x)$ is

$$\Delta y = f(a + dx) - f(a).$$

The corresponding change in the tangent line L is

$$\begin{aligned}
\Delta L &= L(a + dx) - L(a) \\
&= \underbrace{f(a) + f'(a)\big[(a + dx) - a\big]}_{L(a + dx)} - \underbrace{f(a)}_{L(a)} \\
&= f'(a)\, dx.
\end{aligned}$$

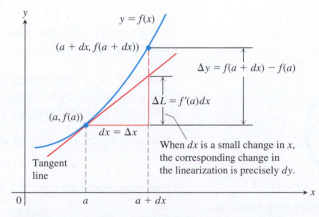

FIGURE 3.56   Geometrically, the differential dy is the change ΔL in the linearization of f when $x = a$ changes by an amount $dx = \Delta x$.

That is, the change in the linearization of f is precisely the value of the differential dy when $x = a$ and $dx = \Delta x$. Therefore, dy represents the amount the tangent line rises or falls when x changes by an amount $dx = \Delta x$.

If $dx \neq 0$, then the quotient of the differential dy by the differential dx is equal to the derivative $f'(x)$ because

$$dy \div dx = \frac{f'(x)\, dx}{dx} = f'(x) = \frac{dy}{dx}.$$

We sometimes write

$$df = f'(x)\, dx$$

in place of $dy = f'(x)\, dx$, calling df the **differential of f.** For instance, if $f(x) = 3x^2 - 6$, then

$$df = d(3x^2 - 6) = 6x\, dx.$$

Every differentiation formula like

$$\frac{d(u + v)}{dx} = \frac{du}{dx} + \frac{dv}{dx} \qquad \text{or} \qquad \frac{d(\sin u)}{dx} = \cos u \frac{du}{dx}$$

has a corresponding differential form like

$$d(u + v) = du + dv \qquad \text{or} \qquad d(\sin u) = \cos u\, du.$$

EXAMPLE 5   We can use the Chain Rule and other differentiation rules to find differentials of functions.

(a) $d(\tan 2x) = \sec^2(2x)\, d(2x) = 2\sec^2 2x\, dx$

(b) $d\left(\dfrac{x}{x + 1}\right) = \dfrac{(x + 1)\, dx - x\, d(x + 1)}{(x + 1)^2} = \dfrac{x\,dx + dx - x\, dx}{(x + 1)^2} = \dfrac{dx}{(x + 1)^2}$ ∎

Estimating with Differentials

Suppose we know the value of a differentiable function $f(x)$ at a point a and want to estimate how much this value will change if we move to a nearby point $a + dx$. If $dx = \Delta x$ is small, then we can see from Figure 3.56 that Δy is approximately equal to the differential dy. Since

$$f(a + dx) = f(a) + \Delta y, \qquad \color{blue}{\Delta x = dx}$$

the differential approximation gives

$$f(a + dx) \approx f(a) + dy$$

when $dx = \Delta x$. Thus the approximation $\Delta y \approx dy$ can be used to estimate $f(a + dx)$ when $f(a)$ is known, dx is small, and $dy = f'(a)\, dx$.

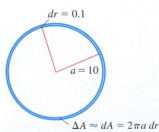

FIGURE 3.57   When dr is small compared with a, the differential dA gives the estimate $A(a + dr) = \pi a^2 + dA$ (Example 6).

EXAMPLE 6   The radius r of a circle increases from $a = 10$ m to 10.1 m (Figure 3.57). Use dA to estimate the increase in the circle's area A. Estimate the area of the enlarged circle and compare your estimate to the true area found by direct calculation.

Solution   Since $A = \pi r^2$, the estimated increase is

$$dA = A'(a)\, dr = 2\pi a\, dr = 2\pi(10)(0.1) = 2\pi \text{ m}^2.$$

Thus, since $A(r + \Delta r) \approx A(r) + dA$, we have

$$A(10 + 0.1) \approx A(10) + 2\pi$$
$$= \pi(10)^2 + 2\pi = 102\pi.$$

The area of a circle of radius 10.1 m is approximately 102π m^2.

The true area is

$$A(10.1) = \pi(10.1)^2$$
$$= 102.01\pi \text{ m}^2.$$

The error in our estimate is $0.01\pi \text{ m}^2$, which is the difference $\Delta A - dA$. ∎

When using differentials to estimate functions, our goal is to choose a nearby point $x = a$ where both $f(a)$ and the derivative $f'(a)$ are easy to evaluate.

EXAMPLE 7   Use differentials to estimate

(a) $7.97^{1/3}$ **(b)** $\sin(\pi/6 + 0.01)$.

Solution

(a) The differential associated with the cube root function $y = x^{1/3}$ is

$$dy = \frac{1}{3x^{2/3}}dx.$$

We set $a = 8$, the closest number near 7.97 where we can easily compute $f(a)$ and $f'(a)$. To arrange that $a + dx = 7.97$, we choose $dx = -0.03$. Approximating with the differential gives

$$f(7.97) = f(a + dx) \approx f(a) + dy$$

$$= 8^{1/3} + \frac{1}{3(8)^{2/3}}(-0.03)$$

$$= 2 + \frac{1}{12}(-0.03) = 1.9975$$

This gives an approximation to the true value of $7.97^{1/3}$, which is 1.997497 to 6 decimals.

(b) The differential associated with $y = \sin x$ is

$$dy = \cos x\, dx.$$

To estimate $\sin(\pi/6 + 0.01)$, we set $a = \pi/6$ and $dx = 0.01$. Then

$$f(\pi/6 + 0.01) = f(a + dx) \approx f(a) + dy$$

$$\sin(a + dx) \approx \sin a + (\cos a)dx$$

$$= \sin\frac{\pi}{6} + \left(\cos\frac{\pi}{6}\right)(0.01)$$

$$= \frac{1}{2} + \frac{\sqrt{3}}{2}(0.01) \approx 0.5087$$

For comparison, the true value of $\sin(\pi/6 + 0.01)$ to 6 decimals is 0.508635. ∎

The method in part (b) of Example 7 can be used in computer algorithms to give values of trigonometric functions. The algorithms store a large table of sine and cosine values between 0 and $\pi/4$. Values between these stored values are computed using differentials as in Example 7b. Values outside of $[0, \pi/4]$ are computed from values in this interval using trigonometric identities.

Error in Differential Approximation

Let $f(x)$ be differentiable at $x = a$ and suppose that $dx = \Delta x$ is an increment of x. We have two ways to describe the change in f as x changes from a to $a + \Delta x$:

The true change: $\Delta f = f(a + \Delta x) - f(a)$

The differential estimate: $df = f'(a)\,\Delta x.$

How well does df approximate Δf?

We measure the approximation error by subtracting df from Δf:

$$\text{Approximation error} = \Delta f - df$$
$$= \Delta f - f'(a)\Delta x$$
$$= \underbrace{f(a + \Delta x) - f(a)}_{\Delta f} - f'(a)\Delta x$$
$$= \underbrace{\left(\frac{f(a + \Delta x) - f(a)}{\Delta x} - f'(a)\right)}_{\text{Call this part } \varepsilon.} \cdot \Delta x$$
$$= \varepsilon \cdot \Delta x.$$

As $\Delta x \rightarrow 0$, the difference quotient

$$\frac{f(a + \Delta x) - f(a)}{\Delta x}$$

approaches $f'(a)$ (remember the definition of $f'(a)$), so the quantity in parentheses becomes a very small number (which is why we called it ε). In fact, $\varepsilon \rightarrow 0$ as $\Delta x \rightarrow 0$. When Δx is small, the approximation error $\varepsilon \Delta x$ is smaller still.

$$\underbrace{\Delta f}_{\substack{\text{true} \\ \text{change}}} = \underbrace{f'(a)\Delta x}_{\substack{\text{estimated} \\ \text{change}}} + \underbrace{\varepsilon \Delta x}_{\text{error}}$$

Although we do not know the exact size of the error, it is the product $\varepsilon \cdot \Delta x$ of two small quantities that both approach zero as $\Delta x \rightarrow 0$. For many common functions, whenever Δx is small, the error is still smaller.

Change in $y = f(x)$ near $x = a$

If $y = f(x)$ is differentiable at $x = a$ and x changes from a to $a + \Delta x$, the change Δy in f is given by

$$\Delta y = f'(a)\, \Delta x + \varepsilon\, \Delta x \tag{1}$$

in which $\varepsilon \rightarrow 0$ as $\Delta x \rightarrow 0$.

In Example 6 we found that

$$\Delta A = \pi(10.1)^2 - \pi(10)^2 = (102.01 - 100)\pi = (\underbrace{2\pi}_{dA} + \underbrace{0.01\pi}_{\text{error}})\ \text{m}^2$$

so the approximation error is $\Delta A - dA = \varepsilon\Delta r = 0.01\pi$ and $\varepsilon = 0.01\pi/\Delta r = 0.01\pi/0.1 = 0.1\pi$ m.

Proof of the Chain Rule

Equation (1) enables us to give a complete proof of the Chain Rule. Our goal is to show that if $f(u)$ is a differentiable function of u and $u = g(x)$ is a differentiable function of x, then the composition $y = f(g(x))$ is a differentiable function of x. Since a function is differentiable if and only if it has a derivative at each point in its domain, we must show that whenever g is differentiable at x_0 and f is differentiable at $g(x_0)$, then the composition is differentiable at x_0 and the derivative of the composition satisfies the equation

$$\left.\frac{dy}{dx}\right|_{x=x_0} = f'(g(x_0)) \cdot g'(x_0).$$

Let Δx be an increment in x and let Δu and Δy be the corresponding increments in u and y. Applying Equation (1) we have

$$\Delta u = g'(x_0)\Delta x + \varepsilon_1 \, \Delta x = (g'(x_0) + \varepsilon_1)\Delta x,$$

where $\varepsilon_1 \to 0$ as $\Delta x \to 0$. Similarly,

$$\Delta y = f'(u_0)\Delta u + \varepsilon_2 \, \Delta u = (f'(u_0) + \varepsilon_2)\Delta u,$$

where $\varepsilon_2 \to 0$ as $\Delta u \to 0$. Notice also that $\Delta u \to 0$ as $\Delta x \to 0$. Combining the equations for Δu and Δy gives

$$\Delta y = (f'(u_0) + \varepsilon_2)(g'(x_0) + \varepsilon_1)\Delta x,$$

so

$$\frac{\Delta y}{\Delta x} = f'(u_0)g'(x_0) + \varepsilon_2 g'(x_0) + f'(u_0)\varepsilon_1 + \varepsilon_2\varepsilon_1.$$

Since ε_1 and ε_2 go to zero as Δx goes to zero, the last three terms on the right vanish in the limit, leaving

$$\frac{dy}{dx}\bigg|_{x=x_0} = \lim_{\Delta x \to 0} \frac{\Delta y}{\Delta x} = f'(u_0)g'(x_0) = f'(g(x_0)) \cdot g'(x_0). \qquad \blacksquare$$

Sensitivity to Change

The equation $df = f'(x)\,dx$ tells how *sensitive* the output of f is to a change in input at different values of x. The larger the value of f' at x, the greater the effect of a given change dx. As we move from a to a nearby point $a + dx$, we can describe the change in f in three ways: absolute, relative, and percentage.

	True	**Estimated**
Absolute change	$\Delta f = f(a + dx) - f(a)$	$df = f'(a)\,dx$
Relative change	$\dfrac{\Delta f}{f(a)}$	$\dfrac{df}{f(a)}$
Percentage change	$\dfrac{\Delta f}{f(a)} \times 100$	$\dfrac{df}{f(a)} \times 100$

EXAMPLE 8   You want to calculate the depth of a well from the equation $s = 16t^2$ by timing how long it takes a heavy stone you drop to splash into the water below. How sensitive will your calculations be to a 0.1-sec error in measuring the time?

Solution   The size of ds in the equation

$$ds = 32t \, dt$$

depends on how big t is. If $t = 2$ sec, the change caused by $dt = 0.1$ is about

$$ds = 32(2)(0.1) = 6.4 \text{ ft}.$$

Three seconds later at $t = 5$ sec, the change caused by the same dt is

$$ds = 32(5)(0.1) = 16 \text{ ft}.$$

For a fixed error in the time measurement, the error in using ds to estimate the depth is larger when it takes a longer time before the stone splashes into the water. That is, the estimate is more sensitive to the effect of the error for larger values of t. $\blacksquare$

EXAMPLE 9   Newton's second law,

$$F = \frac{d}{dt}(mv) = m\frac{dv}{dt} = ma,$$

is stated with the assumption that mass is constant, but we know this is not strictly true because the mass of an object increases with velocity. In Einstein's corrected formula, mass has the value

$$m = \frac{m_0}{\sqrt{1 - v^2/c^2}},$$

where the "rest mass" m_0 represents the mass of an object that is not moving and c is the speed of light, which is about 300,000 km/sec. Use the approximation

$$\frac{1}{\sqrt{1 - x^2}} \approx 1 + \frac{1}{2}x^2 \qquad (2)$$

to estimate the increase Δm in mass resulting from the added velocity v.

Solution   When v is very small compared with c, v^2/c^2 is close to zero and it is safe to use the approximation

$$\frac{1}{\sqrt{1 - v^2/c^2}} \approx 1 + \frac{1}{2}\left(\frac{v^2}{c^2}\right) \qquad \text{Eq. (2) with } x = \frac{v}{c}$$

to obtain

$$m = \frac{m_0}{\sqrt{1 - v^2/c^2}} \approx m_0\left[1 + \frac{1}{2}\left(\frac{v^2}{c^2}\right)\right] = m_0 + \frac{1}{2}m_0v^2\left(\frac{1}{c^2}\right),$$

or

$$m \approx m_0 + \frac{1}{2}m_0v^2\left(\frac{1}{c^2}\right). \qquad (3)$$

Equation (3) expresses the increase in mass that results from the added velocity v.   ■

Converting Mass to Energy

Equation (3) derived in Example 9 has an important interpretation. In Newtonian physics, $(1/2)m_0v^2$ is the kinetic energy (KE) of the object, and if we rewrite Equation (3) in the form

$$(m - m_0)c^2 \approx \frac{1}{2}m_0v^2,$$

we see that

$$(m - m_0)c^2 \approx \frac{1}{2}m_0v^2 = \frac{1}{2}m_0v^2 - \frac{1}{2}m_0(0)^2 = \Delta(\text{KE}),$$

or

$$(\Delta m)c^2 \approx \Delta(\text{KE}).$$

So the change in kinetic energy $\Delta(\text{KE})$ in going from velocity 0 to velocity v is approximately equal to $(\Delta m)c^2$, the change in mass times the square of the speed of light. Using $c \approx 3 \times 10^8$ m/sec, we see that a small change in mass can create a large change in energy.

EXERCISES 3.11

Finding Linearizations

In Exercises 1–5, find the linearization $L(x)$ of $f(x)$ at $x = a$.

1. $f(x) = x^3 - 2x + 3, \quad a = 2$

2. $f(x) = \sqrt{x^2 + 9}, \quad a = -4$

3. $f(x) = x + \dfrac{1}{x}, \quad a = 1$

4. $f(x) = \sqrt[3]{x}, \quad a = -8$

5. $f(x) = \tan x, \quad a = \pi$

6. Common linear approximations at $x = 0$ Find the linearizations of the following functions at $x = 0$.

    a. $\sin x$ **b.** $\cos x$ **c.** $\tan x$ **d.** e^x **e.** $\ln (1 + x)$

Linearization for Approximation

In Exercises 7–14, find a linearization at a suitably chosen integer near a at which the given function and its derivative are easy to evaluate.

7. $f(x) = x^2 + 2x, \quad a = 0.1$

8. $f(x) = x^{-1}, \quad a = 0.9$

9. $f(x) = 2x^2 + 3x - 3, \quad a = -0.9$

10. $f(x) = 1 + x, \quad a = 8.1$

11. $f(x) = \sqrt[3]{x}, \quad a = 8.5$

12. $f(x) = \dfrac{x}{x + 1}, \quad a = 1.3$

13. $f(x) = e^{-x}, \quad a = -0.1$

14. $f(x) = \sin^{-1} x, \quad a = \pi/12$

15. Show that the linearization of $f(x) = (1 + x)^k$ at $x = 0$ is $L(x) = 1 + kx$.

16. Use the linear approximation $(1 + x)^k \approx 1 + kx$ to find an approximation for the function $f(x)$ for values of x near zero.

    a. $f(x) = (1 - x)^6$ **b.** $f(x) = \dfrac{2}{1 - x}$

    c. $f(x) = \dfrac{1}{\sqrt{1 + x}}$ **d.** $f(x) = \sqrt{2 + x^2}$

    e. $f(x) = (4 + 3x)^{1/3}$ **f.** $f(x) = \sqrt[3]{\left(1 - \dfrac{x}{2 + x}\right)^2}$

17. Faster than a calculator Use the approximation $(1 + x)^k \approx 1 + kx$ to estimate the following.

    a. $(1.0002)^{50}$ **b.** $\sqrt[3]{1.009}$

18. Find the linearization of $f(x) = \sqrt{x + 1} + \sin x$ at $x = 0$. How is it related to the individual linearizations of $\sqrt{x + 1}$ and $\sin x$ at $x = 0$?

Derivatives in Differential Form

In Exercises 19–38, find dy.

19. $y = x^3 - 3\sqrt{x}$ **20.** $y = x\sqrt{1 - x^2}$

21. $y = \dfrac{2x}{1 + x^2}$ **22.** $y = \dfrac{2\sqrt{x}}{3(1 + \sqrt{x})}$

23. $2y^{3/2} + xy - x = 0$ **24.** $xy^2 - 4x^{3/2} - y = 0$

25. $y = \sin (5\sqrt{x})$ **26.** $y = \cos (x^2)$

27. $y = 4\tan (x^3/3)$ **28.** $y = \sec (x^2 - 1)$

29. $y = 3 \csc \left(1 - 2\sqrt{x}\right)$ **30.** $y = 2 \cot \left(\dfrac{1}{\sqrt{x}}\right)$

31. $y = e^{\sqrt{x}}$ **32.** $y = xe^{-x}$

33. $y = \ln (1 + x^2)$ **34.** $y = \ln \left(\dfrac{x + 1}{\sqrt{x - 1}}\right)$

35. $y = \tan^{-1}(e^{x^2})$ **36.** $y = \cot^{-1}\left(\dfrac{1}{x^2}\right) + \cos^{-1} 2x$

37. $y = \sec^{-1}(e^{-x})$ **38.** $y = e^{\tan^{-1}\sqrt{x^2+1}}$

Approximation Error

In Exercises 39–44, each function $f(x)$ changes value when x changes from x_0 to $x_0 + dx$. Find

    a. the change $\Delta f = f(x_0 + dx) - f(x_0)$;

    b. the value of the estimate $df = f'(x_0)\, dx$; and

    c. the approximation error $|\Delta f - df|$.

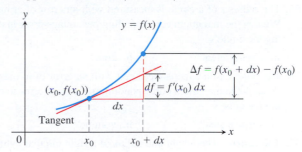

39. $f(x) = x^2 + 2x, \quad x_0 = 1, \quad dx = 0.1$

40. $f(x) = 2x^2 + 4x - 3, \quad x_0 = -1, \quad dx = 0.1$

41. $f(x) = x^3 - x, \quad x_0 = 1, \quad dx = 0.1$

42. $f(x) = x^4, \quad x_0 = 1, \quad dx = 0.1$

43. $f(x) = x^{-1}, \quad x_0 = 0.5, \quad dx = 0.1$

44. $f(x) = x^3 - 2x + 3, \quad x_0 = 2, \quad dx = 0.1$

Differential Estimates of Change

In Exercises 45–50, write a differential formula that estimates the given change in volume or surface area.

45. The change in the volume $V = (4/3)\pi r^3$ of a sphere when the radius changes from r_0 to $r_0 + dr$

46. The change in the volume $V = x^3$ of a cube when the edge lengths change from x_0 to $x_0 + dx$

47. The change in the surface area $S = 6x^2$ of a cube when the edge lengths change from x_0 to $x_0 + dx$

48. The change in the lateral surface area $S = \pi r\sqrt{r^2 + h^2}$ of a right circular cone when the radius changes from r_0 to $r_0 + dr$ and the height does not change

49. The change in the volume $V = \pi r^2 h$ of a right circular cylinder when the radius changes from r_0 to $r_0 + dr$ and the height does not change

50. The change in the lateral surface area $S = 2\pi rh$ of a right circular cylinder when the height changes from h_0 to $h_0 + dh$ and the radius does not change

Applications

51. The radius of a circle is increased from 2.00 to 2.02 m.

    a. Estimate the resulting change in area.

    b. Express the estimate as a percentage of the circle's original area.

52. The diameter of a tree was 10 in. During the following year, the circumference increased 2 in. About how much did the tree's diameter increase? The tree's cross-sectional area?

53. Estimating volume Estimate the volume of material in a cylindrical shell with length 30 in., radius 6 in., and shell thickness 0.5 in.

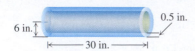

54. Estimating height of a building A surveyor, standing 30 ft from the base of a building, measures the angle of elevation to the top of the building to be 75°. How accurately must the angle be measured for the percentage error in estimating the height of the building to be less than 4%?

55. The radius r of a circle is measured with an error of at most 2%. What is the maximum corresponding percentage error in computing the circle's

    a. circumference?   **b.** area?

56. The edge x of a cube is measured with an error of at most 0.5%. What is the maximum corresponding percentage error in computing the cube's

    a. surface area?   **b.** volume?

57. Tolerance The height and radius of a right circular cylinder are equal, so the cylinder's volume is $V = \pi h^3$. The volume is to be calculated with an error of no more than 1% of the true value. Find approximately the greatest error that can be tolerated in the measurement of h, expressed as a percentage of h.

58. Tolerance

    a. About how accurately must the interior diameter of a 10-m-high cylindrical storage tank be measured to calculate the tank's volume to within 1% of its true value?

    b. About how accurately must the tank's exterior diameter be measured to calculate the amount of paint it will take to paint the side of the tank to within 5% of the true amount?

59. The diameter of a sphere is measured as 100 ± 1 cm and the volume is calculated from this measurement. Estimate the percentage error in the volume calculation.

60. Estimate the allowable percentage error in measuring the diameter D of a sphere if the volume is to be calculated correctly to within 3%.

61. The effect of flight maneuvers on the heart The amount of work done by the heart's main pumping chamber, the left ventricle, is given by the equation

$$W = PV + \frac{V\delta v^2}{2g},$$

where W is the work per unit time, P is the average blood pressure, V is the volume of blood pumped out during the unit of time, δ ("delta") is the weight density of the blood, v is the average velocity of the exiting blood, and g is the acceleration of gravity.

When P, V, δ, and v remain constant, W becomes a function of g, and the equation takes the simplified form

$$W = a + \frac{b}{g} \ (a, b \text{ constant}).$$

As a member of NASA's medical team, you want to know how sensitive W is to apparent changes in g caused by flight maneuvers, and this depends on the initial value of g. As part of your investigation, you decide to compare the effect on W of a given change dg on the moon, where $g = 5.2 \text{ ft/sec}^2$, with the effect the same change dg would have on Earth, where $g = 32 \text{ ft/sec}^2$. Use the simplified equation above to find the ratio of dW_{moon} to dW_{Earth}.

62. Drug concentration The concentration C in (mg/ml) milligrams per milliliter (mg/ml) of a certain drug in a person's bloodstream t hrs after a pill is swallowed is modeled by

$$C(t) = 1 + \frac{4t}{1 + t^3} - e^{-0.06t}.$$

Estimate the change in concentration when t changes from 20 to 30 min.

63. Unclogging arteries The formula $V = kr^4$, discovered by the physiologist Jean Poiseuille (1797–1869), allows us to predict how much the radius of a partially clogged artery has to be expanded in order to restore normal blood flow. The formula says that the volume V of blood flowing through the artery in a unit of time at a fixed pressure is a constant k times the radius of the artery to the fourth power. How will a 10% increase in r affect V?

64. Measuring acceleration of gravity When the length L of a clock pendulum is held constant by controlling its temperature, the pendulum's period T depends on the acceleration of gravity g. The period will therefore vary slightly as the clock is moved from place to place on the earth's surface, depending on the change in g. By keeping track of ΔT, we can estimate the variation in g from the equation $T = 2\pi(L/g)^{1/2}$ that relates T, g, and L.

    a. With L held constant and g as the independent variable, calculate dT and use it to answer parts (b) and (c).

    b. If g increases, will T increase or decrease? Will a pendulum clock speed up or slow down? Explain.

    c. A clock with a 100-cm pendulum is moved from a location where $g = 980 \text{ cm/sec}^2$ to a new location. This increases the period by $dT = 0.001$ sec. Find dg and estimate the value of g at the new location.

65. Quadratic approximations

    a. Let $Q(x) = b_0 + b_1(x - a) + b_2(x - a)^2$ be a quadratic approximation to $f(x)$ at $x = a$ with the properties:

       i) $Q(a) = f(a)$

       ii) $Q'(a) = f'(a)$

       iii) $Q''(a) = f''(a)$.

      Determine the coefficients b_0, b_1, and b_2.

    b. Find the quadratic approximation to $f(x) = 1/(1 - x)$ at $x = 0$.

    T **c.** Graph $f(x) = 1/(1 - x)$ and its quadratic approximation at $x = 0$. Then zoom in on the two graphs at the point $(0, 1)$. Comment on what you see.

    T **d.** Find the quadratic approximation to $g(x) = 1/x$ at $x = 1$. Graph g and its quadratic approximation together. Comment on what you see.

e. Find the quadratic approximation to $h(x) = \sqrt{1 + x}$ at $x = 0$. Graph h and its quadratic approximation together. Comment on what you see.

f. What are the linearizations of f, g, and h at the respective points in parts (b), (d), and (e)?

66. The linearization is the best linear approximation Suppose that $y = f(x)$ is differentiable at $x = a$ and that $g(x) = g(x) = m(x - a) + c$ is a linear function in which m and c are constants. If the error $E(x) = f(x) - g(x)$ were small enough near $x = a$, we might think of using g as a linear approximation of f instead of the linearization $L(x) = f(a) + f'(a)(x - a)$. Show that if we impose on g the conditions

1. $E(a) = 0$ The approximation error is zero at $x = a$.

2. $\displaystyle\lim_{x \to a} \frac{E(x)}{x - a} = 0$ The error is negligible when compared with $x - a$.

then $g(x) = f(a) + f'(a)(x - a)$. Thus, the linearization $L(x)$ gives the only linear approximation whose error is both zero at $x = a$ and negligible in comparison with $x - a$.

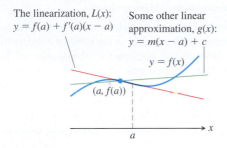

The linearization, $L(x)$:
$y = f(a) + f'(a)(x - a)$

Some other linear approximation, $g(x)$:
$y = m(x - a) + c$

$y = f(x)$

$(a, f(a))$

a

67. The linearization of 2^x

a. Find the linearization of $f(x) = 2^x$ at $x = 0$. Then round its coefficients to two decimal places.

b. Graph the linearization and function together for $-3 \le x \le 3$ and $-1 \le x \le 1$.

68. The linearization of $\log_3 x$

a. Find the linearization of $f(x) = \log_3 x$ at $x = 3$. Then round its coefficients to two decimal places.

b. Graph the linearization and function together in the window $0 \le x \le 8$ and $2 \le x \le 4$.

COMPUTER EXPLORATIONS

In Exercises 69–74, use a CAS to estimate the magnitude of the error in using the linearization in place of the function over a specified interval I. Perform the following steps:

a. Plot the function f over I.

b. Find the linearization L of the function at the point a.

c. Plot f and L together on a single graph.

d. Plot the absolute error $|f(x) - L(x)|$ over I and find its maximum value.

e. From your graph in part (d), estimate as large a $\delta > 0$ as you can, satisfying

$$|x - a| < \delta \quad \Rightarrow \quad |f(x) - L(x)| < \varepsilon$$

for $\varepsilon = 0.5, 0.1$, and 0.01. Then check graphically to see if your δ-estimate holds true.

69. $f(x) = x^3 + x^2 - 2x, \quad [-1, 2], \quad a = 1$

70. $f(x) = \dfrac{x - 1}{4x^2 + 1}, \quad \left[-\dfrac{3}{4}, 1\right], \quad a = \dfrac{1}{2}$

71. $f(x) = x^{2/3}(x - 2), \quad [-2, 3], \quad a = 2$

72. $f(x) = \sqrt{x} - \sin x, \quad [0, 2\pi], \quad a = 2$

73. $f(x) = x2^x, \quad [0, 2], \quad a = 1$

74. $f(x) = \sqrt{x}\sin^{-1}x, \quad [0, 1], \quad a = \dfrac{1}{2}$

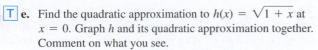

CHAPTER 3   Questions to Guide Your Review

1. What is the derivative of a function f? How is its domain related to the domain of f? Give examples.

2. What role does the derivative play in defining slopes, tangent lines, and rates of change?

3. How can you sometimes graph the derivative of a function when all you have is a table of the function's values?

4. What does it mean for a function to be differentiable on an open interval? On a closed interval?

5. How are derivatives and one-sided derivatives related?

6. Describe geometrically when a function typically does *not* have a derivative at a point.

7. How is a function's differentiability at a point related to its continuity there, if at all?

8. What rules do you know for calculating derivatives? Give some examples.

9. Explain how the three formulas

  a. $\dfrac{d}{dx}(x^n) = nx^{n-1}$ **b.** $\dfrac{d}{dx}(cu) = c\dfrac{du}{dx}$

  c. $\dfrac{d}{dx}(u_1 + u_2 + \cdots + u_n) = \dfrac{du_1}{dx} + \dfrac{du_2}{dx} + \cdots + \dfrac{du_n}{dx}$

  enable us to differentiate any polynomial.

10. What formula do we need, in addition to the three listed in Question 9, to differentiate rational functions?

11. What is a second derivative? A third derivative? How many derivatives do the functions you know have? Give examples.

12. What is the derivative of the exponential function e^x? How does the domain of the derivative compare with the domain of the function?

13. What is the relationship between a function's average and instantaneous rates of change? Give an example.

14. How do derivatives arise in the study of motion? What can you learn about an object's motion along a line by examining the derivatives of the object's position function? Give examples.

15. How can derivatives arise in economics?

16. Give examples of still other applications of derivatives.

17. What do the limits $\lim_{h\to 0}((\sin h)/h)$ and $\lim_{h\to 0}((\cos h - 1)/h)$ have to do with the derivatives of the sine and cosine functions? What *are* the derivatives of these functions?

18. Once you know the derivatives of $\sin x$ and $\cos x$, how can you find the derivatives of $\tan x$, $\cot x$, $\sec x$, and $\csc x$? What *are* the derivatives of these functions?

19. At what points are the six basic trigonometric functions continuous? How do you know?

20. What is the rule for calculating the derivative of a composition of two differentiable functions? How is such a derivative evaluated? Give examples.

21. If u is a differentiable function of x, how do you find $(d/dx)(u^n)$ if n is an integer? If n is a real number? Give examples.

22. What is implicit differentiation? When do you need it? Give examples.

23. What is the derivative of the natural logarithm function $\ln x$? How does the domain of the derivative compare with the domain of the function?

24. What is the derivative of the exponential function a^x, $a > 0$ and $a \neq 1$? What is the geometric significance of the limit of $(a^h - 1)/h$ as $h \to 0$? What is the limit when a is the number e?

25. What is the derivative of $\log_a x$? Are there any restrictions on a?

26. What is logarithmic differentiation? Give an example.

27. How can you write any real power of x as a power of e? Are there any restrictions on x? How does this lead to the Power Rule for differentiating arbitrary real powers?

28. What is one way of expressing the special number e as a limit? What is an approximate numerical value of e correct to 7 decimal places?

29. What are the derivatives of the inverse trigonometric functions? How do the domains of the derivatives compare with the domains of the functions?

30. How do related rates problems arise? Give examples.

31. Outline a strategy for solving related rates problems. Illustrate with an example.

32. What is the linearization $L(x)$ of a function $f(x)$ at a point $x = a$? What is required of f at a for the linearization to exist? How are linearizations used? Give examples.

33. If x moves from a to a nearby value $a + dx$, how do you estimate the corresponding change in the value of a differentiable function $f(x)$? How do you estimate the relative change? The percentage change? Give an example.

CHAPTER 3 Practice Exercises

Derivatives of Functions

Find the derivatives of the functions in Exercises 1–64.

1. $y = x^5 - 0.125x^2 + 0.25x$
2. $y = 3 - 0.7x^3 + 0.3x^7$
3. $y = x^3 - 3(x^2 + \pi^2)$
4. $y = x^7 + \sqrt{7}x - \dfrac{1}{\pi + 1}$
5. $y = (x + 1)^2(x^2 + 2x)$
6. $y = (2x - 5)(4 - x)^{-1}$
7. $y = (\theta^2 + \sec \theta + 1)^3$
8. $y = \left(-1 - \dfrac{\csc \theta}{2} - \dfrac{\theta^2}{4}\right)^2$
9. $s = \dfrac{\sqrt{t}}{1 + \sqrt{t}}$
10. $s = \dfrac{1}{\sqrt{t} - 1}$
11. $y = 2\tan^2 x - \sec^2 x$
12. $y = \dfrac{1}{\sin^2 x} - \dfrac{2}{\sin x}$
13. $s = \cos^4(1 - 2t)$
14. $s = \cot^3\left(\dfrac{2}{t}\right)$
15. $s = (\sec t + \tan t)^5$
16. $s = \csc^5(1 - t + 3t^2)$
17. $r = \sqrt{2\theta \sin \theta}$
18. $r = 2\theta\sqrt{\cos \theta}$
19. $r = \sin\sqrt{2\theta}$
20. $r = \sin(\theta + \sqrt{\theta + 1})$
21. $y = \dfrac{1}{2}x^2 \csc \dfrac{2}{x}$
22. $y = 2\sqrt{x} \sin \sqrt{x}$
23. $y = x^{-1/2}\sec(2x)^2$
24. $y = \sqrt{x}\csc(x + 1)^3$
25. $y = 5\cot x^2$
26. $y = x^2\cot 5x$
27. $y = x^2\sin^2(2x^2)$
28. $y = x^{-2}\sin^2(x^3)$

29. $s = \left(\dfrac{4t}{t + 1}\right)^{-2}$
30. $s = \dfrac{-1}{15(15t - 1)^3}$
31. $y = \left(\dfrac{\sqrt{x}}{1 + x}\right)^2$
32. $y = \left(\dfrac{2\sqrt{x}}{2\sqrt{x} + 1}\right)^2$
33. $y = \sqrt{\dfrac{x^2 + x}{x^2}}$
34. $y = 4x\sqrt{x + \sqrt{x}}$
35. $r = \left(\dfrac{\sin \theta}{\cos \theta - 1}\right)^2$
36. $r = \left(\dfrac{1 + \sin \theta}{1 - \cos \theta}\right)^2$
37. $y = (2x + 1)\sqrt{2x + 1}$
38. $y = 20(3x - 4)^{1/4}(3x - 4)^{-1/5}$
39. $y = \dfrac{3}{(5x^2 + \sin 2x)^{3/2}}$
40. $y = (3 + \cos^3 3x)^{-1/3}$
41. $y = 10e^{-x/5}$
42. $y = \sqrt{2}e^{\sqrt{2}x}$
43. $y = \dfrac{1}{4}xe^{4x} - \dfrac{1}{16}e^{4x}$
44. $y = x^2e^{-2/x}$
45. $y = \ln(\sin^2 \theta)$
46. $y = \ln(\sec^2 \theta)$
47. $y = \log_2(x^2/2)$
48. $y = \log_5(3x - 7)$
49. $y = 8^{-t}$
50. $y = 9^{2t}$
51. $y = 5x^{3.6}$
52. $y = \sqrt{2}x^{-\sqrt{2}}$
53. $y = (x + 2)^{x+2}$
54. $y = 2(\ln x)^{x/2}$
55. $y = \sin^{-1}\sqrt{1 - u^2}, \quad 0 < u < 1$
56. $y = \sin^{-1}\left(\dfrac{1}{\sqrt{v}}\right), \quad v > 1$

57. $y = \ln \cos^{-1} x$

58. $y = z \cos^{-1} z - \sqrt{1 - z^2}$

59. $y = t \tan^{-1} t - \dfrac{1}{2} \ln t$

60. $y = (1 + t^2) \cot^{-1} 2t$

61. $y = z \sec^{-1} z - \sqrt{z^2 - 1}, \quad z > 1$

62. $y = 2\sqrt{x - 1} \sec^{-1} \sqrt{x}$

63. $y = \csc^{-1}(\sec \theta), \quad 0 < \theta < \pi/2$

64. $y = (1 + x^2) e^{\tan^{-1} x}$

Implicit Differentiation

In Exercises 65–78, find dy/dx by implicit differentiation.

65. $xy + 2x + 3y = 1$

66. $x^2 + xy + y^2 - 5x = 2$

67. $x^3 + 4xy - 3y^{4/3} = 2x$

68. $5x^{4/5} + 10y^{6/5} = 15$

69. $\sqrt{xy} = 1$

70. $x^2 y^2 = 1$

71. $y^2 = \dfrac{x}{x + 1}$

72. $y^2 = \sqrt{\dfrac{1 + x}{1 - x}}$

73. $e^{x + 2y} = 1$

74. $y^2 = 2e^{-1/x}$

75. $\ln(x/y) = 1$

76. $x \sin^{-1} y = 1 + x^2$

77. $y e^{\tan^{-1} x} = 2$

78. $x^y = \sqrt{2}$

In Exercises 79 and 80, find dp/dq.

79. $p^3 + 4pq - 3q^2 = 2$

80. $q = (5p^2 + 2p)^{-3/2}$

In Exercises 81 and 82, find dr/ds.

81. $r \cos 2s + \sin^2 s = \pi$

82. $2rs - r - s + s^2 = -3$

83. Find d^2y/dx^2 by implicit differentiation:

    a. $x^3 + y^3 = 1$

    b. $y^2 = 1 - \dfrac{2}{x}$

84. a. By differentiating $x^2 - y^2 = 1$ implicitly, show that $dy/dx = x/y$.

    b. Then show that $d^2y/dx^2 = -1/y^3$.

Numerical Values of Derivatives

85. Suppose that functions $f(x)$ and $g(x)$ and their first derivatives have the following values at $x = 0$ and $x = 1$.

x	$f(x)$	$g(x)$	$f'(x)$	$g'(x)$
0	1	1	-3	$1/2$
1	3	5	$1/2$	-4

Find the first derivatives of the following combinations at the given value of x.

    a. $6f(x) - g(x), \quad x = 1$

    b. $f(x)g^2(x), \quad x = 0$

    c. $\dfrac{f(x)}{g(x) + 1}, \quad x = 1$

    d. $f(g(x)), \quad x = 0$

    e. $g(f(x)), \quad x = 0$

    f. $(x + f(x))^{3/2}, \quad x = 1$

    g. $f(x + g(x)), \quad x = 0$

86. Suppose that the function $f(x)$ and its first derivative have the following values at $x = 0$ and $x = 1$.

x	$f(x)$	$f'(x)$
0	9	-2
1	-3	$1/5$

Find the first derivatives of the following combinations at the given value of x.

    a. $\sqrt{x}\, f(x), \quad x = 1$

    b. $\sqrt{f(x)}, \quad x = 0$

    c. $f(\sqrt{x}), \quad x = 1$

    d. $f(1 - 5 \tan x), \quad x = 0$

    e. $\dfrac{f(x)}{2 + \cos x}, \quad x = 0$

    f. $10 \sin\left(\dfrac{\pi x}{2}\right) f^2(x), \quad x = 1$

87. Find the value of dy/dt at $t = 0$ if $y = 3 \sin 2x$ and $x = t^2 + \pi$.

88. Find the value of ds/du at $u = 2$ if $s = t^2 + 5t$ and $t = (u^2 + 2u)^{1/3}$.

89. Find the value of dw/ds at $s = 0$ if $w = \sin\left(e^{\sqrt{r}}\right)$ and $r = 3 \sin(s + \pi/6)$.

90. Find the value of dr/dt at $t = 0$ if $r = (\theta^2 + 7)^{1/3}$ and $\theta^2 t + \theta = 1$.

91. If $y^3 + y = 2 \cos x$, find the value of d^2y/dx^2 at the point $(0, 1)$.

92. If $x^{1/3} + y^{1/3} = 4$, find d^2y/dx^2 at the point $(8, 8)$.

Applying the Derivative Definition

In Exercises 93 and 94, find the derivative using the definition.

93. $f(t) = \dfrac{1}{2t + 1}$

94. $g(x) = 2x^2 + 1$

95. a. Graph the function

$$f(x) = \begin{cases} x^2, & -1 \le x < 0 \\ -x^2, & 0 \le x \le 1. \end{cases}$$

    b. Is f continuous at $x = 0$?

    c. Is f differentiable at $x = 0$?

Give reasons for your answers.

96. a. Graph the function

$$f(x) = \begin{cases} x, & -1 \le x < 0 \\ \tan x, & 0 \le x \le \pi/4. \end{cases}$$

    b. Is f continuous at $x = 0$?

    c. Is f differentiable at $x = 0$?

Give reasons for your answers.

97. a. Graph the function

$$f(x) = \begin{cases} x, & 0 \le x \le 1 \\ 2 - x, & 1 < x \le 2. \end{cases}$$

    b. Is f continuous at $x = 1$?

    c. Is f differentiable at $x = 1$?

Give reasons for your answers.

98. For what value or values of the constant m, if any, is

$$f(x) = \begin{cases} \sin 2x, & x \le 0 \\ mx, & x > 0 \end{cases}$$

   a. continuous at $x = 0$?

   b. differentiable at $x = 0$?

   Give reasons for your answers.

Slopes, Tangents, and Normals

99. Tangent lines with specified slope Are there any points on the curve $y = (x/2) + 1/(2x - 4)$ where the slope is $-3/2$? If so, find them.

100. Tangent lines with specified slope Are there any points on the curve $y = x - e^{-x}$ where the slope is 2? If so, find them.

101. Horizontal tangent lines Find the points on the curve $y = 2x^3 - 3x^2 - 12x + 20$ where the tangent line is parallel to the x-axis.

102. Tangent intercepts Find the x- and y-intercepts of the line that is tangent to the curve $y = x^3$ at the point $(-2, -8)$.

103. Tangent lines perpendicular or parallel to lines Find the points on the curve $y = 2x^3 - 3x^2 - 12x + 20$ where the tangent line is

   a. perpendicular to the line $y = 1 - (x/24)$.

   b. parallel to the line $y = \sqrt{2} - 12x$.

104. Intersecting tangent lines Show that the tangent lines to the curve $y = (\pi \sin x)/x$ at $x = \pi$ and $x = -\pi$ intersect at right angles.

105. Normal lines parallel to a line Find the points on the curve $y = \tan x, -\pi/2 < x < \pi/2$, where the normal line is parallel to the line $y = -x/2$. Sketch the curve and normal lines together, labeling each with its equation.

106. Tangent lines and normal lines Find equations for the tangent and normal lines to the curve $y = 1 + \cos x$ at the point $(\pi/2, 1)$. Sketch the curve, tangent line, and normal line together, labeling each with its equation.

107. Tangent parabola The parabola $y = x^2 + C$ is to be tangent to the line $y = x$. Find C.

108. Slope of a tangent line Show that the tangent line to the curve $y = x^3$ at any point (a, a^3) meets the curve again at a point where the slope is four times the slope at (a, a^3).

109. Tangent curve For what value of c is the curve $y = c/(x + 1)$ tangent to the line through the points $(0, 3)$ and $(5, -2)$?

110. Normal lines to a circle Show that the normal line at any point of the circle $x^2 + y^2 = a^2$ passes through the origin.

In Exercises 111–116, find equations for the lines that are tangent and normal to the curve at the given point.

111. $x^2 + 2y^2 = 9, \quad (1, 2)$

112. $e^x + y^2 = 2, \quad (0, 1)$

113. $xy + 2x - 5y = 2, \quad (3, 2)$

114. $(y - x)^2 = 2x + 4, \quad (6, 2)$

115. $x + \sqrt{xy} = 6, \quad (4, 1)$

116. $x^{3/2} + 2y^{3/2} = 17, \quad (1, 4)$

117. Find the slope of the curve $x^3 y^3 + y^2 = x + y$ at the points $(1, 1)$ and $(1, -1)$.

118. The graph shown suggests that the curve $y = \sin(x - \sin x)$ might have horizontal tangent lines at the x-axis. Does it? Give reasons for your answer.

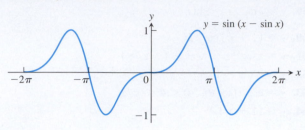

Analyzing Graphs

Each of the figures in Exercises 119 and 120 shows two graphs, the graph of a function $y = f(x)$ together with the graph of its derivative $f'(x)$. Which graph is which? How do you know?

119.   **120.**

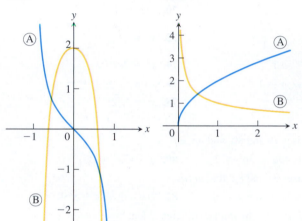

121. Use the following information to graph the function $y = f(x)$ for $-1 \le x \le 6$.

   i) The graph of f is made of line segments joined end to end.

   ii) The graph starts at the point $(-1, 2)$.

   iii) The derivative of f, where defined, agrees with the step function shown here.

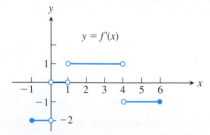

122. Repeat Exercise 121, supposing that the graph starts at $(-1, 0)$ instead of $(-1, 2)$.

Exercises 123 and 124 are about the accompanying graphs. The graphs in part (a) show the numbers of rabbits and foxes in a small arctic population. They are plotted as functions of time for 200 days. The number of rabbits increases at first, as the rabbits reproduce. But the foxes prey on rabbits and, as the number of foxes increases, the rabbit population levels off and then drops. Part (b) shows the graph of the derivative of the rabbit population, made by plotting slopes.

123. a. What is the value of the derivative of the rabbit population when the number of rabbits is largest? Smallest?

b. What is the size of the rabbit population when its derivative is largest? Smallest (negative value)?

124. In what units should the slopes of the rabbit and fox population curves be measured?

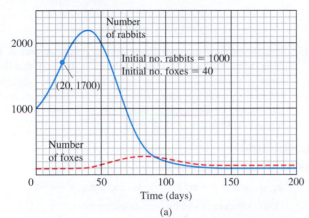

(a)

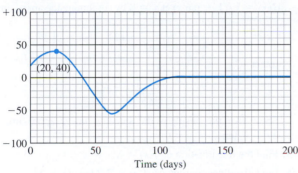

Derivative of the rabbit population

(b)

Source: NCPMF "Differentiation" by W.U. Walton et al., Project CALC. Reprinted by permission of Educational Development Center, Inc.

Trigonometric Limits

Find the limits in Exercises 125–132.

125. $\lim\limits_{x \to 0} \dfrac{\sin x}{2x^2 - x}$

126. $\lim\limits_{x \to 0} \dfrac{3x - \tan 7x}{2x}$

127. $\lim\limits_{r \to 0} \dfrac{\sin r}{\tan 2r}$

128. $\lim\limits_{\theta \to 0} \dfrac{\sin (\sin \theta)}{\theta}$

129. $\lim\limits_{\theta \to (\pi/2)^-} \dfrac{4\tan^2 \theta + \tan \theta + 1}{\tan^2 \theta + 5}$

130. $\lim\limits_{\theta \to 0^+} \dfrac{1 - 2\cot^2 \theta}{5\cot^2 \theta - 7 \cot \theta - 8}$

131. $\lim\limits_{x \to 0} \dfrac{x \sin x}{2 - 2 \cos x}$

132. $\lim\limits_{\theta \to 0} \dfrac{1 - \cos \theta}{\theta^2}$

Show how to extend the functions in Exercises 133 and 134 to be continuous at the origin.

133. $g(x) = \dfrac{\tan (\tan x)}{\tan x}$

134. $f(x) = \dfrac{\tan (\tan x)}{\sin (\sin x)}$

Logarithmic Differentiation

In Exercises 135–140, use logarithmic differentiation to find the derivative of y with respect to the appropriate variable.

135. $y = \dfrac{2(x^2 + 1)}{\sqrt{\cos 2x}}$

136. $y = \sqrt[10]{\dfrac{3x + 4}{2x - 4}}$

137. $y = \left(\dfrac{(t + 1)(t - 1)}{(t - 2)(t + 3)} \right)^5, \quad t > 2$

138. $y = \dfrac{2u2^u}{\sqrt{u^2 + 1}}$

139. $y = (\sin \theta)^{\sqrt{\theta}}$

140. $y = (\ln x)^{1/(\ln x)}$

Related Rates

141. Right circular cylinder The total surface area S of a right circular cylinder is related to the base radius r and height h by the equation $S = 2\pi r^2 + 2\pi rh$.

a. How is dS/dt related to dr/dt if h is constant?

b. How is dS/dt related to dh/dt if r is constant?

c. How is dS/dt related to dr/dt and dh/dt if neither r nor h is constant?

d. How is dr/dt related to dh/dt if S is constant?

142. Right circular cone The lateral surface area S of a right circular cone is related to the base radius r and height h by the equation $S = \pi r \sqrt{r^2 + h^2}$.

a. How is dS/dt related to dr/dt if h is constant?

b. How is dS/dt related to dh/dt if r is constant?

c. How is dS/dt related to dr/dt and dh/dt if neither r nor h is constant?

143. Circle's changing area The radius of a circle is changing at the rate of $-2/\pi$ m/sec. At what rate is the circle's area changing when $r = 10$ m?

144. Cube's changing edges The volume of a cube is increasing at the rate of 1200 cm^3/min at the instant its edges are 20 cm long. At what rate are the lengths of the edges changing at that instant?

145. Resistors connected in parallel If two resistors of R_1 and R_2 ohms are connected in parallel in an electric circuit to make an R-ohm resistor, the value of R can be found from the equation

$$\frac{1}{R} = \frac{1}{R_1} + \frac{1}{R_2}.$$

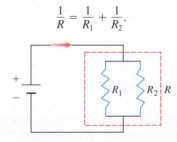

If R_1 is decreasing at the rate of 1 ohm/sec and R_2 is increasing at the rate of 0.5 ohm/sec, at what rate is R changing when $R_1 = 75$ ohms and $R_2 = 50$ ohms?

146. Impedance in a series circuit The impedance Z (ohms) in a series circuit is related to the resistance R (ohms) and reactance X (ohms) by the equation $Z = \sqrt{R^2 + X^2}$. If R is increasing at 3 ohms/sec and X is decreasing at 2 ohms/sec, at what rate is Z changing when $R = 10$ ohms and $X = 20$ ohms?

147. Speed of moving particle The coordinates of a particle moving in the metric xy-plane are differentiable functions of time t with $dx/dt = 10$ m/sec and $dy/dt = 5$ m/sec. How fast is the particle moving away from the origin as it passes through the point $(3, -4)$?

148. Motion of a particle A particle moves along the curve $y = x^{3/2}$ in the first quadrant in such a way that its distance from the origin increases at the rate of 11 units per second. Find dx/dt when $x = 3$.

149. Draining a tank Water drains from the conical tank shown in the accompanying figure at the rate of 5 ft^3/min.

   a. What is the relation between the variables h and r in the figure?

   b. How fast is the water level dropping when $h = 6$ ft?

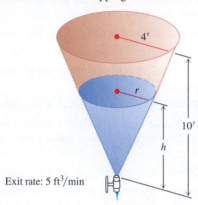

4'

r

10'

h

Exit rate: 5 ft³/min

150. Rotating spool As television cable is pulled from a large spool to be strung from the telephone poles along a street, it unwinds from the spool in layers of constant radius (see accompanying figure). If the truck pulling the cable moves at a steady 6 ft/sec (a touch over 4 mph), use the equation $s = r\theta$ to find how fast (radians per second) the spool is turning when the layer of radius 1.2 ft is being unwound.

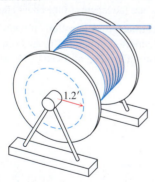

1.2'

151. Moving searchlight beam The figure shows a boat 1 km offshore, sweeping the shore with a searchlight. The light turns at a constant rate, $d\theta/dt = -0.6$ rad/sec.

   a. How fast is the light moving along the shore when it reaches point A?

   b. How many revolutions per minute is 0.6 rad/sec?

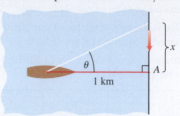

x

θ

A

1 km

152. Points moving on coordinate axes Points A and B move along the x- and y-axes, respectively, in such a way that the distance r (meters) along the perpendicular from the origin to the line AB remains constant. How fast is OA changing, and is it increasing, or decreasing, when $OB = 2r$ and B is moving toward O at the rate of 0.3r m/sec?

Linearization

153. Find the linearizations of

   a. $\tan x$ at $x = -\pi/4$ **b.** $\sec x$ at $x = -\pi/4$.

   Graph the curves and linearizations together.

154. We can obtain a useful linear approximation of the function $f(x) = 1/(1 + \tan x)$ at $x = 0$ by combining the approximations

$$\frac{1}{1 + x} \approx 1 - x \quad \text{and} \quad \tan x \approx x$$

to get

$$\frac{1}{1 + \tan x} \approx 1 - x.$$

Show that this result is the standard linear approximation of $1/(1 + \tan x)$ at $x = 0$.

155. Find the linearization of $f(x) = \sqrt{1 + x} + \sin x - 0.5$ at $x = 0$.

156. Find the linearization of $f(x) = 2/(1 - x) + \sqrt{1 + x} - 3.1$ at $x = 0$.

Differential Estimates of Change

157. Surface area of a cone Write a formula that estimates the change that occurs in the lateral surface area of a right circular cone when the height changes from h_0 to $h_0 + dh$ and the radius does not change.

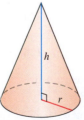

h

r

$V = \frac{1}{3}\pi r^2 h$

$S = \pi r \sqrt{r^2 + h^2}$

(Lateral surface area)

158. Controlling error

    a. How accurately should you measure the edge of a cube to be reasonably sure of calculating the cube's surface area with an error of no more than 2%?

    b. Suppose that the edge is measured with the accuracy required in part (a). About how accurately can the cube's volume be calculated from the edge measurement? To find out, estimate the percentage error in the volume calculation that might result from using the edge measurement.

159. Compounding error The circumference of the equator of a sphere is measured as 10 cm with a possible error of 0.4 cm. This measurement is used to calculate the radius. The radius is then used to calculate the surface area and volume of the sphere. Estimate the percentage errors in the calculated values of

    a. the radius.

    b. the surface area.

    c. the volume.

160. Finding height To find the height of a lamppost (see accompanying figure), you stand a 6 ft pole 20 ft from the lamp and measure the length a of its shadow, finding it to be 15 ft, give or take an inch. Calculate the height of the lamppost using the value $a = 15$ and estimate the possible error in the result.

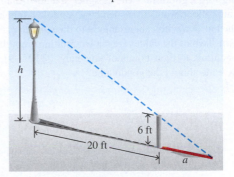

CHAPTER 3   Additional and Advanced Exercises

1. An equation like $\sin^2\theta + \cos^2\theta = 1$ is called an **identity** because it holds for all values of θ. An equation like $\sin\theta = 0.5$ is not an identity because it holds only for selected values of θ, not all. If you differentiate both sides of a trigonometric identity in θ with respect to θ, the resulting new equation will also be an identity.

    Differentiate the following to show that the resulting equations hold for all θ.

    a. $\sin 2\theta = 2\sin\theta\cos\theta$

    b. $\cos 2\theta = \cos^2\theta - \sin^2\theta$

2. If the identity $\sin(x + a) = \sin x \cos a + \cos x \sin a$ is differentiated with respect to x, is the resulting equation also an identity? Does this principle apply to the equation $x^2 - 2x - 8 = 0$? Explain.

3. a. Find values for the constants a, b, and c that will make

$$f(x) = \cos x \quad \text{and} \quad g(x) = a + bx + cx^2$$

    satisfy the conditions

$$f(0) = g(0), \quad f'(0) = g'(0), \quad \text{and} \quad f''(0) = g''(0).$$

    b. Find values for b and c that will make

$$f(x) = \sin(x + a) \quad \text{and} \quad g(x) = b \sin x + c \cos x$$

    satisfy the conditions

$$f(0) = g(0) \quad \text{and} \quad f'(0) = g'(0).$$

    c. For the determined values of a, b, and c, what happens for the third and fourth derivatives of f and g in each of parts (a) and (b)?

4. Solutions to differential equations

    a. Show that $y = \sin x$, $y = \cos x$, and $y = a \cos x + b \sin x$ (a and b constants) all satisfy the equation

$$y'' + y = 0.$$

    b. How would you modify the functions in part (a) to satisfy the equation

$$y'' + 4y = 0?$$

    Generalize this result.

5. An osculating circle Find the values of h, k, and a that make the circle $(x - h)^2 + (y - k)^2 = a^2$ tangent to the parabola $y = x^2 + 1$ at the point $(1, 2)$ and that also make the second derivatives d^2y/dx^2 have the same value on both curves there. Circles like this one that are tangent to a curve and have the same second derivative as the curve at the point of tangency are called *osculating circles* (from the Latin *osculari*, meaning "to kiss"). We encounter them again in Chapter 13.

6. Marginal revenue A bus will hold 60 people. The number x of people per trip who use the bus is related to the fare charged (p dollars) by the law $p = [3 - (x/40)]^2$. Write an expression for the total revenue $r(x)$ per trip received by the bus company. What number of people per trip will make the marginal revenue dr/dx equal to zero? What is the corresponding fare? (This fare is the one that maximizes the revenue.)

7. Industrial production

    a. Economists often use the expression "rate of growth" in relative rather than absolute terms. For example, let $u = f(t)$ be the number of people in the labor force at time t in a given industry. (We treat this function as though it were differentiable even though it is an integer-valued step function.)

        Let $v = g(t)$ be the average production per person in the labor force at time t. The total production is then $y = uv$. If the labor force is growing at the rate of 4% per year ($du/dt = 0.04u$) and the production per worker is growing at the rate of 5% per year ($dv/dt = 0.05v$), find the rate of growth of the total production, y.

b. Suppose that the labor force in part (a) is decreasing at the rate of 2% per year while the production per person is increasing at the rate of 3% per year. Is the total production increasing, or is it decreasing, and at what rate?

8. Designing a gondola The designer of a 30-ft-diameter spherical hot air balloon wants to suspend the gondola 8 ft below the bottom of the balloon with cables tangent to the surface of the balloon, as shown. Two of the cables are shown running from the top edges of the gondola to their points of tangency, $(-12, -9)$ and $(12, -9)$. How wide should the gondola be?

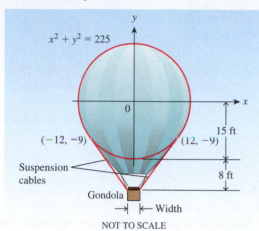

NOT TO SCALE

9. Pisa by parachute On August 5, 1988, Mike McCarthy of London jumped from the top of the Tower of Pisa. He then opened his parachute in what he said was a world record low-level parachute jump of 179 ft. Make a rough sketch to show the shape of the graph of his speed during the jump. (*Source: Boston Globe, Aug. 6, 1988.*)

10. Motion of a particle The position at time $t \geq 0$ of a particle moving along a coordinate line is

$$s = 10\cos(t + \pi/4).$$

a. What is the particle's starting position ($t = 0$)?

b. What are the points farthest to the left and right of the origin reached by the particle?

c. Find the particle's velocity and acceleration at the points in part (b).

d. When does the particle first reach the origin? What are its velocity, speed, and acceleration then?

11. Shooting a paper clip On Earth, you can easily shoot a paper clip 64 ft straight up into the air with a rubber band. In t sec after firing, the paper clip is $s = 64t - 16t^2$ ft above your hand.

a. How long does it take the paper clip to reach its maximum height? With what velocity does it leave your hand?

b. On the moon, the same acceleration will send the paper clip to a height of $s = 64t - 2.6t^2$ ft in t sec. About how long will it take the paper clip to reach its maximum height, and how high will it go?

12. Velocities of two particles At time t sec, the positions of two particles on a coordinate line are $s_1 = 3t^3 - 12t^2 + 18t + 5$ m and $s_2 = -t^3 + 9t^2 - 12t$ m. When do the particles have the same velocities?

13. Velocity of a particle A particle of constant mass m moves along the x-axis. Its velocity v and position x satisfy the equation

$$\tfrac{1}{2}m(v^2 - v_0^2) = \tfrac{1}{2}k(x_0^2 - x^2),$$

where k, v_0, and x_0 are constants. Show that whenever $v \neq 0$,

$$m\frac{dv}{dt} = -kx.$$

In Exercises 14 and 15, use implicit differentiation to find $\dfrac{dy}{dx}$.

14. $y^{\ln x} = x^{x^y}$
15. $y^{e^x} = x^y + 1$

16. Average and instantaneous velocity

a. Show that if the position x of a moving point is given by a quadratic function of t, $x = At^2 + Bt + C$, then the average velocity over any time interval $[t_1, t_2]$ is equal to the instantaneous velocity at the midpoint of the time interval.

b. What is the geometric significance of the result in part (a)?

17. Find all values of the constants m and b for which the function

$$y = \begin{cases} \sin x, & x < \pi \\ mx + b, & x \geq \pi \end{cases}$$

is

a. continuous at $x = \pi$.

b. differentiable at $x = \pi$.

18. Does the function

$$f(x) = \begin{cases} \dfrac{1 - \cos x}{x}, & x \neq 0 \\ 0, & x = 0 \end{cases}$$

have a derivative at $x = 0$? Explain.

19. a. For what values of a and b will

$$f(x) = \begin{cases} ax, & x < 2 \\ ax^2 - bx + 3, & x \geq 2 \end{cases}$$

be differentiable for all values of x?

b. Discuss the geometry of the resulting graph of f.

20. a. For what values of a and b will

$$g(x) = \begin{cases} ax + b, & x \leq -1 \\ ax^3 + x + 2b, & x > -1 \end{cases}$$

be differentiable for all values of x?

b. Discuss the geometry of the resulting graph of g.

21. Odd differentiable functions Is there anything special about the derivative of an odd differentiable function of x? Give reasons for your answer.

22. Even differentiable functions Is there anything special about the derivative of an even differentiable function of x? Give reasons for your answer.

23. Suppose that the functions f and g are defined throughout an open interval containing the point x_0, that f is differentiable at x_0, that $f(x_0) = 0$, and that g is continuous at x_0. Show that the product fg is differentiable at x_0. This process shows, for example, that although $|x|$ is not differentiable at $x = 0$, the product $x|x|$ *is* differentiable at $x = 0$.

24. (*Continuation of Exercise 23.*) Use the result of Exercise 23 to show that the following functions are differentiable at $x = 0$.

a. $|x| \sin x$ **b.** $x^{2/3} \sin x$ **c.** $\sqrt[3]{x}\,(1 - \cos x)$

d. $h(x) = \begin{cases} x^2 \sin(1/x), & x \neq 0 \\ 0, & x = 0 \end{cases}$

25. Is the derivative of

$$h(x) = \begin{cases} x^2 \sin(1/x), & x \neq 0 \\ 0, & x = 0 \end{cases}$$

continuous at $x = 0$? How about the derivative of $k(x) = xh(x)$? Give reasons for your answers.

26. Let $f(x) = \begin{cases} x^2, & x \text{ is rational} \\ 0, & x \text{ is irrational.} \end{cases}$

Show that f is differentiable at $x = 0$.

27. Point B moves from point A to point C at 2 cm/sec in the accompanying diagram. At what rate is θ changing when $x = 4$ cm?

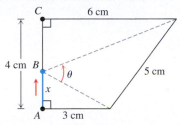

28. Suppose that a function f satisfies the following conditions for all real values of x and y:

i) $f(x + y) = f(x) \cdot f(y)$.

ii) $f(x) = 1 + xg(x)$, where $\lim_{x \to 0} g(x) = 1$.

iii) Show that the derivative $f'(x)$ exists at every value of x and that $f'(x) = f(x)$.

29. The generalized product rule   Use mathematical induction to prove that if $y = u_1 u_2 \cdots u_n$ is a finite product of differentiable functions, then y is differentiable on their common domain and

$$\frac{dy}{dx} = \frac{du_1}{dx} u_2 \cdots u_n + u_1 \frac{du_2}{dx} \cdots u_n + \cdots + u_1 u_2 \cdots u_{n-1} \frac{du_n}{dx}.$$

30. Leibniz's rule for higher-order derivatives of products   Leibniz's rule for higher-order derivatives of products of differentiable functions says that

a. $\dfrac{d^2(uv)}{dx^2} = \dfrac{d^2u}{dx^2} v + 2 \dfrac{du}{dx} \dfrac{dv}{dx} + u \dfrac{d^2v}{dx^2}.$

b. $\dfrac{d^3(uv)}{dx^3} = \dfrac{d^3u}{dx^3} v + 3 \dfrac{d^2u}{dx^2} \dfrac{dv}{dx} + 3 \dfrac{du}{dx} \dfrac{d^2v}{dx^2} + u \dfrac{d^3v}{dx^3}.$

c. $\dfrac{d^n(uv)}{dx^n} = \dfrac{d^nu}{dx^n} v + n \dfrac{d^{n-1}u}{dx^{n-1}} \dfrac{dv}{dx} + \cdots$

$$+ \frac{n(n-1)\cdots(n-k+1)}{k!} \frac{d^{n-k}u}{dx^{n-k}} \frac{d^k v}{dx^k}$$

$$+ \cdots + u \frac{d^n v}{dx^n}.$$

The equations in parts (a) and (b) are special cases of the equation in part (c). Derive the equation in part (c) by mathematical induction, using

$$\binom{m}{k} + \binom{m}{k+1} = \frac{m!}{k!(m-k)!} + \frac{m!}{(k+1)!(m-k-1)!}.$$

31. The period of a clock pendulum   The period T of a clock pendulum (time for one full swing and back) is given by the formula $T^2 = 4\pi^2 L/g$, where T is measured in seconds, $g = 32.2$ ft/sec^2, and L, the length of the pendulum, is measured in feet. Find approximately

a. the length of a clock pendulum whose period is $T = 1$ sec.

b. the change dT in T if the pendulum in part (a) is lengthened 0.01 ft.

c. the amount the clock gains or loses in a day as a result of the period's changing by the amount dT found in part (b).

32. The melting ice cube   Assume that an ice cube retains its cubical shape as it melts. If we call its edge length s, its volume is $V = s^3$ and its surface area is $6s^2$. We assume that V and s are differentiable functions of time t. We assume also that the cube's volume decreases at a rate that is proportional to its surface area. (This latter assumption seems reasonable enough when we think that the melting takes place at the surface: Changing the amount of surface changes the amount of ice exposed to melt.) In mathematical terms,

$$\frac{dV}{dt} = -k(6s^2), \qquad k > 0.$$

The minus sign indicates that the volume is decreasing. We assume that the proportionality factor k is constant. (It probably depends on many things, such as the relative humidity of the surrounding air, the air temperature, and the incidence or absence of sunlight, to name only a few.) Assume a particular set of conditions in which the cube lost $1/4$ of its volume during the first hour, and that the volume is V_0 when $t = 0$. How long will it take the ice cube to melt?

CHAPTER 3   Technology Application Projects

Mathematica/Maple Projects

Projects can be found at pearsonhighered.com/Thomas or within MyMathLab.

- *Convergence of Secant Slopes to the Derivative Function*
  You will visualize the secant line between successive points on a curve and observe what happens as the distance between them becomes small. The function, sample points, and secant lines are plotted on a single graph, while a second graph compares the slopes of the secant lines with the derivative function.

- *Derivatives, Slopes, Tangent Lines, and Making Movies*
  Parts I–III. You will visualize the derivative at a point, the linearization of a function, and the derivative of a function. You will learn how to plot the function and selected tangent lines on the same graph.
  Part IV (Plotting Many Tangent Lines)
  Part V (Making Movies). Parts IV and V of the module can be used to animate tangent lines as one moves along the graph of a function.

- *Convergence of Secant Slopes to the Derivative Function*
  You will visualize right-hand and left-hand derivatives.

- *Motion Along a Straight Line: Position → Velocity → Acceleration*
  Observe dramatic animated visualizations of the derivative relations among the position, velocity, and acceleration functions. Figures in the text can be animated.

4

Applications of Derivatives

OVERVIEW One of the most important applications of the derivative is its use as a tool for finding the optimal (best) solutions to problems. Optimization problems abound in mathematics, physical science and engineering, business and economics, and biology and medicine. For example, what are the height and diameter of the cylinder of largest volume that can be inscribed in a given sphere? What are the dimensions of the strongest rectangular wooden beam that can be cut from a cylindrical log of given diameter? Based on production costs and sales revenue, how many items should a manufacturer produce to maximize profit? How much does the trachea (windpipe) contract to expel air at maximum speed during a cough? What is the branching angle at which blood vessels minimize the energy loss due to friction as blood flows through the branches?

In this chapter we apply derivatives to find extreme values of functions, to determine and analyze the shapes of graphs, and to solve equations numerically. We also introduce the idea of recovering a function from its derivative. The key to many of these applications is the Mean Value Theorem, which connects the derivative and the average change of a function.

4.1 Extreme Values of Functions on Closed Intervals

This section shows how to locate and identify extreme (maximum or minimum) values of a function from its derivative. Once we can do this, we can solve a variety of optimization problems (see Section 4.6). The domains of the functions we consider are intervals or unions of separate intervals.

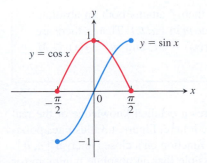

FIGURE 4.1 Absolute extrema for the sine and cosine functions on $[-\pi/2, \pi/2]$. These values can depend on the domain of a function.

DEFINITIONS Let f be a function with domain D. Then f has an **absolute maximum** value on D at a point c if

$$f(x) \leq f(c) \qquad \text{for all } x \text{ in } D$$

and an **absolute minimum** value on D at c if

$$f(x) \geq f(c) \qquad \text{for all } x \text{ in } D.$$

Maximum and minimum values are called **extreme values** of the function f. Absolute maxima or minima are also referred to as **global** maxima or minima.

For example, on the closed interval $[-\pi/2, \pi/2]$ the function $f(x) = \cos x$ takes on an absolute maximum value of 1 (once) and an absolute minimum value of 0 (twice). On the same interval, the function $g(x) = \sin x$ takes on a maximum value of 1 and a minimum value of -1 (Figure 4.1).

Functions defined by the same equation or formula can have different extrema (maximum or minimum values), depending on the domain. A function might not have a maximum or minimum if the domain is unbounded or fails to contain an endpoint. We see this in the following example.

EXAMPLE 1   The absolute extrema of the following functions on their domains can be seen in Figure 4.2. Each function has the same defining equation, $y = x^2$, but the domains vary.

Function rule	Domain D	Absolute extrema on D
(a) $y = x^2$	$(-\infty, \infty)$	No absolute maximum Absolute minimum of 0 at $x = 0$
(b) $y = x^2$	$[0, 2]$	Absolute maximum of 4 at $x = 2$ Absolute minimum of 0 at $x = 0$
(c) $y = x^2$	$(0, 2]$	Absolute maximum of 4 at $x = 2$ No absolute minimum
(d) $y = x^2$	$(0, 2)$	No absolute extrema

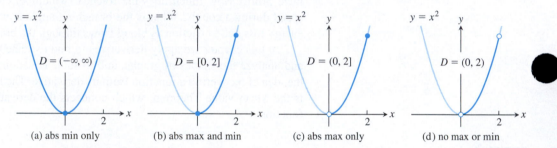

(a) abs min only   (b) abs max and min   (c) abs max only   (d) no max or min

FIGURE 4.2   Graphs for Example 1.

HISTORICAL BIOGRAPHY

Daniel Bernoulli

(1700–1789)

www.goo.gl/JYed90

Some of the functions in Example 1 do not have a maximum or a minimum value. The following theorem asserts that a function which is *continuous* over (or on) a finite *closed* interval $[a, b]$ has an absolute maximum and an absolute minimum value on the interval. We look for these extreme values when we graph a function.

THEOREM 1—The Extreme Value Theorem

If f is continuous on a closed interval $[a, b]$, then f attains both an absolute maximum value M and an absolute minimum value m in $[a, b]$. That is, there are numbers x_1 and x_2 in $[a, b]$ with $f(x_1) = m$, $f(x_2) = M$, and $m \leq f(x) \leq M$ for every other x in $[a, b]$.

The proof of the Extreme Value Theorem requires a detailed knowledge of the real number system (see Appendix 7) and we will not give it here. Figure 4.3 illustrates possible locations for the absolute extrema of a continuous function on a closed interval $[a, b]$. As we observed for the function $y = \cos x$, it is possible that an absolute minimum (or absolute maximum) may occur at two or more different points of the interval.

The requirements in Theorem 1 that the interval be closed and finite, and that the function be continuous, are essential. Without them, the conclusion of the theorem need not hold. Example 1 shows that an absolute extreme value may not exist if the interval fails

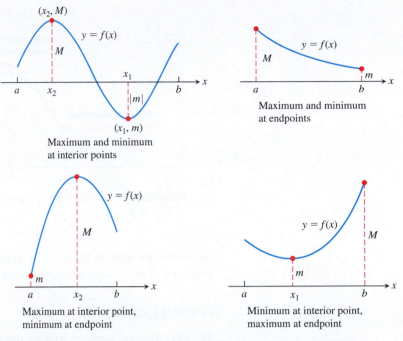

FIGURE 4.3   Some possibilities for a continuous function's maximum and minimum on a closed interval $[a, b]$.

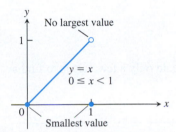

FIGURE 4.4   Even a single point of discontinuity can keep a function from having either a maximum or a minimum value on a closed interval. The function

$$y = \begin{cases} x, & 0 \le x < 1 \\ 0, & x = 1 \end{cases}$$

is continuous at every point of $[0, 1]$ except $x = 1$, yet its graph over $[0, 1]$ does not have a highest point.

to be both closed and finite. The exponential function $y = e^x$ over $(-\infty, \infty)$ shows that neither extreme value need exist on an infinite interval. Figure 4.4 shows that the continuity requirement cannot be omitted.

Local (Relative) Extreme Values

Figure 4.5 shows a graph with five points where a function has extreme values on its domain $[a, b]$. The function's absolute minimum occurs at a even though at e the function's value is smaller than at any other point *nearby*. The curve rises to the left and falls to the right around c, making $f(c)$ a maximum locally. The function attains its absolute maximum at d. We now define what we mean by local extrema.

> **DEFINITIONS**   A function f has a **local maximum** value at a point c within its domain D if $f(x) \le f(c)$ for all $x \in D$ lying in some open interval containing c.
>
> A function f has a **local minimum** value at a point c within its domain D if $f(x) \ge f(c)$ for all $x \in D$ lying in some open interval containing c.

If the domain of f is the closed interval $[a, b]$, then f has a local maximum at the endpoint $x = a$ if $f(x) \le f(a)$ for all x in some half-open interval $[a, a + \delta)$, $\delta > 0$. Likewise, f has a local maximum at an interior point $x = c$ if $f(x) \le f(c)$ for all x in some open interval $(c - \delta, c + \delta)$, $\delta > 0$, and a local maximum at the endpoint $x = b$ if $f(x) \le f(b)$ for all x in some half-open interval $(b - \delta, b]$, $\delta > 0$. The inequalities are reversed for local minimum values. In Figure 4.5, the function f has local maxima at c and d and local minima at a, e, and b. Local extrema are also called **relative extrema**. Some functions can have infinitely many local extrema, even over a finite interval. One example is the function $f(x) = \sin(1/x)$ on the interval $(0, 1]$. (We graphed this function in Figure 2.40.)

An absolute maximum is also a local maximum. Being the largest value overall, it is also the largest value in its immediate neighborhood. Hence, *a list of all local maxima will*

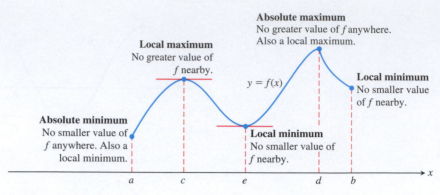

FIGURE 4.5 How to identify types of maxima and minima for a function with domain $a \leq x \leq b$.

automatically include the absolute maximum if there is one. Similarly, *a list of all local minima will include the absolute minimum if there is one.*

Finding Extrema

The next theorem explains why we usually need to investigate only a few values to find a function's extrema.

> **THEOREM 2—The First Derivative Theorem for Local Extreme Values**
> If f has a local maximum or minimum value at an interior point c of its domain, and if f' is defined at c, then
> $$f'(c) = 0.$$

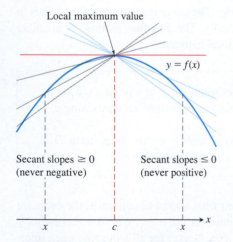

FIGURE 4.6 A curve with a local maximum value. The slope at c, simultaneously the limit of nonpositive numbers and nonnegative numbers, is zero.

Proof To prove that $f'(c)$ is zero at a local extremum, we show first that $f'(c)$ cannot be positive and second that $f'(c)$ cannot be negative. The only number that is neither positive nor negative is zero, so that is what $f'(c)$ must be.

To begin, suppose that f has a local maximum value at $x = c$ (Figure 4.6) so that $f(x) - f(c) \leq 0$ for all values of x near enough to c. Since c is an interior point of f's domain, $f'(c)$ is defined by the two-sided limit

$$\lim_{x \to c} \frac{f(x) - f(c)}{x - c}.$$

This means that the right-hand and left-hand limits both exist at $x = c$ and equal $f'(c)$. When we examine these limits separately, we find that

$$f'(c) = \lim_{x \to c^+} \frac{f(x) - f(c)}{x - c} \leq 0. \qquad \text{\color{blue}Because } (x - c) > 0 \text{ and } f(x) \leq f(c) \qquad (1)$$

Similarly,

$$f'(c) = \lim_{x \to c^-} \frac{f(x) - f(c)}{x - c} \geq 0. \qquad \text{\color{blue}Because } (x - c) < 0 \text{ and } f(x) \leq f(c) \qquad (2)$$

Together, Equations (1) and (2) imply $f'(c) = 0$.

This proves the theorem for local maximum values. To prove it for local minimum values, we simply use $f(x) \geq f(c)$, which reverses the inequalities in Equations (1) and (2). ∎

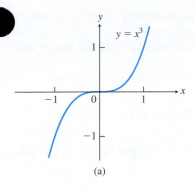

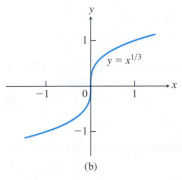

FIGURE 4.7 Critical points without extreme values. (a) $y' = 3x^2$ is 0 at $x = 0$, but $y = x^3$ has no extremum there. (b) $y' = (1/3)x^{-2/3}$ is undefined at $x = 0$, but $y = x^{1/3}$ has no extremum there.

Theorem 2 says that a function's first derivative is always zero at an interior point where the function has a local extreme value and the derivative is defined. If we recall that all the domains we consider are intervals or unions of separate intervals, the only places where a function f can possibly have an extreme value (local or global) are

1. interior points where $f' = 0$,   At $x = c$ and $x = e$ in Fig. 4.5
2. interior points where f' is undefined,   At $x = d$ in Fig. 4.5
3. endpoints of the domain of f.   At $x = a$ and $x = b$ in Fig. 4.5

The following definition helps us to summarize these results.

> **DEFINITION**   An interior point of the domain of a function f where f' is zero or undefined is a **critical point** of f.

Thus the only domain points where a function can assume extreme values are critical points and endpoints. However, be careful not to misinterpret what is being said here. A function may have a critical point at $x = c$ without having a local extreme value there. For instance, both of the functions $y = x^3$ and $y = x^{1/3}$ have critical points at the origin, but neither function has a local extreme value at the origin. Instead, each function has a *point of inflection* there (see Figure 4.7). We define and explore inflection points in Section 4.4.

Most problems that ask for extreme values call for finding the extrema of a continuous function on a closed and finite interval. Theorem 1 assures us that such values exist; Theorem 2 tells us that they are taken on only at critical points and endpoints. Often we can simply list these points and calculate the corresponding function values to find what the largest and smallest values are, and where they are located. However, if the interval is not closed or not finite (such as $a < x < b$ or $a < x < \infty$), we have seen that absolute extrema need not exist. When an absolute maximum or minimum value does exist, it must occur at a critical point or at a right- or left-hand endpoint of the interval.

> **Finding the Absolute Extrema of a Continuous Function f on a Finite Closed Interval**
>
> 1. Find all critical points of f on the interval.
> 2. Evaluate f at all critical points and endpoints.
> 3. Take the largest and smallest of these values.

EXAMPLE 2   Find the absolute maximum and minimum values of $f(x) = x^2$ on $[-2, 1]$.

Solution   The function is differentiable over its entire domain, so the only critical point is where $f'(x) = 2x = 0$, namely $x = 0$. We need to check the function's values at $x = 0$ and at the endpoints $x = -2$ and $x = 1$:

Critical point value: $f(0) = 0$

Endpoint values: $f(-2) = 4$

$f(1) = 1.$

The function has an absolute maximum value of 4 at $x = -2$ and an absolute minimum value of 0 at $x = 0$.   ∎

EXAMPLE 3   Find the absolute maximum and minimum values of $f(x) = 10x(2 - \ln x)$ on the interval $[1, e^2]$.

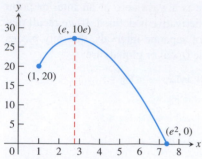

FIGURE 4.8 The extreme values of $f(x) = 10x(2 - \ln x)$ on $[1, e^2]$ occur at $x = e$ and $x = e^2$ (Example 3).

Solution Figure 4.8 suggests that f has its absolute maximum value near $x = 3$ and its absolute minimum value of 0 at $x = e^2$. Let's verify this observation.

We evaluate the function at the critical points and endpoints and take the largest and smallest of the resulting values.

The first derivative is

$$f'(x) = 10(2 - \ln x) - 10x\left(\frac{1}{x}\right) = 10(1 - \ln x).$$

The only critical point in the domain $[1, e^2]$ is the point $x = e$, where $\ln x = 1$. The values of f at this one critical point and at the endpoints are

Critical point value: $f(e) = 10e$

Endpoint values: $f(1) = 10(2 - \ln 1) = 20$

$$f(e^2) = 10e^2(2 - 2\ln e) = 0.$$

We can see from this list that the function's absolute maximum value is $10e \approx 27.2$; it occurs at the critical interior point $x = e$. The absolute minimum value is 0 and occurs at the right endpoint $x = e^2$. ∎

EXAMPLE 4 Find the absolute maximum and minimum values of $f(x) = x^{2/3}$ on the interval $[-2, 3]$.

Solution We evaluate the function at the critical points and endpoints and take the largest and smallest of the resulting values.

The first derivative

$$f'(x) = \frac{2}{3}x^{-1/3} = \frac{2}{3\sqrt[3]{x}}$$

has no zeros but is undefined at the interior point $x = 0$. The values of f at this one critical point and at the endpoints are

Critical point value: $f(0) = 0$

Endpoint values: $f(-2) = (-2)^{2/3} = \sqrt[3]{4}$

$$f(3) = (3)^{2/3} = \sqrt[3]{9}.$$

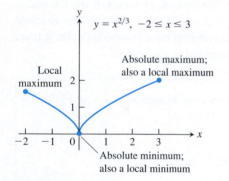

FIGURE 4.9 The extreme values of $f(x) = x^{2/3}$ on $[-2, 3]$ occur at $x = 0$ and $x = 3$ (Example 4).

We can see from this list that the function's absolute maximum value is $\sqrt[3]{9} \approx 2.08$, and it occurs at the right endpoint $x = 3$. The absolute minimum value is 0, and it occurs at the interior point $x = 0$ where the graph has a cusp (Figure 4.9). ∎

Theorem 1 gives a method to find the absolute maxima and absolute minima of a differentiable function on a finite closed interval. On more general domains, such as $(0, 1)$, $[2, 5)$, $[1, \infty)$, and $(-\infty, \infty)$, absolute maxima and minima may or may not exist. To determine if they exist, and to locate them when they do, we will develop methods to sketch the graph of a differentiable function. With knowledge of the asymptotes of the function, as well as the local maxima and minima, we can deduce the locations of the absolute maxima and minima, if any. For now we can find the absolute maxima and the absolute minima of a function on a finite closed interval by comparing the values of the function at its critical points and at the endpoints of the interval. For a differentiable function, these are the only points where the extrema have the potential to occur.

EXERCISES 4.1

Finding Extrema from Graphs

In Exercises 1–6, determine from the graph whether the function has any absolute extreme values on $[a, b]$. Then explain how your answer is consistent with Theorem 1.

1.
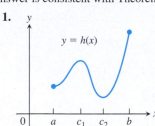

2.
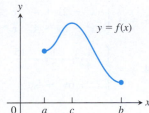

3.
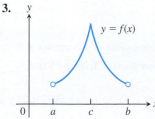

4.
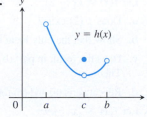

5.
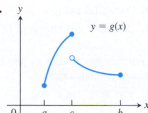

6.
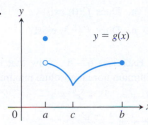

In Exercises 7–10, find the absolute extreme values and where they occur.

7.
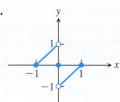

8.
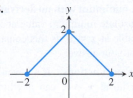

9.

10.
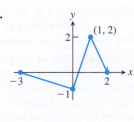

In Exercises 11–14, match the table with a graph.

11.

x	$f'(x)$
a	0
b	0
c	5

12.

x	$f'(x)$
a	0
b	0
c	-5

13.

x	$f'(x)$
a	does not exist
b	0
c	-2

14.

x	$f'(x)$
a	does not exist
b	does not exist
c	-1.7

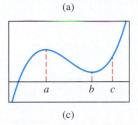

(a)

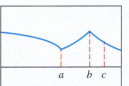

(b)

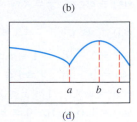

(c)

(d)

In Exercises 15–20, sketch the graph of each function and determine whether the function has any absolute extreme values on its domain. Explain how your answer is consistent with Theorem 1.

15. $f(x) = |x|, \quad -1 < x < 2$

16. $y = \dfrac{6}{x^2 + 2}, \quad -1 < x < 1$

17. $g(x) = \begin{cases} -x, & 0 \le x < 1 \\ x - 1, & 1 \le x \le 2 \end{cases}$

18. $h(x) = \begin{cases} \dfrac{1}{x}, & -1 \le x < 0 \\ \sqrt{x}, & 0 \le x \le 4 \end{cases}$

19. $y = 3 \sin x, \quad 0 < x < 2\pi$

20. $f(x) = \begin{cases} x + 1, & -1 \le x < 0 \\ \cos x, & 0 < x \le \dfrac{\pi}{2} \end{cases}$

Absolute Extrema on Finite Closed Intervals

In Exercises 21–40, find the absolute maximum and minimum values of each function on the given interval. Then graph the function. Identify the points on the graph where the absolute extrema occur, and include their coordinates.

21. $f(x) = \frac{2}{3}x - 5, \quad -2 \le x \le 3$

22. $f(x) = -x - 4, \quad -4 \le x \le 1$

23. $f(x) = x^2 - 1, \quad -1 \le x \le 2$

24. $f(x) = 4 - x^3, \quad -2 \le x \le 1$

25. $F(x) = -\frac{1}{x^2}, \quad 0.5 \le x \le 2$

26. $F(x) = -\frac{1}{x}, \quad -2 \le x \le -1$

27. $h(x) = \sqrt[3]{x}, \quad -1 \le x \le 8$

28. $h(x) = -3x^{2/3}, \quad -1 \le x \le 1$

29. $g(x) = \sqrt{4 - x^2}, \quad -2 \le x \le 1$

30. $g(x) = -\sqrt{5 - x^2}, \quad -\sqrt{5} \le x \le 0$

31. $f(\theta) = \sin \theta, \quad -\frac{\pi}{2} \le \theta \le \frac{5\pi}{6}$

32. $f(\theta) = \tan \theta, \quad -\frac{\pi}{3} \le \theta \le \frac{\pi}{4}$

33. $g(x) = \csc x, \quad \frac{\pi}{3} \le x \le \frac{2\pi}{3}$

34. $g(x) = \sec x, \quad -\frac{\pi}{3} \le x \le \frac{\pi}{6}$

35. $f(t) = 2 - |t|, \quad -1 \le t \le 3$

36. $f(t) = |t - 5|, \quad 4 \le t \le 7$

37. $g(x) = xe^{-x}, \quad -1 \le x \le 1$

38. $h(x) = \ln(x + 1), \quad 0 \le x \le 3$

39. $f(x) = \frac{1}{x} + \ln x, \quad 0.5 \le x \le 4$

40. $g(x) = e^{-x^2}, \quad -2 \le x \le 1$

In Exercises 41–44, find the function's absolute maximum and minimum values and say where they occur.

41. $f(x) = x^{4/3}, \quad -1 \le x \le 8$

42. $f(x) = x^{5/3}, \quad -1 \le x \le 8$

43. $g(\theta) = \theta^{3/5}, \quad -32 \le \theta \le 1$

44. $h(\theta) = 3\theta^{2/3}, \quad -27 \le \theta \le 8$

Finding Critical Points

In Exercises 45–56, determine critical points and domain endpoints for each function.

45. $y = x^2 - 6x + 7$ **46.** $f(x) = 6x^2 - x^3$

47. $f(x) = x(4 - x)^3$ **48.** $g(x) = (x - 1)^2(x - 3)^2$

49. $y = x^2 + \frac{2}{x}$ **50.** $f(x) = \frac{x^2}{x - 2}$

51. $y = x^2 - 32\sqrt{x}$ **52.** $g(x) = \sqrt{2x - x^2}$

53. $y = \ln(x + 1) - \tan^{-1} x$ **54.** $y = 2\sqrt{1 - x^2} + \sin^{-1} x$

55. $y = x^3 + 3x^2 - 24x + 7$ **56.** $y = x - 3x^{2/3}$

Local Extrema and Critical Points

In Exercises 57–64, find the critical points and domain endpoints for each function. Then find the value of the function at each of these points and identify extreme values (absolute and local).

57. $y = x^{2/3}(x + 2)$ **58.** $y = x^{2/3}(x^2 - 4)$

59. $y = x\sqrt{4 - x^2}$ **60.** $y = x^2\sqrt{3 - x}$

61. $y = \begin{cases} 4 - 2x, & x \le 1 \\ x + 1, & x > 1 \end{cases}$

62. $y = \begin{cases} 3 - x, & x < 0 \\ 3 + 2x - x^2, & x \ge 0 \end{cases}$

63. $y = \begin{cases} -x^2 - 2x + 4, & x \le 1 \\ -x^2 + 6x - 4, & x > 1 \end{cases}$

64. $y = \begin{cases} -\frac{1}{4}x^2 - \frac{1}{2}x + \frac{15}{4}, & x \le 1 \\ x^3 - 6x^2 + 8x, & x > 1 \end{cases}$

In Exercises 65 and 66, give reasons for your answers.

65. Let $f(x) = (x - 2)^{2/3}$.

   a. Does $f'(2)$ exist?

   b. Show that the only local extreme value of f occurs at $x = 2$.

   c. Does the result in part (b) contradict the Extreme Value Theorem?

   d. Repeat parts (a) and (b) for $f(x) = (x - a)^{2/3}$, replacing 2 by a.

66. Let $f(x) = |x^3 - 9x|$.

   a. Does $f'(0)$ exist?   **b.** Does $f'(3)$ exist?

   c. Does $f'(-3)$ exist?   **d.** Determine all extrema of f.

In Exercises 67–70, show that the function has neither an absolute minimum nor an absolute maximum on its natural domain.

67. $y = x^{11} + x^3 + x - 5$ **68.** $y = 3x + \tan x$

69. $y = \frac{1 - e^x}{e^x + 1}$ **70.** $y = 2x - \sin 2x$

Theory and Examples

71. A minimum with no derivative   The function $f(x) = |x|$ has an absolute minimum value at $x = 0$ even though f is not differentiable at $x = 0$. Is this consistent with Theorem 2? Give reasons for your answer.

72. Even functions   If an even function $f(x)$ has a local maximum value at $x = c$, can anything be said about the value of f at $x = -c$? Give reasons for your answer.

73. Odd functions   If an odd function $g(x)$ has a local minimum value at $x = c$, can anything be said about the value of g at $x = -c$? Give reasons for your answer.

74. No critical points or endpoints exist   We know how to find the extreme values of a continuous function $f(x)$ by investigating its values at critical points and endpoints. But what if there *are* no critical points or endpoints? What happens then? Do such functions really exist? Give reasons for your answers.

75. The function

$$V(x) = x(10 - 2x)(16 - 2x), \quad 0 < x < 5,$$

models the volume of a box.

   a. Find the extreme values of V.

   b. Interpret any values found in part (a) in terms of the volume of the box.

76. Cubic functions Consider the cubic function

$$f(x) = ax^3 + bx^2 + cx + d.$$

   a. Show that f can have 0, 1, or 2 critical points. Give examples and graphs to support your argument.

   b. How many local extreme values can f have?

77. Maximum height of a vertically moving body The height of a body moving vertically is given by

$$s = -\frac{1}{2}gt^2 + v_0 t + s_0, \qquad g > 0,$$

with s in meters and t in seconds. Find the body's maximum height.

78. Peak alternating current Suppose that at any given time t (in seconds) the current i (in amperes) in an alternating current circuit is $i = 2 \cos t + 2 \sin t$. What is the peak current for this circuit (largest magnitude)?

T Graph the functions in Exercises 79–82. Then find the extreme values of the function on the interval and say where they occur.

79. $f(x) = |x - 2| + |x + 3|, \quad -5 \leq x \leq 5$

80. $g(x) = |x - 1| - |x - 5|, \quad -2 \leq x \leq 7$

81. $h(x) = |x + 2| - |x - 3|, \quad -\infty < x < \infty$

82. $k(x) = |x + 1| + |x - 3|, \quad -\infty < x < \infty$

COMPUTER EXPLORATIONS

In Exercises 83–90, you will use a CAS to help find the absolute extrema of the given function over the specified closed interval. Perform the following steps.

   a. Plot the function over the interval to see its general behavior there.

   b. Find the interior points where $f' = 0$. (In some exercises, you may have to use the numerical equation solver to approximate a solution.) You may want to plot f' as well.

   c. Find the interior points where f' does not exist.

   d. Evaluate the function at all points found in parts (b) and (c) and at the endpoints of the interval.

   e. Find the function's absolute extreme values on the interval and identify where they occur.

83. $f(x) = x^4 - 8x^2 + 4x + 2, \quad [-20/25, 64/25]$

84. $f(x) = -x^4 + 4x^3 - 4x + 1, \quad [-3/4, 3]$

85. $f(x) = x^{2/3}(3 - x), \quad [-2, 2]$

86. $f(x) = 2 + 2x - 3x^{2/3}, \quad [-1, 10/3]$

87. $f(x) = \sqrt{x} + \cos x, \quad [0, 2\pi]$

88. $f(x) = x^{3/4} - \sin x + \frac{1}{2}, \quad [0, 2\pi]$

89. $f(x) = \pi x^2 e^{-3x/2}, \quad [0, 5]$

90. $f(x) = \ln(2x + x \sin x), \quad [1, 15]$

4.2 The Mean Value Theorem

We know that constant functions have zero derivatives, but could there be a more complicated function whose derivative is always zero? If two functions have identical derivatives over an interval, how are the functions related? We answer these and other questions in this chapter by applying the Mean Value Theorem. First we introduce a special case, known as Rolle's Theorem, which is used to prove the Mean Value Theorem.

Rolle's Theorem

As suggested by its graph, if a differentiable function crosses a horizontal line at two different points, there is at least one point between them where the tangent to the graph is horizontal and the derivative is zero (Figure 4.10). We now state and prove this result.

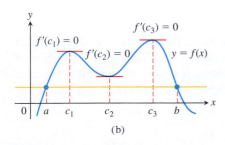

FIGURE 4.10 Rolle's Theorem says that a differentiable curve has at least one horizontal tangent between any two points where it crosses a horizontal line. It may have just one (a), or it may have more (b).

> **THEOREM 3—Rolle's Theorem**
>
> Suppose that $y = f(x)$ is continuous over the closed interval $[a, b]$ and differentiable at every point of its interior (a, b). If $f(a) = f(b)$, then there is at least one number c in (a, b) at which $f'(c) = 0$.

Proof Being continuous, f assumes absolute maximum and minimum values on $[a, b]$ by Theorem 1. These can occur only

1. at interior points where f' is zero,

2. at interior points where f' does not exist,

3. at endpoints of the function's domain, in this case a and b.

HISTORICAL BIOGRAPHY

Michel Rolle
(1652–1719)
www.goo.gl/BfgcNr

By hypothesis, f has a derivative at every interior point. That rules out possibility (2), leaving us with interior points where $f' = 0$ and with the two endpoints a and b.

If either the maximum or the minimum occurs at a point c between a and b, then $f'(c) = 0$ by Theorem 2 in Section 4.1, and we have found a point for Rolle's Theorem.

If both the absolute maximum and the absolute minimum occur at the endpoints, then because $f(a) = f(b)$ it must be the case that f is a constant function with $f(x) = f(a) = f(b)$ for every $x \in [a, b]$. Therefore $f'(x) = 0$ and the point c can be taken anywhere in the interior (a, b). ∎

The hypotheses of Theorem 3 are essential. If they fail at even one point, the graph may not have a horizontal tangent (Figure 4.11).

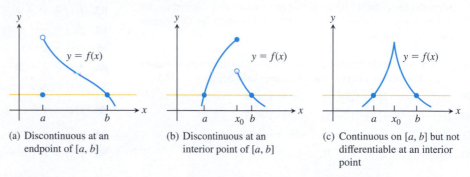

(a) Discontinuous at an endpoint of $[a, b]$

(b) Discontinuous at an interior point of $[a, b]$

(c) Continuous on $[a, b]$ but not differentiable at an interior point

FIGURE 4.11 There may be no horizontal tangent if the hypotheses of Rolle's Theorem do not hold.

Rolle's Theorem may be combined with the Intermediate Value Theorem to show when there is only one real solution of an equation $f(x) = 0$, as we illustrate in the next example.

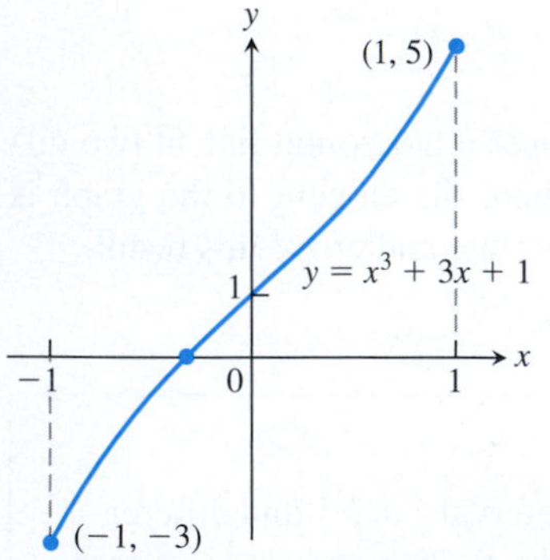

EXAMPLE 1 Show that the equation

$$x^3 + 3x + 1 = 0$$

has exactly one real solution.

Solution We define the continuous function

$$f(x) = x^3 + 3x + 1.$$

Since $f(-1) = -3$ and $f(0) = 1$, the Intermediate Value Theorem tells us that the graph of f crosses the x-axis somewhere in the open interval $(-1, 0)$. (See Figure 4.12.) Now, if there were even two points $x = a$ and $x = b$ where $f(x)$ was zero, Rolle's Theorem would guarantee the existence of a point $x = c$ in between them where f' was zero. However, the derivative

$$f'(x) = 3x^2 + 3$$

is never zero (because it is always positive). Therefore, f has no more than one zero. ∎

Our main use of Rolle's Theorem is in proving the Mean Value Theorem.

FIGURE 4.12 The only real zero of the polynomial $y = x^3 + 3x + 1$ is the one shown here where the curve crosses the x-axis between -1 and 0 (Example 1).

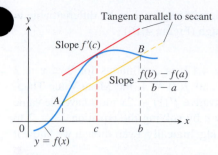

FIGURE 4.13 Geometrically, the Mean Value Theorem says that somewhere between a and b the curve has at least one tangent line parallel to the secant line that joins A and B.

HISTORICAL BIOGRAPHY

Joseph-Louis Lagrange
(1736–1813)
www.goo.gl/WLub9z

The Mean Value Theorem

The Mean Value Theorem, which was first stated by Joseph-Louis Lagrange, is a slanted version of Rolle's Theorem (Figure 4.13). The Mean Value Theorem guarantees that there is a point where the tangent line is parallel to the secant line that joins A and B.

THEOREM 4—The Mean Value Theorem

Suppose $y = f(x)$ is continuous over a closed interval $[a, b]$ and differentiable on the interval's interior (a, b). Then there is at least one point c in (a, b) at which

$$\frac{f(b) - f(a)}{b - a} = f'(c). \qquad (1)$$

Proof We picture the graph of f and draw a line through the points $A(a, f(a))$ and $B(b, f(b))$. (See Figure 4.14.) The secant line is the graph of the function

$$g(x) = f(a) + \frac{f(b) - f(a)}{b - a}(x - a) \qquad (2)$$

(point-slope equation). The vertical difference between the graphs of f and g at x is

$$h(x) = f(x) - g(x)$$
$$= f(x) - f(a) - \frac{f(b) - f(a)}{b - a}(x - a). \qquad (3)$$

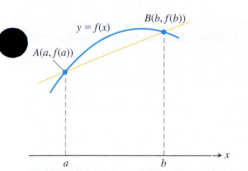

FIGURE 4.14 The graph of f and the secant AB over the interval $[a, b]$.

Figure 4.15 shows the graphs of f, g, and h together.

The function h satisfies the hypotheses of Rolle's Theorem on $[a, b]$. It is continuous on $[a, b]$ and differentiable on (a, b) because both f and g are. Also, $h(a) = h(b) = 0$ because the graphs of f and g both pass through A and B. Therefore $h'(c) = 0$ at some point $c \in (a, b)$. This is the point we want for Equation (1) in the theorem.

To verify Equation (1), we differentiate both sides of Equation (3) with respect to x and then set $x = c$:

$$h'(x) = f'(x) - \frac{f(b) - f(a)}{b - a} \qquad \text{Derivative of Eq. (3)}$$

$$h'(c) = f'(c) - \frac{f(b) - f(a)}{b - a} \qquad \text{Evaluated at } x = c$$

$$0 = f'(c) - \frac{f(b) - f(a)}{b - a} \qquad h'(c) = 0$$

$$f'(c) = \frac{f(b) - f(a)}{b - a}, \qquad \text{Rearranged}$$

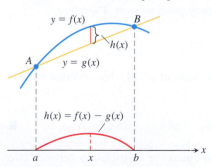

FIGURE 4.15 The secant AB is the graph of the function $g(x)$. The function $h(x) = f(x) - g(x)$ gives the vertical distance between the graphs of f and g at x.

which is what we set out to prove.

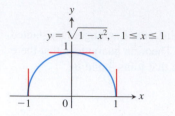

FIGURE 4.16 The function $f(x) = \sqrt{1 - x^2}$ satisfies the hypotheses (and conclusion) of the Mean Value Theorem on $[-1, 1]$ even though f is not differentiable at -1 and 1.

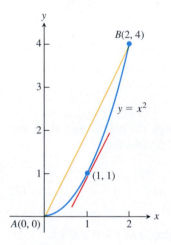

FIGURE 4.17 As we find in Example 2, $c = 1$ is where the tangent is parallel to the secant line.

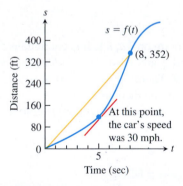

FIGURE 4.18 Distance versus elapsed time for the car in Example 3.

The hypotheses of the Mean Value Theorem do not require f to be differentiable at either a or b. One-sided continuity at a and b is enough (Figure 4.16).

EXAMPLE 2   The function $f(x) = x^2$ (Figure 4.17) is continuous for $0 \leq x \leq 2$ and differentiable for $0 < x < 2$. Since $f(0) = 0$ and $f(2) = 4$, the Mean Value Theorem says that at some point c in the interval, the derivative $f'(x) = 2x$ must have the value $(4 - 0)/(2 - 0) = 2$. In this case we can identify c by solving the equation $2c = 2$ to get $c = 1$. However, it is not always easy to find c algebraically, even though we know it always exists. ■

A Physical Interpretation

We can think of the number $(f(b) - f(a))/(b - a)$ as the average change in f over $[a, b]$ and $f'(c)$ as an instantaneous change. Then the Mean Value Theorem says that the instantaneous change at some interior point is equal to the average change over the entire interval.

EXAMPLE 3   If a car accelerating from zero takes 8 sec to go 352 ft, its average velocity for the 8-sec interval is $352/8 = 44$ ft/sec. The Mean Value Theorem says that at some point during the acceleration the speedometer must read exactly 30 mph (44 ft/sec) (Figure 4.18). ■

Mathematical Consequences

At the beginning of the section, we asked what kind of function has a zero derivative over an interval. The first corollary of the Mean Value Theorem provides the answer that only constant functions have zero derivatives.

COROLLARY 1 If $f'(x) = 0$ at each point x of an open interval (a, b), then $f(x) = C$ for all $x \in (a, b)$, where C is a constant.

Proof   We want to show that f has a constant value on the interval (a, b). We do so by showing that if x_1 and x_2 are any two points in (a, b) with $x_1 < x_2$, then $f(x_1) = f(x_2)$. Now f satisfies the hypotheses of the Mean Value Theorem on $[x_1, x_2]$: It is differentiable at every point of $[x_1, x_2]$ and hence continuous at every point as well. Therefore,

$$\frac{f(x_2) - f(x_1)}{x_2 - x_1} = f'(c)$$

at some point c between x_1 and x_2. Since $f' = 0$ throughout (a, b), this equation implies successively that

$$\frac{f(x_2) - f(x_1)}{x_2 - x_1} = 0, \qquad f(x_2) - f(x_1) = 0, \qquad \text{and} \qquad f(x_1) = f(x_2). \blacksquare$$

At the beginning of this section, we also asked about the relationship between two functions that have identical derivatives over an interval. The next corollary tells us that their values on the interval have a constant difference.

COROLLARY 2 If $f'(x) = g'(x)$ at each point x in an open interval (a, b), then there exists a constant C such that $f(x) = g(x) + C$ for all $x \in (a, b)$. That is, $f - g$ is a constant function on (a, b).

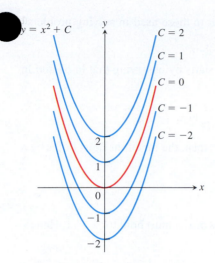

FIGURE 4.19 From a geometric point of view, Corollary 2 of the Mean Value Theorem says that the graphs of functions with identical derivatives on an interval can differ only by a vertical shift. The graphs of the functions with derivative $2x$ are the parabolas $y = x^2 + C$, shown here for several values of C.

Proof   At each point $x \in (a, b)$ the derivative of the difference function $h = f - g$ is

$$h'(x) = f'(x) - g'(x) = 0.$$

Thus, $h(x) = C$ on (a, b) by Corollary 1. That is, $f(x) - g(x) = C$ on (a, b), so $f(x) = g(x) + C$.   ∎

Corollaries 1 and 2 are also true if the open interval (a, b) fails to be finite. That is, they remain true if the interval is (a, ∞), $(-\infty, b)$, or $(-\infty, \infty)$.

Corollary 2 will play an important role when we discuss antiderivatives in Section 4.8. It tells us, for instance, that since the derivative of $f(x) = x^2$ on $(-\infty, \infty)$ is $2x$, any other function with derivative $2x$ on $(-\infty, \infty)$ must have the formula $x^2 + C$ for some value of C (Figure 4.19).

EXAMPLE 4   Find the function $f(x)$ whose derivative is $\sin x$ and whose graph passes through the point $(0, 2)$.

Solution   Since the derivative of $g(x) = -\cos x$ is $g'(x) = \sin x$, we see that f and g have the same derivative. Corollary 2 then says that $f(x) = -\cos x + C$ for some constant C. Since the graph of f passes through the point $(0, 2)$, the value of C is determined from the condition that $f(0) = 2$:

$$f(0) = -\cos(0) + C = 2, \qquad \text{so} \qquad C = 3.$$

The function is $f(x) = -\cos x + 3$.   ■

Finding Velocity and Position from Acceleration

We can use Corollary 2 to find the velocity and position functions of an object moving along a vertical line. Assume the object or body is falling freely from rest with acceleration 9.8 m/sec^2. We assume the position $s(t)$ of the body is measured positive downward from the rest position (so the vertical coordinate line points *downward*, in the direction of the motion, with the rest position at 0).

We know that the velocity $v(t)$ is some function whose derivative is 9.8. We also know that the derivative of $g(t) = 9.8t$ is 9.8. By Corollary 2,

$$v(t) = 9.8t + C$$

for some constant C. Since the body falls from rest, $v(0) = 0$. Thus

$$9.8(0) + C = 0, \qquad \text{and} \qquad C = 0.$$

The velocity function must be $v(t) = 9.8t$. What about the position function $s(t)$?

We know that $s(t)$ is some function whose derivative is $9.8t$. We also know that the derivative of $f(t) = 4.9t^2$ is $9.8t$. By Corollary 2,

$$s(t) = 4.9t^2 + C$$

for some constant C. Since $s(0) = 0$,

$$4.9(0)^2 + C = 0, \qquad \text{and} \qquad C = 0.$$

The position function is $s(t) = 4.9t^2$ until the body hits the ground.

The ability to find functions from their rates of change is one of the very powerful tools of calculus. As we will see, it lies at the heart of the mathematical developments in Chapter 5.

Proofs of the Laws of Logarithms

The algebraic properties of logarithms were stated in Section 1.6. These properties follow from Corollary 2 of the Mean Value Theorem and the formula for the derivative of the logarithm function. The steps in the proofs are similar to those used in solving problems involving logarithms.

Proof that $\ln bx = \ln b + \ln x$ The argument starts by observing that $\ln bx$ and $\ln x$ have the same derivative:

$$\frac{d}{dx}\ln(bx) = \frac{b}{bx} = \frac{1}{x} = \frac{d}{dx}\ln x.$$

According to Corollary 2 of the Mean Value Theorem, then, the functions must differ by a constant, which means that

$$\ln bx = \ln x + C$$

for some C.

Since this last equation holds for all positive values of x, it must hold for $x = 1$. Hence,

$$\ln(b \cdot 1) = \ln 1 + C$$
$$\ln b = 0 + C \qquad \textcolor{blue}{\ln 1 = 0}$$
$$C = \ln b.$$

By substituting, we conclude

$$\ln bx = \ln b + \ln x. \qquad \blacksquare$$

Proof that $\ln x^r = r \ln x$ We use the same-derivative argument again. For all positive values of x,

$$\frac{d}{dx}\ln x^r = \frac{1}{x^r}\frac{d}{dx}(x^r) \qquad \textcolor{blue}{\text{Chain Rule}}$$

$$= \frac{1}{x^r}rx^{r-1} \qquad \textcolor{blue}{\text{Derivative Power Rule}}$$

$$= r \cdot \frac{1}{x} = \frac{d}{dx}(r \ln x).$$

Since $\ln x^r$ and $r \ln x$ have the same derivative,

$$\ln x^r = r \ln x + C$$

for some constant C. Taking x to be 1 identifies C as zero, and we're done. $\blacksquare$

You are asked to prove the Quotient Rule for logarithms,

$$\ln\left(\frac{b}{x}\right) = \ln b - \ln x,$$

in Exercise 75. The Reciprocal Rule, $\ln(1/x) = -\ln x$, is a special case of the Quotient Rule, obtained by taking $b = 1$ and noting that $\ln 1 = 0$.

Laws of Exponents

The laws of exponents for the natural exponential e^x are consequences of the algebraic properties of $\ln x$. They follow from the inverse relationship between these functions.

Laws of Exponents for e^x

For all numbers x, x_1, and x_2, the natural exponential e^x obeys the following laws:

1. $e^{x_1} \cdot e^{x_2} = e^{x_1 + x_2}$

2. $e^{-x} = \dfrac{1}{e^x}$

3. $\dfrac{e^{x_1}}{e^{x_2}} = e^{x_1 - x_2}$

4. $(e^{x_1})^{x_2} = e^{x_1 x_2} = (e^{x_2})^{x_1}$

Proof of Law 1   Let

$$y_1 = e^{x_1} \quad \text{and} \quad y_2 = e^{x_2}. \tag{4}$$

Then

$$x_1 = \ln y_1 \quad \text{and} \quad x_2 = \ln y_2 \qquad \text{Take logs of both sides of Eqs. (4).}$$
$$x_1 + x_2 = \ln y_1 + \ln y_2$$
$$= \ln y_1 y_2 \qquad \text{Product Rule for logarithms}$$
$$e^{x_1 + x_2} = e^{\ln y_1 y_2} \qquad \text{Exponentiate.}$$
$$= y_1 y_2 \qquad e^{\ln u} = u$$
$$= e^{x_1} e^{x_2}.$$

The proof of Law 4 is similar. Laws 2 and 3 follow from Law 1 (Exercises 77 and 78).

EXERCISES 4.2

Checking the Mean Value Theorem

Find the value or values of c that satisfy the equation

$$\frac{f(b) - f(a)}{b - a} = f'(c)$$

in the conclusion of the Mean Value Theorem for the functions and intervals in Exercises 1–8.

1. $f(x) = x^2 + 2x - 1, \quad [0, 1]$

2. $f(x) = x^{2/3}, \quad [0, 1]$

3. $f(x) = x + \dfrac{1}{x}, \quad \left[\dfrac{1}{2}, 2\right]$

4. $f(x) = \sqrt{x - 1}, \quad [1, 3]$

5. $f(x) = \sin^{-1} x, \quad [-1, 1]$

6. $f(x) = \ln(x - 1), \quad [2, 4]$

7. $f(x) = x^3 - x^2, \quad [-1, 2]$

8. $g(x) = \begin{cases} x^3, & -2 \le x \le 0 \\ x^2, & 0 < x \le 2 \end{cases}$

Which of the functions in Exercises 9–14 satisfy the hypotheses of the Mean Value Theorem on the given interval, and which do not? Give reasons for your answers.

9. $f(x) = x^{2/3}, \quad [-1, 8]$

10. $f(x) = x^{4/5}, \quad [0, 1]$

11. $f(x) = \sqrt{x(1 - x)}, \quad [0, 1]$

12. $f(x) = \begin{cases} \dfrac{\sin x}{x}, & -\pi \le x < 0 \\ 0, & x = 0 \end{cases}$

13. $f(x) = \begin{cases} x^2 - x, & -2 \le x \le -1 \\ 2x^2 - 3x - 3, & -1 < x \le 0 \end{cases}$

14. $f(x) = \begin{cases} 2x - 3, & 0 \le x \le 2 \\ 6x - x^2 - 7, & 2 < x \le 3 \end{cases}$

15. The function

$$f(x) = \begin{cases} x, & 0 \le x < 1 \\ 0, & x = 1 \end{cases}$$

is zero at $x = 0$ and $x = 1$ and differentiable on $(0, 1)$, but its derivative on $(0, 1)$ is never zero. How can this be? Doesn't Rolle's Theorem say the derivative has to be zero somewhere in $(0, 1)$? Give reasons for your answer.

16. For what values of a, m, and b does the function

$$f(x) = \begin{cases} 3, & x = 0 \\ -x^2 + 3x + a, & 0 < x < 1 \\ mx + b, & 1 \le x \le 2 \end{cases}$$

satisfy the hypotheses of the Mean Value Theorem on the interval $[0, 2]$?

Roots (Zeros)

17. **a.** Plot the zeros of each polynomial on a line together with the zeros of its first derivative.

   i) $y = x^2 - 4$

   ii) $y = x^2 + 8x + 15$

   iii) $y = x^3 - 3x^2 + 4 = (x + 1)(x - 2)^2$

   iv) $y = x^3 - 33x^2 + 216x = x(x - 9)(x - 24)$

   b. Use Rolle's Theorem to prove that between every two zeros of $x^n + a_{n-1}x^{n-1} + \cdots + a_1 x + a_0$ there lies a zero of

   $$nx^{n-1} + (n - 1)a_{n-1}x^{n-2} + \cdots + a_1.$$

18. Suppose that f'' is continuous on $[a, b]$ and that f has three zeros in the interval. Show that f'' has at least one zero in (a, b). Generalize this result.

19. Show that if $f'' > 0$ throughout an interval $[a, b]$, then f' has at most one zero in $[a, b]$. What if $f'' < 0$ throughout $[a, b]$ instead?

20. Show that a cubic polynomial can have at most three real zeros.

Show that the functions in Exercises 21–28 have exactly one zero in the given interval.

21. $f(x) = x^4 + 3x + 1, \quad [-2, -1]$

22. $f(x) = x^3 + \dfrac{4}{x^2} + 7, \quad (-\infty, 0)$

23. $g(t) = \sqrt{t} + \sqrt{1 + t} - 4, \quad (0, \infty)$

24. $g(t) = \dfrac{1}{1 - t} + \sqrt{1 + t} - 3.1, \quad (-1, 1)$

25. $r(\theta) = \theta + \sin^2\left(\dfrac{\theta}{3}\right) - 8, \quad (-\infty, \infty)$

26. $r(\theta) = 2\theta - \cos^2\theta + \sqrt{2}, \quad (-\infty, \infty)$

27. $r(\theta) = \sec\theta - \dfrac{1}{\theta^3} + 5, \quad (0, \pi/2)$

28. $r(\theta) = \tan\theta - \cot\theta - \theta, \quad (0, \pi/2)$

Finding Functions from Derivatives

29. Suppose that $f(-1) = 3$ and that $f'(x) = 0$ for all x. Must $f(x) = 3$ for all x? Give reasons for your answer.

30. Suppose that $f(0) = 5$ and that $f'(x) = 2$ for all x. Must $f(x) = 2x + 5$ for all x? Give reasons for your answer.

31. Suppose that $f'(x) = 2x$ for all x. Find $f(2)$ if

  a. $f(0) = 0$ **b.** $f(1) = 0$ **c.** $f(-2) = 3$.

32. What can be said about functions whose derivatives are constant? Give reasons for your answer.

In Exercises 33–38, find all possible functions with the given derivative.

33. a. $y' = x$ **b.** $y' = x^2$ **c.** $y' = x^3$

34. a. $y' = 2x$ **b.** $y' = 2x - 1$ **c.** $y' = 3x^2 + 2x - 1$

35. a. $y' = -\dfrac{1}{x^2}$ **b.** $y' = 1 - \dfrac{1}{x^2}$ **c.** $y' = 5 + \dfrac{1}{x^2}$

36. a. $y' = \dfrac{1}{2\sqrt{x}}$ **b.** $y' = \dfrac{1}{\sqrt{x}}$ **c.** $y' = 4x - \dfrac{1}{\sqrt{x}}$

37. a. $y' = \sin 2t$ **b.** $y' = \cos\dfrac{t}{2}$ **c.** $y' = \sin 2t + \cos\dfrac{t}{2}$

38. a. $y' = \sec^2\theta$ **b.** $y' = \sqrt{\theta}$ **c.** $y' = \sqrt{\theta} - \sec^2\theta$

In Exercises 39–42, find the function with the given derivative whose graph passes through the point P.

39. $f'(x) = 2x - 1, \quad P(0, 0)$

40. $g'(x) = \dfrac{1}{x^2} + 2x, \quad P(-1, 1)$

41. $f'(x) = e^{2x}, \quad P\left(0, \dfrac{3}{2}\right)$

42. $r'(t) = \sec t \tan t - 1, \quad P(0, 0)$

Finding Position from Velocity or Acceleration

Exercises 43–46 give the velocity $v = ds/dt$ and initial position of an object moving along a coordinate line. Find the object's position at time t.

43. $v = 9.8t + 5, \quad s(0) = 10$ **44.** $v = 32t - 2, \quad s(0.5) = 4$

45. $v = \sin \pi t, \quad s(0) = 0$ **46.** $v = \dfrac{2}{\pi}\cos\dfrac{2t}{\pi}, \quad s(\pi^2) = 1$

Exercises 47–50 give the acceleration $a = d^2s/dt^2$, initial velocity, and initial position of an object moving on a coordinate line. Find the object's position at time t.

47. $a = e^t, \quad v(0) = 20, \quad s(0) = 5$

48. $a = 9.8, \quad v(0) = -3, \quad s(0) = 0$

49. $a = -4\sin 2t, \quad v(0) = 2, \quad s(0) = -3$

50. $a = \dfrac{9}{\pi^2}\cos\dfrac{3t}{\pi}, \quad v(0) = 0, \quad s(0) = -1$

Applications

51. Temperature change It took 14 sec for a mercury thermometer to rise from $-19°C$ to $100°C$ when it was taken from a freezer and placed in boiling water. Show that somewhere along the way the mercury was rising at the rate of $8.5°C/\text{sec}$.

52. A trucker handed in a ticket at a toll booth showing that in 2 hours she had covered 159 mi on a toll road with speed limit 65 mph. The trucker was cited for speeding. Why?

53. Classical accounts tell us that a 170-oar trireme (ancient Greek or Roman warship) once covered 184 sea miles in 24 hours. Explain why at some point during this feat the trireme's speed exceeded 7.5 knots (sea or nautical miles per hour).

54. A marathoner ran the 26.2-mi New York City Marathon in 2.2 hours. Show that at least twice the marathoner was running at exactly 11 mph, assuming the initial and final speeds are zero.

55. Show that at some instant during a 2-hour automobile trip the car's speedometer reading will equal the average speed for the trip.

56. Free fall on the moon On our moon, the acceleration of gravity is 1.6 m/sec^2. If a rock is dropped into a crevasse, how fast will it be going just before it hits bottom 30 sec later?

Theory and Examples

57. The geometric mean of a and b The *geometric mean* of two positive numbers a and b is the number $\sqrt{ab}$. Show that the value of c in the conclusion of the Mean Value Theorem for $f(x) = 1/x$ on an interval of positive numbers $[a, b]$ is $c = \sqrt{ab}$.

58. The arithmetic mean of a and b The *arithmetic mean* of two numbers a and b is the number $(a + b)/2$. Show that the value of c in the conclusion of the Mean Value Theorem for $f(x) = x^2$ on any interval $[a, b]$ is $c = (a + b)/2$.

T **59.** Graph the function

$$f(x) = \sin x \sin (x + 2) - \sin^2 (x + 1).$$

What does the graph do? Why does the function behave this way? Give reasons for your answers.

60. Rolle's Theorem

  a. Construct a polynomial $f(x)$ that has zeros at $x = -2, -1, 0,$ 1, and 2.

  b. Graph f and its derivative f' together. How is what you see related to Rolle's Theorem?

  c. Do $g(x) = \sin x$ and its derivative g' illustrate the same phenomenon as f and f'?

61. Unique solution Assume that f is continuous on $[a, b]$ and differentiable on (a, b). Also assume that $f(a)$ and $f(b)$ have opposite signs and that $f' \neq 0$ between a and b. Show that $f(x) = 0$ exactly once between a and b.

62. Parallel tangents Assume that f and g are differentiable on $[a, b]$ and that $f(a) = g(a)$ and $f(b) = g(b)$. Show that there is at least one point between a and b where the tangents to the graphs of f and g are parallel or the same line. Illustrate with a sketch.

63. Suppose that $f'(x) \leq 1$ for $1 \leq x \leq 4$. Show that $f(4) - f(1) \leq 3$.

64. Suppose that $0 < f'(x) < 1/2$ for all x-values. Show that $f(-1) < f(1) < 2 + f(-1)$.

65. Show that $|\cos x - 1| \le |x|$ for all x-values. (*Hint:* Consider $f(t) = \cos t$ on $[0, x]$.)

66. Show that for any numbers a and b, the sine inequality $|\sin b - \sin a| \le |b - a|$ is true.

67. If the graphs of two differentiable functions $f(x)$ and $g(x)$ start at the same point in the plane and the functions have the same rate of change at every point, do the graphs have to be identical? Give reasons for your answer.

68. If $|f(w) - f(x)| \le |w - x|$ for all values w and x and f is a differentiable function, show that $-1 \le f'(x) \le 1$ for all x-values.

69. Assume that f is differentiable on $a \le x \le b$ and that $f(b) < f(a)$. Show that f' is negative at some point between a and b.

70. Let f be a function defined on an interval $[a, b]$. What conditions could you place on f to guarantee that

$$\min f' \le \frac{f(b) - f(a)}{b - a} \le \max f',$$

where $\min f'$ and $\max f'$ refer to the minimum and maximum values of f' on $[a, b]$? Give reasons for your answers.

T 71. Use the inequalities in Exercise 70 to estimate $f(0.1)$ if $f'(x) = 1/(1 + x^4 \cos x)$ for $0 \le x \le 0.1$ and $f(0) = 1$.

T 72. Use the inequalities in Exercise 70 to estimate $f(0.1)$ if $f'(x) = 1/(1 - x^4)$ for $0 \le x \le 0.1$ and $f(0) = 2$.

73. Let f be differentiable at every value of x and suppose that $f(1) = 1$, that $f' < 0$ on $(-\infty, 1)$, and that $f' > 0$ on $(1, \infty)$.
   a. Show that $f(x) \ge 1$ for all x.
   b. Must $f'(1) = 0$? Explain.

74. Let $f(x) = px^2 + qx + r$ be a quadratic function defined on a closed interval $[a, b]$. Show that there is exactly one point c in (a, b) at which f satisfies the conclusion of the Mean Value Theorem.

75. Use the same-derivative argument, as was done to prove the Product and Power Rules for logarithms, to prove the Quotient Rule property.

76. Use the same-derivative argument to prove the identities

   a. $\tan^{-1} x + \cot^{-1} x = \frac{\pi}{2}$ **b.** $\sec^{-1} x + \csc^{-1} x = \frac{\pi}{2}$

77. Starting with the equation $e^{x_1}e^{x_2} = e^{x_1 + x_2}$, derived in the text, show that $e^{-x} = 1/e^x$ for any real number x. Then show that $e^{x_1}/e^{x_2} = e^{x_1 - x_2}$ for any numbers x_1 and x_2.

78. Show that $(e^{x_1})^{x_2} = e^{x_1 x_2} = (e^{x_2})^{x_1}$ for any numbers x_1 and x_2.

4.3 Monotonic Functions and the First Derivative Test

In sketching the graph of a differentiable function, it is useful to know where it increases (rises from left to right) and where it decreases (falls from left to right) over an interval. This section gives a test to determine where it increases and where it decreases. We also show how to test the critical points of a function to identify whether local extreme values are present.

Increasing Functions and Decreasing Functions

As another corollary to the Mean Value Theorem, we show that functions with positive derivatives are increasing functions and functions with negative derivatives are decreasing functions. A function that is increasing or decreasing on an interval is said to be **monotonic** on the interval.

COROLLARY 3   Suppose that f is continuous on $[a, b]$ and differentiable on (a, b).

If $f'(x) > 0$ at each point $x \in (a, b)$, then f is increasing on $[a, b]$.
If $f'(x) < 0$ at each point $x \in (a, b)$, then f is decreasing on $[a, b]$.

Proof   Let x_1 and x_2 be any two points in $[a, b]$ with $x_1 < x_2$. The Mean Value Theorem applied to f on $[x_1, x_2]$ says that

$$f(x_2) - f(x_1) = f'(c)(x_2 - x_1)$$

for some c between x_1 and x_2. The sign of the right-hand side of this equation is the same as the sign of $f'(c)$ because $x_2 - x_1$ is positive. Therefore, $f(x_2) > f(x_1)$ if f' is positive on (a, b) and $f(x_2) < f(x_1)$ if f' is negative on (a, b). ∎

Corollary 3 tells us that $f(x) = \sqrt{x}$ is increasing on the interval $[0, b]$ for any $b > 0$ because $f'(x) = 1/2\sqrt{x}$ is positive on $(0, b)$. The derivative does not exist at $x = 0$, but Corollary 3 still applies. The corollary is valid for infinite as well as finite intervals, so $f(x) = \sqrt{x}$ is increasing on $[0, \infty)$.

To find the intervals where a function f is increasing or decreasing, we first find all of the critical points of f. If $a < b$ are two critical points for f, and if the derivative f' is continuous but never zero on the interval (a, b), then by the Intermediate Value Theorem applied to f', the derivative must be everywhere positive on (a, b), or everywhere negative there. One way we can determine the sign of f' on (a, b) is simply by evaluating the derivative at a single point c in (a, b). If $f'(c) > 0$, then $f'(x) > 0$ for all x in (a, b) so f is increasing on $[a, b]$ by Corollary 3; if $f'(c) < 0$, then f is decreasing on $[a, b]$. It doesn't matter which point c we choose in (a, b), since the sign of $f'(c)$ is the same for all choices. Usually we pick c to be a point where it is easy to evaluate $f'(c)$. The next example illustrates how we use this procedure.

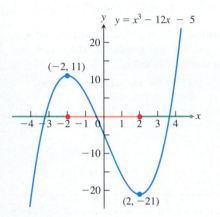

FIGURE 4.20 The function $f(x) = x^3 - 12x - 5$ is monotonic on three separate intervals (Example 1).

EXAMPLE 1 Find the critical points of $f(x) = x^3 - 12x - 5$ and identify the open intervals on which f is increasing and on which f is decreasing.

Solution The function f is everywhere continuous and differentiable. The first derivative

$$f'(x) = 3x^2 - 12 = 3(x^2 - 4)$$
$$= 3(x + 2)(x - 2)$$

is zero at $x = -2$ and $x = 2$. These critical points subdivide the domain of f to create nonoverlapping open intervals $(-\infty, -2)$, $(-2, 2)$, and $(2, \infty)$ on which f' is either positive or negative. We determine the sign of f' by evaluating f' at a convenient point in each subinterval. We evaluate f' at $x = -3$ in the first interval, $x = 0$ in the second interval and $x = 3$ in the third, since f' is relatively easy to compute at these points. The behavior of f is determined by then applying Corollary 3 to each subinterval. The results are summarized in the following table, and the graph of f is given in Figure 4.20.

Interval	$-\infty < x < -2$	$-2 < x < 2$	$2 < x < \infty$
f' evaluated	$f'(-3) = 15$	$f'(0) = -12$	$f'(3) = 15$
Sign of f'	$+$	$-$	$+$
Behavior of f	increasing	decreasing	increasing

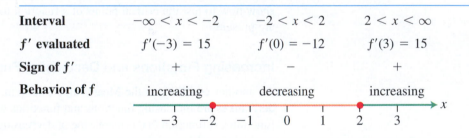

We used "strict" less-than inequalities to identify the intervals in the summary table for Example 1, since open intervals were specified. Corollary 3 says that we could use $\leq$ inequalities as well. That is, the function f in the example is increasing on $-\infty < x \leq -2$, decreasing on $-2 \leq x \leq 2$, and increasing on $2 \leq x < \infty$. We do not talk about whether a function is increasing or decreasing at a single point.

First Derivative Test for Local Extrema

In Figure 4.21, at the points where f has a minimum value, $f' < 0$ immediately to the left and $f' > 0$ immediately to the right. (If the point is an endpoint, there is only one side to consider.) Thus, the function is decreasing on the left of the minimum value and it is increasing on its right. Similarly, at the points where f has a maximum value, $f' > 0$ immediately to the left and $f' < 0$ immediately to the right. Thus, the function is increasing on the left of the maximum value and decreasing on its right. In summary, at a local extreme point, the sign of $f'(x)$ changes.

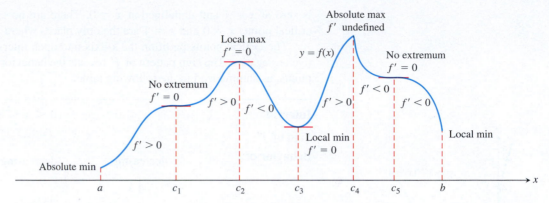

FIGURE 4.21 The critical points of a function locate where it is increasing and where it is decreasing. The first derivative changes sign at a critical point where a local extremum occurs.

These observations lead to a test for the presence and nature of local extreme values of differentiable functions.

First Derivative Test for Local Extrema

Suppose that c is a critical point of a continuous function f, and that f is differentiable at every point in some interval containing c except possibly at c itself. Moving across this interval from left to right,

1. if f' changes from negative to positive at c, then f has a local minimum at c;
2. if f' changes from positive to negative at c, then f has a local maximum at c;
3. if f' does not change sign at c (that is, f' is positive on both sides of c or negative on both sides), then f has no local extremum at c.

The test for local extrema at endpoints is similar, but there is only one side to consider in determining whether f is increasing or decreasing, based on the sign of f'.

Proof of the First Derivative Test   Part (1). Since the sign of f' changes from negative to positive at c, there are numbers a and b such that $a < c < b$, $f' < 0$ on (a, c), and $f' > 0$ on (c, b). If $x \in (a, c)$, then $f(c) < f(x)$ because $f' < 0$ implies that f is decreasing on $[a, c]$. If $x \in (c, b)$, then $f(c) < f(x)$ because $f' > 0$ implies that f is increasing on $[c, b]$. Therefore, $f(x) \geq f(c)$ for every $x \in (a, b)$. By definition, f has a local minimum at c.

Parts (2) and (3) are proved similarly.   ∎

EXAMPLE 2   Find the critical points of

$$f(x) = x^{1/3}(x - 4) = x^{4/3} - 4x^{1/3}.$$

Identify the open intervals on which f is increasing and decreasing. Find the function's local and absolute extreme values.

Solution   The function f is continuous at all x since it is the product of two continuous functions, $x^{1/3}$ and $(x - 4)$. The first derivative

$$f'(x) = \frac{d}{dx}(x^{4/3} - 4x^{1/3}) = \frac{4}{3}x^{1/3} - \frac{4}{3}x^{-2/3}$$

$$= \frac{4}{3}x^{-2/3}(x - 1) = \frac{4(x - 1)}{3x^{2/3}}$$

is zero at $x = 1$ and undefined at $x = 0$. There are no endpoints in the domain, so the critical points $x = 0$ and $x = 1$ are the only places where f might have an extreme value.

The critical points partition the x-axis into open intervals on which f' is either positive or negative. The sign pattern of f' reveals the behavior of f between and at the critical points, as summarized in the following table.

Interval	$x < 0$	$0 < x < 1$	$x > 1$
Sign of f'	$-$	$-$	$+$
Behavior of f	decreasing	decreasing	increasing

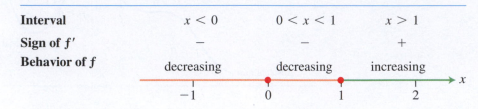

Corollary 3 to the Mean Value Theorem implies that f decreases on $(-\infty, 0)$, decreases on $(0, 1)$, and increases on $(1, \infty)$. The First Derivative Test for Local Extrema tells us that f does not have an extreme value at $x = 0$ (f' does not change sign) and that f has a local minimum at $x = 1$ (f' changes from negative to positive).

The value of the local minimum is $f(1) = 1^{1/3}(1 - 4) = -3$. This is also an absolute minimum since f is decreasing on $(-\infty, 1)$ and increasing on $(1, \infty)$. Figure 4.22 shows this value in relation to the function's graph.

Note that $\lim_{x \to 0} f'(x) = -\infty$, so the graph of f has a vertical tangent at the origin. ∎

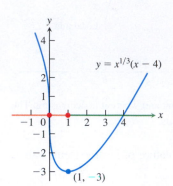

FIGURE 4.22 The function $f(x) = x^{1/3}(x - 4)$ decreases when $x < 1$ and increases when $x > 1$ (Example 2).

EXAMPLE 3 Find the critical points of

$$f(x) = (x^2 - 3)e^x.$$

Identify the open intervals on which f is increasing and decreasing. Find the function's local and absolute extreme values.

Solution The function f is continuous and differentiable for all real numbers, so the critical points occur only at the zeros of f'.

Using the Derivative Product Rule, we find the derivative

$$f'(x) = (x^2 - 3) \cdot \frac{d}{dx} e^x + \frac{d}{dx}(x^2 - 3) \cdot e^x$$

$$= (x^2 - 3)e^x + (2x)e^x$$

$$= (x^2 + 2x - 3)e^x.$$

Since e^x is never zero, the first derivative is zero if and only if

$$x^2 + 2x - 3 = 0$$

$$(x + 3)(x - 1) = 0.$$

The zeros $x = -3$ and $x = 1$ partition the x-axis into open intervals as follows.

Interval	$x < -3$	$-3 < x < 1$	$1 < x$
Sign of f'	$+$	$-$	$+$
Behavior of f	increasing	decreasing	increasing

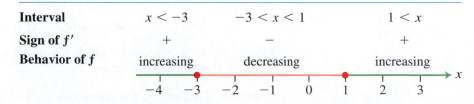

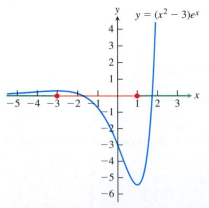

FIGURE 4.23 The graph of $f(x) = (x^2 - 3)e^x$ (Example 3).

We can see from the table that there is a local maximum (about 0.299) at $x = -3$ and a local minimum (about -5.437) at $x = 1$. The local minimum value is also an absolute minimum because $f(x) > 0$ for $|x| > \sqrt{3}$. There is no absolute maximum. The function increases on $(-\infty, -3)$ and $(1, \infty)$ and decreases on $(-3, 1)$. Figure 4.23 shows the graph. ∎

EXERCISES 4.3

Analyzing Functions from Derivatives

Answer the following questions about the functions whose derivatives are given in Exercises 1–14:

   a. What are the critical points of f?

   b. On what open intervals is f increasing or decreasing?

   c. At what points, if any, does f assume local maximum and minimum values?

1. $f'(x) = x(x - 1)$ **2.** $f'(x) = (x - 1)(x + 2)$

3. $f'(x) = (x - 1)^2(x + 2)$ **4.** $f'(x) = (x - 1)^2(x + 2)^2$

5. $f'(x) = (x - 1)e^{-x}$

6. $f'(x) = (x - 7)(x + 1)(x + 5)$

7. $f'(x) = \dfrac{x^2(x - 1)}{x + 2}, \quad x \neq -2$

8. $f'(x) = \dfrac{(x - 2)(x + 4)}{(x + 1)(x - 3)}, \quad x \neq -1, 3$

9. $f'(x) = 1 - \dfrac{4}{x^2}, \quad x \neq 0$ **10.** $f'(x) = 3 - \dfrac{6}{\sqrt{x}}, \quad x \neq 0$

11. $f'(x) = x^{-1/3}(x + 2)$ **12.** $f'(x) = x^{-1/2}(x - 3)$

13. $f'(x) = (\sin x - 1)(2 \cos x + 1), 0 \leq x \leq 2\pi$

14. $f'(x) = (\sin x + \cos x)(\sin x - \cos x), 0 \leq x \leq 2\pi$

Identifying Extrema

In Exercises 15–46:

   a. Find the open intervals on which the function is increasing and decreasing.

   b. Identify the function's local and absolute extreme values, if any, saying where they occur.

15.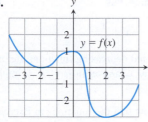

16.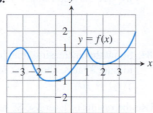

17.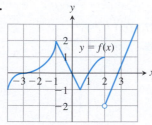

18.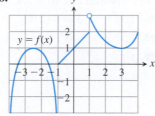

19. $g(t) = -t^2 - 3t + 3$ **20.** $g(t) = -3t^2 + 9t + 5$

21. $h(x) = -x^3 + 2x^2$ **22.** $h(x) = 2x^3 - 18x$

23. $f(\theta) = 3\theta^2 - 4\theta^3$ **24.** $f(\theta) = 6\theta - \theta^3$

25. $f(r) = 3r^3 + 16r$ **26.** $h(r) = (r + 7)^3$

27. $f(x) = x^4 - 8x^2 + 16$ **28.** $g(x) = x^4 - 4x^3 + 4x^2$

29. $H(t) = \dfrac{3}{2}t^4 - t^6$ **30.** $K(t) = 15t^3 - t^5$

31. $f(x) = x - 6\sqrt{x - 1}$ **32.** $g(x) = 4\sqrt{x} - x^2 + 3$

33. $g(x) = x\sqrt{8 - x^2}$ **34.** $g(x) = x^2\sqrt{5 - x}$

35. $f(x) = \dfrac{x^2 - 3}{x - 2}, \quad x \neq 2$ **36.** $f(x) = \dfrac{x^3}{3x^2 + 1}$

37. $f(x) = x^{1/3}(x + 8)$ **38.** $g(x) = x^{2/3}(x + 5)$

39. $h(x) = x^{1/3}(x^2 - 4)$ **40.** $k(x) = x^{2/3}(x^2 - 4)$

41. $f(x) = e^{2x} + e^{-x}$ **42.** $f(x) = e^{\sqrt{x}}$

43. $f(x) = x \ln x$ **44.** $f(x) = x^2 \ln x$

45. $g(x) = x(\ln x)^2$ **46.** $g(x) = x^2 - 2x - 4 \ln x$

In Exercises 47–58:

   a. Identify the function's local extreme values in the given domain, and say where they occur.

   b. Which of the extreme values, if any, are absolute?

   ☐T **c.** Support your findings with a graphing calculator or computer grapher.

47. $f(x) = 2x - x^2, \quad -\infty < x \leq 2$

48. $f(x) = (x + 1)^2, \quad -\infty < x \leq 0$

49. $g(x) = x^2 - 4x + 4, \quad 1 \leq x < \infty$

50. $g(x) = -x^2 - 6x - 9, \quad -4 \leq x < \infty$

51. $f(t) = 12t - t^3, \quad -3 \leq t < \infty$

52. $f(t) = t^3 - 3t^2, \quad -\infty < t \leq 3$

53. $h(x) = \dfrac{x^3}{3} - 2x^2 + 4x, \quad 0 \leq x < \infty$

54. $k(x) = x^3 + 3x^2 + 3x + 1, \quad -\infty < x \leq 0$

55. $f(x) = \sqrt{25 - x^2}, \quad -5 \leq x \leq 5$

56. $f(x) = \sqrt{x^2 - 2x - 3}, \quad 3 \leq x < \infty$

57. $g(x) = \dfrac{x - 2}{x^2 - 1}, \quad 0 \leq x < 1$

58. $g(x) = \dfrac{x^2}{4 - x^2}, \quad -2 < x \leq 1$

In Exercises 59–66:

   a. Find the local extrema of each function on the given interval, and say where they occur.

   ☐T **b.** Graph the function and its derivative together. Comment on the behavior of f in relation to the signs and values of f'.

59. $f(x) = \sin 2x, \quad 0 \leq x \leq \pi$

60. $f(x) = \sin x - \cos x, \quad 0 \leq x \leq 2\pi$

61. $f(x) = \sqrt{3} \cos x + \sin x, \quad 0 \leq x \leq 2\pi$

62. $f(x) = -2x + \tan x, \quad \dfrac{-\pi}{2} < x < \dfrac{\pi}{2}$

63. $f(x) = \dfrac{x}{2} - 2 \sin \dfrac{x}{2}, \quad 0 \leq x \leq 2\pi$

64. $f(x) = -2 \cos x - \cos^2 x, \quad -\pi \leq x \leq \pi$

65. $f(x) = \csc^2 x - 2 \cot x, \quad 0 < x < \pi$

66. $f(x) = \sec^2 x - 2 \tan x, \quad \dfrac{-\pi}{2} < x < \dfrac{\pi}{2}$

In Exercises 67 and 68, the graph of f' is given. Assume that f is continuous and determine the x-values corresponding to local minima and local maxima.

67.   **68.**

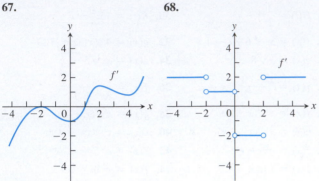

Theory and Examples

Show that the functions in Exercises 69 and 70 have local extreme values at the given values of θ, and say which kind of local extreme the function has.

69. $h(\theta) = 3 \cos \dfrac{\theta}{2}$, $0 \le \theta \le 2\pi$, at $\theta = 0$ and $\theta = 2\pi$

70. $h(\theta) = 5 \sin \dfrac{\theta}{2}$, $0 \le \theta \le \pi$, at $\theta = 0$ and $\theta = \pi$

71. Sketch the graph of a differentiable function $y = f(x)$ through the point $(1, 1)$ if $f'(1) = 0$ and

   a. $f'(x) > 0$ for $x < 1$ and $f'(x) < 0$ for $x > 1$;

   b. $f'(x) < 0$ for $x < 1$ and $f'(x) > 0$ for $x > 1$;

   c. $f'(x) > 0$ for $x \ne 1$;

   d. $f'(x) < 0$ for $x \ne 1$.

72. Sketch the graph of a differentiable function $y = f(x)$ that has

   a. a local minimum at $(1, 1)$ and a local maximum at $(3, 3)$;

   b. a local maximum at $(1, 1)$ and a local minimum at $(3, 3)$;

   c. local maxima at $(1, 1)$ and $(3, 3)$;

   d. local minima at $(1, 1)$ and $(3, 3)$.

73. Sketch the graph of a continuous function $y = g(x)$ such that

   a. $g(2) = 2, 0 < g' < 1$ for $x < 2, g'(x) \to 1^-$ as $x \to 2^-$, $-1 < g' < 0$ for $x > 2$, and $g'(x) \to -1^+$ as $x \to 2^+$;

   b. $g(2) = 2, g' < 0$ for $x < 2, g'(x) \to -\infty$ as $x \to 2^-$, $g' > 0$ for $x > 2$, and $g'(x) \to \infty$ as $x \to 2^+$.

74. Sketch the graph of a continuous function $y = h(x)$ such that

   a. $h(0) = 0, -2 \le h(x) \le 2$ for all $x, h'(x) \to \infty$ as $x \to 0^-$, and $h'(x) \to \infty$ as $x \to 0^+$;

   b. $h(0) = 0, -2 \le h(x) \le 0$ for all $x, h'(x) \to \infty$ as $x \to 0^-$, and $h'(x) \to -\infty$ as $x \to 0^+$.

75. Discuss the extreme-value behavior of the function $f(x) = x \sin (1/x)$, $x \ne 0$. How many critical points does this function have? Where are they located on the x-axis? Does f have an absolute minimum? An absolute maximum? (See Exercise 49 in Section 2.3.)

76. Find the open intervals on which the function $f(x) = ax^2 + bx + c$, $a \ne 0$, is increasing and decreasing. Describe the reasoning behind your answer.

77. Determine the values of constants a and b so that $f(x) = ax^2 + bx$ has an absolute maximum at the point $(1, 2)$.

78. Determine the values of constants a, b, c, and d so that $f(x) = ax^3 + bx^2 + cx + d$ has a local maximum at the point $(0, 0)$ and a local minimum at the point $(1, -1)$.

79. Locate and identify the absolute extreme values of

   a. $\ln (\cos x)$ on $\left[-\pi/4, \pi/3 \right]$,

   b. $\cos (\ln x)$ on $\left[1/2, 2 \right]$.

80. a. Prove that $f(x) = x - \ln x$ is increasing for $x > 1$.

   b. Using part (a), show that $\ln x < x$ if $x > 1$.

81. Find the absolute maximum and minimum values of $f(x) = e^x - 2x$ on $\left[0, 1 \right]$.

82. Where does the periodic function $f(x) = 2e^{\sin (x/2)}$ take on its extreme values and what are these values?

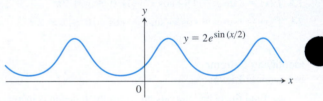

83. Find the absolute maximum value of $f(x) = x^2 \ln (1/x)$ and say where it occurs.

84. a. Prove that $e^x \ge 1 + x$ if $x \ge 0$.

   b. Use the result in part (a) to show that

$$e^x \ge 1 + x + \frac{1}{2}x^2.$$

85. Show that increasing functions and decreasing functions are one-to-one. That is, show that for any x_1 and x_2 in I, $x_2 \ne x_1$ implies $f(x_2) \ne f(x_1)$.

Use the results of Exercise 85 to show that the functions in Exercises 86–90 have inverses over their domains. Find a formula for df^{-1}/dx using Theorem 3, Section 3.8.

86. $f(x) = (1/3)x + (5/6)$ **87.** $f(x) = 27x^3$

88. $f(x) = 1 - 8x^3$ **89.** $f(x) = (1 - x)^3$

90. $f(x) = x^{5/3}$

4.4 Concavity and Curve Sketching

We have seen how the first derivative tells us where a function is increasing, where it is decreasing, and whether a local maximum or local minimum occurs at a critical point. In this section we see that the second derivative gives us information about how the graph of a differentiable function bends or turns. With this knowledge about the first and second derivatives, coupled with our previous understanding of symmetry and asymptotic behavior studied in Sections 1.1 and 2.6, we can now draw an accurate graph of a function. By organizing all of these ideas into a coherent procedure, we give a method for sketching graphs and revealing visually the key features of functions. Identifying and knowing the locations of these features is of major importance in mathematics and its applications to science and engineering, especially in the graphical analysis and interpretation of data. When the domain of a function is not a finite closed interval, sketching a graph helps to determine whether absolute maxima or absolute minima exist and, if they do exist, where they are located.

Concavity

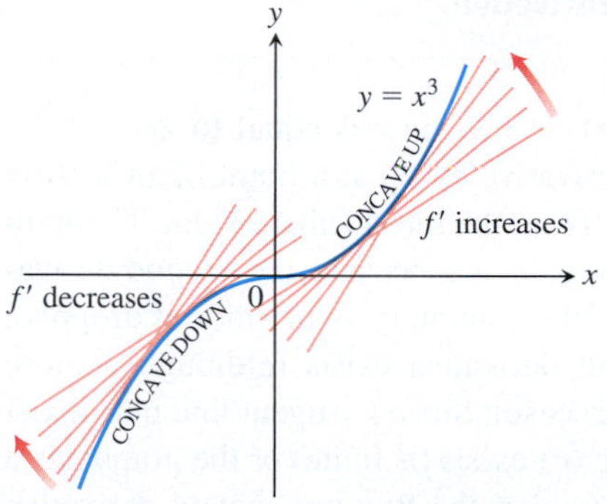

As you can see in Figure 4.24, the curve $y = x^3$ rises as x increases, but the portions defined on the intervals $(-\infty, 0)$ and $(0, \infty)$ turn in different ways. As we approach the origin from the left along the curve, the curve turns to our right and falls below its tangents. The slopes of the tangents are decreasing on the interval $(-\infty, 0)$. As we move away from the origin along the curve to the right, the curve turns to our left and rises above its tangents. The slopes of the tangents are increasing on the interval $(0, \infty)$. This turning or bending behavior defines the *concavity* of the curve.

FIGURE 4.24 The graph of $f(x) = x^3$ is concave down on $(-\infty, 0)$ and concave up on $(0, \infty)$ (Example 1a).

DEFINITION The graph of a differentiable function $y = f(x)$ is

(a) concave up on an open interval I if f' is increasing on I;

(b) concave down on an open interval I if f' is decreasing on I.

A function whose graph is concave up is also often called **convex**.

If $y = f(x)$ has a second derivative, we can apply Corollary 3 of the Mean Value Theorem to the first derivative function. We conclude that f' increases if $f'' > 0$ on I, and decreases if $f'' < 0$.

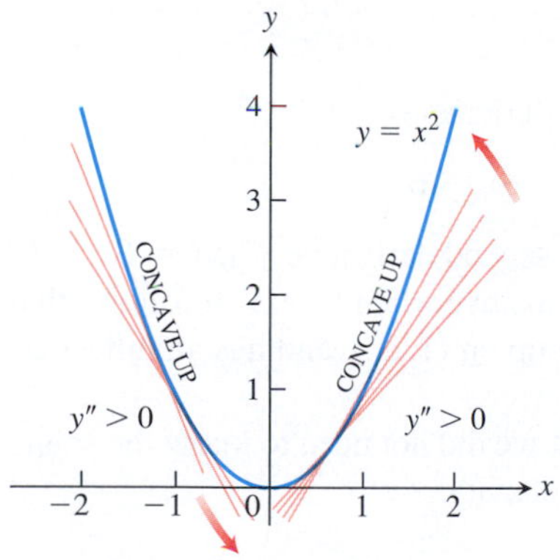

The Second Derivative Test for Concavity
Let $y = f(x)$ be twice-differentiable on an interval I.

1. If $f'' > 0$ on I, the graph of f over I is concave up.

2. If $f'' < 0$ on I, the graph of f over I is concave down.

FIGURE 4.25 The graph of $f(x) = x^2$ is concave up on every interval (Example 1b).

If $y = f(x)$ is twice-differentiable, we will use the notations f'' and y'' interchangeably when denoting the second derivative.

EXAMPLE 1

(a) The curve $y = x^3$ (Figure 4.24) is concave down on $(-\infty, 0)$, where $y'' = 6x < 0$, and concave up on $(0, \infty)$, where $y'' = 6x > 0$.

(b) The curve $y = x^2$ (Figure 4.25) is concave up on $(-\infty, \infty)$ because its second derivative $y'' = 2$ is always positive. ∎

EXAMPLE 2

Determine the concavity of $y = 3 + \sin x$ on $[0, 2\pi]$.

Solution The first derivative of $y = 3 + \sin x$ is $y' = \cos x$, and the second derivative is $y'' = -\sin x$. The graph of $y = 3 + \sin x$ is concave down on $(0, \pi)$, where $y'' = -\sin x$ is negative. It is concave up on $(\pi, 2\pi)$, where $y'' = -\sin x$ is positive (Figure 4.26). ∎

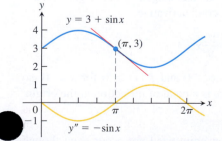

FIGURE 4.26 Using the sign of y'' to determine the concavity of y (Example 2).

Points of Inflection

The curve $y = 3 + \sin x$ in Example 2 changes concavity at the point $(\pi, 3)$. Since the first derivative $y' = \cos x$ exists for all x, we see that the curve has a tangent line of slope -1 at the point $(\pi, 3)$. This point is called a *point of inflection* of the curve. Notice from Figure 4.26 that the graph crosses its tangent line at this point and that the second derivative $y'' = -\sin x$ has value 0 when $x = \pi$. In general, we have the following definition.

> **DEFINITION** A point $(c, f(c))$ where the graph of a function has a tangent line and where the concavity changes is a **point of inflection**.

We observed that the second derivative of $f(x) = 3 + \sin x$ is equal to zero at the inflection point $(\pi, 3)$. Generally, if the second derivative exists at a point of inflection $(c, f(c))$, then $f''(c) = 0$. This follows immediately from the Intermediate Value Theorem whenever f'' is continuous over an interval containing $x = c$ because the second derivative changes sign moving across this interval. Even if the continuity assumption is dropped, it is still true that $f''(c) = 0$, provided the second derivative exists (although a more advanced argument is required in this noncontinuous case). Since a tangent line must exist at the point of inflection, either the first derivative $f'(c)$ exists (is finite) or the graph has a vertical tangent at the point. At a vertical tangent neither the first nor second derivative exists. In summary, one of two things can happen at a point of inflection.

> At a point of inflection $(c, f(c))$, either $f''(c) = 0$ or $f''(c)$ fails to exist.

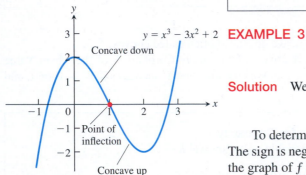

$y = x^3 - 3x^2 + 2$

FIGURE 4.27 The concavity of the graph of f changes from concave down to concave up at the inflection point.

EXAMPLE 3 Determine the concavity and find the inflection points of the function

$$f(x) = x^3 - 3x^2 + 2.$$

Solution We start by computing the first and second derivatives.

$$f'(x) = 3x^2 - 6x, \qquad f''(x) = 6x - 6.$$

To determine concavity, we look at the sign of the second derivative $f''(x) = 6x - 6$. The sign is negative when $x < 1$, is 0 at $x = 1$, and is positive when $x > 1$. It follows that the graph of f is concave down on $(-\infty, 1)$, is concave up on $(1, \infty)$, and has an inflection point at the point $(1, 0)$ where the concavity changes.

The graph of f is shown in Figure 4.27. Notice that we did not need to know the shape of this graph ahead of time in order to determine its concavity. ∎

The next example illustrates that a function can have a point of inflection where the first derivative exists but the second derivative fails to exist.

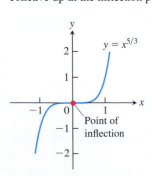

$y = x^{5/3}$

FIGURE 4.28 The graph of $f(x) = x^{5/3}$ has a horizontal tangent at the origin where the concavity changes, although f'' does not exist at $x = 0$ (Example 4).

EXAMPLE 4 The graph of $f(x) = x^{5/3}$ has a horizontal tangent at the origin because $f'(x) = (5/3)x^{2/3} = 0$ when $x = 0$. However, the second derivative

$$f''(x) = \frac{d}{dx}\left(\frac{5}{3}x^{2/3}\right) = \frac{10}{9}x^{-1/3}$$

fails to exist at $x = 0$. Nevertheless, $f''(x) < 0$ for $x < 0$ and $f''(x) > 0$ for $x > 0$, so the second derivative changes sign at $x = 0$ and there is a point of inflection at the origin. The graph is shown in Figure 4.28. ∎

The following example shows that an inflection point need not occur even though both derivatives exist and $f'' = 0$.

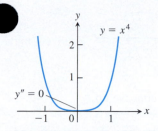

FIGURE 4.29 The graph of $y = x^4$ has no inflection point at the origin, even though $y'' = 0$ there (Example 5).

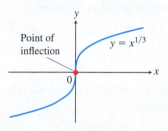

FIGURE 4.30 A point of inflection where y' and y'' fail to exist (Example 6).

EXAMPLE 5 The curve $y = x^4$ has no inflection point at $x = 0$ (Figure 4.29). Even though the second derivative $y'' = 12x^2$ is zero there, it does not change sign. The curve is concave up everywhere. ∎

In the next example a point of inflection occurs at a vertical tangent to the curve where neither the first nor the second derivative exists.

EXAMPLE 6 The graph of $y = x^{1/3}$ has a point of inflection at the origin because the second derivative is positive for $x < 0$ and negative for $x > 0$:

$$y'' = \frac{d^2}{dx^2}\left(x^{1/3}\right) = \frac{d}{dx}\left(\frac{1}{3}x^{-2/3}\right) = -\frac{2}{9}x^{-5/3}.$$

However, both $y' = x^{-2/3}/3$ and y'' fail to exist at $x = 0$, and there is a vertical tangent there. See Figure 4.30. ∎

Caution Example 4 in Section 4.1 (Figure 4.9) shows that the function $f(x) = x^{2/3}$ does not have a second derivative at $x = 0$ and does not have a point of inflection there (there is no change in concavity at $x = 0$). Combined with the behavior of the function in Example 6 above, we see that when the second derivative does not exist at $x = c$, an inflection point may or may not occur there. So we need to be careful about interpreting functional behavior whenever first or second derivatives fail to exist at a point. At such points the graph can have vertical tangents, corners, cusps, or various discontinuities. ●

To study the motion of an object moving along a line as a function of time, we often are interested in knowing when the object's acceleration, given by the second derivative, is positive or negative. The points of inflection on the graph of the object's position function reveal where the acceleration changes sign.

EXAMPLE 7 A particle is moving along a horizontal coordinate line (positive to the right) with position function

$$s(t) = 2t^3 - 14t^2 + 22t - 5, \qquad t \geq 0.$$

Find the velocity and acceleration, and describe the motion of the particle.

Solution The velocity is

$$v(t) = s'(t) = 6t^2 - 28t + 22 = 2(t - 1)(3t - 11),$$

and the acceleration is

$$a(t) = v'(t) = s''(t) = 12t - 28 = 4(3t - 7).$$

When the function $s(t)$ is increasing, the particle is moving to the right; when $s(t)$ is decreasing, the particle is moving to the left.

Notice that the first derivative ($v = s'$) is zero at the critical points $t = 1$ and $t = 11/3$.

Interval	$0 < t < 1$	$1 < t < 11/3$	$11/3 < t$
Sign of $v = s'$	+	−	+
Behavior of s	increasing	decreasing	increasing
Particle motion	right	left	right

The particle is moving to the right in the time intervals $[0, 1)$ and $(11/3, \infty)$, and moving to the left in $(1, 11/3)$. It is momentarily stationary (at rest) at $t = 1$ and $t = 11/3$.

The acceleration $a(t) = s''(t) = 4(3t - 7)$ is zero when $t = 7/3$.

Interval	$0 < t < 7/3$	$7/3 < t$
Sign of $a = s''$	$-$	$+$
Graph of s	concave down	concave up

The particle starts out moving to the right while slowing down, and then reverses and begins moving to the left at $t = 1$ under the influence of the leftward acceleration over the time interval $[0, 7/3]$. The acceleration then changes direction at $t = 7/3$ but the particle continues moving leftward, while slowing down under the rightward acceleration. At $t = 11/3$ the particle reverses direction again: moving to the right in the same direction as the acceleration, so it is speeding up. ∎

Second Derivative Test for Local Extrema

Instead of looking for sign changes in f' at critical points, we can sometimes use the following test to determine the presence and nature of local extrema.

THEOREM 5—Second Derivative Test for Local Extrema
Suppose f'' is continuous on an open interval that contains $x = c$.

1. If $f'(c) = 0$ and $f''(c) < 0$, then f has a local maximum at $x = c$.
2. If $f'(c) = 0$ and $f''(c) > 0$, then f has a local minimum at $x = c$.
3. If $f'(c) = 0$ and $f''(c) = 0$, then the test fails. The function f may have a local maximum, a local minimum, or neither.

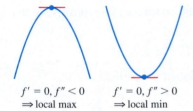

$f' = 0, f'' < 0$ $f' = 0, f'' > 0$
$\Rightarrow$ local max $\Rightarrow$ local min

Proof   Part (1). If $f''(c) < 0$, then $f''(x) < 0$ on some open interval I containing the point c, since f'' is continuous. Therefore, f' is decreasing on I. Since $f'(c) = 0$, the sign of f' changes from positive to negative at c so f has a local maximum at c by the First Derivative Test.

The proof of Part (2) is similar.

For Part (3), consider the three functions $y = x^4$, $y = -x^4$, and $y = x^3$. For each function, the first and second derivatives are zero at $x = 0$. Yet the function $y = x^4$ has a local minimum there, $y = -x^4$ has a local maximum, and $y = x^3$ is increasing in any open interval containing $x = 0$ (having neither a maximum nor a minimum there). Thus the test fails. ∎

This test requires us to know f'' *only at c itself* and not in an interval about c. This makes the test easy to apply. That's the good news. The bad news is that the test is inconclusive if $f'' = 0$ or if f'' does not exist at $x = c$. When this happens, use the First Derivative Test for local extreme values.

Together f' and f'' tell us the shape of the function's graph—that is, where the critical points are located and what happens at a critical point, where the function is increasing and where it is decreasing, and how the curve is turning or bending as defined by its concavity. We use this information to sketch a graph of the function that captures its key features.

EXAMPLE 8   Sketch a graph of the function

$$f(x) = x^4 - 4x^3 + 10$$

using the following steps.

(a) Identify where the extrema of f occur.

(b) Find the intervals on which f is increasing and the intervals on which f is decreasing.

(c) Find where the graph of f is concave up and where it is concave down.

(d) Sketch the general shape of the graph for f.

(e) Plot some specific points, such as local maximum and minimum points, points of inflection, and intercepts. Then sketch the curve.

Solution   The function f is continuous since $f'(x) = 4x^3 - 12x^2$ exists. The domain of f is $(-\infty, \infty)$, and the domain of f' is also $(-\infty, \infty)$. Thus, the critical points of f occur only at the zeros of f'. Since

$$f'(x) = 4x^3 - 12x^2 = 4x^2(x - 3),$$

the first derivative is zero at $x = 0$ and $x = 3$. We use these critical points to define intervals where f is increasing or decreasing.

Interval	$x < 0$	$0 < x < 3$	$3 < x$
Sign of f'	$-$	$-$	$+$
Behavior of f	decreasing	decreasing	increasing

(a) Using the First Derivative Test for local extrema and the table above, we see that there is no extremum at $x = 0$ and a local minimum at $x = 3$.

(b) Using the table above, we see that f is decreasing on $(-\infty, 0]$ and $[0, 3]$, and increasing on $[3, \infty)$.

(c) $f''(x) = 12x^2 - 24x = 12x(x - 2)$ is zero at $x = 0$ and $x = 2$. We use these points to define intervals where f is concave up or concave down.

Interval	$x < 0$	$0 < x < 2$	$2 < x$
Sign of f''	$+$	$-$	$+$
Behavior of f	concave up	concave down	concave up

We see that f is concave up on the intervals $(-\infty, 0)$ and $(2, \infty)$, and concave down on $(0, 2)$.

(d) Summarizing the information in the last two tables, we obtain the following.

$x < 0$	$0 < x < 2$	$2 < x < 3$	$3 < x$
decreasing	decreasing	decreasing	increasing
concave up	concave down	concave up	concave up

The general shape of the curve is shown in the accompanying figure.

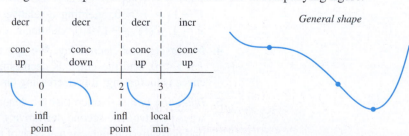

(e) Plot the curve's intercepts (if possible) and the points where y' and y'' are zero. Indicate any local extreme values and inflection points. Use the general shape as a guide to sketch the curve. (Plot additional points as needed.) Figure 4.31 shows the graph of f. ■

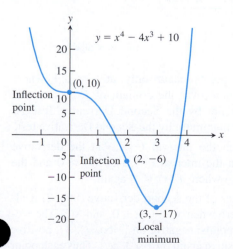

FIGURE 4.31   The graph of $f(x) = x^4 - 4x^3 + 10$ (Example 8).

The steps in Example 8 give a procedure for graphing the key features of a function. Asymptotes were defined and discussed in Section 2.6. We can find them for rational functions, and the methods in the next section give tools to help find them for more general functions.

Procedure for Graphing $y = f(x)$

1. Identify the domain of f and any symmetries the curve may have.
2. Find the derivatives y' and y''.
3. Find the critical points of f, if any, and identify the function's behavior at each one.
4. Find where the curve is increasing and where it is decreasing.
5. Find the points of inflection, if any occur, and determine the concavity of the curve.
6. Identify any asymptotes that may exist.
7. Plot key points, such as the intercepts and the points found in Steps 3–5, and sketch the curve together with any asymptotes that exist.

EXAMPLE 9   Sketch the graph of $f(x) = \dfrac{(x + 1)^2}{1 + x^2}$.

Solution

1. The domain of f is $(-\infty, \infty)$ and there are no symmetries about either axis or the origin (Section 1.1).

2. *Find f' and f''.*

$$f(x) = \frac{(x + 1)^2}{1 + x^2}$$

x-intercept at $x = -1$,
y-intercept at $y = 1$

$$f'(x) = \frac{(1 + x^2) \cdot 2(x + 1) - (x + 1)^2 \cdot 2x}{(1 + x^2)^2}$$

$$= \frac{2(1 - x^2)}{(1 + x^2)^2}$$

Critical points: $x = -1, x = 1$

$$f''(x) = \frac{(1 + x^2)^2 \cdot 2(-2x) - 2(1 - x^2)[2(1 + x^2) \cdot 2x]}{(1 + x^2)^4}$$

$$= \frac{4x(x^2 - 3)}{(1 + x^2)^3}$$

After some algebra

3. *Behavior at critical points.* The critical points occur only at $x = \pm 1$ where $f'(x) = 0$ (Step 2) since f' exists everywhere over the domain of f. At $x = -1$, $f''(-1) = 1 > 0$, yielding a relative minimum by the Second Derivative Test. At $x = 1$, $f''(1) = -1 < 0$, yielding a relative maximum by the Second Derivative test.

4. *Increasing and decreasing.* We see that on the interval $(-\infty, -1)$ the derivative $f'(x) < 0$, and the curve is decreasing. On the interval $(-1, 1)$, $f'(x) > 0$ and the curve is increasing; it is decreasing on $(1, \infty)$ where $f'(x) < 0$ again.

5. *Inflection points.* Notice that the denominator of the second derivative (Step 2) is always positive. The second derivative f'' is zero when $x = -\sqrt{3}, 0$, and $\sqrt{3}$. The second derivative changes sign at each of these points: negative on $\left(-\infty, -\sqrt{3}\right)$, positive on $\left(-\sqrt{3}, 0\right)$, negative on $\left(0, \sqrt{3}\right)$, and positive again on $\left(\sqrt{3}, \infty\right)$. Thus each point is a point of inflection. The curve is concave down on the interval $\left(-\infty, -\sqrt{3}\right)$, concave up on $\left(-\sqrt{3}, 0\right)$, concave down on $\left(0, \sqrt{3}\right)$, and concave up again on $\left(\sqrt{3}, \infty\right)$.

6. *Asymptotes.* Expanding the numerator of $f(x)$ and then dividing both numerator and denominator by x^2 gives

$$f(x) = \frac{(x+1)^2}{1+x^2} = \frac{x^2+2x+1}{1+x^2} \qquad \text{Expanding numerator}$$

$$= \frac{1+(2/x)+(1/x^2)}{(1/x^2)+1}. \qquad \text{Dividing by } x^2$$

We see that $f(x) \to 1^+$ as $x \to \infty$ and that $f(x) \to 1^-$ as $x \to -\infty$. Thus, the line $y=1$ is a horizontal asymptote.

Since f decreases on $(-\infty, -1)$ and then increases on $(-1, 1)$, we know that $f(-1) = 0$ is a local minimum. Although f decreases on $(1, \infty)$, it never crosses the horizontal asymptote $y=1$ on that interval (it approaches the asymptote from above). So the graph never becomes negative, and $f(-1) = 0$ is an absolute minimum as well. Likewise, $f(1) = 2$ is an absolute maximum because the graph never crosses the asymptote $y = 1$ on the interval $(-\infty, -1)$, approaching it from below. Therefore, there are no vertical asymptotes (the range of f is $0 \le y \le 2$).

7. The graph of f is sketched in Figure 4.32. Notice how the graph is concave down as it approaches the horizontal asymptote $y=1$ as $x \to -\infty$, and concave up in its approach to $y=1$ as $x \to \infty$. ∎

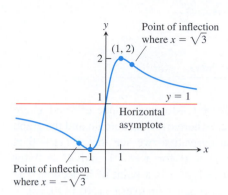

FIGURE 4.32 The graph of $y = \dfrac{(x+1)^2}{1+x^2}$ (Example 9).

EXAMPLE 10 Sketch the graph of $f(x) = \dfrac{x^2+4}{2x}$.

Solution

1. The domain of f is all nonzero real numbers. There are no intercepts because neither x nor $f(x)$ can be zero. Since $f(-x) = -f(x)$, we note that f is an odd function, so the graph of f is symmetric about the origin.

2. We calculate the derivatives of the function, but first rewrite it in order to simplify our computations:

$$f(x) = \frac{x^2+4}{2x} = \frac{x}{2} + \frac{2}{x} \qquad \text{Function simplified for differentiation}$$

$$f'(x) = \frac{1}{2} - \frac{2}{x^2} = \frac{x^2-4}{2x^2} \qquad \text{Combine fractions to solve easily } f'(x) = 0.$$

$$f''(x) = \frac{4}{x^3} \qquad \text{Exists throughout the entire domain of } f$$

3. The critical points occur at $x = \pm 2$ where $f'(x) = 0$. Since $f''(-2) < 0$ and $f''(2) > 0$, we see from the Second Derivative Test that a relative maximum occurs at $x = -2$ with $f(-2) = -2$, and a relative minimum occurs at $x = 2$ with $f(2) = 2$.

4. On the interval $(-\infty, -2)$ the derivative f' is positive because $x^2 - 4 > 0$ so the graph is increasing; on the interval $(-2, 0)$ the derivative is negative and the graph is decreasing. Similarly, the graph is decreasing on the interval $(0, 2)$ and increasing on $(2, \infty)$.

5. There are no points of inflection because $f''(x) < 0$ whenever $x < 0$, $f''(x) > 0$ whenever $x > 0$, and f'' exists everywhere and is never zero throughout the domain of f. The graph is concave down on the interval $(-\infty, 0)$ and concave up on the interval $(0, \infty)$.

6. From the rewritten formula for $f(x)$, we see that

$$\lim_{x \to 0^+} \left(\frac{x}{2} + \frac{2}{x} \right) = +\infty \quad \text{and} \quad \lim_{x \to 0^-} \left(\frac{x}{2} + \frac{2}{x} \right) = -\infty,$$

so the y-axis is a vertical asymptote. Also, as $x \to \infty$ or as $x \to -\infty$, the graph of $f(x)$ approaches the line $y = x/2$. Thus $y = x/2$ is an oblique asymptote.

7. The graph of f is sketched in Figure 4.33. ∎

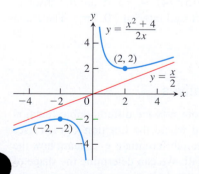

FIGURE 4.33 The graph of $y = \dfrac{x^2+4}{2x}$ (Example 10).

EXAMPLE 11 Sketch the graph of $f(x) = e^{2/x}$.

Solution The domain of f is $(-\infty, 0) \cup (0, \infty)$ and there are no symmetries about either axis or the origin. The derivatives of f are

$$f'(x) = e^{2/x}\left(-\frac{2}{x^2}\right) = -\frac{2e^{2/x}}{x^2}$$

and

$$f''(x) = -\frac{x^2(2e^{2/x})(-2/x^2) - 2e^{2/x}(2x)}{x^4} = \frac{4e^{2/x}(1 + x)}{x^4}.$$

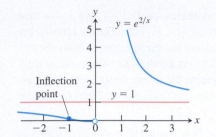

Both derivatives exist everywhere over the domain of f. Moreover, since $e^{2/x}$ and x^2 are both positive for all $x \neq 0$, we see that $f' < 0$ everywhere over the domain and the graph is everywhere decreasing. Examining the second derivative, we see that $f''(x) = 0$ at $x = -1$. Since $e^{2/x} > 0$ and $x^4 > 0$, we have $f'' < 0$ for $x < -1$ and $f'' > 0$ for $x > -1, x \neq 0$. Since f'' changes sign, the point $(-1, e^{-2})$ is a point of inflection. The curve is concave down on the interval $(-\infty, -1)$ and concave up over $(-1, 0) \cup (0, \infty)$.

FIGURE 4.34 The graph of $y = e^{2/x}$ has a point of inflection at $(-1, e^{-2})$. The line $y = 1$ is a horizontal asymptote and $x = 0$ is a vertical asymptote (Example 11).

From Example 7, Section 2.6, we see that $\lim_{x \to 0^-} f(x) = 0$. As $x \to 0^+$, we see that $2/x \to \infty$, so $\lim_{x \to 0^+} f(x) = \infty$ and the y-axis is a vertical asymptote. Also, as $x \to -\infty$ or $x \to \infty$, $2/x \to 0$ and so $\lim_{x \to -\infty} f(x) = \lim_{x \to \infty} f(x) = e^0 = 1$. Therefore, $y = 1$ is a horizontal asymptote. There are no absolute extrema, since f never takes on the value 0 and has no absolute maximum. The graph of f is sketched in Figure 4.34. ∎

EXAMPLE 12 Sketch the graph of $f(x) = \cos x - \dfrac{\sqrt{2}}{2}x$ over $0 \leq x \leq 2\pi$.

Solution The derivatives of f are

$$f'(x) = -\sin x - \frac{\sqrt{2}}{2} \quad \text{and} \quad f''(x) = -\cos x.$$

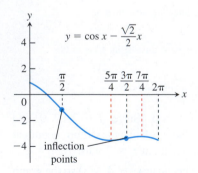

Both derivatives exist everywhere over the interval $(0, 2\pi)$. Within that open interval, the first derivative is zero when $\sin x = -\sqrt{2}/2$, so the critical points are $x = 5\pi/4$ and $x = 7\pi/4$. Since $f''(5\pi/4) = -\cos(5\pi/4) = \sqrt{2}/2 > 0$, the function has a local minimum value of $f(5\pi/4) \approx -3.48$ (evaluated with a calculator) by the Second Derivative Test. Also, $f''(7\pi/4) = -\cos(7\pi/4) = -\sqrt{2}/2 < 0$, so the function has a local maximum value of $f(7\pi/4) \approx -3.18$.

Examining the second derivative, we find that $f'' = 0$ when $x = \pi/2$ or $x = 3\pi/2$. We conclude that $(\pi/2, f(\pi/2)) \approx (\pi/2, -1.11)$ and $(3\pi/2, f(3\pi/2)) \approx (3\pi/2, -3.33)$ are points of inflection.

FIGURE 4.35 The graph of the function in Example 12.

Finally, we evaluate f at the endpoints of the interval to find $f(0) = 1$ and $f(2\pi) \approx -3.44$. Therefore, the values $f(0) = 1$ and $f(5\pi/4) \approx -3.48$ are the absolute maximum and absolute minimum values of f over the closed interval $[0, 2\pi]$. The graph of f is sketched in Figure 4.35. ∎

Graphical Behavior of Functions from Derivatives

As we saw in Examples 8–12, we can learn much about a twice-differentiable function $y = f(x)$ by examining its first derivative. We can find where the function's graph rises and falls and where any local extrema are located. We can differentiate y' to learn how the graph bends as it passes over the intervals of rise and fall. We can determine the shape of the function's graph. Information we cannot get from the derivative is how to place the graph in the xy-plane. But, as we discovered in Section 4.2, the only additional information we need to position the graph is the value of f at one point. Information about the asymptotes is found using limits (Section 2.6). The following figure summarizes how the first derivative and second derivative affect the shape of a graph.

$y = f(x)$	$y = f(x)$	$y = f(x)$
Differentiable $\Rightarrow$ smooth, connected; graph may rise and fall	$y' > 0 \Rightarrow$ rises from left to right; may be wavy	$y' < 0 \Rightarrow$ falls from left to right; may be wavy
or	or	$+$ $-$
$y'' > 0 \Rightarrow$ concave up throughout; no waves; graph may rise or fall or both	$y'' < 0 \Rightarrow$ concave down throughout; no waves; graph may rise or fall or both	y'' changes sign at an inflection point
$+$ $-$ or $-$ $+$		
y' changes sign $\Rightarrow$ graph has local maximum or local minimum	$y' = 0$ and $y'' < 0$ at a point; graph has local maximum	$y' = 0$ and $y'' > 0$ at a point; graph has local minimum

EXERCISES 4.4

Analyzing Functions from Graphs

Identify the inflection points and local maxima and minima of the functions graphed in Exercises 1–8. Identify the intervals on which the functions are concave up and concave down.

1. $y = \dfrac{x^3}{3} - \dfrac{x^2}{2} - 2x + \dfrac{1}{3}$

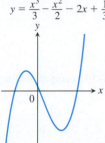

2. $y = \dfrac{x^4}{4} - 2x^2 + 4$

3. $y = \dfrac{3}{4}(x^2 - 1)^{2/3}$

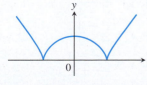

4. $y = \dfrac{9}{14}x^{1/3}(x^2 - 7)$

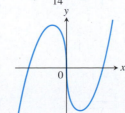

5. $y = x + \sin 2x, \ -\dfrac{2\pi}{3} \le x \le \dfrac{2\pi}{3}$

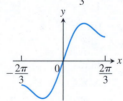

6. $y = \tan x - 4x, \ -\dfrac{\pi}{2} < x < \dfrac{\pi}{2}$

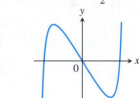

7. $y = \sin |x|, \ -2\pi \le x \le 2\pi$

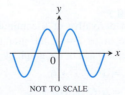

NOT TO SCALE

8. $y = 2 \cos x - \sqrt{2}x, \ -\pi \le x \le \dfrac{3\pi}{2}$

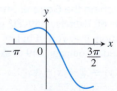

Graphing Functions

In Exercises 9–58, identify the coordinates of any local and absolute extreme points and inflection points. Graph the function.

9. $y = x^2 - 4x + 3$

10. $y = 6 - 2x - x^2$

11. $y = x^3 - 3x + 3$

12. $y = x(6 - 2x)^2$

13. $y = -2x^3 + 6x^2 - 3$

14. $y = 1 - 9x - 6x^2 - x^3$

15. $y = (x - 2)^3 + 1$

16. $y = 1 - (x + 1)^3$

17. $y = x^4 - 2x^2 = x^2(x^2 - 2)$

18. $y = -x^4 + 6x^2 - 4 = x^2(6 - x^2) - 4$

19. $y = 4x^3 - x^4 = x^3(4 - x)$ **20.** $y = x^4 + 2x^3 = x^3(x + 2)$

21. $y = x^5 - 5x^4 = x^4(x - 5)$ **22.** $y = x\left(\dfrac{x}{2} - 5\right)^4$

23. $y = x + \sin x, \quad 0 \le x \le 2\pi$

24. $y = x - \sin x, \quad 0 \le x \le 2\pi$

25. $y = \sqrt{3}x - 2\cos x, \quad 0 \le x \le 2\pi$

26. $y = \dfrac{4}{3}x - \tan x, \quad \dfrac{-\pi}{2} < x < \dfrac{\pi}{2}$

27. $y = \sin x \cos x, \quad 0 \le x \le \pi$

28. $y = \cos x + \sqrt{3}\sin x, \quad 0 \le x \le 2\pi$

29. $y = x^{1/5}$ **30.** $y = x^{2/5}$

31. $y = \dfrac{x}{\sqrt{x^2 + 1}}$ **32.** $y = \dfrac{\sqrt{1 - x^2}}{2x + 1}$

33. $y = 2x - 3x^{2/3}$ **34.** $y = 5x^{2/5} - 2x$

35. $y = x^{2/3}\left(\dfrac{5}{2} - x\right)$ **36.** $y = x^{2/3}(x - 5)$

37. $y = x\sqrt{8 - x^2}$ **38.** $y = (2 - x^2)^{3/2}$

39. $y = \sqrt{16 - x^2}$ **40.** $y = x^2 + \dfrac{2}{x}$

41. $y = \dfrac{x^2 - 3}{x - 2}$ **42.** $y = \sqrt[3]{x^3 + 1}$

43. $y = \dfrac{8x}{x^2 + 4}$ **44.** $y = \dfrac{5}{x^4 + 5}$

45. $y = |x^2 - 1|$ **46.** $y = |x^2 - 2x|$

47. $y = \sqrt{|x|} = \begin{cases} \sqrt{-x}, & x < 0 \\ \sqrt{x}, & x \ge 0 \end{cases}$

48. $y = \sqrt{|x - 4|}$

49. $y = \dfrac{x}{9 - x^2}$ **50.** $y = \dfrac{x^2}{1 - x}$

51. $y = \ln(3 - x^2)$ **52.** $y = (\ln x)^2$

53. $y = e^x - 2e^{-x} - 3x$ **54.** $y = xe^{-x}$

55. $y = \ln(\cos x)$ **56.** $y = \dfrac{\ln x}{\sqrt{x}}$

57. $y = \dfrac{1}{1 + e^{-x}}$ **58.** $y = \dfrac{e^x}{1 + e^x}$

Sketching the General Shape, Knowing y'

Each of Exercises 59–80 gives the first derivative of a continuous function $y = f(x)$. Find y'' and then use Steps 2–4 of the graphing procedure on page 249 to sketch the general shape of the graph of f.

59. $y' = 2 + x - x^2$ **60.** $y' = x^2 - x - 6$

61. $y' = x(x - 3)^2$ **62.** $y' = x^2(2 - x)$

63. $y' = x(x^2 - 12)$ **64.** $y' = (x - 1)^2(2x + 3)$

65. $y' = (8x - 5x^2)(4 - x)^2$ **66.** $y' = (x^2 - 2x)(x - 5)^2$

67. $y' = \sec^2 x, \quad -\dfrac{\pi}{2} < x < \dfrac{\pi}{2}$

68. $y' = \tan x, \quad -\dfrac{\pi}{2} < x < \dfrac{\pi}{2}$

69. $y' = \cot\dfrac{\theta}{2}, \quad 0 < \theta < 2\pi$ **70.** $y' = \csc^2\dfrac{\theta}{2}, \quad 0 < \theta < 2\pi$

71. $y' = \tan^2\theta - 1, \quad -\dfrac{\pi}{2} < \theta < \dfrac{\pi}{2}$

72. $y' = 1 - \cot^2\theta, \quad 0 < \theta < \pi$

73. $y' = \cos t, \quad 0 \le t \le 2\pi$

74. $y' = \sin t, \quad 0 \le t \le 2\pi$

75. $y' = (x + 1)^{-2/3}$ **76.** $y' = (x - 2)^{-1/3}$

77. $y' = x^{-2/3}(x - 1)$ **78.** $y' = x^{-4/5}(x + 1)$

79. $y' = 2|x| = \begin{cases} -2x, & x \le 0 \\ 2x, & x > 0 \end{cases}$

80. $y' = \begin{cases} -x^2, & x \le 0 \\ x^2, & x > 0 \end{cases}$

Sketching y from Graphs of y' and y''

Each of Exercises 81–84 shows the graphs of the first and second derivatives of a function $y = f(x)$. Copy the picture and add to it a sketch of the approximate graph of f, given that the graph passes through the point P.

81.

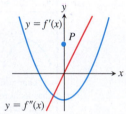

82.

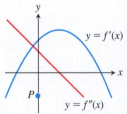

83.

84.

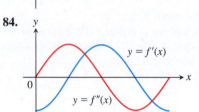

Graphing Rational Functions

Graph the rational functions in Exercises 85–102 using all the steps in the graphing procedure on page 249.

85. $y = \dfrac{2x^2 + x - 1}{x^2 - 1}$ **86.** $y = \dfrac{x^2 - 49}{x^2 + 5x - 14}$

87. $y = \dfrac{x^4 + 1}{x^2}$ **88.** $y = \dfrac{x^2 - 4}{2x}$

89. $y = \dfrac{1}{x^2 - 1}$ **90.** $y = \dfrac{x^2}{x^2 - 1}$

91. $y = -\dfrac{x^2 - 2}{x^2 - 1}$ **92.** $y = \dfrac{x^2 - 4}{x^2 - 2}$

93. $y = \dfrac{x^2}{x + 1}$ **94.** $y = -\dfrac{x^2 - 4}{x + 1}$

95. $y = \dfrac{x^2 - x + 1}{x - 1}$ **96.** $y = -\dfrac{x^2 - x + 1}{x - 1}$

97. $y = \dfrac{x^3 - 3x^2 + 3x - 1}{x^2 + x - 2}$

98. $y = \dfrac{x^3 + x - 2}{x - x^2}$

99. $y = \dfrac{x}{x^2 - 1}$

100. $y = \dfrac{x - 1}{x^2(x - 2)}$

101. $y = \dfrac{8}{x^2 + 4}$ (Agnesi's witch)

102. $y = \dfrac{4x}{x^2 + 4}$ (Newton's serpentine)

Theory and Examples

103. The accompanying figure shows a portion of the graph of a twice-differentiable function $y = f(x)$. At each of the five labeled points, classify y' and y'' as positive, negative, or zero.

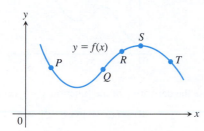

104. Sketch a smooth connected curve $y = f(x)$ with

$f(-2) = 8,$ $f'(2) = f'(-2) = 0,$

$f(0) = 4,$ $f'(x) < 0$ for $|x| < 2,$

$f(2) = 0,$ $f''(x) < 0$ for $x < 0,$

$f'(x) > 0$ for $|x| > 2,$ $f''(x) > 0$ for $x > 0.$

105. Sketch the graph of a twice-differentiable function $y = f(x)$ with the following properties. Label coordinates where possible.

x	y	Derivatives
$x < 2$		$y' < 0, \quad y'' > 0$
2	1	$y' = 0, \quad y'' > 0$
$2 < x < 4$		$y' > 0, \quad y'' > 0$
4	4	$y' > 0, \quad y'' = 0$
$4 < x < 6$		$y' > 0, \quad y'' < 0$
6	7	$y' = 0, \quad y'' < 0$
$x > 6$		$y' < 0, \quad y'' < 0$

106. Sketch the graph of a twice-differentiable function $y = f(x)$ that passes through the points $(-2, 2)$, $(-1, 1)$, $(0, 0)$, $(1, 1)$, and $(2, 2)$ and whose first two derivatives have the following sign patterns.

$$y': \quad \underset{-2}{\overset{+}{\rule{0pt}{0pt}}} \quad \underset{0}{\overset{-}{\rule{0pt}{0pt}}} \quad \underset{2}{\overset{+}{\rule{0pt}{0pt}}} \quad \overset{-}{\rule{0pt}{0pt}}$$

$$y'': \quad \underset{-1}{\overset{-}{\rule{0pt}{0pt}}} \quad \underset{1}{\overset{+}{\rule{0pt}{0pt}}} \quad \overset{-}{\rule{0pt}{0pt}}$$

107. Sketch the graph of a twice-differentiable $y = f(x)$ with the following properties. Label coordinates where possible.

x	y	Derivatives
$x < -2$		$y' > 0, \quad y'' < 0$
-2	-1	$y' = 0, \quad y'' = 0$
$-2 < x < -1$		$y' > 0, \quad y'' > 0$
-1	0	$y' > 0, \quad y'' = 0$
$-1 < x < 0$		$y' > 0, \quad y'' < 0$
0	3	$y' = 0, \quad y'' < 0$
$0 < x < 1$		$y' < 0, \quad y'' < 0$
1	2	$y' < 0, \quad y'' = 0$
$1 < x < 2$		$y' < 0, \quad y'' > 0$
2	0	$y' = 0, \quad y'' > 0$
$x > 2$		$y' > 0, \quad y'' > 0$

108. Sketch the graph of a twice-differentiable function $y = f(x)$ that passes through the points $(-3, -2)$, $(-2, 0)$, $(0, 1)$, $(1, 2)$, and $(2, 3)$ and whose first two derivatives have the following sign patterns.

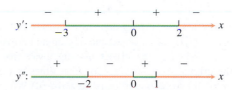

In Exercises 109 and 110, the graph of f' is given. Determine x-values corresponding to inflection points for the graph of f.

109.

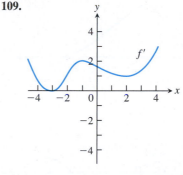

110.

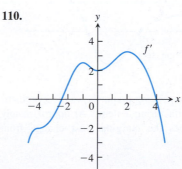

In Exercises 111 and 112, the graph of f' is given. Determine x-values corresponding to local minima, local maxima, and inflection points for the graph of f.

111.

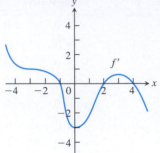

112.

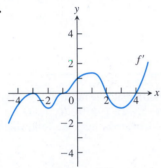

Motion Along a Line The graphs in Exercises 113 and 114 show the position $s = f(t)$ of an object moving up and down on a coordinate line. **(a)** When is the object moving away from the origin? Toward the origin? At approximately what times is the **(b)** velocity equal to zero? **(c)** Acceleration equal to zero? **(d)** When is the acceleration positive? Negative?

113.

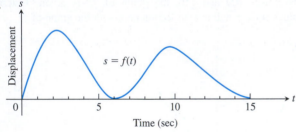

114.

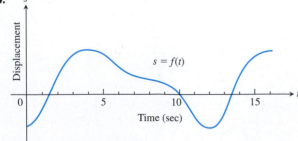

115. Marginal cost The accompanying graph shows the hypothetical cost $c = f(x)$ of manufacturing x items. At approximately what production level does the marginal cost change from decreasing to increasing?

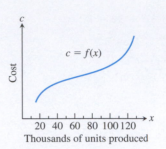

116. The accompanying graph shows the monthly revenue of the Widget Corporation for the past 12 years. During approximately what time intervals was the marginal revenue increasing? Decreasing?

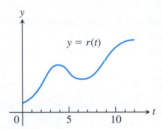

117. Suppose the derivative of the function $y = f(x)$ is

$$y' = (x - 1)^2(x - 2).$$

At what points, if any, does the graph of f have a local minimum, local maximum, or point of inflection? (*Hint:* Draw the sign pattern for y'.)

118. Suppose the derivative of the function $y = f(x)$ is

$$y' = (x - 1)^2(x - 2)(x - 4).$$

At what points, if any, does the graph of f have a local minimum, local maximum, or point of inflection?

119. For $x > 0$, sketch a curve $y = f(x)$ that has $f(1) = 0$ and $f'(x) = 1/x$. Can anything be said about the concavity of such a curve? Give reasons for your answer.

120. Can anything be said about the graph of a function $y = f(x)$ that has a continuous second derivative that is never zero? Give reasons for your answer.

121. If b, c, and d are constants, for what value of b will the curve $y = x^3 + bx^2 + cx + d$ have a point of inflection at $x = 1$? Give reasons for your answer.

122. Parabolas

a. Find the coordinates of the vertex of the parabola $y = ax^2 + bx + c, a \neq 0$.

b. When is the parabola concave up? Concave down? Give reasons for your answers.

123. Quadratic curves What can you say about the inflection points of a quadratic curve $y = ax^2 + bx + c, a \neq 0$? Give reasons for your answer.

124. Cubic curves What can you say about the inflection points of a cubic curve $y = ax^3 + bx^2 + cx + d, a \neq 0$? Give reasons for your answer.

125. Suppose that the second derivative of the function $y = f(x)$ is

$$y'' = (x + 1)(x - 2).$$

For what x-values does the graph of f have an inflection point?

126. Suppose that the second derivative of the function $y = f(x)$ is

$$y'' = x^2(x - 2)^3(x + 3).$$

For what x-values does the graph of f have an inflection point?

127. Find the values of constants a, b, and c so that the graph of $y = ax^3 + bx^2 + cx$ has a local maximum at $x = 3$, local minimum at $x = -1$, and inflection point at $(1, 11)$.

128. Find the values of constants a, b, and c so that the graph of $y = (x^2 + a)/(bx + c)$ has a local minimum at $x = 3$ and a local maximum at $(-1, -2)$.

COMPUTER EXPLORATIONS

In Exercises 129–132, find the inflection points (if any) on the graph of the function and the coordinates of the points on the graph where the function has a local maximum or local minimum value. Then graph the function in a region large enough to show all these points simultaneously. Add to your picture the graphs of the function's first and second derivatives. How are the values at which these graphs intersect the x-axis related to the graph of the function? In what other ways are the graphs of the derivatives related to the graph of the function?

129. $y = x^5 - 5x^4 - 240$ **130.** $y = x^3 - 12x^2$

131. $y = \dfrac{4}{5}x^5 + 16x^2 - 25$

132. $y = \dfrac{x^4}{4} - \dfrac{x^3}{3} - 4x^2 + 12x + 20$

133. Graph $f(x) = 2x^4 - 4x^2 + 1$ and its first two derivatives together. Comment on the behavior of f in relation to the signs and values of f' and f''.

134. Graph $f(x) = x \cos x$ and its second derivative together for $0 \le x \le 2\pi$. Comment on the behavior of the graph of f in relation to the signs and values of f''.

4.5 Indeterminate Forms and L'Hôpital's Rule

Expressions such as "0/0" and "∞/∞" look something like ordinary numbers. We say that they have the *form* of a number. But values cannot be assigned to them in a way that is consistent with the usual rules to add and multiply numbers. We are led to call them "indeterminate forms." Although we must remain careful to remember that they are not numbers, we will see that they can play useful roles in summarizing the limiting behavior of a function.

John (Johann) Bernoulli discovered a rule using derivatives to calculate limits of fractions whose numerators and denominators both approach zero or $+\infty$. The rule is known today as **l'Hôpital's Rule**, after Guillaume de l'Hôpital. He was a French nobleman who wrote the first introductory differential calculus text, where the rule first appeared in print. Limits involving transcendental functions often require some use of the rule.

Indeterminate Form 0/0

If we want to know how the function

$$f(x) = \frac{3x - \sin x}{x}$$

behaves *near* $x = 0$ (where it is undefined), we can examine the limit of $f(x)$ as $x \to 0$. We cannot apply the Quotient Rule for limits (Theorem 1 of Chapter 2) because the limit of the denominator is 0. Moreover, in this case, *both* the numerator and denominator approach 0, and 0/0 is undefined. Such limits may or may not exist in general, but the limit does exist for the function $f(x)$ under discussion by applying l'Hôpital's Rule, as we will see in Example 1d.

If the continuous functions $f(x)$ and $g(x)$ are both zero at $x = a$, then

$$\lim_{x \to a} \frac{f(x)}{g(x)}$$

cannot be found by substituting $x = a$. The substitution produces 0/0, a meaningless expression, which we cannot evaluate. We use 0/0 as a notation for an expression that does not have a numerical value, known as an **indeterminate form**. Other meaningless expressions often occur, such as ∞/∞, $\infty \cdot 0$, $\infty - \infty$, 0^0, and 1^∞, which cannot be evaluated in a consistent way; these are called indeterminate forms as well. Sometimes, but not always, limits that lead to indeterminate forms may be found by cancelation, rearrangement of terms, or other algebraic manipulations. This was our experience in Chapter 2. It took considerable analysis in Section 2.4 to find $\lim_{x\to 0}(\sin x)/x$. But we have had success with the limit

$$f'(a) = \lim_{x\to a} \frac{f(x) - f(a)}{x - a},$$

from which we calculate derivatives and which produces the indeterminant form $0/0$ when we attempt to substitute $x = a$. L'Hôpital's Rule enables us to draw on our success with derivatives to evaluate limits that otherwise lead to indeterminate forms.

THEOREM 6—L'Hôpital's Rule

Suppose that $f(a) = g(a) = 0$, that f and g are differentiable on an open interval I containing a, and that $g'(x) \neq 0$ on I if $x \neq a$. Then

$$\lim_{x\to a} \frac{f(x)}{g(x)} = \lim_{x\to a} \frac{f'(x)}{g'(x)},$$

assuming that the limit on the right side of this equation exists.

We give a proof of Theorem 6 at the end of this section.

Caution

To apply l'Hôpital's Rule to f/g, divide the derivative of f by the derivative of g. Do not fall into the trap of taking the derivative of f/g. The quotient to use is f'/g', not $(f/g)'$.

EXAMPLE 1 The following limits involve $0/0$ indeterminate forms, so we apply l'Hôpital's Rule. In some cases, it must be applied repeatedly.

(a) $\lim\limits_{x\to 0} \dfrac{3x - \sin x}{x} = \lim\limits_{x\to 0} \dfrac{3 - \cos x}{1} = \dfrac{3 - \cos x}{1}\bigg|_{x=0} = 2$

(b) $\lim\limits_{x\to 0} \dfrac{\sqrt{1 + x} - 1}{x} = \lim\limits_{x\to 0} \dfrac{\dfrac{1}{2\sqrt{1 + x}}}{1} = \dfrac{1}{2}$

(c) $\lim\limits_{x\to 0} \dfrac{\sqrt{1 + x} - 1 - x/2}{x^2}$ $\dfrac{0}{0}$; apply l'Hôpital's Rule.

$= \lim\limits_{x\to 0} \dfrac{(1/2)(1 + x)^{-1/2} - 1/2}{2x}$ Still $\dfrac{0}{0}$; apply l'Hôpital's Rule again.

$= \lim\limits_{x\to 0} \dfrac{-(1/4)(1 + x)^{-3/2}}{2} = -\dfrac{1}{8}$ Not $\dfrac{0}{0}$; limit is found.

(d) $\lim\limits_{x\to 0} \dfrac{x - \sin x}{x^3}$ $\dfrac{0}{0}$; apply l'Hôpital's Rule.

$= \lim\limits_{x\to 0} \dfrac{1 - \cos x}{3x^2}$ Still $\dfrac{0}{0}$; apply l'Hôpital's Rule again.

$= \lim\limits_{x\to 0} \dfrac{\sin x}{6x}$ Still $\dfrac{0}{0}$; apply l'Hôpital's Rule again.

$= \lim\limits_{x\to 0} \dfrac{\cos x}{6} = \dfrac{1}{6}$ Not $\dfrac{0}{0}$; limit is found.

Here is a summary of the procedure we followed in Example 1.

Using L'Hôpital's Rule
To find

$$\lim_{x \to a} \frac{f(x)}{g(x)}$$

by l'Hôpital's Rule, we continue to differentiate f and g, so long as we still get the form $0/0$ at $x = a$. But as soon as one or the other of these derivatives is different from zero at $x = a$ we stop differentiating. L'Hôpital's Rule does not apply when either the numerator or denominator has a finite nonzero limit.

EXAMPLE 2   Be careful to apply l'Hôpital's Rule correctly:

$$\lim_{x \to 0} \frac{1 - \cos x}{x + x^2} \qquad \frac{0}{0}$$

$$= \lim_{x \to 0} \frac{\sin x}{1 + 2x} \qquad \text{Not } \frac{0}{0}$$

It is tempting to try to apply l'Hôpital's Rule again, which would result in

$$\lim_{x \to 0} \frac{\cos x}{2} = \frac{1}{2},$$

but this is not the correct limit. l'Hôpital's Rule can be applied only to limits that give indeterminate forms, and $\lim_{x \to 0} (\sin x)/(1 + 2x)$ does not give an indeterminate form. Instead, this limit is $0/1 = 0$, and the correct answer for the original limit is 0.   ■

L'Hôpital's Rule applies to one-sided limits as well.

EXAMPLE 3   In this example the one-sided limits are different.

(a) $\lim\limits_{x \to 0^+} \dfrac{\sin x}{x^2}$ $\qquad\qquad \dfrac{0}{0}$

$\qquad = \lim\limits_{x \to 0^+} \dfrac{\cos x}{2x} = \infty \qquad$ Positive for $x > 0$

(b) $\lim\limits_{x \to 0^-} \dfrac{\sin x}{x^2}$ $\qquad\qquad \dfrac{0}{0}$

$\qquad = \lim\limits_{x \to 0^-} \dfrac{\cos x}{2x} = -\infty \qquad$ Negative for $x < 0$ ■

Indeterminate Forms ∞/∞, $\infty \cdot 0$, $\infty - \infty$

Recall that ∞ and $+\infty$ mean the same thing.

Sometimes when we try to evaluate a limit as $x \to a$ by substituting $x = a$ we get an indeterminant form like ∞/∞, $\infty \cdot 0$, or $\infty - \infty$, instead of $0/0$. We first consider the form ∞/∞.

More advanced treatments of calculus prove that l'Hôpital's Rule applies to the indeterminate form ∞/∞, as well as to $0/0$. If $f(x) \to \pm\infty$ and $g(x) \to \pm\infty$ as $x \to a$, then

$$\lim_{x \to a} \frac{f(x)}{g(x)} = \lim_{x \to a} \frac{f'(x)}{g'(x)}$$

provided the limit on the right exists. In the notation $x \to a$, a may be either finite or infinite. Moreover, $x \to a$ may be replaced by the one-sided limits $x \to a^+$ or $x \to a^-$.

EXAMPLE 4   Find the limits of these ∞/∞ forms:

(a) $\displaystyle\lim_{x\to\pi/2}\frac{\sec x}{1+\tan x}$ (b) $\displaystyle\lim_{x\to\infty}\frac{\ln x}{2\sqrt{x}}$ (c) $\displaystyle\lim_{x\to\infty}\frac{e^x}{x^2}$.

Solution

(a) The numerator and denominator are discontinuous at $x=\pi/2$, so we investigate the one-sided limits there. To apply l'Hôpital's Rule, we can choose I to be any open interval with $x=\pi/2$ as an endpoint.

$$\lim_{x\to(\pi/2)^-}\frac{\sec x}{1+\tan x} \qquad \text{\small $\frac{\infty}{\infty}$ from left, apply l'Hôpital's Rule}$$

$$= \lim_{x\to(\pi/2)^-}\frac{\sec x\tan x}{\sec^2 x} = \lim_{x\to(\pi/2)^-}\sin x = 1$$

The right-hand limit is 1 also, with $(-\infty)/(-\infty)$ as the indeterminate form. Therefore, the two-sided limit is equal to 1.

(b) $\displaystyle\lim_{x\to\infty}\frac{\ln x}{2\sqrt{x}} = \lim_{x\to\infty}\frac{1/x}{1/\sqrt{x}} = \lim_{x\to\infty}\frac{1}{\sqrt{x}} = 0$ $\dfrac{1/x}{1/\sqrt{x}} = \dfrac{\sqrt{x}}{x} = \dfrac{1}{\sqrt{x}}$

(c) $\displaystyle\lim_{x\to\infty}\frac{e^x}{x^2} = \lim_{x\to\infty}\frac{e^x}{2x} = \lim_{x\to\infty}\frac{e^x}{2} = \infty$ ∎

Next we turn our attention to the indeterminate forms $\infty\cdot 0$ and $\infty-\infty$. Sometimes these forms can be handled by using algebra to convert them to a $0/0$ or ∞/∞ form. Here again we do not mean to suggest that $\infty\cdot 0$ or $\infty-\infty$ is a number. They are only notations for functional behaviors when considering limits. Here are examples of how we might work with these indeterminate forms.

EXAMPLE 5   Find the limits of these $\infty\cdot 0$ forms:

(a) $\displaystyle\lim_{x\to\infty}\left(x\sin\frac{1}{x}\right)$ (b) $\displaystyle\lim_{x\to 0^+}\sqrt{x}\ln x$

Solution

(a) $\displaystyle\lim_{x\to\infty}\left(x\sin\frac{1}{x}\right) = \lim_{h\to 0^+}\left(\frac{1}{h}\sin h\right) = \lim_{h\to 0^+}\frac{\sin h}{h} = 1$ $\infty\cdot 0$; let $h=1/x$.

(b) $\displaystyle\lim_{x\to 0^+}\sqrt{x}\ln x = \lim_{x\to 0^+}\frac{\ln x}{1/\sqrt{x}}$ $\infty\cdot 0$ converted to ∞/∞

$$= \lim_{x\to 0^+}\frac{1/x}{-1/2x^{3/2}}$$ l'Hôpital's Rule applied

$$= \lim_{x\to 0^+}\left(-2\sqrt{x}\right) = 0$$ ∎

EXAMPLE 6   Find the limit of this $\infty-\infty$ form:

$$\lim_{x\to 0}\left(\frac{1}{\sin x}-\frac{1}{x}\right).$$

Solution   If $x\to 0^+$, then $\sin x\to 0^+$ and

$$\frac{1}{\sin x}-\frac{1}{x}\to\infty-\infty.$$

Similarly, if $x \to 0^-$, then $\sin x \to 0^-$ and

$$\frac{1}{\sin x} - \frac{1}{x} \to -\infty - (-\infty) = -\infty + \infty.$$

Neither form reveals what happens in the limit. To find out, we first combine the fractions:

$$\frac{1}{\sin x} - \frac{1}{x} = \frac{x - \sin x}{x \sin x}. \qquad \text{\color{blue} Common denominator is } x \sin x.$$

Then we apply l'Hôpital's Rule to the result:

$$\lim_{x \to 0} \left(\frac{1}{\sin x} - \frac{1}{x} \right) = \lim_{x \to 0} \frac{x - \sin x}{x \sin x} \qquad \frac{0}{0}$$

$$= \lim_{x \to 0} \frac{1 - \cos x}{\sin x + x \cos x} \qquad \text{\color{blue}Still } \frac{0}{0}$$

$$= \lim_{x \to 0} \frac{\sin x}{2 \cos x - x \sin x} = \frac{0}{2} = 0. \qquad \blacksquare$$

Indeterminate Powers

Limits that lead to the indeterminate forms 1^∞, 0^0, and ∞^0 can sometimes be handled by first taking the logarithm of the function. We use l'Hôpital's Rule to find the limit of the logarithm expression and then exponentiate the result to find the original function limit. This procedure is justified by the continuity of the exponential function and Theorem 10 in Section 2.5, and it is formulated as follows. (The formula is also valid for one-sided limits.)

If $\lim_{x \to a} \ln f(x) = L$, then

$$\lim_{x \to a} f(x) = \lim_{x \to a} e^{\ln f(x)} = e^L.$$

Here a may be either finite or infinite.

EXAMPLE 7   Apply l'Hôpital's Rule to show that $\lim_{x \to 0^+} (1 + x)^{1/x} = e$.

Solution   The limit leads to the indeterminate form 1^∞. We let $f(x) = (1 + x)^{1/x}$ and find $\lim_{x \to 0^+} \ln f(x)$. Since

$$\ln f(x) = \ln (1 + x)^{1/x} = \frac{1}{x} \ln (1 + x),$$

l'Hôpital's Rule now applies to give

$$\lim_{x \to 0^+} \ln f(x) = \lim_{x \to 0^+} \frac{\ln (1 + x)}{x} \qquad \frac{0}{0}$$

$$= \lim_{x \to 0^+} \frac{\frac{1}{1 + x}}{1} \qquad \text{\color{blue}l'Hôpital's Rule applied}$$

$$= \frac{1}{1} = 1.$$

Therefore, $\lim_{x \to 0^+} (1 + x)^{1/x} = \lim_{x \to 0^+} f(x) = \lim_{x \to 0^+} e^{\ln f(x)} = e^1 = e.$ $\blacksquare$

EXAMPLE 8   Find $\lim_{x \to \infty} x^{1/x}$.

Solution   The limit leads to the indeterminate form ∞^0. We let $f(x) = x^{1/x}$ and find $\lim_{x \to \infty} \ln f(x)$. Since

$$\ln f(x) = \ln x^{1/x} = \frac{\ln x}{x},$$

l'Hôpital's Rule gives

$$\lim_{x \to \infty} \ln f(x) = \lim_{x \to \infty} \frac{\ln x}{x} \qquad \frac{\infty}{\infty}$$

$$= \lim_{x \to \infty} \frac{1/x}{1} \qquad \text{l'Hôpital's Rule applied}$$

$$= \frac{0}{1} = 0.$$

Therefore $\lim_{x \to \infty} x^{1/x} = \lim_{x \to \infty} f(x) = \lim_{x \to \infty} e^{\ln f(x)} = e^0 = 1.$ ∎

Proof of L'Hôpital's Rule

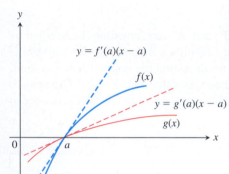

FIGURE 4.36   The two functions in l'Hôpital's Rule, graphed with their linear approximations at $x = a$.

Before we prove l'Hôpital's Rule, we consider a special case to provide some geometric insight for its reasonableness. Consider the two functions $f(x)$ and $g(x)$ having *continuous* derivatives and satisfying $f(a) = g(a) = 0$, $g'(a) \neq 0$. The graphs of $f(x)$ and $g(x)$, together with their linearizations $y = f'(a)(x - a)$ and $y = g'(a)(x - a)$, are shown in Figure 4.36. We know that near $x = a$, the linearizations provide good approximations to the functions. In fact,

$$f(x) = f'(a)(x - a) + \varepsilon_1(x - a) \quad \text{and} \quad g(x) = g'(a)(x - a) + \varepsilon_2(x - a)$$

where $\varepsilon_1 \to 0$ and $\varepsilon_2 \to 0$ as $x \to a$. So, as Figure 4.36 suggests,

$$\lim_{x \to a} \frac{f(x)}{g(x)} = \lim_{x \to a} \frac{f'(a)(x - a) + \varepsilon_1(x - a)}{g'(a)(x - a) + \varepsilon_2(x - a)}$$

$$= \lim_{x \to a} \frac{f'(a) + \varepsilon_1}{g'(a) + \varepsilon_2} = \frac{f'(a)}{g'(a)} \qquad g'(a) \neq 0$$

$$= \lim_{x \to a} \frac{f'(x)}{g'(x)}, \qquad \text{Continuous derivatives}$$

as asserted by l'Hôpital's Rule. We now proceed to a proof of the rule based on the more general assumptions stated in Theorem 6, which do not require that $g'(a) \neq 0$ and that the two functions have *continuous* derivatives.

The proof of l'Hôpital's Rule is based on Cauchy's Mean Value Theorem, an extension of the Mean Value Theorem that involves two functions instead of one. We prove Cauchy's Theorem first and then show how it leads to l'Hôpital's Rule.

When $g(x) = x$, Theorem 7 is the Mean Value Theorem.

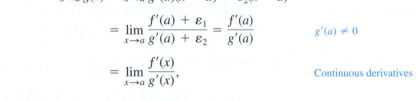

THEOREM 7—Cauchy's Mean Value Theorem
Suppose functions f and g are continuous on $[a, b]$ and differentiable throughout (a, b) and also suppose $g'(x) \neq 0$ throughout (a, b). Then there exists a number c in (a, b) at which

$$\frac{f'(c)}{g'(c)} = \frac{f(b) - f(a)}{g(b) - g(a)}.$$

Proof   We apply the Mean Value Theorem of Section 4.2 twice. First we use it to show that $g(a) \neq g(b)$. For if $g(b)$ did equal $g(a)$, then the Mean Value Theorem would give

$$g'(c) = \frac{g(b) - g(a)}{b - a} = 0$$

for some c between a and b, which cannot happen because $g'(x) \neq 0$ in (a, b).

We next apply the Mean Value Theorem to the function

$$F(x) = f(x) - f(a) - \frac{f(b) - f(a)}{g(b) - g(a)} [g(x) - g(a)].$$

This function is continuous and differentiable where f and g are, and $F(b) = F(a) = 0$. Therefore, there is a number c between a and b for which $F'(c) = 0$. When expressed in terms of f and g, this equation becomes

$$F'(c) = f'(c) - \frac{f(b) - f(a)}{g(b) - g(a)} [g'(c)] = 0$$

so that

$$\frac{f'(c)}{g'(c)} = \frac{f(b) - f(a)}{g(b) - g(a)}.$$ ∎

Cauchy's Mean Value Theorem has a geometric interpretation for a general winding curve C in the plane joining the two points $A = (g(a), f(a))$ and $B = (g(b), f(b))$. In Chapter 11 you will learn how the curve C can be formulated so that there is at least one point P on the curve for which the tangent to the curve at P is parallel to the secant line joining the points A and B. The slope of that tangent line turns out to be the quotient f'/g' evaluated at the number c in the interval (a, b), which is the left-hand side of the equation in Theorem 7. Because the slope of the secant line joining A and B is

$$\frac{f(b) - f(a)}{g(b) - g(a)},$$

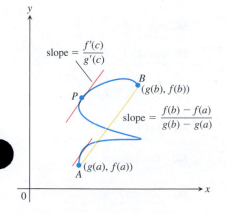

FIGURE 4.37   There is at least one point P on the curve C for which the slope of the tangent to the curve at P is the same as the slope of the secant line joining the points $A(g(a), f(a))$ and $B(g(b), f(b))$.

the equation in Cauchy's Mean Value Theorem says that the slope of the tangent line equals the slope of the secant line. This geometric interpretation is shown in Figure 4.37. Notice from the figure that it is possible for more than one point on the curve C to have a tangent line that is parallel to the secant line joining A and B.

Proof of l'Hôpital's Rule   We first establish the limit equation for the case $x \to a^+$. The method needs almost no change to apply to $x \to a^-$, and the combination of these two cases establishes the result.

Suppose that x lies to the right of a. Then $g'(x) \neq 0$, and we can apply Cauchy's Mean Value Theorem to the closed interval from a to x. This step produces a number c between a and x such that

$$\frac{f'(c)}{g'(c)} = \frac{f(x) - f(a)}{g(x) - g(a)}.$$

But $f(a) = g(a) = 0$, so

$$\frac{f'(c)}{g'(c)} = \frac{f(x)}{g(x)}.$$

As x approaches a, c approaches a because it always lies between a and x. Therefore,

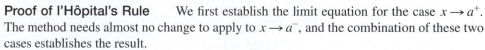

$$\lim_{x \to a^+} \frac{f(x)}{g(x)} = \lim_{c \to a^+} \frac{f'(c)}{g'(c)} = \lim_{x \to a^+} \frac{f'(x)}{g'(x)},$$

which establishes l'Hôpital's Rule for the case where x approaches a from above. The case where x approaches a from below is proved by applying Cauchy's Mean Value Theorem to the closed interval $[x, a]$, $x < a$. ∎

EXERCISES 4.5

Finding Limits in Two Ways

In Exercises 1–6, use l'Hôpital's Rule to evaluate the limit. Then evaluate the limit using a method studied in Chapter 2.

1. $\lim\limits_{x \to -2} \dfrac{x + 2}{x^2 - 4}$

2. $\lim\limits_{x \to 0} \dfrac{\sin 5x}{x}$

3. $\lim\limits_{x \to \infty} \dfrac{5x^2 - 3x}{7x^2 + 1}$

4. $\lim\limits_{x \to 1} \dfrac{x^3 - 1}{4x^3 - x - 3}$

5. $\lim\limits_{x \to 0} \dfrac{1 - \cos x}{x^2}$

6. $\lim\limits_{x \to \infty} \dfrac{2x^2 + 3x}{x^3 + x + 1}$

Applying l'Hôpital's Rule

Use l'Hôpital's rule to find the limits in Exercises 7–50.

7. $\lim\limits_{x \to 2} \dfrac{x - 2}{x^2 - 4}$

8. $\lim\limits_{x \to -5} \dfrac{x^2 - 25}{x + 5}$

9. $\lim\limits_{t \to -3} \dfrac{t^3 - 4t + 15}{t^2 - t - 12}$

10. $\lim\limits_{t \to -1} \dfrac{3t^3 + 3}{4t^3 - t + 3}$

11. $\lim\limits_{x \to \infty} \dfrac{5x^3 - 2x}{7x^3 + 3}$

12. $\lim\limits_{x \to \infty} \dfrac{x - 8x^2}{12x^2 + 5x}$

13. $\lim\limits_{t \to 0} \dfrac{\sin t^2}{t}$

14. $\lim\limits_{t \to 0} \dfrac{\sin 5t}{2t}$

15. $\lim\limits_{x \to 0} \dfrac{8x^2}{\cos x - 1}$

16. $\lim\limits_{x \to 0} \dfrac{\sin x - x}{x^3}$

17. $\lim\limits_{\theta \to \pi/2} \dfrac{2\theta - \pi}{\cos(2\pi - \theta)}$

18. $\lim\limits_{\theta \to -\pi/3} \dfrac{3\theta + \pi}{\sin(\theta + (\pi/3))}$

19. $\lim\limits_{\theta \to \pi/2} \dfrac{1 - \sin \theta}{1 + \cos 2\theta}$

20. $\lim\limits_{x \to 1} \dfrac{x - 1}{\ln x - \sin \pi x}$

21. $\lim\limits_{x \to 0} \dfrac{x^2}{\ln(\sec x)}$

22. $\lim\limits_{x \to \pi/2} \dfrac{\ln(\csc x)}{(x - (\pi/2))^2}$

23. $\lim\limits_{t \to 0} \dfrac{t(1 - \cos t)}{t - \sin t}$

24. $\lim\limits_{t \to 0} \dfrac{t \sin t}{1 - \cos t}$

25. $\lim\limits_{x \to (\pi/2)^-} \left(x - \dfrac{\pi}{2}\right) \sec x$

26. $\lim\limits_{x \to (\pi/2)^-} \left(\dfrac{\pi}{2} - x\right) \tan x$

27. $\lim\limits_{\theta \to 0} \dfrac{3^{\sin \theta} - 1}{\theta}$

28. $\lim\limits_{\theta \to 0} \dfrac{(1/2)^\theta - 1}{\theta}$

29. $\lim\limits_{x \to 0} \dfrac{x 2^x}{2^x - 1}$

30. $\lim\limits_{x \to 0} \dfrac{3^x - 1}{2^x - 1}$

31. $\lim\limits_{x \to \infty} \dfrac{\ln(x + 1)}{\log_2 x}$

32. $\lim\limits_{x \to \infty} \dfrac{\log_2 x}{\log_3(x + 3)}$

33. $\lim\limits_{x \to 0^+} \dfrac{\ln(x^2 + 2x)}{\ln x}$

34. $\lim\limits_{x \to 0^+} \dfrac{\ln(e^x - 1)}{\ln x}$

35. $\lim\limits_{y \to 0} \dfrac{\sqrt{5y + 25} - 5}{y}$

36. $\lim\limits_{y \to 0} \dfrac{\sqrt{ay + a^2} - a}{y}, \quad a > 0$

37. $\lim\limits_{x \to \infty} (\ln 2x - \ln(x + 1))$

38. $\lim\limits_{x \to 0^+} (\ln x - \ln \sin x)$

39. $\lim\limits_{x \to 0^+} \dfrac{(\ln x)^2}{\ln(\sin x)}$

40. $\lim\limits_{x \to 0^+} \left(\dfrac{3x + 1}{x} - \dfrac{1}{\sin x}\right)$

41. $\lim\limits_{x \to 1^+} \left(\dfrac{1}{x - 1} - \dfrac{1}{\ln x}\right)$

42. $\lim\limits_{x \to 0^+} (\csc x - \cot x + \cos x)$

43. $\lim\limits_{\theta \to 0} \dfrac{\cos \theta - 1}{e^\theta - \theta - 1}$

44. $\lim\limits_{h \to 0} \dfrac{e^h - (1 + h)}{h^2}$

45. $\lim\limits_{t \to \infty} \dfrac{e^t + t^2}{e^t - t}$

46. $\lim\limits_{x \to \infty} x^2 e^{-x}$

47. $\lim\limits_{x \to 0} \dfrac{x - \sin x}{x \tan x}$

48. $\lim\limits_{x \to 0} \dfrac{(e^x - 1)^2}{x \sin x}$

49. $\lim\limits_{\theta \to 0} \dfrac{\theta - \sin \theta \cos \theta}{\tan \theta - \theta}$

50. $\lim\limits_{x \to 0} \dfrac{\sin 3x - 3x + x^2}{\sin x \sin 2x}$

Indeterminate Powers and Products

Find the limits in Exercises 51–66.

51. $\lim\limits_{x \to 1^+} x^{1/(1-x)}$

52. $\lim\limits_{x \to 1^+} x^{1/(x-1)}$

53. $\lim\limits_{x \to \infty} (\ln x)^{1/x}$

54. $\lim\limits_{x \to e^+} (\ln x)^{1/(x-e)}$

55. $\lim\limits_{x \to 0^+} x^{-1/\ln x}$

56. $\lim\limits_{x \to \infty} x^{1/\ln x}$

57. $\lim\limits_{x \to \infty} (1 + 2x)^{1/(2 \ln x)}$

58. $\lim\limits_{x \to 0} (e^x + x)^{1/x}$

59. $\lim\limits_{x \to 0^+} x^x$

60. $\lim\limits_{x \to 0^+} \left(1 + \dfrac{1}{x}\right)^x$

61. $\lim\limits_{x \to \infty} \left(\dfrac{x + 2}{x - 1}\right)^x$

62. $\lim\limits_{x \to \infty} \left(\dfrac{x^2 + 1}{x + 2}\right)^{1/x}$

63. $\lim\limits_{x \to 0^+} x^2 \ln x$

64. $\lim\limits_{x \to 0^+} x (\ln x)^2$

65. $\lim\limits_{x \to 0^+} x \tan\left(\dfrac{\pi}{2} - x\right)$

66. $\lim\limits_{x \to 0^+} \sin x \cdot \ln x$

Theory and Applications

L'Hôpital's Rule does not help with the limits in Exercises 67–74. Try it—you just keep on cycling. Find the limits some other way.

67. $\lim\limits_{x \to \infty} \dfrac{\sqrt{9x + 1}}{\sqrt{x + 1}}$

68. $\lim\limits_{x \to 0^+} \dfrac{\sqrt{x}}{\sqrt{\sin x}}$

69. $\lim\limits_{x \to (\pi/2)^-} \dfrac{\sec x}{\tan x}$

70. $\lim\limits_{x \to 0^+} \dfrac{\cot x}{\csc x}$

71. $\lim\limits_{x \to \infty} \dfrac{2^x - 3^x}{3^x + 4^x}$

72. $\lim\limits_{x \to -\infty} \dfrac{2^x + 4^x}{5^x - 2^x}$

73. $\lim\limits_{x \to \infty} \dfrac{e^{x^2}}{xe^x}$

74. $\lim\limits_{x \to 0^+} \dfrac{x}{e^{-1/x}}$

75. Which one is correct, and which one is wrong? Give reasons for your answers.

    a. $\lim\limits_{x \to 3} \dfrac{x - 3}{x^2 - 3} = \lim\limits_{x \to 3} \dfrac{1}{2x} = \dfrac{1}{6}$ **b.** $\lim\limits_{x \to 3} \dfrac{x - 3}{x^2 - 3} = \dfrac{0}{6} = 0$

76. Which one is correct, and which one is wrong? Give reasons for your answers.

    a. $\lim\limits_{x \to 0} \dfrac{x^2 - 2x}{x^2 - \sin x} = \lim\limits_{x \to 0} \dfrac{2x - 2}{2x - \cos x}$

$$= \lim\limits_{x \to 0} \dfrac{2}{2 + \sin x} = \dfrac{2}{2 + 0} = 1$$

    b. $\lim\limits_{x \to 0} \dfrac{x^2 - 2x}{x^2 - \sin x} = \lim\limits_{x \to 0} \dfrac{2x - 2}{2x - \cos x} = \dfrac{-2}{0 - 1} = 2$

77. Only one of these calculations is correct. Which one? Why are the others wrong? Give reasons for your answers.

a. $\lim\limits_{x \to 0^+} x \ln x = 0 \cdot (-\infty) = 0$

b. $\lim\limits_{x \to 0^+} x \ln x = 0 \cdot (-\infty) = -\infty$

c. $\lim\limits_{x \to 0^+} x \ln x = \lim\limits_{x \to 0^+} \dfrac{\ln x}{(1/x)} = \dfrac{-\infty}{\infty} = -1$

d. $\lim\limits_{x \to 0^+} x \ln x = \lim\limits_{x \to 0^+} \dfrac{\ln x}{(1/x)}$

$= \lim\limits_{x \to 0^+} \dfrac{(1/x)}{(-1/x^2)} = \lim\limits_{x \to 0^+} (-x) = 0$

78. Find all values of c that satisfy the conclusion of Cauchy's Mean Value Theorem for the given functions and interval.

a. $f(x) = x, \qquad g(x) = x^2, \qquad (a, b) = (-2, 0)$

b. $f(x) = x, \qquad g(x) = x^2, \qquad (a, b)$ arbitrary

c. $f(x) = x^3/3 - 4x, \qquad g(x) = x^2, \qquad (a, b) = (0, 3)$

79. Continuous extension Find a value of c that makes the function

$$f(x) = \begin{cases} \dfrac{9x - 3 \sin 3x}{5x^3}, & x \neq 0 \\ c, & x = 0 \end{cases}$$

continuous at $x = 0$. Explain why your value of c works.

80. For what values of a and b is

$$\lim\limits_{x \to 0} \left(\dfrac{\tan 2x}{x^3} + \dfrac{a}{x^2} + \dfrac{\sin bx}{x} \right) = 0?$$

T **81.** $\infty - \infty$ **Form**

a. Estimate the value of

$$\lim\limits_{x \to \infty} \left(x - \sqrt{x^2 + x} \right)$$

by graphing $f(x) = x - \sqrt{x^2 + x}$ over a suitably large interval of x-values.

b. Now confirm your estimate by finding the limit with l'Hôpital's Rule. As the first step, multiply $f(x)$ by the fraction $\left(x + \sqrt{x^2 + x} \right)/\left(x + \sqrt{x^2 + x} \right)$ and simplify the new numerator.

82. Find $\lim\limits_{x \to \infty} \left(\sqrt{x^2 + 1} - \sqrt{x} \right)$.

T **83.** **0/0 Form** Estimate the value of

$$\lim\limits_{x \to 1} \dfrac{2x^2 - (3x + 1)\sqrt{x} + 2}{x - 1}$$

by graphing. Then confirm your estimate with l'Hôpital's Rule.

84. This exercise explores the difference between the limit

$$\lim\limits_{x \to \infty} \left(1 + \dfrac{1}{x^2} \right)^x$$

and the limit

$$\lim\limits_{x \to \infty} \left(1 + \dfrac{1}{x} \right)^x = e.$$

a. Use l'Hôpital's Rule to show that

$$\lim\limits_{x \to \infty} \left(1 + \dfrac{1}{x} \right)^x = e.$$

T **b.** Graph

$$f(x) = \left(1 + \dfrac{1}{x^2} \right)^x \quad \text{and} \quad g(x) = \left(1 + \dfrac{1}{x} \right)^x$$

together for $x \geq 0$. How does the behavior of f compare with that of g? Estimate the value of $\lim\limits_{x \to \infty} f(x)$.

c. Confirm your estimate of $\lim\limits_{x \to \infty} f(x)$ by calculating it with l'Hôpital's Rule.

85. Show that

$$\lim\limits_{k \to \infty} \left(1 + \dfrac{r}{k} \right)^k = e^r.$$

86. Given that $x > 0$, find the maximum value, if any, of

a. $x^{1/x}$

b. x^{1/x^2}

c. x^{1/x^n} (n a positive integer)

d. Show that $\lim\limits_{x \to \infty} x^{1/x^n} = 1$ for every positive integer n.

87. Use limits to find horizontal asymptotes for each function.

a. $y = x \tan \left(\dfrac{1}{x} \right)$

b. $y = \dfrac{3x + e^{2x}}{2x + e^{3x}}$

88. Find $f'(0)$ for $f(x) = \begin{cases} e^{-1/x^2}, & x \neq 0 \\ 0, & x = 0. \end{cases}$

T **89. The continuous extension of $(\sin x)^x$ to $[0, \pi]$**

a. Graph $f(x) = (\sin x)^x$ on the interval $0 \leq x \leq \pi$. What value would you assign to f to make it continuous at $x = 0$?

b. Verify your conclusion in part (a) by finding $\lim\limits_{x \to 0^+} f(x)$ with l'Hôpital's Rule.

c. Returning to the graph, estimate the maximum value of f on $[0, \pi]$. About where is max f taken on?

d. Sharpen your estimate in part (c) by graphing f' in the same window to see where its graph crosses the x-axis. To simplify your work, you might want to delete the exponential factor from the expression for f' and graph just the factor that has a zero.

T **90. The function $(\sin x)^{\tan x}$** (*Continuation of Exercise 89.*)

a. Graph $f(x) = (\sin x)^{\tan x}$ on the interval $-7 \leq x \leq 7$. How do you account for the gaps in the graph? How wide are the gaps?

b. Now graph f on the interval $0 \leq x \leq \pi$. The function is not defined at $x = \pi/2$, but the graph has no break at this point. What is going on? What value does the graph appear to give for f at $x = \pi/2$? (*Hint:* Use l'Hôpital's Rule to find $\lim f$ as $x \to (\pi/2)^-$ and $x \to (\pi/2)^+$.)

c. Continuing with the graphs in part (b), find max f and min f as accurately as you can and estimate the values of x at which they are taken on.

4.6 Applied Optimization

What are the dimensions of a rectangle with fixed perimeter having *maximum area*? What are the dimensions for the *least expensive* cylindrical can of a given volume? How many items should be produced for the *most profitable* production run? Each of these questions asks for the best, or optimal, value of a given function. In this section we use derivatives to solve a variety of optimization problems in mathematics, physics, economics, and business.

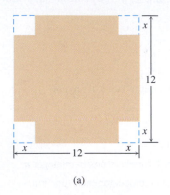

(a)

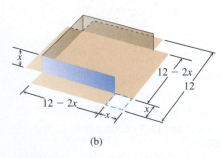

(b)

FIGURE 4.38 An open box made by cutting the corners from a square sheet of tin. What size corners maximize the box's volume (Example 1)?

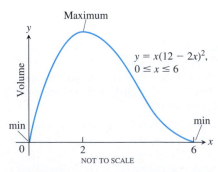

FIGURE 4.39 The volume of the box in Figure 4.38 graphed as a function of *x*.

Solving Applied Optimization Problems

1. *Read the problem*. Read the problem until you understand it. What is given? What is the unknown quantity to be optimized?

2. *Draw a picture*. Label any part that may be important to the problem.

3. *Introduce variables*. List every relation in the picture and in the problem as an equation or algebraic expression, and identify the unknown variable.

4. *Write an equation for the unknown quantity*. If you can, express the unknown as a function of a single variable or in two equations in two unknowns. This may require considerable manipulation.

5. *Test the critical points and endpoints in the domain of the unknown*. Use what you know about the shape of the function's graph. Use the first and second derivatives to identify and classify the function's critical points.

EXAMPLE 1   An open-top box is to be made by cutting small congruent squares from the corners of a 12-in.-by-12-in. sheet of tin and bending up the sides. How large should the squares cut from the corners be to make the box hold as much as possible?

Solution   We start with a picture (Figure 4.38). In the figure, the corner squares are x in. on a side. The volume of the box is a function of this variable:

$$V(x) = x(12 - 2x)^2 = 144x - 48x^2 + 4x^3. \qquad V = hlw$$

Since the sides of the sheet of tin are only 12 in. long, $x \le 6$ and the domain of V is the interval $0 \le x \le 6$.

A graph of V (Figure 4.39) suggests a minimum value of 0 at $x = 0$ and $x = 6$ and a maximum near $x = 2$. To learn more, we examine the first derivative of V with respect to x:

$$\frac{dV}{dx} = 144 - 96x + 12x^2 = 12(12 - 8x + x^2) = 12(2 - x)(6 - x).$$

Of the two zeros, $x = 2$ and $x = 6$, only $x = 2$ lies in the interior of the function's domain and makes the critical-point list. The values of V at this one critical point and two endpoints are

Critical point value: $V(2) = 128$

Endpoint values: $V(0) = 0, \qquad V(6) = 0.$

The maximum volume is 128 in³. The cutout squares should be 2 in. on a side.   ∎

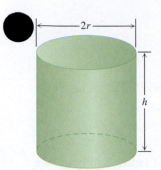

FIGURE 4.40 This one-liter can uses the least material when $h = 2r$ (Example 2).

EXAMPLE 2   You have been asked to design a one-liter can shaped like a right circular cylinder (Figure 4.40). What dimensions will use the least material?

Solution   *Volume of can:* If r and h are measured in centimeters, then the volume of the can in cubic centimeters is

$$\pi r^2 h = 1000. \qquad \text{1 liter} = 1000 \text{ cm}^3$$

Surface area of can: $\quad A = \underbrace{2\pi r^2}_{\substack{\text{circular} \\ \text{ends}}} + \underbrace{2\pi rh}_{\substack{\text{cylindrical} \\ \text{wall}}}$

How can we interpret the phrase "least material"? For a first approximation we can ignore the thickness of the material and the waste in manufacturing. Then we ask for dimensions r and h that make the total surface area as small as possible while satisfying the constraint $\pi r^2 h = 1000 \text{ cm}^3$.

To express the surface area as a function of one variable, we solve for one of the variables in $\pi r^2 h = 1000$ and substitute that expression into the surface area formula. Solving for h is easier:

$$h = \frac{1000}{\pi r^2}.$$

Thus,

$$\begin{aligned} A &= 2\pi r^2 + 2\pi rh \\ &= 2\pi r^2 + 2\pi r\left(\frac{1000}{\pi r^2}\right) \\ &= 2\pi r^2 + \frac{2000}{r}. \end{aligned}$$

Our goal is to find a value of $r > 0$ that minimizes the value of A. Figure 4.41 suggests that such a value exists.

Notice from the graph that for small r (a tall, thin cylindrical container), the term $2000/r$ dominates (see Section 2.6) and A is large. A very thin cylinder containing 1 liter is so tall that its surface area becomes very large. For large r (a short, wide cylindrical container), the term $2\pi r^2$ dominates and A again is large.

Since A is differentiable on $r > 0$, an interval with no endpoints, it can have a minimum value only where its first derivative is zero.

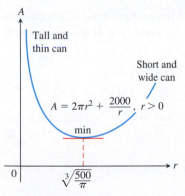

FIGURE 4.41 The graph of $A = 2\pi r^2 + 2000/r$ is concave up.

$$\frac{dA}{dr} = 4\pi r - \frac{2000}{r^2}$$

$$0 = 4\pi r - \frac{2000}{r^2} \qquad \text{Set } dA/dr = 0.$$

$$4\pi r^3 = 2000 \qquad \text{Multiply by } r^2.$$

$$r = \sqrt[3]{\frac{500}{\pi}} \approx 5.42 \qquad \text{Solve for } r.$$

What happens at $r = \sqrt[3]{500/\pi}$?

The second derivative

$$\frac{d^2A}{dr^2} = 4\pi + \frac{4000}{r^3}$$

is positive throughout the domain of A. The graph is therefore everywhere concave up and the value of A at $r = \sqrt[3]{500/\pi}$ is an absolute minimum.

The corresponding value of h (after a little algebra) is

$$h = \frac{1000}{\pi r^2} = 2\sqrt[3]{\frac{500}{\pi}} = 2r.$$

The one-liter can that uses the least material has height equal to twice the radius, here with $r \approx 5.42$ cm and $h \approx 10.84$ cm. ∎

Examples from Mathematics and Physics

EXAMPLE 3   A rectangle is to be inscribed in a semicircle of radius 2. What is the largest area the rectangle can have, and what are its dimensions?

Solution   Let $\left(x, \sqrt{4 - x^2}\right)$ be the coordinates of the corner of the rectangle obtained by placing the circle and rectangle in the coordinate plane (Figure 4.42). The length, height, and area of the rectangle can then be expressed in terms of the position x of the lower right-hand corner:

$$\text{Length: } 2x, \qquad \text{Height: } \sqrt{4 - x^2}, \qquad \text{Area: } 2x\sqrt{4 - x^2}.$$

Notice that the values of x are to be found in the interval $0 \le x \le 2$, where the selected corner of the rectangle lies.

Our goal is to find the absolute maximum value of the function

$$A(x) = 2x\sqrt{4 - x^2}$$

on the domain $[0, 2]$.

The derivative

$$\frac{dA}{dx} = \frac{-2x^2}{\sqrt{4 - x^2}} + 2\sqrt{4 - x^2}$$

is not defined when $x = 2$ and is equal to zero when

$$\frac{-2x^2}{\sqrt{4 - x^2}} + 2\sqrt{4 - x^2} = 0$$
$$-2x^2 + 2(4 - x^2) = 0$$
$$8 - 4x^2 = 0$$
$$x^2 = 2$$
$$x = \pm\sqrt{2}.$$

Of the two zeros, $x = \sqrt{2}$ and $x = -\sqrt{2}$, only $x = \sqrt{2}$ lies in the interior of A's domain and makes the critical-point list. The values of A at the endpoints and at this one critical point are

$$\text{Critical point value: } \quad A\left(\sqrt{2}\right) = 2\sqrt{2}\sqrt{4 - 2} = 4$$
$$\text{Endpoint values: } \qquad A(0) = 0, \qquad A(2) = 0.$$

The area has a maximum value of 4 when the rectangle is $\sqrt{4 - x^2} = \sqrt{2}$ units high and $2x = 2\sqrt{2}$ units long. ∎

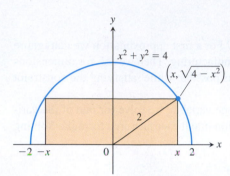

FIGURE 4.42 The rectangle inscribed in the semicircle in Example 3.

HISTORICAL BIOGRAPHY

Willebrord Snell van Royen
(1580–1626)
www.goo.gl/yEeoAi

EXAMPLE 4   The speed of light depends on the medium through which it travels, and is generally slower in denser media.

Fermat's principle in optics states that light travels from one point to another along a path for which the time of travel is a minimum. Describe the path that a ray of light will follow in going from a point A in a medium where the speed of light is c_1 to a point B in a second medium where its speed is c_2.

Solution   Since light traveling from A to B follows the quickest route, we look for a path that will minimize the travel time. We assume that A and B lie in the xy-plane and that the

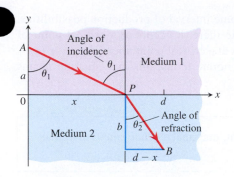

FIGURE 4.43 A light ray refracted (deflected from its path) as it passes from one medium to a denser medium (Example 4).

line separating the two media is the x-axis (Figure 4.43). We place A at coordinates $(0, a)$ and B at coordinates $(d, -b)$ in the xy-plane.

In a uniform medium, where the speed of light remains constant, "shortest time" means "shortest path," and the ray of light will follow a straight line. Thus the path from A to B will consist of a line segment from A to a boundary point P, followed by another line segment from P to B. Distance traveled equals rate times time, so

$$\text{Time} = \frac{\text{distance}}{\text{rate}}.$$

From Figure 4.43, the time required for light to travel from A to P is

$$t_1 = \frac{AP}{c_1} = \frac{\sqrt{a^2 + x^2}}{c_1}.$$

From P to B, the time is

$$t_2 = \frac{PB}{c_2} = \frac{\sqrt{b^2 + (d - x)^2}}{c_2}.$$

The time from A to B is the sum of these:

$$t = t_1 + t_2 = \frac{\sqrt{a^2 + x^2}}{c_1} + \frac{\sqrt{b^2 + (d - x)^2}}{c_2}.$$

This equation expresses t as a differentiable function of x whose domain is $[0, d]$. We want to find the absolute minimum value of t on this closed interval. We find the derivative

$$\frac{dt}{dx} = \frac{x}{c_1 \sqrt{a^2 + x^2}} - \frac{d - x}{c_2 \sqrt{b^2 + (d - x)^2}}$$

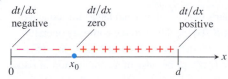

FIGURE 4.44 The sign pattern of dt/dx in Example 4.

and observe that it is continuous. In terms of the angles θ_1 and θ_2 in Figure 4.43,

$$\frac{dt}{dx} = \frac{\sin \theta_1}{c_1} - \frac{\sin \theta_2}{c_2}.$$

The function t has a negative derivative at $x = 0$ and a positive derivative at $x = d$. Since dt/dx is continuous over the interval $[0, d]$, by the Intermediate Value Theorem for continuous functions (Section 2.5), there is a point $x_0 \in [0, d]$ where $dt/dx = 0$ (Figure 4.44). There is only one such point because dt/dx is an increasing function of x (Exercise 62). At this unique point we then have

$$\frac{\sin \theta_1}{c_1} = \frac{\sin \theta_2}{c_2}.$$

This equation is **Snell's Law** or the **Law of Refraction**, and is an important principle in the theory of optics. It describes the path the ray of light follows. ■

Examples from Economics

Suppose that

$$r(x) = \text{the revenue from selling } x \text{ items}$$
$$c(x) = \text{the cost of producing the } x \text{ items}$$
$$p(x) = r(x) - c(x) = \text{the profit from producing and selling } x \text{ items}.$$

Although x is usually an integer in many applications, we can learn about the behavior of these functions by defining them for all nonzero real numbers and by assuming they are differentiable functions. Economists use the terms **marginal revenue**, **marginal cost**, and **marginal profit** to name the derivatives $r'(x)$, $c'(x)$, and $p'(x)$ of the revenue, cost, and profit functions. Let's consider the relationship of the profit p to these derivatives.

If $r(x)$ and $c(x)$ are differentiable for x in some interval of production possibilities, and if $p(x) = r(x) - c(x)$ has a maximum value there, it occurs at a critical point of $p(x)$ or at an endpoint of the interval. If it occurs at a critical point, then $p'(x) = r'(x) - c'(x) = 0$ and we see that $r'(x) = c'(x)$. In economic terms, this last equation means that

> At a production level yielding maximum profit, marginal revenue equals marginal cost (Figure 4.45).

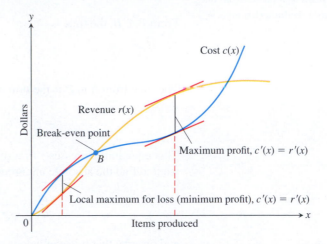

FIGURE 4.45 The graph of a typical cost function starts concave down and later turns concave up. It crosses the revenue curve at the break-even point B. To the left of B, the company operates at a loss. To the right, the company operates at a profit, with the maximum profit occurring where $c'(x) = r'(x)$. Farther to the right, cost exceeds revenue (perhaps because of a combination of rising labor and material costs and market saturation) and production levels become unprofitable again.

EXAMPLE 5   Suppose that $r(x) = 9x$ and $c(x) = x^3 - 6x^2 + 15x$, where x represents millions of MP3 players produced. Is there a production level that maximizes profit? If so, what is it?

Solution   Notice that $r'(x) = 9$ and $c'(x) = 3x^2 - 12x + 15$.

$$3x^2 - 12x + 15 = 9 \qquad \text{Set } c'(x) = r'(x).$$
$$3x^2 - 12x + 6 = 0$$

The two solutions of the quadratic equation are

$$x_1 = \frac{12 - \sqrt{72}}{6} = 2 - \sqrt{2} \approx 0.586 \qquad \text{and}$$

$$x_2 = \frac{12 + \sqrt{72}}{6} = 2 + \sqrt{2} \approx 3.414.$$

The possible production levels for maximum profit are $x \approx 0.586$ million MP3 players or $x \approx 3.414$ million. The second derivative of $p(x) = r(x) - c(x)$ is $p''(x) = -c''(x)$ since $r''(x)$ is everywhere zero. Thus, $p''(x) = 6(2 - x)$, which is negative at $x = 2 + \sqrt{2}$ and positive at $x = 2 - \sqrt{2}$. By the Second Derivative Test, a maximum profit occurs at about $x = 3.414$ (where revenue exceeds costs) and maximum loss occurs at about $x = 0.586$. The graphs of $r(x)$ and $c(x)$ are shown in Figure 4.46. ∎

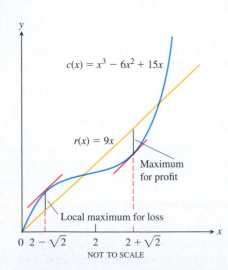

FIGURE 4.46 The cost and revenue curves for Example 5.

EXAMPLE 6   A cabinetmaker uses cherry wood to produce 5 desks each day. Each delivery of one container of wood is $5000, whereas the storage of that material is $10 per day per unit stored, where a unit is the amount of material needed by her to produce 1 desk. How much material should be ordered each time, and how often should the material be delivered, to minimize her average daily cost in the production cycle between deliveries?

Solution   If she asks for a delivery every x days, then she must order $5x$ units to have enough material for that delivery cycle. The *average* amount in storage is approximately one-half of the delivery amount, or $5x/2$. Thus, the cost of delivery and storage for each cycle is approximately

$$\text{Cost per cycle} = \text{delivery costs} + \text{storage costs}$$

$$\text{Cost per cycle} = \underbrace{5000}_{\substack{\text{delivery}\\\text{cost}}} + \underbrace{\left(\frac{5x}{2}\right)}_{\substack{\text{average}\\\text{amount stored}}} \cdot \underbrace{x}_{\substack{\text{number of}\\\text{days stored}}} \cdot \underbrace{10}_{\substack{\text{storage cost}\\\text{per day}}}$$

We compute the *average daily cost* $c(x)$ by dividing the cost per cycle by the number of days x in the cycle (see Figure 4.47).

$$c(x) = \frac{5000}{x} + 25x, \qquad x > 0.$$

As $x \to 0$ and as $x \to \infty$, the average daily cost becomes large. So we expect a minimum to exist, but where? Our goal is to determine the number of days x between deliveries that provides the absolute minimum cost.

We find the critical points by determining where the derivative is equal to zero:

$$c'(x) = -\frac{500}{x^2} + 25 = 0$$

$$x = \pm\sqrt{200} \approx \pm 14.14.$$

Of the two critical points, only $\sqrt{200}$ lies in the domain of $c(x)$. The critical point value of the average daily cost is

$$c\left(\sqrt{200}\right) = \frac{5000}{\sqrt{200}} + 25\sqrt{200} = 500\sqrt{2} \approx \$707.11.$$

We note that $c(x)$ is defined over the open interval $(0, \infty)$ with $c''(x) = 10000/x^3 > 0$. Thus, an absolute minimum exists at $x = \sqrt{200} \approx 14.14$ days.

The cabinetmaker should schedule a delivery of $5(14) = 70$ units of wood every 14 days.   ∎

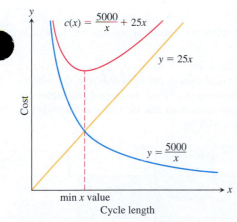

FIGURE 4.47   The average daily cost $c(x)$ is the sum of a hyperbola and a linear function (Example 6).

EXERCISES 4.6

Mathematical Applications

Whenever you are maximizing or minimizing a function of a single variable, we urge you to graph it over the domain that is appropriate to the problem you are solving. The graph will provide insight before you calculate and will furnish a visual context for understanding your answer.

1. **Minimizing perimeter**   What is the smallest perimeter possible for a rectangle whose area is 16 in², and what are its dimensions?

2. Show that among all rectangles with an 8-m perimeter, the one with largest area is a square.

3. The figure shows a rectangle inscribed in an isosceles right triangle whose hypotenuse is 2 units long.

   a. Express the y-coordinate of P in terms of x. (*Hint:* Write an equation for the line AB.)

   b. Express the area of the rectangle in terms of x.

c. What is the largest area the rectangle can have, and what are its dimensions?

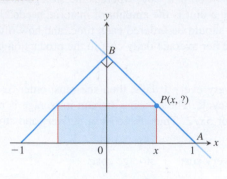

4. A rectangle has its base on the *x*-axis and its upper two vertices on the parabola $y = 12 - x^2$. What is the largest area the rectangle can have, and what are its dimensions?

5. You are planning to make an open rectangular box from an 8-in.-by-15-in. piece of cardboard by cutting congruent squares from the corners and folding up the sides. What are the dimensions of the box of largest volume you can make this way, and what is its volume?

6. You are planning to close off a corner of the first quadrant with a line segment 20 units long running from $(a, 0)$ to $(0, b)$. Show that the area of the triangle enclosed by the segment is largest when $a = b$.

7. The best fencing plan A rectangular plot of farmland will be bounded on one side by a river and on the other three sides by a single-strand electric fence. With 800 m of wire at your disposal, what is the largest area you can enclose, and what are its dimensions?

8. The shortest fence A 216 m² rectangular pea patch is to be enclosed by a fence and divided into two equal parts by another fence parallel to one of the sides. What dimensions for the outer rectangle will require the smallest total length of fence? How much fence will be needed?

9. Designing a tank Your iron works has contracted to design and build a 500 ft³, square-based, open-top, rectangular steel holding tank for a paper company. The tank is to be made by welding thin stainless steel plates together along their edges. As the production engineer, your job is to find dimensions for the base and height that will make the tank weigh as little as possible.

a. What dimensions do you tell the shop to use?

b. Briefly describe how you took weight into account.

10. Catching rainwater A 1125 ft³ open-top rectangular tank with a square base *x* ft on a side and *y* ft deep is to be built with its top flush with the ground to catch runoff water. The costs associated with the tank involve not only the material from which the tank is made but also an excavation charge proportional to the product *xy*.

a. If the total cost is

$$c = 5(x^2 + 4xy) + 10xy,$$

what values of *x* and *y* will minimize it?

b. Give a possible scenario for the cost function in part (a).

11. Designing a poster You are designing a rectangular poster to contain 50 in² of printing with a 4-in. margin at the top and bottom and a 2-in. margin at each side. What overall dimensions will minimize the amount of paper used?

12. Find the volume of the largest right circular cone that can be inscribed in a sphere of radius 3.

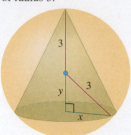

13. Two sides of a triangle have lengths *a* and *b*, and the angle between them is θ. What value of θ will maximize the triangle's area? (*Hint:* $A = (1/2)ab\sin\theta$.)

14. Designing a can What are the dimensions of the lightest open-top right circular cylindrical can that will hold a volume of 1000 cm³? Compare the result here with the result in Example 2.

15. Designing a can You are designing a 1000 cm³ right circular cylindrical can whose manufacture will take waste into account. There is no waste in cutting the aluminum for the side, but the top and bottom of radius *r* will be cut from squares that measure 2*r* units on a side. The total amount of aluminum used up by the can will therefore be

$$A = 8r^2 + 2\pi rh$$

rather than the $A = 2\pi r^2 + 2\pi rh$ in Example 2. In Example 2, the ratio of *h* to *r* for the most economical can was 2 to 1. What is the ratio now?

T **16. Designing a box with a lid** A piece of cardboard measures 10 in. by 15 in. Two equal squares are removed from the corners of a 10-in. side as shown in the figure. Two equal rectangles are removed from the other corners so that the tabs can be folded to form a rectangular box with lid.

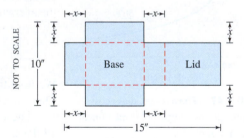

a. Write a formula $V(x)$ for the volume of the box.

b. Find the domain of *V* for the problem situation and graph *V* over this domain.

c. Use a graphical method to find the maximum volume and the value of *x* that gives it.

d. Confirm your result in part (c) analytically.

T **17. Designing a suitcase** A 24-in.-by-36-in. sheet of cardboard is folded in half to form a 24-in.-by-18-in. rectangle as shown in the accompanying figure. Then four congruent squares of side length *x* are cut from the corners of the folded rectangle. The sheet is unfolded, and the six tabs are folded up to form a box with sides and a lid.

a. Write a formula $V(x)$ for the volume of the box.

b. Find the domain of *V* for the problem situation and graph *V* over this domain.

c. Use a graphical method to find the maximum volume and the value of x that gives it.

d. Confirm your result in part (c) analytically.

e. Find a value of x that yields a volume of 1120 in³.

f. Write a paragraph describing the issues that arise in part (b).

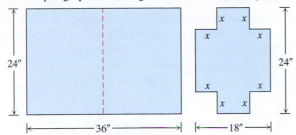

The sheet is then unfolded.

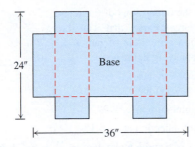

18. A rectangle is to be inscribed under the arch of the curve $y = 4\cos(0.5x)$ from $x = -\pi$ to $x = \pi$. What are the dimensions of the rectangle with largest area, and what is the largest area?

19. Find the dimensions of a right circular cylinder of maximum volume that can be inscribed in a sphere of radius 10 cm. What is the maximum volume?

20. a. The U.S. Postal Service will accept a box for domestic shipment only if the sum of its length and girth (distance around) does not exceed 108 in. What dimensions will give a box with a square end the largest possible volume?

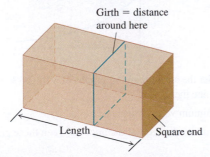

T b. Graph the volume of a 108-in. box (length plus girth equals 108 in.) as a function of its length and compare what you see with your answer in part (a).

21. (*Continuation of Exercise 20.*)

a. Suppose that instead of having a box with square ends you have a box with square sides so that its dimensions are h by h by w and the girth is $2h + 2w$. What dimensions will give the box its largest volume now?

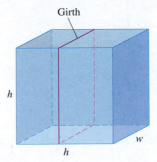

T b. Graph the volume as a function of h and compare what you see with your answer in part (a).

22. A window is in the form of a rectangle surmounted by a semicircle. The rectangle is of clear glass, whereas the semicircle is of tinted glass that transmits only half as much light per unit area as clear glass does. The total perimeter is fixed. Find the proportions of the window that will admit the most light. Neglect the thickness of the frame.

23. A silo (base not included) is to be constructed in the form of a cylinder surmounted by a hemisphere. The cost of construction per square unit of surface area is twice as great for the hemisphere as it is for the cylindrical sidewall. Determine the dimensions to be used if the volume is fixed and the cost of construction is to be kept to a minimum. Neglect the thickness of the silo and waste in construction.

24. The trough in the figure is to be made to the dimensions shown. Only the angle θ can be varied. What value of θ will maximize the trough's volume?

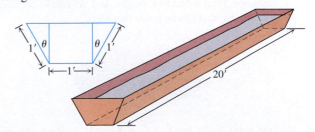

25. Paper folding A rectangular sheet of 8.5-in.-by-11-in. paper is placed on a flat surface. One of the corners is placed on the opposite longer edge, as shown in the figure, and held there as the paper is smoothed flat. The problem is to make the length of the crease as small as possible. Call the length L. Try it with paper.

  a. Show that $L^2 = 2x^3/(2x - 8.5)$.

  b. What value of x minimizes L^2?

  c. What is the minimum value of L?

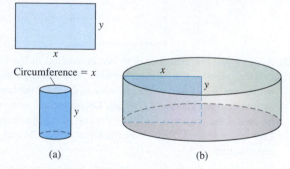

26. Constructing cylinders Compare the answers to the following two construction problems.

  a. A rectangular sheet of perimeter 36 cm and dimensions x cm by y cm is to be rolled into a cylinder as shown in part (a) of the figure. What values of x and y give the largest volume?

  b. The same sheet is to be revolved about one of the sides of length y to sweep out the cylinder as shown in part (b) of the figure. What values of x and y give the largest volume?

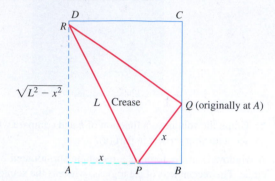

       (a)   (b)

27. Constructing cones A right triangle whose hypotenuse is $\sqrt{3}$ m long is revolved about one of its legs to generate a right circular cone. Find the radius, height, and volume of the cone of greatest volume that can be made this way.

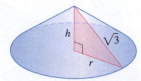

28. Find the point on the line $\dfrac{x}{a} + \dfrac{y}{b} = 1$ that is closest to the origin.

29. Find a positive number for which the sum of it and its reciprocal is the smallest (least) possible.

30. Find a positive number for which the sum of its reciprocal and four times its square is the smallest possible.

31. A wire b m long is cut into two pieces. One piece is bent into an equilateral triangle and the other is bent into a circle. If the sum of the areas enclosed by each part is a minimum, what is the length of each part?

32. Answer Exercise 31 if one piece is bent into a square and the other into a circle.

33. Determine the dimensions of the rectangle of largest area that can be inscribed in the right triangle shown in the accompanying figure.

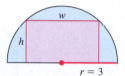

34. Determine the dimensions of the rectangle of largest area that can be inscribed in a semicircle of radius 3. (See accompanying figure.)

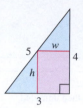

35. What value of a makes $f(x) = x^2 + (a/x)$ have

  a. a local minimum at $x = 2$?

  b. a point of inflection at $x = 1$?

36. What values of a and b make $f(x) = x^3 + ax^2 + bx$ have

  a. a local maximum at $x = -1$ and a local minimum at $x = 3$?

  b. a local minimum at $x = 4$ and a point of inflection at $x = 1$?

37. A right circular cone is circumscribed by a sphere of radius 1. Determine the height h and radius r of the cone of maximum volume.

38. Determine the dimensions of the inscribed rectangle of maximum area.

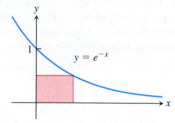

39. Consider the accompanying graphs of $y = 2x + 3$ and $y = \ln x$. Determine the

  a. minimum vertical distance

  b. minimum horizontal distance between these graphs.

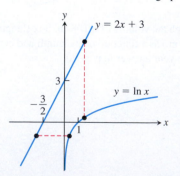

40. Find the point on the graph of $y = 20x^3 + 60x - 3x^5 - 5x^4$ with the largest slope.

41. Among all triangles in the first quadrant formed by the x-axis, the y-axis, and tangent lines to the graph of $y = 3x - x^2$, what is the smallest possible area?

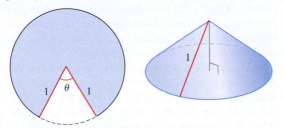

42. A cone is formed from a circular piece of material of radius 1 meter by removing a section of angle θ and then joining the two straight edges. Determine the largest possible volume for the cone.

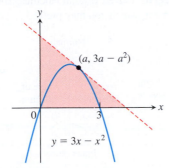

Physical Applications

43. Vertical motion The height above ground of an object moving vertically is given by

$$s = -16t^2 + 96t + 112,$$

with s in feet and t in seconds. Find

a. the object's velocity when $t = 0$;

b. its maximum height and when it occurs;

c. its velocity when $s = 0$.

44. Quickest route Jane is 2 mi offshore in a boat and wishes to reach a coastal village 6 mi down a straight shoreline from the point nearest the boat. She can row 2 mph and can walk 5 mph. Where should she land her boat to reach the village in the least amount of time?

45. Shortest beam The 8-ft wall shown here stands 27 ft from the building. Find the length of the shortest straight beam that will reach to the side of the building from the ground outside the wall.

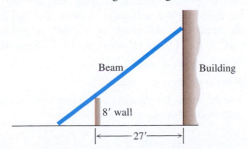

46. Motion on a line The positions of two particles on the s-axis are $s_1 = \sin t$ and $s_2 = \sin (t + \pi/3)$, with s_1 and s_2 in meters and t in seconds.

a. At what time(s) in the interval $0 \leq t \leq 2\pi$ do the particles meet?

b. What is the farthest apart that the particles ever get?

c. When in the interval $0 \leq t \leq 2\pi$ is the distance between the particles changing the fastest?

47. The intensity of illumination at any point from a light source is proportional to the square of the reciprocal of the distance between the point and the light source. Two lights, one having an intensity eight times that of the other, are 6 m apart. How far from the stronger light is the total illumination least?

48. Projectile motion The *range R* of a projectile fired from the origin over horizontal ground is the distance from the origin to the point of impact. If the projectile is fired with an initial velocity v_0 at an angle α with the horizontal, then in Chapter 13 we find that

$$R = \frac{v_0^2}{g} \sin 2\alpha,$$

where g is the downward acceleration due to gravity. Find the angle α for which the range R is the largest possible.

T 49. Strength of a beam The strength S of a rectangular wooden beam is proportional to its width times the square of its depth. (See the accompanying figure.)

a. Find the dimensions of the strongest beam that can be cut from a 12-in.-diameter cylindrical log.

b. Graph S as a function of the beam's width w, assuming the proportionality constant to be $k = 1$. Reconcile what you see with your answer in part (a).

d. On the same screen, graph S as a function of the beam's depth d, again taking $k = 1$. Compare the graphs with one another and with your answer in part (a). What would be the effect of changing to some other value of k? Try it.

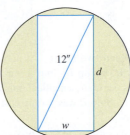

T 50. Stiffness of a beam The stiffness S of a rectangular beam is proportional to its width times the cube of its depth.

a. Find the dimensions of the stiffest beam that can be cut from a 12-in.-diameter cylindrical log.

b. Graph S as a function of the beam's width w, assuming the proportionality constant to be $k = 1$. Reconcile what you see with your answer in part (a).

c. On the same screen, graph S as a function of the beam's depth d, again taking $k = 1$. Compare the graphs with one another and with your answer in part (a). What would be the effect of changing to some other value of k? Try it.

51. Frictionless cart A small frictionless cart, attached to the wall by a spring, is pulled 10 cm from its rest position and released at time $t = 0$ to roll back and forth for 4 sec. Its position at time t is $s = 10 \cos \pi t$.

a. What is the cart's maximum speed? When is the cart moving that fast? Where is it then? What is the magnitude of the acceleration then?

b. Where is the cart when the magnitude of the acceleration is greatest? What is the cart's speed then?

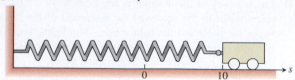

52. Two masses hanging side by side from springs have positions $s_1 = 2 \sin t$ and $s_2 = \sin 2t$, respectively.

a. At what times in the interval $0 < t$ do the masses pass each other? (*Hint:* $\sin 2t = 2 \sin t \cos t.$)

b. When in the interval $0 \le t \le 2\pi$ is the vertical distance between the masses the greatest? What is this distance? (*Hint:* $\cos 2t = 2 \cos^2 t - 1.$)

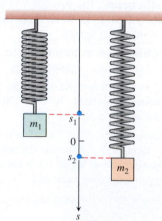

53. Distance between two ships At noon, ship A was 12 nautical miles due north of ship B. Ship A was sailing south at 12 knots (nautical miles per hour; a nautical mile is 2000 yd) and continued to do so all day. Ship B was sailing east at 8 knots and continued to do so all day.

a. Start counting time with $t = 0$ at noon and express the distance s between the ships as a function of t.

b. How rapidly was the distance between the ships changing at noon? One hour later?

c. The visibility that day was 5 nautical miles. Did the ships ever sight each other?

T **d.** Graph s and ds/dt together as functions of t for $-1 \le t \le 3$, using different colors if possible. Compare the graphs and reconcile what you see with your answers in parts (b) and (c).

e. The graph of ds/dt looks as if it might have a horizontal asymptote in the first quadrant. This in turn suggests that ds/dt approaches a limiting value as $t \to \infty$. What is this value? What is its relation to the ships' individual speeds?

54. Fermat's principle in optics Light from a source A is reflected by a plane mirror to a receiver at point B, as shown in the accompanying figure. Show that for the light to obey Fermat's principle, the angle of incidence must equal the angle of reflection, both measured from the line normal to the reflecting surface. (This result can also be derived without calculus. There is a purely geometric argument, which you may prefer.)

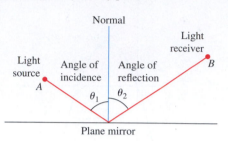

55. Tin pest When metallic tin is kept below 13.2°C, it slowly becomes brittle and crumbles to a gray powder. Tin objects eventually crumble to this gray powder spontaneously if kept in a cold climate for years. The Europeans who saw tin organ pipes in their churches crumble away years ago called the change *tin pest* because it seemed to be contagious, and indeed it was, for the gray powder is a catalyst for its own formation.

A *catalyst* for a chemical reaction is a substance that controls the rate of reaction without undergoing any permanent change in itself. An *autocatalytic reaction* is one whose product is a catalyst for its own formation. Such a reaction may proceed slowly at first if the amount of catalyst present is small and slowly again at the end, when most of the original substance is used up. But in between, when both the substance and its catalyst product are abundant, the reaction proceeds at a faster pace.

In some cases, it is reasonable to assume that the rate $v = dx/dt$ of the reaction is proportional both to the amount of the original substance present and to the amount of product. That is, v may be considered to be a function of x alone, and

$$v = kx(a - x) = kax - kx^2,$$

where

$x = $ the amount of product

$a = $ the amount of substance at the beginning

$k = $ a positive constant.

At what value of x does the rate v have a maximum? What is the maximum value of v?

56. Airplane landing path An airplane is flying at altitude H when it begins its descent to an airport runway that is at horizontal ground distance L from the airplane, as shown in the figure on the next page. Assume that the landing path of the airplane is the graph of a cubic polynomial function $y = ax^3 + bx^2 + cx + d$, where $y(-L) = H$ and $y(0) = 0$.

a. What is dy/dx at $x = 0$?

b. What is dy/dx at $x = -L$?

c. Use the values for dy/dx at $x = 0$ and $x = -L$ together with $y(0) = 0$ and $y(-L) = H$ to show that

$$y(x) = H\left[2\left(\frac{x}{L}\right)^3 + 3\left(\frac{x}{L}\right)^2\right].$$

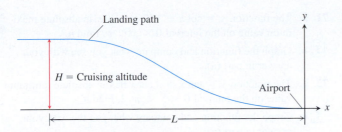

Landing path

H = Cruising altitude

Airport

L

Business and Economics

57. It costs you c dollars each to manufacture and distribute backpacks. If the backpacks sell at x dollars each, the number sold is given by

$$n = \frac{a}{x - c} + b(100 - x),$$

where a and b are positive constants. What selling price will bring a maximum profit?

58. You operate a tour service that offers the following rates:

$200 per person if 50 people (the minimum number to book the tour) go on the tour.

For each additional person, up to a maximum of 80 people total, the rate per person is reduced by $2.

It costs $6000 (a fixed cost) plus $32 per person to conduct the tour. How many people does it take to maximize your profit?

59. Wilson lot size formula One of the formulas for inventory management says that the average weekly cost of ordering, paying for, and holding merchandise is

$$A(q) = \frac{km}{q} + cm + \frac{hq}{2},$$

where q is the quantity you order when things run low (shoes, radios, brooms, or whatever the item might be), k is the cost of placing an order (the same, no matter how often you order), c is the cost of one item (a constant), m is the number of items sold each week (a constant), and h is the weekly holding cost per item (a constant that takes into account things such as space, utilities, insurance, and security).

a. Your job, as the inventory manager for your store, is to find the quantity that will minimize $A(q)$. What is it? (The formula you get for the answer is called the *Wilson lot size formula*.)

b. Shipping costs sometimes depend on order size. When they do, it is more realistic to replace k by $k + bq$, the sum of k and a constant multiple of q. What is the most economical quantity to order now?

60. Production level Prove that the production level (if any) at which average cost is smallest is a level at which the average cost equals marginal cost.

61. Show that if $r(x) = 6x$ and $c(x) = x^3 - 6x^2 + 15x$ are your revenue and cost functions, then the best you can do is break even (have revenue equal cost).

62. Production level Suppose that $c(x) = x^3 - 20x^2 + 20{,}000x$ is the cost of manufacturing x items. Find a production level that will minimize the average cost of making x items.

63. You are to construct an open rectangular box with a square base and a volume of 48 ft^3. If material for the bottom costs $6/ft^2 and material for the sides costs $4/ft^2, what dimensions will result in the least expensive box? What is the minimum cost?

64. The 800-room Mega Motel chain is filled to capacity when the room charge is $50 per night. For each $10 increase in room charge, 40 fewer rooms are filled each night. What charge per room will result in the maximum revenue per night?

Biology

65. Sensitivity to medicine (*Continuation of Exercise 74, Section 3.3.*) Find the amount of medicine to which the body is most sensitive by finding the value of M that maximizes the derivative dR/dM, where

$$R = M^2\left(\frac{C}{2} - \frac{M}{3}\right)$$

and C is a constant.

66. How we cough

a. When we cough, the trachea (windpipe) contracts to increase the velocity of the air going out. This raises the questions of how much it should contract to maximize the velocity and whether it really contracts that much when we cough.

Under reasonable assumptions about the elasticity of the tracheal wall and about how the air near the wall is slowed by friction, the average flow velocity v can be modeled by the equation

$$v = c(r_0 - r)r^2 \text{ cm/sec}, \qquad \frac{r_0}{2} \le r \le r_0,$$

where r_0 is the rest radius of the trachea in centimeters and c is a positive constant whose value depends in part on the length of the trachea.

Show that v is greatest when $r = (2/3)r_0$; that is, when the trachea is about 33% contracted. The remarkable fact is that X-ray photographs confirm that the trachea contracts about this much during a cough.

T **b.** Take r_0 to be 0.5 and c to be 1 and graph v over the interval $0 \le r \le 0.5$. Compare what you see with the claim that v is at a maximum when $r = (2/3)r_0$.

Theory and Examples

67. An inequality for positive integers Show that if a, b, c, and d are positive integers, then

$$\frac{(a^2 + 1)(b^2 + 1)(c^2 + 1)(d^2 + 1)}{abcd} \ge 16.$$

68. The derivative dt/dx in Example 4

   a. Show that

$$f(x) = \frac{x}{\sqrt{a^2 + x^2}}$$

   is an increasing function of x.

   b. Show that

$$g(x) = \frac{d - x}{\sqrt{b^2 + (d - x)^2}}$$

   is a decreasing function of x.

   c. Show that

$$\frac{dt}{dx} = \frac{x}{c_1 \sqrt{a^2 + x^2}} - \frac{d - x}{c_2 \sqrt{b^2 + (d - x)^2}}$$

   is an increasing function of x.

69. Let $f(x)$ and $g(x)$ be the differentiable functions graphed here. Point c is the point where the vertical distance between the curves is the greatest. Is there anything special about the tangents to the two curves at c? Give reasons for your answer.

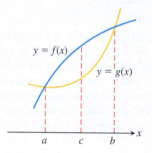

70. You have been asked to determine whether the function $f(x) = 3 + 4\cos x + \cos 2x$ is ever negative.

   a. Explain why you need to consider values of x only in the interval $[0, 2\pi]$.

   b. Is f ever negative? Explain.

71. a. The function $y = \cot x - \sqrt{2} \csc x$ has an absolute maximum value on the interval $0 < x < \pi$. Find it.

   T b. Graph the function and compare what you see with your answer in part (a).

72. a. The function $y = \tan x + 3 \cot x$ has an absolute minimum value on the interval $0 < x < \pi/2$. Find it.

   T b. Graph the function and compare what you see with your answer in part (a).

73. a. How close does the curve $y = \sqrt{x}$ come to the point $(3/2, 0)$? (*Hint:* If you minimize the *square* of the distance, you can avoid square roots.)

   T b. Graph the distance function $D(x)$ and $y = \sqrt{x}$ together and reconcile what you see with your answer in part (a).

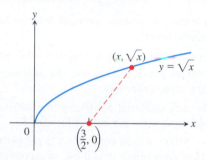

74. a. How close does the semicircle $y = \sqrt{16 - x^2}$ come to the point $(1, \sqrt{3})$?

   T b. Graph the distance function and $y = \sqrt{16 - x^2}$ together and reconcile what you see with your answer in part (a).

4.7 Newton's Method

For thousands of years, one of the main goals of mathematics has been to find solutions to equations. For linear equations $(ax + b = 0)$, and for quadratic equations $(ax^2 + bx + c = 0)$, we can explicitly solve for a solution. However, for most equations there is no simple formula that gives the solutions.

In this section we study a numerical method called *Newton's method* or the *Newton–Raphson method*, which is a technique to approximate the solutions to an equation $f(x) = 0$. Newton's method estimates the solutions using tangent lines of the graph of $y = f(x)$ near the points where f is zero. A value of x where f is zero is called a *root* of the function f and a *solution* of the equation $f(x) = 0$. Newton's method is both powerful and efficient, and it has numerous applications in engineering and other fields where solutions to complicated equations are needed.

Procedure for Newton's Method

The goal of Newton's method for estimating a solution of an equation $f(x) = 0$ is to produce a sequence of approximations that approach the solution. We pick the first number x_0 of

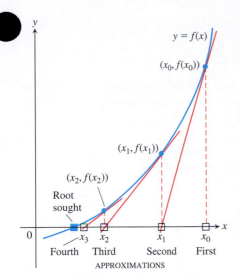

FIGURE 4.48 Newton's method starts with an initial guess x_0 and (under favorable circumstances) improves the guess one step at a time.

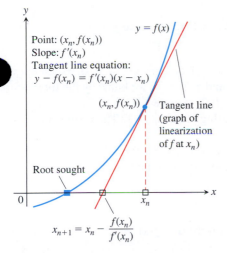

FIGURE 4.49 The geometry of the successive steps of Newton's method. From x_n we go up to the curve and follow the tangent line down to find x_{n+1}.

the sequence. Then, under favorable circumstances, the method moves step by step toward a point where the graph of f crosses the x-axis (Figure 4.48). At each step the method approximates a zero of f with a zero of one of its linearizations. Here is how it works.

The initial estimate, x_0, may be found by graphing or just plain guessing. The method then uses the tangent to the curve $y = f(x)$ at $(x_0, f(x_0))$ to approximate the curve, calling the point x_1 where the tangent meets the x-axis (Figure 4.48). The number x_1 is usually a better approximation to the solution than is x_0. The point x_2 where the tangent to the curve at $(x_1, f(x_1))$ crosses the x-axis is the next approximation in the sequence. We continue, using each approximation to generate the next, until we are close enough to the root to stop.

We can derive a formula for generating the successive approximations in the following way. Given the approximation x_n, the point-slope equation for the tangent to the curve at $(x_n, f(x_n))$ is

$$y = f(x_n) + f'(x_n)(x - x_n).$$

We can find where it crosses the x-axis by setting $y = 0$ (Figure 4.49):

$$0 = f(x_n) + f'(x_n)(x - x_n)$$

$$-\frac{f(x_n)}{f'(x_n)} = x - x_n$$

$$x = x_n - \frac{f(x_n)}{f'(x_n)} \qquad \text{If } f'(x_n) \neq 0$$

This value of x is the next approximation x_{n+1}. Here is a summary of Newton's method.

Newton's Method

1. Guess a first approximation to a solution of the equation $f(x) = 0$. A graph of $y = f(x)$ may help.

2. Use the first approximation to get a second, the second to get a third, and so on, using the formula

$$x_{n+1} = x_n - \frac{f(x_n)}{f'(x_n)}, \qquad \text{if } f'(x_n) \neq 0. \tag{1}$$

Applying Newton's Method

Applications of Newton's method generally involve many numerical computations, making them well suited for computers or calculators. Nevertheless, even when the calculations are done by hand (which may be very tedious), they give a powerful way to find solutions of equations.

In our first example, we find decimal approximations to $\sqrt{2}$ by estimating the positive root of the equation $f(x) = x^2 - 2 = 0$.

EXAMPLE 1   Approximate the positive root of the equation

$$f(x) = x^2 - 2 = 0.$$

Solution   With $f(x) = x^2 - 2$ and $f'(x) = 2x$, Equation (1) becomes

$$x_{n+1} = x_n - \frac{x_n{}^2 - 2}{2x_n}$$

$$= x_n - \frac{x_n}{2} + \frac{1}{x_n}$$

$$= \frac{x_n}{2} + \frac{1}{x_n}.$$

The equation

$$x_{n+1} = \frac{x_n}{2} + \frac{1}{x_n}$$

enables us to go from each approximation to the next with just a few keystrokes. With the starting value $x_0 = 1$, we get the results in the first column of the following table. (To five decimal places, or, equivalently, to six digits, $\sqrt{2} = 1.41421$.)

	Error	Number of correct digits
$x_0 = 1$	-0.41421	1
$x_1 = 1.5$	0.08579	1
$x_2 = 1.41667$	0.00246	3
$x_3 = 1.41422$	0.00001	5

Newton's method is the method used by most software applications to calculate roots because it converges so fast (more about this later). If the arithmetic in the table in Example 1 had been carried to 13 decimal places instead of 5, then going one step further would have given $\sqrt{2}$ correctly to more than 10 decimal places.

EXAMPLE 2 Find the x-coordinate of the point where the curve $y = x^3 - x$ crosses the horizontal line $y = 1$.

Solution The curve crosses the line when $x^3 - x = 1$ or $x^3 - x - 1 = 0$. When does $f(x) = x^3 - x - 1$ equal zero? Since $f(1) = -1$ and $f(2) = 5$, we know by the Intermediate Value Theorem there is a root in the interval $(1, 2)$ (Figure 4.50).

We apply Newton's method to f with the starting value $x_0 = 1$. The results are displayed in Table 4.1 and Figure 4.51.

At $n = 5$, we come to the result $x_6 = x_5 = 1.3247\ 17957$. When $x_{n+1} = x_n$, Equation (1) shows that $f(x_n) = 0$, up to the accuracy of our computation. We have found a solution of $f(x) = 0$ to nine decimals.

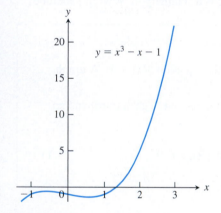

FIGURE 4.50 The graph of $f(x) = x^3 - x - 1$ crosses the x-axis once; this is the root we want to find (Example 2).

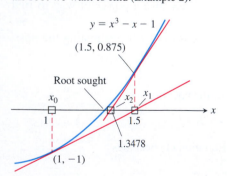

FIGURE 4.51 The first three x-values in Table 4.1 (four decimal places).

TABLE 4.1 The result of applying Newton's method to $f(x) = x^3 - x - 1$ with $x_0 = 1$

n	x_n	$f(x_n)$	$f'(x_n)$	$x_{n+1} = x_n - \dfrac{f(x_n)}{f'(x_n)}$
0	1	-1	2	1.5
1	1.5	0.875	5.75	1.3478 26087
2	1.3478 26087	0.1006 82173	4.4499 05482	1.3252 00399
3	1.3252 00399	0.0020 58362	4.2684 68292	1.3247 18174
4	1.3247 18174	0.0000 00924	4.2646 34722	1.3247 17957
5	1.3247 17957	$-1.8672\text{E-}13$	4.2646 32999	1.3247 17957

In Figure 4.52 we have indicated that the process in Example 2 might have started at the point $B_0(3, 23)$ on the curve, with $x_0 = 3$. Point B_0 is quite far from the x-axis, but the tangent at B_0 crosses the x-axis at about $(2.12, 0)$, so x_1 is still an improvement over x_0. If

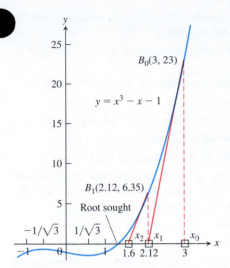

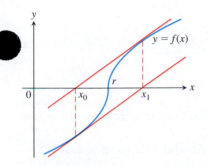

FIGURE 4.52 Any starting value x_0 to the right of $x = 1/\sqrt{3}$ will lead to the root in Example 2.

FIGURE 4.53 Newton's method fails to converge. You go from x_0 to x_1 and back to x_0, never getting any closer to r.

we use Equation (1) repeatedly as before, with $f(x) = x^3 - x - 1$ and $f'(x) = 3x^2 - 1$, we obtain the nine-place solution $x_7 = x_6 = 1.3247\ 17957$ in seven steps.

Convergence of the Approximations

In Chapter 10 we define precisely the idea of *convergence* for the approximations x_n in Newton's method. Intuitively, we mean that as the number n of approximations increases without bound, the values x_n get arbitrarily close to the desired root r. (This notion is similar to the idea of the limit of a function $g(t)$ as t approaches infinity, as defined in Section 2.6.)

In practice, Newton's method usually gives convergence with impressive speed, but this is not guaranteed. One way to test convergence is to begin by graphing the function to estimate a good starting value for x_0. You can test that you are getting closer to a zero of the function by checking that $|f(x_n)|$ is approaching zero, and you can check that the approximations are converging by evaluating $|x_n - x_{n+1}|$.

Newton's method does not always converge. For instance, if

$$f(x) = \begin{cases} -\sqrt{r - x}, & x < r \\ \sqrt{x - r}, & x \geq r, \end{cases}$$

the graph will be like the one in Figure 4.53. If we begin with $x_0 = r - h$, we get $x_1 = r + h$, and successive approximations go back and forth between these two values. No amount of iteration brings us closer to the root than our first guess.

If Newton's method does converge, it converges to a root. Be careful, however. There are situations in which the method appears to converge but no root is there. Fortunately, such situations are rare.

When Newton's method converges to a root, it may not be the root you have in mind. Figure 4.54 shows two ways this can happen.

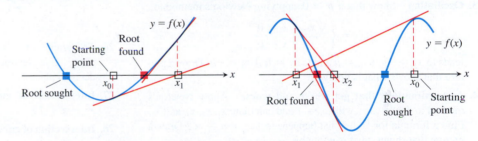

FIGURE 4.54 If you start too far away, Newton's method may miss the root you want.

EXERCISES 4.7

Root Finding

1. Use Newton's method to estimate the solutions of the equation $x^2 + x - 1 = 0$. Start with $x_0 = -1$ for the left-hand solution and with $x_0 = 1$ for the solution on the right. Then, in each case, find x_2.

2. Use Newton's method to estimate the one real solution of $x^3 + 3x + 1 = 0$. Start with $x_0 = 0$ and then find x_2.

3. Use Newton's method to estimate the two zeros of the function $f(x) = x^4 + x - 3$. Start with $x_0 = -1$ for the left-hand zero and with $x_0 = 1$ for the zero on the right. Then, in each case, find x_2.

4. Use Newton's method to estimate the two zeros of the function $f(x) = 2x - x^2 + 1$. Start with $x_0 = 0$ for the left-hand zero

and with $x_0 = 2$ for the zero on the right. Then, in each case, find x_2.

5. Use Newton's method to find the positive fourth root of 2 by solving the equation $x^4 - 2 = 0$. Start with $x_0 = 1$ and find x_2.

6. Use Newton's method to find the negative fourth root of 2 by solving the equation $x^4 - 2 = 0$. Start with $x_0 = -1$ and find x_2.

7. Use Newton's method to find an approximate solution of $3 - x = \ln x$. Start with $x_0 = 2$ and find x_2.

8. Use Newton's method to find an approximate solution of $x - 1 = \tan^{-1}x$. Start with $x_0 = 1$ and find x_2.

9. Use Newton's method to find an approximate solution of $xe^x = 1$. Start with $x_0 = 0$ and find x_2.

Dependence on Initial Point

10. Using the function shown in the figure, and for each initial estimate x_0, determine graphically what happens to the sequence of Newton's method approximations

a. $x_0 = 0$ **b.** $x_0 = 1$

c. $x_0 = 2$ **d.** $x_0 = 4$

e. $x_0 = 5.5$

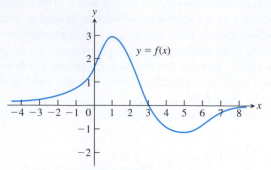

11. Guessing a root Suppose that your first guess is lucky, in the sense that x_0 is a root of $f(x) = 0$. Assuming that $f'(x_0)$ is defined and not 0, what happens to x_1 and later approximations?

12. Estimating pi You plan to estimate $\pi/2$ to five decimal places by using Newton's method to solve the equation $\cos x = 0$. Does it matter what your starting value is? Give reasons for your answer.

Theory and Examples

13. Oscillation Show that if $h > 0$, applying Newton's method to

$$f(x) = \begin{cases} \sqrt{x}, & x \geq 0 \\ \sqrt{-x}, & x < 0 \end{cases}$$

leads to $x_1 = -h$ if $x_0 = h$ and to $x_1 = h$ if $x_0 = -h$. Draw a picture that shows what is going on.

14. Approximations that get worse and worse Apply Newton's method to $f(x) = x^{1/3}$ with $x_0 = 1$ and calculate x_1, x_2, x_3, and x_4. Find a formula for $|x_n|$. What happens to $|x_n|$ as $n \to \infty$? Draw a picture that shows what is going on.

15. Explain why the following four statements ask for the same information:

 i) Find the roots of $f(x) = x^3 - 3x - 1$.

 ii) Find the x-coordinates of the intersections of the curve $y = x^3$ with the line $y = 3x + 1$.

 iii) Find the x-coordinates of the points where the curve $y = x^3 - 3x$ crosses the horizontal line $y = 1$.

 iv) Find the values of x where the derivative of $g(x) = (1/4)x^4 - (3/2)x^2 - x + 5$ equals zero.

16. Locating a planet To calculate a planet's space coordinates, we have to solve equations like $x = 1 + 0.5 \sin x$. Graphing the function $f(x) = x - 1 - 0.5 \sin x$ suggests that the function has a root near $x = 1.5$. Use one application of Newton's method to improve this estimate. That is, start with $x_0 = 1.5$ and find x_1. (The value of the root is 1.49870 to five decimal places.) Remember to use radians.

T **17. Intersecting curves** The curve $y = \tan x$ crosses the line $y = 2x$ between $x = 0$ and $x = \pi/2$. Use Newton's method to find where.

T **18. Real solutions of a quartic** Use Newton's method to find the two real solutions of the equation $x^4 - 2x^3 - x^2 - 2x + 2 = 0$.

T **19. a.** How many solutions does the equation $\sin 3x = 0.99 - x^2$ have?

 b. Use Newton's method to find them.

20. Intersection of curves

 a. Does $\cos 3x$ ever equal x? Give reasons for your answer.

 b. Use Newton's method to find where.

21. Find the four real zeros of the function $f(x) = 2x^4 - 4x^2 + 1$.

T **22. Estimating pi** Estimate π to as many decimal places as your calculator will display by using Newton's method to solve the equation $\tan x = 0$ with $x_0 = 3$.

23. Intersection of curves At what value(s) of x does $\cos x = 2x$?

24. Intersection of curves At what value(s) of x does $\cos x = -x$?

25. The graphs of $y = x^2(x + 1)$ and $y = 1/x$ $(x > 0)$ intersect at one point $x = r$. Use Newton's method to estimate the value of r to four decimal places.

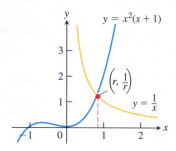

26. The graphs of $y = \sqrt{x}$ and $y = 3 - x^2$ intersect at one point $x = r$. Use Newton's method to estimate the value of r to four decimal places.

27. Intersection of curves At what value(s) of x does $e^{-x^2} = x^2 - x + 1$?

28. Intersection of curves At what value(s) of x does $\ln(1 - x^2) = x - 1$?

29. Use the Intermediate Value Theorem from Section 2.5 to show that $f(x) = x^3 + 2x - 4$ has a root between $x = 1$ and $x = 2$. Then find the root to five decimal places.

30. Factoring a quartic Find the approximate values of r_1 through r_4 in the factorization

$$8x^4 - 14x^3 - 9x^2 + 11x - 1 = 8(x - r_1)(x - r_2)(x - r_3)(x - r_4).$$

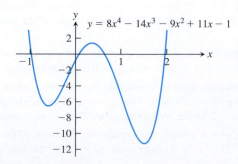

31. Converging to different zeros Use Newton's method to find the zeros of $f(x) = 4x^4 - 4x^2$ using the given starting values.

a. $x_0 = -2$ and $x_0 = -0.8$, lying in $\left(-\infty, -\sqrt{2}/2\right)$

b. $x_0 = -0.5$ and $x_0 = 0.25$, lying in $\left(-\sqrt{21}/7, \sqrt{21}/7\right)$

c. $x_0 = 0.8$ and $x_0 = 2$, lying in $\left(\sqrt{2}/2, \infty\right)$

d. $x_0 = -\sqrt{21}/7$ and $x_0 = \sqrt{21}/7$

32. The sonobuoy problem In submarine location problems, it is often necessary to find a submarine's closest point of approach (CPA) to a sonobuoy (sound detector) in the water. Suppose that the submarine travels on the parabolic path $y = x^2$ and that the buoy is located at the point $(2, -1/2)$.

a. Show that the value of x that minimizes the distance between the submarine and the buoy is a solution of the equation $x = 1/(x^2 + 1)$.

b. Solve the equation $x = 1/(x^2 + 1)$ with Newton's method.

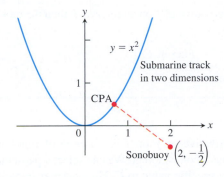

T **33. Curves that are nearly flat at the root** Some curves are so flat that, in practice, Newton's method stops too far from the root to

give a useful estimate. Try Newton's method on $f(x) = (x - 1)^{40}$ with a starting value of $x_0 = 2$ to see how close your machine comes to the root $x = 1$. See the accompanying graph.

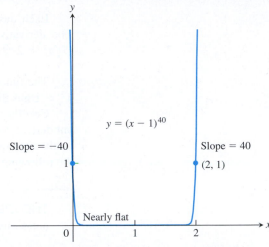

34. The accompanying figure shows a circle of radius r with a chord of length 2 and an arc s of length 3. Use Newton's method to solve for r and θ (radians) to four decimal places. Assume $0 < \theta < \pi$.

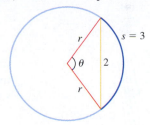

4.8 Antiderivatives

Many problems require that we recover a function from its derivative, or from its rate of change. For instance, the laws of physics tell us the acceleration of an object falling from an initial height, and we can use this to compute its velocity and its height at any time. More generally, starting with a function f, we want to find a function F whose derivative is f. If such a function F exists, it is called an *antiderivative* of f. Antiderivatives are the link connecting the two major elements of calculus: derivatives and definite integrals.

Finding Antiderivatives

> **DEFINITION** A function F is an **antiderivative** of f on an interval I if $F'(x) = f(x)$ for all x in I.

The process of recovering a function $F(x)$ from its derivative $f(x)$ is called *antidifferentiation*. We use capital letters such as F to represent an antiderivative of a function f, G to represent an antiderivative of g, and so forth.

EXAMPLE 1   Find an antiderivative for each of the following functions.

(a) $f(x) = 2x$ **(b)** $g(x) = \cos x$ **(c)** $h(x) = \dfrac{1}{x} + 2e^{2x}$

Solution   We need to think backward here: What function do we know has a derivative equal to the given function?

(a) $F(x) = x^2$ **(b)** $G(x) = \sin x$ **(c)** $H(x) = \ln |x| + e^{2x}$

Each answer can be checked by differentiating. The derivative of $F(x) = x^2$ is $2x$. The derivative of $G(x) = \sin x$ is $\cos x$, and the derivative of $H(x) = \ln |x| + e^{2x}$ is $(1/x) + 2e^{2x}$.   ■

The function $F(x) = x^2$ is not the only function whose derivative is $2x$. The function $x^2 + 1$ has the same derivative. So does $x^2 + C$ for any constant C. Are there others?

Corollary 2 of the Mean Value Theorem in Section 4.2 gives the answer: Any two antiderivatives of a function differ by a constant. So the functions $x^2 + C$, where C is an **arbitrary constant**, form *all* the antiderivatives of $f(x) = 2x$. More generally, we have the following result.

THEOREM 8   If F is an antiderivative of f on an interval I, then the most general antiderivative of f on I is

$$F(x) + C$$

where C is an arbitrary constant.

Thus the most general antiderivative of f on I is a *family* of functions $F(x) + C$ whose graphs are vertical translations of one another. We can select a particular antiderivative from this family by assigning a specific value to C. Here is an example showing how such an assignment might be made.

EXAMPLE 2   Find an antiderivative of $f(x) = 3x^2$ that satisfies $F(1) = -1$.

Solution   Since the derivative of x^3 is $3x^2$, the general antiderivative

$$F(x) = x^3 + C$$

gives all the antiderivatives of $f(x)$. The condition $F(1) = -1$ determines a specific value for C. Substituting $x = 1$ into $f(x) = x^3 + C$ gives

$$F(1) = (1)^3 + C = 1 + C.$$

Since $F(1) = -1$, solving $1 + C = -1$ for C gives $C = -2$. So

$$F(x) = x^3 - 2$$

is the antiderivative satisfying $F(1) = -1$. Notice that this assignment for C selects the particular curve from the family of curves $y = x^3 + C$ that passes through the point $(1, -1)$ in the plane (Figure 4.55).   ■

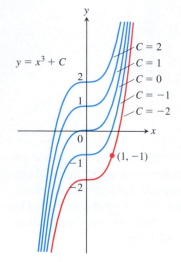

FIGURE 4.55   The curves $y = x^3 + C$ fill the coordinate plane without overlapping. In Example 2, we identify the curve $y = x^3 - 2$ as the one that passes through the given point $(1, -1)$.

By working backward from assorted differentiation rules, we can derive formulas and rules for antiderivatives. In each case there is an arbitrary constant C in the general expression representing all antiderivatives of a given function. Table 4.2 gives antiderivative formulas for a number of important functions.

The rules in Table 4.2 are easily verified by differentiating the general antiderivative formula to obtain the function to its left. For example, the derivative of $(\tan kx)/k + C$ is $\sec^2 kx$, whatever the value of the constants C or $k \neq 0$, and this verifies that Formula 4 gives the general antiderivative of $\sec^2 kx$.

TABLE 4.2 Antiderivative formulas, k a nonzero constant

Function	General antiderivative	Function	General antiderivative
1. x^n	$\frac{1}{n+1}x^{n+1} + C, \quad n \neq -1$	8. e^{kx}	$\frac{1}{k}e^{kx} + C$
2. $\sin kx$	$-\frac{1}{k}\cos kx + C$	9. $\frac{1}{x}$	$\ln|x| + C, \quad x \neq 0$
3. $\cos kx$	$\frac{1}{k}\sin kx + C$	10. $\frac{1}{\sqrt{1 - k^2x^2}}$	$\frac{1}{k}\sin^{-1} kx + C$
4. $\sec^2 kx$	$\frac{1}{k}\tan kx + C$	11. $\frac{1}{1 + k^2x^2}$	$\frac{1}{k}\tan^{-1} kx + C$
5. $\csc^2 kx$	$-\frac{1}{k}\cot kx + C$	12. $\frac{1}{x\sqrt{k^2x^2 - 1}}$	$\sec^{-1} kx + C, kx > 1$
6. $\sec kx \tan kx$	$\frac{1}{k}\sec kx + C$	13. a^{kx}	$\left(\frac{1}{k \ln a}\right)a^{kx} + C, a > 0, a \neq 1$
7. $\csc kx \cot kx$	$-\frac{1}{k}\csc kx + C$		

EXAMPLE 3   Find the general antiderivative of each of the following functions.

(a) $f(x) = x^5$ (b) $g(x) = \frac{1}{\sqrt{x}}$ (c) $h(x) = \sin 2x$

(d) $i(x) = \cos\frac{x}{2}$ (e) $j(x) = e^{-3x}$ (f) $k(x) = 2^x$

Solution   In each case, we can use one of the formulas listed in Table 4.2.

(a) $F(x) = \frac{x^6}{6} + C$ Formula 1 with $n = 5$

(b) $g(x) = x^{-1/2}$, so

$G(x) = \frac{x^{1/2}}{1/2} + C = 2\sqrt{x} + C$ Formula 1 with $n = -1/2$

(c) $H(x) = \frac{-\cos 2x}{2} + C$ Formula 2 with $k = 2$

(d) $I(x) = \frac{\sin(x/2)}{1/2} + C = 2\sin\frac{x}{2} + C$ Formula 3 with $k = 1/2$

(e) $J(x) = -\frac{1}{3}e^{-3x} + C$ Formula 8 with $k = -3$

(f) $K(x) = \left(\frac{1}{\ln 2}\right)2^x + C$ Formula 13 with $a = 2, 5 = 1$

Other derivative rules also lead to corresponding antiderivative rules. We can add and subtract antiderivatives and multiply them by constants.

TABLE 4.3 Antiderivative linearity rules

	Function	General antiderivative
1. *Constant Multiple Rule*:	$kf(x)$	$kF(x) + C, \quad k$ a constant
2. *Sum or Difference Rule*:	$f(x) \pm g(x)$	$F(x) \pm G(x) + C$

The formulas in Table 4.3 are easily proved by differentiating the antiderivatives and verifying that the result agrees with the original function.

EXAMPLE 4   Find the general antiderivative of

$$f(x) = \frac{3}{\sqrt{x}} + \sin 2x.$$

Solution   We have that $f(x) = 3g(x) + h(x)$ for the functions g and h in Example 3. Since $G(x) = 2\sqrt{x}$ is an antiderivative of $g(x)$ from Example 3b, it follows from the Constant Multiple Rule for antiderivatives that $3G(x) = 3 \cdot 2\sqrt{x} = 6\sqrt{x}$ is an antiderivative of $3g(x) = 3/\sqrt{x}$. Likewise, from Example 3c we know that $H(x) = (-1/2)\cos 2x$ is an antiderivative of $h(x) = \sin 2x$. From the Sum Rule for antiderivatives, we then get that

$$F(x) = 3G(x) + H(x) + C$$

$$= 6\sqrt{x} - \frac{1}{2}\cos 2x + C$$

is the general antiderivative formula for $f(x)$, where C is an arbitrary constant.   ■

Initial Value Problems and Differential Equations

Antiderivatives play several important roles in mathematics and its applications. Methods and techniques for finding them are a major part of calculus, and we take up that study in Chapter 8. Finding an antiderivative for a function $f(x)$ is the same problem as finding a function $y(x)$ that satisfies the equation

$$\frac{dy}{dx} = f(x).$$

This is called a **differential equation**, since it is an equation involving an unknown function y that is being differentiated. To solve it, we need a function $y(x)$ that satisfies the equation. This function is found by taking the antiderivative of $f(x)$. We can fix the arbitrary constant arising in the antidifferentiation process by specifying an initial condition

$$y(x_0) = y_0.$$

This condition means the function $y(x)$ has the value y_0 when $x = x_0$. The combination of a differential equation and an initial condition is called an **initial value problem**. Such problems play important roles in all branches of science.

The most general antiderivative $F(x) + C$ of the function $f(x)$ (such as $x^3 + C$ for the function $3x^2$ in Example 2) gives the **general solution** $y = F(x) + C$ of the differential equation $dy/dx = f(x)$. The general solution gives *all* the solutions of the equation (there are infinitely many, one for each value of C). We **solve** the differential equation by finding its general solution. We then solve the initial value problem by finding the **particular solution** that satisfies the initial condition $y(x_0) = y_0$. In Example 2, the function $y = x^3 - 2$ is the particular solution of the differential equation $dy/dx = 3x^2$ satisfying the initial condition $y(1) = -1$.

Antiderivatives and Motion

We have seen that the derivative of the position function of an object gives its velocity, and the derivative of its velocity function gives its acceleration. If we know an object's acceleration, then by finding an antiderivative we can recover the velocity, and from an antiderivative of the velocity we can recover its position function. This procedure was used as an application of Corollary 2 in Section 4.2. Now that we have a terminology and conceptual framework in terms of antiderivatives, we revisit the problem from the point of view of differential equations.

EXAMPLE 5   A hot-air balloon ascending at the rate of 12 ft/sec is at a height 80 ft above the ground when a package is dropped. How long does it take the package to reach the ground?

Solution Let $v(t)$ denote the velocity of the package at time t, and let $s(t)$ denote its height above the ground. The acceleration of gravity near the surface of the earth is 32 ft/sec^2. Assuming no other forces act on the dropped package, we have

$$\frac{dv}{dt} = -32. \qquad \text{Negative because gravity acts in the direction of decreasing } s$$

This leads to the following initial value problem (Figure 4.56):

Differential equation: $\dfrac{dv}{dt} = -32$

Initial condition: $v(0) = 12.$ Balloon initially rising

This is our mathematical model for the package's motion. We solve the initial value problem to obtain the velocity of the package.

1. *Solve the differential equation*: The general formula for an antiderivative of -32 is

$$v = -32t + C.$$

Having found the general solution of the differential equation, we use the initial condition to find the particular solution that solves our problem.

2. *Evaluate C*:

$$12 = -32(0) + C \qquad \text{Initial condition } v(0) = 12$$
$$C = 12.$$

The solution of the initial value problem is

$$v = -32t + 12.$$

Since velocity is the derivative of height, and the height of the package is 80 ft at time $t = 0$ when it is dropped, we now have a second initial value problem:

Differential equation: $\dfrac{ds}{dt} = -32t + 12$ Set $v = ds/dt$ in the previous equation.

Initial condition: $s(0) = 80.$

We solve this initial value problem to find the height as a function of t.

1. *Solve the differential equation*: Finding the general antiderivative of $-32t + 12$ gives

$$s = -16t^2 + 12t + C.$$

2. *Evaluate C*:

$$80 = -16(0)^2 + 12(0) + C \qquad \text{Initial condition } s(0) = 80$$
$$C = 80.$$

The package's height above ground at time t is

$$s = -16t^2 + 12t + 80.$$

Use the solution: To find how long it takes the package to reach the ground, we set s equal to 0 and solve for t:

$$-16t^2 + 12t + 80 = 0$$
$$-4t^2 + 3t + 20 = 0$$
$$t = \frac{-3 \pm \sqrt{329}}{-8} \qquad \text{Quadratic formula}$$
$$t \approx -1.89, \qquad t \approx 2.64.$$

The package hits the ground about 2.64 sec after it is dropped from the balloon. (The negative root has no physical meaning.) ∎

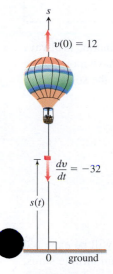

FIGURE 4.56 A package dropped from a rising hot-air balloon (Example 5).

s

$v(0) = 12$

$\dfrac{dv}{dt} = -32$

$s(t)$

$0 \quad$ ground

Indefinite Integrals

A special symbol is used to denote the collection of all antiderivatives of a function f.

DEFINITION The collection of all antiderivatives of f is called the **indefinite integral** of f with respect to x, and is denoted by

$$\int f(x)\, dx.$$

The symbol $\int$ is an **integral sign**. The function f is the **integrand** of the integral, and x is the **variable of integration**.

After the integral sign in the notation we just defined, the integrand function is always followed by a differential to indicate the variable of integration. We will have more to say about why this is important in Chapter 5. Using this notation, we restate the solutions of Example 1, as follows:

$$\int 2x\, dx = x^2 + C,$$

$$\int \cos x\, dx = \sin x + C,$$

$$\int \left(\frac{1}{x} + 2e^{2x}\right) dx = \ln|x| + e^{2x} + C.$$

This notation is related to the main application of antiderivatives, which will be explored in Chapter 5. Antiderivatives play a key role in computing limits of certain infinite sums, an unexpected and wonderfully useful role that is described in a central result of Chapter 5, the Fundamental Theorem of Calculus.

EXAMPLE 6 Evaluate

$$\int (x^2 - 2x + 5)\, dx.$$

Solution If we recognize that $(x^3/3) - x^2 + 5x$ is an antiderivative of $x^2 - 2x + 5$, we can evaluate the integral as

$$\int (x^2 - 2x + 5)\, dx = \overbrace{\frac{x^3}{3} - x^2 + 5x}^{\text{antiderivative}} + \underbrace{C.}_{\text{arbitrary constant}}$$

If we do not recognize the antiderivative right away, we can generate it term-by-term with the Sum, Difference, and Constant Multiple Rules:

$$\int (x^2 - 2x + 5)\, dx = \int x^2\, dx - \int 2x\, dx + \int 5\, dx$$

$$= \int x^2\, dx - 2\int x\, dx + 5\int 1\, dx$$

$$= \left(\frac{x^3}{3} + C_1\right) - 2\left(\frac{x^2}{2} + C_2\right) + 5(x + C_3)$$

$$= \frac{x^3}{3} + C_1 - x^2 - 2C_2 + 5x + 5C_3.$$

This formula is more complicated than it needs to be. If we combine C_1, $-2C_2$, and $5C_3$ into a single arbitrary constant $C = C_1 - 2C_2 + 5C_3$, the formula simplifies to

$$\frac{x^3}{3} - x^2 + 5x + C$$

and *still* gives all the possible antiderivatives there are. For this reason, we recommend that you go right to the final form even if you elect to integrate term-by-term. Write

$$\int (x^2 - 2x + 5)\,dx = \int x^2\,dx - \int 2x\,dx + \int 5\,dx$$

$$= \frac{x^3}{3} - x^2 + 5x + C.$$

Find the simplest antiderivative you can for each part and add the arbitrary constant of integration at the end. ∎

EXERCISES 4.8

Finding Antiderivatives

In Exercises 1–24, find an antiderivative for each function. Do as many as you can mentally. Check your answers by differentiation.

1. a. $2x$　　**b.** x^2　　**c.** $x^2 - 2x + 1$

2. a. $6x$　　**b.** x^7　　**c.** $x^7 - 6x + 8$

3. a. $-3x^{-4}$　　**b.** x^{-4}　　**c.** $x^{-4} + 2x + 3$

4. a. $2x^{-3}$　　**b.** $\dfrac{x^{-3}}{2} + x^2$　　**c.** $-x^{-3} + x - 1$

5. a. $\dfrac{1}{x^2}$　　**b.** $\dfrac{5}{x^2}$　　**c.** $2 - \dfrac{5}{x^2}$

6. a. $-\dfrac{2}{x^3}$　　**b.** $\dfrac{1}{2x^3}$　　**c.** $x^3 - \dfrac{1}{x^3}$

7. a. $\dfrac{3}{2}\sqrt{x}$　　**b.** $\dfrac{1}{2\sqrt{x}}$　　**c.** $\sqrt{x} + \dfrac{1}{\sqrt{x}}$

8. a. $\dfrac{4}{3}\sqrt[3]{x}$　　**b.** $\dfrac{1}{3\sqrt[3]{x}}$　　**c.** $\sqrt[3]{x} + \dfrac{1}{\sqrt[3]{x}}$

9. a. $\dfrac{2}{3}x^{-1/3}$　　**b.** $\dfrac{1}{3}x^{-2/3}$　　**c.** $-\dfrac{1}{3}x^{-4/3}$

10. a. $\dfrac{1}{2}x^{-1/2}$　　**b.** $-\dfrac{1}{2}x^{-3/2}$　　**c.** $-\dfrac{3}{2}x^{-5/2}$

11. a. $\dfrac{1}{x}$　　**b.** $\dfrac{7}{x}$　　**c.** $1 - \dfrac{5}{x}$

12. a. $\dfrac{1}{3x}$　　**b.** $\dfrac{2}{5x}$　　**c.** $1 + \dfrac{4}{3x} - \dfrac{1}{x^2}$

13. a. $-\pi \sin \pi x$　　**b.** $3 \sin x$　　**c.** $\sin \pi x - 3 \sin 3x$

14. a. $\pi \cos \pi x$　　**b.** $\dfrac{\pi}{2} \cos \dfrac{\pi x}{2}$　　**c.** $\cos \dfrac{\pi x}{2} + \pi \cos x$

15. a. $\sec^2 x$　　**b.** $\dfrac{2}{3} \sec^2 \dfrac{x}{3}$　　**c.** $-\sec^2 \dfrac{3x}{2}$

16. a. $\csc^2 x$　　**b.** $-\dfrac{3}{2} \csc^2 \dfrac{3x}{2}$　　**c.** $1 - 8 \csc^2 2x$

17. a. $\csc x \cot x$　　**b.** $-\csc 5x \cot 5x$　　**c.** $-\pi \csc \dfrac{\pi x}{2} \cot \dfrac{\pi x}{2}$

18. a. $\sec x \tan x$　　**b.** $4 \sec 3x \tan 3x$　　**c.** $\sec \dfrac{\pi x}{2} \tan \dfrac{\pi x}{2}$

19. a. e^{3x}　　**b.** e^{-x}　　**c.** $e^{x/2}$

20. a. e^{-2x}　　**b.** $e^{4x/3}$　　**c.** $e^{-x/5}$

21. a. 3^x　　**b.** 2^{-x}　　**c.** $\left(\dfrac{5}{3}\right)^x$

22. a. $x^{\sqrt{3}}$　　**b.** x^{π}　　**c.** $x^{\sqrt{2}-1}$

23. a. $\dfrac{2}{\sqrt{1-x^2}}$　　**b.** $\dfrac{1}{2(x^2+1)}$　　**c.** $\dfrac{1}{1+4x^2}$

24. a. $x - \left(\dfrac{1}{2}\right)^x$　　**b.** $x^2 + 2^x$　　**c.** $\pi^x - x^{-1}$

Finding Indefinite Integrals

In Exercises 25–70, find the most general antiderivative or indefinite integral. You may need to try a solution and then adjust your guess. Check your answers by differentiation.

25. $\displaystyle\int (x + 1)\,dx$

26. $\displaystyle\int (5 - 6x)\,dx$

27. $\displaystyle\int \left(3t^2 + \dfrac{t}{2}\right) dt$

28. $\displaystyle\int \left(\dfrac{t^2}{2} + 4t^3\right) dt$

29. $\displaystyle\int (2x^3 - 5x + 7)\,dx$

30. $\displaystyle\int (1 - x^2 - 3x^5)\,dx$

31. $\displaystyle\int \left(\dfrac{1}{x^2} - x^2 - \dfrac{1}{3}\right) dx$

32. $\displaystyle\int \left(\dfrac{1}{5} - \dfrac{2}{x^3} + 2x\right) dx$

33. $\displaystyle\int x^{-1/3}\,dx$

34. $\displaystyle\int x^{-5/4}\,dx$

35. $\displaystyle\int \left(\sqrt{x} + \sqrt[3]{x}\right) dx$

36. $\displaystyle\int \left(\dfrac{\sqrt{x}}{2} + \dfrac{2}{\sqrt{x}}\right) dx$

37. $\displaystyle\int \left(8y - \dfrac{2}{y^{1/4}}\right) dy$

38. $\displaystyle\int \left(\dfrac{1}{7} - \dfrac{1}{y^{5/4}}\right) dy$

39. $\displaystyle\int 2x(1 - x^{-3})\,dx$

40. $\displaystyle\int x^{-3}(x + 1)\,dx$

41. $\displaystyle\int \dfrac{t\sqrt{t} + \sqrt{t}}{t^2}\,dt$

42. $\displaystyle\int \dfrac{4 + \sqrt{t}}{t^3}\,dt$

43. $\int (-2 \cos t)\, dt$

44. $\int (-5 \sin t)\, dt$

45. $\int 7 \sin \dfrac{\theta}{3}\, d\theta$

46. $\int 3 \cos 5\theta\, d\theta$

47. $\int (-3 \csc^2 x)\, dx$

48. $\int \left(-\dfrac{\sec^2 x}{3} \right) dx$

49. $\int \dfrac{\csc \theta \cot \theta}{2}\, d\theta$

50. $\int \dfrac{2}{5} \sec \theta \tan \theta\, d\theta$

51. $\int (e^{3x} + 5e^{-x})\, dx$

52. $\int (2e^x - 3e^{-2x})\, dx$

53. $\int (e^{-x} + 4^x)\, dx$

54. $\int (1.3)^x\, dx$

55. $\int (4 \sec x \tan x - 2 \sec^2 x)\, dx$

56. $\int \dfrac{1}{2}(\csc^2 x - \csc x \cot x)\, dx$

57. $\int (\sin 2x - \csc^2 x)\, dx$

58. $\int (2 \cos 2x - 3 \sin 3x)\, dx$

59. $\int \dfrac{1 + \cos 4t}{2}\, dt$

60. $\int \dfrac{1 - \cos 6t}{2}\, dt$

61. $\int \left(\dfrac{1}{x} - \dfrac{5}{x^2 + 1} \right) dx$

62. $\int \left(\dfrac{2}{\sqrt{1 - y^2}} - \dfrac{1}{y^{1/4}} \right) dy$

63. $\int 3x^{\sqrt{3}}\, dx$

64. $\int x^{\sqrt{2} - 1}\, dx$

65. $\int (1 + \tan^2 \theta)\, d\theta$

66. $\int (2 + \tan^2 \theta)\, d\theta$

(*Hint:* $1 + \tan^2 \theta = \sec^2 \theta$)

67. $\int \cot^2 x\, dx$

68. $\int (1 - \cot^2 x)\, dx$

(*Hint:* $1 + \cot^2 x = \csc^2 x$)

69. $\int \cos \theta (\tan \theta + \sec \theta)\, d\theta$ **70.** $\int \dfrac{\csc \theta}{\csc \theta - \sin \theta}\, d\theta$

Checking Antiderivative Formulas
Verify the formulas in Exercises 71–82 by differentiation.

71. $\int (7x - 2)^3\, dx = \dfrac{(7x - 2)^4}{28} + C$

72. $\int (3x + 5)^{-2}\, dx = -\dfrac{(3x + 5)^{-1}}{3} + C$

73. $\int \sec^2 (5x - 1)\, dx = \dfrac{1}{5} \tan (5x - 1) + C$

74. $\int \csc^2 \left(\dfrac{x - 1}{3} \right) dx = -3 \cot \left(\dfrac{x - 1}{3} \right) + C$

75. $\int \dfrac{1}{(x + 1)^2}\, dx = -\dfrac{1}{x + 1} + C$

76. $\int \dfrac{1}{(x + 1)^2}\, dx = \dfrac{x}{x + 1} + C$

77. $\int \dfrac{1}{x + 1}\, dx = \ln |x + 1| + C, \quad x \ne -1$

78. $\int xe^x\, dx = xe^x - e^x + C$

79. $\int \dfrac{dx}{a^2 + x^2} = \dfrac{1}{a} \tan^{-1} \left(\dfrac{x}{a} \right) + C$

80. $\int \dfrac{dx}{\sqrt{a^2 - x^2}} = \sin^{-1} \left(\dfrac{x}{a} \right) + C$

T 81. $\int \dfrac{\tan^{-1} x}{x^2}\, dx = \ln x - \dfrac{1}{2} \ln (1 + x^2) - \dfrac{\tan^{-1} x}{x} + C$

82. $\int (\sin^{-1} x)^2\, dx = x(\sin^{-1} x)^2 - 2x + 2\sqrt{1 - x^2} \sin^{-1} x + C$

T 83. Right, or wrong? Say which for each formula and give a brief reason for each answer.

  a. $\int x \sin x\, dx = \dfrac{x^2}{2} \sin x + C$

  b. $\int x \sin x\, dx = -x \cos x + C$

  c. $\int x \sin x\, dx = -x \cos x + \sin x + C$

84. Right, or wrong? Say which for each formula and give a brief reason for each answer.

  a. $\int \tan \theta \sec^2 \theta\, d\theta = \dfrac{\sec^3 \theta}{3} + C$

  b. $\int \tan \theta \sec^2 \theta\, d\theta = \dfrac{1}{2} \tan^2 \theta + C$

  c. $\int \tan \theta \sec^2 \theta\, d\theta = \dfrac{1}{2} \sec^2 \theta + C$

85. Right, or wrong? Say which for each formula and give a brief reason for each answer.

  a. $\int (2x + 1)^2\, dx = \dfrac{(2x + 1)^3}{3} + C$

  b. $\int 3(2x + 1)^2\, dx = (2x + 1)^3 + C$

  c. $\int 6(2x + 1)^2\, dx = (2x + 1)^3 + C$

86. Right, or wrong? Say which for each formula and give a brief reason for each answer.

  a. $\int \sqrt{2x + 1}\, dx = \sqrt{x^2 + x} + C$

  b. $\int \sqrt{2x + 1}\, dx = \sqrt{x^2 + x} + C$

  c. $\int \sqrt{2x + 1}\, dx = \dfrac{1}{3}\left(\sqrt{2x + 1} \right)^3 + C$

87. Right, or wrong? Give a brief reason why.

$$\int \frac{-15(x + 3)^2}{(x - 2)^4} \, dx = \left(\frac{x + 3}{x - 2}\right)^3 + C$$

88. Right, or wrong? Give a brief reason why.

$$\int \frac{x \cos(x^2) - \sin(x^2)}{x^2} \, dx = \frac{\sin(x^2)}{x} + C$$

Initial Value Problems

89. Which of the following graphs shows the solution of the initial value problem

$$\frac{dy}{dx} = 2x, \quad y = 4 \text{ when } x = 1?$$

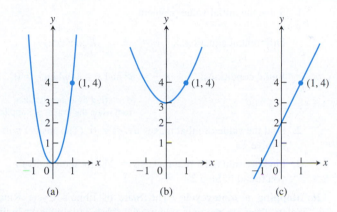

Give reasons for your answer.

90. Which of the following graphs shows the solution of the initial value problem

$$\frac{dy}{dx} = -x, \quad y = 1 \text{ when } x = -1?$$

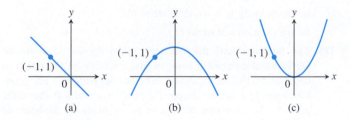

Give reasons for your answer.

Solve the initial value problems in Exercises 91–112.

91. $\dfrac{dy}{dx} = 2x - 7, \quad y(2) = 0$

92. $\dfrac{dy}{dx} = 10 - x, \quad y(0) = -1$

93. $\dfrac{dy}{dx} = \dfrac{1}{x^2} + x, \quad x > 0; \quad y(2) = 1$

94. $\dfrac{dy}{dx} = 9x^2 - 4x + 5, \quad y(-1) = 0$

95. $\dfrac{dy}{dx} = 3x^{-2/3}, \quad y(-1) = -5$

96. $\dfrac{dy}{dx} = \dfrac{1}{2\sqrt{x}}, \quad y(4) = 0$

97. $\dfrac{ds}{dt} = 1 + \cos t, \quad s(0) = 4$

98. $\dfrac{ds}{dt} = \cos t + \sin t, \quad s(\pi) = 1$

99. $\dfrac{dr}{d\theta} = -\pi \sin \pi\theta, \quad r(0) = 0$

100. $\dfrac{dr}{d\theta} = \cos \pi\theta, \quad r(0) = 1$

101. $\dfrac{dv}{dt} = \dfrac{1}{2} \sec t \tan t, \quad v(0) = 1$

102. $\dfrac{dv}{dt} = 8t + \csc^2 t, \quad v\left(\dfrac{\pi}{2}\right) = -7$

103. $\dfrac{dv}{dt} = \dfrac{3}{t\sqrt{t^2 - 1}}, \quad t > 1, v(2) = 0$

104. $\dfrac{dv}{dt} = \dfrac{8}{1 + t^2} + \sec^2 t, \quad v(0) = 1$

105. $\dfrac{d^2y}{dx^2} = 2 - 6x; \quad y'(0) = 4, \quad y(0) = 1$

106. $\dfrac{d^2y}{dx^2} = 0; \quad y'(0) = 2, \quad y(0) = 0$

107. $\dfrac{d^2r}{dt^2} = \dfrac{2}{t^3}; \quad \dfrac{dr}{dt}\bigg|_{t=1} = 1, \quad r(1) = 1$

108. $\dfrac{d^2s}{dt^2} = \dfrac{3t}{8}; \quad \dfrac{ds}{dt}\bigg|_{t=4} = 3, \quad s(4) = 4$

109. $\dfrac{d^3y}{dx^3} = 6; \quad y''(0) = -8, \quad y'(0) = 0, \quad y(0) = 5$

110. $\dfrac{d^3\theta}{dt^3} = 0; \quad \theta''(0) = -2, \quad \theta'(0) = -\dfrac{1}{2}, \quad \theta(0) = \sqrt{2}$

111. $y^{(4)} = -\sin t + \cos t;$
$y'''(0) = 7, \quad y''(0) = y'(0) = -1, \quad y(0) = 0$

112. $y^{(4)} = -\cos x + 8 \sin 2x;$
$y'''(0) = 0, \quad y''(0) = y'(0) = 1, \quad y(0) = 3$

113. Find the curve $y = f(x)$ in the xy-plane that passes through the point $(9, 4)$ and whose slope at each point is $3\sqrt{x}$.

114. a. Find a curve $y = f(x)$ with the following properties:

   i) $\dfrac{d^2y}{dx^2} = 6x$

   ii) Its graph passes through the point $(0, 1)$ and has a horizontal tangent there.

  b. How many curves like this are there? How do you know?

In Exercises 115–118, the graph of f' is given. Assume that $f(0) = 1$ and sketch a possible continuous graph of f.

115.   **116.**

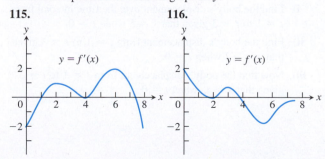

117.

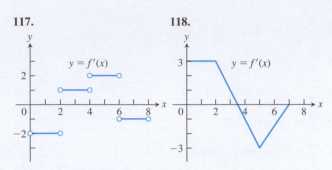

118.

Solution (Integral) Curves

Exercises 119–122 show solution curves of differential equations. In each exercise, find an equation for the curve through the labeled point.

119.

$$\frac{dy}{dx} = 1 - \frac{4}{3}x^{1/3}$$

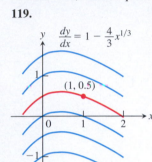

120.

$$\frac{dy}{dx} = x - 1$$

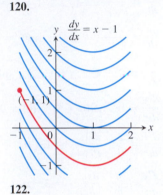

121.

$$\frac{dy}{dx} = \sin x - \cos x$$

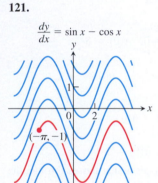

122.

$$\frac{dy}{dx} = \frac{1}{2\sqrt{x}} + \pi \sin \pi x$$

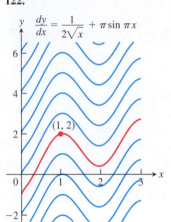

Applications

123. Finding displacement from an antiderivative of velocity

  a. Suppose that the velocity of a body moving along the s-axis is

$$\frac{ds}{dt} = v = 9.8t - 3.$$

  i) Find the body's displacement over the time interval from $t = 1$ to $t = 3$ given that $s = 5$ when $t = 0$.

  ii) Find the body's displacement from $t = 1$ to $t = 3$ given that $s = -2$ when $t = 0$.

  iii) Now find the body's displacement from $t = 1$ to $t = 3$ given that $s = s_0$ when $t = 0$.

  b. Suppose that the position s of a body moving along a coordinate line is a differentiable function of time t. Is it true that once you know an antiderivative of the velocity function ds/dt you can find the body's displacement from $t = a$ to $t = b$ even if you do not know the body's exact position at either of those times? Give reasons for your answer.

124. Liftoff from Earth A rocket lifts off the surface of Earth with a constant acceleration of 20 m/sec^2. How fast will the rocket be going 1 min later?

125. Stopping a car in time You are driving along a highway at a steady 60 mph (88 ft/sec) when you see an accident ahead and slam on the brakes. What constant deceleration is required to stop your car in 242 ft? To find out, carry out the following steps.

  1. Solve the initial value problem

   Differential equation: $\dfrac{d^2s}{dt^2} = -k$ (k constant)

   Initial conditions: $\dfrac{ds}{dt} = 88$ and $s = 0$ when $t = 0$.

   Measuring time and distance from when the brakes are applied

  2. Find the value of t that makes $ds/dt = 0$. (The answer will involve k.)

  3. Find the value of k that makes $s = 242$ for the value of t you found in Step 2.

126. Stopping a motorcycle The State of Illinois Cycle Rider Safety Program requires motorcycle riders to be able to brake from 30 mph (44 ft/sec) to 0 in 45 ft. What constant deceleration does it take to do that?

127. Motion along a coordinate line A particle moves on a coordinate line with acceleration $a = d^2s/dt^2 = 15\sqrt{t} - (3/\sqrt{t})$, subject to the conditions that $ds/dt = 4$ and $s = 0$ when $t = 1$. Find

  a. the velocity $v = ds/dt$ in terms of t.

  b. the position s in terms of t.

T **128. The hammer and the feather** When *Apollo 15* astronaut David Scott dropped a hammer and a feather on the moon to demonstrate that in a vacuum all bodies fall with the same (constant) acceleration, he dropped them from about 4 ft above the ground. The television footage of the event shows the hammer and the feather falling more slowly than on Earth, where, in a vacuum, they would have taken only half a second to fall the 4 ft. How long did it take the hammer and feather to fall 4 ft on the moon? To find out, solve the following initial value problem for s as a function of t. Then find the value of t that makes s equal to 0.

   Differential equation: $\dfrac{d^2s}{dt^2} = -5.2 \text{ ft/sec}^2$

   Initial conditions: $\dfrac{ds}{dt} = 0$ and $s = 4$ when $t = 0$

129. Motion with constant acceleration The standard equation for the position s of a body moving with a constant acceleration a along a coordinate line is

$$s = \frac{a}{2}t^2 + v_0 t + s_0, \qquad (1)$$

where v_0 and s_0 are the body's velocity and position at time $t = 0$. Derive this equation by solving the initial value problem

Differential equation: $\quad \dfrac{d^2 s}{dt^2} = a$

Initial conditions: $\quad \dfrac{ds}{dt} = v_0$ and $s = s_0$ when $t = 0$.

130. Free fall near the surface of a planet For free fall near the surface of a planet where the acceleration due to gravity has a constant magnitude of g length-units/sec^2, Equation (1) in Exercise 129 takes the form

$$s = -\frac{1}{2}gt^2 + v_0 t + s_0, \qquad (2)$$

where s is the body's height above the surface. The equation has a minus sign because the acceleration acts downward, in the direction of decreasing s. The velocity v_0 is positive if the object is rising at time $t = 0$ and negative if the object is falling.

Instead of using the result of Exercise 129, you can derive Equation (2) directly by solving an appropriate initial value problem. What initial value problem? Solve it to be sure you have the right one, explaining the solution steps as you go along.

131. Suppose that

$$f(x) = \frac{d}{dx}\left(1 - \sqrt{x}\right) \quad \text{and} \quad g(x) = \frac{d}{dx}(x + 2).$$

Find:

a. $\displaystyle\int f(x)\,dx$ \quad b. $\displaystyle\int g(x)\,dx$

c. $\displaystyle\int [-f(x)]\,dx$ \quad d. $\displaystyle\int [-g(x)]\,dx$

e. $\displaystyle\int [f(x) + g(x)]\,dx$ \quad f. $\displaystyle\int_i [f(x) - g(x)]\,dx$

132. Uniqueness of solutions If differentiable functions $y = F(x)$ and $y = g(x)$ both solve the initial value problem

$$\frac{dy}{dx} = f(x), \qquad y(x_0) = y_0,$$

on an interval I, must $F(x) = G(x)$ for every x in I? Give reasons for your answer.

COMPUTER EXPLORATIONS

Use a CAS to solve the initial value problems in Exercises 133–136. Plot the solution curves.

133. $y' = \cos^2 x + \sin x, \quad y(\pi) = 1$

134. $y' = \dfrac{1}{x} + x, \quad y(1) = -1$

135. $y' = \dfrac{1}{\sqrt{4 - x^2}}, \quad y(0) = 2$

136. $y'' = \dfrac{2}{x} + \sqrt{x}, \quad y(1) = 0, \quad y'(1) = 0$

| CHAPTER 4 | **Questions to Guide Your Review** |

1. What can be said about the extreme values of a function that is continuous on a closed interval?

2. What does it mean for a function to have a local extreme value on its domain? An absolute extreme value? How are local and absolute extreme values related, if at all? Give examples.

3. How do you find the absolute extrema of a continuous function on a closed interval? Give examples.

4. What are the hypotheses and conclusion of Rolle's Theorem? Are the hypotheses really necessary? Explain.

5. What are the hypotheses and conclusion of the Mean Value Theorem? What physical interpretations might the theorem have?

6. State the Mean Value Theorem's three corollaries.

7. How can you sometimes identify a function $f(x)$ by knowing f' and knowing the value of f at a point $x = x_0$? Give an example.

8. What is the First Derivative Test for Local Extreme Values? Give examples of how it is applied.

9. How do you test a twice-differentiable function to determine where its graph is concave up or concave down? Give examples.

10. What is an inflection point? Give an example. What physical significance do inflection points sometimes have?

11. What is the Second Derivative Test for Local Extreme Values? Give examples of how it is applied.

12. What do the derivatives of a function tell you about the shape of its graph?

13. List the steps you would take to graph a polynomial function. Illustrate with an example.

14. What is a cusp? Give examples.

15. List the steps you would take to graph a rational function. Illustrate with an example.

16. Outline a general strategy for solving max-min problems. Give examples.

17. Describe l'Hôpital's Rule. How do you know when to use the rule and when to stop? Give an example.

18. How can you sometimes handle limits that lead to indeterminate forms ∞/∞, $\infty \cdot 0$, and $\infty - \infty$? Give examples.

19. How can you sometimes handle limits that lead to indeterminate forms 1^∞, 0^0, and ∞^∞? Give examples.

20. Describe Newton's method for solving equations. Give an example. What is the theory behind the method? What are some of the things to watch out for when you use the method?

21. Can a function have more than one antiderivative? If so, how are the antiderivatives related? Explain.

22. What is an indefinite integral? How do you evaluate one? What general formulas do you know for finding indefinite integrals?

23. How can you sometimes solve a differential equation of the form $dy/dx = f(x)$?

24. What is an initial value problem? How do you solve one? Give an example.

25. If you know the acceleration of a body moving along a coordinate line as a function of time, what more do you need to know to find the body's position function? Give an example.

CHAPTER 4   Practice Exercises

Finding Extreme Values

In Exercises 1–16, find the extreme values (absolute and local) of the function over its natural domain, and where they occur.

1. $y = 2x^2 - 8x + 9$

2. $y = x^3 - 2x + 4$

3. $y = x^3 + x^2 - 8x + 5$

4. $y = x^3(x - 5)^2$

5. $y = \sqrt{x^2 - 1}$

6. $y = x - 4\sqrt{x}$

7. $y = \dfrac{1}{\sqrt[3]{1 - x^2}}$

8. $y = \sqrt{3 + 2x - x^2}$

9. $y = \dfrac{x}{x^2 + 1}$

10. $y = \dfrac{x + 1}{x^2 + 2x + 2}$

11. $y = e^x + e^{-x}$

12. $y = e^x - e^{-x}$

13. $y = x \ln x$

14. $y = x^2 \ln x$

15. $y = \cos^{-1}(x^2)$

16. $y = \sin^{-1}(e^x)$

Extreme Values

17. Does $f(x) = x^3 + 2x + \tan x$ have any local maximum or minimum values? Give reasons for your answer.

18. Does $g(x) = \csc x + 2 \cot x$ have any local maximum values? Give reasons for your answer.

19. Does $f(x) = (7 + x)(11 - 3x)^{1/3}$ have an absolute minimum value? An absolute maximum? If so, find them or give reasons why they fail to exist. List all critical points of f.

20. Find values of a and b such that the function

$$f(x) = \frac{ax + b}{x^2 - 1}$$

has a local extreme value of 1 at $x = 3$. Is this extreme value a local maximum, or a local minimum? Give reasons for your answer.

21. Does $g(x) = e^x - x$ have an absolute minimum value? An absolute maximum? If so, find them or give reasons why they fail to exist. List all critical points of g.

22. Does $f(x) = 2e^x/(1 + x^2)$ have an absolute minimum value? An absolute maximum? If so, find them or give reasons why they fail to exist. List all critical points of f.

In Exercises 23 and 24, find the absolute maximum and absolute minimum values of f over the interval.

23. $f(x) = x - 2 \ln x, \quad 1 \le x \le 3$

24. $f(x) = (4/x) + \ln x^2, \quad 1 \le x \le 4$

25. The greatest integer function $f(x) = \lfloor x \rfloor$, defined for all values of x, assumes a local maximum value of 0 at each point of $[0, 1)$. Could any of these local maximum values also be local minimum values of f? Give reasons for your answer.

26. a. Give an example of a differentiable function f whose first derivative is zero at some point c even though f has neither a local maximum nor a local minimum at c.

b. How is this consistent with Theorem 2 in Section 4.1? Give reasons for your answer.

27. The function $y = 1/x$ does not take on either a maximum or a minimum on the interval $0 < x < 1$ even though the function is continuous on this interval. Does this contradict the Extreme Value Theorem for continuous functions? Why?

28. What are the maximum and minimum values of the function $y = |x|$ on the interval $-1 \le x < 1$? Notice that the interval is not closed. Is this consistent with the Extreme Value Theorem for continuous functions? Why?

T **29.** A graph that is large enough to show a function's global behavior may fail to reveal important local features. The graph of $f(x) = (x^8/8) - (x^6/2) - x^5 + 5x^3$ is a case in point.

a. Graph f over the interval $-2.5 \le x \le 2.5$. Where does the graph appear to have local extreme values or points of inflection?

b. Now factor $f'(x)$ and show that f has a local maximum at $x = \sqrt[5]{5} \approx 1.70998$ and local minima at $x = \pm\sqrt{3} \approx \pm 1.73205$.

c. Zoom in on the graph to find a viewing window that shows the presence of the extreme values at $x = \sqrt[5]{5}$ and $x = \sqrt{3}$.

The moral here is that without calculus the existence of two of the three extreme values would probably have gone unnoticed. On any normal graph of the function, the values would lie close enough together to fall within the dimensions of a single pixel on the screen.

(*Source: Uses of Technology in the Mathematics Curriculum*, by Benny Evans and Jerry Johnson, Oklahoma State University, published in 1990 under a grant from the National Science Foundation, USE-8950044.)

T **30.** (*Continuation of Exercise 29.*)

a. Graph $f(x) = (x^8/8) - (2/5)x^5 - 5x - (5/x^2) + 11$ over the interval $-2 \le x \le 2$. Where does the graph appear to have local extreme values or points of inflection?

b. Show that f has a local maximum value at $x = \sqrt[7]{5} \approx 1.2585$ and a local minimum value at $x = \sqrt[3]{2} \approx 1.2599$.

c. Zoom in to find a viewing window that shows the presence of the extreme values at $x = \sqrt[7]{5}$ and $x = \sqrt[3]{2}$.

The Mean Value Theorem

31. a. Show that $g(t) = \sin^2 t - 3t$ decreases on every interval in its domain.

    b. How many solutions does the equation $\sin^2 t - 3t = 5$ have? Give reasons for your answer.

32. a. Show that $y = \tan \theta$ increases on every open interval in its domain.

    b. If the conclusion in part (a) is really correct, how do you explain the fact that $\tan \pi = 0$ is less than $\tan (\pi/4) = 1$?

33. a. Show that the equation $x^4 + 2x^2 - 2 = 0$ has exactly one solution on $[0, 1]$.

    T **b.** Find the solution to as many decimal places as you can.

34. a. Show that $f(x) = x/(x + 1)$ increases on every open interval in its domain.

    b. Show that $f(x) = x^3 + 2x$ has no local maximum or minimum values.

35. Water in a reservoir As a result of a heavy rain, the volume of water in a reservoir increased by 1400 acre-ft in 24 hours. Show that at some instant during that period the reservoir's volume was increasing at a rate in excess of 225,000 gal/min. (An acre-foot is 43,560 ft^3, the volume that would cover 1 acre to the depth of 1 ft. A cubic foot holds 7.48 gal.)

36. The formula $F(x) = 3x + C$ gives a different function for each value of C. All of these functions, however, have the same derivative with respect to x, namely $F'(x) = 3$. Are these the only differentiable functions whose derivative is 3? Could there be any others? Give reasons for your answers.

37. Show that

$$\frac{d}{dx}\left(\frac{x}{x + 1}\right) = \frac{d}{dx}\left(-\frac{1}{x + 1}\right)$$

even though

$$\frac{x}{x + 1} \neq -\frac{1}{x + 1}.$$

Doesn't this contradict Corollary 2 of the Mean Value Theorem? Give reasons for your answer.

38. Calculate the first derivatives of $f(x) = x^2/(x^2 + 1)$ and $g(x) = -1/(x^2 + 1)$. What can you conclude about the graphs of these functions?

Analyzing Graphs

In Exercises 39 and 40, use the graph to answer the questions.

39. Identify any global extreme values of f and the values of x at which they occur.

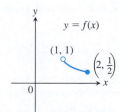

40. Estimate the open intervals on which the function $y = f(x)$ is

    a. increasing.

    b. decreasing.

c. Use the given graph of f' to indicate where any local extreme values of the function occur, and whether each extreme is a relative maximum or minimum.

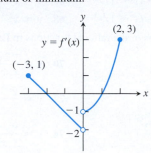

Each of the graphs in Exercises 41 and 42 is the graph of the position function $s = f(t)$ of an object moving on a coordinate line (t represents time). At approximately what times (if any) is each object's **(a)** velocity equal to zero? **(b)** Acceleration equal to zero? During approximately what time intervals does the object move **(c)** forward? **(d)** Backward?

41.

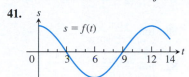

42.

Graphs and Graphing

Graph the curves in Exercises 43–58.

43. $y = x^2 - (x^3/6)$ **44.** $y = x^3 - 3x^2 + 3$

45. $y = -x^3 + 6x^2 - 9x + 3$

46. $y = (1/8)(x^3 + 3x^2 - 9x - 27)$

47. $y = x^3(8 - x)$ **48.** $y = x^2(2x^2 - 9)$

49. $y = x - 3x^{2/3}$ **50.** $y = x^{1/3}(x - 4)$

51. $y = x\sqrt{3 - x}$ **52.** $y = x\sqrt{4 - x^2}$

53. $y = (x - 3)^2 e^x$ **54.** $y = xe^{-x^2}$

55. $y = \ln(x^2 - 4x + 3)$ **56.** $y = \ln(\sin x)$

57. $y = \sin^{-1}\left(\frac{1}{x}\right)$ **58.** $y = \tan^{-1}\left(\frac{1}{x}\right)$

Each of Exercises 59–64 gives the first derivative of a function $y = f(x)$. **(a)** At what points, if any, does the graph of f have a local maximum, local minimum, or inflection point? **(b)** Sketch the general shape of the graph.

59. $y' = 16 - x^2$

60. $y' = x^2 - x - 6$

61. $y' = 6x(x + 1)(x - 2)$

62. $y' = x^2(6 - 4x)$

63. $y' = x^4 - 2x^2$

64. $y' = 4x^2 - x^4$

In Exercises 65–68, graph each function. Then use the function's first derivative to explain what you see.

65. $y = x^{2/3} + (x - 1)^{1/3}$ **66.** $y = x^{2/3} + (x - 1)^{2/3}$

67. $y = x^{1/3} + (x - 1)^{1/3}$ **68.** $y = x^{2/3} - (x - 1)^{1/3}$

Sketch the graphs of the rational functions in Exercises 69–76.

69. $y = \dfrac{x + 1}{x - 3}$ **70.** $y = \dfrac{2x}{x + 5}$

71. $y = \dfrac{x^2 + 1}{x}$ **72.** $y = \dfrac{x^2 - x + 1}{x}$

73. $y = \dfrac{x^3 + 2}{2x}$ **74.** $y = \dfrac{x^4 - 1}{x^2}$

75. $y = \dfrac{x^2 - 4}{x^2 - 3}$ **76.** $y = \dfrac{x^2}{x^2 - 4}$

Using L'Hôpital's Rule
Use l'Hôpital's Rule to find the limits in Exercises 77–88.

77. $\lim\limits_{x \to 1} \dfrac{x^2 + 3x - 4}{x - 1}$ **78.** $\lim\limits_{x \to 1} \dfrac{x^a - 1}{x^b - 1}$

79. $\lim\limits_{x \to \pi} \dfrac{\tan x}{x}$ **80.** $\lim\limits_{x \to 0} \dfrac{\tan x}{x + \sin x}$

81. $\lim\limits_{x \to 0} \dfrac{\sin^2 x}{\tan(x^2)}$ **82.** $\lim\limits_{x \to 0} \dfrac{\sin mx}{\sin nx}$

83. $\lim\limits_{x \to \pi/2^-} \sec 7x \cos 3x$ **84.** $\lim\limits_{x \to 0^+} \sqrt{x} \sec x$

85. $\lim\limits_{x \to 0} (\csc x - \cot x)$ **86.** $\lim\limits_{x \to 0}\left(\dfrac{1}{x^4} - \dfrac{1}{x^2}\right)$

87. $\lim\limits_{x \to \infty} \left(\sqrt{x^2 + x + 1} - \sqrt{x^2 - x} \right)$

88. $\lim\limits_{x \to \infty}\left(\dfrac{x^3}{x^2 - 1} - \dfrac{x^3}{x^2 + 1}\right)$

Find the limits in Exercises 89–102.

89. $\lim\limits_{x \to 0} \dfrac{10^x - 1}{x}$

90. $\lim\limits_{\theta \to 0} \dfrac{3^\theta - 1}{\theta}$

91. $\lim\limits_{x \to 0} \dfrac{2^{\sin x} - 1}{e^x - 1}$

92. $\lim\limits_{x \to 0} \dfrac{2^{-\sin x} - 1}{e^x - 1}$

93. $\lim\limits_{x \to 0} \dfrac{5 - 5 \cos x}{e^x - x - 1}$

94. $\lim\limits_{x \to 0} \dfrac{4 - 4e^x}{xe^x}$

95. $\lim\limits_{t \to 0^+} \dfrac{t - \ln(1 + 2t)}{t^2}$

96. $\lim\limits_{x \to 4} \dfrac{\sin^2(\pi x)}{e^{x-4} + 3 - x}$

97. $\lim\limits_{t \to 0^+} \left(\dfrac{e^t}{t} - \dfrac{1}{t}\right)$

98. $\lim\limits_{y \to 0^+} e^{-1/y} \ln y$

99. $\lim\limits_{x \to \infty} \left(1 + \dfrac{b}{x}\right)^{kx}$

100. $\lim\limits_{x \to \infty} \left(1 + \dfrac{2}{x} + \dfrac{7}{x^2}\right)$

101. $\lim\limits_{x \to 0} \dfrac{\cos 2x - 1 - \sqrt{1 - \cos x}}{\sin^2 x}$

102. $\lim\limits_{x \to 0} \dfrac{\sqrt{1 + \tan x} - \sqrt{1 + \sin x}}{x^3}$

Optimization

103. The sum of two nonnegative numbers is 36. Find the numbers if

    a. the difference of their square roots is to be as large as possible.

    b. the sum of their square roots is to be as large as possible.

104. The sum of two nonnegative numbers is 20. Find the numbers

    a. if the product of one number and the square root of the other is to be as large as possible.

    b. if one number plus the square root of the other is to be as large as possible.

105. An isosceles triangle has its vertex at the origin and its base parallel to the x-axis with the vertices above the axis on the curve $y = 27 - x^2$. Find the largest area the triangle can have.

106. A customer has asked you to design an open-top rectangular stainless steel vat. It is to have a square base and a volume of 32 ft^3, to be welded from quarter-inch plate, and to weigh no more than necessary. What dimensions do you recommend?

107. Find the height and radius of the largest right circular cylinder that can be put in a sphere of radius $\sqrt{3}$.

108. The figure here shows two right circular cones, one upside down inside the other. The two bases are parallel, and the vertex of the smaller cone lies at the center of the larger cone's base. What values of r and h will give the smaller cone the largest possible volume?

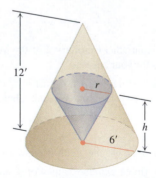

109. Manufacturing tires Your company can manufacture x hundred grade A tires and y hundred grade B tires a day, where $0 \le x \le 4$ and

$$y = \dfrac{40 - 10x}{5 - x}.$$

Your profit on a grade A tire is twice your profit on a grade B tire. What is the most profitable number of each kind to make?

110. Particle motion The positions of two particles on the s-axis are $s_1 = \cos t$ and $s_2 = \cos(t + \pi/4)$.

    a. What is the farthest apart the particles ever get?

    b. When do the particles collide?

111. Open-top box An open-top rectangular box is constructed from a 10-in.-by-16-in. piece of cardboard by cutting squares of equal side length from the corners and folding up the sides. Find analytically the dimensions of the box of largest volume and the maximum volume. Support your answers graphically.

112. The ladder problem What is the approximate length (in feet) of the longest ladder you can carry horizontally around the corner of the corridor shown here? Round your answer down to the nearest foot.

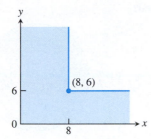

Newton's Method

113. Let $f(x) = 3x - x^3$. Show that the equation $f(x) = -4$ has a solution in the interval $[2, 3]$ and use Newton's method to find it.

114. Let $f(x) = x^4 - x^3$. Show that the equation $f(x) = 75$ has a solution in the interval $[3, 4]$ and use Newton's method to find it.

Finding Indefinite Integrals

Find the indefinite integrals (most general antiderivatives) in Exercises 115–138. You may need to try a solution and then adjust your guess. Check your answers by differentiation.

115. $\int (x^3 + 5x - 7)\, dx$

116. $\int \left(8t^3 - \dfrac{t^2}{2} + t\right) dt$

117. $\int \left(3\sqrt{t} + \dfrac{4}{t^2}\right) dt$

118. $\int \left(\dfrac{1}{2\sqrt{t}} - \dfrac{3}{t^4}\right) dt$

119. $\int \dfrac{dr}{(r+5)^2}$

120. $\int \dfrac{6\, dr}{(r - \sqrt{2})^3}$

121. $\int 3\theta\sqrt{\theta^2 + 1}\, d\theta$

122. $\int \dfrac{\theta}{\sqrt{7 + \theta^2}}\, d\theta$

123. $\int x^3(1 + x^4)^{-1/4}\, dx$

124. $\int (2 - x)^{3/5}\, dx$

125. $\int \sec^2 \dfrac{s}{10}\, ds$

126. $\int \csc^2 \pi s\, ds$

127. $\int \csc \sqrt{2}\theta \cot \sqrt{2}\theta\, d\theta$

128. $\int \sec \dfrac{\theta}{3} \tan \dfrac{\theta}{3}\, d\theta$

129. $\int \sin^2 \dfrac{x}{4}\, dx$ $\left(\text{Hint: } \sin^2\theta = \dfrac{1 - \cos 2\theta}{2}\right)$

130. $\int \cos^2 \dfrac{x}{2}\, dx$

131. $\int \left(\dfrac{3}{x} - x\right) dx$

132. $\int \left(\dfrac{5}{x^2} + \dfrac{2}{x^2 + 1}\right) dx$

133. $\int \left(\dfrac{1}{2}e^t - e^{-t}\right) dt$

134. $\int (5^s + s^5)\, ds$

135. $\int \theta^{1-\pi}\, d\theta$

136. $\int 2^{\pi + r}\, dr$

137. $\int \dfrac{3}{2x\sqrt{x^2 - 1}}\, dx$

138. $\int \dfrac{d\theta}{\sqrt{16 - \theta^2}}$

Initial Value Problems

Solve the initial value problems in Exercises 139–142.

139. $\dfrac{dy}{dx} = \dfrac{x^2 + 1}{x^2}, \quad y(1) = -1$

140. $\dfrac{dy}{dx} = \left(x + \dfrac{1}{x}\right)^2, \quad y(1) = 1$

141. $\dfrac{d^2r}{dt^2} = 15\sqrt{t} + \dfrac{3}{\sqrt{t}}; \quad r'(1) = 8, \quad r(1) = 0$

142. $\dfrac{d^3r}{dt^3} = -\cos t; \quad r''(0) = r'(0) = 0, \quad r(0) = -1$

Applications and Examples

143. Can the integrations in (a) and (b) both be correct? Explain.

a. $\int \dfrac{dx}{\sqrt{1 - x^2}} = \sin^{-1} x + C$

b. $\int \dfrac{dx}{\sqrt{1 - x^2}} = -\int -\dfrac{dx}{\sqrt{1 - x^2}} = -\cos^{-1} x + C$

144. Can the integrations in (a) and (b) both be correct? Explain.

a. $\int \dfrac{dx}{\sqrt{1 - x^2}} = -\int -\dfrac{dx}{\sqrt{1 - x^2}} = -\cos^{-1} x + C$

b. $\int \dfrac{dx}{\sqrt{1 - x^2}} = \int \dfrac{-du}{\sqrt{1 - (-u)^2}}$ $\begin{array}{l} x = -u \\ dx = -du \end{array}$

$\quad = \int \dfrac{-du}{\sqrt{1 - u^2}}$

$\quad = \cos^{-1} u + C$

$\quad = \cos^{-1}(-x) + C$ $\quad u = -x$

145. The rectangle shown here has one side on the positive y-axis, one side on the positive x-axis, and its upper right-hand vertex on the curve $y = e^{-x^2}$. What dimensions give the rectangle its largest area, and what is that area?

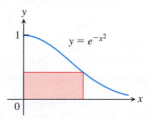

146. The rectangle shown here has one side on the positive y-axis, one side on the positive x-axis, and its upper right-hand vertex on the curve $y = (\ln x)/x^2$. What dimensions give the rectangle its largest area, and what is that area?

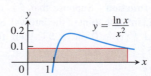

In Exercises 147 and 148, find the absolute maximum and minimum values of each function on the given interval.

147. $y = x \ln 2x - x$, $\left[\frac{1}{2e}, \frac{e}{2} \right]$

148. $y = 10x(2 - \ln x)$, $(0, e^2]$

In Exercises 149 and 150, find the absolute maxima and minima of the functions and give the x-coordinates where they occur.

149. $f(x) = e^{x/\sqrt{x^4+1}}$

150. $g(x) = e^{\sqrt{3-2x-x^2}}$

T **151.** Graph the following functions and use what you see to locate and estimate the extreme values, identify the coordinates of the inflection points, and identify the intervals on which the graphs are concave up and concave down. Then confirm your estimates by working with the functions' derivatives.

 a. $y = (\ln x)/\sqrt{x}$ **b.** $y = e^{-x^2}$

 c. $y = (1 + x)e^{-x}$

T **152.** Graph $f(x) = x \ln x$. Does the function appear to have an absolute minimum value? Confirm your answer with calculus.

T **153.** Graph $f(x) = (\sin x)^{\sin x}$ over $[0, 3\pi]$. Explain what you see.

154. A round underwater transmission cable consists of a core of copper wires surrounded by nonconducting insulation. If x denotes the ratio of the radius of the core to the thickness of the insulation, it is known that the speed of the transmission signal is given by the equation $v = x^2 \ln(1/x)$. If the radius of the core is 1 cm, what insulation thickness h will allow the greatest transmission speed?

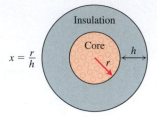

$$x = \frac{r}{h}$$

CHAPTER 4   Additional and Advanced Exercises

Functions and Derivatives

1. What can you say about a function whose maximum and minimum values on an interval are equal? Give reasons for your answer.

2. Is it true that a discontinuous function cannot have both an absolute maximum and an absolute minimum value on a closed interval? Give reasons for your answer.

3. Can you conclude anything about the extreme values of a continuous function on an open interval? On a half-open interval? Give reasons for your answer.

4. Local extrema   Use the sign pattern for the derivative

$$\frac{df}{dx} = 6(x - 1)(x - 2)^2(x - 3)^3(x - 4)^4$$

to identify the points where f has local maximum and minimum values.

5. Local extrema

 a. Suppose that the first derivative of $y = f(x)$ is

$$y' = 6(x + 1)(x - 2)^2.$$

 At what points, if any, does the graph of f have a local maximum, local minimum, or point of inflection?

 b. Suppose that the first derivative of $y = f(x)$ is

$$y' = 6x(x + 1)(x - 2).$$

 At what points, if any, does the graph of f have a local maximum, local minimum, or point of inflection?

6. If $f'(x) \leq 2$ for all x, what is the most the values of f can increase on $[0, 6]$? Give reasons for your answer.

7. Bounding a function   Suppose that f is continuous on $[a, b]$ and that c is an interior point of the interval. Show that if $f'(x) \leq 0$ on $[a, c]$ and $f'(x) \geq 0$ on $(c, b]$, then $f(x)$ is never less than $f(c)$ on $[a, b]$.

8. An inequality

 a. Show that $-1/2 \leq x/(1 + x^2) \leq 1/2$ for every value of x.

 b. Suppose that f is a function whose derivative is $f'(x) = x/(1 + x^2)$. Use the result in part (a) to show that

$$|f(b) - f(a)| \leq \frac{1}{2}|b - a|$$

 for any a and b.

9. The derivative of $f(x) = x^2$ is zero at $x = 0$, but f is not a constant function. Doesn't this contradict the corollary of the Mean Value Theorem that says that functions with zero derivatives are constant? Give reasons for your answer.

10. Extrema and inflection points   Let $h = fg$ be the product of two differentiable functions of x.

 a. If f and g are positive, with local maxima at $x = a$, and if f' and g' change sign at a, does h have a local maximum at a?

 b. If the graphs of f and g have inflection points at $x = a$, does the graph of h have an inflection point at a?

In either case, if the answer is yes, give a proof. If the answer is no, give a counterexample.

11. Finding a function   Use the following information to find the values of a, b, and c in the formula $f(x) = (x + a)/(bx^2 + cx + 2)$.

 a. The values of a, b, and c are either 0 or 1.

 b. The graph of f passes through the point $(-1, 0)$.

 c. The line $y = 1$ is an asymptote of the graph of f.

12. Horizontal tangent   For what value or values of the constant k will the curve $y = x^3 + kx^2 + 3x - 4$ have exactly one horizontal tangent?

Optimization

13. Largest inscribed triangle   Points A and B lie at the ends of a diameter of a unit circle and point C lies on the circumference. Is it true that the area of triangle ABC is largest when the triangle is isosceles? How do you know?

14. Proving the second derivative test   The Second Derivative Test for Local Maxima and Minima (Section 4.4) says:

a. f has a local maximum value at $x = c$ if $f'(c) = 0$ and $f''(c) < 0$

b. f has a local minimum value at $x = c$ if $f'(c) = 0$ and $f''(c) > 0$.

To prove statement (a), let $\varepsilon = (1/2)|f''(c)|$. Then use the fact that

$$f''(c) = \lim_{h \to 0} \frac{f'(c + h) - f'(c)}{h} = \lim_{h \to 0} \frac{f'(c + h)}{h}$$

to conclude that for some $\delta > 0$,

$$0 < |h| < \delta \quad \Rightarrow \quad \frac{f'(c + h)}{h} < f''(c) + \varepsilon < 0.$$

Thus, $f'(c + h)$ is positive for $-\delta < h < 0$ and negative for $0 < h < \delta$. Prove statement (b) in a similar way.

15. Hole in a water tank You want to bore a hole in the side of the tank shown here at a height that will make the stream of water coming out hit the ground as far from the tank as possible. If you drill the hole near the top, where the pressure is low, the water will exit slowly but spend a relatively long time in the air. If you drill the hole near the bottom, the water will exit at a higher velocity but have only a short time to fall. Where is the best place, if any, for the hole? (*Hint:* How long will it take an exiting droplet of water to fall from height y to the ground?)

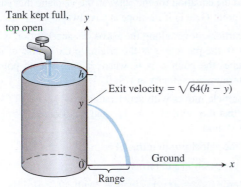

16. Kicking a field goal An American football player wants to kick a field goal with the ball being on a right hash mark. Assume that the goal posts are b feet apart and that the hash mark line is a distance $a > 0$ feet from the right goal post. (See the accompanying figure.) Find the distance h from the goal post line that gives the kicker his largest angle β. Assume that the football field is flat.

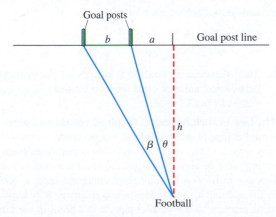

17. A max-min problem with a variable answer Sometimes the solution of a max-min problem depends on the proportions of the shapes involved. As a case in point, suppose that a right circular cylinder of radius r and height h is inscribed in a right circular cone of radius R and height H, as shown here. Find the value of r (in terms of R and H) that maximizes the total surface area of the cylinder (including top and bottom). As you will see, the solution depends on whether $H \leq 2R$ or $H > 2R$.

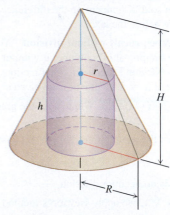

18. Minimizing a parameter Find the smallest value of the positive constant m that will make $mx - 1 + (1/x)$ greater than or equal to zero for all positive values of x.

19. Determine the dimensions of the rectangle of largest area that can be inscribed in the right triangle in the accompanying figure.

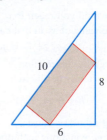

20. A rectangular box with a square base is inscribed in a right circular cone of height 4 and base radius 3. If the base of the box sits on the base of the cone, what is the largest possible volume of the box?

Limits

21. Evaluate the following limits.

a. $\lim_{x \to 0} \dfrac{2 \sin 5x}{3x}$

b. $\lim_{x \to 0} \sin 5x \cot 3x$

c. $\lim_{x \to 0} x \csc^2 \sqrt{2x}$

d. $\lim_{x \to \pi/2} (\sec x - \tan x)$

e. $\lim_{x \to 0} \dfrac{x - \sin x}{x - \tan x}$

f. $\lim_{x \to 0} \dfrac{\sin x^2}{x \sin x}$

g. $\lim_{x \to 0} \dfrac{\sec x - 1}{x^2}$

h. $\lim_{x \to 2} \dfrac{x^3 - 8}{x^2 - 4}$

22. L'Hôpital's Rule does not help with the following limits. Find them some other way.

a. $\lim_{x \to \infty} \dfrac{\sqrt{x + 5}}{\sqrt{x} + 5}$

b. $\lim_{x \to \infty} \dfrac{2x}{x + 7\sqrt{x}}$

Theory and Examples

23. Suppose that it costs a company $y = a + bx$ dollars to produce x units per week. It can sell x units per week at a price of $P = c - ex$ dollars per unit. Each of a, b, c, and e represents a positive constant. **(a)** What production level maximizes the profit? **(b)** What is the corresponding price? **(c)** What is the weekly profit at this level of production? **(d)** At what price should each item be sold to maximize profits if the government imposes a tax of t dollars per item sold? Comment on the difference between this price and the price before the tax.

24. Estimating reciprocals without division You can estimate the value of the reciprocal of a number a without ever dividing by a if you apply Newton's method to the function $f(x) = (1/x) - a$. For example, if $a = 3$, the function involved is $f(x) = (1/x) - 3$.

a. Graph $y = (1/x) - 3$. Where does the graph cross the x-axis?

b. Show that the recursion formula in this case is

$$x_{n+1} = x_n(2 - 3x_n),$$

so there is no need for division.

25. To find $x = \sqrt[q]{a}$, we apply Newton's method to $f(x) = x^q - a$. Here we assume that a is a positive real number and q is a positive integer. Show that x_1 is a "weighted average" of x_0 and a/x_0^{q-1}, and find the coefficients m_0, m_1 such that

$$x_1 = m_0 x_0 + m_1 \left(\frac{a}{x_0^{q-1}}\right), \quad \begin{array}{l} m_0 > 0, m_1 > 0, \\ m_0 + m_1 = 1. \end{array}$$

What conclusion would you reach if x_0 and a/x_0^{q-1} were equal? What would be the value of x_1 in that case?

26. The family of straight lines $y = ax + b$ (a, b arbitrary constants) can be characterized by the relation $y'' = 0$. Find a similar relation satisfied by the family of all circles

$$(x - h)^2 + (y - h)^2 = r^2,$$

where h and r are arbitrary constants. (*Hint:* Eliminate h and r from the set of three equations including the given one and two obtained by successive differentiation.)

27. Free fall in the fourteenth century In the middle of the fourteenth century, Albert of Saxony (1316–1390) proposed a model of free fall that assumed that the velocity of a falling body was proportional to the distance fallen. It seemed reasonable to think that a body that had fallen 20 ft might be moving twice as fast as a body that had fallen 10 ft. And besides, none of the instruments in use at the time were accurate enough to prove otherwise. Today we can see just how far off Albert of Saxony's model was by solving the initial value problem implicit in his model. Solve the problem and compare your solution graphically with the equation $s = 16t^2$. You will see that it describes a motion that starts too slowly at first and then becomes too fast too soon to be realistic.

T **28. Group blood testing** During World War II it was necessary to administer blood tests to large numbers of recruits. There are two standard ways to administer a blood test to N people. In method 1, each person is tested separately. In method 2, the blood samples of x people are pooled and tested as one large sample. If the test is negative, this one test is enough for all x people. If the test is positive, then each of the x people is tested separately, requiring a total of $x + 1$ tests. Using the second method and some probability

theory it can be shown that, on the average, the total number of tests y will be

$$y = N\left(1 - q^x + \frac{1}{x}\right).$$

With $q = 0.99$ and $N = 1000$, find the integer value of x that minimizes y. Also find the integer value of x that maximizes y. (This second result is not important to the real-life situation.) The group testing method was used in World War II with a savings of 80% over the individual testing method, but not with the given value of q.

29. Assume that the brakes of an automobile produce a constant deceleration of k ft/sec^2. **(a)** Determine what k must be to bring an automobile traveling 60 mi/hr (88 ft/sec) to rest in a distance of 100 ft from the point where the brakes are applied. **(b)** With the same k, how far would a car traveling 30 mi/hr go before being brought to a stop?

30. Let $f(x)$, $g(x)$ be two continuously differentiable functions satisfying the relationships $f'(x) = g(x)$ and $f''(x) = -f(x)$. Let $h(x) = f^2(x) + g^2(x)$. If $h(0) = 5$, find $h(10)$.

31. Can there be a curve satisfying the following conditions? d^2y/dx^2 is everywhere equal to zero and, when $x = 0$, $y = 0$ and $dy/dx = 1$. Give a reason for your answer.

32. Find the equation for the curve in the xy-plane that passes through the point $(1, -1)$ if its slope at x is always $3x^2 + 2$.

33. A particle moves along the x-axis. Its acceleration is $a = -t^2$. At $t = 0$, the particle is at the origin. In the course of its motion, it reaches the point $x = b$, where $b > 0$, but no point beyond b. Determine its velocity at $t = 0$.

34. A particle moves with acceleration $a = \sqrt{t} - \left(1/\sqrt{t}\right)$. Assuming that the velocity $v = 4/3$ and the position $s = -4/15$ when $t = 0$, find

a the velocity v in terms of t.

b the position s in terms of t.

35. Given $f(x) = ax^2 + 2bx + c$ with $a > 0$. By considering the minimum, prove that $f(x) \geq 0$ for all real x if and only if $b^2 - ac \leq 0$.

36. Schwarz's inequality

a. In Exercise 35, let

$$f(x) = (a_1x + b_1)^2 + (a_2x + b_2)^2 + \cdots + (a_nx + b_n)^2,$$

and deduce Schwarz's inequality:

$$(a_1b_1 + a_2b_2 + \cdots + a_nb_n)^2$$
$$\leq \left(a_1^2 + a_2^2 + \cdots + a_n^2\right)\left(b_1^2 + b_2^2 + \cdots + b_n^2\right).$$

b. Show that equality holds in Schwarz's inequality only if there exists a real number x that makes a_ix equal $-b_i$ for every value of i from 1 to n.

37. The best branching angles for blood vessels and pipes When a smaller pipe branches off from a larger one in a flow system, we may want it to run off at an angle that is best from some energy-saving point of view. We might require, for instance, that energy loss due to friction be minimized along the section AOB shown in the accompanying figure. In this diagram, B is a given point to be reached by the smaller pipe, A is a point in the larger pipe upstream from B, and O is the point where the branching occurs.

A law due to Poiseuille states that the loss of energy due to friction in nonturbulent flow is proportional to the length of the path and inversely proportional to the fourth power of the radius. Thus, the loss along AO is $(kd_1)/R^4$ and along OB is $(kd_2)/r^4$, where k is a constant, d_1 is the length of AO, d_2 is the length of OB, R is the radius of the larger pipe, and r is the radius of the smaller pipe. The angle θ is to be chosen to minimize the sum of these two losses:

$$L = k\frac{d_1}{R^4} + k\frac{d_2}{r^4}.$$

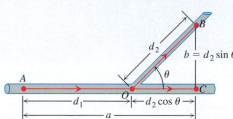

In our model, we assume that $AC = a$ and $BC = b$ are fixed. Thus we have the relations

$$d_1 + d_2 \cos\theta = a \quad d_2 \sin\theta = b,$$

so that

$$d_2 = b\csc\theta,$$

$$d_1 = a - d_2\cos\theta = a - b\cot\theta.$$

We can express the total loss L as a function of θ:

$$L = k\left(\frac{a - b\cot\theta}{R^4} + \frac{b\csc\theta}{r^4}\right).$$

a. Show that the critical value of θ for which $dL/d\theta$ equals zero is

$$\theta_c = \cos^{-1}\frac{r^4}{R^4}.$$

b. If the ratio of the pipe radii is $r/R = 5/6$, estimate to the nearest degree the optimal branching angle given in part (a).

38. Consider point (a, b) on the graph of $y = \ln x$ and triangle ABC formed by the tangent line at (a, b), the y-axis, and the line $y = b$. Show that

$$\text{(area triangle } ABC) = \frac{a}{2}.$$

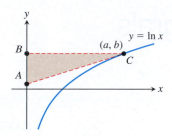

39. Consider the unit circle centered at the origin and with a vertical tangent line passing through point A in the accompanying figure. Assume that the lengths of segments AB and AC are equal, and let point D be the intersection of the x-axis with the line passing through points B and C. Find the limit of t as B approaches A.

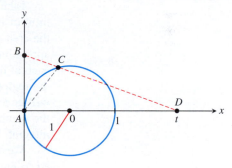

CHAPTER 4   Technology Application Projects

Mathematica/Maple Projects

Projects can be found within MyMathLab.

- ***Motion Along a Straight Line: Position → Velocity → Acceleration***
  You will observe the shape of a graph through dramatic animated visualizations of the derivative relations among the position, velocity, and acceleration. Figures in the text can be animated.

- ***Newton's Method: Estimate π to How Many Places?***
  Plot a function, observe a root, pick a starting point near the root, and use Newton's Iteration Procedure to approximate the root to a desired accuracy. The numbers π, e, and $\sqrt{2}$ are approximated.

5

Integrals

OVERVIEW A great achievement of classical geometry was obtaining formulas for the areas and volumes of triangles, spheres, and cones. In this chapter we develop a method, called *integration*, to calculate the areas and volumes of more general shapes. The *definite integral* is the key tool in calculus for defining and calculating areas and volumes. We also use it to compute quantities such as the lengths of curved paths, probabilities, averages, energy consumption, the mass of an object, and the force against a dam's floodgates, to name only a few.

Like the derivative, the definite integral is defined as a limit. The definite integral is a limit of increasingly fine approximations. The idea is to approximate a quantity (such as the area of a curvy region) by dividing it into many small pieces, each of which we can approximate by something simple (such as a rectangle). Summing the contributions of each of the simple pieces gives us an approximation to the original quantity. As we divide the region into more and more pieces, the approximation given by the sum of the pieces will generally improve, converging to the quantity we are measuring. We take a limit as the number of terms increases to infinity, and when the limit exists, the result is a definite integral. We develop this idea in Section 5.3.

We also show that the process of computing these definite integrals is closely connected to finding antiderivatives. This is one of the most important relationships in calculus; it gives us an efficient way to compute definite integrals, providing a simple and powerful method that eliminates the difficulty of directly computing limits of approximations. This connection is captured in the Fundamental Theorem of Calculus.

5.1 Area and Estimating with Finite Sums

The basis for formulating definite integrals is the construction of approximations by finite sums. In this section we consider three examples of this process: finding the area under a graph, the distance traveled by a moving object, and the average value of a function. Although we have yet to define precisely what we mean by the area of a general region in the plane, or the average value of a function over a closed interval, we do have intuitive ideas of what these notions mean. We begin our approach to integration by *approximating* these quantities with simpler finite sums related to these intuitive ideas. We then consider what happens when we take more and more terms in the summation process. In subsequent sections we look at taking the limit of these sums as the number of terms goes to infinity, which leads to a precise definition of the definite integral.

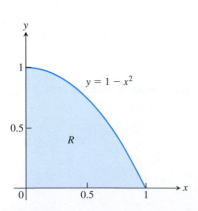

FIGURE 5.1 The area of a region R cannot be found by a simple formula.

Area

Suppose we want to find the area of the shaded region R that lies above the x-axis, below the graph of $y = 1 - x^2$, and between the vertical lines $x = 0$ and $x = 1$ (see Figure 5.1).

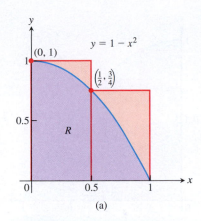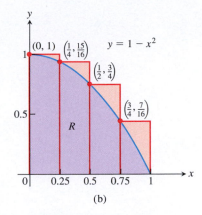

FIGURE 5.2   (a) We get an upper estimate of the area of R by using two rectangles containing R. (b) Four rectangles give a better upper estimate. Both estimates overshoot the true value for the area by the amount shaded in light red.

Unfortunately, there is no simple geometric formula for calculating the areas of general shapes having curved boundaries like the region R. How, then, can we find the area of R?

While we do not yet have a method for determining the exact area of R, we can approximate it in a simple way. Figure 5.2a shows two rectangles that together contain the region R. Each rectangle has width $1/2$ and they have heights 1 and $3/4$ (left to right). The height of each rectangle is the maximum value of the function f in each subinterval. Because the function f is decreasing, the height is its value at the left endpoint of the subinterval of $[0, 1]$ that forms the base of the rectangle. The total area of the two rectangles approximates the area A of the region R:

$$A \approx 1 \cdot \frac{1}{2} + \frac{3}{4} \cdot \frac{1}{2} = \frac{7}{8} = 0.875.$$

This estimate is larger than the true area A since the two rectangles contain R. We say that 0.875 is an **upper sum** because it is obtained by taking the height of the rectangle corresponding to the maximum (uppermost) value of $f(x)$ over points x lying in the base of each rectangle. In Figure 5.2b, we improve our estimate by using four thinner rectangles, each of width $1/4$, which taken together contain the region R. These four rectangles give the approximation

$$A \approx 1 \cdot \frac{1}{4} + \frac{15}{16} \cdot \frac{1}{4} + \frac{3}{4} \cdot \frac{1}{4} + \frac{7}{16} \cdot \frac{1}{4} = \frac{25}{32} = 0.78125,$$

which is still greater than A since the four rectangles contain R.

Suppose instead we use four rectangles contained *inside* the region R to estimate the area, as in Figure 5.3a. Each rectangle has width $1/4$ as before, but the rectangles are shorter and lie entirely beneath the graph of f. The function $f(x) = 1 - x^2$ is decreasing on $[0, 1]$, so the height of each of these rectangles is given by the value of f at the right endpoint of the subinterval forming its base. The fourth rectangle has zero height and therefore contributes no area. Summing these rectangles, whose heights are the minimum value of $f(x)$ over points x in the rectangle's base, gives a **lower sum** approximation to the area:

$$A \approx \frac{15}{16} \cdot \frac{1}{4} + \frac{3}{4} \cdot \frac{1}{4} + \frac{7}{16} \cdot \frac{1}{4} + 0 \cdot \frac{1}{4} = \frac{17}{32} = 0.53125.$$

This estimate is smaller than the area A since the rectangles all lie inside of the region R. The true value of A lies somewhere between these lower and upper sums:

$$0.53125 < A < 0.78125.$$

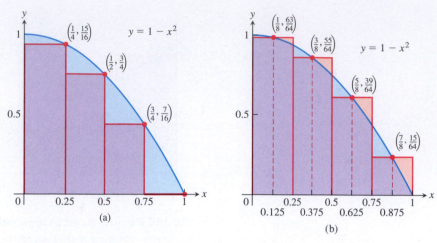

(a)   (b)

FIGURE 5.3   (a) Rectangles contained in R give an estimate for the area that under-shoots the true value by the amount shaded in light blue. (b) The midpoint rule uses rectangles whose height is the value of $y = f(x)$ at the midpoints of their bases. The estimate appears closer to the true value of the area because the light red overshoot areas roughly balance the light blue undershoot areas.

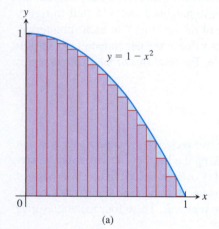

(a)

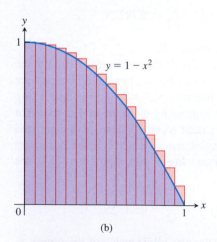

(b)

FIGURE 5.4   (a) A lower sum using 16 rectangles of equal width $\Delta x = 1/16$. (b) An upper sum using 16 rectangles.

Considering both lower and upper sum approximations gives us estimates for the area and a bound on the size of the possible error in these estimates, since the true value of the area lies somewhere between them. Here the error cannot be greater than the difference $0.78125 - 0.53125 = 0.25$.

Yet another estimate can be obtained by using rectangles whose heights are the values of f at the midpoints of the bases of the rectangles (Figure 5.3b). This method of estimation is called the **midpoint rule** for approximating the area. The midpoint rule gives an estimate that is between a lower sum and an upper sum, but it is not quite so clear whether it overestimates or underestimates the true area. With four rectangles of width $1/4$ as before, the midpoint rule estimates the area of R to be

$$A \approx \frac{63}{64} \cdot \frac{1}{4} + \frac{55}{64} \cdot \frac{1}{4} + \frac{39}{64} \cdot \frac{1}{4} + \frac{15}{64} \cdot \frac{1}{4} = \frac{172}{64} \cdot \frac{1}{4} = 0.671875.$$

In each of the sums that we computed, the interval $[a, b]$ over which the function f is defined was subdivided into n subintervals of equal width (or length) $\Delta x = (b - a)/n$, and f was evaluated at a point in each subinterval: c_1 in the first subinterval, c_2 in the second subinterval, and so on. For the upper sum we chose c_k so that $f(c_k)$ was the maximum value of f in the kth subinterval, for the lower sum we chose it so that $f(c_k)$ was the minimum, and for the midpoint rule we chose c_k to be the midpoint of the kth subinterval. In each case the finite sums have the form

$$f(c_1) \, \Delta x + f(c_2) \, \Delta x + f(c_3) \, \Delta x + \cdots + f(c_n) \, \Delta x.$$

By taking more and more rectangles, with each rectangle thinner than before, it appears that these finite sums give better and better approximations to the true area of the region R.

Figure 5.4a shows a lower sum approximation for the area of R using 16 rectangles of equal width. The sum of their areas is 0.634765625, which appears close to the true area, but is still smaller since the rectangles lie inside R.

Figure 5.4b shows an upper sum approximation using 16 rectangles of equal width. The sum of their areas is 0.697265625, which is somewhat larger than the true area because the rectangles taken together contain R. The midpoint rule for 16 rectangles gives a total area approximation of 0.6669921875, but it is not immediately clear whether this estimate is larger or smaller than the true area.

Table 5.1 shows the values of upper and lower sum approximations to the area of R, using up to 1000 rectangles. The values of these approximations appear to be approaching $2/3$. In Section 5.2 we will see how to get an exact value of the area of regions such as R by taking a limit as the base width of each rectangle goes to zero and the number of rectangles goes to infinity. With the techniques developed there, we will be able to show that the area of R is exactly $2/3$.

TABLE 5.1 Finite approximations for the area of R

Number of subintervals	Lower sum	Midpoint sum	Upper sum
2	0.375	0.6875	0.875
4	0.53125	0.671875	0.78125
16	0.634765625	0.6669921875	0.697265625
50	0.6566	0.6667	0.6766
100	0.66165	0.666675	0.67165
1000	0.6661665	0.66666675	0.6671665

Distance Traveled

Suppose we know the velocity function $v(t)$ of a car that moves straight down a highway without changing direction, and we want to know how far it traveled between times $t = a$ and $t = b$. The position function $s(t)$ of the car has derivative $v(t)$. If we can find an antiderivative $F(t)$ of $v(t)$ then we can find the car's position function $s(t)$ by setting $s(t) = F(t) + C$. The distance traveled can then be found by calculating the change in position, $s(b) - s(a) = F(b) - F(a)$. However, if the velocity is known only by the readings at various times of a speedometer on the car, then we have no formula from which to obtain an antiderivative for the velocity. So what do we do in this situation?

When we don't know an antiderivative for the velocity $v(t)$, we can approximate the distance traveled by using finite sums in a way similar to the area estimates that we discussed before. We subdivide the interval $[a, b]$ into short time intervals and assume that the velocity on each subinterval is fairly constant. Then we approximate the distance traveled on each time subinterval with the usual distance formula

$$\text{distance} = \text{velocity} \times \text{time}$$

and add the results across $[a, b]$.

Suppose the subdivided interval looks like

with the subintervals all of equal length Δt. Pick a number t_1 in the first interval. If Δt is so small that the velocity barely changes over a short time interval of duration Δt, then the distance traveled in the first time interval is about $v(t_1)\,\Delta t$. If t_2 is a number in the second interval, the distance traveled in the second time interval is about $v(t_2)\,\Delta t$. The sum of the distances traveled over all the time intervals is

$$D \approx v(t_1)\,\Delta t + v(t_2)\,\Delta t + \cdots + v(t_n)\,\Delta t,$$

where n is the total number of subintervals. This sum is only an approximation to the true distance D, but the approximation increases in accuracy as we take more and more subintervals.

EXAMPLE 1   The velocity function of a projectile fired straight into the air is $f(t) = 160 - 9.8t$ m/sec. Use the summation technique just described to estimate how far the projectile rises during the first 3 sec. How close do the sums come to the exact value of 435.9 m? (You will learn how to compute the exact value of this and similar quantities in Section 5.4.)

Solution   We explore the results for different numbers of subintervals and different choices of evaluation points. Notice that $f(t)$ is decreasing, so choosing left endpoints gives an upper sum estimate; choosing right endpoints gives a lower sum estimate.

(a) *Three subintervals of length 1, with f evaluated at left endpoints giving an upper sum:*

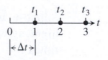

With f evaluated at $t = 0, 1$, and 2, we have

$$D \approx f(t_1)\,\Delta t + f(t_2)\,\Delta t + f(t_3)\,\Delta t$$
$$= [160 - 9.8(0)](1) + [160 - 9.8(1)](1) + [160 - 9.8(2)](1)$$
$$= 450.6.$$

(b) *Three subintervals of length 1, with f evaluated at right endpoints giving a lower sum:*

With f evaluated at $t = 1, 2$, and 3, we have

$$D \approx f(t_1)\,\Delta t + f(t_2)\,\Delta t + f(t_3)\,\Delta t$$
$$= [160 - 9.8(1)](1) + [160 - 9.8(2)](1) + [160 - 9.8(3)](1)$$
$$= 421.2.$$

(c) *With six subintervals of length 1/2, we get*

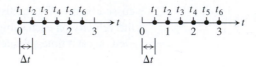

These estimates give an upper sum using left endpoints: $D \approx 443.25$; and a lower sum using right endpoints: $D \approx 428.55$. These six-interval estimates are somewhat closer than the three-interval estimates. The results improve as the subintervals get shorter.

As we can see in Table 5.2, the left-endpoint upper sums approach the true value 435.9 from above, whereas the right-endpoint lower sums approach it from below. The true value lies between these upper and lower sums. The magnitude of the error in the closest entry is 0.23, a small percentage of the true value.

$$\text{Error magnitude} = |\text{true value} - \text{calculated value}|$$
$$= |435.9 - 435.67| = 0.23.$$
$$\text{Error percentage} = \frac{0.23}{435.9} \approx 0.05\%.$$

It would be reasonable to conclude from the table's last entries that the projectile rose about 436 m during its first 3 sec of flight.   ■

TABLE 5.2 Travel-distance estimates

Number of subintervals	Length of each subinterval	Upper sum	Lower sum
3	1	450.6	421.2
6	1/2	443.25	428.55
12	1/4	439.58	432.23
24	1/8	437.74	434.06
48	1/16	436.82	434.98
96	1/32	436.36	435.44
192	1/64	436.13	435.67

Displacement Versus Distance Traveled

If an object with position function $s(t)$ moves along a coordinate line without changing direction, we can calculate the total distance it travels from $t = a$ to $t = b$ by summing the distance traveled over small intervals, as in Example 1. If the object reverses direction one or more times during the trip, then we need to use the object's *speed* $|v(t)|$, which is the absolute value of its velocity function, $v(t)$, to find the total distance traveled. Using the velocity itself, as in Example 1, gives instead an estimate to the object's **displacement**, $s(b) - s(a)$, the difference between its initial and final positions. To see the difference, think about what happens when you walk a mile from your home and then walk back. The total distance traveled is two miles, but your displacement is zero, because you end up back where you started.

To see why using the velocity function in the summation process gives an estimate to the displacement, partition the time interval $[a, b]$ into small enough equal subintervals Δt so that the object's velocity does not change very much from time t_{k-1} to t_k. Then $v(t_k)$ gives a good approximation of the velocity throughout the interval. Accordingly, the change in the object's position coordinate, which is its displacement during the time interval, is about

$$v(t_k) \, \Delta t.$$

The change is positive if $v(t_k)$ is positive and negative if $v(t_k)$ is negative.

In either case, the distance traveled by the object during the subinterval is about

$$|v(t_k)| \, \Delta t.$$

The **total distance traveled** over the time interval is approximately the sum

$$|v(t_1)|\Delta t + |v(t_2)|\Delta t + \cdots + |v(t_n)| \, \Delta t.$$

We will revisit these ideas in Section 5.4.

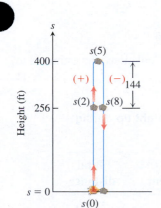

FIGURE 5.5   The rock in Example 2. The height $s = 256$ ft is reached at $t = 2$ and $t = 8$ sec. The rock falls 144 ft from its maximum height when $t = 8$.

EXAMPLE 2   In Example 4 in Section 3.4, we analyzed the motion of a heavy rock blown straight up by a dynamite blast. In that example, we found the velocity of the rock at time t was $v(t) = 160 - 32t$ ft/sec. The rock was 256 ft above the ground 2 sec after the explosion, continued upward to reach a maximum height of 400 ft at 5 sec after the explosion, and then fell back down a distance of 144 ft to reach the height of 256 ft again at $t = 8$ sec after the explosion. (See Figure 5.5.) The total distance traveled in these 8 seconds is $400 + 144 = 544$ ft.

If we follow a procedure like the one presented in Example 1, using the velocity function $v(t)$ in the summation process from $t = 0$ to $t = 8$, we obtain an estimate of the rock's height above the ground at time $t = 8$. Starting at time $t = 0$, the rock traveled upward a total of $256 + 144 = 400$ ft, but then it peaked and traveled downward

TABLE 5.3 Velocity function

t	$v(t)$	t	$v(t)$
0	160	4.5	16
0.5	144	5.0	0
1.0	128	5.5	-16
1.5	112	6.0	-32
2.0	96	6.5	-48
2.5	80	7.0	-64
3.0	64	7.5	-80
3.5	48	8.0	-96
4.0	32		

144 ft, ending at a height of 256 ft at time $t = 8$. The velocity $v(t)$ is positive during the upward travel, but negative while the rock falls back down. When we compute the sum $v(t_1)\Delta t + v(t_2)\Delta t + \cdots + v(t_n)\Delta t$, part of the upward positive distance change is canceled by the negative downward movement, giving in the end an approximation of the displacement from the initial position, equal to a positive change of 256 ft.

On the other hand, if we use the speed $|v(t)|$, which is the absolute value of the velocity function, then distances traveled while moving up and distances traveled while moving down are both counted positively. Both the total upward motion of 400 ft and the downward motion of 144 ft are now counted as positive distances traveled, so the sum $|v(t_1)|\Delta t + |v(t_2)|\Delta t + \cdots + |v(t_n)|\Delta t$ gives us an approximation of 544 ft, the total distance that the rock traveled from time $t = 0$ to time $t = 8$.

As an illustration of our discussion, we subdivide the interval $[0, 8]$ into sixteen subintervals of length $\Delta t = 1/2$ and take the right endpoint of each subinterval as the value of t_k. Table 5.3 shows the values of the velocity function at these endpoints.

Using $v(t)$ in the summation process, we estimate the displacement at $t = 8$:

$$(144 + 128 + 112 + 96 + 80 + 64 + 48 + 32 + 16$$

$$+ 0 - 16 - 32 - 48 - 64 - 80 - 96) \cdot \frac{1}{2} = 192$$

$$\text{Error magnitude} = 256 - 192 = 64$$

Using $|v(t)|$ in the summation process, we estimate the total distance traveled over the time interval $[0, 8]$:

$$(144 + 128 + 112 + 96 + 80 + 64 + 48 + 32 + 16$$

$$+ 0 + 16 + 32 + 48 + 64 + 80 + 96) \cdot \frac{1}{2} = 528$$

$$\text{Error magnitude} = 544 - 528 = 16$$

If we take more and more subintervals of $[0, 8]$ in our calculations, the estimates to the heights 256 ft and 544 ft improve, as shown in Table 5.4. ∎

TABLE 5.4 Travel estimates for a rock blown straight up during the time interval [0, 8]

Number of subintervals	Length of each subinterval	Displacement	Total distance
16	1/2	192.0	528.0
32	1/4	224.0	536.0
64	1/8	240.0	540.0
128	1/16	248.0	542.0
256	1/32	252.0	543.0
512	1/64	254.0	543.5

Average Value of a Nonnegative Continuous Function

The average value of a collection of n numbers $x_1, x_2, \ldots, x_n$ is obtained by adding them together and dividing by n. But what is the average value of a continuous function f on an interval $[a, b]$? Such a function can assume infinitely many values. For example, the temperature at a certain location in a town is a continuous function that goes up and down each day. What does it mean to say that the average temperature in the town over the course of a day is 73 degrees?

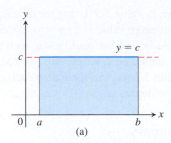
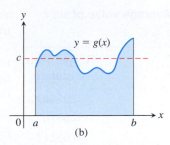

FIGURE 5.6 (a) The average value of $f(x) = c$ on $[a, b]$ is the area of the rectangle divided by $b - a$. (b) The average value of $g(x)$ on $[a, b]$ is the area beneath its graph divided by $b - a$.

When a function is constant, this question is easy to answer. A function with constant value c on an interval $[a, b]$ has average value c. When c is positive, its graph over $[a, b]$ gives a rectangle of height c. The average value of the function can then be interpreted geometrically as the area of this rectangle divided by its width $b - a$ (see Figure 5.6a).

What if we want to find the average value of a nonconstant function, such as the function g in Figure 5.6b? We can think of this graph as a snapshot of the height of some water that is sloshing around in a tank between enclosing walls at $x = a$ and $x = b$. As the water moves, its height over each point changes, but its average height remains the same. To get the average height of the water, we let it settle down until it is level and its height is constant. The resulting height c equals the area under the graph of g divided by $b - a$. We are led to *define* the average value of a nonnegative function on an interval $[a, b]$ to be the area under its graph divided by $b - a$. For this definition to be valid, we need a precise understanding of what is meant by the area under a graph. This will be obtained in Section 5.3, but for now we look at an example.

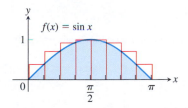

FIGURE 5.7 Approximating the area under $f(x) = \sin x$ between 0 and π to compute the average value of $\sin x$ over $[0, \pi]$, using eight rectangles (Example 3).

EXAMPLE 3 Estimate the average value of the function $f(x) = \sin x$ on the interval $[0, \pi]$.

Solution Looking at the graph of $\sin x$ between 0 and π in Figure 5.7, we can see that its average height is somewhere between 0 and 1. To find the average, we need to calculate the area A under the graph and then divide this area by the length of the interval, $\pi - 0 = \pi$.

We do not have a simple way to determine the area, so we approximate it with finite sums. To get an upper sum approximation, we add the areas of eight rectangles of equal width $\pi/8$ that together contain the region that is beneath the graph of $y = \sin x$ and above the x-axis on $[0, \pi]$. We choose the heights of the rectangles to be the largest value of $\sin x$ on each subinterval. Over a particular subinterval, this largest value may occur at the left endpoint, the right endpoint, or somewhere between them. We evaluate $\sin x$ at this point to get the height of the rectangle for an upper sum. The sum of the rectangular areas then gives an estimate of the total area (Figure 5.7):

$$A \approx \left(\sin\frac{\pi}{8} + \sin\frac{\pi}{4} + \sin\frac{3\pi}{8} + \sin\frac{\pi}{2} + \sin\frac{\pi}{2} + \sin\frac{5\pi}{8} + \sin\frac{3\pi}{4} + \sin\frac{7\pi}{8} \right) \cdot \frac{\pi}{8}$$

$$\approx (.38 + .71 + .92 + 1 + 1 + .92 + .71 + .38) \cdot \frac{\pi}{8} = (6.02) \cdot \frac{\pi}{8} \approx 2.364.$$

To estimate the average value of $\sin x$ on $[0, \pi]$ we divide the estimated area by the length π of the interval and obtain the approximation $2.364/\pi \approx 0.753$.

Since we used an upper sum to approximate the area, this estimate is greater than the actual average value of $\sin x$ over $[0, \pi]$. If we use more and more rectangles, with each rectangle getting thinner and thinner, we get closer and closer to the exact average value, as

TABLE 5.5 Average value of sin x on $0 \le x \le \pi$

Number of subintervals	Upper sum estimate
8	0.75342
16	0.69707
32	0.65212
50	0.64657
100	0.64161
1000	0.63712

shown in Table 5.5. Using the techniques covered in Section 5.3, we will later show that the true average value is $2/\pi \approx 0.63662$.

As before, we could just as well have used rectangles lying under the graph of $y = \sin x$ and calculated a lower sum approximation, or we could have used the midpoint rule. In Section 5.3 we will see that in each case, the approximations are close to the true area if all the rectangles are sufficiently thin. ∎

Summary

The area under the graph of a positive function, the distance traveled by a moving object that doesn't change direction, and the average value of a nonnegative function f over an interval can all be approximated by finite sums constructed in a certain way. First we subdivide the interval into subintervals, treating f as if it were constant over each subinterval. Then we multiply the width of each subinterval by the value of f at some point within it, and add these products together. If the interval $[a, b]$ is subdivided into n subintervals of equal widths $\Delta x = (b - a)/n$, and if $f(c_k)$ is the value of f at the chosen point c_k in the kth subinterval, this process gives a finite sum of the form

$$f(c_1)\,\Delta x + f(c_2)\,\Delta x + f(c_3)\,\Delta x + \cdots + f(c_n)\,\Delta x.$$

The choices for the c_k could maximize or minimize the value of f in the kth subinterval, or give some value in between. The true value lies somewhere between the approximations given by upper sums and lower sums. In the examples that we looked at, the finite sum approximations improved as we took more subintervals of thinner width.

EXERCISES 5.1

Area

In Exercises 1–4, use finite approximations to estimate the area under the graph of the function using

  a. a lower sum with two rectangles of equal width.

  b. a lower sum with four rectangles of equal width.

  c. an upper sum with two rectangles of equal width.

  d. an upper sum with four rectangles of equal width.

1. $f(x) = x^2$ between $x = 0$ and $x = 1$.

2. $f(x) = x^3$ between $x = 0$ and $x = 1$.

3. $f(x) = 1/x$ between $x = 1$ and $x = 5$.

4. $f(x) = 4 - x^2$ between $x = -2$ and $x = 2$.

    Using rectangles each of whose height is given by the value of the function at the midpoint of the rectangle's base (*the midpoint rule*), estimate the area under the graphs of the following functions, using first two and then four rectangles.

5. $f(x) = x^2$ between $x = 0$ and $x = 1$.

6. $f(x) = x^3$ between $x = 0$ and $x = 1$.

7. $f(x) = 1/x$ between $x = 1$ and $x = 5$.

8. $f(x) = 4 - x^2$ between $x = -2$ and $x = 2$.

Distance

9. Distance traveled The accompanying table shows the velocity of a model train engine moving along a track for 10 sec. Estimate

the distance traveled by the engine using 10 subintervals of length 1 with

  a. left-endpoint values.

  b. right-endpoint values.

Time (sec)	Velocity (cm / sec)	Time (sec)	Velocity (cm / sec)
0	0	6	28
1	30	7	15
2	56	8	5
3	25	9	15
4	38	10	0
5	33		

10. Distance traveled upstream You are sitting on the bank of a tidal river watching the incoming tide carry a bottle upstream. You record the velocity of the flow every 5 minutes for an hour, with the results shown in the accompanying table. About how far upstream did the bottle travel during that hour? Find an estimate using 12 subintervals of length 5 with

a. left-endpoint values.

b. right-endpoint values.

Time (min)	Velocity (m/sec)	Time (min)	Velocity (m/sec)
0	1	35	1.2
5	1.2	40	1.0
10	1.7	45	1.8
15	2.0	50	1.5
20	1.8	55	1.2
25	1.6	60	0
30	1.4		

11. Length of a road You and a companion are about to drive a twisty stretch of dirt road in a car whose speedometer works but whose odometer (mileage counter) is broken. To find out how long this particular stretch of road is, you record the car's velocity at 10-sec intervals, with the results shown in the accompanying table. Estimate the length of the road using

a. left-endpoint values.

b. right-endpoint values.

Time (sec)	Velocity (converted to ft/sec) (30 mi/h = 44 ft/sec)	Time (sec)	Velocity (converted to ft/sec) (30 mi/h = 44 ft/sec)
0	0	70	15
10	44	80	22
20	15	90	35
30	35	100	44
40	30	110	30
50	44	120	35
60	35		

12. Distance from velocity data The accompanying table gives data for the velocity of a vintage sports car accelerating from 0 to 142 mi/h in 36 sec (10 thousandths of an hour).

Time (h)	Velocity (mi/h)	Time (h)	Velocity (mi/h)
0.0	0	0.006	116
0.001	40	0.007	125
0.002	62	0.008	132
0.003	82	0.009	137
0.004	96	0.010	142
0.005	108		

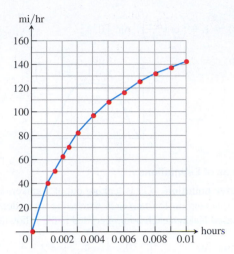

a. Use rectangles to estimate how far the car traveled during the 36 sec it took to reach 142 mi/h.

b. Roughly how many seconds did it take the car to reach the halfway point? About how fast was the car going then?

13. Free fall with air resistance An object is dropped straight down from a helicopter. The object falls faster and faster but its acceleration (rate of change of its velocity) decreases over time because of air resistance. The acceleration is measured in ft/sec^2 and recorded every second after the drop for 5 sec, as shown:

t	0	1	2	3	4	5
a	32.00	19.41	11.77	7.14	4.33	2.63

a. Find an upper estimate for the speed when $t = 5$.

b. Find a lower estimate for the speed when $t = 5$.

c. Find an upper estimate for the distance fallen when $t = 3$.

14. Distance traveled by a projectile An object is shot straight upward from sea level with an initial velocity of 400 ft/sec.

a. Assuming that gravity is the only force acting on the object, give an upper estimate for its velocity after 5 sec have elapsed. Use $g = 32$ ft/sec^2 for the gravitational acceleration.

b. Find a lower estimate for the height attained after 5 sec.

Average Value of a Function

In Exercises 15–18, use a finite sum to estimate the average value of f on the given interval by partitioning the interval into four subintervals of equal length and evaluating f at the subinterval midpoints.

15. $f(x) = x^3$ on $[0, 2]$

16. $f(x) = 1/x$ on $[1, 9]$

17. $f(t) = (1/2) + \sin^2 \pi t$ on $[0, 2]$

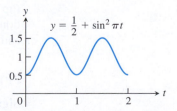

18. $f(t) = 1 - \left(\cos\frac{\pi t}{4}\right)^4$ on $[0, 4]$

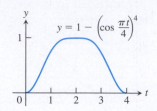

Month	Jul	Aug	Sep	Oct	Nov	Dec
Pollutant release rate (tons/day)	0.63	0.70	0.81	0.85	0.89	0.95

a. Assuming a 30-day month and that new scrubbers allow only 0.05 ton/day to be released, give an upper estimate of the total tonnage of pollutants released by the end of June. What is a lower estimate?

b. In the best case, approximately when will a total of 125 tons of pollutants have been released into the atmosphere?

21. Inscribe a regular n-sided polygon inside a circle of radius 1 and compute the area of the polygon for the following values of n:

    a. 4 (square)   **b.** 8 (octagon)   **c.** 16

    d. Compare the areas in parts (a), (b), and (c) with the area of the circle.

22. (*Continuation of Exercise 21.*)

    a. Inscribe a regular n-sided polygon inside a circle of radius 1 and compute the area of one of the n congruent triangles formed by drawing radii to the vertices of the polygon.

    b. Compute the limit of the area of the inscribed polygon as $n \to \infty$.

    c. Repeat the computations in parts (a) and (b) for a circle of radius r.

Examples of Estimations

19. Water pollution Oil is leaking out of a tanker damaged at sea. The damage to the tanker is worsening as evidenced by the increased leakage each hour, recorded in the following table.

Time (h)	0	1	2	3	4
Leakage (gal/h)	50	70	97	136	190

Time (h)	5	6	7	8
Leakage (gal/h)	265	369	516	720

a. Give an upper and a lower estimate of the total quantity of oil that has escaped after 5 hours.

b. Repeat part (a) for the quantity of oil that has escaped after 8 hours.

c. The tanker continues to leak 720 gal/h after the first 8 hours. If the tanker originally contained 25,000 gal of oil, approximately how many more hours will elapse in the worst case before all the oil has spilled? In the best case?

20. Air pollution A power plant generates electricity by burning oil. Pollutants produced as a result of the burning process are removed by scrubbers in the smokestacks. Over time, the scrubbers become less efficient and eventually they must be replaced when the amount of pollution released exceeds government standards. Measurements are taken at the end of each month determining the rate at which pollutants are released into the atmosphere, recorded as follows.

Month	Jan	Feb	Mar	Apr	May	Jun
Pollutant release rate (tons/day)	0.20	0.25	0.27	0.34	0.45	0.52

COMPUTER EXPLORATIONS
In Exercises 23–26, use a CAS to perform the following steps.

    a. Plot the functions over the given interval.

    b. Subdivide the interval into $n = 100$, 200, and 1000 subintervals of equal length and evaluate the function at the midpoint of each subinterval.

    c. Compute the average value of the function values generated in part (b).

    d. Solve the equation $f(x) = $ (average value) for x using the average value calculated in part (c) for the $n = 1000$ partitioning.

23. $f(x) = \sin x$ on $[0, \pi]$ **24.** $f(x) = \sin^2 x$ on $[0, \pi]$

25. $f(x) = x \sin\frac{1}{x}$ on $\left[\frac{\pi}{4}, \pi\right]$ **26.** $f(x) = x \sin^2\frac{1}{x}$ on $\left[\frac{\pi}{4}, \pi\right]$

5.2 Sigma Notation and Limits of Finite Sums

While estimating with finite sums in Section 5.1, we encountered sums that had many terms (up to 1000 in Table 5.1, for instance). In this section we introduce a more convenient notation for working with sums that have a large number of terms. After describing this notation and its properties, we consider what happens as the number of terms approaches infinity.

Finite Sums and Sigma Notation

Sigma notation enables us to write a sum with many terms in the compact form

$$\sum_{k=1}^{n} a_k = a_1 + a_2 + a_3 + \cdots + a_{n-1} + a_n.$$

Σ is the capital Greek letter Sigma

The Greek letter Σ (capital sigma, corresponding to our letter S), stands for "sum." The **index of summation** k tells us where the sum begins (at the number below the Σ symbol) and where it ends (at the number above Σ). Any letter can be used to denote the index, but the letters i, j, k, and n are customary.

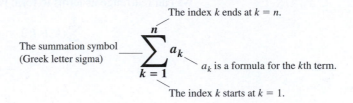

The index k ends at $k = n$.

The summation symbol (Greek letter sigma)

$$\sum_{k=1}^{n} a_k$$

a_k is a formula for the kth term.

The index k starts at $k = 1$.

Thus we can write the squares of the numbers 1 through 11 as

$$1^2 + 2^2 + 3^2 + 4^2 + 5^2 + 6^2 + 7^2 + 8^2 + 9^2 + 10^2 + 11^2 = \sum_{k=1}^{11} k^2,$$

and the sum of $f(i)$ for integers i from 1 to 100 as

$$f(1) + f(2) + f(3) + \cdots + f(100) = \sum_{i=1}^{100} f(i).$$

The starting index does not have to be 1; it can be any integer.

EXAMPLE 1

A sum in sigma notation	The sum written out, one term for each value of k	The value of the sum
$\displaystyle\sum_{k=1}^{5} k$	$1 + 2 + 3 + 4 + 5$	15
$\displaystyle\sum_{k=1}^{3} (-1)^k k$	$(-1)^1(1) + (-1)^2(2) + (-1)^3(3)$	$-1 + 2 - 3 = -2$
$\displaystyle\sum_{k=1}^{2} \frac{k}{k+1}$	$\dfrac{1}{1+1} + \dfrac{2}{2+1}$	$\dfrac{1}{2} + \dfrac{2}{3} = \dfrac{7}{6}$
$\displaystyle\sum_{k=4}^{5} \frac{k^2}{k-1}$	$\dfrac{4^2}{4-1} + \dfrac{5^2}{5-1}$	$\dfrac{16}{3} + \dfrac{25}{4} = \dfrac{139}{12}$

EXAMPLE 2   Express the sum $1 + 3 + 5 + 7 + 9$ in sigma notation.

Solution   The formula generating the terms depends on what we choose the lower limit of summation to be, but the terms generated remain the same. It is often simplest to choose the starting index to be $k = 0$ or $k = 1$, but we can start with any integer.

Starting with $k = 0$: $1 + 3 + 5 + 7 + 9 = \displaystyle\sum_{k=0}^{4} (2k + 1)$

Starting with $k = 1$: $1 + 3 + 5 + 7 + 9 = \displaystyle\sum_{k=1}^{5} (2k - 1)$

Starting with $k = 2$: $1 + 3 + 5 + 7 + 9 = \displaystyle\sum_{k=2}^{6} (2k - 3)$

Starting with $k = -3$: $1 + 3 + 5 + 7 + 9 = \displaystyle\sum_{k=-3}^{1} (2k + 7)$

When we have a sum such as

$$\sum_{k=1}^{3} (k + k^2)$$

we can rearrange its terms to form two sums:

$$\sum_{k=1}^{3} (k + k^2) = (1 + 1^2) + (2 + 2^2) + (3 + 3^2)$$

$$= (1 + 2 + 3) + (1^2 + 2^2 + 3^2) \qquad \text{Regroup terms.}$$

$$= \sum_{k=1}^{3} k + \sum_{k=1}^{3} k^2.$$

This illustrates a general rule for finite sums:

$$\sum_{k=1}^{n} (a_k + b_k) = \sum_{k=1}^{n} a_k + \sum_{k=1}^{n} b_k.$$

This and three other rules are given below. Proofs of these rules can be obtained using mathematical induction (see Appendix 2).

Algebra Rules for Finite Sums

1. *Sum Rule:* $\displaystyle\sum_{k=1}^{n} (a_k + b_k) = \sum_{k=1}^{n} a_k + \sum_{k=1}^{n} b_k$

2. *Difference Rule:* $\displaystyle\sum_{k=1}^{n} (a_k - b_k) = \sum_{k=1}^{n} a_k - \sum_{k=1}^{n} b_k$

3. *Constant Multiple Rule:* $\displaystyle\sum_{k=1}^{n} c a_k = c \cdot \sum_{k=1}^{n} a_k$ (Any number c)

4. *Constant Value Rule:* $\displaystyle\sum_{k=1}^{n} c = n \cdot c$ (Any number c)

EXAMPLE 3 We demonstrate the use of the algebra rules.

(a) $\displaystyle\sum_{k=1}^{n} (3k - k^2) = 3\sum_{k=1}^{n} k - \sum_{k=1}^{n} k^2$ Difference Rule and Constant Multiple Rule

(b) $\displaystyle\sum_{k=1}^{n} (-a_k) = \sum_{k=1}^{n} (-1) \cdot a_k = -1 \cdot \sum_{k=1}^{n} a_k = -\sum_{k=1}^{n} a_k$ Constant Multiple Rule

(c) $\displaystyle\sum_{k=1}^{3} (k + 4) = \sum_{k=1}^{3} k + \sum_{k=1}^{3} 4$ Sum Rule

$$= (1 + 2 + 3) + (3 \cdot 4) \qquad \text{Constant Value Rule}$$

$$= 6 + 12 = 18$$

(d) $\displaystyle\sum_{k=1}^{n} \frac{1}{n} = n \cdot \frac{1}{n} = 1$ Constant Value Rule ($1/n$ is constant)

HISTORICAL BIOGRAPHY

Carl Friedrich Gauss
(1777–1855)
www.goo.gl/LZMP1A

Over the years people have discovered a variety of formulas for the values of finite sums. The most famous of these are the formula for the sum of the first n integers (Gauss is said to have discovered it at age 8) and the formulas for the sums of the squares and cubes of the first n integers.

EXAMPLE 4   Show that the sum of the first n integers is

$$\sum_{k=1}^{n} k = \frac{n(n+1)}{2}.$$

Solution   The formula tells us that the sum of the first 4 integers is

$$\frac{(4)(5)}{2} = 10.$$

Addition verifies this prediction:

$$1 + 2 + 3 + 4 = 10.$$

To prove the formula in general, we write out the terms in the sum twice, once forward and once backward.

$$
\begin{array}{ccccccccc}
1 & + & 2 & + & 3 & + & \cdots & + & n \\
n & + & (n-1) & + & (n-2) & + & \cdots & + & 1
\end{array}
$$

If we add the two terms in the first column we get $1 + n = n + 1$. Similarly, if we add the two terms in the second column we get $2 + (n - 1) = n + 1$. The two terms in any column sum to $n + 1$. When we add the n columns together we get n terms, each equal to $n + 1$, for a total of $n(n + 1)$. Since this is twice the desired quantity, the sum of the first n integers is $n(n + 1)/2$.   ■

Formulas for the sums of the squares and cubes of the first n integers are proved using mathematical induction (see Appendix 2). We state them here.

The first n squares: $\displaystyle\sum_{k=1}^{n} k^2 = \frac{n(n+1)(2n+1)}{6}$

The first n cubes: $\displaystyle\sum_{k=1}^{n} k^3 = \left(\frac{n(n+1)}{2}\right)^2$

Limits of Finite Sums

The finite sum approximations that we considered in Section 5.1 became more accurate as the number of terms increased and the subinterval widths (lengths) narrowed. The next example shows how to calculate a limiting value as the widths of the subintervals go to zero and the number of subintervals grows to infinity.

EXAMPLE 5   Find the limiting value of lower sum approximations to the area of the region R below the graph of $y = 1 - x^2$ and above the interval $[0, 1]$ on the x-axis using equal-width rectangles whose widths approach zero and whose number approaches infinity. (See Figure 5.4a.)

Solution   We compute a lower sum approximation using n rectangles of equal width $\Delta x = (1 - 0)/n$, and then we see what happens as $n \rightarrow \infty$. We start by subdividing $[0, 1]$ into n equal width subintervals

$$\left[0, \frac{1}{n}\right], \left[\frac{1}{n}, \frac{2}{n}\right], \ldots, \left[\frac{n-1}{n}, \frac{n}{n}\right].$$

Each subinterval has width $1/n$. The function $1 - x^2$ is decreasing on $[0, 1]$, and its smallest value in a subinterval occurs at the subinterval's right endpoint. So a lower sum is constructed with rectangles whose height over the subinterval $[(k-1)/n, k/n]$ is $f(k/n) = 1 - (k/n)^2$, giving the sum

$$f\left(\frac{1}{n}\right) \cdot \frac{1}{n} + f\left(\frac{2}{n}\right) \cdot \frac{1}{n} + \cdots + f\left(\frac{k}{n}\right) \cdot \frac{1}{n} + \cdots + f\left(\frac{n}{n}\right) \cdot \frac{1}{n}.$$

We write this in sigma notation and simplify,

$$\sum_{k=1}^{n} f\left(\frac{k}{n}\right) \cdot \frac{1}{n} = \sum_{k=1}^{n}\left(1 - \left(\frac{k}{n}\right)^2\right)\frac{1}{n}$$

$$= \sum_{k=1}^{n}\left(\frac{1}{n} - \frac{k^2}{n^3}\right)$$

$$= \sum_{k=1}^{n}\frac{1}{n} - \sum_{k=1}^{n}\frac{k^2}{n^3} \qquad \text{Difference Rule}$$

$$= n \cdot \frac{1}{n} - \frac{1}{n^3}\sum_{k=1}^{n}k^2 \qquad \begin{array}{l}\text{Constant Value and}\\ \text{Constant Multiple Rules}\end{array}$$

$$= 1 - \left(\frac{1}{n^3}\right)\frac{n(n+1)(2n+1)}{6} \qquad \text{Sum of the First } n \text{ Squares}$$

$$= 1 - \frac{2n^3 + 3n^2 + n}{6n^3}. \qquad \text{Numerator expanded}$$

We have obtained an expression for the lower sum that holds for any n. Taking the limit of this expression as $n \to \infty$, we see that the lower sums converge as the number of subintervals increases and the subinterval widths approach zero:

$$\lim_{n\to\infty}\left(1 - \frac{2n^3 + 3n^2 + n}{6n^3}\right) = 1 - \frac{2}{6} = \frac{2}{3}.$$

The lower sum approximations converge to $2/3$. A similar calculation shows that the upper sum approximations also converge to $2/3$. Any finite sum approximation $\sum_{k=1}^{n} f(c_k)(1/n)$ also converges to the same value, $2/3$. This is because it is possible to show that any finite sum approximation is trapped between the lower and upper sum approximations. For this reason we are led to *define* the area of the region R as this limiting value. In Section 5.3 we study the limits of such finite approximations in a general setting. ∎

Riemann Sums

HISTORICAL BIOGRAPHY

Georg Friedrich Bernhard Riemann
(1826–1866)
www.goo.gl/hPFV65

The theory of limits of finite approximations was made precise by the German mathematician Bernhard Riemann. We now introduce the notion of a *Riemann sum*, which underlies the theory of the definite integral that will be presented in the next section.

We begin with an arbitrary bounded function f defined on a closed interval $[a, b]$. Like the function pictured in Figure 5.8, f may have negative as well as positive values. We subdivide the interval $[a, b]$ into subintervals, not necessarily of equal widths (or lengths), and form sums in the same way as for the finite approximations in Section 5.1. To do so, we choose $n - 1$ points $\{x_1, x_2, x_3, \ldots, x_{n-1}\}$ between a and b that are in increasing order, so that

$$a < x_1 < x_2 < \cdots < x_{n-1} < b.$$

To make the notation consistent, we set $x_0 = a$ and $x_n = b$, so that

$$a = x_0 < x_1 < x_2 < \cdots < x_{n-1} < x_n = b.$$

The set of all of these points,

$$P = \{x_0, x_1, x_2, \ldots, x_{n-1}, x_n\},$$

is called a **partition** of $[a, b]$.

The partition P divides $[a, b]$ into the n closed subintervals

$$[x_0, x_1], [x_1, x_2], \ldots, [x_{n-1}, x_n].$$

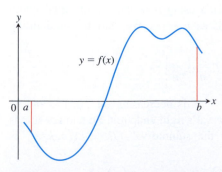

FIGURE 5.8 A typical continuous function $y = f(x)$ over a closed interval $[a, b]$.

The first of these subintervals is $[x_0, x_1]$, the second is $[x_1, x_2]$, and the **kth subinterval** is $[x_{k-1}, x_k]$ (where k is an integer between 1 and n).

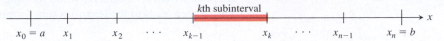

The width of the first subinterval $[x_0, x_1]$ is denoted Δx_1, the width of the second $[x_1, x_2]$ is denoted Δx_2, and the width of the kth subinterval is $\Delta x_k = x_k - x_{k-1}$.

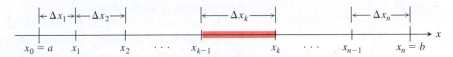

If all n subintervals have equal width, then their common width, which we call Δx, is equal to $(b - a)/n$.

In each subinterval we select some point. The point chosen in the kth subinterval $[x_{k-1}, x_k]$ is called c_k. Then on each subinterval we stand a vertical rectangle that stretches from the x-axis to touch the curve at $(c_k, f(c_k))$. These rectangles can be above or below the x-axis, depending on whether $f(c_k)$ is positive or negative, or on the x-axis if $f(c_k) = 0$ (see Figure 5.9).

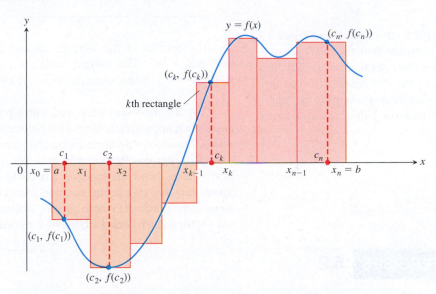

FIGURE 5.9   The rectangles approximate the region between the graph of the function $y = f(x)$ and the x-axis. Figure 5.8 has been repeated and enlarged, the partition of $[a, b]$ and the points c_k have been added, and the corresponding rectangles with heights $f(c_k)$ are shown.

On each subinterval we form the product $f(c_k) \cdot \Delta x_k$. This product is positive, negative, or zero, depending on the sign of $f(c_k)$. When $f(c_k) > 0$, the product $f(c_k) \cdot \Delta x_k$ is the area of a rectangle with height $f(c_k)$ and width Δx_k. When $f(c_k) < 0$, the product $f(c_k) \cdot \Delta x_k$ is a negative number, the negative of the area of a rectangle of width Δx_k that drops from the x-axis to the negative number $f(c_k)$.

Finally we sum all these products to get

$$S_P = \sum_{k=1}^{n} f(c_k)\, \Delta x_k.$$

The sum S_P is called a **Riemann sum for f on the interval $[a, b]$**. There are many such sums, depending on the partition P we choose, and the choices of the points c_k in the

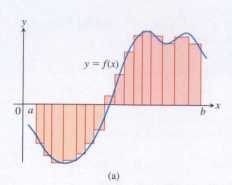

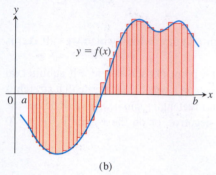

FIGURE 5.10 The curve of Figure 5.9 with rectangles from finer partitions of $[a, b]$. Finer partitions create collections of rectangles with thinner bases that approximate the region between the graph of f and the x-axis with increasing accuracy.

subintervals. For instance, we could choose n subintervals all having equal width $\Delta x = (b - a)/n$ to partition $[a, b]$, and then choose the point c_k to be the right-hand endpoint of each subinterval when forming the Riemann sum (as we did in Example 5). This choice leads to the Riemann sum formula

$$S_n = \sum_{k=1}^{n} f\left(a + k\frac{(b - a)}{n}\right) \cdot \left(\frac{b - a}{n}\right).$$

Similar formulas can be obtained if instead we choose c_k to be the left-hand endpoint, or the midpoint, of each subinterval.

In the cases in which the subintervals all have equal width $\Delta x = (b - a)/n$, we can make them thinner by simply increasing their number n. When a partition has subintervals of varying widths, we can ensure they are all thin by controlling the width of a widest (longest) subinterval. We define the **norm** of a partition P, written $\|P\|$, to be the largest of all the subinterval widths. If $\|P\|$ is a small number, then all of the subintervals in the partition P have a small width.

EXAMPLE 6 The set $P = \{0, 0.2, 0.6, 1, 1.5, 2\}$ is a partition of $[0, 2]$. There are five subintervals of P: $[0, 0.2]$, $[0.2, 0.6]$, $[0.6, 1]$, $[1, 1.5]$, and $[1.5, 2]$:

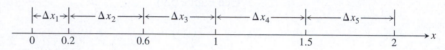

The lengths of the subintervals are $\Delta x_1 = 0.2$, $\Delta x_2 = 0.4$, $\Delta x_3 = 0.4$, $\Delta x_4 = 0.5$, and $\Delta x_5 = 0.5$. The longest subinterval length is 0.5, so the norm of the partition is $\|P\| = 0.5$. In this example, there are two subintervals of this length. ■

Any Riemann sum associated with a partition of a closed interval $[a, b]$ defines rectangles that approximate the region between the graph of a continuous function f and the x-axis. Partitions with norm approaching zero lead to collections of rectangles that approximate this region with increasing accuracy, as suggested by Figure 5.10. We will see in the next section that if the function f is continuous over the closed interval $[a, b]$, then no matter how we choose the partition P and the points c_k in its subintervals, the Riemann sums corresponding to these choices will approach a single limiting value as the subinterval widths (which are controlled by the norm of the partition) approach zero.

EXERCISES 5.2

Sigma Notation

Write the sums in Exercises 1–6 without sigma notation. Then evaluate them.

1. $\sum_{k=1}^{2} \frac{6k}{k + 1}$

2. $\sum_{k=1}^{3} \frac{k - 1}{k}$

3. $\sum_{k=1}^{4} \cos k\pi$

4. $\sum_{k=1}^{5} \sin k\pi$

5. $\sum_{k=1}^{3} (-1)^{k+1} \sin \frac{\pi}{k}$

6. $\sum_{k=1}^{4} (-1)^k \cos k\pi$

7. Which of the following express $1 + 2 + 4 + 8 + 16 + 32$ in sigma notation?

a. $\sum_{k=1}^{6} 2^{k-1}$ b. $\sum_{k=0}^{5} 2^k$ c. $\sum_{k=-1}^{4} 2^{k+1}$

8. Which of the following express $1 - 2 + 4 - 8 + 16 - 32$ in sigma notation?

a. $\sum_{k=1}^{6} (-2)^{k-1}$ b. $\sum_{k=0}^{5} (-1)^k 2^k$ c. $\sum_{k=-2}^{3} (-1)^{k+1} 2^{k+2}$

9. Which formula is not equivalent to the other two?

a. $\sum_{k=2}^{4} \frac{(-1)^{k-1}}{k - 1}$ b. $\sum_{k=0}^{2} \frac{(-1)^k}{k + 1}$ c. $\sum_{k=-1}^{1} \frac{(-1)^k}{k + 2}$

10. Which formula is not equivalent to the other two?

a. $\sum_{k=1}^{4} (k - 1)^2$ b. $\sum_{k=-1}^{3} (k + 1)^2$ c. $\sum_{k=-3}^{-1} k^2$

Express the sums in Exercises 11–16 in sigma notation. The form of your answer will depend on your choice for the starting index.

11. $1 + 2 + 3 + 4 + 5 + 6$ 12. $1 + 4 + 9 + 16$

13. $\frac{1}{2} + \frac{1}{4} + \frac{1}{8} + \frac{1}{16}$ 14. $2 + 4 + 6 + 8 + 10$

15. $1 - \dfrac{1}{2} + \dfrac{1}{3} - \dfrac{1}{4} + \dfrac{1}{5}$ **16.** $-\dfrac{1}{5} + \dfrac{2}{5} - \dfrac{3}{5} + \dfrac{4}{5} - \dfrac{5}{5}$

Values of Finite Sums

17. Suppose that $\displaystyle\sum_{k=1}^{n} a_k = -5$ and $\displaystyle\sum_{k=1}^{n} b_k = 6$. Find the values of

   a. $\displaystyle\sum_{k=1}^{n} 3a_k$ **b.** $\displaystyle\sum_{k=1}^{n} \dfrac{b_k}{6}$ **c.** $\displaystyle\sum_{k=1}^{n} (a_k + b_k)$

   d. $\displaystyle\sum_{k=1}^{n} (a_k - b_k)$ **e.** $\displaystyle\sum_{k=1}^{n} (b_k - 2a_k)$

18. Suppose that $\displaystyle\sum_{k=1}^{n} a_k = 0$ and $\displaystyle\sum_{k=1}^{n} b_k = 1$. Find the values of

   a. $\displaystyle\sum_{k=1}^{n} 8a_k$ **b.** $\displaystyle\sum_{k=1}^{n} 250b_k$

   c. $\displaystyle\sum_{k=1}^{n} (a_k + 1)$ **d.** $\displaystyle\sum_{k=1}^{n} (b_k - 1)$

Evaluate the sums in Exercises 19–32.

19. a. $\displaystyle\sum_{k=1}^{10} k$ **b.** $\displaystyle\sum_{k=1}^{10} k^2$ **c.** $\displaystyle\sum_{k=1}^{10} k^3$

20. a. $\displaystyle\sum_{k=1}^{13} k$ **b.** $\displaystyle\sum_{k=1}^{13} k^2$ **c.** $\displaystyle\sum_{k=1}^{13} k^3$

21. $\displaystyle\sum_{k=1}^{7} (-2k)$ **22.** $\displaystyle\sum_{k=1}^{5} \dfrac{\pi k}{15}$

23. $\displaystyle\sum_{k=1}^{6} (3 - k^2)$ **24.** $\displaystyle\sum_{k=1}^{6} (k^2 - 5)$

25. $\displaystyle\sum_{k=1}^{5} k(3k + 5)$ **26.** $\displaystyle\sum_{k=1}^{7} k(2k + 1)$

27. $\displaystyle\sum_{k=1}^{5} \dfrac{k^3}{225} + \left(\sum_{k=1}^{5} k\right)^3$ **28.** $\left(\displaystyle\sum_{k=1}^{7} k\right)^2 - \displaystyle\sum_{k=1}^{7} \dfrac{k^3}{4}$

29. a. $\displaystyle\sum_{k=1}^{7} 3$ **b.** $\displaystyle\sum_{k=1}^{500} 7$ **c.** $\displaystyle\sum_{k=3}^{264} 10$

30. a. $\displaystyle\sum_{k=9}^{36} k$ **b.** $\displaystyle\sum_{k=3}^{17} k^2$ **c.** $\displaystyle\sum_{k=18}^{71} k(k - 1)$

31. a. $\displaystyle\sum_{k=1}^{n} 4$ **b.** $\displaystyle\sum_{k=1}^{n} c$ **c.** $\displaystyle\sum_{k=1}^{n} (k - 1)$

32. a. $\displaystyle\sum_{k=1}^{n} \left(\dfrac{1}{n} + 2n\right)$ **b.** $\displaystyle\sum_{k=1}^{n} \dfrac{c}{n}$ **c.** $\displaystyle\sum_{k=1}^{n} \dfrac{k}{n^2}$

33. $\displaystyle\sum_{k=1}^{50} [(k + 1)^2 - k^2]$ **34.** $\displaystyle\sum_{k=2}^{20} [\sin(k - 1) - \sin k]$

35. $\displaystyle\sum_{k=7}^{30} \left(\sqrt{k - 4} - \sqrt{k - 3}\right)$

36. $\displaystyle\sum_{k=1}^{40} \dfrac{1}{k(k + 1)}$ $\left(Hint: \ \dfrac{1}{k(k + 1)} = \dfrac{1}{k} - \dfrac{1}{k + 1}\right)$

Riemann Sums

In Exercises 37–42, graph each function $f(x)$ over the given interval. Partition the interval into four subintervals of equal length. Then add to your sketch the rectangles associated with the Riemann sum $\sum_{k=1}^{4} f(c_k) \, \Delta x_k$, given that c_k is the **(a)** left-hand endpoint, **(b)** right-hand endpoint, **(c)** midpoint of the kth subinterval. (Make a separate sketch for each set of rectangles.)

37. $f(x) = x^2 - 1, \quad [0, 2]$ **38.** $f(x) = -x^2, \quad [0, 1]$

39. $f(x) = \sin x, \quad [-\pi, \pi]$

40. $f(x) = \sin x + 1, \quad [-\pi, \pi]$

41. Find the norm of the partition $P = \{0, 1.2, 1.5, 2.3, 2.6, 3\}$.

42. Find the norm of the partition $P = \{-2, -1.6, -0.5, 0, 0.8, 1\}$.

Limits of Riemann Sums

For the functions in Exercises 43–50, find a formula for the Riemann sum obtained by dividing the interval $[a, b]$ into n equal subintervals and using the right-hand endpoint for each c_k. Then take a limit of these sums as $n \to \infty$ to calculate the area under the curve over $[a, b]$.

43. $f(x) = 1 - x^2$ over the interval $[0, 1]$.

44. $f(x) = 2x$ over the interval $[0, 3]$.

45. $f(x) = x^2 + 1$ over the interval $[0, 3]$.

46. $f(x) = 3x^2$ over the interval $[0, 1]$.

47. $f(x) = x + x^2$ over the interval $[0, 1]$.

48. $f(x) = 3x + 2x^2$ over the interval $[0, 1]$.

49. $f(x) = 2x^3$ over the interval $[0, 1]$.

50. $f(x) = x^2 - x^3$ over the interval $[-1, 0]$.

5.3   The Definite Integral

In this section we consider the limit of general Riemann sums as the norm of the partitions of a closed interval $[a, b]$ approaches zero. This limiting process leads us to the definition of the *definite integral* of a function over a closed interval $[a, b]$.

Definition of the Definite Integral

The definition of the definite integral is based on the fact that for some functions, as the norm of the partitions of $[a, b]$ approaches zero, the values of the corresponding Riemann sums approach a limiting value J. We introduce the symbol ε as a small positive number that specifies how close to J the Riemann sum must be, and the symbol δ as a second small positive number that specifies how small the norm of a partition must be in order for convergence to happen. We now define this limit precisely.

DEFINITION Let $f(x)$ be a function defined on a closed interval $[a, b]$. We say that a number J is the **definite integral of f over $[a, b]$** and that J is the limit of the Riemann sums $\sum_{k=1}^{n} f(c_k) \, \Delta x_k$ if the following condition is satisfied:

Given any number $\varepsilon > 0$ there is a corresponding number $\delta > 0$ such that for every partition $P = \{x_0, x_1, \ldots, x_n\}$ of $[a, b]$ with $\|P\| < \delta$ and any choice of c_k in $[x_{k-1}, x_k]$, we have

$$\left| \sum_{k=1}^{n} f(c_k) \, \Delta x_k - J \right| < \varepsilon.$$

The definition involves a limiting process in which the norm of the partition goes to zero.

We have many choices for a partition P with norm going to zero, and many choices of points c_k for each partition. The definite integral exists when we always get the same limit J, no matter what choices are made. When the limit exists we write

$$J = \lim_{\|P\| \to 0} \sum_{k=1}^{n} f(c_k) \, \Delta x_k,$$

and we say that the *definite integral exists*. The limit of any Riemann sum is always taken as the norm of the partitions approaches zero and the number of subintervals goes to infinity, and furthermore the same limit J must be obtained no matter what choices we make for the points c_k.

Leibniz introduced a notation for the definite integral that captures its construction as a limit of Riemann sums. He envisioned the finite sums $\sum_{k=1}^{n} f(c_k) \, \Delta x_k$ becoming an infinite sum of function values $f(x)$ multiplied by "infinitesimal" subinterval widths dx. The sum symbol $\sum$ is replaced in the limit by the integral symbol $\int$, whose origin is in the letter "S" (for sum). The function values $f(c_k)$ are replaced by a continuous selection of function values $f(x)$. The subinterval widths Δx_k become the differential dx. It is as if we are summing all products of the form $f(x) \cdot dx$ as x goes from a to b. While this notation captures the process of constructing an integral, it is Riemann's definition that gives a precise meaning to the definite integral.

If the definite integral exists, then instead of writing J we write

$$\int_{a}^{b} f(x) \, dx.$$

We read this as "the integral from a to b of f of x dee x" or sometimes as "the integral from a to b of f of x with respect to x." The component parts in the integral symbol also have names:

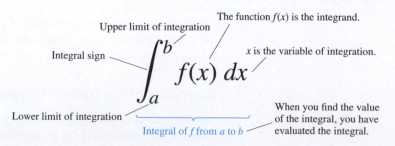

The function $f(x)$ is the integrand.

Upper limit of integration

Integral sign

x is the variable of integration.

$$\int_{a}^{b} f(x) \, dx$$

Lower limit of integration

Integral of f from a to b

When you find the value of the integral, you have evaluated the integral.

When the definite integral exists, we say that the Riemann sums of f on $[a, b]$ **converge** to the definite integral $J = \int_a^b f(x)\, dx$ and that f is **integrable** over $[a, b]$.

In the cases where the subintervals all have equal width $\Delta x = (b - a)/n$, the Riemann sums have the form

$$S_n = \sum_{k=1}^{n} f(c_k)\, \Delta x_k = \sum_{k=1}^{n} f(c_k)\left(\frac{b - a}{n}\right), \qquad \Delta x_k = \Delta x = (b - a)/n \text{ for all } k$$

where c_k is chosen in the kth subinterval. If the definite integral exists, then these Riemann sums converge to the definite integral of f over $[a, b]$, so

$$J = \int_a^b f(x)\, dx = \lim_{n \to \infty} \sum_{k=1}^{n} f(c_k)\left(\frac{b - a}{n}\right). \qquad \text{For equal-width subintervals,} \; \|P\| \to 0 \text{ is the same as } n \to \infty.$$

If we pick the point c_k to be the right endpoint of the kth subinterval, so that $c_k = a + k\,\Delta x = a + k(b - a)/n$, then the formula for the definite integral becomes

---

A Formula for the Riemann Sum with Equal-Width Subintervals

$$\int_a^b f(x)\, dx = \lim_{n \to \infty} \sum_{k=1}^{n} f\left(a + k\,\frac{b - a}{n}\right)\left(\frac{b - a}{n}\right) \qquad (1)$$

Equation (1) gives one explicit formula that can be used to compute definite integrals. As long as the definite integral exists, the Riemann sums corresponding to other choices of partitions and locations of points c_k will have the same limit as $n \to \infty$, provided that the norm of the partition approaches zero.

The value of the definite integral of a function over any particular interval depends on the function, not on the letter we choose to represent its independent variable. If we decide to use t or u instead of x, we simply write the integral as

$$\int_a^b f(t)\, dt \qquad \text{or} \qquad \int_a^b f(u)\, du \qquad \text{instead of} \qquad \int_a^b f(x)\, dx.$$

No matter how we write the integral, it is still the same number, the limit of the Riemann sums as the norm of the partition approaches zero. Since it does not matter what letter we use, the variable of integration is called a **dummy variable**. In the three integrals given above, the dummy variables are t, u, and x.

Integrable and Nonintegrable Functions

Not every function defined over a closed interval $[a, b]$ is integrable even if the function is bounded. That is, the Riemann sums for some functions might not converge to the same limiting value, or to any value at all. A full development of exactly which functions defined over $[a, b]$ are integrable requires advanced mathematical analysis, but fortunately most functions that commonly occur in applications are integrable. In particular, every *continuous* function over $[a, b]$ is integrable over this interval, and so is every function that has no more than a finite number of jump discontinuities on $[a, b]$. (See Figures 1.9 and 1.10. The latter functions are called *piecewise-continuous functions*, and they are defined in Additional Exercises 11–18 at the end of this chapter.) The following theorem, which is proved in more advanced courses, establishes these results.

THEOREM 1 — Integrability of Continuous Functions

If a function f is continuous over the interval $[a, b]$, or if f has at most finitely many jump discontinuities there, then the definite integral $\int_a^b f(x)\, dx$ exists and f is integrable over $[a, b]$.

The idea behind Theorem 1 for continuous functions is given in Exercises 86 and 87. Briefly, when f is continuous we can choose each c_k so that $f(c_k)$ gives the maximum value of f on the subinterval $[x_{k-1}, x_k]$, resulting in an upper sum. Likewise, we can choose c_k to give the minimum value of f on $[x_{k-1}, x_k]$ to obtain a lower sum. The upper and lower sums can be shown to converge to the same limiting value as the norm of the partition P tends to zero. Moreover, every Riemann sum is trapped between the values of the upper and lower sums, so every Riemann sum converges to the same limit as well. Therefore, the number J in the definition of the definite integral exists, and the continuous function f is integrable over $[a, b]$.

For integrability to fail, a function needs to be sufficiently discontinuous that the region between its graph and the x-axis cannot be approximated well by increasingly thin rectangles. Our first example shows a function that is not integrable over a closed interval.

EXAMPLE 1   The function

$$f(x) = \begin{cases} 1, & \text{if } x \text{ is rational,} \\ 0, & \text{if } x \text{ is irrational,} \end{cases}$$

has no Riemann integral over $[0, 1]$. Underlying this is the fact that between any two numbers there is both a rational number and an irrational number. Thus the function jumps up and down too erratically over $[0, 1]$ to allow the region beneath its graph and above the x-axis to be approximated by rectangles, no matter how thin they are. In fact, we will show that upper sum approximations and lower sum approximations converge to different limiting values.

If we choose a partition P of $[0, 1]$, then the lengths of the intervals in the partition sum to 1; that is, $\sum_{k=1}^{n} \Delta x_k = 1$. In each subinterval $[x_{k-1}, x_k]$ there is a rational point, say c_k. Because c_k is rational, $f(c_k) = 1$. Since 1 is the maximum value that f can take anywhere, the upper sum approximation for this choice of c_k's is

$$U = \sum_{k=1}^{n} f(c_k) \, \Delta x_k = \sum_{k=1}^{n} (1) \, \Delta x_k = 1.$$

As the norm of the partition approaches 0, these upper sum approximations converge to 1 (because each approximation is equal to 1).

On the other hand, we could pick the c_k's differently and get a different result. Each subinterval $[x_{k-1}, x_k]$ also contains an irrational point c_k, and for this choice $f(c_k) = 0$. Since 0 is the minimum value that f can take anywhere, this choice of c_k gives us the minimum value of f on the subinterval. The corresponding lower sum approximation is

$$L = \sum_{k=1}^{n} f(c_k) \, \Delta x_k = \sum_{k=1}^{n} (0) \, \Delta x_k = 0.$$

These lower sum approximations converge to 0 as the norm of the partition converges to 0 (because they each equal 0).

Thus making different choices for the points c_k results in different limits for the corresponding Riemann sums. We conclude that the definite integral of f over the interval $[0, 1]$ does not exist, and that f is not integrable over $[0, 1]$.   ■

Theorem 1 says nothing about how to *calculate* definite integrals. A method of calculation will be developed in Section 5.4, through a connection to antiderivatives.

Properties of Definite Integrals

In defining $\int_a^b f(x) \, dx$ as a limit of sums $\sum_{k=1}^{n} f(c_k) \, \Delta x_k$, we moved from left to right across the interval $[a, b]$. What would happen if we instead move right to left, starting with $x_0 = b$ and ending at $x_n = a$? Each Δx_k in the Riemann sum would change its sign, with $x_k - x_{k-1}$ now negative instead of positive. With the same choices of c_k in each sub-interval, the sign of any Riemann sum would change, as would the sign of the limit, the integral $\int_b^a f(x)\, dx$. Since we have not previously given a meaning to integrating backward, we are led to define

$$\int_b^a f(x)\, dx = -\int_a^b f(x)\, dx. \qquad \text{\color{blue}{a and b interchanged}}$$

Although we have only defined the integral over intervals $[a, b]$ with $a < b$, it is convenient to have a definition for the integral over $[a, b]$ when $a = b$, that is, for the integral over an interval of zero width. Since $a = b$ gives $\Delta x = 0$, whenever $f(a)$ exists we define

$$\int_a^a f(x)\, dx = 0. \qquad \text{\color{blue}{a is both the lower and the upper limit of integration.}}$$

Theorem 2 states some basic properties of integrals, including the two just discussed. These properties, listed in Table 5.6, are very useful for computing integrals. We will refer to them repeatedly to simplify our calculations. Rules 2 through 7 have geometric interpretations, which are shown in Figure 5.11. The graphs in these figures show only positive functions, but the rules apply to general integrable functions, which could take both positive and negative values.

> **THEOREM 2**   When f and g are integrable over the interval $[a, b]$, the definite integral satisfies the rules listed in Table 5.6.

While Rules 1 and 2 are definitions, Rules 3 to 7 of Table 5.6 must be proved. Below we give a proof of Rule 6. Similar proofs can be given to verify the other properties in Table 5.6.

TABLE 5.6   **Rules satisfied by definite integrals**

1. *Order of Integration:* $\displaystyle\int_b^a f(x)\, dx = -\int_a^b f(x)\, dx$ {\color{blue}A definition}

2. *Zero Width Interval:* $\displaystyle\int_a^a f(x)\, dx = 0$ {\color{blue}A definition when $f(a)$ exists}

3. *Constant Multiple:* $\displaystyle\int_a^b k f(x)\, dx = k\int_a^b f(x)\, dx$ {\color{blue}Any constant k}

4. *Sum and Difference:* $\displaystyle\int_a^b (f(x) \pm g(x))\, dx = \int_a^b f(x)\, dx \pm \int_a^b g(x)\, dx$

5. *Additivity:* $\displaystyle\int_a^b f(x)\, dx + \int_b^c f(x)\, dx = \int_a^c f(x)\, dx$

6. *Max-Min Inequality:*   If f has maximum value max f and minimum value min f on $[a, b]$, then

$$(\min f)\cdot(b - a) \le \int_a^b f(x)\, dx \le (\max f)\cdot(b - a).$$

7. *Domination:* If $f(x) \ge g(x)$ on $[a, b]$ then $\displaystyle\int_a^b f(x)\, dx \ge \int_a^b g(x)\, dx.$

If $f(x) \ge 0$ on $[a, b]$ then $\displaystyle\int_a^b f(x)\, dx \ge 0.$ {\color{blue}Special case}

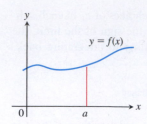

(a) *Zero Width Interval:*

$$\int_a^a f(x)\,dx = 0$$

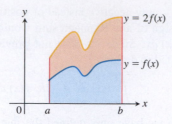

(b) *Constant Multiple: (k = 2)*

$$\int_a^b kf(x)\,dx = k\int_a^b f(x)\,dx$$

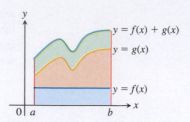

(c) *Sum: (areas add)*

$$\int_a^b (f(x) + g(x))\,dx = \int_a^b f(x)\,dx + \int_a^b g(x)\,dx$$

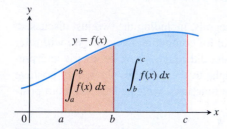

(d) *Additivity for Definite Integrals:*

$$\int_a^b f(x)\,dx + \int_b^c f(x)\,dx = \int_a^c f(x)\,dx$$

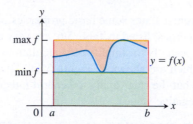

(e) *Max-Min Inequality:*

$$(\min f)\cdot(b - a) \le \int_a^b f(x)\,dx$$
$$\le (\max f)\cdot(b - a)$$

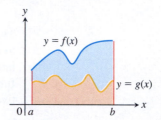

(f) *Domination:*

If $f(x) \ge g(x)$ on $[a, b]$ then

$$\int_a^b f(x)\,dx \ge \int_a^b g(x)\,dx$$

FIGURE 5.11   Geometric interpretations of Rules 2–7 in Table 5.6.

Proof of Rule 6   Rule 6 says that the integral of f over $[a, b]$ is never smaller than the minimum value of f times the length of the interval and never larger than the maximum value of f times the length of the interval. The reason is that for every partition of $[a, b]$ and for every choice of the points c_k,

$$(\min f)\cdot(b - a) = (\min f)\cdot\sum_{k=1}^{n}\Delta x_k \qquad \sum_{k=1}^{n}\Delta x_k = b - a$$

$$= \sum_{k=1}^{n}(\min f)\cdot\Delta x_k \qquad \text{Constant Multiple Rule}$$

$$\le \sum_{k=1}^{n}f(c_k)\,\Delta x_k \qquad \min f \le f(c_k)$$

$$\le \sum_{k=1}^{n}(\max f)\cdot\Delta x_k \qquad f(c_k) \le \max f$$

$$= (\max f)\cdot\sum_{k=1}^{n}\Delta x_k \qquad \text{Constant Multiple Rule}$$

$$= (\max f)\cdot(b - a).$$

In short, all Riemann sums for f on $[a, b]$ satisfy the inequalities

$$(\min f)\cdot(b - a) \le \sum_{k=1}^{n}f(c_k)\,\Delta x_k \le (\max f)\cdot(b - a).$$

Hence their limit, which is the integral, satisfies the same inequalities.   ∎

EXAMPLE 2   To illustrate some of the rules, we suppose that

$$\int_{-1}^{1} f(x)\, dx = 5, \qquad \int_{1}^{4} f(x)\, dx = -2, \quad \text{and} \quad \int_{-1}^{1} h(x)\, dx = 7.$$

Then

1. $\displaystyle \int_{4}^{1} f(x)\, dx = -\int_{1}^{4} f(x)\, dx = -(-2) = 2$ Rule 1

2. $\displaystyle \int_{-1}^{1} \left[2f(x) + 3h(x) \right]\, dx = 2\int_{-1}^{1} f(x)\, dx + 3\int_{-1}^{1} h(x)\, dx$ Rules 3 and 4

$$= 2(5) + 3(7) = 31$$

3. $\displaystyle \int_{-1}^{4} f(x)\, dx = \int_{-1}^{1} f(x)\, dx + \int_{1}^{4} f(x)\, dx = 5 + (-2) = 3$ Rule 5   ■

EXAMPLE 3   Show that the value of $\int_{0}^{1} \sqrt{1 + \cos x}\, dx$ is less than or equal to $\sqrt{2}$.

Solution   The Max-Min Inequality for definite integrals (Rule 6) says that $(\min f) \cdot (b - a)$ is a *lower bound* for the value of $\int_{a}^{b} f(x)\, dx$ and that $(\max f) \cdot (b - a)$ is an *upper bound*. The maximum value of $\sqrt{1 + \cos x}$ on $[0, 1]$ is $\sqrt{1 + 1} = \sqrt{2}$, so

$$\int_{0}^{1} \sqrt{1 + \cos x}\, dx \le \sqrt{2} \cdot (1 - 0) = \sqrt{2}.$$ ■

Area Under the Graph of a Nonnegative Function

We now return to the problem that started this chapter, which is defining what we mean by the *area* of a region having a curved boundary. In Section 5.1 we approximated the area under the graph of a nonnegative continuous function using several types of finite sums of areas of rectangles that approximate the region—upper sums, lower sums, and sums using the midpoints of each subinterval—all of which are Riemann sums constructed in special ways. Theorem 1 guarantees that all of these Riemann sums converge to a single definite integral as the norm of the partitions approaches zero and the number of subintervals goes to infinity. As a result, we can now *define* the area under the graph of a nonnegative integrable function to be the value of that definite integral.

> **DEFINITION**   If $y = f(x)$ is nonnegative and integrable over a closed interval $[a, b]$, then the **area under the curve $y = f(x)$ over $[a, b]$** is the integral of f from a to b,
>
> $$A = \int_{a}^{b} f(x)\, dx.$$

For the first time we have a rigorous definition for the area of a region whose boundary is the graph of a continuous function. We now apply this to a simple example, the area under a straight line, and we verify that our new definition agrees with our previous notion of area.

EXAMPLE 4   Compute $\int_{0}^{b} x\, dx$ and find the area A under $y = x$ over the interval $[0, b]$, $b > 0$.

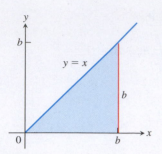

FIGURE 5.12 The region in Example 4 is a triangle.

Solution The region of interest is a triangle (Figure 5.12). We compute the area in two ways.

(a) To compute the definite integral as the limit of Riemann sums, we calculate $\lim_{\|P\|\to 0} \sum_{k=1}^{n} f(c_k)\,\Delta x_k$ for partitions whose norms go to zero. Theorem 1 tells us that it does not matter how we choose the partitions or the points c_k as long as the norms approach zero. All choices give the exact same limit. So we consider the partition P that subdivides the interval $[0, b]$ into n subintervals of equal width $\Delta x = (b - 0)/n = b/n$, and we choose c_k to be the right endpoint in each subinterval. The partition is $P = \left\{ 0, \dfrac{b}{n}, \dfrac{2b}{n}, \dfrac{3b}{n}, \cdots, \dfrac{nb}{n} \right\}$ and $c_k = \dfrac{kb}{n}$. So

$$
\begin{aligned}
\sum_{k=1}^{n} f(c_k)\,\Delta x &= \sum_{k=1}^{n} \frac{kb}{n} \cdot \frac{b}{n} && f(c_k) = c_k \\
&= \sum_{k=1}^{n} \frac{kb^2}{n^2} \\
&= \frac{b^2}{n^2} \sum_{k=1}^{n} k && \text{Constant Multiple Rule} \\
&= \frac{b^2}{n^2} \cdot \frac{n(n + 1)}{2} && \text{Sum of First } n \text{ Integers} \\
&= \frac{b^2}{2}\left(1 + \frac{1}{n}\right).
\end{aligned}
$$

As $n \to \infty$ and $\|P\| \to 0$, this last expression on the right has the limit $b^2/2$. Therefore,

$$
\int_{0}^{b} x\,dx = \frac{b^2}{2}.
$$

(b) Since the area equals the definite integral for a nonnegative function, we can quickly derive the definite integral by using the formula for the area of a triangle having base length b and height $y = b$. The area is $A = (1/2)\,b \cdot b = b^2/2$. Again we conclude that $\int_{0}^{b} x\,dx = b^2/2$. ∎

Example 4 can be generalized to integrate $f(x) = x$ over any closed interval $[a, b], 0 < a < b$.

$$
\begin{aligned}
\int_{a}^{b} x\,dx &= \int_{a}^{0} x\,dx + \int_{0}^{b} x\,dx && \text{Rule 5} \\
&= -\int_{0}^{a} x\,dx + \int_{0}^{b} x\,dx && \text{Rule 1} \\
&= -\frac{a^2}{2} + \frac{b^2}{2}. && \text{Example 4}
\end{aligned}
$$

In conclusion, we have the following rule for integrating $f(x) = x$:

$$
\int_{a}^{b} x\,dx = \frac{b^2}{2} - \frac{a^2}{2}, \qquad a < b \tag{2}
$$

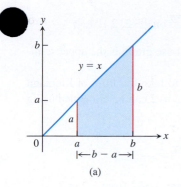

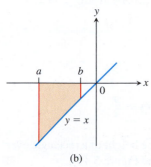

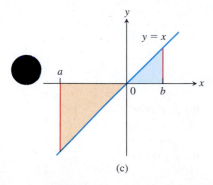

FIGURE 5.13 (a) The area of this trapezoidal region is $A = (b^2 - a^2)/2$. (b) The definite integral in Equation (2) gives the negative of the area of this trapezoidal region. (c) The definite integral in Equation (2) gives the area of the blue triangular region added to the negative of the area of the tan triangular region.

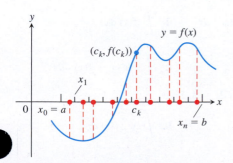

FIGURE 5.14 A sample of values of a function on an interval $[a, b]$.

This computation gives the area of the trapezoid in Figure 5.13a. Equation (2) remains valid when a and b are negative, but the interpretation of the definite integral changes. When $a < b < 0$, the definite integral value $(b^2 - a^2)/2$ is a negative number, the negative of the area of a trapezoid dropping down to the line $y = x$ below the x-axis (Figure 5.13b). When $a < 0$ and $b > 0$, Equation (2) is still valid and the definite integral gives the difference between two areas, the area under the graph and above $[0, b]$ minus the area below $[a, 0]$ and over the graph (Figure 5.13c).

The following results can also be established by using a Riemann sum calculation similar to the one that we used in Example 4 (Exercises 63 and 65).

$$\int_a^b c \, dx = c(b - a), \qquad c \text{ any constant} \tag{3}$$

$$\int_a^b x^2 \, dx = \frac{b^3}{3} - \frac{a^3}{3}, \qquad a < b \tag{4}$$

Average Value of a Continuous Function Revisited

In Section 5.1 we informally introduced the average value of a nonnegative continuous function f over an interval $[a, b]$, leading us to define this average as the area under the graph of $y = f(x)$ divided by $b - a$. In integral notation we write this as

$$\text{Average} = \frac{1}{b - a} \int_a^b f(x) \, dx.$$

This formula gives us a precise definition of the average value of a continuous (or integrable) function, whether it is positive, negative, or both.

Alternatively, we justify this formula through the following reasoning. We start with the idea from arithmetic that the average of n numbers is their sum divided by n. A continuous function f on $[a, b]$ may have infinitely many values, but we can still sample them in an orderly way. We divide $[a, b]$ into n subintervals of equal width $\Delta x = (b - a)/n$ and evaluate f at a point c_k in each (Figure 5.14). The average of the n sampled values is

$$\frac{f(c_1) + f(c_2) + \cdots + f(c_n)}{n} = \frac{1}{n} \sum_{k=1}^{n} f(c_k)$$

$$= \frac{\Delta x}{b - a} \sum_{k=1}^{n} f(c_k) \qquad \Delta x = \frac{b - a}{n}, \text{ so } \frac{1}{n} = \frac{\Delta x}{b - a}$$

$$= \frac{1}{b - a} \sum_{k=1}^{n} f(c_k) \, \Delta x. \qquad \text{Constant Multiple Rule}$$

The average of the samples is obtained by dividing a Riemann sum for f on $[a, b]$ by $(b - a)$. As we increase the number of samples and let the norm of the partition approach zero, the average approaches $(1/(b - a)) \int_a^b f(x) \, dx$. Both points of view lead us to the following definition.

> **DEFINITION** If f is integrable on $[a, b]$, then its **average value on $[a, b]$**, which is also called its **mean**, is
>
> $$\text{av}(f) = \frac{1}{b - a} \int_a^b f(x) \, dx.$$

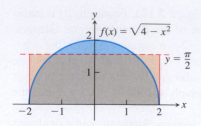

FIGURE 5.15 The average value of $f(x) = \sqrt{4 - x^2}$ on $[-2, 2]$ is $\pi/2$ (Example 5). The area of the rectangle shown here is $4 \cdot (\pi/2) = 2\pi$, which is also the area of the semicircle.

EXAMPLE 5 Find the average value of $f(x) = \sqrt{4 - x^2}$ on $[-2, 2]$.

Solution We recognize $f(x) = \sqrt{4 - x^2}$ as the function whose graph is the upper semicircle of radius 2 centered at the origin (Figure 5.15).

Since we know the area inside a circle, we do not need to take the limit of Riemann sums. The area between the semicircle and the x-axis from -2 to 2 can be computed using the geometry formula

$$\text{Area} = \frac{1}{2} \cdot \pi r^2 = \frac{1}{2} \cdot \pi (2)^2 = 2\pi.$$

Because f is nonnegative, the area is also the value of the integral of f from -2 to 2,

$$\int_{-2}^{2} \sqrt{4 - x^2} \, dx = 2\pi.$$

Therefore, the average value of f is

$$\text{av}(f) = \frac{1}{2 - (-2)} \int_{-2}^{2} \sqrt{4 - x^2} \, dx = \frac{1}{4}(2\pi) = \frac{\pi}{2}.$$

Notice that the average value of f over $[-2, 2]$ is the same as the height of a rectangle over $[-2, 2]$ whose area equals the area of the upper semicircle (see Figure 5.15). ■

EXERCISES 5.3

Interpreting Limits of Sums as Integrals
Express the limits in Exercises 1–8 as definite integrals.

1. $\displaystyle\lim_{\|P\| \to 0} \sum_{k=1}^{n} c_k^2 \, \Delta x_k$, where P is a partition of $[0, 2]$

2. $\displaystyle\lim_{\|P\| \to 0} \sum_{k=1}^{n} 2c_k^3 \, \Delta x_k$, where P is a partition of $[-1, 0]$

3. $\displaystyle\lim_{\|P\| \to 0} \sum_{k=1}^{n} (c_k^2 - 3c_k) \, \Delta x_k$, where P is a partition of $[-7, 5]$

4. $\displaystyle\lim_{\|P\| \to 0} \sum_{k=1}^{n} \left(\frac{1}{c_k}\right) \Delta x_k$, where P is a partition of $[1, 4]$

5. $\displaystyle\lim_{\|P\| \to 0} \sum_{k=1}^{n} \frac{1}{1 - c_k} \, \Delta x_k$, where P is a partition of $[2, 3]$

6. $\displaystyle\lim_{\|P\| \to 0} \sum_{k=1}^{n} \sqrt{4 - c_k^2} \, \Delta x_k$, where P is a partition of $[0, 1]$

7. $\displaystyle\lim_{\|P\| \to 0} \sum_{k=1}^{n} (\sec c_k) \, \Delta x_k$, where P is a partition of $[-\pi/4, 0]$

8. $\displaystyle\lim_{\|P\| \to 0} \sum_{k=1}^{n} (\tan c_k) \, \Delta x_k$, where P is a partition of $[0, \pi/4]$

Using the Definite Integral Rules

9. Suppose that f and g are integrable and that

$$\int_{1}^{2} f(x) \, dx = -4, \quad \int_{1}^{5} f(x) \, dx = 6, \quad \int_{1}^{5} g(x) \, dx = 8.$$

Use the rules in Table 5.6 to find

a. $\displaystyle\int_{2}^{2} g(x) \, dx$ b. $\displaystyle\int_{5}^{1} g(x) \, dx$

c. $\displaystyle\int_{1}^{2} 3f(x) \, dx$ d. $\displaystyle\int_{2}^{5} f(x) \, dx$

e. $\displaystyle\int_{1}^{5} [f(x) - g(x)] \, dx$ f. $\displaystyle\int_{1}^{5} [4f(x) - g(x)] \, dx$

10. Suppose that f and h are integrable and that

$$\int_{1}^{9} f(x) \, dx = -1, \quad \int_{7}^{9} f(x) \, dx = 5, \quad \int_{7}^{9} h(x) \, dx = 4.$$

Use the rules in Table 5.6 to find

a. $\displaystyle\int_{1}^{9} -2f(x) \, dx$ b. $\displaystyle\int_{7}^{9} [f(x) + h(x)] \, dx$

c. $\displaystyle\int_{7}^{9} [2f(x) - 3h(x)] \, dx$ d. $\displaystyle\int_{9}^{1} f(x) \, dx$

e. $\displaystyle\int_{1}^{7} f(x) \, dx$ f. $\displaystyle\int_{9}^{7} [h(x) - f(x)] \, dx$

11. Suppose that $\int_{1}^{2} f(x) \, dx = 5$. Find

a. $\displaystyle\int_{1}^{2} f(u) \, du$ b. $\displaystyle\int_{1}^{2} \sqrt{3} f(z) \, dz$

c. $\displaystyle\int_{2}^{1} f(t) \, dt$ d. $\displaystyle\int_{1}^{2} [-f(x)] \, dx$

12. Suppose that $\int_{-3}^{0} g(t)\, dt = \sqrt{2}$. Find

   a. $\displaystyle\int_{0}^{-3} g(t)\, dt$ **b.** $\displaystyle\int_{-3}^{0} g(u)\, du$

   c. $\displaystyle\int_{-3}^{0} [-g(x)]\, dx$ **d.** $\displaystyle\int_{-3}^{0} \frac{g(r)}{\sqrt{2}}\, dr$

13. Suppose that f is integrable and that $\int_{0}^{3} f(z)\, dz = 3$ and $\int_{0}^{4} f(z)\, dz = 7$. Find

   a. $\displaystyle\int_{3}^{4} f(z)\, dz$ **b.** $\displaystyle\int_{4}^{3} f(t)\, dt$

14. Suppose that h is integrable and that $\int_{-1}^{1} h(r)\, dr = 0$ and $\int_{-1}^{3} h(r)\, dr = 6$. Find

   a. $\displaystyle\int_{1}^{3} h(r)\, dr$ **b.** $-\displaystyle\int_{3}^{1} h(u)\, du$

Using Known Areas to Find Integrals

In Exercises 15–22, graph the integrands and use known area formulas to evaluate the integrals.

15. $\displaystyle\int_{-2}^{4} \left(\frac{x}{2} + 3\right) dx$ **16.** $\displaystyle\int_{1/2}^{3/2} (-2x + 4)\, dx$

17. $\displaystyle\int_{-3}^{3} \sqrt{9 - x^2}\, dx$ **18.** $\displaystyle\int_{-4}^{0} \sqrt{16 - x^2}\, dx$

19. $\displaystyle\int_{-2}^{1} |x|\, dx$ **20.** $\displaystyle\int_{-1}^{1} (1 - |x|)\, dx$

21. $\displaystyle\int_{-1}^{1} (2 - |x|)\, dx$ **22.** $\displaystyle\int_{-1}^{1} \left(1 + \sqrt{1 - x^2}\right) dx$

Use known area formulas to evaluate the integrals in Exercises 23–28.

23. $\displaystyle\int_{0}^{b} \frac{x}{2}\, dx, \quad b > 0$ **24.** $\displaystyle\int_{0}^{b} 4x\, dx, \quad b > 0$

25. $\displaystyle\int_{a}^{b} 2s\, ds, \quad 0 < a < b$ **26.** $\displaystyle\int_{a}^{b} 3t\, dt, \quad 0 < a < b$

27. $f(x) = \sqrt{4 - x^2}$ on   **a.** $[-2, 2]$, **b.** $[0, 2]$

28. $f(x) = 3x + \sqrt{1 - x^2}$ on   **a.** $[-1, 0]$, **b.** $[-1, 1]$

Evaluating Definite Integrals

Use the results of Equations (2) and (4) to evaluate the integrals in Exercises 29–40.

29. $\displaystyle\int_{1}^{\sqrt{2}} x\, dx$ **30.** $\displaystyle\int_{0.5}^{2.5} x\, dx$ **31.** $\displaystyle\int_{\pi}^{2\pi} \theta\, d\theta$

32. $\displaystyle\int_{\sqrt{2}}^{5\sqrt{2}} r\, dr$ **33.** $\displaystyle\int_{0}^{\sqrt[3]{7}} x^2\, dx$ **34.** $\displaystyle\int_{0}^{0.3} s^2\, ds$

35. $\displaystyle\int_{0}^{1/2} t^2\, dt$ **36.** $\displaystyle\int_{0}^{\pi/2} \theta^2\, d\theta$ **37.** $\displaystyle\int_{a}^{2a} x\, dx$

38. $\displaystyle\int_{a}^{\sqrt{3}} x\, dx$ **39.** $\displaystyle\int_{0}^{\sqrt[3]{b}} x^2\, dx$ **40.** $\displaystyle\int_{0}^{3b} x^2\, dx$

Use the rules in Table 5.6 and Equations (2)–(4) to evaluate the integrals in Exercises 41–50.

41. $\displaystyle\int_{3}^{1} 7\, dx$ **42.** $\displaystyle\int_{0}^{2} 5x\, dx$

43. $\displaystyle\int_{0}^{2} (2t - 3)\, dt$ **44.** $\displaystyle\int_{0}^{\sqrt{2}} \left(t - \sqrt{2}\right) dt$

45. $\displaystyle\int_{2}^{1} \left(1 + \frac{z}{2}\right) dz$ **46.** $\displaystyle\int_{3}^{0} (2z - 3)\, dz$

47. $\displaystyle\int_{1}^{2} 3u^2\, du$ **48.** $\displaystyle\int_{1/2}^{1} 24u^2\, du$

49. $\displaystyle\int_{0}^{2} (3x^2 + x - 5)\, dx$ **50.** $\displaystyle\int_{1}^{0} (3x^2 + x - 5)\, dx$

Finding Area by Definite Integrals

In Exercises 51–54, use a definite integral to find the area of the region between the given curve and the x-axis on the interval $[0, b]$.

51. $y = 3x^2$ **52.** $y = \pi x^2$

53. $y = 2x$ **54.** $y = \dfrac{x}{2} + 1$

Finding Average Value

In Exercises 55–62, graph the function and find its average value over the given interval.

55. $f(x) = x^2 - 1$ on $[0, \sqrt{3}]$

56. $f(x) = -\dfrac{x^2}{2}$ on $[0, 3]$

57. $f(x) = -3x^2 - 1$ on $[0, 1]$

58. $f(x) = 3x^2 - 3$ on $[0, 1]$

59. $f(t) = (t - 1)^2$ on $[0, 3]$

60. $f(t) = t^2 - t$ on $[-2, 1]$

61. $g(x) = |x| - 1$ on   **a.** $[-1, 1]$, **b.** $[1, 3]$, and **c.** $[-1, 3]$

62. $h(x) = -|x|$ on   **a.** $[-1, 0]$, **b.** $[0, 1]$, and **c.** $[-1, 1]$

Definite Integrals as Limits of Sums

Use the method of Example 4a or Equation (1) to evaluate the definite integrals in Exercises 63–70.

63. $\displaystyle\int_{a}^{b} c\, dx$ **64.** $\displaystyle\int_{0}^{2} (2x + 1)\, dx$

65. $\displaystyle\int_{a}^{b} x^2\, dx, \quad a < b$ **66.** $\displaystyle\int_{-1}^{0} (x - x^2)\, dx$

67. $\displaystyle\int_{-1}^{2} (3x^2 - 2x + 1)\, dx$ **68.** $\displaystyle\int_{-1}^{1} x^3\, dx$

69. $\displaystyle\int_{a}^{b} x^3\, dx, \quad a < b$ **70.** $\displaystyle\int_{0}^{1} (3x - x^3)\, dx$

Theory and Examples

71. What values of a and b maximize the value of

$$\int_{a}^{b} (x - x^2)\, dx?$$

(*Hint:* Where is the integrand positive?)

72. What values of a and b minimize the value of

$$\int_a^b (x^4 - 2x^2)\, dx?$$

73. Use the Max-Min Inequality to find upper and lower bounds for the value of

$$\int_0^1 \frac{1}{1 + x^2}\, dx.$$

74. (*Continuation of Exercise 73.*) Use the Max-Min Inequality to find upper and lower bounds for

$$\int_0^{0.5} \frac{1}{1 + x^2}\, dx \quad \text{and} \quad \int_{0.5}^1 \frac{1}{1 + x^2}\, dx.$$

Add these to arrive at an improved estimate of

$$\int_0^1 \frac{1}{1 + x^2}\, dx.$$

75. Show that the value of $\int_0^1 \sin(x^2)\, dx$ cannot possibly be 2.

76. Show that the value of $\int_0^1 \sqrt{x + 8}\, dx$ lies between $2\sqrt{2} \approx 2.8$ and 3.

77. Integrals of nonnegative functions Use the Max-Min Inequality to show that if f is integrable then

$$f(x) \geq 0 \quad \text{on} \quad [a, b] \quad \Rightarrow \quad \int_a^b f(x)\, dx \geq 0.$$

78. Integrals of nonpositive functions Show that if f is integrable then

$$f(x) \leq 0 \quad \text{on} \quad [a, b] \quad \Rightarrow \quad \int_a^b f(x)\, dx \leq 0.$$

79. Use the inequality $\sin x \leq x$, which holds for $x \geq 0$, to find an upper bound for the value of $\int_0^1 \sin x\, dx$.

80. The inequality $\sec x \geq 1 + (x^2/2)$ holds on $(-\pi/2, \pi/2)$. Use it to find a lower bound for the value of $\int_0^1 \sec x\, dx$.

81. If $\text{av}(f)$ really is a typical value of the integrable function $f(x)$ on $[a, b]$, then the constant function $\text{av}(f)$ should have the same integral over $[a, b]$ as f. Does it? That is, does

$$\int_a^b \text{av}(f)\, dx = \int_a^b f(x)\, dx?$$

Give reasons for your answer.

82. It would be nice if average values of integrable functions obeyed the following rules on an interval $[a, b]$.

a. $\text{av}(f + g) = \text{av}(f) + \text{av}(g)$

b. $\text{av}(kf) = k\, \text{av}(f)$ \quad (any number k)

c. $\text{av}(f) \leq \text{av}(g)$ \quad if \quad $f(x) \leq g(x)$ \quad on \quad $[a, b]$.

Do these rules ever hold? Give reasons for your answers.

83. Upper and lower sums for increasing functions

a. Suppose the graph of a continuous function $f(x)$ rises steadily as x moves from left to right across an interval $[a, b]$. Let P be a partition of $[a, b]$ into n subintervals of equal length

$\Delta x = (b - a)/n$. Show by referring to the accompanying figure that the difference between the upper and lower sums for f on this partition can be represented graphically as the area of a rectangle R whose dimensions are $[f(b) - f(a)]$ by Δx. (*Hint:* The difference $U - L$ is the sum of areas of rectangles whose diagonals $Q_0 Q_1, Q_1 Q_2, \ldots, Q_{n-1} Q_n$ lie approximately along the curve. There is no overlapping when these rectangles are shifted horizontally onto R.)

b. Suppose that instead of being equal, the lengths Δx_k of the subintervals of the partition of $[a, b]$ vary in size. Show that

$$U - L \leq |f(b) - f(a)|\, \Delta x_{\max},$$

where $\Delta x_{\max}$ is the norm of P, and hence that $\lim_{\|P\| \to 0} (U - L) = 0$.

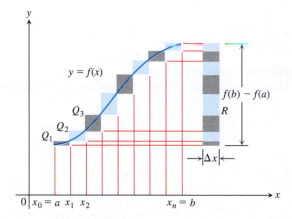

84. Upper and lower sums for decreasing functions (*Continuation of Exercise 83.*)

a. Draw a figure like the one in Exercise 83 for a continuous function $f(x)$ whose values decrease steadily as x moves from left to right across the interval $[a, b]$. Let P be a partition of $[a, b]$ into subintervals of equal length. Find an expression for $U - L$ that is analogous to the one you found for $U - L$ in Exercise 83a.

b. Suppose that instead of being equal, the lengths Δx_k of the subintervals of P vary in size. Show that the inequality

$$U - L \leq |f(b) - f(a)|\, \Delta x_{\max}$$

of Exercise 83b still holds and hence that $\lim_{\|P\| \to 0} (U - L) = 0$.

85. Use the formula

$$\sin h + \sin 2h + \sin 3h + \cdots + \sin mh$$
$$= \frac{\cos(h/2) - \cos((m + (1/2))h)}{2 \sin(h/2)}$$

to find the area under the curve $y = \sin x$ from $x = 0$ to $x = \pi/2$ in two steps:

a. Partition the interval $[0, \pi/2]$ into n subintervals of equal length and calculate the corresponding upper sum U; then

b. Find the limit of U as $n \to \infty$ and $\Delta x = (b - a)/n \to 0$.

86. Suppose that f is continuous and nonnegative over $[a, b]$, as in the accompanying figure. By inserting points

$$x_1, x_2, \ldots, x_{k-1}, x_k, \ldots, x_{n-1}$$

as shown, divide $[a, b]$ into n subintervals of lengths $\Delta x_1 = x_1 - a$, $\Delta x_2 = x_2 - x_1, \ldots, \Delta x_n = b - x_{n-1}$, which need not be equal.

a. If $m_k = \min \{ f(x) \text{ for } x \text{ in the } k\text{th subinterval} \}$, explain the connection between the **lower sum**

$$L = m_1 \, \Delta x_1 + m_2 \, \Delta x_2 + \cdots + m_n \, \Delta x_n$$

and the shaded regions in the first part of the figure.

b. If $M_k = \max \{ f(x) \text{ for } x \text{ in the } k\text{th subinterval} \}$, explain the connection between the **upper sum**

$$U = M_1 \, \Delta x_1 + M_2 \, \Delta x_2 + \cdots + M_n \, \Delta x_n$$

and the shaded regions in the second part of the figure.

c. Explain the connection between $U - L$ and the shaded regions along the curve in the third part of the figure.

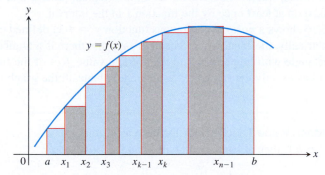

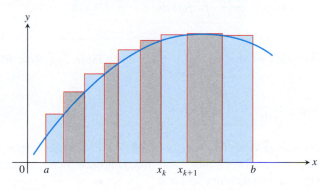

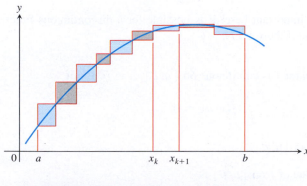

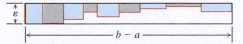

87. We say f is **uniformly continuous** on $[a, b]$ if given any $\varepsilon > 0$, there is a $\delta > 0$ such that if x_1, x_2 are in $[a, b]$ and $|x_1 - x_2| < \delta$, then $|f(x_1) - f(x_2)| < \varepsilon$. It can be shown that a continuous function on $[a, b]$ is uniformly continuous. Use this and the figure for Exercise 86 to show that if f is continuous and $\varepsilon > 0$ is given, it is possible to make $U - L \le \varepsilon \cdot (b - a)$ by making the largest of the Δx_k's sufficiently small.

88. If you average 30 mi/h on a 150-mi trip and then return over the same 150 mi at the rate of 50 mi/h, what is your average speed for the trip? Give reasons for your answer.

COMPUTER EXPLORATIONS

If your CAS can draw rectangles associated with Riemann sums, use it to draw rectangles associated with Riemann sums that converge to the integrals in Exercises 89–94. Use $n = 4, 10, 20$, and 50 subintervals of equal length in each case.

89. $\displaystyle \int_0^1 (1 - x) \, dx = \frac{1}{2}$

90. $\displaystyle \int_0^1 (x^2 + 1) \, dx = \frac{4}{3}$

91. $\displaystyle \int_{-\pi}^{\pi} \cos x \, dx = 0$

92. $\displaystyle \int_0^{\pi/4} \sec^2 x \, dx = 1$

93. $\displaystyle \int_{-1}^{1} |x| \, dx = 1$

94. $\displaystyle \int_1^2 \frac{1}{x} \, dx$ (The integral's value is about 0.693.)

In Exercises 95–102, use a CAS to perform the following steps:

a. Plot the functions over the given interval.

b. Partition the interval into $n = 100, 200$, and 1000 subintervals of equal length, and evaluate the function at the midpoint of each subinterval.

c. Compute the average value of the function values generated in part (b).

d. Solve the equation $f(x) = $ (average value) for x using the average value calculated in part (c) for the $n = 1000$ partitioning.

95. $f(x) = \sin x$ on $[0, \pi]$

96. $f(x) = \sin^2 x$ on $[0, \pi]$

97. $f(x) = x \sin \dfrac{1}{x}$ on $\left[\dfrac{\pi}{4}, \pi \right]$

98. $f(x) = x \sin^2 \dfrac{1}{x}$ on $\left[\dfrac{\pi}{4}, \pi \right]$

99. $f(x) = xe^{-x}$ on $[0, 1]$

100. $f(x) = e^{-x^2}$ on $[0, 1]$

101. $f(x) = \dfrac{\ln x}{x}$ on $[2, 5]$

102. $f(x) = \dfrac{1}{\sqrt{1 - x^2}}$ on $\left[0, \dfrac{1}{2} \right]$

5.4 The Fundamental Theorem of Calculus

In this section we present the Fundamental Theorem of Calculus, which is the central theorem of integral calculus. It connects integration and differentiation, enabling us to compute integrals by using an antiderivative of the integrand function rather than by taking limits of Riemann sums as we did in Section 5.3. Leibniz and Newton exploited this relationship and started mathematical developments that fueled the scientific revolution for the next 200 years.

Along the way, we will present an integral version of the Mean Value Theorem, which is another important theorem of integral calculus and is used to prove the Fundamental Theorem. We also find that the net change of a function over an interval is the integral of its rate of change, as suggested by Example 2 in Section 5.1.

Mean Value Theorem for Definite Integrals

In the previous section we defined the average value of a continuous function over a closed interval $[a, b]$ to be the definite integral $\int_a^b f(x)\, dx$ divided by the length or width $b - a$ of the interval. The Mean Value Theorem for Definite Integrals asserts that this average value is *always* taken on at least once by the function f in the interval.

The graph in Figure 5.16 shows a *positive* continuous function $y = f(x)$ defined over the interval $[a, b]$. Geometrically, the Mean Value Theorem says that there is a number c in $[a, b]$ such that the rectangle with height equal to the average value $f(c)$ of the function and base width $b - a$ has exactly the same area as the region beneath the graph of f from a to b.

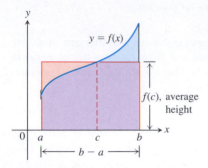

FIGURE 5.16 The value $f(c)$ in the Mean Value Theorem is, in a sense, the average (or *mean*) height of f on $[a, b]$. When $f \geq 0$, the area of the rectangle is the area under the graph of f from a to b,

$$f(c)(b - a) = \int_a^b f(x)\, dx.$$

THEOREM 3—The Mean Value Theorem for Definite Integrals
If f is continuous on $[a, b]$, then at some point c in $[a, b]$,

$$f(c) = \frac{1}{b - a}\int_a^b f(x)\, dx.$$

Proof If we divide both sides of the Max-Min Inequality (Table 5.6, Rule 6) by $(b - a)$, we obtain

$$\min f \leq \frac{1}{b - a}\int_a^b f(x)\, dx \leq \max f.$$

Since f is continuous, the Intermediate Value Theorem for Continuous Functions (Section 2.5) says that f must assume every value between min f and max f. It must therefore assume the value $(1/(b - a)) \int_a^b f(x)\, dx$ at some point c in $[a, b]$. ∎

The continuity of f is important here. It is possible for a discontinuous function to never equal its average value (Figure 5.17).

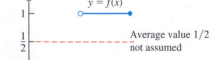

FIGURE 5.17 A discontinuous function need not assume its average value.

EXAMPLE 1 Show that if f is continuous on $[a, b]$, $a \neq b$, and if

$$\int_a^b f(x)\, dx = 0,$$

then $f(x) = 0$ at least once in $[a, b]$.

Solution The average value of f on $[a, b]$ is

$$\mathrm{av}(f) = \frac{1}{b - a}\int_a^b f(x)\, dx = \frac{1}{b - a}\cdot 0 = 0.$$

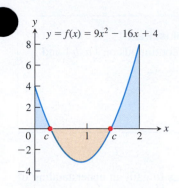

FIGURE 5.18 The function $f(x) = 9x^2 - 16x + 4$ satisfies $\int_0^2 f(x)\, dx = 0$, and there are two values of c in the interval $[0, 2]$ where $f(c) = 0$.

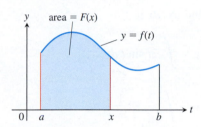

FIGURE 5.19 The function $F(x)$ defined by Equation (1) gives the area under the graph of f from a to x when f is nonnegative and $x > a$.

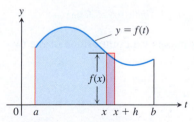

FIGURE 5.20 In Equation (1), $F(x)$ is the area to the left of x. Also, $F(x + h)$ is the area to the left of $x + h$. The difference quotient $[F(x + h) - F(x)]/h$ is then approximately equal to $f(x)$, the height of the rectangle shown here.

By the Mean Value Theorem, f assumes this value at some point $c \in [a, b]$. This is illustrated in Figure 5.18 for the function $f(x) = 9x^2 - 16x + 4$ on the interval $[0, 2]$. ∎

Fundamental Theorem, Part 1

It can be very difficult to compute definite integrals by taking the limit of Riemann sums. We now develop a powerful new method for evaluating definite integrals, based on using antiderivatives. This method combines the two strands of calculus. One strand involves the idea of taking the limits of finite sums to obtain a definite integral, and the other strand contains derivatives and antiderivatives. They come together in the Fundamental Theorem of Calculus. We begin by considering how to differentiate a certain type of function that is described as an integral.

If $f(t)$ is an integrable function over a finite interval I, then the integral from any fixed number $a \in I$ to another number $x \in I$ defines a new function F whose value at x is

$$F(x) = \int_a^x f(t)\, dt. \tag{1}$$

For example, if f is nonnegative and x lies to the right of a, then $F(x)$ is the area under the graph from a to x (Figure 5.19). The variable x is the upper limit of integration of an integral, but F is just like any other real-valued function of a real variable. For each value of the input x, there is a single numerical output, in this case the definite integral of f from a to x.

Equation (1) gives a useful way to define new functions (as we will see in Section 7.1), but its key importance is the connection that it makes between integrals and derivatives. If f is a continuous function, then the Fundamental Theorem asserts that F is a differentiable function of x whose derivative is f itself. That is, at each x in the interval $[a, b]$ we have

$$F'(x) = f(x).$$

To gain some insight into why this holds, we look at the geometry behind it.

If $f \geq 0$ on $[a, b]$, then to compute $F'(x)$ from the definition of the derivative we must take the limit as $h \to 0$ of the difference quotient

$$\frac{F(x + h) - F(x)}{h}.$$

If $h > 0$, then $F(x + h)$ is the area under the graph of f from a to $x + h$, while $F(x)$ is the area under the graph of f from a to x. Subtracting the two gives us the area under the graph of f between x and $x + h$ (see Figure 5.20). As shown in Figure 5.20, if h is small, the area under the graph of f from x to $x + h$ is approximated by the area of the rectangle whose height is $f(x)$ and whose base is the interval $[x, x + h]$. That is,

$$F(x + h) - F(x) \approx h f(x).$$

Dividing both sides by h, we see that the value of the difference quotient is very close to the value of $f(x)$:

$$\frac{F(x + h) - F(x)}{h} \approx f(x).$$

This approximation improves as h approaches 0. It is reasonable to expect that $F'(x)$, which is the limit of this difference quotient as $h \to 0$, equals $f(x)$, so that

$$F'(x) = \lim_{h \to 0} \frac{F(x + h) - F(x)}{h} = f(x).$$

This equation is true even if the function f is not positive, and it forms the first part of the Fundamental Theorem of Calculus.

> **THEOREM 4—The Fundamental Theorem of Calculus, Part 1**
> If f is continuous on $[a, b]$, then $F(x) = \int_a^x f(t)\, dt$ is continuous on $[a, b]$ and differentiable on (a, b) and its derivative is $f(x)$:
>
> $$F'(x) = \frac{d}{dx}\int_a^x f(t)\, dt = f(x). \tag{2}$$

Before proving Theorem 4, we look at several examples to gain an understanding of what it says. In each of these examples, notice that the independent variable x appears in either the upper or the lower limit of integration (either as part of a formula or by itself). The independent variable on which y depends in these examples is x, while t is merely a dummy variable in the integral.

EXAMPLE 2   Use the Fundamental Theorem to find dy/dx if

(a) $y = \displaystyle\int_a^x (t^3 + 1)\, dt$ **(b)** $y = \displaystyle\int_x^5 3t \sin t\, dt$

(c) $y = \displaystyle\int_1^{x^2} \cos t\, dt$ **(d)** $y = \displaystyle\int_{1+3x^2}^4 \frac{1}{2 + e^t}\, dt$

Solution   We calculate the derivatives with respect to the independent variable x.

(a) $\dfrac{dy}{dx} = \dfrac{d}{dx}\displaystyle\int_a^x (t^3 + 1)\, dt = x^3 + 1$ Eq. (2) with $f(t) = t^3 + 1$

(b) $\dfrac{dy}{dx} = \dfrac{d}{dx}\displaystyle\int_x^5 3t \sin t\, dt = \dfrac{d}{dx}\left(-\displaystyle\int_5^x 3t \sin t\, dt\right)$ Table 5.6, Rule 1

$\qquad\qquad = -\dfrac{d}{dx}\displaystyle\int_5^x 3t \sin t\, dt$

$\qquad\qquad = -3x \sin x$ Eq. (2) with $f(t) = 3t \sin t$

(c) The upper limit of integration is not x but x^2. This makes y a composition of the two functions

$$y = \int_1^u \cos t\, dt \qquad \text{and} \qquad u = x^2.$$

We must therefore apply the Chain Rule to find dy/dx:

$$\frac{dy}{dx} = \frac{dy}{du} \cdot \frac{du}{dx}$$

$$= \left(\frac{d}{du}\int_1^u \cos t\, dt\right) \cdot \frac{du}{dx}$$

$$= \cos u \cdot \frac{du}{dx} \qquad\qquad \text{Eq. (2) with } f(t) = \cos t$$

$$= \cos(x^2) \cdot 2x$$

$$= 2x \cos x^2$$

(d) $\displaystyle \frac{d}{dx} \int_{1+3x^2}^{4} \frac{1}{2+e^t}\, dt = \frac{d}{dx}\left(-\int_{4}^{1+3x^2} \frac{1}{2+e^t}\, dt\right)$ Rule 1

$$= -\frac{d}{dx} \int_{4}^{1+3x^2} \frac{1}{2+e^t}\, dt$$

$$= -\frac{1}{2+e^{(1+3x^2)}} \frac{d}{dx}\left(1+3x^2\right) \qquad \text{\color{blue}Eq. (2) and the Chain Rule}$$

$$= -\frac{6x}{2+e^{(1+3x^2)}} \qquad\qquad \blacksquare$$

Proof of Theorem 4   We prove the Fundamental Theorem, Part 1, by applying the definition of the derivative directly to the function $F(x)$, when x and $x+h$ are in (a, b). This means writing out the difference quotient

$$\frac{F(x+h)-F(x)}{h} \tag{3}$$

and showing that its limit as $h \to 0$ is the number $f(x)$. Doing so, we find that

$$F'(x) = \lim_{h\to 0} \frac{F(x+h)-F(x)}{h}$$

$$= \lim_{h\to 0} \frac{1}{h}\left[\int_{a}^{x+h} f(t)\, dt - \int_{a}^{x} f(t)\, dt\right]$$

$$= \lim_{h\to 0} \frac{1}{h} \int_{x}^{x+h} f(t)\, dt. \qquad \text{\color{blue}Table 5.6, Rule 5}$$

According to the Mean Value Theorem for Definite Integrals, there is some point c between x and $x+h$ where $f(c)$ equals the average value of f on the interval $[x, x+h]$. That is, there is some number c in $[x, x+h]$ such that

$$\frac{1}{h}\int_{x}^{x+h} f(t)\, dt = f(c). \tag{4}$$

As $h \to 0$, $x+h$ approaches x, which forces c to approach x also (because c is trapped between x and $x+h$). Since f is continuous at x, $f(c)$ therefore approaches $f(x)$:

$$\lim_{h\to 0} f(c) = f(x). \tag{5}$$

Hence we have shown that, for any x in (a, b),

$$F'(x) = \lim_{h\to 0} \frac{1}{h} \int_{x}^{x+h} f(t)\, dt$$

$$= \lim_{h\to 0} f(c) \qquad \text{\color{blue}Eq. (4)}$$

$$= f(x), \qquad \text{\color{blue}Eq. (5)}$$

and therefore F is differentiable at x. Since differentiability implies continuity, this also shows that F is continuous on the open interval (a, b). To complete the proof, we just have to show that F is also continuous at $x = a$ and $x = b$. To do this, we make a very similar argument, except that at $x = a$ we need only consider the one-sided limit as $h \to 0^+$, and similarly at $x = b$ we need only consider $h \to 0^-$. This shows that F has a one-sided derivative at $x = a$ and at $x = b$, and therefore Theorem 1 in Section 3.2 implies that F is continuous at those two points. $\blacksquare$

Fundamental Theorem, Part 2 (The Evaluation Theorem)

We now come to the second part of the Fundamental Theorem of Calculus. This part describes how to evaluate definite integrals without having to calculate limits of Riemann sums. Instead we find and evaluate an antiderivative at the upper and lower limits of integration.

THEOREM 4 (Continued)—The Fundamental Theorem of Calculus, Part 2

If f is continuous over $[a, b]$ and F is any antiderivative of f on $[a, b]$, then

$$\int_a^b f(x) \, dx = F(b) - F(a).$$

Proof   Part 1 of the Fundamental Theorem tells us that an antiderivative of f exists, namely

$$G(x) = \int_a^x f(t) \, dt.$$

Thus, if F is *any* antiderivative of f, then $F(x) = G(x) + C$ for some constant C for $a < x < b$ (by Corollary 2 of the Mean Value Theorem for Derivatives, Section 4.2). Since both F and G are continuous on $[a, b]$, we see that the equality $F(x) = G(x) + C$ also holds when $x = a$ and $x = b$ by taking one-sided limits (as $x \to a^+$ and $x \to b^-$).

Evaluating $F(b) - F(a)$, we have

$$
\begin{aligned}
F(b) - F(a) &= [G(b) + C] - [G(a) + C] \\
&= G(b) - G(a) \\
&= \int_a^b f(t) \, dt - \int_a^a f(t) \, dt \\
&= \int_a^b f(t) \, dt - 0 \\
&= \int_a^b f(t) \, dt.
\end{aligned}
$$
∎

The Evaluation Theorem is important because it says that to calculate the definite integral of f over an interval $[a, b]$ we need do only two things:

1. Find an antiderivative F of f, and
2. Calculate the number $F(b) - F(a)$, which is equal to $\int_a^b f(x) \, dx$.

This process is much easier than using a Riemann sum computation. The power of the theorem follows from the realization that the definite integral, which is defined by a complicated process involving all of the values of the function f over $[a, b]$, can be found by knowing the values of *any* antiderivative F at only the two endpoints a and b. The usual notation for the difference $F(b) - F(a)$ is

$$F(x) \Bigg]_a^b \quad \text{or} \quad \left[F(x) \right]_a^b,$$

depending on whether F has one or more terms.

EXAMPLE 3   We calculate several definite integrals using the Evaluation Theorem, rather than by taking limits of Riemann sums.

(a) $\displaystyle \int_0^\pi \cos x \, dx = \sin x \Bigg]_0^\pi$ $\dfrac{d}{dx} \sin x = \cos x$

$$= \sin \pi - \sin 0 = 0 - 0 = 0$$

(b) $\displaystyle\int_{-\pi/4}^{0} \sec x \tan x \, dx = \sec x \Big]_{-\pi/4}^{0}$ $\quad\frac{d}{dx}\sec x = \sec x \tan x$

$$= \sec 0 - \sec\left(-\frac{\pi}{4}\right) = 1 - \sqrt{2}$$

(c) $\displaystyle\int_{1}^{4}\left(\frac{3}{2}\sqrt{x} - \frac{4}{x^2}\right)dx = \left[x^{3/2} + \frac{4}{x}\right]_{1}^{4}$ $\quad\frac{d}{dx}\left(x^{3/2} + \frac{4}{x}\right) = \frac{3}{2}x^{1/2} - \frac{4}{x^2}$

$$= \left[(4)^{3/2} + \frac{4}{4}\right] - \left[(1)^{3/2} + \frac{4}{1}\right]$$

$$= [8 + 1] - [5] = 4$$

(d) $\displaystyle\int_{0}^{1}\frac{dx}{x + 1} = \ln|x + 1|\Big]_{0}^{1}$ $\quad\frac{d}{dx}\ln|x + 1| = \frac{1}{x + 1}$

$$= \ln 2 - \ln 1 = \ln 2$$

(e) $\displaystyle\int_{0}^{1}\frac{dx}{x^2 + 1} = \tan^{-1}x\Big]_{0}^{1}$ $\quad\frac{d}{dx}\tan^{-1}x = \ = \frac{1}{x^2 + 1}$

$$= \tan^{-1}1 - \tan^{-1}0 = \frac{\pi}{4} - 0 = \frac{\pi}{4}. \qquad\blacksquare$$

Exercise 82 offers another proof of the Evaluation Theorem, bringing together the ideas of Riemann sums, the Mean Value Theorem, and the definition of the definite integral.

The Integral of a Rate

We can interpret Part 2 of the Fundamental Theorem in another way. If F is any antiderivative of f, then $F' = f$. The equation in the theorem can then be rewritten as

$$\int_{a}^{b} F'(x)\, dx = F(b) - F(a).$$

Now $F'(x)$ represents the rate of change of the function $F(x)$ with respect to x, so the last equation asserts that the integral of F' is just the *net change* in F as x changes from a to b. Formally, we have the following result.

THEOREM 5—The Net Change Theorem

The net change in a differentiable function $F(x)$ over an interval $a \le x \le b$ is the integral of its rate of change:

$$F(b) - F(a) = \int_{a}^{b} F'(x)\, dx. \qquad (6)$$

EXAMPLE 4   Here are several interpretations of the Net Change Theorem.

(a) If $c(x)$ is the cost of producing x units of a certain commodity, then $c'(x)$ is the marginal cost (Section 3.4). From Theorem 5,

$$\int_{x_1}^{x_2} c'(x)\, dx = c(x_2) - c(x_1),$$

which is the cost of increasing production from x_1 units to x_2 units.

(b) If an object with position function $s(t)$ moves along a coordinate line, its velocity is $v(t) = s'(t)$. Theorem 5 says that

$$\int_{t_1}^{t_2} v(t) \, dt = s(t_2) - s(t_1),$$

so the integral of velocity is the **displacement** over the time interval $t_1 \leq t \leq t_2$. On the other hand, the integral of the speed $|v(t)|$ is the **total distance traveled** over the time interval. This is consistent with our discussion in Section 5.1. ∎

If we rearrange Equation (6) as

$$F(b) = F(a) + \int_a^b F'(x) \, dx,$$

we see that the Net Change Theorem also says that the final value of a function $F(x)$ over an interval $[a, b]$ equals its initial value $F(a)$ plus its net change over the interval. So if $v(t)$ represents the velocity function of an object moving along a coordinate line, this means that the object's final position $s(t_2)$ over a time interval $t_1 \leq t \leq t_2$ is its initial position $s(t_1)$ plus its net change in position along the line (see Example 4b).

EXAMPLE 5   Consider again our analysis of a heavy rock blown straight up from the ground by a dynamite blast (Example 2, Section 5.1). The velocity of the rock at any time t during its motion was given as $v(t) = 160 - 32t$ ft/sec.

(a) Find the displacement of the rock during the time period $0 \leq t \leq 8$.

(b) Find the total distance traveled during this time period.

Solution

(a) From Example 4b, the displacement is the integral

$$\int_0^8 v(t) \, dt = \int_0^8 (160 - 32t) \, dt = \left[160t - 16t^2 \right]_0^8$$

$$= (160)(8) - (16)(64) = 256.$$

This means that the height of the rock is 256 ft above the ground 8 sec after the explosion, which agrees with our conclusion in Example 2, Section 5.1.

(b) As we noted in Table 5.3, the velocity function $v(t)$ is positive over the time interval $[0, 5]$ and negative over the interval $[5, 8]$. Therefore, from Example 4b, the total distance traveled is the integral

$$\int_0^8 |v(t)| \, dt = \int_0^5 |v(t)| \, dt + \int_5^8 |v(t)| \, dt$$

$$= \int_0^5 (160 - 32t) \, dt - \int_5^8 (160 - 32t) \, dt \qquad |v(t)| = -(160 - 32t) \text{ over } [5, 8]$$

$$= \left[160t - 16t^2 \right]_0^5 - \left[160t - 16t^2 \right]_5^8$$

$$= \left[(160)(5) - (16)(25) \right] - \left[(160)(8) - (16)(64) - ((160)(5) - (16)(25)) \right]$$

$$= 400 - (-144) = 544.$$

Again, this calculation agrees with our conclusion in Example 2, Section 5.1. That is, the total distance of 544 ft traveled by the rock during the time period $0 \leq t \leq 8$ is (i) the maximum height of 400 ft it reached over the time interval $[0, 5]$ plus (ii) the additional distance of 144 ft the rock fell over the time interval $[5, 8]$. ∎

The Relationship Between Integration and Differentiation

The conclusions of the Fundamental Theorem tell us several things. Equation (2) can be rewritten as

$$\frac{d}{dx}\int_{a}^{x} f(t)\, dt = f(x),$$

which says that if you first integrate the function f and then differentiate the result, you get the function f back again. Likewise, replacing b by x and x by t in Equation (6) gives

$$\int_{a}^{x} F'(t)\, dt = F(x) - F(a),$$

so that if you first differentiate the function F and then integrate the result, you get the function F back (adjusted by an integration constant). In a sense, the processes of integration and differentiation are "inverses" of each other. The Fundamental Theorem also says that every continuous function f has an antiderivative F. It shows the importance of finding antiderivatives in order to evaluate definite integrals easily. Furthermore, it says that the differential equation $dy/dx = f(x)$ has a solution (namely, any of the functions $y = F(x) + C$) when f is a continuous function.

Total Area

Area is always a nonnegative quantity. The Riemann sum approximations contain terms such as $f(c_k)\,\Delta x_k$ that give the area of a rectangle when $f(c_k)$ is positive. When $f(c_k)$ is negative, then the product $f(c_k)\,\Delta x_k$ is the negative of the rectangle's area. When we add up such terms for a negative function, we get the negative of the area between the curve and the x-axis. If we then take the absolute value, we obtain the correct positive area.

EXAMPLE 6   Figure 5.21 shows the graph of $f(x) = x^2 - 4$ and its mirror image $g(x) = 4 - x^2$ reflected across the x-axis. For each function, compute

(a) the definite integral over the interval $[-2, 2]$, and

(b) the area between the graph and the x-axis over $[-2, 2]$.

Solution

(a) $$\int_{-2}^{2} f(x)\, dx = \left[\frac{x^3}{3} - 4x\right]_{-2}^{2} = \left(\frac{8}{3} - 8\right) - \left(-\frac{8}{3} + 8\right) = -\frac{32}{3},$$

and

$$\int_{-2}^{2} g(x)\, dx = \left[4x - \frac{x^3}{3}\right]_{-2}^{2} = \frac{32}{3}.$$

(b) In both cases, the area between the curve and the x-axis over $[-2, 2]$ is $32/3$ square units. Although the definite integral of $f(x)$ is negative, the area is still positive.   ■

To compute the area of the region bounded by the graph of a function $y = f(x)$ and the x-axis when the function takes on both positive and negative values, we must be careful to break up the interval $[a, b]$ into subintervals on which the function doesn't change sign. Otherwise we might get cancelation between positive and negative signed areas, leading to an incorrect total. The correct total area is obtained by adding the absolute value of the definite integral over each subinterval where $f(x)$ does not change sign. The term "area" will be taken to mean this *total area*.

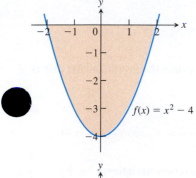

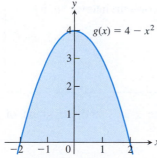

FIGURE 5.21 These graphs enclose the same amount of area with the x-axis, but the definite integrals of the two functions over $[-2, 2]$ differ in sign (Example 6).

EXAMPLE 7   Figure 5.22 shows the graph of the function $f(x) = \sin x$ between $x = 0$ and $x = 2\pi$. Compute

(a) the definite integral of $f(x)$ over $[0, 2\pi]$,

(b) the area between the graph of $f(x)$ and the x-axis over $[0, 2\pi]$.

Solution

(a) The definite integral for $f(x) = \sin x$ is given by

$$\int_0^{2\pi} \sin x \, dx = -\cos x \Big]_0^{2\pi} = -[\cos 2\pi - \cos 0] = -[1 - 1] = 0.$$

The definite integral is zero because the portions of the graph above and below the x-axis make canceling contributions.

(b) The area between the graph of $f(x)$ and the x-axis over $[0, 2\pi]$ is calculated by breaking up the domain of $\sin x$ into two pieces: the interval $[0, \pi]$ over which it is nonnegative and the interval $[\pi, 2\pi]$ over which it is nonpositive.

$$\int_0^{\pi} \sin x \, dx = -\cos x \Big]_0^{\pi} = -[\cos \pi - \cos 0] = -[-1 - 1] = 2$$

$$\int_{\pi}^{2\pi} \sin x \, dx = -\cos x \Big]_{\pi}^{2\pi} = -[\cos 2\pi - \cos \pi] = -[1 - (-1)] = -2$$

The second integral gives a negative value. The area between the graph and the axis is obtained by adding the absolute values,

$$\text{Area} = |2| + |-2| = 4.$$

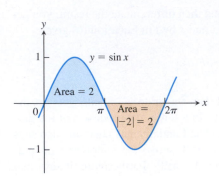

FIGURE 5.22   The total area between $y = \sin x$ and the x-axis for $0 \le x \le 2\pi$ is the sum of the absolute values of two integrals (Example 7).

Summary:
To find the area between the graph of $y = f(x)$ and the x-axis over the interval $[a, b]$:

1. Subdivide $[a, b]$ at the zeros of f.

2. Integrate f over each subinterval.

3. Add the absolute values of the integrals.

EXAMPLE 8   Find the area of the region between the x-axis and the graph of $f(x) = x^3 - x^2 - 2x$, $-1 \le x \le 2$.

Solution   First find the zeros of f. Since

$$f(x) = x^3 - x^2 - 2x = x(x^2 - x - 2) = x(x + 1)(x - 2),$$

the zeros are $x = 0, -1$, and 2 (Figure 5.23). The zeros subdivide $[-1, 2]$ into two subintervals: $[-1, 0]$, on which $f \ge 0$, and $[0, 2]$, on which $f \le 0$. We integrate f over each subinterval and add the absolute values of the calculated integrals.

$$\int_{-1}^{0} (x^3 - x^2 - 2x) \, dx = \left[\frac{x^4}{4} - \frac{x^3}{3} - x^2 \right]_{-1}^{0} = 0 - \left[\frac{1}{4} + \frac{1}{3} - 1 \right] = \frac{5}{12}$$

$$\int_{0}^{2} (x^3 - x^2 - 2x) \, dx = \left[\frac{x^4}{4} - \frac{x^3}{3} - x^2 \right]_{0}^{2} = \left[4 - \frac{8}{3} - 4 \right] - 0 = -\frac{8}{3}$$

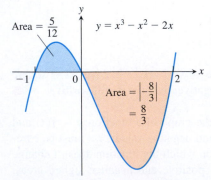

FIGURE 5.23   The region between the curve $y = x^3 - x^2 - 2x$ and the x-axis (Example 8).

The total enclosed area is obtained by adding the absolute values of the calculated integrals.

$$\text{Total enclosed area} = \frac{5}{12} + \left| -\frac{8}{3} \right| = \frac{37}{12}$$

EXERCISES 5.4

Evaluating Integrals

Evaluate the integrals in Exercises 1–34.

1. $\int_0^2 x(x - 3)\, dx$

2. $\int_{-1}^1 (x^2 - 2x + 3)\, dx$

3. $\int_{-2}^2 \dfrac{3}{(x + 3)^4}\, dx$

4. $\int_{-1}^1 x^{299}\, dx$

5. $\int_1^4 \left(3x^2 - \dfrac{x^3}{4}\right) dx$

6. $\int_{-2}^3 (x^3 - 2x + 3)\, dx$

7. $\int_0^1 \left(x^2 + \sqrt{x}\right) dx$

8. $\int_1^{32} x^{-6/5}\, dx$

9. $\int_0^{\pi/3} 2 \sec^2 x\, dx$

10. $\int_0^\pi (1 + \cos x)\, dx$

11. $\int_{\pi/4}^{3\pi/4} \csc \theta \cot \theta\, d\theta$

12. $\int_0^{\pi/3} 4 \dfrac{\sin u}{\cos^2 u}\, du$

13. $\int_{\pi/2}^0 \dfrac{1 + \cos 2t}{2}\, dt$

14. $\int_{-\pi/3}^{\pi/3} \sin^2 t\, dt$

15. $\int_0^{\pi/4} \tan^2 x\, dx$

16. $\int_0^{\pi/6} (\sec x + \tan x)^2\, dx$

17. $\int_0^{\pi/8} \sin 2x\, dx$

18. $\int_{-\pi/3}^{-\pi/4} \left(4 \sec^2 t + \dfrac{\pi}{t^2}\right) dt$

19. $\int_1^{-1} (r + 1)^2\, dr$

20. $\int_{-\sqrt{3}}^{\sqrt{3}} (t + 1)(t^2 + 4)\, dt$

21. $\int_{\sqrt{2}}^1 \left(\dfrac{u^7}{2} - \dfrac{1}{u^5}\right) du$

22. $\int_{-3}^{-1} \dfrac{y^5 - 2y}{y^3}\, dy$

23. $\int_1^{\sqrt{2}} \dfrac{s^2 + \sqrt{s}}{s^2}\, ds$

24. $\int_1^8 \dfrac{(x^{1/3} + 1)(2 - x^{2/3})}{x^{1/3}}\, dx$

25. $\int_{\pi/2}^\pi \dfrac{\sin 2x}{2 \sin x}\, dx$

26. $\int_0^{\pi/3} (\cos x + \sec x)^2\, dx$

27. $\int_{-4}^4 |x|\, dx$

28. $\int_0^\pi \dfrac{1}{2}(\cos x + |\cos x|)\, dx$

29. $\int_0^{\ln 2} e^{3x}\, dx$

30. $\int_1^2 \left(\dfrac{1}{x} - e^{-x}\right) dx$

31. $\int_0^{1/2} \dfrac{4}{\sqrt{1 - x^2}}\, dx$

32. $\int_0^{1/\sqrt{3}} \dfrac{dx}{1 + 4x^2}$

33. $\int_2^4 x^{\pi - 1}\, dx$

34. $\int_{-1}^0 \pi^{x - 1}\, dx$

In Exercises 35–38, guess an antiderivative for the integrand function. Validate your guess by differentiation and then evaluate the given definite integral. (*Hint:* Keep the Chain Rule in mind when trying to guess an antiderivative. You will learn how to find such antiderivatives in the next section.)

35. $\int_0^1 xe^{x^2}\, dx$

36. $\int_1^2 \dfrac{\ln x}{x}\, dx$

37. $\int_2^5 \dfrac{x\, dx}{\sqrt{1 + x^2}}$

38. $\int_0^{\pi/3} \sin^2 x \cos x\, dx$

Derivatives of Integrals

Find the derivatives in Exercises 39–44.

   a. by evaluating the integral and differentiating the result.

   b. by differentiating the integral directly.

39. $\dfrac{d}{dx} \int_0^{\sqrt{x}} \cos t\, dt$

40. $\dfrac{d}{dx} \int_1^{\sin x} 3t^2\, dt$

41. $\dfrac{d}{dt} \int_0^{t^4} \sqrt{u}\, du$

42. $\dfrac{d}{d\theta} \int_0^{\tan \theta} \sec^2 y\, dy$

43. $\dfrac{d}{dx} \int_0^{x^3} e^{-t}\, dt$

44. $\dfrac{d}{dt} \int_0^{\sqrt{t}} \left(x^4 + \dfrac{3}{\sqrt{1 - x^2}}\right) dx$

Find dy/dx in Exercises 45–56.

45. $y = \int_0^x \sqrt{1 + t^2}\, dt$

46. $y = \int_1^x \dfrac{1}{t}\, dt, \quad x > 0$

47. $y = \int_{\sqrt{x}}^0 \sin (t^2)\, dt$

48. $y = x \int_2^{x^2} \sin (t^3)\, dt$

49. $y = \int_{-1}^x \dfrac{t^2}{t^2 + 4}\, dt - \int_3^x \dfrac{t^2}{t^2 + 4}\, dt$

50. $y = \left(\int_0^x (t^3 + 1)^{10}\, dt\right)^3$

51. $y = \int_0^{\sin x} \dfrac{dt}{\sqrt{1 - t^2}}, \quad |x| < \dfrac{\pi}{2}$

52. $y = \int_{\tan x}^0 \dfrac{dt}{1 + t^2}$

53. $y = \int_0^{e^x} \dfrac{1}{\sqrt{t}}\, dt$

54. $y = \int_{2^x}^1 \sqrt[3]{t}\, dt$

55. $y = \int_0^{\sin^{-1} x} \cos t\, dt$

56. $y = \int_{-1}^{x^{1/\pi}} \sin^{-1} t\, dt$

Area

In Exercises 57–60, find the total area between the region and the x-axis.

57. $y = -x^2 - 2x$, $-3 \le x \le 2$

58. $y = 3x^2 - 3$, $-2 \le x \le 2$

59. $y = x^3 - 3x^2 + 2x$, $0 \le x \le 2$

60. $y = x^{1/3} - x$, $-1 \le x \le 8$

Find the areas of the shaded regions in Exercises 61–64.

61.

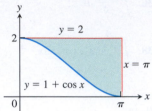

62.

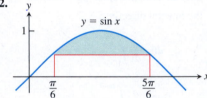

63.

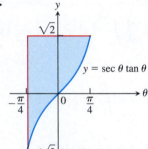

64.

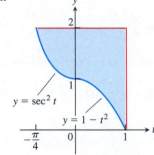

Initial Value Problems

Each of the following functions solves one of the initial value problems in Exercises 65–68. Which function solves which problem? Give brief reasons for your answers.

a. $y = \int_1^x \frac{1}{t} dt - 3$ **b.** $y = \int_0^x \sec t \, dt + 4$

c. $y = \int_{-1}^x \sec t \, dt + 4$ **d.** $y = \int_\pi^x \frac{1}{t} dt - 3$

65. $\frac{dy}{dx} = \frac{1}{x}$, $y(\pi) = -3$ **66.** $y' = \sec x$, $y(-1) = 4$

67. $y' = \sec x$, $y(0) = 4$ **68.** $y' = \frac{1}{x}$, $y(1) = -3$

Express the solutions of the initial value problems in Exercises 69 and 70 in terms of integrals.

69. $\frac{dy}{dx} = \sec x$, $y(2) = 3$ **70.** $\frac{dy}{dx} = \sqrt{1 + x^2}$, $y(1) = -2$

For exercises 71 and 72 find a function f satisfying each equation.

71. $\int_2^x \sqrt{f(t)} \, dt = x \ln x$ **72.** $f(x) = e^2 + \int_1^x f(t) \, dt$

Theory and Examples

73. Archimedes' area formula for parabolic arches Archimedes (287–212 B.C.), inventor, military engineer, physicist, and the greatest mathematician of classical times in the Western world, discovered that the area under a parabolic arch is two-thirds the base times the height. Sketch the parabolic arch $y = h - (4h/b^2)x^2$, $-b/2 \le x \le b/2$, assuming that h and b are positive. Then use calculus to find the area of the region enclosed between the arch and the x-axis.

74. Show that if k is a positive constant, then the area between the x-axis and one arch of the curve $y = \sin kx$ is $2/k$.

75. Cost from marginal cost The marginal cost of printing a poster when x posters have been printed is

$$\frac{dc}{dx} = \frac{1}{2\sqrt{x}}$$

dollars. Find $c(100) - c(1)$, the cost of printing posters 2–100.

76. Revenue from marginal revenue Suppose that a company's marginal revenue from the manufacture and sale of eggbeaters is

$$\frac{dr}{dx} = 2 - 2/(x + 1)^2,$$

where r is measured in thousands of dollars and x in thousands of units. How much money should the company expect from a production run of $x = 3$ thousand eggbeaters? To find out, integrate the marginal revenue from $x = 0$ to $x = 3$.

77. The temperature $T(°F)$ of a room at time t minutes is given by

$$T = 85 - 3\sqrt{25 - t} \quad \text{for} \quad 0 \le t \le 25.$$

a. Find the room's temperature when $t = 0$, $t = 16$, and $t = 25$.

b. Find the room's average temperature for $0 \le t \le 25$.

78. The height $H(\text{ft})$ of a palm tree after growing for t years is given by

$$H = \sqrt{t + 1} + 5t^{1/3} \quad \text{for} \quad 0 \le t \le 8.$$

a. Find the tree's height when $t = 0$, $t = 4$, and $t = 8$.

b. Find the tree's average height for $0 \le t \le 8$.

79. Suppose that $\int_1^x f(t) \, dt = x^2 - 2x + 1$. Find $f(x)$.

80. Find $f(4)$ if $\int_0^x f(t) \, dt = x \cos \pi x$.

81. Find the linearization of

$$f(x) = 2 - \int_2^{x+1} \frac{9}{1 + t} dt$$

at $x = 1$.

82. Find the linearization of

$$g(x) = 3 + \int_1^{x^2} \sec(t - 1) \, dt$$

at $x = -1$.

83. Suppose that f has a positive derivative for all values of x and that $f(1) = 0$. Which of the following statements must be true of the function

$$g(x) = \int_0^x f(t)\,dt?$$

Give reasons for your answers.

a. g is a differentiable function of x.

b. g is a continuous function of x.

c. The graph of g has a horizontal tangent at $x = 1$.

d. g has a local maximum at $x = 1$.

e. g has a local minimum at $x = 1$.

f. The graph of g has an inflection point at $x = 1$.

g. The graph of dg/dx crosses the x-axis at $x = 1$.

84. Another proof of the Evaluation Theorem

a. Let $a = x_0 < x_1 < x_2 \cdots < x_n = b$ be any partition of $[a, b]$, and let F be any antiderivative of f. Show that

$$F(b) - F(a) = \sum_{i=1}^{n} \left[F(x_i) - F(x_{i-1}) \right].$$

b. Apply the Mean Value Theorem to each term to show that $F(x_i) - F(x_{i-1}) = f(c_i)(x_i - x_{i-1})$ for some c_i in the interval (x_{i-1}, x_i). Then show that $F(b) - F(a)$ is a Riemann sum for f on $[a, b]$.

c. From part (b) and the definition of the definite integral, show that

$$F(b) - F(a) = \int_a^b f(x)\,dx.$$

85. Suppose that f is the differentiable function shown in the accompanying graph and that the position at time t (sec) of a particle moving along a coordinate axis is

$$s = \int_0^t f(x)\,dx$$

meters. Use the graph to answer the following questions. Give reasons for your answers.

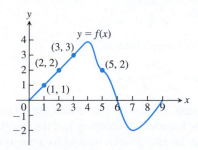

a. What is the particle's velocity at time $t = 5$?

b. Is the acceleration of the particle at time $t = 5$ positive, or negative?

c. What is the particle's position at time $t = 3$?

d. At what time during the first 9 sec does s have its largest value?

e. Approximately when is the acceleration zero?

f. When is the particle moving toward the origin? Away from the origin?

g. On which side of the origin does the particle lie at time $t = 9$?

86. Find $\displaystyle \lim_{x \to \infty} \frac{1}{\sqrt{x}} \int_1^x \frac{dt}{\sqrt{t}}$.

COMPUTER EXPLORATIONS

In Exercises 87–90, let $F(x) = \int_a^x f(t)\,dt$ for the specified function f and interval $[a, b]$. Use a CAS to perform the following steps and answer the questions posed.

a. Plot the functions f and F together over $[a, b]$.

b. Solve the equation $F'(x) = 0$. What can you see to be true about the graphs of f and F at points where $F'(x) = 0$? Is your observation borne out by Part 1 of the Fundamental Theorem coupled with information provided by the first derivative? Explain your answer.

c. Over what intervals (approximately) is the function F increasing and decreasing? What is true about f over those intervals?

d. Calculate the derivative f' and plot it together with F. What can you see to be true about the graph of F at points where $f'(x) = 0$? Is your observation borne out by Part 1 of the Fundamental Theorem? Explain your answer.

87. $f(x) = x^3 - 4x^2 + 3x,\quad [0, 4]$

88. $f(x) = 2x^4 - 17x^3 + 46x^2 - 43x + 12,\quad \left[0, \dfrac{9}{2}\right]$

89. $f(x) = \sin 2x \cos \dfrac{x}{3},\quad [0, 2\pi]$

90. $f(x) = x \cos \pi x,\quad [0, 2\pi]$

In Exercises 91–94, let $F(x) = \int_a^{u(x)} f(t)\,dt$ for the specified a, u, and f. Use a CAS to perform the following steps and answer the questions posed.

a. Find the domain of F.

b. Calculate $F'(x)$ and determine its zeros. For what points in its domain is F increasing? Decreasing?

c. Calculate $F''(x)$ and determine its zero. Identify the local extrema and the points of inflection of F.

d. Using the information from parts (a)–(c), draw a rough hand-sketch of $y = F(x)$ over its domain. Then graph $F(x)$ on your CAS to support your sketch.

91. $a = 1,\quad u(x) = x^2,\quad f(x) = \sqrt{1 - x^2}$

92. $a = 0,\quad u(x) = x^2,\quad f(x) = \sqrt{1 - x^2}$

93. $a = 0,\quad u(x) = 1 - x,\quad f(x) = x^2 - 2x - 3$

94. $a = 0,\quad u(x) = 1 - x^2,\quad f(x) = x^2 - 2x - 3$

In Exercises 95 and 96, assume that f is continuous and $u(x)$ is twice-differentiable.

95. Calculate $\dfrac{d}{dx} \displaystyle\int_a^{u(x)} f(t)\,dt$ and check your answer using a CAS.

96. Calculate $\dfrac{d^2}{dx^2} \displaystyle\int_a^{u(x)} f(t)\,dt$ and check your answer using a CAS.

5.5 Indefinite Integrals and the Substitution Method

The Fundamental Theorem of Calculus says that a definite integral of a continuous function can be computed directly if we can find an antiderivative of the function. In Section 4.8 we defined the **indefinite integral** of the function f with respect to x as the set of *all* antiderivatives of f, symbolized by $\int f(x)\,dx$. Since any two antiderivatives of f differ by a constant, the indefinite integral $\int$ notation means that for any antiderivative F of f,

$$\int f(x)\,dx = F(x) + C,$$

where C is any arbitrary constant. The connection between antiderivatives and the definite integral stated in the Fundamental Theorem now explains this notation:

$$\int_a^b f(x)\,dx = F(b) - F(a) = [F(b) + C] - [F(a) + C]$$

$$= [F(x) + C]_a^b = \left[\int f(x)\,dx\right]_a^b.$$

When finding the indefinite integral of a function f, remember that it always includes an arbitrary constant C.

We must distinguish carefully between definite and indefinite integrals. A definite integral $\int_a^b f(x)\,dx$ is a *number*. An indefinite integral $\int f(x)\,dx$ is a *function* plus an arbitrary constant C.

So far, we have only been able to find antiderivatives of functions that are clearly recognizable as derivatives. In this section we begin to develop more general techniques for finding antiderivatives of functions we can't easily recognize as derivatives.

Substitution: Running the Chain Rule Backwards

If u is a differentiable function of x and n is any number different from -1, the Chain Rule tells us that

$$\frac{d}{dx}\left(\frac{u^{n+1}}{n+1}\right) = u^n \frac{du}{dx}.$$

From another point of view, this same equation says that $u^{n+1}/(n+1)$ is one of the antiderivatives of the function $u^n(du/dx)$. Therefore,

$$\int u^n \frac{du}{dx}\,dx = \frac{u^{n+1}}{n+1} + C. \tag{1}$$

The integral in Equation (1) is equal to the simpler integral

$$\int u^n\,du = \frac{u^{n+1}}{n+1} + C,$$

which suggests that the simpler expression du can be substituted for $(du/dx)\,dx$ when computing an integral. Leibniz, one of the founders of calculus, had the insight that indeed this substitution could be done, leading to the *substitution method* for computing integrals. As with differentials, when computing integrals we have

$$du = \frac{du}{dx}\,dx.$$

EXAMPLE 1 Find the integral $\displaystyle\int (x^3 + x)^5(3x^2 + 1)\,dx.$

Solution We set $u = x^3 + x$. Then

$$du = \frac{du}{dx}\,dx = (3x^2 + 1)\,dx,$$

so that by substitution we have

$$\int (x^3 + x)^5 (3x^2 + 1)\, dx = \int u^5\, du \qquad \text{Let } u = x^3 + x, du = (3x^2 + 1)\, dx.$$

$$= \frac{u^6}{6} + C \qquad \text{Integrate with respect to } u.$$

$$= \frac{(x^3 + x)^6}{6} + C \qquad \text{Substitute } x^3 + x \text{ for } u. \qquad \blacksquare$$

EXAMPLE 2   Find $\displaystyle\int \sqrt{2x + 1}\, dx$.

Solution   The integral does not fit the formula

$$\int u^n\, du,$$

with $u = 2x + 1$ and $n = 1/2$, because

$$du = \frac{du}{dx}\, dx = 2\, dx,$$

which is not precisely dx. The constant factor 2 is missing from the integral. However, we can introduce this factor after the integral sign if we compensate for it by introducing a factor of $1/2$ in front of the integral sign. So we write

$$\int \sqrt{2x + 1}\, dx = \frac{1}{2}\int \underbrace{\sqrt{2x + 1}}_{u} \cdot \underbrace{2\, dx}_{du}$$

$$= \frac{1}{2}\int u^{1/2}\, du \qquad \text{Let } u = 2x + 1, du = 2\, dx.$$

$$= \frac{1}{2}\frac{u^{3/2}}{3/2} + C \qquad \text{Integrate with respect to } u.$$

$$= \frac{1}{3}(2x + 1)^{3/2} + C \qquad \text{Substitute } 2x + 1 \text{ for } u. \qquad \blacksquare$$

The substitutions in Examples 1 and 2 are instances of the following general rule.

THEOREM 6—The Substitution Rule

If $u = g(x)$ is a differentiable function whose range is an interval I, and f is continuous on I, then

$$\int f(g(x)) \cdot g'(x)\, dx = \int f(u)\, du.$$

Proof   By the Chain Rule, $F(g(x))$ is an antiderivative of $f(g(x)) \cdot g'(x)$ whenever F is an antiderivative of f, because

$$\frac{d}{dx}F(g(x)) = F'(g(x)) \cdot g'(x) \qquad \text{Chain Rule}$$

$$= f(g(x)) \cdot g'(x). \qquad F' = f$$

If we make the substitution $u = g(x)$, then

$$\int f(g(x)) g'(x)\, dx = \int \frac{d}{dx} F(g(x))\, dx$$

$$= F(g(x)) + C \qquad \text{Theorem 8 in Chapter 4}$$

$$= F(u) + C \qquad u = g(x)$$

$$= \int F'(u)\, du \qquad \text{Theorem 8 in Chapter 4}$$

$$= \int f(u)\, du. \qquad F' = f \qquad \blacksquare$$

The use of the variable u in the Substitution Rule is traditional (sometimes it is referred to as u-substitution), but any letter can be used, such as v, t, θ and so forth. The rule provides a method for evaluating an integral of the form $\int f(g(x)) g'(x)\, dx$ given that the conditions of Theorem 6 are satisfied. The primary challenge is deciding what expression involving x to substitute for in the integrand. The following examples give helpful ideas.

The Substitution Method to evaluate $\int f(g(x)) g'(x)\, dx$

1. Substitute $u = g(x)$ and $du = (du/dx)\, dx = g'(x)\, dx$ to obtain $\int f(u)\, du$.
2. Integrate with respect to u.
3. Replace u by $g(x)$.

EXAMPLE 3 Find $\displaystyle\int \sec^2(5x + 1) \cdot 5\, dx$

Solution We substitute $u = 5x + 1$ and $du = 5\, dx$. Then,

$$\int \sec^2(5x + 1) \cdot 5\, dx = \int \sec^2 u\, du \qquad \text{Let } u = 5x + 1, du = 5\, dx.$$

$$= \tan u + C \qquad \frac{d}{du} \tan u = \sec^2 u$$

$$= \tan(5x + 1) + C. \qquad \text{Substitute } 5x + 1 \text{ for } u. \qquad \blacksquare$$

EXAMPLE 4 Find $\displaystyle\int \cos(7\theta + 3)\, d\theta$.

Solution We let $u = 7\theta + 3$ so that $du = 7\, d\theta$. The constant factor 7 is missing from the $d\theta$ term in the integral. We can compensate for it by multiplying and dividing by 7, using the same procedure as in Example 2. Then,

$$\int \cos(7\theta + 3)\, d\theta = \frac{1}{7} \int \cos(7\theta + 3) \cdot 7\, d\theta \qquad \text{Place factor } 1/7 \text{ in front of integral.}$$

$$= \frac{1}{7} \int \cos u\, du \qquad \text{Let } u = 7\theta + 3, du = 7\, d\theta.$$

$$= \frac{1}{7} \sin u + C \qquad \text{Integrate.}$$

$$= \frac{1}{7} \sin(7\theta + 3) + C. \qquad \text{Substitute } 7\theta + 3 \text{ for } u.$$

There is another approach to this problem. With $u = 7\theta + 3$ and $du = 7\, d\theta$ as before, we solve for $d\theta$ to obtain $d\theta = (1/7)\, du$. Then the integral becomes

$$\int \cos\,(7\theta + 3)\, d\theta = \int \cos u \cdot \frac{1}{7}\, du \qquad \text{Let } u = 7\theta + 3,\, du = 7\, d\theta,\text{ and } d\theta = (1/7)\, du.$$

$$= \frac{1}{7} \sin u + C \qquad\qquad \text{Integrate.}$$

$$= \frac{1}{7} \sin\,(7\theta + 3) + C. \qquad \text{Substitute } 7\theta + 3 \text{ for } u.$$

We can verify this solution by differentiating and checking that we obtain the original function $\cos\,(7\theta + 3)$. ■

EXAMPLE 5   Sometimes we observe that a power of x appears in the integrand that is one less than the power of x appearing in the argument of a function we want to integrate. This observation immediately suggests we try a substitution for the higher power of x. For example, in the integral below we see that x^3 appears as the exponent of one factor, and this factor is multiplied by x^2. This suggests trying the substitution $u = x^3$.

$$\int x^2 e^{x^3}\, dx = \int e^{x^3} \cdot x^2\, dx$$

$$= \int e^u \cdot \frac{1}{3}\, du \qquad \text{Let } u = x^3,\, du = 3x^2\, dx,$$
$$\qquad\qquad\qquad\qquad (1/3)\, du = x^2\, dx.$$

$$= \frac{1}{3} \int e^u\, du$$

$$= \frac{1}{3} e^u + C \qquad \text{Integrate with respect to } u.$$

$$= \frac{1}{3} e^{x^3} + C \qquad \text{Replace } u \text{ by } x^3.$$ ■

HISTORICAL BIOGRAPHY
George David Birkhoff
(1884–1944)
www.goo.gl/0YjM2t

It may happen that an extra factor of x appears in the integrand when we try a substitution $u = g(x)$. In that case, it may be possible to solve the equation $u = g(x)$ for x in terms of u. Replacing the extra factor of x with that expression may then result in an integral that we can evaluate. Here is an example of this situation.

EXAMPLE 6   Evaluate $\displaystyle\int x\sqrt{2x + 1}\, dx$.

Solution   Our previous experience with the integral in Example 2 suggests the substitution $u = 2x + 1$ with $du = 2\, dx$. Then

$$\sqrt{2x + 1}\, dx = \frac{1}{2}\sqrt{u}\, du.$$

However, in this example the integrand contains an extra factor of x that multiplies the term $\sqrt{2x + 1}$. To adjust for this, we solve the substitution equation $u = 2x + 1$ for x to obtain $x = (u - 1)/2$, and find that

$$x\sqrt{2x + 1}\, dx = \frac{1}{2}(u - 1) \cdot \frac{1}{2}\sqrt{u}\, du.$$

The integration now becomes

$$\int x\sqrt{2x+1}\,dx = \frac{1}{4}\int (u-1)\sqrt{u}\,du = \frac{1}{4}\int (u-1)u^{1/2}\,du \qquad \text{Substitute.}$$

$$= \frac{1}{4}\int (u^{3/2}-u^{1/2})\,du \qquad \text{Multiply terms.}$$

$$= \frac{1}{4}\left(\frac{2}{5}u^{5/2}-\frac{2}{3}u^{3/2}\right)+C \qquad \text{Integrate.}$$

$$= \frac{1}{10}(2x+1)^{5/2}-\frac{1}{6}(2x+1)^{3/2}+C. \qquad \text{Replace } u \text{ by } 2x+1. \ \blacksquare$$

EXAMPLE 7 Sometimes we can use trigonometric identities to transform integrals we do not know how to evaluate into ones we can evaluate using the Substitution Rule.

(a) $\displaystyle\int \sin^2 x\,dx = \int \frac{1-\cos 2x}{2}\,dx \qquad \sin^2 x = \frac{1-\cos 2x}{2}$

$$= \frac{1}{2}\int (1-\cos 2x)\,dx$$

$$= \frac{1}{2}x - \frac{1}{2}\frac{\sin 2x}{2} + C = \frac{x}{2}-\frac{\sin 2x}{4}+C$$

(b) $\displaystyle\int \cos^2 x\,dx = \int \frac{1+\cos 2x}{2}\,dx = \frac{x}{2}+\frac{\sin 2x}{4}+C \qquad \cos^2 x = \frac{1+\cos 2x}{2}$

(c) $\displaystyle\int \tan x\,du = \int \frac{\sin x}{\cos x}\,dx = \int \frac{-du}{u} \qquad u = \cos x,\, du = -\sin x\,dx$

$$= -\ln|u| + C = -\ln|\cos x| + C$$

$$= \ln \frac{1}{|\cos x|} + C = \ln|\sec x| + C \qquad \text{Reciprocal Rule} \qquad \blacksquare$$

EXAMPLE 8 An integrand may require some algebraic manipulation before the substitution method can be applied. This example gives two integrals for which we simplify by multiplying the integrand by an algebraic form equal to 1 before attempting a substitution.

(a) $\displaystyle\int \frac{dx}{e^x + e^{-x}} = \int \frac{e^x\,dx}{e^{2x}+1} \qquad \text{Multiply by } (e^x/e^x) = 1.$

$$= \int \frac{du}{u^2+1} \qquad \text{Let } u = e^x,\, u^2 = e^{2x},\; du = e^x\,dx.$$

$$= \tan^{-1}u + C \qquad \text{Integrate with respect to } u.$$

$$= \tan^{-1}(e^x) + C \qquad \text{Replace } u \text{ by } e^x.$$

(b) $\displaystyle\int \sec x\,dx = \int (\sec x)(1)\,dx = \int \sec x \cdot \frac{\sec x + \tan x}{\sec x + \tan x}\,dx \qquad \frac{\sec x + \tan x}{\sec x + \tan x} \text{ is equal to } 1.$

$$= \int \frac{\sec^2 x + \sec x \tan x}{\sec x + \tan x}\,dx$$

$$= \int \frac{du}{u} \qquad u = \tan x + \sec x,\; du = (\sec^2 + \sec x \tan x)\,dx$$

$$= \ln|u| + C = \ln|\sec x + \tan x| + C. \qquad \blacksquare$$

The integrals of cot x and csc x are computed in a way similar to how we found the integrals of tan x and sec x in Examples 7c and 8b (see Exercises 71 and 72). We summarize the results for these four basic trigonometric integrals here.

Integrals of the tangent, cotangent, secant, and cosecant functions

$$\int \tan x \, dx = \ln|\sec x| + C \qquad\qquad \int \sec x \, dx = \ln|\sec x + \tan x| + C$$

$$\int \cot x \, dx = \ln|\sin x| + C \qquad\qquad \int \csc x \, dx = -\ln|\csc x + \cot x| + C$$

Trying Different Substitutions

The success of the substitution method depends on finding a substitution that changes an integral we cannot evaluate directly into one that we can. Finding the right substitution gets easier with practice and experience. If your first substitution fails, try another substitution, possibly coupled with other algebraic or trigonometric simplifications to the integrand. Several more complicated types of substitutions will be studied in Chapter 8.

EXAMPLE 9   Evaluate $\displaystyle\int \frac{2z\,dz}{\sqrt[3]{z^2 + 1}}$.

Solution   We will use the substitution method of integration as an exploratory tool: We substitute for the most troublesome part of the integrand and see how things work out. For the integral here, we might try $u = z^2 + 1$ or we might even press our luck and take u to be the entire cube root. In this example both substitutions turn out to be successful, but that is not always the case. If one substitution does not help, a different substitution may work instead.

Method 1:   Substitute $u = z^2 + 1$.

$$\int \frac{2z\,dz}{\sqrt[3]{z^2 + 1}} = \int \frac{du}{u^{1/3}} \qquad\qquad \text{Let } u = z^2 + 1, \\ du = 2z\,dz.$$

$$= \int u^{-1/3}\,du \qquad\qquad \text{In the form } \int u^n\,du$$

$$= \frac{u^{2/3}}{2/3} + C \qquad\qquad \text{Integrate.}$$

$$= \frac{3}{2}u^{2/3} + C$$

$$= \frac{3}{2}(z^2 + 1)^{2/3} + C \qquad\qquad \text{Replace } u \text{ by } z^2 + 1.$$

Method 2:   Substitute $u = \sqrt[3]{z^2 + 1}$ instead.

$$\int \frac{2z\,dz}{\sqrt[3]{z^2 + 1}} = \int \frac{3u^2\,du}{u} \qquad\qquad \text{Let } u = \sqrt[3]{z^2 + 1}, \\ u^3 = z^2 + 1,\ 3u^2\,du = 2z\,dz.$$

$$= 3\int u\,du$$

$$= 3 \cdot \frac{u^2}{2} + C \qquad\qquad \text{Integrate.}$$

$$= \frac{3}{2}(z^2 + 1)^{2/3} + C \qquad\qquad \text{Replace } u \text{ by } (z^2 + 1)^{1/3}. \qquad\blacksquare$$

EXERCISES 5.5

Evaluating Indefinite Integrals

Evaluate the indefinite integrals in Exercises 1–16 by using the given substitutions to reduce the integrals to standard form.

1. $\displaystyle \int 2(2x + 4)^5 \, dx, \quad u = 2x + 4$

2. $\displaystyle \int 7\sqrt{7x - 1} \, dx, \quad u = 7x - 1$

3. $\displaystyle \int 2x(x^2 + 5)^{-4} \, dx, \quad u = x^2 + 5$

4. $\displaystyle \int \frac{4x^3}{(x^4 + 1)^2} \, dx, \quad u = x^4 + 1$

5. $\displaystyle \int (3x + 2)(3x^2 + 4x)^4 \, dx, \quad u = 3x^2 + 4x$

6. $\displaystyle \int \frac{\left(1 + \sqrt{x}\right)^{1/3}}{\sqrt{x}} \, dx, \quad u = 1 + \sqrt{x}$

7. $\displaystyle \int \sin 3x \, dx, \quad u = 3x$ **8.** $\displaystyle \int x \sin (2x^2) \, dx, \quad u = 2x^2$

9. $\displaystyle \int \sec 2t \tan 2t \, dt, \quad u = 2t$

10. $\displaystyle \int \left(1 - \cos \frac{t}{2}\right)^2 \sin \frac{t}{2} \, dt, \quad u = 1 - \cos \frac{t}{2}$

11. $\displaystyle \int \frac{9r^2 \, dr}{\sqrt{1 - r^3}}, \quad u = 1 - r^3$

12. $\displaystyle \int 12(y^4 + 4y^2 + 1)^2(y^3 + 2y) \, dy, \quad u = y^4 + 4y^2 + 1$

13. $\displaystyle \int \sqrt{x} \sin^2 (x^{3/2} - 1) \, dx, \quad u = x^{3/2} - 1$

14. $\displaystyle \int \frac{1}{x^2} \cos^2 \left(\frac{1}{x}\right) \, dx, \quad u = -\frac{1}{x}$

15. $\displaystyle \int \csc^2 2\theta \cot 2\theta \, d\theta$

    a. Using $u = \cot 2\theta$ **b.** Using $u = \csc 2\theta$

16. $\displaystyle \int \frac{dx}{\sqrt{5x + 8}}$

    a. Using $u = 5x + 8$ **b.** Using $u = \sqrt{5x + 8}$

Evaluate the integrals in Exercises 17–66.

17. $\displaystyle \int \sqrt{3 - 2s} \, ds$ **18.** $\displaystyle \int \frac{1}{\sqrt{5s + 4}} \, ds$

19. $\displaystyle \int \theta \sqrt[4]{1 - \theta^2} \, d\theta$ **20.** $\displaystyle \int 3y\sqrt{7 - 3y^2} \, dy$

21. $\displaystyle \int \frac{1}{\sqrt{x}\left(1 + \sqrt{x}\right)^2} \, dx$ **22.** $\displaystyle \int \sqrt{\sin x} \cos^3 x \, dx$

23. $\displaystyle \int \sec^2 (3x + 2) \, dx$ **24.** $\displaystyle \int \tan^2 x \sec^2 x \, dx$

25. $\displaystyle \int \sin^5 \frac{x}{3} \cos \frac{x}{3} \, dx$ **26.** $\displaystyle \int \tan^7 \frac{x}{2} \sec^2 \frac{x}{2} \, dx$

27. $\displaystyle \int r^2 \left(\frac{r^3}{18} - 1\right)^5 \, dr$ **28.** $\displaystyle \int r^4 \left(7 - \frac{r^5}{10}\right)^3 \, dr$

29. $\displaystyle \int x^{1/2} \sin (x^{3/2} + 1) \, dx$

30. $\displaystyle \int \csc \left(\frac{v - \pi}{2}\right) \cot \left(\frac{v - \pi}{2}\right) \, dv$

31. $\displaystyle \int \frac{\sin (2t + 1)}{\cos^2 (2t + 1)} \, dt$ **32.** $\displaystyle \int \frac{\sec z \tan z}{\sqrt{\sec z}} \, dz$

33. $\displaystyle \int \frac{1}{t^2} \cos \left(\frac{1}{t} - 1\right) \, dt$ **34.** $\displaystyle \int \frac{1}{\sqrt{t}} \cos \left(\sqrt{t} + 3\right) \, dt$

35. $\displaystyle \int \frac{1}{\theta^2} \sin \frac{1}{\theta} \cos \frac{1}{\theta} \, d\theta$ **36.** $\displaystyle \int \frac{\cos \sqrt{\theta}}{\sqrt{\theta} \sin^2 \sqrt{\theta}} \, d\theta$

37. $\displaystyle \int \frac{x}{\sqrt{1 + x}} \, dx$ **38.** $\displaystyle \int \sqrt{\frac{x - 1}{x^5}} \, dx$

39. $\displaystyle \int \frac{1}{x^2} \sqrt{2 - \frac{1}{x}} \, dx$ **40.** $\displaystyle \int \frac{1}{x^3} \sqrt{\frac{x^2 - 1}{x^2}} \, dx$

41. $\displaystyle \int \sqrt{\frac{x^3 - 3}{x^{11}}} \, dx$ **42.** $\displaystyle \int \sqrt{\frac{x^4}{x^3 - 1}} \, dx$

43. $\displaystyle \int x(x - 1)^{10} \, dx$ **44.** $\displaystyle \int x\sqrt{4 - x} \, dx$

45. $\displaystyle \int (x + 1)^2(1 - x)^5 \, dx$ **46.** $\displaystyle \int (x + 5)(x - 5)^{1/3} \, dx$

47. $\displaystyle \int x^3\sqrt{x^2 + 1} \, dx$ **48.** $\displaystyle \int 3x^5\sqrt{x^3 + 1} \, dx$

49. $\displaystyle \int \frac{x}{(x^2 - 4)^3} \, dx$ **50.** $\displaystyle \int \frac{x}{(2x - 1)^{2/3}} \, dx$

51. $\displaystyle \int (\cos x) \, e^{\sin x} \, dx$ **52.** $\displaystyle \int (\sin 2\theta) \, e^{\sin^2 \theta} \, d\theta$

53. $\displaystyle \int \frac{1}{\sqrt{x} e^{-\sqrt{x}}} \sec^2(e^{\sqrt{x}} + 1) \, dx$

54. $\displaystyle \int \frac{1}{x^2} e^{1/x} \sec (1 + e^{1/x}) \tan (1 + e^{1/x}) \, dx$

55. $\displaystyle \int \frac{dx}{x \ln x}$ **56.** $\displaystyle \int \frac{\ln \sqrt{t}}{t} \, dt$

57. $\displaystyle \int \frac{dz}{1 + e^z}$ **58.** $\displaystyle \int \frac{dx}{x\sqrt{x^4 - 1}}$

59. $\displaystyle \int \frac{5}{9 + 4r^2} \, dr$ **60.** $\displaystyle \int \frac{1}{\sqrt{e^{2\theta} - 1}} \, d\theta$

61. $\int \dfrac{e^{\sin^{-1}x}\,dx}{\sqrt{1-x^2}}$

62. $\int \dfrac{e^{\cos^{-1}x}\,dx}{\sqrt{1-x^2}}$

63. $\int \dfrac{(\sin^{-1}x)^2\,dx}{\sqrt{1-x^2}}$

64. $\int \dfrac{\sqrt{\tan^{-1}x}\,dx}{1+x^2}$

65. $\int \dfrac{dy}{(\tan^{-1}y)(1+y^2)}$

66. $\int \dfrac{dy}{(\sin^{-1}y)\sqrt{1-y^2}}$

If you do not know what substitution to make, try reducing the integral step by step, using a trial substitution to simplify the integral a bit and then another to simplify it some more. You will see what we mean if you try the sequences of substitutions in Exercises 67 and 68.

67. $\int \dfrac{18\tan^2 x\sec^2 x}{(2+\tan^3 x)^2}\,dx$

    a. $u = \tan x$, followed by $v = u^3$, then by $w = 2 + v$

    b. $u = \tan^3 x$, followed by $v = 2 + u$

    c. $u = 2 + \tan^3 x$

68. $\int \sqrt{1+\sin^2(x-1)}\,\sin(x-1)\cos(x-1)\,dx$

    a. $u = x - 1$, followed by $v = \sin u$, then by $w = 1 + v^2$

    b. $u = \sin(x - 1)$, followed by $v = 1 + u^2$

    c. $u = 1 + \sin^2(x - 1)$

Evaluate the integrals in Exercises 69 and 70.

69. $\int \dfrac{(2r-1)\cos\sqrt{3(2r-1)^2+6}}{\sqrt{3(2r-1)^2+6}}\,dr$

70. $\int \dfrac{\sin\sqrt{\theta}}{\sqrt{\theta}\cos^3\sqrt{\theta}}\,d\theta$

71. Find the integral of $\cot x$ using a substitution like that in Example 7c.

72. Find the integral of $\csc x$ by multiplying by an appropriate form equal to 1, as in Example 8b.

Initial Value Problems

Solve the initial value problems in Exercises 73–78.

73. $\dfrac{ds}{dt} = 12t\,(3t^2 - 1)^3, \quad s(1) = 3$

74. $\dfrac{dy}{dx} = 4x\,(x^2 + 8)^{-1/3}, \quad y(0) = 0$

75. $\dfrac{ds}{dt} = 8\sin^2\!\left(t + \dfrac{\pi}{12}\right), \quad s(0) = 8$

76. $\dfrac{dr}{d\theta} = 3\cos^2\!\left(\dfrac{\pi}{4} - \theta\right), \quad r(0) = \dfrac{\pi}{8}$

77. $\dfrac{d^2s}{dt^2} = -4\sin\!\left(2t - \dfrac{\pi}{2}\right), \quad s'(0) = 100, \quad s(0) = 0$

78. $\dfrac{d^2y}{dx^2} = 4\sec^2 2x\tan 2x, \quad y'(0) = 4, \quad y(0) = -1$

79. The velocity of a particle moving back and forth on a line is $v = ds/dt = 6\sin 2t$ m/sec for all t. If $s = 0$ when $t = 0$, find the value of s when $t = \pi/2$ sec.

80. The acceleration of a particle moving back and forth on a line is $a = d^2s/dt^2 = \pi^2\cos\pi t$ m/sec^2 for all t. If $s = 0$ and $v = 8$ m/sec when $t = 0$, find s when $t = 1$ sec.

5.6   Definite Integral Substitutions and the Area Between Curves

There are two methods for evaluating a definite integral by substitution. One method is to find an antiderivative using substitution and then to evaluate the definite integral by applying the Evaluation Theorem. The other method extends the process of substitution directly to *definite* integrals by changing the limits of integration. We will use the new formula that we introduce here to compute the area between two curves.

The Substitution Formula

The following formula shows how the limits of integration change when we apply a substitution to an integral.

> **THEOREM 7—Substitution in Definite Integrals**
> If g' is continuous on the interval $[a, b]$ and f is continuous on the range of $g(x) = u$, then
> $$\int_a^b f(g(x)) \cdot g'(x)\,dx = \int_{g(a)}^{g(b)} f(u)\,du.$$

Proof   Let F denote any antiderivative of f. Then,

$$\int_a^b f(g(x)) \cdot g'(x)\, dx = F(g(x)) \Big]_{x=a}^{x=b}$$

$$\frac{d}{dx} F(g(x))$$
$$= F'(g(x))g'(x)$$
$$= f(g(x))g'(x)$$

$$= F(g(b)) - F(g(a))$$

$$= F(u) \Big]_{u=g(a)}^{u=g(b)}$$

$$= \int_{g(a)}^{g(b)} f(u)\, du. \qquad \text{Fundamental Theorem, Part 2} \qquad \blacksquare$$

To use Theorem 7, make the same u-substitution $u = g(x)$ and $du = g'(x)\, dx$ that you would use to evaluate the corresponding indefinite integral. Then integrate the transformed integral with respect to u from the value $g(a)$ (the value of u at $x = a$) to the value $g(b)$ (the value of u at $x = b$).

EXAMPLE 1   Evaluate $\displaystyle\int_{-1}^{1} 3x^2 \sqrt{x^3 + 1}\, dx$.

Solution   We will show how to evaluate the integral using Theorem 7, and how to evaluate it using the original limits of integration.

Method 1:   Transform the integral and evaluate the transformed integral with the transformed limits given in Theorem 7.

$$\int_{-1}^{1} 3x^2 \sqrt{x^3 + 1}\, dx \qquad \begin{array}{l} \text{Let } u = x^3 + 1,\, du = 3x^2\, dx. \\ \text{When } x = -1,\, u = (-1)^3 + 1 = 0. \\ \text{When } x = 1,\, u = (1)^3 + 1 = 2. \end{array}$$

$$= \int_0^2 \sqrt{u}\, du$$

$$= \frac{2}{3} u^{3/2} \Big]_0^2 \qquad \text{Evaluate the new definite integral.}$$

$$= \frac{2}{3}\left[2^{3/2} - 0^{3/2} \right] = \frac{2}{3}\left[2\sqrt{2} \right] = \frac{4\sqrt{2}}{3}$$

Method 2:   Transform the integral as an indefinite integral, integrate, change back to x, and use the original x-limits.

$$\int 3x^2 \sqrt{x^3 + 1}\, dx = \int \sqrt{u}\, du \qquad \text{Let } u = x^3 + 1,\, du = 3x^2\, dx.$$

$$= \frac{2}{3} u^{3/2} + C \qquad \text{Integrate with respect to } u.$$

$$= \frac{2}{3}(x^3 + 1)^{3/2} + C \qquad \text{Replace } u \text{ by } x^3 + 1.$$

$$\int_{-1}^{1} 3x^2 \sqrt{x^3 + 1}\, dx = \frac{2}{3}(x^3 + 1)^{3/2} \Big]_{-1}^{1} \qquad \begin{array}{l} \text{Use the integral just found, with} \\ \text{limits of integration for } x. \end{array}$$

$$= \frac{2}{3}\left[((1)^3 + 1)^{3/2} - ((-1)^3 + 1)^{3/2} \right]$$

$$= \frac{2}{3}\left[2^{3/2} - 0^{3/2} \right] = \frac{2}{3}\left[2\sqrt{2} \right] = \frac{4\sqrt{2}}{3}$$

Which method is better—evaluating the transformed definite integral with transformed limits using Theorem 7, or transforming the integral, integrating, and transforming back to use the original limits of integration? In Example 1, the first method seems easier, but that is not always the case. Generally, it is best to know both methods and to use whichever one seems better at the time.

EXAMPLE 2   We use the method of transforming the limits of integration.

(a) $\displaystyle\int_{\pi/4}^{\pi/2} \cot\theta\,\csc^2\theta\,d\theta = \int_1^0 u\cdot(-du)$ Let $u = \cot\theta$, $du = -\csc^2\theta\,d\theta$,
$-du = \csc^2\theta\,d\theta$.
When $\theta = \pi/4$, $u = \cot(\pi/4) = 1$.
When $\theta = \pi/2$, $u = \cot(\pi/2) = 0$.

$$= -\int_1^0 u\,du$$

$$= -\left[\frac{u^2}{2}\right]_1^0$$

$$= -\left[\frac{(0)^2}{2} - \frac{(1)^2}{2}\right] = \frac{1}{2}$$

(b) $\displaystyle\int_{-\pi/4}^{\pi/4} \tan x\,dx = \int_{-\pi/4}^{\pi/4} \frac{\sin x}{\cos x}\,dx$

$$= -\int_{\sqrt{2}/2}^{\sqrt{2}/2} \frac{du}{u}$$ Let $u = \cos x$, $du = -\sin x\,dx$.
When $x = -\pi/4$, $u = \sqrt{2}/2$.
When $x = \pi/4$, $u = \sqrt{2}/2$.

$$= -\ln|u|\Big]_{\sqrt{2}/2}^{\sqrt{2}/2} = 0$$ Integrate, zero width interval   ■

Definite Integrals of Symmetric Functions

The Substitution Formula in Theorem 7 simplifies the calculation of definite integrals of even and odd functions (Section 1.1) over a symmetric interval $[-a, a]$ (Figure 5.24).

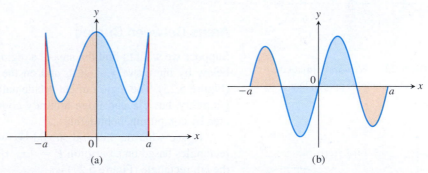

FIGURE 5.24   (a) For f an even function, the integral from $-a$ to a is twice the integral from 0 to a. (b) For f an odd function, the integral from $-a$ to a equals 0.

THEOREM 8   Let f be continuous on the symmetric interval $[-a, a]$.

(a)   If f is even, then $\displaystyle\int_{-a}^{a} f(x)\,dx = 2\int_0^a f(x)\,dx$.

(b)   If f is odd, then $\displaystyle\int_{-a}^{a} f(x)\,dx = 0$.

Proof of Part (a)

$$\int_{-a}^{a} f(x)\,dx = \int_{-a}^{0} f(x)\,dx + \int_{0}^{a} f(x)\,dx \qquad \text{Additivity Rule for Definite Integrals}$$

$$= -\int_{0}^{-a} f(x)\,dx + \int_{0}^{a} f(x)\,dx \qquad \text{Order of Integration Rule}$$

$$= -\int_{0}^{a} f(-u)(-du) + \int_{0}^{a} f(x)\,dx \qquad \begin{array}{l}\text{Let } u = -x, du = -dx. \\ \text{When } x = 0, u = 0. \\ \text{When } x = -a, u = a.\end{array}$$

$$= \int_{0}^{a} f(-u)\,du + \int_{0}^{a} f(x)\,dx$$

$$= \int_{0}^{a} f(u)\,du + \int_{0}^{a} f(x)\,dx \qquad \begin{array}{l}f \text{ is even, so} \\ f(-u) = f(u).\end{array}$$

$$= 2\int_{0}^{a} f(x)\,dx$$

The proof of part (b) is entirely similar and you are asked to give it in Exercise 116. ∎

EXAMPLE 3   Evaluate $\displaystyle\int_{-2}^{2} (x^4 - 4x^2 + 6)\,dx$.

Solution   Since $f(x) = x^4 - 4x^2 + 6$ satisfies $f(-x) = f(x)$, it is even on the symmetric interval $[-2, 2]$, so

$$\int_{-2}^{2} (x^4 - 4x^2 + 6)\,dx = 2\int_{0}^{2} (x^4 - 4x^2 + 6)\,dx$$

$$= 2\left[\frac{x^5}{5} - \frac{4}{3}x^3 + 6x\right]_{0}^{2}$$

$$= 2\left(\frac{32}{5} - \frac{32}{3} + 12\right) = \frac{232}{15}. \quad\blacksquare$$

Areas Between Curves

Suppose we want to find the area of a region that is bounded above by the curve $y = f(x)$, below by the curve $y = g(x)$, and on the left and right by the lines $x = a$ and $x = b$ (Figure 5.25). The region might accidentally have a shape whose area we could find with geometry, but if f and g are arbitrary continuous functions, we usually have to find the area by computing an integral.

To see what the integral should be, we first approximate the region with n vertical rectangles based on a partition $P = \{x_0, x_1, \ldots, x_n\}$ of $[a, b]$ (Figure 5.26). The area of the kth rectangle (Figure 5.27) is

$$\Delta A_k = \text{height} \times \text{width} = [f(c_k) - g(c_k)]\,\Delta x_k.$$

We then approximate the area of the region by adding the areas of the n rectangles:

$$A \approx \sum_{k=1}^{n} \Delta A_k = \sum_{k=1}^{n} [f(c_k) - g(c_k)]\,\Delta x_k. \qquad \text{Riemann sum}$$

As $\|P\| \to 0$, the sums on the right approach the limit $\int_{a}^{b} [f(x) - g(x)]\,dx$ because f and g are continuous. The area of the region is defined to be the value of this integral. That is,

$$A = \lim_{\|P\| \to 0} \sum_{k=1}^{n} [f(c_k) - g(c_k)]\,\Delta x_k = \int_{a}^{b} [f(x) - g(x)]\,dx.$$

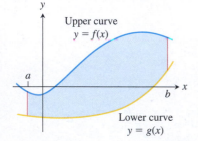

FIGURE 5.25   The region between the curves $y = f(x)$ and $y = g(x)$ and the lines $x = a$ and $x = b$.

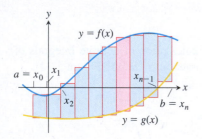

FIGURE 5.26   We approximate the region with rectangles perpendicular to the x-axis.

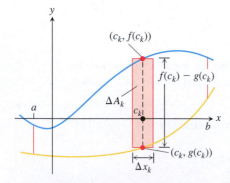

FIGURE 5.27   The area ΔA_k of the kth rectangle is the product of its height, $f(c_k) - g(c_k)$, and its width, Δx_k.

When applying this definition it is usually helpful to graph the curves. The graph reveals which curve is the upper curve f and which is the lower curve g. It also helps you find the limits of integration if they are not given. You may need to find where the curves intersect to determine the limits of integration, and this may involve solving the equation $f(x) = g(x)$ for values of x. Then you can integrate the function $f - g$ for the area between the intersections.

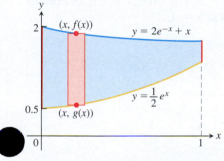

FIGURE 5.28 The region in Example 4 with a typical approximating rectangle.

EXAMPLE 4 Find the area of the region bounded above by the curve $y = 2e^{-x} + x$, below by the curve $y = e^x/2$, on the left by $x = 0$, and on the right by $x = 1$.

Solution Figure 5.28 displays the graphs of the curves and the region whose area we want to find. The area between the curves over the interval $0 \leq x \leq 1$ is

$$A = \int_0^1 \left[(2e^{-x} + x) - \frac{1}{2}e^x\right] dx = \left[-2e^{-x} + \frac{1}{2}x^2 - \frac{1}{2}e^x\right]_0^1$$

$$= \left(-2e^{-1} + \frac{1}{2} - \frac{1}{2}e\right) - \left(-2 + 0 - \frac{1}{2}\right)$$

$$= 3 - \frac{2}{e} - \frac{e}{2} \approx 0.9051. \qquad \blacksquare$$

EXAMPLE 5 Find the area of the region enclosed by the parabola $y = 2 - x^2$ and the line $y = -x$.

Solution First we sketch the two curves (Figure 5.29). The limits of integration are found by solving $y = 2 - x^2$ and $y = -x$ simultaneously for x.

$$2 - x^2 = -x \qquad \text{Equate } f(x) \text{ and } g(x).$$
$$x^2 - x - 2 = 0 \qquad \text{Rewrite.}$$
$$(x + 1)(x - 2) = 0 \qquad \text{Factor.}$$
$$x = -1, \qquad x = 2. \qquad \text{Solve.}$$

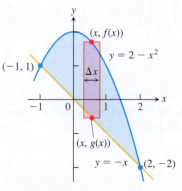

FIGURE 5.29 The region in Example 5 with a typical approximating rectangle from a Riemann sum.

The region runs from $x = -1$ to $x = 2$. The limits of integration are $a = -1$, $b = 2$. The area between the curves is

$$A = \int_a^b [f(x) - g(x)] \, dx = \int_{-1}^2 [(2 - x^2) - (-x)] \, dx$$

$$= \int_{-1}^2 (2 + x - x^2) \, dx = \left[2x + \frac{x^2}{2} - \frac{x^3}{3}\right]_{-1}^2$$

$$= \left(4 + \frac{4}{2} - \frac{8}{3}\right) - \left(-2 + \frac{1}{2} + \frac{1}{3}\right) = \frac{9}{2}. \qquad \blacksquare$$

If the formula for a bounding curve changes at one or more points, we subdivide the region into subregions that correspond to the formula changes and apply the formula for the area between curves to each subregion.

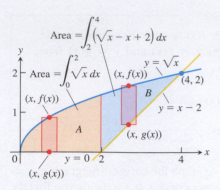

FIGURE 5.30 When the formula for a bounding curve changes, the area integral changes to become the sum of integrals to match, one integral for each of the shaded regions shown here for Example 6.

EXAMPLE 6 Find the area of the region in the first quadrant that is bounded above by $y = \sqrt{x}$ and below by the x-axis and the line $y = x - 2$.

Solution The sketch (Figure 5.30) shows that the region's upper boundary is the graph of $f(x) = \sqrt{x}$. The lower boundary changes from $g(x) = 0$ for $0 \le x \le 2$ to $g(x) = x - 2$ for $2 \le x \le 4$ (both formulas agree at $x = 2$). We subdivide the region at $x = 2$ into subregions A and B, shown in Figure 5.30.

The limits of integration for region A are $a = 0$ and $b = 2$. The left-hand limit for region B is $a = 2$. To find the right-hand limit, we solve the equations $y = \sqrt{x}$ and $y = x - 2$ simultaneously for x:

$$\sqrt{x} = x - 2 \qquad \text{Equate } f(x) \text{ and } g(x).$$
$$x = (x - 2)^2 = x^2 - 4x + 4 \qquad \text{Square both sides.}$$
$$x^2 - 5x + 4 = 0 \qquad \text{Rewrite.}$$
$$(x - 1)(x - 4) = 0 \qquad \text{Factor.}$$
$$x = 1, \quad x = 4. \qquad \text{Solve.}$$

Only the value $x = 4$ satisfies the equation $\sqrt{x} = x - 2$. The value $x = 1$ is an extraneous root introduced by squaring. The right-hand limit is $b = 4$.

For $0 \le x \le 2$: $\quad f(x) - g(x) = \sqrt{x} - 0 = \sqrt{x}$
For $2 \le x \le 4$: $\quad f(x) - g(x) = \sqrt{x} - (x - 2) = \sqrt{x} - x + 2$

We add the areas of subregions A and B to find the total area:

$$\text{Total area} = \underbrace{\int_0^2 \sqrt{x}\, dx}_{\text{area of } A} + \underbrace{\int_2^4 \left(\sqrt{x} - x + 2 \right) dx}_{\text{area of } B}$$

$$= \left[\frac{2}{3} x^{3/2} \right]_0^2 + \left[\frac{2}{3} x^{3/2} - \frac{x^2}{2} + 2x \right]_2^4$$

$$= \frac{2}{3}(2)^{3/2} - 0 + \left(\frac{2}{3}(4)^{3/2} - 8 + 8 \right) - \left(\frac{2}{3}(2)^{3/2} - 2 + 4 \right)$$

$$= \frac{2}{3}(8) - 2 = \frac{10}{3}.$$

Integration with Respect to y

If a region's bounding curves are described by functions of y, the approximating rectangles are horizontal instead of vertical and the basic formula has y in place of x.

For regions like these:

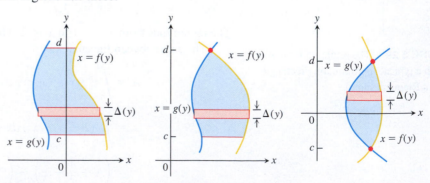

use the formula

$$A = \int_c^d [\, f(y) - g(y) \,] \, dy.$$

In this equation f always denotes the right-hand curve and g the left-hand curve, so $f(y) - g(y)$ is nonnegative.

EXAMPLE 7   Find the area of the region in Example 6 by integrating with respect to y.

Solution   We first sketch the region and a typical *horizontal* rectangle based on a partition of an interval of y-values (Figure 5.31). The region's right-hand boundary is the line $x = y + 2$, so $f(y) = y + 2$. The left-hand boundary is the curve $x = y^2$, so $g(y) = y^2$. The lower limit of integration is $y = 0$. We find the upper limit by solving $x = y + 2$ and $x = y^2$ simultaneously for y:

$$y + 2 = y^2 \qquad \text{Equate } f(y) = y + 2 \text{ and } g(y) = y^2.$$
$$y^2 - y - 2 = 0 \qquad \text{Rewrite.}$$
$$(y + 1)(y - 2) = 0 \qquad \text{Factor.}$$
$$y = -1, \qquad y = 2 \qquad \text{Solve.}$$

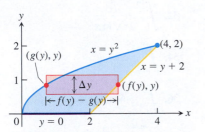

FIGURE 5.31   It takes two integrations to find the area of this region if we integrate with respect to x. It takes only one if we integrate with respect to y (Example 7).

The upper limit of integration is $b = 2$. (The value $y = -1$ gives a point of intersection *below* the x-axis.)

The area of the region is

$$A = \int_c^d [\, f(y) - g(y)\,]\, dy = \int_0^2 [\, y + 2 - y^2\,]\, dy$$

$$= \int_0^2 [\, 2 + y - y^2\,]\, dy$$

$$= \left[2y + \frac{y^2}{2} - \frac{y^3}{3} \right]_0^2$$

$$= 4 + \frac{4}{2} - \frac{8}{3} = \frac{10}{3}.$$

This is the result of Example 6, found with less work.   ∎

Although it was easier to find the area in Example 6 by integrating with respect to y rather than x (just as we did in Example 7), there is an easier way yet. Looking at Figure 5.32, we see that the area we want is the area between the curve $y = \sqrt{x}$ and the x-axis for $0 \le x \le 4$, *minus* the area of an isosceles triangle of base and height equal to 2. So by combining calculus with some geometry, we find

$$\text{Area} = \int_0^4 \sqrt{x}\, dx - \frac{1}{2}(2)(2)$$

$$= \frac{2}{3} x^{3/2} \Big]_0^4 - 2$$

$$= \frac{2}{3}(8) - 0 - 2 = \frac{10}{3}.$$

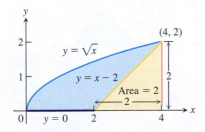

FIGURE 5.32   The area of the blue region is the area under the parabola $y = \sqrt{x}$ minus the area of the triangle.

EXERCISES **5.6**

Evaluating Definite Integrals

Use the Substitution Formula in Theorem 7 to evaluate the integrals in Exercises 1–48.

1. a. $\displaystyle\int_0^3 \sqrt{y + 1}\, dy$ **b.** $\displaystyle\int_{-1}^0 \sqrt{y + 1}\, dy$

2. a. $\displaystyle\int_0^1 r\sqrt{1 - r^2}\, dr$ **b.** $\displaystyle\int_{-1}^1 r\sqrt{1 - r^2}\, dr$

3. a. $\displaystyle\int_0^{\pi/4} \tan x \sec^2 x\, dx$ **b.** $\displaystyle\int_{-\pi/4}^0 \tan x \sec^2 x\, dx$

4. a. $\displaystyle\int_0^\pi 3\cos^2 x \sin x \, dx$ **b.** $\displaystyle\int_{2\pi}^{3\pi} 3\cos^2 x \sin x \, dx$

5. a. $\displaystyle\int_0^1 t^3(1+t^4)^3 \, dt$ **b.** $\displaystyle\int_{-1}^1 t^3(1+t^4)^3 \, dt$

6. a. $\displaystyle\int_0^{\sqrt 7} t(t^2+1)^{1/3} \, dt$ **b.** $\displaystyle\int_{-\sqrt 7}^0 t(t^2+1)^{1/3} \, dt$

7. a. $\displaystyle\int_{-1}^1 \frac{5r}{(4+r^2)^2} dr$ **b.** $\displaystyle\int_0^1 \frac{5r}{(4+r^2)^2} dr$

8. a. $\displaystyle\int_0^1 \frac{10\sqrt v}{(1+v^{3/2})^2} dv$ **b.** $\displaystyle\int_1^4 \frac{10\sqrt v}{(1+v^{3/2})^2} dv$

9. a. $\displaystyle\int_0^{\sqrt 3} \frac{4x}{\sqrt{x^2+1}} dx$ **b.** $\displaystyle\int_{-\sqrt 3}^{\sqrt 3} \frac{4x}{\sqrt{x^2+1}} dx$

10. a. $\displaystyle\int_0^1 \frac{x^3}{\sqrt{x^4+9}} dx$ **b.** $\displaystyle\int_{-1}^0 \frac{x^3}{\sqrt{x^4+9}} dx$

11. a. $\displaystyle\int_0^1 t\sqrt{4+5t} \, dt$ **b.** $\displaystyle\int_1^9 t\sqrt{4+5t} \, dt$

12. a. $\displaystyle\int_0^{\pi/6} (1-\cos 3t)\sin 3t \, dt$

    b. $\displaystyle\int_{\pi/6}^{\pi/3} (1-\cos 3t)\sin 3t \, dt$

13. a. $\displaystyle\int_0^{2\pi} \frac{\cos z}{\sqrt{4+3\sin z}} dz$ **b.** $\displaystyle\int_{-\pi}^{\pi} \frac{\cos z}{\sqrt{4+3\sin z}} dz$

14. a. $\displaystyle\int_{-\pi/2}^0 \left(2+\tan\frac t2\right)\sec^2\frac t2 dt$

    b. $\displaystyle\int_{-\pi/2}^{\pi/2} \left(2+\tan\frac t2\right)\sec^2\frac t2 \, dt$

15. $\displaystyle\int_0^1 \sqrt{t^5+2t}(5t^4+2) \, dt$ **16.** $\displaystyle\int_1^4 \frac{dy}{2\sqrt y(1+\sqrt y)^2}$

17. $\displaystyle\int_0^{\pi/6} \cos^{-3} 2\theta \sin 2\theta \, d\theta$ **18.** $\displaystyle\int_{\pi}^{3\pi/2} \cot^5\left(\frac\theta6\right)\sec^2\left(\frac\theta6\right) d\theta$

19. $\displaystyle\int_0^{\pi} 5(5-4\cos t)^{1/4}\sin t \, dt$ **20.** $\displaystyle\int_0^{\pi/4} (1-\sin 2t)^{3/2}\cos 2t \, dt$

21. $\displaystyle\int_0^1 (4y-y^2+4y^3+1)^{-2/3}(12y^2-2y+4) \, dy$

22. $\displaystyle\int_0^1 (y^3+6y^2-12y+9)^{-1/2}(y^2+4y-4) \, dy$

23. $\displaystyle\int_0^{\sqrt[3]{\pi^2}} \sqrt\theta \cos^2(\theta^{3/2}) \, d\theta$ **24.** $\displaystyle\int_{-1}^{-1/2} t^{-2}\sin^2\left(1+\frac1t\right) dt$

25. $\displaystyle\int_0^{\pi/4} (1+e^{\tan\theta})\sec^2\theta \, d\theta$ **26.** $\displaystyle\int_{\pi/4}^{\pi/2} (1+e^{\cot\theta})\csc^2\theta \, d\theta$

27. $\displaystyle\int_0^{\pi} \frac{\sin t}{2-\cos t} dt$ **28.** $\displaystyle\int_0^{\pi/3} \frac{4\sin\theta}{1-4\cos\theta} d\theta$

29. $\displaystyle\int_1^2 \frac{2\ln x}{x} dx$ **30.** $\displaystyle\int_2^4 \frac{dx}{x\ln x}$

31. $\displaystyle\int_2^4 \frac{dx}{x(\ln x)^2}$ **32.** $\displaystyle\int_2^{16} \frac{dx}{2x\sqrt{\ln x}}$

33. $\displaystyle\int_0^{\pi/2} \tan\frac x2 dx$ **34.** $\displaystyle\int_{\pi/4}^{\pi/2} \cot t \, dt$

35. $\displaystyle\int_0^{\pi/3} \tan^2\theta \cos\theta \, d\theta$ **36.** $\displaystyle\int_0^{\pi/12} 6\tan 3x \, dx$

37. $\displaystyle\int_{-\pi/2}^{\pi/2} \frac{2\cos\theta \, d\theta}{1+(\sin\theta)^2}$ **38.** $\displaystyle\int_{\pi/6}^{\pi/4} \frac{\csc^2 x \, dx}{1+(\cot x)^2}$

39. $\displaystyle\int_0^{\ln\sqrt 3} \frac{e^x \, dx}{1+e^{2x}}$ **40.** $\displaystyle\int_1^{e^{\pi/4}} \frac{4 \, dt}{t(1+\ln^2 t)}$

41. $\displaystyle\int_0^1 \frac{4 \, ds}{\sqrt{4-s^2}}$ **42.** $\displaystyle\int_0^{\sqrt[3]2/4} \frac{ds}{\sqrt{9-4s^2}}$

43. $\displaystyle\int_{\sqrt 2}^2 \frac{\sec^2(\sec^{-1}x) \, dx}{x\sqrt{x^2-1}}$ **44.** $\displaystyle\int_{2/\sqrt 3}^2 \frac{\cos(\sec^{-1}x) \, dx}{x\sqrt{x^2-1}}$

45. $\displaystyle\int_{-1}^{-\sqrt 2/2} \frac{dy}{y\sqrt{4y^2-1}}$ **46.** $\displaystyle\int_0^3 \frac{y \, dy}{\sqrt{5y+1}}$

47. $\displaystyle\int_0^1 \frac{\tan^{-1}x}{1+x^2} dx$ **48.** $\displaystyle\int_{-\sqrt 3}^{1/\sqrt 3} \frac{\cos(\tan^{-1}3x)}{1+9x^2} dx$

Area

Find the total areas of the shaded regions in Exercises 49–64.

49.

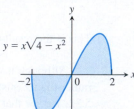

$y = x\sqrt{4-x^2}$

50.

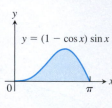

$y = (1-\cos x)\sin x$

51.

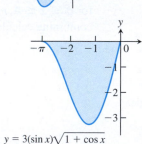

$y = 3(\sin x)\sqrt{1+\cos x}$

52. $y = \dfrac{\pi}{2}(\cos x)(\sin(\pi+\pi\sin x))$

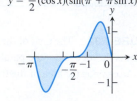

53.

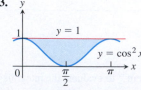

$y = 1$

$y = \cos^2 x$

54.

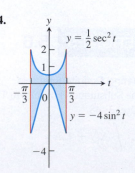

$y = \dfrac12 \sec^2 t$

$y = -4\sin^2 t$

55.

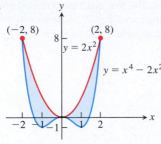

NOT TO SCALE

56.

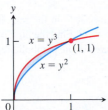

57.

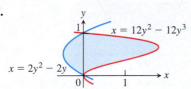

58.

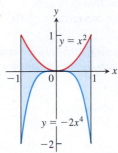

59.

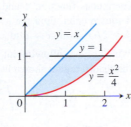

60.

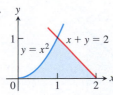

61.   **62.**

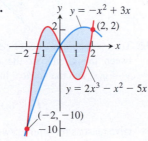

63.   **64.**

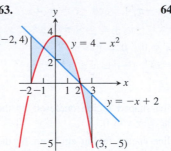

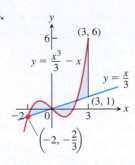

Find the areas of the regions enclosed by the lines and curves in Exercises 65–74.

65. $y = x^2 - 2$ and $y = 2$ **66.** $y = 2x - x^2$ and $y = -3$

67. $y = x^4$ and $y = 8x$ **68.** $y = x^2 - 2x$ and $y = x$

69. $y = x^2$ and $y = -x^2 + 4x$

70. $y = 7 - 2x^2$ and $y = x^2 + 4$

71. $y = x^4 - 4x^2 + 4$ and $y = x^2$

72. $y = x\sqrt{a^2 - x^2}$, $a > 0$, and $y = 0$

73. $y = \sqrt{|x|}$ and $5y = x + 6$ (How many intersection points are there?)

74. $y = |x^2 - 4|$ and $y = (x^2/2) + 4$

Find the areas of the regions enclosed by the lines and curves in Exercises 75–82.

75. $x = 2y^2$, $x = 0$, and $y = 3$

76. $x = y^2$ and $x = y + 2$

77. $y^2 - 4x = 4$ and $4x - y = 16$

78. $x - y^2 = 0$ and $x + 2y^2 = 3$

79. $x + y^2 = 0$ and $x + 3y^2 = 2$

80. $x - y^{2/3} = 0$ and $x + y^4 = 2$

81. $x = y^2 - 1$ and $x = |y|\sqrt{1 - y^2}$

82. $x = y^3 - y^2$ and $x = 2y$

Find the areas of the regions enclosed by the curves in Exercises 83–86.

83. $4x^2 + y = 4$ and $x^4 - y = 1$

84. $x^3 - y = 0$ and $3x^2 - y = 4$

85. $x + 4y^2 = 4$ and $x + y^4 = 1$, for $x \geq 0$

86. $x + y^2 = 3$ and $4x + y^2 = 0$

Find the areas of the regions enclosed by the lines and curves in Exercises 87–94.

87. $y = 2 \sin x$ and $y = \sin 2x$, $0 \leq x \leq \pi$

88. $y = 8 \cos x$ and $y = \sec^2 x$, $-\pi/3 \leq x \leq \pi/3$

89. $y = \cos(\pi x/2)$ and $y = 1 - x^2$

90. $y = \sin(\pi x/2)$ and $y = x$

91. $y = \sec^2 x,$ $y = \tan^2 x,$ $x = -\pi/4,$ and $x = \pi/4$

92. $x = \tan^2 y$ and $x = -\tan^2 y,$ $-\pi/4 \le y \le \pi/4$

93. $x = 3 \sin y \sqrt{\cos y}$ and $x = 0,$ $0 \le y \le \pi/2$

94. $y = \sec^2 (\pi x/3)$ and $y = x^{1/3},$ $-1 \le x \le 1$

Area Between Curves

95. Find the area of the propeller-shaped region enclosed by the curve $x - y^3 = 0$ and the line $x - y = 0$.

96. Find the area of the propeller-shaped region enclosed by the curves $x - y^{1/3} = 0$ and $x - y^{1/5} = 0$.

97. Find the area of the region in the first quadrant bounded by the line $y = x$, the line $x = 2$, the curve $y = 1/x^2$, and the x-axis.

98. Find the area of the "triangular" region in the first quadrant bounded on the left by the y-axis and on the right by the curves $y = \sin x$ and $y = \cos x$.

99. Find the area between the curves $y = \ln x$ and $y = \ln 2x$ from $x = 1$ to $x = 5$.

100. Find the area between the curve $y = \tan x$ and the x-axis from $x = -\pi/4$ to $x = \pi/3$.

101. Find the area of the "triangular" region in the first quadrant that is bounded above by the curve $y = e^{2x}$, below by the curve $y = e^x$, and on the right by the line $x = \ln 3$.

102. Find the area of the "triangular" region in the first quadrant that is bounded above by the curve $y = e^{x/2}$, below by the curve $y = e^{-x/2}$, and on the right by the line $x = 2 \ln 2$.

103. Find the area of the region between the curve $y = 2x/(1 + x^2)$ and the interval $-2 \le x \le 2$ of the x-axis.

104. Find the area of the region between the curve $y = 2^{1-x}$ and the interval $-1 \le x \le 1$ of the x-axis.

105. The region bounded below by the parabola $y = x^2$ and above by the line $y = 4$ is to be partitioned into two subsections of equal area by cutting across it with the horizontal line $y = c$.

   a. Sketch the region and draw a line $y = c$ across it that looks about right. In terms of c, what are the coordinates of the points where the line and parabola intersect? Add them to your figure.

   b. Find c by integrating with respect to y. (This puts c in the limits of integration.)

   c. Find c by integrating with respect to x. (This puts c into the integrand as well.)

106. Find the area of the region between the curve $y = 3 - x^2$ and the line $y = -1$ by integrating with respect to   **a.** x,   **b.** y.

107. Find the area of the region in the first quadrant bounded on the left by the y-axis, below by the line $y = x/4$, above left by the curve $y = 1 + \sqrt{x}$, and above right by the curve $y = 2/\sqrt{x}$.

108. Find the area of the region in the first quadrant bounded on the left by the y-axis, below by the curve $x = 2\sqrt{y}$, above left by the curve $x = (y - 1)^2$, and above right by the line $x = 3 - y$.

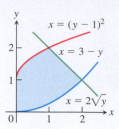

109. The figure here shows triangle AOC inscribed in the region cut from the parabola $y = x^2$ by the line $y = a^2$. Find the limit of the ratio of the area of the triangle to the area of the parabolic region as a approaches zero.

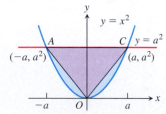

110. Suppose the area of the region between the graph of a positive continuous function f and the x-axis from $x = a$ to $x = b$ is 4 square units. Find the area between the curves $y = f(x)$ and $y = 2f(x)$ from $x = a$ to $x = b$.

111. Which of the following integrals, if either, calculates the area of the shaded region shown here? Give reasons for your answer.

   a. $\displaystyle \int_{-1}^{1} (x - (-x)) \, dx = \int_{-1}^{1} 2x \, dx$

   b. $\displaystyle \int_{-1}^{1} (-x - (x)) \, dx = \int_{-1}^{1} -2x \, dx$

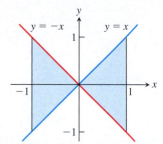

112. True, sometimes true, or never true? The area of the region between the graphs of the continuous functions $y = f(x)$ and $y = g(x)$ and the vertical lines $x = a$ and $x = b$ ($a < b$) is

$$\int_{a}^{b} [f(x) - g(x)] \, dx.$$

Give reasons for your answer.

Theory and Examples

113. Suppose that $F(x)$ is an antiderivative of $f(x) = (\sin x)/x$, $x > 0$. Express

$$\int_1^3 \frac{\sin 2x}{x} dx$$

in terms of F.

114. Show that if f is continuous, then

$$\int_0^1 f(x) \, dx = \int_0^1 f(1 - x) \, dx.$$

115. Suppose that

$$\int_0^1 f(x) \, dx = 3.$$

Find

$$\int_{-1}^0 f(x) \, dx$$

if   **a.** f is odd,   **b.** f is even.

116. a. Show that if f is odd on $[-a, a]$, then

$$\int_{-a}^a f(x) \, dx = 0.$$

b. Test the result in part (a) with $f(x) = \sin x$ and $a = \pi/2$.

117. If f is a continuous function, find the value of the integral

$$I = \int_0^a \frac{f(x) \, dx}{f(x) + f(a - x)}$$

by making the substitution $u = a - x$ and adding the resulting integral to I.

118. By using a substitution, prove that for all positive numbers x and y,

$$\int_x^{xy} \frac{1}{t} dt = \int_1^y \frac{1}{t} dt.$$

The Shift Property for Definite Integrals   A basic property of definite integrals is their invariance under translation, as expressed by the equation

$$\int_a^b f(x) \, dx = \int_{a-c}^{b-c} f(x + c) \, dx. \qquad (1)$$

The equation holds whenever f is integrable and defined for the necessary values of x. For example in the accompanying figure, show that

$$\int_{-2}^{-1} (x + 2)^3 \, dx = \int_0^1 x^3 \, dx$$

because the areas of the shaded regions are congruent.

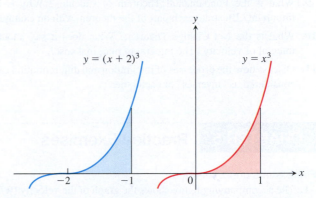

119. Use a substitution to verify Equation (1).

120. For each of the following functions, graph $f(x)$ over $[a, b]$ and $f(x + c)$ over $[a - c, b - c]$ to convince yourself that Equation (1) is reasonable.

a. $f(x) = x^2$, $a = 0$, $b = 1$, $c = 1$

b. $f(x) = \sin x$, $a = 0$, $b = \pi$, $c = \pi/2$

c. $f(x) = \sqrt{x - 4}$, $a = 4$, $b = 8$, $c = 5$

COMPUTER EXPLORATIONS

In Exercises 121–124, you will find the area between curves in the plane when you cannot find their points of intersection using simple algebra. Use a CAS to perform the following steps:

a. Plot the curves together to see what they look like and how many points of intersection they have.

b. Use the numerical equation solver in your CAS to find all the points of intersection.

c. Integrate $|f(x) - g(x)|$ over consecutive pairs of intersection values.

d. Sum together the integrals found in part (c).

121. $f(x) = \dfrac{x^3}{3} - \dfrac{x^2}{2} - 2x + \dfrac{1}{3}$, $g(x) = x - 1$

122. $f(x) = \dfrac{x^4}{2} - 3x^3 + 10$, $g(x) = 8 - 12x$

123. $f(x) = x + \sin(2x)$, $g(x) = x^3$

124. $f(x) = x^2 \cos x$, $g(x) = x^3 - x$

| CHAPTER 5 | **Questions to Guide Your Review** |

1. How can you sometimes estimate quantities like distance traveled, area, and average value with finite sums? Why might you want to do so?

2. What is sigma notation? What advantage does it offer? Give examples.

3. What is a Riemann sum? Why might you want to consider such a sum?

4. What is the norm of a partition of a closed interval?

5. What is the definite integral of a function f over a closed interval $[a, b]$? When can you be sure it exists?

6. What is the relation between definite integrals and area? Describe some other interpretations of definite integrals.

7. What is the average value of an integrable function over a closed interval? Must the function assume its average value? Explain.

8. Describe the rules for working with definite integrals (Table 5.6). Give examples.

9. What is the Fundamental Theorem of Calculus? Why is it so important? Illustrate each part of the theorem with an example.

10. What is the Net Change Theorem? What does it say about the integral of velocity? The integral of marginal cost?

11. Discuss how the processes of integration and differentiation can be considered as "inverses" of each other.

12. How does the Fundamental Theorem provide a solution to the initial value problem $dy/dx = f(x)$, $y(x_0) = y_0$, when f is continuous?

13. How is integration by substitution related to the Chain Rule?

14. How can you sometimes evaluate indefinite integrals by substitution? Give examples.

15. How does the method of substitution work for definite integrals? Give examples.

16. How do you define and calculate the area of the region between the graphs of two continuous functions? Give an example.

CHAPTER 5 Practice Exercises

Finite Sums and Estimates

1. The accompanying figure shows the graph of the velocity (ft/sec) of a model rocket for the first 8 sec after launch. The rocket accelerated straight up for the first 2 sec and then coasted to reach its maximum height at $t = 8$ sec.

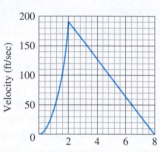

Time after launch (sec)

a. Assuming that the rocket was launched from ground level, about how high did it go? (This is the rocket in Section 3.3, Exercise 17, but you do not need to do Exercise 17 to do the exercise here.)

b. Sketch a graph of the rocket's height above ground as a function of time for $0 \le t \le 8$.

2. a. The accompanying figure shows the velocity (m/sec) of a body moving along the s-axis during the time interval from $t = 0$ to $t = 10$ sec. About how far did the body travel during those 10 sec?

b. Sketch a graph of s as a function of t for $0 \le t \le 10$, assuming $s(0) = 0$.

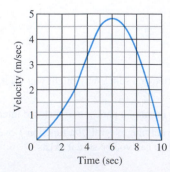

Time (sec)

3. Suppose that $\sum_{k=1}^{10} a_k = -2$ and $\sum_{k=1}^{10} b_k = 25$. Find the value of

a. $\sum_{k=1}^{10} \frac{a_k}{4}$

b. $\sum_{k=1}^{10} (b_k - 3a_k)$

c. $\sum_{k=1}^{10} (a_k + b_k - 1)$

d. $\sum_{k=1}^{10} \left(\frac{5}{2} - b_k\right)$

4. Suppose that $\sum_{k=1}^{20} a_k = 0$ and $\sum_{k=1}^{20} b_k = 7$. Find the values of

a. $\sum_{k=1}^{20} 3a_k$

b. $\sum_{k=1}^{20} (a_k + b_k)$

c. $\sum_{k=1}^{20} \left(\frac{1}{2} - \frac{2b_k}{7}\right)$

d. $\sum_{k=1}^{20} (a_k - 2)$

Definite Integrals

In Exercises 5–8, express each limit as a definite integral. Then evaluate the integral to find the value of the limit. In each case, P is a partition of the given interval and the numbers c_k are chosen from the subintervals of P.

5. $\lim_{\|P\| \to 0} \sum_{k=1}^{n} (2c_k - 1)^{-1/2} \Delta x_k$, where P is a partition of $[1, 5]$

6. $\lim_{\|P\| \to 0} \sum_{k=1}^{n} c_k(c_k^2 - 1)^{1/3} \Delta x_k$, where P is a partition of $[1, 3]$

7. $\lim_{\|P\| \to 0} \sum_{k=1}^{n} \left(\cos\left(\frac{c_k}{2}\right)\right) \Delta x_k$, where P is a partition of $[-\pi, 0]$

8. $\lim_{\|P\| \to 0} \sum_{k=1}^{n} (\sin c_k)(\cos c_k) \Delta x_k$, where P is a partition of $[0, \pi/2]$

9. If $\int_{-2}^{2} 3f(x)\, dx = 12$, $\int_{-2}^{5} f(x)\, dx = 6$, and $\int_{-2}^{5} g(x)\, dx = 2$, find the values of the following.

a. $\int_{-2}^{2} f(x)\, dx$

b. $\int_{2}^{5} f(x)\, dx$

c. $\int_{5}^{-2} g(x)\, dx$

d. $\int_{-2}^{5} (-\pi g(x))\, dx$

e. $\int_{-2}^{5} \left(\frac{f(x) + g(x)}{5}\right) dx$

10. If $\int_0^2 f(x)\, dx = \pi$, $\int_0^2 7g(x)\, dx = 7$, and $\int_0^1 g(x)\, dx = 2$, find the values of the following.

a. $\displaystyle\int_0^2 g(x)\, dx$

b. $\displaystyle\int_1^2 g(x)\, dx$

c. $\displaystyle\int_2^0 f(x)\, dx$

d. $\displaystyle\int_0^2 \sqrt{2}\, f(x)\, dx$

e. $\displaystyle\int_0^2 (g(x) - 3f(x))\, dx$

Area

In Exercises 11–14, find the total area of the region between the graph of f and the x-axis.

11. $f(x) = x^2 - 4x + 3, \quad 0 \le x \le 3$

12. $f(x) = 1 - (x^2/4), \quad -2 \le x \le 3$

13. $f(x) = 5 - 5x^{2/3}, \quad -1 \le x \le 8$

14. $f(x) = 1 - \sqrt{x}, \quad 0 \le x \le 4$

Find the areas of the regions enclosed by the curves and lines in Exercises 15–26.

15. $y = x, \quad y = 1/x^2, \quad x = 2$

16. $y = x, \quad y = 1/\sqrt{x}, \quad x = 2$

17. $\sqrt{x} + \sqrt{y} = 1, \quad x = 0, \quad y = 0$

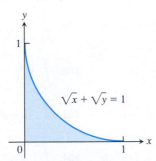

$\sqrt{x} + \sqrt{y} = 1$

18. $x^3 + \sqrt{y} = 1, \quad x = 0, \quad y = 0, \quad \text{for} \quad 0 \le x \le 1$

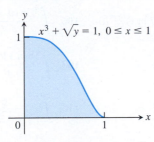

$x^3 + \sqrt{y} = 1, \ 0 \le x \le 1$

19. $x = 2y^2, \quad x = 0, \quad y = 3$ **20.** $x = 4 - y^2, \quad x = 0$

21. $y^2 = 4x, \quad y = 4x - 2$

22. $y^2 = 4x + 4, \quad y = 4x - 16$

23. $y = \sin x, \quad y = x, \quad 0 \le x \le \pi/4$

24. $y = |\sin x|, \quad y = 1, \quad -\pi/2 \le x \le \pi/2$

25. $y = 2 \sin x, \quad y = \sin 2x, \quad 0 \le x \le \pi$

26. $y = 8 \cos x, \quad y = \sec^2 x, \quad -\pi/3 \le x \le \pi/3$

27. Find the area of the "triangular" region bounded on the left by $x + y = 2$, on the right by $y = x^2$, and above by $y = 2$.

28. Find the area of the "triangular" region bounded on the left by $y = \sqrt{x}$, on the right by $y = 6 - x$, and below by $y = 1$.

29. Find the extreme values of $f(x) = x^3 - 3x^2$ and find the area of the region enclosed by the graph of f and the x-axis.

30. Find the area of the region cut from the first quadrant by the curve $x^{1/2} + y^{1/2} = a^{1/2}$.

31. Find the total area of the region enclosed by the curve $x = y^{2/3}$ and the lines $x = y$ and $y = -1$.

32. Find the total area of the region between the curves $y = \sin x$ and $y = \cos x$ for $0 \le x \le 3\pi/2$.

33. Find the area between the curve $y = 2(\ln x)/x$ and the x-axis from $x = 1$ to $x = e$.

34. a. Show that the area between the curve $y = 1/x$ and the x-axis from $x = 10$ to $x = 20$ is the same as the area between the curve and the x-axis from $x = 1$ to $x = 2$.

    b. Show that the area between the curve $y = 1/x$ and the x-axis from ka to kb is the same as the area between the curve and the x-axis from $x = a$ to $x = b$ $(0 < a < b, k > 0)$.

Initial Value Problems

35. Show that $y = x^2 + \int_1^x \dfrac{1}{t}\, dt$ solves the initial value problem

$$\frac{d^2 y}{dx^2} = 2 - \frac{1}{x^2}; \quad y'(1) = 3, \quad y(1) = 1.$$

36. Show that $y = \int_0^x \left(1 + 2\sqrt{\sec t}\right) dt$ solves the initial value problem

$$\frac{d^2 y}{dx^2} = \sqrt{\sec x} \tan x; \quad y'(0) = 3, \quad y(0) = 0.$$

Express the solutions of the initial value problems in Exercises 37 and 38 in terms of integrals.

37. $\dfrac{dy}{dx} = \dfrac{\sin x}{x}, \quad y(5) = -3$

38. $\dfrac{dy}{dx} = \sqrt{2 - \sin^2 x}, \quad y(-1) = 2$

Solve the initial value problems in Exercises 39–42.

39. $\dfrac{dy}{dx} = \dfrac{1}{\sqrt{1 - x^2}}, \quad y(0) = 0$

40. $\dfrac{dy}{dx} = \dfrac{1}{x^2 + 1} - 1, \quad y(0) = 1$

41. $\dfrac{dy}{dx} = \dfrac{1}{x\sqrt{x^2 - 1}}, \quad x > 1; \quad y(2) = \pi$

42. $\dfrac{dy}{dx} = \dfrac{1}{1 + x^2} - \dfrac{2}{\sqrt{1 - x^2}}, \quad y(0) = 2$

For Exercises 43 and 44, find a function f that satisfies each equation.

43. $f(x) = 1 + \displaystyle\int_1^x t\, f(t)\, dt$ **44.** $f(x) = \displaystyle\int_0^x \left(1 + f(t)^2\right) dt$

Evaluating Indefinite Integrals
Evaluate the integrals in Exercises 45–76.

45. $\int 2(\cos x)^{-1/2} \sin x \, dx$

46. $\int (\tan x)^{-3/2} \sec^2 x \, dx$

47. $\int (2\theta + 1 + 2\cos(2\theta + 1)) \, d\theta$

48. $\int \left(\dfrac{1}{\sqrt{2\theta - \pi}} + 2\sec^2(2\theta - \pi) \right) d\theta$

49. $\int \left(t - \dfrac{2}{t} \right)\left(t + \dfrac{2}{t} \right) dt$

50. $\int \dfrac{(t + 1)^2 - 1}{t^4} \, dt$

51. $\int \sqrt{t} \sin(2t^{3/2}) \, dt$

52. $\int (\sec \theta \tan \theta) \sqrt{1 + \sec \theta} \, d\theta$

53. $\int e^x \sec^2(e^x - 7) \, dx$

54. $\int e^y \csc(e^y + 1) \cot(e^y + 1) \, dy$

55. $\int (\sec^2 x) e^{\tan x} \, dx$

56. $\int (\csc^2 x) e^{\cot x} \, dx$

57. $\int_{-1}^{1} \dfrac{dx}{3x - 4}$

58. $\int_{1}^{e} \dfrac{\sqrt{\ln x}}{x} \, dx$

59. $\int_{0}^{4} \dfrac{2t}{t^2 - 25} \, dt$

60. $\int \dfrac{\tan(\ln v)}{v} \, dv$

61. $\int \dfrac{(\ln x)^{-3}}{x} \, dx$

62. $\int \dfrac{1}{r} \csc^2(1 + \ln r) \, dr$

63. $\int x 3^{x^2} \, dx$

64. $\int 2^{\tan x} \sec^2 x \, dx$

65. $\int \dfrac{3 \, dr}{\sqrt{1 - 4(r - 1)^2}}$

66. $\int \dfrac{6 \, dr}{\sqrt{4 - (r + 1)^2}}$

67. $\int \dfrac{dx}{2 + (x - 1)^2}$

68. $\int \dfrac{dx}{1 + (3x + 1)^2}$

69. $\int \dfrac{dx}{(2x - 1)\sqrt{(2x - 1)^2 - 4}}$

70. $\int \dfrac{dx}{(x + 3)\sqrt{(x + 3)^2 - 25}}$

71. $\int \dfrac{e^{\sin^{-1}\sqrt{x}} \, dx}{2\sqrt{x - x^2}}$

72. $\int \dfrac{\sqrt{\sin^{-1} x} \, dx}{\sqrt{1 - x^2}}$

73. $\int \dfrac{dy}{\sqrt{\tan^{-1} y}\,(1 + y^2)}$

74. $\int \dfrac{(\tan^{-1} x)^2 \, dx}{1 + x^2}$

75. $\int \dfrac{\sin 2\theta - \cos 2\theta}{(\sin 2\theta + \cos 2\theta)^3} \, d\theta$

76. $\int \cos \theta \cdot \sin(\sin \theta) \, d\theta$

Evaluating Definite Integrals
Evaluate the integrals in Exercises 77–116.

77. $\int_{-1}^{1} (3x^2 - 4x + 7) \, dx$

78. $\int_{0}^{1} (8s^3 - 12s^2 + 5) \, ds$

79. $\int_{1}^{2} \dfrac{4}{v^2} \, dv$

80. $\int_{1}^{27} x^{-4/3} \, dx$

81. $\int_{1}^{4} \dfrac{dt}{t\sqrt{t}}$

82. $\int_{1}^{4} \dfrac{(1 + \sqrt{u})^{1/2}}{\sqrt{u}} \, du$

83. $\int_{0}^{1} \dfrac{36 \, dx}{(2x + 1)^3}$

84. $\int_{0}^{1} \dfrac{dr}{\sqrt[3]{(7 - 5r)^2}}$

85. $\int_{1/8}^{1} x^{-1/3}(1 - x^{2/3})^{3/2} \, dx$

86. $\int_{0}^{1/2} x^3(1 + 9x^4)^{-3/2} \, dx$

87. $\int_{0}^{\pi} \sin^2 5r \, dr$

88. $\int_{0}^{\pi/4} \cos^2 \left(4t - \dfrac{\pi}{4} \right) dt$

89. $\int_{0}^{\pi/3} \sec^2 \theta \, d\theta$

90. $\int_{\pi/4}^{3\pi/4} \csc^2 x \, dx$

91. $\int_{\pi}^{3\pi} \cot^2 \dfrac{x}{6} \, dx$

92. $\int_{0}^{\pi} \tan^2 \dfrac{\theta}{3} \, d\theta$

93. $\int_{-\pi/3}^{0} \sec x \tan x \, dx$

94. $\int_{\pi/4}^{3\pi/4} \csc z \cot z \, dz$

95. $\int_{0}^{\pi/2} 5(\sin x)^{3/2} \cos x \, dx$

96. $\int_{-\pi/2}^{\pi/2} 15 \sin^4 3x \cos 3x \, dx$

97. $\int_{0}^{\pi/2} \dfrac{3 \sin x \cos x}{\sqrt{1 + 3\sin^2 x}} \, dx$

98. $\int_{0}^{\pi/4} \dfrac{\sec^2 x}{(1 + 7\tan x)^{2/3}} \, dx$

99. $\int_{1}^{4} \left(\dfrac{x}{8} + \dfrac{1}{2x} \right) dx$

100. $\int_{1}^{8} \left(\dfrac{2}{3x} - \dfrac{8}{x^2} \right) dx$

101. $\int_{-2}^{-1} e^{-(x+1)} \, dx$

102. $\int_{-\ln 2}^{0} e^{2w} \, dw$

103. $\int_{0}^{\ln 5} e^r (3e^r + 1)^{-3/2} \, dr$

104. $\int_{0}^{\ln 9} e^\theta (e^\theta - 1)^{1/2} \, d\theta$

105. $\int_{1}^{e} \dfrac{1}{x}(1 + 7\ln x)^{-1/3} \, dx$

106. $\int_{1}^{3} \dfrac{(\ln(v + 1))^2}{v + 1} \, dv$

107. $\int_{1}^{8} \dfrac{\log_4 \theta}{\theta} \, d\theta$

108. $\int_{1}^{e} \dfrac{8 \ln 3 \log_3 \theta}{\theta} \, d\theta$

109. $\int_{-3/4}^{3/4} \dfrac{6 \, dx}{\sqrt{9 - 4x^2}}$

110. $\int_{-1/5}^{1/5} \dfrac{6 \, dx}{\sqrt{4 - 25x^2}}$

111. $\int_{-2}^{2} \dfrac{3 \, dt}{4 + 3t^2}$

112. $\int_{\sqrt{3}}^{3} \dfrac{dt}{3 + t^2}$

113. $\int_{1/\sqrt{3}}^{1} \dfrac{dy}{y\sqrt{4y^2 - 1}}$

114. $\int_{4\sqrt{2}}^{8} \dfrac{24 \, dy}{y\sqrt{y^2 - 16}}$

115. $\int_{\sqrt{2}/3}^{2/3} \dfrac{dy}{|y|\sqrt{9y^2 - 1}}$

116. $\int_{-2/\sqrt{5}}^{-\sqrt{6}/\sqrt{5}} \dfrac{dy}{|y|\sqrt{5y^2 - 3}}$

Average Values

117. Find the average value of $f(x) = mx + b$

   a. over $[-1, 1]$

   b. over $[-k, k]$

118. Find the average value of

   a. $y = \sqrt{3x}$ over $[0, 3]$

   b. $y = \sqrt{ax}$ over $[0, a]$

119. Let f be a function that is differentiable on $[a, b]$. In Chapter 2 we defined the average rate of change of f over $[a, b]$ to be

$$\frac{f(b) - f(a)}{b - a}$$

and the instantaneous rate of change of f at x to be $f'(x)$. In this chapter we defined the average value of a function. For the new definition of average to be consistent with the old one, we should have

$$\frac{f(b) - f(a)}{b - a} = \text{average value of } f' \text{ on } [a, b].$$

Is this the case? Give reasons for your answer.

120. Is it true that the average value of an integrable function over an interval of length 2 is half the function's integral over the interval? Give reasons for your answer.

121. a. Verify that $\int \ln x \, dx = x \ln x - x + C$.

   b. Find the average value of $\ln x$ over $[1, e]$.

122. Find the average value of $f(x) = 1/x$ on $[1, 2]$.

123. Compute the average value of the temperature function

$$f(x) = 37 \sin\left(\frac{2\pi}{365}(x - 101)\right) + 25$$

for a 365-day year. (See Exercise 100, Section 3.6.) This is one way to estimate the annual mean air temperature in Fairbanks, Alaska. The National Weather Service's official figure, a numerical average of the daily normal mean air temperatures for the year, is 25.7°F, which is slightly higher than the average value of $f(x)$.

T 124. Specific heat of a gas Specific heat C_v is the amount of heat required to raise the temperature of one mole (gram molecule) of a gas with constant volume by 1°C. The specific heat of oxygen depends on its temperature T and satisfies the formula

$$C_v = 8.27 + 10^{-5}(26T - 1.87T^2).$$

Find the average value of C_v for $20° \leq T \leq 675°C$ and the temperature at which it is attained.

Differentiating Integrals

In Exercises 125–132, find dy/dx.

125. $y = \int_2^x \sqrt{2 + \cos^3 t} \, dt$

126. $y = \int_2^{7x^2} \sqrt{2 + \cos^3 t} \, dt$

127. $y = \int_x^1 \frac{6}{3 + t^4} dt$

128. $y = \int_{\sec x}^2 \frac{1}{t^2 + 1} dt$

129. $y = \int_{\ln x^2}^0 e^{\cos t} \, dt$

130. $y = \int_1^{e^{\sqrt{x}}} \ln(t^2 + 1) \, dt$

131. $y = \int_0^{\sin^{-1} x} \frac{dt}{\sqrt{1 - 2t^2}}$

132. $y = \int_{\tan^{-1} x}^{\pi/4} e^{\sqrt{t}} \, dt$

Theory and Examples

133. Is it true that every function $y = f(x)$ that is differentiable on $[a, b]$ is itself the derivative of some function on $[a, b]$? Give reasons for your answer.

134. Suppose that $F(x)$ is an antiderivative of $f(x) = \sqrt{1 + x^4}$. Express $\int_0^1 \sqrt{1 + x^4} \, dx$ in terms of F and give a reason for your answer.

135. Find dy/dx if $y = \int_x^1 \sqrt{1 + t^2} \, dt$. Explain the main steps in your calculation.

136. Find dy/dx if $y = \int_{\cos x}^0 (1/(1 - t^2)) \, dt$. Explain the main steps in your calculation.

137. A new parking lot To meet the demand for parking, your town has allocated the area shown here. As the town engineer, you have been asked by the town council to find out if the lot can be built for $10,000. The cost to clear the land will be $0.10 a square foot, and the lot will cost $2.00 a square foot to pave. Can the job be done for $10,000? Use a lower sum estimate to see. (Answers may vary slightly, depending on the estimate used.)

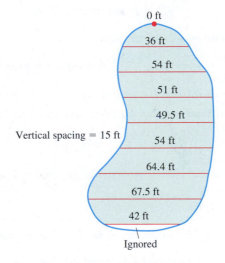

138. Skydivers A and B are in a helicopter hovering at 6400 ft. Skydiver A jumps and descends for 4 sec before opening her parachute. The helicopter then climbs to 7000 ft and hovers there. Forty-five seconds after A leaves the aircraft, B jumps and descends for 13 sec before opening his parachute. Both skydivers descend at 16 ft/sec with parachutes open. Assume that the skydivers fall freely (no effective air resistance) before their parachutes open.

   a. At what altitude does A's parachute open?

   b. At what altitude does B's parachute open?

   c. Which skydiver lands first?

CHAPTER 5   Additional and Advanced Exercises

Theory and Examples

1. a. If $\int_0^1 7f(x)\,dx = 7$, does $\int_0^1 f(x)\,dx = 1$?

  b. If $\int_0^1 f(x)\,dx = 4$ and $f(x) \geq 0$, does

$$\int_0^1 \sqrt{f(x)}\,dx = \sqrt{4} = 2?$$

Give reasons for your answers.

2. Suppose $\int_{-2}^2 f(x)\,dx = 4$, $\int_2^5 f(x)\,dx = 3$, $\int_{-2}^5 g(x)\,dx = 2$.

Which, if any, of the following statements are true?

  a. $\int_5^2 f(x)\,dx = -3$

  b. $\int_{-2}^5 (f(x) + g(x)) = 9$

  c. $f(x) \leq g(x)$ on the interval $-2 \leq x \leq 5$

3. Initial value problem Show that

$$y = \frac{1}{a}\int_0^x f(t)\sin a(x - t)\,dt$$

solves the initial value problem

$$\frac{d^2y}{dx^2} + a^2 y = f(x), \qquad \frac{dy}{dx} = 0 \text{ and } y = 0 \text{ when } x = 0.$$

(*Hint:* $\sin(ax - at) = \sin ax \cos at - \cos ax \sin at$.)

4. Proportionality Suppose that x and y are related by the equation

$$x = \int_0^y \frac{1}{\sqrt{1 + 4t^2}}\,dt.$$

Show that d^2y/dx^2 is proportional to y and find the constant of proportionality.

5. Find $f(4)$ if

  a. $\int_0^{x^2} f(t)\,dt = x \cos \pi x$ **b.** $\int_0^{f(x)} t^2\,dt = x \cos \pi x.$

6. Find $f(\pi/2)$ from the following information.

  i) f is positive and continuous.

  ii) The area under the curve $y = f(x)$ from $x = 0$ to $x = a$ is

$$\frac{a^2}{2} + \frac{a}{2}\sin a + \frac{\pi}{2}\cos a.$$

7. The area of the region in the xy-plane enclosed by the x-axis, the curve $y = f(x)$, $f(x) \geq 0$, and the lines $x = 1$ and $x = b$ is equal to $\sqrt{b^2 + 1} - \sqrt{2}$ for all $b > 1$. Find $f(x)$.

8. Prove that

$$\int_0^x \left(\int_0^u f(t)\,dt \right) du = \int_0^x f(u)(x - u)\,du.$$

(*Hint:* Express the integral on the right-hand side as the difference of two integrals. Then show that both sides of the equation have the same derivative with respect to x.)

9. Finding a curve Find the equation for the curve in the xy-plane that passes through the point $(1, -1)$ if its slope at x is always $3x^2 + 2$.

10. Shoveling dirt You sling a shovelful of dirt up from the bottom of a hole with an initial velocity of 32 ft/sec. The dirt must rise 17 ft above the release point to clear the edge of the hole. Is that enough speed to get the dirt out, or had you better duck?

Piecewise Continuous Functions

Although we are mainly interested in continuous functions, many functions in applications are piecewise continuous. A function $f(x)$ is **piecewise continuous on a closed interval** I if f has only finitely many discontinuities in I, the limits

$$\lim_{x \to c^-} f(x) \qquad \text{and} \qquad \lim_{x \to c^+} f(x)$$

exist and are finite at every interior point of I, and the appropriate one-sided limits exist and are finite at the endpoints of I. All piecewise continuous functions are integrable. The points of discontinuity subdivide I into open and half-open subintervals on which f is continuous, and the limit criteria above guarantee that f has a continuous extension to the closure of each subinterval. To integrate a piecewise continuous function, we integrate the individual extensions and add the results. The integral of

$$f(x) = \begin{cases} 1 - x, & -1 \leq x < 0 \\ x^2, & 0 \leq x < 2 \\ -1, & 2 \leq x \leq 3 \end{cases}$$

(Figure 5.33) over $[-1, 3]$ is

$$\int_{-1}^3 f(x)\,dx = \int_{-1}^0 (1 - x)\,dx + \int_0^2 x^2\,dx + \int_2^3 (-1)\,dx$$

$$= \left[x - \frac{x^2}{2} \right]_{-1}^0 + \left[\frac{x^3}{3} \right]_0^2 + \left[-x \right]_2^3$$

$$= \frac{3}{2} + \frac{8}{3} - 1 = \frac{19}{6}.$$

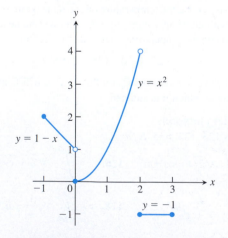

FIGURE 5.33 Piecewise continuous functions like this are integrated piece by piece.

The Fundamental Theorem applies to piecewise continuous functions with the restriction that $(d/dx) \int_a^x f(t)\, dt$ is expected to equal $f(x)$ only at values of x at which f is continuous. There is a similar restriction on Leibniz's Rule (see Exercises 31–38).

Graph the functions in Exercises 11–16 and integrate them over their domains.

11. $f(x) = \begin{cases} x^{2/3}, & -8 \le x < 0 \\ -4, & 0 \le x \le 3 \end{cases}$

12. $f(x) = \begin{cases} \sqrt{-x}, & -4 \le x < 0 \\ x^2 - 4, & 0 \le x \le 3 \end{cases}$

13. $g(t) = \begin{cases} t, & 0 \le t < 1 \\ \sin \pi t, & 1 \le t \le 2 \end{cases}$

14. $h(z) = \begin{cases} \sqrt{1-z}, & 0 \le z < 1 \\ (7z - 6)^{-1/3}, & 1 \le z \le 2 \end{cases}$

15. $f(x) = \begin{cases} 1, & -2 \le x < -1 \\ 1 - x^2, & -1 \le x < 1 \\ 2, & 1 \le x \le 2 \end{cases}$

16. $h(r) = \begin{cases} r, & -1 \le r < 0 \\ 1 - r^2, & 0 \le r < 1 \\ 1, & 1 \le r \le 2 \end{cases}$

17. Find the average value of the function graphed in the accompanying figure.

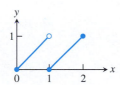

18. Find the average value of the function graphed in the accompanying figure.

Limits
Find the limits in Exercises 19–22.

19. $\lim\limits_{b \to 1^-} \int_0^b \dfrac{dx}{\sqrt{1 - x^2}}$

20. $\lim\limits_{x \to \infty} \dfrac{1}{x} \int_0^x \tan^{-1} t\, dt$

21. $\lim\limits_{n \to \infty} \left(\dfrac{1}{n + 1} + \dfrac{1}{n + 2} + \cdots + \dfrac{1}{2n} \right)$

22. $\lim\limits_{n \to \infty} \dfrac{1}{n} \left(e^{1/n} + e^{2/n} + \cdots + e^{(n-1)/n} + e^{n/n} \right)$

Approximating Finite Sums with Integrals
In many applications of calculus, integrals are used to approximate finite sums—the reverse of the usual procedure of using finite sums to approximate integrals.

For example, let's estimate the sum of the square roots of the first n positive integers, $\sqrt{1} + \sqrt{2} + \cdots + \sqrt{n}$. The integral

$$\int_0^1 \sqrt{x}\, dx = \frac{2}{3} x^{3/2} \Big]_0^1 = \frac{2}{3}$$

is the limit of the upper sums

$$S_n = \sqrt{\frac{1}{n}} \cdot \frac{1}{n} + \sqrt{\frac{2}{n}} \cdot \frac{1}{n} + \cdots + \sqrt{\frac{n}{n}} \cdot \frac{1}{n}$$

$$= \frac{\sqrt{1} + \sqrt{2} + \cdots + \sqrt{n}}{n^{3/2}}.$$

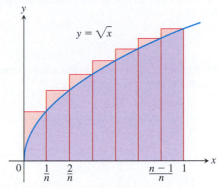

Therefore, when n is large, S_n will be close to $2/3$ and we will have

$$\text{Root sum} = \sqrt{1} + \sqrt{2} + \cdots + \sqrt{n} = S_n \cdot n^{3/2} \approx \frac{2}{3} n^{3/2}.$$

The following table shows how good the approximation can be.

n	Root sum	$(2/3)n^{3/2}$	Relative error
10	22.468	21.082	$1.386/22.468 \approx 6\%$
50	239.04	235.70	1.4%
100	671.46	666.67	0.7%
1000	21,097	21,082	0.07%

23. Evaluate

$$\lim_{n \to \infty} \frac{1^5 + 2^5 + 3^5 + \cdots + n^5}{n^6}$$

by showing that the limit is

$$\int_0^1 x^5\, dx$$

and evaluating the integral.

24. See Exercise 23. Evaluate

$$\lim_{n \to \infty} \frac{1}{n^4} (1^3 + 2^3 + 3^3 + \cdots + n^3).$$

25. Let $f(x)$ be a continuous function. Express

$$\lim_{n \to \infty} \frac{1}{n} \left[f\left(\frac{1}{n}\right) + f\left(\frac{2}{n}\right) + \cdots + f\left(\frac{n}{n}\right) \right]$$

as a definite integral.

26. Use the result of Exercise 25 to evaluate

a. $\lim\limits_{n \to \infty} \dfrac{1}{n^2} (2 + 4 + 6 + \cdots + 2n),$

b. $\lim\limits_{n \to \infty} \dfrac{1}{n^{16}} (1^{15} + 2^{15} + 3^{15} + \cdots + n^{15}),$

c. $\lim\limits_{n \to \infty} \dfrac{1}{n} \left(\sin \dfrac{\pi}{n} + \sin \dfrac{2\pi}{n} + \sin \dfrac{3\pi}{n} + \cdots + \sin \dfrac{n\pi}{n} \right).$

What can be said about the following limits?

d. $\lim_{n \to \infty} \dfrac{1}{n^{17}}(1^{15} + 2^{15} + 3^{15} + \cdots + n^{15})$

e. $\lim_{n \to \infty} \dfrac{1}{n^{15}}(1^{15} + 2^{15} + 3^{15} + \cdots + n^{15})$

27. a. Show that the area A_n of an n-sided regular polygon in a circle of radius r is

$$A_n = \frac{nr^2}{2}\sin\frac{2\pi}{n}.$$

b. Find the limit of A_n as $n \to \infty$. Is this answer consistent with what you know about the area of a circle?

28. Let

$$S_n = \frac{1^2}{n^3} + \frac{2^2}{n^3} + \cdots + \frac{(n-1)^2}{n^3}.$$

To calculate $\lim_{n\to\infty} S_n$, show that

$$S_n = \frac{1}{n}\left[\left(\frac{1}{n}\right)^2 + \left(\frac{2}{n}\right)^2 + \cdots + \left(\frac{n-1}{n}\right)^2\right]$$

and interpret S_n as an approximating sum of the integral

$$\int_0^1 x^2\, dx.$$

(*Hint:* Partition $[0, 1]$ into n intervals of equal length and write out the approximating sum for inscribed rectangles.)

Defining Functions Using the Fundamental Theorem

29. A function defined by an integral The graph of a function f consists of a semicircle and two line segments as shown. Let $g(x) = \int_1^x f(t)\, dt$.

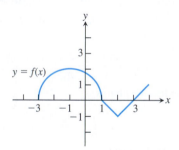

a. Find $g(1)$.   **b.** Find $g(3)$.   **c.** Find $g(-1)$.

d. Find all values of x on the open interval $(-3, 4)$ at which g has a relative maximum.

e. Write an equation for the line tangent to the graph of g at $x = -1$.

f. Find the x-coordinate of each point of inflection of the graph of g on the open interval $(-3, 4)$.

g. Find the range of g.

30. A differential equation Show that both of the following conditions are satisfied by $y = \sin x + \int_x^\pi \cos 2t\, dt + 1$:

i) $y'' = -\sin x + 2 \sin 2x$

ii) $y = 1$ and $y' = -2$ when $x = \pi$.

Leibniz's Rule In applications, we sometimes encounter functions defined by integrals that have variable upper limits of integration and variable lower limits of integration at the same time. We can find the derivative of such an integral by a formula called **Leibniz's Rule**.

Leibniz's Rule

If f is continuous on $[a, b]$ and if $u(x)$ and $v(x)$ are differentiable functions of x whose values lie in $[a, b]$, then

$$\frac{d}{dx}\int_{u(x)}^{v(x)} f(t)\, dt = f(v(x))\frac{dv}{dx} - f(u(x))\frac{du}{dx}.$$

To prove the rule, let F be an antiderivative of f on $[a, b]$. Then

$$\int_{u(x)}^{v(x)} f(t)\, dt = F(v(x)) - F(u(x)).$$

Differentiating both sides of this equation with respect to x gives the equation we want:

$$\frac{d}{dx}\int_{u(x)}^{v(x)} f(t)\, dt = \frac{d}{dx}\left[F(v(x)) - F(u(x))\right]$$

$$= F'(v(x))\frac{dv}{dx} - F'(u(x))\frac{du}{dx} \qquad \text{Chain Rule}$$

$$= f(v(x))\frac{dv}{dx} - f(u(x))\frac{du}{dx}.$$

Use Leibniz's Rule to find the derivatives of the functions in Exercises 31–38.

31. $f(x) = \displaystyle\int_{1/x}^x \frac{1}{t}\, dt$

32. $f(x) = \displaystyle\int_{\cos x}^{\sin x} \frac{1}{1 - t^2}\, dt$

33. $g(y) = \displaystyle\int_{\sqrt{y}}^{2\sqrt{y}} \sin t^2\, dt$

34. $g(y) = \displaystyle\int_{\sqrt{y}}^{y^2} \frac{e^t}{t}\, dt$

35. $y = \displaystyle\int_{x^2/2}^{x^2} \ln\sqrt{t}\, dt$

36. $y = \displaystyle\int_{\sqrt{x}}^{\sqrt[3]{x}} \ln t\, dt$

37. $y = \displaystyle\int_0^{\ln x} \sin e^t\, dt$

38. $y = \displaystyle\int_{e^{4\sqrt{x}}}^{e^{2x}} \ln t\, dt$

Theory and Examples

39. Use Leibniz's Rule to find the value of x that maximizes the value of the integral

$$\int_x^{x+3} t(5 - t)\, dt.$$

40. For what $x > 0$ does $x^{(x^x)} = (x^x)^x$? Give reasons for your answer.

41. Find the areas between the curves $y = 2(\log_2 x)/x$ and $y = 2(\log_4 x)/x$ and the x-axis from $x = 1$ to $x = e$. What is the ratio of the larger area to the smaller?

42. a. Find df/dx if

$$f(x) = \int_1^{e^x} \frac{2\ln t}{t}\, dt.$$

b. Find $f(0)$.

c. What can you conclude about the graph of f? Give reasons for your answer.

43. Find $f'(2)$ if $f(x) = e^{g(x)}$ and $g(x) = \displaystyle\int_2^x \frac{t}{1 + t^4}\, dt$.

44. Use the accompanying figure to show that

$$\int_0^{\pi/2} \sin x \, dx = \frac{\pi}{2} - \int_0^1 \sin^{-1} x \, dx.$$

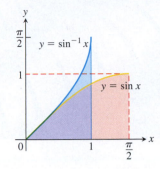

45. Napier's inequality   Here are two pictorial proofs that

$$b > a > 0 \quad \Rightarrow \quad \frac{1}{b} < \frac{\ln b - \ln a}{b - a} < \frac{1}{a}.$$

Explain what is going on in each case.

a.

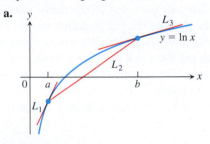

b.

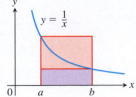

(*Source:* Roger B. Nelson, *College Mathematics Journal*, Vol. 24, No. 2, March 1993, p. 165.)

46. Bound on an integral   Let f be a continuously differentiable function on $[a, b]$ satisfying $\int_a^b f(x) \, dx = 0$.

a. If $c = (a + b)/2$, show that

$$\int_a^b x f(x) \, dx = \int_a^c (x - c) f(x) \, dx + \int_c^b (x - c) f(x) \, dx.$$

b. Let $t = |x - c|$ and $\ell = (b - a)/2$. Show that

$$\int_a^b x f(x) \, dx = \int_0^\ell t(f(c + t) - f(c - t)) \, dt.$$

c. Apply the Mean Value Theorem from Section 4.2 to part (b) to prove that

$$\left| \int_a^b x f(x) \, dx \right| \le \frac{(b - a)^3}{12} M,$$

where M is the absolute maximum of f' on $[a, b]$.

CHAPTER 5   Technology Application Projects

Mathematica/Maple Projects

Projects can be found within MyMathLab.

- *Using Riemann Sums to Estimate Areas, Volumes, and Lengths of Curves*
  Visualize and approximate areas and volumes in Part I.

- *Riemann Sums, Definite Integrals, and the Fundamental Theorem of Calculus*
  Parts I, II, and III develop Riemann sums and definite integrals. Part IV continues the development of the Riemann sum and definite integral using the Fundamental Theorem to solve problems previously investigated.

- *Rain Catchers, Elevators, and Rockets*
  Part I illustrates that the area under a curve is the same as the area of an appropriate rectangle for examples taken from the chapter. You will compute the amount of water accumulating in basins of different shapes as the basin is filled and drained.

- *Motion Along a Straight Line, Part II*
  You will observe the shape of a graph through dramatic animated visualizations of the derivative relations among position, velocity, and acceleration. Figures in the text can be animated using this software.

- *Bending of Beams*
  Study bent shapes of beams, determine their maximum deflections, concavity, and inflection points, and interpret the results in terms of a beams's compression and tension.

6

Applications of Definite Integrals

OVERVIEW In Chapter 5 we saw that a continuous function over a closed interval has a definite integral, which is the limit of Riemann sum approximations for the function. We found a way to evaluate definite integrals using the Fundamental Theorem of Calculus. We saw that the area under a curve and the area between two curves could be defined and computed as definite integrals. In this chapter we will see some of the many additional applications of definite integrals. We will use the definite integral to define and find volumes, lengths of plane curves, and areas of surfaces of revolution. We will see how integrals are used to solve physical problems involving the work done by a force, and how they give the location of an object's center of mass. The integral arises in these and other applications in which we can approximate a desired quantity by Riemann sums. The limit of those Riemann sums, which is the quantity we seek, is given by a definite integral.

6.1 Volumes Using Cross-Sections

In this section we define volumes of solids by using the areas of their cross-sections. A **cross-section** of a solid S is the planar region formed by intersecting S with a plane (Figure 6.1). We present three different methods for obtaining the cross-sections appropriate to finding the volume of a particular solid: the method of slicing, the disk method, and the washer method.

Suppose that we want to find the volume of a solid S like the one pictured in Figure 6.1. At each point x in the interval $[a, b]$ we form a cross-section $S(x)$ by intersecting S with a plane perpendicular to the x-axis through the point x, which gives a planar region whose area is $A(x)$. We will show that if A is a continuous function of x, then the volume of the solid S is the definite integral of $A(x)$. This method of computing volumes is known as the **method of slicing**.

Before showing how this method works, we need to extend the definition of a cylinder from the usual cylinders of classical geometry (which have circular, square, or other regular bases) to cylindrical solids that have more general bases. As shown in Figure 6.2, if the

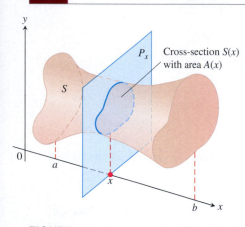

FIGURE 6.1 A cross-section $S(x)$ of the solid S formed by intersecting S with a plane P_x perpendicular to the x-axis through the point x in the interval $[a, b]$.

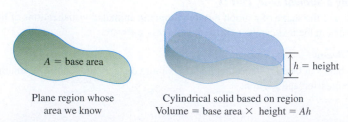

Plane region whose area we know

Cylindrical solid based on region
Volume = base area × height = Ah

FIGURE 6.2 The volume of a cylindrical solid is always defined to be its base area times its height.

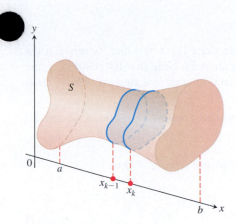

FIGURE 6.3 A typical thin slab in the solid S.

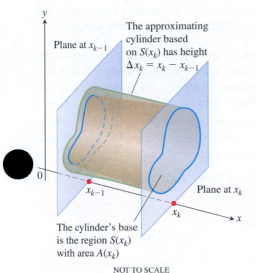

Plane at x_{k-1}

The approximating cylinder based on $S(x_k)$ has height $\Delta x_k = x_k - x_{k-1}$

Plane at x_k

The cylinder's base is the region $S(x_k)$ with area $A(x_k)$

NOT TO SCALE

FIGURE 6.4 The solid thin slab in Figure 6.3 is shown enlarged here. It is approximated by the cylindrical solid with base $S(x_k)$ having area $A(x_k)$ and height $\Delta x_k = x_k - x_{k-1}$.

cylindrical solid has a base whose area is A and its height is h, then the volume of the cylindrical solid is

$$\text{Volume} = \text{area} \times \text{height} = A \cdot h.$$

In the method of slicing, the base will be the cross-section of S that has area $A(x)$, and the height will correspond to the width Δx_k of subintervals formed by partitioning the interval $[a, b]$ into finitely many subintervals $[x_{k-1}, x_k]$.

Slicing by Parallel Planes

We partition $[a, b]$ into subintervals of width (length) Δx_k and slice the solid, as we would a loaf of bread, by planes perpendicular to the x-axis at the partition points $a = x_0 < x_1 < \cdots < x_n = b$. These planes slice S into thin "slabs" (like thin slices of a loaf of bread). A typical slab is shown in Figure 6.3. We approximate the slab between the plane at x_{k-1} and the plane at x_k by a cylindrical solid with base area $A(x_k)$ and height $\Delta x_k = x_k - x_{k-1}$ (Figure 6.4). The volume V_k of this cylindrical solid is $A(x_k) \cdot \Delta x_k$, which is approximately the same volume as that of the slab:

$$\text{Volume of the } k\text{th slab} \approx V_k = A(x_k) \, \Delta x_k.$$

The volume V of the entire solid S is therefore approximated by the sum of these cylindrical volumes,

$$V \approx \sum_{k=1}^{n} V_k = \sum_{k=1}^{n} A(x_k) \, \Delta x_k.$$

This is a Riemann sum for the function $A(x)$ on $[a, b]$. The approximation given by this Riemann sum converges to the definite integral of $A(x)$ as $n \to \infty$:

$$\lim_{n \to \infty} \sum_{k=1}^{n} A(x_k) \, \Delta x_k = \int_a^b A(x) \, dx.$$

Therefore, we define this definite integral to be the volume of the solid S.

DEFINITION The **volume** of a solid of integrable cross-sectional area $A(x)$ from $x = a$ to $x = b$ is the integral of A from a to b,

$$V = \int_a^b A(x) \, dx.$$

This definition applies whenever $A(x)$ is integrable, and in particular when $A(x)$ is continuous. To apply this definition to calculate the volume of a solid using cross-sections perpendicular to the x-axis, take the following steps:

Calculating the Volume of a Solid

1. *Sketch the solid and a typical cross-section.*
2. *Find a formula for $A(x)$, the area of a typical cross-section.*
3. *Find the limits of integration.*
4. *Integrate $A(x)$ to find the volume.*

EXAMPLE 1 A pyramid 3 m high has a square base that is 3 m on a side. The cross-section of the pyramid perpendicular to the altitude x m down from the vertex is a square x m on a side. Find the volume of the pyramid.

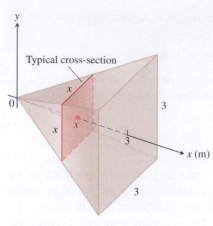

FIGURE 6.5 The cross-sections of the pyramid in Example 1 are squares.

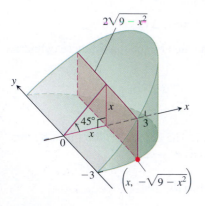

FIGURE 6.6 The wedge of Example 2, sliced perpendicular to the x-axis. The cross-sections are rectangles.

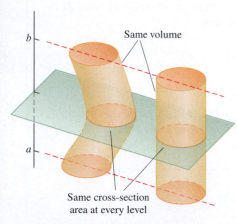

FIGURE 6.7 *Cavalieri's principle:* These solids have the same volume (imagine each solid as a stack of coins).

HISTORICAL BIOGRAPHY

Bonaventura Cavalieri
(1598–1647)
www.goo.gl/kPgDdQ

Solution

1. *A sketch.* We draw the pyramid with its altitude along the x-axis and its vertex at the origin and include a typical cross-section (Figure 6.5). Note that by positioning the pyramid in this way, we have vertical cross-sections that are squares, whose areas are easy to calculate.

2. *A formula for A(x).* The cross-section at x is a square x meters on a side, so its area is

$$A(x) = x^2.$$

3. *The limits of integration.* The squares lie on the planes from $x = 0$ to $x = 3$.

4. *Integrate to find the volume:*

$$V = \int_0^3 A(x)\, dx = \int_0^3 x^2\, dx = \frac{x^3}{3}\Big]_0^3 = 9 \text{ m}^3.$$ ∎

EXAMPLE 2   A curved wedge is cut from a circular cylinder of radius 3 by two planes. One plane is perpendicular to the axis of the cylinder. The second plane crosses the first plane at a 45° angle at the center of the cylinder. Find the volume of the wedge.

Solution   We draw the wedge and sketch a typical cross-section perpendicular to the x-axis (Figure 6.6). The base of the wedge in the figure is the semicircle with $x \geq 0$ that is cut from the circle $x^2 + y^2 = 9$ by the 45° plane when it intersects the y-axis. For any x in the interval $[0, 3]$, the y-values in this semicircular base vary from $y = -\sqrt{9 - x^2}$ to $y = \sqrt{9 - x^2}$. When we slice through the wedge by a plane perpendicular to the x-axis, we obtain a cross-section at x which is a rectangle of height x whose width extends across the semicircular base. The area of this cross-section is

$$A(x) = (\text{height})(\text{width}) = (x)\left(2\sqrt{9 - x^2}\right)$$
$$= 2x\sqrt{9 - x^2}.$$

The rectangles run from $x = 0$ to $x = 3$, so we have

$$V = \int_a^b A(x)\, dx = \int_0^3 2x\sqrt{9 - x^2}\, dx$$

$$= -\frac{2}{3}(9 - x^2)^{3/2}\Big]_0^3 \qquad \begin{array}{l}\text{Let } u = 9 - x^2, \\ du = -2x\, dx, \text{ integrate,} \\ \text{and substitute back.}\end{array}$$

$$= 0 + \frac{2}{3}(9)^{3/2}$$

$$= 18.$$ ∎

EXAMPLE 3   Cavalieri's principle says that solids with equal altitudes and identical cross-sectional areas at each height have the same volume (Figure 6.7). This follows immediately from the definition of volume, because the cross-sectional area function $A(x)$ and the interval $[a, b]$ are the same for both solids. ∎

Solids of Revolution: The Disk Method

The solid generated by rotating (or revolving) a planar region about an axis in its plane is called a **solid of revolution**. To find the volume of a solid like the one shown in Figure 6.8, we first observe that the cross-sectional area $A(x)$ is the area of a disk of radius $R(x)$, where $R(x)$ is the distance from the axis of revolution to the planar region's boundary. The area is then

$$A(x) = \pi(\text{radius})^2 = \pi[R(x)]^2.$$

Therefore, the definition of volume gives us the following formula.

> **Volume by Disks for Rotation About the x-Axis**
>
> $$V = \int_a^b A(x)\,dx = \int_a^b \pi\,[\,R(x)\,]^2\,dx.$$

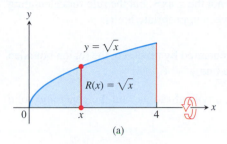

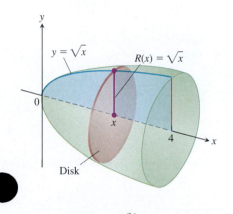

FIGURE 6.8 The region (a) and solid of revolution (b) in Example 4.

This method for calculating the volume of a solid of revolution is often called the **disk method** because a cross-section is a circular disk of radius $R(x)$.

EXAMPLE 4   The region between the curve $y = \sqrt{x}$, $0 \le x \le 4$, and the x-axis is revolved about the x-axis to generate a solid. Find its volume.

Solution   We draw figures showing the region, a typical radius, and the generated solid (Figure 6.8). The volume is

$$V = \int_a^b \pi\,[\,R(x)\,]^2\,dx$$

$$= \int_0^4 \pi\,[\,\sqrt{x}\,]^2\,dx \qquad \text{\color{blue}{Radius } } R(x) = \sqrt{x} \text{ \color{blue}{for}}$$
$$\text{\color{blue}{rotation around } } x\text{\color{blue}{-axis.}}$$

$$= \pi \int_0^4 x\,dx = \pi \frac{x^2}{2}\bigg]_0^4 = \pi \frac{(4)^2}{2} = 8\pi. \qquad \blacksquare$$

EXAMPLE 5   The circle

$$x^2 + y^2 = a^2$$

is rotated about the x-axis to generate a sphere. Find its volume.

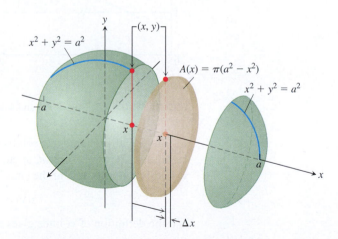

FIGURE 6.9 The sphere generated by rotating the circle $x^2 + y^2 = a^2$ about the x-axis. The radius is $R(x) = y = \sqrt{a^2 - x^2}$ (Example 5).

Solution   We imagine the sphere cut into thin slices by planes perpendicular to the x-axis (Figure 6.9). The cross-sectional area at a typical point x between $-a$ and a is

$$A(x) = \pi y^2 = \pi(a^2 - x^2). \qquad \text{\color{blue}{$R(x) = \sqrt{a^2 - x^2}$ for}}$$
$$\text{\color{blue}{rotation around } } x\text{\color{blue}{-axis.}}$$

Therefore, the volume is

$$V = \int_{-a}^{a} A(x)\,dx = \int_{-a}^{a} \pi(a^2 - x^2)\,dx = \pi \left[a^2 x - \frac{x^3}{3} \right]_{-a}^{a} = \frac{4}{3}\pi a^3. \qquad \blacksquare$$

The axis of revolution in the next example is not the x-axis, but the rule for calculating the volume is the same: Integrate $\pi(\text{radius})^2$ between appropriate limits.

EXAMPLE 6   Find the volume of the solid generated by revolving the region bounded by $y = \sqrt{x}$ and the lines $y = 1$, $x = 4$ about the line $y = 1$.

Solution   We draw figures showing the region, a typical radius, and the generated solid (Figure 6.10). The volume is

$$V = \int_{1}^{4} \pi [R(x)]^2\,dx$$

$$= \int_{1}^{4} \pi [\sqrt{x} - 1]^2\,dx \qquad \text{Radius } R(x) = \sqrt{x} - 1 \text{ for rotation around } y = 1.$$

$$= \pi \int_{1}^{4} [x - 2\sqrt{x} + 1]\,dx \qquad \text{Expand integrand.}$$

$$= \pi \left[\frac{x^2}{2} - 2 \cdot \frac{2}{3} x^{3/2} + x \right]_{1}^{4} = \frac{7\pi}{6}. \qquad \text{Integrate.}$$

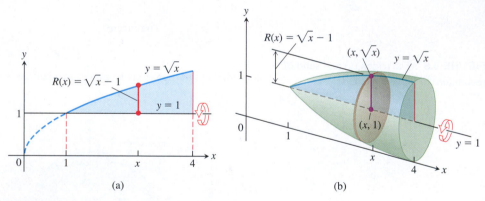

FIGURE 6.10   The region (a) and solid of revolution (b) in Example 6.   ■

To find the volume of a solid generated by revolving a region between the y-axis and a curve $x = R(y)$, $c \le y \le d$, about the y-axis, we use the same method with x replaced by y. In this case, the area of the circular cross-section is

$$A(y) = \pi[\text{radius}]^2 = \pi[R(y)]^2,$$

and the definition of volume gives us the following formula.

Volume by Disks for Rotation About the y-Axis

$$V = \int_{c}^{d} A(y)\,dy = \int_{c}^{d} \pi[R(y)]^2\,dy.$$

EXAMPLE 7   Find the volume of the solid generated by revolving the region between the y-axis and the curve $x = 2/y$, $1 \le y \le 4$, about the y-axis.

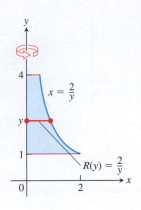

(a)

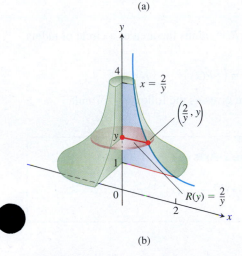

(b)

FIGURE 6.11 The region (a) and part of the solid of revolution (b) in Example 7.

Solution We draw figures showing the region, a typical radius, and the generated solid (Figure 6.11). The volume is

$$V = \int_1^4 \pi [R(y)]^2 \, dy$$

$$= \int_1^4 \pi \left(\frac{2}{y}\right)^2 dy \qquad \text{Radius } R(y) = \frac{2}{y} \text{ for rotation around } y\text{-axis}$$

$$= \pi \int_1^4 \frac{4}{y^2} \, dy = 4\pi \left[-\frac{1}{y}\right]_1^4 = 4\pi \left[\frac{3}{4}\right] = 3\pi. \qquad \blacksquare$$

EXAMPLE 8 Find the volume of the solid generated by revolving the region between the parabola $x = y^2 + 1$ and the line $x = 3$ about the line $x = 3$.

Solution We draw figures showing the region, a typical radius, and the generated solid (Figure 6.12). Note that the cross-sections are perpendicular to the line $x = 3$ and have y-coordinates from $y = -\sqrt{2}$ to $y = \sqrt{2}$. The volume is

$$V = \int_{-\sqrt{2}}^{\sqrt{2}} \pi [R(y)]^2 \, dy \qquad y = \pm\sqrt{2} \text{ when } x = 3$$

$$= \int_{-\sqrt{2}}^{\sqrt{2}} \pi [2 - y^2]^2 \, dy \qquad \begin{array}{l} \text{Radius } R(y) = 3 - (y^2 + 1) \\ \text{for rotation around axis } x = 3. \end{array}$$

$$= \pi \int_{-\sqrt{2}}^{\sqrt{2}} [4 - 4y^2 + y^4] \, dy \qquad \text{Expand integrand.}$$

$$= \pi \left[4y - \frac{4}{3}y^3 + \frac{y^5}{5}\right]_{-\sqrt{2}}^{\sqrt{2}} \qquad \text{Integrate.}$$

$$= \frac{64\pi\sqrt{2}}{15}. \qquad \blacksquare$$

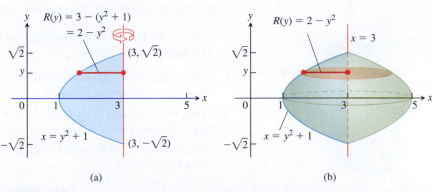

(a)   **(b)**

FIGURE 6.12 The region (a) and solid of revolution (b) in Example 8.   ■

Solids of Revolution: The Washer Method

If the region we revolve to generate a solid does not border on or cross the axis of revolution, then the solid has a hole in it (Figure 6.13). The cross-sections perpendicular to the axis of revolution are *washers* (the purplish circular surface in Figure 6.13) instead of disks. The dimensions of a typical washer are

Outer radius: $R(x)$

Inner radius: $r(x)$

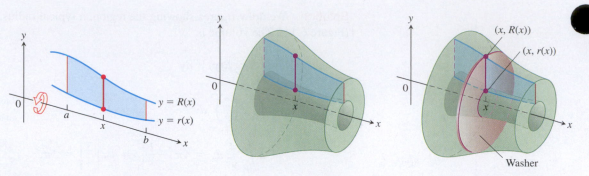

FIGURE 6.13 The cross-sections of the solid of revolution generated here are washers, not disks, so the integral $\int_a^b A(x)\,dx$ leads to a slightly different formula.

The washer's area is the area of a circle of radius $R(x)$ minus the area of a circle of radius $r(x)$:

$$A(x) = \pi\,[\,R(x)\,]^2 - \pi\,[\,r(x)\,]^2 = \pi\big(\,[\,R(x)\,]^2 - [\,r(x)\,]^2\big).$$

Consequently, the definition of volume in this case gives us the following formula.

> **Volume by Washers for Rotation About the x-Axis**
>
> $$V = \int_a^b A(x)\,dx = \int_a^b \pi\big(\,[\,R(x)\,]^2 - [\,r(x)\,]^2\big)\,dx.$$

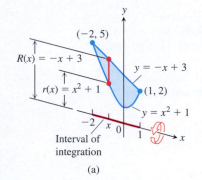

Interval of integration

(a)

This method for calculating the volume of a solid of revolution is called the **washer method** because a thin slab of the solid resembles a circular washer with outer radius $R(x)$ and inner radius $r(x)$.

EXAMPLE 9 The region bounded by the curve $y = x^2 + 1$ and the line $y = -x + 3$ is revolved about the x-axis to generate a solid. Find the volume of the solid.

Solution We use the same four steps for calculating the volume of a solid that were discussed earlier in this section.

1. Draw the region and sketch a line segment across it perpendicular to the axis of revolution (the red segment in Figure 6.14a).

2. Find the outer and inner radii of the washer that would be swept out by the line segment if it were revolved about the x-axis along with the region.

   These radii are the distances of the ends of the line segment from the axis of revolution (see Figure 6.14).

   $$\text{Outer radius:}\quad R(x) = -x + 3$$
   $$\text{Inner radius:}\quad r(x) = x^2 + 1$$

Washer cross-section
Outer radius: $R(x) = -x + 3$
Inner radius: $r(x) = x^2 + 1$

(b)

FIGURE 6.14 (a) The region in Example 9 spanned by a line segment perpendicular to the axis of revolution. (b) When the region is revolved about the x-axis, the line segment generates a washer.

3. Find the limits of integration by finding the x-coordinates of the intersection points of the curve and line in Figure 6.14a.

$$x^2 + 1 = -x + 3$$
$$x^2 + x - 2 = 0$$
$$(x + 2)(x - 1) = 0$$
$$x = -2, \quad x = 1 \qquad \text{Limits of integration}$$

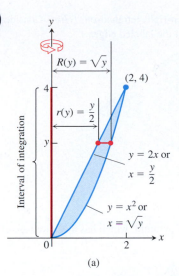

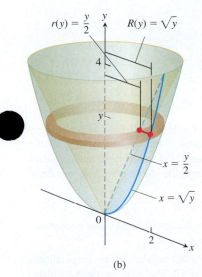

FIGURE 6.15 (a) The region being rotated about the y-axis, the washer radii, and limits of integration in Example 10. (b) The washer swept out by the line segment in part (a).

4. Evaluate the volume integral.

$$V = \int_a^b \pi \left([R(x)]^2 - [r(x)]^2 \right) dx \qquad \text{Rotation around } x\text{-axis}$$

$$= \int_{-2}^1 \pi \left((-x + 3)^2 - (x^2 + 1)^2 \right) dx \qquad \text{Values from Steps 2 and 3}$$

$$= \pi \int_{-2}^1 (8 - 6x - x^2 - x^4) \, dx \qquad \text{Simplify algebraically.}$$

$$= \pi \left[8x - 3x^2 - \frac{x^3}{3} - \frac{x^5}{5} \right]_{-2}^1 = \frac{117\pi}{5} \qquad \text{Integrate.} \qquad \blacksquare$$

To find the volume of a solid formed by revolving a region about the y-axis, we use the same procedure as in Example 9, but integrate with respect to y instead of x. In this situation the line segment sweeping out a typical washer is perpendicular to the y-axis (the axis of revolution), and the outer and inner radii of the washer are functions of y.

EXAMPLE 10 The region bounded by the parabola $y = x^2$ and the line $y = 2x$ in the first quadrant is revolved about the y-axis to generate a solid. Find the volume of the solid.

Solution First we sketch the region and draw a line segment across it perpendicular to the axis of revolution (the y-axis). See Figure 6.15a.

The radii of the washer swept out by the line segment are $R(y) = \sqrt{y}$, $r(y) = y/2$ (Figure 6.15).

The line and parabola intersect at $y = 0$ and $y = 4$, so the limits of integration are $c = 0$ and $d = 4$. We integrate to find the volume:

$$V = \int_c^d \pi \left([R(y)]^2 - [r(y)]^2 \right) dy \qquad \text{Rotation around } y\text{-axis}$$

$$= \int_0^4 \pi \left([\sqrt{y}]^2 - \left[\frac{y}{2} \right]^2 \right) dy \qquad \begin{array}{l}\text{Substitute for radii and}\\ \text{limits of integration.}\end{array}$$

$$= \pi \int_0^4 \left(y - \frac{y^2}{4} \right) dy = \pi \left[\frac{y^2}{2} - \frac{y^3}{12} \right]_0^4 = \frac{8}{3}\pi \qquad \blacksquare$$

EXERCISES 6.1

Volumes by Slicing

Find the volumes of the solids in Exercises 1–10.

1. The solid lies between planes perpendicular to the x-axis at $x = 0$ and $x = 4$. The cross-sections perpendicular to the axis on the interval $0 \le x \le 4$ are squares whose diagonals run from the parabola $y = -\sqrt{x}$ to the parabola $y = \sqrt{x}$.

2. The solid lies between planes perpendicular to the x-axis at $x = -1$ and $x = 1$. The cross-sections perpendicular to the x-axis are circular disks whose diameters run from the parabola $y = x^2$ to the parabola $y = 2 - x^2$.

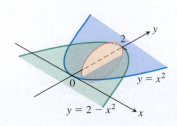

3. The solid lies between planes perpendicular to the x-axis at $x = -1$ and $x = 1$. The cross-sections perpendicular to the x-axis between these planes are squares whose bases run from the semicircle $y = -\sqrt{1 - x^2}$ to the semicircle $y = \sqrt{1 - x^2}$.

4. The solid lies between planes perpendicular to the x-axis at $x = -1$ and $x = 1$. The cross-sections perpendicular to the x-axis between these planes are squares whose diagonals run from the semicircle $y = -\sqrt{1 - x^2}$ to the semicircle $y = \sqrt{1 - x^2}$.

5. The base of a solid is the region between the curve $y = 2\sqrt{\sin x}$ and the interval $[0, \pi]$ on the x-axis. The cross-sections perpendicular to the x-axis are

    a. equilateral triangles with bases running from the x-axis to the curve as shown in the accompanying figure.

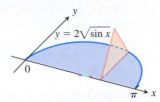

    b. squares with bases running from the x-axis to the curve.

6. The solid lies between planes perpendicular to the x-axis at $x = -\pi/3$ and $x = \pi/3$. The cross-sections perpendicular to the x-axis are

    a. circular disks with diameters running from the curve $y = \tan x$ to the curve $y = \sec x$.

    b. squares whose bases run from the curve $y = \tan x$ to the curve $y = \sec x$.

7. The base of a solid is the region bounded by the graphs of $y = 3x$, $y = 6$, and $x = 0$. The cross-sections perpendicular to the x-axis are

    a. rectangles of height 10.

    b. rectangles of perimeter 20.

8. The base of a solid is the region bounded by the graphs of $y = \sqrt{x}$ and $y = x/2$. The cross-sections perpendicular to the x-axis are

    a. isosceles triangles of height 6.

    b. semicircles with diameters running across the base of the solid.

9. The solid lies between planes perpendicular to the y-axis at $y = 0$ and $y = 2$. The cross-sections perpendicular to the y-axis are circular disks with diameters running from the y-axis to the parabola $x = \sqrt{5}y^2$.

10. The base of the solid is the disk $x^2 + y^2 \leq 1$. The cross-sections by planes perpendicular to the y-axis between $y = -1$ and $y = 1$ are isosceles right triangles with one leg in the disk.

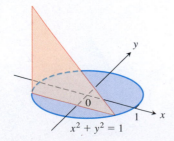

11. Find the volume of the given right tetrahedron. (*Hint:* Consider slices perpendicular to one of the labeled edges.)

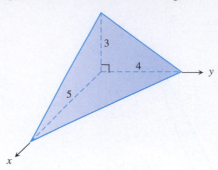

12. Find the volume of the given pyramid, which has a square base of area 9 and height 5.

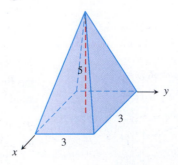

13. **A twisted solid** A square of side length s lies in a plane perpendicular to a line L. One vertex of the square lies on L. As this square moves a distance h along L, the square turns one revolution about L to generate a corkscrew-like column with square cross-sections.

    a. Find the volume of the column.

    b. What will the volume be if the square turns twice instead of once? Give reasons for your answer.

14. **Cavalieri's principle** A solid lies between planes perpendicular to the x-axis at $x = 0$ and $x = 12$. The cross-sections by planes perpendicular to the x-axis are circular disks whose diameters run from the line $y = x/2$ to the line $y = x$ as shown in the accompanying figure. Explain why the solid has the same volume as a right circular cone with base radius 3 and height 12.

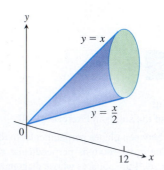

15. Intersection of two half-cylinders Two half-cylinders of diameter 2 meet at a right angle in the accompanying figure. Find the volume of the solid region common to both half-cylinders. (*Hint*: Consider slices parallel to the base of the solid.)

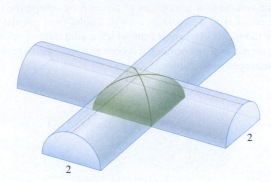

16. Gasoline in a tank A gasoline tank is in the shape of a right circular cylinder (lying on its side) of length 10 ft and radius 4 ft. Set up an integral that represents the volume of the gas in the tank if it is filled to a depth of 6 ft. You will learn how to compute this integral in Chapter 8 (or you may use geometry to find its value).

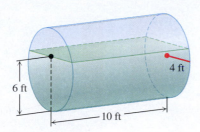

Volumes by the Disk Method

In Exercises 17–20, find the volume of the solid generated by revolving the shaded region about the given axis.

17. About the *x*-axis

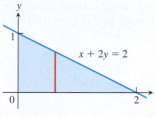

$x + 2y = 2$

18. About the *y*-axis

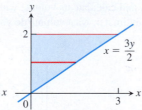

$x = \dfrac{3y}{2}$

19. About the *y*-axis

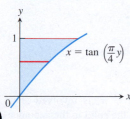

$x = \tan\left(\dfrac{\pi}{4}y\right)$

20. About the *x*-axis

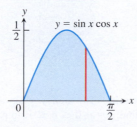

$y = \sin x \cos x$

Find the volumes of the solids generated by revolving the regions bounded by the lines and curves in Exercises 21–30 about the *x*-axis.

21. $y = x^2$, $y = 0$, $x = 2$
22. $y = x^3$, $y = 0$, $x = 2$
23. $y = \sqrt{9 - x^2}$, $y = 0$
24. $y = x - x^2$, $y = 0$
25. $y = \sqrt{\cos x}$, $0 \le x \le \pi/2$, $y = 0$, $x = 0$
26. $y = \sec x$, $y = 0$, $x = -\pi/4$, $x = \pi/4$
27. $y = e^{-x}$, $y = 0$, $x = 0$, $x = 1$
28. The region between the curve $y = \sqrt{\cot x}$ and the *x*-axis from $x = \pi/6$ to $x = \pi/2$
29. The region between the curve $y = 1/(2\sqrt{x})$ and the *x*-axis from $x = 1/4$ to $x = 4$
30. $y = e^{x-1}$, $y = 0$, $x = 1$, $x = 3$

In Exercises 31 and 32, find the volume of the solid generated by revolving the region about the given line.

31. The region in the first quadrant bounded above by the line $y = \sqrt{2}$, below by the curve $y = \sec x \tan x$, and on the left by the *y*-axis, about the line $y = \sqrt{2}$

32. The region in the first quadrant bounded above by the line $y = 2$, below by the curve $y = 2 \sin x$, $0 \le x \le \pi/2$, and on the left by the *y*-axis, about the line $y = 2$

Find the volumes of the solids generated by revolving the regions bounded by the lines and curves in Exercises 33–38 about the *y*-axis.

33. The region enclosed by $x = \sqrt{5}y^2$, $x = 0$, $y = -1$, $y = 1$
34. The region enclosed by $x = y^{3/2}$, $x = 0$, $y = 2$
35. The region enclosed by $x = \sqrt{2\sin 2y}$, $0 \le y \le \pi/2$, $x = 0$
36. The region enclosed by $x = \sqrt{\cos(\pi y/4)}$, $-2 \le y \le 0$, $x = 0$
37. $x = 2/\sqrt{y + 1}$, $x = 0$, $y = 0$, $y = 3$
38. $x = \sqrt{2y}/(y^2 + 1)$, $x = 0$, $y = 1$

Volumes by the Washer Method

Find the volumes of the solids generated by revolving the shaded regions in Exercises 39 and 40 about the indicated axes.

39. The *x*-axis

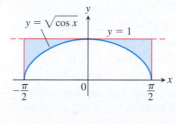

$y = \sqrt{\cos x}$, $y = 1$

40. The *y*-axis

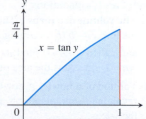

$x = \tan y$

Find the volumes of the solids generated by revolving the regions bounded by the lines and curves in Exercises 41–46 about the *x*-axis.

41. $y = x$, $y = 1$, $x = 0$
42. $y = 2\sqrt{x}$, $y = 2$, $x = 0$
43. $y = x^2 + 1$, $y = x + 3$
44. $y = 4 - x^2$, $y = 2 - x$
45. $y = \sec x$, $y = \sqrt{2}$, $-\pi/4 \le x \le \pi/4$
46. $y = \sec x$, $y = \tan x$, $x = 0$, $x = 1$

In Exercises 47–50, find the volume of the solid generated by revolving each region about the y-axis.

47. The region enclosed by the triangle with vertices $(1, 0)$, $(2, 1)$, and $(1, 1)$

48. The region enclosed by the triangle with vertices $(0, 1)$, $(1, 0)$, and $(1, 1)$

49. The region in the first quadrant bounded above by the parabola $y = x^2$, below by the x-axis, and on the right by the line $x = 2$

50. The region in the first quadrant bounded on the left by the circle $x^2 + y^2 = 3$, on the right by the line $x = \sqrt{3}$, and above by the line $y = \sqrt{3}$

In Exercises 51 and 52, find the volume of the solid generated by revolving each region about the given axis.

51. The region in the first quadrant bounded above by the curve $y = x^2$, below by the x-axis, and on the right by the line $x = 1$, about the line $x = -1$

52. The region in the second quadrant bounded above by the curve $y = -x^3$, below by the x-axis, and on the left by the line $x = -1$, about the line $x = -2$

Volumes of Solids of Revolution

53. Find the volume of the solid generated by revolving the region bounded by $y = \sqrt{x}$ and the lines $y = 2$ and $x = 0$ about

  a. the x-axis.   **b.** the y-axis.

  c. the line $y = 2$.   **d.** the line $x = 4$.

54. Find the volume of the solid generated by revolving the triangular region bounded by the lines $y = 2x$, $y = 0$, and $x = 1$ about

  a. the line $x = 1$.   **b.** the line $x = 2$.

55. Find the volume of the solid generated by revolving the region bounded by the parabola $y = x^2$ and the line $y = 1$ about

  a. the line $y = 1$.   **b.** the line $y = 2$.

  c. the line $y = -1$.

56. By integration, find the volume of the solid generated by revolving the triangular region with vertices $(0, 0)$, $(b, 0)$, $(0, h)$ about

  a. the x-axis.   **b.** the y-axis.

Theory and Applications

57. The volume of a torus The disk $x^2 + y^2 \le a^2$ is revolved about the line $x = b$ $(b > a)$ to generate a solid shaped like a doughnut and called a *torus*. Find its volume. (*Hint:* $\int_{-a}^{a} \sqrt{a^2 - y^2}\, dy = \pi a^2/2$, since it is the area of a semicircle of radius a.)

58. Volume of a bowl A bowl has a shape that can be generated by revolving the graph of $y = x^2/2$ between $y = 0$ and $y = 5$ about the y-axis.

  a. Find the volume of the bowl.

  b. Related rates If we fill the bowl with water at a constant rate of 3 cubic units per second, how fast will the water level in the bowl be rising when the water is 4 units deep?

59. Volume of a bowl

  a. A hemispherical bowl of radius a contains water to a depth h. Find the volume of water in the bowl.

  b. Related rates Water runs into a sunken concrete hemispherical bowl of radius 5 m at the rate of 0.2 m³/sec. How fast is the water level in the bowl rising when the water is 4 m deep?

60. Explain how you could estimate the volume of a solid of revolution by measuring the shadow cast on a table parallel to its axis of revolution by a light shining directly above it.

61. Volume of a hemisphere Derive the formula $V = (2/3)\pi R^3$ for the volume of a hemisphere of radius R by comparing its cross-sections with the cross-sections of a solid right circular cylinder of radius R and height R from which a solid right circular cone of base radius R and height R has been removed, as suggested by the accompanying figure.

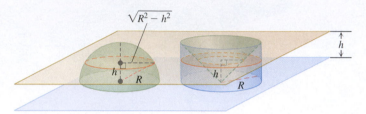

62. Designing a plumb bob Having been asked to design a brass plumb bob that will weigh in the neighborhood of 190 g, you decide to shape it like the solid of revolution shown here. Find the plumb bob's volume. If you specify a brass that weighs 8.5 g/cm³, how much will the plumb bob weigh (to the nearest gram)?

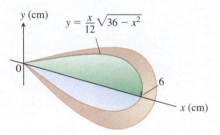

63. Designing a wok You are designing a wok frying pan that will be shaped like a spherical bowl with handles. A bit of experimentation at home persuades you that you can get one that holds about 3 L if you make it 9 cm deep and give the sphere a radius of 16 cm. To be sure, you picture the wok as a solid of revolution, as shown here, and calculate its volume with an integral. To the nearest cubic centimeter, what volume do you really get? $(1 \text{ L} = 1000 \text{ cm}^3)$

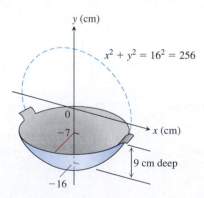

64. Max-min The arch $y = \sin x$, $0 \le x \le \pi$, is revolved about the line $y = c$, $0 \le c \le 1$, to generate the solid in the accompanying figure.

  a. Find the value of c that minimizes the volume of the solid. What is the minimum volume?

b. What value of c in $[0, 1]$ maximizes the volume of the solid?

T **c.** Graph the solid's volume as a function of c, first for $0 \leq c \leq 1$ and then on a larger domain. What happens to the volume of the solid as c moves away from $[0, 1]$? Does this make sense physically? Give reasons for your answers.

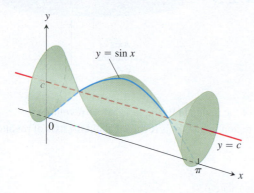

65. Consider the region R bounded by the graphs of $y = f(x) > 0$, $x = a > 0$, $x = b > a$, and $y = 0$ (see accompanying figure). If the volume of the solid formed by revolving R about the x-axis is 4π, and the volume of the solid formed by revolving R about the line $y = -1$ is 8π, find the area of R.

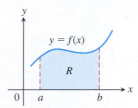

66. Consider the region R given in Exercise 65. If the volume of the solid formed by revolving R around the x-axis is 6π, and the volume of the solid formed by revolving R around the line $y = -2$ is 10π, find the area of R.

6.2 Volumes Using Cylindrical Shells

In Section 6.1 we defined the volume of a solid to be the definite integral $V = \int_a^b A(x)\, dx$, where $A(x)$ is an integrable cross-sectional area of the solid from $x = a$ to $x = b$. The area $A(x)$ was obtained by slicing through the solid with a plane perpendicular to the x-axis. However, this method of slicing is sometimes awkward to apply, as we will illustrate in our first example. To overcome this difficulty, we use the same integral definition for volume, but obtain the area by slicing through the solid in a different way.

Slicing with Cylinders

Suppose we slice through the solid using circular cylinders of increasing radii, like cookie cutters. We slice straight down through the solid so that the axis of each cylinder is parallel to the y-axis. The vertical axis of each cylinder is always the same line, but the radii of the cylinders increase with each slice. In this way the solid is sliced up into thin cylindrical shells of constant thickness that grow outward from their common axis, like circular tree rings. Unrolling a cylindrical shell shows that its volume is approximately that of a rectangular slab with area $A(x)$ and thickness Δx. This slab interpretation allows us to apply the same integral definition for volume as before. The following example provides some insight.

EXAMPLE 1   The region enclosed by the x-axis and the parabola $y = f(x) = 3x - x^2$ is revolved about the vertical line $x = -1$ to generate a solid (see Figure 6.16). Find the volume of the solid.

Solution   Using the washer method from Section 6.1 would be awkward here because we would need to express the x-values of the left and right sides of the parabola in Figure 6.16a in terms of y. (These x-values are the inner and outer radii for a typical washer, requiring us to solve $y = 3x - x^2$ for x, which leads to a complicated formula for x.) Instead of rotating a horizontal strip of thickness Δy, we rotate a *vertical strip* of thickness Δx. This rotation produces a *cylindrical shell* of height y_k above a point x_k within the base of the vertical strip and of thickness Δx. An example of a cylindrical shell is shown as the orange-shaded region in Figure 6.17. We can think of the cylindrical shell shown in the figure as approximating a slice of the solid obtained by cutting straight down through it, parallel to the axis of revolution, all the way around close to the inside hole. We then cut

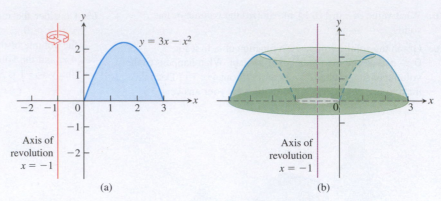

FIGURE 6.16 (a) The graph of the region in Example 1, before revolution. (b) The solid formed when the region in part (a) is revolved about the axis of revolution $x = -1$.

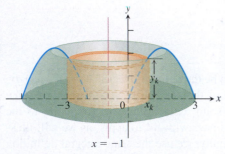

FIGURE 6.17 A cylindrical shell of height y_k obtained by rotating a vertical strip of thickness Δx_k about the line $x = -1$. The outer radius of the cylinder occurs at x_k, where the height of the parabola is $y_k = 3x_k - x_k^2$ (Example 1).

another cylindrical slice around the enlarged hole, then another, and so on, obtaining n cylinders. The radii of the cylinders gradually increase, and the heights of the cylinders follow the contour of the parabola: shorter to taller, then back to shorter (Figure 6.16a). The sum of the volumes of the shells is a Riemann sum that approximates the volume of the entire solid.

Each shell sits over a subinterval $[x_{k-1}, x_k]$ in the x-axis. The thickness of the shell is $\Delta x_k = x_k - x_{k-1}$. Because the parabola is rotated around the line $x = -1$, the outer radius of the shell is $1 + x_k$. The height of the shell is the height of the parabola at some point in the interval $[x_{k-1}, x_k]$, or approximately $y_k = f(x_k) = 3x_k - x_k^2$. If we unroll this cylinder and flatten it out, it becomes (approximately) a rectangular slab with thickness Δx_k (see Figure 6.18). The height of the rectangular slab is approximately $y_k = 3x_k - x_k^2$, and its length is the circumference of the shell, which is approximately $2\pi \cdot \text{radius} = 2\pi(1 + x_k)$. Hence the volume of the shell is approximately the volume of the rectangular slab, which is

$$\Delta V_k = \text{circumference} \times \text{height} \times \text{thickness}$$
$$= 2\pi(1 + x_k) \cdot \left(3x_k - x_k^2\right) \cdot \Delta x_k.$$

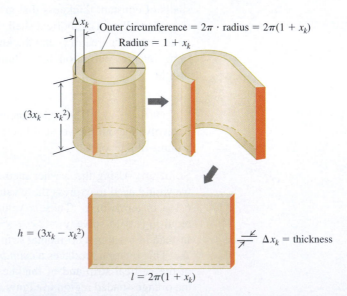

FIGURE 6.18 Cutting and unrolling a cylindrical shell gives a nearly rectangular solid (Example 1).

Summing together the volumes ΔV_k of the individual cylindrical shells over the interval $[0, 3]$ gives the Riemann sum

$$\sum_{k=1}^{n} \Delta V_k = \sum_{k=1}^{n} 2\pi(x_k + 1)\left(3x_k - x_k^2\right)\Delta x_k.$$

Taking the limit as the thickness $\Delta x_k \to 0$ and $n \to \infty$ gives the volume integral

$$V = \lim_{n \to \infty} \sum_{k=1}^{n} 2\pi(x_k + 1)\left(3x_k - x_k^2\right)\Delta x_k$$

$$= \int_0^3 2\pi(x + 1)(3x - x^2)\,dx$$

$$= \int_0^3 2\pi(3x^2 + 3x - x^3 - x^2)\,dx$$

$$= 2\pi\int_0^3 (2x^2 + 3x - x^3)\,dx$$

$$= 2\pi\left[\frac{2}{3}x^3 + \frac{3}{2}x^2 - \frac{1}{4}x^4\right]_0^3 = \frac{45\pi}{2}.$$

We now generalize this procedure to more general solids.

The Shell Method

Suppose that the region bounded by the graph of a nonnegative continuous function $y = f(x)$ and the x-axis over the finite closed interval $[a, b]$ lies to the right of the vertical line $x = L$ (see Figure 6.19a). We assume $a \geq L$, so the vertical line may touch the region but cannot pass through it. We generate a solid S by rotating this region about the vertical line L.

Let P be a partition of the interval $[a, b]$ by the points $a = x_0 < x_1 < \cdots < x_n = b$. As usual, we choose a point c_k in each subinterval $[x_{k-1}, x_k]$. In Example 1 we chose c_k to be the endpoint x_k, but now it will be more convenient to let c_k be the midpoint of the subinterval $[x_{k-1}, x_k]$. We approximate the region in Figure 6.19a with rectangles based on this partition of $[a, b]$. A typical approximating rectangle has height $f(c_k)$ and width

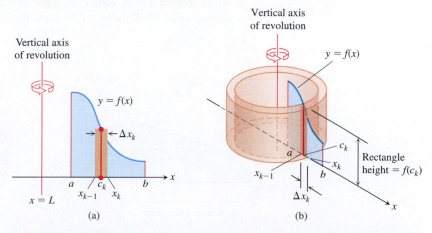

FIGURE 6.19 When the region shown in (a) is revolved about the vertical line $x = L$, a solid is produced which can be sliced into cylindrical shells. A typical shell is shown in (b).

> The volume of a cylindrical shell of height h with inner radius r and outer radius R is
>
> $$\pi R^2 h - \pi r^2 h = 2\pi\left(\frac{R+r}{2}\right)(h)(R-r).$$

$\Delta x_k = x_k - x_{k-1}$. If this rectangle is rotated about the vertical line $x = L$, then a shell is swept out, as in Figure 6.19b. A formula from geometry tells us that the volume of the shell swept out by the rectangle is

$$\Delta V_k = 2\pi \times \text{average shell radius} \times \text{shell height} \times \text{thickness}$$
$$= 2\pi \cdot (c_k - L) \cdot f(c_k) \cdot \Delta x_k. \qquad R = x_k - L \text{ and } r = x_{k-1} - L$$

We approximate the volume of the solid S by summing the volumes of the shells swept out by the n rectangles:

$$V \approx \sum_{k=1}^{n} \Delta V_k.$$

The limit of this Riemann sum as each $\Delta x_k \to 0$ and $n \to \infty$ gives the volume of the solid as a definite integral:

$$V = \lim_{n \to \infty} \sum_{k-1}^{n} \Delta V_k = \int_{a}^{b} 2\pi(\text{shell radius})(\text{shell height})\, dx$$

$$= \int_{a}^{b} 2\pi(x - L)f(x)\, dx.$$

We refer to the variable of integration, here x, as the **thickness variable**. To emphasize the *process* of the shell method, we state the general formula in terms of the shell radius and shell height. This will allow for rotations about a horizontal line L as well.

Shell Formula for Revolution About a Vertical Line

The volume of the solid generated by revolving the region between the x-axis and the graph of a continuous function $y = f(x) \geq 0, L \leq a \leq x \leq b$, about a vertical line $x = L$ is

$$V = \int_{a}^{b} 2\pi \binom{\text{shell}}{\text{radius}}\binom{\text{shell}}{\text{height}} dx.$$

EXAMPLE 2   The region bounded by the curve $y = \sqrt{x}$, the x-axis, and the line $x = 4$ is revolved about the y-axis to generate a solid. Find the volume of the solid.

Solution   Sketch the region and draw a line segment across it *parallel* to the axis of revolution (Figure 6.20a). Label the segment's height (shell height) and distance from the axis of revolution (shell radius). (We drew the shell in Figure 6.20b, but you need not do that.)

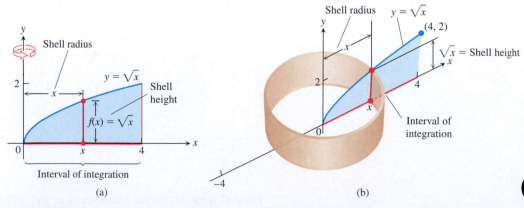

FIGURE 6.20   (a) The region, shell dimensions, and interval of integration in Example 2. (b) The shell swept out by the vertical segment in part (a) with a width Δx.

The shell thickness variable is x, so the limits of integration for the shell formula are $a = 0$ and $b = 4$ (Figure 6.20). The volume is

$$V = \int_a^b 2\pi \binom{\text{shell}}{\text{radius}}\binom{\text{shell}}{\text{height}} dx$$

$$= \int_0^4 2\pi(x)\left(\sqrt{x}\right) dx$$

$$= 2\pi \int_0^4 x^{3/2}\, dx = 2\pi\left[\frac{2}{5}x^{5/2}\right]_0^4 = \frac{128\pi}{5}.$$ ∎

So far, we have used vertical axes of revolution. For horizontal axes, we replace the x's with y's.

EXAMPLE 3   The region bounded by the curve $y = \sqrt{x}$, the x-axis, and the line $x = 4$ is revolved about the x-axis to generate a solid. Find the volume of the solid by the shell method.

Solution   This is the solid whose volume was found by the disk method in Example 4 of Section 6.1. Now we find its volume by the shell method. First, sketch the region and draw a line segment across it *parallel* to the axis of revolution (Figure 6.21a). Label the segment's length (shell height) and distance from the axis of revolution (shell radius). (We drew the shell in Figure 6.21b, but you need not do that.)

In this case, the shell thickness variable is y, so the limits of integration for the shell formula method are $a = 0$ and $b = 2$ (along the y-axis in Figure 6.21). The volume of the solid is

$$V = \int_a^b 2\pi \binom{\text{shell}}{\text{radius}}\binom{\text{shell}}{\text{height}} dy$$

$$= \int_0^2 2\pi(y)(4 - y^2)\, dy$$

$$= 2\pi \int_0^2 (4y - y^3)\, dy$$

$$= 2\pi\left[2y^2 - \frac{y^4}{4}\right]_0^2 = 8\pi.$$

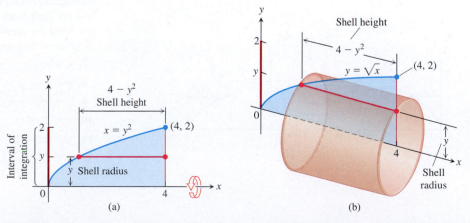

FIGURE 6.21   (a) The region, shell dimensions, and interval of integration in Example 3. (b) The shell swept out by the horizontal segment in part (a) with a width Δy. ∎

> **Summary of the Shell Method**
>
> Regardless of the position of the axis of revolution (horizontal or vertical), the steps for implementing the shell method are these.
>
> 1. *Draw the region and sketch a line segment* across it *parallel* to the axis of revolution. *Label* the segment's height or length (shell height) and distance from the axis of revolution (shell radius).
> 2. *Find the limits of integration* for the thickness variable.
> 3. *Integrate* the product 2π (shell radius) (shell height) with respect to the thickness variable (x or y) to find the volume.

The shell method gives the same answer as the washer method when both are used to calculate the volume of a region. We do not prove that result here, but it is illustrated in Exercises 37 and 38. (Exercise 45 outlines a proof.) Both volume formulas are actually special cases of a general volume formula we will look at when studying double and triple integrals in Chapter 15. That general formula also allows for computing volumes of solids other than those swept out by regions of revolution.

EXERCISES 6.2

Revolution About the Axes

In Exercises 1–6, use the shell method to find the volumes of the solids generated by revolving the shaded region about the indicated axis.

1.

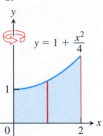

$y = 1 + \dfrac{x^2}{4}$

2.

$y = 2 - \dfrac{x^2}{4}$

3.

$y = \sqrt{2}$

$x = y^2$

4.

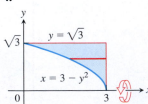

$y = \sqrt{3}$

$x = 3 - y^2$

5. The y-axis

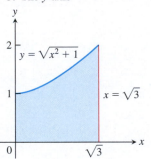

$y = \sqrt{x^2 + 1}$

$x = \sqrt{3}$

6. The y-axis

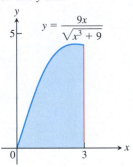

$y = \dfrac{9x}{\sqrt{x^3 + 9}}$

Revolution About the y-Axis

Use the shell method to find the volumes of the solids generated by revolving the regions bounded by the curves and lines in Exercises 7–12 about the y-axis.

7. $y = x$, $\quad y = -x/2$, $\quad x = 2$

8. $y = 2x$, $\quad y = x/2$, $\quad x = 1$

9. $y = x^2$, $\quad y = 2 - x$, $\quad x = 0$, $\quad$ for $x \geq 0$

10. $y = 2 - x^2$, $\quad y = x^2$, $\quad x = 0$

11. $y = 2x - 1$, $\quad y = \sqrt{x}$, $\quad x = 0$

12. $y = 3/(2\sqrt{x})$, $\quad y = 0$, $\quad x = 1$, $\quad x = 4$

13. Let $f(x) = \begin{cases} (\sin x)/x, & 0 < x \leq \pi \\ 1, & x = 0 \end{cases}$

a. Show that $x f(x) = \sin x, 0 \leq x \leq \pi$.

b. Find the volume of the solid generated by revolving the shaded region about the y-axis in the accompanying figure.

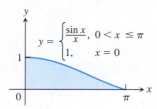

14. Let $g(x) = \begin{cases} (\tan x)^2/x, & 0 < x \leq \pi/4 \\ 0, & x = 0 \end{cases}$

a. Show that $x\, g(x) = (\tan x)^2, 0 \leq x \leq \pi/4$.

b. Find the volume of the solid generated by revolving the shaded region about the y-axis in the accompanying figure.

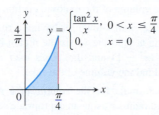

Revolution About the x-Axis

Use the shell method to find the volumes of the solids generated by revolving the regions bounded by the curves and lines in Exercises 15–22 about the x-axis.

15. $x = \sqrt{y}, \quad x = -y, \quad y = 2$

16. $x = y^2, \quad x = -y, \quad y = 2, \quad y \geq 0$

17. $x = 2y - y^2, \quad x = 0$ **18.** $x = 2y - y^2, \quad x = y$

19. $y = |x|, \quad y = 1$ **20.** $y = x, \quad y = 2x, \quad y = 2$

21. $y = \sqrt{x}, \quad y = 0, \quad y = x - 2$

22. $y = \sqrt{x}, \quad y = 0, \quad y = 2 - x$

Revolution About Horizontal and Vertical Lines

In Exercises 23–26, use the shell method to find the volumes of the solids generated by revolving the regions bounded by the given curves about the given lines.

23. $y = 3x, \quad y = 0, \quad x = 2$

    a. The y-axis   **b.** The line $x = 4$

    c. The line $x = -1$ **d.** The x-axis

    e. The line $y = 7$ **f.** The line $y = -2$

24. $y = x^3, \quad y = 8, \quad x = 0$

    a. The y-axis   **b.** The line $x = 3$

    c. The line $x = -2$ **d.** The x-axis

    e. The line $y = 8$ **f.** The line $y = -1$

25. $y = x + 2, \quad y = x^2$

    a. The line $x = 2$ **b.** The line $x = -1$

    c. The x-axis   **d.** The line $y = 4$

26. $y = x^4, \quad y = 4 - 3x^2$

    a. The line $x = 1$ **b.** The x-axis

In Exercises 27 and 28, use the shell method to find the volumes of the solids generated by revolving the shaded regions about the indicated axes.

27. a. The x-axis   **b.** The line $y = 1$

     c. The line $y = 8/5$ **d.** The line $y = -2/5$

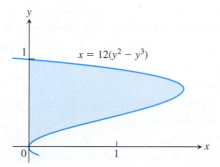

28. a. The x-axis   **b.** The line $y = 2$

     c. The line $y = 5$ **d.** The line $y = -5/8$

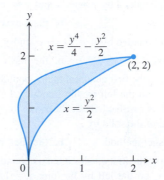

Choosing the Washer Method or Shell Method

For some regions, both the washer and shell methods work well for the solid generated by revolving the region about the coordinate axes, but this is not always the case. When a region is revolved about the y-axis, for example, and washers are used, we must integrate with respect to y. It may not be possible, however, to express the integrand in terms of y. In such a case, the shell method allows us to integrate with respect to x instead. Exercises 29 and 30 provide some insight.

29. Compute the volume of the solid generated by revolving the region bounded by $y = x$ and $y = x^2$ about each coordinate axis using

    a. the shell method.   **b.** the washer method.

30. Compute the volume of the solid generated by revolving the triangular region bounded by the lines $2y = x + 4$, $y = x$, and $x = 0$ about

  a. the x-axis using the washer method.

  b. the y-axis using the shell method.

  c. the line $x = 4$ using the shell method.

  d. the line $y = 8$ using the washer method.

In Exercises 31–36, find the volumes of the solids generated by revolving the regions about the given axes. If you think it would be better to use washers in any given instance, feel free to do so.

31. The triangle with vertices $(1, 1)$, $(1, 2)$, and $(2, 2)$ about

  a. the x-axis   **b.** the y-axis

  c. the line $x = 10/3$ **d.** the line $y = 1$

32. The region bounded by $y = \sqrt{x}$, $y = 2$, $x = 0$ about

  a. the x-axis   **b.** the y-axis

  c. the line $x = 4$ **d.** the line $y = 2$

33. The region in the first quadrant bounded by the curve $x = y - y^3$ and the y-axis about

  a. the x-axis   **b.** the line $y = 1$

34. The region in the first quadrant bounded by $x = y - y^3$, $x = 1$, and $y = 1$ about

  a. the x-axis   **b.** the y-axis

  c. the line $x = 1$ **d.** the line $y = 1$

35. The region bounded by $y = \sqrt{x}$ and $y = x^2/8$ about

  a. the x-axis   **b.** the y-axis

36. The region bounded by $y = 2x - x^2$ and $y = x$ about

  a. the y-axis   **b.** the line $x = 1$

37. The region in the first quadrant that is bounded above by the curve $y = 1/x^{1/4}$, on the left by the line $x = 1/16$, and below by the line $y = 1$ is revolved about the x-axis to generate a solid. Find the volume of the solid by

  a. the washer method.   **b.** the shell method.

38. The region in the first quadrant that is bounded above by the curve $y = 1/\sqrt{x}$, on the left by the line $x = 1/4$, and below by the line $y = 1$ is revolved about the y-axis to generate a solid. Find the volume of the solid by

  a. the washer method.   **b.** the shell method.

Theory and Examples

39. The region shown here is to be revolved about the x-axis to generate a solid. Which of the methods (disk, washer, shell) could you use to find the volume of the solid? How many integrals would be required in each case? Explain.

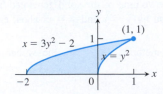

40. The region shown here is to be revolved about the y-axis to generate a solid. Which of the methods (disk, washer, shell) could you use to find the volume of the solid? How many integrals would be required in each case? Give reasons for your answers.

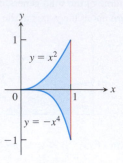

41. A bead is formed from a sphere of radius 5 by drilling through a diameter of the sphere with a drill bit of radius 3.

  a. Find the volume of the bead.

  b. Find the volume of the removed portion of the sphere.

42. A Bundt cake, well known for having a ringed shape, is formed by revolving around the y-axis the region bounded by the graph of $y = \sin(x^2 - 1)$ and the x-axis over the interval $1 \le x \le \sqrt{1 + \pi}$. Find the volume of the cake.

43. Derive the formula for the volume of a right circular cone of height h and radius r using an appropriate solid of revolution.

44. Derive the equation for the volume of a sphere of radius r using the shell method.

45. **Equivalence of the washer and shell methods for finding volume** Let f be differentiable and increasing on the interval $a \le x \le b$, with $a > 0$, and suppose that f has a differentiable inverse, f^{-1}. Revolve about the y-axis the region bounded by the graph of f and the lines $x = a$ and $y = f(b)$ to generate a solid. Then the values of the integrals given by the washer and shell methods for the volume have identical values:

$$\int_{f(a)}^{f(b)} \pi((f^{-1}(y))^2 - a^2)\, dy = \int_{a}^{b} 2\pi x(f(b) - f(x))\, dx.$$

To prove this equality, define

$$W(t) = \int_{f(a)}^{f(t)} \pi((f^{-1}(y))^2 - a^2)\, dy$$

$$S(t) = \int_{a}^{t} 2\pi x(f(t) - f(x))\, dx.$$

Then show that the functions W and S agree at a point of $[a, b]$ and have identical derivatives on $[a, b]$. As you saw in Section 4.8, Exercise 132, this will guarantee $W(t) = S(t)$ for all t in $[a, b]$. In particular, $W(b) = S(b)$. (*Source:* "Disks and Shells Revisited" by Walter Carlip, in *American Mathematical Monthly*, Feb. 1991, vol. 98, no. 2, pp. 154–156.)

46. The region between the curve $y = \sec^{-1} x$ and the x-axis from $x = 1$ to $x = 2$ (shown here) is revolved about the y-axis to generate a solid. Find the volume of the solid.

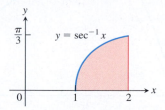

47. Find the volume of the solid generated by revolving the region enclosed by the graphs of $y = e^{-x^2}$, $y = 0$, $x = 0$, and $x = 1$ about the y-axis.

48. Find the volume of the solid generated by revolving the region enclosed by the graphs of $y = e^{x/2}$, $y = 1$, and $x = \ln 3$ about the x-axis.

49. Consider the region R bounded by the graphs of $y = f(x) > 0$, $x = a > 0$, and $x = b > a$. If the volume of the solid formed by revolving R about the y-axis is 2π, and the volume formed by revolving R about the line $x = -2$ is 10π, find the area of R.

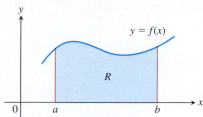

50. Consider the region R given in Exercise 49. If the area of region R is 1, and the volume of the solid formed by revolving R about the line $x = -3$ is 10π, find the volume of the solid formed by revolving R about the y-axis.

6.3 Arc Length

We know what is meant by the length of a straight-line segment, but without calculus, we have no precise definition of the length of a general winding curve. If the curve is the graph of a continuous function defined over an interval, then we can find the length of the curve using a procedure similar to that we used for defining the area between the curve and the x-axis. We divide the curve into many pieces, and we approximate each piece by a straight-line segment. The sum of the lengths of these segments is an approximation to the total curve length that we seek. The total length of the curve is the limiting value of these approximations as the number of segments goes to infinity.

Length of a Curve $y = f(x)$

Suppose the curve whose length we want to find is the graph of the function $y = f(x)$ from $x = a$ to $x = b$. In order to derive an integral formula for the length of the curve, we assume that f has a continuous derivative at every point of $[a, b]$. Such a function is called **smooth**, and its graph is a **smooth curve** because it does not have any breaks, corners, or cusps.

We partition the interval $[a, b]$ into n subintervals with $a = x_0 < x_1 < x_2 < \cdots < x_n = b$. If $y_k = f(x_k)$, then the corresponding point $P_k(x_k, y_k)$ lies on the curve. Next we connect successive points P_{k-1} and P_k with straight-line segments that, taken together, form a polygonal path whose length approximates the length of the curve (Figure 6.22). If we set $\Delta x_k = x_k - x_{k-1}$ and $\Delta y_k = y_k - y_{k-1}$, then a representative line segment in the path has length (see Figure 6.23)

$$L_k = \sqrt{(\Delta x_k)^2 + (\Delta y_k)^2},$$

so the length of the curve is approximated by the sum

$$\sum_{k=1}^{n} L_k = \sum_{k=1}^{n} \sqrt{(\Delta x_k)^2 + (\Delta y_k)^2}. \tag{1}$$

We expect the approximation to improve as the partition of $[a, b]$ becomes finer. In order to evaluate this limit, we use the Mean Value Theorem, which tells us that there is a point c_k, with $x_{k-1} < c_k < x_k$, such that

$$\Delta y_k = f'(c_k) \, \Delta x_k.$$

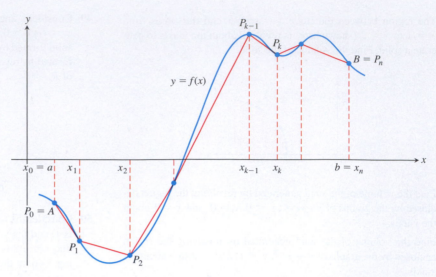

FIGURE 6.22 The length of the polygonal path $P_0P_1P_2\cdots P_n$ approximates the length of the curve $y = f(x)$ from point A to point B.

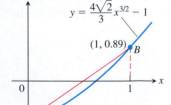

FIGURE 6.23 The arc $P_{k-1}P_k$ of the curve $y = f(x)$ is approximated by the straight-line segment shown here, which has length $L_k = \sqrt{(\Delta x_k)^2 + (\Delta y_k)^2}$.

Substituting this for Δy_k, the sums in Equation (1) take the form

$$\sum_{k=1}^{n} L_k = \sum_{k=1}^{n} \sqrt{(\Delta x_k)^2 + (f'(c_k)\Delta x_k)^2} = \sum_{k=1}^{n} \sqrt{1 + [f'(c_k)]^2}\,\Delta x_k. \qquad (2)$$

This is a Riemann sum whose limit we can evaluate. Because $\sqrt{1 + [f'(x)]^2}$ is continuous on $[a, b]$, the limit of the Riemann sum on the right-hand side of Equation (2) exists and has the value

$$\lim_{n\to\infty} \sum_{k=1}^{n} L_k = \lim_{n\to\infty} \sum_{k=1}^{n} \sqrt{1 + [f'(c_k)]^2}\,\Delta x_k = \int_a^b \sqrt{1 + [f'(x)]^2}\,dx.$$

We define the length of the curve to be this integral.

DEFINITION If f' is continuous on $[a, b]$, then the **length (arc length)** of the curve $y = f(x)$ from the point $A = (a, f(a))$ to the point $B = (b, f(b))$ is the value of the integral

$$L = \int_a^b \sqrt{1 + [f'(x)]^2}\,dx = \int_a^b \sqrt{1 + \left(\frac{dy}{dx}\right)^2}\,dx. \qquad (3)$$

EXAMPLE 1 Find the length of the curve shown in Figure 6.24, which is the graph of the function

$$y = \frac{4\sqrt{2}}{3}x^{3/2} - 1, \qquad 0 \le x \le 1.$$

Solution We use Equation (3) with $a = 0$, $b = 1$, and

$$y = \frac{4\sqrt{2}}{3}x^{3/2} - 1 \qquad \text{If } x = 1, \text{ then } y \approx 0.89$$

$$\frac{dy}{dx} = \frac{4\sqrt{2}}{3} \cdot \frac{3}{2}x^{1/2} = 2\sqrt{2}\,x^{1/2}$$

$$\left(\frac{dy}{dx}\right)^2 = \left(2\sqrt{2}\,x^{1/2}\right)^2 = 8x.$$

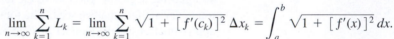

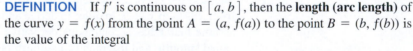

FIGURE 6.24 The length of the curve is slightly larger than the length of the line segment joining points A and B (Example 1).

The length of the curve over $x = 0$ to $x = 1$ is

$$L = \int_0^1 \sqrt{1 + \left(\frac{dy}{dx}\right)^2}\, dx = \int_0^1 \sqrt{1 + 8x}\, dx \qquad \text{Eq. (3) with } a = 0, b = 1.$$

$$= \frac{2}{3} \cdot \frac{1}{8}(1 + 8x)^{3/2}\Big]_0^1 = \frac{13}{6} \approx 2.17. \qquad \begin{array}{l}\text{Let } u = 1 + 8x, \text{ integrate,}\\ \text{and replace } u \text{ by } 1 + 8x.\end{array}$$

Notice that the length of the curve is slightly larger than the length of the straight-line segment joining the points $A = (0, -1)$ and $B = \left(1, 4\sqrt{2}/3 - 1\right)$ on the curve (see Figure 6.24):

$$2.17 > \sqrt{1^2 + (1.89)^2} \approx 2.14. \qquad \text{Decimal approximations} \qquad \blacksquare$$

EXAMPLE 2 Find the length of the graph of

$$f(x) = \frac{x^3}{12} + \frac{1}{x}, \qquad 1 \le x \le 4.$$

Solution A graph of the function is shown in Figure 6.25. To use Equation (3), we find

$$f'(x) = \frac{x^2}{4} - \frac{1}{x^2}$$

so

$$1 + [f'(x)]^2 = 1 + \left(\frac{x^2}{4} - \frac{1}{x^2}\right)^2 = 1 + \left(\frac{x^4}{16} - \frac{1}{2} + \frac{1}{x^4}\right)$$

$$= \frac{x^4}{16} + \frac{1}{2} + \frac{1}{x^4} = \left(\frac{x^2}{4} + \frac{1}{x^2}\right)^2.$$

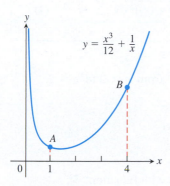

FIGURE 6.25 The curve in Example 2, where $A = (1, 13/12)$ and $B = (4, 67/12)$.

The length of the graph over $[1, 4]$ is

$$L = \int_1^4 \sqrt{1 + [f'(x)]^2}\, dx = \int_1^4 \left(\frac{x^2}{4} + \frac{1}{x^2}\right) dx$$

$$= \left[\frac{x^3}{12} - \frac{1}{x}\right]_1^4 = \left(\frac{64}{12} - \frac{1}{4}\right) - \left(\frac{1}{12} - 1\right) = \frac{72}{12} = 6. \qquad \blacksquare$$

EXAMPLE 3 Find the length of the curve

$$y = \frac{1}{2}(e^x + e^{-x}), \qquad 0 \le x \le 2.$$

Solution We use Equation (3) with $a = 0$, $b = 2$, and

$$y = \frac{1}{2}(e^x + e^{-x})$$

$$\frac{dy}{dx} = \frac{1}{2}(e^x - e^{-x})$$

$$\left(\frac{dy}{dx}\right)^2 = \frac{1}{4}(e^{2x} - 2 + e^{-2x})$$

$$1 + \left(\frac{dy}{dx}\right)^2 = \frac{1}{4}(e^{2x} + 2 + e^{-2x}) = \left[\frac{1}{2}(e^x + e^{-x})\right]^2.$$

The length of the curve from $x = 0$ to $x = 2$ is

$$L = \int_0^2 \sqrt{1 + \left(\frac{dy}{dx}\right)^2}\, dx = \int_0^2 \frac{1}{2}(e^x + e^{-x})\, dx \qquad \text{Eq. (3) with } a = 0, b = 2.$$

$$= \frac{1}{2}\left[e^x - e^{-x}\right]_0^2 = \frac{1}{2}(e^2 - e^{-2}) \approx 3.63. \qquad \blacksquare$$

Dealing with Discontinuities in dy/dx

Even if the derivative dy/dx does not exist at some point on a curve, it is possible that dx/dy could exist. This can happen, for example, when a curve has a vertical tangent. In this case, we may be able to find the curve's length by expressing x as a function of y and applying the following analogue of Equation (3):

Formula for the Length of $x = g(y)$, $c \leq y \leq d$

If g' is continuous on $[c, d]$, the length of the curve $x = g(y)$ from $A = (g(c), c)$ to $B = (g(d), d)$ is

$$L = \int_c^d \sqrt{1 + \left(\frac{dx}{dy}\right)^2}\, dy = \int_c^d \sqrt{1 + [g'(y)]^2}\, dy. \qquad (4)$$

EXAMPLE 4 Find the length of the curve $y = (x/2)^{2/3}$ from $x = 0$ to $x = 2$.

Solution The derivative

$$\frac{dy}{dx} = \frac{2}{3}\left(\frac{x}{2}\right)^{-1/3}\left(\frac{1}{2}\right) = \frac{1}{3}\left(\frac{2}{x}\right)^{1/3}$$

is not defined at $x = 0$, so we cannot find the curve's length with Equation (3).

We therefore rewrite the equation to express x in terms of y:

$$y = \left(\frac{x}{2}\right)^{2/3}$$

$$y^{3/2} = \frac{x}{2} \qquad \text{Raise both sides to the power } 3/2.$$

$$x = 2y^{3/2}. \qquad \text{Solve for } x.$$

From this we see that the curve whose length we want is also the graph of $x = 2y^{3/2}$ from $y = 0$ to $y = 1$ (see Figure 6.26).

The derivative

$$\frac{dx}{dy} = 2\left(\frac{3}{2}\right)y^{1/2} = 3y^{1/2}$$

is continuous on $[0, 1]$. We may therefore use Equation (4) to find the curve's length:

$$L = \int_c^d \sqrt{1 + \left(\frac{dx}{dy}\right)^2}\, dy = \int_0^1 \sqrt{1 + 9y}\, dy \qquad \text{Eq. (4) with } c = 0, d = 1.$$

$$= \frac{1}{9} \cdot \frac{2}{3}(1 + 9y)^{3/2}\Big]_0^1 \qquad \begin{array}{l}\text{Let } u = 1 + 9y, \, du/9 = dy, \\ \text{integrate, and substitute back.}\end{array}$$

$$= \frac{2}{27}\left(10\sqrt{10} - 1\right) \approx 2.27. \qquad \blacksquare$$

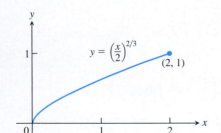

FIGURE 6.26 The graph of $y = (x/2)^{2/3}$ from $x = 0$ to $x = 2$ is also the graph of $x = 2y^{3/2}$ from $y = 0$ to $y = 1$ (Example 4).

The Differential Formula for Arc Length

If $y = f(x)$ and if f' is continuous on $[a, b]$, then by the Fundamental Theorem of Calculus we can define a new function

$$s(x) = \int_a^x \sqrt{1 + [f'(t)]^2}\, dt. \qquad (5)$$

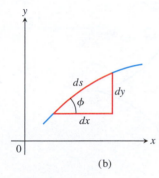

FIGURE 6.27 Diagrams for remembering the equation $ds = \sqrt{dx^2 + dy^2}$.

From Equation (3) and Figure 6.22, we see that this function $s(x)$ is continuous and measures the length along the curve $y = f(x)$ from the initial point $P_0(a, f(a))$ to the point $Q(x, f(x))$ for each $x \in [a, b]$. The function s is called the **arc length function** for $y = f(x)$. From the Fundamental Theorem, the function s is differentiable on (a, b) and

$$\frac{ds}{dx} = \sqrt{1 + [f'(x)]^2} = \sqrt{1 + \left(\frac{dy}{dx}\right)^2}.$$

Then the differential of arc length is

$$ds = \sqrt{1 + \left(\frac{dy}{dx}\right)^2}\, dx. \tag{6}$$

A useful way to remember Equation (6) is to write

$$ds = \sqrt{dx^2 + dy^2}, \tag{7}$$

which can be integrated between appropriate limits to give the total length of a curve. From this point of view, all the arc length formulas are simply different expressions for the equation $L = \int ds$. Figure 6.27a gives the exact interpretation of ds corresponding to Equation (7). Figure 6.27b is not strictly accurate, but it can be thought of as a simplified approximation of Figure 6.27a. That is, $ds \approx \Delta s$.

EXAMPLE 5   Find the arc length function for the curve in Example 2, taking $A = (1, 13/12)$ as the starting point (see Figure 6.25).

Solution   In the solution to Example 2, we found that

$$1 + [f'(x)]^2 = \left(\frac{x^2}{4} + \frac{1}{x^2}\right)^2.$$

Therefore the arc length function is given by

$$s(x) = \int_1^x \sqrt{1 + [f'(t)]^2}\, dt = \int_1^x \left(\frac{t^2}{4} + \frac{1}{t^2}\right) dt$$

$$= \left[\frac{t^3}{12} - \frac{1}{t}\right]_1^x = \frac{x^3}{12} - \frac{1}{x} + \frac{11}{12}.$$

To compute the arc length along the curve from $A = (1, 13/12)$ to $B = (4, 67/12)$, for instance, we simply calculate

$$s(4) = \frac{4^3}{12} - \frac{1}{4} + \frac{11}{12} = 6.$$

This is the same result we obtained in Example 2.   ∎

EXERCISES 6.3

Finding Lengths of Curves

Find the lengths of the curves in Exercises 1–16. If you have graphing software, you may want to graph these curves to see what they look like.

1. $y = (1/3)(x^2 + 2)^{3/2}$ from $x = 0$ to $x = 3$

2. $y = x^{3/2}$ from $x = 0$ to $x = 4$

3. $x = (y^3/3) + 1/(4y)$ from $y = 1$ to $y = 3$

4. $x = (y^{3/2}/3) - y^{1/2}$ from $y = 1$ to $y = 9$

5. $x = (y^4/4) + 1/(8y^2)$ from $y = 1$ to $y = 2$

6. $x = (y^3/6) + 1/(2y)$ from $y = 2$ to $y = 3$

7. $y = (3/4)x^{4/3} - (3/8)x^{2/3} + 5$, $1 \le x \le 8$

8. $y = (x^3/3) + x^2 + x + 1/(4x + 4)$, $0 \le x \le 2$

9. $y = \ln x - \dfrac{x^2}{8}$ from $x = 1$ to $x = 2$

10. $y = \dfrac{x^2}{2} - \dfrac{\ln x}{4}$ from $x = 1$ to $x = 3$

11. $y = \dfrac{x^3}{3} + \dfrac{1}{4x}$, $1 \le x \le 3$

12. $y = \dfrac{x^5}{5} + \dfrac{1}{12x^3}$, $\dfrac{1}{2} \le x \le 1$

13. $y = \dfrac{3}{2}x^{2/3} + 1$, $\dfrac{1}{8} \le x \le 1$

14. $y = \dfrac{1}{2}(e^x + e^{-x})$, $-1 \le x \le 1$

15. $x = \displaystyle\int_0^y \sqrt{\sec^4 t - 1} \, dt$, $-\pi/4 \le y \le \pi/4$

16. $y = \displaystyle\int_{-2}^x \sqrt{3t^4 - 1} \, dt$, $-2 \le x \le -1$

T **Finding Integrals for Lengths of Curves**
In Exercises 17–24, do the following.

    a. Set up an integral for the length of the curve.

    b. Graph the curve to see what it looks like.

    c. Use your grapher's or computer's integral evaluator to find the curve's length numerically.

17. $y = x^2$, $-1 \le x \le 2$

18. $y = \tan x$, $-\pi/3 \le x \le 0$

19. $x = \sin y$, $0 \le y \le \pi$

20. $x = \sqrt{1 - y^2}$, $-1/2 \le y \le 1/2$

21. $y^2 + 2y = 2x + 1$ from $(-1, -1)$ to $(7, 3)$

22. $y = \sin x - x \cos x$, $0 \le x \le \pi$

23. $y = \displaystyle\int_0^x \tan t \, dt$, $0 \le x \le \pi/6$

24. $x = \displaystyle\int_0^y \sqrt{\sec^2 t - 1} \, dt$, $-\pi/3 \le y \le \pi/4$

Theory and Examples

25. a. Find a curve with a positive derivative through the point $(1, 1)$ whose length integral (Equation 3) is
$$L = \int_1^4 \sqrt{1 + \frac{1}{4x}} \, dx.$$

    b. How many such curves are there? Give reasons for your answer.

26. a. Find a curve with a positive derivative through the point $(0, 1)$ whose length integral (Equation 4) is
$$L = \int_1^2 \sqrt{1 + \frac{1}{y^4}} \, dy.$$

    b. How many such curves are there? Give reasons for your answer.

27. Find the length of the curve
$$y = \int_0^x \sqrt{\cos 2t} \, dt$$
from $x = 0$ to $x = \pi/4$.

28. The length of an astroid The graph of the equation $x^{2/3} + y^{2/3} = 1$ is one of a family of curves called *astroids* (not "asteroids") because of their starlike appearance (see the accompanying figure). Find the length of this particular astroid by finding the length of half the first-quadrant portion, $y = (1 - x^{2/3})^{3/2}$, $\sqrt{2}/4 \le x \le 1$, and multiplying by 8.

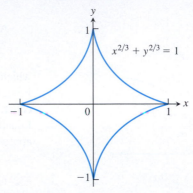

29. Length of a line segment Use the arc length formula (Equation 3) to find the length of the line segment $y = 3 - 2x$, $0 \le x \le 2$. Check your answer by finding the length of the segment as the hypotenuse of a right triangle.

30. Circumference of a circle Set up an integral to find the circumference of a circle of radius r centered at the origin. You will learn how to evaluate the integral in Section 8.3.

31. If $9x^2 = y(y - 3)^2$, show that
$$ds^2 = \frac{(y + 1)^2}{4y} \, dy^2.$$

32. If $4x^2 - y^2 = 64$, show that
$$ds^2 = \frac{4}{y^2} (5x^2 - 16) \, dx^2.$$

33. Is there a smooth (continuously differentiable) curve $y = f(x)$ whose length over the interval $0 \le x \le a$ is always $\sqrt{2}a$? Give reasons for your answer.

34. Using tangent fins to derive the length formula for curves Assume that f is smooth on $[a, b]$ and partition the interval $[a, b]$ in the usual way. In each subinterval $[x_{k-1}, x_k]$, construct the *tangent fin* at the point $(x_{k-1}, f(x_{k-1}))$, as shown in the accompanying figure.

    a. Show that the length of the kth tangent fin over the interval $[x_{k-1}, x_k]$ equals $\sqrt{(\Delta x_k)^2 + (f'(x_{k-1}) \, \Delta x_k)^2}$.

    b. Show that
$$\lim_{n \to \infty} \sum_{k=1}^n (\text{length of } k\text{th tangent fin}) = \int_a^b \sqrt{1 + (f'(x))^2} \, dx,$$

which is the length L of the curve $y = f(x)$ from a to b.

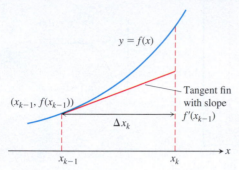

35. Approximate the arc length of one-quarter of the unit circle (which is $\pi/2$) by computing the length of the polygonal approximation with $n = 4$ segments (see accompanying figure).

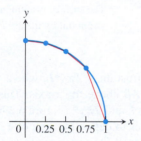

36. Distance between two points Assume that the two points (x_1, y_1) and (x_2, y_2) lie on the graph of the straight line $y = mx + b$. Use the arc length formula (Equation 3) to find the distance between the two points.

37. Find the arc length function for the graph of $f(x) = 2x^{3/2}$ using $(0, 0)$ as the starting point. What is the length of the curve from $(0, 0)$ to $(1, 2)$?

38. Find the arc length function for the curve in Exercise 8, using $(0, 1/4)$ as the starting point. What is the length of the curve from $(0, 1/4)$ to $(1, 59/24)$?

COMPUTER EXPLORATIONS

In Exercises 39–44, use a CAS to perform the following steps for the given graph of the function over the closed interval.

   a. Plot the curve together with the polygonal path approximations for $n = 2, 4, 8$ partition points over the interval. (See Figure 6.22.)

   b. Find the corresponding approximation to the length of the curve by summing the lengths of the line segments.

   c. Evaluate the length of the curve using an integral. Compare your approximations for $n = 2, 4, 8$ with the actual length given by the integral. How does the actual length compare with the approximations as n increases? Explain your answer.

39. $f(x) = \sqrt{1 - x^2}, \quad -1 \le x \le 1$

40. $f(x) = x^{1/3} + x^{2/3}, \quad 0 \le x \le 2$

41. $f(x) = \sin(\pi x^2), \quad 0 \le x \le \sqrt{2}$

42. $f(x) = x^2 \cos x, \quad 0 \le x \le \pi$

43. $f(x) = \dfrac{x - 1}{4x^2 + 1}, \quad -\dfrac{1}{2} \le x \le 1$

44. $f(x) = x^3 - x^2, \quad -1 \le x \le 1$

6.4 Areas of Surfaces of Revolution

When you jump rope, the rope sweeps out a surface in the space around you similar to what is called a *surface of revolution*. The surface surrounds a volume of revolution, and many applications require that we know the area of the surface rather than the volume it encloses. In this section we define areas of surfaces of revolution. More general surfaces are treated in Chapter 16.

Defining Surface Area

If you revolve a region in the plane that is bounded by the graph of a function over an interval, it sweeps out a solid of revolution, as we saw earlier in the chapter. However, if you revolve only the bounding curve itself, it does not sweep out any interior volume but rather a surface that surrounds the solid and forms part of its boundary. Just as we were interested in defining and finding the length of a curve in the last section, we are now interested in defining and finding the area of a surface generated by revolving a curve about an axis.

Before considering general curves, we begin by rotating horizontal and slanted line segments about the x-axis. If we rotate the horizontal line segment AB having length Δx about the x-axis (Figure 6.28a), we generate a cylinder with surface area $2\pi y \Delta x$. This area is the same as that of a rectangle with side lengths Δx and $2\pi y$ (Figure 6.28b). The length $2\pi y$ is the circumference of the circle of radius y generated by rotating the point (x, y) on the line AB about the x-axis.

Suppose the line segment AB has length L and is slanted rather than horizontal. Now when AB is rotated about the x-axis, it generates a frustum of a cone (Figure 6.29a). From

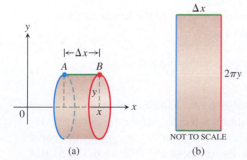

FIGURE 6.28 (a) A cylindrical surface generated by rotating the horizontal line segment AB of length Δx about the x-axis has area $2\pi y \Delta x$. (b) The cut and rolled-out cylindrical surface as a rectangle.

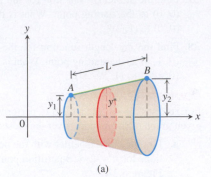

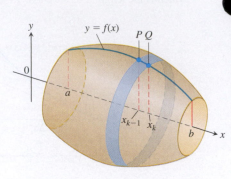

NOT TO SCALE

(a)

(b)

FIGURE 6.29 (a) The frustum of a cone generated by rotating the slanted line segment AB of length L about the x-axis has area $2\pi y^* L$. (b) The area of the rectangle for $y^* = \dfrac{y_1 + y_2}{2}$, the average height of AB above the x-axis.

FIGURE 6.30 The surface generated by revolving the graph of a nonnegative function $y = f(x)$, $a \leq x \leq b$, about the x-axis. The surface is a union of bands like the one swept out by the arc PQ.

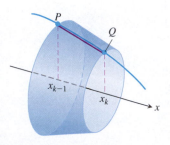

FIGURE 6.31 The line segment joining P and Q sweeps out a frustum of a cone.

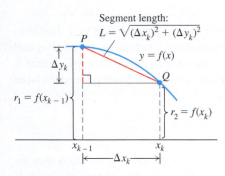

FIGURE 6.32 Dimensions associated with the arc and line segment PQ.

classical geometry, the surface area of this frustum is $2\pi y^* L$, where $y^* = (y_1 + y_2)/2$ is the average height of the slanted segment AB above the x-axis. This surface area is the same as that of a rectangle with side lengths L and $2\pi y^*$ (Figure 6.29b).

Let's build on these geometric principles to define the area of a surface swept out by revolving more general curves about the x-axis. Suppose we want to find the area of the surface swept out by revolving the graph of a nonnegative continuous function $y = f(x)$, $a \leq x \leq b$, about the x-axis. We partition the closed interval $[a, b]$ in the usual way and use the points in the partition to subdivide the graph into short arcs. Figure 6.30 shows a typical arc PQ and the band it sweeps out as part of the graph of f.

As the arc PQ revolves about the x-axis, the line segment joining P and Q sweeps out a frustum of a cone whose axis lies along the x-axis (Figure 6.31). The surface area of this frustum approximates the surface area of the band swept out by the arc PQ. The surface area of the frustum of the cone shown in Figure 6.31 is $2\pi y^* L$, where y^* is the average height of the line segment joining P and Q, and L is its length (just as before). Since $f \geq 0$, from Figure 6.32 we see that the average height of the line segment is $y^* = (f(x_{k-1}) + f(x_k))/2$, and the slant length is $L = \sqrt{(\Delta x_k)^2 + (\Delta y_k)^2}$. Therefore,

$$\text{Frustum surface area} = 2\pi \cdot \frac{f(x_{k-1}) + f(x_k)}{2} \cdot \sqrt{(\Delta x_k)^2 + (\Delta y_k)^2}$$

$$= \pi(f(x_{k-1}) + f(x_k))\sqrt{(\Delta x_k)^2 + (\Delta y_k)^2}.$$

The area of the original surface, being the sum of the areas of the bands swept out by arcs like arc PQ, is approximated by the frustum area sum

$$\sum_{k=1}^{n} \pi(f(x_{k-1}) + f(x_k))\sqrt{(\Delta x_k)^2 + (\Delta y_k)^2}. \tag{1}$$

We expect the approximation to improve as the partition of $[a, b]$ becomes finer. To find the limit, we first need to find an appropriate substitution for Δy_k. If the function f is differentiable, then by the Mean Value Theorem, there is a point $(c_k, f(c_k))$ on the curve between P and Q where the tangent is parallel to the segment PQ (Figure 6.33). At this point,

$$f'(c_k) = \frac{\Delta y_k}{\Delta x_k},$$

$$\Delta y_k = f'(c_k)\, \Delta x_k.$$

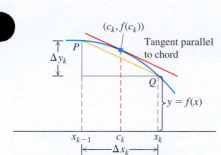

FIGURE 6.33 If f is smooth, the Mean Value Theorem guarantees the existence of a point c_k where the tangent is parallel to segment PQ.

With this substitution for Δy_k, the sums in Equation (1) take the form

$$\sum_{k=1}^{n} \pi(f(x_{k-1}) + f(x_k))\sqrt{(\Delta x_k)^2 + (f'(c_k)\,\Delta x_k)^2}$$

$$= \sum_{k=1}^{n} \pi(f(x_{k-1}) + f(x_k))\sqrt{1 + (f'(c_k))^2}\,\Delta x_k. \qquad (2)$$

These sums are not the Riemann sums of any function because the points $x_{k-1}, x_k,$ and c_k are not the same. However, the points $x_{k-1}, x_k,$ and c_k are very close to each other, and so we expect (and it can be proved) that as the norm of the partition of $[a, b]$ goes to zero, the sums in Equation (2) converge to the integral

$$\int_{a}^{b} 2\pi f(x)\sqrt{1 + (f'(x))^2}\,dx.$$

We therefore define this integral to be the area of the surface swept out by the graph of f from a to b.

DEFINITION   If the function $f(x) \geq 0$ is continuously differentiable on $[a, b]$, the **area of the surface** generated by revolving the graph of $y = f(x)$ about the x-axis is

$$S = \int_{a}^{b} 2\pi y \sqrt{1 + \left(\frac{dy}{dx}\right)^2}\,dx = \int_{a}^{b} 2\pi f(x)\sqrt{1 + (f'(x))^2}\,dx. \qquad (3)$$

Note that the square root in Equation (3) is similar to the one that appears in the formula for the arc length differential of the generating curve in Equation (6) of Section 6.3.

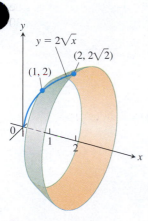

FIGURE 6.34 In Example 1 we calculate the area of this surface.

EXAMPLE 1   Find the area of the surface generated by revolving the curve $y = 2\sqrt{x}$, $1 \leq x \leq 2$, about the x-axis (Figure 6.34).

Solution   We evaluate the formula

$$S = \int_{a}^{b} 2\pi y \sqrt{1 + \left(\frac{dy}{dx}\right)^2}\,dx \qquad \text{Eq. (3)}$$

with

$$a = 1, \qquad b = 2, \qquad y = 2\sqrt{x}, \qquad \frac{dy}{dx} = \frac{1}{\sqrt{x}}.$$

First, we perform some algebraic manipulation on the radical in the integrand to transform it into an expression that is easier to integrate.

$$\sqrt{1 + \left(\frac{dy}{dx}\right)^2} = \sqrt{1 + \left(\frac{1}{\sqrt{x}}\right)^2}$$

$$= \sqrt{1 + \frac{1}{x}} = \sqrt{\frac{x + 1}{x}} = \frac{\sqrt{x + 1}}{\sqrt{x}}$$

With these substitutions, we have

$$S = \int_{1}^{2} 2\pi \cdot 2\sqrt{x}\,\frac{\sqrt{x + 1}}{\sqrt{x}}\,dx = 4\pi \int_{1}^{2} \sqrt{x + 1}\,dx$$

$$= 4\pi \cdot \frac{2}{3}(x + 1)^{3/2}\Big]_{1}^{2} = \frac{8\pi}{3}\left(3\sqrt{3} - 2\sqrt{2}\right).$$ ∎

Revolution About the *y*-Axis

For revolution about the *y*-axis, we interchange *x* and *y* in Equation (3).

Surface Area for Revolution About the *y*-Axis

If $x = g(y) \geq 0$ is continuously differentiable on $[c, d]$, the area of the surface generated by revolving the graph of $x = g(y)$ about the *y*-axis is

$$S = \int_c^d 2\pi x \sqrt{1 + \left(\frac{dx}{dy}\right)^2}\, dy = \int_c^d 2\pi g(y)\sqrt{1 + (g'(y))^2}\, dy. \qquad (4)$$

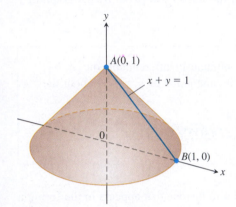

FIGURE 6.35 Revolving line segment *AB* about the *y*-axis generates a cone whose lateral surface area we can now calculate in two different ways (Example 2).

EXAMPLE 2   The line segment $x = 1 - y, 0 \leq y \leq 1$, is revolved about the *y*-axis to generate the cone in Figure 6.35. Find its lateral surface area (which excludes the base area).

Solution   Here we have a calculation we can check with a formula from geometry:

$$\text{Lateral surface area} = \frac{\text{base circumference}}{2} \times \text{slant height} = \pi\sqrt{2}.$$

To see how Equation (4) gives the same result, we take

$$c = 0, \quad d = 1, \quad x = 1 - y, \quad \frac{dx}{dy} = -1,$$

$$\sqrt{1 + \left(\frac{dx}{dy}\right)^2} = \sqrt{1 + (-1)^2} = \sqrt{2}$$

and calculate

$$S = \int_c^d 2\pi x \sqrt{1 + \left(\frac{dx}{dy}\right)^2}\, dy = \int_0^1 2\pi (1 - y)\sqrt{2}\, dy$$

$$= 2\pi\sqrt{2}\left[y - \frac{y^2}{2}\right]_0^1 = 2\pi\sqrt{2}\left(1 - \frac{1}{2}\right)$$

$$= \pi\sqrt{2}.$$

The results agree, as they should. ∎

EXERCISES 6.4

Finding Integrals for Surface Area

In Exercises 1–8:

  a. Set up an integral for the area of the surface generated by revolving the given curve about the indicated axis.

  T **b.** Graph the curve to see what it looks like. If you can, graph the surface too.

  T **c.** Use your utility's integral evaluator to find the surface's area numerically.

1. $y = \tan x, \quad 0 \leq x \leq \pi/4; \quad x$-axis

2. $y = x^2, \quad 0 \leq x \leq 2; \quad x$-axis

3. $xy = 1, \quad 1 \leq y \leq 2; \quad y$-axis

4. $x = \sin y, \quad 0 \leq y \leq \pi; \quad y$-axis

5. $x^{1/2} + y^{1/2} = 3$ from $(4, 1)$ to $(1, 4); \quad x$-axis

6. $y + 2\sqrt{y} = x, \quad 1 \leq y \leq 2; \quad y$-axis

7. $x = \int_0^y \tan t\, dt, \quad 0 \leq y \leq \pi/3; \quad y$-axis

8. $y = \int_1^x \sqrt{t^2 - 1}\, dt, \quad 1 \leq x \leq \sqrt{5}; \quad x$-axis

Finding Surface Area

9. Find the lateral (side) surface area of the cone generated by revolving the line segment $y = x/2, 0 \leq x \leq 4$, about the x-axis. Check your answer with the geometry formula

$$\text{Lateral surface area} = \frac{1}{2} \times \text{base circumference} \times \text{slant height.}$$

10. Find the lateral surface area of the cone generated by revolving the line segment $y = x/2, 0 \leq x \leq 4$, about the y-axis. Check your answer with the geometry formula

$$\text{Lateral surface area} = \frac{1}{2} \times \text{base circumference} \times \text{slant height.}$$

11. Find the surface area of the cone frustum generated by revolving the line segment $y = (x/2) + (1/2), 1 \leq x \leq 3$, about the x-axis. Check your result with the geometry formula

$$\text{Frustum surface area} = \pi(r_1 + r_2) \times \text{slant height.}$$

12. Find the surface area of the cone frustum generated by revolving the line segment $y = (x/2) + (1/2), 1 \leq x \leq 3$, about the y-axis. Check your result with the geometry formula

$$\text{Frustum surface area} = \pi(r_1 + r_2) \times \text{slant height.}$$

Find the areas of the surfaces generated by revolving the curves in Exercises 13–23 about the indicated axes. If you have a grapher, you may want to graph these curves to see what they look like.

13. $y = x^3/9, \quad 0 \leq x \leq 2; \quad x$-axis

14. $y = \sqrt{x}, \quad 3/4 \leq x \leq 15/4; \quad x$-axis

15. $y = \sqrt{2x - x^2}, \quad 0.5 \leq x \leq 1.5; \quad x$-axis

16. $y = \sqrt{x + 1}, \quad 1 \leq x \leq 5; \quad x$-axis

17. $x = y^3/3, \quad 0 \leq y \leq 1; \quad y$-axis

18. $x = (1/3)y^{3/2} - y^{1/2}, \quad 1 \leq y \leq 3; \quad y$-axis

19. $x = 2\sqrt{4 - y}, \quad 0 \leq y \leq 15/4; \quad y$-axis

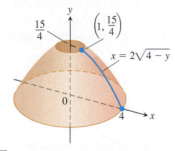

20. $x = \sqrt{2y - 1}, \quad 5/8 \leq y \leq 1; \quad y$-axis

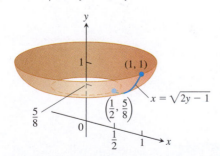

21. $x = (e^y + e^{-y})/2, \quad 0 \leq y \leq \ln 2; \quad y$-axis

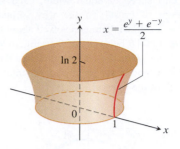

22. $y = (1/3)(x^2 + 2)^{3/2}, \quad 0 \leq x \leq \sqrt{2}; \quad y$-axis (*Hint:* Express $ds = \sqrt{dx^2 + dy^2}$ in terms of dx, and evaluate the integral $S = \int 2\pi x \, ds$ with appropriate limits.)

23. $x = (y^4/4) + 1/(8y^2), \quad 1 \leq y \leq 2; \quad x$-axis (*Hint:* Express $ds = \sqrt{dx^2 + dy^2}$ in terms of dy, and evaluate the integral $S = \int 2\pi y \, ds$ with appropriate limits.)

24. Write an integral for the area of the surface generated by revolving the curve $y = \cos x, -\pi/2 \leq x \leq \pi/2$, about the x-axis. In Section 8.4 we will see how to evaluate such integrals.

25. Testing the new definition Show that the surface area of a sphere of radius a is still $4\pi a^2$ by using Equation (3) to find the area of the surface generated by revolving the curve $y = \sqrt{a^2 - x^2}$, $-a \leq x \leq a$, about the x-axis.

26. Testing the new definition The lateral (side) surface area of a cone of height h and base radius r should be $\pi r \sqrt{r^2 + h^2}$, the semiperimeter of the base times the slant height. Show that this is still the case by finding the area of the surface generated by revolving the line segment $y = (r/h)x, 0 \leq x \leq h$, about the x-axis.

27. Enameling woks Your company decided to put out a deluxe version of a wok you designed. The plan is to coat it inside with white enamel and outside with blue enamel. Each enamel will be sprayed on 0.5 mm thick before baking. (See accompanying figure.) Your manufacturing department wants to know how much enamel to have on hand for a production run of 5000 woks. What do you tell them? (Neglect waste and unused material and give your answer in liters. Remember that $1 \, \text{cm}^3 = 1 \, \text{mL}$, so $1 \, \text{L} = 1000 \, \text{cm}^3$.)

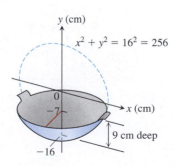

28. Slicing bread Did you know that if you cut a spherical loaf of bread into slices of equal width, each slice will have the same amount of crust? To see why, suppose the semicircle $y = \sqrt{r^2 - x^2}$ shown here is revolved about the x-axis to generate a sphere. Let AB be an arc of the semicircle that lies above an interval of length h on the x-axis. Show that the area swept out by AB does not depend on the location of the interval. (It does depend on the length of the interval.)

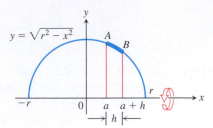

29. The shaded band shown here is cut from a sphere of radius R by parallel planes h units apart. Show that the surface area of the band is $2\pi Rh$.

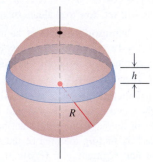

30. Here is a schematic drawing of the 90-ft dome used by the U.S. National Weather Service to house radar in Bozeman, Montana.

  a. How much outside surface is there to paint (not counting the bottom)?

T **b.** Express the answer to the nearest square foot.

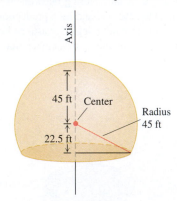

31. An alternative derivation of the surface area formula Assume f is smooth on $[a, b]$ and partition $[a, b]$ in the usual way. In the kth subinterval $[x_{k-1}, x_k]$, construct the tangent line to the curve at the midpoint $m_k = (x_{k-1} + x_k)/2$, as in the accompanying figure.

  a. Show that

$$r_1 = f(m_k) - f'(m_k)\frac{\Delta x_k}{2} \quad \text{and} \quad r_2 = f(m_k) + f'(m_k)\frac{\Delta x_k}{2}.$$

  b. Show that the length L_k of the tangent line segment in the kth subinterval is $L_k = \sqrt{(\Delta x_k)^2 + (f'(m_k)\,\Delta x_k)^2}$.

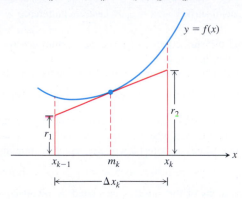

  c. Show that the lateral surface area of the frustum of the cone swept out by the tangent line segment as it revolves about the x-axis is $2\pi f(m_k)\sqrt{1 + (f'(m_k))^2}\,\Delta x_k$.

  d. Show that the area of the surface generated by revolving $y = f(x)$ about the x-axis over $[a, b]$ is

$$\lim_{n\to\infty} \sum_{k=1}^{n} \binom{\text{lateral surface area}}{\text{of } k\text{th frustum}} = \int_a^b 2\pi f(x)\sqrt{1 + (f'(x))^2}\,dx.$$

32. The surface of an astroid Find the area of the surface generated by revolving about the x-axis the portion of the astroid $x^{2/3} + y^{2/3} = 1$ shown in the accompanying figure.

(*Hint:* Revolve the first-quadrant portion $y = (1 - x^{2/3})^{3/2}$, $0 \le x \le 1$, about the x-axis and double your result.)

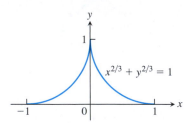

6.5 Work and Fluid Forces

In everyday life, *work* means an activity that requires muscular or mental effort. In science, the term refers specifically to a force acting on an object and the object's subsequent displacement. This section shows how to calculate work. The applications run from compressing railroad car springs and emptying subterranean tanks to forcing subatomic particles to collide and lifting satellites into orbit.

Work Done by a Constant Force

When an object moves a distance d along a straight line as a result of being acted on by a force of constant magnitude F in the direction of motion, we define the **work** W done by the force on the object with the formula

$$W = Fd \qquad \text{(Constant-force formula for work).} \tag{1}$$

From Equation (1) we see that the unit of work in any system is the unit of force multiplied by the unit of distance. In SI units (SI stands for *Système International*, or International System), the unit of force is a newton, the unit of distance is a meter, and the unit of work is a newton-meter ($N \cdot m$). This combination appears so often it has a special name, the **joule**. Taking gravitational acceleration at sea level to be 9.8 m/sec^2, to lift one kilogram one meter requires work of 9.8 joules. This is seen by multiplying the force of 9.8 newtons exerted on one kilogram by the one-meter distance moved. In the British system, the unit of work is the foot-pound, a unit sometimes used in applications. It requires one foot-pound of work to lift a one pound weight a distance of one foot.

Joules

The joule, abbreviated J, is named after the English physicist James Prescott Joule (1818–1889). The defining equation is

1 joule = (1 newton)(1 meter).

In symbols, $1 \text{ J} = 1 \text{ N} \cdot \text{m}$.

EXAMPLE 1   Suppose you jack up the side of a 2000-lb car 1.25 ft to change a tire. The jack applies a constant vertical force of about 1000 lb in lifting the side of the car (but because of the mechanical advantage of the jack, the force you apply to the jack itself is only about 30 lb). The total work performed by the jack on the car is $1000 \times 1.25 = 1250$ ft-lb. In SI units, the jack has applied a force of 4448 N through a distance of 0.381 m to do $4448 \times 0.381 \approx 1695$ J of work. ∎

Work Done by a Variable Force Along a Line

If the force you apply varies along the way, as it will if you are stretching or compressing a spring, the formula $W = Fd$ has to be replaced by an integral formula that takes the variation in F into account.

Suppose that the force performing the work acts on an object moving along a straight line, which we take to be the x-axis. We assume that the magnitude of the force is a continuous function F of the object's position x. We want to find the work done over the interval from $x = a$ to $x = b$. We partition $[a, b]$ in the usual way and choose an arbitrary point c_k in each subinterval $[x_{k-1}, x_k]$. If the subinterval is short enough, the continuous function F will not vary much from x_{k-1} to x_k. The amount of work done across the interval will be about $F(c_k)$ times the distance Δx_k, the same as it would be if F were constant and we could apply Equation (1). The total work done from a to b is therefore approximated by the Riemann sum

$$\text{Work} \approx \sum_{k=1}^{n} F(c_k) \, \Delta x_k.$$

We expect the approximation to improve as the norm of the partition goes to zero, so we define the work done by the force from a to b to be the integral of F from a to b:

$$\lim_{n \to \infty} \sum_{k=1}^{n} F(c_k) \, \Delta x_k = \int_{a}^{b} F(x) \, dx.$$

DEFINITION The **work** done by a variable force $F(x)$ in moving an object along the x-axis from $x = a$ to $x = b$ is

$$W = \int_a^b F(x)\,dx. \tag{2}$$

The units of the integral are joules if F is in newtons and x is in meters, and foot-pounds if F is in pounds and x is in feet. So the work done by a force of $F(x) = 1/x^2$ newtons in moving an object along the x-axis from $x = 1$ m to $x = 10$ m is

$$W = \int_1^{10} \frac{1}{x^2}\,dx = -\frac{1}{x}\bigg]_1^{10} = -\frac{1}{10} + 1 = 0.9 \text{ J}.$$

Hooke's Law for Springs: $F = kx$

One calculation for work arises in finding the work required to stretch or compress a spring. **Hooke's Law** says that the force required to hold a stretched or compressed spring x units from its natural (unstressed) length is proportional to x. In symbols,

$$F = kx. \tag{3}$$

The constant k, measured in force units per unit length, is a characteristic of the spring, called the **force constant** (or **spring constant**) of the spring. Hooke's Law, Equation (3), gives good results as long as the force doesn't distort the metal in the spring. We assume that the forces in this section are too small to do that.

EXAMPLE 2 Find the work required to compress a spring from its natural length of 1 ft to a length of 0.75 ft if the force constant is $k = 16$ lb/ft.

Solution We picture the uncompressed spring laid out along the x-axis with its movable end at the origin and its fixed end at $x = 1$ ft (Figure 6.36). This enables us to describe the force required to compress the spring from 0 to x with the formula $F = 16x$. To compress the spring from 0 to 0.25 ft, the force must increase from

$$F(0) = 16 \cdot 0 = 0 \text{ lb} \qquad \text{to} \qquad F(0.25) = 16 \cdot 0.25 = 4 \text{ lb}.$$

The work done by F over this interval is

$$W = \int_0^{0.25} 16x\,dx = 8x^2\bigg]_0^{0.25} = 0.5 \text{ ft-lb.} \qquad \begin{array}{l}\text{Eq. (2) with}\\ a = 0, b = 0.25,\\ F(x) = 16x\end{array} \qquad \blacksquare$$

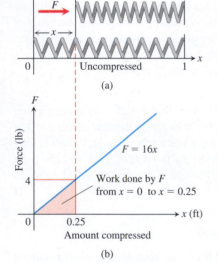

FIGURE 6.36 The force F needed to hold a spring under compression increases linearly as the spring is compressed (Example 2).

EXAMPLE 3 A spring has a natural length of 1 m. A force of 24 N holds the spring stretched to a total length of 1.8 m.

(a) Find the force constant k.

(b) How much work will it take to stretch the spring 2 m beyond its natural length?

(c) How far will a 45-N force stretch the spring?

Solution

(a) *The force constant.* We find the force constant from Equation (3). A force of 24 N maintains the spring at a position where it is stretched 0.8 m from its natural length, so

$$24 = k(0.8) \qquad \qquad \text{Eq. (3) with } F = 24, x = 0.8$$
$$k = 24/0.8 = 30 \text{ N/m}.$$

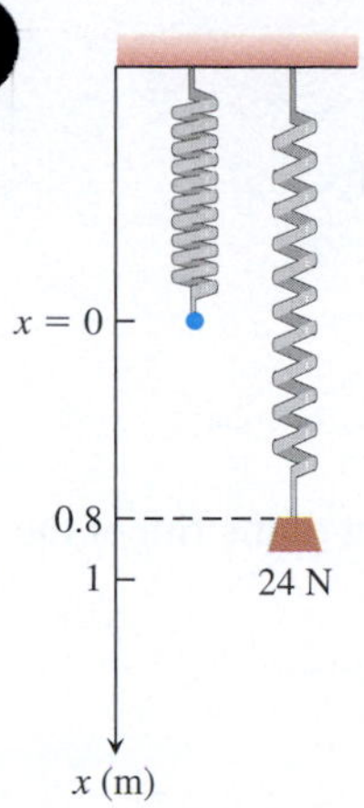

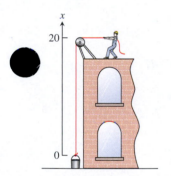

$x = 0$

0.8
1 24 N

x (m)

FIGURE 6.37 A 24-N weight stretches this spring 0.8 m beyond its unstressed length (Example 3).

(b) *The work to stretch the spring* 2 m. We imagine the unstressed spring hanging along the x-axis with its free end at $x = 0$ (Figure 6.37). The force required to stretch the spring x m beyond its natural length is the force required to hold the free end of the spring x units from the origin. Hooke's Law with $k = 30$ says that this force is

$$F(x) = 30x.$$

The work done by F on the spring from $x = 0$ m to $x = 2$ m is

$$W = \int_0^2 30x \, dx = 15x^2 \Big]_0^2 = 60 \text{ J}.$$

(c) *How far will a 45-N force stretch the spring?* We substitute $F = 45$ in the equation $F = 30x$ to find

$$45 = 30x, \quad \text{or} \quad x = 1.5 \text{ m}.$$

A 45-N force will keep the spring stretched 1.5 m beyond its natural length. ■

Lifting Objects and Pumping Liquids from Containers

The work integral is useful for calculating the work done in lifting objects whose weights vary with their elevation.

EXAMPLE 4 A 5-kg bucket is lifted from the ground into the air by pulling in 20 m of rope at a constant speed (Figure 6.38). The rope weighs 0.08 kg/m. How much work was spent lifting the bucket and rope?

Solution The weight of the bucket is obtained by multiplying the mass (5 kg) and the acceleration due to gravity, approximately 9.8 m/s². So the bucket's weight is $(5)(9.8) = 49$ N, and the work done lifting it alone is weight × distance $= (49)(20) = 980$ J.

The weight of the rope varies with the bucket's elevation, because less of it is freely hanging as the bucket is raised. When the bucket is x m off the ground, the remaining portion of the rope still being lifted weighs $(0.08)(20 - x)(9.8)$ N. So the work in lifting the rope is

$$\text{Work on rope} = \int_0^{20} (0.08)(20 - x)(9.8) \, dx = \int_0^{20} (15.68 - 0.784x) \, dx$$

$$= \left[15.68x - 0.392x^2 \right]_0^{20} = 313.6 - 156.8 = 156.8 \text{ J}.$$

The total work for the bucket and rope combined is

$$980 + 156.8 = 1136.8 \text{ J}.$$ ■

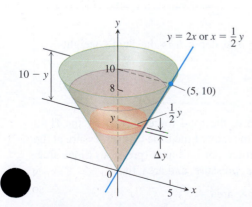

x
20

0

FIGURE 6.38 Lifting the bucket in Example 4.

How much work does it take to pump all or part of the liquid from a container? Engineers often need to know the answer in order to design or choose the right pump, or to compute the cost to transport water or some other liquid from one place to another. To find out how much work is required to pump the liquid, we imagine lifting the liquid out one thin horizontal slab at a time and applying the equation $W = Fd$ to each slab. We then evaluate the integral that this leads to as the slabs become thinner and more numerous.

EXAMPLE 5 The conical tank in Figure 6.39 is filled to within 2 ft of the top with olive oil weighing 57 lb/ft³. How much work does it take to pump the oil to the rim of the tank?

y
$y = 2x \text{ or } x = \frac{1}{2} y$
$10 - y$
10
8 (5, 10)
$\frac{1}{2} y$
y
Δy
0
5 x

FIGURE 6.39 The olive oil and tank in Example 5.

Solution We imagine the oil divided into thin slabs by planes perpendicular to the y-axis at the points of a partition of the interval $[0, 8]$.

The typical slab between the planes at y and $y + \Delta y$ has a volume of about

$$\Delta V = \pi(\text{radius})^2(\text{thickness}) = \pi\left(\frac{1}{2}y\right)^2 \Delta y = \frac{\pi}{4}y^2 \, \Delta y \, \text{ft}^3.$$

The force $F(y)$ required to lift this slab is equal to its weight,

$$F(y) = 57 \, \Delta V = \frac{57\pi}{4}y^2 \, \Delta y \, \text{lb}. \qquad \text{Weight} = (\text{weight per unit volume}) \times \text{volume}$$

The distance through which $F(y)$ must act to lift this slab to the level of the rim of the cone is about $(10 - y)$ ft, so the work done lifting the slab is about

$$\Delta W = \frac{57\pi}{4}(10 - y)y^2 \, \Delta y \, \text{ft-lb}.$$

Assuming there are n slabs associated with the partition of $[0, 8]$, and that $y = y_k$ denotes the plane associated with the kth slab of thickness Δy_k, we can approximate the work done lifting all of the slabs with the Riemann sum

$$W \approx \sum_{k=1}^{n} \frac{57\pi}{4}(10 - y_k)y_k^2 \, \Delta y_k \, \text{ft-lb}.$$

The work of pumping the oil to the rim is the limit of these sums as the norm of the partition goes to zero and the number of slabs tends to infinity:

$$W = \lim_{n \to \infty} \sum_{k=1}^{n} \frac{57\pi}{4}(10 - y_k)y_k^2 \, \Delta y_k = \int_0^8 \frac{57\pi}{4}(10 - y)y^2 \, dy$$

$$= \frac{57\pi}{4}\int_0^8 (10y^2 - y^3) \, dy$$

$$= \frac{57\pi}{4}\left[\frac{10y^3}{3} - \frac{y^4}{4}\right]_0^8 \approx 30{,}561 \, \text{ft-lb}. \quad \blacksquare$$

Fluid Pressure and Forces

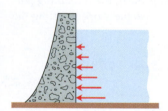

FIGURE 6.40 To withstand the increasing pressure, dams are built thicker as they go down.

Dams are built thicker at the bottom than at the top (Figure 6.40) because the pressure against them increases with depth. The pressure at any point on a dam depends only on how far below the surface the point is and not on how much the surface of the dam happens to be tilted at that point. The pressure, in pounds per square foot at a point h feet below the surface, is always $62.4h$. The number 62.4 is the weight-density of freshwater in pounds per cubic foot. The pressure h feet below the surface of any fluid is the fluid's *weight-density* times h.

The Pressure-Depth Equation

In a fluid that is standing still, the pressure p at depth h is the fluid's weight-density w times h:

$$p = wh. \tag{4}$$

Weight-density

A fluid's weight-density w is its weight per unit volume. Typical values (lb/ft³) are listed below.

Gasoline	42
Mercury	849
Milk	64.5
Molasses	100
Olive oil	57
Seawater	64
Freshwater	62.4

In a container of fluid with a flat horizontal base, the total force exerted by the fluid against the base can be calculated by multiplying the area of the base by the pressure at the base. We can do this because total force equals force per unit area (pressure) times area. (See Figure 6.41.) If F, p, and A are the total force, pressure, and area, then

$$F = \text{total force} = \text{force per unit area} \times \text{area}$$

$$= \text{pressure} \times \text{area} = pA$$

$$= whA. \qquad p = wh \text{ from Eq. (4)}$$

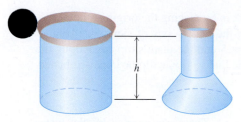

FIGURE 6.41 These containers are filled with water to the same depth and have the same base area. The total force is therefore the same on the bottom of each container. The containers' shapes do not matter here.

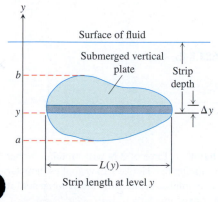

FIGURE 6.42 The force exerted by a fluid against one side of a thin, flat horizontal strip is about ΔF = pressure × area = $w \times$ (strip depth) × $L(y) \Delta y$.

Fluid Force on a Constant-Depth Surface

$$F = pA = whA \qquad (5)$$

For example, the weight-density of freshwater is $62.4 \text{ lb}/\text{ft}^3$, so the fluid force at the bottom of a 10 ft × 20 ft rectangular swimming pool 3 ft deep is

$$F = whA = (62.4 \text{ lb}/\text{ft}^3)(3 \text{ ft})(10 \cdot 20 \text{ ft}^2)$$
$$= 37{,}440 \text{ lb}.$$

For a flat plate submerged *horizontally*, like the bottom of the swimming pool just discussed, the downward force acting on its upper face due to liquid pressure is given by Equation (5). If the plate is submerged *vertically*, however, then the pressure against it will be different at different depths and Equation (5) no longer is usable in that form (because h varies).

Suppose we want to know the force exerted by a fluid against one side of a vertical plate submerged in a fluid of weight-density w. To find it, we model the plate as a region extending from $y = a$ to $y = b$ in the xy-plane (Figure 6.42). We partition $[a, b]$ in the usual way and imagine the region to be cut into thin horizontal strips by planes perpendicular to the y-axis at the partition points. The typical strip from y to $y + \Delta y$ is Δy units wide by $L(y)$ units long. We assume $L(y)$ to be a continuous function of y.

The pressure varies across the strip from top to bottom. If the strip is narrow enough, however, the pressure will remain close to its bottom-edge value of $w \times$ (strip depth). The force exerted by the fluid against one side of the strip will be about

$$\Delta F = (\text{pressure along bottom edge}) \times (\text{area})$$
$$= w \cdot (\text{strip depth}) \cdot L(y) \Delta y.$$

Assume there are n strips associated with the partition of $a \leq y \leq b$ and that y_k is the bottom edge of the kth strip having length $L(y_k)$ and width Δy_k. The force against the entire plate is approximated by summing the forces against each strip, giving the Riemann sum

$$F \approx \sum_{k=1}^{n} w \cdot (\text{strip depth})_k \cdot L(y_k) \Delta y_k. \qquad (6)$$

The sum in Equation (6) is a Riemann sum for a continuous function on $[a, b]$, and we expect the approximations to improve as the norm of the partition goes to zero. The force against the plate is the limit of these sums:

$$\lim_{n \to \infty} \sum_{k=1}^{n} w \cdot (\text{strip depth})_k \cdot L(y_k) \Delta y_k = \int_a^b w \cdot (\text{strip depth}) \cdot L(y) \, dy.$$

The Integral for Fluid Force Against a Vertical Flat Plate

Suppose that a plate submerged vertically in fluid of weight-density w runs from $y = a$ to $y = b$ on the y-axis. Let $L(y)$ be the length of the horizontal strip measured from left to right along the surface of the plate at level y. Then the force exerted by the fluid against one side of the plate is

$$F = \int_a^b w \cdot (\text{strip depth}) \cdot L(y) \, dy. \qquad (7)$$

EXAMPLE 6   A flat isosceles right-triangular plate with base 6 ft and height 3 ft is submerged vertically, base up, 2 ft below the surface of a swimming pool. Find the force exerted by the water against one side of the plate.

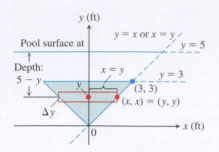

FIGURE 6.43 To find the force on one side of the submerged plate in Example 6, we can use a coordinate system like the one here.

Solution We establish a coordinate system to work in by placing the origin at the plate's bottom vertex and running the y-axis upward along the plate's axis of symmetry (Figure 6.43). The surface of the pool lies along the line $y = 5$ and the plate's top edge along the line $y = 3$. The plate's right-hand edge lies along the line $y = x$, with the upper-right vertex at $(3, 3)$. The length of a thin strip at level y is

$$L(y) = 2x = 2y.$$

The depth of the strip beneath the surface is $(5 - y)$. The force exerted by the water against one side of the plate is therefore

$$F = \int_a^b w \cdot \left(\begin{array}{c} \text{strip} \\ \text{depth} \end{array} \right) \cdot L(y)\, dy \qquad \text{Eq. (7)}$$

$$= \int_0^3 62.4\,(5 - y)\,2y\, dy$$

$$= 124.8 \int_0^3 (5y - y^2)\, dy$$

$$= 124.8 \left[\frac{5}{2}y^2 - \frac{y^3}{3} \right]_0^3 = 1684.8 \text{ lb.} \quad \blacksquare$$

EXERCISES 6.5

Springs

The graphs of force functions (in newtons) are given in Exercises 1 and 2. How much work is done by each force in moving an object 10 m?

1.

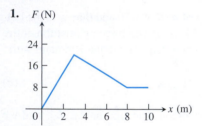

2.

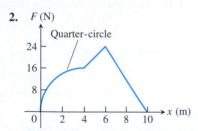

3. Spring constant It took 1800 J of work to stretch a spring from its natural length of 2 m to a length of 5 m. Find the spring's force constant.

4. Stretching a spring A spring has a natural length of 10 in. An 800-lb force stretches the spring to 14 in.

   a. Find the force constant.

   b. How much work is done in stretching the spring from 10 in. to 12 in.?

   c. How far beyond its natural length will a 1600-lb force stretch the spring?

5. Stretching a rubber band A force of 2 N will stretch a rubber band 2 cm (0.02 m). Assuming that Hooke's Law applies, how far will a 4-N force stretch the rubber band? How much work does it take to stretch the rubber band this far?

6. Stretching a spring If a force of 90 N stretches a spring 1 m beyond its natural length, how much work does it take to stretch the spring 5 m beyond its natural length?

7. Subway car springs It takes a force of 21,714 lb to compress a coil spring assembly on a New York City Transit Authority subway car from its free height of 8 in. to its fully compressed height of 5 in.

   a. What is the assembly's force constant?

   b. How much work does it take to compress the assembly the first half inch? the second half inch? Answer to the nearest in.-lb.

8. Bathroom scale A bathroom scale is compressed 1/16 in. when a 150-lb person stands on it. Assuming that the scale behaves like a spring that obeys Hooke's Law, how much does someone who compresses the scale 1/8 in. weigh? How much work is done compressing the scale 1/8 in.?

Work Done by a Variable Force

9. Lifting a rope A mountain climber is about to haul up a 50-m length of hanging rope. How much work will it take if the rope weighs 0.624 N/m?

10. Leaky sandbag A bag of sand originally weighing 144 lb was lifted at a constant rate. As it rose, sand also leaked out at a constant rate. The sand was half gone by the time the bag had been lifted to 18 ft. How much work was done lifting the sand this far? (Neglect the weight of the bag and lifting equipment.)

11. **Lifting an elevator cable** An electric elevator with a motor at the top has a multistrand cable weighing 4.5 lb/ft. When the car is at the first floor, 180 ft of cable are paid out, and effectively 0 ft are out when the car is at the top floor. How much work does the motor do just lifting the cable when it takes the car from the first floor to the top?

12. **Force of attraction** When a particle of mass m is at $(x, 0)$, it is attracted toward the origin with a force whose magnitude is k/x^2. If the particle starts from rest at $x = b$ and is acted on by no other forces, find the work done on it by the time it reaches $x = a$, $0 < a < b$.

13. **Leaky bucket** Assume the bucket in Example 4 is leaking. It starts with 5 liters of water (5 kg) and leaks at a constant rate. It finishes draining just as it reaches the top. How much work was spent lifting the water alone? (*Hint:* Do not include the rope and bucket, and find the proportion of water left at elevation x m.)

14. (*Continuation of Exercise 13.*) The workers in Example 4 and Exercise 13 changed to a larger bucket that held 10 liters (10 kg) of water, but the new bucket had an even larger leak so that it, too, was empty by the time it reached the top. Assuming that the water leaked out at a steady rate, how much work was done lifting the water alone? (Do not include the rope and bucket.)

Pumping Liquids from Containers

15. **Pumping water** The rectangular tank shown here, with its top at ground level, is used to catch runoff water. Assume that the water weighs 62.4 lb/ft³.

    a. How much work does it take to empty the tank by pumping the water back to ground level once the tank is full?

    b. If the water is pumped to ground level with a (5/11)-horsepower (hp) motor (work output 250 ft-lb/sec), how long will it take to empty the full tank (to the nearest minute)?

    c. Show that the pump in part (b) will lower the water level 10 ft (halfway) during the first 25 min of pumping.

    d. **The weight of water** What are the answers to parts (a) and (b) in a location where water weighs 62.26 lb/ft³? 62.59 lb/ft³?

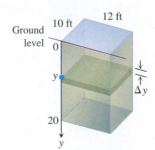

16. **Emptying a cistern** The rectangular cistern (storage tank for rainwater) shown has its top 10 ft below ground level. The cistern, currently full, is to be emptied for inspection by pumping its contents to ground level.

    a. How much work will it take to empty the cistern?

    b. How long will it take a 1/2-hp pump, rated at 275 ft-lb/sec, to pump the tank dry?

c. How long will it take the pump in part (b) to empty the tank halfway? (It will be less than half the time required to empty the tank completely.)

d. **The weight of water** What are the answers to parts (a) through (c) in a location where water weighs 62.26 lb/ft³? 62.59 lb/ft³?

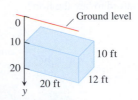

17. **Pumping oil** How much work would it take to pump oil from the tank in Example 5 to the level of the top of the tank if the tank were completely full?

18. **Pumping a half-full tank** Suppose that, instead of being full, the tank in Example 5 is only half full. How much work does it take to pump the remaining oil to a level 4 ft above the top of the tank?

19. **Emptying a tank** A vertical right-circular cylindrical tank measures 30 ft high and 20 ft in diameter. It is full of kerosene weighing 51.2 lb/ft³. How much work does it take to pump the kerosene to the level of the top of the tank?

20. a. **Pumping milk** Suppose that the conical container in Example 5 contains milk (weighing 64.5 lb/ft³) instead of olive oil. How much work will it take to pump the contents to the rim?

    b. **Pumping oil** How much work will it take to pump the oil in Example 5 to a level 3 ft above the cone's rim?

21. The graph of $y = x^2$ on $0 \le x \le 2$ is revolved about the y-axis to form a tank that is then filled with salt water from the Dead Sea (weighing approximately 73 lb/ft³). How much work does it take to pump all of the water to the top of the tank?

22. A right-circular cylindrical tank of height 10 ft and radius 5 ft is lying horizontally and is full of diesel fuel weighing 53 lb/ft³. How much work is required to pump all of the fuel to a point 15 ft above the top of the tank?

23. **Emptying a water reservoir** We model pumping from spherical containers the way we do from other containers, with the axis of integration along the vertical axis of the sphere. Use the figure here to find how much work it takes to empty a full hemispherical water reservoir of radius 5 m by pumping the water to a height of 4 m above the top of the reservoir. Water weighs 9800 N/m³.

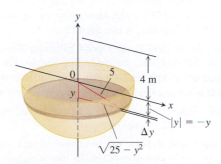

24. You are in charge of the evacuation and repair of the storage tank shown here. The tank is a hemisphere of radius 10 ft and is full of benzene weighing 56 lb/ft³. A firm you contacted says it can empty the tank for 1/2¢ per foot-pound of work. Find the work required to empty the tank by pumping the benzene to an outlet 2 ft above the top of the tank. If you have $5000 budgeted for the job, can you afford to hire the firm?

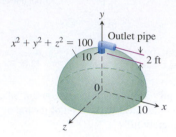

Work and Kinetic Energy

25. Kinetic energy If a variable force of magnitude $F(x)$ moves an object of mass m along the x-axis from x_1 to x_2, the object's velocity v can be written as dx/dt (where t represents time). Use Newton's second law of motion $F = m(dv/dt)$ and the Chain Rule

$$\frac{dv}{dt} = \frac{dv}{dx}\frac{dx}{dt} = v\frac{dv}{dx}$$

to show that the net work done by the force in moving the object from x_1 to x_2 is

$$W = \int_{x_1}^{x_2} F(x)\,dx = \frac{1}{2}mv_2^2 - \frac{1}{2}mv_1^2,$$

where v_1 and v_2 are the object's velocities at x_1 and x_2. In physics, the expression $(1/2)mv^2$ is called the *kinetic energy* of an object of mass m moving with velocity v. Therefore, *the work done by the force equals the change in the object's kinetic energy*, and we can find the work by calculating this change.

In Exercises 26–30, use the result of Exercise 25.

26. Tennis A 2-oz tennis ball was served at 160 ft/sec (about 109 mph). How much work was done on the ball to make it go this fast? (To find the ball's mass from its weight, express the weight in pounds and divide by 32 ft/sec², the acceleration of gravity.)

27. Baseball How many foot-pounds of work does it take to throw a baseball 90 mph? A baseball weighs 5 oz, or 0.3125 lb.

28. Golf A 1.6-oz golf ball is driven off the tee at a speed of 280 ft/sec (about 191 mph). How many foot-pounds of work are done on the ball getting it into the air?

29. Tennis At the 2012 Busan Open Challenger Tennis Tournament is Busan, South Korea, the Australian Samuel Groth hit a serve measured at 263 kph (163.4 mph). How much work was required by Groth to serve a 0.056699-kg (2-oz) tennis ball at that speed?

30. Softball How much work has to be performed on a 6.5-oz softball to pitch it 132 ft/sec (90 mph)?

31. Drinking a milkshake The truncated conical container shown here is full of strawberry milkshake that weighs 4/9 oz/in³. As you can see, the container is 7 in. deep, 2.5 in. across at the base, and 3.5 in. across at the top (a standard size at Brigham's in Boston). The straw sticks up an inch above the top. About how much work

does it take to suck up the milkshake through the straw (neglecting friction)? Answer in inch-ounces.

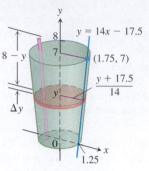

Dimensions in inches

32. Water tower Your town has decided to drill a well to increase its water supply. As the town engineer, you have determined that a water tower will be necessary to provide the pressure needed for distribution, and you have designed the system shown here. The water is to be pumped from a 300-ft well through a vertical 4-in. pipe into the base of a cylindrical tank 20 ft in diameter and 25 ft high. The base of the tank will be 60 ft above ground. The pump is a 3-hp pump, rated at 1650 ft · lb/sec. To the nearest hour, how long will it take to fill the tank the first time? (Include the time it takes to fill the pipe.) Assume that water weighs 62.4 lb/ft³.

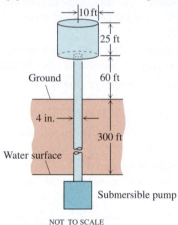

NOT TO SCALE

33. Putting a satellite in orbit The strength of Earth's gravitational field varies with the distance r from Earth's center, and the magnitude of the gravitational force experienced by a satellite of mass m during and after launch is

$$F(r) = \frac{mMG}{r^2}.$$

Here, $M = 5.975 \times 10^{24}$ kg is Earth's mass, $G = 6.6720 \times 10^{-11}$ N · m² kg⁻² is the universal gravitational constant, and r is measured in meters. The work it takes to lift a 1000-kg satellite from Earth's surface to a circular orbit 35,780 km above Earth's center is therefore given by the integral

$$\text{Work} = \int_{6,370,000}^{35,780,000} \frac{1000MG}{r^2}\,dr \text{ joules}.$$

Evaluate the integral. The lower limit of integration is Earth's radius in meters at the launch site. (This calculation does not take into account energy spent lifting the launch vehicle or energy spent bringing the satellite to orbit velocity.)

34. Forcing electrons together Two electrons r meters apart repel each other with a force of

$$F = \frac{23 \times 10^{-29}}{r^2} \text{ newtons.}$$

a. Suppose one electron is held fixed at the point $(1, 0)$ on the x-axis (units in meters). How much work does it take to move a second electron along the x-axis from the point $(-1, 0)$ to the origin?

b. Suppose an electron is held fixed at each of the points $(-1, 0)$ and $(1, 0)$. How much work does it take to move a third electron along the x-axis from $(5, 0)$ to $(3, 0)$?

Finding Fluid Forces

35. Triangular plate Calculate the fluid force on one side of the plate in Example 6 using the coordinate system shown here.

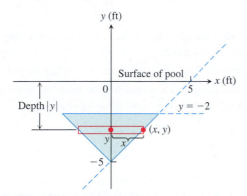

36. Triangular plate Calculate the fluid force on one side of the plate in Example 6 using the coordinate system shown here.

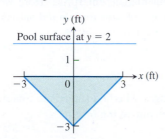

37. Rectangular plate In a pool filled with water to a depth of 10 ft, calculate the fluid force on one side of a 3 ft by 4 ft rectangular plate if the plate rests vertically at the bottom of the pool

a. on its 4-ft edge.   **b.** on its 3-ft edge.

38. Semicircular plate Calculate the fluid force on one side of a semicircular plate of radius 5 ft that rests vertically on its diameter at the bottom of a pool filled with water to a depth of 6 ft.

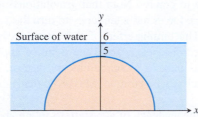

39. Triangular plate The isosceles triangular plate shown here is submerged vertically 1 ft below the surface of a freshwater lake.

a. Find the fluid force against one face of the plate.

b. What would be the fluid force on one side of the plate if the water were seawater instead of freshwater?

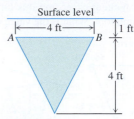

40. Rotated triangular plate The plate in Exercise 39 is revolved 180° about line AB so that part of the plate sticks out of the lake, as shown here. What force does the water exert on one face of the plate now?

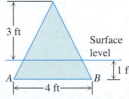

41. New England Aquarium The viewing portion of the rectangular glass window in a typical fish tank at the New England Aquarium in Boston is 63 in. wide and runs from 0.5 in. below the water's surface to 33.5 in. below the surface. Find the fluid force against this portion of the window. The weight-density of seawater is 64 lb/ft³. (In case you were wondering, the glass is 3/4 in. thick and the tank walls extend 4 in. above the water to keep the fish from jumping out.)

42. Semicircular plate A semicircular plate 2 ft in diameter sticks straight down into freshwater with the diameter along the surface. Find the force exerted by the water on one side of the plate.

43. Tilted plate Calculate the fluid force on one side of a 5 ft by 5 ft square plate if the plate is at the bottom of a pool filled with water to a depth of 8 ft and

a. lying flat on its 5 ft by 5 ft face.

b. resting vertically on a 5-ft edge.

c. resting on a 5-ft edge and tilted at 45° to the bottom of the pool.

44. Tilted plate Calculate the fluid force on one side of a right-triangular plate with edges 3 ft, 4 ft, and 5 ft if the plate sits at the bottom of a pool filled with water to a depth of 6 ft on its 3-ft edge and tilted at 60° to the bottom of the pool.

45. The cubical metal tank shown here has a parabolic gate held in place by bolts and designed to withstand a fluid force of 160 lb without rupturing. The liquid you plan to store has a weight-density of 50 lb/ft³.

a. What is the fluid force on the gate when the liquid is 2 ft deep?

b. What is the maximum height to which the container can be filled without exceeding the gate's design limitation?

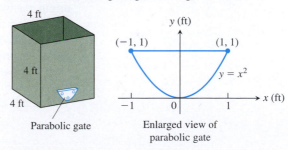

Parabolic gate   Enlarged view of parabolic gate

46. The end plates of the trough shown here were designed to withstand a fluid force of 6667 lb. How many cubic feet of water can the tank hold without exceeding this limitation? Round down to the nearest cubic foot. What is the value of h?

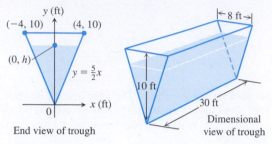

End view of trough

Dimensional
view of trough

47. A vertical rectangular plate a units long by b units wide is submerged in a fluid of weight-density w with its long edges parallel to the fluid's surface. Find the average value of the pressure along the vertical dimension of the plate. Explain your answer.

48. (*Continuation of Exercise 47.*) Show that the force exerted by the fluid on one side of the plate is the average value of the pressure (found in Exercise 47) times the area of the plate.

49. Water pours into the tank shown here at the rate of 4 ft³/min. The tank's cross-sections are 4-ft-diameter semicircles. One end of the tank is movable, but moving it to increase the volume compresses

a spring. The spring constant is $k = 100$ lb/ft. If the end of the tank moves 5 ft against the spring, the water will drain out of a safety hole in the bottom at the rate of 5 ft³/min. Will the movable end reach the hole before the tank overflows?

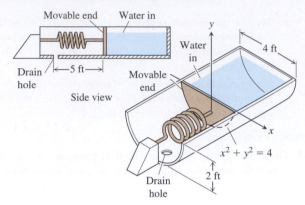

50. Watering trough   The vertical ends of a watering trough are squares 3 ft on a side.

  a. Find the fluid force against the ends when the trough is full.

  b. How many inches do you have to lower the water level in the trough to reduce the fluid force by 25%?

6.6 Moments and Centers of Mass

Many structures and mechanical systems behave as if their masses were concentrated at a single point, called the *center of mass* (Figure 6.44). It is important to know how to locate this point, and doing so is basically a mathematical enterprise. Here we consider masses distributed along a line or region in the plane. Masses distributed across a region or curve in three-dimensional space are treated in Chapters 15 and 16.

Masses Along a Line

We develop our mathematical model in stages. The first stage is to imagine masses m_1, m_2, and m_3 on a rigid x-axis supported by a fulcrum at the origin.

The resulting system might balance, or it might not, depending on how large the masses are and how they are arranged along the x-axis.

    Each mass m_k exerts a downward force $m_k g$ (the weight of m_k) equal to the magnitude of the mass times the acceleration due to gravity. Note that gravitational acceleration is downward, hence negative. Each of these forces has a tendency to turn the x-axis about the origin, the way a child turns a seesaw. This turning effect, called a **torque**, is measured by multiplying the force $m_k g$ by the signed distance x_k from the point of application to the origin. By convention, a positive torque induces a counterclockwise turn. Masses to the left of the origin exert positive (counterclockwise) torque. Masses to the right of the origin exert negative (clockwise) torque.

    The sum of the torques measures the tendency of a system to rotate about the origin. This sum is called the **system torque**.

$$\text{System torque} = m_1 g x_1 + m_2 g x_2 + m_3 g x_3 \tag{1}$$

The system will balance if and only if its torque is zero.

If we factor out the g in Equation (1), we see that the system torque is

$$\underbrace{g}_{\substack{\text{a feature of the}\\\text{environment}}} \cdot \underbrace{(m_1x_1 + m_2x_2 + m_3x_3)}_{\substack{\text{a feature of}\\\text{the system}}}.$$

Thus, the torque is the product of the gravitational acceleration g, which is a feature of the environment in which the system happens to reside, and the number $(m_1x_1 + m_2x_2 + m_3x_3)$, which is a feature of the system itself.

The number $(m_1x_1 + m_2x_2 + m_3x_3)$ is called the **moment of the system about the origin**. It is the sum of the **moments** m_1x_1, m_2x_2, m_3x_3 of the individual masses.

$$M_0 = \text{Moment of system about origin} = \sum m_k x_k$$

(We shift to sigma notation here to allow for sums with more terms.)

We usually want to know where to place the fulcrum to make the system balance; that is, we want to know at what point $\bar{x}$ to place the fulcrum to make the torques add to zero.

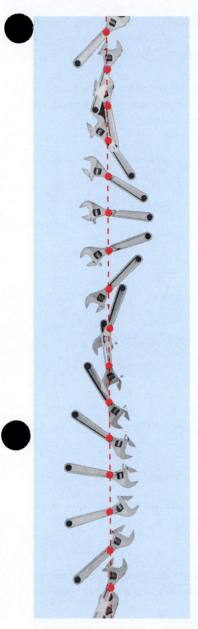

The torque of each mass about the fulcrum in this special location is

$$\text{Torque of } m_k \text{ about } \bar{x} = \left(\begin{array}{c}\text{signed distance}\\\text{of } m_k \text{ from } \bar{x}\end{array}\right)\left(\begin{array}{c}\text{downward}\\\text{force}\end{array}\right)$$

$$= (x_k - \bar{x})m_k g.$$

When we write the equation that says that the sum of these torques is zero, we get an equation we can solve for $\bar{x}$:

$$\sum (x_k - \bar{x})m_k g = 0 \qquad \text{Sum of the torques equals zero.}$$

$$\bar{x} = \frac{\sum m_k x_k}{\sum m_k}. \qquad \text{Solved for } \bar{x}$$

This last equation tells us to find $\bar{x}$ by dividing the system's moment about the origin by the system's total mass:

$$\bar{x} = \frac{\sum m_k x_k}{\sum m_k} = \frac{\text{system moment about origin}}{\text{system mass}}. \tag{2}$$

The point $\bar{x}$ is called the system's **center of mass**.

Thin Wires

Instead of a discrete set of masses arranged in a line, suppose that we have a straight wire or rod located on interval $[a, b]$ on the x-axis. Suppose further that this wire is not homogeneous, but rather the density varies continuously from point to point. If a short segment of a rod containing the point x with length Δx has mass Δm, then the density at x is given by

$$\delta(x) = \lim_{\Delta x \to 0} \Delta m / \Delta x.$$

We often write this formula in one of the alternative forms $\delta = dm/dx$ and $dm = \delta\, dx$.

Partition the interval $[a, b]$ into finitely many subintervals $[x_{k-1}, x_k]$. If we take n subintervals and replace the portion of a wire along a subinterval of length Δx_k containing x_k by a point mass located at x_k with mass $\Delta m_k = \delta(x_k)\,\Delta x_k$, then we obtain a collection of point masses that have approximately the same total mass and same moment as the wire.

FIGURE 6.44 A wrench gliding on ice turning about its center of mass as the center glides in a vertical line. (*Source: PSSC Physics*, 2nd ed., Reprinted by permission of Education Development Center, Inc.)

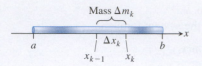

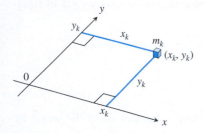

FIGURE 6.45 A rod of varying density can be modeled by a finite number of point masses of mass $\Delta m_k = \delta(x_k)\,\Delta x_k$ located at points x_k along the rod.

The mass M of the wire and the moment M_0 are approximated by the Riemann sums

$$M \approx \sum_{k=1}^{n} \Delta m_k = \sum_{k=1}^{n} \delta(x_k)\,\Delta x_k, \qquad M_0 \approx \sum_{k=1}^{n} x_k\,\Delta m_k = \sum_{k=1}^{n} x_k\,\delta(x_k)\,\Delta x_k.$$

By taking a limit of these Riemann sums as the length of the intervals in the partition approaches zero, we get integral formulas for the mass and the moment of the wire about the origin. The mass M, moment about the origin M_0, and center of mass $\bar{x}$ are

$$M = \int_a^b \delta(x)\,dx, \qquad M_0 = \int_a^b x\,\delta(x)\,dx, \qquad \bar{x} = \frac{M_0}{M} = \frac{\displaystyle\int_a^b x\,\delta(x)\,dx}{\displaystyle\int_a^b \delta(x)\,dx}.$$

EXAMPLE 1   Find the mass M and the center of mass $\bar{x}$ of a rod lying on the x-axis over the interval $[1, 2]$ whose density is given by $\delta(x) = 2 + 3x^2$.

Solution   The mass of the rod is obtained by integrating the density,

$$M = \int_1^2 (2 + 3x^2)\,dx = \left[2x + x^3\right]_1^2 = (4 + 8) - (2 + 1) = 9,$$

and the center of mass is

$$\bar{x} = \frac{M_0}{M} = \frac{\displaystyle\int_1^2 x(2 + 3x^2)\,dx}{9} = \frac{\left[x^2 + \dfrac{3x^4}{4}\right]_1^2}{9} = \frac{19}{12}.$$

Masses Distributed over a Plane Region

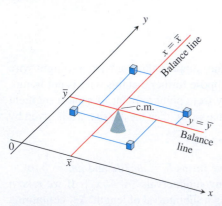

FIGURE 6.46 Each mass m_k has a moment about each axis.

Suppose that we have a finite collection of masses located in the plane, with mass m_k at the point (x_k, y_k) (see Figure 6.46). The mass of the system is

$$\text{System mass:} \qquad M = \sum m_k.$$

Each mass m_k has a moment about each axis. Its moment about the x-axis is $m_k y_k$, and its moment about the y-axis is $m_k x_k$. The moments of the entire system about the two axes are

$$\text{Moment about } x\text{-axis:} \qquad M_x = \sum m_k y_k,$$
$$\text{Moment about } y\text{-axis:} \qquad M_y = \sum m_k x_k.$$

The x-coordinate of the system's center of mass is defined to be

$$\bar{x} = \frac{M_y}{M} = \frac{\sum m_k x_k}{\sum m_k}. \tag{3}$$

With this choice of $\bar{x}$, as in the one-dimensional case, the system balances about the line $x = \bar{x}$ (Figure 6.47).

The y-coordinate of the system's center of mass is defined to be

$$\bar{y} = \frac{M_x}{M} = \frac{\sum m_k y_k}{\sum m_k}. \tag{4}$$

FIGURE 6.47 A two-dimensional array of masses balances on its center of mass.

With this choice of $\bar{y}$, the system balances about the line $y = \bar{y}$ as well. The torques exerted by the masses about the line $y = \bar{y}$ cancel out. Thus, as far as balance is concerned, the system behaves as if all its mass were at the single point $(\bar{x}, \bar{y})$. We call this point the system's **center of mass**.

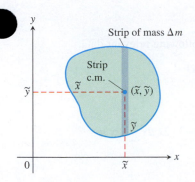

FIGURE 6.48 A plate cut into thin strips parallel to the y-axis. The moment exerted by a typical strip about each axis is the moment its mass Δm would exert if concentrated at the strip's center of mass $(\tilde{x}, \tilde{y})$.

Thin, Flat Plates

In many applications, we need to find the center of mass of a thin, flat plate: a disk of aluminum, say, or a triangular sheet of steel. In such cases, we assume the distribution of mass to be continuous, and the formulas we use to calculate $\bar{x}$ and $\bar{y}$ contain integrals instead of finite sums. The integrals arise in the following way.

Imagine that the plate occupying a region in the xy-plane is cut into thin strips parallel to one of the axes (in Figure 6.48, the y-axis). The center of mass of a typical strip is $(\tilde{x}, \tilde{y})$. We treat the strip's mass Δm as if it were concentrated at $(\tilde{x}, \tilde{y})$. The moment of the strip about the y-axis is then $\tilde{x}\,\Delta m$. The moment of the strip about the x-axis is $\tilde{y}\,\Delta m$. Equations (3) and (4) then become

$$\bar{x} = \frac{M_y}{M} = \frac{\sum \tilde{x}\,\Delta m}{\sum \Delta m}, \qquad \bar{y} = \frac{M_x}{M} = \frac{\sum \tilde{y}\,\Delta m}{\sum \Delta m}.$$

These sums are Riemann sums for integrals, and they approach these integrals in the limit as the strips become narrower and narrower. We write these integrals symbolically as

$$\bar{x} = \frac{\int \tilde{x}\,dm}{\int dm} \qquad \text{and} \qquad \bar{y} = \frac{\int \tilde{y}\,dm}{\int dm}.$$

Moments, Mass, and Center of Mass of a Thin Plate Covering a Region in the xy-Plane

Moment about the x-axis: $\quad M_x = \displaystyle\int \tilde{y}\,dm$

Moment about the y-axis: $\quad M_y = \displaystyle\int \tilde{x}\,dm$ (5)

Mass: $\quad M = \displaystyle\int dm$

Center of mass: $\quad \bar{x} = \dfrac{M_y}{M}, \quad \bar{y} = \dfrac{M_x}{M}$

Density of a plate

A material's density is its mass per unit area. For wires, rods, and narrow strips, the density is given by mass per unit length.

The differential dm in these integrals is the mass of the strip. For this section, we assume the density δ of the plate is a constant or a continuous function of x. Then $dm = \delta\,dA$, which is the mass per unit area δ times the area dA of the strip.

To evaluate the integrals in Equations (5), we picture the plate in the coordinate plane and sketch a strip of mass parallel to one of the coordinate axes. We then express the strip's mass dm and the coordinates $(\tilde{x}, \tilde{y})$ of the strip's center of mass in terms of x or y. Finally, we integrate $\tilde{y}\,dm$, $\tilde{x}\,dm$, and dm between limits of integration determined by the plate's location in the plane.

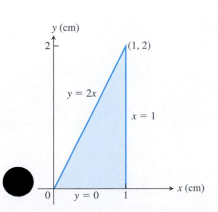

FIGURE 6.49 The plate in Example 2.

EXAMPLE 2   The triangular plate shown in Figure 6.49 has a constant density of $\delta = 3$ g/cm². Find

(a) the plate's moment M_y about the y-axis.

(b) the plate's mass M.

(c) the x-coordinate of the plate's center of mass (c.m.).

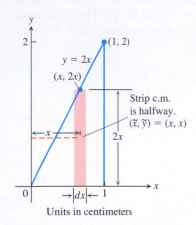

FIGURE 6.50 Modeling the plate in Example 2 with vertical strips.

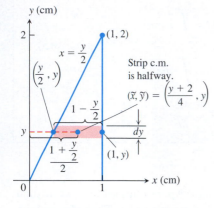

FIGURE 6.51 Modeling the plate in Example 2 with horizontal strips.

Solution   Method 1: Vertical Strips   (Figure 6.50)

(a) The moment M_y: The typical vertical strip has the following relevant data.

center of mass (c.m.): $(\widetilde{x}, \widetilde{y}) = (x, x)$
length: $2x$
width: dx
area: $dA = 2x\,dx$
mass: $dm = \delta\,dA = 3 \cdot 2x\,dx = 6x\,dx$
distance of c.m. from y-axis: $\widetilde{x} = x$

The moment of the strip about the y-axis is

$$\widetilde{x}\,dm = x \cdot 6x\,dx = 6x^2\,dx.$$

The moment of the plate about the y-axis is therefore

$$M_y = \int \widetilde{x}\,dm = \int_0^1 6x^2\,dx = 2x^3 \Big]_0^1 = 2\,\text{g}\cdot\text{cm}.$$

(b) The plate's mass:

$$M = \int dm = \int_0^1 6x\,dx = 3x^2 \Big]_0^1 = 3\,\text{g}.$$

(c) The x-coordinate of the plate's center of mass:

$$\overline{x} = \frac{M_y}{M} = \frac{2\,\text{g}\cdot\text{cm}}{3\,\text{g}} = \frac{2}{3}\,\text{cm}.$$

By a similar computation, we could find M_x and $\overline{y} = M_x/M$.

Method 2: Horizontal Strips   (Figure 6.51)

(a) The moment M_y: The y-coordinate of the center of mass of a typical horizontal strip is y (see the figure), so

$$\widetilde{y} = y.$$

The x-coordinate is the x-coordinate of the point halfway across the triangle. This makes it the average of $y/2$ (the strip's left-hand x-value) and 1 (the strip's right-hand x-value):

$$\widetilde{x} = \frac{(y/2) + 1}{2} = \frac{y}{4} + \frac{1}{2} = \frac{y+2}{4}.$$

We also have

length: $1 - \dfrac{y}{2} = \dfrac{2-y}{2}$
width: dy
area: $dA = \dfrac{2-y}{2}\,dy$
mass: $dm = \delta\,dA = 3 \cdot \dfrac{2-y}{2}\,dy$
distance of c.m. to y-axis: $\widetilde{x} = \dfrac{y+2}{4}.$

The moment of the strip about the y-axis is

$$\widetilde{x}\,dm = \frac{y+2}{4} \cdot 3 \cdot \frac{2-y}{2}\,dy = \frac{3}{8}(4 - y^2)\,dy.$$

The moment of the plate about the y-axis is

$$M_y = \int \tilde{x} \, dm = \int_0^2 \frac{3}{8}(4 - y^2) \, dy = \frac{3}{8}\left[4y - \frac{y^3}{3}\right]_0^2 = \frac{3}{8}\left(\frac{16}{3}\right) = 2 \text{ g} \cdot \text{cm}.$$

(b) The plate's mass:

$$M = \int dm = \int_0^2 \frac{3}{2}(2 - y) \, dy = \frac{3}{2}\left[2y - \frac{y^2}{2}\right]_0^2 = \frac{3}{2}(4 - 2) = 3 \text{ g}.$$

(c) The x-coordinate of the plate's center of mass:

$$\bar{x} = \frac{M_y}{M} = \frac{2 \text{ g} \cdot \text{cm}}{3 \text{ g}} = \frac{2}{3} \text{ cm}.$$

By a similar computation, we could find M_x and $\bar{y}$. ∎

If the distribution of mass in a thin, flat plate has an axis of symmetry, the center of mass will lie on this axis. If there are two axes of symmetry, the center of mass will lie at their intersection. These facts often help to simplify our work.

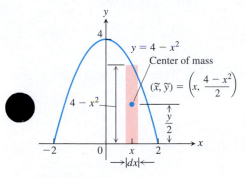

FIGURE 6.52 Modeling the plate in Example 3 with vertical strips.

EXAMPLE 3   Find the center of mass of a thin plate covering the region bounded above by the parabola $y = 4 - x^2$ and below by the x-axis (Figure 6.52). Assume the density of the plate at the point (x, y) is $\delta = 2x^2$, which is twice the square of the distance from the point to the y-axis.

Solution   The mass distribution is symmetric about the y-axis, so $\bar{x} = 0$. We model the distribution of mass with vertical strips, since the density is given as a function of the variable x. The typical vertical strip (see Figure 6.52) has the following relevant data.

$$\text{center of mass (c.m.):} \quad (\tilde{x}, \tilde{y}) = \left(x, \frac{4 - x^2}{2}\right)$$

$$\begin{aligned}
\text{length:} &\quad 4 - x^2 \\
\text{width:} &\quad dx \\
\text{area:} &\quad dA = (4 - x^2) \, dx \\
\text{mass:} &\quad dm = \delta \, dA = \delta(4 - x^2) \, dx \\
\text{distance from c.m. to } x\text{-axis:} &\quad \tilde{y} = \frac{4 - x^2}{2}
\end{aligned}$$

The moment of the strip about the x-axis is

$$\tilde{y} \, dm = \frac{4 - x^2}{2} \cdot \delta(4 - x^2) \, dx = \frac{\delta}{2}(4 - x^2)^2 \, dx.$$

The moment of the plate about the x-axis is

$$M_x = \int \tilde{y} \, dm = \int_{-2}^2 \frac{\delta}{2}(4 - x^2)^2 \, dx = \int_{-2}^2 x^2(4 - x^2)^2 \, dx$$

$$= \int_{-2}^2 (16x^2 - 8x^4 + x^6) \, dx = \frac{2048}{105}.$$

The mass of the plate is

$$M = \int dm = \int_{-2}^2 \delta(4 - x^2) \, dx = \int_{-2}^2 2x^2(4 - x^2) \, dx$$

$$= \int_{-2}^2 (8x^2 - 2x^4) \, dx = \frac{256}{15}.$$

Therefore,

$$\bar{y} = \frac{M_x}{M} = \frac{2048}{105} \cdot \frac{15}{256} = \frac{8}{7}.$$

The plate's center of mass is

$$(\bar{x}, \bar{y}) = \left(0, \frac{8}{7}\right).$$

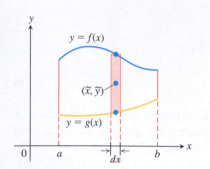

FIGURE 6.53 Modeling the plate bounded by two curves with vertical strips. The strip c.m. is halfway, so $\tilde{y} = \frac{1}{2}[f(x) + g(x)]$.

Plates Bounded by Two Curves

Suppose a plate covers a region that lies between two curves $y = g(x)$ and $y = f(x)$, where $f(x) \geq g(x)$ and $a \leq x \leq b$. The typical vertical strip (see Figure 6.53) has

center of mass (c.m.): $(\tilde{x}, \tilde{y}) = \left(x, \frac{1}{2}[f(x) + g(x)]\right)$
length: $f(x) - g(x)$
width: dx
area: $dA = [f(x) - g(x)]\, dx$
mass: $dm = \delta\, dA = \delta[f(x) - g(x)]\, dx.$

The moment of the plate about the y-axis is

$$M_y = \int x\, dm = \int_a^b x\delta[f(x) - g(x)]\, dx,$$

and the moment about the x-axis is

$$M_x = \int y\, dm = \int_a^b \frac{1}{2}[f(x) + g(x)] \cdot \delta[f(x) - g(x)]\, dx$$

$$= \int_a^b \frac{\delta}{2}[f^2(x) - g^2(x)]\, dx.$$

These moments give us the following formulas.

$$\bar{x} = \frac{1}{M}\int_a^b \delta x[f(x) - g(x)]\, dx \qquad (6)$$

$$\bar{y} = \frac{1}{M}\int_a^b \frac{\delta}{2}[f^2(x) - g^2(x)]\, dx \qquad (7)$$

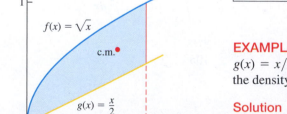

FIGURE 6.54 The region in Example 4.

EXAMPLE 4 Find the center of mass for the thin plate bounded by the curves $g(x) = x/2$ and $f(x) = \sqrt{x}, 0 \leq x \leq 1$ (Figure 6.54), using Equations (6) and (7) with the density function $\delta(x) = x^2$.

Solution We first compute the mass of the plate, using $dm = \delta[f(x) - g(x)]\, dx$:

$$M = \int_0^1 x^2\left(\sqrt{x} - \frac{x}{2}\right) dx = \int_0^1 \left(x^{5/2} - \frac{x^3}{2}\right) dx = \left[\frac{2}{7}x^{7/2} - \frac{1}{8}x^4\right]_0^1 = \frac{9}{56}.$$

Then from Equations (6) and (7) we get

$$\bar{x} = \frac{56}{9} \int_0^1 x^2 \cdot x \left(\sqrt{x} - \frac{x}{2} \right) dx$$

$$= \frac{56}{9} \int_0^1 \left(x^{7/2} - \frac{x^4}{2} \right) dx$$

$$= \frac{56}{9} \left[\frac{2}{9} x^{9/2} - \frac{1}{10} x^5 \right]_0^1 = \frac{308}{405},$$

and

$$\bar{y} = \frac{56}{9} \int_0^1 \frac{x^2}{2} \left(x - \frac{x^2}{4} \right) dx$$

$$= \frac{28}{9} \int_0^1 \left(x^3 - \frac{x^4}{4} \right) dx$$

$$= \frac{28}{9} \left[\frac{1}{4} x^4 - \frac{1}{20} x^5 \right]_0^1 = \frac{252}{405}.$$

The center of mass is shown in Figure 6.54.

Centroids

The center of mass in Example 4 is not located at the geometric center of the region. This is due to the region's nonuniform density. When the density function is constant, it cancels out of the numerator and denominator of the formulas for $\bar{x}$ and $\bar{y}$. Thus, when the density is constant, the location of the center of mass is a feature of the geometry of the object and not of the material from which it is made. In such cases, engineers may call the center of mass the **centroid** of the shape, as in "Find the centroid of a triangle or a solid cone." To do so, just set δ equal to 1 and proceed to find $\bar{x}$ and $\bar{y}$ as before, by dividing moments by masses.

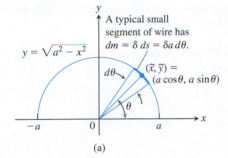

EXAMPLE 5   Find the center of mass (centroid) of a thin wire of constant density δ shaped like a semicircle of radius a.

Solution   We model the wire with the semicircle $y = \sqrt{a^2 - x^2}$ (Figure 6.55). The distribution of mass is symmetric about the y-axis, so $\bar{x} = 0$. To find $\bar{y}$, we imagine the wire divided into short subarc segments. If $(\tilde{x}, \tilde{y})$ is the center of mass of a subarc and θ is the angle between the x-axis and the radial line joining the origin to $(\tilde{x}, \tilde{y})$, then $\tilde{y} = a \sin \theta$ is a function of the angle θ measured in radians (see Figure 6.55a). The length ds of the subarc containing $(\tilde{x}, \tilde{y})$ subtends an angle of $d\theta$ radians, so $ds = a\, d\theta$. Thus a typical subarc segment has these relevant data for calculating $\bar{y}$:

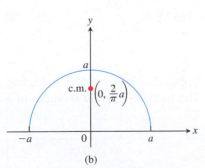

FIGURE 6.55   The semicircular wire in Example 5. (a) The dimensions and variables used in finding the center of mass. (b) The center of mass does not lie on the wire.

$$\text{length:} \quad ds = a\, d\theta$$
$$\text{mass:} \quad dm = \delta\, ds = \delta a\, d\theta \qquad \text{\textcolor{blue}{Mass per unit length times length}}$$
$$\text{distance of c.m. to } x\text{-axis:} \quad \tilde{y} = a \sin \theta.$$

Hence,

$$\bar{y} = \frac{\int \tilde{y}\, dm}{\int dm} = \frac{\int_0^\pi a \sin \theta \cdot \delta a\, d\theta}{\int_0^\pi \delta a\, d\theta} = \frac{\delta a^2 \left[-\cos \theta \right]_0^\pi}{\delta a \pi} = \frac{2}{\pi} a.$$

The center of mass lies on the axis of symmetry at the point $(0, 2a/\pi)$, about two-thirds of the way up from the origin (Figure 6.55b). Notice how δ cancels in the equation for $\bar{y}$, so we could have set $\delta = 1$ everywhere and obtained the same value for $\bar{y}$.

In Example 5 we found the center of mass of a thin wire lying along the graph of a differentiable function in the *xy*-plane. In Chapter 16 we will learn how to find the center of mass of a wire lying along a more general smooth curve in the plane or in space.

Fluid Forces and Centroids

Surface level of fluid

$\bar{h}$ = centroid depth

Plate centroid

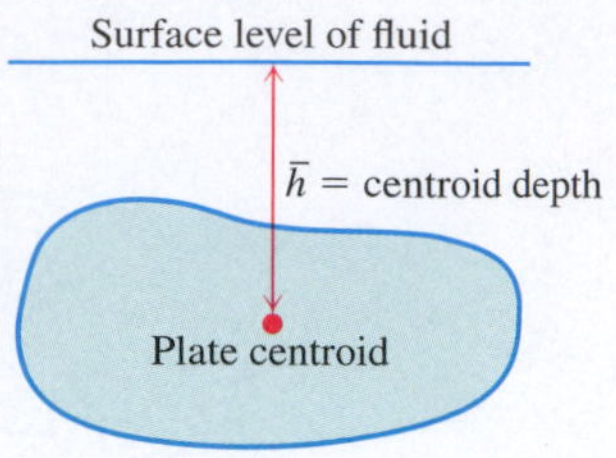

FIGURE 6.56 The force against one side of the plate is $w \cdot \bar{h} \cdot$ plate area.

If we know the location of the centroid of a submerged flat vertical plate (Figure 6.56), we can take a shortcut to find the force against one side of the plate. From Equation (7) in Section 6.5, and the definition of the moment about the *x*-axis, we have

$$F = \int_a^b w \times (\text{strip depth}) \times L(y) \, dy$$

$$= w \int_a^b (\text{strip depth}) \times L(y) \, dy$$

$$= w \times (\text{moment about surface level line of region occupied by plate})$$

$$= w \times (\text{depth of plate's centroid}) \times (\text{area of plate}).$$

> ### Fluid Forces and Centroids
> The force of a fluid of weight-density w against one side of a submerged flat vertical plate is the product of w, the distance $\bar{h}$ from the plate's centroid to the fluid surface, and the plate's area:
>
> $$F = w\bar{h}A. \tag{8}$$

EXAMPLE 6   A flat isosceles triangular plate with base 6 ft and height 3 ft is submerged vertically, base up with its vertex at the origin, so that the base is 2 ft below the surface of a swimming pool. (This is Example 6, Section 6.5.) Use Equation (8) to find the force exerted by the water against one side of the plate.

Solution   The centroid of the triangle (Figure 6.43) lies on the *y*-axis, one-third of the way from the base to the vertex, so $\bar{h} = 3$ (where $y = 2$), since the pool's surface is $y = 5$. The triangle's area is

$$A = \frac{1}{2} (\text{base})(\text{height}) = \frac{1}{2} (6)(3) = 9.$$

Hence,

$$F = w\bar{h}A = (62.4)(3)(9) = 1684.8 \text{ lb.} \qquad \blacksquare$$

The Theorems of Pappus

In the fourth century, an Alexandrian Greek named Pappus discovered two formulas that relate centroids to surfaces and solids of revolution. The formulas provide shortcuts to a number of otherwise lengthy calculations.

> ### THEOREM 1—Pappus's Theorem for Volumes
> If a plane region is revolved once about a line in the plane that does not cut through the region's interior, then the volume of the solid it generates is equal to the region's area times the distance traveled by the region's centroid during the revolution. If ρ is the distance from the axis of revolution to the centroid, then
>
> $$V = 2\pi\rho A. \tag{9}$$

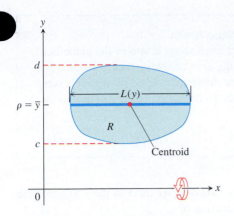

FIGURE 6.57 The region R is to be revolved (once) about the x-axis to generate a solid. A 1700-year-old theorem says that the solid's volume can be calculated by multiplying the region's area by the distance traveled by its centroid during the revolution.

Proof We draw the axis of revolution as the x-axis with the region R in the first quadrant (Figure 6.57). We let $L(y)$ denote the length of the cross-section of R perpendicular to the y-axis at y. We assume $L(y)$ to be continuous.

By the method of cylindrical shells, the volume of the solid generated by revolving the region about the x-axis is

$$V = \int_c^d 2\pi(\text{shell radius})(\text{shell height}) \, dy = 2\pi \int_c^d y \, L(y) \, dy. \qquad (10)$$

The y-coordinate of R's centroid is

$$\bar{y} = \frac{\int_c^d \tilde{y} \, dA}{A} = \frac{\int_c^d y \, L(y) \, dy}{A}, \qquad \tilde{y} = y, \, dA = L(y) \, dy$$

so that

$$\int_c^d y \, L(y) \, dy = A\bar{y}.$$

Substituting $A\bar{y}$ for the last integral in Equation (10) gives $V = 2\pi\bar{y}A$. With ρ equal to $\bar{y}$, we have $V = 2\pi\rho A$. ∎

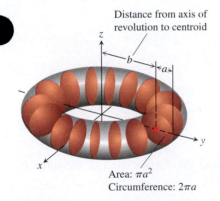

FIGURE 6.58 With Pappus's first theorem, we can find the volume of a torus without having to integrate (Example 7).

EXAMPLE 7 Find the volume of the torus (doughnut) generated by revolving a circular disk of radius a about an axis in its plane at a distance $b \geq a$ from its center (Figure 6.58).

Solution We apply Pappus's Theorem for volumes. The centroid of a disk is located at its center, the area is $A = \pi a^2$, and $\rho = b$ is the distance from the centroid to the axis of revolution (see Figure 6.58). Substituting these values into Equation (9), we find the volume of the torus to be

$$V = 2\pi(b)(\pi a^2) = 2\pi^2 b a^2. \qquad ■$$

The next example shows how we can use Equation (9) in Pappus's Theorem to find one of the coordinates of the centroid of a plane region of known area A when we also know the volume V of the solid generated by revolving the region about the other coordinate axis. That is, if $\bar{y}$ is the coordinate we want to find, we revolve the region around the x-axis so that $\bar{y} = \rho$ is the distance from the centroid to the axis of revolution. The idea is that the rotation generates a solid of revolution whose volume V is an already known quantity. Then we can solve Equation (9) for ρ, which is the value of the centroid's coordinate $\bar{y}$.

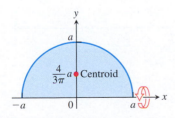

FIGURE 6.59 With Pappus's first theorem, we can locate the centroid of a semicircular region without having to integrate (Example 8).

EXAMPLE 8 Locate the centroid of a semicircular region of radius a.

Solution We consider the region between the semicircle $y = \sqrt{a^2 - x^2}$ (Figure 6.59) and the x-axis and imagine revolving the region about the x-axis to generate a solid sphere. By symmetry, the x-coordinate of the centroid is $\bar{x} = 0$. With $\bar{y} = \rho$ in Equation (9), we have

$$\bar{y} = \frac{V}{2\pi A} = \frac{(4/3)\pi a^3}{2\pi(1/2)\pi a^2} = \frac{4}{3\pi}a. \qquad ■$$

> **THEOREM 2—Pappus's Theorem for Surface Areas**
> If an arc of a smooth plane curve is revolved once about a line in the plane that does not cut through the arc's interior, then the area of the surface generated by the arc equals the length L of the arc times the distance traveled by the arc's centroid during the revolution. If ρ is the distance from the axis of revolution to the centroid, then
>
> $$S = 2\pi\rho L. \qquad (11)$$

The proof we give assumes that we can model the axis of revolution as the x-axis and the arc as the graph of a continuously differentiable function of x.

Proof   We draw the axis of revolution as the x-axis with the arc extending from $x = a$ to $x = b$ in the first quadrant (Figure 6.60). The area of the surface generated by the arc is

$$S = \int_{x=a}^{x=b} 2\pi y \, ds = 2\pi \int_{x=a}^{x=b} y \, ds. \qquad (12)$$

The y-coordinate of the arc's centroid is

$$\bar{y} = \frac{\displaystyle\int_{x=a}^{x=b} \tilde{y} \, ds}{\displaystyle\int_{x=a}^{x=b} ds} = \frac{\displaystyle\int_{x=a}^{x=b} y \, ds}{L}. \qquad \begin{array}{l} L = \int ds \text{ is the arc's} \\ \text{length and } \tilde{y} = y. \end{array}$$

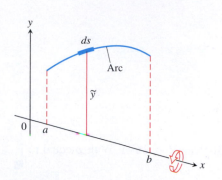

FIGURE 6.60 Figure for proving Pappus's Theorem for surface area. The arc length differential ds is given by Equation (6) in Section 6.3.

Hence

$$\int_{x=a}^{x=b} y \, ds = \bar{y}L.$$

Substituting $\bar{y}L$ for the last integral in Equation (12) gives $S = 2\pi\bar{y}L$. With ρ equal to $\bar{y}$, we have $S = 2\pi\rho L$. ∎

EXAMPLE 9   Use Pappus's area theorem to find the surface area of the torus in Example 7.

Solution   From Figure 6.58, the surface of the torus is generated by revolving a circle of radius a about the z-axis, and $b \geq a$ is the distance from the centroid to the axis of revolution. The arc length of the smooth curve generating this surface of revolution is the circumference of the circle, so $L = 2\pi a$. Substituting these values into Equation (11), we find the surface area of the torus to be

$$S = 2\pi(b)(2\pi a) = 4\pi^2 ba. \qquad ■$$

EXERCISES 6.6

Mass of a wire

In Exercises 1–6, find the mass M and center of mass $\bar{x}$ of the linear wire covering the given interval and having the given density $\delta(x)$.

1. $1 \leq x \leq 4, \quad \delta(x) = \sqrt{x}$

2. $-3 \leq x \leq 3, \quad \delta(x) = 1 + 3x^2$

3. $0 \leq x \leq 3, \quad \delta(x) = \dfrac{1}{x+1}$

4. $1 \leq x \leq 2, \quad \delta(x) = \dfrac{8}{x^3}$

5. $\delta(x) = \begin{cases} 4, & 0 \leq x \leq 2 \\ 5, & 2 < x \leq 3 \end{cases}$

6. $\delta(x) = \begin{cases} 2 - x, & 0 \leq x < 1 \\ x, & 1 \leq x \leq 2 \end{cases}$

Thin Plates with Constant Density

In Exercises 7–20, find the center of mass of a thin plate of constant density δ covering the given region.

7. The region bounded by the parabola $y = x^2$ and the line $y = 4$

8. The region bounded by the parabola $y = 25 - x^2$ and the x-axis

9. The region bounded by the parabola $y = x - x^2$ and the line $y = -x$

10. The region enclosed by the parabolas $y = x^2 - 3$ and $y = -2x^2$

11. The region bounded by the y-axis and the curve $x = y - y^3$, $0 \le y \le 1$

12. The region bounded by the parabola $x = y^2 - y$ and the line $y = x$

13. The region bounded by the x-axis and the curve $y = \cos x$, $-\pi/2 \le x \le \pi/2$

14. The region between the curve $y = \sec^2 x$, $-\pi/4 \le x \le \pi/4$ and the x-axis

T 15. The region between the curve $y = 1/x$ and the x-axis from $x = 1$ to $x = 2$. Give the coordinates to two decimal places.

16. **a.** The region cut from the first quadrant by the circle $x^2 + y^2 = 9$

    b. The region bounded by the x-axis and the semicircle $y = \sqrt{9 - x^2}$

Compare your answer in part (b) with the answer in part (a).

17. The region in the first and fourth quadrants enclosed by the curves $y = 1/(1 + x^2)$ and $y = -1/(1 + x^2)$ and by the lines $x = 0$ and $x = 1$

18. The region bounded by the parabolas $y = 2x^2 - 4x$ and $y = 2x - x^2$

19. The region between the curve $y = 1/\sqrt{x}$ and the x-axis from $x = 1$ to $x = 16$

20. The region bounded above by the curve $y = 1/x^3$, below by the curve $y = -1/x^3$, and on the left and right by the lines $x = 1$ and $x = a > 1$. Also, find $\lim_{a \to \infty} \bar{x}$.

21. Consider a region bounded by the graphs of $y = x^4$ and $y = x^5$. Show that the center of mass lies outside the region.

22. Consider a thin plate of constant density δ lies in the region bounded by the graphs of $y = \sqrt{x}$ and $x = 2y$. Find the plate's

    a. moment about the x-axis.

    b. moment about the y-axis.

    c. moment about the line $x = 5$.

    d. moment about the line $x = -1$.

    e. moment about the line $y = 2$.

    f. moment about the line $y = -3$.

    g. mass.

    h. center of mass.

Thin Plates with Varying Density

23. Find the center of mass of a thin plate covering the region between the x-axis and the curve $y = 2/x^2$, $1 \le x \le 2$, if the plate's density at the point (x, y) is $\delta(x) = x^2$.

24. Find the center of mass of a thin plate covering the region bounded below by the parabola $y = x^2$ and above by the line $y = x$ if the plate's density at the point (x, y) is $\delta(x) = 12x$.

25. The region bounded by the curves $y = \pm 4/\sqrt{x}$ and the lines $x = 1$ and $x = 4$ is revolved about the y-axis to generate a solid.

    a. Find the volume of the solid.

    b. Find the center of mass of a thin plate covering the region if the plate's density at the point (x, y) is $\delta(x) = 1/x$.

    c. Sketch the plate and show the center of mass in your sketch.

26. The region between the curve $y = 2/x$ and the x-axis from $x = 1$ to $x = 4$ is revolved about the x-axis to generate a solid.

    a. Find the volume of the solid.

    b. Find the center of mass of a thin plate covering the region if the plate's density at the point (x, y) is $\delta(x) = \sqrt{x}$.

    c. Sketch the plate and show the center of mass in your sketch.

Centroids of Triangles

27. **The centroid of a triangle lies at the intersection of the triangle's medians** You may recall that the point inside a triangle that lies one-third of the way from each side toward the opposite vertex is the point where the triangle's three medians intersect. Show that the centroid lies at the intersection of the medians by showing that it too lies one-third of the way from each side toward the opposite vertex. To do so, take the following steps.

    i) Stand one side of the triangle on the x-axis as in part (b) of the accompanying figure. Express dm in terms of L and dy.

    ii) Use similar triangles to show that $L = (b/h)(h - y)$. Substitute this expression for L in your formula for dm.

    iii) Show that $\bar{y} = h/3$.

    iv) Extend the argument to the other sides.

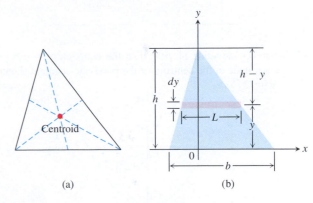

(a) (b)

Use the result in Exercise 27 to find the centroids of the triangles whose vertices appear in Exercises 28–32. Assume $a, b > 0$.

28. $(-1, 0), (1, 0), (0, 3)$ 29. $(0, 0), (1, 0), (0, 1)$

30. $(0, 0), (a, 0), (0, a)$ 31. $(0, 0), (a, 0), (0, b)$

32. $(0, 0), (a, 0), (a/2, b)$

Thin Wires

33. **Constant density** Find the moment about the x-axis of a wire of constant density that lies along the curve $y = \sqrt{x}$ from $x = 0$ to $x = 2$.

34. **Constant density** Find the moment about the x-axis of a wire of constant density that lies along the curve $y = x^3$ from $x = 0$ to $x = 1$.

35. Variable density Suppose that the density of the wire in Example 5 is $\delta = k \sin \theta$ (k constant). Find the center of mass.

36. Variable density Suppose that the density of the wire in Example 5 is $\delta = 1 + k|\cos \theta|$ (k constant). Find the center of mass.

Plates Bounded by Two Curves

In Exercises 37–40, find the centroid of the thin plate bounded by the graphs of the given functions. Use Equations (6) and (7) with $\delta = 1$ and $M = $ area of the region covered by the plate.

37. $g(x) = x^2$ and $f(x) = x + 6$

38. $g(x) = x^2(x + 1)$, $f(x) = 2$, and $x = 0$

39. $g(x) = x^2(x - 1)$ and $f(x) = x^2$

40. $g(x) = 0$, $f(x) = 2 + \sin x$, $x = 0$, and $x = 2\pi$

(*Hint:* $\int x \sin x \, dx = \sin x - x \cos x + C$.)

Theory and Examples

Verify the statements and formulas in Exercises 41 and 42.

41. The coordinates of the centroid of a differentiable plane curve are

$$\bar{x} = \frac{\int x \, ds}{\text{length}}, \qquad \bar{y} = \frac{\int y \, ds}{\text{length}}.$$

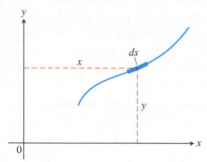

42. Whatever the value of $p > 0$ in the equation $y = x^2/(4p)$, the y-coordinate of the centroid of the parabolic segment shown here is $\bar{y} = (3/5)a$.

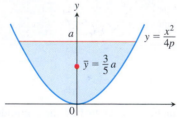

The Theorems of Pappus

43. The square region with vertices $(0, 2)$, $(2, 0)$, $(4, 2)$, and $(2, 4)$ is revolved about the x-axis to generate a solid. Find the volume and surface area of the solid.

44. Use a theorem of Pappus to find the volume generated by revolving about the line $x = 5$ the triangular region bounded by the coordinate axes and the line $2x + y = 6$ (see Exercise 27).

45. Find the volume of the torus generated by revolving the circle $(x - 2)^2 + y^2 = 1$ about the y-axis.

46. Use the theorems of Pappus to find the lateral surface area and the volume of a right-circular cone.

47. Use Pappus's Theorem for surface area and the fact that the surface area of a sphere of radius a is $4\pi a^2$ to find the centroid of the semicircle $y = \sqrt{a^2 - x^2}$.

48. As found in Exercise 47, the centroid of the semicircle $y = \sqrt{a^2 - x^2}$ lies at the point $(0, 2a/\pi)$. Find the area of the surface swept out by revolving the semicircle about the line $y = a$.

49. The area of the region R enclosed by the semiellipse $y = (b/a)\sqrt{a^2 - x^2}$ and the x-axis is $(1/2)\pi ab$, and the volume of the ellipsoid generated by revolving R about the x-axis is $(4/3)\pi ab^2$. Find the centroid of R. Notice that the location is independent of a.

50. As found in Example 8, the centroid of the region enclosed by the x-axis and the semicircle $y = \sqrt{a^2 - x^2}$ lies at the point $(0, 4a/3\pi)$. Find the volume of the solid generated by revolving this region about the line $y = -a$.

51. The region of Exercise 50 is revolved about the line $y = x - a$ to generate a solid. Find the volume of the solid.

52. As found in Exercise 47, the centroid of the semicircle $y = \sqrt{a^2 - x^2}$ lies at the point $(0, 2a/\pi)$. Find the area of the surface generated by revolving the semicircle about the line $y = x - a$.

In Exercises 53 and 54, use a theorem of Pappus to find the centroid of the given triangle. Use the fact that the volume of a cone of radius r and height h is $V = \frac{1}{3}\pi r^2 h$.

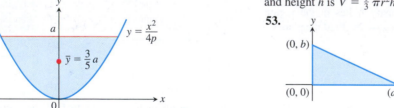

CHAPTER 6 **Questions to Guide Your Review**

1. How do you define and calculate the volumes of solids by the method of slicing? Give an example.

2. How are the disk and washer methods for calculating volumes derived from the method of slicing? Give examples of volume calculations by these methods.

3. Describe the method of cylindrical shells. Give an example.

4. How do you find the length of the graph of a smooth function over a closed interval? Give an example. What about functions that do not have continuous first derivatives?

5. How do you define and calculate the area of the surface swept out by revolving the graph of a smooth function $y = f(x)$, $a \le x \le b$, about the x-axis? Give an example.

6. How do you define and calculate the work done by a variable force directed along a portion of the *x*-axis? How do you calculate the work it takes to pump a liquid from a tank? Give examples.

7. How do you calculate the force exerted by a liquid against a portion of a flat vertical wall? Give an example.

8. What is a center of mass? a centroid?

9. How do you locate the center of mass of a thin flat plate of material? Give an example.

10. How do you locate the center of mass of a thin plate bounded by two curves $y = f(x)$ and $y = g(x)$ over $a \le x \le b$?

CHAPTER 6 Practice Exercises

Volumes

Find the volumes of the solids in Exercises 1–18.

1. The solid lies between planes perpendicular to the *x*-axis at $x = 0$ and $x = 1$. The cross-sections perpendicular to the *x*-axis between these planes are circular disks whose diameters run from the parabola $y = x^2$ to the parabola $y = \sqrt{x}$.

2. The base of the solid is the region in the first quadrant between the line $y = x$ and the parabola $y = 2\sqrt{x}$. The cross-sections of the solid perpendicular to the *x*-axis are equilateral triangles whose bases stretch from the line to the curve.

3. The solid lies between planes perpendicular to the *x*-axis at $x = \pi/4$ and $x = 5\pi/4$. The cross-sections between these planes are circular disks whose diameters run from the curve $y = 2\cos x$ to the curve $y = 2\sin x$.

4. The solid lies between planes perpendicular to the *x*-axis at $x = 0$ and $x = 6$. The cross-sections between these planes are squares whose bases run from the *x*-axis up to the curve $x^{1/2} + y^{1/2} = \sqrt{6}$.

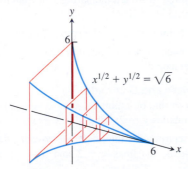

$x^{1/2} + y^{1/2} = \sqrt{6}$

5. The solid lies between planes perpendicular to the *x*-axis at $x = 0$ and $x = 4$. The cross-sections of the solid perpendicular to the *x*-axis between these planes are circular disks whose diameters run from the curve $x^2 = 4y$ to the curve $y^2 = 4x$.

6. The base of the solid is the region bounded by the parabola $y^2 = 4x$ and the line $x = 1$ in the *xy*-plane. Each cross-section perpendicular to the *x*-axis is an equilateral triangle with one edge in the plane. (The triangles all lie on the same side of the plane.)

7. Find the volume of the solid generated by revolving the region bounded by the *x*-axis, the curve $y = 3x^4$, and the lines $x = 1$ and $x = -1$ about **(a)** the *x*-axis; **(b)** the *y*-axis; **(c)** the line $x = 1$; **(d)** the line $y = 3$.

8. Find the volume of the solid generated by revolving the "triangular" region bounded by the curve $y = 4/x^3$ and the lines $x = 1$ and $y = 1/2$ about **(a)** the *x*-axis; **(b)** the *y*-axis; **(c)** the line $x = 2$; **(d)** the line $y = 4$.

9. Find the volume of the solid generated by revolving the region bounded on the left by the parabola $x = y^2 + 1$ and on the right by the line $x = 5$ about **(a)** the *x*-axis; **(b)** the *y*-axis; **(c)** the line $x = 5$.

10. Find the volume of the solid generated by revolving the region bounded by the parabola $y^2 = 4x$ and the line $y = x$ about **(a)** the *x*-axis; **(b)** the *y*-axis; **(c)** the line $x = 4$; **(d)** the line $y = 4$.

11. Find the volume of the solid generated by revolving the "triangular" region bounded by the *x*-axis, the line $x = \pi/3$, and the curve $y = \tan x$ in the first quadrant about the *x*-axis.

12. Find the volume of the solid generated by revolving the region bounded by the curve $y = \sin x$ and the lines $x = 0$, $x = \pi$, and $y = 2$ about the line $y = 2$.

13. Find the volume of the solid generated by revolving the region bounded by the curve $x = e^{y^2}$ and the lines $y = 0$, $x = 0$, and $y = 1$ about the *x*-axis.

14. Find the volume of the solid generated by revolving about the *x*-axis the region bounded by $y = 2\tan x$, $y = 0$, $x = -\pi/4$, and $x = \pi/4$. (The region lies in the first and third quadrants and resembles a skewed bowtie.)

15. **Volume of a solid sphere hole** A round hole of radius $\sqrt{3}$ ft is bored through the center of a solid sphere of a radius 2 ft. Find the volume of material removed from the sphere.

16. **Volume of a football** The profile of a football resembles the ellipse shown here. Find the football's volume to the nearest cubic inch.

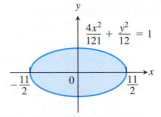

17. Find the volume of the given circular frustum of height *h* and radii *a* and *b*.

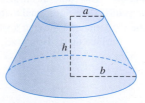

18. The graph of $x^{2/3} + y^{2/3} = 1$ is called an astroid and is given below. Find the volume of the solid formed by revolving the region enclosed by the astroid about the *x*-axis.

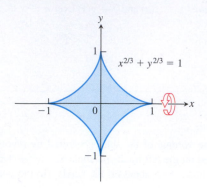

Lengths of Curves

Find the lengths of the curves in Exercises 19–22.

19. $y = x^{1/2} - (1/3)x^{3/2}, \quad 1 \le x \le 4$

20. $x = y^{2/3}, \quad 1 \le y \le 8$

21. $y = x^2 - (\ln x)/8, \quad 1 \le x \le 2$

22. $x = (y^3/12) + (1/y), \quad 1 \le y \le 2$

23. $y = \sin^{-1} x - \sqrt{1 - x^2}, \quad 0 \le x \le \dfrac{3}{4}$

24. $y = \dfrac{2}{3}x^{3/2} - 1, \quad 0 \le x \le 1$

Areas of Surfaces of Revolution

In Exercises 25–28, find the areas of the surfaces generated by revolving the curves about the given axes.

25. $y = \sqrt{2x + 1}, \quad 0 \le x \le 3; \quad x$-axis

26. $y = x^3/3, \quad 0 \le x \le 1; \quad x$-axis

27. $x = \sqrt{4y - y^2}, \quad 1 \le y \le 2; \quad y$-axis

28. $x = \sqrt{y}, \quad 2 \le y \le 6; \quad y$-axis

Work

29. Lifting equipment A rock climber is about to haul up 100 N (about 22.5 lb) of equipment that has been hanging beneath her on 40 m of rope that weighs 0.8 newton per meter. How much work will it take? (*Hint:* Solve for the rope and equipment separately, then add.)

30. Leaky tank truck You drove an 800-gal tank truck of water from the base of Mt. Washington to the summit and discovered on arrival that the tank was only half full. You started with a full tank, climbed at a steady rate, and accomplished the 4750-ft elevation change in 50 min. Assuming that the water leaked out at a steady rate, how much work was spent in carrying water to the top? Do not count the work done in getting yourself and the truck there. Water weighs 8 lb/U.S. gal.

31. Earth's attraction The force of attraction on an object below Earth's surface is directly proportional to its distance from Earth's center. Find the work done in moving a weight of *w* lb located *a* mi below Earth's surface up to the surface itself. Assume Earth's radius is a constant *r* mi.

32. Garage door spring A force of 200 N will stretch a garage door spring 0.8 m beyond its unstressed length. How far will a 300-N force stretch the spring? How much work does it take to stretch the spring this far from its unstressed length?

33. Pumping a reservoir A reservoir shaped like a right-circular cone, point down, 20 ft across the top and 8 ft deep, is full of water. How much work does it take to pump the water to a level 6 ft above the top?

34. Pumping a reservoir (*Continuation of Exercise 33.*) The reservoir is filled to a depth of 5 ft, and the water is to be pumped to the same level as the top. How much work does it take?

35. Pumping a conical tank A right-circular conical tank, point down, with top radius 5 ft and height 10 ft is filled with a liquid whose weight-density is 60 lb/ft³. How much work does it take to pump the liquid to a point 2 ft above the tank? If the pump is driven by a motor rated at 275 ft-lb/sec (1/2 hp), how long will it take to empty the tank?

36. Pumping a cylindrical tank A storage tank is a right-circular cylinder 20 ft long and 8 ft in diameter with its axis horizontal. If the tank is half full of olive oil weighing 57 lb/ft³, find the work done in emptying it through a pipe that runs from the bottom of the tank to an outlet that is 6 ft above the top of the tank.

37. Assume that a spring does not follow Hooke's Law. Instead, the force required to stretch the spring *x* ft from its natural length is $F(x) = 10x^{3/2}$ lb. How much work does it take to

  a. stretch the spring 4 ft from its natural length?

  b. stretch the spring from an initial 1 ft past its natural length to 5 ft past its natural length?

38. Assume that a spring does not follow Hooke's Law. Instead, the force required to stretch the spring *x* m from its natural length is $F(x) = k\sqrt{5 + x^2}$ N.

  a. If a 3-N force stretches the spring 2 m, find the value of *k*.

  b. How much work is required to stretch the spring 1 m from its natural length?

Centers of Mass and Centroids

39. Find the centroid of a thin, flat plate covering the region enclosed by the parabolas $y = 2x^2$ and $y = 3 - x^2$.

40. Find the centroid of a thin, flat plate covering the region enclosed by the *x*-axis, the lines $x = 2$ and $x = -2$, and the parabola $y = x^2$.

41. Find the centroid of a thin, flat plate covering the "triangular" region in the first quadrant bounded by the *y*-axis, the parabola $y = x^2/4$, and the line $y = 4$.

42. Find the centroid of a thin, flat plate covering the region enclosed by the parabola $y^2 = x$ and the line $x = 2y$.

43. Find the center of mass of a thin, flat plate covering the region enclosed by the parabola $y^2 = x$ and the line $x = 2y$ if the density function is $\delta(y) = 1 + y$. (Use horizontal strips.)

44. a. Find the center of mass of a thin plate of constant density covering the region between the curve $y = 3/x^{3/2}$ and the *x*-axis from $x = 1$ to $x = 9$.

  b. Find the plate's center of mass if, instead of being constant, the density is $\delta(x) = x$. (Use vertical strips.)

Fluid Force

45. Trough of water   The vertical triangular plate shown here is the end plate of a trough full of water ($w = 62.4$). What is the fluid force against the plate?

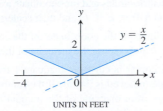

UNITS IN FEET

46. Trough of maple syrup   The vertical trapezoidal plate shown here is the end plate of a trough full of maple syrup weighing 75 lb/ft^3. What is the force exerted by the syrup against the end plate of the trough when the syrup is 10 in. deep?

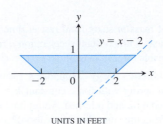

UNITS IN FEET

47. Force on a parabolic gate   A flat vertical gate in the face of a dam is shaped like the parabolic region between the curve $y = 4x^2$ and the line $y = 4$, with measurements in feet. The top of the gate lies 5 ft below the surface of the water. Find the force exerted by the water against the gate ($w = 62.4$).

T **48.** You plan to store mercury ($w = 849 \text{ lb/ft}^3$) in a vertical rectangular tank with a 1 ft square base side whose interior side wall can withstand a total fluid force of 40,000 lb. About how many cubic feet of mercury can you store in the tank at any one time?

CHAPTER 6 Additional and Advanced Exercises

Volume and Length

1. A solid is generated by revolving about the x-axis the region bounded by the graph of the positive continuous function $y = f(x)$, the x-axis, the fixed line $x = a$, and the variable line $x = b, b > a$. Its volume, for all b, is $b^2 - ab$. Find $f(x)$.

2. A solid is generated by revolving about the x-axis the region bounded by the graph of the positive continuous function $y = f(x)$, the x-axis, and the lines $x = 0$ and $x = a$. Its volume, for all $a > 0$, is $a^2 + a$. Find $f(x)$.

3. Suppose that the increasing function $f(x)$ is smooth for $x \geq 0$ and that $f(0) = a$. Let $s(x)$ denote the length of the graph of f from $(0, a)$ to $(x, f(x))$, $x > 0$. Find $f(x)$ if $s(x) = Cx$ for some constant C. What are the allowable values for C?

4. a. Show that for $0 < \alpha \leq \pi/2$,

$$\int_0^\alpha \sqrt{1 + \cos^2\theta} \, d\theta > \sqrt{\alpha^2 + \sin^2\alpha}.$$

b. Generalize the result in part (a).

5. Find the volume of the solid formed by revolving the region bounded by the graphs of $y = x$ and $y = x^2$ about the line $y = x$.

6. Consider a right-circular cylinder of diameter 1. Form a wedge by making one slice parallel to the base of the cylinder completely through the cylinder, and another slice at an angle of 45° to the first slice and intersecting the first slice at the opposite edge of the cylinder (see accompanying diagram). Find the volume of the wedge.

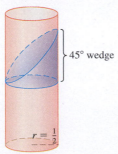

45° wedge

$r = \frac{1}{2}$

Surface Area

7. At points on the curve $y = 2\sqrt{x}$, line segments of length $h = y$ are drawn perpendicular to the xy-plane. (See accompanying figure.) Find the area of the surface formed by these perpendiculars from $(0, 0)$ to $(3, 2\sqrt{3})$.

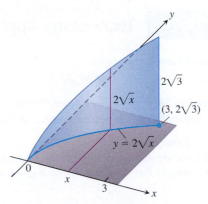

8. At points on a circle of radius a, line segments are drawn perpendicular to the plane of the circle, the perpendicular at each point P being of length ks, where s is the length of the arc of the circle measured counterclockwise from $(a, 0)$ to P and k is a positive constant, as shown here. Find the area of the surface formed by the perpendiculars along the arc beginning at $(a, 0)$ and extending once around the circle.

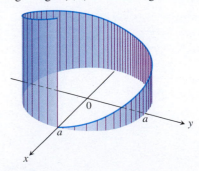

Work

9. A particle of mass m starts from rest at time $t = 0$ and is moved along the x-axis with constant acceleration a from $x = 0$ to $x = h$ against a variable force of magnitude $F(t) = t^2$. Find the work done.

10. Work and kinetic energy Suppose a 1.6-oz golf ball is placed on a vertical spring with force constant $k = 2\ \text{lb}/\text{in}$. The spring is compressed 6 in. and released. About how high does the ball go (measured from the spring's rest position)?

Centers of Mass

11. Find the centroid of the region bounded below by the x-axis and above by the curve $y = 1 - x^n$, n an even positive integer. What is the limiting position of the centroid as $n \to \infty$?

12. If you haul a telephone pole on a two-wheeled carriage behind a truck, you want the wheels to be 3 ft or so behind the pole's center of mass to provide an adequate "tongue" weight. The 40-ft wooden telephone poles used by Verizon have a 27-in. circumference at the top and a 43.5-in. circumference at the base. About how far from the top is the center of mass?

13. Suppose that a thin metal plate of area A and constant density δ occupies a region R in the xy-plane, and let M_y be the plate's moment about the y-axis. Show that the plate's moment about the line $x = b$ is

a. $M_y - b\delta A$ if the plate lies to the right of the line, and

b. $b\delta A - M_y$ if the plate lies to the left of the line.

14. Find the center of mass of a thin plate covering the region bounded by the curve $y^2 = 4ax$ and the line $x = a$, $a =$ positive constant, if the density at (x, y) is directly proportional to (**a**) x, (**b**) $|y|$.

15. a. Find the centroid of the region in the first quadrant bounded by two concentric circles and the coordinate axes, if the circles have radii a and b, $0 < a < b$, and their centers are at the origin.

b. Find the limits of the coordinates of the centroid as a approaches b and discuss the meaning of the result.

16. A triangular corner is cut from a square 1 ft on a side. The area of the triangle removed is 36 in^2. If the centroid of the remaining region is 7 in. from one side of the original square, how far is it from the remaining sides?

Fluid Force

17. A triangular plate ABC is submerged in water with its plane vertical. The side AB, 4 ft long, is 6 ft below the surface of the water, while the vertex C is 2 ft below the surface. Find the force exerted by the water on one side of the plate.

18. A vertical rectangular plate is submerged in a fluid with its top edge parallel to the fluid's surface. Show that the force exerted by the fluid on one side of the plate equals the average value of the pressure up and down the plate times the area of the plate.

CHAPTER 6	**Technology Application Projects**

Mathematica/Maple Projects

Projects can be found within MyMathLab.

- *Using Riemann Sums to Estimate Areas, Volumes, and Lengths of Curves*
  Visualize and approximate areas and volumes in **Part I** and **Part II**: Volumes of Revolution; and **Part III**: Lengths of Curves.

- *Modeling a Bungee Cord Jump*
  Collect data (or use data previously collected) to build and refine a model for the force exerted by a jumper's bungee cord. Use the work-energy theorem to compute the distance fallen for a given jumper and a given length of bungee cord.

7

Integrals and Transcendental Functions

OVERVIEW Our treatment of the logarithmic and exponential functions has been rather informal. In this chapter, we give a rigorous analytic approach to the definitions and properties of these functions. We also introduce the hyperbolic functions and their inverses. Like the trigonometric functions, all of these functions belong to the class of transcendental functions.

7.1 The Logarithm Defined as an Integral

In Chapter 1, we introduced the natural logarithm function $\ln x$ as the inverse of the exponential function e^x. The function e^x was chosen as that function in the family of general exponential functions $a^x, a > 0$, whose graph has slope 1 as it crosses the y-axis. The function a^x was presented intuitively, however, based on its graph at rational values of x.

In this section we recreate the theory of logarithmic and exponential functions from an entirely different point of view. Here we define these functions analytically and recover their behaviors. To begin, we use the Fundamental Theorem of Calculus to define the natural logarithm function $\ln x$ as an integral. We quickly develop its properties, including the algebraic, geometric, and analytic properties seen before. Next we introduce the function e^x as the inverse function of $\ln x$, and establish its previously seen properties. Defining $\ln x$ as an integral and e^x as its inverse is an indirect approach. While it may at first seem strange, it gives an elegant and powerful way to obtain and validate the key properties of logarithmic and exponential functions.

Definition of the Natural Logarithm Function

The natural logarithm of a positive number x, written as $\ln x$, is the value of an integral. The appropriate integral is suggested by our earlier results in Chapter 5.

DEFINITION The **natural logarithm** is the function given by

$$\ln x = \int_1^x \frac{1}{t}\, dt, \qquad x > 0.$$

From the Fundamental Theorem of Calculus, we know that $\ln x$ is a continuous function. Geometrically, if $x > 1$, then $\ln x$ is the area under the curve $y = 1/t$ from $t = 1$ to $t = x$ (Figure 7.1). For $0 < x < 1$, $\ln x$ gives the negative of the area under the curve from x to 1, and the function is not defined for $x \le 0$. From the Zero Width Interval Rule for definite integrals, we also have

$$\ln 1 = \int_1^1 \frac{1}{t}\, dt = 0.$$

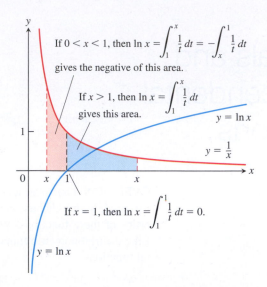

FIGURE 7.1   The graph of $y = \ln x$ and its relation to the function $y = 1/x, x > 0$. The graph of the logarithm rises above the x-axis as x moves from 1 to the right, and it falls below the axis as x moves from 1 to the left.

TABLE 7.1 Typical 2-place values of $\ln x$

x	$\ln x$
0	undefined
0.05	-3.00
0.5	-0.69
1	0
2	0.69
3	1.10
4	1.39
10	2.30

Notice that we show the graph of $y = 1/x$ in Figure 7.1 but use $y = 1/t$ in the integral. Using x for everything would have us writing

$$\ln x = \int_1^x \frac{1}{x}\, dx,$$

with x meaning two different things. So we change the variable of integration to t.

By using rectangles to obtain finite approximations of the area under the graph of $y = 1/t$ and over the interval between $t = 1$ and $t = x$, as in Section 5.1, we can approximate the values of the function $\ln x$. Several values are given in Table 7.1. There is an important number between $x = 2$ and $x = 3$ whose natural logarithm equals 1. This number, which we now define, exists because $\ln x$ is a continuous function and therefore satisfies the Intermediate Value Theorem on the interval $[2, 3]$.

DEFINITION   The **number e** is the number in the domain of the natural logarithm that satisfies

$$\ln(e) = \int_1^e \frac{1}{t}\, dt = 1.$$

Interpreted geometrically, the number e corresponds to the point on the x-axis for which the area under the graph of $y = 1/t$ and above the interval $[1, e]$ equals the area of the unit square. That is, the area of the region shaded blue in Figure 7.1 is 1 sq unit when $x = e$. We will see further on that this is the same number $e \approx 2.718281828$ we have encountered before.

The Derivative of $y = \ln x$

By the first part of the Fundamental Theorem of Calculus (Section 5.4),

$$\frac{d}{dx}\ln x = \frac{d}{dx}\int_1^x \frac{1}{t}\,dt = \frac{1}{x}.$$

For every positive value of x, we have

$$\frac{d}{dx}\ln x = \frac{1}{x}. \tag{1}$$

Therefore, the function $y = \ln x$ is a solution to the initial value problem $dy/dx = 1/x$, $x > 0$, with $y(1) = 0$. Notice that the derivative is always positive.

If u is a differentiable function of x whose values are positive, so that $\ln u$ is defined, then by applying the Chain Rule we obtain

$$\frac{d}{dx}\ln u = \frac{1}{u}\frac{du}{dx}, \qquad u > 0. \tag{2}$$

The derivative of $\ln|x|$ can be found just as in Example 3(c) of Section 3.8, giving

$$\frac{d}{dx}\ln|x| = \frac{1}{x}, \qquad x \neq 0. \tag{3}$$

Moreover, if b is any constant with $bx > 0$, Equation (2) gives

$$\frac{d}{dx}\ln bx = \frac{1}{bx}\cdot\frac{d}{dx}(bx) = \frac{1}{bx}(b) = \frac{1}{x}.$$

The Graph and Range of $\ln x$

The derivative $d(\ln x)/dx = 1/x$ is positive for $x > 0$, so $\ln x$ is an increasing function of x. The second derivative, $-1/x^2$, is negative, so the graph of $\ln x$ is concave down. (See Figure 7.2a.)

The function $\ln x$ has the following familiar algebraic properties, which we stated in Section 1.6. In Section 4.2 we showed these properties are a consequence of Corollary 2 of the Mean Value Theorem, and those derivations still apply.

1. $\ln bx = \ln b + \ln x$	**2.** $\ln\dfrac{b}{x} = \ln b - \ln x$
3. $\ln\dfrac{1}{x} = -\ln x$	**4.** $\ln x^r = r\ln x,\ r$ rational

We can estimate the value of $\ln 2$ by considering the area under the graph of $y = 1/x$ and above the interval $[1, 2]$. In Figure 7.2(b) a rectangle of height $1/2$ over the interval

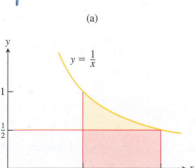

(a)

(b)

FIGURE 7.2 (a) The graph of the natural logarithm. (b) The rectangle of height $y = 1/2$ fits beneath the graph of $y = 1/x$ for the interval $1 \le x \le 2$.

$[1, 2]$ fits under the graph. Therefore the area under the graph, which is $\ln 2$, is greater than the area, $1/2$, of the rectangle. So $\ln 2 > 1/2$. Knowing this we have

$$\ln 2^n = n \ln 2 > n\left(\frac{1}{2}\right) = \frac{n}{2}.$$

This result shows that $\ln(2^n) \to \infty$ as $n \to \infty$. Since $\ln x$ is an increasing function,

$$\lim_{x \to \infty} \ln x = \infty. \qquad \text{\small \color{blue} ln x is increasing and not bounded above}$$

We also have

$$\lim_{x \to 0^+} \ln x = \lim_{t \to \infty} \ln t^{-1} = \lim_{t \to \infty} (-\ln t) = -\infty. \qquad {\small \color{blue} x = 1/t = t^{-1}}$$

We defined $\ln x$ for $x > 0$, so the domain of $\ln x$ is the set of positive real numbers. The above discussion and the Intermediate Value Theorem show that its range is the entire real line, giving the familiar graph of $y = \ln x$ shown in Figure 7.2(a).

The Integral $\int 1/u \, du$

Equation (3) leads to the following integral formula.

If u is a differentiable function that is never zero, then

$$\int \frac{1}{u} \, du = \ln |u| + C. \tag{4}$$

Equation (4) applies anywhere on the domain of $1/u$, which is the set of points where $u \neq 0$. It says that integrals that have the form $\int \frac{du}{u}$ lead to logarithms. Whenever $u = f(x)$ is a differentiable function that is never zero, we have that $du = f'(x) \, dx$ and

$$\int \frac{f'(x)}{f(x)} \, dx = \ln |f(x)| + C.$$

EXAMPLE 1   Here we recognize an integral of the form $\int \frac{du}{u}$.

$$\int_{-\pi/2}^{\pi/2} \frac{4 \cos \theta}{3 + 2 \sin \theta} \, d\theta = \int_1^5 \frac{2}{u} \, du \qquad {\small \color{blue} \begin{aligned} u &= 3 + 2 \sin \theta, \quad du = 2 \cos \theta \, d\theta, \\ u(-\pi/2) &= 1, \quad u(\pi/2) = 5 \end{aligned}}$$

$$= 2 \ln |u| \, \Big]_1^5$$

$$= 2 \ln |5| - 2 \ln |1| = 2 \ln 5$$

Note that $u = 3 + 2 \sin \theta$ is always positive on $[-\pi/2, \pi/2]$, so Equation (4) applies. ∎

The Inverse of $\ln x$ and the Number e

The function $\ln x$, being an increasing function of x with domain $(0, \infty)$ and range $(-\infty, \infty)$, has an inverse $\ln^{-1} x$ with domain $(-\infty, \infty)$ and range $(0, \infty)$. The graph of $\ln^{-1} x$ is the graph of $\ln x$ reflected across the line $y = x$. As you can see in Figure 7.3,

$$\lim_{x \to \infty} \ln^{-1} x = \infty \qquad \text{and} \qquad \lim_{x \to -\infty} \ln^{-1} x = 0.$$

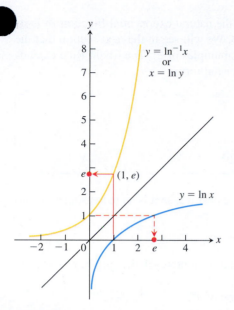

FIGURE 7.3 The graphs of $y = \ln x$ and $y = \ln^{-1} x = \exp x$. The number e is $\ln^{-1} 1 = \exp(1)$.

The notations $\ln^{-1} x$, $\exp x$, and e^x all refer to the natural exponential function.

Typical values of e^x

x	e^x (rounded)
-1	0.37
0	1
1	2.72
2	7.39
10	22026
100	2.6881×10^{43}

The inverse function $\ln^{-1} x$ is also denoted by $\exp x$. We have not yet established that $\exp x$ is an exponential function, only that $\exp x$ is the inverse of the function $\ln x$. We will now show that $\ln^{-1} x = \exp x$ is, in fact, the exponential function with base e.

The number e was defined to satisfy the equation $\ln(e) = 1$, so $e = \exp(1)$. We can raise the number e to a rational power r using algebra:

$$e^2 = e \cdot e, \quad e^{-2} = \frac{1}{e^2}, \quad e^{1/2} = \sqrt{e}, \quad e^{2/3} = \sqrt[3]{e^2},$$

and so on. Since e is positive, e^r is positive too. Thus, e^r has a logarithm. When we take the logarithm, we find that for r rational

$$\ln e^r = r \ln e = r \cdot 1 = r.$$

Then applying the function $\ln^{-1}$ to both sides of the equation $\ln e^r = r$, we find that

$$e^r = \exp r \quad \text{for } r \text{ rational}. \qquad\qquad \text{exp is } \ln^{-1}. \qquad (5)$$

Thus $\exp r$ coincides with the exponential function e^r for all rational values of r. While we have not yet found a way to give an exact meaning to e^x for x irrational, we can use Equation (5) to do so. The function $\exp x = \ln^{-1} x$ has domain $(-\infty, \infty)$, so it is defined for every x. We have $\exp r = e^r$ for r rational by Equation (5), and we now define e^x to equal $\exp x$ for all x.

> **DEFINITION** For every real number x, we define the **natural exponential function** to be $e^x = \exp x$.

For the first time we have a precise meaning for a number raised to an irrational power. Usually the exponential function is denoted by e^x rather than $\exp x$. Since $\ln x$ and e^x are inverses of one another, we have the following relations.

> **Inverse Equations for e^x and ln x**
>
> $$e^{\ln x} = x \qquad \text{(all } x > 0)$$
> $$\ln(e^x) = x \qquad \text{(all } x)$$

The Derivative and Integral of e^x

The exponential function is differentiable because it is the inverse of a differentiable function whose derivative is never zero. We calculate its derivative using Theorem 3 of Section 3.8 and our knowledge of the derivative of $\ln x$. Let

$$f(x) = \ln x \quad \text{and} \quad y = e^x = \ln^{-1} x = f^{-1}(x).$$

Then,

$$\frac{dy}{dx} = \frac{d}{dx} e^x = \frac{d}{dx} \ln^{-1} x$$

$$= \frac{d}{dx} f^{-1}(x)$$

$$= \frac{1}{f'(f^{-1}(x))} \qquad \text{Theorem 3, Section 3.8}$$

$$= \frac{1}{f'(e^x)} \qquad f^{-1}(x) = e^x$$

$$= \frac{1}{\left(\dfrac{1}{e^x}\right)} \qquad f'(z) = \frac{1}{z} \text{ with } z = e^x$$

$$= e^x.$$

That is, for $y = e^x$, we find that $dy/dx = e^x$ so the natural exponential function e^x is its own derivative, just as we claimed in Section 3.3. We will see in the next section that the only functions that behave this way are constant multiples of e^x. The Chain Rule extends the derivative result in the usual way to a more general form.

If u is any differentiable function of x, then

$$\frac{d}{dx} e^u = e^u \frac{du}{dx}. \tag{6}$$

Since $e^x > 0$, its derivative is also everywhere positive, so it is an increasing and continuous function for all x, having limits

$$\lim_{x \to -\infty} e^x = 0 \quad \text{and} \quad \lim_{x \to \infty} e^x = \infty.$$

It follows that the x-axis ($y = 0$) is a horizontal asymptote of the graph $y = e^x$ (see Figure 7.3).

Equation (6) also tells us the indefinite integral of e^u.

$$\int e^u \, du = e^u + C$$

If $f(x) = e^x$, then from Equation (6), $f'(0) = e^0 = 1$. That is, the exponential function e^x has slope 1 as it crosses the y-axis at $x = 0$. This agrees with our assertion for the natural exponential in Section 3.3.

Laws of Exponents

Even though e^x is defined in a seemingly roundabout way as $\ln^{-1} x$, it obeys the familiar laws of exponents from algebra. Theorem 1 shows us that these laws are consequences of the definitions of $\ln x$ and e^x. We proved the laws in Section 4.2, and the proofs are still valid here because they are based on the inverse relationship between $\ln x$ and e^x.

THEOREM 1—Laws of Exponents for e^x
For all numbers x, x_1, and x_2, the natural exponential e^x obeys the following laws:

1. $e^{x_1} \cdot e^{x_2} = e^{x_1 + x_2}$ **2.** $e^{-x} = \dfrac{1}{e^x}$

3. $\dfrac{e^{x_1}}{e^{x_2}} = e^{x_1 - x_2}$ **4.** $(e^{x_1})^{x_2} = e^{x_1 x_2} = (e^{x_2})^{x_1}$

The General Exponential Function a^x

Since $a = e^{\ln a}$ for any positive number a, we can express a^x as $(e^{\ln a})^x = e^{x \ln a}$. We therefore make the following definition, consistent with what we stated in Section 1.6.

DEFINITION For any numbers $a > 0$ and x, the exponential function with base a is given by

$$a^x = e^{x \ln a}.$$

The General Power Function

x^r is the function $e^{r\ln x}$

When $a = e$, the definition gives $a^x = e^{x\ln a} = e^{x\ln e} = e^{x\cdot 1} = e^x$. Similarly, the power function $f(x) = x^r$ is defined for positive x by the formula $x^r = e^{r\ln x}$, which holds for any real number r, rational or irrational.

Theorem 1 is also valid for a^x, the exponential function with base a. For example,

$$
\begin{aligned}
a^{x_1} \cdot a^{x_2} &= e^{x_1\ln a} \cdot e^{x_2\ln a} && \text{Definition of } a^x \\
&= e^{x_1\ln a + x_2\ln a} && \text{Law 1} \\
&= e^{(x_1+x_2)\ln a} && \text{Factor } \ln a \\
&= a^{x_1+x_2}. && \text{Definition of } a^x
\end{aligned}
$$

Starting with the definition $a^x = e^{x\ln a}$, $a > 0$, we get the derivative

$$\frac{d}{dx}a^x = \frac{d}{dx}e^{x\ln a} = (\ln a)e^{x\ln a} = (\ln a)a^x,$$

so

$$\frac{d}{dx}a^x = a^x\ln a.$$

Alternatively, we get the same derivative rule by applying logarithmic differentiation:

$$
\begin{aligned}
y &= a^x \\
\ln y &= x\ln a && \text{Taking logarithms} \\
\frac{1}{y}\frac{dy}{dx} &= \ln a && \text{Differentiating with respect to } x \\
\frac{dy}{dx} &= y\ln a = a^x\ln a.
\end{aligned}
$$

With the Chain Rule, we get a more general form, as in Section 3.8.

If $a > 0$ and u is a differentiable function of x, then a^u is a differentiable function of x and

$$\frac{d}{dx}a^u = a^u\ln a\,\frac{du}{dx}.$$

The integral equivalent of this last result is

$$\int a^u\,du = \frac{a^u}{\ln a} + C.$$

Logarithms with Base a

If a is any positive number other than 1, the function a^x is one-to-one and has a nonzero derivative at every point. It therefore has a differentiable inverse.

DEFINITION For any positive number $a \ne 1$, the **logarithm of x with base a**, denoted by $\log_a x$, is the inverse function of a^x.

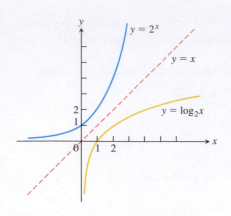

FIGURE 7.4 The graph of 2^x and its inverse, $\log_2 x$.

The graph of $y = \log_a x$ can be obtained by reflecting the graph of $y = a^x$ across the 45° line $y = x$ (Figure 7.4). When $a = e$, we have $\log_e x =$ inverse of $e^x = \ln x$. Since $\log_a x$ and a^x are inverses of one another, composing them in either order gives the identity function.

Inverse Equations for a^x and $\log_a x$

$$a^{\log_a x} = x \qquad (x > 0)$$
$$\log_a(a^x) = x \qquad (\text{all } x)$$

As stated in Section 1.6, the function $\log_a x$ is just a numerical multiple of $\ln x$. We see this from the following derivation:

$y = \log_a x$	Defining equation for y
$a^y = x$	Equivalent equation
$\ln a^y = \ln x$	Natural log of both sides
$y \ln a = \ln x$	Algebra Rule 4 for natural log
$y = \dfrac{\ln x}{\ln a}$	Solve for y.
$\log_a x = \dfrac{\ln x}{\ln a}$	Substitute for y.

It then follows easily that the arithmetic rules satisfied by $\log_a x$ are the same as the ones for $\ln x$. These rules, given in Table 7.2, can be proved by dividing the corresponding rules for the natural logarithm function by $\ln a$. For example,

$$\ln xy = \ln x + \ln y \qquad \text{Rule 1 for natural logarithms} \ldots$$
$$\frac{\ln xy}{\ln a} = \frac{\ln x}{\ln a} + \frac{\ln y}{\ln a} \qquad \ldots \text{divided by } \ln a \ldots$$
$$\log_a xy = \log_a x + \log_a y. \qquad \ldots \text{gives Rule 1 for base } a \text{ logarithms.}$$

TABLE 7.2 Rules for base a logarithms

For any numbers $x > 0$ and $y > 0$,

1. *Product Rule:*

   $\log_a xy = \log_a x + \log_a y$

2. *Quotient Rule:*

   $\log_a \dfrac{x}{y} = \log_a x - \log_a y$

3. *Reciprocal Rule:*

   $\log_a \dfrac{1}{y} = -\log_a y$

4. *Power Rule:*

   $\log_a x^y = y \log_a x$

Derivatives and Integrals Involving $\log_a x$

To find derivatives or integrals involving base a logarithms, we convert them to natural logarithms. If u is a positive differentiable function of x, then

$$\frac{d}{dx}(\log_a u) = \frac{d}{dx}\left(\frac{\ln u}{\ln a}\right) = \frac{1}{\ln a}\frac{d}{dx}(\ln u) = \frac{1}{\ln a} \cdot \frac{1}{u}\frac{du}{dx}.$$

$$\frac{d}{dx}(\log_a u) = \frac{1}{\ln a} \cdot \frac{1}{u}\frac{du}{dx}$$

EXAMPLE 2 We illustrate the derivative and integral results.

(a) $\dfrac{d}{dx}\log_{10}(3x + 1) = \dfrac{1}{\ln 10} \cdot \dfrac{1}{3x + 1}\dfrac{d}{dx}(3x + 1) = \dfrac{3}{(\ln 10)(3x + 1)}$

Transcendental Numbers and Transcendental Functions

Numbers that are solutions of polynomial equations with rational coefficients are called **algebraic**: -2 is algebraic because it satisfies the equation $x + 2 = 0$, and $\sqrt{3}$ is algebraic because it satisfies the equation $x^2 - 3 = 0$. Numbers such as e and π that are not algebraic are called **transcendental**.

We call a function $y = f(x)$ algebraic if it satisfies an equation of the form

$$P_n y^n + \cdots + P_1 y + P_0 = 0$$

in which the P's are polynomials in x with rational coefficients. The function $y = 1/\sqrt{x + 1}$ is algebraic because it satisfies the equation $(x + 1)y^2 - 1 = 0$. Here the polynomials are $P_2 = x + 1$, $P_1 = 0$, and $P_0 = -1$. Functions that are not algebraic are called transcendental.

(b)
$$\int \frac{\log_2 x}{x}\,dx = \frac{1}{\ln 2}\int \frac{\ln x}{x}\,dx \qquad \log_2 x = \frac{\ln x}{\ln 2}$$

$$= \frac{1}{\ln 2}\int u\,du \qquad u = \ln x, \quad du = \frac{1}{x}dx$$

$$= \frac{1}{\ln 2}\frac{u^2}{2} + C = \frac{1}{\ln 2}\frac{(\ln x)^2}{2} + C = \frac{(\ln x)^2}{2\ln 2} + C \qquad \blacksquare$$

Summary

In this section we used calculus to give precise definitions of the logarithmic and exponential functions. This approach is somewhat different from our earlier treatments of the polynomial, rational, and trigonometric functions. There we first defined the function and then we studied its derivatives and integrals. Here we started with an integral from which the functions of interest were obtained. The motivation behind this approach was to address mathematical difficulties that arise when we attempt to define functions such as a^x for any real number x, rational or irrational. Defining $\ln x$ as the integral of the function $1/t$ from $t = 1$ to $t = x$ enabled us to define all of the exponential and logarithmic functions, and then derive their key algebraic and analytic properties.

EXERCISES 7.1

Integration

Evaluate the integrals in Exercises 1–46.

1. $\displaystyle\int_{-3}^{-2} \frac{dx}{x}$

2. $\displaystyle\int_{-1}^{0} \frac{3\,dx}{3x - 2}$

3. $\displaystyle\int \frac{2y\,dy}{y^2 - 25}$

4. $\displaystyle\int \frac{8r\,dr}{4r^2 - 5}$

5. $\displaystyle\int \frac{3\sec^2 t}{6 + 3\tan t}\,dt$

6. $\displaystyle\int \frac{\sec y \tan y}{2 + \sec y}\,dy$

7. $\displaystyle\int \frac{dx}{2\sqrt{x} + 2x}$

8. $\displaystyle\int \frac{\sec x\,dx}{\sqrt{\ln(\sec x + \tan x)}}$

9. $\displaystyle\int_{\ln 2}^{\ln 3} e^x\,dx$

10. $\displaystyle\int 8e^{(x+1)}\,dx$

11. $\displaystyle\int_{1}^{4} \frac{(\ln x)^3}{2x}\,dx$

12. $\displaystyle\int \frac{\ln(\ln x)}{x \ln x}\,dx$

13. $\displaystyle\int_{\ln 4}^{\ln 9} e^{x/2}\,dx$

14. $\displaystyle\int \tan x \ln(\cos x)\,dx$

15. $\displaystyle\int \frac{e^{\sqrt{r}}}{\sqrt{r}}\,dr$

16. $\displaystyle\int \frac{e^{-\sqrt{r}}}{\sqrt{r}}\,dr$

17. $\displaystyle\int 2t\,e^{-t^2}\,dt$

18. $\displaystyle\int \frac{\ln x\,dx}{x\sqrt{\ln^2 x + 1}}$

19. $\displaystyle\int \frac{e^{1/x}}{x^2}\,dx$

20. $\displaystyle\int \frac{e^{-1/x^2}}{x^3}\,dx$

21. $\displaystyle\int e^{\sec \pi t}\sec \pi t \tan \pi t\,dt$

22. $\displaystyle\int e^{\csc(\pi+t)}\csc(\pi + t)\cot(\pi + t)\,dt$

23. $\displaystyle\int_{\ln(\pi/6)}^{\ln(\pi/2)} 2e^v \cos e^v\,dv$

24. $\displaystyle\int_{0}^{\sqrt{\ln \pi}} 2xe^{x^2}\cos(e^{x^2})\,dx$

25. $\displaystyle\int \frac{e^r}{1 + e^r}\,dr$

26. $\displaystyle\int \frac{dx}{1 + e^x}$

27. $\displaystyle\int_{0}^{1} 2^{-\theta}\,d\theta$

28. $\displaystyle\int_{-2}^{0} 5^{-\theta}\,d\theta$

29. $\displaystyle\int_{1}^{\sqrt{2}} x2^{(x^2)}\,dx$

30. $\displaystyle\int_{1}^{4} \frac{2^{\sqrt{x}}}{\sqrt{x}}\,dx$

31. $\displaystyle\int_{0}^{\pi/2} 7^{\cos t}\sin t\,dt$

32. $\displaystyle\int_{0}^{\pi/4} \left(\frac{1}{3}\right)^{\tan t}\sec^2 t\,dt$

33. $\displaystyle\int_{2}^{4} x^{2x}(1 + \ln x)\,dx$

34. $\displaystyle\int_{1}^{2} \frac{2^{\ln x}}{x}\,dx$

35. $\displaystyle\int_{0}^{3} \left(\sqrt{2} + 1\right)x^{\sqrt{2}}\,dx$

36. $\displaystyle\int_{1}^{e} x^{(\ln 2)-1}\,dx$

37. $\displaystyle\int \frac{\log_{10} x}{x}\,dx$

38. $\displaystyle\int_{1}^{4} \frac{\log_2 x}{x}\,dx$

39. $\displaystyle\int_{1}^{4} \frac{\ln 2 \log_2 x}{x}\,dx$

40. $\displaystyle\int_{1}^{e} \frac{2\ln 10 \log_{10} x}{x}\,dx$

41. $\displaystyle\int_{0}^{2} \frac{\log_2(x + 2)}{x + 2}\,dx$

42. $\displaystyle\int_{1/10}^{10} \frac{\log_{10}(10x)}{x}\,dx$

43. $\displaystyle\int_{0}^{9} \frac{2\log_{10}(x + 1)}{x + 1}\,dx$

44. $\displaystyle\int_{2}^{3} \frac{2\log_2(x - 1)}{x - 1}\,dx$

45. $\displaystyle\int \frac{dx}{x \log_{10} x}$

46. $\displaystyle\int \frac{dx}{x (\log_8 x)^2}$

Initial Value Problems

Solve the initial value problems in Exercises 47–52.

47. $\dfrac{dy}{dt} = e^t \sin(e^t - 2), \quad y(\ln 2) = 0$

48. $\dfrac{dy}{dt} = e^{-t} \sec^2(\pi e^{-t}), \quad y(\ln 4) = 2/\pi$

49. $\dfrac{d^2 y}{dx^2} = 2e^{-x}, \quad y(0) = 1 \quad$ and $\quad y'(0) = 0$

50. $\dfrac{d^2 y}{dt^2} = 1 - e^{2t}, \quad y(1) = -1 \quad$ and $\quad y'(1) = 0$

51. $\dfrac{dy}{dx} = 1 + \dfrac{1}{x}, \quad y(1) = 3$

52. $\dfrac{d^2 y}{dx^2} = \sec^2 x, \quad y(0) = 0 \quad$ and $\quad y'(0) = 1$

Theory and Applications

53. The region between the curve $y = 1/x^2$ and the x-axis from $x = 1/2$ to $x = 2$ is revolved about the y-axis to generate a solid. Find the volume of the solid.

54. In Section 6.2, Exercise 6, we revolved about the y-axis the region between the curve $y = 9x/\sqrt{x^3 + 9}$ and the x-axis from $x = 0$ to $x = 3$ to generate a solid of volume 36π. What volume do you get if you revolve the region about the x-axis instead? (See Section 6.2, Exercise 6, for a graph.)

Find the lengths of the curves in Exercises 55 and 56.

55. $y = (x^2/8) - \ln x, \quad 4 \le x \le 8$

56. $x = (y/4)^2 - 2\ln(y/4), \quad 4 \le y \le 12$

T **57. The linearization of $\ln(1 + x)$ at $x = 0$** Instead of approximating $\ln x$ near $x = 1$, we approximate $\ln(1 + x)$ near $x = 0$. We get a simpler formula this way.

  a. Derive the linearization $\ln(1 + x) \approx x$ at $x = 0$.

  b. Estimate to five decimal places the error involved in replacing $\ln(1 + x)$ by x on the interval $[0, 0.1]$.

  c. Graph $\ln(1 + x)$ and x together for $0 \le x \le 0.5$. Use different colors, if available. At what points does the approximation of $\ln(1 + x)$ seem best? Least good? By reading coordinates from the graphs, find as good an upper bound for the error as your grapher will allow.

58. The linearization of e^x at $x = 0$

  a. Derive the linear approximation $e^x \approx 1 + x$ at $x = 0$.

  T **b.** Estimate to five decimal places the magnitude of the error involved in replacing e^x by $1 + x$ on the interval $[0, 0.2]$.

  T **c.** Graph e^x and $1 + x$ together for $-2 \le x \le 2$. Use different colors, if available. On what intervals does the approximation appear to overestimate e^x? Underestimate e^x?

59. Show that for any number $a > 1$

$$\int_1^a \ln x \, dx + \int_0^{\ln a} e^y \, dy = a \ln a,$$

as suggested by the accompanying figure.

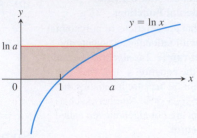

60. The geometric, logarithmic, and arithmetic mean inequality

  a. Show that the graph of e^x is concave up over every interval of x-values.

  b. Show, by reference to the accompanying figure, that if $0 < a < b$ then

$$e^{(\ln a + \ln b)/2} \cdot (\ln b - \ln a) < \int_{\ln a}^{\ln b} e^x \, dx < \frac{e^{\ln a} + e^{\ln b}}{2} \cdot (\ln b - \ln a).$$

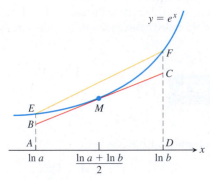

NOT TO SCALE

  c. Use the inequality in part (b) to conclude that

$$\sqrt{ab} < \frac{b - a}{\ln b - \ln a} < \frac{a + b}{2}.$$

This inequality says that the geometric mean of two positive numbers is less than their logarithmic mean, which in turn is less than their arithmetic mean.

61. Use Figure 7.1 and appropriate areas to show that

$$\frac{1}{2} + \frac{1}{3} + \frac{1}{4} + \cdots + \frac{1}{n} < \ln n < 1 + \frac{1}{2} + \frac{1}{3} + \cdots + \frac{1}{n - 1}.$$

62. Partition the interval $[1, 2]$ into n equal parts. Then use Figure 7.1 and appropriate partition points and areas to show that

$$\frac{1}{n + 1} + \frac{1}{n + 2} + \frac{1}{n + 3} + \cdots + \frac{1}{2n} < \ln 2 < \frac{1}{n} + \frac{1}{n + 1}$$

$$+ \frac{1}{n + 2} + \cdots + \frac{1}{2n - 1}.$$

Grapher Explorations

63. Graph $\ln x$, $\ln 2x$, $\ln 4x$, $\ln 8x$, and $\ln 16x$ (as many as you can) together for $0 < x \le 10$. What is going on? Explain.

64. Graph $y = \ln|\sin x|$ in the window $0 \le x \le 22, -2 \le y \le 0$. Explain what you see. How could you change the formula to turn the arches upside down?

65. a. Graph $y = \sin x$ and the curves $y = \ln(a + \sin x)$ for $a = 2, 4, 8, 20,$ and 50 together for $0 \leq x \leq 23$.

  b. Why do the curves flatten as a increases? (*Hint:* Find an a-dependent upper bound for $|y'|$.)

66. Does the graph of $y = \sqrt{x} - \ln x, x > 0$, have an inflection point? Try to answer the question **(a)** by graphing, **(b)** by using calculus.

T **67.** The equation $x^2 = 2^x$ has three solutions: $x = 2, x = 4,$ and one other. Estimate the third solution as accurately as you can by graphing.

T **68.** Could $x^{\ln 2}$ possibly be the same as $2^{\ln x}$ for some $x > 0$? Graph the two functions and explain what you see.

T **69. Which is bigger, π^e or e^π?** Calculators have taken some of the mystery out of this once-challenging question. (Go ahead and check; you will see that it is a surprisingly close call.) You can answer the question without a calculator, though.

  a. Find an equation for the line through the origin tangent to the graph of $y = \ln x$.

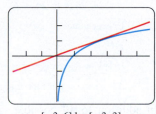

[−3, 6] by [−3, 3]

  b. Give an argument based on the graphs of $y = \ln x$ and the tangent line to explain why $\ln x < x/e$ for all positive $x \neq e$.

  c. Show that $\ln(x^e) < x$ for all positive $x \neq e$.

  d. Conclude that $x^e < e^x$ for all positive $x \neq e$.

  e. So which is bigger, π^e or e^π?

T **70. A decimal representation of e** Find e to as many decimal places as you can by solving the equation $\ln x = 1$ using Newton's method in Section 4.7.

Calculations with Other Bases

T **71.** Most scientific calculators have keys for $\log_{10} x$ and $\ln x$. To find logarithms to other bases, we use the equation $\log_a x = (\ln x)/(\ln a)$.

Find the following logarithms to five decimal places.

  a. $\log_3 8$

  b. $\log_7 0.5$

  c. $\log_{20} 17$

  d. $\log_{0.5} 7$

  e. $\ln x$, given that $\log_{10} x = 2.3$

  f. $\ln x$, given that $\log_2 x = 1.4$

  g. $\ln x$, given that $\log_2 x = -1.5$

  h. $\ln x$, given that $\log_{10} x = -0.7$

72. Conversion factors

  a. Show that the equation for converting base 10 logarithms to base 2 logarithms is

$$\log_2 x = \frac{\ln 10}{\ln 2} \log_{10} x.$$

  b. Show that the equation for converting base a logarithms to base b logarithms is

$$\log_b x = \frac{\ln a}{\ln b} \log_a x.$$

73. Alternative proof that $\lim\limits_{x \to \infty} \left(1 + \dfrac{1}{x}\right)^x = e$:

  a. Let $x > 0$ be given, and use Figure 7.1 to show that

$$\frac{1}{x+1} < \int_x^{x+1} \frac{1}{t}\, dt < \frac{1}{x}.$$

  b. Conclude from part (a) that

$$\frac{1}{x+1} < \ln\left(1 + \frac{1}{x}\right) < \frac{1}{x}.$$

  c. Conclude from part (b) that

$$e^{\frac{x}{x+1}} < \left(1 + \frac{1}{x}\right)^x < e.$$

  d. Conclude from part (c) that

$$\lim_{x \to \infty} \left(1 + \frac{1}{x}\right)^x = e.$$

7.2 Exponential Change and Separable Differential Equations

Exponential functions increase or decrease very rapidly with changes in the independent variable. They describe growth or decay in many natural and industrial situations. The variety of models based on these functions partly accounts for their importance.

Exponential Change

In modeling many real-world situations, a quantity y increases or decreases at a rate proportional to its size at a given time t. Examples of such quantities include the size of a population, the amount of a decaying radioactive material, and the temperature difference between a hot object and its surrounding medium. Such quantities are said to undergo **exponential change**.

If the amount present at time $t = 0$ is called y_0, then we can find y as a function of by solving the following initial value problem:

Differential equation: $\dfrac{dy}{dt} = ky$ (1a)

Initial condition: $y = y_0$ when $t = 0$. (1b)

If y is positive and increasing, then k is positive, and we use Equation (1a) to say that the rate of growth is proportional to what has already been accumulated. If y is positive and decreasing, then k is negative, and we use Equation (1a) to say that the rate of decay is proportional to the amount still left.

We see right away that the constant function $y = 0$ is a solution of Equation (1a) if $y_0 = 0$. To find the nonzero solutions, we divide Equation (1a) by y:

$$\frac{1}{y} \cdot \frac{dy}{dt} = k \qquad y \neq 0$$

$$\int \frac{1}{y} \frac{dy}{dt}\, dt = \int k\, dt \qquad \text{Integrate with respect to } t;$$

$$\ln|y| = kt + C \qquad \textstyle\int (1/u)\, du = \ln|u| + C.$$
$$|y| = e^{kt+C} \qquad \text{Exponentiate.}$$
$$|y| = e^{C} \cdot e^{kt} \qquad e^{a+b} = e^{a} \cdot e^{b}$$
$$y = \pm e^{C} e^{kt} \qquad \text{If } |y| = r, \text{ then } y = \pm r.$$
$$y = A e^{kt}. \qquad A \text{ is a shorter name for } \pm e^{C}.$$

By allowing A to take on the value 0 in addition to all possible values $\pm e^{C}$, we can include the solution $y = 0$ in the formula.

We find the value of A for the initial value problem by solving for A when $y = y_0$ and $t = 0$:

$$y_0 = A e^{k \cdot 0} = A.$$

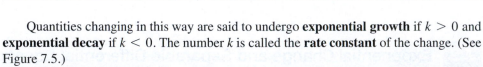

The solution of the initial value problem

$$\frac{dy}{dt} = ky, \qquad y(0) = y_0$$

is

$$y = y_0 e^{kt}. \tag{2}$$

Quantities changing in this way are said to undergo **exponential growth** if $k > 0$ and **exponential decay** if $k < 0$. The number k is called the **rate constant** of the change. (See Figure 7.5.)

The derivation of Equation (2) shows also that the only functions that are their own derivatives ($k = 1$) are constant multiples of the exponential function.

Before presenting several examples of exponential change, let us consider the process we used to derive it.

Separable Differential Equations

Exponential change is modeled by a differential equation of the form $dy/dx = ky$, where k is a nonzero constant. More generally, suppose we have a differential equation of the form

$$\frac{dy}{dx} = f(x, y), \tag{3}$$

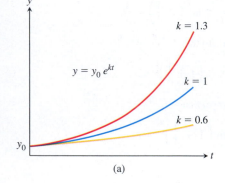

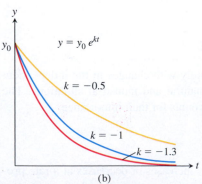

FIGURE 7.5 Graphs of (a) exponential growth and (b) exponential decay. As $|k|$ increases, the growth ($k > 0$) or decay ($k < 0$) intensifies.

where f is a function of *both* the independent and dependent variables. A **solution** of the equation is a differentiable function $y = y(x)$ defined on an interval of x-values (perhaps infinite) such that

$$\frac{d}{dx} y(x) = f(x, y(x))$$

on that interval. That is, when $y(x)$ and its derivative $y'(x)$ are substituted into the differential equation, the resulting equation is true for all x in the solution interval. The **general solution** is a solution $y(x)$ that contains all possible solutions and it always contains an arbitrary constant.

Equation (3) is **separable** if f can be expressed as a product of a function of x and a function of y. The differential equation then has the form

$$\frac{dy}{dx} = g(x) H(y). \qquad \begin{array}{l} g \text{ is a function of } x; \\ H \text{ is a function of } y. \end{array}$$

Then collect all y terms with dy and all x terms with dx:

$$\frac{1}{H(y)} dy = g(x) dx.$$

Now we simply integrate both sides of this equation:

$$\int \frac{1}{H(y)} dy = \int g(x) dx. \qquad (4)$$

After completing the integrations, we obtain the solution y defined implicitly as a function of x.

The justification that we can integrate both sides in Equation (4) in this way is based on the Substitution Rule (Section 5.5):

$$\int \frac{1}{H(y)} dy = \int \frac{1}{H(y(x))} \frac{dy}{dx} dx$$

$$= \int \frac{1}{H(y(x))} H(y(x)) g(x) dx \qquad \frac{dy}{dx} = H(y(x)) g(x)$$

$$= \int g(x) dx.$$

EXAMPLE 1   Solve the differential equation

$$\frac{dy}{dx} = (1 + y)e^x, \quad y > -1.$$

Solution   Since $1 + y$ is never zero for $y > -1$, we can solve the equation by separating the variables.

$$\frac{dy}{dx} = (1 + y)e^x$$

$$dy = (1 + y)e^x dx \qquad \begin{array}{l} \text{Treat } dy/dx \text{ as a quotient of differentials} \\ \text{and multiply both sides by } dx. \end{array}$$

$$\frac{dy}{1 + y} = e^x dx \qquad \text{Divide by } (1 + y).$$

$$\int \frac{dy}{1 + y} = \int e^x dx \qquad \text{Integrate both sides.}$$

$$\ln(1 + y) = e^x + C \qquad \begin{array}{l} C \text{ represents the combined constants} \\ \text{of integration.} \end{array}$$

The last equation gives y as an implicit function of x. ∎

EXAMPLE 2   Solve the equation $y(x + 1)\dfrac{dy}{dx} = x(y^2 + 1)$.

Solution   We change to differential form, separate the variables, and integrate:

$$y(x + 1)\, dy = x(y^2 + 1)\, dx$$

$$\frac{y\, dy}{y^2 + 1} = \frac{x\, dx}{x + 1} \qquad\qquad x \neq -1$$

$$\int \frac{y\, dy}{1 + y^2} = \int \left(1 - \frac{1}{x + 1}\right) dx \qquad \text{Divide } x \text{ by } x + 1.$$

$$\frac{1}{2}\ln(1 + y^2) = x - \ln|x + 1| + C.$$

The last equation gives the solution y as an implicit function of x.   ∎

The initial value problem

$$\frac{dy}{dt} = ky, \qquad y(0) = y_0$$

involves a separable differential equation, and the solution $y = y_0 e^{kt}$ expresses exponential change. We now present several examples of such change.

Unlimited Population Growth

Strictly speaking, the number of individuals in a population (of people, plants, animals, or bacteria, for example) is a discontinuous function of time because it takes on discrete values. However, when the number of individuals becomes large enough, the population can be approximated by a continuous function. Differentiability of the approximating function is another reasonable hypothesis in many settings, allowing for the use of calculus to model and predict population sizes.

If we assume that the proportion of reproducing individuals remains constant and assume a constant fertility, then at any instant t the birth rate is proportional to the number $y(t)$ of individuals present. Let's assume, too, that the death rate of the population is stable and proportional to $y(t)$. If, further, we neglect departures and arrivals, the growth rate dy/dt is the birth rate minus the death rate, which is the difference of the two proportionalities under our assumptions. In other words, $dy/dt = ky$ so that $y = y_0 e^{kt}$, where y_0 is the size of the population at time $t = 0$. As with all kinds of growth, there may be limitations imposed by the surrounding environment, but we will not go into these here. (We treat one model imposing such limitations in Section 9.4.) When k is positive, the proportionality $dy/dt = ky$ models *unlimited population growth*. (See Figure 7.6.)

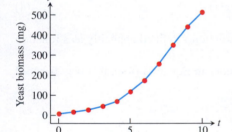

FIGURE 7.6   Graph of the growth of a yeast population over a 10-hour period, based on the data in Table 7.3.

TABLE 7.3 Population of yeast

Time (hr)	Yeast biomass (mg)
0	9.6
1	18.3
2	29.0
3	47.2
4	71.1
5	119.1
6	174.6
7	257.3
8	350.7
9	441.0
10	513.3

EXAMPLE 3   The biomass of a yeast culture in an experiment is initially 29 grams. After 30 minutes the mass is 37 grams. Assuming that the equation for unlimited population growth gives a good model for the growth of the yeast when the mass is below 100 grams, how long will its take for the mass to double from its initial value?

Solution   Let $y(t)$ be the yeast biomass after t minutes. We use the exponential growth model $dy/dt = ky$ for unlimited population growth, with solution $y = y_0 e^{kt}$.
We have $y_0 = y(0) = 29$. We are also told that,

$$y(30) = 29e^{k(30)} = 37.$$

Solving this equation for k, we find

$$e^{k(30)} = \frac{37}{29}$$

$$30k = \ln\left(\frac{37}{29}\right)$$

$$k = \frac{1}{30}\ln\left(\frac{37}{29}\right) \approx 0.008118.$$

Then the mass of the yeast in grams after t minutes is given by the equation

$$y = 29e^{(0.008118)t}.$$

To solve the problem we find the time t for which $y(t) = 58$, which is twice the initial amount.

$$29e^{(0.008118)t} = 58$$

$$(0.008118)t = \ln\left(\frac{58}{29}\right)$$

$$t = \frac{\ln 2}{0.008118} \approx 85.38$$

It takes about 85 minutes for the yeast population to double.   ■

In the next example we model the number of people within a given population who are infected by a disease which is being eradicated from the population. Here the constant of proportionality k is negative, and the model describes an exponentially decaying number of infected individuals.

EXAMPLE 4   One model for the way diseases die out when properly treated assumes that the rate dy/dt at which the number of infected people changes is proportional to the number y. The number of people cured is proportional to the number y that are infected with the disease. Suppose that in the course of any given year the number of cases of a disease is reduced by 20%. If there are 10,000 cases today, how many years will it take to reduce the number to 1000?

Solution   We use the equation $y = y_0 e^{kt}$. There are three things to find: the value of y_0, the value of k, and the time t when $y = 1000$.

The value of y_0. We are free to count time beginning anywhere we want. If we count from today, then $y = 10,000$ when $t = 0$, so $y_0 = 10,000$. Our equation is now

$$y = 10,000e^{kt}. \tag{5}$$

The value of k. When $t = 1$ year, the number of cases will be 80% of its present value, or 8000. Hence,

$$8000 = 10,000e^{k(1)} \qquad \text{Eq. (5) with } t = 1 \text{ and } y = 8000$$

$$e^k = 0.8$$

$$\ln(e^k) = \ln 0.8 \qquad \text{Logs of both sides}$$

$$k = \ln 0.8 < 0. \qquad \ln 0.8 \approx -0.223$$

At any given time t,

$$y = 10,000e^{(\ln 0.8)t}. \tag{6}$$

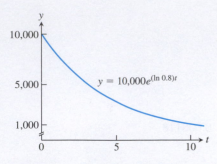

FIGURE 7.7   A graph of the number of people infected by a disease exhibits exponential decay (Example 4).

The value of t that makes y = 1000. We set y equal to 1000 in Equation (6) and solve for t:

$$1000 = 10{,}000e^{(\ln 0.8)t}$$

$$e^{(\ln 0.8)t} = 0.1$$

$$(\ln 0.8)t = \ln 0.1 \qquad \text{Logs of both sides}$$

$$t = \frac{\ln 0.1}{\ln 0.8} \approx 10.32 \text{ years}.$$

It will take a little more than 10 years to reduce the number of cases to 1000. (See Figure 7.7.)   ■

Radioactivity

Some atoms are unstable and can spontaneously emit mass or radiation. This process is called **radioactive decay**, and an element whose atoms go spontaneously through this process is called **radioactive**. Sometimes when an atom emits some of its mass through this process of radioactivity, the remainder of the atom re-forms to make an atom of some new element. For example, radioactive carbon-14 decays into nitrogen; radium, through a number of intermediate radioactive steps, decays into lead.

For radon-222 gas, t is measured in days and $k = 0.18$. For radium-226, which used to be painted on watch dials to make them glow at night (a dangerous practice), t is measured in years and $k = 4.3 \times 10^{-4}$.

Experiments have shown that at any given time the rate at which a radioactive element decays (as measured by the number of nuclei that change per unit time) is approximately proportional to the number of radioactive nuclei present. Thus, the decay of a radioactive element is described by the equation $dy/dt = -ky, k > 0$. It is conventional to use $-k$, with $k > 0$, to emphasize that y is decreasing. If y_0 is the number of radioactive nuclei present at time zero, the number still present at any later time t will be

$$y = y_0e^{-kt}, \qquad k > 0.$$

In Section 1.6, we defined the **half-life** of a radioactive element to be the time required for half of the radioactive nuclei present in a sample to decay. It is an interesting fact that the half-life is a constant that does not depend on the number of radioactive nuclei initially present in the sample, but only on the radioactive substance. We found the half-life is given by

$$\text{Half-life} = \frac{\ln 2}{k} \tag{7}$$

For example, the half-life for radon-222 is

$$\text{half-life} = \frac{\ln 2}{0.18} \approx 3.9 \text{ days}.$$

Carbon-14 dating uses the half-life of 5730 years.

EXAMPLE 5   The decay of radioactive elements can sometimes be used to date events from Earth's past. In a living organism, the ratio of radioactive carbon, carbon-14, to ordinary carbon stays fairly constant during the lifetime of the organism, being approximately equal to the ratio in the organism's atmosphere at the time. After the organism's death, however, no new carbon is ingested, and the proportion of carbon-14 in the organism's remains decreases as the carbon-14 decays.

Scientists who do carbon-14 dating often use a figure of 5730 years for its half-life. Find the age of a sample in which 10% of the radioactive nuclei originally present have decayed.

Solution   We use the decay equation $y = y_0e^{-kt}$. There are two things to find: the value of k and the value of t when y is $0.9y_0$ (90% of the radioactive nuclei are still present). That is, find t when $y_0e^{-kt} = 0.9y_0$, or $e^{-kt} = 0.9$.

The value of k. We use the half-life Equation (7):

$$k = \frac{\ln 2}{\text{half-life}} = \frac{\ln 2}{5730} \qquad \text{(about } 1.2 \times 10^{-4}\text{).}$$

The value of t that makes $e^{-kt} = 0.9.$

$$e^{-kt} = 0.9$$

$$e^{-(\ln 2/5730)t} = 0.9$$

$$-\frac{\ln 2}{5730}t = \ln 0.9 \qquad\qquad\qquad \text{Logs of both sides}$$

$$t = -\frac{5730 \ln 0.9}{\ln 2} \approx 871 \text{ years}$$

The sample is about 871 years old. ■

Heat Transfer: Newton's Law of Cooling

Hot soup left in a tin cup cools to the temperature of the surrounding air. A hot silver bar immersed in a large tub of water cools to the temperature of the surrounding water. In situations like these, the rate at which an object's temperature is changing at any given time is roughly proportional to the difference between its temperature and the temperature of the surrounding medium. This observation is called *Newton's Law of Cooling*, although it applies to warming as well.

If H is the temperature of the object at time t and H_S is the constant surrounding temperature, then the differential equation is

$$\frac{dH}{dt} = -k(H - H_S). \tag{8}$$

If we substitute y for $(H - H_S)$, then

$$\frac{dy}{dt} = \frac{d}{dt}(H - H_S) = \frac{dH}{dt} - \frac{d}{dt}(H_S)$$

$$= \frac{dH}{dt} - 0 \qquad\qquad\qquad H_S \text{ is a constant.}$$

$$= \frac{dH}{dt}$$

$$= -k(H - H_S) \qquad\qquad \text{Eq. (8)}$$

$$= -ky. \qquad\qquad\qquad H - H_S = y$$

We know that the solution of the equation $dy/dt = -ky$ is $y = y_0 e^{-kt}$, where $y(0) = y_0$. Substituting $(H - H_S)$ for y, this says that

$$H - H_S = (H_0 - H_S)e^{-kt}, \tag{9}$$

where H_0 is the temperature at $t = 0$. This equation is the solution to Newton's Law of Cooling.

EXAMPLE 6   A hard-boiled egg at 98°C is put in a sink of 18°C water. After 5 min, the egg's temperature is 38°C. Assuming that the water has not warmed appreciably, how much longer will it take the egg to reach 20°C?

Solution   We find how long it would take the egg to cool from 98°C to 20°C and subtract the 5 min that have already elapsed. Using Equation (9) with $H_S = 18$ and $H_0 = 98$, the egg's temperature t min after it is put in the sink is

$$H = 18 + (98 - 18)e^{-kt} = 18 + 80e^{-kt}.$$

To find k, we use the information that $H = 38$ when $t = 5$:

$$38 = 18 + 80e^{-5k}$$

$$e^{-5k} = \frac{1}{4}$$

$$-5k = \ln\frac{1}{4} = -\ln 4$$

$$k = \frac{1}{5}\ln 4 = 0.2 \ln 4 \qquad \text{(about 0.28)}.$$

The egg's temperature at time t is $H = 18 + 80e^{-(0.2\ln 4)t}$. Now find the time t when $H = 20$:

$$20 = 18 + 80e^{-(0.2\ln 4)t}$$

$$80e^{-(0.2\ln 4)t} = 2$$

$$e^{-(0.2\ln 4)t} = \frac{1}{40}$$

$$-(0.2\ln 4)t = \ln\frac{1}{40} = -\ln 40$$

$$t = \frac{\ln 40}{0.2\ln 4} \approx 13 \text{ min}.$$

The egg's temperature will reach 20°C about 13 min after it is put in the water to cool. Since it took 5 min to reach 38°C, it will take about 8 min more to reach 20°C. ∎

EXERCISES 7.2

Verifying Solutions

In Exercises 1–4, show that each function $y = f(x)$ is a solution of the accompanying differential equation.

1. $2y' + 3y = e^{-x}$

  a. $y = e^{-x}$ **b.** $y = e^{-x} + e^{-(3/2)x}$

  c. $y = e^{-x} + Ce^{-(3/2)x}$

2. $y' = y^2$

  a. $y = -\dfrac{1}{x}$ **b.** $y = -\dfrac{1}{x+3}$ **c.** $y = -\dfrac{1}{x+C}$

3. $y = \dfrac{1}{x}\displaystyle\int_1^x \dfrac{e^t}{t}\,dt, \quad x^2y' + xy = e^x$

4. $y = \dfrac{1}{\sqrt{1+x^4}}\displaystyle\int_1^x \sqrt{1+t^4}\,dt, \quad y' + \dfrac{2x^3}{1+x^4}y = 1$

Initial Value Problems

In Exercises 5–8, show that each function is a solution of the given initial value problem.

Differential equation	Initial equation	Solution candidate
5. $y' + y = \dfrac{2}{1+4e^{2x}}$	$y(-\ln 2) = \dfrac{\pi}{2}$	$y = e^{-x}\tan^{-1}(2e^x)$
6. $y' = e^{-x^2} - 2xy$	$y(2) = 0$	$y = (x-2)e^{-x^2}$
7. $xy' + y = -\sin x,$ $x > 0$	$y\left(\dfrac{\pi}{2}\right) = 0$	$y = \dfrac{\cos x}{x}$
8. $x^2y' = xy - y^2,$ $x > 1$	$y(e) = e$	$y = \dfrac{x}{\ln x}$

Separable Differential Equations

Solve the differential equation in Exercises 9–22.

9. $2\sqrt{xy}\dfrac{dy}{dx} = 1, \quad x, y > 0$ **10.** $\dfrac{dy}{dx} = x^2\sqrt{y}, \quad y > 0$

11. $\dfrac{dy}{dx} = e^{x-y}$ **12.** $\dfrac{dy}{dx} = 3x^2 e^{-y}$

13. $\dfrac{dy}{dx} = \sqrt{y}\cos^2\sqrt{y}$ **14.** $\sqrt{2xy}\dfrac{dy}{dx} = 1$

15. $\sqrt{x}\dfrac{dy}{dx} = e^{y+\sqrt{x}}, \quad x > 0$ **16.** $(\sec x)\dfrac{dy}{dx} = e^{y+\sin x}$

17. $\dfrac{dy}{dx} = 2x\sqrt{1-y^2}, \quad -1 < y < 1$

18. $\dfrac{dy}{dx} = \dfrac{e^{2x-y}}{e^{x+y}}$

19. $y^2\dfrac{dy}{dx} = 3x^2y^3 - 6x^2$ **20.** $\dfrac{dy}{dx} = xy + 3x - 2y - 6$

21. $\dfrac{1}{x}\dfrac{dy}{dx} = ye^{x^2} + 2\sqrt{y}\,e^{x^2}$ **22.** $\dfrac{dy}{dx} = e^{x-y} + e^x + e^{-y} + 1$

Applications and Examples

The answers to most of the following exercises are in terms of logarithms and exponentials. A calculator can be helpful, enabling you to express the answers in decimal form.

23. Human evolution continues The analysis of tooth shrinkage by C. Loring Brace and colleagues at the University of Michigan's

Museum of Anthropology indicates that human tooth size is continuing to decrease and that the evolutionary process has not yet come to a halt. In northern Europeans, for example, tooth size reduction now has a rate of 1% per 1000 years.

a. If t represents time in years and y represents tooth size, use the condition that $y = 0.99y_0$ when $t = 1000$ to find the value of k in the equation $y = y_0e^{kt}$. Then use this value of k to answer the following questions.

b. In about how many years will human teeth be 90% of their present size?

c. What will be our descendants' tooth size 20,000 years from now (as a percentage of our present tooth size)?

24. **Atmospheric pressure** The earth's atmospheric pressure p is often modeled by assuming that the rate dp/dh at which p changes with the altitude h above sea level is proportional to p. Suppose that the pressure at sea level is 1013 millibars (about 14.7 pounds per square inch) and that the pressure at an altitude of 20 km is 90 millibars.

a. Solve the initial value problem

Differential equation: $dp/dh = kp$ (k a constant)
Initial condition: $p = p_0$ when $h = 0$

to express p in terms of h. Determine the values of p_0 and k from the given altitude-pressure data.

b. What is the atmospheric pressure at $h = 50$ km?

c. At what altitude does the pressure equal 900 millibars?

25. **First-order chemical reactions** In some chemical reactions, the rate at which the amount of a substance changes with time is proportional to the amount present. For the change of δ-glucono lactone into gluconic acid, for example,

$$\frac{dy}{dt} = -0.6y$$

when t is measured in hours. If there are 100 grams of δ-glucono lactone present when $t = 0$, how many grams will be left after the first hour?

26. **The inversion of sugar** The processing of raw sugar has a step called "inversion" that changes the sugar's molecular structure. Once the process has begun, the rate of change of the amount of raw sugar is proportional to the amount of raw sugar remaining. If 1000 kg of raw sugar reduces to 800 kg of raw sugar during the first 10 hours, how much raw sugar will remain after another 14 hours?

27. **Working underwater** The intensity $L(x)$ of light x feet beneath the surface of the ocean satisfies the differential equation

$$\frac{dL}{dx} = -kL.$$

As a diver, you know from experience that diving to 18 ft in the Caribbean Sea cuts the intensity in half. You cannot work without artificial light when the intensity falls below one-tenth of the surface value. About how deep can you expect to work without artificial light?

28. **Voltage in a discharging capacitor** Suppose that electricity is draining from a capacitor at a rate that is proportional to the voltage V across its terminals and that, if t is measured in seconds,

$$\frac{dV}{dt} = -\frac{1}{40}V.$$

Solve this equation for V, using V_0 to denote the value of V when $t = 0$. How long will it take the voltage to drop to 10% of its original value?

29. **Cholera bacteria** Suppose that the bacteria in a colony can grow unchecked, by the law of exponential change. The colony starts with 1 bacterium and doubles every half-hour. How many bacteria will the colony contain at the end of 24 hours? (Under favorable laboratory conditions, the number of cholera bacteria can double every 30 min. In an infected person, many bacteria are destroyed, but this example helps explain why a person who feels well in the morning may be dangerously ill by evening.)

30. **Growth of bacteria** A colony of bacteria is grown under ideal conditions in a laboratory so that the population increases exponentially with time. At the end of 3 hours there are 10,000 bacteria. At the end of 5 hours there are 40,000. How many bacteria were present initially?

31. **The incidence of a disease** (*Continuation of Example 4.*) Suppose that in any given year the number of cases can be reduced by 25% instead of 20%.

a. How long will it take to reduce the number of cases to 1000?

b. How long will it take to eradicate the disease, that is, reduce the number of cases to less than 1?

32. **Drug concentration** An antibiotic is administered intravenously into the bloodstream at a constant rate r. As the drug flows through the patient's system and acts on the infection that is present, it is removed from the bloodstream at a rate proportional to the amount in the bloodstream at that time. Since the amount of blood in the patient is constant, this means that the concentration $y = y(t)$ of the antibiotic in the bloodstream can be modeled by the differential equation

$$\frac{dy}{dt} = r - ky, \quad k > 0 \text{ and constant.}$$

a. If $y(0) = y_0$, find the concentration $y(t)$ at any time t.

b. Assume that $y_0 < (r/k)$ and find $\lim_{y \to \infty} y(t)$. Sketch the solution curve for the concentration.

33. **Endangered species** Biologists consider a species of animal or plant to be endangered if it is expected to become extinct within 20 years. If a certain species of wildlife is counted to have 1147 members at the present time, and the population has been steadily declining exponentially at an annual rate averaging 39% over the past 7 years, do you think the species is endangered? Explain your answer.

34. **The U.S. population** The U.S. Census Bureau keeps a running clock totaling the U.S. population. On September 20, 2012, the total was increasing at the rate of 1 person every 12 sec. The population figure for 8:11 P.M. EST on that day was 314,419,198.

a. Assuming exponential growth at a constant rate, find the rate constant for the population's growth (people per 365-day year).

b. At this rate, what will the U.S. population be at 8:11 P.M. EST on September 20, 2019?

35. **Oil depletion** Suppose the amount of oil pumped from one of the canyon wells in Whittier, California, decreases at the continuous rate of 10% per year. When will the well's output fall to one-fifth of its present value?

36. Continuous price discounting To encourage buyers to place 100-unit orders, your firm's sales department applies a continuous discount that makes the unit price a function $p(x)$ of the number of units x ordered. The discount decreases the price at the rate of $0.01 per unit ordered. The price per unit for a 100-unit order is $p(100) = \$20.09$.

a. Find $p(x)$ by solving the following initial value problem:

Differential equation: $\dfrac{dp}{dx} = -\dfrac{1}{100}p$

Initial condition: $p(100) = 20.09$.

b. Find the unit price $p(10)$ for a 10-unit order and the unit price $p(90)$ for a 90-unit order.

c. The sales department has asked you to find out if it is discounting so much that the firm's revenue, $r(x) = x \cdot p(x)$, will actually be less for a 100-unit order than, say, for a 90-unit order. Reassure them by showing that r has its maximum value at $x = 100$.

d. Graph the revenue function $r(x) = xp(x)$ for $0 \le x \le 200$.

37. Plutonium-239 The half-life of the plutonium isotope is 24,360 years. If 10 g of plutonium is released into the atmosphere by a nuclear accident, how many years will it take for 80% of the isotope to decay?

38. Polonium-210 The half-life of polonium is 139 days, but your sample will not be useful to you after 95% of the radioactive nuclei present on the day the sample arrives has disintegrated. For about how many days after the sample arrives will you be able to use the polonium?

39. The mean life of a radioactive nucleus Physicists using the radioactivity equation $y = y_0 e^{-kt}$ call the number $1/k$ the *mean life* of a radioactive nucleus. The mean life of a radon nucleus is about $1/0.18 = 5.6$ days. The mean life of a carbon-14 nucleus is more than 8000 years. Show that 95% of the radioactive nuclei originally present in a sample will disintegrate within three mean lifetimes, i.e., by time $t = 3/k$. Thus, the mean life of a nucleus gives a quick way to estimate how long the radioactivity of a sample will last.

40. Californium-252 What costs $60 million per gram and can be used to treat brain cancer, analyze coal for its sulfur content, and detect explosives in luggage? The answer is californium-252, a radioactive isotope so rare that only 8 g of it have been made in the Western world since its discovery by Glenn Seaborg in 1950. The half-life of the isotope is 2.645 years—long enough for a useful service life and short enough to have a high radioactivity per unit mass. One microgram of the isotope releases 170 million neutrons per minute.

a. What is the value of k in the decay equation for this isotope?

b. What is the isotope's mean life? (See Exercise 39.)

c. How long will it take 95% of a sample's radioactive nuclei to disintegrate?

41. Cooling soup Suppose that a cup of soup cooled from 90°C to 60°C after 10 min in a room whose temperature was 20°C. Use Newton's Law of Cooling to answer the following questions.

a. How much longer would it take the soup to cool to 35°C?

b. Instead of being left to stand in the room, the cup of 90°C soup is put in a freezer whose temperature is −15°C. How long will it take the soup to cool from 90°C to 35°C?

42. A beam of unknown temperature An aluminum beam was brought from the outside cold into a machine shop where the temperature was held at 65°F. After 10 min, the beam warmed to 35°F and after another 10 min it was 50°F. Use Newton's Law of Cooling to estimate the beam's initial temperature.

43. Surrounding medium of unknown temperature A pan of warm water (46°C) was put in a refrigerator. Ten minutes later, the water's temperature was 39°C; 10 min after that, it was 33°C. Use Newton's Law of Cooling to estimate how cold the refrigerator was.

44. Silver cooling in air The temperature of an ingot of silver is 60°C above room temperature right now. Twenty minutes ago, it was 70°C above room temperature. How far above room temperature will the silver be

a. 15 min from now?

b. 2 hours from now?

c. When will the silver be 10°C above room temperature?

45. The age of Crater Lake The charcoal from a tree killed in the volcanic eruption that formed Crater Lake in Oregon contained 44.5% of the carbon-14 found in living matter. About how old is Crater Lake?

46. The sensitivity of carbon-14 dating to measurement To see the effect of a relatively small error in the estimate of the amount of carbon-14 in a sample being dated, consider this hypothetical situation:

a. A bone fragment found in central Illinois in the year 2000 contains 17% of its original carbon-14 content. Estimate the year the animal died.

b. Repeat part (a), assuming 18% instead of 17%.

c. Repeat part (a), assuming 16% instead of 17%.

47. Carbon-14 The oldest known frozen human mummy, discovered in the Schnalstal glacier of the Italian Alps in 1991 and called *Otzi*, was found wearing straw shoes and a leather coat with goat fur, and holding a copper ax and stone dagger. It was estimated that Otzi died 5000 years before he was discovered in the melting glacier. How much of the original carbon-14 remained in Otzi at the time of his discovery?

48. Art forgery A painting attributed to Vermeer (1632–1675), which should contain no more than 96.2% of its original carbon-14, contains 99.5% instead. About how old is the forgery?

49. Lascaux Cave paintings Prehistoric cave paintings of animals were found in the Lascaux Cave in France in 1940. Scientific analysis revealed that only 15% of the original carbon-14 in the paintings remained. What is an estimate of the age of the paintings?

50. Incan mummy The frozen remains of a young Incan woman were discovered by archeologist Johan Reinhard on Mt. Ampato in Peru during an expedition in 1995.

a. How much of the original carbon-14 was present if the estimated age of the "Ice Maiden" was 500 years?

b. If a 1% error can occur in the carbon-14 measurement, what is the oldest possible age for the Ice Maiden?

7.3 Hyperbolic Functions

The hyperbolic functions are formed by taking combinations of the two exponential functions e^x and e^{-x}. The hyperbolic functions simplify many mathematical expressions and occur frequently in mathematical and engineering applications.

Definitions and Identities

The hyperbolic sine and hyperbolic cosine functions are defined by the equations

$$\sinh x = \frac{e^x - e^{-x}}{2} \quad \text{and} \quad \cosh x = \frac{e^x + e^{-x}}{2}.$$

We pronounce $\sinh x$ as "cinch x," rhyming with "pinch x," and $\cosh x$ as "kosh x," rhyming with "gosh x." From this basic pair, we define the hyperbolic tangent, cotangent, secant, and cosecant functions. The defining equations and graphs of these functions are shown in Table 7.4. We will see that the hyperbolic functions bear many similarities to the trigonometric functions after which they are named.

Hyperbolic functions satisfy the identities in Table 7.5. Except for differences in sign, these resemble identities we know for the trigonometric functions. The identities are proved directly from the definitions, as we show here for the second one:

$$2 \sinh x \cosh x = 2 \left(\frac{e^x - e^{-x}}{2} \right) \left(\frac{e^x + e^{-x}}{2} \right)$$

$$= \frac{e^{2x} - e^{-2x}}{2} \qquad \text{Simplify.}$$

$$= \sinh 2x. \qquad \text{Definition of sinh}$$

TABLE 7.4 The six basic hyperbolic functions

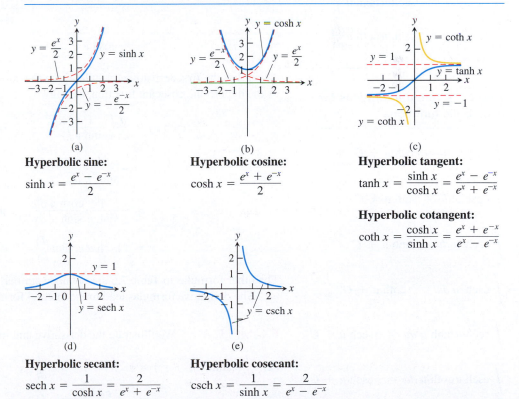

(a)

Hyperbolic sine:
$$\sinh x = \frac{e^x - e^{-x}}{2}$$

(b)

Hyperbolic cosine:
$$\cosh x = \frac{e^x + e^{-x}}{2}$$

(c)

Hyperbolic tangent:
$$\tanh x = \frac{\sinh x}{\cosh x} = \frac{e^x - e^{-x}}{e^x + e^{-x}}$$

Hyperbolic cotangent:
$$\coth x = \frac{\cosh x}{\sinh x} = \frac{e^x + e^{-x}}{e^x - e^{-x}}$$

(d)

Hyperbolic secant:
$$\operatorname{sech} x = \frac{1}{\cosh x} = \frac{2}{e^x + e^{-x}}$$

(e)

Hyperbolic cosecant:
$$\operatorname{csch} x = \frac{1}{\sinh x} = \frac{2}{e^x - e^{-x}}$$

TABLE 7.5 Identities for hyperbolic functions

$$\cosh^2 x - \sinh^2 x = 1$$

$$\sinh 2x = 2 \sinh x \cosh x$$

$$\cosh 2x = \cosh^2 x + \sinh^2 x$$

$$\cosh^2 x = \frac{\cosh 2x + 1}{2}$$

$$\sinh^2 x = \frac{\cosh 2x - 1}{2}$$

$$\tanh^2 x = 1 - \operatorname{sech}^2 x$$

$$\coth^2 x = 1 + \operatorname{csch}^2 x$$

TABLE 7.6 Derivatives of hyperbolic functions

$$\frac{d}{dx}(\sinh u) = \cosh u \frac{du}{dx}$$

$$\frac{d}{dx}(\cosh u) = \sinh u \frac{du}{dx}$$

$$\frac{d}{dx}(\tanh u) = \operatorname{sech}^2 u \frac{du}{dx}$$

$$\frac{d}{dx}(\coth u) = -\operatorname{csch}^2 u \frac{du}{dx}$$

$$\frac{d}{dx}(\operatorname{sech} u) = -\operatorname{sech} u \tanh u \frac{du}{dx}$$

$$\frac{d}{dx}(\operatorname{csch} u) = -\operatorname{csch} u \coth u \frac{du}{dx}$$

TABLE 7.7 Integral formulas for hyperbolic functions

$$\int \sinh u \, du = \cosh u + C$$

$$\int \cosh u \, du = \sinh u + C$$

$$\int \operatorname{sech}^2 u \, du = \tanh u + C$$

$$\int \operatorname{csch}^2 u \, du = -\coth u + C$$

$$\int \operatorname{sech} u \tanh u \, du = -\operatorname{sech} u + C$$

$$\int \operatorname{csch} u \coth u \, du = -\operatorname{csch} u + C$$

The other identities are obtained similarly, by substituting in the definitions of the hyperbolic functions and using algebra.

For any real number u, we know the point with coordinates $(\cos u, \sin u)$ lies on the unit circle $x^2 + y^2 = 1$. So the trigonometric functions are sometimes called the *circular* functions. Because of the first identity

$$\cosh^2 u - \sinh^2 u = 1,$$

with u substituted for x in Table 7.5, the point having coordinates $(\cosh u, \sinh u)$ lies on the right-hand branch of the hyperbola $x^2 - y^2 = 1$. This is where the *hyperbolic* functions get their names (see Exercise 86).

Hyperbolic functions are useful in finding integrals, which we will see in Chapter 8. They play an important role in science and engineering as well. The hyperbolic cosine describes the shape of a hanging cable or wire that is strung between two points at the same height and hanging freely (see Exercise 83). The shape of the St. Louis Arch is an inverted hyperbolic cosine. The hyperbolic tangent occurs in the formula for the velocity of an ocean wave moving over water having a constant depth, and the inverse hyperbolic tangent describes how relative velocities sum according to Einstein's Law in the Special Theory of Relativity.

Derivatives and Integrals of Hyperbolic Functions

The six hyperbolic functions, being rational combinations of the differentiable functions e^x and e^{-x}, have derivatives at every point at which they are defined (Table 7.6). Again, there are similarities with trigonometric functions.

The derivative formulas are derived from the derivative of e^u:

$$\frac{d}{dx}(\sinh u) = \frac{d}{dx}\left(\frac{e^u - e^{-u}}{2}\right) \qquad \text{Definition of } \sinh u$$

$$= \frac{e^u \, du/dx + e^{-u} \, du/dx}{2} \qquad \text{Derivative of } e^u$$

$$= \cosh u \frac{du}{dx}. \qquad \text{Definition of } \cosh u$$

This gives the first derivative formula. From the definition, we can calculate the derivative of the hyperbolic cosecant function, as follows:

$$\frac{d}{dx}(\operatorname{csch} u) = \frac{d}{dx}\left(\frac{1}{\sinh u}\right) \qquad \text{Definition of } \operatorname{csch} u$$

$$= -\frac{\cosh u}{\sinh^2 u}\frac{du}{dx} \qquad \text{Quotient Rule for derivatives}$$

$$= -\frac{1}{\sinh u}\frac{\cosh u}{\sinh u}\frac{du}{dx} \qquad \text{Rearrange terms.}$$

$$= -\operatorname{csch} u \coth u \frac{du}{dx} \qquad \text{Definitions of } \operatorname{csch} u \text{ and } \coth u$$

The other formulas in Table 7.6 are obtained similarly.

The derivative formulas lead to the integral formulas in Table 7.7.

EXAMPLE 1 We illustrate the derivative and integral formulas.

(a) $\dfrac{d}{dt}\left(\tanh \sqrt{1 + t^2}\right) = \operatorname{sech}^2 \sqrt{1 + t^2} \cdot \dfrac{d}{dt}\left(\sqrt{1 + t^2}\right)$

$$= \frac{t}{\sqrt{1 + t^2}}\operatorname{sech}^2 \sqrt{1 + t^2}$$

(b) $\displaystyle\int \coth 5x\, dx = \int \frac{\cosh 5x}{\sinh 5x}\, dx = \frac{1}{5}\int \frac{du}{u}$ $u = \sinh 5x,$
$du = 5 \cosh 5x\, dx$

$$= \frac{1}{5}\ln|u| + C = \frac{1}{5}\ln|\sinh 5x| + C$$

(c) $\displaystyle\int_0^1 \sinh^2 x\, dx = \int_0^1 \frac{\cosh 2x - 1}{2}\, dx$ Table 7.5

$$= \frac{1}{2}\int_0^1 (\cosh 2x - 1)\, dx = \frac{1}{2}\left[\frac{\sinh 2x}{2} - x\right]_0^1$$

$$= \frac{\sinh 2}{4} - \frac{1}{2} \approx 0.40672$$ Evaluate with a calculator.

(d) $\displaystyle\int_0^{\ln 2} 4e^x \sinh x\, dx = \int_0^{\ln 2} 4e^x \frac{e^x - e^{-x}}{2}\, dx = \int_0^{\ln 2} (2e^{2x} - 2)\, dx$

$$= \left[e^{2x} - 2x\right]_0^{\ln 2} = (e^{2\ln 2} - 2\ln 2) - (1 - 0)$$

$$= 4 - 2\ln 2 - 1 \approx 1.6137$$ ∎

Inverse Hyperbolic Functions

The inverses of the six basic hyperbolic functions are very useful in integration (see Chapter 8). Since $d(\sinh x)/dx = \cosh x > 0$, the hyperbolic sine is an increasing function of x. We denote its inverse by

$$y = \sinh^{-1}x.$$

For every value of x in the interval $-\infty < x < \infty$, the value of $y = \sinh^{-1}x$ is the number whose hyperbolic sine is x. The graphs of $y = \sinh x$ and $y = \sinh^{-1}x$ are shown in Figure 7.8a.

The function $y = \cosh x$ is not one-to-one because its graph in Table 7.4 does not pass the horizontal line test. The restricted function $y = \cosh x, x \geq 0$, however, is one-to-one and therefore has an inverse, denoted by

$$y = \cosh^{-1}x.$$

For every value of $x \geq 1, y = \cosh^{-1}x$ is the number in the interval $0 \leq y < \infty$ whose hyperbolic cosine is x. The graphs of $y = \cosh x, x \geq 0$, and $y = \cosh^{-1}x$ are shown in Figure 7.8b.

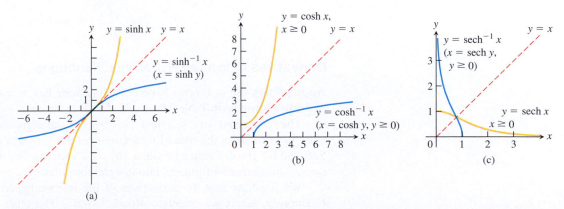

FIGURE 7.8   The graphs of the inverse hyperbolic sine, cosine, and secant of x. Notice the symmetries about the line $y = x$.

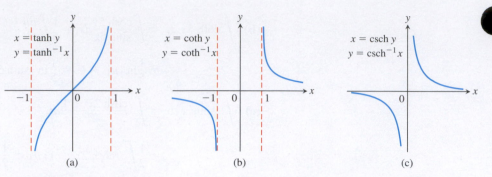

FIGURE 7.9   The graphs of the inverse hyperbolic tangent, cotangent, and cosecant of x.

Like $y = \cosh x$, the function $y = \operatorname{sech} x = 1/\cosh x$ fails to be one-to-one, but its restriction to nonnegative values of x does have an inverse, denoted by

$$y = \operatorname{sech}^{-1} x.$$

For every value of x in the interval $(0, 1\,]$, $y = \operatorname{sech}^{-1} x$ is the nonnegative number whose hyperbolic secant is x. The graphs of $y = \operatorname{sech} x$, $x \geq 0$, and $y = \operatorname{sech}^{-1} x$ are shown in Figure 7.8c.

The hyperbolic tangent, cotangent, and cosecant are one-to-one on their domains and therefore have inverses, denoted by

$$y = \tanh^{-1} x, \qquad y = \coth^{-1} x, \qquad y = \operatorname{csch}^{-1} x.$$

These functions are graphed in Figure 7.9.

Useful Identities

TABLE 7.8 Identities for inverse hyperbolic functions

$$\operatorname{sech}^{-1} x = \cosh^{-1} \frac{1}{x}$$

$$\operatorname{csch}^{-1} x = \sinh^{-1} \frac{1}{x}$$

$$\coth^{-1} x = \tanh^{-1} \frac{1}{x}$$

We can use the identities in Table 7.8 to express $\operatorname{sech}^{-1} x$, $\operatorname{csch}^{-1} x$, and $\coth^{-1} x$ in terms of $\cosh^{-1} x$, $\sinh^{-1} x$, and $\tanh^{-1} x$. These identities are direct consequences of the definitions. For example, if $0 < x \leq 1$, then

$$\operatorname{sech}\left(\cosh^{-1}\left(\frac{1}{x}\right)\right) = \frac{1}{\cosh\left(\cosh^{-1}\left(\frac{1}{x}\right)\right)} = \frac{1}{\left(\frac{1}{x}\right)} = x.$$

We also know that $\operatorname{sech}(\operatorname{sech}^{-1} x) = x$, so because the hyperbolic secant is one-to-one on $(0, 1\,]$, we have

$$\cosh^{-1}\left(\frac{1}{x}\right) = \operatorname{sech}^{-1} x.$$

Derivatives of Inverse Hyperbolic Functions

An important use of inverse hyperbolic functions lies in antiderivatives that reverse the derivative formulas in Table 7.9.

The restrictions $|u| < 1$ and $|u| > 1$ on the derivative formulas for $\tanh^{-1} u$ and $\coth^{-1} u$ come from the natural restrictions on the values of these functions. (See Figure 7.9a and b.) The distinction between $|u| < 1$ and $|u| > 1$ becomes important when we convert the derivative formulas into integral formulas.

We illustrate how the derivatives of the inverse hyperbolic functions are found in Example 2, where we calculate $d(\cosh^{-1} u)/dx$. The other derivatives are obtained by similar calculations.

TABLE 7.9 Derivatives of inverse hyperbolic functions

$$\frac{d(\sinh^{-1} u)}{dx} = \frac{1}{\sqrt{1 + u^2}}\frac{du}{dx}$$

$$\frac{d(\cosh^{-1} u)}{dx} = \frac{1}{\sqrt{u^2 - 1}}\frac{du}{dx}, \qquad u > 1$$

$$\frac{d(\tanh^{-1} u)}{dx} = \frac{1}{1 - u^2}\frac{du}{dx}, \qquad |u| < 1$$

$$\frac{d(\coth^{-1} u)}{dx} = \frac{1}{1 - u^2}\frac{du}{dx}, \qquad |u| > 1$$

$$\frac{d(\operatorname{sech}^{-1} u)}{dx} = -\frac{1}{u\sqrt{1 - u^2}}\frac{du}{dx}, \qquad 0 < u < 1$$

$$\frac{d(\operatorname{csch}^{-1} u)}{dx} = -\frac{1}{|u|\sqrt{1 + u^2}}\frac{du}{dx}, \qquad u \neq 0$$

EXAMPLE 2 Show that if u is a differentiable function of x whose values are greater than 1, then

$$\frac{d}{dx}(\cosh^{-1} u) = \frac{1}{\sqrt{u^2 - 1}}\frac{du}{dx}.$$

Solution First we find the derivative of $y = \cosh^{-1} x$ for $x > 1$ by applying Theorem 3 of Section 3.8 with $f(x) = \cosh x$ and $f^{-1}(x) = \cosh^{-1} x$. Theorem 3 can be applied because the derivative of $\cosh x$ is positive when $x > 0$.

$$(f^{-1})'(x) = \frac{1}{f'(f^{-1}(x))} \qquad \text{Theorem 3, Section 3.8}$$

$$= \frac{1}{\sinh(\cosh^{-1} x)} \qquad f'(u) = \sinh u$$

$$= \frac{1}{\sqrt{\cosh^2(\cosh^{-1} x) - 1}} \qquad \begin{array}{l}\cosh^2 u - \sinh^2 u = 1, \\ \sinh u = \sqrt{\cosh^2 u - 1}\end{array}$$

$$= \frac{1}{\sqrt{x^2 - 1}} \qquad \cosh(\cosh^{-1} x) = x$$

The Chain Rule gives the final result:

$$\frac{d}{dx}(\cosh^{-1} u) = \frac{1}{\sqrt{u^2 - 1}}\frac{du}{dx}. \qquad \blacksquare$$

HISTORICAL BIOGRAPHY
Sonya Kovalevsky
(1850–1891)
www.goo.gl/6TLoDz

With appropriate substitutions, the derivative formulas in Table 7.9 lead to the integration formulas in Table 7.10. Each of the formulas in Table 7.10 can be verified by differentiating the expression on the right-hand side.

EXAMPLE 3 Evaluate

$$\int_0^1 \frac{2\, dx}{\sqrt{3 + 4x^2}}.$$

TABLE 7.10 Integrals leading to inverse hyperbolic functions

1. $\displaystyle\int \frac{du}{\sqrt{a^2 + u^2}} = \sinh^{-1}\left(\frac{u}{a}\right) + C, \qquad a > 0$

2. $\displaystyle\int \frac{du}{\sqrt{u^2 - a^2}} = \cosh^{-1}\left(\frac{u}{a}\right) + C, \qquad u > a > 0$

3. $\displaystyle\int \frac{du}{a^2 - u^2} = \begin{cases} \dfrac{1}{a}\tanh^{-1}\left(\dfrac{u}{a}\right) + C, & u^2 < a^2 \\[2mm] \dfrac{1}{a}\coth^{-1}\left(\dfrac{u}{a}\right) + C, & u^2 > a^2 \end{cases}$

4. $\displaystyle\int \frac{du}{u\sqrt{a^2 - u^2}} = -\frac{1}{a}\operatorname{sech}^{-1}\left(\frac{u}{a}\right) + C, \qquad 0 < u < a$

5. $\displaystyle\int \frac{du}{u\sqrt{a^2 + u^2}} = -\frac{1}{a}\operatorname{csch}^{-1}\left|\frac{u}{a}\right| + C, \qquad u \neq 0 \text{ and } a > 0$

Solution The indefinite integral is

$$\int \frac{2\,dx}{\sqrt{3 + 4x^2}} = \int \frac{du}{\sqrt{a^2 + u^2}} \qquad u = 2x, \quad du = 2\,dx, \quad a = \sqrt{3}$$

$$= \sinh^{-1}\left(\frac{u}{a}\right) + C \qquad \text{Formula from Table 7.10}$$

$$= \sinh^{-1}\left(\frac{2x}{\sqrt{3}}\right) + C.$$

Therefore,

$$\int_0^1 \frac{2\,dx}{\sqrt{3 + 4x^2}} = \sinh^{-1}\left(\frac{2x}{\sqrt{3}}\right)\bigg]_0^1 = \sinh^{-1}\left(\frac{2}{\sqrt{3}}\right) - \sinh^{-1}(0)$$

$$= \sinh^{-1}\left(\frac{2}{\sqrt{3}}\right) - 0 \approx 0.98665. \qquad \blacksquare$$

EXERCISES 7.3

Values and Identities

Each of Exercises 1–4 gives a value of $\sinh x$ or $\cosh x$. Use the definitions and the identity $\cosh^2 x - \sinh^2 x = 1$ to find the values of the remaining five hyperbolic functions.

1. $\sinh x = -\dfrac{3}{4}$

2. $\sinh x = \dfrac{4}{3}$

3. $\cosh x = \dfrac{17}{15}, \quad x > 0$

4. $\cosh x = \dfrac{13}{5}, \quad x > 0$

Rewrite the expressions in Exercises 5–10 in terms of exponentials and simplify the results as much as you can.

5. $2\cosh(\ln x)$

6. $\sinh(2\ln x)$

7. $\cosh 5x + \sinh 5x$

8. $\cosh 3x - \sinh 3x$

9. $(\sinh x + \cosh x)^4$

10. $\ln(\cosh x + \sinh x) + \ln(\cosh x - \sinh x)$

11. Prove the identities

$$\sinh(x + y) = \sinh x \cosh y + \cosh x \sinh y,$$
$$\cosh(x + y) = \cosh x \cosh y + \sinh x \sinh y.$$

Then use them to show that

a. $\sinh 2x = 2\sinh x \cosh x.$

b. $\cosh 2x = \cosh^2 x + \sinh^2 x.$

12. Use the definitions of $\cosh x$ and $\sinh x$ to show that

$$\cosh^2 x - \sinh^2 x = 1.$$

Finding Derivatives

In Exercises 13–24, find the derivative of y with respect to the appropriate variable.

13. $y = 6 \sinh \dfrac{x}{3}$

14. $y = \dfrac{1}{2} \sinh (2x + 1)$

15. $y = 2\sqrt{t}\, \tanh \sqrt{t}$

16. $y = t^2 \tanh \dfrac{1}{t}$

17. $y = \ln (\sinh z)$

18. $y = \ln (\cosh z)$

19. $y = (\operatorname{sech} \theta)(1 - \ln \operatorname{sech} \theta)$ **20.** $y = (\operatorname{csch} \theta)(1 - \ln \operatorname{csch} \theta)$

21. $y = \ln \cosh v - \dfrac{1}{2} \tanh^2 v$ **22.** $y = \ln \sinh v - \dfrac{1}{2} \coth^2 v$

23. $y = (x^2 + 1) \operatorname{sech} (\ln x)$

(*Hint:* Before differentiating, express in terms of exponentials and simplify.)

24. $y = (4x^2 - 1) \operatorname{csch} (\ln 2x)$

In Exercises 25–36, find the derivative of y with respect to the appropriate variable.

25. $y = \sinh^{-1} \sqrt{x}$

26. $y = \cosh^{-1} 2\sqrt{x + 1}$

27. $y = (1 - \theta) \tanh^{-1} \theta$

28. $y = (\theta^2 + 2\theta) \tanh^{-1} (\theta + 1)$

29. $y = (1 - t) \coth^{-1} \sqrt{t}$

30. $y = (1 - t^2) \coth^{-1} t$

31. $y = \cos^{-1} x - x \operatorname{sech}^{-1} x$

32. $y = \ln x + \sqrt{1 - x^2}\, \operatorname{sech}^{-1} x$

33. $y = \operatorname{csch}^{-1} \left(\dfrac{1}{2} \right)^{\theta}$

34. $y = \operatorname{csch}^{-1} 2^{\theta}$

35. $y = \sinh^{-1} (\tan x)$

36. $y = \cosh^{-1} (\sec x), \quad 0 < x < \pi/2$

Integration Formulas

Verify the integration formulas in Exercises 37–40.

37. a. $\displaystyle \int \operatorname{sech} x \, dx = \tan^{-1} (\sinh x) + C$

    b. $\displaystyle \int \operatorname{sech} x \, dx = \sin^{-1} (\tanh x) + C$

38. $\displaystyle \int x \operatorname{sech}^{-1} x \, dx = \dfrac{x^2}{2} \operatorname{sech}^{-1} x - \dfrac{1}{2}\sqrt{1 - x^2} + C$

39. $\displaystyle \int x \coth^{-1} x \, dx = \dfrac{x^2 - 1}{2} \coth^{-1} x + \dfrac{x}{2} + C$

40. $\displaystyle \int \tanh^{-1} x \, dx = x \tanh^{-1} x + \dfrac{1}{2} \ln (1 - x^2) + C$

Evaluating Integrals

Evaluate the integrals in Exercises 41–60.

41. $\displaystyle \int \sinh 2x \, dx$

42. $\displaystyle \int \sinh \dfrac{x}{5} \, dx$

43. $\displaystyle \int 6 \cosh \left(\dfrac{x}{2} - \ln 3 \right) dx$

44. $\displaystyle \int 4 \cosh (3x - \ln 2) \, dx$

45. $\displaystyle \int \tanh \dfrac{x}{7} \, dx$

46. $\displaystyle \int \coth \dfrac{\theta}{\sqrt{3}} \, d\theta$

47. $\displaystyle \int \operatorname{sech}^2 \left(x - \dfrac{1}{2} \right) dx$

48. $\displaystyle \int \operatorname{csch}^2 (5 - x) \, dx$

49. $\displaystyle \int \dfrac{\operatorname{sech} \sqrt{t}\, \tanh \sqrt{t}\, dt}{\sqrt{t}}$

50. $\displaystyle \int \dfrac{\operatorname{csch} (\ln t)\, \coth (\ln t)\, dt}{t}$

51. $\displaystyle \int_{\ln 2}^{\ln 4} \coth x \, dx$

52. $\displaystyle \int_0^{\ln 2} \tanh 2x \, dx$

53. $\displaystyle \int_{-\ln 4}^{-\ln 2} 2e^{\theta} \cosh \theta \, d\theta$

54. $\displaystyle \int_0^{\ln 2} 4e^{-\theta} \sinh \theta \, d\theta$

55. $\displaystyle \int_{-\pi/4}^{\pi/4} \cosh (\tan \theta) \sec^2 \theta \, d\theta$ **56.** $\displaystyle \int_0^{\pi/2} 2 \sinh (\sin \theta) \cos \theta \, d\theta$

57. $\displaystyle \int_1^2 \dfrac{\cosh (\ln t)}{t} \, dt$

58. $\displaystyle \int_1^4 \dfrac{8 \cosh \sqrt{x}}{\sqrt{x}} \, dx$

59. $\displaystyle \int_{-\ln 2}^0 \cosh^2 \left(\dfrac{x}{2} \right) dx$

60. $\displaystyle \int_0^{\ln 10} 4 \sinh^2 \left(\dfrac{x}{2} \right) dx$

Inverse Hyperbolic Functions and Integrals

Since the hyperbolic functions can be expressed in terms of exponential functions, it is possible to express the inverse hyperbolic functions in terms of logarithms, as shown in the following table.

$$\sinh^{-1} x = \ln \left(x + \sqrt{x^2 + 1} \right), \qquad -\infty < x < \infty$$

$$\cosh^{-1} x = \ln \left(x + \sqrt{x^2 - 1} \right), \qquad x \geq 1$$

$$\tanh^{-1} x = \dfrac{1}{2} \ln \dfrac{1 + x}{1 - x}, \qquad |x| < 1$$

$$\operatorname{sech}^{-1} x = \ln \left(\dfrac{1 + \sqrt{1 - x^2}}{x} \right), \qquad 0 < x \leq 1$$

$$\operatorname{csch}^{-1} x = \ln \left(\dfrac{1}{x} + \dfrac{\sqrt{1 + x^2}}{|x|} \right), \qquad x \neq 0$$

$$\coth^{-1} x = \dfrac{1}{2} \ln \dfrac{x + 1}{x - 1}, \qquad |x| > 1$$

Use these formulas to express the numbers in Exercises 61–66 in terms of natural logarithms.

61. $\sinh^{-1} (-5/12)$

62. $\cosh^{-1} (5/3)$

63. $\tanh^{-1} (-1/2)$

64. $\coth^{-1} (5/4)$

65. $\operatorname{sech}^{-1} (3/5)$

66. $\operatorname{csch}^{-1} \left(-1/\sqrt{3} \right)$

Evaluate the integrals in Exercises 67–74 in terms of

    a. inverse hyperbolic functions.

    b. natural logarithms.

67. $\displaystyle \int_0^{2\sqrt{3}} \dfrac{dx}{\sqrt{4 + x^2}}$

68. $\displaystyle \int_0^{1/3} \dfrac{6 \, dx}{\sqrt{1 + 9x^2}}$

69. $\displaystyle \int_{5/4}^2 \dfrac{dx}{1 - x^2}$

70. $\displaystyle \int_0^{1/2} \dfrac{dx}{1 - x^2}$

71. $\displaystyle \int_{1/5}^{3/13} \dfrac{dx}{x\sqrt{1 - 16x^2}}$

72. $\displaystyle \int_1^2 \dfrac{dx}{x\sqrt{4 + x^2}}$

73. $\displaystyle \int_0^{\pi} \dfrac{\cos x \, dx}{\sqrt{1 + \sin^2 x}}$

74. $\displaystyle \int_1^e \dfrac{dx}{x\sqrt{1 + (\ln x)^2}}$

Applications and Examples

75. Show that if a function f is defined on an interval symmetric about the origin (so that f is defined at $-x$ whenever it is defined at x), then

$$f(x) = \frac{f(x) + f(-x)}{2} + \frac{f(x) - f(-x)}{2}. \qquad (1)$$

Then show that $(f(x) + f(-x))/2$ is even and that $(f(x) - f(-x))/2$ is odd.

76. Derive the formula $\sinh^{-1} x = \ln\left(x + \sqrt{x^2 + 1}\right)$ for all real x. Explain in your derivation why the plus sign is used with the square root instead of the minus sign.

77. Skydiving If a body of mass m falling from rest under the action of gravity encounters an air resistance proportional to the square of the velocity, then the body's velocity t sec into the fall satisfies the differential equation

$$m\frac{dv}{dt} = mg - kv^2,$$

where k is a constant that depends on the body's aerodynamic properties and the density of the air. (We assume that the fall is short enough so that the variation in the air's density will not affect the outcome significantly.)

a. Show that

$$v = \sqrt{\frac{mg}{k}}\tanh\left(\sqrt{\frac{gk}{m}}\, t\right)$$

satisfies the differential equation and the initial condition that $v = 0$ when $t = 0$.

b. Find the body's *limiting velocity*, $\lim_{t\to\infty} v$.

c. For a 160-lb skydiver ($mg = 160$), with time in seconds and distance in feet, a typical value for k is 0.005. What is the diver's limiting velocity?

78. Accelerations whose magnitudes are proportional to displacement Suppose that the position of a body moving along a coordinate line at time t is

a. $s = a \cos kt + b \sin kt$. **b.** $s = a \cosh kt + b \sinh kt$.

Show in both cases that the acceleration d^2s/dt^2 is proportional to s but that in the first case it is directed toward the origin, whereas in the second case it is directed away from the origin.

79. Volume A region in the first quadrant is bounded above by the curve $y = \cosh x$, below by the curve $y = \sinh x$, and on the left and right by the y-axis and the line $x = 2$, respectively. Find the volume of the solid generated by revolving the region about the x-axis.

80. Volume The region enclosed by the curve $y = \operatorname{sech} x$, the x-axis, and the lines $x = \pm \ln \sqrt{3}$ is revolved about the x-axis to generate a solid. Find the volume of the solid.

81. Arc length Find the length of the graph of $y = (1/2) \cosh 2x$ from $x = 0$ to $x = \ln \sqrt{5}$.

82. Use the definitions of the hyperbolic functions to find each of the following limits.

a. $\lim_{x\to\infty} \tanh x$ **b.** $\lim_{x\to-\infty} \tanh x$

c. $\lim_{x\to\infty} \sinh x$ **d.** $\lim_{x\to-\infty} \sinh x$

e. $\lim_{x\to\infty} \operatorname{sech} x$ **f.** $\lim_{x\to\infty} \coth x$

g. $\lim_{x\to0^+} \coth x$ **h.** $\lim_{x\to0^-} \coth x$

i. $\lim_{x\to-\infty} \operatorname{csch} x$

83. Hanging cables Imagine a cable, like a telephone line or TV cable, strung from one support to another and hanging freely. The cable's weight per unit length is a constant w and the horizontal tension at its lowest point is a *vector* of length H. If we choose a coordinate system for the plane of the cable in which the x-axis is horizontal, the force of gravity is straight down, the positive y-axis points straight up, and the lowest point of the cable lies at the point $y = H/w$ on the y-axis (see accompanying figure), then it can be shown that the cable lies along the graph of the hyperbolic cosine

$$y = \frac{H}{w}\cosh\frac{w}{H}x.$$

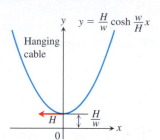

Such a curve is sometimes called a **chain curve** or a **catenary**, the latter deriving from the Latin *catena*, meaning "chain."

a. Let $P(x, y)$ denote an arbitrary point on the cable. The next accompanying figure displays the tension at P as a vector of length (magnitude) T, as well as the tension H at the lowest point A. Show that the cable's slope at P is

$$\tan \phi = \frac{dy}{dx} = \sinh\frac{w}{H}x.$$

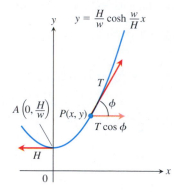

b. Using the result from part (a) and the fact that the horizontal tension at P must equal H (the cable is not moving), show that $T = wy$. Hence, the magnitude of the tension at $P(x, y)$ is exactly equal to the weight of y units of cable.

84. (*Continuation of Exercise 83.*) The length of arc AP in the Exercise 83 figure is $s = (1/a)\sinh ax$, where $a = w/H$. Show that the coordinates of P may be expressed in terms of s as

$$x = \frac{1}{a}\sinh^{-1} as, \qquad y = \sqrt{s^2 + \frac{1}{a^2}}.$$

85. Area Show that the area of the region in the first quadrant enclosed by the curve $y = (1/a)\cosh ax$, the coordinate axes, and the line $x = b$ is the same as the area of a rectangle of height $1/a$ and length s, where s is the length of the curve from $x = 0$ to $x = b$. Draw a figure illustrating this result.

86. The hyperbolic in hyperbolic functions   Just as $x = \cos u$ and $y = \sin u$ are identified with points (x, y) on the unit circle, the functions $x = \cosh u$ and $y = \sinh u$ are identified with points (x, y) on the right-hand branch of the unit hyperbola, $x^2 - y^2 = 1$.

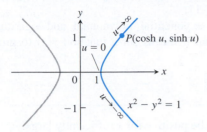

Since $\cosh^2 u - \sinh^2 u = 1$, the point $(\cosh u, \sinh u)$ lies on the right-hand branch of the hyperbola $x^2 - y^2 = 1$ for every value of u.

    Another analogy between hyperbolic and circular functions is that the variable u in the coordinates $(\cosh u, \sinh u)$ for the points of the right-hand branch of the hyperbola $x^2 - y^2 = 1$ is twice the area of the sector AOP pictured in the accompanying figure. To see why this is so, carry out the following steps.

a. Show that the area $A(u)$ of sector AOP is

$$A(u) = \frac{1}{2}\cosh u \sinh u - \int_1^{\cosh u} \sqrt{x^2 - 1}\, dx.$$

b. Differentiate both sides of the equation in part (a) with respect to u to show that

$$A'(u) = \frac{1}{2}.$$

c. Solve this last equation for $A(u)$. What is the value of $A(0)$? What is the value of the constant of integration C in your solution? With C determined, what does your solution say about the relationship of u to $A(u)$?

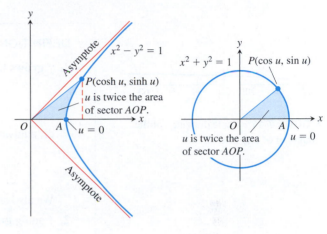

One of the analogies between hyperbolic and circular functions is revealed by these two diagrams (Exercise 86).

7.4   Relative Rates of Growth

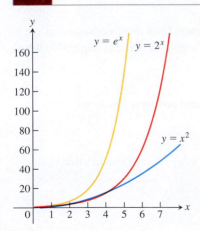

FIGURE 7.10   The graphs of e^x, 2^x, and x^2.

It is often important in mathematics, computer science, and engineering to compare the rates at which functions of x grow as x becomes large. Exponential functions are important in these comparisons because of their very fast growth, and logarithmic functions because of their very slow growth. In this section we introduce the *little-oh* and *big-oh* notation used to describe the results of these comparisons. We restrict our attention to functions whose values eventually become and remain positive as $x \to \infty$.

Growth Rates of Functions

You may have noticed that exponential functions like 2^x and e^x seem to grow more rapidly as x gets large than do polynomials and rational functions. These exponentials certainly grow more rapidly than x itself, and you can see 2^x outgrowing x^2 as x increases in Figure 7.10. In fact, as $x \to \infty$, the functions 2^x and e^x grow faster than any power of x, even $x^{1,000,000}$ (Exercise 19). In contrast, logarithmic functions like $y = \log_2 x$ and $y = \ln x$ grow more slowly as $x \to \infty$ than any positive power of x (Exercise 21).

    To get a feeling for how rapidly the values of $y = e^x$ grow with increasing x, think of graphing the function on a large blackboard, with the axes scaled in centimeters. At $x = 1$ cm, the graph is $e^1 \approx 3$ cm above the x-axis. At $x = 6$ cm, the graph is $e^6 \approx 403$ cm ≈ 4 m high (it is about to go through the ceiling if it hasn't done so already). At $x = 10$ cm, the graph is $e^{10} \approx 22{,}026$ cm ≈ 220 m high, higher than most

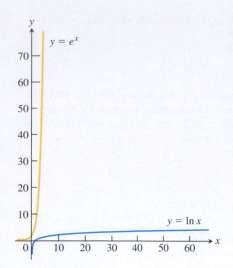

FIGURE 7.11 Scale drawings of the graphs of e^x and $\ln x$.

buildings. At $x = 24$ cm, the graph is more than halfway to the moon, and at $x = 43$ cm from the origin, the graph is high enough to reach past the sun's closest stellar neighbor, the red dwarf star Proxima Centauri. By contrast, with axes scaled in centimeters, you have to go nearly 5 light-years out on the x-axis to find a point where the graph of $y = \ln x$ is even $y = 43$ cm high. See Figure 7.11.

These important comparisons of exponential, polynomial, and logarithmic functions can be made precise by defining what it means for a function $f(x)$ to grow faster than another function $g(x)$ as $x \to \infty$.

DEFINITION Let $f(x)$ and $g(x)$ be positive for x sufficiently large.

1. f **grows faster than** g as $x \to \infty$ if

$$\lim_{x \to \infty} \frac{f(x)}{g(x)} = \infty$$

or, equivalently, if

$$\lim_{x \to \infty} \frac{g(x)}{f(x)} = 0.$$

We also say that g **grows slower than** f as $x \to \infty$.

2. f and g **grow at the same rate** as $x \to \infty$ if

$$\lim_{x \to \infty} \frac{f(x)}{g(x)} = L$$

where L is finite and positive.

According to these definitions, $y = 2x$ does not grow faster than $y = x$. The two functions grow at the same rate because

$$\lim_{x \to \infty} \frac{2x}{x} = \lim_{x \to \infty} 2 = 2,$$

which is a finite, positive limit. The reason for this departure from more colloquial usage is that we want "f grows faster than g" to mean that for large x-values g is negligible when compared with f.

EXAMPLE 1 We compare the growth rates of several common functions.

(a) e^x grows faster than x^2 as $x \to \infty$ because

$$\lim_{x \to \infty} \frac{e^x}{x^2} = \lim_{x \to \infty} \frac{e^x}{2x} = \lim_{x \to \infty} \frac{e^x}{2} = \infty. \qquad \text{Using l'Hôpital's Rule twice}$$
$$\underbrace{\phantom{\lim_{x \to \infty} \frac{e^x}{x^2}}}_{\infty/\infty} \quad \underbrace{\phantom{\lim_{x \to \infty} \frac{e^x}{2x}}}_{\infty/\infty}$$

(b) 3^x grows faster than 2^x as $x \to \infty$ because

$$\lim_{x \to \infty} \frac{3^x}{2^x} = \lim_{x \to \infty} \left(\frac{3}{2}\right)^x = \infty.$$

(c) x^2 grows faster than $\ln x$ as $x \to \infty$ because

$$\lim_{x \to \infty} \frac{x^2}{\ln x} = \lim_{x \to \infty} \frac{2x}{1/x} = \lim_{x \to \infty} 2x^2 = \infty. \qquad \text{l'Hôpital's Rule}$$

(d) $\ln x$ grows slower than $x^{1/n}$ as $x \to \infty$ for any positive integer n because

$$\lim_{x \to \infty} \frac{\ln x}{x^{1/n}} = \lim_{x \to \infty} \frac{1/x}{(1/n)x^{(1/n)-1}} \qquad \text{l'Hôpital's Rule}$$

$$= \lim_{x \to \infty} \frac{n}{x^{1/n}} = 0. \qquad n \text{ is constant.}$$

(e) As part (b) suggests, exponential functions with different bases never grow at the same rate as $x \to \infty$. If $a > b > 0$, then a^x grows faster than b^x. Since $(a/b) > 1$,

$$\lim_{x \to \infty} \frac{a^x}{b^x} = \lim_{x \to \infty} \left(\frac{a}{b}\right)^x = \infty.$$

(f) In contrast to exponential functions, logarithmic functions with different bases $a > 1$ and $b > 1$ always grow at the same rate as $x \to \infty$:

$$\lim_{x \to \infty} \frac{\log_a x}{\log_b x} = \lim_{x \to \infty} \frac{\ln x / \ln a}{\ln x / \ln b} = \frac{\ln b}{\ln a}.$$

The limiting ratio is always finite and never zero. ∎

If f grows at the same rate as g as $x \to \infty$, and g grows at the same rate as h as $x \to \infty$, then f grows at the same rate as h as $x \to \infty$. The reason is that

$$\lim_{x \to \infty} \frac{f}{g} = L_1 \qquad \text{and} \qquad \lim_{x \to \infty} \frac{g}{h} = L_2$$

together imply

$$\lim_{x \to \infty} \frac{f}{h} = \lim_{x \to \infty} \frac{f}{g} \cdot \frac{g}{h} = L_1 L_2.$$

If L_1 and L_2 are finite and nonzero, then so is $L_1 L_2$.

EXAMPLE 2   Show that $\sqrt{x^2 + 5}$ and $(2\sqrt{x} - 1)^2$ grow at the same rate as $x \to \infty$.

Solution   We show that the functions grow at the same rate by showing that they both grow at the same rate as the function $g(x) = x$:

$$\lim_{x \to \infty} \frac{\sqrt{x^2 + 5}}{x} = \lim_{x \to \infty} \sqrt{1 + \frac{5}{x^2}} = 1,$$

$$\lim_{x \to \infty} \frac{(2\sqrt{x} - 1)^2}{x} = \lim_{x \to \infty} \left(\frac{2\sqrt{x} - 1}{\sqrt{x}}\right)^2 = \lim_{x \to \infty} \left(2 - \frac{1}{\sqrt{x}}\right)^2 = 4. \qquad ∎$$

Order and Oh-Notation

The "little-oh" and "big-oh" notation was invented by number theorists a hundred years ago and is now commonplace in mathematical analysis and computer science. According to this definition, saying $f = o(g)$ as $x \to \infty$ is another way to say that f grows slower than g as $x \to \infty$.

DEFINITION   A function f is **of smaller order than** g as $x \to \infty$ if $\lim_{x \to \infty} \dfrac{f(x)}{g(x)} = 0$. We indicate this by writing $\boldsymbol{f = o(g)}$ ("f is little-oh of g").

EXAMPLE 3   Here we use little-oh notation.

(a) $\ln x = o(x)$ as $x \to \infty$ because $\displaystyle \lim_{x \to \infty} \frac{\ln x}{x} = 0$

(b) $x^2 = o(x^3 + 1)$ as $x \to \infty$ because $\displaystyle \lim_{x \to \infty} \frac{x^2}{x^3 + 1} = 0$

DEFINITION   Let $f(x)$ and $g(x)$ be positive for x sufficiently large. Then f is **of at most the order of** g as $x \to \infty$ if there is a positive integer M for which

$$\frac{f(x)}{g(x)} \leq M,$$

for x sufficiently large. We indicate this by writing $\boldsymbol{f = O(g)}$ ("f is big-oh of g").

EXAMPLE 4   Here we use big-oh notation.

(a) $x + \sin x = O(x)$ as $x \to \infty$ because $\dfrac{x + \sin x}{x} \leq 2$ for x sufficiently large.

(b) $e^x + x^2 = O(e^x)$ as $x \to \infty$ because $\dfrac{e^x + x^2}{e^x} \to 1$ as $x \to \infty$.

(c) $x = O(e^x)$ as $x \to \infty$ because $\dfrac{x}{e^x} \to 0$ as $x \to \infty$.

If you look at the definitions again, you will see that $f = o(g)$ implies $f = O(g)$ for functions that are positive for all sufficiently large x. Also, if f and g grow at the same rate, then $f = O(g)$ and $g = O(f)$ (Exercise 11).

Sequential vs. Binary Search

Computer scientists often measure the efficiency of an algorithm by counting the number of steps a computer must take to execute the algorithm. There can be significant differences in how efficiently algorithms perform, even if they are designed to accomplish the same task. These differences are often described using big-oh notation. Here is an example.

Webster's International Dictionary lists about 26,000 words that begin with the letter *a*. One way to look up a word, or to learn if it is not there, is to read through the list one word at a time until you either find the word or determine that it is not there. This method, called **sequential search**, makes no particular use of the words' alphabetical arrangement in the list. You are sure to get an answer, but it might take 26,000 steps.

Another way to find the word or to learn it is not there is to go straight to the middle of the list (give or take a few words). If you do not find the word, then go to the middle of the half that contains it and forget about the half that does not. (You know which half contains it because you know the list is ordered alphabetically.) This method, called a **binary search**, eliminates roughly 13,000 words in a single step. If you do not find the word on the second try, then jump to the middle of the half that contains it. Continue this way until you have either found the word or divided the list in half so many times there are no words left. How many times do you have to divide the list to find the word or learn that it is not there? At most 15, because

$$(26{,}000/2^{15}) < 1.$$

That certainly beats a possible 26,000 steps.

For a list of length n, a sequential search algorithm takes on the order of n steps to find a word or determine that it is not in the list. A binary search, as the second algorithm is called, takes on the order of $\log_2 n$ steps. The reason is that if $2^{m-1} < n \leq 2^m$, then

$m - 1 < \log_2 n \leq m$, and the number of bisections required to narrow the list to one word will be at most $m = \lceil \log_2 n \rceil$, the integer ceiling of the number $\log_2 n$.

Big-oh notation provides a compact way to say all this. The number of steps in a sequential search of an ordered list is $O(n)$; the number of steps in a binary search is $O(\log_2 n)$. In our example, there is a big difference between the two (26,000 vs. 15), and the difference can only increase with n because n grows faster than $\log_2 n$ as $n \to \infty$.

EXERCISES 7.4

Comparisons with the Exponential e^x

1. Which of the following functions grow faster than e^x as $x \to \infty$? Which grow at the same rate as e^x? Which grow slower?

  a. $x - 3$ **b.** $x^3 + \sin^2 x$

  c. $\sqrt{x}$ **d.** 4^x

  e. $(3/2)^x$ **f.** $e^{x/2}$

  g. $e^x/2$ **h.** $\log_{10} x$

2. Which of the following functions grow faster than e^x as $x \to \infty$? Which grow at the same rate as e^x? Which grow slower?

  a. $10x^4 + 30x + 1$ **b.** $x \ln x - x$

  c. $\sqrt{1 + x^4}$ **d.** $(5/2)^x$

  e. e^{-x} **f.** xe^x

  g. $e^{\cos x}$ **h.** e^{x-1}

Comparisons with the Power x^2

3. Which of the following functions grow faster than x^2 as $x \to \infty$? Which grow at the same rate as x^2? Which grow slower?

  a. $x^2 + 4x$ **b.** $x^5 - x^2$

  c. $\sqrt{x^4 + x^3}$ **d.** $(x + 3)^2$

  e. $x \ln x$ **f.** 2^x

  g. $x^3 e^{-x}$ **h.** $8x^2$

4. Which of the following functions grow faster than x^2 as $x \to \infty$? Which grow at the same rate as x^2? Which grow slower?

  a. $x^2 + \sqrt{x}$ **b.** $10x^2$

  c. $x^2 e^{-x}$ **d.** $\log_{10}(x^2)$

  e. $x^3 - x^2$ **f.** $(1/10)^x$

  g. $(1.1)^x$ **h.** $x^2 + 100x$

Comparisons with the Logarithm $\ln x$

5. Which of the following functions grow faster than $\ln x$ as $x \to \infty$? Which grow at the same rate as $\ln x$? Which grow slower?

  a. $\log_3 x$ **b.** $\ln 2x$

  c. $\ln \sqrt{x}$ **d.** $\sqrt{x}$

  e. x **f.** $5 \ln x$

  g. $1/x$ **h.** e^x

6. Which of the following functions grow faster than $\ln x$ as $x \to \infty$? Which grow at the same rate as $\ln x$? Which grow slower?

  a. $\log_2(x^2)$ **b.** $\log_{10} 10x$

  c. $1/\sqrt{x}$ **d.** $1/x^2$

  e. $x - 2 \ln x$ **f.** e^{-x}

  g. $\ln(\ln x)$ **h.** $\ln(2x + 5)$

Ordering Functions by Growth Rates

7. Order the following functions from slowest growing to fastest growing as $x \to \infty$.

  a. e^x **b.** x^x

  c. $(\ln x)^x$ **d.** $e^{x/2}$

8. Order the following functions from slowest growing to fastest growing as $x \to \infty$.

  a. 2^x **b.** x^2

  c. $(\ln 2)^x$ **d.** e^x

Big-oh and Little-oh; Order

9. True, or false? As $x \to \infty$,

  a. $x = o(x)$ **b.** $x = o(x + 5)$

  c. $x = O(x + 5)$ **d.** $x = O(2x)$

  e. $e^x = o(e^{2x})$ **f.** $x + \ln x = O(x)$

  g. $\ln x = o(\ln 2x)$ **h.** $\sqrt{x^2 + 5} = O(x)$

10. True, or false? As $x \to \infty$,

  a. $\dfrac{1}{x + 3} = O\left(\dfrac{1}{x}\right)$ **b.** $\dfrac{1}{x} + \dfrac{1}{x^2} = O\left(\dfrac{1}{x}\right)$

  c. $\dfrac{1}{x} - \dfrac{1}{x^2} = o\left(\dfrac{1}{x}\right)$ **d.** $2 + \cos x = O(2)$

  e. $e^x + x = O(e^x)$ **f.** $x \ln x = o(x^2)$

  g. $\ln(\ln x) = O(\ln x)$ **h.** $\ln(x) = o(\ln(x^2 + 1))$

11. Show that if positive functions $f(x)$ and $g(x)$ grow at the same rate as $x \to \infty$, then $f = O(g)$ and $g = O(f)$.

12. When is a polynomial $f(x)$ of smaller order than a polynomial $g(x)$ as $x \to \infty$? Give reasons for your answer.

13. When is a polynomial $f(x)$ of at most the order of a polynomial $g(x)$ as $x \to \infty$? Give reasons for your answer.

14. What do the conclusions we drew in Section 2.8 about the limits of rational functions tell us about the relative growth of polynomials as $x \to \infty$?

Other Comparisons

T **15.** Investigate

$$\lim_{x \to \infty} \frac{\ln(x + 1)}{\ln x} \quad \text{and} \quad \lim_{x \to \infty} \frac{\ln(x + 999)}{\ln x}.$$

Then use l'Hôpital's Rule to explain what you find.

16. (*Continuation of Exercise 15.*) Show that the value of

$$\lim_{x \to \infty} \frac{\ln(x + a)}{\ln x}$$

is the same no matter what value you assign to the constant a. What does this say about the relative rates at which the functions $f(x) = \ln(x + a)$ and $g(x) = \ln x$ grow?

17. Show that $\sqrt{10x + 1}$ and $\sqrt{x + 1}$ grow at the same rate as $x \to \infty$ by showing that they both grow at the same rate as $\sqrt{x}$ as $x \to \infty$.

18. Show that $\sqrt{x^4 + x}$ and $\sqrt{x^4 - x^3}$ grow at the same rate as $x \to \infty$ by showing that they both grow at the same rate as x^2 as $x \to \infty$.

19. Show that e^x grows faster as $x \to \infty$ than x^n for any positive integer n, even $x^{1,000,000}$. (*Hint:* What is the nth derivative of x^n?)

20. The function e^x outgrows any polynomial Show that e^x grows faster as $x \to \infty$ than any polynomial

$$a_n x^n + a_{n-1} x^{n-1} + \cdots + a_1 x + a_0.$$

21. a. Show that $\ln x$ grows slower as $x \to \infty$ than $x^{1/n}$ for any positive integer n, even $x^{1/1,000,000}$.

 T b. Although the values of $x^{1/1,000,000}$ eventually overtake the values of $\ln x$, you have to go way out on the x-axis before this happens. Find a value of x greater than 1 for which $x^{1/1,000,000} > \ln x$. You might start by observing that when $x > 1$ the equation $\ln x = x^{1/1,000,000}$ is equivalent to the equation $\ln(\ln x) = (\ln x)/1,000,000$.

 T c. Even $x^{1/10}$ takes a long time to overtake $\ln x$. Experiment with a calculator to find the value of x at which the graphs of $x^{1/10}$ and $\ln x$ cross, or, equivalently, at which $\ln x = 10 \ln(\ln x)$. Bracket the crossing point between powers of 10 and then close in by successive halving.

 T d. (*Continuation of part (c).*) The value of x at which $\ln x = 10 \ln(\ln x)$ is too far out for some graphers and root finders to identify. Try it on the equipment available to you and see what happens.

22. The function $\ln x$ grows slower than any polynomial Show that $\ln x$ grows slower as $x \to \infty$ than any nonconstant polynomial.

Algorithms and Searches

23. a. Suppose you have three different algorithms for solving the same problem and each algorithm takes a number of steps that is of the order of one of the functions listed here:

$$n \log_2 n, \quad n^{3/2}, \quad n(\log_2 n)^2.$$

 Which of the algorithms is the most efficient in the long run? Give reasons for your answer.

 T b. Graph the functions in part (a) together to get a sense of how rapidly each one grows.

24. Repeat Exercise 23 for the functions

$$n, \quad \sqrt{n} \log_2 n, \quad (\log_2 n)^2.$$

T 25. Suppose you are looking for an item in an ordered list one million items long. How many steps might it take to find that item with a sequential search? A binary search?

T 26. You are looking for an item in an ordered list 450,000 items long (the length of *Webster's Third New International Dictionary*). How many steps might it take to find the item with a sequential search? A binary search?

CHAPTER 7 **Questions to Guide Your Review**

1. How is the natural logarithm function defined as an integral? What are its domain, range, and derivative? What arithmetic properties does it have? Comment on its graph.

2. What integrals lead to logarithms? Give examples.

3. What are the integrals of $\tan x$ and $\cot x$? $\sec x$ and $\csc x$?

4. How is the exponential function e^x defined? What are its domain, range, and derivative? What laws of exponents does it obey? Comment on its graph.

5. How are the functions a^x and $\log_a x$ defined? Are there any restrictions on a? How is the graph of $\log_a x$ related to the graph of $\ln x$? What truth is there in the statement that there is really only one exponential function and one logarithmic function?

6. How do you solve separable first-order differential equations?

7. What is the law of exponential change? How can it be derived from an initial value problem? What are some of the applications of the law?

8. What are the six basic hyperbolic functions? Comment on their domains, ranges, and graphs. What are some of the identities relating them?

9. What are the derivatives of the six basic hyperbolic functions? What are the corresponding integral formulas? What similarities do you see here with the six basic trigonometric functions?

10. How are the inverse hyperbolic functions defined? Comment on their domains, ranges, and graphs. How can you find values of $\text{sech}^{-1} x$, $\text{csch}^{-1} x$, and $\coth^{-1} x$ using a calculator's keys for $\cosh^{-1} x$, $\sinh^{-1} x$, and $\tanh^{-1} x$?

11. What integrals lead naturally to inverse hyperbolic functions?

12. How do you compare the growth rates of positive functions as $x \to \infty$?

13. What roles do the functions e^x and $\ln x$ play in growth comparisons?

14. Describe big-oh and little-oh notation. Give examples.

15. Which is more efficient—a sequential search or a binary search? Explain.

CHAPTER 7   Practice Exercises

Integration

Evaluate the integrals in Exercises 1–12.

1. $\displaystyle\int e^x \sin(e^x)\,dx$

2. $\displaystyle\int e^t \cos(3e^t - 2)\,dt$

3. $\displaystyle\int_0^\pi \tan\frac{x}{3}\,dx$

4. $\displaystyle\int_{1/6}^{1/4} 2\cot \pi x\,dx$

5. $\displaystyle\int_{-\pi/2}^{\pi/6} \frac{\cos t}{1 - \sin t}\,dt$

6. $\displaystyle\int e^x \sec e^x\,dx$

7. $\displaystyle\int \frac{\ln(x-5)}{x-5}\,dx$

8. $\displaystyle\int \frac{\cos(1 - \ln v)}{v}\,dv$

9. $\displaystyle\int_1^7 \frac{3}{x}\,dx$

10. $\displaystyle\int_1^{32} \frac{1}{5x}\,dx$

11. $\displaystyle\int_e^{e^2} \frac{1}{x\sqrt{\ln x}}\,dx$

12. $\displaystyle\int_2^4 (1 + \ln t)\,t \ln t\,dt$

Solving Equations with Logarithmic or Exponential Terms

In Exercises 13–18, solve for y.

13. $3^y = 2^{y+1}$

14. $4^{-y} = 3^{y+2}$

15. $9e^{2y} = x^2$

16. $3^y = 3\ln x$

17. $\ln(y - 1) = x + \ln y$

18. $\ln(10\ln y) = \ln 5x$

Comparing Growth Rates of Functions

19. Does f grow faster, slower, or at the same rate as g as $x \to \infty$? Give reasons for your answers.

  a. $f(x) = \log_2 x$, $g(x) = \log_3 x$

  b. $f(x) = x$, $g(x) = x + \dfrac{1}{x}$

  c. $f(x) = x/100$, $g(x) = xe^{-x}$

  d. $f(x) = x$, $g(x) = \tan^{-1}x$

  e. $f(x) = \csc^{-1}x$, $g(x) = 1/x$

  f. $f(x) = \sinh x$, $g(x) = e^x$

20. Does f grow faster, slower, or at the same rate as g as $x \to \infty$? Give reasons for your answers.

  a. $f(x) = 3^{-x}$, $g(x) = 2^{-x}$

  b. $f(x) = \ln 2x$, $g(x) = \ln x^2$

  c. $f(x) = 10x^3 + 2x^2$, $g(x) = e^x$

  d. $f(x) = \tan^{-1}(1/x)$, $g(x) = 1/x$

  e. $f(x) = \sin^{-1}(1/x)$, $g(x) = 1/x^2$

  f. $f(x) = \text{sech}\,x$, $g(x) = e^{-x}$

21. True, or false? Give reasons for your answers.

  a. $\dfrac{1}{x^2} + \dfrac{1}{x^4} = O\left(\dfrac{1}{x^2}\right)$ **b.** $\dfrac{1}{x^2} + \dfrac{1}{x^4} = O\left(\dfrac{1}{x^4}\right)$

  c. $x = o(x + \ln x)$ **d.** $\ln(\ln x) = o(\ln x)$

  e. $\tan^{-1}x = O(1)$ **f.** $\cosh x = O(e^x)$

22. True, or false? Give reasons for your answers.

  a. $\dfrac{1}{x^4} = o\left(\dfrac{1}{x^2} + \dfrac{1}{x^4}\right)$ **b.** $\dfrac{1}{x^4} = o\left(\dfrac{1}{x^2} + \dfrac{1}{x^4}\right)$

  c. $\ln x = o(x + 1)$ **d.** $\ln 2x = O(\ln x)$

  e. $\sec^{-1}x = O(1)$ **f.** $\sinh x = O(e^x)$

Theory and Applications

23. The function $f(x) = e^x + x$, being differentiable and one-to-one, has a differentiable inverse $f^{-1}(x)$. Find the value of df^{-1}/dx at the point $f(\ln 2)$.

24. Find the inverse of the function $f(x) = 1 + (1/x)$, $x \ne 0$. Then show that $f^{-1}(f(x)) = f(f^{-1}(x)) = x$ and that

$$\left.\frac{df^{-1}}{dx}\right|_{f(x)} = \frac{1}{f'(x)}.$$

25. A particle is traveling upward and to the right along the curve $y = \ln x$. Its x-coordinate is increasing at the rate $(dx/dt) = \sqrt{x}$ m/sec. At what rate is the y-coordinate changing at the point $(e^2, 2)$?

26. A girl is sliding down a slide shaped like the curve $y = 9e^{-x/3}$. Her y-coordinate is changing at the rate $dy/dt = (-1/4)\sqrt{9 - y}$ ft/sec. At approximately what rate is her x-coordinate changing when she reaches the bottom of the slide at $x = 9$ ft? (Take e^3 to be 20 and round your answer to the nearest ft/sec.)

27. The functions $f(x) = \ln 5x$ and $g(x) = \ln 3x$ differ by a constant. What constant? Give reasons for your answer.

28. a. If $(\ln x)/x = (\ln 2)/2$, must $x = 2$?

  b. If $(\ln x)/x = -2\ln 2$, must $x = 1/2$?

  Give reasons for your answers.

29. The quotient $(\log_4 x)/(\log_2 x)$ has a constant value. What value? Give reasons for your answer.

T 30. $\log_x(2)$ vs. $\log_2(x)$ How does $f(x) = \log_x(2)$ compare with $g(x) = \log_2(x)$? Here is one way to find out.

  a. Use the equation $\log_a b = (\ln b)/(\ln a)$ to express $f(x)$ and $g(x)$ in terms of natural logarithms.

  b. Graph f and g together. Comment on the behavior of f in relation to the signs and values of g.

In Exercises 31–34, solve the differential equation.

31. $\dfrac{dy}{dx} = \sqrt{y}\cos^2\sqrt{y}$

32. $y' = \dfrac{3y(x + 1)^2}{y - 1}$

33. $yy' = \sec y^2 \sec^2 x$

34. $y\cos^2 x\,dy + \sin x\,dx = 0$

In Exercises 35–38, solve the initial value problem.

35. $\dfrac{dy}{dx} = e^{-x-y-2}$, $y(0) = -2$

36. $\dfrac{dy}{dx} = \dfrac{y \ln y}{1 + x^2}$, $y(0) = e^2$

37. $x\,dy - \left(y + \sqrt{y}\right)dx = 0$, $y(1) = 1$

38. $y^{-2} \dfrac{dx}{dy} = \dfrac{e^x}{e^{2x} + 1}$, $y(0) = 1$

39. What is the age of a sample of charcoal in which 90% of the carbon-14 originally present has decayed?

40. **Cooling a pie** A deep-dish apple pie, whose internal temperature was 220°F when removed from the oven, was set out on a breezy 40°F porch to cool. Fifteen minutes later, the pie's internal temperature was 180°F. How long did it take the pie to cool from there to 70°F?

41. Find the length of the curve $y = \ln(e^x - 1) - \ln(e^x + 1)$, $\ln 2 \le x \le \ln 3$.

42. The Austrian biologist Ludwig von Bentalanffy derived and published the von Bertalanffy growth equation in 1934, which continues to be widely used and is especially important in fisheries studies. Let $L(t)$ denote the length of a fish at time t and assume $L(0) = L_0$. The von Bertalanffy equation is

$$\frac{dL}{dt} = k(A - L),$$

where $A = \lim_{t \to \infty} L(t)$ is the asymptotic length of the fish and k is a proportionality constant.

Assume that $L(t)$ is the length in meters of a shark of age t years. In addition, assume $A = 3$, $L(0) = 0.5$ m, and $L(5) = 1.75$ m.

a. Solve the von Bertalanffy differential equation.

b. What is $L(10)$?

c. When does $L = 2.5$ m?

CHAPTER 7 Additional and Advanced Exercises

1. Let $A(t)$ be the area of the region in the first quadrant enclosed by the coordinate axes, the curve $y = e^{-x}$, and the vertical line $x = t$, $t > 0$. Let $V(t)$ be the volume of the solid generated by revolving the region about the x-axis. Find the following limits.

    a. $\lim_{t \to \infty} A(t)$ b. $\lim_{t \to \infty} V(t)/A(t)$ c. $\lim_{t \to 0^+} V(t)/A(t)$

2. **Varying a logarithm's base**

    a. Find $\lim \log_a 2$ as $a \to 0^+$, 1^-, 1^+, and ∞.

    T b. Graph $y = \log_a 2$ as a function of a over the interval $0 < a \le 4$.

T 3. Graph $f(x) = \tan^{-1}x + \tan^{-1}(1/x)$ for $-5 \le x \le 5$. Then use calculus to explain what you see. How would you expect f to behave beyond the interval $[-5, 5]$? Give reasons for your answer.

T 4. Graph $f(x) = (\sin x)^{\sin x}$ over $[0, 3\pi]$. Explain what you see.

5. **Even-odd decompositions**

    a. Suppose that g is an even function of x and h is an odd function of x. Show that if $g(x) + h(x) = 0$ for all x then $g(x) = 0$ for all x and $h(x) = 0$ for all x.

    b. Use the result in part (a) to show that if $f(x) = f_E(x) + f_O(x)$ is the sum of an even function $f_E(x)$ and an odd function $f_O(x)$, then

$$f_E(x) = (f(x) + f(-x))/2 \quad \text{and} \quad f_O(x) = (f(x) - f(-x))/2.$$

    c. What is the significance of the result in part (b)?

6. Let g be a function that is differentiable throughout an open interval containing the origin. Suppose g has the following properties:

    i. $g(x + y) = \dfrac{g(x) + g(y)}{1 - g(x)g(y)}$ for all real numbers x, y, and $x + y$ in the domain of g.

    ii. $\lim_{h \to 0} g(h) = 0$

    iii. $\lim_{h \to 0} \dfrac{g(h)}{h} = 1$

    a. Show that $g(0) = 0$.

    b. Show that $g'(x) = 1 + [g(x)]^2$.

    c. Find $g(x)$ by solving the differential equation in part (b).

7. **Center of mass** Find the center of mass of a thin plate of constant density covering the region in the first and fourth quadrants enclosed by the curves $y = 1/(1 + x^2)$ and $y = -1/(1 + x^2)$ and by the lines $x = 0$ and $x = 1$.

8. **Solid of revolution** The region between the curve $y = 1/(2\sqrt{x})$ and the x-axis from $x = 1/4$ to $x = 4$ is revolved about the x-axis to generate a solid.

    a. Find the volume of the solid.

    b. Find the centroid of the region.

9. **The Rule of 70** If you use the approximation $\ln 2 \approx 0.70$ (in place of $0.69314 \ldots$), you can derive a rule of thumb that says, "To estimate how many years it will take an amount of money to double when invested at r percent compounded continuously, divide r into 70." For instance, an amount of money invested at 5% will double in about $70/5 = 14$ years. If you want it to double in 10 years instead, you have to invest it at $70/10 = 7\%$. Show how the Rule of 70 is derived. (A similar "Rule of 72" uses 72 instead of 70, because 72 has more integer factors.)

T 10. **Urban gardening** A vegetable garden 50 ft wide is to be grown between two buildings, which are 500 ft apart along an east-west line. If the buildings are 200 ft and 350 ft tall, where should the garden be placed in order to receive the maximum number of hours of sunlight exposure? (*Hint:* Determine the value of x in the accompanying figure that maximizes sunlight exposure for the garden.)

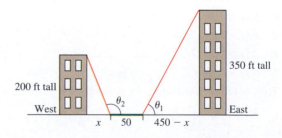

8

Techniques of Integration

OVERVIEW The Fundamental Theorem tells us how to evaluate a definite integral once we have an antiderivative for the integrand function. However, finding antiderivatives (or indefinite integrals) is not as straightforward as finding derivatives. In this chapter we study a number of important techniques that apply to finding integrals for specialized classes of functions such as trigonometric functions, products of certain functions, and rational functions. Since we cannot always find an antiderivative, we develop numerical methods for calculating definite integrals. We also study integrals whose domain or range are infinite, called *improper integrals*.

8.1 Using Basic Integration Formulas

Table 8.1 summarizes the indefinite integrals of many of the functions we have studied so far, and the substitution method helps us use the table to evaluate more complicated functions involving these basic ones. In this section we combine the Substitution Rules (studied in Chapter 5) with algebraic methods and trigonometric identities to help us use Table 8.1. A more extensive Table of Integrals is given at the back of the chapter, and we discuss its use in Section 8.6.

Sometimes we have to rewrite an integral to match it to a standard form of the type displayed in Table 8.1. We start with an example of this procedure.

EXAMPLE 1   Evaluate the integral

$$\int_3^5 \frac{2x - 3}{\sqrt{x^2 - 3x + 1}} \, dx.$$

Solution   We rewrite the integral and apply the Substitution Rule for Definite Integrals presented in Section 5.6, to find

$$\int_3^5 \frac{2x - 3}{\sqrt{x^2 - 3x + 1}} \, dx = \int_1^{11} \frac{du}{\sqrt{u}} \qquad \begin{array}{l} u = x^2 - 3x + 1, \ du = (2x - 3) \, dx; \\ u = 1 \text{ when } x = 3, \ u = 11 \text{ when } x = 5 \end{array}$$

$$= \int_1^{11} u^{-1/2} \, du$$

$$= 2\sqrt{u}\,\Big]_1^{11} = 2\left(\sqrt{11} - 1\right) \approx 4.63. \qquad \text{Table 8.1, Formula 2} \quad \blacksquare$$

TABLE 8.1 Basic integration formulas

1. $\int k \, dx = kx + C$ (any number k)

12. $\int \tan x \, dx = \ln |\sec x| + C$

2. $\int x^n \, dx = \dfrac{x^{n+1}}{n+1} + C$ $(n \neq -1)$

13. $\int \cot x \, dx = \ln |\sin x| + C$

3. $\int \dfrac{dx}{x} = \ln |x| + C$

14. $\int \sec x \, dx = \ln |\sec x + \tan x| + C$

4. $\int e^x \, dx = e^x + C$

15. $\int \csc x \, dx = -\ln |\csc x + \cot x| + C$

5. $\int a^x \, dx = \dfrac{a^x}{\ln a} + C$ $(a > 0, a \neq 1)$

16. $\int \sinh x \, dx = \cosh x + C$

6. $\int \sin x \, dx = -\cos x + C$

17. $\int \cosh x \, dx = \sinh x + C$

7. $\int \cos x \, dx = \sin x + C$

18. $\int \dfrac{dx}{\sqrt{a^2 - x^2}} = \sin^{-1}\left(\dfrac{x}{a}\right) + C$

8. $\int \sec^2 x \, dx = \tan x + C$

19. $\int \dfrac{dx}{a^2 + x^2} = \dfrac{1}{a} \tan^{-1}\left(\dfrac{x}{a}\right) + C$

8. $\int \csc^2 x \, dx = -\cot x + C$

20. $\int \dfrac{dx}{x\sqrt{x^2 - a^2}} = \dfrac{1}{a} \sec^{-1}\left|\dfrac{x}{a}\right| + C$

10. $\int \sec x \tan x \, dx = \sec x + C$

21. $\int \dfrac{dx}{\sqrt{a^2 + x^2}} = \sinh^{-1}\left(\dfrac{x}{a}\right) + C$ $(a > 0)$

11. $\int \csc x \cot x \, dx = -\csc x + C$

22. $\int \dfrac{dx}{\sqrt{x^2 - a^2}} = \cosh^{-1}\left(\dfrac{x}{a}\right) + C$ $(x > a > 0)$

EXAMPLE 2   Complete the square to evaluate

$$\int \frac{dx}{\sqrt{8x - x^2}}.$$

Solution   We complete the square to simplify the denominator:

$$8x - x^2 = -(x^2 - 8x) = -(x^2 - 8x + 16 - 16)$$
$$= -(x^2 - 8x + 16) + 16 = 16 - (x - 4)^2.$$

Then

$$\int \frac{dx}{\sqrt{8x - x^2}} = \int \frac{dx}{\sqrt{16 - (x - 4)^2}}$$

$$= \int \frac{du}{\sqrt{a^2 - u^2}} \qquad a = 4, u = (x - 4),$$
$$\qquad\qquad\qquad du = dx$$

$$= \sin^{-1}\left(\frac{u}{a}\right) + C \qquad \text{Table 8.1, Formula 18}$$

$$= \sin^{-1}\left(\frac{x - 4}{4}\right) + C.$$

EXAMPLE 3   Evaluate the integral

$$\int (\cos x \sin 2x + \sin x \cos 2x)\, dx.$$

Solution   We can replace the integrand with an equivalent trigonometric expression using the Sine Addition Formula to obtain a simple substitution:

$$\int (\cos x \sin 2x + \sin x \cos 2x)\, dx = \int (\sin(x + 2x))\, dx$$

$$= \int \sin 3x\, dx$$

$$= \int \frac{1}{3} \sin u\, du \qquad\qquad u = 3x,\ du = 3\,dx$$

$$= -\frac{1}{3} \cos 3x + C. \qquad\qquad \text{Table 8.1, Formula 6} \quad\blacksquare$$

In Section 5.5 we found the indefinite integral of the secant function by multiplying it by a fractional form identically equal to one, and then integrating the equivalent result. We can use that same procedure in other instances as well, as we illustrate next.

EXAMPLE 4   Find $\displaystyle\int_0^{\pi/4} \frac{dx}{1 - \sin x}.$

Solution   We multiply the numerator and denominator of the integrand by $1 + \sin x$. This procedure transforms the integral into one we can evaluate:

$$\int_0^{\pi/4} \frac{dx}{1 - \sin x} = \int_0^{\pi/4} \frac{1}{1 - \sin x} \cdot \frac{1 + \sin x}{1 + \sin x}\, dx \qquad \begin{array}{l}\text{Multiply and divide}\\ \text{by conjugate.}\end{array}$$

$$= \int_0^{\pi/4} \frac{1 + \sin x}{1 - \sin^2 x}\, dx \qquad \text{Simplify.}$$

$$= \int_0^{\pi/4} \frac{1 + \sin x}{\cos^2 x}\, dx \qquad 1 - \sin^2 x = \cos^2 x$$

$$= \int_0^{\pi/4} (\sec^2 x + \sec x \tan x)\, dx \qquad \begin{array}{l}\text{Use Table 8.1,}\\ \text{Formulas 8 and 10}\end{array}$$

$$= \Big[\tan x + \sec x\Big]_0^{\pi/4} = \left(1 + \sqrt{2} - (0 + 1)\right) = \sqrt{2}. \qquad \blacksquare$$

EXAMPLE 5   Evaluate

$$\int \frac{3x^2 - 7x}{3x + 2}\, dx.$$

Solution   The integrand is an improper fraction since the degree of the numerator is greater than the degree of the denominator. To integrate it, we perform long division to obtain a quotient plus a remainder that is a proper fraction:

$$\frac{3x^2 - 7x}{3x + 2} = x - 3 + \frac{6}{3x + 2}.$$

$$\begin{array}{r} x - 3 \\ 3x + 2\overline{\smash{)}\,3x^2 - 7x} \\ \underline{3x^2 + 2x} \\ -9x \\ \underline{-9x - 6} \\ +\,6 \end{array}$$

Therefore,

$$\int \frac{3x^2 - 7x}{3x + 2}\, dx = \int \left(x - 3 + \frac{6}{3x + 2} \right) dx = \frac{x^2}{2} - 3x + 2 \ln |3x + 2| + C. \quad \blacksquare$$

Reducing an improper fraction by long division (Example 5) does not always lead to an expression we can integrate directly. We see what to do about that in Section 8.5.

EXAMPLE 6   Evaluate

$$\int \frac{3x + 2}{\sqrt{1 - x^2}}\, dx.$$

Solution   We first separate the integrand to get

$$\int \frac{3x + 2}{\sqrt{1 - x^2}}\, dx = 3 \int \frac{x\, dx}{\sqrt{1 - x^2}} + 2 \int \frac{dx}{\sqrt{1 - x^2}}.$$

In the first of these new integrals, we substitute

$$u = 1 - x^2, \qquad du = -2x\, dx, \qquad \text{so} \qquad x\, dx = -\frac{1}{2}\, du.$$

Then we obtain

$$3 \int \frac{x\, dx}{\sqrt{1 - x^2}} = 3 \int \frac{(-1/2)\, du}{\sqrt{u}} = -\frac{3}{2} \int u^{-1/2}\, du$$

$$= -\frac{3}{2} \cdot \frac{u^{1/2}}{1/2} + C_1 = -3\sqrt{1 - x^2} + C_1.$$

The second of the new integrals is a standard form,

$$2 \int \frac{dx}{\sqrt{1 - x^2}} = 2 \sin^{-1} x + C_2. \qquad \text{\color{blue}Table 8.1, Formula 18}$$

Combining these results and renaming $C_1 + C_2$ as C gives

$$\int \frac{3x + 2}{\sqrt{1 - x^2}}\, dx = -3\sqrt{1 - x^2} + 2 \sin^{-1} x + C. \quad \blacksquare$$

The question of what to substitute for in an integrand is not always quite so clear. Sometimes we simply proceed by trial-and-error, and if nothing works out, we then try another method altogether. The next several sections of the text present some of these new methods, but substitution works in the following example.

EXAMPLE 7   Evaluate

$$\int \frac{dx}{\left(1 + \sqrt{x}\right)^3}.$$

Solution   We might try substituting for the term $\sqrt{x}$, but the derivative factor $1/\sqrt{x}$ is missing from the integrand, so this substitution will not help. The other possibility is to substitute for $\left(1 + \sqrt{x}\right)$, and it turns out this works:

$$\int \frac{dx}{\left(1 + \sqrt{x}\right)^3} = \int \frac{2(u - 1)\, du}{u^3} \qquad \color{blue}{u = 1 + \sqrt{x},\ du = \frac{1}{2\sqrt{x}}\, dx;}$$
$$\color{blue}{dx = 2\sqrt{x}\, du = 2(u - 1)\, du}$$

$$= \int \left(\frac{2}{u^2} - \frac{2}{u^3} \right) du$$

$$= -\frac{2}{u} + \frac{1}{u^2} + C$$

$$= \frac{1 - 2u}{u^2} + C$$

$$= \frac{1 - 2(1 + \sqrt{x})}{(1 + \sqrt{x})^2} + C$$

$$= C - \frac{1 + 2\sqrt{x}}{(1 + \sqrt{x})^2}. \qquad \blacksquare$$

When evaluating definite integrals, a property of the integrand may help us in calculating the result.

EXAMPLE 8 Evaluate $\displaystyle\int_{-\pi/2}^{\pi/2} x^3 \cos x \, dx$.

Solution No substitution or algebraic manipulation is clearly helpful here. But we observe that the interval of integration is the symmetric interval $[-\pi/2, \pi/2]$. Moreover, the factor x^3 is an odd function, and $\cos x$ is an even function, so their product is odd. Therefore,

$$\int_{-\pi/2}^{\pi/2} x^3 \cos x \, dx = 0. \qquad \text{Theorem 8, Section 5.6} \qquad \blacksquare$$

EXERCISES 8.1

Assorted Integrations

The integrals in Exercises 1–44 are in no particular order. Evaluate each integral using any algebraic method or trigonometric identity you think is appropriate. When necessary, use a substitution to reduce it to a standard form.

1. $\displaystyle\int_0^1 \frac{16x}{8x^2 + 2} \, dx$

2. $\displaystyle\int \frac{x^2}{x^2 + 1} \, dx$

3. $\displaystyle\int (\sec x - \tan x)^2 \, dx$

4. $\displaystyle\int_{\pi/4}^{\pi/3} \frac{dx}{\cos^2 x \tan x}$

5. $\displaystyle\int \frac{1 - x}{\sqrt{1 - x^2}} \, dx$

6. $\displaystyle\int \frac{dx}{x - \sqrt{x}}$

7. $\displaystyle\int \frac{e^{-\cot z}}{\sin^2 z} \, dz$

8. $\displaystyle\int \frac{2^{\ln z^3}}{16z} \, dz$

9. $\displaystyle\int \frac{dz}{e^z + e^{-z}}$

10. $\displaystyle\int_1^2 \frac{8 \, dx}{x^2 - 2x + 2}$

11. $\displaystyle\int_{-1}^0 \frac{4 \, dx}{1 + (2x + 1)^2}$

12. $\displaystyle\int_{-1}^3 \frac{4x^2 - 7}{2x + 3} \, dx$

13. $\displaystyle\int \frac{dt}{1 - \sec t}$

14. $\displaystyle\int \csc t \sin 3t \, dt$

15. $\displaystyle\int_0^{\pi/4} \frac{1 + \sin \theta}{\cos^2 \theta} \, d\theta$

16. $\displaystyle\int \frac{d\theta}{\sqrt{2\theta - \theta^2}}$

17. $\displaystyle\int \frac{\ln y}{y + 4y \ln^2 y} \, dy$

18. $\displaystyle\int \frac{2^{\sqrt{y}} \, dy}{2\sqrt{y}}$

19. $\displaystyle\int \frac{d\theta}{\sec \theta + \tan \theta}$

20. $\displaystyle\int \frac{dt}{t\sqrt{3 + t^2}}$

21. $\displaystyle\int \frac{4t^3 - t^2 + 16t}{t^2 + 4} \, dt$

22. $\displaystyle\int \frac{x + 2\sqrt{x} - 1}{2x\sqrt{x} - 1} \, dx$

23. $\displaystyle\int_0^{\pi/2} \sqrt{1 - \cos \theta} \, d\theta$

24. $\displaystyle\int (\sec t + \cot t)^2 \, dt$

25. $\displaystyle\int \frac{dy}{\sqrt{e^{2y} - 1}}$

26. $\displaystyle\int \frac{6 \, dy}{\sqrt{y}(1 + y)}$

27. $\displaystyle\int \frac{2 \, dx}{x\sqrt{1 - 4 \ln^2 x}}$

28. $\displaystyle\int \frac{dx}{(x - 2)\sqrt{x^2 - 4x + 3}}$

29. $\displaystyle\int (\csc x - \sec x)(\sin x + \cos x) \, dx$

30. $\displaystyle\int 3 \sinh \left(\frac{x}{2} + \ln 5 \right) dx$

31. $\displaystyle\int_{\sqrt{2}}^3 \frac{2x^3}{x^2 - 1} \, dx$

32. $\displaystyle\int_{-1}^1 \sqrt{1 + x^2} \sin x \, dx$

33. $\displaystyle\int_{-1}^0 \sqrt{\frac{1 + y}{1 - y}} \, dy$

34. $\displaystyle\int e^{z + e^z} \, dz$

35. $\displaystyle\int \frac{7\,dx}{(x-1)\sqrt{x^2-2x-48}}$

36. $\displaystyle\int \frac{dx}{(2x+1)\sqrt{4x+4x^2}}$

37. $\displaystyle\int \frac{2\theta^3-7\theta^2+7\theta}{2\theta-5}\,d\theta$

38. $\displaystyle\int \frac{d\theta}{\cos\theta-1}$

39. $\displaystyle\int \frac{dx}{1+e^x}$

40. $\displaystyle\int \frac{\sqrt{x}}{1+x^3}\,dx$

Hint: Use long division.

Hint: Let $u=x^{3/2}$.

41. $\displaystyle\int \frac{e^{3x}}{e^x+1}\,dx$

42. $\displaystyle\int \frac{2^x-1}{3^x}\,dx$

43. $\displaystyle\int \frac{1}{\sqrt{x}\,(1+x)}\,dx$

44. $\displaystyle\int \frac{\tan\theta+3}{\sin\theta}\,d\theta$

Theory and Examples

45. Area Find the area of the region bounded above by $y=2\cos x$ and below by $y=\sec x$, $-\pi/4 \le x \le \pi/4$.

46. Volume Find the volume of the solid generated by revolving the region in Exercise 45 about the x-axis.

47. Arc length Find the length of the curve $y=\ln(\cos x)$, $0 \le x \le \pi/3$.

48. Arc length Find the length of the curve $y=\ln(\sec x)$, $0 \le x \le \pi/4$.

49. Centroid Find the centroid of the region bounded by the x-axis, the curve $y=\sec x$, and the lines $x=-\pi/4$, $x=\pi/4$.

50. Centroid Find the centroid of the region bounded by the x-axis, the curve $y=\csc x$, and the lines $x=\pi/6$, $x=5\pi/6$.

51. The functions $y=e^{x^3}$ and $y=x^3e^{x^3}$ do not have elementary antiderivatives, but $y=(1+3x^3)e^{x^3}$ does. Evaluate

$$\int (1+3x^3)e^{x^3}\,dx.$$

52. Use the substitution $u=\tan x$ to evaluate the integral

$$\int \frac{dx}{1+\sin^2 x}.$$

53. Use the substitution $u=x^4+1$ to evaluate the integral

$$\int x^7\sqrt{x^4+1}\,dx.$$

54. Using different substitutions Show that the integral

$$\int ((x^2-1)(x+1))^{-2/3}dx$$

can be evaluated with any of the following substitutions.

a. $u=1/(x+1)$

b. $u=((x-1)/(x+1))^k$ for $k=1,\ 1/2,\ 1/3,\ -1/3,\ -2/3,$ and -1

c. $u=\tan^{-1}x$ **d.** $u=\tan^{-1}\sqrt{x}$

e. $u=\tan^{-1}((x-1)/2)$ **f.** $u=\cos^{-1}x$

g. $u=\cosh^{-1}x$

What is the value of the integral?

8.2 Integration by Parts

Integration by parts is a technique for simplifying integrals of the form

$$\int u(x)\,v'(x)\,dx.$$

It is useful when u can be differentiated repeatedly and v' can be integrated repeatedly without difficulty. The integrals

$$\int x\cos x\,dx \qquad \text{and} \qquad \int x^2 e^x\,dx$$

are such integrals because $u(x)=x$ or $u(x)=x^2$ can be differentiated repeatedly to become zero, and $v'(x)=\cos x$ or $v'(x)=e^x$ can be integrated repeatedly without difficulty. Integration by parts also applies to integrals like

$$\int \ln x\,dx \qquad \text{and} \qquad \int e^x\cos x\,dx.$$

In the first case, the integrand $\ln x$ can be rewritten as $(\ln x)(1)$, and $u(x)=\ln x$ is easy to differentiate while $v'(x)=1$ easily integrates to x. In the second case, each part of the integrand appears again after repeated differentiation or integration.

Product Rule in Integral Form

If u and v are differentiable functions of x, the Product Rule says that

$$\frac{d}{dx}\big[u(x)\,v(x)\big]=u'(x)\,v(x)+u(x)\,v'(x).$$

In terms of indefinite integrals, this equation becomes

$$\int \frac{d}{dx}[u(x)v(x)]\,dx = \int [u'(x)v(x) + u(x)v'(x)]\,dx$$

or

$$\int \frac{d}{dx}[u(x)v(x)]\,dx = \int u'(x)v(x)\,dx + \int u(x)v'(x)\,dx.$$

Rearranging the terms of this last equation, we get

$$\int u(x)v'(x)\,dx = \int \frac{d}{dx}[u(x)v(x)]\,dx - \int v(x)u'(x)\,dx,$$

leading to the **integration by parts** formula

Integration by Parts Formula

$$\int u(x)v'(x)\,dx = u(x)v(x) - \int v(x)u'(x)\,dx \tag{1}$$

This formula allows us to exchange the problem of computing the integral $\int u(x)v'(x)\,dx$ with the problem of computing a different integral, $\int v(x)u'(x)\,dx$. In many cases, we can choose the functions u and v so that the second integral is easier to compute than the first. There can be many choices for u and v, and it is not always clear which choice works best, so sometimes we need to try several.

The formula is often given in differential form. With $v'(x)\,dx = dv$ and $u'(x)\,dx = du$, the integration by parts formula becomes

---

Integration by Parts Formula—Differential Version

$$\int u\,dv = uv - \int v\,du \tag{2}$$

The next examples illustrate the technique.

EXAMPLE 1   Find

$$\int x \cos x\,dx.$$

Solution   There is no obvious antiderivative of $x \cos x$, so we use the integration by parts formula

$$\int u(x)v'(x)\,dx = u(x)v(x) - \int v(x)u'(x)\,dx$$

to change this expression to one that is easier to integrate. We first decide how to choose the functions $u(x)$ and $v(x)$. In this case we factor the expression $x \cos x$ into

$$u(x) = x \quad \text{and} \quad v'(x) = \cos x.$$

Next we differentiate $u(x)$ and find an antiderivative of $v'(x)$,

$$u'(x) = 1 \quad \text{and} \quad v(x) = \sin x.$$

When finding an antiderivative for $v'(x)$ we have a choice of how to pick a constant of integration C. We choose the constant $C = 0$, since that makes this antiderivative as simple as possible. We now apply the integration by parts formula:

$$\int \underset{\substack{| \ | \\ u(x) \ v'(x)}}{x \cos x \ dx} = \underset{\substack{| \quad \ \backslash \\ u(x) \quad v(x)}}{x \sin x} - \int \underset{\substack{| \quad \ | \\ v(x) \ u'(x)}}{\sin x \ (1) \ dx} \qquad \text{Integration by parts formula}$$

$$= x \sin x + \cos x + C \qquad \text{Integrate and simplify.} \qquad ■$$

and we have found the integral of the original function.

There are four apparent choices available for $u(x)$ and $v'(x)$ in Example 1:

1. Let $u(x) = 1$ and $v'(x) = x \cos x$.
2. Let $u(x) = x$ and $v'(x) = \cos x$.
3. Let $u(x) = x \cos x$ and $v'(x) = 1$.
4. Let $u(x) = \cos x$ and $v'(x) = x$.

Choice 2 was used in Example 1. The other three choices lead to integrals we don't know how to integrate. For instance, Choice 3, with $u'(x) = \cos x - x \sin x$, leads to the integral

$$\int (x \cos x - x^2 \sin x) \, dx.$$

The goal of integration by parts is to go from an integral $\int u(x) v'(x) \, dx$ that we don't see how to evaluate to an integral $\int v(x) u'(x) \, dx$ that we can evaluate. Generally, you choose $v'(x)$ first to be as much of the integrand as we can readily integrate; $u(x)$ is the leftover part. When finding $v(x)$ from $v'(x)$, any antiderivative will work, and we usually pick the simplest one; no arbitrary constant of integration is needed in $v(x)$ because it would simply cancel out of the right-hand side of Equation (2).

EXAMPLE 2 Find $\int \ln x \, dx$.

Solution We have not yet seen how to find an antiderivative for $\ln x$. If we set $u(x) = \ln x$, then $u'(x)$ is the simpler function $1/x$. It may not appear that a second function $v'(x)$ is multiplying $\ln x$, but we can choose $v'(x)$ to be the constant function $v'(x) = 1$. We use the integration by parts formula Equation (1) with

$$u(x) = \ln x \quad \text{and} \quad v'(x) = 1.$$

We differentiate $u(x)$ and find an antiderivative of $v'(x)$,

$$u'(x) = \frac{1}{x} \quad \text{and} \quad v(x) = x.$$

Then

$$\int \underset{\substack{| \quad | \\ u(x) \ v'(x)}}{\ln x \cdot 1 \ dx} = \underset{\substack{| \quad \ | \\ u(x) \ v(x)}}{(\ln x) \ x} - \int \underset{\substack{| \quad \backslash \\ v(x) \ u'(x)}}{x \frac{1}{x} \ dx} \qquad \text{Integration by parts formula}$$

$$= x \ln x - x + C \qquad \text{Simplify and integrate.} \qquad ■$$

In the following examples we use the differential form to indicate the process of integration by parts. The computations are the same, with du and dv providing shorter expressions for $u'(x) \, dx$ and $v'(x) \, dx$. Sometimes we have to use integration by parts more than once, as in the next example.

EXAMPLE 3   Evaluate

$$\int x^2 e^x \, dx.$$

Solution   We use the integration by parts formula Equation (1) with

$$u(x) = x^2 \quad \text{and} \quad v'(x) = e^x.$$

We differentiate $u(x)$ and find an antiderivative of $v'(x)$,

$$u'(x) = 2x \quad \text{and} \quad v(x) = e^x.$$

We summarize this choice by setting $du = u'(x) \, dx$ and $dv = v'(x) \, dx$, so

$$du = 2x \, dx \quad \text{and} \quad dv = e^x \, dx.$$

We then have

$$\int \underbrace{x^2}_{u} \, \underbrace{e^x \, dx}_{dv} = \underbrace{x^2}_{u} \underbrace{e^x}_{v} - \int \underbrace{e^x}_{v} \, \underbrace{2x \, dx}_{du}.$$ Integration by parts formula

The new integral is less complicated than the original because the exponent on x is reduced by one. To evaluate the integral on the right, we integrate by parts again with $u = x, dv = e^x \, dx$. Then $du = dx, v = e^x$, and

$$\int \underbrace{x}_{u} \, \underbrace{e^x \, dx}_{dv} = \underbrace{x}_{u} \underbrace{e^x}_{v} - \int \underbrace{e^x}_{v} \, \underbrace{dx}_{du} = xe^x - e^x + C.$$

Integration by parts Equation (2)
$u = x, dv = e^x \, dx$
$v = e^x, \quad du = dx$

Using this last evaluation, we then obtain

$$\int x^2 e^x \, dx = x^2 e^x - 2 \int xe^x \, dx$$

$$= x^2 e^x - 2xe^x + 2e^x + C,$$

where the constant of integration is renamed after substituting for the integral on the right. ∎

    The technique of Example 3 works for any integral $\int x^n e^x \, dx$ in which n is a positive integer, because differentiating x^n will eventually lead to zero and integrating e^x is easy.

    Integrals like the one in the next example occur in electrical engineering. Their evaluation requires two integrations by parts, followed by solving for the unknown integral.

EXAMPLE 4   Evaluate

$$\int e^x \cos x \, dx.$$

Solution   Let $u = e^x$ and $dv = \cos x \, dx$. Then $du = e^x \, dx, v = \sin x$, and

$$\int e^x \cos x \, dx = e^x \sin x - \int e^x \sin x \, dx.$$ $u(x) = e^x, \quad v(x) = \sin x$

The second integral is like the first except that it has $\sin x$ in place of $\cos x$. To evaluate it, we use integration by parts with

$$u = e^x, \quad dv = \sin x \, dx, \quad v = -\cos x, \quad du = e^x \, dx.$$

Then

$$\int e^x \cos x \, dx = e^x \sin x - \left(-e^x \cos x - \int (-\cos x)(e^x \, dx)\right) \qquad u(x) = e^x, \quad v(x) = -\cos x$$

$$= e^x \sin x + e^x \cos x - \int e^x \cos x \, dx.$$

The unknown integral now appears on both sides of the equation, but with opposite signs. Adding the integral to both sides and adding the constant of integration give

$$2\int e^x \cos x \, dx = e^x \sin x + e^x \cos x + C_1.$$

Dividing by 2 and renaming the constant of integration give

$$\int e^x \cos x \, dx = \frac{e^x \sin x + e^x \cos x}{2} + C. \qquad \blacksquare$$

EXAMPLE 5   Obtain a formula that expresses the integral

$$\int \cos^n x \, dx$$

in terms of an integral of a lower power of $\cos x$.

Solution   We may think of $\cos^n x$ as $\cos^{n-1} x \cdot \cos x$. Then we let

$$u = \cos^{n-1} x \qquad \text{and} \qquad dv = \cos x \, dx,$$

so that

$$du = (n-1)\cos^{n-2} x \, (-\sin x \, dx) \qquad \text{and} \qquad v = \sin x.$$

Integration by parts then gives

$$\int \cos^n x \, dx = \cos^{n-1} x \sin x + (n-1)\int \sin^2 x \cos^{n-2} x \, dx$$

$$= \cos^{n-1} x \sin x + (n-1)\int (1 - \cos^2 x) \cos^{n-2} x \, dx$$

$$= \cos^{n-1} x \sin x + (n-1)\int \cos^{n-2} x \, dx - (n-1)\int \cos^n x \, dx.$$

If we add

$$(n-1)\int \cos^n x \, dx$$

to both sides of this equation, we obtain

$$n\int \cos^n x \, dx = \cos^{n-1} x \sin x + (n-1)\int \cos^{n-2} x \, dx.$$

We then divide through by n, and the final result is

$$\int \cos^n x \, dx = \frac{\cos^{n-1} x \sin x}{n} + \frac{n-1}{n}\int \cos^{n-2} x \, dx. \qquad \blacksquare$$

The formula found in Example 5 is called a **reduction formula** because it replaces an integral containing some power of a function with an integral of the same form having the power reduced. When n is a positive integer, we may apply the formula repeatedly until the remaining integral is easy to evaluate. For example, the result in Example 5 tells us that

$$\int \cos^3 x \, dx = \frac{\cos^2 x \sin x}{3} + \frac{2}{3} \int \cos x \, dx$$

$$= \frac{1}{3} \cos^2 x \sin x + \frac{2}{3} \sin x + C.$$

Evaluating Definite Integrals by Parts

The integration by parts formula in Equation (1) can be combined with Part 2 of the Fundamental Theorem in order to evaluate definite integrals by parts. Assuming that both u' and v' are continuous over the interval $[a, b]$, Part 2 of the Fundamental Theorem gives

---

Integration by Parts Formula for Definite Integrals

$$\int_a^b u(x) \, v'(x) \, dx = u(x) v(x) \Big]_a^b - \int_a^b v(x) \, u'(x) \, dx \qquad (3)$$

EXAMPLE 6   Find the area of the region bounded by the curve $y = xe^{-x}$ and the x-axis from $x = 0$ to $x = 4$.

Solution   The region is shaded in Figure 8.1. Its area is

$$\int_0^4 xe^{-x} \, dx.$$

Let $u = x$, $dv = e^{-x} dx$, $v = -e^{-x}$, and $du = dx$. Then,

$$\int_0^4 xe^{-x} \, dx = -xe^{-x} \Big]_0^4 - \int_0^4 (-e^{-x}) \, dx \qquad \text{Integration by parts Formula (3)}$$

$$= \left[-4e^{-4} - (-0e^{-0}) \right] + \int_0^4 e^{-x} \, dx$$

$$= -4e^{-4} - e^{-x} \Big]_0^4$$

$$= -4e^{-4} - (e^{-4} - e^{-0}) = 1 - 5e^{-4} \approx 0.91. \qquad \blacksquare$$

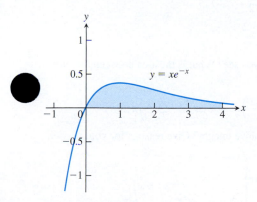

FIGURE 8.1   The region in Example 6.

EXERCISES 8.2

Integration by Parts
Evaluate the integrals in Exercises 1–24 using integration by parts.

1. $\displaystyle\int x \sin \frac{x}{2} \, dx$

2. $\displaystyle\int \theta \cos \pi\theta \, d\theta$

3. $\displaystyle\int t^2 \cos t \, dt$

4. $\displaystyle\int x^2 \sin x \, dx$

5. $\displaystyle\int_1^2 x \ln x \, dx$

6. $\displaystyle\int_1^e x^3 \ln x \, dx$

7. $\displaystyle\int xe^x \, dx$

8. $\displaystyle\int xe^{3x} \, dx$

9. $\displaystyle\int x^2 e^{-x} \, dx$

10. $\displaystyle\int (x^2 - 2x + 1)e^{2x} \, dx$

11. $\displaystyle\int \tan^{-1} y \, dy$

12. $\displaystyle\int \sin^{-1} y \, dy$

13. $\displaystyle\int x \sec^2 x \, dx$

14. $\displaystyle\int 4x \sec^2 2x \, dx$

15. $\displaystyle\int x^3 e^x \, dx$

16. $\displaystyle\int p^4 e^{-p} \, dp$

17. $\displaystyle\int (x^2 - 5x)e^x \, dx$

18. $\displaystyle\int (r^2 + r + 1)e^r \, dr$

19. $\int x^5 e^x \, dx$

20. $\int t^2 e^{4t} \, dt$

21. $\int e^\theta \sin \theta \, d\theta$

22. $\int e^{-y} \cos y \, dy$

23. $\int e^{2x} \cos 3x \, dx$

24. $\int e^{-2x} \sin 2x \, dx$

Using Substitution

Evaluate the integrals in Exercises 25–30 by using a substitution prior to integration by parts.

25. $\int e^{\sqrt{3s+9}} \, ds$

26. $\int_0^1 x\sqrt{1-x} \, dx$

27. $\int_0^{\pi/3} x \tan^2 x \, dx$

28. $\int \ln (x + x^2) \, dx$

29. $\int \sin (\ln x) \, dx$

30. $\int z(\ln z)^2 \, dz$

Evaluating Integrals

Evaluate the integrals in Exercises 31–56. Some integrals do not require integration by parts.

31. $\int x \sec x^2 \, dx$

32. $\int \frac{\cos \sqrt{x}}{\sqrt{x}} \, dx$

33. $\int x (\ln x)^2 \, dx$

34. $\int \frac{1}{x (\ln x)^2} \, dx$

35. $\int \frac{\ln x}{x^2} \, dx$

36. $\int \frac{(\ln x)^3}{x} \, dx$

37. $\int x^3 e^{x^4} \, dx$

38. $\int x^5 e^{x^3} \, dx$

39. $\int x^3 \sqrt{x^2 + 1} \, dx$

40. $\int x^2 \sin x^3 \, dx$

41. $\int \sin 3x \cos 2x \, dx$

42. $\int \sin 2x \cos 4x \, dx$

43. $\int \sqrt{x} \ln x \, dx$

44. $\int \frac{e^{\sqrt{x}}}{\sqrt{x}} \, dx$

45. $\int \cos \sqrt{x} \, dx$

46. $\int \sqrt{x} \, e^{\sqrt{x}} \, dx$

47. $\int_0^{\pi/2} \theta^2 \sin 2\theta \, d\theta$

48. $\int_0^{\pi/2} x^3 \cos 2x \, dx$

49. $\int_{2/\sqrt{3}}^2 t \sec^{-1} t \, dt$

50. $\int_0^{1/\sqrt{2}} 2x \sin^{-1} (x^2) \, dx$

51. $\int x \tan^{-1} x \, dx$

52. $\int x^2 \tan^{-1} \frac{x}{2} \, dx$

53. $\int (1 + 2x^2) e^{x^2} \, dx$

54. $\int \frac{xe^x}{(x + 1)^2} \, dx$

55. $\int \sqrt{x} \left(\sin^{-1} \sqrt{x} \right) dx$

56. $\int \frac{(\sin^{-1} x)^2}{\sqrt{1 - x^2}} \, dx$

Theory and Examples

57. Finding area Find the area of the region enclosed by the curve $y = x \sin x$ and the x-axis (see the accompanying figure) for

a. $0 \leq x \leq \pi$.

b. $\pi \leq x \leq 2\pi$.

c. $2\pi \leq x \leq 3\pi$.

d. What pattern do you see here? What is the area between the curve and the x-axis for $n\pi \leq x \leq (n + 1)\pi$, n an arbitrary nonnegative integer? Give reasons for your answer.

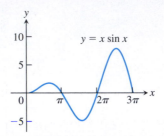

58. Finding area Find the area of the region enclosed by the curve $y = x \cos x$ and the x-axis (see the accompanying figure) for

a. $\pi/2 \leq x \leq 3\pi/2$.

b. $3\pi/2 \leq x \leq 5\pi/2$.

c. $5\pi/2 \leq x \leq 7\pi/2$.

d. What pattern do you see? What is the area between the curve and the x-axis for

$$\left(\frac{2n - 1}{2}\right)\pi \leq x \leq \left(\frac{2n + 1}{2}\right)\pi,$$

n an arbitrary positive integer? Give reasons for your answer.

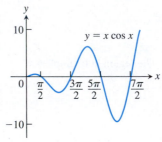

59. Finding volume Find the volume of the solid generated by revolving the region in the first quadrant bounded by the coordinate axes, the curve $y = e^x$, and the line $x = \ln 2$ about the line $x = \ln 2$.

60. Finding volume Find the volume of the solid generated by revolving the region in the first quadrant bounded by the coordinate axes, the curve $y = e^{-x}$, and the line $x = 1$

a. about the y-axis.

b. about the line $x = 1$.

61. Finding volume Find the volume of the solid generated by revolving the region in the first quadrant bounded by the coordinate axes and the curve $y = \cos x$, $0 \leq x \leq \pi/2$, about

a. the y-axis.

b. the line $x = \pi/2$.

62. Finding volume Find the volume of the solid generated by revolving the region bounded by the x-axis and the curve $y = x \sin x, 0 \le x \le \pi$, about

  a. the y-axis.

  b. the line $x = \pi$.

  (See Exercise 57 for a graph.)

63. Consider the region bounded by the graphs of $y = \ln x, y = 0$, and $x = e$.

  a. Find the area of the region.

  b. Find the volume of the solid formed by revolving this region about the x-axis.

  c. Find the volume of the solid formed by revolving this region about the line $x = -2$.

  d. Find the centroid of the region.

64. Consider the region bounded by the graphs of $y = \tan^{-1} x, y = 0$, and $x = 1$.

  a. Find the area of the region.

  b. Find the volume of the solid formed by revolving this region about the y-axis.

65. Average value A retarding force, symbolized by the dashpot in the accompanying figure, slows the motion of the weighted spring so that the mass's position at time t is

$$y = 2e^{-t} \cos t, \qquad t \ge 0.$$

Find the average value of y over the interval $0 \le t \le 2\pi$.

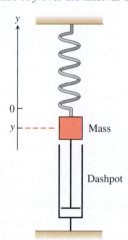

66. Average value In a mass-spring-dashpot system like the one in Exercise 65, the mass's position at time t is

$$y = 4e^{-t}(\sin t - \cos t), \qquad t \ge 0.$$

Find the average value of y over the interval $0 \le t \le 2\pi$.

Reduction Formulas

In Exercises 67–71, use integration by parts to establish the reduction formula.

67. $\displaystyle \int x^n \cos x \, dx = x^n \sin x - n \int x^{n-1} \sin x \, dx$

68. $\displaystyle \int x^n \sin x \, dx = -x^n \cos x + n \int x^{n-1} \cos x \, dx$

69. $\displaystyle \int x^n e^{ax} \, dx = \frac{x^n e^{ax}}{a} - \frac{n}{a} \int x^{n-1} e^{ax} \, dx, \quad a \ne 0$

70. $\displaystyle \int (\ln x)^n \, dx = x(\ln x)^n - n \int (\ln x)^{n-1} \, dx$

71. $\displaystyle \int x^m (\ln x)^n \, dx = \frac{x^{m+1}}{m+1} (\ln x)^n$
$$- \frac{n}{m+1} \int x^m (\ln x)^{n-1} \, dx, \quad m \ne -1$$

72. $\displaystyle \int x^n \sqrt{x+1} \, dx = \frac{2x^n}{2n+3}(x+1)^{3/2}$
$$- \frac{2n}{2n+3} \int x^{n-1} \sqrt{x+1} \, dx$$

73. $\displaystyle \int \frac{x^n}{\sqrt{x+1}} \, dx = \frac{2x^n}{2n+1} \sqrt{x+1}$
$$- \frac{2n}{2n+1} \int \frac{x^{n-1}}{\sqrt{x+1}} \, dx$$

74. Use Example 5 to show that

$$\int_0^{\pi/2} \sin^n x \, dx = \int_0^{\pi/2} \cos^n x \, dx$$
$$= \begin{cases} \left(\dfrac{\pi}{2}\right)\dfrac{1 \cdot 3 \cdot 5 \cdots (n-1)}{2 \cdot 4 \cdot 6 \cdots n}, & n \text{ even} \\[3ex] \dfrac{2 \cdot 4 \cdot 6 \cdots (n-1)}{1 \cdot 3 \cdot 5 \cdots n}, & n \text{ odd} \end{cases}$$

75. Show that

$$\int_a^b \left(\int_x^b f(t) \, dt \right) dx = \int_a^b (x-a)f(x) \, dx.$$

76. Use integration by parts to obtain the formula

$$\int \sqrt{1-x^2} \, dx = \frac{1}{2}x\sqrt{1-x^2} + \frac{1}{2}\int \frac{1}{\sqrt{1-x^2}} \, dx.$$

Integrating Inverses of Functions

Integration by parts leads to a rule for integrating inverses that usually gives good results:

$$\int f^{-1}(x) \, dx = \int yf'(y) \, dy \qquad \begin{array}{l} y = f^{-1}(x), \quad x = f(y) \\ dx = f'(y) \, dy \end{array}$$

$$= yf(y) - \int f(y) \, dy \qquad \begin{array}{l} \text{Integration by parts with} \\ u = y, \, dv = f'(y) \, dy \end{array}$$

$$= xf^{-1}(x) - \int f(y) \, dy$$

The idea is to take the most complicated part of the integral, in this case $f^{-1}(x)$, and simplify it first. For the integral of $\ln x$, we get

$$\int \ln x \, dx = \int y e^{y} \, dy \qquad \begin{array}{l} y = \ln x, \quad x = e^{y} \\ dx = e^{y} \, dy \end{array}$$

$$= y e^{y} - e^{y} + C$$

$$= x \ln x - x + C.$$

For the integral of $\cos^{-1} x$ we get

$$\int \cos^{-1} x \, dx = x \cos^{-1} x - \int \cos y \, dy \qquad y = \cos^{-1} x$$

$$= x \cos^{-1} x - \sin y + C$$

$$= x \cos^{-1} x - \sin (\cos^{-1} x) + C.$$

Use the formula

$$\int f^{-1}(x) \, dx = x f^{-1}(x) - \int f(y) \, dy \qquad y = f^{-1}(x) \quad (4)$$

to evaluate the integrals in Exercises 77–80. Express your answers in terms of x.

77. $\displaystyle\int \sin^{-1} x \, dx$

78. $\displaystyle\int \tan^{-1} x \, dx$

79. $\displaystyle\int \sec^{-1} x \, dx$

80. $\displaystyle\int \log_2 x \, dx$

Another way to integrate $f^{-1}(x)$ (when f^{-1} is integrable) is to use integration by parts with $u = f^{-1}(x)$ and $dv = dx$ to rewrite the integral of f^{-1} as

$$\int f^{-1}(x) \, dx = x f^{-1}(x) - \int x \left(\frac{d}{dx} f^{-1}(x) \right) dx. \qquad (5)$$

Exercises 81 and 82 compare the results of using Equations (4) and (5).

81. Equations (4) and (5) give different formulas for the integral of $\cos^{-1} x$:

  a. $\displaystyle\int \cos^{-1} x \, dx = x \cos^{-1} x - \sin (\cos^{-1} x) + C$ Eq. (4)

  b. $\displaystyle\int \cos^{-1} x \, dx = x \cos^{-1} x - \sqrt{1 - x^2} + C$ Eq. (5)

  Can both integrations be correct? Explain.

82. Equations (4) and (5) lead to different formulas for the integral of $\tan^{-1} x$:

  a. $\displaystyle\int \tan^{-1} x \, dx = x \tan^{-1} x - \ln \sec (\tan^{-1} x) + C$ Eq. (4)

  b. $\displaystyle\int \tan^{-1} x \, dx = x \tan^{-1} x - \ln \sqrt{1 + x^2} + C$ Eq. (5)

  Can both integrations be correct? Explain.

Evaluate the integrals in Exercises 83 and 84 with (**a**) Eq. (4) and (**b**) Eq. (5). In each case, check your work by differentiating your answer with respect to x.

83. $\displaystyle\int \sinh^{-1} x \, dx$

84. $\displaystyle\int \tanh^{-1} x \, dx$

8.3   Trigonometric Integrals

Trigonometric integrals involve algebraic combinations of the six basic trigonometric functions. In principle, we can always express such integrals in terms of sines and cosines, but it is often simpler to work with other functions, as in the integral

$$\int \sec^2 x \, dx = \tan x + C.$$

The general idea is to use identities to transform the integrals we have to find into integrals that are easier to work with.

Products of Powers of Sines and Cosines

We begin with integrals of the form

$$\int \sin^m x \cos^n x \, dx,$$

where m and n are nonnegative integers (positive or zero). We can divide the appropriate substitution into three cases according to m and n being odd or even.

Case 1   If *m* **is odd**, we write *m* as $2k + 1$ and use the identity $\sin^2 x = 1 - \cos^2 x$ to obtain

$$\sin^m x = \sin^{2k+1} x = (\sin^2 x)^k \sin x = (1 - \cos^2 x)^k \sin x. \qquad (1)$$

Then we combine the single $\sin x$ with dx in the integral and set $\sin x \, dx$ equal to $-d(\cos x)$.

Case 2   If *n* **is odd** in $\int \sin^m x \cos^n x \, dx$, we write *n* as $2k + 1$ and use the identity $\cos^2 x = 1 - \sin^2 x$ to obtain

$$\cos^n x = \cos^{2k+1} x = (\cos^2 x)^k \cos x = (1 - \sin^2 x)^k \cos x.$$

We then combine the single $\cos x$ with dx and set $\cos x \, dx$ equal to $d(\sin x)$.

Case 3   If **both *m* and *n* are even** in $\int \sin^m x \cos^n x \, dx$, we substitute

$$\sin^2 x = \frac{1 - \cos 2x}{2}, \qquad \cos^2 x = \frac{1 + \cos 2x}{2} \qquad (2)$$

to reduce the integrand to one in lower powers of $\cos 2x$.

Here are some examples illustrating each case.

EXAMPLE 1   Evaluate

$$\int \sin^3 x \cos^2 x \, dx.$$

Solution   This is an example of Case 1.

$$\int \sin^3 x \cos^2 x \, dx = \int \sin^2 x \cos^2 x \sin x \, dx \qquad \textit{m is odd.}$$

$$= \int (1 - \cos^2 x)(\cos^2 x)(-d(\cos x)) \qquad \sin x \, dx = -d(\cos x)$$

$$= \int (1 - u^2)(u^2)(-du) \qquad u = \cos x$$

$$= \int (u^4 - u^2) \, du \qquad \text{Multiply terms.}$$

$$= \frac{u^5}{5} - \frac{u^3}{3} + C = \frac{\cos^5 x}{5} - \frac{\cos^3 x}{3} + C \qquad \blacksquare$$

EXAMPLE 2   Evaluate

$$\int \cos^5 x \, dx.$$

Solution   This is an example of Case 2, where $m = 0$ is even and $n = 5$ is odd.

$$\int \cos^5 x \, dx = \int \cos^4 x \cos x \, dx = \int (1 - \sin^2 x)^2 \, d(\sin x) \qquad \cos x \, dx = d(\sin x)$$

$$= \int (1 - u^2)^2 \, du \qquad u = \sin x$$

$$= \int (1 - 2u^2 + u^4) \, du \qquad \text{Square } 1 - u^2.$$

$$= u - \frac{2}{3}u^3 + \frac{1}{5}u^5 + C = \sin x - \frac{2}{3}\sin^3 x + \frac{1}{5}\sin^5 x + C \qquad \blacksquare$$

EXAMPLE 3   Evaluate

$$\int \sin^2 x \cos^4 x \, dx.$$

Solution   This is an example of Case 3.

$$\int \sin^2 x \cos^4 x \, dx = \int \left(\frac{1 - \cos 2x}{2} \right) \left(\frac{1 + \cos 2x}{2} \right)^2 dx \qquad \text{\textit{m} and \textit{n} both even}$$

$$= \frac{1}{8} \int (1 - \cos 2x)(1 + 2 \cos 2x + \cos^2 2x) \, dx$$

$$= \frac{1}{8} \int (1 + \cos 2x - \cos^2 2x - \cos^3 2x) \, dx$$

$$= \frac{1}{8} \left[x + \frac{1}{2} \sin 2x - \int (\cos^2 2x + \cos^3 2x) \, dx \right]$$

For the term involving $\cos^2 2x$, we use

$$\int \cos^2 2x \, dx = \frac{1}{2} \int (1 + \cos 4x) \, dx$$

$$= \frac{1}{2} \left(x + \frac{1}{4} \sin 4x \right). \qquad \text{Omit constant of integration until final result.}$$

For the $\cos^3 2x$ term, we have

$$\int \cos^3 2x \, dx = \int (1 - \sin^2 2x) \cos 2x \, dx \qquad u = \sin 2x, \ du = 2 \cos 2x \, dx$$

$$= \frac{1}{2} \int (1 - u^2) \, du = \frac{1}{2} \left(\sin 2x - \frac{1}{3} \sin^3 2x \right). \qquad \text{Again omit } C.$$

Combining everything and simplifying, we get

$$\int \sin^2 x \cos^4 x \, dx = \frac{1}{16} \left(x - \frac{1}{4} \sin 4x + \frac{1}{3} \sin^3 2x \right) + C. \qquad \blacksquare$$

Eliminating Square Roots

In the next example, we use the identity $\cos^2 \theta = (1 + \cos 2\theta)/2$ to eliminate a square root.

EXAMPLE 4   Evaluate

$$\int_0^{\pi/4} \sqrt{1 + \cos 4x} \, dx.$$

Solution   To eliminate the square root, we use the identity

$$\cos^2 \theta = \frac{1 + \cos 2\theta}{2} \qquad \text{or} \qquad 1 + \cos 2\theta = 2 \cos^2 \theta.$$

With $\theta = 2x$, this becomes

$$1 + \cos 4x = 2 \cos^2 2x.$$

Therefore,

$$\int_0^{\pi/4} \sqrt{1 + \cos 4x}\, dx = \int_0^{\pi/4} \sqrt{2 \cos^2 2x}\, dx = \int_0^{\pi/4} \sqrt{2}\sqrt{\cos^2 2x}\, dx$$

$$= \sqrt{2} \int_0^{\pi/4} |\cos 2x|\, dx = \sqrt{2} \int_0^{\pi/4} \cos 2x\, dx \qquad \begin{matrix} \cos 2x \ge 0 \text{ on} \\ [0, \pi/4] \end{matrix}$$

$$= \sqrt{2} \left[\frac{\sin 2x}{2} \right]_0^{\pi/4} = \frac{\sqrt{2}}{2} [1 - 0] = \frac{\sqrt{2}}{2}. \qquad \blacksquare$$

Integrals of Powers of tan x and sec x

We know how to integrate the tangent and secant functions and their squares. To integrate higher powers, we use the identities $\tan^2 x = \sec^2 x - 1$ and $\sec^2 x = \tan^2 x + 1$, and integrate by parts when necessary to reduce the higher powers to lower powers.

EXAMPLE 5   Evaluate

$$\int \tan^4 x\, dx.$$

Solution

$$\int \tan^4 x\, dx = \int \tan^2 x \cdot \tan^2 x\, dx = \int \tan^2 x \cdot (\sec^2 x - 1)\, dx$$

$$= \int \tan^2 x \sec^2 x\, dx - \int \tan^2 x\, dx$$

$$= \int \tan^2 x \sec^2 x\, dx - \int (\sec^2 x - 1)\, dx$$

$$= \int \tan^2 x \sec^2 x\, dx - \int \sec^2 x\, dx + \int dx$$

In the first integral, we let

$$u = \tan x, \qquad du = \sec^2 x\, dx$$

and have

$$\int u^2\, du = \frac{1}{3} u^3 + C_1.$$

The remaining integrals are standard forms, so

$$\int \tan^4 x\, dx = \frac{1}{3} \tan^3 x - \tan x + x + C. \qquad \blacksquare$$

EXAMPLE 6   Evaluate

$$\int \sec^3 x\, dx.$$

Solution   We integrate by parts using

$$u = \sec x, \qquad dv = \sec^2 x\, dx, \qquad v = \tan x, \qquad du = \sec x \tan x\, dx.$$

Then

$$\int \sec^3 x \, dx = \sec x \tan x - \int (\tan x)(\sec x \tan x \, dx)$$

$$= \sec x \tan x - \int (\sec^2 x - 1) \sec x \, dx \qquad \tan^2 x = \sec^2 x - 1$$

$$= \sec x \tan x + \int \sec x \, dx - \int \sec^3 x \, dx.$$

Combining the two secant-cubed integrals gives

$$2 \int \sec^3 x \, dx = \sec x \tan x + \int \sec x \, dx$$

and

$$\int \sec^3 x \, dx = \frac{1}{2} \sec x \tan x + \frac{1}{2} \ln |\sec x + \tan x| + C. \qquad \blacksquare$$

EXAMPLE 7   Evaluate

$$\int \tan^4 x \sec^4 x \, dx.$$

Solution

$$\int (\tan^4 x)(\sec^4 x) \, dx = \int (\tan^4 x)(1 + \tan^2 x)(\sec^2 x) \, dx \qquad \sec^2 x = 1 + \tan^2 x$$

$$= \int (\tan^4 x + \tan^6 x)(\sec^2 x) \, dx$$

$$= \int (\tan^4 x)(\sec^2 x) \, dx + \int (\tan^6 x)(\sec^2 x) \, dx$$

$$= \int u^4 \, du + \int u^6 \, du = \frac{u^5}{5} + \frac{u^7}{7} + C \qquad \begin{matrix} u = \tan x, \\ du = \sec^2 x \, dx \end{matrix}$$

$$= \frac{\tan^5 x}{5} + \frac{\tan^7 x}{7} + C \qquad \blacksquare$$

Products of Sines and Cosines

The integrals

$$\int \sin mx \sin nx \, dx, \qquad \int \sin mx \cos nx \, dx, \qquad \text{and} \qquad \int \cos mx \cos nx \, dx$$

arise in many applications involving periodic functions. We can evaluate these integrals through integration by parts, but two such integrations are required in each case. It is simpler to use the identities

$$\sin mx \sin nx = \frac{1}{2} [\cos (m - n)x - \cos (m + n)x], \qquad (3)$$

$$\sin mx \cos nx = \frac{1}{2} [\sin (m - n)x + \sin (m + n)x], \qquad (4)$$

$$\cos mx \cos nx = \frac{1}{2} [\cos (m - n)x + \cos (m + n)x]. \qquad (5)$$

These identities come from the angle sum formulas for the sine and cosine functions (Section 1.3). They give functions whose antiderivatives are easily found.

EXAMPLE 8   Evaluate

$$\int \sin 3x \cos 5x \, dx.$$

Solution   From Equation (4) with $m = 3$ and $n = 5$, we get

$$\int \sin 3x \cos 5x \, dx = \frac{1}{2} \int [\sin(-2x) + \sin 8x] \, dx$$

$$= \frac{1}{2} \int (\sin 8x - \sin 2x) \, dx$$

$$= -\frac{\cos 8x}{16} + \frac{\cos 2x}{4} + C.$$

EXERCISES 8.3

Powers of Sines and Cosines
Evaluate the integrals in Exercises 1–22.

1. $\displaystyle\int \cos 2x \, dx$

2. $\displaystyle\int_0^\pi 3 \sin \frac{x}{3} \, dx$

3. $\displaystyle\int \cos^3 x \sin x \, dx$

4. $\displaystyle\int \sin^4 2x \cos 2x \, dx$

5. $\displaystyle\int \sin^3 x \, dx$

6. $\displaystyle\int \cos^3 4x \, dx$

7. $\displaystyle\int \sin^5 x \, dx$

8. $\displaystyle\int_0^\pi \sin^5 \frac{x}{2} \, dx$

9. $\displaystyle\int \cos^3 x \, dx$

10. $\displaystyle\int_0^{\pi/6} 3 \cos^5 3x \, dx$

11. $\displaystyle\int \sin^3 x \cos^3 x \, dx$

12. $\displaystyle\int \cos^3 2x \sin^5 2x \, dx$

13. $\displaystyle\int \cos^2 x \, dx$

14. $\displaystyle\int_0^{\pi/2} \sin^2 x \, dx$

15. $\displaystyle\int_0^{\pi/2} \sin^7 y \, dy$

16. $\displaystyle\int 7 \cos^7 t \, dt$

17. $\displaystyle\int_0^\pi 8 \sin^4 x \, dx$

18. $\displaystyle\int 8 \cos^4 2\pi x \, dx$

19. $\displaystyle\int 16 \sin^2 x \cos^2 x \, dx$

20. $\displaystyle\int_0^\pi 8 \sin^4 y \cos^2 y \, dy$

21. $\displaystyle\int 8 \cos^3 2\theta \sin 2\theta \, d\theta$

22. $\displaystyle\int_0^{\pi/2} \sin^2 2\theta \cos^3 2\theta \, d\theta$

Integrating Square Roots
Evaluate the integrals in Exercises 23–32.

23. $\displaystyle\int_0^{2\pi} \sqrt{\frac{1 - \cos x}{2}} \, dx$

24. $\displaystyle\int_0^\pi \sqrt{1 - \cos 2x} \, dx$

25. $\displaystyle\int_0^\pi \sqrt{1 - \sin^2 t} \, dt$

26. $\displaystyle\int_0^\pi \sqrt{1 - \cos^2 \theta} \, d\theta$

27. $\displaystyle\int_{\pi/3}^{\pi/2} \frac{\sin^2 x}{\sqrt{1 - \cos x}} \, dx$

28. $\displaystyle\int_0^{\pi/6} \sqrt{1 + \sin x} \, dx$

$$\left(\text{Hint: Multiply by } \sqrt{\frac{1 - \sin x}{1 - \sin x}}. \right)$$

29. $\displaystyle\int_{5\pi/6}^\pi \frac{\cos^4 x}{\sqrt{1 - \sin x}} \, dx$

30. $\displaystyle\int_{\pi/2}^{3\pi/4} \sqrt{1 - \sin 2x} \, dx$

31. $\displaystyle\int_0^{\pi/2} \theta \sqrt{1 - \cos 2\theta} \, d\theta$

32. $\displaystyle\int_{-\pi}^\pi (1 - \cos^2 t)^{3/2} \, dt$

Powers of Tangents and Secants
Evaluate the integrals in Exercises 33–50.

33. $\displaystyle\int \sec^2 x \tan x \, dx$

34. $\displaystyle\int \sec x \tan^2 x \, dx$

35. $\displaystyle\int \sec^3 x \tan x \, dx$

36. $\displaystyle\int \sec^3 x \tan^3 x \, dx$

37. $\displaystyle\int \sec^2 x \tan^2 x \, dx$

38. $\displaystyle\int \sec^4 x \tan^2 x \, dx$

39. $\displaystyle\int_{-\pi/3}^0 2 \sec^3 x \, dx$

40. $\displaystyle\int e^x \sec^3 e^x \, dx$

41. $\int \sec^4 \theta \, d\theta$

42. $\int 3 \sec^4 3x \, dx$

43. $\int_{\pi/4}^{\pi/2} \csc^4 \theta \, d\theta$

44. $\int \sec^6 x \, dx$

45. $\int 4 \tan^3 x \, dx$

46. $\int_{-\pi/4}^{\pi/4} 6 \tan^4 x \, dx$

47. $\int \tan^5 x \, dx$

48. $\int \cot^6 2x \, dx$

49. $\int_{\pi/6}^{\pi/3} \cot^3 x \, dx$

50. $\int 8 \cot^4 t \, dt$

Products of Sines and Cosines

Evaluate the integrals in Exercises 51–56.

51. $\int \sin 3x \cos 2x \, dx$

52. $\int \sin 2x \cos 3x \, dx$

53. $\int_{-\pi}^{\pi} \sin 3x \sin 3x \, dx$

54. $\int_{0}^{\pi/2} \sin x \cos x \, dx$

55. $\int \cos 3x \cos 4x \, dx$

56. $\int_{-\pi/2}^{\pi/2} \cos x \cos 7x \, dx$

Exercises 57–62 require the use of various trigonometric identities before you evaluate the integrals.

57. $\int \sin^2 \theta \cos 3\theta \, d\theta$

58. $\int \cos^2 2\theta \sin \theta \, d\theta$

59. $\int \cos^3 \theta \sin 2\theta \, d\theta$

60. $\int \sin^3 \theta \cos 2\theta \, d\theta$

61. $\int \sin \theta \cos \theta \cos 3\theta \, d\theta$

62. $\int \sin \theta \sin 2\theta \sin 3\theta \, d\theta$

Assorted Integrations

Use any method to evaluate the integrals in Exercises 63–68.

63. $\int \frac{\sec^3 x}{\tan x} \, dx$

64. $\int \frac{\sin^3 x}{\cos^4 x} \, dx$

65. $\int \frac{\tan^2 x}{\csc x} \, dx$

66. $\int \frac{\cot x}{\cos^2 x} \, dx$

67. $\int x \sin^2 x \, dx$

68. $\int x \cos^3 x \, dx$

Applications

69. Arc length Find the length of the curve

$$y = \ln (\sin x), \quad \frac{\pi}{6} \le x \le \frac{\pi}{2}$$

70. Center of gravity Find the center of gravity of the region bounded by the x-axis, the curve $y = \sec x$, and the lines $x = -\pi/4, x = \pi/4$.

71. Volume Find the volume generated by revolving one arch of the curve $y = \sin x$ about the x-axis.

72. Area Find the area between the x-axis and the curve $y = \sqrt{1 + \cos 4x}, 0 \le x \le \pi$.

73. Centroid Find the centroid of the region bounded by the graphs of $y = x + \cos x$ and $y = 0$ for $0 \le x \le 2\pi$.

74. Volume Find the volume of the solid formed by revolving the region bounded by the graphs of $y = \sin x + \sec x, y = 0, x = 0,$ and $x = \pi/3$ about the x-axis.

75. Volume Find the volume of the solid formed by revolving the region bounded by the graphs of $y = \tan^{-1}x, x = 0,$ and $y = \pi/4$ about the y-axis.

76. Average Value Find the average value of the function $f(x) = \dfrac{1}{1 - \sin \theta}$ on $[0, \pi/6]$.

8.4 Trigonometric Substitutions

Trigonometric substitutions occur when we replace the variable of integration by a trigonometric function. The most common substitutions are $x = a \tan \theta, x = a \sin \theta,$ and $x = a \sec \theta$. These substitutions are effective in transforming integrals involving $\sqrt{a^2 + x^2}, \sqrt{a^2 - x^2},$ and $\sqrt{x^2 - a^2}$ into integrals we can evaluate directly since they come from the reference right triangles in Figure 8.2.

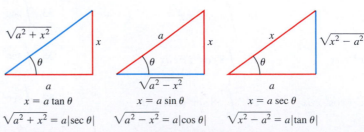

FIGURE 8.2 Reference triangles for the three basic substitutions identifying the sides labeled x and a for each substitution.

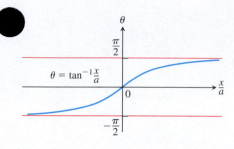

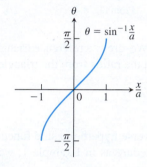

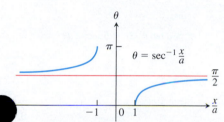

FIGURE 8.3 The arctangent, arcsine, and arcsecant of x/a, graphed as functions of x/a.

With $x = a \tan \theta$,

$$a^2 + x^2 = a^2 + a^2 \tan^2 \theta = a^2(1 + \tan^2 \theta) = a^2 \sec^2 \theta.$$

With $x = a \sin \theta$,

$$a^2 - x^2 = a^2 - a^2 \sin^2 \theta = a^2(1 - \sin^2 \theta) = a^2 \cos^2 \theta.$$

With $x = a \sec \theta$,

$$x^2 - a^2 = a^2 \sec^2 \theta - a^2 = a^2(\sec^2 \theta - 1) = a^2 \tan^2 \theta.$$

We want any substitution we use in an integration to be reversible so that we can change back to the original variable afterward. For example, if $x = a \tan \theta$, we want to be able to set $\theta = \tan^{-1}(x/a)$ after the integration takes place. If $x = a \sin \theta$, we want to be able to set $\theta = \sin^{-1}(x/a)$ when we're done, and similarly for $x = a \sec \theta$.

As we know from Section 1.6, the functions in these substitutions have inverses only for selected values of θ (Figure 8.3). For reversibility,

$$x = a \tan \theta \quad \text{requires} \quad \theta = \tan^{-1}\left(\frac{x}{a}\right) \quad \text{with} \quad -\frac{\pi}{2} < \theta < \frac{\pi}{2},$$

$$x = a \sin \theta \quad \text{requires} \quad \theta = \sin^{-1}\left(\frac{x}{a}\right) \quad \text{with} \quad -\frac{\pi}{2} \le \theta \le \frac{\pi}{2},$$

$$x = a \sec \theta \quad \text{requires} \quad \theta = \sec^{-1}\left(\frac{x}{a}\right) \quad \text{with} \quad \begin{cases} 0 \le \theta < \dfrac{\pi}{2} & \text{if } \dfrac{x}{a} \ge 1, \\[2mm] \dfrac{\pi}{2} < \theta \le \pi & \text{if } \dfrac{x}{a} \le -1. \end{cases}$$

To simplify calculations with the substitution $x = a \sec \theta$, we will restrict its use to integrals in which $x/a \ge 1$. This will place θ in $[0, \pi/2)$ and make $\tan \theta \ge 0$. We will then have $\sqrt{x^2 - a^2} = \sqrt{a^2 \tan^2 \theta} = |a \tan \theta| = a \tan \theta$, free of absolute values, provided $a > 0$.

Procedure for a Trigonometric Substitution

1. Write down the substitution for x, calculate the differential dx, and specify the selected values of θ for the substitution.

2. Substitute the trigonometric expression and the calculated differential into the integrand, and then simplify the results algebraically.

3. Integrate the trigonometric integral, keeping in mind the restrictions on the angle θ for reversibility.

4. Draw an appropriate reference triangle to reverse the substitution in the integration result and convert it back to the original variable x.

EXAMPLE 1　　Evaluate

$$\int \frac{dx}{\sqrt{4 + x^2}}.$$

Solution　　We set

$$x = 2 \tan \theta, \qquad dx = 2 \sec^2 \theta \, d\theta, \qquad -\frac{\pi}{2} < \theta < \frac{\pi}{2},$$

$$4 + x^2 = 4 + 4 \tan^2 \theta = 4(1 + \tan^2 \theta) = 4 \sec^2 \theta.$$

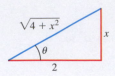

FIGURE 8.4 Reference triangle for $x = 2 \tan \theta$ (Example 1):

$$\tan \theta = \frac{x}{2}$$

and

$$\sec \theta = \frac{\sqrt{4 + x^2}}{2}.$$

Then

$$\int \frac{dx}{\sqrt{4 + x^2}} = \int \frac{2 \sec^2 \theta \, d\theta}{\sqrt{4 \sec^2 \theta}} = \int \frac{\sec^2 \theta \, d\theta}{|\sec \theta|} \qquad \sqrt{\sec^2 \theta} = |\sec \theta|$$

$$= \int \sec \theta \, d\theta \qquad \sec \theta > 0 \text{ for } -\frac{\pi}{2} < \theta < \frac{\pi}{2}$$

$$= \ln |\sec \theta + \tan \theta| + C$$

$$= \ln \left| \frac{\sqrt{4 + x^2}}{2} + \frac{x}{2} \right| + C. \qquad \text{From Fig. 8.4}$$

Notice how we expressed $\ln |\sec \theta + \tan \theta|$ in terms of x: We drew a reference triangle for the original substitution $x = 2 \tan \theta$ (Figure 8.4) and read the ratios from the triangle. ∎

EXAMPLE 2 Here we find an expression for the inverse hyperbolic sine function in terms of the natural logarithm. Following the same procedure as in Example 1, we find that

$$\int \frac{dx}{\sqrt{a^2 + x^2}} = \int \sec \theta \, d\theta \qquad x = a \tan \theta, \, dx = a \sec^2 \theta \, d\theta$$

$$= \ln |\sec \theta + \tan \theta| + C$$

$$= \ln \left| \frac{\sqrt{a^2 + x^2}}{a} + \frac{x}{a} \right| + C \qquad \text{Fig. 8.2}$$

From Table 7.9, $\sinh^{-1}(x/a)$ is also an antiderivative of $1/\sqrt{a^2 + x^2}$, so the two antiderivatives differ by a constant, giving

$$\sinh^{-1} \frac{x}{a} = \ln \left| \frac{\sqrt{a^2 + x^2}}{a} + \frac{x}{a} \right| + C.$$

Setting $x = 0$ in this last equation, we find $0 = \ln |1| + C$, so $C = 0$. Since $\sqrt{a^2 + x^2} > |x|$, we conclude that

$$\sinh^{-1} \frac{x}{a} = \ln \left(\frac{\sqrt{a^2 + x^2}}{a} + \frac{x}{a} \right)$$

(See also Exercise 76 in Section 7.3.) ∎

EXAMPLE 3 Evaluate

$$\int \frac{x^2 \, dx}{\sqrt{9 - x^2}}.$$

Solution We set

$$x = 3 \sin \theta, \qquad dx = 3 \cos \theta \, d\theta, \qquad -\frac{\pi}{2} < \theta < \frac{\pi}{2}$$

$$9 - x^2 = 9 - 9 \sin^2 \theta = 9(1 - \sin^2 \theta) = 9 \cos^2 \theta.$$

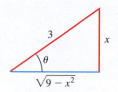

FIGURE 8.5 Reference triangle for $x = 3 \sin \theta$ (Example 3):

$$\sin \theta = \frac{x}{3}$$

and

$$\cos \theta = \frac{\sqrt{9 - x^2}}{3}.$$

Then

$$\int \frac{x^2 \, dx}{\sqrt{9 - x^2}} = \int \frac{9 \sin^2 \theta \cdot 3 \cos \theta \, d\theta}{|3 \cos \theta|}$$

$$= 9 \int \sin^2 \theta \, d\theta \qquad \cos \theta > 0 \text{ for } -\frac{\pi}{2} < \theta < \frac{\pi}{2}$$

$$= 9 \int \frac{1 - \cos 2\theta}{2} \, d\theta$$

$$= \frac{9}{2} \left(\theta - \frac{\sin 2\theta}{2} \right) + C$$

$$= \frac{9}{2} (\theta - \sin \theta \cos \theta) + C \qquad \sin 2\theta = 2 \sin \theta \cos \theta$$

$$= \frac{9}{2} \left(\sin^{-1} \frac{x}{3} - \frac{x}{3} \cdot \frac{\sqrt{9 - x^2}}{3} \right) + C \qquad \text{From Fig. 8.5}$$

$$= \frac{9}{2} \sin^{-1} \frac{x}{3} - \frac{x}{2} \sqrt{9 - x^2} + C.$$

EXAMPLE 4   Evaluate

$$\int \frac{dx}{\sqrt{25x^2 - 4}}, \qquad x > \frac{2}{5}.$$

Solution   We first rewrite the radical as

$$\sqrt{25x^2 - 4} = \sqrt{25 \left(x^2 - \frac{4}{25} \right)}$$

$$= 5 \sqrt{x^2 - \left(\frac{2}{5} \right)^2} \qquad \sqrt{x^2 - a^2} \text{ with } a = \frac{2}{5}$$

to put the radicand in the form $x^2 - a^2$. We then substitute

$$x = \frac{2}{5} \sec \theta, \qquad dx = \frac{2}{5} \sec \theta \tan \theta \, d\theta, \qquad 0 < \theta < \frac{\pi}{2}.$$

We then get

$$x^2 - \left(\frac{2}{5} \right)^2 = \frac{4}{25} \sec^2 \theta - \frac{4}{25} = \frac{4}{25} (\sec^2 \theta - 1) = \frac{4}{25} \tan^2 \theta$$

and

$$\sqrt{x^2 - \left(\frac{2}{5} \right)^2} = \frac{2}{5} |\tan \theta| = \frac{2}{5} \tan \theta. \qquad \begin{array}{l} \tan \theta > 0 \text{ for} \\ 0 < \theta < \pi/2 \end{array}$$

With these substitutions, we have

$$\int \frac{dx}{\sqrt{25x^2 - 4}} = \int \frac{dx}{5\sqrt{x^2 - (4/25)}} = \int \frac{(2/5) \sec \theta \tan \theta \, d\theta}{5 \cdot (2/5) \tan \theta}$$

$$= \frac{1}{5} \int \sec \theta \, d\theta = \frac{1}{5} \ln |\sec \theta + \tan \theta| + C$$

$$= \frac{1}{5} \ln \left| \frac{5x}{2} + \frac{\sqrt{25x^2 - 4}}{2} \right| + C. \qquad \text{From Fig. 8.6}$$

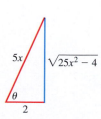

FIGURE 8.6   If $x = (2/5)\sec \theta$, $0 < \theta < \pi/2$, then $\theta = \sec^{-1}(5x/2)$, and we can read the values of the other trigonometric functions of θ from this right triangle (Example 4).

EXERCISES 8.4

Using Trigonometric Substitutions

Evaluate the integrals in Exercises 1–14.

1. $\displaystyle\int \frac{dx}{\sqrt{9 + x^2}}$

2. $\displaystyle\int \frac{3\,dx}{\sqrt{1 + 9x^2}}$

3. $\displaystyle\int_{-2}^{2} \frac{dx}{4 + x^2}$

4. $\displaystyle\int_{0}^{2} \frac{dx}{8 + 2x^2}$

5. $\displaystyle\int_{0}^{3/2} \frac{dx}{\sqrt{9 - x^2}}$

6. $\displaystyle\int_{0}^{1/2\sqrt{2}} \frac{2\,dx}{\sqrt{1 - 4x^2}}$

7. $\displaystyle\int \sqrt{25 - t^2}\,dt$

8. $\displaystyle\int \sqrt{1 - 9t^2}\,dt$

9. $\displaystyle\int \frac{dx}{\sqrt{4x^2 - 49}}, \quad x > \frac{7}{2}$

10. $\displaystyle\int \frac{5\,dx}{\sqrt{25x^2 - 9}}, \quad x > \frac{3}{5}$

11. $\displaystyle\int \frac{\sqrt{y^2 - 49}}{y}\,dy, \quad y > 7$

12. $\displaystyle\int \frac{\sqrt{y^2 - 25}}{y^3}\,dy, \quad y > 5$

13. $\displaystyle\int \frac{dx}{x^2\sqrt{x^2 - 1}}, \quad x > 1$

14. $\displaystyle\int \frac{2\,dx}{x^3\sqrt{x^2 - 1}}, \quad x > 1$

Assorted Integrations

Use any method to evaluate the integrals in Exercises 15–34. Most will require trigonometric substitutions, but some can be evaluated by other methods.

15. $\displaystyle\int \frac{x}{\sqrt{9 - x^2}}\,dx$

16. $\displaystyle\int \frac{x^2}{4 + x^2}\,dx$

17. $\displaystyle\int \frac{x^3\,dx}{\sqrt{x^2 + 4}}$

18. $\displaystyle\int \frac{dx}{x^2\sqrt{x^2 + 1}}$

19. $\displaystyle\int \frac{8\,dw}{w^2\sqrt{4 - w^2}}$

20. $\displaystyle\int \frac{\sqrt{9 - w^2}}{w^2}\,dw$

21. $\displaystyle\int \sqrt{\frac{x + 1}{1 - x}}\,dx$

22. $\displaystyle\int x\sqrt{x^2 - 4}\,dx$

23. $\displaystyle\int_{0}^{\sqrt{3}/2} \frac{4x^2\,dx}{(1 - x^2)^{3/2}}$

24. $\displaystyle\int_{0}^{1} \frac{dx}{(4 - x^2)^{3/2}}$

25. $\displaystyle\int \frac{dx}{(x^2 - 1)^{3/2}}, \quad x > 1$

26. $\displaystyle\int \frac{x^2\,dx}{(x^2 - 1)^{5/2}}, \quad x > 1$

27. $\displaystyle\int \frac{(1 - x^2)^{3/2}}{x^6}\,dx$

28. $\displaystyle\int \frac{(1 - x^2)^{1/2}}{x^4}\,dx$

29. $\displaystyle\int \frac{8\,dx}{(4x^2 + 1)^2}$

30. $\displaystyle\int \frac{6\,dt}{(9t^2 + 1)^2}$

31. $\displaystyle\int \frac{x^3\,dx}{x^2 - 1}$

32. $\displaystyle\int \frac{x\,dx}{25 + 4x^2}$

33. $\displaystyle\int \frac{v^2\,dv}{(1 - v^2)^{5/2}}$

34. $\displaystyle\int \frac{(1 - r^2)^{5/2}}{r^8}\,dr$

In Exercises 35–48, use an appropriate substitution and then a trigonometric substitution to evaluate the integrals.

35. $\displaystyle\int_{0}^{\ln 4} \frac{e^t\,dt}{\sqrt{e^{2t} + 9}}$

36. $\displaystyle\int_{\ln(3/4)}^{\ln(4/3)} \frac{e^t\,dt}{(1 + e^{2t})^{3/2}}$

37. $\displaystyle\int_{1/12}^{1/4} \frac{2\,dt}{\sqrt{t} + 4t\sqrt{t}}$

38. $\displaystyle\int_{1}^{e} \frac{dy}{y\sqrt{1 + (\ln y)^2}}$

39. $\displaystyle\int \frac{dx}{x\sqrt{x^2 - 1}}$

40. $\displaystyle\int \frac{dx}{1 + x^2}$

41. $\displaystyle\int \frac{x\,dx}{\sqrt{x^2 - 1}}$

42. $\displaystyle\int \frac{dx}{\sqrt{1 - x^2}}$

43. $\displaystyle\int \frac{x\,dx}{\sqrt{1 + x^4}}$

44. $\displaystyle\int \frac{\sqrt{1 - (\ln x)^2}}{x \ln x}\,dx$

45. $\displaystyle\int \sqrt{\frac{4 - x}{x}}\,dx$

(*Hint:* Let $x = u^2$.)

46. $\displaystyle\int \sqrt{\frac{x}{1 - x^3}}\,dx$

(*Hint:* Let $u = x^{3/2}$.)

47. $\displaystyle\int \sqrt{x}\,\sqrt{1 - x}\,dx$

48. $\displaystyle\int \frac{\sqrt{x - 2}}{\sqrt{x - 1}}\,dx$

Complete the Square Before Using Trigonometric Substitutions

For Exercises 49–52, complete the square before using an appropriate trigonometric substitution.

49. $\displaystyle\int \sqrt{8 - 2x - x^2}\,dx$

50. $\displaystyle\int \frac{1}{\sqrt{x^2 - 2x + 5}}\,dx$

51. $\displaystyle\int \frac{\sqrt{x^2 + 4x + 3}}{x + 2}\,dx$

52. $\displaystyle\int \frac{\sqrt{x^2 + 2x + 2}}{x^2 + 2x + 1}\,dx$

Initial Value Problems

Solve the initial value problems in Exercises 53–56 for y as a function of x.

53. $x\dfrac{dy}{dx} = \sqrt{x^2 - 4}, \quad x \geq 2, \quad y(2) = 0$

54. $\sqrt{x^2 - 9}\dfrac{dy}{dx} = 1, \quad x > 3, \quad y(5) = \ln 3$

55. $(x^2 + 4)\dfrac{dy}{dx} = 3, \quad y(2) = 0$

56. $(x^2 + 1)^2\dfrac{dy}{dx} = \sqrt{x^2 + 1}, \quad y(0) = 1$

Applications and Examples

57. Area Find the area of the region in the first quadrant that is enclosed by the coordinate axes and the curve $y = \sqrt{9 - x^2}/3$.

58. Area Find the area enclosed by the ellipse

$$\frac{x^2}{a^2} + \frac{y^2}{b^2} = 1.$$

59. Consider the region bounded by the graphs of $y = \sin^{-1} x$, $y = 0$, and $x = 1/2$.

    a. Find the area of the region.

    b. Find the centroid of the region.

60. Consider the region bounded by the graphs of $y = \sqrt{x} \tan^{-1} x$ and $y = 0$ for $0 \le x \le 1$. Find the volume of the solid formed by revolving this region about the x-axis (see accompanying figure).

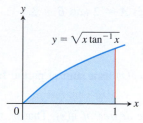

61. Evaluate $\int x^3 \sqrt{1 - x^2}\, dx$ using

    a. integration by parts.

    b. a u-substitution.

    c. a trigonometric substitution.

62. Path of a water skier Suppose that a boat is positioned at the origin with a water skier tethered to the boat at the point $(30, 0)$ on a rope 30 ft long. As the boat travels along the positive y-axis, the skier is pulled behind the boat along an unknown path $y = f(x)$, as shown in the accompanying figure.

    a. Show that $f'(x) = \dfrac{-\sqrt{900 - x^2}}{x}$.

    (*Hint*: Assume that the skier is always pointed directly at the boat and the rope is on a line tangent to the path $y = f(x)$.)

    b. Solve the equation in part (a) for $f(x)$, using $f(30) = 0$.

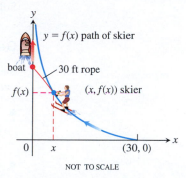

NOT TO SCALE

63. Find the average value of $f(x) = \dfrac{\sqrt{x + 1}}{\sqrt{x}}$ on the interval $[1, 3]$.

64. Find the length of the curve $y = 1 - e^{-x}$, $0 \le x \le 1$.

8.5 Integration of Rational Functions by Partial Fractions

This section shows how to express a rational function (a quotient of polynomials) as a sum of simpler fractions, called *partial fractions*, which are easily integrated. For instance, the rational function $(5x - 3)/(x^2 - 2x - 3)$ can be rewritten as

$$\frac{5x - 3}{x^2 - 2x - 3} = \frac{2}{x + 1} + \frac{3}{x - 3}.$$

You can verify this equation algebraically by placing the fractions on the right side over a common denominator $(x + 1)(x - 3)$. The skill acquired in writing rational functions as such a sum is useful in other settings as well (for instance, when using certain transform methods to solve differential equations). To integrate the rational function $(5x - 3)/(x^2 - 2x - 3)$ on the left side of our previous expression, we simply sum the integrals of the fractions on the right side:

$$\int \frac{5x - 3}{(x + 1)(x - 3)}\, dx = \int \frac{2}{x + 1}\, dx + \int \frac{3}{x - 3}\, dx$$

$$= 2 \ln|x + 1| + 3 \ln|x - 3| + C.$$

The method for rewriting rational functions as a sum of simpler fractions is called **the method of partial fractions**. In the case of the preceding example, it consists of finding constants A and B such that

$$\frac{5x - 3}{x^2 - 2x - 3} = \frac{A}{x + 1} + \frac{B}{x - 3}. \tag{1}$$

(Pretend for a moment that we do not know that $A = 2$ and $B = 3$ will work.) We call the fractions $A/(x + 1)$ and $B/(x - 3)$ **partial fractions** because their denominators are only part of the original denominator $x^2 - 2x - 3$. We call A and B **undetermined coefficients** until suitable values for them have been found.

To find A and B, we first clear Equation (1) of fractions and regroup in powers of x, obtaining

$$5x - 3 = A(x - 3) + B(x + 1) = (A + B)x - 3A + B.$$

This will be an identity in x if and only if the coefficients of like powers of x on the two sides are equal:

$$A + B = 5, \qquad -3A + B = -3.$$

Solving these equations simultaneously gives $A = 2$ and $B = 3$.

General Description of the Method

Success in writing a rational function $f(x)/g(x)$ as a sum of partial fractions depends on two things:

- *The degree of $f(x)$ must be less than the degree of $g(x)$.* That is, the fraction must be proper. If it isn't, divide $f(x)$ by $g(x)$ and work with the remainder term. Example 3 of this section illustrates such a case.

- We must know the factors of $g(x)$. In theory, any polynomial with real coefficients can be written as a product of real linear factors and real quadratic factors. In practice, the factors may be hard to find.

Here is how we find the partial fractions of a proper fraction $f(x)/g(x)$ when the factors of g are known. A quadratic polynomial (or factor) is **irreducible** if it cannot be written as the product of two linear factors with real coefficients. That is, the polynomial has no real roots.

Method of Partial Fractions When $f(x)/g(x)$ Is Proper

1. Let $x - r$ be a linear factor of $g(x)$. Suppose that $(x - r)^m$ is the highest power of $x - r$ that divides $g(x)$. Then, to this factor, assign the sum of the m partial fractions:

$$\frac{A_1}{(x - r)} + \frac{A_2}{(x - r)^2} + \cdots + \frac{A_m}{(x - r)^m}.$$

Do this for each distinct linear factor of $g(x)$.

2. Let $x^2 + px + q$ be an irreducible quadratic factor of $g(x)$ so that $x^2 + px + q$ has no real roots. Suppose that $(x^2 + px + q)^n$ is the highest power of this factor that divides $g(x)$. Then, to this factor, assign the sum of the n partial fractions:

$$\frac{B_1 x + C_1}{(x^2 + px + q)} + \frac{B_2 x + C_2}{(x^2 + px + q)^2} + \cdots + \frac{B_n x + C_n}{(x^2 + px + q)^n}.$$

Do this for each distinct quadratic factor of $g(x)$.

3. Set the original fraction $f(x)/g(x)$ equal to the sum of all these partial fractions. Clear the resulting equation of fractions and arrange the terms in decreasing powers of x.

4. Equate the coefficients of corresponding powers of x and solve the resulting equations for the undetermined coefficients.

EXAMPLE 1   Use partial fractions to evaluate

$$\int \frac{x^2 + 4x + 1}{(x - 1)(x + 1)(x + 3)}\, dx.$$

Solution   Note that each of the factors $(x - 1)$, $(x + 1)$, and $(x + 3)$ is raised only to the first power. Therefore, the partial fraction decomposition has the form

$$\frac{x^2 + 4x + 1}{(x - 1)(x + 1)(x + 3)} = \frac{A}{x - 1} + \frac{B}{x + 1} + \frac{C}{x + 3}.$$

To find the values of the undetermined coefficients A, B, and C, we clear fractions and get

$$\begin{aligned}
x^2 + 4x + 1 &= A(x + 1)(x + 3) + B(x - 1)(x + 3) + C(x - 1)(x + 1) \\
&= A(x^2 + 4x + 3) + B(x^2 + 2x - 3) + C(x^2 - 1) \\
&= (A + B + C)x^2 + (4A + 2B)x + (3A - 3B - C).
\end{aligned}$$

The polynomials on both sides of the above equation are identical, so we equate coefficients of like powers of x, obtaining

$$\begin{aligned}
\text{Coefficient of } x^2: &\quad A + B + C = 1 \\
\text{Coefficient of } x^1: &\quad 4A + 2B = 4 \\
\text{Coefficient of } x^0: &\quad 3A - 3B - C = 1
\end{aligned}$$

There are several ways of solving such a system of linear equations for the unknowns A, B, and C, including elimination of variables or the use of a calculator or computer. The solution is $A = 3/4$, $B = 1/2$, and $C = -1/4$. Hence we have

$$\int \frac{x^2 + 4x + 1}{(x - 1)(x + 1)(x + 3)} \, dx = \int \left[\frac{3}{4} \frac{1}{x - 1} + \frac{1}{2} \frac{1}{x + 1} - \frac{1}{4} \frac{1}{x + 3} \right] dx$$

$$= \frac{3}{4} \ln |x - 1| + \frac{1}{2} \ln |x + 1| - \frac{1}{4} \ln |x + 3| + K,$$

where K is the arbitrary constant of integration (we call it K here to avoid confusion with the undetermined coefficient we labeled as C). ∎

EXAMPLE 2   Use partial fractions to evaluate

$$\int \frac{6x + 7}{(x + 2)^2} \, dx.$$

Solution   First we express the integrand as a sum of partial fractions with undetermined coefficients.

$$\frac{6x + 7}{(x + 2)^2} = \frac{A}{x + 2} + \frac{B}{(x + 2)^2} \qquad \text{Two terms because } (x + 2) \text{ is squared}$$

$$6x + 7 = A(x + 2) + B \qquad \text{Multiply both sides by } (x + 2)^2.$$

$$= Ax + (2A + B)$$

Equating coefficients of corresponding powers of x gives

$$A = 6 \quad \text{and} \quad 2A + B = 12 + B = 7, \quad \text{or} \quad A = 6 \quad \text{and} \quad B = -5.$$

Therefore,

$$\int \frac{6x + 7}{(x + 2)^2} \, dx = \int \left(\frac{6}{x + 2} - \frac{5}{(x + 2)^2} \right) dx$$

$$= 6 \int \frac{dx}{x + 2} - 5 \int (x + 2)^{-2} \, dx$$

$$= 6 \ln |x + 2| + 5(x + 2)^{-1} + C. \qquad ∎$$

The next example shows how to handle the case when $f(x)/g(x)$ is an improper fraction. It is a case where the degree of f is larger than the degree of g.

EXAMPLE 3   Use partial fractions to evaluate

$$\int \frac{2x^3 - 4x^2 - x - 3}{x^2 - 2x - 3}\, dx.$$

Solution   First we divide the denominator into the numerator to get a polynomial plus a proper fraction.

$$
\begin{array}{r}
2x \\
x^2 - 2x - 3 \overline{\smash{)}\ 2x^3 - 4x^2 - x - 3} \\
\underline{2x^3 - 4x^2 - 6x - 3} \\
5x - 3
\end{array}
$$

Then we write the improper fraction as a polynomial plus a proper fraction.

$$\frac{2x^3 - 4x^2 - x - 3}{x^2 - 2x - 3} = 2x + \frac{5x - 3}{x^2 - 2x - 3}$$

We found the partial fraction decomposition of the fraction on the right in the opening example, so

$$\int \frac{2x^3 - 4x^2 - x - 3}{x^2 - 2x - 3}\, dx = \int 2x\, dx + \int \frac{5x - 3}{x^2 - 2x - 3}\, dx$$

$$= \int 2x\, dx + \int \frac{2}{x + 1}\, dx + \int \frac{3}{x - 3}\, dx$$

$$= x^2 + 2 \ln |x + 1| + 3 \ln |x - 3| + C. \qquad \blacksquare$$

EXAMPLE 4   Use partial fractions to evaluate

$$\int \frac{-2x + 4}{(x^2 + 1)(x - 1)^2}\, dx.$$

Solution   The denominator has an irreducible quadratic factor $x^2 + 1$ as well as a repeated linear factor $(x - 1)^2$, so we write

$$\frac{-2x + 4}{(x^2 + 1)(x - 1)^2} = \frac{Ax + B}{x^2 + 1} + \frac{C}{x - 1} + \frac{D}{(x - 1)^2}. \qquad (2)$$

Clearing the equation of fractions gives

$$-2x + 4 = (Ax + B)(x - 1)^2 + C(x - 1)(x^2 + 1) + D(x^2 + 1)$$

$$= (A + C)x^3 + (-2A + B - C + D)x^2$$

$$+ (A - 2B + C)x + (B - C + D).$$

Equating coefficients of like terms gives

$$
\begin{array}{ll}
\text{Coefficients of } x^3: & 0 = A + C \\
\text{Coefficients of } x^2: & 0 = -2A + B - C + D \\
\text{Coefficients of } x^1: & -2 = A - 2B + C \\
\text{Coefficients of } x^0: & 4 = B - C + D
\end{array}
$$

We solve these equations simultaneously to find the values of A, B, C, and D:

$$-4 = -2A, \quad A = 2 \qquad \text{\color{blue}{Subtract fourth equation from second.}}$$

$$C = -A = -2 \qquad \text{\color{blue}{From the first equation}}$$

$$B = (A + C + 2)/2 = 1 \qquad \text{\color{blue}{From the third equation and } C = -A}$$

$$D = 4 - B + C = 1. \qquad \text{\color{blue}{From the fourth equation}}$$

We substitute these values into Equation (2), obtaining

$$\frac{-2x + 4}{(x^2 + 1)(x - 1)^2} = \frac{2x + 1}{x^2 + 1} - \frac{2}{x - 1} + \frac{1}{(x - 1)^2}.$$

Finally, using the expansion above we can integrate:

$$\int \frac{-2x + 4}{(x^2 + 1)(x - 1)^2} dx = \int \left(\frac{2x + 1}{x^2 + 1} - \frac{2}{x - 1} + \frac{1}{(x - 1)^2} \right) dx$$

$$= \int \left(\frac{2x}{x^2 + 1} + \frac{1}{x^2 + 1} - \frac{2}{x - 1} + \frac{1}{(x - 1)^2} \right) dx$$

$$= \ln (x^2 + 1) + \tan^{-1} x - 2 \ln |x - 1| - \frac{1}{x - 1} + C. \quad \blacksquare$$

EXAMPLE 5 Use partial fractions to evaluate

$$\int \frac{dx}{x(x^2 + 1)^2}.$$

Solution The form of the partial fraction decomposition is

$$\frac{1}{x(x^2 + 1)^2} = \frac{A}{x} + \frac{Bx + C}{x^2 + 1} + \frac{Dx + E}{(x^2 + 1)^2}.$$

Multiplying by $x(x^2 + 1)^2$, we have

$$1 = A(x^2 + 1)^2 + (Bx + C)x(x^2 + 1) + (Dx + E)x$$
$$= A(x^4 + 2x^2 + 1) + B(x^4 + x^2) + C(x^3 + x) + Dx^2 + Ex$$
$$= (A + B)x^4 + Cx^3 + (2A + B + D)x^2 + (C + E)x + A.$$

If we equate coefficients, we get the system

$$A + B = 0, \quad C = 0, \quad 2A + B + D = 0, \quad C + E = 0, \quad A = 1.$$

Solving this system gives $A = 1, B = -1, C = 0, D = -1,$ and $E = 0.$ Thus,

$$\int \frac{dx}{x(x^2 + 1)^2} = \int \left[\frac{1}{x} + \frac{-x}{x^2 + 1} + \frac{-x}{(x^2 + 1)^2} \right] dx$$

$$= \int \frac{dx}{x} - \int \frac{x \, dx}{x^2 + 1} - \int \frac{x \, dx}{(x^2 + 1)^2}$$

$$= \int \frac{dx}{x} - \frac{1}{2} \int \frac{du}{u} - \frac{1}{2} \int \frac{du}{u^2} \qquad \begin{array}{l} u = x^2 + 1, \\ du = 2x \, dx \end{array}$$

$$= \ln |x| - \frac{1}{2} \ln |u| + \frac{1}{2u} + K$$

$$= \ln |x| - \frac{1}{2} \ln (x^2 + 1) + \frac{1}{2(x^2 + 1)} + K$$

$$= \ln \frac{|x|}{\sqrt{x^2 + 1}} + \frac{1}{2(x^2 + 1)} + K. \quad \blacksquare$$

HISTORICAL BIOGRAPHY
Oliver Heaviside
(1850–1925)
www.goo.gl/5rnavZ

When the degree of the polynomial $f(x)$ is less than the degree of $g(x)$ and

$$g(x) = (x - r_1)(x - r_2) \cdots (x - r_n)$$

is a product of n distinct linear factors, each raised to the first power, there is a quick way to expand $f(x)/g(x)$ by partial fractions.

EXAMPLE 6 Find A, B, and C in the partial fraction expansion

$$\frac{x^2 + 1}{(x - 1)(x - 2)(x - 3)} = \frac{A}{x - 1} + \frac{B}{x - 2} + \frac{C}{x - 3}. \tag{3}$$

Solution If we multiply both sides of Equation (3) by $(x - 1)$ to get

$$\frac{x^2 + 1}{(x - 2)(x - 3)} = A + \frac{B(x - 1)}{x - 2} + \frac{C(x - 1)}{x - 3}$$

and set $x = 1$, the resulting equation gives the value of A:

$$\frac{(1)^2 + 1}{(1 - 2)(1 - 3)} = A + 0 + 0,$$

$$A = 1.$$

In exactly the same way, we can multiply both sides by $(x - 2)$ and then substitute in $x = 2$. This gives

$$\frac{(2)^2 + 1}{(2 - 1)(2 - 3)} = B.$$

So $B = -5$. Finally, we multiply both sides by $(x - 3)$ and then substitute in $x = 3$, which yields

$$\frac{(3)^2 + 1}{(3 - 1)(3 - 2)} = C,$$

and $C = 5$. ∎

Other Ways to Determine the Coefficients

Another way to determine the constants that appear in partial fractions is to differentiate, as in the next example. Still another is to assign selected numerical values to x.

EXAMPLE 7 Find A, B, and C in the equation

$$\frac{x - 1}{(x + 1)^3} = \frac{A}{x + 1} + \frac{B}{(x + 1)^2} + \frac{C}{(x + 1)^3}$$

by clearing fractions, differentiating the result, and substituting $x = -1$.

Solution We first clear fractions:

$$x - 1 = A(x + 1)^2 + B(x + 1) + C.$$

Substituting $x = -1$ shows $C = -2$. We then differentiate both sides with respect to x, obtaining

$$1 = 2A(x + 1) + B.$$

Substituting $x = -1$ shows $B = 1$. We differentiate again to get $0 = 2A$, which shows $A = 0$. Hence,

$$\frac{x - 1}{(x + 1)^3} = \frac{1}{(x + 1)^2} - \frac{2}{(x + 1)^3}.$$

In some problems, assigning small values to x, such as $x = 0, \pm 1, \pm 2$, to get equations in A, B, and C provides a fast alternative to other methods.

EXAMPLE 8   Find A, B, and C in the expression

$$\frac{x^2 + 1}{(x - 1)(x - 2)(x - 3)} = \frac{A}{x - 1} + \frac{B}{x - 2} + \frac{C}{x - 3}$$

by assigning numerical values to x.

Solution   Clear fractions to get

$$x^2 + 1 = A(x - 2)(x - 3) + B(x - 1)(x - 3) + C(x - 1)(x - 2).$$

Then let $x = 1, 2, 3$ successively to find A, B, and C:

$$x = 1: \quad (1)^2 + 1 = A(-1)(-2) + B(0) + C(0)$$
$$2 = 2A$$
$$A = 1$$

$$x = 2: \quad (2)^2 + 1 = A(0) + B(1)(-1) + C(0)$$
$$5 = -B$$
$$B = -5$$

$$x = 3: \quad (3)^2 + 1 = A(0) + B(0) + C(2)(1)$$
$$10 = 2C$$
$$C = 5.$$

Conclusion:

$$\frac{x^2 + 1}{(x - 1)(x - 2)(x - 3)} = \frac{1}{x - 1} - \frac{5}{x - 2} + \frac{5}{x - 3}.$$

EXERCISES 8.5

Expanding Quotients into Partial Fractions
Expand the quotients in Exercises 1–8 by partial fractions.

1. $\dfrac{5x - 13}{(x - 3)(x - 2)}$

2. $\dfrac{5x - 7}{x^2 - 3x + 2}$

3. $\dfrac{x + 4}{(x + 1)^2}$

4. $\dfrac{2x + 2}{x^2 - 2x + 1}$

5. $\dfrac{z + 1}{z^2(z - 1)}$

6. $\dfrac{z}{z^3 - z^2 - 6z}$

7. $\dfrac{t^2 + 8}{t^2 - 5t + 6}$

8. $\dfrac{t^4 + 9}{t^4 + 9t^2}$

Nonrepeated Linear Factors
In Exercises 9–16, express the integrand as a sum of partial fractions and evaluate the integrals.

9. $\displaystyle\int \frac{dx}{1 - x^2}$

10. $\displaystyle\int \frac{dx}{x^2 + 2x}$

11. $\displaystyle\int \frac{x + 4}{x^2 + 5x - 6}\,dx$

12. $\displaystyle\int \frac{2x + 1}{x^2 - 7x + 12}\,dx$

13. $\displaystyle\int_4^8 \frac{y\,dy}{y^2 - 2y - 3}$

14. $\displaystyle\int_{1/2}^1 \frac{y + 4}{y^2 + y}\,dy$

15. $\displaystyle\int \frac{dt}{t^3 + t^2 - 2t}$

16. $\displaystyle\int \frac{x + 3}{2x^3 - 8x}\,dx$

Repeated Linear Factors
In Exercises 17–20, express the integrand as a sum of partial fractions and evaluate the integrals.

17. $\displaystyle\int_0^1 \frac{x^3\,dx}{x^2 + 2x + 1}$

18. $\displaystyle\int_{-1}^0 \frac{x^3\,dx}{x^2 - 2x + 1}$

19. $\displaystyle\int \frac{dx}{(x^2 - 1)^2}$

20. $\displaystyle\int \frac{x^2\,dx}{(x - 1)(x^2 + 2x + 1)}$

Irreducible Quadratic Factors
In Exercises 21–32, express the integrand as a sum of partial fractions and evaluate the integrals.

21. $\displaystyle\int_0^1 \frac{dx}{(x + 1)(x^2 + 1)}$

22. $\displaystyle\int_1^{\sqrt{3}} \frac{3t^2 + t + 4}{t^3 + t}\,dt$

23. $\displaystyle\int \frac{y^2 + 2y + 1}{(y^2 + 1)^2}\,dy$

24. $\displaystyle\int \frac{8x^2 + 8x + 2}{(4x^2 + 1)^2}\,dx$

25. $\displaystyle\int \frac{2s + 2}{(s^2 + 1)(s - 1)^3}\,ds$

26. $\displaystyle\int \frac{s^4 + 81}{s(s^2 + 9)^2}\,ds$

27. $\displaystyle\int \frac{x^2 - x + 2}{x^3 - 1}\,dx$

28. $\displaystyle\int \frac{1}{x^4 + x}\,dx$

29. $\displaystyle\int \frac{x^2}{x^4 - 1}\,dx$

30. $\displaystyle\int \frac{x^2 + x}{x^4 - 3x^2 - 4}\,dx$

31. $\displaystyle\int \frac{2\theta^3 + 5\theta^2 + 8\theta + 4}{(\theta^2 + 2\theta + 2)^2}\,d\theta$

32. $\displaystyle\int \frac{\theta^4 - 4\theta^3 + 2\theta^2 - 3\theta + 1}{(\theta^2 + 1)^3}\,d\theta$

Improper Fractions

In Exercises 33–38, perform long division on the integrand, write the proper fraction as a sum of partial fractions, and then evaluate the integral.

33. $\displaystyle\int \frac{2x^3 - 2x^2 + 1}{x^2 - x}\, dx$

34. $\displaystyle\int \frac{x^4}{x^2 - 1}\, dx$

35. $\displaystyle\int \frac{9x^3 - 3x + 1}{x^3 - x^2}\, dx$

36. $\displaystyle\int \frac{16x^3}{4x^2 - 4x + 1}\, dx$

37. $\displaystyle\int \frac{y^4 + y^2 - 1}{y^3 + y}\, dy$

38. $\displaystyle\int \frac{2y^4}{y^3 - y^2 + y - 1}\, dy$

Evaluating Integrals

Evaluate the integrals in Exercises 39–54.

39. $\displaystyle\int \frac{e^t\, dt}{e^{2t} + 3e^t + 2}$

40. $\displaystyle\int \frac{e^{4t} + 2e^{2t} - e^t}{e^{2t} + 1}\, dt$

41. $\displaystyle\int \frac{\cos y\, dy}{\sin^2 y + \sin y - 6}$

42. $\displaystyle\int \frac{\sin\theta\, d\theta}{\cos^2\theta + \cos\theta - 2}$

43. $\displaystyle\int \frac{(x-2)^2 \tan^{-1}(2x) - 12x^3 - 3x}{(4x^2 + 1)(x - 2)^2}\, dx$

44. $\displaystyle\int \frac{(x+1)^2 \tan^{-1}(3x) + 9x^3 + x}{(9x^2 + 1)(x + 1)^2}\, dx$

45. $\displaystyle\int \frac{1}{x^{3/2} - \sqrt{x}}\, dx$

46. $\displaystyle\int \frac{1}{(x^{1/3} - 1)\sqrt{x}}\, dx$
(*Hint:* Let $x = u^6$.)

47. $\displaystyle\int \frac{\sqrt{x+1}}{x}\, dx$
(*Hint:* Let $x + 1 = u^2$.)

48. $\displaystyle\int \frac{1}{x\sqrt{x+9}}\, dx$

49. $\displaystyle\int \frac{1}{x(x^4 + 1)}\, dx$
$\left(\text{\textit{Hint:} Multiply by }\dfrac{x^3}{x^3}.\right)$

50. $\displaystyle\int \frac{1}{x^6(x^5 + 4)}\, dx$

51. $\displaystyle\int \frac{1}{\cos 2\theta \sin\theta}\, d\theta$

52. $\displaystyle\int \frac{1}{\cos\theta + \sin 2\theta}\, d\theta$

53. $\displaystyle\int \frac{\sqrt{1 + \sqrt{x}}}{x}\, dx$

54. $\displaystyle\int \frac{\sqrt{x}}{\sqrt{2 - \sqrt{x}} + \sqrt{x}}\, dx$

Use any method to evaluate the integrals in Exercises 55–66.

55. $\displaystyle\int \frac{x^3 - 2x^2 - 3x}{x + 2}\, dx$

56. $\displaystyle\int \frac{x + 2}{x^3 - 2x^2 - 3x}\, dx$

57. $\displaystyle\int \frac{2^x - 2^{-x}}{2^x + 2^{-x}}\, dx$

58. $\displaystyle\int \frac{2^x}{2^{2x} + 2^x - 2}\, dx$

59. $\displaystyle\int \frac{1}{x^4 - 1}\, dx$

60. $\displaystyle\int \frac{x^4 - 1}{x^5 - 5x + 1}\, dx$

61. $\displaystyle\int \frac{\ln x + 2}{x(\ln x + 1)(\ln x + 3)}\, dx$

62. $\displaystyle\int \frac{2}{x(\ln x - 2)^3}\, dx$

63. $\displaystyle\int \frac{1}{\sqrt{x^2 - 1}}\, dx$

64. $\displaystyle\int \frac{x}{x + \sqrt{x^2 + 2}}\, dx$

65. $\displaystyle\int x^5 \sqrt{x^3 + 1}\, dx$

66. $\displaystyle\int x^2 \sqrt{1 - x^2}\, dx$

Initial Value Problems

Solve the initial value problems in Exercises 67–70 for x as a function of t.

67. $(t^2 - 3t + 2)\dfrac{dx}{dt} = 1 \quad (t > 2), \quad x(3) = 0$

68. $(3t^4 + 4t^2 + 1)\dfrac{dx}{dt} = 2\sqrt{3}, \quad x(1) = -\pi\sqrt{3}/4$

69. $(t^2 + 2t)\dfrac{dx}{dt} = 2x + 2 \quad (t, x > 0), \quad x(1) = 1$

70. $(t + 1)\dfrac{dx}{dt} = x^2 + 1 \quad (t > -1), \quad x(0) = 0$

Applications and Examples

In Exercises 71 and 72, find the volume of the solid generated by revolving the shaded region about the indicated axis.

71. The x-axis

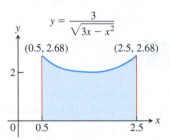

72. The y-axis

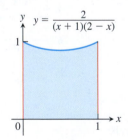

73. Find the length of the curve $y = \ln(1 - x^2), 0 \le x \le \dfrac{1}{2}$.

74. Integrate $\displaystyle\int \sec\theta\, d\theta$ by

a. multiplying by $\dfrac{\sec\theta + \tan\theta}{\sec\theta + \tan\theta}$ and then using a u-substitution.

b. writing the integral as $\displaystyle\int \frac{1}{\cos\theta}\, d\theta$. Then multiply by $\dfrac{\cos\theta}{\cos\theta}$, use a trigonometric identity and a u-substitution, and finally integrate using partial fractions.

75. Find, to two decimal places, the x-coordinate of the centroid of the region in the first quadrant bounded by the x-axis, the curve $y = \tan^{-1} x$, and the line $x = \sqrt{3}$.

T 76. Find the x-coordinate of the centroid of this region to two decimal places.

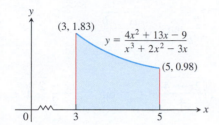

$(3, 1.83)$

$y = \dfrac{4x^2 + 13x - 9}{x^3 + 2x^2 - 3x}$

$(5, 0.98)$

T 77. **Social diffusion** Sociologists sometimes use the phrase "social diffusion" to describe the way information spreads through a population. The information might be a rumor, a cultural fad, or news about a technical innovation. In a sufficiently large population, the number of people x who have the information is treated as a differentiable function of time t, and the rate of diffusion, dx/dt, is assumed to be proportional to the number of people who have the information times the number of people who do not. This leads to the equation

$$\frac{dx}{dt} = kx(N - x),$$

where N is the number of people in the population.

Suppose t is in days, $k = 1/250$, and two people start a rumor at time $t = 0$ in a population of $N = 1000$ people.

a. Find x as a function of t.

b. When will half the population have heard the rumor? (This is when the rumor will be spreading the fastest.)

T 78. **Second-order chemical reactions** Many chemical reactions are the result of the interaction of two molecules that undergo a change to produce a new product. The rate of the reaction typically depends on the concentrations of the two kinds of molecules. If a is the amount of substance A and b is the amount of substance B at time $t = 0$, and if x is the amount of product at time t, then the rate of formation of x may be given by the differential equation

$$\frac{dx}{dt} = k(a - x)(b - x),$$

or

$$\frac{1}{(a - x)(b - x)} \frac{dx}{dt} = k,$$

where k is a constant for the reaction. Integrate both sides of this equation to obtain a relation between x and t **(a)** if $a = b$, and **(b)** if $a \neq b$. Assume in each case that $x = 0$ when $t = 0$.

Integral Tables and Computer Algebra Systems

In this section we discuss how to use tables and computer algebra systems (CAS) to evaluate integrals.

Integral Tables

A Brief Table of Integrals is provided at the back of the text, after the index. (More extensive tables appear in compilations such as *CRC Mathematical Tables*, which contain thousands of integrals.) The integration formulas are stated in terms of constants a, b, c, m, n, and so on. These constants can usually assume any real value and need not be integers. Occasional limitations on their values are stated with the formulas. Formula 21 requires $n \neq -1$, for example, and Formula 27 requires $n \neq -2$.

The formulas also assume that the constants do not take on values that require dividing by zero or taking even roots of negative numbers. For example, Formula 24 assumes that $a \neq 0$, and Formulas 29a and 29b cannot be used unless b is positive.

EXAMPLE 1   Find

$$\int x(2x + 5)^{-1} \, dx.$$

Solution   We use Formula 24 at the back of the book (not 22, which requires $n \neq -1$):

$$\int x(ax + b)^{-1} \, dx = \frac{x}{a} - \frac{b}{a^2} \ln |ax + b| + C.$$

With $a = 2$ and $b = 5$, we have

$$\int x(2x + 5)^{-1} \, dx = \frac{x}{2} - \frac{5}{4} \ln |2x + 5| + C. \qquad \blacksquare$$

EXAMPLE 2   Find

$$\int \frac{dx}{x\sqrt{2x - 4}}.$$

Solution   We use Formula 29b:

$$\int \frac{dx}{x\sqrt{ax - b}} = \frac{2}{\sqrt{b}} \tan^{-1} \sqrt{\frac{ax - b}{b}} + C.$$

With $a = 2$ and $b = 4$, we have

$$\int \frac{dx}{x\sqrt{2x - 4}} = \frac{2}{\sqrt{4}} \tan^{-1} \sqrt{\frac{2x - 4}{4}} + C = \tan^{-1} \sqrt{\frac{x - 2}{2}} + C. \qquad \blacksquare$$

EXAMPLE 3   Find

$$\int x \sin^{-1} x \, dx.$$

Solution   We begin by using Formula 106:

$$\int x^n \sin^{-1} ax \, dx = \frac{x^{n+1}}{n + 1} \sin^{-1} ax - \frac{a}{n + 1} \int \frac{x^{n+1} \, dx}{\sqrt{1 - a^2 x^2}}, \qquad n \neq -1.$$

With $n = 1$ and $a = 1$, we have

$$\int x \sin^{-1} x \, dx = \frac{x^2}{2} \sin^{-1} x - \frac{1}{2} \int \frac{x^2 \, dx}{\sqrt{1 - x^2}}.$$

Next we use Formula 49 to find the integral on the right:

$$\int \frac{x^2}{\sqrt{a^2 - x^2}} \, dx = \frac{a^2}{2} \sin^{-1} \left(\frac{x}{a} \right) - \frac{1}{2} x \sqrt{a^2 - x^2} + C.$$

With $a = 1$,

$$\int \frac{x^2 \, dx}{\sqrt{1 - x^2}} = \frac{1}{2} \sin^{-1} x - \frac{1}{2} x \sqrt{1 - x^2} + C.$$

The combined result is

$$\int x \sin^{-1} x \, dx = \frac{x^2}{2} \sin^{-1} x - \frac{1}{2} \left(\frac{1}{2} \sin^{-1} x - \frac{1}{2} x \sqrt{1 - x^2} + C \right)$$

$$= \left(\frac{x^2}{2} - \frac{1}{4} \right) \sin^{-1} x + \frac{1}{4} x \sqrt{1 - x^2} + C'. \qquad \blacksquare$$

Reduction Formulas

The time required for repeated integrations by parts can sometimes be shortened by applying reduction formulas like

$$\int \tan^n x \, dx = \frac{1}{n - 1} \tan^{n-1} x - \int \tan^{n-2} x \, dx \qquad (1)$$

$$\int (\ln x)^n \, dx = x(\ln x)^n - n \int (\ln x)^{n-1} \, dx \qquad (2)$$

$$\int \sin^n x \cos^m x \, dx = -\frac{\sin^{n-1} x \cos^{m+1} x}{m + n} + \frac{n - 1}{m + n} \int \sin^{n-2} x \cos^m x \, dx \qquad (n \neq -m). \qquad (3)$$

By applying such a formula repeatedly, we can eventually express the original integral in terms of a power low enough to be evaluated directly. The next example illustrates this procedure.

EXAMPLE 4   Find

$$\int \tan^5 x \, dx.$$

Solution   We apply Equation (1) with $n = 5$ to get

$$\int \tan^5 x \, dx = \frac{1}{4} \tan^4 x - \int \tan^3 x \, dx.$$

We then apply Equation (1) again, with $n = 3$, to evaluate the remaining integral:

$$\int \tan^3 x \, dx = \frac{1}{2} \tan^2 x - \int \tan x \, dx = \frac{1}{2} \tan^2 x + \ln |\cos x| + C.$$

The combined result is

$$\int \tan^5 x \, dx = \frac{1}{4} \tan^4 x - \frac{1}{2} \tan^2 x - \ln |\cos x| + C'. \qquad \blacksquare$$

As their form suggests, reduction formulas are derived using integration by parts. (See Example 5 in Section 8.2.)

Integration with a CAS

A powerful capability of computer algebra systems is their ability to integrate symbolically. This is performed with the **integrate command** specified by the particular system (for example, **int** in Maple, **Integrate** in Mathematica).

EXAMPLE 5   Suppose that you want to evaluate the indefinite integral of the function

$$f(x) = x^2 \sqrt{a^2 + x^2}.$$

Using Maple, you first define or name the function:

$$> f := x\text{\textasciicircum}2 * \text{sqrt}(a\text{\textasciicircum}2 + x\text{\textasciicircum}2);$$

Then you use the integrate command on f, identifying the variable of integration:

$$> \text{int}(f, x);$$

Maple returns the answer

$$\frac{1}{4} x(a^2 + x^2)^{3/2} - \frac{1}{8} a^2 x \sqrt{a^2 + x^2} - \frac{1}{8} a^4 \ln \left(x + \sqrt{a^2 + x^2} \right).$$

If you want to see whether the answer can be simplified, enter

$$> \text{simplify}(\%);$$

Maple returns

$$\frac{1}{8} a^2 x \sqrt{a^2 + x^2} + \frac{1}{4} x^3 \sqrt{a^2 + x^2} - \frac{1}{8} a^4 \ln \left(x + \sqrt{a^2 + x^2} \right).$$

If you want the definite integral for $0 \le x \le \pi/2$, you can use the format

$$> \text{int}(f, x = 0..\text{Pi}/2);$$

Maple will return the expression

$$\frac{1}{64}\pi(4a^2 + \pi^2)^{(3/2)} - \frac{1}{32}a^2\pi\sqrt{4a^2 + \pi^2} + \frac{1}{8}a^4\ln(2)$$

$$- \frac{1}{8}a^4\ln\left(\pi + \sqrt{4a^2 + \pi^2}\right) + \frac{1}{16}a^4\ln(a^2).$$

You can also find the definite integral for a particular value of the constant a:

$$> a:= 1;$$
$$> \text{int}(f, x = 0..1);$$

Maple returns the numerical answer

$$\frac{3}{8}\sqrt{2} + \frac{1}{8}\ln\left(\sqrt{2} - 1\right).$$ ■

EXAMPLE 6   Use a CAS to find

$$\int \sin^2 x\cos^3 x\,dx.$$

Solution   With Maple, we have the entry

$$> \text{int}\,((\sin^\wedge 2)(x) * (\cos^\wedge 3)(x), x);$$

with the immediate return

$$-\frac{1}{5}\sin(x)\cos(x)^4 + \frac{1}{15}\cos(x)^2\sin(x) + \frac{2}{15}\sin(x).$$ ■

Computer algebra systems vary in how they process integrations. We used Maple in Examples 5 and 6. Mathematica would have returned somewhat different results:

1. In Example 5, given

$$In[1]:= \text{Integrate}[x^\wedge 2 * \text{Sqrt}[a^\wedge 2 + x^\wedge 2], x]$$

Mathematica returns

$$Out[1] = \sqrt{a^2 + x^2}\left(\frac{a^2 x}{8} + \frac{x^3}{4}\right) - \frac{1}{8}a^4\,\text{Log}\left[x + \sqrt{a^2 + x^2}\right]$$

without having to simplify an intermediate result. The answer is close to Formula 22 in the integral tables.

2. The Mathematica answer to the integral

$$In[2]:= \text{Integrate}[\,\text{Sin}[x]^\wedge 2 * \text{Cos}[x]^\wedge 3, x]$$

in Example 6 is

$$Out[2] = \frac{\text{Sin}[x]}{8} - \frac{1}{48}\text{Sin}[3x] - \frac{1}{80}\text{Sin}[5x]$$

differing from the Maple answer. Both answers are correct.

   Although a CAS is very powerful and can aid us in solving difficult problems, each CAS has its own limitations. There are even situations where a CAS may further complicate a problem (in the sense of producing an answer that is extremely difficult to use or interpret). Note, too, that neither Maple nor Mathematica returns an arbitrary constant $+C$. On the other hand, a little mathematical thinking on your part may reduce the problem to one that is quite easy to handle. We provide an example in Exercise 67.

Many hardware devices have an availability to integration applications, based on software (like Maple or Mathematica), that provide for symbolic input of the integrand to return symbolic output of the indefinite integral. Many of these software applications calculate definite integrals as well. These applications give another tool for finding integrals, aside from using integral tables. However, in some instances, the integration software may not provide an output answer at all.

Nonelementary Integrals

Many functions have antiderivatives that cannot be expressed using the standard functions that we have encountered, such as polynomials, trigonometric functions, and exponential functions. Integrals of functions that do not have elementary antiderivatives are called **nonelementary** integrals. These integrals can sometimes be expressed with infinite series (Chapter 10) or approximated using numerical methods (Section 8.7). Examples of nonelementary integrals include the error function (which measures the probability of random errors)

$$\text{erf}(x) = \frac{2}{\sqrt{\pi}} \int_0^x e^{-t^2}\, dt$$

and integrals such as

$$\int \sin x^2\, dx \qquad \text{and} \qquad \int \sqrt{1 + x^4}\, dx$$

that arise in engineering and physics. These and a number of others, such as

$$\int \frac{e^x}{x}\, dx, \qquad \int e^{(e^x)}\, dx, \qquad \int \frac{1}{\ln x}\, dx, \qquad \int \ln(\ln x)\, dx, \qquad \int \frac{\sin x}{x}\, dx,$$

$$\int \sqrt{1 - k^2 \sin^2 x}\, dx, \qquad 0 < k < 1,$$

look so easy they tempt us to try them just to see how they turn out. It can be proved, however, that there is no way to express any of these integrals as finite combinations of elementary functions. The same applies to integrals that can be changed into these by substitution. The functions in these integrals all have antiderivatives, as a consequence of the Fundamental Theorem of Calculus, Part 1, because they are continuous. However, none of the antiderivatives are elementary. The integrals you are asked to evaluate in this chapter have elementary antiderivatives.

EXERCISES 8.6

Using Integral Tables

Use the table of integrals at the back of the book to evaluate the integrals in Exercises 1–26.

1. $\displaystyle\int \frac{dx}{x\sqrt{x - 3}}$

2. $\displaystyle\int \frac{dx}{x\sqrt{x + 4}}$

3. $\displaystyle\int \frac{x\, dx}{\sqrt{x - 2}}$

4. $\displaystyle\int \frac{x\, dx}{(2x + 3)^{3/2}}$

5. $\displaystyle\int x\sqrt{2x - 3}\, dx$

6. $\displaystyle\int x(7x + 5)^{3/2}\, dx$

7. $\displaystyle\int \frac{\sqrt{9 - 4x}}{x^2}\, dx$

8. $\displaystyle\int \frac{dx}{x^2\sqrt{4x - 9}}$

9. $\displaystyle\int x\sqrt{4x - x^2}\, dx$

10. $\displaystyle\int \frac{\sqrt{x - x^2}}{x}\, dx$

11. $\displaystyle\int \frac{dx}{x\sqrt{7 + x^2}}$

12. $\displaystyle\int \frac{dx}{x\sqrt{7 - x^2}}$

13. $\displaystyle\int \frac{\sqrt{4 - x^2}}{x}\, dx$

14. $\displaystyle\int \frac{\sqrt{x^2 - 4}}{x}\, dx$

15. $\displaystyle\int e^{2t}\cos 3t\, dt$

16. $\displaystyle\int e^{-3t}\sin 4t\, dt$

17. $\displaystyle\int x\cos^{-1} x\, dx$

18. $\displaystyle\int x\tan^{-1} x\, dx$

19. $\int x^2 \tan^{-1} x \, dx$

20. $\int \dfrac{\tan^{-1} x}{x^2} \, dx$

21. $\int \sin 3x \cos 2x \, dx$

22. $\int \sin 2x \cos 3x \, dx$

23. $\int 8 \sin 4t \sin \dfrac{t}{2} \, dt$

24. $\int \sin \dfrac{t}{3} \sin \dfrac{t}{6} \, dt$

25. $\int \cos \dfrac{\theta}{3} \cos \dfrac{\theta}{4} \, d\theta$

26. $\int \cos \dfrac{\theta}{2} \cos 7\theta \, d\theta$

Substitution and Integral Tables

In Exercises 27–40, use a substitution to change the integral into one you can find in the table. Then evaluate the integral.

27. $\int \dfrac{x^3 + x + 1}{(x^2 + 1)^2} \, dx$

28. $\int \dfrac{x^2 + 6x}{(x^2 + 3)^2} \, dx$

29. $\int \sin^{-1} \sqrt{x} \, dx$

30. $\int \dfrac{\cos^{-1} \sqrt{x}}{\sqrt{x}} \, dx$

31. $\int \dfrac{\sqrt{x}}{\sqrt{1 - x}} \, dx$

32. $\int \dfrac{\sqrt{2 - x}}{\sqrt{x}} \, dx$

33. $\int \cot t \sqrt{1 - \sin^2 t} \, dt, \quad 0 < t < \pi/2$

34. $\int \dfrac{dt}{\tan t \sqrt{4 - \sin^2 t}}$

35. $\int \dfrac{dy}{y \sqrt{3 + (\ln y)^2}}$

36. $\int \tan^{-1} \sqrt{y} \, dy$

37. $\int \dfrac{1}{\sqrt{x^2 + 2x + 5}} \, dx$

38. $\int \dfrac{x^2}{\sqrt{x^2 - 4x + 5}} \, dx$

(*Hint*: Complete the square.)

39. $\int \sqrt{5 - 4x - x^2} \, dx$

40. $\int x^2 \sqrt{2x - x^2} \, dx$

Using Reduction Formulas

Use reduction formulas to evaluate the integrals in Exercises 41–50.

41. $\int \sin^5 2x \, dx$

42. $\int 8 \cos^4 2\pi t \, dt$

43. $\int \sin^2 2\theta \cos^3 2\theta \, d\theta$

44. $\int 2 \sin^2 t \sec^4 t \, dt$

45. $\int 4 \tan^3 2x \, dx$

46. $\int 8 \cot^4 t \, dt$

47. $\int 2 \sec^3 \pi x \, dx$

48. $\int 3 \sec^4 3x \, dx$

49. $\int \csc^5 x \, dx$

50. $\int 16x^3 (\ln x)^2 \, dx$

Evaluate the integrals in Exercises 51–56 by making a substitution (possibly trigonometric) and then applying a reduction formula.

51. $\int e^t \sec^3 (e^t - 1) \, dt$

52. $\int \dfrac{\csc^3 \sqrt{\theta}}{\sqrt{\theta}} \, d\theta$

53. $\int_0^1 2\sqrt{x^2 + 1} \, dx$

54. $\int_0^{\sqrt{3}/2} \dfrac{dy}{(1 - y^2)^{5/2}}$

55. $\int_1^2 \dfrac{(r^2 - 1)^{3/2}}{r} \, dr$

56. $\int_0^{1/\sqrt{3}} \dfrac{dt}{(t^2 + 1)^{7/2}}$

Applications

57. Surface area Find the area of the surface generated by revolving the curve $y = \sqrt{x^2 + 2}$, $0 \le x \le \sqrt{2}$, about the x-axis.

58. Arc length Find the length of the curve $y = x^2$, $0 \le x \le \sqrt{3}/2$.

59. Centroid Find the centroid of the region cut from the first quadrant by the curve $y = 1/\sqrt{x + 1}$ and the line $x = 3$.

60. Moment about y-axis A thin plate of constant density $\delta = 1$ occupies the region enclosed by the curve $y = 36/(2x + 3)$ and the line $x = 3$ in the first quadrant. Find the moment of the plate about the y-axis.

T 61. Use the integral table and a calculator to find to two decimal places the area of the surface generated by revolving the curve $y = x^2$, $-1 \le x \le 1$, about the x-axis.

62. Volume The head of your firm's accounting department has asked you to find a formula she can use in a computer program to calculate the year-end inventory of gasoline in the company's tanks. A typical tank is shaped like a right circular cylinder of radius r and length L, mounted horizontally, as shown in the accompanying figure. The data come to the accounting office as depth measurements taken with a vertical measuring stick marked in centimeters.

a. Show, in the notation of the figure, that the volume of gasoline that fills the tank to a depth d is

$$V = 2L \int_{-r}^{-r+d} \sqrt{r^2 - y^2} \, dy.$$

b. Evaluate the integral.

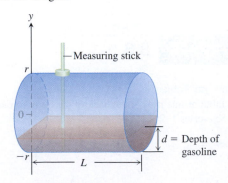

63. What is the largest value

$$\int_a^b \sqrt{x - x^2} \, dx$$

can have for any a and b? Give reasons for your answer.

64. What is the largest value

$$\int_a^b x\sqrt{2x - x^2}\, dx$$

can have for any a and b? Give reasons for your answer.

COMPUTER EXPLORATIONS

In Exercises 65 and 66, use a CAS to perform the integrations.

65. Evaluate the integrals

a. $\displaystyle\int x \ln x\, dx$ **b.** $\displaystyle\int x^2 \ln x\, dx$ **c.** $\displaystyle\int x^3 \ln x\, dx.$

d. What pattern do you see? Predict the formula for $\int x^4 \ln x\, dx$ and then see if you are correct by evaluating it with a CAS.

e. What is the formula for $\int x^n \ln x\, dx, n \geq 1$? Check your answer using a CAS.

66. Evaluate the integrals

a. $\displaystyle\int \frac{\ln x}{x^2}\, dx$ **b.** $\displaystyle\int \frac{\ln x}{x^3}\, dx$ **c.** $\displaystyle\int \frac{\ln x}{x^4}\, dx.$

d. What pattern do you see? Predict the formula for

$$\int \frac{\ln x}{x^5}\, dx$$

and then see if you are correct by evaluating it with a CAS.

e. What is the formula for

$$\int \frac{\ln x}{x^n}\, dx, \quad n \geq 2?$$

Check your answer using a CAS.

67. a. Use a CAS to evaluate

$$\int_0^{\pi/2} \frac{\sin^n x}{\sin^n x + \cos^n x}\, dx$$

where n is an arbitrary positive integer. Does your CAS find the result?

b. In succession, find the integral when $n = 1, 2, 3, 5,$ and 7. Comment on the complexity of the results.

c. Now substitute $x = (\pi/2) - u$ and add the new and old integrals. What is the value of

$$\int_0^{\pi/2} \frac{\sin^n x}{\sin^n x + \cos^n x}\, dx?$$

This exercise illustrates how a little mathematical ingenuity solves a problem not immediately amenable to solution by a CAS.

8.7 Numerical Integration

The antiderivatives of some functions, like $\sin(x^2)$, $1/\ln x$, and $\sqrt{1 + x^4}$, have no elementary formulas. When we cannot find a workable antiderivative for a function f that we have to integrate, we can partition the interval of integration, replace f by a closely fitting polynomial on each subinterval, integrate the polynomials, and add the results to approximate the definite integral of f. This procedure is an example of numerical integration. In this section we study two such methods, the *Trapezoidal Rule* and *Simpson's Rule*. A key goal in our analysis is to control the possible error that is introduced when computing an approximation to an integral.

Trapezoidal Approximations

The Trapezoidal Rule for the value of a definite integral is based on approximating the region between a curve and the x-axis with trapezoids instead of rectangles, as in Figure 8.7. It is not necessary for the subdivision points $x_0, x_1, x_2, \ldots, x_n$ in the figure to be evenly spaced, but the resulting formula is simpler if they are. We therefore assume that the length of each subinterval is

$$\Delta x = \frac{b - a}{n}.$$

The length $\Delta x = (b - a)/n$ is called the **step size** or **mesh size**. The area of the trapezoid that lies above the ith subinterval is

$$\Delta x \left(\frac{y_{i-1} + y_i}{2} \right) = \frac{\Delta x}{2} (y_{i-1} + y_i),$$

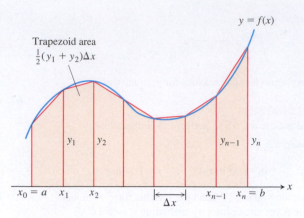

FIGURE 8.7 The Trapezoidal Rule approximates short stretches of the curve $y = f(x)$ with line segments. To approximate the integral of f from a to b, we add the areas of the trapezoids made by joining the ends of the segments to the x-axis.

where $y_{i-1} = f(x_{i-1})$ and $y_i = f(x_i)$. (See Figure 8.7.) The area below the curve $y = f(x)$ and above the x-axis is then approximated by adding the areas of all the trapezoids:

$$T = \frac{1}{2}(y_0 + y_1)\Delta x + \frac{1}{2}(y_1 + y_2)\Delta x + \cdots$$

$$+ \frac{1}{2}(y_{n-2} + y_{n-1})\Delta x + \frac{1}{2}(y_{n-1} + y_n)\Delta x$$

$$= \Delta x \left(\frac{1}{2}y_0 + y_1 + y_2 + \cdots + y_{n-1} + \frac{1}{2}y_n \right)$$

$$= \frac{\Delta x}{2}(y_0 + 2y_1 + 2y_2 + \cdots + 2y_{n-1} + y_n),$$

where

$$y_0 = f(a), \qquad y_1 = f(x_1), \ldots, \qquad y_{n-1} = f(x_{n-1}), \qquad y_n = f(b).$$

The Trapezoidal Rule says: Use T to estimate the integral of f from a to b.

The Trapezoidal Rule

To approximate $\int_a^b f(x)\, dx$, use

$$T = \frac{\Delta x}{2}\left(y_0 + 2y_1 + 2y_2 + \cdots + 2y_{n-1} + y_n \right).$$

The y's are the values of f at the partition points

$$x_0 = a, \, x_1 = a + \Delta x, \, x_2 = a + 2\Delta x, \, \ldots, \, x_{n-1} = a + (n - 1)\Delta x, \, x_n = b,$$

where $\Delta x = (b - a)/n$.

EXAMPLE 1 Use the Trapezoidal Rule with $n = 4$ to estimate $\int_1^2 x^2\, dx$. Compare the estimate with the exact value.

Solution Partition $[\,1, 2\,]$ into four subintervals of equal length (Figure 8.8). Then evaluate $y = x^2$ at each partition point (Table 8.2).

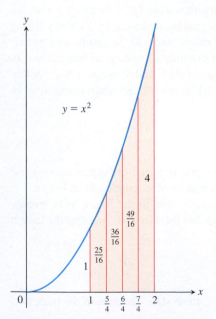

FIGURE 8.8 The trapezoidal approximation of the area under the graph of $y = x^2$ from $x = 1$ to $x = 2$ is a slight overestimate (Example 1).

TABLE 8.2

x	$y = x^2$
1	1
$\dfrac{5}{4}$	$\dfrac{25}{16}$
$\dfrac{6}{4}$	$\dfrac{36}{16}$
$\dfrac{7}{4}$	$\dfrac{49}{16}$
2	4

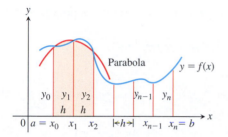

FIGURE 8.9 Simpson's Rule approximates short stretches of the curve with parabolas.

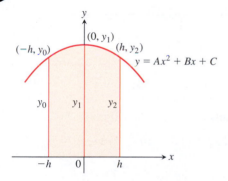

FIGURE 8.10 By integrating from $-h$ to h, we find the shaded area to be

$$\frac{h}{3}(y_0 + 4y_1 + y_2).$$

Using these y-values, $n = 4$, and $\Delta x = (2 - 1)/4 = 1/4$ in the Trapezoidal Rule, we have

$$T = \frac{\Delta x}{2}\left(y_0 + 2y_1 + 2y_2 + 2y_3 + y_4\right)$$

$$= \frac{1}{8}\left(1 + 2\left(\frac{25}{16}\right) + 2\left(\frac{36}{16}\right) + 2\left(\frac{49}{16}\right) + 4\right)$$

$$= \frac{75}{32} = 2.34375.$$

Since the parabola is concave *up*, the approximating segments lie above the curve, giving each trapezoid slightly more area than the corresponding strip under the curve. The exact value of the integral is

$$\int_1^2 x^2\, dx = \frac{x^3}{3}\Bigg]_1^2 = \frac{8}{3} - \frac{1}{3} = \frac{7}{3}.$$

The T approximation overestimates the integral by about half a percent of its true value of $7/3$. The percentage error is $(2.34375 - 7/3)/(7/3) \approx 0.00446$, or 0.446%. ∎

Simpson's Rule: Approximations Using Parabolas

Another rule for approximating the definite integral of a continuous function results from using parabolas instead of the straight-line segments that produced trapezoids. As before, we partition the interval $[a, b]$ into n subintervals of equal length $h = \Delta x = (b - a)/n$, but this time we require that n be an *even* number. On each consecutive pair of intervals we approximate the curve $y = f(x) \geq 0$ by a parabola, as shown in Figure 8.9. A typical parabola passes through three consecutive points (x_{i-1}, y_{i-1}), (x_i, y_i), and (x_{i+1}, y_{i+1}) on the curve.

Let's calculate the shaded area beneath a parabola passing through three consecutive points. To simplify our calculations, we first take the case where $x_0 = -h$, $x_1 = 0$, and $x_2 = h$ (Figure 8.10), where $h = \Delta x = (b - a)/n$. The area under the parabola will be the same if we shift the y-axis to the left or right. The parabola has an equation of the form

$$y = Ax^2 + Bx + C,$$

so the area under it from $x = -h$ to $x = h$ is

$$A_p = \int_{-h}^{h} (Ax^2 + Bx + C)\, dx$$

$$= \frac{Ax^3}{3} + \frac{Bx^2}{2} + Cx\Bigg]_{-h}^{h}$$

$$= \frac{2Ah^3}{3} + 2Ch = \frac{h}{3}(2Ah^2 + 6C).$$

Since the curve passes through the three points $(-h, y_0)$, $(0, y_1)$, and (h, y_2), we also have

$$y_0 = Ah^2 - Bh + C, \qquad y_1 = C, \qquad y_2 = Ah^2 + Bh + C,$$

from which we obtain

$$C = y_1,$$
$$Ah^2 - Bh = y_0 - y_1,$$
$$Ah^2 + Bh = y_2 - y_1,$$
$$2Ah^2 = y_0 + y_2 - 2y_1.$$

Hence, expressing the area A_p in terms of the ordinates y_0, y_1, and y_2, we have

$$A_p = \frac{h}{3}(2Ah^2 + 6C) = \frac{h}{3}((y_0 + y_2 - 2y_1) + 6y_1) = \frac{h}{3}(y_0 + 4y_1 + y_2).$$

Now shifting the parabola horizontally to its shaded position in Figure 8.9 does not change the area under it. Thus the area under the parabola through (x_0, y_0), (x_1, y_1), and (x_2, y_2) in Figure 8.9 is still

$$\frac{h}{3}(y_0 + 4y_1 + y_2).$$

Similarly, the area under the parabola through the points (x_2, y_2), (x_3, y_3), and (x_4, y_4) is

$$\frac{h}{3}(y_2 + 4y_3 + y_4).$$

Computing the areas under all the parabolas and adding the results gives the approximation

$$\int_a^b f(x)\,dx \approx \frac{h}{3}(y_0 + 4y_1 + y_2) + \frac{h}{3}(y_2 + 4y_3 + y_4) + \cdots$$

$$+ \frac{h}{3}(y_{n-2} + 4y_{n-1} + y_n)$$

$$= \frac{h}{3}(y_0 + 4y_1 + 2y_2 + 4y_3 + 2y_4 + \cdots + 2y_{n-2} + 4y_{n-1} + y_n).$$

HISTORICAL BIOGRAPHY

Thomas Simpson
(1720–1761)
www.goo.gl/idqvuc

The result is known as Simpson's Rule. The function need not be positive, as in our derivation, but the number n of subintervals must be even to apply the rule because each parabolic arc uses two subintervals.

Simpson's Rule
To approximate $\int_a^b f(x)\,dx$, use

$$S = \frac{\Delta x}{3}(y_0 + 4y_1 + 2y_2 + 4y_3 + \cdots + 2y_{n-2} + 4y_{n-1} + y_n).$$

The y's are the values of f at the partition points

$$x_0 = a, x_1 = a + \Delta x, x_2 = a + 2\Delta x, \ldots, x_{n-1} = a + (n - 1)\Delta x, x_n = b.$$

The number n is even, and $\Delta x = (b - a)/n$.

Note the pattern of the coefficients in the above rule: $1, 4, 2, 4, 2, 4, 2, \ldots, 4, 1$.

EXAMPLE 2 Use Simpson's Rule with $n = 4$ to approximate $\int_0^2 5x^4\,dx$.

Solution Partition $[0, 2]$ into four subintervals and evaluate $y = 5x^4$ at the partition points (Table 8.3). Then apply Simpson's Rule with $n = 4$ and $\Delta x = 1/2$:

$$S = \frac{\Delta x}{3}\left(y_0 + 4y_1 + 2y_2 + 4y_3 + y_4\right)$$

$$= \frac{1}{6}\left(0 + 4\left(\frac{5}{16}\right) + 2(5) + 4\left(\frac{405}{16}\right) + 80\right)$$

$$= 32\frac{1}{12}.$$

This estimate differs from the exact value (32) by only $1/12$, a percentage error of less than three-tenths of one percent, and this was with just four subintervals. ∎

TABLE 8.3

x	$y = 5x^4$
0	0
$\frac{1}{2}$	$\frac{5}{16}$
1	5
$\frac{3}{2}$	$\frac{405}{16}$
2	80

Error Analysis

Whenever we use an approximation technique, the issue arises as to how accurate the approximation might be. The following theorem gives formulas for estimating the errors when using the Trapezoidal Rule and Simpson's Rule. The **error** is the difference between the approximation obtained by the rule and the actual value of the definite integral $\int_a^b f(x)\,dx$.

THEOREM 1—Error Estimates in the Trapezoidal and Simpson's Rules

If f'' is continuous and M is any upper bound for the values of $|f''|$ on $[a, b]$, then the error E_T in the trapezoidal approximation of the integral of f from a to b for n steps satisfies the inequality

$$|E_T| \le \frac{M(b-a)^3}{12n^2}. \qquad \text{Trapezoidal Rule}$$

If $f^{(4)}$ is continuous and M is any upper bound for the values of $|f^{(4)}|$ on $[a, b]$, then the error E_S in the Simpson's Rule approximation of the integral of f from a to b for n steps satisfies the inequality

$$|E_S| \le \frac{M(b-a)^5}{180n^4}. \qquad \text{Simpson's Rule}$$

To see why Theorem 1 is true in the case of the Trapezoidal Rule, we begin with a result from advanced calculus, which says that if f'' is continuous on the interval $[a, b]$, then

$$\int_a^b f(x)\,dx = T - \frac{b-a}{12} \cdot f''(c)(\Delta x)^2$$

for some number c between a and b. Thus, as Δx approaches zero, the error defined by

$$E_T = -\frac{b-a}{12} \cdot f''(c)(\Delta x)^2$$

approaches *zero* as the *square* of Δx.

The inequality

$$|E_T| \le \frac{b-a}{12} \max|f''(x)|(\Delta x)^2,$$

where max refers to the interval $[a, b]$, gives an upper bound for the magnitude of the error. In practice, we usually cannot find the exact value of $\max|f''(x)|$ and have to estimate an upper bound or "worst case" value for it instead. If M is any upper bound for the values of $|f''(x)|$ on $[a, b]$, so that $|f''(x)| \le M$ on $[a, b]$, then

$$|E_T| \le \frac{b-a}{12} M(\Delta x)^2.$$

If we substitute $(b - a)/n$ for Δx, we get

$$|E_T| \le \frac{M(b-a)^3}{12n^2}.$$

To estimate the error in Simpson's Rule, we start with a result from advanced calculus that says that if the fourth derivative $f^{(4)}$ is continuous, then

$$\int_a^b f(x)\,dx = S - \frac{b-a}{180} \cdot f^{(4)}(c)(\Delta x)^4$$

for some point c between a and b. Thus, as Δx approaches zero, the error,

$$E_S = -\frac{b-a}{180} \cdot f^{(4)}(c)(\Delta x)^4,$$

approaches zero as the *fourth power* of Δx. (This helps to explain why Simpson's Rule is likely to give better results than the Trapezoidal Rule.)

The inequality

$$|E_S| \le \frac{b-a}{180} \max |f^{(4)}(x)| \, (\Delta x)^4,$$

where max refers to the interval $[a, b]$, gives an upper bound for the magnitude of the error. As with $\max |f''|$ in the error formula for the Trapezoidal Rule, we usually cannot find the exact value of $\max |f^{(4)}(x)|$ and have to replace it with an upper bound. If M is any upper bound for the values of $|f^{(4)}|$ on $[a, b]$, then

$$|E_S| \le \frac{b-a}{180} M(\Delta x)^4.$$

Substituting $(b-a)/n$ for Δx in this last expression gives

$$|E_S| \le \frac{M(b-a)^5}{180 n^4}.$$

EXAMPLE 3   Find an upper bound for the error in estimating $\int_0^2 5x^4 \, dx$ using Simpson's Rule with $n = 4$ (Example 2).

Solution   To estimate the error, we first find an upper bound M for the magnitude of the fourth derivative of $f(x) = 5x^4$ on the interval $0 \le x \le 2$. Since the fourth derivative has the constant value $f^{(4)}(x) = 120$, we take $M = 120$. With $b - a = 2$ and $n = 4$, the error estimate for Simpson's Rule gives

$$|E_S| \le \frac{M(b-a)^5}{180 n^4} = \frac{120 \, (2)^5}{180 \cdot 4^4} = \frac{1}{12}.$$

This estimate is consistent with the result of Example 2.   ∎

Theorem 1 can also be used to estimate the number of subintervals required when using the Trapezoidal or Simpson's Rule if we specify a certain tolerance for the error.

EXAMPLE 4   Estimate the minimum number of subintervals needed to approximate the integral in Example 3 using Simpson's Rule with an error of magnitude less than 10^{-4}.

Solution   Using the inequality in Theorem 1, if we choose the number of subintervals n to satisfy

$$\frac{M(b-a)^5}{180 n^4} < 10^{-4},$$

then the error E_S in Simpson's Rule satisfies $|E_S| < 10^{-4}$ as required.

From the solution in Example 3, we have $M = 120$ and $b - a = 2$, so we want n to satisfy

$$\frac{120(2)^5}{180 n^4} < \frac{1}{10^4}$$

or, equivalently,

$$n^4 > \frac{64 \cdot 10^4}{3}.$$

It follows that

$$n > 10\left(\frac{64}{3}\right)^{1/4} \approx 21.5.$$

Since n must be even in Simpson's Rule, we estimate the minimum number of subintervals required for the error tolerance to be $n = 22$. ■

EXAMPLE 5   As we saw in Chapter 7, the value of ln 2 can be calculated from the integral

$$\ln 2 = \int_1^2 \frac{1}{x}\,dx.$$

Table 8.4 shows T and S values for approximations of $\int_1^2 (1/x)\,dx$ using various values of n. Notice how Simpson's Rule dramatically improves over the Trapezoidal Rule.

TABLE 8.4 Trapezoidal Rule approximations (T_n) and Simpson's Rule approximations (S_n) of ln 2 $= \int_1^2 (1/x)\,dx$

n	T_n	|Error| less than ...	S_n	|Error| less than ...
10	0.6937714032	0.0006242227	0.6931502307	0.0000030502
20	0.6933033818	0.0001562013	0.6931473747	0.0000001942
30	0.6932166154	0.0000694349	0.6931472190	0.0000000385
40	0.6931862400	0.0000390595	0.6931471927	0.0000000122
50	0.6931721793	0.0000249988	0.6931471856	0.0000000050
100	0.6931534305	0.0000062500	0.6931471809	0.0000000004

In particular, notice that when we double the value of n (thereby halving the value of $h = \Delta x$), the T error is divided by 2 *squared*, whereas the S error is divided by 2 *to the fourth*.

This has a dramatic effect as $\Delta x = (2 - 1)/n$ gets very small. The Simpson approximation for $n = 50$ rounds accurately to seven places and for $n = 100$ agrees to nine decimal places (billionths)! ■

If $f(x)$ is a polynomial of degree less than four, then its fourth derivative is zero, and

$$E_S = -\frac{b-a}{180} f^{(4)}(c)(\Delta x)^4 = -\frac{b-a}{180}(0)(\Delta x)^4 = 0.$$

Thus, there will be no error in the Simpson approximation of any integral of f. In other words, if f is a constant, a linear function, or a quadratic or cubic polynomial, Simpson's Rule will give the value of any integral of f exactly, whatever the number of subdivisions. Similarly, if f is a constant or a linear function, then its second derivative is zero, and

$$E_T = -\frac{b-a}{12} f''(c)(\Delta x)^2 = -\frac{b-a}{12}(0)(\Delta x)^2 = 0.$$

The Trapezoidal Rule will therefore give the exact value of any integral of f. This is no surprise, for the trapezoids fit the graph perfectly.

Although decreasing the step size Δx reduces the error in the Simpson and Trapezoidal approximations in theory, it may fail to do so in practice. When Δx is very small, say $\Delta x = 10^{-8}$, computer or calculator round-off errors in the arithmetic required to evaluate S and T may accumulate to such an extent that the error formulas no longer describe what is going on. Shrinking Δx below a certain size can actually make things worse. You should consult a text on numerical analysis for more sophisticated methods if you are having problems with round-off error using the rules discussed in this section.

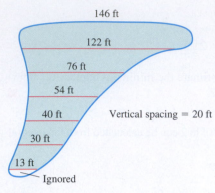

146 ft

122 ft

76 ft

54 ft

40 ft   Vertical spacing = 20 ft

30 ft

13 ft

Ignored

FIGURE 8.11 The dimensions of the swamp in Example 6.

EXAMPLE 6   A town wants to drain and fill a small polluted swamp (Figure 8.11). The swamp averages 5 ft deep. About how many cubic yards of dirt will it take to fill the area after the swamp is drained?

Solution   To calculate the volume of the swamp, we estimate the surface area and multiply by 5. To estimate the area, we use Simpson's Rule with $\Delta x = 20$ ft and the y's equal to the distances measured across the swamp, as shown in Figure 8.11.

$$S = \frac{\Delta x}{3}(y_0 + 4y_1 + 2y_2 + 4y_3 + 2y_4 + 4y_5 + y_6)$$

$$= \frac{20}{3}(146 + 488 + 152 + 216 + 80 + 120 + 13) = 8100$$

The volume is about $(8100)(5) = 40{,}500$ ft^3 or 1500 yd^3. ∎

EXERCISES 8.7

Estimating Definite Integrals

The instructions for the integrals in Exercises 1–10 have two parts, one for the Trapezoidal Rule and one for Simpson's Rule.

I. Using the Trapezoidal Rule

  a. Estimate the integral with $n = 4$ steps and find an upper bound for $|E_T|$.

  b. Evaluate the integral directly and find $|E_T|$.

  c. Use the formula $(|E_T|/(\text{true value})) \times 100$ to express $|E_T|$ as a percentage of the integral's true value.

II. Using Simpson's Rule

  a. Estimate the integral with $n = 4$ steps and find an upper bound for $|E_S|$.

  b. Evaluate the integral directly and find $|E_S|$.

  c. Use the formula $(|E_S|/(\text{true value})) \times 100$ to express $|E_S|$ as a percentage of the integral's true value.

1. $\displaystyle\int_1^2 x\, dx$

2. $\displaystyle\int_1^3 (2x - 1)\, dx$

3. $\displaystyle\int_{-1}^1 (x^2 + 1)\, dx$

4. $\displaystyle\int_{-2}^0 (x^2 - 1)\, dx$

5. $\displaystyle\int_0^2 (t^3 + t)\, dt$

6. $\displaystyle\int_{-1}^1 (t^3 + 1)\, dt$

7. $\displaystyle\int_1^2 \frac{1}{s^2}\, ds$

8. $\displaystyle\int_2^4 \frac{1}{(s-1)^2}\, ds$

9. $\displaystyle\int_0^\pi \sin t\, dt$

10. $\displaystyle\int_0^1 \sin \pi t\, dt$

Estimating the Number of Subintervals

In Exercises 11–22, estimate the minimum number of subintervals needed to approximate the integrals with an error of magnitude less than 10^{-4} by **(a)** the Trapezoidal Rule and **(b)** Simpson's Rule. (The integrals in Exercises 11–18 are the integrals from Exercises 1–8.)

11. $\displaystyle\int_1^2 x\, dx$

12. $\displaystyle\int_1^3 (2x - 1)\, dx$

13. $\displaystyle\int_{-1}^1 (x^2 + 1)\, dx$

14. $\displaystyle\int_{-2}^0 (x^2 - 1)\, dx$

15. $\displaystyle\int_0^2 (t^3 + t)\, dt$

16. $\displaystyle\int_{-1}^1 (t^3 + 1)\, dt$

17. $\displaystyle\int_1^2 \frac{1}{s^2}\, ds$

18. $\displaystyle\int_2^4 \frac{1}{(s-1)^2}\, ds$

19. $\displaystyle\int_0^3 \sqrt{x + 1}\, dx$

20. $\displaystyle\int_0^3 \frac{1}{\sqrt{x + 1}}\, dx$

21. $\displaystyle\int_0^2 \sin (x + 1)\, dx$

22. $\displaystyle\int_{-1}^1 \cos (x + \pi)\, dx$

Estimates with Numerical Data

23. Volume of water in a swimming pool   A rectangular swimming pool is 30 ft wide and 50 ft long. The accompanying table shows the depth $h(x)$ of the water at 5-ft intervals from one end of the pool to the other. Estimate the volume of water in the pool using the Trapezoidal Rule with $n = 10$ applied to the integral

$$V = \int_0^{50} 30 \cdot h(x)\, dx.$$

Position (ft) x	Depth (ft) $h(x)$	Position (ft) x	Depth (ft) $h(x)$
0	6.0	30	11.5
5	8.2	35	11.9
10	9.1	40	12.3
15	9.9	45	12.7
20	10.5	50	13.0
25	11.0		

24. Distance traveled The accompanying table shows time-to-speed data for a sports car accelerating from rest to 130 mph. How far had the car traveled by the time it reached this speed? (Use trapezoids to estimate the area under the velocity curve, but be careful: The time intervals vary in length.)

Speed change	Time (sec)
Zero to 30 mph	2.2
40 mph	3.2
50 mph	4.5
60 mph	5.9
70 mph	7.8
80 mph	10.2
90 mph	12.7
100 mph	16.0
110 mph	20.6
120 mph	26.2
130 mph	37.1

25. Wing design The design of a new airplane requires a gasoline tank of constant cross-sectional area in each wing. A scale drawing of a cross-section is shown here. The tank must hold 5000 lb of gasoline, which has a density of 42 lb/ft³. Estimate the length of the tank by Simpson's Rule.

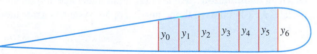

$y_0 = 1.5$ ft, $y_1 = 1.6$ ft, $y_2 = 1.8$ ft, $y_3 = 1.9$ ft, $y_4 = 2.0$ ft, $y_5 = y_6 = 2.1$ ft   Horizontal spacing $= 1$ ft

26. Oil consumption on Pathfinder Island A diesel generator runs continuously, consuming oil at a gradually increasing rate until it must be temporarily shut down to have the filters replaced. Use the Trapezoidal Rule to estimate the amount of oil consumed by the generator during that week.

Day	Oil consumption rate (liters / h)
Sun	0.019
Mon	0.020
Tue	0.021
Wed	0.023
Thu	0.025
Fri	0.028
Sat	0.031
Sun	0.035

Theory and Examples

27. Usable values of the sine-integral function *The sine-integral function,*

$$Si(x) = \int_0^x \frac{\sin t}{t}\, dt, \qquad \text{"Sine integral of x"}$$

is one of the many functions in engineering whose formulas cannot be simplified. There is no elementary formula for the antiderivative of $(\sin t)/t$. The values of Si(x), however, are readily estimated by numerical integration.

Although the notation does not show it explicitly, the function being integrated is

$$f(t) = \begin{cases} \dfrac{\sin t}{t}, & t \neq 0 \\ 1, & t = 0, \end{cases}$$

the continuous extension of $(\sin t)/t$ to the interval $[0, x]$. The function has derivatives of all orders at every point of its domain. Its graph is smooth, and you can expect good results from Simpson's Rule.

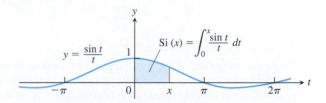

a. Use the fact that $|f^{(4)}| \leq 1$ on $[0, \pi/2]$ to give an upper bound for the error that will occur if

$$Si\left(\frac{\pi}{2}\right) = \int_0^{\pi/2} \frac{\sin t}{t}\, dt$$

is estimated by Simpson's Rule with $n = 4$.

b. Estimate Si($\pi/2$) by Simpson's Rule with $n = 4$.

c. Express the error bound you found in part (a) as a percentage of the value you found in part (b).

28. The error function *The error function,*

$$erf(x) = \frac{2}{\sqrt{\pi}} \int_0^x e^{-t^2}\, dt,$$

important in probability and in the theories of heat flow and signal transmission, must be evaluated numerically because there is no elementary expression for the antiderivative of e^{-t^2}.

a. Use Simpson's Rule with $n = 10$ to estimate erf(1).

b. In $[0, 1]$,

$$\left| \frac{d^4}{dt^4}\left(e^{-t^2} \right) \right| \leq 12.$$

Give an upper bound for the magnitude of the error of the estimate in part (a).

29. Prove that the sum T in the Trapezoidal Rule for $\int_a^b f(x)\, dx$ is a Riemann sum for f continuous on $[a, b]$. (*Hint:* Use the Intermediate Value Theorem to show the existence of c_k in the subinterval $[x_{k-1}, x_k]$ satisfying $f(c_k) = (f(x_{k-1}) + f(x_k))/2$.)

30. Prove that the sum S in Simpson's Rule for $\int_a^b f(x)\, dx$ is a Riemann sum for f continuous on $[a, b]$. (See Exercise 29.)

T 31. Elliptic integrals The length of the ellipse

$$\frac{x^2}{a^2} + \frac{y^2}{b^2} = 1$$

turns out to be

$$\text{Length} = 4a \int_0^{\pi/2} \sqrt{1 - e^2 \cos^2 t}\, dt,$$

where $e = \sqrt{a^2 - b^2}/a$ is the ellipse's eccentricity. The integral in this formula, called an *elliptic integral*, is nonelementary except when $e = 0$ or 1.

a. Use the Trapezoidal Rule with $n = 10$ to estimate the length of the ellipse when $a = 1$ and $e = 1/2$.

b. Use the fact that the absolute value of the second derivative of $f(t) = \sqrt{1 - e^2 \cos^2 t}$ is less than 1 to find an upper bound for the error in the estimate you obtained in part (a).

Applications

T **32.** The length of one arch of the curve $y = \sin x$ is given by

$$L = \int_0^{\pi} \sqrt{1 + \cos^2 x}\, dx.$$

Estimate L by Simpson's Rule with $n = 8$.

T **33.** Your metal fabrication company is bidding for a contract to make sheets of corrugated iron roofing like the one shown here. The cross-sections of the corrugated sheets are to conform to the curve

$$y = \sin \frac{3\pi}{20} x, \quad 0 \le x \le 20 \text{ in.}$$

If the roofing is to be stamped from flat sheets by a process that does not stretch the material, how wide should the original material be? To find out, use numerical integration to approximate the length of the sine curve to two decimal places.

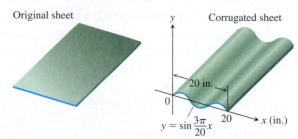

T **34.** Your engineering firm is bidding for the contract to construct the tunnel shown here. The tunnel is 300 ft long and 50 ft wide at the base. The cross-section is shaped like one arch of the curve $y = 25 \cos(\pi x/50)$. Upon completion, the tunnel's inside surface (excluding the roadway) will be treated with a waterproof

sealer that costs \$2.35 per square foot to apply. How much will it cost to apply the sealer? (*Hint:* Use numerical integration to find the length of the cosine curve.)

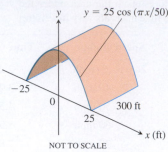

NOT TO SCALE

Find, to two decimal places, the areas of the surfaces generated by revolving the curves in Exercises 35 and 36 about the *x*-axis.

35. $y = \sin x, \quad 0 \le x \le \pi$ **36.** $y = x^2/4, \quad 0 \le x \le 2$

37. Use numerical integration to estimate the value of

$$\sin^{-1} 0.6 = \int_0^{0.6} \frac{dx}{\sqrt{1 - x^2}}.$$

For reference, $\sin^{-1} 0.6 = 0.64350$ to five decimal places.

38. Use numerical integration to estimate the value of

$$\pi = 4 \int_0^1 \frac{1}{1 + x^2}\, dx.$$

39. Drug assimilation An average adult under age 60 years assimilates a 12-hr cold medicine into his or her system at a rate modeled by

$$\frac{dy}{dt} = 6 - \ln(2t^2 - 3t + 3),$$

where y is measured in milligrams and t is the time in hours since the medication was taken. What amount of medicine is absorbed into a person's system over a 12-hr period?

40. Effects of an antihistamine The concentration of an antihistamine in the bloodstream of a healthy adult is modeled by

$$C = 12.5 - 4 \ln(t^2 - 3t + 4),$$

where C is measured in grams per liter and t is the time in hours since the medication was taken. What is the average level of concentration in the bloodstream over a 6-hr period?

8.8 Improper Integrals

Up to now, we have required definite integrals to satisfy two properties. First, the domain of integration $[a, b]$ must be finite. Second, the range of the integrand must be finite on this domain. In practice, we may encounter problems that fail to meet one or both of these conditions. The integral for the area under the curve $y = (\ln x)/x^2$ from $x = 1$ to $x = \infty$ is an example for which the domain is infinite (Figure 8.12a). The integral for the area under the curve of $y = 1/\sqrt{x}$ between $x = 0$ and $x = 1$ is an example for which the range of the integrand is infinite (Figure 8.12b). In either case, the integrals are said to be *improper* and are calculated as limits. We will see in Section 8.9 that improper integrals play an important role in probability. They are also useful when investigating the convergence of certain infinite series in Chapter 10.

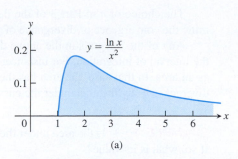

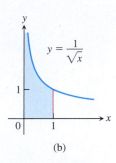

FIGURE 8.12 Are the areas under these infinite curves finite? We will see that the answer is yes for both curves.

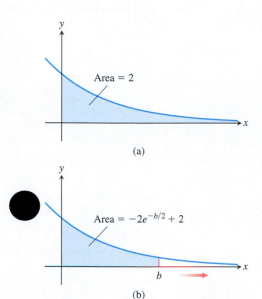

(a)

(b)

FIGURE 8.13 (a) The area in the first quadrant under the curve $y = e^{-x/2}$. (b) The area is an improper integral of the first type.

Infinite Limits of Integration

Consider the infinite region (unbounded on the right) that lies under the curve $y = e^{-x/2}$ in the first quadrant (Figure 8.13a). You might think this region has infinite area, but we will see that the value is finite. We assign a value to the area in the following way. First find the area $A(b)$ of the portion of the region that is bounded on the right by $x = b$ (Figure 8.13b).

$$A(b) = \int_0^b e^{-x/2} \, dx = -2e^{-x/2} \Big]_0^b = -2e^{-b/2} + 2$$

Then find the limit of $A(b)$ as $b \to \infty$

$$\lim_{b \to \infty} A(b) = \lim_{b \to \infty} \left(-2e^{-b/2} + 2 \right) = 2.$$

The value we assign to the area under the curve from 0 to ∞ is

$$\int_0^\infty e^{-x/2} \, dx = \lim_{b \to \infty} \int_0^b e^{-x/2} \, dx = 2.$$

DEFINITION   Integrals with infinite limits of integration are **improper integrals of Type I**.

1. If $f(x)$ is continuous on $[a, \infty)$, then

$$\int_a^\infty f(x) \, dx = \lim_{b \to \infty} \int_a^b f(x) \, dx.$$

2. If $f(x)$ is continuous on $(-\infty, b\,]$, then

$$\int_{-\infty}^b f(x) \, dx = \lim_{a \to -\infty} \int_a^b f(x) \, dx.$$

3. If $f(x)$ is continuous on $(-\infty, \infty)$, then

$$\int_{-\infty}^\infty f(x) \, dx = \int_{-\infty}^c f(x) \, dx + \int_c^\infty f(x) \, dx,$$

where c is any real number.

In each case, if the limit exists and is finite, we say that the improper integral **converges** and that the limit is the **value** of the improper integral. If the limit fails to exist, the improper integral **diverges**.

The choice of c in Part 3 of the definition is unimportant. We can evaluate or determine the convergence or divergence of $\int_{-\infty}^{\infty} f(x) \, dx$ with any convenient choice.

Any of the integrals in the above definition can be interpreted as an area if $f \geq 0$ on the interval of integration. For instance, we interpreted the improper integral in Figure 8.13 as an area. In that case, the area has the finite value 2. If $f \geq 0$ and the improper integral diverges, we say the area under the curve is **infinite**.

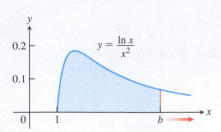

FIGURE 8.14 The area under this curve is an improper integral (Example 1).

EXAMPLE 1   Is the area under the curve $y = (\ln x)/x^2$ from $x = 1$ to $x = \infty$ finite? If so, what is its value?

Solution   We find the area under the curve from $x = 1$ to $x = b$ and examine the limit as $b \to \infty$. If the limit is finite, we take it to be the area under the curve (Figure 8.14). The area from 1 to b is

$$\int_1^b \frac{\ln x}{x^2} \, dx = \left[(\ln x)\left(-\frac{1}{x}\right) \right]_1^b - \int_1^b \left(-\frac{1}{x}\right)\left(\frac{1}{x}\right) dx$$

Integration by parts with $u = \ln x$, $dv = dx/x^2$, $du = dx/x$, $v = -1/x$

$$= -\frac{\ln b}{b} - \left[\frac{1}{x}\right]_1^b$$

$$= -\frac{\ln b}{b} - \frac{1}{b} + 1.$$

The limit of the area as $b \to \infty$ is

$$\int_1^\infty \frac{\ln x}{x^2} \, dx = \lim_{b \to \infty} \int_1^b \frac{\ln x}{x^2} \, dx$$

$$= \lim_{b \to \infty} \left[-\frac{\ln b}{b} - \frac{1}{b} + 1 \right]$$

$$= -\left[\lim_{b \to \infty} \frac{\ln b}{b} \right] - 0 + 1$$

$$= -\left[\lim_{b \to \infty} \frac{1/b}{1} \right] + 1 = 0 + 1 = 1. \qquad \text{l'Hôpital's Rule}$$

Thus, the improper integral converges and the area has finite value 1.   ∎

EXAMPLE 2   Evaluate

$$\int_{-\infty}^{\infty} \frac{dx}{1 + x^2}.$$

HISTORICAL BIOGRAPHY

Lejeune Dirichlet
(1805–1859)
www.goo.gl/QGwXLL

Solution   According to the definition (Part 3), we can choose $c = 0$ and write

$$\int_{-\infty}^{\infty} \frac{dx}{1 + x^2} = \int_{-\infty}^{0} \frac{dx}{1 + x^2} + \int_{0}^{\infty} \frac{dx}{1 + x^2}.$$

Next we evaluate each improper integral on the right side of the equation above.

$$\int_{-\infty}^{0} \frac{dx}{1 + x^2} = \lim_{a \to -\infty} \int_{a}^{0} \frac{dx}{1 + x^2}$$

$$= \lim_{a \to -\infty} \tan^{-1} x \Big]_a^0$$

$$= \lim_{a \to -\infty} (\tan^{-1} 0 - \tan^{-1} a) = 0 - \left(-\frac{\pi}{2}\right) = \frac{\pi}{2}$$

$$\int_0^\infty \frac{dx}{1+x^2} = \lim_{b\to\infty} \int_0^b \frac{dx}{1+x^2}$$

$$= \lim_{b\to\infty} \tan^{-1} x \Big]_0^b$$

$$= \lim_{b\to\infty} (\tan^{-1} b - \tan^{-1} 0) = \frac{\pi}{2} - 0 = \frac{\pi}{2}$$

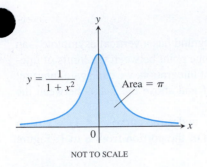

$y = \dfrac{1}{1+x^2}$ Area $= \pi$

NOT TO SCALE

FIGURE 8.15 The area under this curve is finite (Example 2).

Thus,

$$\int_{-\infty}^\infty \frac{dx}{1+x^2} = \frac{\pi}{2} + \frac{\pi}{2} = \pi.$$

Since $1/(1+x^2) > 0$, the improper integral can be interpreted as the (finite) area beneath the curve and above the x-axis (Figure 8.15). ■

The Integral $\displaystyle\int_1^\infty \frac{dx}{x^p}$

The function $y = 1/x$ is the boundary between the convergent and divergent improper integrals with integrands of the form $y = 1/x^p$. As the next example shows, the improper integral converges if $p > 1$ and diverges if $p \le 1$.

EXAMPLE 3   For what values of p does the integral $\int_1^\infty dx/x^p$ converge? When the integral does converge, what is its value?

Solution   If $p \ne 1$,

$$\int_1^b \frac{dx}{x^p} = \frac{x^{-p+1}}{-p+1} \Big]_1^b = \frac{1}{1-p}(b^{-p+1} - 1) = \frac{1}{1-p}\left(\frac{1}{b^{p-1}} - 1\right).$$

Thus,

$$\int_1^\infty \frac{dx}{x^p} = \lim_{b\to\infty} \int_1^b \frac{dx}{x^p}$$

$$= \lim_{b\to\infty} \left[\frac{1}{1-p}\left(\frac{1}{b^{p-1}} - 1\right)\right] = \begin{cases} \dfrac{1}{p-1}, & p > 1 \\ \infty, & p < 1 \end{cases}$$

because

$$\lim_{b\to\infty} \frac{1}{b^{p-1}} = \begin{cases} 0, & p > 1 \\ \infty, & p < 1. \end{cases}$$

Therefore, the integral converges to the value $1/(p-1)$ if $p > 1$ and it diverges if $p < 1$.

If $p = 1$, the integral also diverges:

$$\int_1^\infty \frac{dx}{x^p} = \int_1^\infty \frac{dx}{x}$$

$$= \lim_{b\to\infty} \int_1^b \frac{dx}{x}$$

$$= \lim_{b\to\infty} \ln x \Big]_1^b$$

$$= \lim_{b\to\infty} (\ln b - \ln 1) = \infty.$$

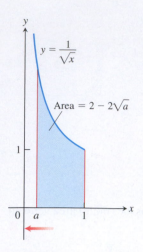

FIGURE 8.16 The area under this curve is an example of an improper integral of the second kind.

Integrands with Vertical Asymptotes

Another type of improper integral arises when the integrand has a vertical asymptote—an infinite discontinuity—at a limit of integration or at some point between the limits of integration. If the integrand f is positive over the interval of integration, we can again interpret the improper integral as the area under the graph of f and above the x-axis between the limits of integration.

Consider the region in the first quadrant that lies under the curve $y = 1/\sqrt{x}$ from $x = 0$ to $x = 1$ (Figure 8.12b). First we find the area of the portion from a to 1 (Figure 8.16):

$$\int_a^1 \frac{dx}{\sqrt{x}} = 2\sqrt{x}\,\Big]_a^1 = 2 - 2\sqrt{a}.$$

Then we find the limit of this area as $a \to 0^+$:

$$\lim_{a \to 0^+} \int_a^1 \frac{dx}{\sqrt{x}} = \lim_{a \to 0}\left(2 - 2\sqrt{a}\right) = 2.$$

Therefore the area under the curve from 0 to 1 is finite and is defined to be

$$\int_0^1 \frac{dx}{\sqrt{x}} = \lim_{a \to 0^+} \int_a^1 \frac{dx}{\sqrt{x}} = 2.$$

> **DEFINITION**   Integrals of functions that become infinite at a point within the interval of integration are **improper integrals of Type II**.
>
> **1.** If $f(x)$ is continuous on $(a, b]$ and discontinuous at a, then
>
> $$\int_a^b f(x)\, dx = \lim_{c \to a^+} \int_c^b f(x)\, dx.$$
>
> **2.** If $f(x)$ is continuous on $[a, b)$ and discontinuous at b, then
>
> $$\int_a^b f(x)\, dx = \lim_{c \to b^-} \int_a^c f(x)\, dx.$$
>
> **3.** If $f(x)$ is discontinuous at c, where $a < c < b$, and continuous on $[a, c) \cup (c, b]$, then
>
> $$\int_a^b f(x)\, dx = \int_a^c f(x)\, dx + \int_c^b f(x)\, dx.$$
>
> In each case, if the limit exists and is finite, we say the improper integral **converges** and that the limit is the **value** of the improper integral. If the limit does not exist, the integral **diverges**.

In Part 3 of the definition, the integral on the left side of the equation converges if *both* integrals on the right side converge; otherwise it diverges.

EXAMPLE 4   Investigate the convergence of

$$\int_0^1 \frac{1}{1 - x}\, dx.$$

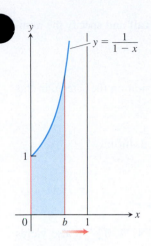

FIGURE 8.17 The area beneath the curve and above the x-axis for $[0, 1)$ is not a real number (Example 4).

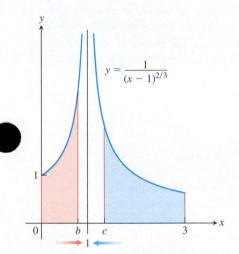

FIGURE 8.18 Example 5 shows that the area under the curve exists (so it is a real number).

Solution The integrand $f(x) = 1/(1 - x)$ is continuous on $[0, 1)$ but is discontinuous at $x = 1$ and becomes infinite as $x \to 1^-$ (Figure 8.17). We evaluate the integral as

$$\lim_{b \to 1^-} \int_0^b \frac{1}{1 - x}\, dx = \lim_{b \to 1^-} \left[-\ln |1 - x| \right]_0^b$$

$$= \lim_{b \to 1^-} \left[-\ln (1 - b) + 0 \right] = \infty.$$

The limit is infinite, so the integral diverges. ■

EXAMPLE 5   Evaluate

$$\int_0^3 \frac{dx}{(x - 1)^{2/3}}.$$

Solution The integrand has a vertical asymptote at $x = 1$ and is continuous on $[0, 1)$ and $(1, 3]$ (Figure 8.18). Thus, by Part 3 of the definition above,

$$\int_0^3 \frac{dx}{(x - 1)^{2/3}} = \int_0^1 \frac{dx}{(x - 1)^{2/3}} + \int_1^3 \frac{dx}{(x - 1)^{2/3}}.$$

Next, we evaluate each improper integral on the right-hand side of this equation.

$$\int_0^1 \frac{dx}{(x - 1)^{2/3}} = \lim_{b \to 1^-} \int_0^b \frac{dx}{(x - 1)^{2/3}}$$

$$= \lim_{b \to 1^-} 3(x - 1)^{1/3} \Big]_0^b$$

$$= \lim_{b \to 1^-} \left[3(b - 1)^{1/3} + 3 \right] = 3$$

$$\int_1^3 \frac{dx}{(x - 1)^{2/3}} = \lim_{c \to 1^+} \int_c^3 \frac{dx}{(x - 1)^{2/3}}$$

$$= \lim_{c \to 1^+} 3(x - 1)^{1/3} \Big]_c^3$$

$$= \lim_{c \to 1^+} \left[3(3 - 1)^{1/3} - 3(c - 1)^{1/3} \right] = 3\sqrt[3]{2}$$

We conclude that

$$\int_0^3 \frac{dx}{(x - 1)^{2/3}} = 3 + 3\sqrt[3]{2}.$$ ■

Improper Integrals with a CAS

Computer algebra systems can evaluate many convergent improper integrals. To evaluate the integral

$$\int_2^\infty \frac{x + 3}{(x - 1)(x^2 + 1)}\, dx$$

(which converges) using Maple, enter

> $f := (x + 3)/((x - 1) * (x^2 + 1));$

Then use the integration command

> $\text{int}(f, x = 2..\text{infinity});$

Maple returns the answer

$$-\frac{1}{2}\pi + \ln (5) + \arctan (2).$$

To obtain a numerical result, use the evaluation command **evalf** and specify the number of digits as follows:

$$> \text{evalf}(\%, 6);$$

The symbol % instructs the computer to evaluate the last expression on the screen, in this case $(-1/2)\pi + \ln(5) + \arctan(2)$. Maple returns 1.14579.

Using Mathematica, entering

$$In[1] := \text{Integrate}[(x + 3)/((x - 1)(x\text{^}2 + 1)), \{x, 2, \text{Infinity}\}]$$

returns

$$Out[1] = -\frac{\pi}{2} + \text{ArcTan}[2] + \text{Log}[5].$$

To obtain a numerical result with six digits, use the command "N[%, 6]"; it also yields 1.14579.

Tests for Convergence and Divergence

When we cannot evaluate an improper integral directly, we try to determine whether it converges or diverges. If the integral diverges, that's the end of the story. If it converges, we can use numerical methods to approximate its value. The principal tests for convergence or divergence are the Direct Comparison Test and the Limit Comparison Test.

EXAMPLE 6 Does the integral $\int_1^\infty e^{-x^2}\,dx$ converge?

Solution By definition,

$$\int_1^\infty e^{-x^2}\,dx = \lim_{b\to\infty} \int_1^b e^{-x^2}\,dx.$$

We cannot evaluate this integral directly because it is nonelementary. But we *can* show that its limit as $b \to \infty$ is finite. We know that $\int_1^b e^{-x^2}\,dx$ is an increasing function of b because the area under the curve increases as b increases. Therefore either it becomes infinite as $b \to \infty$ or it has a finite limit as $b \to \infty$. For our function it does not become infinite: For every value of $x \geq 1$, we have $e^{-x^2} \leq e^{-x}$ (Figure 8.19) so that

$$\int_1^b e^{-x^2}\,dx \leq \int_1^b e^{-x}\,dx = -e^{-b} + e^{-1} < e^{-1} \approx 0.36788.$$

Hence,

$$\int_1^\infty e^{-x^2}\,dx = \lim_{b\to\infty} \int_1^b e^{-x^2}\,dx$$

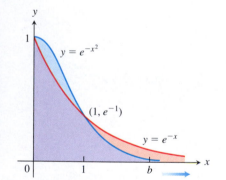

FIGURE 8.19 The graph of e^{-x^2} lies below the graph of e^{-x} for $x > 1$ (Example 6).

converges to some finite value. We do not know exactly what the value is except that it is something positive and less than 0.37. Here we are relying on the completeness property of the real numbers, discussed in Appendix 6. ∎

The comparison of e^{-x^2} and e^{-x} in Example 6 is a special case of the following test.

THEOREM 2—Direct Comparison Test

Let f and g be continuous on $[a, \infty)$ with $0 \leq f(x) \leq g(x)$ for all $x \geq a$. Then

1. If $\displaystyle\int_a^\infty g(x)\,dx$ converges, then $\displaystyle\int_a^\infty f(x)\,dx$ also converges.

2. If $\displaystyle\int_a^\infty f(x)\,dx$ diverges, then $\displaystyle\int_a^\infty g(x)\,dx$ also diverges.

Proof   The reasoning behind the argument establishing Theorem 2 is similar to that in Example 6. If $0 \le f(x) \le g(x)$ for $x \ge a$, then from Rule 7 in Theorem 2 of Section 5.3 we have

$$\int_a^b f(x)\,dx \le \int_a^b g(x)\,dx, \qquad b > a.$$

From this it can be argued, as in Example 6, that

$$\int_a^\infty f(x)\,dx \qquad \text{converges if} \qquad \int_a^\infty g(x)\,dx \qquad \text{converges.}$$

Turning this around to its contrapositive form, this says that

$$\int_a^\infty g(x)\,dx \qquad \text{diverges if} \qquad \int_a^\infty f(x)\,dx \qquad \text{diverges.} \qquad \blacksquare$$

Although the theorem is stated for Type I improper integrals, a similar result is true for integrals of Type II as well.

EXAMPLE 7   These examples illustrate how we use Theorem 2.

HISTORICAL BIOGRAPHY
Karl Weierstrass
(1815–1897)
www.goo.gl/3RH2r0

(a) $\displaystyle\int_1^\infty \frac{\sin^2 x}{x^2}\,dx$ converges because

$$0 \le \frac{\sin^2 x}{x^2} \le \frac{1}{x^2} \quad \text{on} \quad [1,\infty) \quad \text{and} \quad \int_1^\infty \frac{1}{x^2}\,dx \quad \text{converges.} \qquad \text{Example 3}$$

(b) $\displaystyle\int_1^\infty \frac{1}{\sqrt{x^2 - 0.1}}\,dx$ diverges because

$$\frac{1}{\sqrt{x^2 - 0.1}} \ge \frac{1}{x} \quad \text{on} \quad [1,\infty) \quad \text{and} \quad \int_1^\infty \frac{1}{x}\,dx \quad \text{diverges.} \qquad \text{Example 3}$$

(c) $\displaystyle\int_0^{\pi/2} \frac{\cos x}{\sqrt{x}}\,dx$ converges because

$$0 \le \frac{\cos x}{\sqrt{x}} \le \frac{1}{\sqrt{x}} \quad \text{on} \quad \left[0, \frac{\pi}{2}\right], \qquad 0 \le \cos x \le 1 \text{ on } \left[0, \frac{\pi}{2}\right]$$

and

$$\int_0^{\pi/2} \frac{dx}{\sqrt{x}} = \lim_{a \to 0^+} \int_a^{\pi/2} \frac{dx}{\sqrt{x}}$$

$$= \lim_{a \to 0^+} \sqrt{4x}\,\Big]_a^{\pi/2} \qquad 2\sqrt{x} = \sqrt{4x}$$

$$= \lim_{a \to 0^+} \left(\sqrt{2\pi} - \sqrt{4a}\right) = \sqrt{2\pi} \qquad \text{converges.} \qquad \blacksquare$$

THEOREM 3—Limit Comparison Test

If the positive functions f and g are continuous on $[a, \infty)$, and if

$$\lim_{x \to \infty} \frac{f(x)}{g(x)} = L, \qquad 0 < L < \infty,$$

then

$$\int_a^\infty f(x)\,dx \qquad \text{and} \qquad \int_a^\infty g(x)\,dx$$

either *both converge* or *both diverge*.

We omit the proof of Theorem 3, which is similar to that of Theorem 2.

Although the improper integrals of two functions from a to ∞ may both converge, this does not mean that their integrals necessarily have the same value, as the next example shows.

EXAMPLE 8 Show that

$$\int_1^\infty \frac{dx}{1 + x^2}$$

converges by comparison with $\int_1^\infty (1/x^2)\, dx$. Find and compare the two integral values.

Solution The functions $f(x) = 1/x^2$ and $g(x) = 1/(1 + x^2)$ are positive and continuous on $[1, \infty)$. Also,

$$\lim_{x \to \infty} \frac{f(x)}{g(x)} = \lim_{x \to \infty} \frac{1/x^2}{1/(1 + x^2)} = \lim_{x \to \infty} \frac{1 + x^2}{x^2}$$

$$= \lim_{x \to \infty} \left(\frac{1}{x^2} + 1 \right) = 0 + 1 = 1,$$

which is a positive finite limit (Figure 8.20). Therefore, $\int_1^\infty \frac{dx}{1 + x^2}$ converges because $\int_1^\infty \frac{dx}{x^2}$ converges.

The integrals converge to different values, however:

$$\int_1^\infty \frac{dx}{x^2} = \frac{1}{2 - 1} = 1 \qquad \text{Example 3}$$

and

$$\int_1^\infty \frac{dx}{1 + x^2} = \lim_{b \to \infty} \int_1^b \frac{dx}{1 + x^2} = \lim_{b \to \infty} \left[\tan^{-1} b - \tan^{-1} 1 \right] = \frac{\pi}{2} - \frac{\pi}{4} = \frac{\pi}{4}. \quad \blacksquare$$

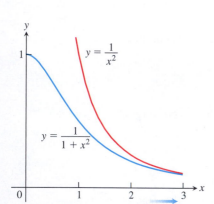

FIGURE 8.20 The functions in Example 8.

EXAMPLE 9 Investigate the convergence of $\int_1^\infty \frac{1 - e^{-x}}{x}\, dx$.

Solution The integrand suggests a comparison of $f(x) = (1 - e^{-x})/x$ with $g(x) = 1/x$. However, we cannot use the Direct Comparison Test because $f(x) \le g(x)$ and the integral of $g(x)$ *diverges*. On the other hand, using the Limit Comparison Test we find that

$$\lim_{x \to \infty} \frac{f(x)}{g(x)} = \lim_{x \to \infty} \left(\frac{1 - e^{-x}}{x} \right) \left(\frac{x}{1} \right) = \lim_{x \to \infty} (1 - e^{-x}) = 1,$$

which is a positive finite limit. Therefore, $\int_1^\infty \frac{1 - e^{-x}}{x}\, dx$ diverges because $\int_1^\infty \frac{dx}{x}$ diverges. Approximations to the improper integral are given in Table 8.5. Note that the values do not appear to approach any fixed limiting value as $b \to \infty$. $\blacksquare$

TABLE 8.5

b	$\int_1^b \frac{1 - e^{-x}}{x}\, dx$
2	0.5226637569
5	1.3912002736
10	2.0832053156
100	4.3857862516
1000	6.6883713446
10000	8.9909564376
100000	11.2935415306

EXERCISES 8.8

Evaluating Improper Integrals

The integrals in Exercises 1–34 converge. Evaluate the integrals without using tables.

1. $\displaystyle\int_0^\infty \frac{dx}{x^2 + 1}$

2. $\displaystyle\int_1^\infty \frac{dx}{x^{1.001}}$

3. $\displaystyle\int_0^1 \frac{dx}{\sqrt{x}}$

4. $\displaystyle\int_0^4 \frac{dx}{\sqrt{4 - x}}$

5. $\displaystyle\int_{-1}^1 \frac{dx}{x^{2/3}}$

6. $\displaystyle\int_{-8}^1 \frac{dx}{x^{1/3}}$

7. $\displaystyle\int_0^1 \frac{dx}{\sqrt{1 - x^2}}$

8. $\displaystyle\int_0^1 \frac{dr}{r^{0.999}}$

9. $\displaystyle\int_{-\infty}^{-2} \frac{2\,dx}{x^2 - 1}$

10. $\displaystyle\int_{-\infty}^2 \frac{2\,dx}{x^2 + 4}$

11. $\displaystyle\int_2^\infty \frac{2}{v^2 - v}\,dv$

12. $\displaystyle\int_2^\infty \frac{2\,dt}{t^2 - 1}$

13. $\displaystyle\int_{-\infty}^\infty \frac{2x\,dx}{(x^2 + 1)^2}$

14. $\displaystyle\int_{-\infty}^\infty \frac{x\,dx}{(x^2 + 4)^{3/2}}$

15. $\displaystyle\int_0^1 \frac{\theta + 1}{\sqrt{\theta^2 + 2\theta}}\,d\theta$

16. $\displaystyle\int_0^2 \frac{s + 1}{\sqrt{4 - s^2}}\,ds$

17. $\displaystyle\int_0^\infty \frac{dx}{(1 + x)\sqrt{x}}$

18. $\displaystyle\int_1^\infty \frac{1}{x\sqrt{x^2 - 1}}\,dx$

19. $\displaystyle\int_0^\infty \frac{dv}{(1 + v^2)(1 + \tan^{-1} v)}$

20. $\displaystyle\int_0^\infty \frac{16\tan^{-1} x}{1 + x^2}\,dx$

21. $\displaystyle\int_{-\infty}^0 \theta e^\theta\,d\theta$

22. $\displaystyle\int_0^\infty 2e^{-\theta}\sin\theta\,d\theta$

23. $\displaystyle\int_{-\infty}^\infty e^{-|x|}\,dx$

24. $\displaystyle\int_{-\infty}^\infty 2xe^{-x^2}\,dx$

25. $\displaystyle\int_0^1 x\ln x\,dx$

26. $\displaystyle\int_0^1 (-\ln x)\,dx$

27. $\displaystyle\int_0^2 \frac{ds}{\sqrt{4 - s^2}}$

28. $\displaystyle\int_0^1 \frac{4r\,dr}{\sqrt{1 - r^4}}$

29. $\displaystyle\int_1^2 \frac{ds}{s\sqrt{s^2 - 1}}$

30. $\displaystyle\int_2^4 \frac{dt}{t\sqrt{t^2 - 4}}$

31. $\displaystyle\int_{-1}^4 \frac{dx}{\sqrt{|x|}}$

32. $\displaystyle\int_0^2 \frac{dx}{\sqrt{|x - 1|}}$

33. $\displaystyle\int_{-1}^\infty \frac{d\theta}{\theta^2 + 5\theta + 6}$

34. $\displaystyle\int_0^\infty \frac{dx}{(x + 1)(x^2 + 1)}$

Testing for Convergence

In Exercises 35–68, use integration, the Direct Comparison Test, or the Limit Comparison Test to test the integrals for convergence. If more than one method applies, use whatever method you prefer.

35. $\displaystyle\int_{1/2}^2 \frac{dx}{x\ln x}$

36. $\displaystyle\int_{-1}^1 \frac{d\theta}{\theta^2 - 2\theta}$

37. $\displaystyle\int_{1/2}^\infty \frac{dx}{x(\ln x)^3}$

38. $\displaystyle\int_0^\infty \frac{d\theta}{\theta^2 - 1}$

39. $\displaystyle\int_0^{\pi/2} \tan\theta\,d\theta$

40. $\displaystyle\int_0^{\pi/2} \cot\theta\,d\theta$

41. $\displaystyle\int_0^1 \frac{\ln x}{x^2}\,dx$

42. $\displaystyle\int_1^2 \frac{dx}{x\ln x}$

43. $\displaystyle\int_0^{\ln 2} x^{-2}e^{-1/x}\,dx$

44. $\displaystyle\int_0^1 \frac{e^{-\sqrt{x}}}{\sqrt{x}}\,dx$

45. $\displaystyle\int_0^\pi \frac{dt}{\sqrt{t + \sin t}}$

46. $\displaystyle\int_0^1 \frac{dt}{t - \sin t}$ (Hint: $t \geq \sin t$ for $t \geq 0$)

47. $\displaystyle\int_0^2 \frac{dx}{1 - x^2}$

48. $\displaystyle\int_0^2 \frac{dx}{1 - x}$

49. $\displaystyle\int_{-1}^1 \ln|x|\,dx$

50. $\displaystyle\int_{-1}^1 -x\ln|x|\,dx$

51. $\displaystyle\int_1^\infty \frac{dx}{x^3 + 1}$

52. $\displaystyle\int_4^\infty \frac{dx}{\sqrt{x} - 1}$

53. $\displaystyle\int_2^\infty \frac{dv}{\sqrt{v - 1}}$

54. $\displaystyle\int_0^\infty \frac{d\theta}{1 + e^\theta}$

55. $\displaystyle\int_0^\infty \frac{dx}{\sqrt{x^6 + 1}}$

56. $\displaystyle\int_2^\infty \frac{dx}{\sqrt{x^2 - 1}}$

57. $\displaystyle\int_1^\infty \frac{\sqrt{x + 1}}{x^2}\,dx$

58. $\displaystyle\int_2^\infty \frac{x\,dx}{\sqrt{x^4 - 1}}$

59. $\displaystyle\int_\pi^\infty \frac{2 + \cos x}{x}\,dx$

60. $\displaystyle\int_\pi^\infty \frac{1 + \sin x}{x^2}\,dx$

61. $\displaystyle\int_4^\infty \frac{2\,dt}{t^{3/2} - 1}$

62. $\displaystyle\int_2^\infty \frac{1}{\ln x}\,dx$

63. $\displaystyle\int_1^\infty \frac{e^x}{x}\,dx$

64. $\displaystyle\int_{e^e}^\infty \ln(\ln x)\,dx$

65. $\displaystyle\int_1^\infty \frac{1}{\sqrt{e^x - x}}\,dx$

66. $\displaystyle\int_1^\infty \frac{1}{e^x - 2^x}\,dx$

67. $\displaystyle\int_{-\infty}^\infty \frac{dx}{\sqrt{x^4 + 1}}$

68. $\displaystyle\int_{-\infty}^\infty \frac{dx}{e^x + e^{-x}}$

Theory and Examples

69. Find the values of p for which each integral converges.

a. $\displaystyle\int_1^2 \frac{dx}{x(\ln x)^p}$

b. $\displaystyle\int_2^\infty \frac{dx}{x(\ln x)^p}$

70. $\int_{-\infty}^{\infty} f(x)\,dx$ **may not equal** $\lim\limits_{b \to \infty} \int_{-b}^{b} f(x)\,dx$ Show that

$$\int_{0}^{\infty} \frac{2x\,dx}{x^2 + 1}$$

diverges and hence that

$$\int_{-\infty}^{\infty} \frac{2x\,dx}{x^2 + 1}$$

diverges. Then show that

$$\lim_{b \to \infty} \int_{-b}^{b} \frac{2x\,dx}{x^2 + 1} = 0.$$

Exercises 71–74 are about the infinite region in the first quadrant between the curve $y = e^{-x}$ and the x-axis.

71. Find the area of the region.

72. Find the centroid of the region.

73. Find the volume of the solid generated by revolving the region about the y-axis.

74. Find the volume of the solid generated by revolving the region about the x-axis.

75. Find the area of the region that lies between the curves $y = \sec x$ and $y = \tan x$ from $x = 0$ to $x = \pi/2$.

76. The region in Exercise 75 is revolved about the x-axis to generate a solid.

    a. Find the volume of the solid.

    b. Show that the inner and outer surfaces of the solid have infinite area.

77. Consider the infinite region in the first quadrant bounded by the graphs of $y = \dfrac{1}{x^2}$, $y = 0$, and $x = 1$.

    a. Find the area of the region.

    b. Find the volume of the solid formed by revolving the region (i) about the x-axis; (ii) about the y-axis.

78. Consider the infinite region in the first quadrant bounded by the graphs of $y = \dfrac{1}{\sqrt{x}}$, $y = 0$, $x = 0$, and $x = 1$.

    a. Find the area of the region.

    b. Find the volume of the solid formed by revolving the region (i) about the x-axis; (ii) about the y-axis.

79. Evaluate the integrals.

    a. $\displaystyle\int_{0}^{1} \frac{dt}{\sqrt{t}(1 + t)}$ **b.** $\displaystyle\int_{0}^{\infty} \frac{dt}{\sqrt{t}(1 + t)}$

80. Evaluate $\displaystyle\int_{3}^{\infty} \frac{dx}{x\sqrt{x^2 - 9}}$.

81. Estimating the value of a convergent improper integral whose domain is infinite

    a. Show that

$$\int_{3}^{\infty} e^{-3x}\,dx = \frac{1}{3}e^{-9} < 0.000042,$$

and hence that $\int_{3}^{\infty} e^{-x^2}\,dx < 0.000042$. Explain why this means that $\int_{0}^{\infty} e^{-x^2}\,dx$ can be replaced by $\int_{0}^{3} e^{-x^2}\,dx$ without introducing an error of magnitude greater than 0.000042.

    T **b.** Evaluate $\int_{0}^{3} e^{-x^2}\,dx$ numerically.

82. The infinite paint can or Gabriel's horn   As Example 3 shows, the integral $\int_{1}^{\infty}(dx/x)$ diverges. This means that the integral

$$\int_{1}^{\infty} 2\pi \frac{1}{x}\sqrt{1 + \frac{1}{x^4}}\,dx,$$

which measures the *surface area* of the solid of revolution traced out by revolving the curve $y = 1/x$, $1 \le x$, about the x-axis, diverges also. By comparing the two integrals, we see that, for every finite value $b > 1$,

$$\int_{1}^{b} 2\pi \frac{1}{x}\sqrt{1 + \frac{1}{x^4}}\,dx > 2\pi \int_{1}^{b} \frac{1}{x}\,dx.$$

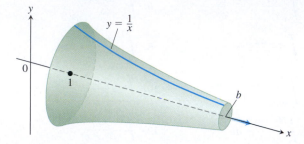

However, the integral

$$\int_{1}^{\infty} \pi\left(\frac{1}{x}\right)^2 dx$$

for the *volume* of the solid converges.

    a. Calculate it.

    b. This solid of revolution is sometimes described as a can that does not hold enough paint to cover its own interior. Think about that for a moment. It is common sense that a finite amount of paint cannot cover an infinite surface. But if we fill the horn with paint (a finite amount), then we *will* have covered an infinite surface. Explain the apparent contradiction.

83. Sine-integral function   The integral

$$\text{Si}(x) = \int_{0}^{x} \frac{\sin t}{t}\,dt,$$

called the *sine-integral function*, has important applications in optics.

    T **a.** Plot the integrand $(\sin t)/t$ for $t > 0$. Is the sine-integral function everywhere increasing or decreasing? Do you think Si $(x) = 0$ for $x \ge 0$? Check your answers by graphing the function Si (x) for $0 \le x \le 25$.

    b. Explore the convergence of

$$\int_{0}^{\infty} \frac{\sin t}{t}\,dt.$$

If it converges, what is its value?

84. Error function   The function

$$\text{erf}(x) = \int_{0}^{x} \frac{2e^{-t^2}}{\sqrt{\pi}}\,dt,$$

called the *error function*, has important applications in probability and statistics.

    T **a.** Plot the error function for $0 \le x \le 25$.

b. Explore the convergence of

$$\int_0^\infty \frac{2e^{-t^2}}{\sqrt{\pi}}\, dt.$$

If it converges, what appears to be its value? You will see how to confirm your estimate in Section 15.4, Exercise 41.

85. Normal probability distribution   The function

$$f(x) = \frac{1}{\sigma\sqrt{2\pi}}\, e^{-\frac{1}{2}\left(\frac{x-\mu}{\sigma}\right)^2}$$

is called the *normal probability density function* with mean μ and standard deviation σ. The number μ tells where the distribution is centered, and σ measures the "scatter" around the mean. (See Section 8.9.)

From the theory of probability, it is known that

$$\int_{-\infty}^\infty f(x)\, dx = 1.$$

In what follows, let $\mu = 0$ and $\sigma = 1$.

T **a.** Draw the graph of f. Find the intervals on which f is increasing, the intervals on which f is decreasing, and any local extreme values and where they occur.

b. Evaluate

$$\int_{-n}^n f(x)\, dx$$

for $n = 1, 2,$ and 3.

c. Give a convincing argument that

$$\int_{-\infty}^\infty f(x)\, dx = 1.$$

(*Hint:* Show that $0 < f(x) < e^{-x/2}$ for $x > 1$, and for $b > 1$,

$$\int_b^\infty e^{-x/2}\, dx \to 0 \quad \text{as} \quad b \to \infty.)$$

86. Show that if $f(x)$ is integrable on every interval of real numbers and a and b are real numbers with $a < b$, then

a. $\int_{-\infty}^a f(x)\, dx$ and $\int_a^\infty f(x)\, dx$ both converge if and only if $\int_{-\infty}^b f(x)\, dx$ and $\int_b^\infty f(x)\, dx$ both converge.

b. $\int_{-\infty}^a f(x)\, dx + \int_a^\infty f(x)\, dx = \int_{-\infty}^b f(x)\, dx + \int_b^\infty f(x)\, dx$ when the integrals involved converge.

COMPUTER EXPLORATIONS

In Exercises 87–90, use a CAS to explore the integrals for various values of p (include noninteger values). For what values of p does the integral converge? What is the value of the integral when it does converge? Plot the integrand for various values of p.

87. $\int_0^e x^p \ln x\, dx$

88. $\int_e^\infty x^p \ln x\, dx$

89. $\int_0^\infty x^p \ln x\, dx$

90. $\int_{-\infty}^\infty x^p \ln |x|\, dx$

Use a CAS to evaluate the integrals.

91. $\int_0^{2/\pi} \sin \frac{1}{x}\, dx$

92. $\int_0^{2/\pi} x \sin \frac{1}{x}\, dx$

8.9 Probability

The outcome of some events, such as a heavy rock falling from a great height, can be modeled so that we can predict with high accuracy what will happen. On the other hand, many events have more than one possible outcome and which one of them will occur is uncertain. If we toss a coin, a head or a tail will result with each outcome being equally likely, but we do not know in advance which one it will be. If we randomly select and then weigh a person from a large population, there are many possible weights the person might have, and it is not certain whether the weight will be between 180 and 190 lb. We are told it is highly likely, but not known for sure, that an earthquake of magnitude 6.0 or greater on the Richter scale will occur near a major population area in California within the next one hundred years. Events having more than one possible outcome are *probabilistic* in nature, and when modeling them we assign a *probability* to the likelihood that a particular outcome may occur. In this section we show how calculus plays a central role in making predictions with probabilistic models.

Random Variables

We begin our discussion with some familiar examples of uncertain events for which the collection of all possible outcomes is finite.

EXAMPLE 1

(a) If we toss a coin once, there are two possible outcomes $\{H, T\}$, where H represents the coin landing head face up and T a tail landing face up. If we toss a coin three times, there are eight possible outcomes, taking into account the order in which a head or tail occurs. The set of outcomes is $\{HHH, HHT, HTH, THH, HTT, THT, TTH, TTT\}$.

(b) If we roll a six-sided die once, the set of possible outcomes is $\{1, 2, 3, 4, 5, 6\}$ representing the six faces of the die.

(c) If we select at random two cards from a 52-card deck, there are 52 possible outcomes for the first card drawn and then 51 possibilities for the second card. Since the order of the cards does not matter, there are $(52 \cdot 51)/2 = 1{,}326$ possible outcomes altogether. ∎

It is customary to refer to the set of all possible outcomes as the *sample space* for an event. With an uncertain event we are usually interested in which outcomes, if any, are more likely to occur than others, and to how large an extent. In tossing a coin three times, is it more likely that two heads or that one head will result? To answer such questions, we need a way to quantify the outcomes.

DEFINITION A **random variable** is a function X that assigns a numerical value to each outcome in a sample space.

Random variables that have only finitely many values are called **discrete** random variables. A **continuous random variable** can take on values in an entire interval, and it is associated with a *distribution function*, which we explain later.

EXAMPLE 2

(a) Suppose we toss a coin three times giving the possible outcomes $\{HHH, HHT, HTH, THH, HTT, THT, TTH, TTT\}$. Define the random variable X to be the number of heads that appear. So $X(HHT) = 2$, $X(THT) = 1$, and so forth. Since X can only assume the values 0, 1, 2, or 3, it is a discrete random variable.

(b) We spin an arrow anchored by a pin located at the origin. The arrow can wind up pointing in any possible direction and we define the random variable X as the radian angle the arrow makes with the positive x-axis, measured counterclockwise. In this case, X is a continuous random variable that can take on any value in the interval $[0, 2\pi)$.

(c) The weight of a randomly selected person in a given population is a continuous random variable W. The cholesterol level of a randomly chosen person, and the waiting time for service of a person in a queue at a bank, are also continuous random variables.

(d) The scores on the national ACT Examination for college admissions in a particular year are described by a discrete random variable S taking on integer values between 1 and 36. If the number of outcomes is large, or for reasons involving statistical analysis, discrete random variables such as test scores are often modeled as continuous random variables (Example 13).

(e) We roll a pair of dice and define the random variable X to be the sum of the numbers on the top faces. This sum can only assume the integer values from 2 through 12, so X is a discrete random variable.

(f) A tire company produces tires for mid-sized sedans. The tires are guaranteed to last for 30,000 miles, but some will fail sooner and some will last many more miles beyond 30,000. The lifetime in miles of a tire is described by a continuous random variable L. ∎

Probability Distributions

A *probability distribution* describes the probabilistic behavior of a random variable. Our chief interest is in probability distributions associated with continuous random variables, but to gain some perspective we first consider a distribution for a discrete random variable.

Suppose we toss a coin three times, with each side H or T equally likely to occur on a given toss. We define the discrete random variable X that assigns the number of heads appearing in each outcome, giving

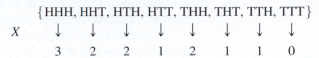

$$\{\text{HHH, HHT, HTH, HTT, THH, THT, TTH, TTT}\}$$

Next we count the *frequency* or number of times a specific value of X occurs. Because each of the eight outcomes is equally likely to occur, we can calculate the probability of the random variable X by dividing the frequency of each value by the total number of outcomes. We summarize our results as follows:

Value of X	0	1	2	3
Frequency	1	3	3	1
P(X)	1/8	3/8	3/8	1/8

We display this information in a probability bar graph of the discrete random variable X, as shown in Figure 8.21. The values of X are portrayed by intervals of length 1 on the x-axis so the area of each bar in the graph is the probability of the corresponding outcome. For instance, the probability that exactly two heads occurs in the three tosses of the coin is the area of the bar associated with the value $X = 2$, which is $3/8$. Similarly, the probability that two or more heads occurs is the sum of areas of the bars associated with the values $X = 2$ and $X = 3$, or $4/8$. The probability that either zero or three heads occurs is $\frac{1}{8} + \frac{1}{8} = \frac{1}{4}$, and so forth. Note that the total area of all the bars in the graph is 1, which is the sum of all the probabilities for X.

With a continuous random variable, even when the outcomes are equally likely, we cannot simply count the number of outcomes in the sample space or the frequencies of outcomes that lead to a specific value of X. In fact, the probability that X takes on any particular one of its values is zero. What *is* meaningful to ask is how probable it is that the random variable takes on a value within some specified *interval* of values.

We capture the information we need about the probabilities of X in a function whose graph behaves much like the bar graph in Figure 8.21. That is, we take a nonnegative function f defined over the range of the random variable with the property that the total area beneath the graph of f is 1. The probability that a value of the random variable X lies within some specified interval $[c, d]$ is then the area under the graph of f over that interval. The following definition assumes the range of the continuous random variable X is any real value, but the definition is general enough to account for random variables having a range of finite length.

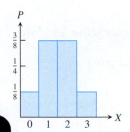

FIGURE 8.21 Probability bar graph for the random variable X when tossing a fair coin three times.

DEFINITIONS A **probability density function** for a continuous random variable is a function f defined over $(-\infty, \infty)$ and having the following properties:

1. f is continuous, except possibly at a finite number of points.

2. f is nonnegative, so $f \geq 0$.

3. $\displaystyle\int_{-\infty}^{\infty} f(x)\,dx = 1.$

If X is a continuous random variable with probability density function f, the **probability** that X assumes a value in the interval between $X = c$ and $X = d$ is given by the integral

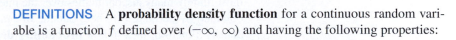

$$P(c \leq X \leq d) = \int_{c}^{d} f(x)\,dx.$$

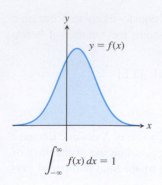

$$\int_{-\infty}^{\infty} f(x)\,dx = 1$$

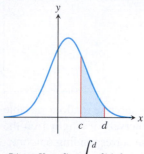

$$P(c \le X \le d) = \int_c^d f(x)\,dx$$

FIGURE 8.22 A probability density function for the continuous random variable X.

The probability that a continuous random variable X assumes a particular real value c is $P(X = c) = \int_c^c f(x)\,dx = 0$, consistent with our previous assertion. Since the area under the graph of f over the interval $[c, d]$ is only a portion of the total area beneath the graph, the probability $P(c \le X \le d)$ is always a number between zero and one. Figure 8.22 illustrates a probability density function.

A probability density function for a random variable X resembles the density function for a wire of varying density. To obtain the mass of a segment of the wire, we integrate the density of the wire over an interval. To obtain the probability that a random variable has values in a particular interval, we integrate the probability density function over that interval.

EXAMPLE 3   Let $f(x) = 2e^{-2x}$ if $0 \le x < \infty$ and $f(x) = 0$ for all negative values of x.

(a) Verify that f is a probability density function.

(b) The time T in hours until a car passes a spot on a remote road is described by the probability density function f. Find the probability $P(T \le 1)$ that a hitchhiker at that spot will see a car within one hour.

(c) Find the probability $P(T = 1)$ that a car passes by the spot after precisely one hour.

Solution

(a) The function f is continuous except at $x = 0$, and is everywhere nonnegative. Moreover,

$$\int_{-\infty}^{\infty} f(x)\,dx = \int_0^{\infty} 2e^{-2x}\,dx = \lim_{b \to \infty} \int_0^b 2e^{-2x}\,dx = \lim_{b \to \infty} \left(1 - e^{-2b}\right) = 1.$$

So all of the conditions are satisfied and we have shown that f is a probability density function.

(b) The probability that a car comes after a time lapse between zero and one hour is given by integrating the probability density function over the interval $[0, 1]$. So

$$P(T \le 1) = \int_0^1 2e^{-2t}\,dt = -e^{-2t}\Big]_0^1 = 1 - e^{-2} \approx 0.865.$$

This result can be interpreted to mean that if 100 people were to hitchhike at that spot, about 87 of them can expect to see a car within one hour.

(c) This probability is the integral $\int_1^1 f(t)\,dt$, which equals zero. We interpret this to mean that a sufficiently accurate measurement of the time until a car comes by the spot would have no possibility of being precisely equal to one hour. It might be very close, perhaps, but it would not be exactly one hour. ∎

We can extend the definition to finite intervals. If f is a nonnegative function with at most finitely many discontinuities over the interval $[a, b]$, and its extension F to $(-\infty, \infty)$, obtained by defining F to be 0 outside of $[a, b]$, satisfies the definition for a probability density function, then f is a **probability density function for $[a, b]$**. This means that $\int_a^b f(x)\,dx = 1$. Similar definitions can be made for the intervals (a, b), $(a, b]$, and $[a, b)$.

EXAMPLE 4   Show that $f(x) = \dfrac{4}{27}x^2(3 - x)$ is a probability density function over the interval $[0, 3]$.

Solution   The function f is continuous and nonnegative over $[0, 3]$. Also,

$$\int_0^3 \frac{4}{27}x^2(3 - x)\,dx = \frac{4}{27}\left[x^3 - \frac{1}{4}x^4\right]_0^3 = \frac{4}{27}\left(27 - \frac{81}{4}\right) = 1.$$

We conclude that f is a probability density function over $[0, 3]$. ∎

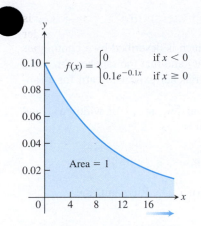

$$f(x) = \begin{cases} 0 & \text{if } x < 0 \\ 0.1e^{-0.1x} & \text{if } x \geq 0 \end{cases}$$

Area = 1

FIGURE 8.23 An exponentially decreasing probability density function.

Exponentially Decreasing Distributions

The distribution in Example 3 is called an *exponentially decreasing probability density function*. These probability density functions always take on the form

$$f(x) = \begin{cases} 0 & \text{if } x < 0 \\ ce^{-cx} & \text{if } x \geq 0 \end{cases}$$

(see Exercise 23). Exponential density functions can provide models for describing random variables such as the lifetimes of light bulbs, radioactive particles, tooth crowns, and many kinds of electronic components. They also model the amount of time until some specific event occurs, such as the time until a pollinator arrives at a flower, the arrival times of a bus at a stop, the time between individuals joining a queue, the waiting time between phone calls at a help desk, and even the lengths of the phone calls themselves. A graph of an exponential density function is shown in Figure 8.23.

Random variables with exponential distributions are *memoryless*. If we think of X as describing the lifetime of some object, then the probability that the object survives for at least $s + t$ hours, given that it has survived t hours, is the same as the initial probability that it survives for at least s hours. For instance, the current age t of a radioactive particle does not change the probability that it will survive for at least another time period of length s. Sometimes the exponential distribution is used as a model when the memoryless principle is violated, because it provides reasonable approximations that are good enough for their intended use. For instance, this might be the case when predicting the lifetime of an artificial hip replacement or heart valve for a particular patient. Here is an application illustrating the exponential distribution.

EXAMPLE 5   An electronics company models the lifetime T in years of a chip they manufacture with the exponential density function

$$f(t) = \begin{cases} 0 & \text{if } t < 0 \\ 0.1e^{-0.1t} & \text{if } t \geq 0. \end{cases}$$

Using this model,

(a) Find the probability $P(T > 2)$ that a chip will last for more than two years.

(b) Find the probability $P(4 \leq T \leq 5)$ that a chip will fail in the fifth year.

(c) If 1000 chips are shipped to a customer, how many can be expected to fail within three years?

Solution

(a) The probability that a chip lasts at least two years is

$$P(T > 2) = \int_2^{\infty} 0.1e^{-0.1t}\, dt = \lim_{b \to \infty} \int_2^b 0.1e^{-0.1t}\, dt$$

$$= \lim_{b \to \infty} \left[e^{-0.2} - e^{-0.1b} \right] = e^{-0.2} \approx 0.819.$$

That is, about 82% of the chips last more than two years.

(b) The probability is

$$P(4 \leq T \leq 5) = \int_4^5 0.1e^{-0.1t}\, dt = -e^{-0.1t} \Big]_4^5 = e^{-0.4} - e^{-0.5} \approx 0.064$$

which means that about 6% of the chips fail during the fifth year.

(c) We want the probability

$$P(0 \leq T \leq 3) = \int_0^3 0.1e^{-0.1t}\, dt = -e^{-0.1t} \Big]_0^3 = 1 - e^{-0.3} \approx 0.259.$$

We can expect that about 259 of the 1000 chips will fail within three years.

Expected Values, Means, and Medians

Suppose the weight in lbs of a steer raised on a cattle ranch is described by a continuous random variable W with probability density function $f(w)$ and that the rancher can sell a steer of weight w for $g(w)$ dollars. How much can the rancher expect to earn for a randomly chosen steer on the ranch?

To answer this question, we consider a small interval $[w_i, w_{i+1}]$ of width Δw_i and note that the probability a steer has weight in this interval is

$$\int_{w_i}^{w_{i+1}} f(w)\, dw \approx f(w_i)\, \Delta w_i.$$

The earning on a steer in this interval is approximately $g(w_i)$. The Riemann sum

$$\sum g(w_i)\, f(w_i)\, \Delta w_i$$

then approximates the amount the rancher would receive for a steer. We assume that steers have a maximum weight, so f is zero outside some finite interval $[0, b]$. Then taking the limit of the Riemann sum as the width of each interval approaches zero gives the integral

$$\int_{-\infty}^{\infty} g(w)\, f(w)\, dw.$$

This integral estimates how much the rancher can expect to earn for a typical steer on the ranch and is the *expected value of the function g*.

The expected values of certain functions of a random variable X have particular importance in probability and statistics. One of the most important of these functions is the expected value of the function $g(x) = x$.

DEFINITION   The **expected value** or **mean** of a continuous random variable X with probability density function f is the number

$$\mu = \mathrm{E}(X) = \int_{-\infty}^{\infty} x f(x)\, dx.$$

The expected value $\mathrm{E}(X)$ can be thought of as a weighted average of the random variable X, where each value of X is weighted by $f(X)$. The mean can also be interpreted as the long-run average value of the random variable X, and it is one measure of the centrality of the random variable X.

EXAMPLE 6   Find the mean of the random variable X with exponential probability density function

$$f(x) = \begin{cases} 0 & \text{if } x < 0 \\ ce^{-cx} & \text{if } x \geq 0. \end{cases}$$

Solution   From the definition we have

$$\mu = \int_{-\infty}^{\infty} x f(x)\, dx = \int_{0}^{\infty} x c e^{-cx}\, dx$$

$$= \lim_{b \to \infty} \int_{0}^{b} x c e^{-cx}\, dx = \lim_{b \to \infty} \left(-x e^{-cx} \Big]_{0}^{b} + \int_{0}^{b} e^{-cx}\, dx \right)$$

$$= \lim_{b \to \infty} \left(-b e^{-cb} - \frac{1}{c} e^{-cb} + \frac{1}{c} \right) = \frac{1}{c}. \qquad \text{l'Hôpital's Rule on first term}$$

Therefore, the mean is $\mu = 1/c$. ∎

From the result in Example 6, knowing the mean or expected value μ of a random variable X having an exponential density function allows us to write its entire formula.

Exponential Density Function for a Random Variable X with Mean μ

$$f(x) = \begin{cases} 0 & \text{if } x < 0 \\ \mu^{-1}e^{-x/\mu} & \text{if } x \geq 0 \end{cases}$$

EXAMPLE 7   Suppose the time T before a chip fails in Example 5 is modeled instead by the exponential density function with a mean of eight years. Find the probability that a chip will fail within five years.

Solution   The exponential density function with mean $\mu = 8$ is

$$f(t) = \begin{cases} 0 & \text{if } t < 0 \\ \dfrac{1}{8}e^{-t/8} & \text{if } t \geq 0 \end{cases}$$

Then the probability a chip will fail within five years is the definite integral

$$P(0 \leq T \leq 5) = \int_0^5 0.125e^{-0.125t}\,dt = -e^{-0.125t}\Big]_0^5 = 1 - e^{-0.625} \approx 0.465$$

so about 47% of the chips can be expected to fail within five years.   ■

EXAMPLE 8   Find the expected value for the random variable X with probability density function given by Example 4.

Solution   The expected value is

$$\mu = \mathrm{E}(X) = \int_0^3 \frac{4}{27}x^3(3-x)\,dx = \frac{4}{27}\left[\frac{3}{4}x^4 - \frac{1}{5}x^5\right]_0^3$$

$$= \frac{4}{27}\left(\frac{243}{4} - \frac{243}{5}\right) = 1.8$$

From Figure 8.24, you can see that this expected value is reasonable because the region beneath the probability density function appears to be balanced about the vertical line $x = 1.8$. That is, the horizontal coordinate of the centroid of a plate described by the region is $\bar{x} = 1.8$.   ■

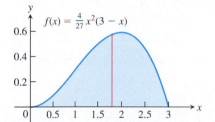

FIGURE 8.24   The expected value of a random variable with this probability density function is $\mu = 1.8$ (Example 8).

There are other ways to measure the centrality of a random variable with a given probability density function.

DEFINITION   The **median** of a continuous random variable X with probability density function f is the number m for which

$$\int_{-\infty}^m f(x)\,dx = \frac{1}{2} \qquad \text{and} \qquad \int_m^\infty f(x)\,dx = \frac{1}{2}.$$

The definition of the median means that there is an equal likelihood that the random variable X will be smaller than m or larger than m.

EXAMPLE 9   Find the median of a random variable X with exponential probability density function

$$f(x) = \begin{cases} 0 & \text{if } x < 0 \\ ce^{-cx} & \text{if } x \geq 0. \end{cases}$$

Solution   The median m must satisfy

$$\frac{1}{2} = \int_0^m ce^{-cx}\, dx = -e^{-cx}\Big]_0^m = 1 - e^{-cm}.$$

It follows that

$$e^{-cm} = \frac{1}{2} \qquad \text{or} \qquad m = \frac{1}{c}\ln 2.$$

Also,

$$\frac{1}{2} = \lim_{b\to\infty}\int_m^b ce^{-cx}\, dx = \lim_{b\to\infty}\left[-e^{-cx}\right]_m^b = \lim_{b\to\infty}\left(e^{-cm} - e^{-cb}\right) = e^{-cm}$$

giving the same value for m. Since $1/c$ is the mean μ of X with an exponential distribution, we conclude that the median is $m = \mu \ln 2$. The mean and median differ because the probability density function is skewed and spreads toward the right.   ■

Variance and Standard Deviation

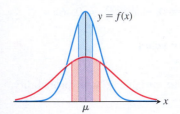

$y = f(x)$

μ

x

FIGURE 8.25   Probability density functions with the same mean can have different spreads in relation to the mean. The blue and red regions under the curves have equal area.

Random variables with exactly the same mean μ but different distributions can behave very differently (see Figure 8.25). The *variance* of a random variable X measures how spread out the values of X are in relation to the mean, and we measure this dispersion by the expected value of $(X - \mu)^2$. Since the variance measures the expected square of the difference from the mean, we often work instead with its square root.

DEFINITIONS   The **variance** of a random variable X with probability density function f is the expected value of $(X - \mu)^2$:

$$\text{Var}(X) = \int_{-\infty}^{\infty} (x - \mu)^2 f(x)\, dx$$

The **standard deviation** of X is

$$\sigma_X = \sqrt{\text{Var}(X)} = \sqrt{\int_{-\infty}^{\infty} (x - \mu)^2 f(x)\, dx}.$$

EXAMPLE 10   Find the standard deviation of the random variable T in Example 5, and find the probability that T lies within one standard deviation of the mean.

Solution   The probability density function is the exponential density function with mean $\mu = 10$ by Example 6. To find the standard deviation we first calculate the variance integral:

$$\int_{-\infty}^{\infty} (t - \mu)^2 f(t)\, dt = \int_0^{\infty} (t - 10)^2 \left(0.1e^{-0.1t}\right) dt$$

$$= \lim_{b\to\infty}\int_0^b (t - 10)^2 \left(0.1e^{-0.1t}\right) dt$$

$$= \lim_{b \to \infty} \left[\left(-(t - 10)^2 - 20(t - 10) \right) e^{-0.1t} \right]_0^b$$

$$+ \lim_{b \to \infty} \int_0^b 20 e^{-0.1t} \, dt \qquad \text{Integrating by parts}$$

$$= \left[0 + (-10)^2 + 20(-10) \right] - 20 \lim_{b \to \infty} \left(10 e^{-0.1t} \right) \Big]_0^b$$

$$= -100 - 200 \lim_{b \to \infty} \left(e^{-0.1b} - 1 \right) = 100.$$

The standard deviation is the square root of the variance, so $\sigma = 10.0$.

To find the probability that T lies within one standard deviation of the mean, we find the probability $P(\mu - \sigma \le T \le \mu + \sigma)$. For this example, we have

$$P(10 - 10 \le T \le 10 + 10) = \int_0^{20} 0.1 e^{-0.1t} \, dt = -e^{-0.1t} \Big]_0^{20} = 1 - e^{-2} \approx 0.865$$

This means that about 87% of the chips will fail within twenty years. ■

Uniform Distributions

The **uniform distribution** is very simple, but it occurs commonly in applications. The probability density function for this distribution on the interval $[a, b]$ is

$$f(x) = \frac{1}{b - a}, \quad a \le x \le b.$$

If each outcome in the sample space is equally likely to occur, then the random variable X has a uniform distribution. Since f is constant on $[a, b]$, a random variable with a uniform distribution is just as likely to be in one subinterval of a fixed length as in any other of the same length. The probability that X assumes a value in a subinterval of $[a, b]$ is the length of that subinterval divided by $(b - a)$.

EXAMPLE 11 An anchored arrow is spun around the origin, and the random variable X is the radian angle the arrow makes with the positive x-axis, measured within the interval $[0, 2\pi)$. Assuming there is equal probability for the arrow pointing in any direction, find the probability density function and the probability that the arrow ends up pointing between North and East.

Solution We model the probability density function with the uniform distribution $f(x) = 1/2\pi$, $0 \le x < 2\pi$, and $f(x) = 0$ elsewhere.

The probability that the arrow ends up pointing between North and East is given by

$$P\left(0 \le X \le \frac{\pi}{2} \right) = \int_0^{\pi/2} \frac{1}{2\pi} \, dx = \frac{1}{4}.$$ ■

Normal Distributions

Numerous applications use the **normal distribution,** which is defined by the probability density function

$$f(x) = \frac{1}{\sigma \sqrt{2\pi}} e^{-(x-\mu)^2 / 2\sigma^2}.$$

It can be shown that the mean of a random variable X with this probability density function is μ and its standard deviation is σ. In applications the values of μ and σ are often estimated using large sets of data. The function is graphed in Figure 8.26, and the graph is

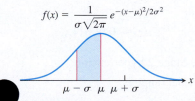

$$f(x) = \frac{1}{\sigma \sqrt{2\pi}} e^{-(x-\mu)^2/2\sigma^2}$$

FIGURE 8.26 The normal probability density function with mean μ and standard deviation σ.

sometimes called a *bell curve* because of its shape. Since the curve is symmetric about the mean, the median for X is the same as its mean. It is often observed in practice that many random variables have approximately a normal distribution. Some examples illustrating this phenomenon are the height of a man, the annual rainfall in a certain region, an individual's blood pressure, the serum cholesterol level in the blood, the brain weights in a certain population of adults, and the amount of growth in a given period for a population of sunflower seeds.

The normal probability density function does not have an antiderivative expressible in terms of familiar functions. Once μ and σ are fixed, however, an integral involving the normal probability density function can be computed using numerical integration methods. Usually we use the numerical integration capability of a computer or calculator to estimate the values of these integrals. Such computations show that for any normal distribution, we get the following values for the probability that the random variable X lies within $k = 1, 2, 3,$ or 4 standard deviations of the mean:

$$P(\mu - \sigma < X < \mu + \sigma) \approx 0.68269$$
$$P(\mu - 2\sigma < X < \mu + 2\sigma) \approx 0.95450$$
$$P(\mu - 3\sigma < X < \mu + 3\sigma) \approx 0.99730$$
$$P(\mu - 4\sigma < X < \mu + 4\sigma) \approx 0.99994$$

This means, for instance, that the random variable X will take on a value within two standard deviations of the mean about 95% of the time. About 68% of the time, X will lie within one standard deviation of the mean (see Figure 8.27).

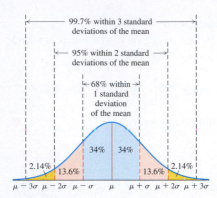

FIGURE 8.27 Probabilities of the normal distribution within its standard deviation bands.

EXAMPLE 12 An individual's blood pressure is an important indicator of overall health. A medical study of healthy individuals between 14 and 70 years of age modeled their systolic blood pressure using a normal distribution with mean 119.7 mm Hg and standard deviation 10.9 mm Hg.

(a) Using this model, what percentage of the population has a systolic blood pressure between 140 and 160 mm Hg, the levels set by the American Heart Association for Stage 1 hypertension?

(b) What percentage has a blood pressure between 160 and 180 mm Hg, the levels set by the American Heart Association for Stage 2 hypertension?

(c) What percentage has a blood pressure in the normal range of 90–120, as set by the American Heart Association?

Solution

(a) Since we cannot find an antiderivative, we use a computer to evaluate the probability integral of the normal probability density function with $\mu = 119.7$ and $\sigma = 10.9$:

$$P(140 \le X \le 160) = \int_{140}^{160} \frac{1}{10.9\sqrt{2\pi}} e^{-(x-119.7)^2/2(10.9)^2} \, dx \approx 0.03117.$$

This means that about 3% of the population in the studied age range have Stage 1 hypertension.

(b) Again we use a computer to calculate the probability that the blood pressure is between 160 and 180 mm Hg:

$$P(160 \le X \le 180) = \int_{160}^{180} \frac{1}{10.9\sqrt{2\pi}} e^{-(x-119.7)^2/2(10.9)^2} \, dx \approx 0.00011.$$

We conclude that about 0.011% of the population has Stage 2 hypertension.

(c) The probability that the blood pressure falls in the normal range is

$$P(90 \le X \le 120) = \int_{90}^{120} \frac{1}{10.9\sqrt{2\pi}} e^{-(x-119.7)^2/2(10.9)^2} \, dx \approx 0.50776.$$

That is, about 51% of the population has a normal systolic blood pressure.

Many national tests are standardized using the normal distribution. The following example illustrates modeling the discrete random variable for scores on a test using the normal distribution function for a continuous random variable.

EXAMPLE 13   The ACT is a standardized test taken by high school students seeking admission to many colleges and universities. The test measures knowledge skills and proficiency in the areas of English, math, and science, with scores ranging over the interval $[1, 36]$. Nearly 1.5 million high school students took the test in 2009, and the composite mean score across the academic areas was $\mu = 21.1$ with standard deviation $\sigma = 5.1$.

(a) What percentage of the population had an ACT score between 18 and 24?

(b) What is the ranking of a student who scored 27 on the test?

(c) What is the minimal integer score a student needed to get in order to be in the top 8% of the scoring population?

Solution

(a) We use a computer to evaluate the probability integral of the normal probability density function with $\mu = 21.1$ and $\sigma = 5.1$:

$$P(18 \leq X \leq 24) = \int_{18}^{24} \frac{1}{5.1\sqrt{2\pi}} e^{-(x-21.1)^2/2(5.1)^2} \, dx \approx 0.44355.$$

This means that about 44% of the students had an ACT score between 18 and 24.

(b) Again we use a computer to calculate the probability of a student getting a score lower than 27 on the test:

$$P(1 \leq X < 27) = \int_{1}^{27} \frac{1}{5.1\sqrt{2\pi}} e^{-(x-21.1)^2/2(5.1)^2} \, dx \approx 0.87630.$$

We conclude that about 88% of the students scored below a score of 27, so the student ranked in the top 12% of the population.

(c) We look at how many students had a mark higher than 28:

$$P(28 < X \leq 36) = \int_{28}^{36} \frac{1}{5.1\sqrt{2\pi}} e^{-(x-21.1)^2/2(5.1)^2} \, dx \approx 0.0863.$$

Since this number gives more than 8% of the students, we look at the next higher integer score:

$$P(29 < X \leq 36) = \int_{29}^{36} \frac{1}{5.1\sqrt{2\pi}} e^{-(x-21.1)^2/2(5.1)^2} \, dx \approx 0.0595.$$

Therefore, 29 is the lowest integer score a student could get in order to score in the top 8% of the population (and actually scoring here in the top 6%). ∎

The simplest form for a normal distribution of X occurs when its mean is zero and its standard deviation is one. The *standard normal probability density function f* giving mean $\mu = 0$ and standard deviation $\sigma = 1$ is

$$f(x) = \frac{1}{\sqrt{2\pi}} e^{-x^2/2}.$$

Note that the substitution $z = (x - \mu)/\sigma$ gives the equivalent integrals

$$\int_{a}^{b} \frac{1}{\sigma\sqrt{2\pi}} e^{-((x-\mu)/\sigma)^2/2} \, dx = \int_{\alpha}^{\beta} \frac{1}{\sqrt{2\pi}} e^{-z^2/2} \, dz,$$

where $\alpha = (a - \mu)/\sigma$ and $\beta = (b - \mu)/\sigma$. So we can convert random variable values to the "z-values" to standardize a normal distribution, and then use the integral on the right-hand side of the last equation to calculate probabilities for the original random variable normal distribution with mean μ and standard deviation σ. In a normal distribution, we know that 95.5% of the population lies within two standard deviations of the mean, so a random variable X converted to a z-value has more than a 95% chance of occurring in the interval $[-2, 2]$.

EXERCISES 8.9

Probability Density Functions

In Exercises 1–8, determine which are probability density functions and justify your answer.

1. $f(x) = \dfrac{1}{18}x$ over $[4, 8]$

2. $f(x) = \dfrac{1}{2}(2 - x)$ over $[0, 2]$

3. $f(x) = 2^x$ over $\left[0, \dfrac{\ln(1 + \ln 2)}{\ln 2}\right]$

4. $f(x) = x - 1$ over $[0, 1 + \sqrt{3}]$

5. $f(x) = \begin{cases} \dfrac{1}{x^2} & x \geq 1 \\ 0 & x < 1 \end{cases}$

6. $f(x) = \begin{cases} \dfrac{8}{\pi(4 + x^2)} & x \geq 0 \\ 0 & x < 0 \end{cases}$

7. $f(x) = 2\cos 2x$ over $\left[0, \dfrac{\pi}{4}\right]$

8. $f(x) = \dfrac{1}{x}$ over $(0, e]$

9. Let f be the probability density function for the random variable L in Example 2f. Explain the meaning of each integral.

a. $\displaystyle\int_{25,000}^{32,000} f(l)\, dl$ **b.** $\displaystyle\int_{30,000}^{\infty} f(l)\, dl$

c. $\displaystyle\int_{0}^{20,000} f(l)\, dl$ **d.** $\displaystyle\int_{-\infty}^{15,000} f(l)\, dl$

10. Let $f(x)$ be the uniform distribution for the random variable X in Example 11. Express the following probabilities as integrals.

a. The probability that the arrow points either between South and West or between North and West.

b. The probability that the arrow makes an angle of at least 2 radians.

Verify that the functions in Exercises 11–16 are probability density functions for a continuous random variable X over the given interval. Determine the specified probability.

11. $f(x) = xe^{-x}$ over $[0, \infty)$, $P(1 \leq X \leq 3)$

T **12.** $f(x) = \dfrac{\ln x}{x^2}$ over $[1, \infty)$, $P(2 < X < 15)$

13. $f(x) = \dfrac{3}{2}x(2 - x)$ over $[0, 1]$, $P(0.5 > X)$

T **14.** $f(x) = \dfrac{\sin^2 \pi x}{\pi x^2}$ over $\left[\dfrac{200}{1059}, \infty\right)$, $P(X < \pi/6)$

15. $f(x) = \begin{cases} \dfrac{2}{x^3} & x > 1 \\ 0 & x \leq 1 \end{cases}$ over $(-\infty, \infty)$, $P(4 \leq X < 9)$

16. $f(x) = \sin x$ over $[0, \pi/2]$, $P\left(\dfrac{\pi}{6} < X \leq \dfrac{\pi}{4}\right)$

In Exercises 17–20, find the value of the constant c so that the given function is a probability density function for a random variable over the specified interval.

17. $f(x) = \dfrac{1}{6}x$ over $[3, c]$ **18.** $f(x) = \dfrac{1}{x}$ over $[c, c + 1]$

19. $f(x) = 4e^{-2x}$ over $[0, c]$

20. $f(x) = cx\sqrt{25 - x^2}$ over $[0, 5]$

21. Let $f(x) = \dfrac{c}{1 + x^2}$. Find the value of c so that f is a probability density function. If f is a probability density function for the random variable X, find the probability $P(1 \leq X < 2)$.

22. Find the value of c so that $f(x) = c\sqrt{x}(1 - x)$ is a probability density function for the random variable X over $[0, 1]$, and find the probability $P(0.25 \leq X \leq 0.5)$.

23. Show that if the exponentially decreasing function

$$f(x) = \begin{cases} 0 & \text{if } x < 0 \\ Ae^{-cx} & \text{if } x \geq 0 \end{cases}$$

is a probability density function, then $A = c$.

24. Suppose f is a probability density function for the random variable X with mean μ. Show that its variance satisfies

$$\text{Var}(X) = \int_{-\infty}^{\infty} x^2 f(x)\, dx - \mu^2.$$

Compute the mean and median for a random variable with the probability density functions in Exercises 25–28.

25. $f(x) = \dfrac{1}{8}x$ over $[0, 4]$ **26.** $f(x) = \dfrac{1}{9}x^2$ over $[0, 3]$

27. $f(x) = \begin{cases} \dfrac{2}{x^3} & x \geq 1 \\ 0 & x < 1 \end{cases}$ **28.** $f(x) = \begin{cases} \dfrac{1}{x} & 1 \leq x \leq e \\ 0 & \text{Otherwise} \end{cases}$

Exponential Distributions

29. Digestion time The digestion time in hours of a fixed amount of food is exponentially distributed with a mean of 1 hour. What is the probability that the food is digested in less than 30 minutes?

30. Pollinating flowers   A biologist models the time in minutes until a bee arrives at a flowering plant with an exponential distribution having a mean of 4 minutes. If 1000 flowers are in a field, how many can be expected to be pollinated within 5 minutes?

31. Lifetime of light bulbs   A manufacturer of light bulbs finds that the mean lifetime of a bulb is 1200 hours. Assume the life of a bulb is exponentially distributed.

   a. Find the probability that a bulb will last less than its guaranteed lifetime of 1000 hours.

   b. In a batch of light bulbs, what is the expected time until half the light bulbs in the batch fail?

32. Lifetime of an electronic component   The life expectancy in years of a component in a microcomputer is exponentially distributed, and $1/3$ of the components fail in the first 3 years. The company that manufactures the component offers a 1 year warranty. What is the probability that a component will fail during the warranty period?

33. Lifetime of an organism   A *hydra* is a small fresh-water animal, and studies have shown that its probability of dying does not increase with the passage of time. The lack of influence of age on mortality rates for this species indicates that an exponential distribution is an appropriate model for the mortality of hydra. A biologist studies a population of 500 hydra and observes that 200 of them die within the first 2 years. How many of the hydra would you expect to die within the first six months?

34. Car accidents   The number of days that elapse between the beginning of a calendar year and the moment a high-risk driver is involved in an accident is exponentially distributed. Based on historical data, an insurance company expects that 30% of high-risk drivers will be involved in an accident during the first 50 days of the calendar year. In a group of 100 high-risk drivers, how many do you expect to be involved in an accident during the first 80 days of the calendar year?

35. Customer service time   The mean waiting time to get served after walking into a bakery is 30 seconds. Assume that an exponential density function describes the waiting times.

   a. What is the probability a customer waits 15 seconds or less?

   b. What is the probability a customer waits longer than one minute?

   c. What is the probability a customer waits exactly 5 minutes?

   d. If 200 customers come to the bakery in a day, how many are likely to be served within three minutes?

36. Airport waiting time   According to the U.S. Customs and Border Protection Agency, the average airport wait time at Chicago's O'Hare International airport is 16 minutes for a traveler arriving during the hours 7–8 A.M., and 32 minutes for arrival during the hours 4–5 P.M. The wait time is defined as the total processing time from arrival at the airport until the completion of a passenger's security screening. Assume the wait time is exponentially distributed.

   a. What is the probability of waiting between 10 and 30 minutes for a traveler arriving during the 7–8 A.M. hour?

   b. What is the probability of waiting more than 25 minutes for a traveler arriving during the 7–8 P.M. hour?

   c. What is the probability of waiting between 35 and 50 minutes for a traveler arriving during the 4–5 P.M. hour?

   d. What is the probability of waiting less than 20 minutes for a traveler arriving during the 4–5 P.M. hour?

37. Printer lifetime   The lifetime of a $200 printer is exponentially distributed with a mean of 2 years. The manufacturer agrees to pay a full refund to a buyer if the printer fails during the first year following its purchase, and a one-half refund if it fails during the second year. If the manufacturer sells 100 printers, how much should it expect to pay in refunds?

38. Failure time   The time between failures of a photocopier is exponentially distributed. Half of the copiers at a university require service during the first 2 years of operations. If the university purchased 150 copiers, how many do you expect to require service during the first year of their operation?

T Normal Distributions

39. Cholesterol levels   The serum cholesterol levels of children aged 12 to 14 years follows a normal distribution with mean $\mu = 162$ mg/dl and standard deviation $\sigma = 28$ mg/dl. In a population of 1000 of these children, how many would you expect to have serum cholesterol levels between 165 and 193? between 148 and 167?

40. Annual rainfall   The annual rainfall in inches for San Francisco, California, is approximately a normal random variable with mean 20.11 in. and standard deviation 4.7 in. What is the probability that next year's rainfall will exceed 17 in.?

41. Manufacturing time   The assembly time in minutes for a component at an electronic manufacturing plant is normally distributed with a mean of $\mu = 55$ and standard deviation $\sigma = 4$. What is the probability that a component will be made in less than one hour?

42. Lifetime of a tire   Assume the random variable L in Example 2f is normally distributed with mean $\mu = 22,000$ miles and $\sigma = 4,000$ miles.

   a. In a batch of 4000 tires, how many can be expected to last for at least 18,000 miles?

   b. What is the minimum number of miles you would expect to find as the lifetime for 90% of the tires?

43. Height   The average height of American females aged 18–24 is normally distributed with mean $\mu = 65.5$ inches and $\sigma = 2.5$ inches.

   a. What percentage of females are taller than 68 inches?

   b. What is the probability a female is between 5′1″ and 5′4″ tall?

44. Life expectancy   At birth, a French citizen has an average life expectancy of 82 years with a standard deviation of 7 years. If 100 newly born French babies are selected at random, how many would you expect to live between 75 and 85 years? Assume life expectancy is normally distributed.

45. Length of pregnancy   A team of medical practitioners determines that in a population of 1000 females with ages ranging from 20 to 35 years, the length of pregnancy from conception to birth is approximately normally distributed with a mean of 266 days and a standard deviation of 16 days. How many of these females would you expect to have a pregnancy lasting from 36 weeks to 40 weeks?

46. Brain weights In a population of 500 adult Swedish males, medical researchers find their brain weights to be approximately normally distributed with mean $\mu = 1400$ gm and standard deviation $\sigma = 100$ gm.

  a. What percentage of brain weights are between 1325 and 1450 gm?

  b. How many males in the population would you expect to have a brain weight exceeding 1480 gm?

47. Blood pressure Diastolic blood pressure in adults is normally distributed with $\mu = 80$ mm Hg and $\sigma = 12$ mm Hg. In a random sample of 300 adults, how many would be expected to have a diastolic blood pressure below 70 mm Hg?

48. Albumin levels Serum albumin in healthy 20-year-old males is normally distributed with $\mu = 4.4$ and $\sigma = 0.2$. How likely is it for a healthy 20-year-old male to have a level in the range 4.3 to 4.45?

49. Quality control A manufacturer of generator shafts finds that it needs to add additional weight to its shafts in order to achieve proper static and dynamic balance. Based on experimental tests, the average weight it needs to add is $\mu = 35$ gm with $\sigma = 9$ gm. Assuming a normal distribution, from 1000 randomly selected shafts, how many would be expected to need an added weight in excess of 40 gm?

50. Miles driven A taxicab company in New York City analyzed the daily number of miles driven by each of its drivers. It found the average distance was 200 mi with a standard deviation of 30 mi. Assuming a normal distribution, what prediction can we make about the percentage of drivers who will log in either more than 260 mi or less than 170 mi?

51. Germination of sunflower seeds The germination rate of a particular seed is the percentage of seeds in the batch which successfully emerge as plants. Assume that the germination rate for a batch of sunflower seeds is 80%, and that among a large population of n seeds the number of successful germinations is normally distributed with mean $\mu = 0.8n$ and $\sigma = 0.4\sqrt{n}$.

  a. In a batch of $n = 2500$ seeds, what is the probability that at least 1960 will successfully germinate?

  b. In a batch of $n = 2500$ seeds, what is the probability that at most 1980 will successfully germinate?

  c. In a batch of $n = 2500$ seeds, what is the probability that between 1940 and 2020 will successfully germinate?

52. Suppose you toss a fair coin n times and record the number of heads that land. Assume that n is large and approximate the discrete random variable X with a continuous random variable that is normally distributed with $\mu = n/2$ and $\sigma = \sqrt{n}/2$. If $n = 400$, find the given probabilities.

  a. $P(190 \le X < 210)$ b. $P(X < 170)$

  c. $P(X > 220)$ d. $P(X = 300)$

Discrete Random Variables

53. A fair coin is tossed four times and the random variable X assigns the number of tails that appear in each outcome.

  a. Determine the set of possible outcomes.

  b. Find the value of X for each outcome.

  c. Create a probability bar graph for X, as in Figure 8.21. What is the probability that at least two heads appear in the four tosses of the coin?

54. You roll a pair of six-sided dice, and the random variable X assigns to each outcome the sum of the number of dots showing on each face, as in Example 2e.

  a. Find the set of possible outcomes.

  b. Create a probability bar graph for X.

  c. What is the probability that $X = 8$?

  d. What is the probability that $X \le 5$? $X > 9$?

55. Three people are asked their opinion in a poll about a particular brand of a common product found in grocery stores. They can answer in one of three ways: "Like the product brand" (L), "Dislike the product brand" (D), or "Undecided" (U). For each outcome, the random variable X assigns the number of L's that appear.

  a. Find the set of possible outcomes and the range of X.

  b. Create a probability bar graph for X.

  c. What is the probability that at least two people like the product brand?

  d. What is the probability that no more than one person dislikes the product brand?

56. Spacecraft components A component of a spacecraft has both a main system and a backup system operating throughout a flight. The probability that both systems fail sometime during the flight is 0.0148. Assuming that each system separately has the same failure rate, what is the probability that the main system fails during the flight?

CHAPTER 8 **Questions to Guide Your Review**

1. What is the formula for integration by parts? Where does it come from? Why might you want to use it?

2. When applying the formula for integration by parts, how do you choose the u and dv? How can you apply integration by parts to an integral of the form $\int f(x)\, dx$?

3. If an integrand is a product of the form $\sin^n x \cos^m x$, where m and n are nonnegative integers, how do you evaluate the integral? Give a specific example of each case.

4. What substitutions are made to evaluate integrals of $\sin mx \sin nx$, $\sin mx \cos nx$, and $\cos mx \cos nx$? Give an example of each case.

5. What substitutions are sometimes used to transform integrals involving $\sqrt{a^2 - x^2}$, $\sqrt{a^2 + x^2}$, and $\sqrt{x^2 - a^2}$ into integrals that can be evaluated directly? Give an example of each case.

6. What restrictions can you place on the variables involved in the three basic trigonometric substitutions to make sure the substitutions are reversible (have inverses)?

7. What is the goal of the method of partial fractions?

8. When the degree of a polynomial $f(x)$ is less than the degree of a polynomial $g(x)$, how do you write $f(x)/g(x)$ as a sum of partial fractions if $g(x)$

   a. is a product of distinct linear factors?

   b. consists of a repeated linear factor?

   a. contains an irreducible quadratic factor?

   What do you do if the degree of f is *not* less than the degree of g?

9. How are integral tables typically used? What do you do if a particular integral you want to evaluate is not listed in the table?

10. What is a reduction formula? How are reduction formulas used? Give an example.

11. How would you compare the relative merits of Simpson's Rule and the Trapezoidal Rule?

12. What is an improper integral of Type I? Type II? How are the values of various types of improper integrals defined? Give examples.

13. What tests are available for determining the convergence and divergence of improper integrals that cannot be evaluated directly? Give examples of their use.

14. What is a random variable? What is a continuous random variable? Give some specific examples.

15. What is a probability density function? What is the probability that a continuous random variable has a value in the interval $[c, d]$?

16. What is an exponentially decreasing probability density function? What are some typical events that might be modeled by this distribution? What do we mean when we say such distributions are *memoryless*?

17. What is the expected value of a continuous random variable? What is the expected value of an exponentially distributed random variable?

18. What is the median of a continuous random variable? What is the median of an exponential distribution?

19. What does the variance of a random variable measure? What is the standard deviation of a continuous random variable X?

20. What probability density function describes the normal distribution? What are some examples typically modeled by a normal distribution? How do we usually calculate probabilities for a normal distribution?

21. In a normal distribution, what percentage of the population lies within 1 standard deviation of the mean? Within 2 standard deviations?

CHAPTER 8   Practice Exercises

Integration by Parts
Evaluate the integrals in Exercises 1–8 using integration by parts.

1. $\displaystyle\int \ln (x + 1)\, dx$

2. $\displaystyle\int x^2 \ln x\, dx$

3. $\displaystyle\int \tan^{-1} 3x\, dx$

4. $\displaystyle\int \cos^{-1}\left(\frac{x}{2}\right) dx$

5. $\displaystyle\int (x + 1)^2 e^x\, dx$

6. $\displaystyle\int x^2 \sin (1 - x)\, dx$

7. $\displaystyle\int e^x \cos 2x\, dx$

8. $\displaystyle\int x \sin x \cos x\, dx$

Partial Fractions
Evaluate the integrals in Exercises 9–28. It may be necessary to use a substitution first.

9. $\displaystyle\int \frac{x\, dx}{x^2 - 3x + 2}$

10. $\displaystyle\int \frac{x\, dx}{x^2 + 4x + 3}$

11. $\displaystyle\int \frac{dx}{x(x + 1)^2}$

12. $\displaystyle\int \frac{x + 1}{x^2(x - 1)}\, dx$

13. $\displaystyle\int \frac{\sin \theta\, d\theta}{\cos^2 \theta + \cos \theta - 2}$

14. $\displaystyle\int \frac{\cos \theta\, d\theta}{\sin^2 \theta + \sin \theta - 6}$

15. $\displaystyle\int \frac{3x^2 + 4x + 4}{x^3 + x}\, dx$

16. $\displaystyle\int \frac{4x\, dx}{x^3 + 4x}$

17. $\displaystyle\int \frac{v + 3}{2v^3 - 8v}\, dv$

18. $\displaystyle\int \frac{(3v - 7)\, dv}{(v - 1)(v - 2)(v - 3)}$

19. $\displaystyle\int \frac{dt}{t^4 + 4t^2 + 3}$

20. $\displaystyle\int \frac{t\, dt}{t^4 - t^2 - 2}$

21. $\displaystyle\int \frac{x^3 + x^2}{x^2 + x - 2}\, dx$

22. $\displaystyle\int \frac{x^3 + 1}{x^3 - x}\, dx$

23. $\displaystyle\int \frac{x^3 + 4x^2}{x^2 + 4x + 3}\, dx$

24. $\displaystyle\int \frac{2x^3 + x^2 - 21x + 24}{x^2 + 2x - 8}\, dx$

25. $\displaystyle\int \frac{dx}{x(3\sqrt{x} + 1)}$

26. $\displaystyle\int \frac{dx}{x(1 + \sqrt[3]{x})}$

27. $\displaystyle\int \frac{ds}{e^s - 1}$

28. $\displaystyle\int \frac{ds}{\sqrt{e^s + 1}}$

Trigonometric Substitutions
Evaluate the integrals in Exercises 29–32 **(a)** without using a trigonometric substitution, **(b)** using a trigonometric substitution.

29. $\displaystyle\int \frac{y\, dy}{\sqrt{16 - y^2}}$

30. $\displaystyle\int \frac{x\, dx}{\sqrt{4 + x^2}}$

31. $\displaystyle\int \frac{x\, dx}{4 - x^2}$

32. $\displaystyle\int \frac{t\, dt}{\sqrt{4t^2 - 1}}$

Evaluate the integrals in Exercises 33–36.

33. $\displaystyle\int \frac{x\, dx}{9 - x^2}$

34. $\displaystyle\int \frac{dx}{x(9 - x^2)}$

35. $\displaystyle\int \frac{dx}{9 - x^2}$

36. $\displaystyle\int \frac{dx}{\sqrt{9 - x^2}}$

Trigonometric Integrals

Evaluate the integrals in Exercises 37–44.

37. $\displaystyle\int \sin^3 x \cos^4 x \, dx$

38. $\displaystyle\int \cos^5 x \sin^5 x \, dx$

39. $\displaystyle\int \tan^4 x \sec^2 x \, dx$

40. $\displaystyle\int \tan^3 x \sec^3 x \, dx$

41. $\displaystyle\int \sin 5\theta \cos 6\theta \, d\theta$

42. $\displaystyle\int \sec^2 \theta \sin^3 \theta \, d\theta$

43. $\displaystyle\int \sqrt{1 + \cos{(t/2)}} \, dt$

44. $\displaystyle\int e^t \sqrt{\tan^2 e^t + 1} \, dt$

Numerical Integration

45. According to the error-bound formula for Simpson's Rule, how many subintervals should you use to be sure of estimating the value of

$$\ln 3 = \int_1^3 \frac{1}{x} \, dx$$

by Simpson's Rule with an error of no more than 10^{-4} in absolute value? (Remember that for Simpson's Rule, the number of subintervals has to be even.)

46. A brief calculation shows that if $0 \le x \le 1$, then the second derivative of $f(x) = \sqrt{1 + x^4}$ lies between 0 and 8. Based on this, about how many subdivisions would you need to estimate the integral of f from 0 to 1 with an error no greater than 10^{-3} in absolute value using the Trapezoidal Rule?

47. A direct calculation shows that

$$\int_0^\pi 2 \sin^2 x \, dx = \pi.$$

How close do you come to this value by using the Trapezoidal Rule with $n = 6$? Simpson's Rule with $n = 6$? Try them and find out.

48. You are planning to use Simpson's Rule to estimate the value of the integral

$$\int_1^2 f(x) \, dx$$

with an error magnitude less than 10^{-5}. You have determined that $|f^{(4)}(x)| \le 3$ throughout the interval of integration. How many subintervals should you use to ensure the required accuracy? (Remember that for Simpson's Rule the number has to be even.)

T **49. Mean temperature** Use Simpson's Rule to approximate the average value of the temperature function

$$f(x) = 37 \sin\left(\frac{2\pi}{365}(x - 101)\right) + 25$$

for a 365-day year. This is one way to estimate the annual mean air temperature in Fairbanks, Alaska. The National Weather Service's official figure, a numerical average of the daily normal mean air temperatures for the year, is 25.7°F, which is slightly higher than the average value of $f(x)$.

50. Heat capacity of a gas Heat capacity C_v is the amount of heat required to raise the temperature of a given mass of gas with constant volume by 1°C, measured in units of cal/deg-mol (calories

per degree gram molecular weight). The heat capacity of oxygen depends on its temperature T and satisfies the formula

$$C_v = 8.27 + 10^{-5}\,(26T - 1.87T^2).$$

Use Simpson's Rule to find the average value of C_v and the temperature at which it is attained for $20° \le T \le 675°C$.

51. Fuel efficiency An automobile computer gives a digital readout of fuel consumption in gallons per hour. During a trip, a passenger recorded the fuel consumption every 5 min for a full hour of travel.

Time	Gal/h	Time	Gal/h
0	2.5	35	2.5
5	2.4	40	2.4
10	2.3	45	2.3
15	2.4	50	2.4
20	2.4	55	2.4
25	2.5	60	2.3
30	2.6		

a. Use the Trapezoidal Rule to approximate the total fuel consumption during the hour.

b. If the automobile covered 60 mi in the hour, what was its fuel efficiency (in miles per gallon) for that portion of the trip?

52. A new parking lot To meet the demand for parking, your town has allocated the area shown here. As the town engineer, you have been asked by the town council to find out if the lot can be built for $11,000. The cost to clear the land will be $0.10 a square foot, and the lot will cost $2.00 a square foot to pave. Use Simpson's Rule to find out if the job can be done for $11,000.

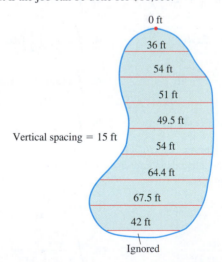

0 ft

36 ft

54 ft

51 ft

49.5 ft

Vertical spacing = 15 ft

54 ft

64.4 ft

67.5 ft

42 ft

Ignored

Improper Integrals

Evaluate the improper integrals in Exercises 53–62.

53. $\displaystyle\int_0^3 \frac{dx}{\sqrt{9 - x^2}}$

54. $\displaystyle\int_0^1 \ln x \, dx$

55. $\displaystyle\int_0^2 \frac{dy}{(y - 1)^{2/3}}$

56. $\displaystyle\int_{-2}^0 \frac{d\theta}{(\theta + 1)^{3/5}}$

57. $\displaystyle\int_{3}^{\infty} \frac{2\,du}{u^2 - 2u}$

58. $\displaystyle\int_{1}^{\infty} \frac{3v - 1}{4v^3 - v^2}\,dv$

59. $\displaystyle\int_{0}^{\infty} x^2 e^{-x}\,dx$

60. $\displaystyle\int_{-\infty}^{0} x e^{3x}\,dx$

61. $\displaystyle\int_{-\infty}^{\infty} \frac{dx}{4x^2 + 9}$

62. $\displaystyle\int_{-\infty}^{\infty} \frac{4\,dx}{x^2 + 16}$

Which of the improper integrals in Exercises 63–68 converge and which diverge?

63. $\displaystyle\int_{6}^{\infty} \frac{d\theta}{\sqrt{\theta^2 + 1}}$

64. $\displaystyle\int_{0}^{\infty} e^{-u}\cos u\,du$

65. $\displaystyle\int_{1}^{\infty} \frac{\ln z}{z}\,dz$

66. $\displaystyle\int_{1}^{\infty} \frac{e^{-t}}{\sqrt{t}}\,dt$

67. $\displaystyle\int_{-\infty}^{\infty} \frac{2\,dx}{e^x + e^{-x}}$

68. $\displaystyle\int_{-\infty}^{\infty} \frac{dx}{x^2(1 + e^x)}$

Assorted Integrations

Evaluate the integrals in Exercises 69–136. The integrals are listed in random order so you need to decide which integration technique to use.

69. $\displaystyle\int x e^{2x}\,dx$

70. $\displaystyle\int_{0}^{1} x^2 e^{x^3}\,dx$

71. $\displaystyle\int (\tan^2 x + \sec^2 x)\,dx$

72. $\displaystyle\int_{0}^{\pi/4} \cos^2 2x\,dx$

73. $\displaystyle\int x \sec^2 x\,dx$

74. $\displaystyle\int x \sec^2 (x^2)\,dx$

75. $\displaystyle\int \sin x \cos^2 x\,dx$

76. $\displaystyle\int \sin 2x \sin (\cos 2x)\,dx$

77. $\displaystyle\int_{-1}^{0} \frac{e^x}{e^x + e^{-x}}\,dx$

78. $\displaystyle\int (e^{2x} + e^{-x})^2\,dx$

79. $\displaystyle\int \frac{x + 1}{x^4 - x^3}\,dx$

80. $\displaystyle\int \frac{e^x + 1}{e^x(e^{2x} - 4)}\,dx$

81. $\displaystyle\int \frac{e^x + e^{3x}}{e^{2x}}\,dx$

82. $\displaystyle\int (e^x - e^{-x})(e^x + e^{-x})^3\,dx$

83. $\displaystyle\int_{0}^{\pi/3} \tan^3 x \sec^2 x\,dx$

84. $\displaystyle\int \tan^4 x \sec^4 x\,dx$

85. $\displaystyle\int_{0}^{3} (x + 2)\sqrt{x + 1}\,dx$

86. $\displaystyle\int (x + 1)\sqrt{x^2 + 2x}\,dx$

87. $\displaystyle\int \cot x \csc^3 x\,dx$

88. $\displaystyle\int \sin x (\tan x - \cot x)^2\,dx$

89. $\displaystyle\int \frac{x\,dx}{1 + \sqrt{x}}$

90. $\displaystyle\int \frac{x^3 + 2}{4 - x^2}\,dx$

91. $\displaystyle\int \sqrt{2x - x^2}\,dx$

92. $\displaystyle\int \frac{dx}{\sqrt{-2x - x^2}}$

93. $\displaystyle\int \frac{2 - \cos x + \sin x}{\sin^2 x}\,dx$

94. $\displaystyle\int \sin^2 \theta \cos^5 \theta\,d\theta$

95. $\displaystyle\int \frac{9\,dv}{81 - v^4}$

96. $\displaystyle\int_{2}^{\infty} \frac{dx}{(x - 1)^2}$

97. $\displaystyle\int \theta \cos (2\theta + 1)\,d\theta$

98. $\displaystyle\int \frac{x^3\,dx}{x^2 - 2x + 1}$

99. $\displaystyle\int \frac{\sin 2\theta\,d\theta}{(1 + \cos 2\theta)^2}$

100. $\displaystyle\int_{\pi/4}^{\pi/2} \sqrt{1 + \cos 4x}\,dx$

101. $\displaystyle\int \frac{x\,dx}{\sqrt{2 - x}}$

102. $\displaystyle\int \frac{\sqrt{1 - v^2}}{v^2}\,dv$

103. $\displaystyle\int \frac{dy}{y^2 - 2y + 2}$

104. $\displaystyle\int \frac{x\,dx}{\sqrt{8 - 2x^2 - x^4}}$

105. $\displaystyle\int \frac{z + 1}{z^2(z^2 + 4)}\,dz$

106. $\displaystyle\int x^2(x - 1)^{1/3}\,dx$

107. $\displaystyle\int \frac{t\,dt}{\sqrt{9 - 4t^2}}$

108. $\displaystyle\int \frac{\tan^{-1} x}{x^2}\,dx$

109. $\displaystyle\int \frac{e^t\,dt}{e^{2t} + 3e^t + 2}$

110. $\displaystyle\int \tan^3 t\,dt$

111. $\displaystyle\int_{1}^{\infty} \frac{\ln y}{y^3}\,dy$

112. $\displaystyle\int y^{3/2}(\ln y)^2\,dy$

113. $\displaystyle\int e^{\ln \sqrt{x}}\,dx$

114. $\displaystyle\int e^{\theta}\sqrt{3 + 4e^{\theta}}\,d\theta$

115. $\displaystyle\int \frac{\sin 5t\,dt}{1 + (\cos 5t)^2}$

116. $\displaystyle\int \frac{dv}{\sqrt{e^{2v} - 1}}$

117. $\displaystyle\int \frac{dr}{1 + \sqrt{r}}$

118. $\displaystyle\int \frac{4x^3 - 20x}{x^4 - 10x^2 + 9}\,dx$

119. $\displaystyle\int \frac{x^3}{1 + x^2}\,dx$

120. $\displaystyle\int \frac{x^2}{1 + x^3}\,dx$

121. $\displaystyle\int \frac{1 + x^2}{1 + x^3}\,dx$

122. $\displaystyle\int \frac{1 + x^2}{(1 + x)^3}\,dx$

123. $\displaystyle\int \sqrt{x} \cdot \sqrt{1 + \sqrt{x}}\,dx$

124. $\displaystyle\int \sqrt{1 + \sqrt{1 + x}}\,dx$

125. $\displaystyle\int \frac{1}{\sqrt{x} \cdot \sqrt{1 + x}}\,dx$

126. $\displaystyle\int_{0}^{1/2} \sqrt{1 + \sqrt{1 - x^2}}\,dx$

127. $\displaystyle\int \frac{\ln x}{x + x \ln x}\,dx$

128. $\displaystyle\int \frac{1}{x \cdot \ln x \cdot \ln (\ln x)}\,dx$

129. $\displaystyle\int \frac{x^{\ln x} \ln x}{x}\,dx$

130. $\displaystyle\int (\ln x)^{\ln x}\left[\frac{1}{x} + \frac{\ln (\ln x)}{x}\right]dx$

131. $\displaystyle\int \frac{1}{x\sqrt{1 - x^4}}\,dx$

132. $\displaystyle\int \frac{\sqrt{1 - x}}{x}\,dx$

133. a. Show that $\displaystyle\int_{0}^{a} f(x)\,dx = \int_{0}^{a} f(a - x)\,dx$.

b. Use part (a) to evaluate

$$\int_{0}^{\pi/2} \frac{\sin x}{\sin x + \cos x}\,dx.$$

134. $\displaystyle\int \frac{\sin x}{\sin x + \cos x}\,dx$

135. $\displaystyle\int \frac{\sin^2 x}{1 + \sin^2 x}\,dx$

136. $\displaystyle\int \frac{1 - \cos x}{1 + \cos x}\,dx$

CHAPTER 8 Additional and Advanced Exercises

Evaluating Integrals

Evaluate the integrals in Exercises 1–6.

1. $\displaystyle\int (\sin^{-1} x)^2 \, dx$

2. $\displaystyle\int \frac{dx}{x(x + 1)(x + 2)\cdots(x + m)}$

3. $\displaystyle\int x \sin^{-1} x \, dx$

4. $\displaystyle\int \sin^{-1} \sqrt{y} \, dy$

5. $\displaystyle\int \frac{dt}{t - \sqrt{1 - t^2}}$

6. $\displaystyle\int \frac{dx}{x^4 + 4}$

Evaluate the limits in Exercise 7 and 8.

7. $\displaystyle\lim_{x \to \infty} \int_{-x}^{x} \sin t \, dt$

8. $\displaystyle\lim_{x \to 0^+} x \int_{x}^{1} \frac{\cos t}{t^2} \, dt$

Evaluate the limits in Exercise 9 and 10 by identifying them with definite integrals and evaluating the integrals.

9. $\displaystyle\lim_{n \to \infty} \sum_{k=1}^{n} \ln \sqrt[n]{1 + \frac{k}{n}}$

10. $\displaystyle\lim_{n \to \infty} \sum_{k=0}^{n-1} \frac{1}{\sqrt{n^2 - k^2}}$

Applications

11. Finding arc length Find the length of the curve

$$y = \int_{0}^{x} \sqrt{\cos 2t} \, dt, \quad 0 \le x \le \pi/4.$$

12. Finding arc length Find the length of the graph of the function $y = \ln(1 - x^2)$, $0 \le x \le 1/2$.

13. Finding volume The region in the first quadrant that is enclosed by the x-axis and the curve $y = 3x\sqrt{1 - x}$ is revolved about the y-axis to generate a solid. Find the volume of the solid.

14. Finding volume The region in the first quadrant that is enclosed by the x-axis, the curve $y = 5/(x\sqrt{5 - x})$, and the lines $x = 1$ and $x = 4$ is revolved about the x-axis to generate a solid. Find the volume of the solid.

15. Finding volume The region in the first quadrant enclosed by the coordinate axes, the curve $y = e^x$, and the line $x = 1$ is revolved about the y-axis to generate a solid. Find the volume of the solid.

16. Finding volume The region in the first quadrant that is bounded above by the curve $y = e^x - 1$, below by the x-axis, and on the right by the line $x = \ln 2$ is revolved about the line $x = \ln 2$ to generate a solid. Find the volume of the solid.

17. Finding volume Let R be the "triangular" region in the first quadrant that is bounded above by the line $y = 1$, below by the curve $y = \ln x$, and on the left by the line $x = 1$. Find the volume of the solid generated by revolving R about

a. the x-axis. **b.** the line $y = 1$.

18. Finding volume (*Continuation of Exercise 17.*) Find the volume of the solid generated by revolving the region R about

a. the y-axis. **b.** the line $x = 1$.

19. Finding volume The region between the x-axis and the curve

$$y = f(x) = \begin{cases} 0, & x = 0 \\ x \ln x, & 0 < x \le 2 \end{cases}$$

is revolved about the x-axis to generate the solid shown here.

a. Show that f is continuous at $x = 0$.

b. Find the volume of the solid.

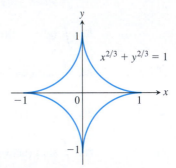

20. Finding volume The infinite region bounded by the coordinate axes and the curve $y = -\ln x$ in the first quadrant is revolved about the x-axis to generate a solid. Find the volume of the solid.

21. Centroid of a region Find the centroid of the region in the first quadrant that is bounded below by the x-axis, above by the curve $y = \ln x$, and on the right by the line $x = e$.

22. Centroid of a region Find the centroid of the region in the plane enclosed by the curves $y = \pm(1 - x^2)^{-1/2}$ and the lines $x = 0$ and $x = 1$.

23. Length of a curve Find the length of the curve $y = \ln x$ from $x = 1$ to $x = e$.

24. Finding surface area Find the area of the surface generated by revolving the curve in Exercise 23 about the y-axis.

25. The surface generated by an astroid The graph of the equation $x^{2/3} + y^{2/3} = 1$ is an *astroid* (see accompanying figure). Find the area of the surface generated by revolving the curve about the x-axis.

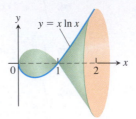

26. Length of a curve Find the length of the curve

$$y = \int_{1}^{x} \sqrt{\sqrt{t} - 1} \, dt, \quad 1 \le x \le 16.$$

27. For what value or values of a does

$$\int_{1}^{\infty} \left(\frac{ax}{x^2 + 1} - \frac{1}{2x} \right) dx$$

converge? Evaluate the corresponding integral(s).

28. For each $x > 0$, let $G(x) = \int_0^\infty e^{-xt}\, dt$. Prove that $xG(x) = 1$ for each $x > 0$.

29. Infinite area and finite volume What values of p have the following property: The area of the region between the curve $y = x^{-p}$, $1 \le x < \infty$, and the x-axis is infinite but the volume of the solid generated by revolving the region about the x-axis is finite.

30. Infinite area and finite volume What values of p have the following property: The area of the region in the first quadrant enclosed by the curve $y = x^{-p}$, the y-axis, the line $x = 1$, and the interval $[0, 1]$ on the x-axis is infinite but the volume of the solid generated by revolving the region about one of the coordinate axes is finite.

31. Integrating the square of the derivative If f is continuously differentiable on $[0, 1]$ and $f(1) = f(0) = -1/6$, prove that

$$\int_0^1 (f'(x))^2\, dx \ge 2\int_0^1 f(x)\, dx + \frac{1}{4}.$$

Hint: Consider the inequality $0 \le \int_0^1 \left(f'(x) + x - \frac{1}{2}\right)^2 dx.$

Source: Mathematics Magazine, vol. 84, no. 4, Oct. 2011.

32. (*Continuation of Exercise 31.*) If f is continuously differentiable on $[0, a]$ for $a > 0$, and $f(a) = f(0) = b$, prove that

$$\int_0^a (f'(x))^2\, dx \ge 2\int_0^a f(x)\, dx - \left(2ab + \frac{a^3}{12}\right).$$

Hint: Consider the inequality $0 \le \int_0^a \left(f'(x) + x - \frac{a}{2}\right)^2 dx.$

Source: Mathematics Magazine, vol. 84, no. 4, Oct. 2011.

The Substitution $z = \tan(x/2)$

The substitution

$$z = \tan\frac{x}{2} \tag{1}$$

reduces the problem of integrating a rational expression in $\sin x$ and $\cos x$ to a problem of integrating a rational function of z. This in turn can be integrated by partial fractions.

From the accompanying figure

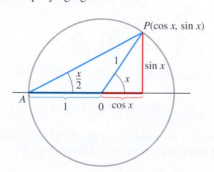

we can read the relation

$$\tan\frac{x}{2} = \frac{\sin x}{1 + \cos x}.$$

To see the effect of the substitution, we calculate

$$\cos x = 2\cos^2\left(\frac{x}{2}\right) - 1 = \frac{2}{\sec^2(x/2)} - 1$$

$$= \frac{2}{1 + \tan^2(x/2)} - 1 = \frac{2}{1 + z^2} - 1$$

$$\cos x = \frac{1 - z^2}{1 + z^2}, \tag{2}$$

and

$$\sin x = 2\sin\frac{x}{2}\cos\frac{x}{2} = 2\frac{\sin(x/2)}{\cos(x/2)} \cdot \cos^2\left(\frac{x}{2}\right)$$

$$= 2\tan\frac{x}{2} \cdot \frac{1}{\sec^2(x/2)} = \frac{2\tan(x/2)}{1 + \tan^2(x/2)}$$

$$\sin x = \frac{2z}{1 + z^2}. \tag{3}$$

Finally, $x = 2\tan^{-1} z$, so

$$dx = \frac{2\,dz}{1 + z^2}. \tag{4}$$

Examples

a. $\displaystyle\int \frac{1}{1 + \cos x}\, dx = \int \frac{1 + z^2}{2}\, \frac{2\,dz}{1 + z^2}$

$\displaystyle = \int dz = z + C$

$\displaystyle = \tan\left(\frac{x}{2}\right) + C$

b. $\displaystyle\int \frac{1}{2 + \sin x}\, dx = \int \frac{1 + z^2}{2 + 2z + 2z^2}\, \frac{2\,dz}{1 + z^2}$

$\displaystyle = \int \frac{dz}{z^2 + z + 1} = \int \frac{dz}{(z + (1/2))^2 + 3/4}$

$\displaystyle = \int \frac{du}{u^2 + a^2}$

$\displaystyle = \frac{1}{a}\tan^{-1}\left(\frac{u}{a}\right) + C$

$\displaystyle = \frac{2}{\sqrt{3}}\tan^{-1}\frac{2z + 1}{\sqrt{3}} + C$

$\displaystyle = \frac{2}{\sqrt{3}}\tan^{-1}\frac{1 + 2\tan(x/2)}{\sqrt{3}} + C$

Use the substitutions in Equations (1)–(4) to evaluate the integrals in Exercises 33–40. Integrals like these arise in calculating the average angular velocity of the output shaft of a universal joint when the input and output shafts are not aligned.

33. $\displaystyle\int \frac{dx}{1 - \sin x}$ **34.** $\displaystyle\int \frac{dx}{1 + \sin x + \cos x}$

35. $\displaystyle\int_0^{\pi/2} \frac{dx}{1 + \sin x}$ **36.** $\displaystyle\int_{\pi/3}^{\pi/2} \frac{dx}{1 - \cos x}$

37. $\displaystyle\int_0^{\pi/2} \frac{d\theta}{2 + \cos\theta}$

38. $\displaystyle\int_{\pi/2}^{2\pi/3} \frac{\cos\theta\, d\theta}{\sin\theta\cos\theta + \sin\theta}$

39. $\displaystyle\int \frac{dt}{\sin t - \cos t}$

40. $\displaystyle\int \frac{\cos t\, dt}{1 - \cos t}$

Use the substitution $z = \tan(\theta/2)$ to evaluate the integrals in Exercises 41 and 42.

41. $\displaystyle\int \sec\theta\, d\theta$

42. $\displaystyle\int \csc\theta\, d\theta$

The Gamma Function and Stirling's Formula

Euler's gamma function $\Gamma(x)$ ("gamma of x"; Γ is a Greek capital g) uses an integral to extend the factorial function from the nonnegative integers to other real values. The formula is

$$\Gamma(x) = \int_0^\infty t^{x-1} e^{-t}\, dt, \quad x > 0.$$

For each positive x, the number $\Gamma(x)$ is the integral of $t^{x-1} e^{-t}$ with respect to t from 0 to ∞. Figure 8.28 shows the graph of Γ near the origin. You will see how to calculate $\Gamma(1/2)$ if you do Additional Exercise 23 in Chapter 15.

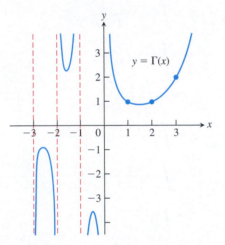

FIGURE 8.28 Euler's gamma function $\Gamma(x)$ is a continuous function of x whose value at each positive integer $n + 1$ is $n!$. The defining integral formula for Γ is valid only for $x > 0$, but we can extend Γ to negative noninteger values of x with the formula $\Gamma(x) = (\Gamma(x + 1))/x$, which is the subject of Exercise 43.

43. If n is a nonnegative integer, $\Gamma(n + 1) = n!$

a. Show that $\Gamma(1) = 1$.

b. Then apply integration by parts to the integral for $\Gamma(x + 1)$ to show that $\Gamma(x + 1) = x\Gamma(x)$. This gives

$$\Gamma(2) = 1\Gamma(1) = 1$$
$$\Gamma(3) = 2\Gamma(2) = 2$$
$$\Gamma(4) = 3\Gamma(3) = 6$$
$$\vdots$$
$$\Gamma(n + 1) = n\,\Gamma(n) = n! \tag{5}$$

c. Use mathematical induction to verify Equation (5) for every nonnegative integer n.

44. Stirling's formula Scottish mathematician James Stirling (1692–1770) showed that

$$\lim_{x\to\infty} \left(\frac{e}{x}\right)^x \sqrt{\frac{x}{2\pi}}\, \Gamma(x) = 1,$$

so, for large x,

$$\Gamma(x) = \left(\frac{x}{e}\right)^x \sqrt{\frac{2\pi}{x}}\,(1 + \varepsilon(x)), \qquad \varepsilon(x) \to 0 \text{ as } x \to \infty. \tag{6}$$

Dropping $\varepsilon(x)$ leads to the approximation

$$\Gamma(x) \approx \left(\frac{x}{e}\right)^x \sqrt{\frac{2\pi}{x}} \qquad \textbf{(Stirling's formula).} \tag{7}$$

a. Stirling's approximation for $n!$ Use Equation (7) and the fact that $n! = n\Gamma(n)$ to show that

$$n! \approx \left(\frac{n}{e}\right)^n \sqrt{2n\pi} \qquad \textbf{(Stirling's approximation).} \tag{8}$$

As you will see if you do Exercise 114 in Section 10.1, Equation (8) leads to the approximation

$$\sqrt[n]{n!} \approx \frac{n}{e}. \tag{9}$$

T **b.** Compare your calculator's value for $n!$ with the value given by Stirling's approximation for $n = 10, 20, 30, \ldots$, as far as your calculator can go.

T **c.** A refinement of Equation (6) gives

$$\Gamma(x) = \left(\frac{x}{e}\right)^x \sqrt{\frac{2\pi}{x}}\, e^{1/(12x)}(1 + \varepsilon(x))$$

or

$$\Gamma(x) \approx \left(\frac{x}{e}\right)^x \sqrt{\frac{2\pi}{x}}\, e^{1/(12x)},$$

which tells us that

$$n! \approx \left(\frac{n}{e}\right)^n \sqrt{2n\pi}\, e^{1/(12n)}. \tag{10}$$

Compare the values given for 10! by your calculator, Stirling's approximation, and Equation (10).

| CHAPTER 8 | Technology Application Projects |

Mathematica/Maple Projects

Projects can be found within MyMathLab.

- *Riemann, Trapezoidal, and Simpson Approximations*
  Part I: Visualize the error involved in using Riemann sums to approximate the area under a curve.
  Part II: Build a table of values and compute the relative magnitude of the error as a function of the step size Δx.
  Part III: Investigate the effect of the derivative function on the error.
  Parts IV and V: Trapezoidal Rule approximations.
  Part VI: Simpson's Rule approximations.

- *Games of Chance: Exploring the Monte Carlo Probabilistic Technique for Numerical Integration*
  Graphically explore the Monte Carlo method for approximating definite integrals.

- *Computing Probabilities with Improper Integrals*
  More explorations of the Monte Carlo method for approximating definite integrals.

9

First-Order Differential Equations

OVERVIEW Many real-world problems, when formulated mathematically, lead to differential equations. We encountered a number of these equations in previous chapters when studying phenomena such as the motion of an object along a straight line, the decay of a radioactive material, the growth of a population, and the cooling of a heated object placed within a medium of lower temperature.

In Section 4.8 we introduced differential equations of the form $dy/dx = f(x)$, where f is given and y is an unknown function of x. When f is continuous over some interval, we learned that the general solution $y(x)$ was found directly by integration, $y = \int f(x)\, dx$. Next, in Section 7.2, we investigated differential equations of the form $dy/dx = f(x, y)$, where f is a function of both the independent variable x and the dependent variable y. There we learned how to find the general solution for the special case when the differential equation is separable. In this chapter we further extend our study to include other commonly occurring *first-order* differential equations. These differential equations involve only first derivatives of the unknown function $y(x)$, and they model phenomena varying from simple electrical circuits to the concentration of a chemical in a container. Differential equations involving second derivatives are studied in Chapter 17.

Chapter 17 is available online.
www.goo.gl/MgDXPY

9.1 Solutions, Slope Fields, and Euler's Method

We begin this section by defining general differential equations involving first derivatives. We then look at slope fields, which give a geometric picture of the solutions to such equations. Many differential equations cannot be solved by obtaining an explicit formula for the solution. However, we can often find numerical approximations to solutions. We present one such method here, called *Euler's method*, which is the basis for many other numerical methods as well.

General First-Order Differential Equations and Solutions

A **first-order differential equation** is an equation

$$\frac{dy}{dx} = f(x, y) \tag{1}$$

in which $f(x, y)$ is a function of two variables defined on a region in the xy-plane. The equation is of *first order* because it involves only the first derivative dy/dx (and not higher-order derivatives). In a typical situation y represents an unknown function of x, and $f(x, y)$ is a known function. Some examples of first-order differential equations are:

$y' = x + y$, $y' = y/x$, and $y' = 3xy$. In all cases we should think of y as an unknown function of x whose derivative is given by $f(x, y)$. The equations

$$y' = f(x, y) \qquad \text{and} \qquad \frac{d}{dx}y = f(x, y)$$

are equivalent to Equation (1) and all three forms will be used interchangeably in the text.

A **solution** of Equation (1) is a differentiable function $y = y(x)$ defined on an interval I of x-values (perhaps infinite) such that

$$\frac{d}{dx}y(x) = f(x, y(x))$$

on that interval. That is, when $y(x)$ and its derivative $y'(x)$ are substituted into Equation (1), the resulting equation is true for all x over the interval I. The **general solution** to a first-order differential equation is a solution that contains all possible solutions. The general solution always contains an arbitrary constant, but having this property doesn't mean a solution is the general solution. That is, a solution may contain an arbitrary constant without being the general solution. Establishing that a solution *is* the general solution may require deeper results from the theory of differential equations and is left to a more advanced course.

EXAMPLE 1   Show that every member of the family of functions

$$y = \frac{C}{x} + 2$$

is a solution of the first-order differential equation

$$\frac{dy}{dx} = \frac{1}{x}(2 - y)$$

on the interval $(0, \infty)$, where C is any constant.

Solution   Differentiating $y = C/x + 2$ gives

$$\frac{dy}{dx} = C\frac{d}{dx}\left(\frac{1}{x}\right) + 0 = -\frac{C}{x^2}.$$

We need to show that the differential equation is satisfied when we substitute into it the expressions $(C/x) + 2$ for y, and $-C/x^2$ for dy/dx. That is, we need to verify that for all $x \in (0, \infty)$,

$$-\frac{C}{x^2} = \frac{1}{x}\left[2 - \left(\frac{C}{x} + 2\right)\right].$$

This last equation follows immediately by expanding the expression on the right-hand side:

$$\frac{1}{x}\left[2 - \left(\frac{C}{x} + 2\right)\right] = \frac{1}{x}\left(-\frac{C}{x}\right) = -\frac{C}{x^2}.$$

Therefore, for every value of C, the function $y = C/x + 2$ is a solution of the differential equation.   ■

As with antiderivatives, we often need a *particular* rather than the general solution to a first-order differential equation $y' = f(x, y)$. The **particular solution** satisfying the initial condition $y(x_0) = y_0$ is the solution $y = y(x)$ whose value is y_0 when $x = x_0$. Thus the graph of the particular solution passes through the point (x_0, y_0) in the xy-plane. A **first-order initial value problem** is a differential equation $y' = f(x, y)$ whose solution must satisfy an initial condition $y(x_0) = y_0$.

EXAMPLE 2   Show that the function

$$y = (x + 1) - \frac{1}{3}e^x$$

is a solution to the first-order initial value problem

$$\frac{dy}{dx} = y - x, \qquad y(0) = \frac{2}{3}.$$

Solution   The equation

$$\frac{dy}{dx} = y - x$$

is a first-order differential equation with $f(x, y) = y - x$.

On the left side of the equation:

$$\frac{dy}{dx} = \frac{d}{dx}\left(x + 1 - \frac{1}{3}e^x\right) = 1 - \frac{1}{3}e^x.$$

On the right side of the equation:

$$y - x = (x + 1) - \frac{1}{3}e^x - x = 1 - \frac{1}{3}e^x.$$

The function satisfies the initial condition because

$$y(0) = \left[(x + 1) - \frac{1}{3}e^x\right]_{x=0} = 1 - \frac{1}{3} = \frac{2}{3}.$$

The graph of the function is shown in Figure 9.1.

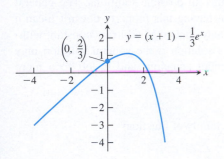

FIGURE 9.1 Graph of the solution to the initial value problem in Example 2.

Slope Fields: Viewing Solution Curves

Each time we specify an initial condition $y(x_0) = y_0$ for the solution of a differential equation $y' = f(x, y)$, the **solution curve** (graph of the solution) is required to pass through the point (x_0, y_0) and to have slope $f(x_0, y_0)$ there. We can picture these slopes graphically by drawing short line segments of slope $f(x, y)$ at selected points (x, y) in the region of the xy-plane that constitutes the domain of f. Each segment has the same slope as the solution curve through (x, y) and so is tangent to the curve there. The resulting picture is called a **slope field** (or **direction field**) and gives a visualization of the general shape of the solution curves. Figure 9.2a shows a slope field, with a particular solution sketched into it in Figure 9.2b. We see how these line segments indicate the direction the solution curve takes at each point it passes through.

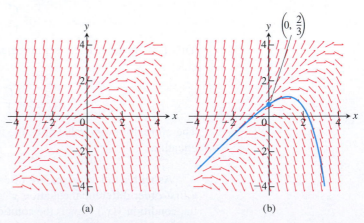

(a)   (b)

FIGURE 9.2   (a) Slope field for $\dfrac{dy}{dx} = y - x$. (b) The particular solution curve through the point $\left(0, \dfrac{2}{3}\right)$ (Example 2).

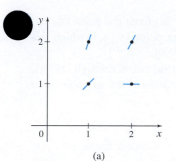

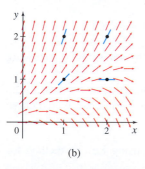

(a)

(b)

FIGURE 9.3 (a) The slope field for $y' = 2y - x$ is shown at four points. (b) The slope field at several hundred additional points in the plane.

EXAMPLE 3 For the differential equation

$$y' = 2y - x$$

draw line segments representing the slope field at the points $(1, 2)$; $(1, 1)$; $(2, 2)$, and $(2, 1)$.

Solution At $(1, 2)$ the slope is $2(2) - (1) = 3$. Similarly the slope at $(2, 2)$ is 2, at $(2, 1)$ is 0, and the slope at $(1, 1)$ is 1. We indicate this on the graph by drawing short line segments of the given slope through each point, as in Figure 9.3. ∎

Figure 9.4 shows three slope fields and we see how the solution curves behave by following the tangent line segments in these fields. Slope fields are useful because they display the overall behavior of the family of solution curves for a given differential equation. For instance, the slope field in Figure 9.4b reveals that every solution $y(x)$ to the differential equation specified in the figure satisfies $\lim_{x \to \pm\infty} y(x) = 0$. We will see that knowing the overall behavior of the solution curves is often critical to understanding and predicting outcomes in a real-world system modeled by a differential equation.

Constructing a slope field with pencil and paper can be quite tedious. All our examples were generated by computer software.

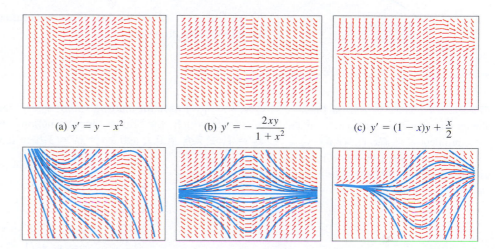

(a) $y' = y - x^2$ (b) $y' = -\dfrac{2xy}{1 + x^2}$ (c) $y' = (1 - x)y + \dfrac{x}{2}$

FIGURE 9.4 Slope fields (top row) and selected solution curves (bottom row). In computer renditions, slope segments are sometimes portrayed with arrows, as they are here, but they should be considered as just tangent line segments.

Euler's Method

If we do not require or cannot immediately find an *exact* solution giving an explicit formula for an initial value problem $y' = f(x, y)$, $y(x_0) = y_0$, we can often use a computer to generate a table of approximate numerical values of y for values of x in an appropriate interval. Such a table is called a **numerical solution** of the problem, and the method by which we generate the table is called a **numerical method**.

Given a differential equation $dy/dx = f(x, y)$ and an initial condition $y(x_0) = y_0$, we can approximate the solution $y = y(x)$ by its linearization

$$L(x) = y(x_0) + y'(x_0)(x - x_0) \quad \text{or} \quad L(x) = y_0 + f(x_0, y_0)(x - x_0).$$

The function $L(x)$ gives a good approximation to the solution $y(x)$ in a short interval about x_0 (Figure 9.5). The basis of Euler's method is to patch together a string of linearizations to approximate the curve over a longer stretch. Here is how the method works.

We know the point (x_0, y_0) lies on the solution curve. Suppose that we specify a new value for the independent variable to be $x_1 = x_0 + dx$. (Recall that $dx = \Delta x$ in the definition of differentials.) If the increment dx is small, then

$$y_1 = L(x_1) = y_0 + f(x_0, y_0)\, dx$$

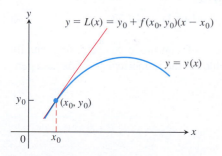

FIGURE 9.5 The linearization $L(x)$ of $y = y(x)$ at $x = x_0$.

$$y = L(x) = y_0 + f(x_0, y_0)(x - x_0)$$

$$y = y(x)$$

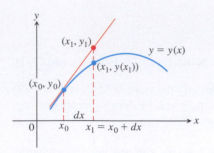

FIGURE 9.6 The first Euler step approximates $y(x_1)$ with $y_1 = L(x_1)$.

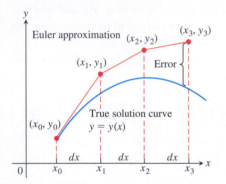

FIGURE 9.7 Three steps in the Euler approximation to the solution of the initial value problem $y' = f(x, y)$, $y(x_0) = y_0$. As we take more steps, the errors involved usually accumulate, but not in the exaggerated way shown here.

is a good approximation to the exact solution value $y = y(x_1)$. So from the point (x_0, y_0), which lies *exactly* on the solution curve, we have obtained the point (x_1, y_1), which lies very close to the point $(x_1, y(x_1))$ on the solution curve (Figure 9.6).

Using the point (x_1, y_1) and the slope $f(x_1, y_1)$ of the solution curve through (x_1, y_1), we take a second step. Setting $x_2 = x_1 + dx$, we use the linearization of the solution curve through (x_1, y_1) to calculate

$$y_2 = y_1 + f(x_1, y_1)\, dx.$$

This gives the next approximation (x_2, y_2) to values along the solution curve $y = y(x)$ (Figure 9.7). Continuing in this fashion, we take a third step from the point (x_2, y_2) with slope $f(x_2, y_2)$ to obtain the third approximation

$$y_3 = y_2 + f(x_2, y_2)\, dx,$$

and so on. We are literally building an approximation to one of the solutions by following the direction of the slope field of the differential equation.

The steps in Figure 9.7 are drawn large to illustrate the construction process, so the approximation looks crude. In practice, dx would be small enough to make the red curve hug the blue one and give a good approximation throughout.

EXAMPLE 4 Find the first three approximations y_1, y_2, y_3 using Euler's method for the initial value problem

$$y' = 1 + y, \qquad y(0) = 1,$$

starting at $x_0 = 0$ with $dx = 0.1$.

Solution We have the starting values $x_0 = 0$ and $y_0 = 1$. Next we determine the values of x at which the Euler approximations will take place: $x_1 = x_0 + dx = 0.1$, $x_2 = x_0 + 2\,dx = 0.2$, and $x_3 = x_0 + 3\,dx = 0.3$. Then we find

$$\begin{aligned}
\textit{First:} \quad y_1 &= y_0 + f(x_0, y_0)\, dx \\
&= y_0 + (1 + y_0)\, dx \\
&= 1 + (1 + 1)(0.1) = 1.2
\end{aligned}$$

$$\begin{aligned}
\textit{Second:} \quad y_2 &= y_1 + f(x_1, y_1)\, dx \\
&= y_1 + (1 + y_1)\, dx \\
&= 1.2 + (1 + 1.2)(0.1) = 1.42
\end{aligned}$$

$$\begin{aligned}
\textit{Third:} \quad y_3 &= y_2 + f(x_2, y_2)\, dx \\
&= y_2 + (1 + y_2)\, dx \\
&= 1.42 + (1 + 1.42)(0.1) = 1.662 \qquad \blacksquare
\end{aligned}$$

The step-by-step process used in Example 4 can be continued easily. Using equally spaced values for the independent variable in the table for the numerical solution, and generating n of them, set

$$\begin{aligned}
x_1 &= x_0 + dx \\
x_2 &= x_1 + dx \\
&\ \ \vdots \\
x_n &= x_{n-1} + dx.
\end{aligned}$$

Then calculate the approximations to the solution,

$$\begin{aligned}
y_1 &= y_0 + f(x_0, y_0)\, dx \\
y_2 &= y_1 + f(x_1, y_1)\, dx \\
&\ \ \vdots \\
y_n &= y_{n-1} + f(x_{n-1}, y_{n-1})\, dx.
\end{aligned}$$

HISTORICAL BIOGRAPHY

Leonhard Euler
(1703–1783)
www.goo.gl/v2nvlA

The number of steps n can be as large as we like, but errors can accumulate if n is too large.

Euler's method is easy to implement on a computer or calculator. The software program generates a table of numerical solutions to an initial value problem, allowing us to input x_0 and y_0, the number of steps n, and the step size dx. It then calculates the approximate solution values $y_1, y_2, \ldots, y_n$ in iterative fashion, as just described.

Solving the separable equation in Example 4, we find that the exact solution to the initial value problem is $y = 2e^x - 1$. We use this information in Example 5.

EXAMPLE 5   Use Euler's method to solve

$$y' = 1 + y, \qquad y(0) = 1,$$

on the interval $0 \leq x \leq 1$, starting at $x_0 = 0$ and taking **(a)** $dx = 0.1$ and **(b)** $dx = 0.05$. Compare the approximations with the values of the exact solution $y = 2e^x - 1$.

Solution

(a) We used a computer to generate the approximate values in Table 9.1. The "error" column is obtained by subtracting the unrounded Euler values from the unrounded values found using the exact solution. All entries are then rounded to four decimal places.

TABLE 9.1 Euler solution of $y' = 1 + y, y(0) = 1$,
step size $dx = 0.1$

x	y (Euler)	y (exact)	Error
0	1	1	0
0.1	1.2	1.2103	0.0103
0.2	1.42	1.4428	0.0228
0.3	1.662	1.6997	0.0377
0.4	1.9282	1.9836	0.0554
0.5	2.2210	2.2974	0.0764
0.6	2.5431	2.6442	0.1011
0.7	2.8974	3.0275	0.1301
0.8	3.2872	3.4511	0.1639
0.9	3.7159	3.9192	0.2033
1.0	4.1875	4.4366	0.2491

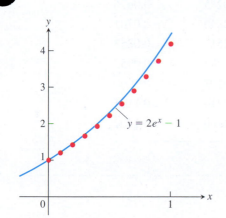

FIGURE 9.8 The graph of $y = 2e^x - 1$ superimposed on a scatterplot of the Euler approximations shown in Table 9.1 (Example 5).

By the time we reach $x = 1$ (after 10 steps), the error is about 5.6% of the exact solution. A plot of the exact solution curve with the scatterplot of Euler solution points from Table 9.1 is shown in Figure 9.8.

(b) One way to try to reduce the error is to decrease the step size. Table 9.2 shows the results and their comparisons with the exact solutions when we decrease the step size to 0.05, doubling the number of steps to 20. As in Table 9.1, all computations are performed before rounding. This time when we reach $x = 1$, the relative error is only about 2.9%. ∎

It might be tempting to reduce the step size even further in Example 5 to obtain greater accuracy. Each additional calculation, however, not only requires additional computer time but more importantly adds to the buildup of round-off errors due to the approximate representations of numbers inside the computer.

The analysis of error and the investigation of methods to reduce it when making numerical calculations are important but are appropriate for a more advanced course. There are numerical methods more accurate than Euler's method, usually presented in a further study of differential equations or in a numerical analysis course.

TABLE 9.2 Euler solution of $y' = 1 + y$, $y(0) = 1$, step size $dx = 0.05$

x	y (Euler)	y (exact)	Error
0	1	1	0
0.05	1.1	1.1025	0.0025
0.10	1.205	1.2103	0.0053
0.15	1.3153	1.3237	0.0084
0.20	1.4310	1.4428	0.0118
0.25	1.5526	1.5681	0.0155
0.30	1.6802	1.6997	0.0195
0.35	1.8142	1.8381	0.0239
0.40	1.9549	1.9836	0.0287
0.45	2.1027	2.1366	0.0340
0.50	2.2578	2.2974	0.0397
0.55	2.4207	2.4665	0.0458
0.60	2.5917	2.6442	0.0525
0.65	2.7713	2.8311	0.0598
0.70	2.9599	3.0275	0.0676
0.75	3.1579	3.2340	0.0761
0.80	3.3657	3.4511	0.0853
0.85	3.5840	3.6793	0.0953
0.90	3.8132	3.9192	0.1060
0.95	4.0539	4.1714	0.1175
1.00	4.3066	4.4366	0.1300

EXERCISES 9.1

Slope Fields

In Exercises 1–4, match the differential equations with their slope fields, graphed here.

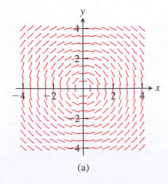

(a)

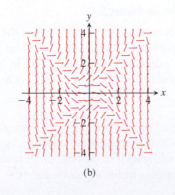

(b)

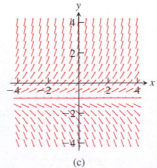

(c)

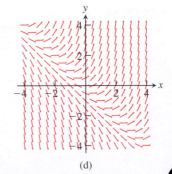

(d)

1. $y' = x + y$

2. $y' = y + 1$

3. $y' = -\dfrac{x}{y}$

4. $y' = y^2 - x^2$

In Exercises 5 and 6, copy the slope fields and sketch in some of the solution curves.

5. $y' = (y + 2)(y - 2)$

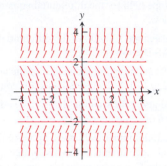

6. $y' = y(y + 1)(y - 1)$

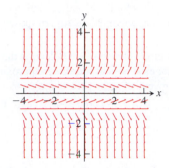

Integral Equations

In Exercises 7–12, write an equivalent first-order differential equation and initial condition for y.

7. $y = -1 + \int_{1}^{x} (t - y(t))\, dt$ **8.** $y = \int_{1}^{x} \frac{1}{t}\, dt$

9. $y = 2 - \int_{0}^{x} (1 + y(t)) \sin t\, dt$

10. $y = 1 + \int_{0}^{x} y(t)\, dt$ **11.** $y = x + 4 + \int_{-2}^{x} t e^{y(t)}\, dt$

12. $y = \ln x + \int_{x}^{e} \sqrt{t^2 + (y(t))^2}\, dt$

In Exercises 13 and 14, consider the differential equation $y' = f(y)$ and the given graph of f. Make a rough sketch of a direction field for each differential equation.

13.   **14.**

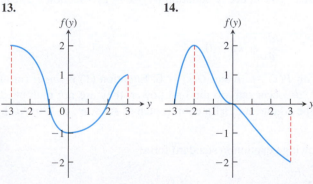

Using Euler's Method

In Exercises 15–20, use Euler's method to calculate the first three approximations to the given initial value problem for the specified increment size. Calculate the exact solution and investigate the accuracy of your approximations. Round your results to four decimal places.

15. $y' = \dfrac{2y}{x}$, $y(1) = -1$, $dx = 0.5$

16. $y' = x(1 - y)$, $y(1) = 0$, $dx = 0.2$

17. $y' = 2xy + 2y$, $y(0) = 3$, $dx = 0.2$

18. $y' = y^2(1 + 2x)$, $y(-1) = 1$, $dx = 0.5$

T **19.** $y' = 2xe^{x^2}$, $y(0) = 2$, $dx = 0.1$

T **20.** $y' = ye^x$, $y(0) = 2$, $dx = 0.5$

21. Use the Euler method with $dx = 0.2$ to estimate $y(1)$ if $y' = y$ and $y(0) = 1$. What is the exact value of $y(1)$?

22. Use the Euler method with $dx = 0.2$ to estimate $y(2)$ if $y' = y/x$ and $y(1) = 2$. What is the exact value of $y(2)$?

23. Use the Euler method with $dx = 0.5$ to estimate $y(5)$ if $y' = y^2/\sqrt{x}$ and $y(1) = -1$. What is the exact value of $y(5)$?

24. Use the Euler method with $dx = 1/3$ to estimate $y(2)$ if $y' = x \sin y$ and $y(0) = 1$. What is the exact value of $y(2)$?

25. Show that the solution of the initial value problem

$$y' = x + y, \qquad y(x_0) = y_0$$

is

$$y = -1 - x + (1 + x_0 + y_0)\, e^{x - x_0}.$$

26. What integral equation is equivalent to the initial value problem $y' = f(x), y(x_0) = y_0$?

COMPUTER EXPLORATIONS

In Exercises 27–32, obtain a slope field and add to it graphs of the solution curves passing through the given points.

27. $y' = y$ with
   a. $(0, 1)$ **b.** $(0, 2)$ **c.** $(0, -1)$

28. $y' = 2(y - 4)$ with
   a. $(0, 1)$ **b.** $(0, 4)$ **c.** $(0, 5)$

29. $y' = y(x + y)$ with
   a. $(0, 1)$ **b.** $(0, -2)$ **c.** $(0, 1/4)$ **d.** $(-1, -1)$

30. $y' = y^2$ with
   a. $(0, 1)$ **b.** $(0, 2)$ **c.** $(0, -1)$ **d.** $(0, 0)$

31. $y' = (y - 1)(x + 2)$ with
   a. $(0, -1)$ **b.** $(0, 1)$ **c.** $(0, 3)$ **d.** $(1, -1)$

32. $y' = \dfrac{xy}{x^2 + 4}$ with
   a. $(0, 2)$ **b.** $(0, -6)$ **c.** $\left(-2\sqrt{3}, -4\right)$

In Exercises 33 and 34, obtain a slope field and graph the particular solution over the specified interval. Use your CAS DE solver to find the general solution of the differential equation.

33. A **logistic equation** $y' = y(2 - y)$, $y(0) = 1/2$; $0 \le x \le 4$, $0 \le y \le 3$

34. $y' = (\sin x)(\sin y)$, $y(0) = 2$; $-6 \le x \le 6$, $-6 \le y \le 6$

Exercises 35 and 36 have no explicit solution in terms of elementary functions. Use a CAS to explore graphically each of the differential equations.

35. $y' = \cos(2x - y)$, $y(0) = 2$; $0 \le x \le 5$, $0 \le y \le 5$

36. A **Gompertz equation** $y' = y(1/2 - \ln y)$, $y(0) = 1/3$; $0 \le x \le 4$, $0 \le y \le 3$

37. Use a CAS to find the solutions of $y' + y = f(x)$ subject to the initial condition $y(0) = 0$, if $f(x)$ is

a. $2x$ **b.** $\sin 2x$ **c.** $3e^{x/2}$ **d.** $2e^{-x/2}\cos 2x$.

Graph all four solutions over the interval $-2 \le x \le 6$ to compare the results.

38. a. Use a CAS to plot the slope field of the differential equation

$$y' = \frac{3x^2 + 4x + 2}{2(y - 1)}$$

over the region $-3 \le x \le 3$ and $-3 \le y \le 3$.

b. Separate the variables and use a CAS integrator to find the general solution in implicit form.

c. Using a CAS implicit function grapher, plot solution curves for the arbitrary constant values $C = -6, -4, -2, 0, 2, 4, 6$.

d. Find and graph the solution that satisfies the initial condition $y(0) = -1$.

In Exercises 39–42, use Euler's method with the specified step size to estimate the value of the solution at the given point x^*. Find the value of the exact solution at x^*.

39. $y' = 2xe^{x^2}$, $y(0) = 2$, $dx = 0.1$, $x^* = 1$

40. $y' = 2y^2(x - 1)$, $y(2) = -1/2$, $dx = 0.1$, $x^* = 3$

41. $y' = \sqrt{x}/y$, $y > 0$, $y(0) = 1$, $dx = 0.1$, $x^* = 1$

42. $y' = 1 + y^2$, $y(0) = 0$, $dx = 0.1$, $x^* = 1$

Use a CAS to explore graphically each of the differential equations in Exercises 43–46. Perform the following steps to help with your explorations.

a. Plot a slope field for the differential equation in the given xy-window.

b. Find the general solution of the differential equation using your CAS DE solver.

c. Graph the solutions for the values of the arbitrary constant $C = -2, -1, 0, 1, 2$ superimposed on your slope field plot.

d. Find and graph the solution that satisfies the specified initial condition over the interval $[0, b]$.

e. Find the Euler numerical approximation to the solution of the initial value problem with 4 subintervals of the x-interval and plot the Euler approximation superimposed on the graph produced in part (d).

f. Repeat part (e) for 8, 16, and 32 subintervals. Plot these three Euler approximations superimposed on the graph from part (e).

g. Find the error $(y(\text{exact}) - y(\text{Euler}))$ at the specified point $x = b$ for each of your four Euler approximations. Discuss the improvement in the percentage error.

43. $y' = x + y$, $y(0) = -7/10$; $-4 \le x \le 4$, $-4 \le y \le 4$; $b = 1$

44. $y' = -x/y$, $y(0) = 2$; $-3 \le x \le 3$, $-3 \le y \le 3$; $b = 2$

45. $y' = y(2 - y)$, $y(0) = 1/2$; $0 \le x \le 4$, $0 \le y \le 3$; $b = 3$

46. $y' = (\sin x)(\sin y)$, $y(0) = 2$; $-6 \le x \le 6$, $-6 \le y \le 6$; $b = 3\pi/2$

9.2 First-Order Linear Equations

A first-order **linear** differential equation is one that can be written in the form

$$\frac{dy}{dx} + P(x)y = Q(x), \tag{1}$$

where P and Q are continuous functions of x. Equation (1) is the linear equation's **standard form**. Since the exponential growth/decay equation $dy/dx = ky$ (Section 7.2) can be put in the standard form

$$\frac{dy}{dx} - ky = 0,$$

we see it is a linear equation with $P(x) = -k$ and $Q(x) = 0$. Equation (1) is *linear* (in y) because y and its derivative dy/dx occur only to the first power, they are not multiplied together, nor do they appear as the argument of a function $\left(\text{such as } \sin y, \, e^y, \text{ or } \sqrt{dy/dx}\right)$.

EXAMPLE 1   Put the following equation in standard form:

$$x\frac{dy}{dx} = x^2 + 3y, \qquad x > 0.$$

Solution

$$x\frac{dy}{dx} = x^2 + 3y$$

$$\frac{dy}{dx} = x + \frac{3}{x}y \qquad \text{Divide by } x.$$

$$\frac{dy}{dx} - \frac{3}{x}y = x \qquad \begin{array}{l}\text{Standard form with } P(x) = -3/x \\ \text{and } Q(x) = x\end{array}$$

Notice that $P(x)$ is $-3/x$, not $+3/x$. The standard form is $y' + P(x)y = Q(x)$, so the minus sign is part of the formula for $P(x)$.   ■

Solving Linear Equations

We solve the equation

$$\frac{dy}{dx} + P(x)y = Q(x)$$

by multiplying both sides by a *positive* function $v(x)$ that transforms the left-hand side into the derivative of the product $v(x) \cdot y$. We will show how to find v in a moment, but first we want to show how, once found, it provides the solution we seek.

Here is why multiplying by $v(x)$ works:

$$\frac{dy}{dx} + P(x)y = Q(x) \qquad \begin{array}{l}\text{Original equation is} \\ \text{in standard form.}\end{array}$$

$$v(x)\frac{dy}{dx} + P(x)v(x)y = v(x)Q(x) \qquad \text{Multiply by positive } v(x).$$

$$\frac{d}{dx}(v(x)\cdot y) = v(x)Q(x) \qquad \begin{array}{l}v(x) \text{ is chosen to make} \\ v\dfrac{dy}{dx} + Pvy = \dfrac{d}{dx}(v\cdot y).\end{array}$$

$$v(x)\cdot y = \int v(x)Q(x)\,dx \qquad \begin{array}{l}\text{Integrate with respect} \\ \text{to } x.\end{array}$$

$$y = \frac{1}{v(x)}\int v(x)Q(x)\,dx \qquad \text{Divide by } v(x). \qquad (2)$$

Equation (2) expresses the solution of Equation (1) in terms of the functions $v(x)$ and $Q(x)$. We call $v(x)$ an **integrating factor** for Equation (1) because its presence makes the equation integrable.

Why doesn't the formula for $P(x)$ appear in the solution as well? It does, but indirectly, in the construction of the positive function $v(x)$. We have

$$\frac{d}{dx}(vy) = v\frac{dy}{dx} + Pvy \qquad \text{Condition imposed on } v$$

$$v\frac{dy}{dx} + y\frac{dv}{dx} = v\frac{dy}{dx} + Pvy \qquad \text{Derivative Product Rule}$$

$$y\frac{dv}{dx} = Pvy \qquad \text{The terms } v\dfrac{dy}{dx} \text{ cancel.}$$

This last equation will hold if

$$\frac{dv}{dx} = Pv$$

$$\frac{dv}{v} = P\,dx \qquad \text{Variables separated, } v > 0$$

$$\int \frac{dv}{v} = \int P\,dx \qquad \text{Integrate both sides.}$$

$$\ln v = \int P\,dx \qquad \text{Since } v > 0, \text{ we do not need absolute value signs in } \ln v.$$

$$e^{\ln v} = e^{\int P\,dx} \qquad \text{Exponentiate both sides to solve for } v. \qquad (3)$$

$$v = e^{\int P\,dx}$$

Thus a formula for the general solution to Equation (1) is given by Equation (2), where $v(x)$ is given by Equation (3). However, rather than memorizing the formula, just remember how to find the integrating factor once you have the standard form so $P(x)$ is correctly identified. Any antiderivative of P works for Equation (3).

> To solve the linear equation $y' + P(x)y = Q(x)$, multiply both sides by the integrating factor $v(x) = e^{\int P(x)\,dx}$ and integrate both sides.

When you integrate the product on the left-hand side in this procedure, you always obtain the product $v(x)y$ of the integrating factor and solution function y because of the way v is defined.

EXAMPLE 2   Solve the equation

$$x\frac{dy}{dx} = x^2 + 3y, \qquad x > 0.$$

Solution   First we put the equation in standard form (Example 1):

$$\frac{dy}{dx} - \frac{3}{x}y = x,$$

so $P(x) = -3/x$ is identified.

The integrating factor is

$$
\begin{aligned}
v(x) = e^{\int P(x)\,dx} &= e^{\int (-3/x)\,dx} \\
&= e^{-3\ln|x|} \qquad \text{Constant of integration is 0,} \\
&\qquad\qquad\quad \text{so } v \text{ is as simple as possible.} \\
&= e^{-3\ln x} \qquad x > 0 \\
&= e^{\ln x^{-3}} = \frac{1}{x^3}.
\end{aligned}
$$

Next we multiply both sides of the standard form by $v(x)$ and integrate:

$$\frac{1}{x^3} \cdot \left(\frac{dy}{dx} - \frac{3}{x}y \right) = \frac{1}{x^3} \cdot x$$

$$\frac{1}{x^3}\frac{dy}{dx} - \frac{3}{x^4}y = \frac{1}{x^2}$$

$$\frac{d}{dx}\left(\frac{1}{x^3}y \right) = \frac{1}{x^2} \qquad \text{Left-hand side is } \frac{d}{dx}(v \cdot y).$$

$$\frac{1}{x^3}y = \int \frac{1}{x^2}\,dx \qquad \text{Integrate both sides.}$$

$$\frac{1}{x^3}y = -\frac{1}{x} + C.$$

Solving this last equation for y gives the general solution:

$$y = x^3\left(-\frac{1}{x} + C \right) = -x^2 + Cx^3, \qquad x > 0.$$

EXAMPLE 3   Find the particular solution of

$$3xy' - y = \ln x + 1, \qquad x > 0,$$

satisfying $y(1) = -2$.

Solution   With $x > 0$, we write the equation in standard form:

$$y' - \frac{1}{3x} y = \frac{\ln x + 1}{3x}.$$

Then the integrating factor is given by

$$v = e^{\int -dx/3x} = e^{(-1/3)\ln x} = x^{-1/3}. \qquad x > 0$$

Thus

$$x^{-1/3} y = \frac{1}{3} \int (\ln x + 1) x^{-4/3} \, dx. \qquad \text{Left-hand side is } vy.$$

Integration by parts of the right-hand side gives

$$x^{-1/3} y = -x^{-1/3} (\ln x + 1) + \int x^{-4/3} \, dx + C.$$

Therefore

$$x^{-1/3} y = -x^{-1/3} (\ln x + 1) - 3x^{-1/3} + C$$

or, solving for y,

$$y = -(\ln x + 4) + Cx^{1/3}.$$

When $x = 1$ and $y = -2$ this last equation becomes

$$-2 = -(0 + 4) + C,$$

so

$$C = 2.$$

Substitution into the equation for y gives the particular solution

$$y = 2x^{1/3} - \ln x - 4. \qquad \blacksquare$$

In solving the linear equation in Example 2, we integrated both sides of the equation after multiplying each side by the integrating factor. However, we can shorten the amount of work, as in Example 3, by remembering that the left-hand side *always* integrates into the product $v(x) \cdot y$ of the integrating factor times the solution function. From Equation (2) this means that

$$v(x) y = \int v(x) Q(x) \, dx. \qquad (4)$$

We need only integrate the product of the integrating factor $v(x)$ with $Q(x)$ on the right-hand side of Equation (1) and then equate the result with $v(x)y$ to obtain the general solution. Nevertheless, to emphasize the role of $v(x)$ in the solution process, we sometimes follow the complete procedure as illustrated in Example 2.

Observe that if the function $Q(x)$ is identically zero in the standard form given by Equation (1), the linear equation is separable and can be solved by the method of Section 7.2:

$$\frac{dy}{dx} + P(x) y = Q(x)$$

$$\frac{dy}{dx} + P(x) y = 0 \qquad Q(x) = 0$$

$$\frac{dy}{y} = -P(x) \, dx \qquad \text{Separating the variables}$$

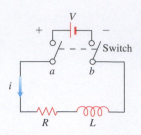

FIGURE 9.9   The *RL* circuit in Example 4.

RL Circuits

The diagram in Figure 9.9 represents an electrical circuit whose total resistance is a constant *R* ohms and whose self-inductance, shown as a coil, is *L* henries, also a constant. There is a switch whose terminals at *a* and *b* can be closed to connect a constant electrical source of *V* volts.

Ohm's Law, $V = RI$, has to be augmented for such a circuit. The correct equation accounting for both resistance and inductance is

$$L\frac{di}{dt} + Ri = V, \tag{5}$$

where *i* is the current in amperes and *t* is the time in seconds. By solving this equation, we can predict how the current will flow after the switch is closed.

EXAMPLE 4   The switch in the *RL* circuit in Figure 9.9 is closed at time $t = 0$. How will the current flow as a function of time?

Solution   Equation (5) is a first-order linear differential equation for *i* as a function of *t*. Its standard form is

$$\frac{di}{dt} + \frac{R}{L}i = \frac{V}{L}, \tag{6}$$

and the corresponding solution, given that $i = 0$ when $t = 0$, is

$$i = \frac{V}{R} - \frac{V}{R}e^{-(R/L)t}. \tag{7}$$

(We leave the calculation of the solution for you to do in Exercise 28.) Since *R* and *L* are positive, $-(R/L)$ is negative and $e^{-(R/L)t} \to 0$ as $t \to \infty$. Thus,

$$\lim_{t\to\infty} i = \lim_{t\to\infty}\left(\frac{V}{R} - \frac{V}{R}e^{-(R/L)t}\right) = \frac{V}{R} - \frac{V}{R}\cdot 0 = \frac{V}{R}.$$

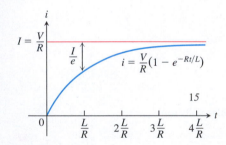

FIGURE 9.10   The growth of the current in the *RL* circuit in Example 4. *I* is the current's steady-state value. The number $t = L/R$ is the time constant of the circuit. The current gets to within 5% of its steady-state value in 3 time constants (Exercise 27).

At any given time, the current is theoretically less than V/R, but as time passes, the current approaches the **steady-state value** V/R. According to the equation

$$L\frac{di}{dt} + Ri = V,$$

$I = V/R$ is the current that will flow in the circuit if either $L = 0$ (no inductance) or $di/dt = 0$ (steady current, $i =$ constant) (Figure 9.10).

Equation (7) expresses the solution of Equation (6) as the sum of two terms: a steady-state solution V/R and a transient solution $-(V/R)e^{-(R/L)t}$ that tends to zero as $t \to \infty$. ∎

EXERCISES 9.2

First-Order Linear Equations

Solve the differential equations in Exercises 1–14.

1. $x\dfrac{dy}{dx} + y = e^x, \quad x > 0$ **2.** $e^x\dfrac{dy}{dx} + 2e^x y = 1$

3. $xy' + 3y = \dfrac{\sin x}{x^2}, \quad x > 0$

4. $y' + (\tan x)y = \cos^2 x, \quad -\pi/2 < x < \pi/2$

5. $x\dfrac{dy}{dx} + 2y = 1 - \dfrac{1}{x}, \quad x > 0$

6. $(1 + x)y' + y = \sqrt{x}$ **7.** $2y' = e^{x/2} + y$

8. $e^{2x}y' + 2e^{2x}y = 2x$ **9.** $xy' - y = 2x\ln x$

10. $x\dfrac{dy}{dx} = \dfrac{\cos x}{x} - 2y, \quad x > 0$

11. $(t - 1)^3 \dfrac{ds}{dt} + 4(t - 1)^2 s = t + 1, \quad t > 1$

12. $(t + 1) \dfrac{ds}{dt} + 2s = 3(t + 1) + \dfrac{1}{(t + 1)^2}, \quad t > -1$

13. $\sin \theta \dfrac{dr}{d\theta} + (\cos \theta)r = \tan \theta, \quad 0 < \theta < \pi/2$

14. $\tan \theta \dfrac{dr}{d\theta} + r = \sin^2 \theta, \quad 0 < \theta < \pi/2$

Solving Initial Value Problems

Solve the initial value problems in Exercises 15–20.

15. $\dfrac{dy}{dt} + 2y = 3, \quad y(0) = 1$

16. $t \dfrac{dy}{dt} + 2y = t^3, \quad t > 0, \quad y(2) = 1$

17. $\theta \dfrac{dy}{d\theta} + y = \sin \theta, \quad \theta > 0, \quad y(\pi/2) = 1$

18. $\theta \dfrac{dy}{d\theta} - 2y = \theta^3 \sec \theta \tan \theta, \quad \theta > 0, \quad y(\pi/3) = 2$

19. $(x + 1) \dfrac{dy}{dx} - 2(x^2 + x)y = \dfrac{e^{x^2}}{x + 1}, \quad x > -1, \quad y(0) = 5$

20. $\dfrac{dy}{dx} + xy = x, \quad y(0) = -6$

21. Solve the exponential growth/decay initial value problem for y as a function of t by thinking of the differential equation as a first-order linear equation with $P(x) = -k$ and $Q(x) = 0$:

$$\dfrac{dy}{dt} = ky \quad (k \text{ constant}), \quad y(0) = y_0$$

22. Solve the following initial value problem for u as a function of t:

$$\dfrac{du}{dt} + \dfrac{k}{m}u = 0 \quad (k \text{ and } m \text{ positive constants}), \quad u(0) = u_0$$

a. as a first-order linear equation.

b. as a separable equation.

Theory and Examples

23. Is either of the following equations correct? Give reasons for your answers.

a. $x \displaystyle\int \dfrac{1}{x} dx = x \ln |x| + C$ **b.** $x \displaystyle\int \dfrac{1}{x} dx = x \ln |x| + Cx$

24. Is either of the following equations correct? Give reasons for your answers.

a. $\dfrac{1}{\cos x} \displaystyle\int \cos x \, dx = \tan x + C$

b. $\dfrac{1}{\cos x} \displaystyle\int \cos x \, dx = \tan x + \dfrac{C}{\cos x}$

25. Current in a closed *RL* circuit How many seconds after the switch in an *RL* circuit is closed will it take the current i to reach half of its steady-state value? Notice that the time depends on R and L and not on how much voltage is applied.

26. Current in an open RL circuit If the switch is thrown open after the current in an *RL* circuit has built up to its steady-state value $I = V/R$, the decaying current (see accompanying figure) obeys the equation

$$L \dfrac{di}{dt} + Ri = 0,$$

which is Equation (5) with $V = 0$.

a. Solve the equation to express i as a function of t.

b. How long after the switch is thrown will it take the current to fall to half its original value?

c. Show that the value of the current when $t = L/R$ is I/e. (The significance of this time is explained in the next exercise.)

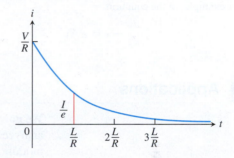

27. Time constants Engineers call the number L/R the *time constant* of the *RL* circuit in Figure 9.10. The significance of the time constant is that the current will reach 95% of its final value within 3 time constants of the time the switch is closed (Figure 9.10). Thus, the time constant gives a built-in measure of how rapidly an individual circuit will reach equilibrium.

a. Find the value of i in Equation (7) that corresponds to $t = 3L/R$ and show that it is about 95% of the steady-state value $I = V/R$.

b. Approximately what percentage of the steady-state current will be flowing in the circuit 2 time constants after the switch is closed (i.e., when $t = 2L/R$)?

28. Derivation of Equation (7) in Example 4

a. Show that the solution of the equation

$$\dfrac{di}{dt} + \dfrac{R}{L}i = \dfrac{V}{L}$$

is

$$i = \dfrac{V}{R} + Ce^{-(R/L)t}.$$

b. Then use the initial condition $i(0) = 0$ to determine the value of C. This will complete the derivation of Equation (7).

c. Show that $i = V/R$ is a solution of Equation (6) and that $i = Ce^{-(R/L)t}$ satisfies the equation

$$\dfrac{di}{dt} + \dfrac{R}{L}i = 0.$$

A **Bernoulli differential equation** is of the form

$$\frac{dy}{dx} + P(x)\,y = Q(x)\,y^n.$$

Observe that, if $n = 0$ or 1, the Bernoulli equation is linear. For other values of n, the substitution $u = y^{1-n}$ transforms the Bernoulli equation into the linear equation

$$\frac{du}{dx} + (1 - n)P(x)u = (1 - n)Q(x).$$

For example, in the equation

$$\frac{dy}{dx} - y = e^{-x} y^2$$

we have $n = 2$, so that $u = y^{1-2} = y^{-1}$ and $du/dx = -y^{-2}\,dy/dx$. Then $dy/dx = -y^2\,du/dx = -u^{-2}\,du/dx$. Substitution into the original equation gives

$$-u^{-2}\frac{du}{dx} - u^{-1} = e^{-x} u^{-2}$$

or, equivalently,

$$\frac{du}{dx} + u = -e^{-x}.$$

This last equation is linear in the (unknown) dependent variable u.

Solve the Bernoulli equations in Exercises 29–32.

29. $y' - y = -y^2$ **30.** $y' - y = xy^2$

31. $xy' + y = y^{-2}$ **32.** $x^2y' + 2xy = y^3$

9.3 Applications

We now look at four applications of first-order differential equations. The first application analyzes an object moving along a straight line while subject to a force opposing its motion. The second is a model of population growth. The third application considers a curve or curves intersecting each curve in a second family of curves *orthogonally* (that is, at right angles). The final application analyzes chemical concentrations entering and leaving a container. The various models involve separable or linear first-order equations.

Motion with Resistance Proportional to Velocity

In some cases it is reasonable to assume that the resistance encountered by a moving object, such as a car coasting to a stop, is proportional to the object's velocity. The faster the object moves, the more its forward progress is resisted by the air through which it passes. Picture the object as a mass m moving along a coordinate line with position function s and velocity v at time t. From Newton's second law of motion, the resisting force opposing the motion is

$$\text{Force} = \text{mass} \times \text{acceleration} = m\frac{dv}{dt}.$$

If the resisting force is proportional to velocity, we have

$$m\frac{dv}{dt} = -kv \qquad \text{or} \qquad \frac{dv}{dt} = -\frac{k}{m}v \qquad (k > 0).$$

This is a separable differential equation representing exponential change. The solution to the equation with initial condition $v = v_0$ at $t = 0$ is (Section 7.2)

$$v = v_0 e^{-(k/m)t}. \tag{1}$$

What can we learn from Equation (1)? For one thing, we can see that if m is something large, like the mass of a 20,000-ton ore boat in Lake Erie, it will take a long time for the velocity to approach zero (because t must be large in the exponent of the equation in order to make kt/m large enough for v to be small). We can learn even more if we integrate Equation (1) to find the position s as a function of time t.

Suppose that an object is coasting to a stop and the only force acting on it is a resistance proportional to its speed. How far will it coast? To find out, we start with Equation (1) and solve the initial value problem

$$\frac{ds}{dt} = v_0 e^{-(k/m)t}, \qquad s(0) = 0.$$

Integrating with respect to t gives

$$s = -\frac{v_0 m}{k} e^{-(k/m)t} + C.$$

Substituting $s = 0$ when $t = 0$ gives

$$0 = -\frac{v_0 m}{k} + C \quad \text{and} \quad C = \frac{v_0 m}{k}.$$

The body's position at time t is therefore

$$s(t) = -\frac{v_0 m}{k} e^{-(k/m)t} + \frac{v_0 m}{k} = \frac{v_0 m}{k} (1 - e^{-(k/m)t}). \tag{2}$$

To find how far the body will coast, we find the limit of $s(t)$ as $t \to \infty$. Since $-(k/m) < 0$, we know that $e^{-(k/m)t} \to 0$ as $t \to \infty$, so that

$$\lim_{t \to \infty} s(t) = \lim_{t \to \infty} \frac{v_0 m}{k} (1 - e^{-(k/m)t}) = \frac{v_0 m}{k} (1 - 0) = \frac{v_0 m}{k}.$$

Thus,

$$\text{Distance coasted} = \frac{v_0 m}{k}. \tag{3}$$

The number $v_0 m/k$ is only an upper bound (albeit a useful one). It is true to life in one respect, at least: If m is large, the body will coast a long way.

In the English system, in which weight is measured in pounds, mass is measured in **slugs**. Thus,

$$\text{Pounds} = \text{slugs} \times 32,$$

assuming that gravitational acceleration is 32 ft/sec².

EXAMPLE 1   For a 192-lb ice skater, the k in Equation (1) is about 1/3 slug/sec and $m = 192/32 = 6$ slugs. How long will it take the skater to coast from 11 ft/sec (7.5 mph) to 1 ft/sec? How far will the skater coast before coming to a complete stop?

Solution   We answer the first question by solving Equation (1) for t:

$$11 e^{-t/18} = 1 \qquad \text{Eq. (1) with } k = 1/3,$$
$$e^{-t/18} = 1/11 \qquad m = 6, v_0 = 11, v = 1$$
$$-t/18 = \ln(1/11) = -\ln 11$$
$$t = 18 \ln 11 \approx 43 \text{ sec}.$$

We answer the second question with Equation (3):

$$\text{Distance coasted} = \frac{v_0 m}{k} = \frac{11 \cdot 6}{1/3} = 198 \text{ ft}. \qquad \blacksquare$$

Inaccuracy of the Exponential Population Growth Model

In Section 7.2 we modeled population growth with the Law of Exponential Change:

$$\frac{dP}{dt} = kP, \qquad P(0) = P_0$$

where P is the population at time t, $k > 0$ is a constant growth rate, and P_0 is the size of the population at time $t = 0$. In Section 7.2 we found the solution $P = P_0 e^{kt}$ to this model.

To assess the model, notice that the exponential growth differential equation says that

$$\frac{dP/dt}{P} = k \tag{4}$$

is constant. This rate is called the **relative growth rate**. Now, Table 9.3 gives the world population at midyear for the years 1980 to 1989. Taking $dt = 1$ and $dP \approx \Delta P$, we see

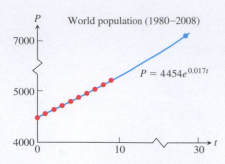

FIGURE 9.11 Notice that the value of the solution $P = 4454e^{0.017t}$ is 7169 when $t = 28$, which is nearly 7% more than the actual population in 2008.

TABLE 9.3 World population (midyear)

Year	Population (millions)	$\Delta P/P$
1980	4454	$76/4454 \approx 0.0171$
1981	4530	$80/4530 \approx 0.0177$
1982	4610	$80/4610 \approx 0.0174$
1983	4690	$80/4690 \approx 0.0171$
1984	4770	$81/4770 \approx 0.0170$
1985	4851	$82/4851 \approx 0.0169$
1986	4933	$85/4933 \approx 0.0172$
1987	5018	$87/5018 \approx 0.0173$
1988	5105	$85/5105 \approx 0.0167$
1989	5190	

Source: U.S. Bureau of the Census (Sept., 2007): www.census .gov/ipc/www/idb.

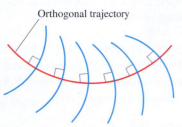

FIGURE 9.12 An orthogonal trajectory intersects the family of curves at right angles, or orthogonally.

from the table that the relative growth rate in Equation (4) is approximately the constant 0.017. Thus, based on the tabled data with $t = 0$ representing 1980, $t = 1$ representing 1981, and so forth, the world population could be modeled by the initial value problem,

$$\frac{dP}{dt} = 0.017P, \qquad P(0) = 4454.$$

The solution to this initial value problem gives the population function $P = 4454e^{0.017t}$. In year 2008 (so $t = 28$), the solution predicts the world population in midyear to be about 7169 million, or 7.2 billion (Figure 9.11), which is more than the actual population of 6707 million from the U.S. Bureau of the Census. A more realistic model would consider environmental and other factors affecting the growth rate, which has been steadily declining to about 0.012 since 1987. We consider one such model in Section 9.4.

Orthogonal Trajectories

An **orthogonal trajectory** of a family of curves is a curve that intersects each curve of the family at right angles, or *orthogonally* (Figure 9.12). For instance, each straight line through the origin is an orthogonal trajectory of the family of circles $x^2 + y^2 = a^2$, centered at the origin (Figure 9.13). Such mutually orthogonal systems of curves are of particular importance in physical problems related to electrical potential, where the curves in one family correspond to strength of an electric field and those in the other family correspond to constant electric potential. They also occur in hydrodynamics and heat-flow problems.

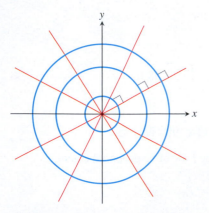

FIGURE 9.13 Every straight line through the origin is orthogonal to the family of circles centered at the origin.

EXAMPLE 2 Find the orthogonal trajectories of the family of curves $xy = a$, where $a \neq 0$ is an arbitrary constant.

Solution The curves $xy = a$ form a family of hyperbolas having the coordinate axes as asymptotes. First we find the slopes of each curve in this family, or their dy/dx values. Differentiating $xy = a$ implicitly gives

$$x\frac{dy}{dx} + y = 0 \qquad \text{or} \qquad \frac{dy}{dx} = -\frac{y}{x}.$$

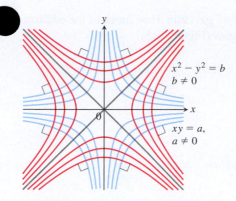

FIGURE 9.14 Each curve is orthogonal to every curve it meets in the other family (Example 2).

Thus the slope of the tangent line at any point (x, y) on one of the hyperbolas $xy = a$ is $y' = -y/x$. On an orthogonal trajectory the slope of the tangent line at this same point must be the negative reciprocal, or x/y. Therefore, the orthogonal trajectories must satisfy the differential equation

$$\frac{dy}{dx} = \frac{x}{y}.$$

This differential equation is separable and we solve it as in Section 7.2:

$$y \, dy = x \, dx \qquad \text{Separate variables.}$$

$$\int y \, dy = \int x \, dx \qquad \text{Integrate both sides.}$$

$$\frac{1}{2}y^2 = \frac{1}{2}x^2 + C$$

$$y^2 - x^2 = b, \tag{5}$$

where $b = 2C$ is an arbitrary constant. The orthogonal trajectories are the family of hyperbolas given by Equation (5) and sketched in Figure 9.14. ∎

Mixture Problems

Suppose a chemical in a liquid solution (or dispersed in a gas) runs into a container holding the liquid (or the gas) with, possibly, a specified amount of the chemical dissolved as well. The mixture is kept uniform by stirring and flows out of the container at a known rate. In this process, it is often important to know the concentration of the chemical in the container at any given time. The differential equation describing the process is based on the formula

$$\begin{matrix} \text{Rate of change} \\ \text{of amount} \\ \text{in container} \end{matrix} = \begin{pmatrix} \text{rate at which} \\ \text{chemical} \\ \text{arrives} \end{pmatrix} - \begin{pmatrix} \text{rate at which} \\ \text{chemical} \\ \text{departs.} \end{pmatrix} \tag{6}$$

If $y(t)$ is the amount of chemical in the container at time t and $V(t)$ is the total volume of liquid in the container at time t, then the departure rate of the chemical at time t is

$$\text{Departure rate} = \frac{y(t)}{V(t)} \cdot (\text{outflow rate})$$

$$= \begin{pmatrix} \text{concentration in} \\ \text{container at time } t \end{pmatrix} \cdot (\text{outflow rate}). \tag{7}$$

Accordingly, Equation (6) becomes

$$\frac{dy}{dt} = (\text{chemical's arrival rate}) - \frac{y(t)}{V(t)} \cdot (\text{outflow rate}). \tag{8}$$

If, say, y is measured in pounds, V in gallons, and t in minutes, the units in Equation (8) are

$$\frac{\text{pounds}}{\text{minutes}} = \frac{\text{pounds}}{\text{minutes}} - \frac{\text{pounds}}{\text{gallons}} \cdot \frac{\text{gallons}}{\text{minutes}}.$$

EXAMPLE 3   In an oil refinery, a storage tank contains 2000 gal of gasoline that initially has 100 lb of an additive dissolved in it. In preparation for winter weather, gasoline containing 2 lb of additive per gallon is pumped into the tank at a rate of 40 gal/min.

The well-mixed solution is pumped out at a rate of 45 gal/min. How much of the additive is in the tank 20 min after the pumping process begins (Figure 9.15)?

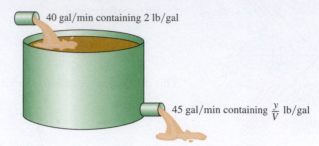

40 gal/min containing 2 lb/gal

45 gal/min containing $\frac{y}{V}$ lb/gal

FIGURE 9.15   The storage tank in Example 3 mixes input liquid with stored liquid to produce an output liquid.

Solution   Let y be the amount (in pounds) of additive in the tank at time t. We know that $y = 100$ when $t = 0$. The number of gallons of gasoline and additive in solution in the tank at any time t is

$$V(t) = 2000 \text{ gal} + \left(40\frac{\text{gal}}{\text{min}} - 45\frac{\text{gal}}{\text{min}}\right)(t \text{ min}) = (2000 - 5t) \text{ gal}.$$

Therefore,

$$\text{Rate out} = \frac{y(t)}{V(t)} \cdot \text{outflow rate} \qquad \text{Eq. (7)}$$

$$= \left(\frac{y}{2000 - 5t}\right)45 \qquad \begin{array}{l}\text{Outflow rate is 45 gal/min}\\\text{and } V = 2000 - 5t.\end{array}$$

$$= \frac{45y}{2000 - 5t}\frac{\text{lb}}{\text{min}}.$$

Also,

$$\text{Rate in} = \left(2\frac{\text{lb}}{\text{gal}}\right)\left(40\frac{\text{gal}}{\text{min}}\right) = 80\frac{\text{lb}}{\text{min}}.$$

The differential equation modeling the mixture process is

$$\frac{dy}{dt} = 80 - \frac{45y}{2000 - 5t} \qquad \text{Eq. (8)}$$

in pounds per minute.

To solve this differential equation, we first write it in standard linear form:

$$\frac{dy}{dt} + \frac{45}{2000 - 5t}y = 80.$$

Thus, $P(t) = 45/(2000 - 5t)$ and $Q(t) = 80$. The integrating factor is

$$v(t) = e^{\int P\, dt} = e^{\int \frac{45}{2000 - 5t}dt}$$

$$= e^{-9 \ln (2000 - 5t)} \qquad 2000 - 5t > 0$$

$$= (2000 - 5t)^{-9}.$$

Multiplying both sides of the standard equation by $v(t)$ and integrating both sides gives

$$(2000 - 5t)^{-9} \cdot \left(\frac{dy}{dt} + \frac{45}{2000 - 5t} y \right) = 80(2000 - 5t)^{-9}$$

$$(2000 - 5t)^{-9} \frac{dy}{dt} + 45(2000 - 5t)^{-10} y = 80(2000 - 5t)^{-9}$$

$$\frac{d}{dt}\left[(2000 - 5t)^{-9} y \right] = 80(2000 - 5t)^{-9}$$

$$(2000 - 5t)^{-9} y = \int 80(2000 - 5t)^{-9} \, dt$$

$$(2000 - 5t)^{-9} y = 80 \cdot \frac{(2000 - 5t)^{-8}}{(-8)(-5)} + C.$$

The general solution is

$$y = 2(2000 - 5t) + C(2000 - 5t)^9.$$

Because $y = 100$ when $t = 0$, we can determine the value of C:

$$100 = 2(2000 - 0) + C(2000 - 0)^9$$

$$C = -\frac{3900}{(2000)^9}.$$

The particular solution of the initial value problem is

$$y = 2(2000 - 5t) - \frac{3900}{(2000)^9} (2000 - 5t)^9.$$

The amount of additive in the tank 20 min after the pumping begins is

$$y(20) = 2[2000 - 5(20)] - \frac{3900}{(2000)^9}[2000 - 5(20)]^9 \approx 1342 \text{ lb.} \quad \blacksquare$$

EXERCISES 9.3

Motion Along a Line

1. Coasting bicycle A 66-kg cyclist on a 7-kg bicycle starts coasting on level ground at 9 m/sec. The k in Equation (1) is about 3.9 kg/sec.

  a. About how far will the cyclist coast before reaching a complete stop?

  b. How long will it take the cyclist's speed to drop to 1 m/sec?

2. Coasting battleship Suppose that an Iowa class battleship has mass around 51,000 metric tons (51,000,000 kg) and a k value in Equation (1) of about 59,000 kg/sec. Assume that the ship loses power when it is moving at a speed of 9 m/sec.

  a. About how far will the ship coast before it is dead in the water?

  b. About how long will it take the ship's speed to drop to 1 m/sec?

3. The data in Table 9.4 were collected with a motion detector and a CBL™ by Valerie Sharritts, then a mathematics teacher at St. Francis DeSales High School in Columbus, Ohio. The table shows the distance s (meters) coasted on inline skates in t sec by her daughter Ashley when she was 10 years old. Find a model for Ashley's position given by the data in Table 9.4 in the form of

TABLE 9.4 Ashley Sharritts skating data

t (sec)	s (m)	t (sec)	s (m)	t (sec)	s (m)
0	0	2.24	3.05	4.48	4.77
0.16	0.31	2.40	3.22	4.64	4.82
0.32	0.57	2.56	3.38	4.80	4.84
0.48	0.80	2.72	3.52	4.96	4.86
0.64	1.05	2.88	3.67	5.12	4.88
0.80	1.28	3.04	3.82	5.28	4.89
0.96	1.50	3.20	3.96	5.44	4.90
1.12	1.72	3.36	4.08	5.60	4.90
1.28	1.93	3.52	4.18	5.76	4.91
1.44	2.09	3.68	4.31	5.92	4.90
1.60	2.30	3.84	4.41	6.08	4.91
1.76	2.53	4.00	4.52	6.24	4.90
1.92	2.73	4.16	4.63	6.40	4.91
2.08	2.89	4.32	4.69	6.56	4.91

Equation (2). Her initial velocity was $v_0 = 2.75$ m/sec, her mass $m = 39.92$ kg (she weighed 88 lb), and her total coasting distance was 4.91 m.

4. Coasting to a stop Table 9.5 shows the distance s (meters) coasted on inline skates in terms of time t (seconds) by Kelly Schmitzer. Find a model for her position in the form of Equation (2). Her initial velocity was $v_0 = 0.80$ m/sec, her mass $m = 49.90$ kg (110 lb), and her total coasting distance was 1.32 m.

TABLE 9.5 Kelly Schmitzer skating data

t (sec)	s (m)	t (sec)	s (m)	t (sec)	s (m)
0	0	1.5	0.89	3.1	1.30
0.1	0.07	1.7	0.97	3.3	1.31
0.3	0.22	1.9	1.05	3.5	1.32
0.5	0.36	2.1	1.11	3.7	1.32
0.7	0.49	2.3	1.17	3.9	1.32
0.9	0.60	2.5	1.22	4.1	1.32
1.1	0.71	2.7	1.25	4.3	1.32
1.3	0.81	2.9	1.28	4.5	1.32

Orthogonal Trajectories

In Exercises 5–10, find the orthogonal trajectories of the family of curves. Sketch several members of each family.

5. $y = mx$ **6.** $y = cx^2$ **7.** $kx^2 + y^2 = 1$

8. $2x^2 + y^2 = c^2$ **9.** $y = ce^{-x}$ **10.** $y = e^{kx}$

11. Show that the curves $2x^2 + 3y^2 = 5$ and $y^2 = x^3$ are orthogonal.

12. Find the family of solutions of the given differential equation and the family of orthogonal trajectories. Sketch both families.

  a. $x\,dx + y\,dy = 0$

  b. $x\,dy - 2y\,dx = 0$

Mixture Problems

13. Salt mixture A tank initially contains 100 gal of brine in which 50 lb of salt are dissolved. A brine containing 2 lb/gal of salt runs into the tank at the rate of 5 gal/min. The mixture is kept uniform by stirring and flows out of the tank at the rate of 4 gal/min.

  a. At what rate (pounds per minute) does salt enter the tank at time t?

  b. What is the volume of brine in the tank at time t?

  c. At what rate (pounds per minute) does salt leave the tank at time t?

  d. Write down and solve the initial value problem describing the mixing process.

  e. Find the concentration of salt in the tank 25 min after the process starts.

14. Mixture problem A 200-gal tank is half full of distilled water. At time $t = 0$, a solution containing 0.5 lb/gal of concentrate enters the tank at the rate of 5 gal/min, and the well-stirred mixture is withdrawn at the rate of 3 gal/min.

  a. At what time will the tank be full?

  b. At the time the tank is full, how many pounds of concentrate will it contain?

15. Fertilizer mixture A tank contains 100 gal of fresh water. A solution containing 1 lb/gal of soluble lawn fertilizer runs into the tank at the rate of 1 gal/min, and the mixture is pumped out of the tank at the rate of 3 gal/min. Find the maximum amount of fertilizer in the tank and the time required to reach the maximum.

16. Carbon monoxide pollution An executive conference room of a corporation contains 4500 ft^3 of air initially free of carbon monoxide. Starting at time $t = 0$, cigarette smoke containing 4% carbon monoxide is blown into the room at the rate of 0.3 ft^3/min. A ceiling fan keeps the air in the room well circulated and the air leaves the room at the same rate of 0.3 ft^3/min. Find the time when the concentration of carbon monoxide in the room reaches 0.01%.

9.4   Graphical Solutions of Autonomous Equations

In Chapter 4 we learned that the sign of the first derivative tells where the graph of a function is increasing and where it is decreasing. The sign of the second derivative tells the concavity of the graph. We can build on our knowledge of how derivatives determine the shape of a graph to solve differential equations graphically. We will see that the ability to discern physical behavior from graphs is a powerful tool in understanding real-world systems. The starting ideas for a graphical solution are the notions of *phase line* and *equilibrium value*. We arrive at these notions by investigating, from a point of view quite different from that studied in Chapter 4, what happens when the derivative of a differentiable function is zero.

Equilibrium Values and Phase Lines

When we differentiate implicitly the equation

$$\frac{1}{5}\ln(5y - 15) = x + 1,$$

we obtain

$$\frac{1}{5}\left(\frac{5}{5y-15}\right)\frac{dy}{dx} = 1.$$

Solving for $y' = dy/dx$ we find $y' = 5y - 15 = 5(y - 3)$. In this case the derivative y' is a function of y only (the dependent variable) and is zero when $y = 3$.

A differential equation for which dy/dx is a function of y only is called an **autonomous** differential equation. Let's investigate what happens when the derivative in an autonomous equation equals zero. We assume any derivatives are continuous.

> **DEFINITION**   If $dy/dx = g(y)$ is an autonomous differential equation, then the values of y for which $dy/dx = 0$ are called **equilibrium values** or **rest points**.

Thus, equilibrium values are those at which no change occurs in the dependent variable, so y is at *rest*. The emphasis is on the value of y where $dy/dx = 0$, not the value of x, as we studied in Chapter 4. For example, the equilibrium values for the autonomous differential equation

$$\frac{dy}{dx} = (y + 1)(y - 2)$$

are $y = -1$ and $y = 2$.

To construct a graphical solution to an autonomous differential equation, we first make a **phase line** for the equation, a plot on the y-axis that shows the equation's equilibrium values along with the intervals where dy/dx and d^2y/dx^2 are positive and negative. Then we know where the solutions are increasing and decreasing, and the concavity of the solution curves. These are the essential features we found in Section 4.4, so we can determine the shapes of the solution curves without having to find formulas for them.

EXAMPLE 1   Draw a phase line for the equation

$$\frac{dy}{dx} = (y + 1)(y - 2)$$

and use it to sketch solutions to the equation.

Solution

1. *Draw a number line for y and mark the equilibrium values $y = -1$ and $y = 2$, where $dy/dx = 0$.*

2. *Identify and label the intervals where $y' > 0$ and $y' < 0$. This step resembles what we did in Section 4.3, only now we are marking the y-axis instead of the x-axis.*

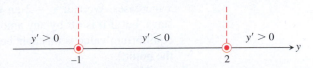

We can encapsulate the information about the sign of y' on the phase line itself. Since $y' > 0$ on the interval to the left of $y = -1$, a solution of the differential equation with a y-value less than -1 will increase from there toward $y = -1$. We display this information by drawing an arrow on the interval pointing to -1.

Similarly, $y' < 0$ between $y = -1$ and $y = 2$, so any solution with a value in this interval will decrease toward $y = -1$.

For $y > 2$, we have $y' > 0$, so a solution with a y-value greater than 2 will increase from there without bound.

In short, solution curves below the horizontal line $y = -1$ in the xy-plane rise toward $y = -1$. Solution curves between the lines $y = -1$ and $y = 2$ fall away from $y = 2$ toward $y = -1$. Solution curves above $y = 2$ rise away from $y = 2$ and keep going up.

3. *Calculate y'' and mark the intervals where $y'' > 0$ and $y'' < 0$.* To find y'', we differentiate y' *with respect to x*, using implicit differentiation.

$$y' = (y + 1)(y - 2) = y^2 - y - 2 \qquad \text{Formula for } y' \dots$$

$$y'' = \frac{d}{dx}(y') = \frac{d}{dx}(y^2 - y - 2) \qquad \text{Substitute for } y'.$$

$$= 2yy' - y' \qquad \begin{array}{l} \text{Differentiate implicitly} \\ \text{with respect to } x. \end{array}$$

$$= (2y - 1)y'$$

$$= (2y - 1)(y + 1)(y - 2).$$

From this formula, we see that y'' changes sign at $y = -1$, $y = 1/2$, and $y = 2$. We add the sign information to the phase line.

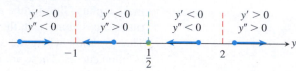

4. *Sketch an assortment of solution curves in the xy-plane.* The horizontal lines $y = -1$, $y = 1/2$, and $y = 2$ partition the plane into horizontal bands in which we know the signs of y' and y''. In each band, this information tells us whether the solution curves rise or fall and how they bend as x increases (Figure 9.16).

The "equilibrium lines" $y = -1$ and $y = 2$ are also solution curves. (The constant functions $y = -1$ and $y = 2$ satisfy the differential equation.) Solution curves that cross the line $y = 1/2$ have an inflection point there. The concavity changes from concave down (above the line) to concave up (below the line).

As predicted in Step 2, solutions in the middle and lower bands approach the equilibrium value $y = -1$ as x increases. Solutions in the upper band rise steadily away from the value $y = 2$. ∎

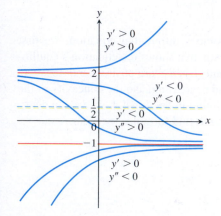

FIGURE 9.16 Graphical solutions from Example 1 include the horizontal lines $y = -1$ and $y = 2$ through the equilibrium values. No two solution curves can ever cross or touch each other.

Stable and Unstable Equilibria

Look at Figure 9.16 once more, in particular at the behavior of the solution curves near the equilibrium values. Once a solution curve has a value near $y = -1$, it tends steadily toward that value; $y = -1$ is a **stable equilibrium**. The behavior near $y = 2$ is just the opposite: All solutions except the equilibrium solution $y = 2$ itself move *away* from it as x increases. We call $y = 2$ an **unstable equilibrium**. If the solution is *at* that value, it stays, but if it is off by any amount, no matter how small, it moves away. (Sometimes an equilibrium value is unstable because a solution moves away from it only on one side of the point.)

Now that we know what to look for, we can already see this behavior on the initial phase line (the second diagram in Step 2 of Example 1). The arrows lead away from $y = 2$ and, once to the left of $y = 2$, toward $y = -1$.

We now present several applied examples for which we can sketch a family of solution curves to the differential equation models using the method in Example 1.

Newton's Law of Cooling

In Section 7.2 we solved analytically the differential equation

$$\frac{dH}{dt} = -k(H - H_S), \qquad k > 0$$

FIGURE 9.17 First step in constructing the phase line for Newton's Law of Cooling. The temperature tends toward the equilibrium (surrounding-medium) value in the long run.

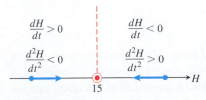

FIGURE 9.18 The complete phase line for Newton's Law of Cooling.

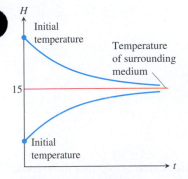

FIGURE 9.19 Temperature versus time. Regardless of initial temperature, the object's temperature $H(t)$ tends toward 15°C, the temperature of the surrounding medium.

modeling Newton's Law of Cooling. Here H is the temperature of an object at time t and H_S is the constant temperature of the surrounding medium.

Suppose that the surrounding medium (say, a room in a house) has a constant Celsius temperature of 15°C. We can then express the difference in temperature as $H(t) - 15$. Assuming H is a differentiable function of time t, by Newton's Law of Cooling, there is a constant of proportionality $k > 0$ such that

$$\frac{dH}{dt} = -k(H - 15) \tag{1}$$

(*minus k to give a negative derivative when* $H > 15$).

Since $dH/dt = 0$ at $H = 15$, the temperature 15°C is an equilibrium value. If $H > 15$, Equation (1) tells us that $(H - 15) > 0$ and $dH/dt < 0$. If the object is hotter than the room, it will get cooler. Similarly, if $H < 15$, then $(H - 15) < 0$ and $dH/dt > 0$. An object cooler than the room will warm up. Thus, the behavior described by Equation (1) agrees with our intuition of how temperature should behave. These observations are captured in the initial phase line diagram in Figure 9.17. The value $H = 15$ is a stable equilibrium.

We determine the concavity of the solution curves by differentiating both sides of Equation (1) with respect to t:

$$\frac{d}{dt}\left(\frac{dH}{dt}\right) = \frac{d}{dt}(-k(H - 15))$$

$$\frac{d^2H}{dt^2} = -k\frac{dH}{dt}.$$

Since $-k$ is negative, we see that d^2H/dt^2 is positive when $dH/dt < 0$ and negative when $dH/dt > 0$. Figure 9.18 adds this information to the phase line.

The completed phase line shows that if the temperature of the object is above the equilibrium value of 15°C, the graph of $H(t)$ will be decreasing and concave upward. If the temperature is below 15°C (the temperature of the surrounding medium), the graph of $H(t)$ will be increasing and concave downward. We use this information to sketch typical solution curves (Figure 9.19).

From the upper solution curve in Figure 9.19, we see that as the object cools down, the rate at which it cools slows down because dH/dt approaches zero. This observation is implicit in Newton's Law of Cooling and contained in the differential equation, but the flattening of the graph as time advances gives an immediate visual representation of the phenomenon.

A Falling Body Encountering Resistance

Newton observed that the rate of change in momentum encountered by a moving object is equal to the net force applied to it. In mathematical terms,

$$F = \frac{d}{dt}(mv), \tag{2}$$

where F is the net force acting on the object, and m and v are the object's mass and velocity. If m varies with time, as it will if the object is a rocket burning fuel, the right-hand side of Equation (2) expands to

$$m\frac{dv}{dt} + v\frac{dm}{dt}$$

using the Derivative Product Rule. In many situations, however, m is constant, $dm/dt = 0$, and Equation (2) takes the simpler form

$$F = m\frac{dv}{dt} \quad \text{or} \quad F = ma, \tag{3}$$

known as *Newton's second law of motion* (see Section 9.3).

In free fall, the constant acceleration due to gravity is denoted by g and the one force propelling the body downward is

$$F_p = mg,$$

the force due to gravity. If, however, we think of a real body falling through the air—say, a penny from a great height or a parachutist from an even greater height—we know that at some point air resistance is a factor in the speed of the fall. A more realistic model of free fall would include air resistance, shown as a force F_r in the schematic diagram in Figure 9.20.

For low speeds well below the speed of sound, physical experiments have shown that F_r is approximately proportional to the body's velocity. The net force on the falling body is therefore

$$F = F_p - F_r,$$

giving

$$m\frac{dv}{dt} = mg - kv$$

$$\frac{dv}{dt} = g - \frac{k}{m}v. \tag{4}$$

We can use a phase line to analyze the velocity functions that solve this differential equation.

The equilibrium point, obtained by setting the right-hand side of Equation (4) equal to zero, is

$$v = \frac{mg}{k}.$$

If the body is initially moving faster than this, dv/dt is negative and the body slows down. If the body is moving at a velocity below mg/k, then $dv/dt > 0$ and the body speeds up. These observations are captured in the initial phase line diagram in Figure 9.21.

We determine the concavity of the solution curves by differentiating both sides of Equation (4) with respect to t:

$$\frac{d^2v}{dt^2} = \frac{d}{dt}\left(g - \frac{k}{m}v\right) = -\frac{k}{m}\frac{dv}{dt}.$$

We see that $d^2v/dt^2 < 0$ when $v < mg/k$ and $d^2v/dt^2 > 0$ when $v > mg/k$. Figure 9.22 adds this information to the phase line. Notice the similarity to the phase line for Newton's Law of Cooling (Figure 9.18). The solution curves are similar as well (Figure 9.23).

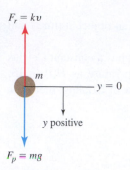

FIGURE 9.20 An object falling under the propulsion due to gravity, with a resistive force assumed to be proportional to the velocity.

FIGURE 9.21 Initial phase line for the falling body encountering resistance.

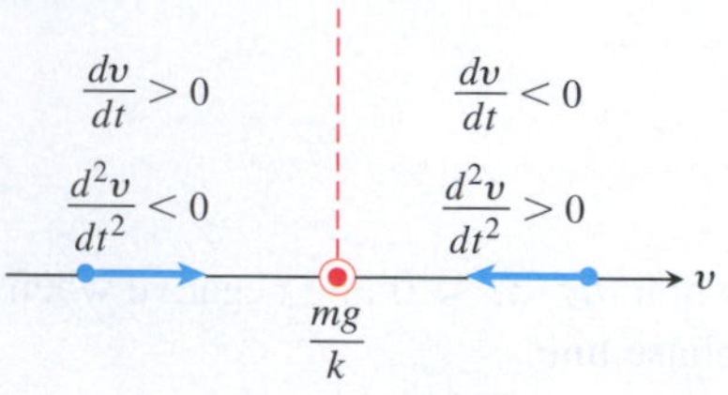

FIGURE 9.22 The completed phase line for the falling body.

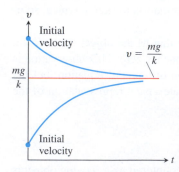

FIGURE 9.23 Typical velocity curves for a falling body encountering resistance. The value $v = mg/k$ is the terminal velocity.

Figure 9.23 shows two typical solution curves. Regardless of the initial velocity, we see the body's velocity tending toward the limiting value $v = mg/k$. This value, a stable equilibrium point, is called the body's **terminal velocity**. Skydivers can vary their terminal velocity from 95 mph to 180 mph by changing the amount of body area opposing the fall, which affects the value of k.

Logistic Population Growth

In Section 9.3 we examined population growth using the model of exponential change. That is, if P represents the number of individuals and we neglect departures and arrivals, then

$$\frac{dP}{dt} = kP, \tag{5}$$

where $k > 0$ is the birth rate minus the death rate per individual per unit time.

Because the natural environment has only a limited number of resources to sustain life, it is reasonable to assume that only a maximum population M can be accommodated. As the population approaches this **limiting population** or **carrying capacity**, resources become less abundant and the growth rate k decreases. A simple relationship exhibiting this behavior is

$$k = r(M - P),$$

where $r > 0$ is a constant. Notice that k decreases as P increases toward M and that k is negative if P is greater than M. Substituting $r(M - P)$ for k in Equation (5) gives the differential equation

$$\frac{dP}{dt} = r(M - P)P = rMP - rP^2. \tag{6}$$

The model given by Equation (6) is referred to as **logistic growth**.

We can forecast the behavior of the population over time by analyzing the phase line for Equation (6). The equilibrium values are $P = M$ and $P = 0$, and we can see that $dP/dt > 0$ if $0 < P < M$ and $dP/dt < 0$ if $P > M$. These observations are recorded on the phase line in Figure 9.24.

FIGURE 9.24 The initial phase line for logistic growth (Equation 6).

We determine the concavity of the population curves by differentiating both sides of Equation (6) with respect to t:

$$\frac{d^2P}{dt^2} = \frac{d}{dt}\left(rMP - rP^2\right)$$

$$= rM\frac{dP}{dt} - 2rP\frac{dP}{dt}$$

$$= r(M - 2P)\frac{dP}{dt}. \tag{7}$$

If $P = M/2$, then $d^2P/dt^2 = 0$. If $P < M/2$, then $(M - 2P)$ and dP/dt are positive and $d^2P/dt^2 > 0$. If $M/2 < P < M$, then $(M - 2P) < 0$, $dP/dt > 0$, and $d^2P/dt^2 < 0$. If $P > M$, then $(M - 2P)$ and dP/dt are both negative and $d^2P/dt^2 > 0$. We add this information to the phase line (Figure 9.25).

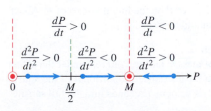

FIGURE 9.25 The completed phase line for logistic growth (Equation 6).

The lines $P = M/2$ and $P = M$ divide the first quadrant of the tP-plane into horizontal bands in which we know the signs of both dP/dt and d^2P/dt^2. In each band, we know how the solution curves rise and fall, and how they bend as time passes. The equilibrium lines $P = 0$ and $P = M$ are both population curves. Population curves crossing the line

$P = M/2$ have an inflection point there, giving them a **sigmoid** shape (curved in two directions like a letter S). Figure 9.26 displays typical population curves. Notice that each population curve approaches the limiting population M as $t \to \infty$.

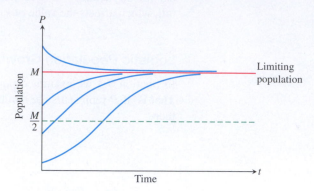

FIGURE 9.26 Population curves for logistic growth.

EXERCISES 9.4

Phase Lines and Solution Curves

In Exercises 1–8,

  a. Identify the equilibrium values. Which are stable and which are unstable?

  b. Construct a phase line. Identify the signs of y' and y''.

  c. Sketch several solution curves.

1. $\dfrac{dy}{dx} = (y + 2)(y - 3)$ 2. $\dfrac{dy}{dx} = y^2 - 4$

3. $\dfrac{dy}{dx} = y^3 - y$ 4. $\dfrac{dy}{dx} = y^2 - 2y$

5. $y' = \sqrt{y}, \quad y > 0$ 6. $y' = y - \sqrt{y}, \quad y > 0$

7. $y' = (y - 1)(y - 2)(y - 3)$ 8. $y' = y^3 - y^2$

Models of Population Growth

The autonomous differential equations in Exercises 9–12 represent models for population growth. For each exercise, use a phase line analysis to sketch solution curves for $P(t)$, selecting different starting values $P(0)$. Which equilibria are stable, and which are unstable?

9. $\dfrac{dP}{dt} = 1 - 2P$ 10. $\dfrac{dP}{dt} = P(1 - 2P)$

11. $\dfrac{dP}{dt} = 2P(P - 3)$ 12. $\dfrac{dP}{dt} = 3P(1 - P)\left(P - \dfrac{1}{2}\right)$

13. **Catastrophic change in logistic growth** Suppose that a healthy population of some species is growing in a limited environment and that the current population P_0 is fairly close to the carrying capacity M_0. You might imagine a population of fish living in a freshwater lake in a wilderness area. Suddenly a catastrophe such as the Mount St. Helens volcanic eruption contaminates the lake and destroys a significant part of the food and oxygen on which the fish depend. The result is a new environment with a carrying capacity M_1 considerably less than M_0 and, in fact, less than the current population P_0. Starting at some time before the catastrophe, sketch a "before-and-after" curve that shows how the fish population responds to the change in environment.

14. **Controlling a population** The fish and game department in a certain state is planning to issue hunting permits to control the deer population (one deer per permit). It is known that if the deer population falls below a certain level m, the deer will become extinct. It is also known that if the deer population rises above the carrying capacity M, the population will decrease back to M through disease and malnutrition.

  a. Discuss the reasonableness of the following model for the growth rate of the deer population as a function of time:

$$\frac{dP}{dt} = rP(M - P)(P - m),$$

  where P is the population of the deer and r is a positive constant of proportionality. Include a phase line.

  b. Explain how this model differs from the logistic model $dP/dt = rP(M - P)$. Is it better or worse than the logistic model?

  c. Show that if $P > M$ for all t, then $\lim_{t \to \infty} P(t) = M$.

  d. What happens if $P < m$ for all t?

  e. Discuss the solutions to the differential equation. What are the equilibrium points of the model? Explain the dependence of the steady-state value of P on the initial values of P. About how many permits should be issued?

Applications and Examples

15. **Skydiving** If a body of mass m falling from rest under the action of gravity encounters an air resistance proportional to the square of velocity, then the body's velocity t seconds into the fall satisfies the equation

$$m\frac{dv}{dt} = mg - kv^2, \quad k > 0$$

  where k is a constant that depends on the body's aerodynamic properties and the density of the air. (We assume that the fall is too short to be affected by changes in the air's density.)

a. Draw a phase line for the equation.

b. Sketch a typical velocity curve.

c. For a 110-lb skydiver ($mg = 110$) and with time in seconds and distance in feet, a typical value of k is 0.005. What is the diver's terminal velocity? Repeat for a 200-lb skydiver.

16. **Resistance proportional to $\sqrt{v}$**   A body of mass m is projected vertically downward with initial velocity v_0. Assume that the resisting force is proportional to the square root of the velocity and find the terminal velocity from a graphical analysis.

17. **Sailing**   A sailboat is running along a straight course with the wind providing a constant forward force of 50 lb. The only other force acting on the boat is resistance as the boat moves through the water. The resisting force is numerically equal to five times the boat's speed, and the initial velocity is 1 ft/sec. What is the maximum velocity in feet per second of the boat under this wind?

18. **The spread of information**   Sociologists recognize a phenomenon called *social diffusion*, which is the spreading of a piece of information, technological innovation, or cultural fad among a population. The members of the population can be divided into two classes: those who have the information and those who do not. In a fixed population whose size is known, it is reasonable to assume that the rate of diffusion is proportional to the number who have the information times the number yet to receive it. If X denotes the number of individuals who have the information in a population of N people, then a mathematical model for social diffusion is given by

$$\frac{dX}{dt} = kX(N - X),$$

where t represents time in days and k is a positive constant.

a. Discuss the reasonableness of the model.

b. Construct a phase line identifying the signs of X' and X''.

c. Sketch representative solution curves.

d. Predict the value of X for which the information is spreading most rapidly. How many people eventually receive the information?

19. **Current in an RL circuit**   The accompanying diagram represents an electrical circuit whose total resistance is a constant R ohms and whose self-inductance, shown as a coil, is L henries, also a constant. There is a switch whose terminals at a and b can be closed to connect a constant electrical source of V volts. From Section 9.2, we have

$$L\frac{di}{dt} + Ri = V,$$

where i is the current in amperes and t is the time in seconds.

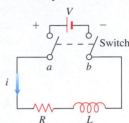

Use a phase line analysis to sketch the solution curve assuming that the switch in the RL circuit is closed at time $t = 0$. What happens to the current as $t \to \infty$? This value is called the *steady-state solution*.

20. **A pearl in shampoo**   Suppose that a pearl is sinking in a thick fluid, like shampoo, subject to a frictional force opposing its fall and proportional to its velocity. Suppose that there is also a resistive buoyant force exerted by the shampoo. According to *Archimedes' principle*, the buoyant force equals the weight of the fluid displaced by the pearl. Using m for the mass of the pearl and P for the mass of the shampoo displaced by the pearl as it descends, complete the following steps.

a. Draw a schematic diagram showing the forces acting on the pearl as it sinks, as in Figure 9.20.

b. Using $v(t)$ for the pearl's velocity as a function of time t, write a differential equation modeling the velocity of the pearl as a falling body.

c. Construct a phase line displaying the signs of v' and v''.

d. Sketch typical solution curves.

e. What is the terminal velocity of the pearl?

9.5 Systems of Equations and Phase Planes

In some situations we are led to consider not one, but several, first-order differential equations. Such a collection is called a **system** of differential equations. In this section we present an approach to understanding systems through a graphical procedure known as a *phase-plane analysis*. We present this analysis in the context of modeling the populations of trout and bass living in a common pond.

Phase Planes

A general system of two first-order differential equations may take the form

$$\frac{dx}{dt} = F(x, y),$$

$$\frac{dy}{dt} = G(x, y).$$

In this system we often think of t as representing time and take $x(t)$ and $y(t)$ to be two functions of t. Such a system of equations is called **autonomous** because dx/dt and dy/dt do not depend on the independent variable time t, but only on the dependent variables x and y. A **solution** of such a system consists of a pair of functions $x(t)$ and $y(t)$ that satisfies both of the differential equations simultaneously for every t over some time interval (finite or infinite).

We cannot look at just one of these equations in isolation to find solutions $x(t)$ or $y(t)$ since each derivative depends on both x and y. To gain insight into the solutions, we look at both dependent variables together by plotting the points $(x(t), y(t))$ in the xy-plane starting at some specified point. Therefore the solution functions define a solution curve through the specified point, called a **trajectory** of the system. The xy-plane itself, in which these trajectories reside, is referred to as the **phase plane**. Thus we consider both solutions together and study the behavior of all the solution trajectories in the phase plane. It can be proved that two trajectories can never cross or touch each other. (Solution trajectories are examples of *parametric curves*, which are studied in detail in Chapter 11.)

A Competitive-Hunter Model

Imagine two species of fish, say trout and bass, competing for the same limited resources (such as food and oxygen) in a certain pond. We let $x(t)$ represent the number of trout and $y(t)$ the number of bass living in the pond at time t. In reality $x(t)$ and $y(t)$ are always integer valued, but we will approximate them with real-valued differentiable functions. This allows us to apply the methods of differential equations.

Several factors affect the rates of change of these populations. As time passes, each species breeds, so we assume its population increases proportionally to its size. Taken by itself, this would lead to exponential growth in each of the two populations. However, there is a countervailing effect from the fact that the two species are in competition. A large number of bass tends to cause a decrease in the number of trout, and vice versa. Our model takes the size of this effect to be proportional to the frequency with which the two species interact, which in turn is proportional to xy, the product of the two populations. These considerations lead to the following model for the growth of the trout and bass in the pond:

$$\frac{dx}{dt} = (a - by)x, \tag{1a}$$

$$\frac{dy}{dt} = (m - nx)y. \tag{1b}$$

Here $x(t)$ represents the trout population, $y(t)$ the bass population, and a, b, m, n are positive constants. A solution of this system then consists of a pair of functions $x(t)$ and $y(t)$ that gives the population of each fish species at time t. Each equation in (1) contains both of the unknown functions x and y, so we are unable to solve them individually. Instead, we will use a graphical analysis to study the solution trajectories of this **competitive-hunter model**.

We now examine the nature of the phase plane in the trout-bass population model. We will be interested in the 1st quadrant of the xy-plane, where $x \geq 0$ and $y \geq 0$, since populations cannot be negative. First, we determine where the bass and trout populations are both constant. Noting that the $(x(t), y(t))$ values remain unchanged when $dx/dt = 0$ and $dy/dt = 0$, Equations (1a and 1b) then become

$$(a - by)x = 0,$$

$$(m - nx)y = 0.$$

This pair of simultaneous equations has two solutions: $(x, y) = (0, 0)$ and $(x, y) = (m/n, a/b)$. At these (x, y) values, called **equilibrium** or **rest points**, the two populations remain at constant values over all time. The point $(0, 0)$ represents a pond containing no members of either fish species; the point $(m/n, a/b)$ corresponds to a pond with an unchanging number of each fish species.

Next, we note that if $y = a/b$, then Equation (1a) implies $dx/dt = 0$, so the trout population $x(t)$ is constant. Similarly, if $x = m/n$, then Equation (1b) implies $dy/dt = 0$, and the bass population $y(t)$ is constant. This information is recorded in Figure 9.27.

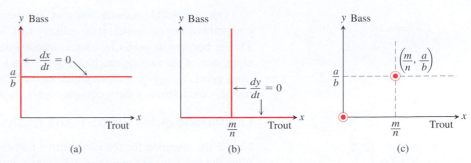

FIGURE 9.27   Rest points in the competitive-hunter model given by Equations (1a) and (1b).

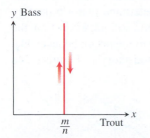

FIGURE 9.28   To the left of the line $x = m/n$ the trajectories move upward, and to the right they move downward.

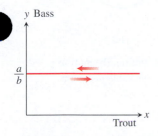

FIGURE 9.29   Above the line $y = a/b$ the trajectories move to the left, and below it they move to the right.

In setting up our competitive-hunter model, precise values of the constants a, b, m, n will not generally be known. Nonetheless, we can analyze the system of Equations (1) to learn the nature of its solution trajectories. We begin by determining the signs of dx/dt and dy/dt throughout the phase plane. Although $x(t)$ represents the number of trout and $y(t)$ the number of bass at time t, we are thinking of the pair of values $(x(t), y(t))$ as a point tracing out a trajectory curve in the phase plane. When dx/dt is positive, $x(t)$ is increasing and the point is moving to the right in the phase plane. If dx/dt is negative, the point is moving to the left. Likewise, the point is moving upward where dy/dt is positive and downward where dy/dt is negative.

We saw that $dy/dt = 0$ along the vertical line $x = m/n$. To the left of this line, dy/dt is positive since $dy/dt = (m - nx)y$ and $x < m/n$. So the trajectories on this side of the line are directed upward. To the right of this line, dy/dt is negative and the trajectories point downward. The directions of the associated trajectories are indicated in Figure 9.28. Similarly, above the horizontal line $y = a/b$, we have $dx/dt < 0$ and the trajectories head leftward; below this line they head rightward, as shown in Figure 9.29. Combining this information gives four distinct regions in the plane A, B, C, D, with their respective trajectory directions shown in Figure 9.30.

Next, we examine what happens near the two equilibrium points. The trajectories near $(0, 0)$ point away from it, upward and to the right. The behavior near the equilibrium point $(m/n, a/b)$ depends on the region in which a trajectory begins. If it starts in region B, for instance, then it will move downward and leftward toward the equilibrium point. Depending on where the trajectory begins, it may move downward into region D, leftward into region A, or perhaps straight into the equilibrium point. If it enters into regions A or D, then it will continue to move away from the rest point. We say that both rest points are **unstable**, meaning (in this setting) there are trajectories near each point that head away from them. These features are indicated in Figure 9.31.

It turns out that in each of the half-planes above and below the line $y = a/b$, there is exactly one trajectory approaching the equilibrium point $(m/n, a/b)$ (see Exercise 7). Above these two trajectories the bass population increases and below them it decreases. The two trajectories approaching the equilibrium point are suggested in Figure 9.32.

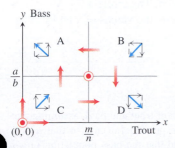

FIGURE 9.30   Composite graphical analysis of the trajectory directions in the four regions determined by $x = m/n$ and $y = a/b$.

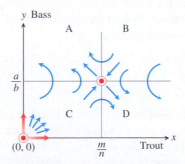

FIGURE 9.31   Motion along the trajectories near the rest points $(0, 0)$ and $(m/n, a/b)$.

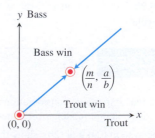

FIGURE 9.32   Qualitative results of analyzing the competitive-hunter model. There are exactly two trajectories approaching the point $(m/n, a/b)$.

Our graphical analysis leads us to conclude that, under the assumptions of the competitive-hunter model, it is unlikely that both species will reach equilibrium levels. This is because it would be almost impossible for the fish populations to move exactly along one of the two approaching trajectories for all time. Furthermore, the initial populations point (x_0, y_0) determines which of the two species is likely to survive over time, and mutual coexistence of the species is highly improbable.

Limitations of the Phase-Plane Analysis Method

FIGURE 9.33 Trajectory direction near the rest point $(0, 0)$.

Unlike the situation for the competitive-hunter model, it is not always possible to determine the behavior of trajectories near a rest point. For example, suppose we know that the trajectories near a rest point, chosen here to be the origin $(0, 0)$, behave as in Figure 9.33. The information provided by Figure 9.33 is not sufficient to distinguish between the three possible trajectories shown in Figure 9.34. Even if we could determine that a trajectory near an equilibrium point resembles that of Figure 9.34c, we would still not know how the other trajectories behave. It could happen that a trajectory closer to the origin behaves like the motions displayed in Figure 9.34a or 9.34b. The spiraling trajectory in Figure 9.34c can never actually reach the rest point in a finite time period.

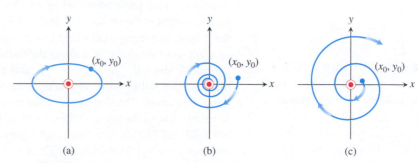

FIGURE 9.34 Three possible trajectory motions: (a) periodic motion, (b) motion toward an asymptotically stable rest point, and (c) motion near an unstable rest point.

Another Type of Behavior

The system

$$\frac{dx}{dt} = y + x - x(x^2 + y^2), \tag{2a}$$

$$\frac{dy}{dt} = -x + y - y(x^2 + y^2) \tag{2b}$$

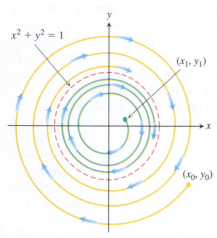

FIGURE 9.35 The solution $x^2 + y^2 = 1$ is a limit cycle.

can be shown to have only one equilibrium point at $(0, 0)$. Yet any trajectory starting on the unit circle traverses it clockwise because, when $x^2 + y^2 = 1$, we have $dy/dx = -x/y$ (see Exercise 2). If a trajectory starts inside the unit circle, it spirals outward, asymptotically approaching the circle as $t \to \infty$. If a trajectory starts outside the unit circle, it spirals inward, again asymptotically approaching the circle as $t \to \infty$. The circle $x^2 + y^2 = 1$ is called a **limit cycle** of the system (Figure 9.35). In this system, the values of x and y eventually become periodic.

EXERCISES 9.5

1. List three important considerations that are ignored in the competitive-hunter model as presented in the text.

2. For the system (2a) and (2b), show that any trajectory starting on the unit circle $x^2 + y^2 = 1$ will traverse the unit circle in a periodic solution. First introduce polar coordinates and rewrite the system as $dr/dt = r(1 - r^2)$ and $-d\theta/dt = -1$.

3. Develop a model for the growth of trout and bass, assuming that in isolation trout demonstrate exponential decay [so that $a < 0$ in Equations (1a) and (1b)] and that the bass population grows logistically with a population limit M. Analyze graphically the motion in the vicinity of the rest points in your model. Is coexistence possible?

4. How might the competitive-hunter model be validated? Include a discussion of how the various constants a, b, m, and n might be estimated. How could state conservation authorities use the model to ensure the survival of both species?

5. Consider another competitive-hunter model defined by

$$\frac{dx}{dt} = a\left(1 - \frac{x}{k_1}\right)x - bxy,$$

$$\frac{dy}{dt} = m\left(1 - \frac{y}{k_2}\right)y - nxy,$$

where x and y represent trout and bass populations, respectively.

a. What assumptions are implicitly being made about the growth of trout and bass in the absence of competition?

b. Interpret the constants a, b, m, n, k_1, and k_2 in terms of the physical problem.

c. Perform a graphical analysis:

   i) Find the possible equilibrium levels.

   ii) Determine whether coexistence is possible.

   iii) Pick several typical starting points and sketch typical trajectories in the phase plane.

   iv) Interpret the outcomes predicted by your graphical analysis in terms of the constants a, b, m, n, k_1, and k_2.

Note: When you get to part (iii), you should realize that five cases exist. You will need to analyze all five cases.

6. An economic model Consider the following economic model. Let P be the price of a single item on the market. Let Q be the quantity of the item available on the market. Both P and Q are functions of time. If one considers price and quantity as two interacting species, the following model might be proposed:

$$\frac{dP}{dt} = aP\left(\frac{b}{Q} - P\right),$$

$$\frac{dQ}{dt} = cQ(fP - Q),$$

where a, b, c, and f are positive constants. Justify and discuss the adequacy of the model.

a. If $a = 1$, $b = 20{,}000$, $c = 1$, and $f = 30$, find the equilibrium points of this system. If possible, classify each equilibrium point with respect to its stability. If a point cannot be readily classified, give some explanation.

b. Perform a graphical stability analysis to determine what will happen to the levels of P and Q as time increases.

c. Give an economic interpretation of the curves that determine the equilibrium points.

7. Two trajectories approach equilibrium Show that the two trajectories leading to $(m/n, a/b)$ shown in Figure 9.32 are unique by carrying out the following steps.

a. From system (1a) and (1b) apply the Chain Rule to derive the following equation:

$$\frac{dy}{dx} = \frac{(m - nx)y}{(a - by)x}.$$

b. Separate the variables, integrate, and exponentiate to obtain

$$y^a e^{-by} = Kx^m e^{-nx},$$

where K is a constant of integration.

c. Let $f(y) = y^a/e^{by}$ and $g(x) = x^m/e^{nx}$. Show that $f(y)$ has a unique maximum of $M_y = (a/eb)^a$ when $y = a/b$ as shown in Figure 9.36. Similarly, show that $g(x)$ has a unique maximum $M_x = (m/en)^m$ when $x = m/n$, also shown in Figure 9.36.

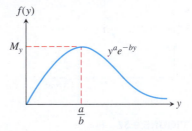

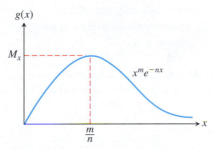

FIGURE 9.36 Graphs of the functions $f(y) = y^a/e^{by}$ and $g(x) = x^m/e^{nx}$.

d. Consider what happens as (x, y) approaches $(m/n, a/b)$. Take limits in part (b) as $x \to m/n$ and $y \to a/b$ to show that either

$$\lim_{\substack{x \to m/n \\ y \to a/b}}\left[\left(\frac{y^a}{e^{by}}\right)\left(\frac{e^{nx}}{x^m}\right)\right] = K$$

or $M_y/M_x = K$. Thus any solution trajectory that approaches $(m/n, a/b)$ must satisfy

$$\frac{y^a}{e^{by}} = \left(\frac{M_y}{M_x}\right)\left(\frac{x^m}{e^{nx}}\right).$$

e. Show that only one trajectory can approach $(m/n, a/b)$ from below the line $y = a/b$. Pick $y_0 < a/b$. From Figure 9.36 you can see that $f(y_0) < M_y$, which implies that

$$\frac{M_y}{M_x}\left(\frac{x^m}{e^{nx}}\right) = y_0{}^a/e^{by_0} < M_y.$$

This in turn implies that

$$\frac{x^m}{e^{nx}} < M_x.$$

Figure 9.36 tells you that for $g(x)$ there is a unique value $x_0 < m/n$ satisfying this last inequality. That is, for each $y < a/b$ there is a unique value of x satisfying the equation in part (d). Thus there can exist only one trajectory solution approaching $(m/n, a/b)$ from below, as shown in Figure 9.37.

f. Use a similar argument to show that the solution trajectory leading to $(m/n, a/b)$ is unique if $y_0 > a/b$.

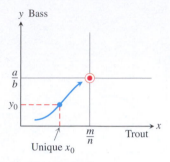

FIGURE 9.37 For any $y < a/b$ only one solution trajectory leads to the rest point $(m/n, a/b)$.

8. Show that the second-order differential equation $y'' = F(x, y, y')$ can be reduced to a system of two first-order differential equations

$$\frac{dy}{dx} = z,$$

$$\frac{dz}{dx} = F(x, y, z).$$

Can something similar be done to the nth-order differential equation $y^{(n)} = F\left(x, y, y', y'', \ldots, y^{(n-1)}\right)$?

Lotka-Volterra Equations for a Predator-Prey Model

In 1925 Lotka and Volterra introduced the *predator-prey* equations, a system of equations that models the populations of two species, one of which preys on the other. Let $x(t)$ represent the number of rabbits living in a region at time t, and $y(t)$ the number of foxes in the same region. As time passes, the number of rabbits increases at a rate proportional to their population, and decreases at a rate proportional to the number of encounters between rabbits and foxes. The foxes, which compete for food, increase in number at a rate proportional to the number of encounters with rabbits but decrease at a rate proportional to the number of foxes. The number of encounters between rabbits and foxes is assumed to be proportional to the product of the two populations. These assumptions lead to the autonomous system

$$\frac{dx}{dt} = (a - by)x$$

$$\frac{dy}{dt} = (-c + dx)y$$

where a, b, c, d are positive constants. The values of these constants vary according to the specific situation being modeled. We can study the nature of the population changes without setting these constants to specific values.

9. What happens to the rabbit population if there are no foxes present?

10. What happens to the fox population if there are no rabbits present?

11. Show that $(0, 0)$ and $(c/d, a/b)$ are equilibrium points. Explain the meaning of each of these points.

12. Show, by differentiating, that the function

$$C(t) = a \ln y(t) - by(t) - dx(t) + c \ln x(t)$$

is constant when $x(t)$ and $y(t)$ are positive and satisfy the predator-prey equations.

While x and y may change over time, $C(t)$ does not. Thus, C is a *conserved quantity* and its existence gives a *conservation law*. A trajectory that begins at a point (x, y) at time $t = 0$ gives a value of C that remains unchanged at future times. Each value of the constant C gives a trajectory for the autonomous system, and these trajectories close up, rather than spiraling inward or outward. The rabbit and fox populations oscillate through repeated cycles along a fixed trajectory. Figure 9.38 shows several trajectories for the predator-prey system.

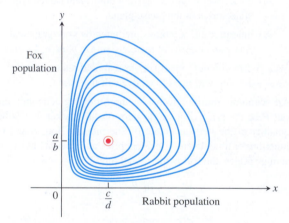

FIGURE 9.38 Some trajectories along which C is conserved.

13. Using a procedure similar to that in the text for the competitive-hunter model, show that each trajectory is traversed in a counterclockwise direction as time t increases.

Along each trajectory, both the rabbit and fox populations fluctuate between their maximum and minimum levels. The maximum and minimum levels for the rabbit population occur where the trajectory intersects the horizontal line $y = a/b$. For the fox population, they occur where the trajectory intersects the vertical line $x = c/d$. When the rabbit population is at its maximum, the fox population is below its maximum value. As the rabbit population declines from this point in time, we move counterclockwise around the trajectory, and the fox population grows until it reaches its maximum value. At this point the rabbit population has declined to $x = c/d$ and is no longer at its peak value. We see that the fox population reaches its maximum value at a later time than the rabbits. The predator population *lags behind* that of the prey in achieving its maximum values. This lag effect is shown in Figure 9.39, which graphs both $x(t)$ and $y(t)$.

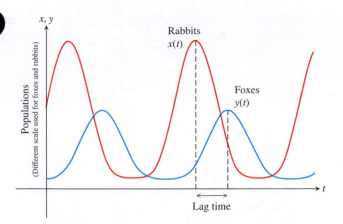

FIGURE 9.39 The fox and rabbit populations oscillate periodically, with the maximum fox population lagging the maximum rabbit population.

14. At some time during a trajectory cycle, a wolf invades the rabbit-fox territory, eats some rabbits, and then leaves. Does this mean that the fox population will from then on have a lower maximum value? Explain your answer.

CHAPTER 9   Questions to Guide Your Review

1. What is a first-order differential equation? When is a function a solution of such an equation?

2. What is a general solution? A particular solution?

3. What is the slope field of a differential equation $y' = f(x, y)$? What can we learn from such fields?

4. Describe Euler's method for solving the initial value problem $y' = f(x, y), y(x_0) = y_0$ numerically. Give an example. Comment on the method's accuracy. Why might you want to solve an initial value problem numerically?

5. How do you solve linear first-order differential equations?

6. What is an orthogonal trajectory of a family of curves? Describe how one is found for a given family of curves.

7. What is an autonomous differential equation? What are its equilibrium values? How do they differ from critical points? What is a stable equilibrium value? Unstable?

8. How do you construct the phase line for an autonomous differential equation? How does the phase line help you produce a graph which qualitatively depicts a solution to the differential equation?

9. Why is the exponential model unrealistic for predicting long-term population growth? How does the logistic model correct for the deficiency in the exponential model for population growth? What is the logistic differential equation? What is the form of its solution? Describe the graph of the logistic solution.

10. What is an autonomous system of differential equations? What is a solution to such a system? What is a trajectory of the system?

CHAPTER 9   Practice Exercises

In Exercises 1–22 solve the differential equation.

1. $y' = xe^y \sqrt{x - 2}$

2. $y' = xye^{x^2}$

3. $\sec x \, dy + x \cos^2 y \, dx = 0$

4. $2x^2 \, dx - 3\sqrt{y} \csc x \, dy = 0$

5. $y' = \dfrac{e^y}{xy}$

6. $y' = xe^{x-y} \csc y$

7. $x(x - 1) \, dy - y \, dx = 0$

8. $y' = (y^2 - 1)x^{-1}$

9. $2y' - y = xe^{x/2}$

10. $\dfrac{y'}{2} + y = e^{-x} \sin x$

11. $xy' + 2y = 1 - x^{-1}$

12. $xy' - y = 2x \ln x$

13. $(1 + e^x) \, dy + (ye^x + e^{-x}) \, dx = 0$

14. $e^{-x} \, dy + (e^{-x}y - 4x) \, dx = 0$

15. $(x + 3y^2) \, dy + y \, dx = 0$ (*Hint:* $d(xy) = y \, dx + x \, dy$)

16. $x \, dy + (3y - x^{-2} \cos x) \, dx = 0, \quad x > 0$

17. $y' = \sin^3 x \cos^2 y$

18. $x \, dy - (x^4 - y) \, dx = 0$

19. $dy + x(2y - e^{x-x^2}) \, dx = 0$

20. $y' + 3x^2y = 7x^2$

21. $y' = xy \ln x \ln y$

22. $xy' + 2y \ln x = \ln x$

Initial Value Problems

In Exercises 23–28 solve the initial value problem.

23. $(x + 1)\dfrac{dy}{dx} + 2y = x, \quad x > -1, \quad y(0) = 1$

24. $x\dfrac{dy}{dx} + 2y = x^2 + 1, \quad x > 0, \quad y(1) = 1$

25. $\dfrac{dy}{dx} + 3x^2y = x^2, \quad y(0) = -1$

26. $x\,dy + (y - \cos x)\,dx = 0, \quad y\left(\dfrac{\pi}{2}\right) = 0$

27. $xy' + (x - 2)y = 3x^3 e^{-x}, \quad y(1) = 0$

28. $y\,dx + (3x - xy + 2)\,dy = 0, \quad y(2) = -1, \quad y < 0$

Euler's Method

In Exercises 29 and 30, use Euler's method to solve the initial value problem on the given interval starting at x_0 with $dx = 0.1$.

T **29.** $y' = y + \cos x, \quad y(0) = 0; \quad 0 \le x \le 2; \quad x_0 = 0$

T **30.** $y' = (2 - y)(2x + 3), \quad y(-3) = 1; \quad -3 \le x \le -1; \quad x_0 = -3$

In Exercises 31 and 32, use Euler's method with $dx = 0.05$ to estimate $y(c)$ where y is the solution to the given initial value problem.

T **31.** $c = 3; \quad \dfrac{dy}{dx} = \dfrac{x - 2y}{x + 1}, \quad y(0) = 1$

T **32.** $c = 4; \quad \dfrac{dy}{dx} = \dfrac{x^2 - 2y + 1}{x}, \quad y(1) = 1$

In Exercises 33 and 34, use Euler's method to solve the initial value problem graphically, starting at $x_0 = 0$ with

    a. $dx = 0.1$.   **b.** $dx = -0.1$.

T **33.** $\dfrac{dy}{dx} = \dfrac{1}{e^{x+y+2}}, \quad y(0) = -2$

T **34.** $\dfrac{dy}{dx} = -\dfrac{x^2 + y}{e^y + x}, \quad y(0) = 0$

Slope Fields

In Exercises 35–38, sketch part of the equation's slope field. Then add to your sketch the solution curve that passes through the point $P(1, -1)$. Use Euler's method with $x_0 = 1$ and $dx = 0.2$ to estimate $y(2)$. Round your answers to four decimal places. Find the exact value of $y(2)$ for comparison.

35. $y' = x$ **36.** $y' = 1/x$

37. $y' = xy$ **38.** $y' = 1/y$

Autonomous Differential Equations and Phase Lines

In Exercises 39 and 40:

    a. Identify the equilibrium values. Which are stable and which are unstable?

    b. Construct a phase line. Identify the signs of y' and y''.

    c. Sketch a representative selection of solution curves.

39. $\dfrac{dy}{dx} = y^2 - 1$ **40.** $\dfrac{dy}{dx} = y - y^2$

Applications

41. Escape velocity The gravitational attraction F exerted by an airless moon on a body of mass m at a distance s from the moon's center is given by the equation $F = -mg\,R^2 s^{-2}$, where g is the acceleration of gravity at the moon's surface and R is the moon's radius (see accompanying figure). The force F is negative because it acts in the direction of decreasing s.

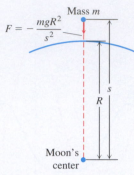

a. If the body is projected vertically upward from the moon's surface with an initial velocity v_0 at time $t = 0$, use Newton's second law, $F = ma$, to show that the body's velocity at position s is given by the equation

$$v^2 = \dfrac{2gR^2}{s} + v_0^2 - 2gR.$$

Thus, the velocity remains positive as long as $v_0 \ge \sqrt{2gR}$. The velocity $v_0 = \sqrt{2gR}$ is the moon's **escape velocity**. A body projected upward with this velocity or a greater one will escape from the moon's gravitational pull.

b. Show that if $v_0 = \sqrt{2gR}$, then

$$s = R\left(1 + \dfrac{3v_0}{2R}t\right)^{2/3}.$$

42. Coasting to a stop Table 9.6 shows the distance s (meters) coasted on inline skates in t sec by Johnathon Krueger. Find a model for his position in the form of Equation (2) of Section 9.3. His initial velocity was $v_0 = 0.86$ m/sec, his mass $m = 30.84$ kg (he weighed 68 lb), and his total coasting distance 0.97 m.

TABLE 9.6 Johnathon Krueger skating data

t (sec)	s (m)	t (sec)	s (m)	t (sec)	s (m)
0	0	0.93	0.61	1.86	0.93
0.13	0.08	1.06	0.68	2.00	0.94
0.27	0.19	1.20	0.74	2.13	0.95
0.40	0.28	1.33	0.79	2.26	0.96
0.53	0.36	1.46	0.83	2.39	0.96
0.67	0.45	1.60	0.87	2.53	0.97
0.80	0.53	1.73	0.90	2.66	0.97

Mixture Problems

In Exercises 43 and 44, let S represent the pounds of salt in a tank at time t minutes. Set up a differential equation representing the given information and the rate at which S changes. Then solve for S and answer the particular questions.

43. A mixture containing $\dfrac{1}{2}$ lb of salt per gallon flows into a tank at the rate of 6 gal/min and the well-stirred mixture flows out of the tank at the rate of 4 gal/min. The tank initially holds 160 gal of solution containing 10 lb of salt.

a. How many gallons of solution are in the tank after 1 minute? after 10 minutes? after 1 hour?

b. How many pounds of salt are in the tank after 1 minute? after 10 minutes? after 1 hour?

44. Pure water flows into a tank at the rate of 4 gal/min, and the well-stirred mixture flows out of the tank at the rate of 5 gal/min. The tank initially holds 200 gal of solution containing 50 pounds of salt.

a. How many gallons of solution are in the tank after 1 minute? after 10 minutes? after 200 minutes?

b. How many pounds of salt are in the tank after 1 minute? after 30 minutes?

c. When will the tank have exactly 5 pounds of salt and how many gallons of solution will be in the tank?

CHAPTER 9   Additional and Advanced Exercises

Theory and Applications

1. **Transport through a cell membrane**   Under some conditions, the result of the movement of a dissolved substance across a cell's membrane is described by the equation

$$\frac{dy}{dt} = k\frac{A}{V}(c - y).$$

In this equation, y is the concentration of the substance inside the cell and dy/dt is the rate at which y changes over time. The letters k, A, V, and c stand for constants, k being the *permeability coefficient* (a property of the membrane), A the surface area of the membrane, V the cell's volume, and c the concentration of the substance outside the cell. The equation says that the rate at which the concentration changes within the cell is proportional to the difference between it and the outside concentration.

a. Solve the equation for $y(t)$, using y_0 to denote $y(0)$.

b. Find the steady-state concentration, $\lim_{t\to\infty} y(t)$.

2. **Height of a rocket**   If an external force F acts upon a system whose mass varies with time, Newton's law of motion is

$$\frac{d(mv)}{dt} = F + (v + u)\frac{dm}{dt}.$$

In this equation, m is the mass of the system at time t, v is its velocity, and $v + u$ is the velocity of the mass that is entering (or leaving) the system at the rate dm/dt. Suppose that a rocket of initial mass m_0 starts from rest, but is driven upward by firing some of its mass directly backward at the constant rate of $dm/dt = -b$ units per second and at constant speed relative to the rocket $u = -c$. The only external force acting on the rocket is $F = -mg$ due to gravity. Under these assumptions, show that the height of the rocket above the ground at the end of t seconds (t small compared to m_0/b) is

$$y = c\left[t + \frac{m_0 - bt}{b}\ln\frac{m_0 - bt}{m_0}\right] - \frac{1}{2}gt^2.$$

3. a. Assume that $P(x)$ and $Q(x)$ are continuous over the interval $[a, b]$. Use the Fundamental Theorem of Calculus, Part 1, to show that any function y satisfying the equation

$$v(x)y = \int v(x)Q(x)\, dx + C$$

for $v(x) = e^{\int P(x)\, dx}$ is a solution to the first-order linear equation

$$\frac{dy}{dx} + P(x)y = Q(x).$$

b. If $C = y_0 v(x_0) - \int_{x_0}^{x} v(t)Q(t)\, dt$, then show that any solution y in part (a) satisfies the initial condition $y(x_0) = y_0$.

4. (*Continuation of Exercise 3.*) Assume the hypotheses of Exercise 3, and assume that $y_1(x)$ and $y_2(x)$ are both solutions to the first-order linear equation satisfying the initial condition $y(x_0) = y_0$.

a. Verify that $y(x) = y_1(x) - y_2(x)$ satisfies the initial value problem

$$y' + P(x)y = 0, \quad y(x_0) = 0.$$

b. For the integrating factor $v(x) = e^{\int P(x)\, dx}$, show that

$$\frac{d}{dx}\left(v(x)[\,y_1(x) - y_2(x)\,]\right) = 0.$$

Conclude that $v(x)[\,y_1(x) - y_2(x)\,] \equiv$ constant.

c. From part (a), we have $y_1(x_0) - y_2(x_0) = 0$. Since $v(x) > 0$ for $a < x < b$, use part (b) to establish that $y_1(x) - y_2(x) \equiv 0$ on the interval (a, b). Thus $y_1(x) = y_2(x)$ for all $a < x < b$.

Homogeneous Equations

A first-order differential equation of the form

$$\frac{dy}{dx} = F\left(\frac{y}{x}\right)$$

is called *homogeneous*. It can be transformed into an equation whose variables are separable by defining the new variable $v = y/x$. Then, $y = vx$ and

$$\frac{dy}{dx} = v + x\frac{dv}{dx}.$$

Substitution into the original differential equation and collecting terms with like variables then gives the separable equation

$$\frac{dx}{x} + \frac{dv}{v - F(v)} = 0.$$

After solving this separable equation, the solution of the original equation is obtained when we replace v by y/x.

Solve the homogeneous equations in Exercises 5–10. First put the equation in the form of a homogeneous equation.

5. $(x^2 + y^2)\, dx + xy\, dy = 0$

6. $x^2\, dy + (y^2 - xy)\, dx = 0$

7. $(xe^{y/x} + y)\, dx - x\, dy = 0$

8. $(x + y)\, dy + (x - y)\, dx = 0$

9. $y' = \dfrac{y}{x} + \cos\dfrac{y - x}{x}$

10. $\left(x\sin\dfrac{y}{x} - y\cos\dfrac{y}{x}\right)dx + x\cos\dfrac{y}{x}\, dy = 0$

CHAPTER 9 Technology Application Projects

Mathematica/Maple Projects

Projects can be found within MyMathLab.

- ***Drug Dosages: Are They Effective? Are They Safe?***
  Formulate and solve an initial value model for the absorption of a drug in the bloodstream.

- ***First-Order Differential Equations and Slope Fields***
  Plot slope fields and solution curves for various initial conditions to selected first-order differential equations.

10

Infinite Sequences and Series

OVERVIEW In this chapter we introduce the topic of *infinite series*. Such series give us precise ways to express many numbers and functions, both familiar and new, as arithmetic sums with infinitely many terms. For example, we will learn that

$$\frac{\pi}{4} = 1 - \frac{1}{3} + \frac{1}{5} - \frac{1}{7} + \frac{1}{9} - \cdots$$

and

$$\cos x = 1 - \frac{x^2}{2} + \frac{x^4}{24} - \frac{x^6}{720} + \frac{x^8}{40,320} - \cdots.$$

We need to develop a method to make sense of such expressions. Everyone knows how to add two numbers together, or even several. But how do you add together infinitely many numbers? Or, when adding together functions, how do you add infinitely many powers of x? In this chapter we answer these questions, which are part of the theory of infinite sequences and series. As with the differential and integral calculus, limits play a major role in the development of infinite series.

One common and important application of series occurs when making computations with complicated functions. A hard-to-compute function is replaced by an expression that looks like an "infinite degree polynomial," an infinite series in powers of x, as we see with the cosine function given above. Using the first few terms of this infinite series can allow for highly accurate approximations of functions by polynomials, enabling us to work with more general functions than those we encountered before. These new functions are commonly obtained as solutions to differential equations arising in important applications of mathematics to science and engineering.

The terms "sequence" and "series" are sometimes used interchangeably in spoken language. In mathematics, however, each has a distinct meaning. A sequence is a type of infinite list, whereas a series is an infinite sum. To understand the infinite sums described by series, we are led to first study infinite sequences.

10.1 Sequences

HISTORICAL ESSAY

Sequences and Series

www.goo.gl/WLjL57

Sequences are fundamental to the study of infinite series and to many aspects of mathematics. We saw one example of a sequence when we studied Newton's Method in Section 4.7. Newton's Method produces a sequence of approximations x_n that become closer and closer to the root of a differentiable function. Now we will explore general sequences of numbers and the conditions under which they converge to a finite number.

Representing Sequences

A sequence is a list of numbers

$$a_1, a_2, a_3, \ldots, a_n, \ldots$$

in a given order. Each of a_1, a_2, a_3 and so on represents a number. These are the **terms** of the sequence. For example, the sequence

$$2, 4, 6, 8, 10, 12, \ldots, 2n, \ldots$$

has first term $a_1 = 2$, second term $a_2 = 4$, and nth term $a_n = 2n$. The integer n is called the **index** of a_n, and indicates where a_n occurs in the list. Order is important. The sequence $2, 4, 6, 8 \ldots$ is not the same as the sequence $4, 2, 6, 8 \ldots$.

We can think of the sequence

$$a_1, a_2, a_3, \ldots, a_n, \ldots$$

as a function that sends 1 to a_1, 2 to a_2, 3 to a_3, and in general sends the positive integer n to the nth term a_n. More precisely, an **infinite sequence** of numbers is a function whose domain is the set of positive integers. For example, the function associated with the sequence

$$2, 4, 6, 8, 10, 12, \ldots, 2n, \ldots$$

sends 1 to $a_1 = 2$, 2 to $a_2 = 4$, and so on. The general behavior of this sequence is described by the formula $a_n = 2n$.

We can change the index to start at any given number n. For example, the sequence

$$12, 14, 16, 18, 20, 22 \ldots$$

is described by the formula $a_n = 10 + 2n$, if we start with $n = 1$. It can also be described by the simpler formula $b_n = 2n$, where the index n starts at 6 and increases. To allow such simpler formulas, we let the first index of the sequence be any appropriate integer. In the sequence above, $\{a_n\}$ starts with a_1 while $\{b_n\}$ starts with b_6.

Sequences can be described by writing rules that specify their terms, such as

$$a_n = \sqrt{n}, \qquad b_n = (-1)^{n+1}\frac{1}{n}, \qquad c_n = \frac{n-1}{n}, \qquad d_n = (-1)^{n+1},$$

or by listing terms:

$$\{a_n\} = \left\{ \sqrt{1}, \sqrt{2}, \sqrt{3}, \ldots, \sqrt{n}, \ldots \right\}$$

$$\{b_n\} = \left\{ 1, -\frac{1}{2}, \frac{1}{3}, -\frac{1}{4}, \ldots, (-1)^{n+1}\frac{1}{n}, \ldots \right\}$$

$$\{c_n\} = \left\{ 0, \frac{1}{2}, \frac{2}{3}, \frac{3}{4}, \frac{4}{5}, \ldots, \frac{n-1}{n}, \ldots \right\}$$

$$\{d_n\} = \left\{ 1, -1, 1, -1, 1, -1, \ldots, (-1)^{n+1}, \ldots \right\}.$$

We also sometimes write a sequence using its rule, as with

$$\{a_n\} = \left\{ \sqrt{n} \right\}_{n=1}^{\infty}$$

and

$$\{b_n\} = \left\{ (-1)^{n+1}\frac{1}{n} \right\}_{n=1}^{\infty}.$$

Figure 10.1 shows two ways to represent sequences graphically. The first marks the first few points from $a_1, a_2, a_3, \ldots, a_n, \ldots$ on the real axis. The second method shows the graph of the function defining the sequence. The function is defined only on integer inputs, and the graph consists of some points in the xy-plane located at $(1, a_1)$, $(2, a_2), \ldots, (n, a_n), \ldots$.

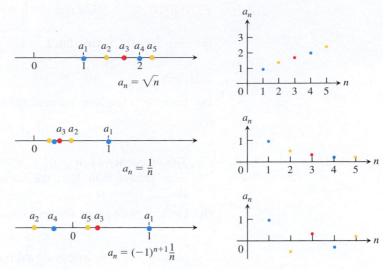

FIGURE 10.1 Sequences can be represented as points on the real line or as points in the plane where the horizontal axis n is the index number of the term and the vertical axis a_n is its value.

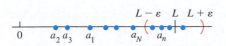

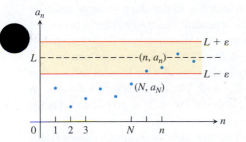

FIGURE 10.2 In the representation of a sequence as points in the plane, $a_n \to L$ if $y = L$ is a horizontal asymptote of the sequence of points $\{(n, a_n)\}$. In this figure, all the a_n's after a_N lie within ε of L.

Convergence and Divergence

Sometimes the numbers in a sequence approach a single value as the index n increases. This happens in the sequence

$$\left\{ 1, \frac{1}{2}, \frac{1}{3}, \frac{1}{4}, \ldots, \frac{1}{n}, \ldots \right\}$$

whose terms approach 0 as n gets large, and in the sequence

$$\left\{ 0, \frac{1}{2}, \frac{2}{3}, \frac{3}{4}, \frac{4}{5}, \ldots, 1 - \frac{1}{n}, \ldots \right\}$$

whose terms approach 1. On the other hand, sequences like

$$\left\{ \sqrt{1}, \sqrt{2}, \sqrt{3}, \ldots, \sqrt{n}, \ldots \right\}$$

have terms that get larger than any number as n increases, and sequences like

$$\left\{ 1, -1, 1, -1, 1, -1, \ldots, (-1)^{n+1}, \ldots \right\}$$

bounce back and forth between 1 and -1, never converging to a single value. The following definition captures the meaning of having a sequence converge to a limiting value. It says that if we go far enough out in the sequence, by taking the index n to be larger than some value N, the difference between a_n and the limit of the sequence becomes less than any preselected number $\varepsilon > 0$.

> **DEFINITIONS** The sequence $\{a_n\}$ **converges** to the number L if for every positive number ε there corresponds an integer N such that
>
> $$|a_n - L| < \varepsilon \qquad \text{whenever} \qquad n > N.$$
>
> If no such number L exists, we say that $\{a_n\}$ **diverges**.
>   If $\{a_n\}$ converges to L, we write $\lim_{n\to\infty} a_n = L$, or simply $a_n \to L$, and call L the **limit** of the sequence (Figure 10.2).

The definition is very similar to the definition of the limit of a function $f(x)$ as x tends to ∞ ($\lim_{x\to\infty} f(x)$ in Section 2.6). We will exploit this connection to calculate limits of sequences.

EXAMPLE 1 Show that

(a) $\lim_{n\to\infty} \dfrac{1}{n} = 0$ **(b)** $\lim_{n\to\infty} k = k$ (any constant k)

Solution

(a) Let $\varepsilon > 0$ be given. We must show that there exists an integer N such that

$$\left| \frac{1}{n} - 0 \right| < \varepsilon \qquad \text{whenever} \qquad n > N.$$

The inequality $|1/n - 0| < \varepsilon$ will hold if $1/n < \varepsilon$ or $n > 1/\varepsilon$. If N is any integer greater than $1/\varepsilon$, the inequality will hold for all $n > N$. This proves that $\lim_{n\to\infty} 1/n = 0$.

(b) Let $\varepsilon > 0$ be given. We must show that there exists an integer N such that

$$|k - k| < \varepsilon \qquad \text{whenever} \qquad n > N.$$

Since $k - k = 0$, we can use any positive integer for N and the inequality $|k - k| < \varepsilon$ will hold. This proves that $\lim_{n\to\infty} k = k$ for any constant k. ■

EXAMPLE 2 Show that the sequence $\{1, -1, 1, -1, 1, -1, \ldots, (-1)^{n+1}, \ldots\}$ diverges.

Solution Suppose the sequence converges to some number L. Then the numbers in the sequence eventually get arbitrarily close to the limit L. This can't happen if they keep oscillating between 1 and -1. We can see this by choosing $\varepsilon = 1/2$ in the definition of the limit. Then all terms a_n of the sequence with index n larger than some N must lie within $\varepsilon = 1/2$ of L. Since the number 1 appears repeatedly as every other term of the sequence, we must have that the number 1 lies within the distance $\varepsilon = 1/2$ of L. It follows that $|L - 1| < 1/2$, or equivalently, $1/2 < L < 3/2$. Likewise, the number -1 appears repeatedly in the sequence with arbitrarily high index. So we must also have that $|L - (-1)| < 1/2$, or equivalently, $-3/2 < L < -1/2$. But the number L cannot lie in both of the intervals $(1/2, 3/2)$ and $(-3/2, -1/2)$ because they have no overlap. Therefore, no such limit L exists and so the sequence diverges.

Note that the same argument works for any positive number ε smaller than 1, not just $1/2$. ■

The sequence $\{\sqrt{n}\}$ also diverges, but for a different reason. As n increases, its terms become larger than any fixed number. We describe the behavior of this sequence by writing

$$\lim_{n\to\infty} \sqrt{n} = \infty.$$

In writing infinity as the limit of a sequence, we are not saying that the differences between the terms a_n and ∞ become small as n increases. Nor are we asserting that there is some number infinity that the sequence approaches. We are merely using a notation that captures the idea that a_n eventually gets and stays larger than any fixed number as n gets large (see Figure 10.3a). The terms of a sequence might also decrease to negative infinity, as in Figure 10.3b.

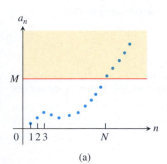

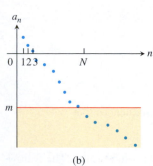

FIGURE 10.3 (a) The sequence diverges to ∞ because no matter what number M is chosen, the terms of the sequence after some index N all lie in the yellow band above M. (b) The sequence diverges to $-\infty$ because all terms after some index N lie below any chosen number m.

> **DEFINITION** The sequence $\{a_n\}$ **diverges to infinity** if for every number M there is an integer N such that for all n larger than N, $a_n > M$. If this condition holds we write
>
> $$\lim_{n\to\infty} a_n = \infty \qquad \text{or} \qquad a_n \to \infty.$$
>
> Similarly, if for every number m there is an integer N such that for all $n > N$ we have $a_n < m$, then we say $\{a_n\}$ **diverges to negative infinity** and write
>
> $$\lim_{n\to\infty} a_n = -\infty \qquad \text{or} \qquad a_n \to -\infty.$$

A sequence may diverge without diverging to infinity or negative infinity, as we saw in Example 2. The sequences $\{1, -2, 3, -4, 5, -6, 7, -8, \ldots\}$ and $\{1, 0, 2, 0, 3, 0, \ldots\}$ are also examples of such divergence.

The convergence or divergence of a sequence is not affected by the values of any number of its initial terms (whether we omit or change the first 10, 1000, or even the first million terms does not matter). From Figure 10.2, we can see that only the part of the sequence that remains after discarding some initial number of terms determines whether the sequence has a limit and the value of that limit when it does exist.

Calculating Limits of Sequences

Since sequences are functions with domain restricted to the positive integers, it is not surprising that the theorems on limits of functions given in Chapter 2 have versions for sequences.

THEOREM 1   Let $\{a_n\}$ and $\{b_n\}$ be sequences of real numbers, and let A and B be real numbers. The following rules hold if $\lim_{n\to\infty} a_n = A$ and $\lim_{n\to\infty} b_n = B$.

1. *Sum Rule:* $\qquad\qquad\qquad$ $\lim_{n\to\infty}(a_n + b_n) = A + B$

2. *Difference Rule:* $\qquad\qquad$ $\lim_{n\to\infty}(a_n - b_n) = A - B$

3. *Constant Multiple Rule:* $\qquad$ $\lim_{n\to\infty}(k \cdot b_n) = k \cdot B$ (any number k)

4. *Product Rule:* $\qquad\qquad$ $\lim_{n\to\infty}(a_n \cdot b_n) = A \cdot B$

5. *Quotient Rule:* $\qquad\qquad$ $\lim_{n\to\infty} \dfrac{a_n}{b_n} = \dfrac{A}{B}$ if $B \neq 0$

The proof is similar to that of Theorem 1 of Section 2.2 and is omitted.

EXAMPLE 3   By combining Theorem 1 with the limits of Example 1, we have:

(a) $\lim\limits_{n\to\infty}\left(-\dfrac{1}{n}\right) = -1 \cdot \lim\limits_{n\to\infty}\dfrac{1}{n} = -1 \cdot 0 = 0$ $\qquad$ Constant Multiple Rule and Example 1a

(b) $\lim\limits_{n\to\infty}\left(\dfrac{n-1}{n}\right) = \lim\limits_{n\to\infty}\left(1 - \dfrac{1}{n}\right) = \lim\limits_{n\to\infty} 1 - \lim\limits_{n\to\infty}\dfrac{1}{n} = 1 - 0 = 1$ $\qquad$ Difference Rule and Example 1a

(c) $\lim\limits_{n\to\infty}\dfrac{5}{n^2} = 5 \cdot \lim\limits_{n\to\infty}\dfrac{1}{n} \cdot \lim\limits_{n\to\infty}\dfrac{1}{n} = 5 \cdot 0 \cdot 0 = 0$ $\qquad$ Product Rule

(d) $\lim\limits_{n\to\infty}\dfrac{4 - 7n^6}{n^6 + 3} = \lim\limits_{n\to\infty}\dfrac{(4/n^6) - 7}{1 + (3/n^6)} = \dfrac{0 - 7}{1 + 0} = -7.$ $\qquad$ Divide numerator and denominator by n^6 and use the Sum and Quotient Rules.

Be cautious in applying Theorem 1. It does not say, for example, that each of the sequences $\{a_n\}$ and $\{b_n\}$ have limits if their sum $\{a_n + b_n\}$ has a limit. For instance, $\{a_n\} = \{1, 2, 3, \ldots\}$ and $\{b_n\} = \{-1, -2, -3, \ldots\}$ both diverge, but their sum $\{a_n + b_n\} = \{0, 0, 0, \ldots\}$ clearly converges to 0.

One consequence of Theorem 1 is that every nonzero multiple of a divergent sequence $\{a_n\}$ diverges. Suppose, to the contrary, that $\{ca_n\}$ converges for some number $c \neq 0$. Then, by taking $k = 1/c$ in the Constant Multiple Rule in Theorem 1, we see that the sequence

$$\left\{\frac{1}{c} \cdot ca_n\right\} = \{a_n\}$$

converges. Thus, $\{ca_n\}$ cannot converge unless $\{a_n\}$ also converges. If $\{a_n\}$ does not converge, then $\{ca_n\}$ does not converge.

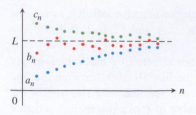

FIGURE 10.4 The terms of sequence $\{b_n\}$ are sandwiched between those of $\{a_n\}$ and $\{c_n\}$, forcing them to the same common limit L.

The next theorem is the sequence version of the Sandwich Theorem in Section 2.2. You are asked to prove the theorem in Exercise 119. (See Figure 10.4.)

> **THEOREM 2—The Sandwich Theorem for Sequences**
>
> Let $\{a_n\}$, $\{b_n\}$, and $\{c_n\}$ be sequences of real numbers. If $a_n \leq b_n \leq c_n$ holds for all n beyond some index N, and if $\lim_{n\to\infty} a_n = \lim_{n\to\infty} c_n = L$, then $\lim_{n\to\infty} b_n = L$ also.

An immediate consequence of Theorem 2 is that, if $|b_n| \leq c_n$ and $c_n \to 0$, then $b_n \to 0$ because $-c_n \leq b_n \leq c_n$. We use this fact in the next example.

EXAMPLE 4   Since $1/n \to 0$, we know that

(a) $\dfrac{\cos n}{n} \to 0$ because $-\dfrac{1}{n} \leq \dfrac{\cos n}{n} \leq \dfrac{1}{n}$;

(b) $\dfrac{1}{2^n} \to 0$ because $0 \leq \dfrac{1}{2^n} \leq \dfrac{1}{n}$;

(c) $(-1)^n \dfrac{1}{n} \to 0$ because $-\dfrac{1}{n} \leq (-1)^n \dfrac{1}{n} \leq \dfrac{1}{n}$.

(d) If $|a_n| \to 0$, then $a_n \to 0$ because $-|a_n| \leq a_n \leq |a_n|$.   ∎

The application of Theorems 1 and 2 is broadened by a theorem stating that applying a continuous function to a convergent sequence produces a convergent sequence. We state the theorem, leaving the proof as an exercise (Exercise 120).

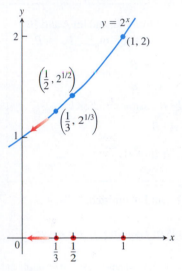

FIGURE 10.5 As $n \to \infty$, $1/n \to 0$ and $2^{1/n} \to 2^0$ (Example 6). The terms of $\{1/n\}$ are shown on the x-axis; the terms of $\{2^{1/n}\}$ are shown as the y-values on the graph of $f(x) = 2^x$.

> **THEOREM 3—The Continuous Function Theorem for Sequences**
>
> Let $\{a_n\}$ be a sequence of real numbers. If $a_n \to L$ and if f is a function that is continuous at L and defined at all a_n, then $f(a_n) \to f(L)$.

EXAMPLE 5   Show that $\sqrt{(n+1)/n} \to 1$.

Solution   We know that $(n+1)/n \to 1$. Taking $f(x) = \sqrt{x}$ and $L = 1$ in Theorem 3 gives $\sqrt{(n+1)/n} \to \sqrt{1} = 1$.   ∎

EXAMPLE 6   The sequence $\{1/n\}$ converges to 0. By taking $a_n = 1/n$, $f(x) = 2^x$, and $L = 0$ in Theorem 3, we see that $2^{1/n} = f(1/n) \to f(L) = 2^0 = 1$. The sequence $\{2^{1/n}\}$ converges to 1 (Figure 10.5).   ∎

Using L'Hôpital's Rule

The next theorem formalizes the connection between $\lim_{n\to\infty} a_n$ and $\lim_{x\to\infty} f(x)$. It enables us to use l'Hôpital's Rule to find the limits of some sequences.

> **THEOREM 4**   Suppose that $f(x)$ is a function defined for all $x \geq n_0$ and that $\{a_n\}$ is a sequence of real numbers such that $a_n = f(n)$ for $n \geq n_0$. Then
>
> $$\lim_{n\to\infty} a_n = L \quad \text{whenever} \quad \lim_{x\to\infty} f(x) = L.$$

Proof Suppose that $\lim_{x \to \infty} f(x) = L$. Then for each positive number ε there is a number M such that

$$|f(x) - L| < \varepsilon \qquad \text{whenever} \qquad x > M.$$

Let N be an integer greater than M and greater than or equal to n_0. Since $a_n = f(n)$, it follows that for all $n > N$ we have

$$|a_n - L| = |f(n) - L| < \varepsilon. \qquad \blacksquare$$

EXAMPLE 7 Show that

$$\lim_{n \to \infty} \frac{\ln n}{n} = 0.$$

Solution The function $(\ln x)/x$ is defined for all $x \geq 1$ and agrees with the given sequence at positive integers. Therefore, by Theorem 4, $\lim_{n \to \infty} (\ln n)/n$ will equal $\lim_{x \to \infty} (\ln x)/x$ if the latter exists. A single application of l'Hôpital's Rule shows that

$$\lim_{x \to \infty} \frac{\ln x}{x} = \lim_{x \to \infty} \frac{1/x}{1} = \frac{0}{1} = 0.$$

We conclude that $\lim_{n \to \infty} (\ln n)/n = 0$. $\qquad \blacksquare$

When we use l'Hôpital's Rule to find the limit of a sequence, we often treat n as a continuous real variable and differentiate directly with respect to n. This saves us from having to rewrite the formula for a_n as we did in Example 7.

EXAMPLE 8 Does the sequence whose nth term is

$$a_n = \left(\frac{n+1}{n-1} \right)^n$$

converge? If so, find $\lim_{n \to \infty} a_n$.

Solution The limit leads to the indeterminate form 1^∞. We can apply l'Hôpital's Rule if we first change the form to $\infty \cdot 0$ by taking the natural logarithm of a_n:

$$\ln a_n = \ln \left(\frac{n+1}{n-1} \right)^n = n \ln \left(\frac{n+1}{n-1} \right).$$

Then,

$$\lim_{n \to \infty} \ln a_n = \lim_{n \to \infty} n \ln \left(\frac{n+1}{n-1} \right) \qquad \color{blue}{\infty \cdot 0 \text{ form}}$$

$$= \lim_{n \to \infty} \frac{\ln \left(\dfrac{n+1}{n-1} \right)}{1/n} \qquad \color{blue}{\frac{0}{0} \text{ form}}$$

$$= \lim_{n \to \infty} \frac{-2/(n^2 - 1)}{-1/n^2} \qquad \color{blue}{\text{L'Hôpital's Rule: differentiate numerator and denominator.}}$$

$$= \lim_{n \to \infty} \frac{2n^2}{n^2 - 1} = 2. \qquad \color{blue}{\text{Simplify and evaluate.}}$$

Since $\ln a_n \to 2$ and $f(x) = e^x$ is continuous, Theorem 3 tells us that

$$a_n = e^{\ln a_n} \to e^2.$$

The sequence $\{a_n\}$ converges to e^2. $\qquad \blacksquare$

Commonly Occurring Limits

The next theorem gives some limits that arise frequently.

Factorial Notation

The notation $n!$ ("n factorial") means the product $1 \cdot 2 \cdot 3 \cdots n$ of the integers from 1 to n. Notice that $(n + 1)! = (n + 1) \cdot n!$. Thus, $4! = 1 \cdot 2 \cdot 3 \cdot 4 = 24$ and $5! = 1 \cdot 2 \cdot 3 \cdot 4 \cdot 5 = 5 \cdot 4! = 120$. We define $0!$ to be 1. Factorials grow even faster than exponentials, as the table suggests. The values in the table are rounded.

n	e^n	$n!$
1	3	1
5	148	120
10	22,026	3,628,800
20	4.9×10^8	2.4×10^{18}

THEOREM 5 The following six sequences converge to the limits listed below:

1. $\displaystyle\lim_{n \to \infty} \frac{\ln n}{n} = 0$

2. $\displaystyle\lim_{n \to \infty} \sqrt[n]{n} = 1$

3. $\displaystyle\lim_{n \to \infty} x^{1/n} = 1 \qquad (x > 0)$

4. $\displaystyle\lim_{n \to \infty} x^n = 0 \qquad (|x| < 1)$

5. $\displaystyle\lim_{n \to \infty} \left(1 + \frac{x}{n}\right)^n = e^x \qquad (\text{any } x)$

6. $\displaystyle\lim_{n \to \infty} \frac{x^n}{n!} = 0 \qquad (\text{any } x)$

In Formulas (3) through (6), x remains fixed as $n \to \infty$.

Proof The first limit was computed in Example 7. The next two can be proved by taking logarithms and applying Theorem 4 (Exercises 117 and 118). The remaining proofs are given in Appendix 5. ∎

EXAMPLE 9 These are examples of the limits in Theorem 5.

(a) $\dfrac{\ln\left(n^2\right)}{n} = \dfrac{2 \ln n}{n} \to 2 \cdot 0 = 0$ ⟶ Formula 1

(b) $\sqrt[n]{n^2} = n^{2/n} = \left(n^{1/n}\right)^2 \to (1)^2 = 1$ ⟶ Formula 2

(c) $\sqrt[n]{3n} = 3^{1/n}\left(n^{1/n}\right) \to 1 \cdot 1 = 1$ ⟶ Formula 3 with $x = 3$ and Formula 2

(d) $\left(-\dfrac{1}{2}\right)^n \to 0$ ⟶ Formula 4 with $x = -\dfrac{1}{2}$

(e) $\left(\dfrac{n - 2}{n}\right)^n = \left(1 + \dfrac{-2}{n}\right)^n \to e^{-2}$ ⟶ Formula 5 with $x = -2$

(f) $\dfrac{100^n}{n!} \to 0$ ⟶ Formula 6 with $x = 100$ ∎

Recursive Definitions

So far, we have calculated each a_n directly from the value of n. But sequences are often defined **recursively** by giving

1. The value(s) of the initial term or terms, and

2. A rule, called a **recursion formula**, for calculating any later term from terms that precede it.

EXAMPLE 10

(a) The statements $a_1 = 1$ and $a_n = a_{n-1} + 1$ for $n > 1$ define the sequence $1, 2, 3, \ldots, n, \ldots$ of positive integers. With $a_1 = 1$, we have $a_2 = a_1 + 1 = 2$, $a_3 = a_2 + 1 = 3$, and so on.

(b) The statements $a_1 = 1$ and $a_n = n \cdot a_{n-1}$ for $n > 1$ define the sequence $1, 2, 6, 24, \ldots, n!, \ldots$ of factorials. With $a_1 = 1$, we have $a_2 = 2 \cdot a_1 = 2$, $a_3 = 3 \cdot a_2 = 6$, $a_4 = 4 \cdot a_3 = 24$, and so on.

(c) The statements $a_1 = 1$, $a_2 = 1$, and $a_{n+1} = a_n + a_{n-1}$ for $n > 2$ define the sequence $1, 1, 2, 3, 5, \ldots$ of **Fibonacci numbers**. With $a_1 = 1$ and $a_2 = 1$, we have $a_3 = 1 + 1 = 2, a_4 = 2 + 1 = 3, a_5 = 3 + 2 = 5$, and so on.

(d) As we can see by applying Newton's method (see Exercise 145), the statements $x_0 = 1$ and $x_{n+1} = x_n - \left[(\sin x_n - x_n{}^2)/(\cos x_n - 2x_n) \right]$ for $n > 0$ define a sequence that, when it converges, gives a solution to the equation $\sin x - x^2 = 0$. ■

Bounded Monotonic Sequences

Two concepts that play a key role in determining the convergence of a sequence are those of a *bounded* sequence and a *monotonic* sequence.

> **DEFINITION** A sequence $\{a_n\}$ is **bounded from above** if there exists a number M such that $a_n \le M$ for all n. The number M is an **upper bound** for $\{a_n\}$. If M is an upper bound for $\{a_n\}$ but no number less than M is an upper bound for $\{a_n\}$, then M is the **least upper bound** for $\{a_n\}$.
>
> A sequence $\{a_n\}$ is **bounded from below** if there exists a number m such that $a_n \ge m$ for all n. The number m is a **lower bound** for $\{a_n\}$. If m is a lower bound for $\{a_n\}$ but no number greater than m is a lower bound for $\{a_n\}$, then m is the **greatest lower bound** for $\{a_n\}$.
>
> If $\{a_n\}$ is bounded from above and below, then $\{a_n\}$ is **bounded**. If $\{a_n\}$ is not bounded, then we say that $\{a_n\}$ is an **unbounded** sequence.

EXAMPLE 11

(a) The sequence $1, 2, 3, \ldots, n, \ldots$ has no upper bound because it eventually surpasses every number M. However, it is bounded below by every real number less than or equal to 1. The number $m = 1$ is the greatest lower bound of the sequence.

(b) The sequence $\dfrac{1}{2}, \dfrac{2}{3}, \dfrac{3}{4}, \ldots, \dfrac{n}{n+1}, \ldots$ is bounded above by every real number greater than or equal to 1. The upper bound $M = 1$ is the least upper bound (Exercise 137). The sequence is also bounded below by every number less than or equal to $\dfrac{1}{2}$, which is its greatest lower bound. ■

Convergent sequences are bounded

If a sequence $\{a_n\}$ converges to the number L, then by definition there is a number N such that $|a_n - L| < 1$ if $n > N$. That is,

$$L - 1 < a_n < L + 1 \quad \text{for } n > N.$$

If M is a number larger than $L + 1$ and all of the finitely many numbers $a_1, a_2, \ldots, a_N$, then for every index n we have $a_n \le M$ so that $\{a_n\}$ is bounded from above. Similarly, if m is a number smaller than $L - 1$ and all of the numbers $a_1, a_2, \ldots, a_N$, then m is a lower bound of the sequence. Therefore, all convergent sequences are bounded.

Although it is true that every convergent sequence is bounded, there are bounded sequences that fail to converge. One example is the bounded sequence $\{(-1)^{n+1}\}$ discussed in Example 2. The problem here is that some bounded sequences bounce around in the band determined by any lower bound m and any upper bound M (Figure 10.6). An important type of sequence that does not behave that way is one for which each term is at least as large, or at least as small, as its predecessor.

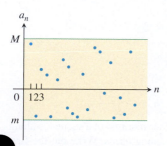

FIGURE 10.6 Some bounded sequences bounce around between their bounds and fail to converge to any limiting value.

> **DEFINITIONS** A sequence $\{a_n\}$ is **nondecreasing** if $a_n \le a_{n+1}$ for all n. That is, $a_1 \le a_2 \le a_3 \le \ldots$. The sequence is **nonincreasing** if $a_n \ge a_{n+1}$ for all n. The sequence $\{a_n\}$ is **monotonic** if it is either nondecreasing or nonincreasing.

EXAMPLE 12

(a) The sequence $1, 2, 3, \ldots, n, \ldots$ is nondecreasing.

(b) The sequence $\dfrac{1}{2}, \dfrac{2}{3}, \dfrac{3}{4}, \ldots, \dfrac{n}{n+1}, \ldots$ is nondecreasing.

(c) The sequence $1, \dfrac{1}{2}, \dfrac{1}{4}, \dfrac{1}{8}, \ldots, \dfrac{1}{2^n}, \ldots$ is nonincreasing.

(d) The constant sequence $3, 3, 3, \ldots, 3, \ldots$ is both nondecreasing and nonincreasing.

(e) The sequence $1, -1, 1, -1, 1, -1, \ldots$ is not monotonic.   ∎

A nondecreasing sequence that is bounded from above always has a least upper bound. Likewise, a nonincreasing sequence bounded from below always has a greatest lower bound. These results are based on the *completeness property* of the real numbers, discussed in Appendix 6. We now prove that if L is the least upper bound of a nondecreasing sequence then the sequence converges to L, and that if L is the greatest lower bound of a nonincreasing sequence then the sequence converges to L.

> **THEOREM 6—The Monotonic Sequence Theorem**
> If a sequence $\{a_n\}$ is both bounded and monotonic, then the sequence converges.

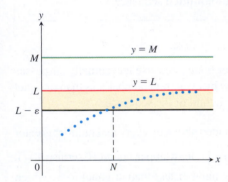

FIGURE 10.7 If the terms of a nondecreasing sequence have an upper bound M, they have a limit $L \leq M$.

Proof   Suppose $\{a_n\}$ is nondecreasing, L is its least upper bound, and we plot the points $(1, a_1), (2, a_2), \ldots, (n, a_n), \ldots$ in the xy-plane. If M is an upper bound of the sequence, all these points will lie on or below the line $y = M$ (Figure 10.7). The line $y = L$ is the lowest such line. None of the points (n, a_n) lies above $y = L$, but some do lie above any lower line $y = L - \varepsilon$, if ε is a positive number (because $L - \varepsilon$ is not an upper bound). The sequence converges to L because

a. $a_n \leq L$ for *all* values of n, and

b. given any $\varepsilon > 0$, there exists at least one integer N for which $a_N > L - \varepsilon$.

The fact that $\{a_n\}$ is nondecreasing tells us further that

$$a_n \geq a_N > L - \varepsilon \qquad \text{for all } n \geq N.$$

Thus, *all* the numbers a_n beyond the Nth number lie within ε of L. This is precisely the condition for L to be the limit of the sequence $\{a_n\}$.

The proof for nonincreasing sequences bounded from below is similar.   ∎

It is important to realize that Theorem 6 does not say that convergent sequences are monotonic. The sequence $\{(-1)^{n+1}/n\}$ converges and is bounded, but it is not monotonic since it alternates between positive and negative values as it tends toward zero. What the theorem does say is that a nondecreasing sequence converges when it is bounded from above, but it diverges to infinity otherwise.

EXERCISES 10.1

Finding Terms of a Sequence

Each of Exercises 1–6 gives a formula for the nth term a_n of a sequence $\{a_n\}$. Find the values of $a_1, a_2, a_3,$ and a_4.

1. $a_n = \dfrac{1-n}{n^2}$

2. $a_n = \dfrac{1}{n!}$

3. $a_n = \dfrac{(-1)^{n+1}}{2n-1}$

4. $a_n = 2 + (-1)^n$

5. $a_n = \dfrac{2^n}{2^{n+1}}$

6. $a_n = \dfrac{2^n - 1}{2^n}$

Each of Exercises 7–12 gives the first term or two of a sequence along with a recursion formula for the remaining terms. Write out the first ten terms of the sequence.

7. $a_1 = 1, \quad a_{n+1} = a_n + (1/2^n)$

8. $a_1 = 1, \quad a_{n+1} = a_n/(n+1)$

9. $a_1 = 2, \quad a_{n+1} = (-1)^{n+1} a_n / 2$

10. $a_1 = -2, \quad a_{n+1} = n a_n / (n + 1)$

11. $a_1 = a_2 = 1, \quad a_{n+2} = a_{n+1} + a_n$

12. $a_1 = 2, \quad a_2 = -1, \quad a_{n+2} = a_{n+1} / a_n$

Finding a Sequence's Formula

In Exercises 13–30, find a formula for the nth term of the sequence.

13. $1, -1, 1, -1, 1, \ldots$ 1's with alternating signs

14. $-1, 1, -1, 1, -1, \ldots$ 1's with alternating signs

15. $1, -4, 9, -16, 25, \ldots$ Squares of the positive integers, with alternating signs

16. $1, -\dfrac{1}{4}, \dfrac{1}{9}, -\dfrac{1}{16}, \dfrac{1}{25}, \ldots$ Reciprocals of squares of the positive integers, with alternating signs

17. $\dfrac{1}{9}, \dfrac{2}{12}, \dfrac{2^2}{15}, \dfrac{2^3}{18}, \dfrac{2^4}{21}, \ldots$ Powers of 2 divided by multiples of 3

18. $-\dfrac{3}{2}, -\dfrac{1}{6}, \dfrac{1}{12}, \dfrac{3}{20}, \dfrac{5}{30}, \ldots$ Integers differing by 2 divided by products of consecutive integers

19. $0, 3, 8, 15, 24, \ldots$ Squares of the positive integers diminished by 1

20. $-3, -2, -1, 0, 1, \ldots$ Integers, beginning with -3

21. $1, 5, 9, 13, 17, \ldots$ Every other odd positive integer

22. $2, 6, 10, 14, 18, \ldots$ Every other even positive integer

23. $\dfrac{5}{1}, \dfrac{8}{2}, \dfrac{11}{6}, \dfrac{14}{24}, \dfrac{17}{120}, \ldots$ Integers differing by 3 divided by factorials

24. $\dfrac{1}{25}, \dfrac{8}{125}, \dfrac{27}{625}, \dfrac{64}{3125}, \dfrac{125}{15,625}, \ldots$ Cubes of positive integers divided by powers of 5

25. $1, 0, 1, 0, 1, \ldots$ Alternating 1's and 0's

26. $0, 1, 1, 2, 2, 3, 3, 4, \ldots$ Each positive integer repeated

27. $\dfrac{1}{2} - \dfrac{1}{3}, \dfrac{1}{3} - \dfrac{1}{4}, \dfrac{1}{4} - \dfrac{1}{5}, \dfrac{1}{5} - \dfrac{1}{6}, \ldots$

28. $\sqrt{5} - \sqrt{4}, \sqrt{6} - \sqrt{5}, \sqrt{7} - \sqrt{6}, \sqrt{8} - \sqrt{7}, \ldots$

29. $\sin\left(\dfrac{\sqrt{2}}{1+4}\right), \sin\left(\dfrac{\sqrt{3}}{1+9}\right), \sin\left(\dfrac{\sqrt{4}}{1+16}\right), \sin\left(\dfrac{\sqrt{5}}{1+25}\right), \ldots$

30. $\sqrt{\dfrac{5}{8}}, \sqrt{\dfrac{7}{11}}, \sqrt{\dfrac{9}{14}}, \sqrt{\dfrac{11}{17}}, \ldots$

Convergence and Divergence

Which of the sequences $\{a_n\}$ in Exercises 31–100 converge, and which diverge? Find the limit of each convergent sequence.

31. $a_n = 2 + (0.1)^n$

32. $a_n = \dfrac{n + (-1)^n}{n}$

33. $a_n = \dfrac{1 - 2n}{1 + 2n}$

34. $a_n = \dfrac{2n + 1}{1 - 3\sqrt{n}}$

35. $a_n = \dfrac{1 - 5n^4}{n^4 + 8n^3}$

36. $a_n = \dfrac{n + 3}{n^2 + 5n + 6}$

37. $a_n = \dfrac{n^2 - 2n + 1}{n - 1}$

38. $a_n = \dfrac{1 - n^3}{70 - 4n^2}$

39. $a_n = 1 + (-1)^n$

40. $a_n = (-1)^n \left(1 - \dfrac{1}{n}\right)$

41. $a_n = \left(\dfrac{n + 1}{2n}\right)\left(1 - \dfrac{1}{n}\right)$

42. $a_n = \left(2 - \dfrac{1}{2^n}\right)\left(3 + \dfrac{1}{2^n}\right)$

43. $a_n = \dfrac{(-1)^{n+1}}{2n - 1}$

44. $a_n = \left(-\dfrac{1}{2}\right)^n$

45. $a_n = \sqrt{\dfrac{2n}{n + 1}}$

46. $a_n = \dfrac{1}{(0.9)^n}$

47. $a_n = \sin\left(\dfrac{\pi}{2} + \dfrac{1}{n}\right)$

48. $a_n = n\pi \cos(n\pi)$

49. $a_n = \dfrac{\sin n}{n}$

50. $a_n = \dfrac{\sin^2 n}{2^n}$

51. $a_n = \dfrac{n}{2^n}$

52. $a_n = \dfrac{3^n}{n^3}$

53. $a_n = \dfrac{\ln(n + 1)}{\sqrt{n}}$

54. $a_n = \dfrac{\ln n}{\ln 2n}$

55. $a_n = 8^{1/n}$

56. $a_n = (0.03)^{1/n}$

57. $a_n = \left(1 + \dfrac{7}{n}\right)^n$

58. $a_n = \left(1 - \dfrac{1}{n}\right)^n$

59. $a_n = \sqrt[n]{10n}$

60. $a_n = \sqrt[n]{n^2}$

61. $a_n = \left(\dfrac{3}{n}\right)^{1/n}$

62. $a_n = (n + 4)^{1/(n+4)}$

63. $a_n = \dfrac{\ln n}{n^{1/n}}$

64. $a_n = \ln n - \ln(n + 1)$

65. $a_n = \sqrt[n]{4^n n}$

66. $a_n = \sqrt[n]{3^{2n+1}}$

67. $a_n = \dfrac{n!}{n^n}$ (*Hint:* Compare with $1/n$.)

68. $a_n = \dfrac{(-4)^n}{n!}$

69. $a_n = \dfrac{n!}{10^{6n}}$

70. $a_n = \dfrac{n!}{2^n \cdot 3^n}$

71. $a_n = \left(\dfrac{1}{n}\right)^{1/(\ln n)}$

72. $a_n = \dfrac{(n + 1)!}{(n + 3)!}$

73. $a_n = \dfrac{(2n + 2)!}{(2n - 1)!}$

74. $a_n = \dfrac{3e^n + e^{-n}}{e^n + 3e^{-n}}$

75. $a_n = \dfrac{e^{-2n} - 2e^{-3n}}{e^{-2n} - e^{-n}}$

76. $a_n = \left(1 - \dfrac{1}{2}\right) + \left(\dfrac{1}{2} - \dfrac{1}{3}\right) + \left(\dfrac{1}{3} - \dfrac{1}{4}\right) + \cdots$
$$+ \left(\dfrac{1}{n - 2} - \dfrac{1}{n - 1}\right) + \left(\dfrac{1}{n - 1} - \dfrac{1}{n}\right)$$

77. $a_n = (\ln 3 - \ln 2) + (\ln 4 - \ln 3) + (\ln 5 - \ln 4) + \cdots$
$$+ (\ln(n - 1) - \ln(n - 2)) + (\ln n - \ln(n - 1))$$

78. $a_n = \ln\left(1 + \dfrac{1}{n}\right)^n$

79. $a_n = \left(\dfrac{3n + 1}{3n - 1}\right)^n$

80. $a_n = \left(\dfrac{n}{n + 1}\right)^n$

81. $a_n = \left(\dfrac{x^n}{2n + 1}\right)^{1/n}, \quad x > 0$

82. $a_n = \left(1 - \dfrac{1}{n^2}\right)^n$

83. $a_n = \dfrac{3^n \cdot 6^n}{2^{-n} \cdot n!}$

84. $a_n = \dfrac{(10/11)^n}{(9/10)^n + (11/12)^n}$

85. $a_n = \tanh n$

86. $a_n = \sinh(\ln n)$

87. $a_n = \dfrac{n^2}{2n - 1}\sin\dfrac{1}{n}$

88. $a_n = n\left(1 - \cos\dfrac{1}{n}\right)$

89. $a_n = \sqrt{n}\,\sin\dfrac{1}{\sqrt{n}}$

90. $a_n = (3^n + 5^n)^{1/n}$

91. $a_n = \tan^{-1} n$

92. $a_n = \dfrac{1}{\sqrt{n}}\tan^{-1} n$

93. $a_n = \left(\dfrac{1}{3}\right)^n + \dfrac{1}{\sqrt{2^n}}$

94. $a_n = \sqrt[n]{n^2 + n}$

95. $a_n = \dfrac{(\ln n)^{200}}{n}$

96. $a_n = \dfrac{(\ln n)^5}{\sqrt{n}}$

97. $a_n = n - \sqrt{n^2 - n}$

98. $a_n = \dfrac{1}{\sqrt{n^2 - 1} - \sqrt{n^2 + n}}$

99. $a_n = \dfrac{1}{n}\displaystyle\int_1^n \dfrac{1}{x}\,dx$

100. $a_n = \displaystyle\int_1^n \dfrac{1}{x^p}\,dx,\quad p > 1$

Recursively Defined Sequences

In Exercises 101–108, assume that each sequence converges and find its limit.

101. $a_1 = 2,\quad a_{n+1} = \dfrac{72}{1 + a_n}$

102. $a_1 = -1,\quad a_{n+1} = \dfrac{a_n + 6}{a_n + 2}$

103. $a_1 = -4,\quad a_{n+1} = \sqrt{8 + 2a_n}$

104. $a_1 = 0,\quad a_{n+1} = \sqrt{8 + 2a_n}$

105. $a_1 = 5,\quad a_{n+1} = \sqrt{5a_n}$

106. $a_1 = 3,\quad a_{n+1} = 12 - \sqrt{a_n}$

107. $2, 2 + \dfrac{1}{2}, 2 + \dfrac{1}{2 + \dfrac{1}{2}}, 2 + \dfrac{1}{2 + \dfrac{1}{2 + \dfrac{1}{2}}}, \ldots$

108. $\sqrt{1}, \sqrt{1 + \sqrt{1}}, \sqrt{1 + \sqrt{1 + \sqrt{1}}},$

$\sqrt{1 + \sqrt{1 + \sqrt{1 + \sqrt{1}}}}, \ldots$

Theory and Examples

109. The first term of a sequence is $x_1 = 1$. Each succeeding term is the sum of all those that come before it:

$$x_{n+1} = x_1 + x_2 + \cdots + x_n.$$

Write out enough early terms of the sequence to deduce a general formula for x_n that holds for $n \geq 2$.

110. A sequence of rational numbers is described as follows:

$$\dfrac{1}{1}, \dfrac{3}{2}, \dfrac{7}{5}, \dfrac{17}{12}, \ldots, \dfrac{a}{b}, \dfrac{a + 2b}{a + b}, \ldots.$$

Here the numerators form one sequence, the denominators form a second sequence, and their ratios form a third sequence. Let x_n and y_n be, respectively, the numerator and the denominator of the nth fraction $r_n = x_n / y_n$.

a. Verify that $x_1^2 - 2y_1^2 = -1$, $x_2^2 - 2y_2^2 = +1$ and, more generally, that if $a^2 - 2b^2 = -1$ or $+1$, then

$$(a + 2b)^2 - 2(a + b)^2 = +1 \quad\text{or}\quad -1,$$

respectively.

b. The fractions $r_n = x_n / y_n$ approach a limit as n increases. What is that limit? (*Hint:* Use part (a) to show that $r_n^2 - 2 = \pm(1/y_n)^2$ and that y_n is not less than n.)

111. Newton's method The following sequences come from the recursion formula for Newton's method,

$$x_{n+1} = x_n - \dfrac{f(x_n)}{f'(x_n)}.$$

Do the sequences converge? If so, to what value? In each case, begin by identifying the function f that generates the sequence.

a. $x_0 = 1,\quad x_{n+1} = x_n - \dfrac{x_n^2 - 2}{2x_n} = \dfrac{x_n}{2} + \dfrac{1}{x_n}$

b. $x_0 = 1,\quad x_{n+1} = x_n - \dfrac{\tan x_n - 1}{\sec^2 x_n}$

c. $x_0 = 1,\quad x_{n+1} = x_n - 1$

112. a. Suppose that $f(x)$ is differentiable for all x in $[0, 1]$ and that $f(0) = 0$. Define sequence $\{a_n\}$ by the rule $a_n = nf(1/n)$. Show that $\lim_{n \to \infty} a_n = f'(0)$. Use the result in part (a) to find the limits of the following sequences $\{a_n\}$.

b. $a_n = n\tan^{-1}\dfrac{1}{n}$

c. $a_n = n(e^{1/n} - 1)$

d. $a_n = n\ln\left(1 + \dfrac{2}{n}\right)$

113. Pythagorean triples A triple of positive integers a, b, and c is called a **Pythagorean triple** if $a^2 + b^2 = c^2$. Let a be an odd positive integer and let

$$b = \left\lfloor \dfrac{a^2}{2} \right\rfloor \quad\text{and}\quad c = \left\lceil \dfrac{a^2}{2} \right\rceil$$

be, respectively, the integer floor and ceiling for $a^2/2$.

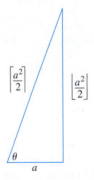

a. Show that $a^2 + b^2 = c^2$. (*Hint:* Let $a = 2n + 1$ and express b and c in terms of n.)

b. By direct calculation, or by appealing to the accompanying figure, find

$$\lim_{a \to \infty} \dfrac{\left\lfloor \dfrac{a^2}{2} \right\rfloor}{\left\lceil \dfrac{a^2}{2} \right\rceil}.$$

114. The *n*th root of *n*!

a. Show that $\lim_{n\to\infty}(2n\pi)^{1/(2n)} = 1$ and hence, using Stirling's approximation (Chapter 8, Additional Exercise 52a), that

$$\sqrt[n]{n!} \approx \frac{n}{e} \quad \text{for large values of } n.$$

T b. Test the approximation in part (a) for $n = 40, 50, 60, \ldots$, as far as your calculator will allow.

115. a. Assuming that $\lim_{n\to\infty}(1/n^c) = 0$ if c is any positive constant, show that

$$\lim_{n\to\infty}\frac{\ln n}{n^c} = 0$$

if c is any positive constant.

b. Prove that $\lim_{n\to\infty}(1/n^c) = 0$ if c is any positive constant. (*Hint:* If $\varepsilon = 0.001$ and $c = 0.04$, how large should N be to ensure that $|1/n^c - 0| < \varepsilon$ if $n > N$?)

116. The zipper theorem Prove the "zipper theorem" for sequences: If $\{a_n\}$ and $\{b_n\}$ both converge to L, then the sequence

$$a_1, b_1, a_2, b_2, \ldots, a_n, b_n, \ldots$$

converges to L.

117. Prove that $\lim_{n\to\infty}\sqrt[n]{n} = 1$.

118. Prove that $\lim_{n\to\infty}x^{1/n} = 1, (x > 0)$.

119. Prove Theorem 2.   **120.** Prove Theorem 3.

In Exercises 121–124, determine if the sequence is monotonic and if it is bounded.

121. $a_n = \dfrac{3n + 1}{n + 1}$ **122.** $a_n = \dfrac{(2n + 3)!}{(n + 1)!}$

123. $a_n = \dfrac{2^n 3^n}{n!}$ **124.** $a_n = 2 - \dfrac{2}{n} - \dfrac{1}{2^n}$

Which of the sequences in Exercises 125–134 converge, and which diverge? Give reasons for your answers.

125. $a_n = 1 - \dfrac{1}{n}$ **126.** $a_n = n - \dfrac{1}{n}$

127. $a_n = \dfrac{2^n - 1}{2^n}$ **128.** $a_n = \dfrac{2^n - 1}{3^n}$

129. $a_n = ((-1)^n + 1)\left(\dfrac{n + 1}{n}\right)$

130. The first term of a sequence is $x_1 = \cos(1)$. The next terms are $x_2 = x_1$ or $\cos(2)$, whichever is larger; and $x_3 = x_2$ or $\cos(3)$, whichever is larger (farther to the right). In general,

$$x_{n+1} = \max\{x_n, \cos(n + 1)\}.$$

131. $a_n = \dfrac{1 + \sqrt{2n}}{\sqrt{n}}$ **132.** $a_n = \dfrac{n + 1}{n}$

133. $a_n = \dfrac{4^{n+1} + 3^n}{4^n}$ **134.** $a_1 = 1, \quad a_{n+1} = 2a_n - 3$

In Exercises 135–136, use the definition of convergence to prove the given limit.

135. $\lim_{n\to\infty}\dfrac{\sin n}{n} = 0$ **136.** $\lim_{n\to\infty}\left(1 - \dfrac{1}{n^2}\right) = 1$

137. The sequence $\{n/(n + 1)\}$ has a least upper bound of 1 Show that if M is a number less than 1, then the terms of $\{n/(n + 1)\}$ eventually exceed M. That is, if $M < 1$ there is an integer N such that $n/(n + 1) > M$ whenever $n > N$. Since $n/(n + 1) < 1$ for every n, this proves that 1 is a least upper bound for $\{n/(n + 1)\}$.

138. Uniqueness of least upper bounds Show that if M_1 and M_2 are least upper bounds for the sequence $\{a_n\}$, then $M_1 = M_2$. That is, a sequence cannot have two different least upper bounds.

139. Is it true that a sequence $\{a_n\}$ of positive numbers must converge if it is bounded from above? Give reasons for your answer.

140. Prove that if $\{a_n\}$ is a convergent sequence, then to every positive number ε there corresponds an integer N such that

$$|a_m - a_n| < \varepsilon \quad \text{whenever} \quad m > N \quad \text{and} \quad n > N.$$

141. Uniqueness of limits Prove that limits of sequences are unique. That is, show that if L_1 and L_2 are numbers such that $a_n \to L_1$ and $a_n \to L_2$, then $L_1 = L_2$.

142. Limits and subsequences If the terms of one sequence appear in another sequence in their given order, we call the first sequence a **subsequence** of the second. Prove that if two sub-sequences of a sequence $\{a_n\}$ have different limits $L_1 \neq L_2$, then $\{a_n\}$ diverges.

143. For a sequence $\{a_n\}$ the terms of even index are denoted by a_{2k} and the terms of odd index by a_{2k+1}. Prove that if $a_{2k} \to L$ and $a_{2k+1} \to L$, then $a_n \to L$.

144. Prove that a sequence $\{a_n\}$ converges to 0 if and only if the sequence of absolute values $\{|a_n|\}$ converges to 0.

145. Sequences generated by Newton's method Newton's method, applied to a differentiable function $f(x)$, begins with a starting value x_0 and constructs from it a sequence of numbers $\{x_n\}$ that under favorable circumstances converges to a zero of f. The recursion formula for the sequence is

$$x_{n+1} = x_n - \frac{f(x_n)}{f'(x_n)}.$$

a. Show that the recursion formula for $f(x) = x^2 - a, a > 0$, can be written as $x_{n+1} = (x_n + a/x_n)/2$.

T b. Starting with $x_0 = 1$ and $a = 3$, calculate successive terms of the sequence until the display begins to repeat. What number is being approximated? Explain.

T **146. A recursive definition of $\pi/2$** If you start with $x_1 = 1$ and define the subsequent terms of $\{x_n\}$ by the rule $x_n = x_{n-1} + \cos x_{n-1}$, you generate a sequence that converges rapidly to $\pi/2$. **(a)** Try it. **(b)** Use the accompanying figure to explain why the convergence is so rapid.

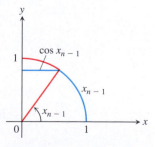

Use a CAS to perform the following steps for the sequences in Exercises 147–158.

a. Calculate and then plot the first 25 terms of the sequence. Does the sequence appear to be bounded from above or below? Does it appear to converge or diverge? If it does converge, what is the limit L?

b. If the sequence converges, find an integer N such that $|a_n - L| \leq 0.01$ for $n \geq N$. How far in the sequence do you have to get for the terms to lie within 0.0001 of L?

147. $a_n = \sqrt[n]{n}$

148. $a_n = \left(1 + \dfrac{0.5}{n}\right)^n$

149. $a_1 = 1, \quad a_{n+1} = a_n + \dfrac{1}{5^n}$

150. $a_1 = 1, \quad a_{n+1} = a_n + (-2)^n$

151. $a_n = \sin n$

152. $a_n = n \sin \dfrac{1}{n}$

153. $a_n = \dfrac{\sin n}{n}$

154. $a_n = \dfrac{\ln n}{n}$

155. $a_n = (0.9999)^n$

156. $a_n = (123456)^{1/n}$

157. $a_n = \dfrac{8^n}{n!}$

158. $a_n = \dfrac{n^{41}}{19^n}$

10.2 Infinite Series

An *infinite series* is the sum of an infinite sequence of numbers

$$a_1 + a_2 + a_3 + \cdots + a_n + \cdots$$

The goal of this section is to understand the meaning of such an infinite sum and to develop methods to calculate it. Since there are infinitely many terms to add in an infinite series, we cannot just keep adding to see what comes out. Instead we look at the result of summing just the first n terms of the sequence. The sum of the first n terms

$$s_n = a_1 + a_2 + a_3 + \cdots + a_n$$

is an ordinary finite sum and can be calculated by normal addition. It is called the *nth partial sum*. As n gets larger, we expect the partial sums to get closer and closer to a limiting value in the same sense that the terms of a sequence approach a limit, as discussed in Section 10.1.

For example, to assign meaning to an expression like

$$1 + \frac{1}{2} + \frac{1}{4} + \frac{1}{8} + \frac{1}{16} + \cdots$$

we add the terms one at a time from the beginning and look for a pattern in how these partial sums grow.

Partial sum		Value	Suggestive expression for partial sum
First:	$s_1 = 1$	1	$2 - 1$
Second:	$s_2 = 1 + \dfrac{1}{2}$	$\dfrac{3}{2}$	$2 - \dfrac{1}{2}$
Third:	$s_3 = 1 + \dfrac{1}{2} + \dfrac{1}{4}$	$\dfrac{7}{4}$	$2 - \dfrac{1}{4}$
$\vdots$	$\vdots$	$\vdots$	$\vdots$
nth:	$s_n = 1 + \dfrac{1}{2} + \dfrac{1}{4} + \cdots + \dfrac{1}{2^{n-1}}$	$\dfrac{2^n - 1}{2^{n-1}}$	$2 - \dfrac{1}{2^{n-1}}$

Indeed there is a pattern. The partial sums form a sequence whose nth term is

$$s_n = 2 - \frac{1}{2^{n-1}}.$$

This sequence of partial sums converges to 2 because $\lim_{n\to\infty}(1/2^{n-1}) = 0$. We say

"the sum of the infinite series $1 + \dfrac{1}{2} + \dfrac{1}{4} + \cdots + \dfrac{1}{2^{n-1}} + \cdots$ is 2."

Is the sum of any finite number of terms in this series equal to 2? No. Can we actually add an infinite number of terms one by one? No. But we can still define their sum by defining it to be the limit of the sequence of partial sums as $n \to \infty$, in this case 2 (Figure 10.8). Our knowledge of sequences and limits enables us to break away from the confines of finite sums.

FIGURE 10.8   As the lengths $1, 1/2, 1/4, 1/8, \ldots$ are added one by one, the sum approaches 2.

DEFINITIONS   Given a sequence of numbers $\{a_n\}$, an expression of the form

$$a_1 + a_2 + a_3 + \cdots + a_n + \cdots$$

is an **infinite series**. The number a_n is the **nth term** of the series. The sequence $\{s_n\}$ defined by

$$s_1 = a_1$$
$$s_2 = a_1 + a_2$$
$$\vdots$$
$$s_n = a_1 + a_2 + \cdots + a_n = \sum_{k=1}^{n} a_k$$
$$\vdots$$

is the **sequence of partial sums** of the series, the number s_n being the **nth partial sum**. If the sequence of partial sums converges to a limit L, we say that the series **converges** and that its **sum** is L. In this case, we also write

$$a_1 + a_2 + \cdots + a_n + \cdots = \sum_{n=1}^{\infty} a_n = L.$$

If the sequence of partial sums of the series does not converge, we say that the series **diverges**.

We can represent each term in an infinite series by the area of a rectangle. If all the terms a_n in the series are positive, then the series converges if the total area is finite, and diverges otherwise. Figure 10.9a shows an example where the series converges and Figure 10.9b shows an example where it diverges. The convergence of the total area is related to the convergence or divergence of improper integrals, as we found in Section 8.8. We make this connection explicit in the next section, where we develop an important test for convergence of series, the Integral Test.

When we begin to study a given series $a_1 + a_2 + \cdots + a_n + \cdots$, we might not know whether it converges or diverges. In either case, it is convenient to use sigma notation to write the series as

$$\sum_{n=1}^{\infty} a_n, \qquad \sum_{k=1}^{\infty} a_k, \qquad \text{or} \qquad \sum a_n$$

A useful shorthand when summation from 1 to ∞ is understood

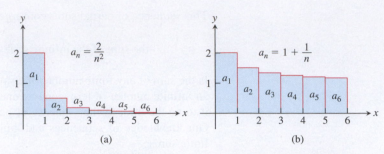

FIGURE 10.9 The sum of a series with positive terms can be interpreted as a total area of an infinite collection of rectangles. The series converges when the total area of the rectangles is finite (a) and diverges when the total area is unbounded (b). Note that the total area can be infinite even if the area of the rectangles is decreasing.

Geometric Series

Geometric series are series of the form

$$a + ar + ar^2 + \cdots + ar^{n-1} + \cdots = \sum_{n=1}^{\infty} ar^{n-1}$$

in which a and r are fixed real numbers and $a \neq 0$. The series can also be written as $\sum_{n=0}^{\infty} ar^n$. The **ratio** r can be positive, as in

$$1 + \frac{1}{2} + \frac{1}{4} + \cdots + \left(\frac{1}{2}\right)^{n-1} + \cdots, \qquad r = 1/2, a = 1$$

or negative, as in

$$1 - \frac{1}{3} + \frac{1}{9} - \cdots + \left(-\frac{1}{3}\right)^{n-1} + \cdots. \qquad r = -1/3, a = 1$$

If $r = 1$, the nth partial sum of the geometric series is

$$s_n = a + a(1) + a(1)^2 + \cdots + a(1)^{n-1} = na,$$

and the series diverges because $\lim_{n\to\infty} s_n = \pm\infty$, depending on the sign of a. If $r = -1$, the series diverges because the nth partial sums alternate between a and 0 and never approach a single limit. If $|r| \neq 1$, we can determine the convergence or divergence of the series in the following way:

$$s_n = a + ar + ar^2 + \cdots + ar^{n-1} \qquad \text{Write the } n\text{th partial sum.}$$
$$rs_n = ar + ar^2 + \cdots + ar^{n-1} + ar^n \qquad \text{Multiply } s_n \text{ by } r.$$
$$s_n - rs_n = a - ar^n \qquad \text{Subtract } rs_n \text{ from } s_n. \text{ Most of the terms on the right cancel.}$$
$$s_n(1 - r) = a(1 - r^n) \qquad \text{Factor.}$$
$$s_n = \frac{a(1 - r^n)}{1 - r}, \qquad (r \neq 1). \qquad \text{We can solve for } s_n \text{ if } r \neq 1.$$

If $|r| < 1$, then $r^n \to 0$ as $n \to \infty$ (as in Section 10.1), so $s_n \to a/(1 - r)$ in this case. On the other hand, if $|r| > 1$, then $|r^n| \to \infty$ and the series diverges.

If $|r| < 1$, the geometric series $a + ar + ar^2 + \cdots + ar^{n-1} + \cdots$ converges to $a/(1 - r)$:

$$\sum_{n=1}^{\infty} ar^{n-1} = \frac{a}{1 - r}, \qquad |r| < 1.$$

If $|r| \geq 1$, the series diverges.

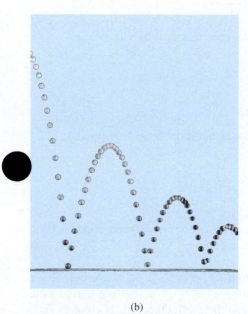

(a)

(b)

FIGURE 10.10 (a) Example 3 shows how to use a geometric series to calculate the total vertical distance traveled by a bouncing ball if the height of each rebound is reduced by the factor r. (b) A stroboscopic photo of a bouncing ball. (*Source: PSSC Physics*, 2nd ed., Reprinted by permission of Educational Development Center, Inc.)

The formula $a/(1 - r)$ for the sum of a geometric series applies *only* when the summation index begins with $n = 1$ in the expression $\sum_{n=1}^{\infty} ar^{n-1}$ (or with the index $n = 0$ if we write the series as $\sum_{n=0}^{\infty} ar^n$).

EXAMPLE 1   The geometric series with $a = 1/9$ and $r = 1/3$ is

$$\frac{1}{9} + \frac{1}{27} + \frac{1}{81} + \cdots = \sum_{n=1}^{\infty} \frac{1}{9}\left(\frac{1}{3}\right)^{n-1} = \frac{1/9}{1 - (1/3)} = \frac{1}{6}.$$

EXAMPLE 2   The series

$$\sum_{n=0}^{\infty} \frac{(-1)^n 5}{4^n} = 5 - \frac{5}{4} + \frac{5}{16} - \frac{5}{64} + \cdots$$

is a geometric series with $a = 5$ and $r = -1/4$. It converges to

$$\frac{a}{1 - r} = \frac{5}{1 + (1/4)} = 4.$$

EXAMPLE 3   You drop a ball from a meters above a flat surface. Each time the ball hits the surface after falling a distance h, it rebounds a distance rh, where r is positive but less than 1. Find the total distance the ball travels up and down (Figure 10.10).

Solution   The total distance is

$$s = a + \underbrace{2ar + 2ar^2 + 2ar^3 + \cdots}_{\text{This sum is } 2ar/(1 - r).} = a + \frac{2ar}{1 - r} = a\frac{1 + r}{1 - r}.$$

If $a = 6$ m and $r = 2/3$, for instance, the distance is

$$s = 6 \cdot \frac{1 + (2/3)}{1 - (2/3)} = 6\left(\frac{5/3}{1/3}\right) = 30 \text{ m.}$$

EXAMPLE 4   Express the repeating decimal $5.232323\ldots$ as the ratio of two integers.

Solution   From the definition of a decimal number, we get a geometric series

$$5.232323\ldots = 5 + \frac{23}{100} + \frac{23}{(100)^2} + \frac{23}{(100)^3} + \cdots$$

$$= 5 + \frac{23}{100}\underbrace{\left(1 + \frac{1}{100} + \left(\frac{1}{100}\right)^2 + \cdots\right)}_{1/(1 - 0.01)} \quad \begin{matrix} a = 1, \\ r = 1/100 \end{matrix}$$

$$= 5 + \frac{23}{100}\left(\frac{1}{0.99}\right) = 5 + \frac{23}{99} = \frac{518}{99}$$

Unfortunately, formulas like the one for the sum of a convergent geometric series are rare and we usually have to settle for an estimate of a series' sum (more about this later). The next example, however, is another case in which we can find the sum exactly.

EXAMPLE 5   Find the sum of the "telescoping" series $\displaystyle\sum_{n=1}^{\infty} \frac{1}{n(n + 1)}$.

Solution   We look for a pattern in the sequence of partial sums that might lead to a formula for s_k. The key observation is the partial fraction decomposition

$$\frac{1}{n(n+1)} = \frac{1}{n} - \frac{1}{n+1},$$

so

$$\sum_{n=1}^{k} \frac{1}{n(n+1)} = \sum_{n=1}^{k} \left(\frac{1}{n} - \frac{1}{n+1} \right)$$

and

$$s_k = \left(\frac{1}{1} - \frac{1}{2} \right) + \left(\frac{1}{2} - \frac{1}{3} \right) + \left(\frac{1}{3} - \frac{1}{4} \right) + \cdots + \left(\frac{1}{k} - \frac{1}{k+1} \right).$$

Removing parentheses and canceling adjacent terms of opposite sign collapses the sum to

$$s_k = 1 - \frac{1}{k+1}.$$

We now see that $s_k \rightarrow 1$ as $k \rightarrow \infty$. The series converges, and its sum is 1:

$$\sum_{n=1}^{\infty} \frac{1}{n(n+1)} = 1. \qquad \blacksquare$$

The nth-Term Test for a Divergent Series

One reason that a series may fail to converge is that its terms don't become small.

EXAMPLE 6   The series

$$\sum_{n=1}^{\infty} \frac{n+1}{n} = \frac{2}{1} + \frac{3}{2} + \frac{4}{3} + \cdots + \frac{n+1}{n} + \cdots$$

diverges because the partial sums eventually outgrow every preassigned number. Each term is greater than 1, so the sum of n terms is greater than n. $\blacksquare$

We now show that $\lim_{n \to \infty} a_n$ must equal zero if the series $\sum_{n=1}^{\infty} a_n$ converges. To see why, let S represent the series' sum and $s_n = a_1 + a_2 + \cdots + a_n$ the nth partial sum. When n is large, both s_n and s_{n-1} are close to S, so their difference, a_n, is close to zero. More formally,

$$a_n = s_n - s_{n-1} \quad \rightarrow \quad S - S = 0. \qquad \text{\color{blue}Difference Rule for sequences}$$

This establishes the following theorem.

Caution

Theorem 7 *does not say* that $\sum_{n=1}^{\infty} a_n$ converges if $a_n \rightarrow 0$. It is possible for a series to diverge when $a_n \rightarrow 0$. (See Example 8.)

> **THEOREM 7**   If $\displaystyle\sum_{n=1}^{\infty} a_n$ converges, then $a_n \rightarrow 0$.

Theorem 7 leads to a test for detecting the kind of divergence that occurred in Example 6.

> **The nth-Term Test for Divergence**
>
> $\displaystyle\sum_{n=1}^{\infty} a_n$ diverges if $\displaystyle\lim_{n \to \infty} a_n$ fails to exist or is different from zero.

EXAMPLE 7   The following are all examples of divergent series.

(a) $\displaystyle\sum_{n=1}^{\infty} n^2$ diverges because $n^2 \rightarrow \infty$.

(b) $\displaystyle\sum_{n=1}^{\infty} \frac{n+1}{n}$ diverges because $\dfrac{n+1}{n} \to 1.$ $\lim_{n\to\infty} a_n \neq 0$

(c) $\displaystyle\sum_{n=1}^{\infty} (-1)^{n+1}$ diverges because $\lim_{n\to\infty}(-1)^{n+1}$ does not exist.

(d) $\displaystyle\sum_{n=1}^{\infty} \frac{-n}{2n+5}$ diverges because $\lim_{n\to\infty} \dfrac{-n}{2n+5} = -\dfrac{1}{2} \neq 0.$ ∎

EXAMPLE 8 The series

$$1 + \underbrace{\frac{1}{2} + \frac{1}{2}}_{2 \text{ terms}} + \underbrace{\frac{1}{4} + \frac{1}{4} + \frac{1}{4} + \frac{1}{4}}_{4 \text{ terms}} + \cdots + \underbrace{\frac{1}{2^n} + \frac{1}{2^n} + \cdots + \frac{1}{2^n}}_{2^n \text{ terms}} + \cdots$$

diverges because the terms can be grouped into infinitely many clusters each of which adds to 1, so the partial sums increase without bound. However, the terms of the series form a sequence that converges to 0. Example 1 of Section 10.3 shows that the harmonic series $\sum 1/n$ also behaves in this manner. ∎

Combining Series

Whenever we have two convergent series, we can add them term by term, subtract them term by term, or multiply them by constants to make new convergent series.

THEOREM 8 If $\sum a_n = A$ and $\sum b_n = B$ are convergent series, then

1. *Sum Rule:* $\sum(a_n + b_n) = \sum a_n + \sum b_n = A + B$

2. *Difference Rule:* $\sum(a_n - b_n) = \sum a_n - \sum b_n = A - B$

3. *Constant Multiple Rule:* $\sum ka_n = k\sum a_n = kA$ (any number k).

Proof The three rules for series follow from the analogous rules for sequences in Theorem 1, Section 10.1. To prove the Sum Rule for series, let

$$A_n = a_1 + a_2 + \cdots + a_n, \quad B_n = b_1 + b_2 + \cdots + b_n.$$

Then the partial sums of $\sum(a_n + b_n)$ are

$$\begin{aligned} s_n &= (a_1 + b_1) + (a_2 + b_2) + \cdots + (a_n + b_n) \\ &= (a_1 + \cdots + a_n) + (b_1 + \cdots + b_n) \\ &= A_n + B_n. \end{aligned}$$

Since $A_n \to A$ and $B_n \to B$, we have $s_n \to A + B$ by the Sum Rule for sequences. The proof of the Difference Rule is similar.

To prove the Constant Multiple Rule for series, observe that the partial sums of $\sum ka_n$ form the sequence

$$s_n = ka_1 + ka_2 + \cdots + ka_n = k(a_1 + a_2 + \cdots + a_n) = kA_n,$$

which converges to kA by the Constant Multiple Rule for sequences. ∎

As corollaries of Theorem 8, we have the following results. We omit the proofs.

---

1. Every nonzero constant multiple of a divergent series diverges.

2. If $\sum a_n$ converges and $\sum b_n$ diverges, then $\sum(a_n + b_n)$ and $\sum(a_n - b_n)$ both diverge.

Caution Remember that $\Sigma(a_n + b_n)$ can converge *even if* both Σa_n and Σb_n diverge. For example, $\Sigma a_n = 1 + 1 + 1 + \cdots$ and $\Sigma b_n = (-1) + (-1) + (-1) + \cdots$ diverge, whereas $\Sigma(a_n + b_n) = 0 + 0 + 0 + \cdots$ converges to 0.

EXAMPLE 9   Find the sums of the following series.

(a) $\displaystyle\sum_{n=1}^{\infty} \frac{3^{n-1} - 1}{6^{n-1}} = \sum_{n=1}^{\infty} \left(\frac{1}{2^{n-1}} - \frac{1}{6^{n-1}} \right)$

$\displaystyle\qquad\qquad = \sum_{n=1}^{\infty} \frac{1}{2^{n-1}} - \sum_{n=1}^{\infty} \frac{1}{6^{n-1}}$ Difference Rule

$\displaystyle\qquad\qquad = \frac{1}{1 - (1/2)} - \frac{1}{1 - (1/6)}$ Geometric series with $a = 1$ and $r = 1/2, 1/6$

$\displaystyle\qquad\qquad = 2 - \frac{6}{5} = \frac{4}{5}$

(b) $\displaystyle\sum_{n=0}^{\infty} \frac{4}{2^n} = 4 \sum_{n=0}^{\infty} \frac{1}{2^n}$ Constant Multiple Rule

$\displaystyle\qquad\qquad = 4 \left(\frac{1}{1 - (1/2)} \right)$ Geometric series with $a = 1, r = 1/2$

$\displaystyle\qquad\qquad = 8$

Adding or Deleting Terms

We can add a finite number of terms to a series or delete a finite number of terms without altering the series' convergence or divergence, although in the case of convergence this will usually change the sum. If $\sum_{n=1}^{\infty} a_n$ converges, then $\sum_{n=k}^{\infty} a_n$ converges for any $k > 1$ and

$$\sum_{n=1}^{\infty} a_n = a_1 + a_2 + \cdots + a_{k-1} + \sum_{n=k}^{\infty} a_n.$$

Conversely, if $\sum_{n=k}^{\infty} a_n$ converges for any $k > 1$, then $\sum_{n=1}^{\infty} a_n$ converges. Thus,

$$\sum_{n=1}^{\infty} \frac{1}{5^n} = \frac{1}{5} + \frac{1}{25} + \frac{1}{125} + \sum_{n=4}^{\infty} \frac{1}{5^n}$$

and

$$\sum_{n=4}^{\infty} \frac{1}{5^n} = \left(\sum_{n=1}^{\infty} \frac{1}{5^n} \right) - \frac{1}{5} - \frac{1}{25} - \frac{1}{125}.$$

The convergence or divergence of a series is not affected by its first few terms. Only the "tail" of the series, the part that remains when we sum beyond some finite number of initial terms, influences whether it converges or diverges.

Reindexing

As long as we preserve the order of its terms, we can reindex any series without altering its convergence. To raise the starting value of the index h units, replace the n in the formula for a_n by $n - h$:

$$\sum_{n=1}^{\infty} a_n = \sum_{n=1+h}^{\infty} a_{n-h} = a_1 + a_2 + a_3 + \cdots.$$

To lower the starting value of the index h units, replace the n in the formula for a_n by $n + h$:

$$\sum_{n=1}^{\infty} a_n = \sum_{n=1-h}^{\infty} a_{n+h} = a_1 + a_2 + a_3 + \cdots.$$

We saw this reindexing in starting a geometric series with the index $n = 0$ instead of the index $n = 1$, but we can use any other starting index value as well. We usually give preference to indexings that lead to simple expressions.

EXAMPLE 10   We can write the geometric series

$$\sum_{n=1}^{\infty} \frac{1}{2^{n-1}} = 1 + \frac{1}{2} + \frac{1}{4} + \cdots$$

as

$$\sum_{n=0}^{\infty} \frac{1}{2^n}, \qquad \sum_{n=5}^{\infty} \frac{1}{2^{n-5}}, \qquad \text{or even} \qquad \sum_{n=-4}^{\infty} \frac{1}{2^{n+4}}.$$

The partial sums remain the same no matter what indexing we choose to use. ∎

EXERCISES 10.2

Finding nth Partial Sums

In Exercises 1–6, find a formula for the nth partial sum of each series and use it to find the series' sum if the series converges.

1. $2 + \frac{2}{3} + \frac{2}{9} + \frac{2}{27} + \cdots + \frac{2}{3^{n-1}} + \cdots$

2. $\frac{9}{100} + \frac{9}{100^2} + \frac{9}{100^3} + \cdots + \frac{9}{100^n} + \cdots$

3. $1 - \frac{1}{2} + \frac{1}{4} - \frac{1}{8} + \cdots + (-1)^{n-1}\frac{1}{2^{n-1}} + \cdots$

4. $1 - 2 + 4 - 8 + \cdots + (-1)^{n-1} 2^{n-1} + \cdots$

5. $\frac{1}{2 \cdot 3} + \frac{1}{3 \cdot 4} + \frac{1}{4 \cdot 5} + \cdots + \frac{1}{(n+1)(n+2)} + \cdots$

6. $\frac{5}{1 \cdot 2} + \frac{5}{2 \cdot 3} + \frac{5}{3 \cdot 4} + \cdots + \frac{5}{n(n+1)} + \cdots$

Series with Geometric Terms

In Exercises 7–14, write out the first eight terms of each series to show how the series starts. Then find the sum of the series or show that it diverges.

7. $\sum_{n=0}^{\infty} \frac{(-1)^n}{4^n}$

8. $\sum_{n=2}^{\infty} \frac{1}{4^n}$

9. $\sum_{n=1}^{\infty} \left(1 - \frac{7}{4^n}\right)$

10. $\sum_{n=0}^{\infty} (-1)^n \frac{5}{4^n}$

11. $\sum_{n=0}^{\infty} \left(\frac{5}{2^n} + \frac{1}{3^n}\right)$

12. $\sum_{n=0}^{\infty} \left(\frac{5}{2^n} - \frac{1}{3^n}\right)$

13. $\sum_{n=0}^{\infty} \left(\frac{1}{2^n} + \frac{(-1)^n}{5^n}\right)$

14. $\sum_{n=0}^{\infty} \left(\frac{2^{n+1}}{5^n}\right)$

In Exercises 15–22, determine if the geometric series converges or diverges. If a series converges, find its sum.

15. $1 + \left(\frac{2}{5}\right) + \left(\frac{2}{5}\right)^2 + \left(\frac{2}{5}\right)^3 + \left(\frac{2}{5}\right)^4 + \cdots$

16. $1 + (-3) + (-3)^2 + (-3)^3 + (-3)^4 + \cdots$

17. $\left(\frac{1}{8}\right) + \left(\frac{1}{8}\right)^2 + \left(\frac{1}{8}\right)^3 + \left(\frac{1}{8}\right)^4 + \left(\frac{1}{8}\right)^5 + \cdots$

18. $\left(\frac{-2}{3}\right)^2 + \left(\frac{-2}{3}\right)^3 + \left(\frac{-2}{3}\right)^4 + \left(\frac{-2}{3}\right)^5 + \left(\frac{-2}{3}\right)^6 + \cdots$

19. $1 - \left(\frac{2}{e}\right) + \left(\frac{2}{e}\right)^2 - \left(\frac{2}{e}\right)^3 + \left(\frac{2}{e}\right)^4 - \cdots$

20. $\left(\frac{1}{3}\right)^{-2} - \left(\frac{1}{3}\right)^{-1} + 1 - \left(\frac{1}{3}\right) + \left(\frac{1}{3}\right)^2 - \cdots$

21. $1 + \left(\frac{10}{9}\right)^2 + \left(\frac{10}{9}\right)^4 + \left(\frac{10}{9}\right)^6 + \left(\frac{10}{9}\right)^8 + \cdots$

22. $\frac{9}{4} - \frac{27}{8} + \frac{81}{16} - \frac{243}{32} + \frac{729}{64} - \cdots$

Repeating Decimals

Express each of the numbers in Exercises 23–30 as the ratio of two integers.

23. $0.\overline{23} = 0.23\ 23\ 23 \ldots$

24. $0.\overline{234} = 0.234\ 234\ 234 \ldots$

25. $0.\overline{7} = 0.7777 \ldots$

26. $0.\overline{d} = 0.dddd \ldots,$ where d is a digit

27. $0.0\overline{6} = 0.06666 \ldots$

28. $1.\overline{414} = 1.414\ 414\ 414 \ldots$

29. $1.24\overline{123} = 1.24\ 123\ 123\ 123 \ldots$

30. $3.\overline{142857} = 3.142857\ 142857 \ldots$

Using the nth-Term Test

In Exercises 31–38, use the nth-Term Test for divergence to show that the series is divergent, or state that the test is inconclusive.

31. $\sum_{n=1}^{\infty} \frac{n}{n+10}$

32. $\sum_{n=1}^{\infty} \frac{n(n+1)}{(n+2)(n+3)}$

33. $\sum_{n=0}^{\infty} \frac{1}{n+4}$

34. $\sum_{n=1}^{\infty} \frac{n}{n^2+3}$

35. $\displaystyle\sum_{n=1}^{\infty} \cos\frac{1}{n}$

36. $\displaystyle\sum_{n=0}^{\infty} \frac{e^n}{e^n + n}$

37. $\displaystyle\sum_{n=1}^{\infty} \ln\frac{1}{n}$

38. $\displaystyle\sum_{n=0}^{\infty} \cos n\pi$

Telescoping Series

In Exercises 39–44, find a formula for the nth partial sum of the series and use it to determine if the series converges or diverges. If a series converges, find its sum.

39. $\displaystyle\sum_{n=1}^{\infty} \left(\frac{1}{n} - \frac{1}{n+1}\right)$

40. $\displaystyle\sum_{n=1}^{\infty} \left(\frac{3}{n^2} - \frac{3}{(n+1)^2}\right)$

41. $\displaystyle\sum_{n=1}^{\infty} \left(\ln\sqrt{n+1} - \ln\sqrt{n}\right)$ **42.** $\displaystyle\sum_{n=1}^{\infty} \left(\tan(n) - \tan(n-1)\right)$

43. $\displaystyle\sum_{n=1}^{\infty} \left(\cos^{-1}\left(\frac{1}{n+1}\right) - \cos^{-1}\left(\frac{1}{n+2}\right)\right)$

44. $\displaystyle\sum_{n=1}^{\infty} \left(\sqrt{n+4} - \sqrt{n+3}\right)$

Find the sum of each series in Exercises 45–52.

45. $\displaystyle\sum_{n=1}^{\infty} \frac{4}{(4n-3)(4n+1)}$

46. $\displaystyle\sum_{n=1}^{\infty} \frac{6}{(2n-1)(2n+1)}$

47. $\displaystyle\sum_{n=1}^{\infty} \frac{40n}{(2n-1)^2(2n+1)^2}$

48. $\displaystyle\sum_{n=1}^{\infty} \frac{2n+1}{n^2(n+1)^2}$

49. $\displaystyle\sum_{n=1}^{\infty} \left(\frac{1}{\sqrt{n}} - \frac{1}{\sqrt{n+1}}\right)$

50. $\displaystyle\sum_{n=1}^{\infty} \left(\frac{1}{2^{1/n}} - \frac{1}{2^{1/(n+1)}}\right)$

51. $\displaystyle\sum_{n=1}^{\infty} \left(\frac{1}{\ln(n+2)} - \frac{1}{\ln(n+1)}\right)$

52. $\displaystyle\sum_{n=1}^{\infty} \left(\tan^{-1}(n) - \tan^{-1}(n+1)\right)$

Convergence or Divergence

Which series in Exercises 53–76 converge, and which diverge? Give reasons for your answers. If a series converges, find its sum.

53. $\displaystyle\sum_{n=0}^{\infty} \left(\frac{1}{\sqrt{2}}\right)^n$

54. $\displaystyle\sum_{n=0}^{\infty} (\sqrt{2})^n$

55. $\displaystyle\sum_{n=1}^{\infty} (-1)^{n+1} \frac{3}{2^n}$

56. $\displaystyle\sum_{n=1}^{\infty} (-1)^{n+1} n$

57. $\displaystyle\sum_{n=0}^{\infty} \cos\left(\frac{n\pi}{2}\right)$

58. $\displaystyle\sum_{n=0}^{\infty} \frac{\cos n\pi}{5^n}$

59. $\displaystyle\sum_{n=0}^{\infty} e^{-2n}$

60. $\displaystyle\sum_{n=1}^{\infty} \ln\frac{1}{3^n}$

61. $\displaystyle\sum_{n=1}^{\infty} \frac{2}{10^n}$

62. $\displaystyle\sum_{n=0}^{\infty} \frac{1}{x^n}, \quad |x| > 1$

63. $\displaystyle\sum_{n=0}^{\infty} \frac{2^n - 1}{3^n}$

64. $\displaystyle\sum_{n=1}^{\infty} \left(1 - \frac{1}{n}\right)^n$

65. $\displaystyle\sum_{n=0}^{\infty} \frac{n!}{1000^n}$

66. $\displaystyle\sum_{n=1}^{\infty} \frac{n^n}{n!}$

67. $\displaystyle\sum_{n=1}^{\infty} \frac{2^n + 3^n}{4^n}$

68. $\displaystyle\sum_{n=1}^{\infty} \frac{2^n + 4^n}{3^n + 4^n}$

69. $\displaystyle\sum_{n=1}^{\infty} \ln\left(\frac{n}{n+1}\right)$

70. $\displaystyle\sum_{n=1}^{\infty} \ln\left(\frac{n}{2n+1}\right)$

71. $\displaystyle\sum_{n=0}^{\infty} \left(\frac{e}{\pi}\right)^n$

72. $\displaystyle\sum_{n=0}^{\infty} \frac{e^{n\pi}}{\pi^{ne}}$

73. $\displaystyle\sum_{n=1}^{\infty} \left(\frac{n}{n+1} - \frac{n+2}{n+3}\right)$

74. $\displaystyle\sum_{n=2}^{\infty} \left(\sin\left(\frac{\pi}{n}\right) - \sin\left(\frac{\pi}{n-1}\right)\right)$

75. $\displaystyle\sum_{n=1}^{\infty} \left(\cos\left(\frac{\pi}{n}\right) + \sin\left(\frac{\pi}{n}\right)\right)$

76. $\displaystyle\sum_{n=0}^{\infty} \left(\ln(4e^n - 1) - \ln(2e^n + 1)\right)$

Geometric Series with a Variable x

In each of the geometric series in Exercises 77–80, write out the first few terms of the series to find a and r, and find the sum of the series. Then express the inequality $|r| < 1$ in terms of x and find the values of x for which the inequality holds and the series converges.

77. $\displaystyle\sum_{n=0}^{\infty} (-1)^n x^n$

78. $\displaystyle\sum_{n=0}^{\infty} (-1)^n x^{2n}$

79. $\displaystyle\sum_{n=0}^{\infty} 3\left(\frac{x-1}{2}\right)^n$

80. $\displaystyle\sum_{n=0}^{\infty} \frac{(-1)^n}{2}\left(\frac{1}{3 + \sin x}\right)^n$

In Exercises 81–86, find the values of x for which the given geometric series converges. Also, find the sum of the series (as a function of x) for those values of x.

81. $\displaystyle\sum_{n=0}^{\infty} 2^n x^n$

82. $\displaystyle\sum_{n=0}^{\infty} (-1)^n x^{-2n}$

83. $\displaystyle\sum_{n=0}^{\infty} (-1)^n (x+1)^n$

84. $\displaystyle\sum_{n=0}^{\infty} \left(-\frac{1}{2}\right)^n (x-3)^n$

85. $\displaystyle\sum_{n=0}^{\infty} \sin^n x$

86. $\displaystyle\sum_{n=0}^{\infty} (\ln x)^n$

Theory and Examples

87. The series in Exercise 5 can also be written as

$$\sum_{n=1}^{\infty} \frac{1}{(n+1)(n+2)} \quad\text{and}\quad \sum_{n=-1}^{\infty} \frac{1}{(n+3)(n+4)}.$$

Write it as a sum beginning with **(a)** $n = -2$, **(b)** $n = 0$, **(c)** $n = 5$.

88. The series in Exercise 6 can also be written as

$$\sum_{n=1}^{\infty} \frac{5}{n(n+1)} \quad\text{and}\quad \sum_{n=0}^{\infty} \frac{5}{(n+1)(n+2)}.$$

Write it as a sum beginning with **(a)** $n = -1$, **(b)** $n = 3$, **(c)** $n = 20$.

89. Make up an infinite series of nonzero terms whose sum is

    a. 1   **b.** -3 **c.** 0.

90. (*Continuation of Exercise 89.*) Can you make an infinite series of nonzero terms that converges to any number you want? Explain.

91. Show by example that $\sum(a_n/b_n)$ may diverge even though $\sum a_n$ and $\sum b_n$ converge and no b_n equals 0.

92. Find convergent geometric series $A = \sum a_n$ and $B = \sum b_n$ that illustrate the fact that $\sum a_n b_n$ may converge without being equal to AB.

93. Show by example that $\sum(a_n/b_n)$ may converge to something other than A/B even when $A = \sum a_n$, $B = \sum b_n \neq 0$, and no b_n equals 0.

94. If $\sum a_n$ converges and $a_n > 0$ for all n, can anything be said about $\sum(1/a_n)$? Give reasons for your answer.

95. What happens if you add a finite number of terms to a divergent series or delete a finite number of terms from a divergent series? Give reasons for your answer.

96. If $\sum a_n$ converges and $\sum b_n$ diverges, can anything be said about their term-by-term sum $\sum(a_n + b_n)$? Give reasons for your answer.

97. Make up a geometric series $\sum ar^{n-1}$ that converges to the number 5 if

    a. $a = 2$ **b.** $a = 13/2$.

98. Find the value of b for which

$$1 + e^b + e^{2b} + e^{3b} + \cdots = 9.$$

99. For what values of r does the infinite series

$$1 + 2r + r^2 + 2r^3 + r^4 + 2r^5 + r^6 + \cdots$$

converge? Find the sum of the series when it converges.

100. The accompanying figure shows the first five of a sequence of squares. The outermost square has an area of 4 m². Each of the other squares is obtained by joining the midpoints of the sides of the squares before it. Find the sum of the areas of all the squares.

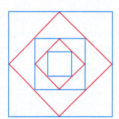

101. Drug dosage A patient takes a 300 mg tablet for the control of high blood pressure every morning at the same time. The concentration of the drug in the patient's system decays exponentially at a constant hourly rate of $k = 0.12$.

    a. How many milligrams of the drug are in the patient's system just before the second tablet is taken? Just before the third tablet is taken?

    b. In the long run, after taking the medication for at least six months, what quantity of drug is in the patient's body just before taking the next regularly scheduled morning tablet?

102. Show that the error $(L - s_n)$ obtained by replacing a convergent geometric series with one of its partial sums s_n is $ar^n/(1 - r)$.

103. The Cantor set To construct this set, we begin with the closed interval $[0, 1]$. From that interval, remove the middle open interval $(1/3, 2/3)$, leaving the two closed intervals $[0, 1/3]$ and $[2/3, 1]$. At the second step we remove the open middle third interval from each of those remaining. From $[0, 1/3]$ we remove the open interval $(1/9, 2/9)$, and from $[2/3, 1]$ we remove $(7/9, 8/9)$, leaving behind the four closed intervals $[0, 1/9]$, $[2/9, 1/3]$, $[2/3, 7/9]$, and $[8/9, 1]$. At the next step, we remove the middle open third interval from each closed interval left behind, so $(1/27, 2/27)$ is removed from $[0, 1/9]$, leaving the closed intervals $[0, 1/27]$ and $[2/27, 1/9]$; $(7/27, 8/27)$ is removed from $[2/9, 1/3]$, leaving behind $[2/9, 7/27]$ and $[8/27, 1/3]$, and so forth. We continue this process repeatedly without stopping, at each step removing the open third interval from every closed interval remaining behind from the preceding step. The numbers remaining in the interval $[0, 1]$, after all open middle third intervals have been removed, are the points in the Cantor set (named after Georg Cantor, 1845–1918). The set has some interesting properties.

    a. The Cantor set contains infinitely many numbers in $[0, 1]$. List 12 numbers that belong to the Cantor set.

    b. Show, by summing an appropriate geometric series, that the total length of all the open middle third intervals that have been removed from $[0, 1]$ is equal to 1.

104. Helga von Koch's snowflake curve Helga von Koch's snowflake is a curve of infinite length that encloses a region of finite area. To see why this is so, suppose the curve is generated by starting with an equilateral triangle whose sides have length 1.

    a. Find the length L_n of the nth curve C_n and show that $\lim_{n \to \infty} L_n = \infty$.

    b. Find the area A_n of the region enclosed by C_n and show that $\lim_{n \to \infty} A_n = (8/5) A_1$.

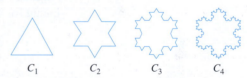

105. The largest circle in the accompanying figure has radius 1. Consider the sequence of circles of maximum area inscribed in semicircles of diminishing size. What is the sum of the areas of all of the circles?

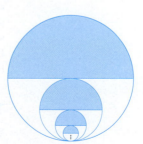

10.3 The Integral Test

The most basic question we can ask about a series is whether it converges. In this section we begin to study this question, starting with series that have nonnegative terms. Such a series converges if its sequence of partial sums is bounded. If we establish that a given series does converge, we generally do not have a formula available for its sum. So to get an estimate for the sum of a convergent series, we investigate the error involved when using a partial sum to approximate the total sum.

Nondecreasing Partial Sums

Suppose that $\sum_{n=1}^{\infty} a_n$ is an infinite series with $a_n \geq 0$ for all n. Then each partial sum is greater than or equal to its predecessor because $s_{n+1} = s_n + a_n$, so

$$s_1 \leq s_2 \leq s_3 \leq \cdots \leq s_n \leq s_{n+1} \leq \cdots.$$

Since the partial sums form a nondecreasing sequence, the Monotonic Sequence Theorem (Theorem 6, Section 10.1) gives the following result.

Corollary of Theorem 6

A series $\sum_{n=1}^{\infty} a_n$ of nonnegative terms converges if and only if its partial sums are bounded from above.

EXAMPLE 1   As an application of the above corollary, consider the **harmonic series**

$$\sum_{n=1}^{\infty} \frac{1}{n} = 1 + \frac{1}{2} + \frac{1}{3} + \cdots + \frac{1}{n} + \cdots.$$

Although the nth term $1/n$ does go to zero, the series diverges because there is no upper bound for its partial sums. To see why, group the terms of the series in the following way:

$$1 + \frac{1}{2} + \underbrace{\left(\frac{1}{3} + \frac{1}{4} \right)}_{> \frac{2}{4} = \frac{1}{2}} + \underbrace{\left(\frac{1}{5} + \frac{1}{6} + \frac{1}{7} + \frac{1}{8} \right)}_{> \frac{4}{8} = \frac{1}{2}} + \underbrace{\left(\frac{1}{9} + \frac{1}{10} + \cdots + \frac{1}{16} \right)}_{> \frac{8}{16} = \frac{1}{2}} + \cdots.$$

The sum of the first two terms is 1.5. The sum of the next two terms is $1/3 + 1/4$, which is greater than $1/4 + 1/4 = 1/2$. The sum of the next four terms is $1/5 + 1/6 + 1/7 + 1/8$, which is greater than $1/8 + 1/8 + 1/8 + 1/8 = 1/2$. The sum of the next eight terms is $1/9 + 1/10 + 1/11 + 1/12 + 1/13 + 1/14 + 1/15 + 1/16$, which is greater than $8/16 = 1/2$. The sum of the next 16 terms is greater than $16/32 = 1/2$, and so on. In general, the sum of 2^n terms ending with $1/2^{n+1}$ is greater than $2^n/2^{n+1} = 1/2$. If $n = 2^k$, the partial sum s_n is greater than $k/2$, so the sequence of partial sums is not bounded from above. The harmonic series diverges. ■

The Integral Test

We now introduce the Integral Test with a series that is related to the harmonic series, but whose nth term is $1/n^2$ instead of $1/n$.

EXAMPLE 2   Does the following series converge?

$$\sum_{n=1}^{\infty} \frac{1}{n^2} = 1 + \frac{1}{4} + \frac{1}{9} + \frac{1}{16} + \cdots + \frac{1}{n^2} + \cdots$$

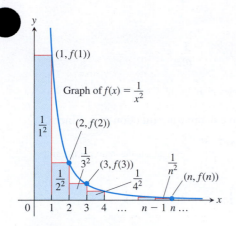

FIGURE 10.11 The sum of the areas of the rectangles under the graph of $f(x) = 1/x^2$ is less than the area under the graph (Example 2).

Solution We determine the convergence of $\sum_{n=1}^{\infty}(1/n^2)$ by comparing it with $\int_{1}^{\infty}(1/x^2)\,dx$. To carry out the comparison, we think of the terms of the series as values of the function $f(x) = 1/x^2$ and interpret these values as the areas of rectangles under the curve $y = 1/x^2$.

As Figure 10.11 shows,

$$s_n = \frac{1}{1^2} + \frac{1}{2^2} + \frac{1}{3^2} + \cdots + \frac{1}{n^2}$$

$$= f(1) + f(2) + f(3) + \cdots + f(n)$$

$$< f(1) + \int_{1}^{n}\frac{1}{x^2}\,dx \qquad \text{Rectangle areas sum to less than area under graph.}$$

$$< 1 + \int_{1}^{\infty}\frac{1}{x^2}\,dx \qquad \int_{1}^{n}(1/x^2)\,dx < \int_{1}^{\infty}(1/x^2)\,dx$$

$$< 1 + 1 = 2. \qquad \text{As in Section 8.8, Example 3,} \\ \int_{1}^{\infty}(1/x^2)\,dx = 1.$$

Thus the partial sums of $\sum_{n=1}^{\infty}(1/n^2)$ are bounded from above (by 2) and the series converges. ∎

Caution

The series and integral need not have the same value in the convergent case. You will see in Example 6 that

$$\sum_{n=1}^{\infty}(1/n^2) \neq \int_{1}^{\infty}(1/x^2)\,dx = 1.$$

> **THEOREM 9—The Integral Test**
>
> Let $\{a_n\}$ be a sequence of positive terms. Suppose that $a_n = f(n)$, where f is a continuous, positive, decreasing function of x for all $x \geq N$ (N a positive integer). Then the series $\sum_{n=N}^{\infty} a_n$ and the integral $\int_{N}^{\infty} f(x)\,dx$ both converge or both diverge.

Proof We establish the test for the case $N = 1$. The proof for general N is similar.

We start with the assumption that f is a decreasing function with $f(n) = a_n$ for every n. This leads us to observe that the rectangles in Figure 10.12a, which have areas $a_1, a_2, \ldots, a_n$, collectively enclose more area than that under the curve $y = f(x)$ from $x = 1$ to $x = n + 1$. That is,

$$\int_{1}^{n+1} f(x)\,dx \leq a_1 + a_2 + \cdots + a_n.$$

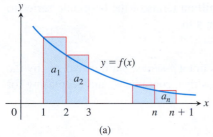

(a)

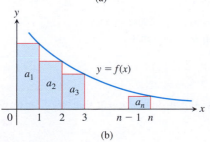

(b)

FIGURE 10.12 Subject to the conditions of the Integral Test, the series $\sum_{n=1}^{\infty} a_n$ and the integral $\int_{1}^{\infty} f(x)\,dx$ both converge or both diverge.

In Figure 10.12b the rectangles have been faced to the left instead of to the right. If we momentarily disregard the first rectangle of area a_1, we see that

$$a_2 + a_3 + \cdots + a_n \leq \int_{1}^{n} f(x)\,dx.$$

If we include a_1, we have

$$a_1 + a_2 + \cdots + a_n \leq a_1 + \int_{1}^{n} f(x)\,dx.$$

Combining these results gives

$$\int_{1}^{n+1} f(x)\,dx \leq a_1 + a_2 + \cdots + a_n \leq a_1 + \int_{1}^{n} f(x)\,dx.$$

These inequalities hold for each n, and continue to hold as $n \to \infty$.

If $\int_{1}^{\infty} f(x)\,dx$ is finite, the right-hand inequality shows that $\sum a_n$ is finite. If $\int_{1}^{\infty} f(x)\,dx$ is infinite, the left-hand inequality shows that $\sum a_n$ is infinite. Hence the series and integral are either both finite or both infinite. ∎

The *p*-series $\sum\limits_{n=1}^{\infty} \dfrac{1}{n^p}$ converges if $p > 1$, diverges if $p \leq 1$.

EXAMPLE 3 Show that the ***p*-series**

$$\sum_{n=1}^{\infty} \frac{1}{n^p} = \frac{1}{1^p} + \frac{1}{2^p} + \frac{1}{3^p} + \cdots + \frac{1}{n^p} + \cdots$$

(p a real constant) converges if $p > 1$, and diverges if $p \leq 1$.

Solution If $p > 1$, then $f(x) = 1/x^p$ is a positive decreasing function of x. Since

$$\int_1^{\infty} \frac{1}{x^p}\,dx = \int_1^{\infty} x^{-p}\,dx = \lim_{b \to \infty} \left[\frac{x^{-p+1}}{-p+1} \right]_1^b \qquad \text{\color{blue}Evaluate the improper integral}$$

$$= \frac{1}{1-p} \lim_{b \to \infty} \left(\frac{1}{b^{p-1}} - 1 \right)$$

$$= \frac{1}{1-p}(0 - 1) = \frac{1}{p-1}, \qquad \begin{array}{l}\text{\color{blue}$b^{p-1} \to \infty$ as $b \to \infty$}\\ \text{\color{blue}because $p - 1 > 0$.}\end{array}$$

the series converges by the Integral Test. We emphasize that the sum of the p-series is *not* $1/(p-1)$. The series converges, but we don't know the value it converges to.

If $p \leq 0$, the series diverges by the nth-term test. If $0 < p < 1$, then $1 - p > 0$ and

$$\int_1^{\infty} \frac{1}{x^p}\,dx = \frac{1}{1-p} \lim_{b \to \infty} (b^{1-p} - 1) = \infty.$$

Therefore, the series diverges by the Integral Test.

If $p = 1$, we have the (divergent) harmonic series

$$1 + \frac{1}{2} + \frac{1}{3} + \cdots + \frac{1}{n} + \cdots.$$

In summary, we have convergence for $p > 1$ but divergence for all other values of p. ∎

The p-series with $p = 1$ is the **harmonic series** (Example 1). The p-Series Test shows that the harmonic series is just *barely* divergent; if we increase p to 1.000000001, for instance, the series converges!

The slowness with which the partial sums of the harmonic series approach infinity is impressive. For instance, it takes more than 178 million terms of the harmonic series to move the partial sums beyond 20. (See also Exercise 49b.)

EXAMPLE 4 The series $\sum_{n=1}^{\infty}(1/(n^2 + 1))$ is not a p-series, but it converges by the Integral Test. The function $f(x) = 1/(x^2 + 1)$ is positive, continuous, and decreasing for $x \geq 1$, and

$$\int_1^{\infty} \frac{1}{x^2 + 1}\,dx = \lim_{b \to \infty} \Big[\arctan x \Big]_1^b$$

$$= \lim_{b \to \infty} \Big[\arctan b - \arctan 1 \Big]$$

$$= \frac{\pi}{2} - \frac{\pi}{4} = \frac{\pi}{4}.$$

The Integral Test tells us that the series converges, but it does *not* say that $\pi/4$ or any other number is the sum of the series. ∎

EXAMPLE 5 Determine the convergence or divergence of the series.

(a) $\sum_{n=1}^{\infty} ne^{-n^2}$ (b) $\sum_{n=1}^{\infty} \frac{1}{2^{\ln n}}$

Solutions

(a) We apply the Integral Test and find that

$$\int_1^\infty \frac{x}{e^{x^2}} \, dx = \frac{1}{2} \int_1^\infty \frac{du}{e^u} \qquad u = x^2, \, du = 2x \, dx$$

$$= \lim_{b\to\infty} \left[-\frac{1}{2} e^{-u} \right]_1^b$$

$$= \lim_{b\to\infty} \left(-\frac{1}{2e^b} + \frac{1}{2e} \right) = \frac{1}{2e}.$$

Since the integral converges, the series also converges.

(b) Again applying the Integral Test,

$$\int_1^\infty \frac{dx}{2^{\ln x}} = \int_0^\infty \frac{e^u \, du}{2^u} \qquad u = \ln x, \, x = e^u, \, dx = e^u \, du$$

$$= \int_0^\infty \left(\frac{e}{2} \right)^u du$$

$$= \lim_{b\to\infty} \frac{1}{\ln \left(\frac{e}{2} \right)} \left(\left(\frac{e}{2} \right)^b - 1 \right) = \infty. \qquad (e/2) > 1$$

The improper integral diverges, so the series diverges also. ∎

Error Estimation

For some convergent series, such as the geometric series or the telescoping series in Example 5 of Section 10.2, we can actually find the total sum of the series. That is, we can find the limiting value S of the sequence of partial sums. For most convergent series, however, we cannot easily find the total sum. Nevertheless, we can *estimate* the sum by adding the first n terms to get s_n, but we need to know how far off s_n is from the total sum S. An approximation to a function or to a number is more useful when it is accompanied by a bound on the size of the worst possible error that could occur. With such an error bound we can try to make an estimate or approximation that is close enough for the problem at hand. Without a bound on the error size, we are just guessing and hoping that we are close to the actual answer. We now show a way to bound the error size using integrals.

Suppose that a series Σa_n with positive terms is shown to be convergent by the Integral Test, and we want to estimate the size of the **remainder** R_n measuring the difference between the total sum S of the series and its nth partial sum s_n. That is, we wish to estimate

$$R_n = S - s_n = a_{n+1} + a_{n+2} + a_{n+3} + \cdots.$$

To get a lower bound for the remainder, we compare the sum of the areas of the rectangles with the area under the curve $y = f(x)$ for $x \geq n$ (see Figure 10.13a). We see that

$$R_n = a_{n+1} + a_{n+2} + a_{n+3} + \cdots \geq \int_{n+1}^\infty f(x) \, dx.$$

Similarly, from Figure 10.13b, we find an upper bound with

$$R_n = a_{n+1} + a_{n+2} + a_{n+3} + \cdots \leq \int_n^\infty f(x) \, dx.$$

These comparisons prove the following result, giving bounds on the size of the remainder.

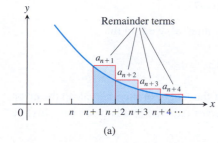

(a)

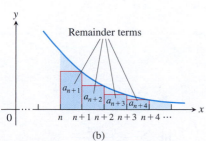

(b)

FIGURE 10.13 The remainder when using n terms is (a) larger than the integral of f over $[n + 1, \infty)$. (b) smaller than the integral of f over $[n, \infty)$.

> ### Bounds for the Remainder in the Integral Test
> Suppose $\{a_k\}$ is a sequence of positive terms with $a_k = f(k)$, where f is a continuous positive decreasing function of x for all $x \geq n$, and that Σa_n converges to S. Then the remainder $R_n = S - s_n$ satisfies the inequalities
>
> $$\int_{n+1}^{\infty} f(x)\, dx \leq R_n \leq \int_n^{\infty} f(x)\, dx. \tag{1}$$

If we add the partial sum s_n to each side of the inequalities in (1), we get

$$s_n + \int_{n+1}^{\infty} f(x)\, dx \leq S \leq s_n + \int_n^{\infty} f(x)\, dx \tag{2}$$

since $s_n + R_n = S$. The inequalities in (2) are useful for estimating the error in approximating the sum of a series known to converge by the Integral Test. The error can be no larger than the length of the interval containing S, with endpoints given by (2).

EXAMPLE 6   Estimate the sum of the series $\Sigma(1/n^2)$ using the inequalities in (2) and $n = 10$.

Solution   We have that

$$\int_n^{\infty} \frac{1}{x^2}\, dx = \lim_{b \to \infty} \left[-\frac{1}{x} \right]_n^b = \lim_{b \to \infty} \left(-\frac{1}{b} + \frac{1}{n} \right) = \frac{1}{n}.$$

Using this result with the inequalities in (2), we get

$$s_{10} + \frac{1}{11} \leq S \leq s_{10} + \frac{1}{10}.$$

Taking $s_{10} = 1 + (1/4) + (1/9) + (1/16) + \cdots + (1/100) \approx 1.54977$, these last inequalities give

$$1.64068 \leq S \leq 1.64977.$$

If we approximate the sum S by the midpoint of this interval, we find that

$$\sum_{n=1}^{\infty} \frac{1}{n^2} \approx 1.6452.$$

The error in this approximation is then less than half the length of the interval, so the error is less than 0.005. Using a trigonometric *Fourier series* (studied in advanced calculus), it can be shown that S is equal to $\pi^2/6 \approx 1.64493$. ∎

The p-series for $p = 2$

$$\sum_{n=1}^{\infty} \frac{1}{n^2} = \frac{\pi^2}{6} \approx 1.64493$$

EXERCISES 10.3

Applying the Integral Test

Use the Integral Test to determine if the series in Exercises 1–12 converge or diverge. Be sure to check that the conditions of the Integral Test are satisfied.

1. $\displaystyle\sum_{n=1}^{\infty} \frac{1}{n^2}$

2. $\displaystyle\sum_{n=1}^{\infty} \frac{1}{n^{0.2}}$

3. $\displaystyle\sum_{n=1}^{\infty} \frac{1}{n^2 + 4}$

4. $\displaystyle\sum_{n=1}^{\infty} \frac{1}{n + 4}$

5. $\displaystyle\sum_{n=1}^{\infty} e^{-2n}$

6. $\displaystyle\sum_{n=2}^{\infty} \frac{1}{n(\ln n)^2}$

7. $\displaystyle\sum_{n=1}^{\infty} \frac{n}{n^2 + 4}$

8. $\displaystyle\sum_{n=2}^{\infty} \frac{\ln(n^2)}{n}$

9. $\displaystyle\sum_{n=1}^{\infty} \frac{n^2}{e^{n/3}}$

10. $\displaystyle\sum_{n=2}^{\infty} \frac{n - 4}{n^2 - 2n + 1}$

11. $\displaystyle\sum_{n=1}^{\infty} \frac{7}{\sqrt{n + 4}}$

12. $\displaystyle\sum_{n=2}^{\infty} \frac{1}{5n + 10\sqrt{n}}$

Determining Convergence or Divergence

Which of the series in Exercises 13–46 converge, and which diverge? Give reasons for your answers. (When you check an answer, remember that there may be more than one way to determine the series' convergence or divergence.)

13. $\displaystyle\sum_{n=1}^{\infty} \frac{1}{10^n}$ **14.** $\displaystyle\sum_{n=1}^{\infty} e^{-n}$ **15.** $\displaystyle\sum_{n=1}^{\infty} \frac{n}{n+1}$

16. $\displaystyle\sum_{n=1}^{\infty} \frac{5}{n+1}$ **17.** $\displaystyle\sum_{n=1}^{\infty} \frac{3}{\sqrt{n}}$ **18.** $\displaystyle\sum_{n=1}^{\infty} \frac{-2}{n\sqrt{n}}$

19. $\displaystyle\sum_{n=1}^{\infty} -\frac{1}{8^n}$ **20.** $\displaystyle\sum_{n=1}^{\infty} \frac{-8}{n}$ **21.** $\displaystyle\sum_{n=2}^{\infty} \frac{\ln n}{n}$

22. $\displaystyle\sum_{n=2}^{\infty} \frac{\ln n}{\sqrt{n}}$ **23.** $\displaystyle\sum_{n=1}^{\infty} \frac{2^n}{3^n}$ **24.** $\displaystyle\sum_{n=1}^{\infty} \frac{5^n}{4^n+3}$

25. $\displaystyle\sum_{n=0}^{\infty} \frac{-2}{n+1}$ **26.** $\displaystyle\sum_{n=1}^{\infty} \frac{1}{2n-1}$ **27.** $\displaystyle\sum_{n=1}^{\infty} \frac{2^n}{n+1}$

28. $\displaystyle\sum_{n=1}^{\infty} \left(1+\frac{1}{n}\right)^n$ **29.** $\displaystyle\sum_{n=2}^{\infty} \frac{\sqrt{n}}{\ln n}$ **30.** $\displaystyle\sum_{n=1}^{\infty} \frac{1}{\sqrt{n}(\sqrt{n}+1)}$

31. $\displaystyle\sum_{n=1}^{\infty} \frac{1}{(\ln 2)^n}$ **32.** $\displaystyle\sum_{n=1}^{\infty} \frac{1}{(\ln 3)^n}$

33. $\displaystyle\sum_{n=3}^{\infty} \frac{(1/n)}{(\ln n)\sqrt{\ln^2 n - 1}}$ **34.** $\displaystyle\sum_{n=1}^{\infty} \frac{1}{n(1+\ln^2 n)}$

35. $\displaystyle\sum_{n=1}^{\infty} n \sin \frac{1}{n}$ **36.** $\displaystyle\sum_{n=1}^{\infty} n \tan \frac{1}{n}$

37. $\displaystyle\sum_{n=1}^{\infty} \frac{e^n}{1+e^{2n}}$ **38.** $\displaystyle\sum_{n=1}^{\infty} \frac{2}{1+e^n}$

39. $\displaystyle\sum_{n=1}^{\infty} \frac{e^n}{10+e^n}$ **40.** $\displaystyle\sum_{n=1}^{\infty} \frac{e^n}{(10+e^n)^2}$

41. $\displaystyle\sum_{n=2}^{\infty} \frac{\sqrt{n+2}-\sqrt{n+1}}{\sqrt{n+1}\sqrt{n+2}}$ **42.** $\displaystyle\sum_{n=3}^{\infty} \frac{7}{\sqrt{n+1}\ln\sqrt{n+1}}$

43. $\displaystyle\sum_{n=1}^{\infty} \frac{8 \tan^{-1} n}{1+n^2}$ **44.** $\displaystyle\sum_{n=1}^{\infty} \frac{n}{n^2+1}$

45. $\displaystyle\sum_{n=1}^{\infty} \operatorname{sech} n$ **46.** $\displaystyle\sum_{n=1}^{\infty} \operatorname{sech}^2 n$

Theory and Examples

For what values of a, if any, do the series in Exercises 47 and 48 converge?

47. $\displaystyle\sum_{n=1}^{\infty} \left(\frac{a}{n+2}-\frac{1}{n+4}\right)$ **48.** $\displaystyle\sum_{n=3}^{\infty} \left(\frac{1}{n-1}-\frac{2a}{n+1}\right)$

49. a. Draw illustrations like those in Figures 10.12a and 10.12b to show that the partial sums of the harmonic series satisfy the inequalities

$$\ln(n+1) = \int_1^{n+1} \frac{1}{x}\,dx \le 1+\frac{1}{2}+\cdots+\frac{1}{n}$$

$$\le 1 + \int_1^n \frac{1}{x}\,dx = 1 + \ln n.$$

T b. There is absolutely no empirical evidence for the divergence of the harmonic series even though we know it diverges. The partial sums just grow too slowly. To see what we mean, suppose you had started with $s_1 = 1$ the day the universe was formed, 13 billion years ago, and added a new term every *second*. About how large would the partial sum s_n be today, assuming a 365-day year?

50. Are there any values of x for which $\sum_{n=1}^{\infty}(1/nx)$ converges? Give reasons for your answer.

51. Is it true that if $\sum_{n=1}^{\infty} a_n$ is a divergent series of positive numbers, then there is also a divergent series $\sum_{n=1}^{\infty} b_n$ of positive numbers with $b_n < a_n$ for every n? Is there a "smallest" divergent series of positive numbers? Give reasons for your answers.

52. (*Continuation of Exercise 51.*) Is there a "largest" convergent series of positive numbers? Explain.

53. $\sum_{n=1}^{\infty} \left(1/\sqrt{n+1}\right)$ **diverges**

a. Use the accompanying graph to show that the partial sum $s_{50} = \sum_{n=1}^{50} \left(1/\sqrt{n+1}\right)$ satisfies

$$\int_1^{51} \frac{1}{\sqrt{x+1}}\,dx < s_{50} < \int_0^{50} \frac{1}{\sqrt{x+1}}\,dx.$$

Conclude that $11.5 < s_{50} < 12.3$.

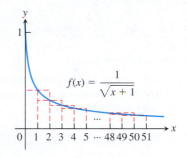

b. What should n be in order that the partial sum

$$s_n = \sum_{i=1}^{n} \left(1/\sqrt{i+1}\right) \text{ satisfy } s_n > 1000?$$

54. $\sum_{n=1}^{\infty} (1/n^4)$ **converges**

a. Use the accompanying graph to find an upper bound for the error if $s_{30} = \sum_{n=1}^{30} (1/n^4)$ is used to estimate the value of $\sum_{n=1}^{\infty} (1/n^4)$.

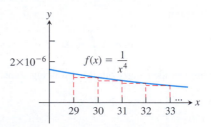

b. Find n so that the partial sum $s_n = \sum_{i=1}^{n} (1/i^4)$ estimates the value of $\sum_{n=1}^{\infty} (1/n^4)$ with an error of at most 0.000001.

55. Estimate the value of $\sum_{n=1}^{\infty} (1/n^3)$ to within 0.01 of its exact value.

56. Estimate the value of $\sum_{n=2}^{\infty} (1/(n^2+4))$ to within 0.1 of its exact value.

57. How many terms of the convergent series $\sum_{n=1}^{\infty} (1/n^{1.1})$ should be used to estimate its value with error at most 0.00001?

58. How many terms of the convergent series $\sum_{n=4}^{\infty} 1/(n(\ln n)^3)$ should be used to estimate its value with error at most 0.01?

59. The Cauchy condensation test The Cauchy condensation test says: Let $\{a_n\}$ be a nonincreasing sequence ($a_n \geq a_{n+1}$ for all n) of positive terms that converges to 0. Then $\sum a_n$ converges if and only if $\sum 2^n a_{2^n}$ converges. For example, $\sum (1/n)$ diverges because $\sum 2^n \cdot (1/2^n) = \sum 1$ diverges. Show why the test works.

60. Use the Cauchy condensation test from Exercise 59 to show that

a. $\displaystyle\sum_{n=2}^{\infty} \frac{1}{n \ln n}$ diverges;

b. $\displaystyle\sum_{n=1}^{\infty} \frac{1}{n^p}$ converges if $p > 1$ and diverges if $p \leq 1$.

61. Logarithmic p-series

a. Show that the improper integral

$$\int_2^{\infty} \frac{dx}{x(\ln x)^p} \quad (p \text{ a positive constant})$$

converges if and only if $p > 1$.

b. What implications does the fact in part (a) have for the convergence of the series

$$\sum_{n=2}^{\infty} \frac{1}{n(\ln n)^p}?$$

Give reasons for your answer.

62. (*Continuation of Exercise 61.*) Use the result in Exercise 61 to determine which of the following series converge and which diverge. Support your answer in each case.

a. $\displaystyle\sum_{n=2}^{\infty} \frac{1}{n(\ln n)}$ **b.** $\displaystyle\sum_{n=2}^{\infty} \frac{1}{n(\ln n)^{1.01}}$

c. $\displaystyle\sum_{n=2}^{\infty} \frac{1}{n \ln(n^3)}$ **d.** $\displaystyle\sum_{n=2}^{\infty} \frac{1}{n(\ln n)^3}$

63. Euler's constant Graphs like those in Figure 10.12 suggest that as n increases there is little change in the difference between the sum

$$1 + \frac{1}{2} + \cdots + \frac{1}{n}$$

and the integral

$$\ln n = \int_1^n \frac{1}{x}\, dx.$$

To explore this idea, carry out the following steps.

a. By taking $f(x) = 1/x$ in the proof of Theorem 9, show that

$$\ln(n + 1) \leq 1 + \frac{1}{2} + \cdots + \frac{1}{n} \leq 1 + \ln n$$

or

$$0 < \ln(n + 1) - \ln n \leq 1 + \frac{1}{2} + \cdots + \frac{1}{n} - \ln n \leq 1.$$

Thus, the sequence

$$a_n = 1 + \frac{1}{2} + \cdots + \frac{1}{n} - \ln n$$

is bounded from below and from above.

b. Show that

$$\frac{1}{n + 1} < \int_n^{n+1} \frac{1}{x}\, dx = \ln(n + 1) - \ln n,$$

and use this result to show that the sequence $\{a_n\}$ in part (a) is decreasing.

Since a decreasing sequence that is bounded from below converges, the numbers a_n defined in part (a) converge:

$$1 + \frac{1}{2} + \cdots + \frac{1}{n} - \ln n \rightarrow \gamma.$$

The number γ, whose value is $0.5772\ldots$, is called *Euler's constant*.

64. Use the Integral Test to show that the series

$$\sum_{n=0}^{\infty} e^{-n^2}$$

converges.

65. a. For the series $\sum(1/n^3)$, use the inequalities in Equation (2) with $n = 10$ to find an interval containing the sum S.

b. As in Example 5, use the midpoint of the interval found in part (a) to approximate the sum of the series. What is the maximum error for your approximation?

66. Repeat Exercise 65 using the series $\sum(1/n^4)$.

67. Area Consider the sequence $\{1/n\}_{n=1}^{\infty}$. On each subinterval $(1/(n + 1), 1/n)$ within the interval $[0, 1]$, erect the rectangle with area a_n having height $1/n$ and width equal to the length of the subinterval. Find the total area $\sum a_n$ of all the rectangles. (*Hint:* Use the result of Example 5 in Section 10.2.)

68. Area Repeat Exercise 67, using trapezoids instead of rectangles. That is, on the subinterval $(1/(n + 1), 1/n)$, let a_n denote the area of the trapezoid having heights $y = 1/(n + 1)$ at $x = 1/(n + 1)$ and $y = 1/n$ at $x = 1/n$.

10.4 Comparison Tests

We have seen how to determine the convergence of geometric series, p-series, and a few others. We can test the convergence of many more series by comparing their terms to those of a series whose convergence is already known.

> **THEOREM 10—Direct Comparison Test**
>
> Let $\sum a_n$ and $\sum b_n$ be two series with $0 \leq a_n \leq b_n$ for all n. Then
>
> **1.** If $\sum b_n$ converges, then $\sum a_n$ also converges.
>
> **2.** If $\sum a_n$ diverges, then $\sum b_n$ also diverges.

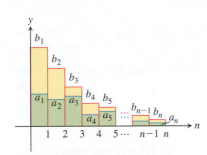

FIGURE 10.14 If the total area $\sum b_n$ of the taller b_n rectangles is finite, then so is the total area $\sum a_n$ of the shorter a_n rectangles.

Proof   The series $\sum a_n$ and $\sum b_n$ have nonnegative terms. The Corollary of Theorem 6 stated in Section 10.3 tells us that the series $\sum a_n$ and $\sum b_n$ converge if and only if their partial sums are bounded from above.

In Part (1) we assume that $\sum b_n$ converges to some number M. The partial sums $\sum_{n=1}^{N} a_n$ are all bounded from above by $M = \sum b_n$, since

$$s_N = a_1 + a_2 + \cdots + a_N \leq b_1 + b_2 + \cdots + b_N \leq \sum_{n=1}^{\infty} b_n = M.$$

Since the partial sums of $\sum a_n$ are bounded from above, the Corollary of Theorem 6 implies that $\sum a_n$ converges. We conclude that when $\sum b_n$ converges, then so does $\sum a_n$. Figure 10.12 illustrates this result, with each term of each series interpreted as the area of a rectangle.

In Part (2), where we assume that $\sum a_n$ diverges, the partial sums of $\sum_{n=1}^{\infty} b_n$ are not bounded from above. If they were, the partial sums for $\sum a_n$ would also be bounded from above, since

$$a_1 + a_2 + \cdots + a_N \leq b_1 + b_2 + \cdots + b_N,$$

and this would mean that $\sum a_n$ converges. We conclude that if $\sum a_n$ diverges, then so does $\sum b_n$. ∎

HISTORICAL BIOGRAPHY

Albert of Saxony

(ca. 1316–1390)

`www.goo.gl/Q2d00w`

EXAMPLE 1   We apply Theorem 10 to several series.

(a) The series

$$\sum_{n=1}^{\infty} \frac{5}{5n - 1}$$

diverges because its nth term

$$\frac{5}{5n - 1} = \frac{1}{n - \frac{1}{5}} > \frac{1}{n}$$

is greater than the nth term of the divergent harmonic series.

(b) The series

$$\sum_{n=0}^{\infty} \frac{1}{n!} = 1 + \frac{1}{1!} + \frac{1}{2!} + \frac{1}{3!} + \cdots$$

converges because its terms are all positive and less than or equal to the corresponding terms of

$$1 + \sum_{n=0}^{\infty} \frac{1}{2^n} = 1 + 1 + \frac{1}{2} + \frac{1}{2^2} + \cdots.$$

The geometric series on the left converges and we have

$$1 + \sum_{n=0}^{\infty} \frac{1}{2^n} = 1 + \frac{1}{1 - (1/2)} = 3.$$

The fact that 3 is an upper bound for the partial sums of $\sum_{n=0}^{\infty} (1/n!)$ does not mean that the series converges to 3. As we will see in Section 10.9, the series converges to e.

(c) The series

$$5 + \frac{2}{3} + \frac{1}{7} + 1 + \frac{1}{2 + \sqrt{1}} + \frac{1}{4 + \sqrt{2}} + \frac{1}{8 + \sqrt{3}} + \cdots + \frac{1}{2^n + \sqrt{n}} + \cdots$$

converges. To see this, we ignore the first three terms and compare the remaining terms with those of the convergent geometric series $\sum_{n=0}^{\infty}(1/2^n)$. The term $1/(2^n + \sqrt{n})$ of the truncated sequence is less than the corresponding term $1/2^n$ of the geometric series. We see that term by term we have the comparison

$$1 + \frac{1}{2 + \sqrt{1}} + \frac{1}{4 + \sqrt{2}} + \frac{1}{8 + \sqrt{3}} + \cdots \leq 1 + \frac{1}{2} + \frac{1}{4} + \frac{1}{8} + \cdots.$$

So the truncated series and the original series converge by an application of the Direct Comparison Test.   ∎

The Limit Comparison Test

We now introduce a comparison test that is particularly useful for series in which a_n is a rational function of n.

THEOREM 11—Limit Comparison Test

Suppose that $a_n > 0$ and $b_n > 0$ for all $n \geq N$ (N an integer).

1. If $\lim\limits_{n\to\infty} \dfrac{a_n}{b_n} = c$ and $c > 0$, then $\sum a_n$ and $\sum b_n$ both converge or both diverge.

2. If $\lim\limits_{n\to\infty} \dfrac{a_n}{b_n} = 0$ and $\sum b_n$ converges, then $\sum a_n$ converges.

3. If $\lim\limits_{n\to\infty} \dfrac{a_n}{b_n} = \infty$ and $\sum b_n$ diverges, then $\sum a_n$ diverges.

Proof   We will prove Part 1. Parts 2 and 3 are left as Exercises 57a and b.
Since $c/2 > 0$, there exists an integer N such that

$$\left| \frac{a_n}{b_n} - c \right| < \frac{c}{2} \qquad \text{whenever} \qquad n > N. \qquad \text{\color{red}{Limit definition with}} \\ \text{\color{red}{$\varepsilon = c/2, L = c$, and}} \\ \text{\color{red}{a_n replaced by a_n/b_n}}$$

Thus, for $n > N$,

$$-\frac{c}{2} < \frac{a_n}{b_n} - c < \frac{c}{2},$$

$$\frac{c}{2} < \frac{a_n}{b_n} < \frac{3c}{2},$$

$$\left(\frac{c}{2}\right) b_n < a_n < \left(\frac{3c}{2}\right) b_n.$$

If $\sum b_n$ converges, then $\sum (3c/2)b_n$ converges and $\sum a_n$ converges by the Direct Comparison Test. If $\sum b_n$ diverges, then $\sum (c/2)b_n$ diverges and $\sum a_n$ diverges by the Direct Comparison Test.   ∎

EXAMPLE 2   Which of the following series converge, and which diverge?

(a) $\dfrac{3}{4} + \dfrac{5}{9} + \dfrac{7}{16} + \dfrac{9}{25} + \cdots = \displaystyle\sum_{n=1}^{\infty} \frac{2n + 1}{(n + 1)^2} = \sum_{n=1}^{\infty} \frac{2n + 1}{n^2 + 2n + 1}$

(b) $\dfrac{1}{1} + \dfrac{1}{3} + \dfrac{1}{7} + \dfrac{1}{15} + \cdots = \displaystyle\sum_{n=1}^{\infty} \dfrac{1}{2^n - 1}$

(c) $\dfrac{1 + 2 \ln 2}{9} + \dfrac{1 + 3 \ln 3}{14} + \dfrac{1 + 4 \ln 4}{21} + \cdots = \displaystyle\sum_{n=2}^{\infty} \dfrac{1 + n \ln n}{n^2 + 5}$

Solution We apply the Limit Comparison Test to each series.

(a) Let $a_n = (2n + 1)/(n^2 + 2n + 1)$. For large n, we expect a_n to behave like $2n/n^2 = 2/n$ since the leading terms dominate for large n, so we let $b_n = 1/n$. Since

$$\sum_{n=1}^{\infty} b_n = \sum_{n=1}^{\infty} \frac{1}{n} \text{ diverges}$$

and

$$\lim_{n\to\infty} \frac{a_n}{b_n} = \lim_{n\to\infty} \frac{2n^2 + n}{n^2 + 2n + 1} = 2,$$

$\sum a_n$ diverges by Part 1 of the Limit Comparison Test. We could just as well have taken $b_n = 2/n$, but $1/n$ is simpler.

(b) Let $a_n = 1/(2^n - 1)$. For large n, we expect a_n to behave like $1/2^n$, so we let $b_n = 1/2^n$. Since

$$\sum_{n=1}^{\infty} b_n = \sum_{n=1}^{\infty} \frac{1}{2^n} \text{ converges}$$

and

$$\lim_{n\to\infty} \frac{a_n}{b_n} = \lim_{n\to\infty} \frac{2^n}{2^n - 1} = \lim_{n\to\infty} \frac{1}{1 - (1/2^n)} = 1,$$

$\sum a_n$ converges by Part 1 of the Limit Comparison Test.

(c) Let $a_n = (1 + n \ln n)/(n^2 + 5)$. For large n, we expect a_n to behave like $(n \ln n)/n^2 = (\ln n)/n$, which is greater than $1/n$ for $n \geq 3$, so we let $b_n = 1/n$. Since

$$\sum_{n=2}^{\infty} b_n = \sum_{n=2}^{\infty} \frac{1}{n} \text{ diverges}$$

and

$$\lim_{n\to\infty} \frac{a_n}{b_n} = \lim_{n\to\infty} \frac{n + n^2 \ln n}{n^2 + 5} = \infty,$$

$\sum a_n$ diverges by Part 3 of the Limit Comparison Test. ■

EXAMPLE 3 Does $\displaystyle\sum_{n=1}^{\infty} \dfrac{\ln n}{n^{3/2}}$ converge?

Solution Because $\ln n$ grows more slowly than n^c for any positive constant c (Section 10.1, Exercise 115), we can compare the series to a convergent p-series. To get the p-series, we see that

$$\frac{\ln n}{n^{3/2}} < \frac{n^{1/4}}{n^{3/2}} = \frac{1}{n^{5/4}}$$

for n sufficiently large. Then taking $a_n = (\ln n)/n^{3/2}$ and $b_n = 1/n^{5/4}$, we have

$$\lim_{n \to \infty} \frac{a_n}{b_n} = \lim_{n \to \infty} \frac{\ln n}{n^{1/4}}$$

$$= \lim_{n \to \infty} \frac{1/n}{(1/4)\,n^{-3/4}} \qquad \text{l'Hôpital's Rule}$$

$$= \lim_{n \to \infty} \frac{4}{n^{1/4}} = 0.$$

Since $\sum b_n = \sum (1/n^{5/4})$ is a p-series with $p > 1$, it converges. Therefore $\sum a_n$ converges by Part 2 of the Limit Comparison Test. ∎

EXERCISES 10.4

Direct Comparison Test
In Exercises 1–8, use the Direct Comparison Test to determine if each series converges or diverges.

1. $\displaystyle\sum_{n=1}^{\infty} \frac{1}{n^2 + 30}$
2. $\displaystyle\sum_{n=1}^{\infty} \frac{n-1}{n^4 + 2}$
3. $\displaystyle\sum_{n=2}^{\infty} \frac{1}{\sqrt{n}-1}$

4. $\displaystyle\sum_{n=2}^{\infty} \frac{n+2}{n^2-n}$
5. $\displaystyle\sum_{n=1}^{\infty} \frac{\cos^2 n}{n^{3/2}}$
6. $\displaystyle\sum_{n=1}^{\infty} \frac{1}{n3^n}$

7. $\displaystyle\sum_{n=1}^{\infty} \sqrt{\frac{n+4}{n^4+4}}$
8. $\displaystyle\sum_{n=1}^{\infty} \frac{\sqrt{n}+1}{\sqrt{n^2+3}}$

Limit Comparison Test
In Exercises 9–16, use the Limit Comparison Test to determine if each series converges or diverges.

9. $\displaystyle\sum_{n=1}^{\infty} \frac{n-2}{n^3 - n^2 + 3}$

(*Hint:* Limit Comparison with $\sum_{n=1}^{\infty} (1/n^2)$)

10. $\displaystyle\sum_{n=1}^{\infty} \sqrt{\frac{n+1}{n^2+2}}$

(*Hint:* Limit Comparison with $\sum_{n=1}^{\infty} (1/\sqrt{n})$)

11. $\displaystyle\sum_{n=2}^{\infty} \frac{n(n+1)}{(n^2+1)(n-1)}$
12. $\displaystyle\sum_{n=1}^{\infty} \frac{2^n}{3+4^n}$

13. $\displaystyle\sum_{n=1}^{\infty} \frac{5^n}{\sqrt{n}\,4^n}$
14. $\displaystyle\sum_{n=1}^{\infty} \left(\frac{2n+3}{5n+4}\right)^n$

15. $\displaystyle\sum_{n=2}^{\infty} \frac{1}{\ln n}$

(*Hint:* Limit Comparison with $\sum_{n=2}^{\infty} (1/n)$)

16. $\displaystyle\sum_{n=1}^{\infty} \ln\left(1 + \frac{1}{n^2}\right)$

(*Hint:* Limit Comparison with $\sum_{n=1}^{\infty} (1/n^2)$)

Determining Convergence or Divergence
Which of the series in Exercises 17–56 converge, and which diverge? Use any method, and give reasons for your answers.

17. $\displaystyle\sum_{n=1}^{\infty} \frac{1}{2\sqrt{n} + \sqrt[3]{n}}$
18. $\displaystyle\sum_{n=1}^{\infty} \frac{3}{n+\sqrt{n}}$
19. $\displaystyle\sum_{n=1}^{\infty} \frac{\sin^2 n}{2^n}$

20. $\displaystyle\sum_{n=1}^{\infty} \frac{1+\cos n}{n^2}$
21. $\displaystyle\sum_{n=1}^{\infty} \frac{2n}{3n-1}$
22. $\displaystyle\sum_{n=1}^{\infty} \frac{n+1}{n^2\sqrt{n}}$

23. $\displaystyle\sum_{n=1}^{\infty} \frac{10n+1}{n(n+1)(n+2)}$
24. $\displaystyle\sum_{n=3}^{\infty} \frac{5n^3 - 3n}{n^2(n-2)(n^2+5)}$

25. $\displaystyle\sum_{n=1}^{\infty} \left(\frac{n}{3n+1}\right)^n$
26. $\displaystyle\sum_{n=1}^{\infty} \frac{1}{\sqrt{n^3+2}}$
27. $\displaystyle\sum_{n=3}^{\infty} \frac{1}{\ln(\ln n)}$

28. $\displaystyle\sum_{n=1}^{\infty} \frac{(\ln n)^2}{n^3}$
29. $\displaystyle\sum_{n=2}^{\infty} \frac{1}{\sqrt{n}\,\ln n}$
30. $\displaystyle\sum_{n=1}^{\infty} \frac{(\ln n)^2}{n^{3/2}}$

31. $\displaystyle\sum_{n=1}^{\infty} \frac{1}{1 + \ln n}$
32. $\displaystyle\sum_{n=2}^{\infty} \frac{\ln(n+1)}{n+1}$
33. $\displaystyle\sum_{n=2}^{\infty} \frac{1}{n\sqrt{n^2-1}}$

34. $\displaystyle\sum_{n=1}^{\infty} \frac{\sqrt{n}}{n^2+1}$
35. $\displaystyle\sum_{n=1}^{\infty} \frac{1-n}{n2^n}$
36. $\displaystyle\sum_{n=1}^{\infty} \frac{n+2^n}{n^2 2^n}$

37. $\displaystyle\sum_{n=1}^{\infty} \frac{1}{3^{n-1}+1}$
38. $\displaystyle\sum_{n=1}^{\infty} \frac{3^{n-1}+1}{3^n}$
39. $\displaystyle\sum_{n=1}^{\infty} \frac{n+1}{n^2+3n}\cdot\frac{1}{5n}$

40. $\displaystyle\sum_{n=1}^{\infty} \frac{2^n + 3^n}{3^n + 4^n}$
41. $\displaystyle\sum_{n=1}^{\infty} \frac{2^n - n}{n2^n}$
42. $\displaystyle\sum_{n=1}^{\infty} \frac{\ln n}{\sqrt{n}\,e^n}$

43. $\displaystyle\sum_{n=2}^{\infty} \frac{1}{n!}$

(*Hint:* First show that $(1/n!) \le (1/n(n-1))$ for $n \ge 2$.)

44. $\displaystyle\sum_{n=1}^{\infty} \frac{(n-1)!}{(n+2)!}$
45. $\displaystyle\sum_{n=1}^{\infty} \sin\frac{1}{n}$
46. $\displaystyle\sum_{n=1}^{\infty} \tan\frac{1}{n}$

47. $\displaystyle\sum_{n=1}^{\infty} \frac{\tan^{-1} n}{n^{1.1}}$
48. $\displaystyle\sum_{n=1}^{\infty} \frac{\sec^{-1} n}{n^{1.3}}$
49. $\displaystyle\sum_{n=1}^{\infty} \frac{\coth n}{n^2}$

50. $\displaystyle\sum_{n=1}^{\infty} \frac{\tanh n}{n^2}$
51. $\displaystyle\sum_{n=1}^{\infty} \frac{1}{n\sqrt[n]{n}}$
52. $\displaystyle\sum_{n=1}^{\infty} \frac{\sqrt[n]{n}}{n^2}$

53. $\displaystyle\sum_{n=1}^{\infty} \frac{1}{1+2+3+\cdots+n}$
54. $\displaystyle\sum_{n=1}^{\infty} \frac{1}{1+2^2+3^2+\cdots+n^2}$

55. $\displaystyle\sum_{n=2}^{\infty} \frac{n}{(\ln n)^2}$
56. $\displaystyle\sum_{n=2}^{\infty} \frac{(\ln n)^2}{n}$

Theory and Examples

57. Prove **(a)** Part 2 and **(b)** Part 3 of the Limit Comparison Test.

58. If $\sum_{n=1}^{\infty} a_n$ is a convergent series of nonnegative numbers, can anything be said about $\sum_{n=1}^{\infty} (a_n/n)$? Explain.

59. Suppose that $a_n > 0$ and $b_n > 0$ for $n \geq N$ (N an integer). If $\lim_{n \to \infty} (a_n/b_n) = \infty$ and $\sum a_n$ converges, can anything be said about $\sum b_n$? Give reasons for your answer.

60. Prove that if $\sum a_n$ is a convergent series of nonnegative terms, then $\sum a_n^2$ converges.

61. Suppose that $a_n > 0$ and $\lim_{n \to \infty} a_n = \infty$. Prove that $\sum a_n$ diverges.

62. Suppose that $a_n > 0$ and $\lim_{n \to \infty} n^2 a_n = 0$. Prove that $\sum a_n$ converges.

63. Show that $\sum_{n=2}^{\infty} ((\ln n)^q/n^p)$ converges for $-\infty < q < \infty$ and $p > 1$.

(*Hint:* Limit Comparison with $\sum_{n=2}^{\infty} 1/n^r$ for $1 < r < p$.)

64. (*Continuation of Exercise 63.*) Show that $\sum_{n=2}^{\infty} ((\ln n)^q/n^p)$ diverges for $-\infty < q < \infty$ and $0 < p < 1$.

(*Hint:* Limit Comparison with an appropriate p-series.)

65. Decimal numbers Any real number in the interval $[0, 1]$ can be represented by a decimal (not necessarily unique) as

$$0.d_1 d_2 d_3 d_4 \ldots = \frac{d_1}{10} + \frac{d_2}{10^2} + \frac{d_3}{10^3} + \frac{d_4}{10^4} + \cdots,$$

where d_i is one of the integers 0, 1, 2, 3, $\ldots$, 9. Prove that the series on the right-hand side always converges.

66. If $\sum a_n$ is a convergent series of positive terms, prove that $\sum \sin(a_n)$ converges.

In Exercises 67–72, use the results of Exercises 63 and 64 to determine if each series converges or diverges.

67. $\sum_{n=2}^{\infty} \dfrac{(\ln n)^3}{n^4}$

68. $\sum_{n=2}^{\infty} \sqrt{\dfrac{\ln n}{n}}$

69. $\sum_{n=2}^{\infty} \dfrac{(\ln n)^{1000}}{n^{1.001}}$

70. $\sum_{n=2}^{\infty} \dfrac{(\ln n)^{1/5}}{n^{0.99}}$

71. $\sum_{n=2}^{\infty} \dfrac{1}{n^{1.1}(\ln n)^3}$

72. $\sum_{n=2}^{\infty} \dfrac{1}{\sqrt{n} \cdot \ln n}$

COMPUTER EXPLORATIONS

73. It is not yet known whether the series

$$\sum_{n=1}^{\infty} \frac{1}{n^3 \sin^2 n}$$

converges or diverges. Use a CAS to explore the behavior of the series by performing the following steps.

a. Define the sequence of partial sums

$$s_k = \sum_{n=1}^{k} \frac{1}{n^3 \sin^2 n}.$$

What happens when you try to find the limit of s_k as $k \to \infty$? Does your CAS find a closed form answer for this limit?

b. Plot the first 100 points (k, s_k) for the sequence of partial sums. Do they appear to converge? What would you estimate the limit to be?

c. Next plot the first 200 points (k, s_k). Discuss the behavior in your own words.

d. Plot the first 400 points (k, s_k). What happens when $k = 355$? Calculate the number $355/113$. Explain from you calculation what happened at $k = 355$. For what values of k would you guess this behavior might occur again?

74. a. Use Theorem 8 to show that

$$S = \sum_{n=1}^{\infty} \frac{1}{n(n+1)} + \sum_{n=1}^{\infty} \left(\frac{1}{n^2} - \frac{1}{n(n+1)} \right)$$

where $S = \sum_{n=1}^{\infty} (1/n^2)$, the sum of a convergent p-series.

b. From Example 5, Section 10.2, show that

$$S = 1 + \sum_{n=1}^{\infty} \frac{1}{n^2(n+1)}.$$

c. Explain why taking the first M terms in the series in part (b) gives a better approximation to S than taking the first M terms in the original series $\sum_{n=1}^{\infty} (1/n^2)$.

d. We know the exact value of S is $\pi^2/6$. Which of the sums

$$\sum_{n=1}^{1000000} \frac{1}{n^2} \quad \text{or} \quad 1 + \sum_{n=1}^{1000} \frac{1}{n^2(n+1)}$$

gives a better approximation to S?

10.5 Absolute Convergence; The Ratio and Root Tests

When some of the terms of a series are positive and others are negative, the series may or may not converge. For example, the geometric series

$$5 - \frac{5}{4} + \frac{5}{16} - \frac{5}{64} + \cdots = \sum_{n=0}^{\infty} 5 \left(\frac{-1}{4} \right)^n \tag{1}$$

converges (since $|r| = \frac{1}{4} < 1$), whereas the different geometric series

$$1 - \frac{5}{4} + \frac{25}{16} - \frac{125}{64} + \cdots = \sum_{n=0}^{\infty} \left(\frac{-5}{4} \right)^n \tag{2}$$

diverges (since $|r| = 5/4 > 1$). In series (1), there is some cancelation in the partial sums, which may be assisting the convergence property of the series. However, if we make all of the terms positive in series (1) to form the new series

$$5 + \frac{5}{4} + \frac{5}{16} + \frac{5}{64} + \cdots = \sum_{n=0}^{\infty} \left| 5\left(\frac{-1}{4}\right)^n \right| = \sum_{n=0}^{\infty} 5\left(\frac{1}{4}\right)^n,$$

we see that it still converges. For a general series with both positive and negative terms, we can apply the tests for convergence studied before to the series of absolute values of its terms. In doing so, we are led naturally to the following concept.

DEFINITION A series $\sum a_n$ **converges absolutely** (is **absolutely convergent**) if the corresponding series of absolute values, $\sum |a_n|$, converges.

So the geometric series (1) is absolutely convergent. We observed, too, that it is also convergent. This situation is always true: An absolutely convergent series is convergent as well, which we now prove.

Caution
Be careful when using Theorem 12. A convergent series need *not* converge absolutely, as you will see in the next section.

THEOREM 12—The Absolute Convergence Test

If $\displaystyle\sum_{n=1}^{\infty} |a_n|$ converges, then $\displaystyle\sum_{n=1}^{\infty} a_n$ converges.

Proof For each n,

$$-|a_n| \leq a_n \leq |a_n|, \qquad \text{so} \qquad 0 \leq a_n + |a_n| \leq 2|a_n|.$$

If $\sum_{n=1}^{\infty} |a_n|$ converges, then $\sum_{n=1}^{\infty} 2|a_n|$ converges and, by the Direct Comparison Test, the nonnegative series $\sum_{n=1}^{\infty} (a_n + |a_n|)$ converges. The equality $a_n = (a_n + |a_n|) - |a_n|$ now lets us express $\sum_{n=1}^{\infty} a_n$ as the difference of two convergent series:

$$\sum_{n=1}^{\infty} a_n = \sum_{n=1}^{\infty} (a_n + |a_n| - |a_n|) = \sum_{n=1}^{\infty} (a_n + |a_n|) - \sum_{n=1}^{\infty} |a_n|.$$

Therefore, $\sum_{n=1}^{\infty} a_n$ converges. ∎

EXAMPLE 1 This example gives two series that converge absolutely.

(a) For $\displaystyle\sum_{n=1}^{\infty} (-1)^{n+1} \frac{1}{n^2} = 1 - \frac{1}{4} + \frac{1}{9} - \frac{1}{16} + \cdots$, the corresponding series of absolute values is the convergent series

$$\sum_{n=1}^{\infty} \frac{1}{n^2} = 1 + \frac{1}{4} + \frac{1}{9} + \frac{1}{16} + \cdots.$$

The original series converges because it converges absolutely.

(b) For $\displaystyle\sum_{n=1}^{\infty} \frac{\sin n}{n^2} = \frac{\sin 1}{1} + \frac{\sin 2}{4} + \frac{\sin 3}{9} + \cdots$, which contains both positive and negative terms, the corresponding series of absolute values is

$$\sum_{n=1}^{\infty} \left| \frac{\sin n}{n^2} \right| = \frac{|\sin 1|}{1} + \frac{|\sin 2|}{4} + \cdots,$$

which converges by comparison with $\sum_{n=1}^{\infty} (1/n^2)$ because $|\sin n| \leq 1$ for every n. The original series converges absolutely; therefore it converges. ∎

The Ratio Test

The Ratio Test measures the rate of growth (or decline) of a series by examining the ratio a_{n+1}/a_n. For a geometric series $\sum ar^n$, this rate is a constant $((ar^{n+1})/(ar^n) = r)$, and the series converges if and only if its ratio is less than 1 in absolute value. The Ratio Test is a powerful rule extending that result.

> **THEOREM 13—The Ratio Test**
>
> Let $\sum a_n$ be any series and suppose that
>
> $$\lim_{n \to \infty} \left| \frac{a_{n+1}}{a_n} \right| = \rho.$$
>
> Then **(a)** the series *converges absolutely* if $\rho < 1$, **(b)** the series *diverges* if $\rho > 1$ or ρ is infinite, **(c)** the test is *inconclusive* if $\rho = 1$.

ρ is the Greek lowercase letter rho, which is pronounced "row."

Proof

(a) $\rho < 1$. Let r be a number between ρ and 1. Then the number $\varepsilon = r - \rho$ is positive. Since

$$\left| \frac{a_{n+1}}{a_n} \right| \to \rho,$$

$|a_{n+1}/a_n|$ must lie within ε of ρ when n is large enough, say, for all $n \geq N$. In particular,

$$\left| \frac{a_{n+1}}{a_n} \right| < \rho + \varepsilon = r, \qquad \text{when } n \geq N.$$

Hence

$$|a_{N+1}| < r|a_N|,$$
$$|a_{N+2}| < r|a_{N+1}| < r^2|a_N|,$$
$$|a_{N+3}| < r|a_{N+2}| < r^3|a_N|,$$
$$\vdots$$
$$|a_{N+m}| < r|a_{N+m-1}| < r^m|a_N|.$$

Therefore,

$$\sum_{m=N}^{\infty} |a_m| = \sum_{m=0}^{\infty} |a_{N+m}| \leq \sum_{m=0}^{\infty} |a_N| \, r^m = |a_N| \sum_{m=0}^{\infty} r^m.$$

The geometric series on the right-hand side converges because $0 < r < 1$, so the series of absolute values $\sum_{m=N}^{\infty} |a_m|$ converges by the Direct Comparison Test. Because adding or deleting finitely many terms in a series does not affect its convergence or divergence property, the series $\sum_{n=1}^{\infty} |a_n|$ also converges. That is, the series $\sum a_n$ is absolutely convergent.

(b) $1 < \rho \leq \infty$. From some index M on,

$$\left| \frac{a_{n+1}}{a_n} \right| > 1 \qquad \text{and} \qquad |a_M| < |a_{M+1}| < |a_{M+2}| < \cdots.$$

The terms of the series do not approach zero as n becomes infinite, and the series diverges by the nth-Term Test.

(c) $\rho = 1$. The two series

$$\sum_{n=1}^{\infty} \frac{1}{n} \qquad \text{and} \qquad \sum_{n=1}^{\infty} \frac{1}{n^2}$$

show that some other test for convergence must be used when $\rho = 1$.

$$\text{For } \sum_{n=1}^{\infty} \frac{1}{n}: \quad \left| \frac{a_{n+1}}{a_n} \right| = \frac{1/(n+1)}{1/n} = \frac{n}{n+1} \rightarrow 1.$$

$$\text{For } \sum_{n=1}^{\infty} \frac{1}{n^2}: \quad \left| \frac{a_{n+1}}{a_n} \right| = \frac{1/(n+1)^2}{1/n^2} = \left(\frac{n}{n+1} \right)^2 \rightarrow 1^2 = 1.$$

In both cases, $\rho = 1$, yet the first series diverges, whereas the second converges. ∎

The Ratio Test is often effective when the terms of a series contain factorials of expressions involving n or expressions raised to a power involving n.

EXAMPLE 2   Investigate the convergence of the following series.

(a) $\displaystyle\sum_{n=0}^{\infty} \frac{2^n + 5}{3^n}$ **(b)** $\displaystyle\sum_{n=1}^{\infty} \frac{(2n)!}{n!n!}$ **(c)** $\displaystyle\sum_{n=1}^{\infty} \frac{4^n n! n!}{(2n)!}$

Solution   We apply the Ratio Test to each series.

(a) For the series $\sum_{n=0}^{\infty} (2^n + 5)/3^n$,

$$\left| \frac{a_{n+1}}{a_n} \right| = \frac{(2^{n+1} + 5)/3^{n+1}}{(2^n + 5)/3^n} = \frac{1}{3} \cdot \frac{2^{n+1} + 5}{2^n + 5} = \frac{1}{3} \cdot \left(\frac{2 + 5 \cdot 2^{-n}}{1 + 5 \cdot 2^{-n}} \right) \rightarrow \frac{1}{3} \cdot \frac{2}{1} = \frac{2}{3}.$$

The series converges absolutely (and thus converges) because $\rho = 2/3$ is less than 1. This does *not* mean that $2/3$ is the sum of the series. In fact,

$$\sum_{n=0}^{\infty} \frac{2^n + 5}{3^n} = \sum_{n=0}^{\infty} \left(\frac{2}{3} \right)^n + \sum_{n=0}^{\infty} \frac{5}{3^n} = \frac{1}{1 - (2/3)} + \frac{5}{1 - (1/3)} = \frac{21}{2}.$$

(b) If $a_n = \dfrac{(2n)!}{n!n!}$, then $a_{n+1} = \dfrac{(2n+2)!}{(n+1)!(n+1)!}$ and

$$\left| \frac{a_{n+1}}{a_n} \right| = \frac{n!n!(2n+2)(2n+1)(2n)!}{(n+1)!(n+1)!(2n)!}$$

$$= \frac{(2n+2)(2n+1)}{(n+1)(n+1)} = \frac{4n+2}{n+1} \rightarrow 4.$$

The series diverges because $\rho = 4$ is greater than 1.

(c) If $a_n = 4^n n! n! / (2n)!$, then

$$\left| \frac{a_{n+1}}{a_n} \right| = \frac{4^{n+1}(n+1)!(n+1)!}{(2n+2)(2n+1)(2n)!} \cdot \frac{(2n)!}{4^n n! n!}$$

$$= \frac{4(n+1)(n+1)}{(2n+2)(2n+1)} = \frac{2(n+1)}{2n+1} \rightarrow 1.$$

Because the limit is $\rho = 1$, we cannot decide from the Ratio Test whether the series converges. However, when we notice that $a_{n+1}/a_n = (2n+2)/(2n+1)$, we conclude that a_{n+1} is always greater than a_n because $(2n+2)/(2n+1)$ is always greater than 1. Therefore, all terms are greater than or equal to $a_1 = 2$, and the nth term does not approach zero as $n \rightarrow \infty$. The series diverges. ∎

The Root Test

The convergence tests we have so far for Σa_n work best when the formula for a_n is relatively simple. However, consider the series with the terms

$$a_n = \begin{cases} n/2^n, & n \text{ odd} \\ 1/2^n, & n \text{ even.} \end{cases}$$

To investigate convergence we write out several terms of the series:

$$\sum_{n=1}^{\infty} a_n = \frac{1}{2^1} + \frac{1}{2^2} + \frac{3}{2^3} + \frac{1}{2^4} + \frac{5}{2^5} + \frac{1}{2^6} + \frac{7}{2^7} + \cdots$$

$$= \frac{1}{2} + \frac{1}{4} + \frac{3}{8} + \frac{1}{16} + \frac{5}{32} + \frac{1}{64} + \frac{7}{128} + \cdots.$$

Clearly, this is not a geometric series. The nth term approaches zero as $n \to \infty$, so the nth-Term Test does not tell us if the series diverges. The Integral Test does not look promising. The Ratio Test produces

$$\left| \frac{a_{n+1}}{a_n} \right| = \begin{cases} \dfrac{1}{2n}, & n \text{ odd} \\ \dfrac{n+1}{2}, & n \text{ even} \end{cases}$$

As $n \to \infty$, the ratio is alternately small and large and therefore has no limit. However, we will see that the following test establishes that the series converges.

THEOREM 14—The Root Test

Let $\sum a_n$ be any series and suppose that

$$\lim_{n \to \infty} \sqrt[n]{|a_n|} = \rho.$$

Then **(a)** the series *converges absolutely* if $\rho < 1$, **(b)** the series *diverges* if $\rho > 1$ or ρ is infinite, **(c)** the test is *inconclusive* if $\rho = 1$.

Proof

(a) $\rho < 1.$ Choose an $\varepsilon > 0$ so small that $\rho + \varepsilon < 1$. Since $\sqrt[n]{|a_n|} \to \rho$, the terms $\sqrt[n]{|a_n|}$ eventually get to within ε of ρ. So there exists an index M such that

$$\sqrt[n]{|a_n|} < \rho + \varepsilon \qquad \text{when } n \geq M.$$

Then it is also true that

$$|a_n| < (\rho + \varepsilon)^n \qquad \text{for } n \geq M.$$

Now, $\sum_{n=M}^{\infty} (\rho + \varepsilon)^n$ is a geometric series with ratio $(\rho + \varepsilon) < 1$ and therefore converges. By the Direct Comparison Test, $\sum_{n=M}^{\infty} |a_n|$ converges. Adding finitely many terms to a series does not affect its convergence or divergence, so the series

$$\sum_{n=1}^{\infty} |a_n| = |a_1| + \cdots + |a_{M-1}| + \sum_{n=M}^{\infty} |a_n|$$

also converges. Therefore, Σa_n converges absolutely.

(b) $1 < \rho \leq \infty.$ For all indices beyond some integer M, we have $\sqrt[n]{|a_n|} > 1$, so that $|a_n| > 1$ for $n > M$. The terms of the series do not converge to zero. The series diverges by the nth-Term Test.

(c) $\rho = 1.$ The series $\sum_{n=1}^{\infty} (1/n)$ and $\sum_{n=1}^{\infty} (1/n^2)$ show that the test is not conclusive when $\rho = 1$. The first series diverges and the second converges, but in both cases $\sqrt[n]{|a_n|} \to 1$.   ∎

EXAMPLE 3   Consider again the series with terms $a_n = \begin{cases} n/2^n, & n \text{ odd} \\ 1/2^n, & n \text{ even}. \end{cases}$

Does Σa_n converge?

Solution   We apply the Root Test, finding that

$$\sqrt[n]{|a_n|} = \begin{cases} \sqrt[n]{n}/2, & n \text{ odd} \\ 1/2, & n \text{ even}. \end{cases}$$

Therefore,

$$\frac{1}{2} \le \sqrt[n]{|a_n|} \le \frac{\sqrt[n]{n}}{2}.$$

Since $\sqrt[n]{n} \to 1$ (Section 10.1, Theorem 5), we have $\lim_{n \to \infty} \sqrt[n]{|a_n|} = 1/2$ by the Sandwich Theorem. The limit is less than 1, so the series converges absolutely by the Root Test. ∎

EXAMPLE 4   Which of the following series converge, and which diverge?

(a) $\displaystyle\sum_{n=1}^{\infty} \frac{n^2}{2^n}$ (b) $\displaystyle\sum_{n=1}^{\infty} \frac{2^n}{n^3}$ (c) $\displaystyle\sum_{n=1}^{\infty} \left(\frac{1}{1+n}\right)^n$

Solution   We apply the Root Test to each series, noting that each series has positive terms.

(a) $\displaystyle\sum_{n=1}^{\infty} \frac{n^2}{2^n}$ converges because $\sqrt[n]{\frac{n^2}{2^n}} = \frac{\sqrt[n]{n^2}}{\sqrt[n]{2^n}} = \frac{(\sqrt[n]{n})^2}{2} \to \frac{1^2}{2} < 1.$

(b) $\displaystyle\sum_{n=1}^{\infty} \frac{2^n}{n^3}$ diverges because $\sqrt[n]{\frac{2^n}{n^3}} = \frac{2}{(\sqrt[n]{n})^3} \to \frac{2}{1^3} > 1.$

(c) $\displaystyle\sum_{n=1}^{\infty} \left(\frac{1}{1+n}\right)^n$ converges because $\sqrt[n]{\left(\frac{1}{1+n}\right)^n} = \frac{1}{1+n} \to 0 < 1.$ ∎

EXERCISES 10.5

Using the Ratio Test
In Exercises 1–8, use the Ratio Test to determine if each series converges absolutely or diverges.

1. $\displaystyle\sum_{n=1}^{\infty} \frac{2^n}{n!}$

2. $\displaystyle\sum_{n=1}^{\infty} (-1)^n \frac{n+2}{3^n}$

3. $\displaystyle\sum_{n=1}^{\infty} \frac{(n-1)!}{(n+1)^2}$

4. $\displaystyle\sum_{n=1}^{\infty} \frac{2^{n+1}}{n3^{n-1}}$

5. $\displaystyle\sum_{n=1}^{\infty} \frac{n^4}{(-4)^n}$

6. $\displaystyle\sum_{n=2}^{\infty} \frac{3^{n+2}}{\ln n}$

7. $\displaystyle\sum_{n=1}^{\infty} (-1)^n \frac{n^2(n+2)!}{n! \, 3^{2n}}$

8. $\displaystyle\sum_{n=1}^{\infty} \frac{n5^n}{(2n+3)\ln(n+1)}$

Using the Root Test
In Exercises 9–16, use the Root Test to determine if each series converges absolutely or diverges.

9. $\displaystyle\sum_{n=1}^{\infty} \frac{7}{(2n+5)^n}$

10. $\displaystyle\sum_{n=1}^{\infty} \frac{4^n}{(3n)^n}$

11. $\displaystyle\sum_{n=1}^{\infty} \left(\frac{4n+3}{3n-5}\right)^n$

12. $\displaystyle\sum_{n=1}^{\infty} \left(-\ln\left(e^2+\frac{1}{n}\right)\right)^{n+1}$

13. $\displaystyle\sum_{n=1}^{\infty} \frac{-8}{(3+(1/n))^{2n}}$

14. $\displaystyle\sum_{n=1}^{\infty} \sin^n\left(\frac{1}{\sqrt{n}}\right)$

15. $\displaystyle\sum_{n=1}^{\infty} (-1)^n \left(1-\frac{1}{n}\right)^{n^2}$

(*Hint:* $\lim_{n \to \infty} (1 + x/n)^n = e^x$)

16. $\displaystyle\sum_{n=2}^{\infty} \frac{(-1)^n}{n^{1+n}}$

Determining Convergence or Divergence
In Exercises 17–46, use any method to determine if the series converges or diverges. Give reasons for your answer.

17. $\displaystyle\sum_{n=1}^{\infty} \frac{n^{\sqrt{2}}}{2^n}$

18. $\displaystyle\sum_{n=1}^{\infty} (-1)^n n^2 e^{-n}$

19. $\displaystyle\sum_{n=1}^{\infty} n!(-e)^{-n}$

20. $\displaystyle\sum_{n=1}^{\infty} \frac{n!}{10^n}$

21. $\displaystyle\sum_{n=1}^{\infty} \frac{n^{10}}{10^n}$

22. $\displaystyle\sum_{n=1}^{\infty} \left(\frac{n-2}{n}\right)^n$

23. $\displaystyle\sum_{n=1}^{\infty} \frac{2 + (-1)^n}{1.25^n}$

24. $\displaystyle\sum_{n=1}^{\infty} \frac{(-2)^n}{3^n}$

25. $\displaystyle\sum_{n=1}^{\infty} (-1)^n \left(1 - \frac{3}{n}\right)^n$

26. $\displaystyle\sum_{n=1}^{\infty} \left(1 - \frac{1}{3n}\right)^n$

27. $\displaystyle\sum_{n=1}^{\infty} \frac{\ln n}{n^3}$

28. $\displaystyle\sum_{n=1}^{\infty} \frac{(-\ln n)^n}{n^n}$

29. $\displaystyle\sum_{n=1}^{\infty} \left(\frac{1}{n} - \frac{1}{n^2}\right)$

30. $\displaystyle\sum_{n=1}^{\infty} \left(\frac{1}{n} - \frac{1}{n^2}\right)^n$

31. $\displaystyle\sum_{n=1}^{\infty} \frac{e^n}{n^e}$

32. $\displaystyle\sum_{n=1}^{\infty} \frac{n \ln n}{(-2)^n}$

33. $\displaystyle\sum_{n=1}^{\infty} \frac{(n + 1)(n + 2)}{n!}$

34. $\displaystyle\sum_{n=1}^{\infty} e^{-n}(n^3)$

35. $\displaystyle\sum_{n=1}^{\infty} \frac{(n + 3)!}{3! n! 3^n}$

36. $\displaystyle\sum_{n=1}^{\infty} \frac{n 2^n (n + 1)!}{3^n n!}$

37. $\displaystyle\sum_{n=1}^{\infty} \frac{n!}{(2n + 1)!}$

38. $\displaystyle\sum_{n=1}^{\infty} \frac{n!}{(-n)^n}$

39. $\displaystyle\sum_{n=2}^{\infty} \frac{-n}{(\ln n)^n}$

40. $\displaystyle\sum_{n=2}^{\infty} \frac{n}{(\ln n)^{(n/2)}}$

41. $\displaystyle\sum_{n=1}^{\infty} \frac{n! \ln n}{n(n + 2)!}$

42. $\displaystyle\sum_{n=1}^{\infty} \frac{(-3)^n}{n^3 2^n}$

43. $\displaystyle\sum_{n=1}^{\infty} \frac{(n!)^2}{(2n)!}$

44. $\displaystyle\sum_{n=1}^{\infty} \frac{(2n + 3)(2^n + 3)}{3^n + 2}$

45. $\displaystyle\sum_{n=3}^{\infty} \frac{2^n}{n^2}$

46. $\displaystyle\sum_{n=3}^{\infty} \frac{2^{n^2}}{n^{2^n}}$

Recursively Defined Terms Which of the series $\sum_{n=1}^{\infty} a_n$ defined by the formulas in Exercises 47–56 converge, and which diverge? Give reasons for your answers.

47. $a_1 = 2, \quad a_{n+1} = \dfrac{1 + \sin n}{n} a_n$

48. $a_1 = 1, \quad a_{n+1} = \dfrac{1 + \tan^{-1} n}{n} a_n$

49. $a_1 = \dfrac{1}{3}, \quad a_{n+1} = \dfrac{3n - 1}{2n + 5} a_n$

50. $a_1 = 3, \quad a_{n+1} = \dfrac{n}{n + 1} a_n$

51. $a_1 = 2, \quad a_{n+1} = \dfrac{2}{n} a_n$

52. $a_1 = 5, \quad a_{n+1} = \dfrac{\sqrt[n]{n}}{2} a_n$

53. $a_1 = 1, \quad a_{n+1} = \dfrac{1 + \ln n}{n} a_n$

54. $a_1 = \dfrac{1}{2}, \quad a_{n+1} = \dfrac{n + \ln n}{n + 10} a_n$

55. $a_1 = \dfrac{1}{3}, \quad a_{n+1} = \sqrt[n]{a_n}$

56. $a_1 = \dfrac{1}{2}, \quad a_{n+1} = (a_n)^{n+1}$

Convergence or Divergence Which of the series in Exercises 57–64 converge, and which diverge? Give reasons for your answers.

57. $\displaystyle\sum_{n=1}^{\infty} \frac{2^n n! n!}{(2n)!}$

58. $\displaystyle\sum_{n=1}^{\infty} \frac{(-1)^n (3n)!}{n! (n + 1)! (n + 2)!}$

59. $\displaystyle\sum_{n=1}^{\infty} \frac{(n!)^n}{(n^n)^2}$

60. $\displaystyle\sum_{n=1}^{\infty} (-1)^n \frac{(n!)^n}{n^{(n^2)}}$

61. $\displaystyle\sum_{n=1}^{\infty} \frac{n^n}{2^{(n^2)}}$

62. $\displaystyle\sum_{n=1}^{\infty} \frac{n^n}{(2^n)^2}$

63. $\displaystyle\sum_{n=1}^{\infty} \frac{1 \cdot 3 \cdot \cdots \cdot (2n - 1)}{4^n 2^n n!}$

64. $\displaystyle\sum_{n=1}^{\infty} \frac{1 \cdot 3 \cdot \cdots \cdot (2n - 1)}{[2 \cdot 4 \cdot \cdots \cdot (2n)](3^n + 1)}$

65. Assume that b_n is a sequence of positive numbers converging to $4/5$. Determine if the following series converge or diverge.

    a. $\displaystyle\sum_{n=1}^{\infty} (b_n)^{1/n}$ **b.** $\displaystyle\sum_{n=1}^{\infty} \left(\frac{5}{4}\right)^n (b_n)$

    c. $\displaystyle\sum_{n=1}^{\infty} (b_n)^n$ **d.** $\displaystyle\sum_{n=1}^{\infty} \frac{1000^n}{n! + b_n}$

66. Assume that b_n is a sequence of positive numbers converging to $1/3$. Determine if the following series converge or diverge.

    a. $\displaystyle\sum_{n=1}^{\infty} \frac{b_{n+1} b_n}{n 4^n}$ **b.** $\displaystyle\sum_{n=1}^{\infty} \frac{n^n}{n! \, b_1^2 b_2^2 \cdots b_n^2}$

Theory and Examples

67. Neither the Ratio Test nor the Root Test helps with p-series. Try them on

$$\sum_{n=1}^{\infty} \frac{1}{n^p}$$

and show that both tests fail to provide information about convergence.

68. Show that neither the Ratio Test nor the Root Test provides information about the convergence of

$$\sum_{n=2}^{\infty} \frac{1}{(\ln n)^p} \qquad (p \text{ constant}).$$

69. Let $a_n = \begin{cases} n/2^n, & \text{if } n \text{ is a prime number} \\ 1/2^n, & \text{otherwise.} \end{cases}$

    Does $\sum a_n$ converge? Give reasons for your answer.

70. Show that $\sum_{n=1}^{\infty} 2^{(n^2)}/n!$ diverges. Recall from the Laws of Exponents that $2^{(n^2)} = (2^n)^n$.

10.6 Alternating Series and Conditional Convergence

A series in which the terms are alternately positive and negative is an **alternating series**. Here are three examples:

$$1 - \frac{1}{2} + \frac{1}{3} - \frac{1}{4} + \frac{1}{5} - \cdots + \frac{(-1)^{n+1}}{n} + \cdots \tag{1}$$

$$-2 + 1 - \frac{1}{2} + \frac{1}{4} - \frac{1}{8} + \cdots + \frac{(-1)^n 4}{2^n} + \cdots \tag{2}$$

$$1 - 2 + 3 - 4 + 5 - 6 + \cdots + (-1)^{n+1} n + \cdots \tag{3}$$

We see from these examples that the nth term of an alternating series is of the form

$$a_n = (-1)^{n+1} u_n \qquad \text{or} \qquad a_n = (-1)^n u_n$$

where $u_n = |a_n|$ is a positive number.

Series (1), called the **alternating harmonic series**, converges, as we will see in a moment. Series (2), a geometric series with ratio $r = -1/2$, converges to $-2/[1 + (1/2)] = -4/3$. Series (3) diverges because the nth term does not approach zero.

We prove the convergence of the alternating harmonic series by applying the Alternating Series Test. This test is for *convergence* of an alternating series and cannot be used to conclude that such a series diverges. If we multiply $(u_1 - u_2 + u_3 - u_4 + \cdots)$ by -1, we see that the test is also valid for the alternating series $-u_1 + u_2 - u_3 + u_4 - \cdots$, as with the one in Series (2) given above.

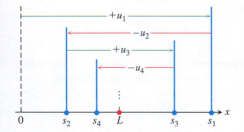

FIGURE 10.15 The partial sums of an alternating series that satisfies the hypotheses of Theorem 15 for $N = 1$ straddle the limit from the beginning.

THEOREM 15—The Alternating Series Test

The series

$$\sum_{n=1}^{\infty} (-1)^{n+1} u_n = u_1 - u_2 + u_3 - u_4 + \cdots$$

converges if the following conditions are satisfied:

1. The u_n's are all positive.

2. The u_n's are eventually nonincreasing: $u_n \geq u_{n+1}$ for all $n \geq N$, for some integer N.

3. $u_n \to 0$.

Proof We look at the case where $u_1, u_2, u_3, u_4, \ldots$ is nonincreasing, so that $N = 1$. If n is an even integer, say $n = 2m$, then the sum of the first n terms is

$$s_{2m} = (u_1 - u_2) + (u_3 - u_4) + \cdots + (u_{2m-1} - u_{2m})$$
$$= u_1 - (u_2 - u_3) - (u_4 - u_5) - \cdots - (u_{2m-2} - u_{2m-1}) - u_{2m}.$$

The first equality shows that s_{2m} is the sum of m nonnegative terms, since each term in parentheses is positive or zero. Hence $s_{2m+2} \geq s_{2m}$, and the sequence $\{s_{2m}\}$ is nondecreasing. The second equality shows that $s_{2m} \leq u_1$. Since $\{s_{2m}\}$ is nondecreasing and bounded from above, it has a limit, say

$$\lim_{m \to \infty} s_{2m} = L. \qquad \text{Theorem 6} \tag{4}$$

If n is an odd integer, say $n = 2m + 1$, then the sum of the first n terms is $s_{2m+1} = s_{2m} + u_{2m+1}$. Since $u_n \to 0$,

$$\lim_{m \to \infty} u_{2m+1} = 0$$

and, as $m \to \infty$,

$$s_{2m+1} = s_{2m} + u_{2m+1} \to L + 0 = L. \tag{5}$$

Combining the results of Equations (4) and (5) gives $\lim_{n \to \infty} s_n = L$ (Section 10.1, Exercise 143). ∎

EXAMPLE 1   The alternating harmonic series

$$\sum_{n=1}^{\infty} (-1)^{n+1} \frac{1}{n} = 1 - \frac{1}{2} + \frac{1}{3} - \frac{1}{4} + \cdots$$

clearly satisfies the three requirements of Theorem 15 with $N = 1$; it therefore converges by the Alternating Series Test. Notice that the test gives no information about what the sum of the series might be. Figure 10.16 shows histograms of the partial sums of the divergent harmonic series and those of the convergent alternating harmonic series. It turns out that the alternating harmonic series converges to $\ln 2$. ∎

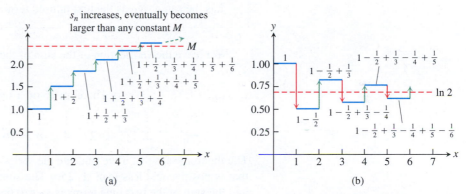

FIGURE 10.16   (a) The harmonic series diverges, with partial sums that eventually exceed any constant. (b) The alternating harmonic series converges to $\ln 2 \approx .693$.

Rather than directly verifying the definition $u_n \geq u_{n+1}$, a second way to show that the sequence $\{u_n\}$ is nonincreasing is to define a differentiable function $f(x)$ satisfying $f(n) = u_n$. That is, the values of f match the values of the sequence at every positive integer n. If $f'(x) \leq 0$ for all x greater than or equal to some positive integer N, then $f(x)$ is nonincreasing for $x \geq N$. It follows that $f(n) \geq f(n + 1)$, or $u_n \geq u_{n+1}$, for $n \geq N$.

EXAMPLE 2   We show that the sequence $u_n = 10n/(n^2 + 16)$ is eventually nonincreasing. Define $f(x) = 10x/(x^2 + 16)$. Then from the Derivative Quotient Rule,

$$f'(x) = \frac{10(16 - x^2)}{(x^2 + 16)^2} \leq 0 \qquad \text{whenever } x \geq 4.$$

It follows that $u_n \geq u_{n+1}$ for $n \geq 4$. That is, the sequence $\{u_n\}$ is nonincreasing for $n \geq 4$. ∎

A graphical interpretation of the partial sums (Figure 10.15) shows how an alternating series converges to its limit L when the three conditions of Theorem 15 are satisfied with $N = 1$. Starting from the origin of the x-axis, we lay off the positive distance $s_1 = u_1$. To find the point corresponding to $s_2 = u_1 - u_2$, we back up a distance equal to u_2. Since $u_2 \leq u_1$, we do not back up any farther than the origin. We continue in this seesaw fashion, backing up or going forward as the signs in the series demand. But for $n \geq N$, each forward or backward step is shorter than (or at most the same size as) the preceding step because $u_{n+1} \leq u_n$. And since the nth term approaches zero as n increases, the size of step we take forward or backward gets smaller and smaller. We oscillate back and forth across the limit L, and the amplitude of oscillation approaches zero. The limit L lies between any two successive sums s_n and s_{n+1} and hence differs from s_n by an amount less than u_{n+1}.

Because

$$|L - s_n| < u_{n+1} \qquad \text{for } n \geq N,$$

we can make useful estimates of the sums of convergent alternating series.

THEOREM 16—The Alternating Series Estimation Theorem

If the alternating series $\sum_{n=1}^{\infty}(-1)^{n+1}u_n$ satisfies the three conditions of Theorem 15, then for $n \geq N$,

$$s_n = u_1 - u_2 + \cdots + (-1)^{n+1}u_n$$

approximates the sum L of the series with an error whose absolute value is less than u_{n+1}, the absolute value of the first unused term. Furthermore, the sum L lies between any two successive partial sums s_n and s_{n+1}, and the remainder, $L - s_n$, has the same sign as the first unused term.

We leave the verification of the sign of the remainder for Exercise 87.

EXAMPLE 3   We try Theorem 16 on a series whose sum we know:

$$\sum_{n=0}^{\infty}(-1)^n \frac{1}{2^n} = 1 - \frac{1}{2} + \frac{1}{4} - \frac{1}{8} + \frac{1}{16} - \frac{1}{32} + \frac{1}{64} - \frac{1}{128} + \frac{1}{256} - \cdots.$$

The theorem says that if we truncate the series after the eighth term, we throw away a total that is positive and less than $1/256$. The sum of the first eight terms is $s_8 = 0.6640625$ and the sum of the first nine terms is $s_9 = 0.66796875$. The sum of the geometric series is

$$\frac{1}{1 - (-1/2)} = \frac{1}{3/2} = \frac{2}{3},$$

and we note that $0.6640625 < (2/3) < 0.66796875$. The difference, $(2/3) - 0.6640625 = 0.0026041666\ldots$, is positive and is less than $(1/256) = 0.00390625$. ∎

Conditional Convergence

If we replace all the negative terms in the alternating series in Example 3, changing them to positive terms instead, we obtain the geometric series $\sum 1/2^n$. The original series and the new series of absolute values both converge (although to different sums). For an absolutely convergent series, changing infinitely many of the negative terms in the series to positive values does not change its property of still being a convergent series. Other convergent series may behave differently. The convergent alternating harmonic series has infinitely many negative terms, but if we change its negative terms to positive values, the resulting series is the divergent harmonic series. So the presence of infinitely many negative terms is essential to the convergence of the alternating harmonic series. The following terminology distinguishes these two types of convergent series.

DEFINITION   A series that is convergent but not absolutely convergent is called **conditionally convergent**.

The alternating harmonic series is conditionally convergent, or **converges conditionally**. The next example extends that result to the alternating p-series.

EXAMPLE 4   If p is a positive constant, the sequence $\{1/n^p\}$ is a decreasing sequence with limit zero. Therefore, the alternating p-series

$$\sum_{n=1}^{\infty} \frac{(-1)^{n-1}}{n^p} = 1 - \frac{1}{2^p} + \frac{1}{3^p} - \frac{1}{4^p} + \cdots, \quad p > 0$$

converges.

If $p > 1$, the series converges absolutely as an ordinary p-series. If $0 < p \leq 1$, the series converges conditionally by the alternating series test. For instance,

Absolute convergence $\left(p = 3/2\right)$: $1 - \dfrac{1}{2^{3/2}} + \dfrac{1}{3^{3/2}} - \dfrac{1}{4^{3/2}} + \cdots$

Conditional convergence $\left(p = 1/2\right)$: $1 - \dfrac{1}{\sqrt{2}} + \dfrac{1}{\sqrt{3}} - \dfrac{1}{\sqrt{4}} + \cdots$ ∎

We need to be careful when using a conditionally convergent series. We have seen with the alternating harmonic series that altering the signs of infinitely many terms of a conditionally convergent series can change its convergence status. Even more, simply changing the order of occurrence of infinitely many of its terms can also have a significant effect, as we now discuss.

Rearranging Series

We can always rearrange the terms of a *finite* collection of numbers without changing their sum. The same result is true for an infinite series that is absolutely convergent (see Exercise 96 for an outline of the proof).

THEOREM 17—The Rearrangement Theorem for Absolutely Convergent Series

If $\sum_{n=1}^{\infty} a_n$ converges absolutely, and $b_1, b_2, \ldots, b_n, \ldots$ is any arrangement of the sequence $\{a_n\}$, then $\sum b_n$ converges absolutely and

$$\sum_{n=1}^{\infty} b_n = \sum_{n=1}^{\infty} a_n.$$

On the other hand, if we rearrange the terms of a conditionally convergent series, we can get different results. In fact, for any real number r, a given conditionally convergent series can be rearranged so its sum is equal to r. (We omit the proof of this fact.) Here's an example of summing the terms of a conditionally convergent series with different orderings, with each ordering giving a different value for the sum.

EXAMPLE 5   We know that the alternating harmonic series $\sum_{n=1}^{\infty} (-1)^{n+1}/n$ converges to some number L. Moreover, by Theorem 16, L lies between the successive partial sums $s_2 = 1/2$ and $s_3 = 5/6$, so $L \neq 0$. If we multiply the series by 2 we obtain

$$2L = 2 \sum_{n=1}^{\infty} \frac{(-1)^{n+1}}{n} = 2\left(1 - \frac{1}{2} + \frac{1}{3} - \frac{1}{4} + \frac{1}{5} - \frac{1}{6} + \frac{1}{7} - \frac{1}{8} + \frac{1}{9} - \frac{1}{10} + \frac{1}{11} - \cdots\right)$$

$$= 2 - 1 + \frac{2}{3} - \frac{1}{2} + \frac{2}{5} - \frac{1}{3} + \frac{2}{7} - \frac{1}{4} + \frac{2}{9} - \frac{1}{5} + \frac{2}{11} - \cdots.$$

Now we change the order of this last sum by grouping each pair of terms with the same odd denominator, but leaving the negative terms with the even denominators as they are placed (so the denominators are the positive integers in their natural order). This rearrangement gives

$$(2 - 1) - \frac{1}{2} + \left(\frac{2}{3} - \frac{1}{3}\right) - \frac{1}{4} + \left(\frac{2}{5} - \frac{1}{5}\right) - \frac{1}{6} + \left(\frac{2}{7} - \frac{1}{7}\right) - \frac{1}{8} + \cdots$$

$$= \left(1 - \frac{1}{2} + \frac{1}{3} - \frac{1}{4} + \frac{1}{5} - \frac{1}{6} + \frac{1}{7} - \frac{1}{8} + \frac{1}{9} - \frac{1}{10} + \frac{1}{11} - \cdots\right)$$

$$= \sum_{n=1}^{\infty} \frac{(-1)^{n+1}}{n} = L.$$

So by rearranging the terms of the conditionally convergent series $\sum_{n=1}^{\infty} 2(-1)^{n+1}/n$, the series becomes $\sum_{n=1}^{\infty} (-1)^{n+1}/n$, which is the alternating harmonic series itself. If the two series are the same, it would imply that $2L = L$, which is clearly false since $L \neq 0$.   ∎

Example 5 shows that we cannot rearrange the terms of a conditionally convergent series and expect the new series to be the same as the original one. When we use a conditionally convergent series, the terms must be added together in the order they are given to obtain a correct result. In contrast, Theorem 17 guarantees that the terms of an absolutely convergent series can be summed in any order without affecting the result.

Summary of Tests to Determine Convergence or Divergence

We have developed a variety of tests to determine convergence or divergence for an infinite series of constants. There are other tests we have not presented which are sometimes given in more advanced courses. Here is a summary of the tests we have considered.

---

1. **The nth-Term Test for Divergence:** Unless $a_n \to 0$, the series diverges.

2. **Geometric series:** $\sum ar^n$ converges if $|r| < 1$; otherwise it diverges.

3. **p-series:** $\sum 1/n^p$ converges if $p > 1$; otherwise it diverges.

4. **Series with nonnegative terms:** Try the Integral Test or try comparing to a known series with the Direct Comparison Test or the Limit Comparison Test. Try the Ratio or Root Test.

5. **Series with some negative terms:** Does $\sum |a_n|$ converge by the Ratio or Root Test, or by another of the tests listed above? Remember, absolute convergence implies convergence.

6. **Alternating series:** $\sum a_n$ converges if the series satisfies the conditions of the Alternating Series Test.

EXERCISES   10.6

Determining Convergence or Divergence

In Exercises 1–14, determine if the alternating series converges or diverges. Some of the series do not satisfy the conditions of the Alternating Series Test.

1. $\displaystyle\sum_{n=1}^{\infty} (-1)^{n+1} \frac{1}{\sqrt{n}}$

2. $\displaystyle\sum_{n=1}^{\infty} (-1)^{n+1} \frac{1}{n^{3/2}}$

3. $\displaystyle\sum_{n=1}^{\infty} (-1)^{n+1} \frac{1}{n3^n}$

4. $\displaystyle\sum_{n=2}^{\infty} (-1)^n \frac{4}{(\ln n)^2}$

5. $\displaystyle\sum_{n=1}^{\infty} (-1)^n \frac{n}{n^2 + 1}$

6. $\displaystyle\sum_{n=1}^{\infty} (-1)^{n+1} \frac{n^2 + 5}{n^2 + 4}$

7. $\displaystyle\sum_{n=1}^{\infty} (-1)^{n+1} \frac{2^n}{n^2}$

8. $\displaystyle\sum_{n=1}^{\infty} (-1)^n \frac{10^n}{(n + 1)!}$

9. $\displaystyle\sum_{n=1}^{\infty} (-1)^{n+1} \left(\frac{n}{10}\right)^n$

10. $\displaystyle\sum_{n=2}^{\infty} (-1)^{n+1} \frac{1}{\ln n}$

11. $\displaystyle\sum_{n=1}^{\infty}(-1)^{n+1}\frac{\ln n}{n}$

12. $\displaystyle\sum_{n=1}^{\infty}(-1)^{n}\ln\left(1+\frac{1}{n}\right)$

13. $\displaystyle\sum_{n=1}^{\infty}(-1)^{n+1}\frac{\sqrt{n}+1}{n+1}$

14. $\displaystyle\sum_{n=1}^{\infty}(-1)^{n+1}\frac{3\sqrt{n}+1}{\sqrt{n}+1}$

Absolute and Conditional Convergence

Which of the series in Exercises 15–48 converge absolutely, which converge, and which diverge? Give reasons for your answers.

15. $\displaystyle\sum_{n=1}^{\infty}(-1)^{n+1}(0.1)^{n}$

16. $\displaystyle\sum_{n=1}^{\infty}(-1)^{n+1}\frac{(0.1)^{n}}{n}$

17. $\displaystyle\sum_{n=1}^{\infty}(-1)^{n}\frac{1}{\sqrt{n}}$

18. $\displaystyle\sum_{n=1}^{\infty}\frac{(-1)^{n}}{1+\sqrt{n}}$

19. $\displaystyle\sum_{n=1}^{\infty}(-1)^{n+1}\frac{n}{n^{3}+1}$

20. $\displaystyle\sum_{n=1}^{\infty}(-1)^{n+1}\frac{n!}{2^{n}}$

21. $\displaystyle\sum_{n=1}^{\infty}(-1)^{n}\frac{1}{n+3}$

22. $\displaystyle\sum_{n=1}^{\infty}(-1)^{n}\frac{\sin n}{n^{2}}$

23. $\displaystyle\sum_{n=1}^{\infty}(-1)^{n+1}\frac{3+n}{5+n}$

24. $\displaystyle\sum_{n=1}^{\infty}\frac{(-2)^{n+1}}{n+5^{n}}$

25. $\displaystyle\sum_{n=1}^{\infty}(-1)^{n+1}\frac{1+n}{n^{2}}$

26. $\displaystyle\sum_{n=1}^{\infty}(-1)^{n+1}\left(\sqrt[n]{10}\right)$

27. $\displaystyle\sum_{n=1}^{\infty}(-1)^{n}n^{2}(2/3)^{n}$

28. $\displaystyle\sum_{n=2}^{\infty}(-1)^{n+1}\frac{1}{n\ln n}$

29. $\displaystyle\sum_{n=1}^{\infty}(-1)^{n}\frac{\tan^{-1}n}{n^{2}+1}$

30. $\displaystyle\sum_{n=1}^{\infty}(-1)^{n}\frac{\ln n}{n-\ln n}$

31. $\displaystyle\sum_{n=1}^{\infty}(-1)^{n}\frac{n}{n+1}$

32. $\displaystyle\sum_{n=1}^{\infty}(-5)^{-n}$

33. $\displaystyle\sum_{n=1}^{\infty}\frac{(-100)^{n}}{n!}$

34. $\displaystyle\sum_{n=1}^{\infty}\frac{(-1)^{n-1}}{n^{2}+2n+1}$

35. $\displaystyle\sum_{n=1}^{\infty}\frac{\cos n\pi}{n\sqrt{n}}$

36. $\displaystyle\sum_{n=1}^{\infty}\frac{\cos n\pi}{n}$

37. $\displaystyle\sum_{n=1}^{\infty}\frac{(-1)^{n}(n+1)^{n}}{(2n)^{n}}$

38. $\displaystyle\sum_{n=1}^{\infty}\frac{(-1)^{n+1}(n!)^{2}}{(2n)!}$

39. $\displaystyle\sum_{n=1}^{\infty}(-1)^{n}\frac{(2n)!}{2^{n}n!n}$

40. $\displaystyle\sum_{n=1}^{\infty}(-1)^{n}\frac{(n!)^{2}3^{n}}{(2n+1)!}$

41. $\displaystyle\sum_{n=1}^{\infty}(-1)^{n}\left(\sqrt{n+1}-\sqrt{n}\right)$
42. $\displaystyle\sum_{n=1}^{\infty}(-1)^{n}\left(\sqrt{n^{2}+n}-n\right)$

43. $\displaystyle\sum_{n=1}^{\infty}(-1)^{n}\left(\sqrt{n+\sqrt{n}}-\sqrt{n}\right)$

44. $\displaystyle\sum_{n=1}^{\infty}\frac{(-1)^{n}}{\sqrt{n}+\sqrt{n+1}}$

45. $\displaystyle\sum_{n=1}^{\infty}(-1)^{n}\operatorname{sech}n$

46. $\displaystyle\sum_{n=1}^{\infty}(-1)^{n}\operatorname{csch}n$

47. $\dfrac{1}{4}-\dfrac{1}{6}+\dfrac{1}{8}-\dfrac{1}{10}+\dfrac{1}{12}-\dfrac{1}{14}+\cdots$

48. $1+\dfrac{1}{4}-\dfrac{1}{9}-\dfrac{1}{16}+\dfrac{1}{25}+\dfrac{1}{36}-\dfrac{1}{49}-\dfrac{1}{64}+\cdots$

Error Estimation

In Exercises 49–52, estimate the magnitude of the error involved in using the sum of the first four terms to approximate the sum of the entire series.

49. $\displaystyle\sum_{n=1}^{\infty}(-1)^{n+1}\frac{1}{n}$

50. $\displaystyle\sum_{n=1}^{\infty}(-1)^{n+1}\frac{1}{10^{n}}$

51. $\displaystyle\sum_{n=1}^{\infty}(-1)^{n+1}\frac{(0.01)^{n}}{n}$ As you will see in Section 10.7, the sum is ln (1.01).

52. $\dfrac{1}{1+t}=\displaystyle\sum_{n=0}^{\infty}(-1)^{n}t^{n},\quad 0<t<1$

In Exercises 53–56, determine how many terms should be used to estimate the sum of the entire series with an error of less than 0.001.

53. $\displaystyle\sum_{n=1}^{\infty}(-1)^{n}\frac{1}{n^{2}+3}$

54. $\displaystyle\sum_{n=1}^{\infty}(-1)^{n+1}\frac{n}{n^{2}+1}$

55. $\displaystyle\sum_{n=1}^{\infty}(-1)^{n+1}\frac{1}{\left(n+3\sqrt{n}\right)^{3}}$

56. $\displaystyle\sum_{n=1}^{\infty}(-1)^{n}\frac{1}{\ln(\ln(n+2))}$

In Exercises 57–82, use any method to determine whether the series converges or diverges. Give reasons for your answer.

57. $\displaystyle\sum_{n=1}^{\infty}\frac{3^{n}}{n^{n}}$

58. $\displaystyle\sum_{n=1}^{\infty}\frac{3^{n}}{n^{3}}$

59. $\displaystyle\sum_{n=1}^{\infty}\left(\frac{1}{n+2}-\frac{1}{n+3}\right)$

60. $\displaystyle\sum_{n=1}^{\infty}\left(\frac{1}{2n+1}-\frac{1}{2n+2}\right)$

61. $\displaystyle\sum_{n=0}^{\infty}(-1)^{n}\frac{(n+2)!}{(2n)!}$

62. $\displaystyle\sum_{n=2}^{\infty}\frac{(3n)!}{(n!)^{3}}$

63. $\displaystyle\sum_{n=1}^{\infty}n^{-2/\sqrt{5}}$

64. $\displaystyle\sum_{n=2}^{\infty}\frac{3}{10+n^{4/3}}$

65. $\displaystyle\sum_{n=1}^{\infty}\left(1-\frac{2}{n}\right)^{n^{2}}$

66. $\displaystyle\sum_{n=0}^{\infty}\left(\frac{n+1}{n+2}\right)^{n}$

67. $\displaystyle\sum_{n=1}^{\infty}\frac{n-2}{n^{2}+3n}\left(-\frac{2}{3}\right)^{n}$

68. $\displaystyle\sum_{n=0}^{\infty}\frac{n+1}{(n+2)!}\left(\frac{3}{2}\right)^{n}$

69. $\dfrac{1}{2}-\dfrac{1}{2}+\dfrac{1}{2}-\dfrac{1}{2}+\dfrac{1}{2}-\dfrac{1}{2}+\cdots$

70. $1-\dfrac{1}{8}+\dfrac{1}{64}-\dfrac{1}{512}+\dfrac{1}{4096}-\cdots$

71. $\displaystyle\sum_{n=3}^{\infty}\sin\left(\frac{1}{\sqrt{n}}\right)$

72. $\displaystyle\sum_{n=1}^{\infty}\tan(n^{1/n})$

73. $\displaystyle\sum_{n=2}^{\infty}\frac{n}{\ln n}$

74. $\displaystyle\sum_{n=2}^{\infty}\frac{1}{n\sqrt{\ln n}}$

75. $\displaystyle\sum_{n=2}^{\infty}\ln\left(\frac{n+2}{n+1}\right)$

76. $\displaystyle\sum_{n=2}^{\infty}\left(\frac{\ln n}{n}\right)^{3}$

77. $\displaystyle\sum_{n=2}^{\infty}\frac{1}{1+2+2^{2}+\cdots+2^{n}}$

78. $\displaystyle\sum_{n=2}^{\infty}\frac{1+3+3^{2}+\cdots+3^{n-1}}{1+2+3+\cdots+n}$

79. $\displaystyle\sum_{n=0}^{\infty} (-1)^n \frac{e^n}{e^n + e^{n^2}}$

80. $\displaystyle\sum_{n=0}^{\infty} \frac{(2n+3)(2^n+3)}{3^n+2}$

81. $\displaystyle\sum_{n=1}^{\infty} \frac{n^2 3^n}{3 \cdot 5 \cdot 7 \cdots (2n+1)}$

82. $\displaystyle\sum_{n=1}^{\infty} \frac{4 \cdot 6 \cdot 8 \cdots (2n)}{5^{n+1}(n+2)!}$

T Approximate the sums in Exercises 83 and 84 with an error of magnitude less than 5×10^{-6}.

83. $\displaystyle\sum_{n=0}^{\infty} (-1)^n \frac{1}{(2n)!}$ As you will see in Section 10.9, the sum is $\cos 1$, the cosine of 1 radian.

84. $\displaystyle\sum_{n=0}^{\infty} (-1)^n \frac{1}{n!}$ As you will see in Section 10.9 the sum is e^{-1}.

Theory and Examples

85. a. The series

$$\frac{1}{3} - \frac{1}{2} + \frac{1}{9} - \frac{1}{4} + \frac{1}{27} - \frac{1}{8} + \cdots + \frac{1}{3^n} - \frac{1}{2^n} + \cdots$$

does not meet one of the conditions of Theorem 14. Which one?

b. Use Theorem 17 to find the sum of the series in part (a).

T **86.** The limit L of an alternating series that satisfies the conditions of Theorem 15 lies between the values of any two consecutive partial sums. This suggests using the average

$$\frac{s_n + s_{n+1}}{2} = s_n + \frac{1}{2}(-1)^{n+2} a_{n+1}$$

to estimate L. Compute

$$s_{20} + \frac{1}{2} \cdot \frac{1}{21}$$

as an approximation to the sum of the alternating harmonic series. The exact sum is $\ln 2 = 0.69314718\ldots$.

87. The sign of the remainder of an alternating series that satisfies the conditions of Theorem 15 Prove the assertion in Theorem 16 that whenever an alternating series satisfying the conditions of Theorem 15 is approximated with one of its partial sums, then the remainder (sum of the unused terms) has the same sign as the first unused term. (*Hint:* Group the remainder's terms in consecutive pairs.)

88. Show that the sum of the first $2n$ terms of the series

$$1 - \frac{1}{2} + \frac{1}{2} - \frac{1}{3} + \frac{1}{3} - \frac{1}{4} + \frac{1}{4} - \frac{1}{5} + \frac{1}{5} - \frac{1}{6} + \cdots$$

is the same as the sum of the first n terms of the series

$$\frac{1}{1 \cdot 2} + \frac{1}{2 \cdot 3} + \frac{1}{3 \cdot 4} + \frac{1}{4 \cdot 5} + \frac{1}{5 \cdot 6} + \cdots.$$

Do these series converge? What is the sum of the first $2n+1$ terms of the first series? If the series converge, what is their sum?

89. Show that if $\sum_{n=1}^{\infty} a_n$ diverges, then $\sum_{n=1}^{\infty} |a_n|$ diverges.

90. Show that if $\sum_{n=1}^{\infty} a_n$ converges absolutely, then

$$\left| \sum_{n=1}^{\infty} a_n \right| \leq \sum_{n=1}^{\infty} |a_n|.$$

91. Show that if $\sum_{n=1}^{\infty} a_n$ and $\sum_{n=1}^{\infty} b_n$ both converge absolutely, then so do the following.

a. $\displaystyle\sum_{n=1}^{\infty} (a_n + b_n)$

b. $\displaystyle\sum_{n=1}^{\infty} (a_n - b_n)$

c. $\displaystyle\sum_{n=1}^{\infty} k a_n$ (k any number)

92. Show by example that $\sum_{n=1}^{\infty} a_n b_n$ may diverge even if $\sum_{n=1}^{\infty} a_n$ and $\sum_{n=1}^{\infty} b_n$ both converge.

93. If $\sum a_n$ converges absolutely, prove that $\sum a_n^2$ converges.

94. Does the series

$$\sum_{n=1}^{\infty} \left(\frac{1}{n} - \frac{1}{n^2} \right)$$

converge or diverge? Justify your answer.

T **95.** In the alternating harmonic series, suppose the goal is to arrange the terms to get a new series that converges to $-1/2$. Start the new arrangement with the first negative term, which is $-1/2$. Whenever you have a sum that is less than or equal to $-1/2$, start introducing positive terms, taken in order, until the new total is greater than $-1/2$. Then add negative terms until the total is less than or equal to $-1/2$ again. Continue this process until your partial sums have been above the target at least three times and finish at or below it. If s_n is the sum of the first n terms of your new series, plot the points (n, s_n) to illustrate how the sums are behaving.

96. Outline of the proof of the Rearrangement Theorem (Theorem 17)

a. Let ε be a positive real number, let $L = \sum_{n=1}^{\infty} a_n$, and let $s_k = \sum_{n=1}^{k} a_n$. Show that for some index N_1 and for some index $N_2 \geq N_1$,

$$\sum_{n=N_1}^{\infty} |a_n| < \frac{\varepsilon}{2} \quad \text{and} \quad |s_{N_2} - L| < \frac{\varepsilon}{2}.$$

Since all the terms $a_1, a_2, \ldots, a_{N_2}$ appear somewhere in the sequence $\{b_n\}$, there is an index $N_3 \geq N_2$ such that if $n \geq N_3$, then $\left(\sum_{k=1}^{n} b_k \right) - s_{N_2}$ is at most a sum of terms a_m with $m \geq N_1$. Therefore, if $n \geq N_3$,

$$\left| \sum_{k=1}^{n} b_k - L \right| \leq \left| \sum_{k=1}^{n} b_k - s_{N_2} \right| + |s_{N_2} - L|$$

$$\leq \sum_{k=N_1}^{\infty} |a_k| + |s_{N_2} - L| < \varepsilon.$$

b. The argument in part (a) shows that if $\sum_{n=1}^{\infty} a_n$ converges absolutely then $\sum_{n=1}^{\infty} b_n$ converges and $\sum_{n=1}^{\infty} b_n = \sum_{n=1}^{\infty} a_n$. Now show that because $\sum_{n=1}^{\infty} a_n$ converges, $\sum_{n=1}^{\infty} b_n$ converges to $\sum_{n=1}^{\infty} a_n$.

10.7 Power Series

Now that we can test many infinite series of numbers for convergence, we can study sums that look like "infinite polynomials." We call these sums *power series* because they are defined as infinite series of powers of some variable, in our case x. Like polynomials, power series can be added, subtracted, multiplied, differentiated, and integrated to give new power series. With power series we can extend the methods of calculus to a vast array of functions, making the techniques of calculus applicable in an even wider setting.

Power Series and Convergence

We begin with the formal definition, which specifies the notation and terminology used for power series.

DEFINITIONS **A power series about $x = 0$** is a series of the form

$$\sum_{n=0}^{\infty} c_n x^n = c_0 + c_1 x + c_2 x^2 + \cdots + c_n x^n + \cdots. \tag{1}$$

A power series about $x = a$ is a series of the form

$$\sum_{n=0}^{\infty} c_n (x - a)^n = c_0 + c_1(x - a) + c_2(x - a)^2 + \cdots + c_n(x - a)^n + \cdots \tag{2}$$

in which the **center** a and the **coefficients** $c_0, c_1, c_2, \ldots, c_n, \ldots$ are constants.

Equation (1) is the special case obtained by taking $a = 0$ in Equation (2). We will see that a power series defines a function $f(x)$ on a certain interval where it converges. Moreover, this function will be shown to be continuous and differentiable over the interior of that interval.

EXAMPLE 1 Taking all the coefficients to be 1 in Equation (1) gives the geometric power series

$$\sum_{n=0}^{\infty} x^n = 1 + x + x^2 + \cdots + x^n + \cdots.$$

This is the geometric series with first term 1 and ratio x. It converges to $1/(1 - x)$ for $|x| < 1$. We express this fact by writing

$$\frac{1}{1 - x} = 1 + x + x^2 + \cdots + x^n + \cdots, \qquad -1 < x < 1. \tag{3}$$

Power Series for $\dfrac{1}{1 - x}$

$$\frac{1}{1 - x} = \sum_{n=0}^{\infty} x^n, \quad |x| < 1$$

Up to now, we have used Equation (3) as a formula for the sum of the series on the right. We now change the focus: We think of the partial sums of the series on the right as polynomials $P_n(x)$ that approximate the function on the left. For values of x near zero, we need take only a few terms of the series to get a good approximation. As we move toward $x = 1$, or -1, we must take more terms. Figure 10.17 shows the graphs of $f(x) = 1/(1 - x)$ and the approximating polynomials $y_n = P_n(x)$ for $n = 0, 1, 2$, and 8. The function $f(x) = 1/(1 - x)$ is not continuous on intervals containing $x = 1$, where it has a vertical asymptote. The approximations do not apply when $x \geq 1$.

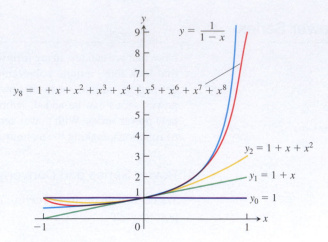

FIGURE 10.17 The graphs of $f(x) = 1/(1 - x)$ in Example 1 and four of its polynomial approximations.

EXAMPLE 2 The power series

$$1 - \frac{1}{2}(x - 2) + \frac{1}{4}(x - 2)^2 + \cdots + \left(-\frac{1}{2}\right)^n (x - 2)^n + \cdots \qquad (4)$$

matches Equation (2) with $a = 2$, $c_0 = 1$, $c_1 = -1/2$, $c_2 = 1/4, \ldots, c_n = (-1/2)^n$. This is a geometric series with first term 1 and ratio $r = -\dfrac{x - 2}{2}$. The series converges for $\left|\dfrac{x - 2}{2}\right| < 1$, which simplifies to $0 < x < 4$. The sum is

$$\frac{1}{1 - r} = \frac{1}{1 + \dfrac{x - 2}{2}} = \frac{2}{x},$$

so

$$\frac{2}{x} = 1 - \frac{(x - 2)}{2} + \frac{(x - 2)^2}{4} - \cdots + \left(-\frac{1}{2}\right)^n (x - 2)^n + \cdots, \qquad 0 < x < 4.$$

Series (4) generates useful polynomial approximations of $f(x) = 2/x$ for values of x near 2:

$$P_0(x) = 1$$

$$P_1(x) = 1 - \frac{1}{2}(x - 2) = 2 - \frac{x}{2}$$

$$P_2(x) = 1 - \frac{1}{2}(x - 2) + \frac{1}{4}(x - 2)^2 = 3 - \frac{3x}{2} + \frac{x^2}{4},$$

and so on (Figure 10.18). ∎

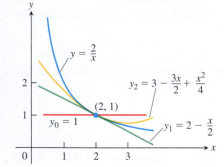

FIGURE 10.18 The graphs of $f(x) = 2/x$ and its first three polynomial approximations (Example 2).

The following example illustrates how we test a power series for convergence by using the Ratio Test to see where it converges and diverges.

EXAMPLE 3 For what values of x do the following power series converge?

(a) $\displaystyle\sum_{n=1}^{\infty} (-1)^{n-1} \frac{x^n}{n} = x - \frac{x^2}{2} + \frac{x^3}{3} - \cdots$

(b) $\displaystyle\sum_{n=1}^{\infty} (-1)^{n-1} \frac{x^{2n-1}}{2n - 1} = x - \frac{x^3}{3} + \frac{x^5}{5} - \cdots$

(c) $\displaystyle\sum_{n=0}^{\infty}\frac{x^{n}}{n!} = 1 + x + \frac{x^{2}}{2!} + \frac{x^{3}}{3!} + \cdots$

(d) $\displaystyle\sum_{n=0}^{\infty}n!x^{n} = 1 + x + 2!x^{2} + 3!x^{3} + \cdots$

Solution   Apply the Ratio Test to the series $\sum |u_{n}|$, where u_{n} is the nth term of the power series in question.

(a) $\left|\dfrac{u_{n+1}}{u_{n}}\right| = \left|\dfrac{x^{n+1}}{n+1}\cdot\dfrac{n}{x}\right| = \dfrac{n}{n+1}|x| \rightarrow |x|.$

By the Ratio Test, the series converges absolutely for $|x| < 1$ and diverges for $|x| > 1$. At $x = 1$, we get the alternating harmonic series $1 - 1/2 + 1/3 - 1/4 + \cdots$, which converges. At $x = -1$, we get $-1 - 1/2 - 1/3 - 1/4 - \cdots$, the negative of the harmonic series, which diverges. Series (a) converges for $-1 < x \leq 1$ and diverges elsewhere.

We will see in Example 6 that this series converges to the function $\ln(1 + x)$ on the interval $(-1, 1]$ (see Figure 10.19).

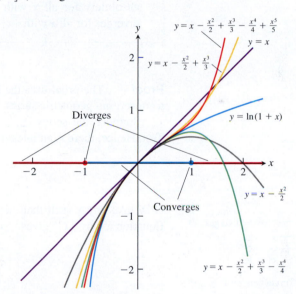

FIGURE 10.19   The power series $x - \dfrac{x^{2}}{2} + \dfrac{x^{3}}{3} - \dfrac{x^{4}}{4} + \cdots$

converges on the interval $(-1, 1]$.

(b) $\left|\dfrac{u_{n+1}}{u_{n}}\right| = \left|\dfrac{x^{2n+1}}{2n+1}\cdot\dfrac{2n-1}{x^{2n-1}}\right| = \dfrac{2n-1}{2n+1}x^{2} \rightarrow x^{2}.$ $\color{blue}2(n+1) - 1 = 2n + 1$

By the Ratio Test, the series converges absolutely for $x^{2} < 1$ and diverges for $x^{2} > 1$. At $x = 1$ the series becomes $1 - 1/3 + 1/5 - 1/7 + \cdots$, which converges by the Alternating Series Theorem. It also converges at $x = -1$ because it is again an alternating series that satisfies the conditions for convergence. The value at $x = -1$ is the negative of the value at $x = 1$. Series (b) converges for $-1 \leq x \leq 1$ and diverges elsewhere.

(c) $\left|\dfrac{u_{n+1}}{u_{n}}\right| = \left|\dfrac{x^{n+1}}{(n+1)!}\cdot\dfrac{n!}{x^{n}}\right| = \dfrac{|x|}{n+1} \rightarrow 0$ for every x. $\color{blue}\dfrac{n!}{(n+1)!} = \dfrac{1\cdot 2\cdot 3\cdots n}{1\cdot 2\cdot 3\cdots n\cdot (n+1)}$

The series converges absolutely for all x.

(d) $\left|\dfrac{u_{n+1}}{u_n}\right| = \left|\dfrac{(n+1)!x^{n+1}}{n!x^n}\right| = (n+1)|x| \rightarrow \infty$ unless $x = 0$.

The series diverges for all values of x except $x = 0$.

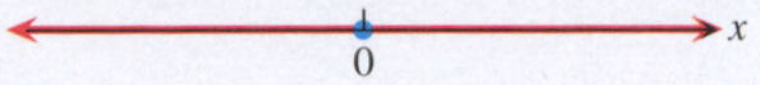

$\xleftarrow{\hspace{3cm}} \underset{0}{\bullet} \xrightarrow{\hspace{3cm}} x$

The previous example illustrated how a power series might converge. The next result shows that if a power series converges at more than one value, then it converges over an entire interval of values. The interval might be finite or infinite and contain one, both, or none of its endpoints. We will see that each endpoint of a finite interval must be tested independently for convergence or divergence.

THEOREM 18—The Convergence Theorem for Power Series

If the power series

$$\sum_{n=0}^{\infty} a_n x^n = a_0 + a_1 x + a_2 x^2 + \cdots \text{ converges at } x = c \neq 0, \text{ then it converges}$$

absolutely for all x with $|x| < |c|$. If the series diverges at $x = d$, then it diverges for all x with $|x| > |d|$.

Proof   The proof uses the Direct Comparison Test, with the given series compared to a converging geometric series.

Suppose the series $\sum_{n=0}^{\infty} a_n c^n$ converges. Then $\lim_{n\to\infty} a_n c^n = 0$ by the nth-Term Test. Hence, there is an integer N such that $|a_n c^n| < 1$ for all $n > N$, so that

$$|a_n| < \frac{1}{|c|^n} \qquad \text{for } n > N. \tag{5}$$

Now take any x such that $|x| < |c|$, so that $|x|/|c| < 1$. Multiplying both sides of Equation (5) by $|x|^n$ gives

$$|a_n||x|^n < \frac{|x|^n}{|c|^n} \qquad \text{for } n > N.$$

Since $|x/c| < 1$, it follows that the geometric series $\sum_{n=0}^{\infty} |x/c|^n$ converges. By the Direct Comparison Test (Theorem 10), the series $\sum_{n=0}^{\infty} |a_n||x^n|$ converges, so the original power series $\sum_{n=0}^{\infty} a_n x^n$ converges absolutely for $-|c| < x < |c|$ as claimed by the theorem. (See Figure 10.20.)

Now suppose that the series $\sum_{n=0}^{\infty} a_n x^n$ diverges at $x = d$. If x is a number with $|x| > |d|$ and the series converges at x, then the first half of the theorem shows that the series also converges at d, contrary to our assumption. So the series diverges for all x with $|x| > |d|$. ∎

To simplify the notation, Theorem 18 deals with the convergence of series of the form $\sum a_n x^n$. For series of the form $\sum a_n(x-a)^n$ we can replace $x-a$ by x' and apply the results to the series $\sum a_n(x')^n$.

The Radius of Convergence of a Power Series

The theorem we have just proved and the examples we have studied lead to the conclusion that a power series $\sum c_n(x-a)^n$ behaves in one of three possible ways. It might converge only at $x = a$, or converge everywhere, or converge on some interval of radius R centered

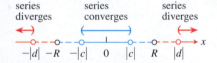

series diverges	series converges	series diverges

$\xleftarrow{\hspace{0.3cm}} \overset{}{\underset{-|d|}{\circ}} \; \overset{}{\underset{-R}{\circ}} \;\; \overset{}{\underset{-|c|}{\circ}} \;\; \overset{}{\underset{0}{}} \;\; \overset{}{\underset{|c|}{\circ}} \;\; \overset{}{\underset{R}{\circ}} \;\; \overset{}{\underset{|d|}{\circ}} \xrightarrow{\hspace{0.3cm}} x$

FIGURE 10.20 Convergence of $\sum a_n x^n$ at $x = c$ implies absolute convergence on the interval $-|c| < x < |c|$; divergence at $x = d$ implies divergence for $|x| > |d|$. The corollary to Theorem 18 asserts the existence of a radius of convergence $R \geq 0$. For $|x| < R$ the series converges absolutely and for $|x| > R$ it diverges.

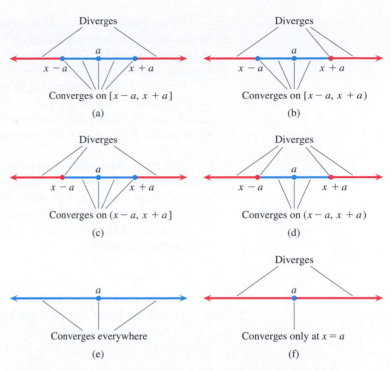

FIGURE 10.21   The six possibilities for an interval of convergence.

at $x = a$. We prove this as a Corollary to Theorem 18. When we also consider the convergence at the endpoints of an interval, there are six different possibilities. These are shown in Figure 10.21.

Corollary to Theorem 18

The convergence of the series $\sum c_n(x - a)^n$ is described by one of the following three cases:

1. There is a positive number R such that the series diverges for x with $|x - a| > R$ but converges absolutely for x with $|x - a| < R$. The series may or may not converge at either of the endpoints $x = a - R$ and $x = a + R$.

2. The series converges absolutely for every x $(R = \infty)$.

3. The series converges at $x = a$ and diverges elsewhere $(R = 0)$.

Proof   We first consider the case where $a = 0$, so that we have a power series $\sum_{n=0}^{\infty} c_n x^n$ centered at 0. If the series converges everywhere we are in Case 2. If it converges only at $x = 0$ then we are in Case 3. Otherwise there is a nonzero number d such that $\sum_{n=0}^{\infty} c_n d^n$ diverges. Let S be the set of values of x for which $\sum_{n=0}^{\infty} c_n x^n$ converges. The set S does not include any x with $|x| > |d|$, since Theorem 18 implies the series diverges at all such values. So the set S is bounded. By the Completeness Property of the Real Numbers (Appendix 6) S has a least upper bound R. (This is the smallest number with the property that all elements of S are less than or equal to R.) Since we are not in Case 3, the series converges at some number $b \neq 0$ and, by Theorem 18, also on the open interval $(-|b|, |b|)$. Therefore, $R > 0$.

If $|x| < R$ then there is a number c in S with $|x| < c < R$, since otherwise R would not be the least upper bound for S. The series converges at c since $c \in S$, so by Theorem 18 the series converges absolutely at x.

Now suppose $|x| > R$. If the series converges at x, then Theorem 18 implies it converges absolutely on the open interval $(-|x|, |x|)$, so that S contains this interval. Since R is an upper bound for S, it follows that $|x| \leq R$, which is a contradiction. So if $|x| > R$ then the series diverges. This proves the theorem for power series centered at $a = 0$.

For a power series centered at an arbitrary point $x = a$, set $x' = x - a$ and repeat the argument above, replacing x with x'. Since $x' = 0$ when $x = a$, convergence of the series $\sum_{n=0}^{\infty} |c_n(x')^n|$ on a radius R open interval centered at $x' = 0$ corresponds to convergence of the series $\sum_{n=0}^{\infty} |c_n(x - a)^n|$ on a radius R open interval centered at $x = a$. ∎

R is called the **radius of convergence** of the power series, and the interval of radius R centered at $x = a$ is called the **interval of convergence**. The interval of convergence may be open, closed, or half-open, depending on the particular series. At points x with $|x - a| < R$, the series converges absolutely. If the series converges for all values of x, we say its radius of convergence is infinite. If it converges only at $x = a$, we say its radius of convergence is zero.

How to Test a Power Series for Convergence

1. Use the Ratio Test (or Root Test) to find the largest open interval where the series converges absolutely,

$$|x - a| < R \quad \text{or} \quad a - R < x < a + R.$$

2. If R is finite, test for convergence or divergence at each endpoint, as in Examples 3a and b. Use a Comparison Test, the Integral Test, or the Alternating Series Test.

3. If R is finite, the series diverges for $|x - a| > R$ (it does not even converge conditionally) because the nth term does not approach zero for those values of x.

Operations on Power Series

On the intersection of their intervals of convergence, two power series can be added and subtracted term by term just like series of constants (Theorem 8). They can be multiplied just as we multiply polynomials, but we often limit the computation of the product to the first few terms, which are the most important. The following result gives a formula for the coefficients in the product, but we omit the proof. (Power series can also be divided in a way similar to division of polynomials, but we do not give a formula for the general coefficient here.)

THEOREM 19—Series Multiplication for Power Series

If $A(x) = \sum_{n=0}^{\infty} a_n x^n$ and $B(x) = \sum_{n=0}^{\infty} b_n x^n$ converge absolutely for $|x| < R$, and

$$c_n = a_0 b_n + a_1 b_{n-1} + a_2 b_{n-2} + \cdots + a_{n-1} b_1 + a_n b_0 = \sum_{k=0}^{n} a_k b_{n-k},$$

then $\sum_{n=0}^{\infty} c_n x^n$ converges absolutely to $A(x)B(x)$ for $|x| < R$:

$$\left(\sum_{n=0}^{\infty} a_n x^n \right) \left(\sum_{n=0}^{\infty} b_n x^n \right) = \sum_{n=0}^{\infty} c_n x^n.$$

Finding the general coefficient c_n in the product of two power series can be very tedious and the term may be unwieldy. The following computation provides an illustration of a product where we find the first few terms by multiplying the terms of the second series by each term of the first series:

$$\left(\sum_{n=0}^{\infty} x^n\right) \cdot \left(\sum_{n=0}^{\infty} (-1)^n \frac{x^{n+1}}{n+1}\right)$$

$$= (1 + x + x^2 + \cdots)\left(x - \frac{x^2}{2} + \frac{x^3}{3} - \cdots\right) \qquad \text{Multiply second series \ldots}$$

$$= \underbrace{\left(x - \frac{x^2}{2} + \frac{x^3}{3} - \cdots\right)}_{\text{by } 1} + \underbrace{\left(x^2 - \frac{x^3}{2} + \frac{x^4}{3} - \cdots\right)}_{\text{by } x} + \underbrace{\left(x^3 - \frac{x^4}{2} + \frac{x^5}{3} - \cdots\right)}_{\text{by } x^2} + \cdots$$

$$= x + \frac{x^2}{2} + \frac{5x^3}{6} + \cdots. \qquad \text{and gather the first three powers.}$$

We can also substitute a function $f(x)$ for x in a convergent power series.

THEOREM 20 If $\sum_{n=0}^{\infty} a_n x^n$ converges absolutely for $|x| < R$ and f is a continuous function, then $\sum_{n=0}^{\infty} a_n (f(x))^n$ converges absolutely on the set of points x where $|f(x)| < R$.

Since $1/(1 - x) = \sum_{n=0}^{\infty} x^n$ converges absolutely for $|x| < 1$, it follows from Theorem 20 that $1/(1 - 4x^2) = \sum_{n=0}^{\infty} (4x^2)^n$ converges absolutely when x satisfies $|4x^2| < 1$ or equivalently when $|x| < 1/2$.

Theorem 21 says that a power series can be differentiated term by term at each interior point of its interval of convergence. A proof is outlined in Exercise 64.

THEOREM 21—Term-by-Term Differentiation

If $\sum c_n(x - a)^n$ has radius of convergence $R > 0$, it defines a function

$$f(x) = \sum_{n=0}^{\infty} c_n(x - a)^n \qquad \text{on the interval} \qquad a - R < x < a + R.$$

This function f has derivatives of all orders inside the interval, and we obtain the derivatives by differentiating the original series term by term:

$$f'(x) = \sum_{n=1}^{\infty} nc_n(x - a)^{n-1},$$

$$f''(x) = \sum_{n=2}^{\infty} n(n - 1)c_n(x - a)^{n-2},$$

and so on. Each of these derived series converges at every point of the interval $a - R < x < a + R$.

EXAMPLE 4 Find series for $f'(x)$ and $f''(x)$ if

$$f(x) = \frac{1}{1 - x} = 1 + x + x^2 + x^3 + x^4 + \cdots + x^n + \cdots$$

$$= \sum_{n=0}^{\infty} x^n, \qquad -1 < x < 1.$$

Solution   We differentiate the power series on the right term by term:

$$f'(x) = \frac{1}{(1-x)^2} = 1 + 2x + 3x^2 + 4x^3 + \cdots + nx^{n-1} + \cdots$$

$$= \sum_{n=1}^{\infty} nx^{n-1}, \qquad -1 < x < 1;$$

$$f''(x) = \frac{2}{(1-x)^3} = 2 + 6x + 12x^2 + \cdots + n(n-1)x^{n-2} + \cdots$$

$$= \sum_{n=2}^{\infty} n(n-1)x^{n-2}, \qquad -1 < x < 1. \qquad \blacksquare$$

Caution   Term-by-term differentiation might not work for other kinds of series. For example, the trigonometric series

$$\sum_{n=1}^{\infty} \frac{\sin(n!x)}{n^2}$$

converges for all x. But if we differentiate term by term we get the series

$$\sum_{n=1}^{\infty} \frac{n! \cos(n!x)}{n^2},$$

which diverges for all x. This is not a power series since it is not a sum of positive integer powers of x.

It is also true that a power series can be integrated term by term throughout its interval of convergence. The proof is outlined in Exercise 65.

THEOREM 22—Term-by-Term Integration
Suppose that

$$f(x) = \sum_{n=0}^{\infty} c_n(x-a)^n$$

converges for $a - R < x < a + R (R > 0)$. Then

$$\sum_{n=0}^{\infty} c_n \frac{(x-a)^{n+1}}{n+1}$$

converges for $a - R < x < a + R$ and

$$\int f(x)\, dx = \sum_{n=0}^{\infty} c_n \frac{(x-a)^{n+1}}{n+1} + C$$

for $a - R < x < a + R$.

EXAMPLE 5   Identify the function

$$f(x) = \sum_{n=0}^{\infty} \frac{(-1)^n x^{2n+1}}{2n+1} = x - \frac{x^3}{3} + \frac{x^5}{5} - \cdots, \qquad -1 \leq x \leq 1.$$

Solution   We differentiate the original series term by term and get

$$f'(x) = 1 - x^2 + x^4 - x^6 + \cdots, \qquad -1 < x < 1. \qquad \text{Theorem 21}$$

This is a geometric series with first term 1 and ratio $-x^2$, so

$$f'(x) = \frac{1}{1 - (-x^2)} = \frac{1}{1 + x^2}.$$

We can now integrate $f'(x) = 1/(1 + x^2)$ to get

$$\int f'(x)\, dx = \int \frac{dx}{1 + x^2} = \tan^{-1} x + C.$$

The series for $f(x)$ is zero when $x = 0$, so $C = 0$. Hence

$$f(x) = x - \frac{x^3}{3} + \frac{x^5}{5} - \frac{x^7}{7} + \cdots = \tan^{-1} x, \qquad -1 < x < 1. \qquad (6)$$

It can be shown that the series also converges to $\tan^{-1} x$ at the endpoints $x = \pm 1$, but we omit the proof. ∎

Notice that the original series in Example 5 converges at both endpoints of the original interval of convergence, but Theorem 22 can only guarantee the convergence of the differentiated series inside the interval.

EXAMPLE 6   The series

$$\frac{1}{1 + t} = 1 - t + t^2 - t^3 + \cdots$$

converges on the open interval $-1 < t < 1$. Therefore,

$$\ln(1 + x) = \int_0^x \frac{1}{1 + t}\, dt = t - \frac{t^2}{2} + \frac{t^3}{3} - \frac{t^4}{4} + \cdots \Big]_0^x \qquad \text{Theorem 22}$$

$$= x - \frac{x^2}{2} + \frac{x^3}{3} - \frac{x^4}{4} + \cdots$$

or

$$\ln(1 + x) = \sum_{n=1}^{\infty} \frac{(-1)^{n-1} x^n}{n}, \qquad -1 < x < 1.$$

It can also be shown that the series converges at $x = 1$ to the number $\ln 2$, but that was not guaranteed by the theorem. A proof of this is outlined in Exercise 61. ∎

The Number π as a Series

$$\frac{\pi}{4} = \tan^{-1} 1 = \sum_{n=0}^{\infty} \frac{(-1)^n}{2n + 1}$$

Alternating Harmonic Series Sum

$$\ln 2 = \sum_{n=1}^{\infty} \frac{(-1)^{n-1}}{n}$$

EXERCISES   10.7

Intervals of Convergence

In Exercises 1–36, **(a)** find the series' radius and interval of convergence. For what values of x does the series converge **(b)** absolutely, **(c)** conditionally?

1. $\displaystyle\sum_{n=0}^{\infty} x^n$

2. $\displaystyle\sum_{n=0}^{\infty} (x + 5)^n$

3. $\displaystyle\sum_{n=0}^{\infty} (-1)^n (4x + 1)^n$

4. $\displaystyle\sum_{n=1}^{\infty} \frac{(3x - 2)^n}{n}$

5. $\displaystyle\sum_{n=0}^{\infty} \frac{(x - 2)^n}{10^n}$

6. $\displaystyle\sum_{n=0}^{\infty} (2x)^n$

7. $\displaystyle\sum_{n=0}^{\infty} \frac{nx^n}{n + 2}$

8. $\displaystyle\sum_{n=1}^{\infty} \frac{(-1)^n (x + 2)^n}{n}$

9. $\displaystyle\sum_{n=1}^{\infty} \frac{x^n}{n\sqrt{n}\, 3^n}$

10. $\displaystyle\sum_{n=1}^{\infty} \frac{(x - 1)^n}{\sqrt{n}}$

11. $\displaystyle\sum_{n=0}^{\infty} \frac{(-1)^n x^n}{n!}$

12. $\displaystyle\sum_{n=0}^{\infty} \frac{3^n x^n}{n!}$

13. $\displaystyle\sum_{n=1}^{\infty} \frac{4^n x^{2n}}{n}$

14. $\displaystyle\sum_{n=1}^{\infty} \frac{(x - 1)^n}{n^3 3^n}$

15. $\displaystyle\sum_{n=0}^{\infty} \frac{x^n}{\sqrt{n^2 + 3}}$

16. $\displaystyle\sum_{n=0}^{\infty} \frac{(-1)^n x^{n+1}}{\sqrt{n + 3}}$

17. $\displaystyle\sum_{n=0}^{\infty} \frac{n(x+3)^n}{5^n}$

18. $\displaystyle\sum_{n=0}^{\infty} \frac{nx^n}{4^n(n^2+1)}$

19. $\displaystyle\sum_{n=0}^{\infty} \frac{\sqrt{n}x^n}{3^n}$

20. $\displaystyle\sum_{n=1}^{\infty} \sqrt[n]{n}(2x+5)^n$

21. $\displaystyle\sum_{n=1}^{\infty} (2+(-1)^n)\cdot(x+1)^{n-1}$

22. $\displaystyle\sum_{n=1}^{\infty} \frac{(-1)^n 3^{2n}(x-2)^n}{3n}$

23. $\displaystyle\sum_{n=1}^{\infty} \left(1+\frac{1}{n}\right)^n x^n$

24. $\displaystyle\sum_{n=1}^{\infty} (\ln n)x^n$

25. $\displaystyle\sum_{n=1}^{\infty} n^n x^n$

26. $\displaystyle\sum_{n=0}^{\infty} n!(x-4)^n$

27. $\displaystyle\sum_{n=1}^{\infty} \frac{(-1)^{n+1}(x+2)^n}{n2^n}$

28. $\displaystyle\sum_{n=0}^{\infty} (-2)^n(n+1)(x-1)^n$

29. $\displaystyle\sum_{n=2}^{\infty} \frac{x^n}{n(\ln n)^2}$ Get the information you need about $\sum 1/(n(\ln n)^2)$ from Section 10.3, Exercise 61.

30. $\displaystyle\sum_{n=2}^{\infty} \frac{x^n}{n\ln n}$ Get the information you need about $\sum 1/(n \ln n)$ from Section 10.3, Exercise 60.

31. $\displaystyle\sum_{n=1}^{\infty} \frac{(4x-5)^{2n+1}}{n^{3/2}}$

32. $\displaystyle\sum_{n=1}^{\infty} \frac{(3x+1)^{n+1}}{2n+2}$

33. $\displaystyle\sum_{n=1}^{\infty} \frac{1}{2\cdot4\cdot6\cdots(2n)}x^n$

34. $\displaystyle\sum_{n=1}^{\infty} \frac{3\cdot5\cdot7\cdots(2n+1)}{n^2\cdot2^n}x^{n+1}$

35. $\displaystyle\sum_{n=1}^{\infty} \frac{1+2+3+\cdots+n}{1^2+2^2+3^2+\cdots+n^2}x^n$

36. $\displaystyle\sum_{n=1}^{\infty} \left(\sqrt{n+1}-\sqrt{n}\right)(x-3)^n$

In Exercises 37–40, find the series' radius of convergence.

37. $\displaystyle\sum_{n=1}^{\infty} \frac{n!}{3\cdot6\cdot9\cdots3n}x^n$

38. $\displaystyle\sum_{n=1}^{\infty} \left(\frac{2\cdot4\cdot6\cdots(2n)}{2\cdot5\cdot8\cdots(3n-1)}\right)^2 x^n$

39. $\displaystyle\sum_{n=1}^{\infty} \frac{(n!)^2}{2^n(2n)!}x^n$

40. $\displaystyle\sum_{n=1}^{\infty} \left(\frac{n}{n+1}\right)^{n^2} x^n$

(*Hint:* Apply the Root Test.)

In Exercises 41–48, use Theorem 20 to find the series' interval of convergence and, within this interval, the sum of the series as a function of x.

41. $\displaystyle\sum_{n=0}^{\infty} 3^n x^n$

42. $\displaystyle\sum_{n=0}^{\infty} (e^x-4)^n$

43. $\displaystyle\sum_{n=0}^{\infty} \frac{(x-1)^{2n}}{4^n}$

44. $\displaystyle\sum_{n=0}^{\infty} \frac{(x+1)^{2n}}{9^n}$

45. $\displaystyle\sum_{n=0}^{\infty} \left(\frac{\sqrt{x}}{2}-1\right)^n$

46. $\displaystyle\sum_{n=0}^{\infty} (\ln x)^n$

47. $\displaystyle\sum_{n=0}^{\infty} \left(\frac{x^2+1}{3}\right)^n$

48. $\displaystyle\sum_{n=0}^{\infty} \left(\frac{x^2-1}{2}\right)^n$

Using the Geometric Series

49. In Example 2 we represented the function $f(x) = 2/x$ as a power series about $x = 2$. Use a geometric series to represent $f(x)$ as a power series about $x = 1$, and find its interval of convergence.

50. Use a geometric series to represent each of the given functions as a power series about $x = 0$, and find their intervals of convergence.

    a. $f(x) = \dfrac{5}{3-x}$ **b.** $g(x) = \dfrac{3}{x-2}$

51. Represent the function $g(x)$ in Exercise 50 as a power series about $x = 5$, and find the interval of convergence.

52. **a.** Find the interval of convergence of the power series

$$\sum_{n=0}^{\infty} \frac{8}{4^{n+2}}x^n.$$

    b. Represent the power series in part (a) as a power series about $x = 3$ and identify the interval of convergence of the new series. (Later in the chapter you will understand why the new interval of convergence does not necessarily include all of the numbers in the original interval of convergence.)

Theory and Examples

53. For what values of x does the series

$$1 - \frac{1}{2}(x-3) + \frac{1}{4}(x-3)^2 + \cdots + \left(-\frac{1}{2}\right)^n(x-3)^n + \cdots$$

converge? What is its sum? What series do you get if you differentiate the given series term by term? For what values of x does the new series converge? What is its sum?

54. If you integrate the series in Exercise 53 term by term, what new series do you get? For what values of x does the new series converge, and what is another name for its sum?

55. The series

$$\sin x = x - \frac{x^3}{3!} + \frac{x^5}{5!} - \frac{x^7}{7!} + \frac{x^9}{9!} - \frac{x^{11}}{11!} + \cdots$$

converges to $\sin x$ for all x.

    a. Find the first six terms of a series for $\cos x$. For what values of x should the series converge?

    b. By replacing x by $2x$ in the series for $\sin x$, find a series that converges to $\sin 2x$ for all x.

    c. Using the result in part (a) and series multiplication, calculate the first six terms of a series for $2 \sin x \cos x$. Compare your answer with the answer in part (b).

56. The series

$$e^x = 1 + x + \frac{x^2}{2!} + \frac{x^3}{3!} + \frac{x^4}{4!} + \frac{x^5}{5!} + \cdots$$

converges to e^x for all x.

    a. Find a series for $(d/dx)e^x$. Do you get the series for e^x? Explain your answer.

b. Find a series for $\int e^x \, dx$. Do you get the series for e^x? Explain your answer.

c. Replace x by $-x$ in the series for e^x to find a series that converges to e^{-x} for all x. Then multiply the series for e^x and e^{-x} to find the first six terms of a series for $e^{-x} \cdot e^x$.

57. The series

$$\tan x = x + \frac{x^3}{3} + \frac{2x^5}{15} + \frac{17x^7}{315} + \frac{62x^9}{2835} + \cdots$$

converges to $\tan x$ for $-\pi/2 < x < \pi/2$.

a. Find the first five terms of the series for $\ln |\sec x|$. For what values of x should the series converge?

b. Find the first five terms of the series for $\sec^2 x$. For what values of x should this series converge?

c. Check your result in part (b) by squaring the series given for $\sec x$ in Exercise 58.

58. The series

$$\sec x = 1 + \frac{x^2}{2} + \frac{5}{24}x^4 + \frac{61}{720}x^6 + \frac{277}{8064}x^8 + \cdots$$

converges to $\sec x$ for $-\pi/2 < x < \pi/2$.

a. Find the first five terms of a power series for the function $\ln |\sec x + \tan x|$. For what values of x should the series converge?

b. Find the first four terms of a series for $\sec x \tan x$. For what values of x should the series converge?

c. Check your result in part (b) by multiplying the series for $\sec x$ by the series given for $\tan x$ in Exercise 57.

59. Uniqueness of convergent power series

a. Show that if two power series $\sum_{n=0}^{\infty} a_n x^n$ and $\sum_{n=0}^{\infty} b_n x^n$ are convergent and equal for all values of x in an open interval $(-c, c)$, then $a_n = b_n$ for every n. (*Hint:* Let $f(x) = \sum_{n=0}^{\infty} a_n x^n = \sum_{n=0}^{\infty} b_n x^n$. Differentiate term by term to show that a_n and b_n both equal $f^{(n)}(0)/(n!)$.)

b. Show that if $\sum_{n=0}^{\infty} a_n x^n = 0$ for all x in an open interval $(-c, c)$, then $a_n = 0$ for every n.

60. The sum of the series $\sum_{n=0}^{\infty} (n^2/2^n)$ To find the sum of this series, express $1/(1 - x)$ as a geometric series, differentiate both sides of the resulting equation with respect to x, multiply both sides of the result by x, differentiate again, multiply by x again, and set x equal to $1/2$. What do you get?

61. The sum of the alternating harmonic series This exercise will show that

$$\sum_{n=1}^{\infty} \frac{(-1)^{n+1}}{n} = \ln 2.$$

Let h_n be the nth partial sum of the harmonic series, and let s_n be the nth partial sum of the alternating harmonic series.

a. Use mathematical induction or algebra to show that

$$s_{2n} = h_{2n} - h_n.$$

b. Use the results in Exercise 63 in Section 10.3 to conclude that

$$\lim_{n \to \infty} (h_n - \ln n) = \gamma$$

and

$$\lim_{n \to \infty} (h_{2n} - \ln 2n) = \gamma,$$

where γ is Euler's constant.

c. Use these facts to show that

$$\sum_{n=1}^{\infty} \frac{(-1)^{n+1}}{n} = \lim_{n \to \infty} s_{2n} = \ln 2.$$

62. Assume that the series $\sum a_n x^n$ converges for $x = 4$ and diverges for $x = 7$. Answer true (T), false (F), or not enough information given (N) for the following statements about the series.

a. Converges absolutely for $x = -4$

b. Diverges for $x = 5$

c. Converges absolutely for $x = -8.5$

d. Converges for $x = -2$

e. Diverges for $x = 8$

f. Diverges for $x = -6$

g. Converges absolutely for $x = 0$

h. Converges absolutely for $x = -7.1$

63. Assume that the series $\sum a_n (x - 2)^n$ converges for $x = -1$ and diverges for $x = 6$. Answer true (T), false (F), or not enough information given (N) for the following statements about the series.

a. Converges absolutely for $x = 1$

b. Diverges for $x = -6$

c. Diverges for $x = 2$

d. Converges for $x = 0$

e. Converges absolutely for $x = 5$

f. Diverges for $x = 4.9$

g. Diverges for $x = 5.1$

h. Converges absolutely for $x = 4$

64. Proof of Theorem 21 Assume that $a = 0$ in Theorem 21 and that $f(x) = \sum_{n=0}^{\infty} c_n x^n$ converges for $-R < x < R$. Let $g(x) = \sum_{n=1}^{\infty} n c_n x^{n-1}$. This exercise will prove that $f'(x) = g(x)$, that is, $\lim_{h \to 0} \dfrac{f(x + h) - f(x)}{h} = g(x)$.

a. Use the Ratio Test to show that $g(x)$ converges for $-R < x < R$.

b. Use the Mean Value Theorem to show that

$$\frac{(x + h)^n - x^n}{h} = n c_n^{n-1}$$

for some c_n between x and $x + h$ for $n = 1, 2, 3, \ldots$.

c. Show that

$$\left| g(x) - \frac{f(x + h) - f(x)}{h} \right| = \left| \sum_{n=2}^{\infty} n a_n \left(x^{n-1} - c_n^{n-1} \right) \right|$$

d. Use the Mean Value Theorem to show that

$$\frac{x^{n-1} - c_n^{n-1}}{x - c_n} = (n - 1) d_n^{n-2}$$

for some d_{n-1} between x and c_n for $n = 2, 3, 4, \ldots$.

e. Explain why $|x - c_n| < h$ and why
$|d_{n-1}| \leq \alpha = \max\{|x|, |x + h|\}$.

f. Show that

$$\left| g(x) - \frac{f(x + h) - f(x)}{h} \right| \leq |h| \sum_{n=2}^{\infty} |n(n - 1)a_n\alpha^{n-2}|$$

g. Show that $\sum_{n=2}^{\infty} n(n - 1)\alpha^{n-2}$ converges for $-R < x < R$.

h. Let $h \to 0$ in part (f) to conclude that

$$\lim_{h \to 0} \frac{f(x + h) - f(x)}{h} = g(x).$$

65. Proof of Theorem 22 Assume that $a = 0$ in Theorem 22 and
that $f(x) = \sum_{n=0}^{\infty} c_n x^n$ converges for $-R < x < R$. Let

$g(x) = \sum_{n=0}^{\infty} \dfrac{c_n}{n + 1} x^{n+1}$. This exercise will prove that $g'(x) = f(x)$.

a. Use the Ratio Test to show that $g(x)$ converges for
$-R < x < R$.

b. Use Theorem 21 to show that $g'(x) = f(x)$, that is,

$$\int f(x)\, dx = g(x) + C.$$

10.8 Taylor and Maclaurin Series

We have seen how geometric series can be used to generate a power series for functions such as $f(x) = 1/(1 - x)$ or $g(x) = 3/(x - 2)$. Now we expand our capability to represent a function with a power series. This section shows how functions that are infinitely differentiable generate power series called *Taylor series*. In many cases, these series provide useful polynomial approximations of the original functions. Because approximation by polynomials is extremely useful to both mathematicians and scientists, Taylor series are an important application of the theory of infinite series.

Series Representations

We know from Theorem 21 that within its interval of convergence I the sum of a power series is a continuous function with derivatives of all orders. But what about the other way around? If a function $f(x)$ has derivatives of all orders on an interval, can it be expressed as a power series on at least part of that interval? And if it can, what are its coefficients?

We can answer the last question readily if we assume that $f(x)$ is the sum of a power series about $x = a$,

$$f(x) = \sum_{n=0}^{\infty} a_n(x - a)^n$$

$$= a_0 + a_1(x - a) + a_2(x - a)^2 + \cdots + a_n(x - a)^n + \cdots$$

with a positive radius of convergence. By repeated term-by-term differentiation within the interval of convergence I, we obtain

$$f'(x) = a_1 + 2a_2(x - a) + 3a_3(x - a)^2 + \cdots + na_n(x - a)^{n-1} + \cdots,$$

$$f''(x) = 1 \cdot 2a_2 + 2 \cdot 3a_3(x - a) + 3 \cdot 4a_4(x - a)^2 + \cdots,$$

$$f'''(x) = 1 \cdot 2 \cdot 3a_3 + 2 \cdot 3 \cdot 4a_4(x - a) + 3 \cdot 4 \cdot 5a_5(x - a)^2 + \cdots,$$

with the nth derivative being

$$f^{(n)}(x) = n!a_n + \text{a sum of terms with } (x - a) \text{ as a factor.}$$

Since these equations all hold at $x = a$, we have

$$f'(a) = a_1, \qquad f''(a) = 1 \cdot 2a_2, \qquad f'''(a) = 1 \cdot 2 \cdot 3a_3,$$

and, in general,

$$f^{(n)}(a) = n!a_n.$$

These formulas reveal a pattern in the coefficients of any power series $\sum_{n=0}^{\infty} a_n(x - a)^n$ that converges to the values of f on I ("represents f on I"). If there *is* such a series (still an open question), then there is only one such series, and its nth coefficient is

$$a_n = \frac{f^{(n)}(a)}{n!}.$$

If f has a series representation, then the series must be

$$f(x) = f(a) + f'(a)(x - a) + \frac{f''(a)}{2!}(x - a)^2$$

$$+ \cdots + \frac{f^{(n)}(a)}{n!}(x - a)^n + \cdots. \tag{1}$$

But if we start with an arbitrary function f that is infinitely differentiable on an interval containing $x = a$ and use it to generate the series in Equation (1), does the series converge to $f(x)$ at each x in the interval of convergence? The answer is maybe—for some functions it will but for other functions it will not (as we will see in Example 4).

HISTORICAL BIOGRAPHIES

Brook Taylor
(1685–1731)
www.goo.gl/5A5Dxl
Colin Maclaurin
(1698–1746)
www.goo.gl/vL7QNQ

Taylor and Maclaurin Series

The series on the right-hand side of Equation (1) is the most important and useful series we will study in this chapter.

DEFINITIONS   Let f be a function with derivatives of all orders throughout some interval containing a as an interior point. Then the **Taylor series generated by f at $x = a$** is

$$\sum_{k=0}^{\infty} \frac{f^{(k)}(a)}{k!}(x - a)^k = f(a) + f'(a)(x - a) + \frac{f''(a)}{2!}(x - a)^2$$

$$+ \cdots + \frac{f^{(n)}(a)}{n!}(x - a)^n + \cdots.$$

The **Maclaurin series of f** is the Taylor series generated by f at $x = 0$, or

$$\sum_{k=0}^{\infty} \frac{f^{(k)}(0)}{k!}x^k = f(0) + f'(0)x + \frac{f''(0)}{2!}x^2 + \cdots + \frac{f^{(n)}(0)}{n!}x^n + \cdots.$$

The Maclaurin series generated by f is often just called the Taylor series of f.

EXAMPLE 1   Find the Taylor series generated by $f(x) = 1/x$ at $a = 2$. Where, if anywhere, does the series converge to $1/x$?

Solution   We need to find $f(2), f'(2), f''(2), \ldots$. Taking derivatives we get

$$f(x) = x^{-1}, \quad f'(x) = -x^{-2}, \quad f''(x) = 2!x^{-3}, \ldots, f^{(n)}(x) = (-1)^n n! x^{-(n+1)},$$

so that

$$f(2) = 2^{-1} = \frac{1}{2}, \quad f'(2) = -\frac{1}{2^2}, \quad \frac{f''(2)}{2!} = 2^{-3} = \frac{1}{2^3}, \ldots, \frac{f^{(n)}(2)}{n!} = \frac{(-1)^n}{2^{n+1}}.$$

The Taylor series is

$$f(2) + f'(2)(x - 2) - \frac{f''(2)}{2!}(x - 2)^2 + \cdots + \frac{f^{(n)}(2)}{n!}(x - 2)^n + \cdots$$

$$= \frac{1}{2} - \frac{(x - 2)}{2^2} + \frac{(x - 2)^2}{2^3} - \cdots + (-1)^n \frac{(x - 2)^n}{2^{n+1}} + \cdots.$$

This is a geometric series with first term $1/2$ and ratio $r = -(x - 2)/2$. It converges absolutely for $|x - 2| < 2$ and its sum is

$$\frac{1/2}{1 + (x - 2)/2} = \frac{1}{2 + (x - 2)} = \frac{1}{x}.$$

In this example the Taylor series generated by $f(x) = 1/x$ at $a = 2$ converges to $1/x$ for $|x - 2| < 2$ or $0 < x < 4$. ∎

Taylor Polynomials

The linearization of a differentiable function f at a point a is the polynomial of degree one given by

$$P_1(x) = f(a) + f'(a)(x - a).$$

In Section 3.11 we used this linearization to approximate $f(x)$ at values of x near a. If f has derivatives of higher order at a, then it has higher-order polynomial approximations as well, one for each available derivative. These polynomials are called the Taylor polynomials of f.

DEFINITION   Let f be a function with derivatives of order k for $k = 1, 2, \dots, N$ in some interval containing a as an interior point. Then for any integer n from 0 through N, the **Taylor polynomial of order n** generated by f at $x = a$ is the polynomial

$$P_n(x) = f(a) + f'(a)(x - a) + \frac{f''(a)}{2!}(x - a)^2 + \cdots$$

$$+ \frac{f^{(k)}(a)}{k!}(x - a)^k + \cdots + \frac{f^{(n)}(a)}{n!}(x - a)^n.$$

We speak of a Taylor polynomial of *order n* rather than *degree n* because $f^{(n)}(a)$ may be zero. The first two Taylor polynomials of $f(x) = \cos x$ at $x = 0$, for example, are $P_0(x) = 1$ and $P_1(x) = 1$. The first-order Taylor polynomial has degree zero, not one.

Just as the linearization of f at $x = a$ provides the best linear approximation of f in the neighborhood of a, the higher-order Taylor polynomials provide the "best" polynomial approximations of their respective degrees. (See Exercise 44.)

EXAMPLE 2   Find the Taylor series and the Taylor polynomials generated by $f(x) = e^x$ at $x = 0$.

Solution   Since $f^{(n)}(x) = e^x$ and $f^{(n)}(0) = 1$ for every $n = 0, 1, 2, \dots$, the Taylor series generated by f at $x = 0$ (see Figure 10.22) is

$$f(0) + f'(0)x + \frac{f''(0)}{2!}x^2 + \cdots + \frac{f^{(n)}(0)}{n!}x^n + \cdots$$

$$= 1 + x + \frac{x^2}{2} + \cdots + \frac{x^n}{n!} + \cdots$$

$$= \sum_{k=0}^{\infty} \frac{x^k}{k!}.$$

This is also the Maclaurin series for e^x. In the next section we will see that the series converges to e^x at every x.

The Taylor polynomial of order n at $x = 0$ is

$$P_n(x) = 1 + x + \frac{x^2}{2} + \cdots + \frac{x^n}{n!}.$$ ∎

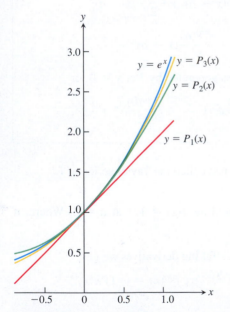

FIGURE 10.22   The graph of $f(x) = e^x$ and its Taylor polynomials

$P_1(x) = 1 + x$

$P_2(x) = 1 + x + (x^2/2!)$

$P_3(x) = 1 + x + (x^2/2!) + (x^3/3!).$

Notice the very close agreement near the center $x = 0$ (Example 2).

EXAMPLE 3   Find the Taylor series and Taylor polynomials generated by $f(x) = \cos x$ at $x = 0$.

Solution   The cosine and its derivatives are

$$f(x) = \cos x, \qquad\qquad f'(x) = -\sin x,$$
$$f''(x) = -\cos x, \qquad\qquad f^{(3)}(x) = \sin x,$$
$$\vdots \qquad\qquad\qquad\qquad \vdots$$
$$f^{(2n)}(x) = (-1)^n \cos x, \qquad f^{(2n+1)}(x) = (-1)^{n+1} \sin x.$$

At $x = 0$, the cosines are 1 and the sines are 0, so

$$f^{(2n)}(0) = (-1)^n, \qquad f^{(2n+1)}(0) = 0.$$

The Taylor series generated by f at 0 is

$$f(0) + f'(0)x + \frac{f''(0)}{2!}x^2 + \frac{f'''(0)}{3!}x^3 + \cdots + \frac{f^{(n)}(0)}{n!}x^n + \cdots$$

$$= 1 + 0 \cdot x - \frac{x^2}{2!} + 0 \cdot x^3 + \frac{x^4}{4!} + \cdots + (-1)^n \frac{x^{2n}}{(2n)!} + \cdots$$

$$= \sum_{k=0}^{\infty} \frac{(-1)^k x^{2k}}{(2k)!}.$$

This is also the Maclaurin series for $\cos x$. Notice that only even powers of x occur in the Taylor series generated by the cosine function, which is consistent with the fact that it is an even function. In Section 10.9, we will see that the series converges to $\cos x$ at every x.

Because $f^{(2n+1)}(0) = 0$, the Taylor polynomials of orders $2n$ and $2n + 1$ are identical:

$$P_{2n}(x) = P_{2n+1}(x) = 1 - \frac{x^2}{2!} + \frac{x^4}{4!} - \cdots + (-1)^n \frac{x^{2n}}{(2n)!}.$$

Figure 10.23 shows how well these polynomials approximate $f(x) = \cos x$ near $x = 0$. Only the right-hand portions of the graphs are given because the graphs are symmetric about the y-axis. ∎

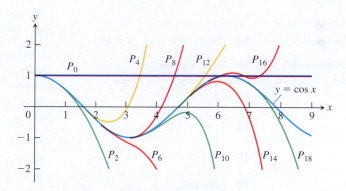

FIGURE 10.23   The polynomials

$$P_{2n}(x) = \sum_{k=0}^{n} \frac{(-1)^k x^{2k}}{(2k)!}$$

converge to $\cos x$ as $n \to \infty$. We can deduce the behavior of $\cos x$ arbitrarily far away solely from knowing the values of the cosine and its derivatives at $x = 0$ (Example 3).

EXAMPLE 4   It can be shown (though not easily) that

$$f(x) = \begin{cases} 0, & x = 0 \\ e^{-1/x^2}, & x \neq 0 \end{cases}$$

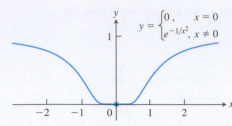

FIGURE 10.24 The graph of the continuous extension of $y = e^{-1/x^2}$ is so flat at the origin that all of its derivatives there are zero (Example 4). Therefore its Taylor series, which is zero everywhere, is not the function itself.

(Figure 10.24) has derivatives of all orders at $x = 0$ and that $f^{(n)}(0) = 0$ for all n. This means that the Taylor series generated by f at $x = 0$ is

$$f(0) + f'(0)x + \frac{f''(0)}{2!}x^2 + \cdots + \frac{f^{(n)}(0)}{n!}x^n + \cdots$$
$$= 0 + 0 \cdot x + 0 \cdot x^2 + \cdots + 0 \cdot x^n + \cdots$$
$$= 0 + 0 + \cdots + 0 + \cdots.$$

The series converges for every x (its sum is 0) but converges to $f(x)$ only at $x = 0$. That is, the Taylor series generated by $f(x)$ in this example is *not* equal to the function $f(x)$ over the entire interval of convergence.   ∎

Two questions still remain.

1. For what values of x can we normally expect a Taylor series to converge to its generating function?

2. How accurately do a function's Taylor polynomials approximate the function on a given interval?

The answers are provided by a theorem of Taylor in the next section.

EXERCISES 10.8

Finding Taylor Polynomials

In Exercises 1–10, find the Taylor polynomials of orders 0, 1, 2, and 3 generated by f at a.

1. $f(x) = e^{2x}, \quad a = 0$

2. $f(x) = \sin x, \quad a = 0$

3. $f(x) = \ln x, \quad a = 1$

4. $f(x) = \ln(1 + x), \quad a = 0$

5. $f(x) = 1/x, \quad a = 2$

6. $f(x) = 1/(x + 2), \quad a = 0$

7. $f(x) = \sin x, \quad a = \pi/4$

8. $f(x) = \tan x, \quad a = \pi/4$

9. $f(x) = \sqrt{x}, \quad a = 4$

10. $f(x) = \sqrt{1 - x}, \quad a = 0$

Finding Taylor Series at $x = 0$ (Maclaurin Series)

Find the Maclaurin series for the functions in Exercises 11–24.

11. e^{-x}

12. xe^x

13. $\dfrac{1}{1 + x}$

14. $\dfrac{2 + x}{1 - x}$

15. $\sin 3x$

16. $\sin \dfrac{x}{2}$

17. $7 \cos(-x)$

18. $5 \cos \pi x$

19. $\cosh x = \dfrac{e^x + e^{-x}}{2}$

20. $\sinh x = \dfrac{e^x - e^{-x}}{2}$

21. $x^4 - 2x^3 - 5x + 4$

22. $\dfrac{x^2}{x + 1}$

23. $x \sin x$

24. $(x + 1) \ln(x + 1)$

Finding Taylor and Maclaurin Series

In Exercises 25–34, find the Taylor series generated by f at $x = a$.

25. $f(x) = x^3 - 2x + 4, \quad a = 2$

26. $f(x) = 2x^3 + x^2 + 3x - 8, \quad a = 1$

27. $f(x) = x^4 + x^2 + 1, \quad a = -2$

28. $f(x) = 3x^5 - x^4 + 2x^3 + x^2 - 2, \quad a = -1$

29. $f(x) = 1/x^2, \quad a = 1$

30. $f(x) = 1/(1 - x)^3, \quad a = 0$

31. $f(x) = e^x, \quad a = 2$

32. $f(x) = 2^x, \quad a = 1$

33. $f(x) = \cos(2x + (\pi/2)), \quad a = \pi/4$

34. $f(x) = \sqrt{x + 1}, \quad a = 0$

In Exercises 35–38, find the first three nonzero terms of the Maclaurin series for each function and the values of x for which the series converges absolutely.

35. $f(x) = \cos x - (2/(1 - x))$

36. $f(x) = (1 - x + x^2)e^x$

37. $f(x) = (\sin x) \ln(1 + x)$

38. $f(x) = x \sin^2 x$

39. $f(x) = x^4 e^{x^2}$

40. $f(x) = \dfrac{x^3}{1 + 2x}$

Theory and Examples

41. Use the Taylor series generated by e^x at $x = a$ to show that

$$e^x = e^a \left[1 + (x - a) + \frac{(x - a)^2}{2!} + \cdots \right].$$

42. (*Continuation of Exercise 41.*) Find the Taylor series generated by e^x at $x = 1$. Compare your answer with the formula in Exercise 41.

43. Let $f(x)$ have derivatives through order n at $x = a$. Show that the Taylor polynomial of order n and its first n derivatives have the same values that f and its first n derivatives have at $x = a$.

44. Approximation properties of Taylor polynomials Suppose that $f(x)$ is differentiable on an interval centered at $x = a$ and that $g(x) = b_0 + b_1(x - a) + \cdots + b_n(x - a)^n$ is a polynomial of degree n with constant coefficients $b_0, \ldots, b_n$. Let $E(x) = f(x) - g(x)$. Show that if we impose on g the conditions

i) $E(a) = 0$ The approximation error is zero at $x = a$.

ii) $\displaystyle\lim_{x \to a} \frac{E(x)}{(x - a)^n} = 0,$ The error is negligible when compared to $(x - a)^n$.

then

$$g(x) = f(a) + f'(a)(x - a) + \frac{f''(a)}{2!}(x - a)^2 + \cdots$$

$$+ \frac{f^{(n)}(a)}{n!}(x - a)^n.$$

Thus, the Taylor polynomial $P_n(x)$ is the only polynomial of degree less than or equal to n whose error is both zero at $x = a$ and negligible when compared with $(x - a)^n$.

Quadratic Approximations The Taylor polynomial of order 2 generated by a twice-differentiable function $f(x)$ at $x = a$ is called the *quadratic approximation* of f at $x = a$. In Exercises 45–50, find the **(a)** linearization (Taylor polynomial of order 1) and **(b)** quadratic approximation of f at $x = 0$.

45. $f(x) = \ln(\cos x)$ **46.** $f(x) = e^{\sin x}$

47. $f(x) = 1/\sqrt{1 - x^2}$ **48.** $f(x) = \cosh x$

49. $f(x) = \sin x$ **50.** $f(x) = \tan x$

10.9 Convergence of Taylor Series

In the last section we asked when a Taylor series for a function can be expected to converge to the function that generates it. The finite-order Taylor polynomials that approximate the Taylor series provide estimates for the generating function. In order for these estimates to be useful, we need a way to control the possible errors we may encounter when approximating a function with its finite-order Taylor polynomials. How do we bound such possible errors? We answer the question in this section with the following theorem.

THEOREM 23—Taylor's Theorem

If f and its first n derivatives f', f'', ..., $f^{(n)}$ are continuous on the closed interval between a and b, and $f^{(n)}$ is differentiable on the open interval between a and b, then there exists a number c between a and b such that

$$f(b) = f(a) + f'(a)(b - a) + \frac{f''(a)}{2!}(b - a)^2 + \cdots$$

$$+ \frac{f^{(n)}(a)}{n!}(b - a)^n + \frac{f^{(n+1)}(c)}{(n + 1)!}(b - a)^{n+1}.$$

Taylor's Theorem is a generalization of the Mean Value Theorem (Exercise 49). There is a proof of Taylor's Theorem at the end of this section.

When we apply Taylor's Theorem, we usually want to hold a fixed and treat b as an independent variable. Taylor's formula is easier to use in circumstances like these if we change b to x. Here is a version of the theorem with this change.

Taylor's Formula

If f has derivatives of all orders in an open interval I containing a, then for each positive integer n and for each x in I,

$$f(x) = f(a) + f'(a)(x - a) + \frac{f''(a)}{2!}(x - a)^2 + \cdots$$

$$+ \frac{f^{(n)}(a)}{n!}(x - a)^n + R_n(x), \tag{1}$$

where

$$R_n(x) = \frac{f^{(n+1)}(c)}{(n + 1)!}(x - a)^{n+1} \qquad \text{for some } c \text{ between } a \text{ and } x. \tag{2}$$

When we state Taylor's theorem this way, it says that for each $x \in I$,

$$f(x) = P_n(x) + R_n(x).$$

The function $R_n(x)$ is determined by the value of the $(n + 1)$st derivative $f^{(n+1)}$ at a point c that depends on both a and x, and that lies somewhere between them. For any value of n we want, the equation gives both a polynomial approximation of f of that order and a formula for the error involved in using that approximation over the interval I.

Equation (1) is called **Taylor's formula**. The function $R_n(x)$ is called the **remainder of order n** or the **error term** for the approximation of f by $P_n(x)$ over I.

> If $R_n(x) \to 0$ as $n \to \infty$ for all $x \in I$, we say that the Taylor series generated by f at $x = a$ **converges** to f on I, and we write
>
> $$f(x) = \sum_{k=0}^{\infty} \frac{f^{(k)}(a)}{k!} (x - a)^k.$$

Often we can estimate R_n without knowing the value of c, as the following example illustrates.

EXAMPLE 1   Show that the Taylor series generated by $f(x) = e^x$ at $x = 0$ converges to $f(x)$ for every real value of x.

Solution   The function has derivatives of all orders throughout the interval $I = (-\infty, \infty)$. Equations (1) and (2) with $f(x) = e^x$ and $a = 0$ give

$$e^x = 1 + x + \frac{x^2}{2!} + \cdots + \frac{x^n}{n!} + R_n(x) \qquad \text{\color{blue}Polynomial from Section 10.8, Example 2}$$

and

$$R_n(x) = \frac{e^c}{(n + 1)!} x^{n+1} \qquad \text{for some } c \text{ between } 0 \text{ and } x.$$

Since e^x is an increasing function of x, e^c lies between $e^0 = 1$ and e^x. When x is negative, so is c, and $e^c < 1$. When x is zero, $e^x = 1$ so that $R_n(x) = 0$. When x is positive, so is c, and $e^c < e^x$. Thus, for $R_n(x)$ given as above,

$$|R_n(x)| \le \frac{|x|^{n+1}}{(n + 1)!} \qquad \text{when } x \le 0, \qquad \text{\color{blue}$e^c < 1$ since $c < 0$}$$

and

$$|R_n(x)| < e^x \frac{x^{n+1}}{(n + 1)!} \qquad \text{when } x > 0. \qquad \text{\color{blue}$e^c < e^x$ since $c < x$}$$

Finally, because

$$\lim_{n \to \infty} \frac{x^{n+1}}{(n + 1)!} = 0 \qquad \text{for every } x, \qquad \text{\color{blue}Section 10.1, Theorem 5}$$

$\lim_{n \to \infty} R_n(x) = 0$, and the series converges to e^x for every x. Thus,

$$e^x = \sum_{k=0}^{\infty} \frac{x^k}{k!} = 1 + x + \frac{x^2}{2!} + \cdots + \frac{x^k}{k!} + \cdots. \tag{3}$$

The Number e as a Series

$$e = \sum_{n=0}^{\infty} \frac{1}{n!}$$

We can use the result of Example 1 with $x = 1$ to write

$$e = 1 + 1 + \frac{1}{2!} + \cdots + \frac{1}{n!} + R_n(1),$$

where for some c between 0 and 1,

$$R_n(1) = e^c \frac{1}{(n+1)!} < \frac{3}{(n+1)!}. \qquad e^c < e^1 < 3$$

Estimating the Remainder

It is often possible to estimate $R_n(x)$ as we did in Example 1. This method of estimation is so convenient that we state it as a theorem for future reference.

THEOREM 24—The Remainder Estimation Theorem
If there is a positive constant M such that $|f^{(n+1)}(t)| \le M$ for all t between x and a, inclusive, then the remainder term $R_n(x)$ in Taylor's Theorem satisfies the inequality

$$|R_n(x)| \le M \frac{|x-a|^{n+1}}{(n+1)!}.$$

If this inequality holds for every n and the other conditions of Taylor's Theorem are satisfied by f, then the series converges to $f(x)$.

The next two examples use Theorem 24 to show that the Taylor series generated by the sine and cosine functions do in fact converge to the functions themselves.

EXAMPLE 2   Show that the Taylor series for $\sin x$ at $x = 0$ converges for all x.

Solution   The function and its derivatives are

$$f(x) = \sin x, \qquad f'(x) = \cos x,$$
$$f''(x) = -\sin x, \qquad f'''(x) = -\cos x,$$
$$\vdots \qquad\qquad \vdots$$
$$f^{(2k)}(x) = (-1)^k \sin x, \qquad f^{(2k+1)}(x) = (-1)^k \cos x,$$

so

$$f^{(2k)}(0) = 0 \quad\text{and}\quad f^{(2k+1)}(0) = (-1)^k.$$

The series has only odd-powered terms and, for $n = 2k+1$, Taylor's Theorem gives

$$\sin x = x - \frac{x^3}{3!} + \frac{x^5}{5!} - \cdots + \frac{(-1)^k x^{2k+1}}{(2k+1)!} + R_{2k+1}(x).$$

All the derivatives of $\sin x$ have absolute values less than or equal to 1, so we can apply the Remainder Estimation Theorem with $M = 1$ to obtain

$$|R_{2k+1}(x)| \le 1 \cdot \frac{|x|^{2k+2}}{(2k+2)!}.$$

From Theorem 5, Rule 6, we have $(|x|^{2k+2}/(2k+2)!) \to 0$ as $k \to \infty$, whatever the value of x, so $R_{2k+1}(x) \to 0$ and the Maclaurin series for $\sin x$ converges to $\sin x$ for every x. Thus,

$$\sin x = x - \frac{x^3}{3!} + \frac{x^5}{5!} - \frac{x^7}{7!} + \cdots$$

$$\sin x = \sum_{k=0}^{\infty} \frac{(-1)^k x^{2k+1}}{(2k+1)!} = x - \frac{x^3}{3!} + \frac{x^5}{5!} - \frac{x^7}{7!} + \cdots. \qquad (4)$$

EXAMPLE 3   Show that the Taylor series for $\cos x$ at $x = 0$ converges to $\cos x$ for every value of x.

Solution   We add the remainder term to the Taylor polynomial for $\cos x$ (Section 10.8, Example 3) to obtain Taylor's formula for $\cos x$ with $n = 2k$:

$$\cos x = 1 - \frac{x^2}{2!} + \frac{x^4}{4!} - \cdots + (-1)^k \frac{x^{2k}}{(2k)!} + R_{2k}(x).$$

Because the derivatives of the cosine have absolute value less than or equal to 1, the Remainder Estimation Theorem with $M = 1$ gives

$$|R_{2k}(x)| \le 1 \cdot \frac{|x|^{2k+1}}{(2k+1)!}.$$

For every value of x, $R_{2k}(x) \to 0$ as $k \to \infty$. Therefore, the series converges to $\cos x$ for every value of x. Thus,

$$\cos x = 1 - \frac{x^2}{2!} + \frac{x^4}{4!} - \frac{x^6}{6!} + \cdots$$

$$\cos x = \sum_{k=0}^{\infty} \frac{(-1)^k x^{2k}}{(2k)!} = 1 - \frac{x^2}{2!} + \frac{x^4}{4!} - \frac{x^6}{6!} + \cdots. \tag{5}$$

Using Taylor Series

Since every Taylor series is a power series, the operations of adding, subtracting, and multiplying Taylor series are all valid on the intersection of their intervals of convergence.

EXAMPLE 4   Using known series, find the first few terms of the Taylor series for the given function by using power series operations.

(a) $\frac{1}{3}(2x + x \cos x)$ **(b)** $e^x \cos x$

Solution

(a) $\frac{1}{3}(2x + x \cos x) = \frac{2}{3}x + \frac{1}{3}x\left(1 - \frac{x^2}{2!} + \frac{x^4}{4!} - \cdots + (-1)^k \frac{x^{2k}}{(2k)!} + \cdots\right)$ Taylor series for $\cos x$

$$= \frac{2}{3}x + \frac{1}{3}x - \frac{x^3}{3!} + \frac{x^5}{3\cdot4!} - \cdots = x - \frac{x^3}{6} + \frac{x^5}{72} - \cdots$$

(b) $e^x \cos x = \left(1 + x + \frac{x^2}{2!} + \frac{x^3}{3!} + \frac{x^4}{4!} + \cdots\right)\left(1 - \frac{x^2}{2!} + \frac{x^4}{4!} - \cdots\right)$ Multiply the first series by each term of the second series.

$$= \left(1 + x + \frac{x^2}{2!} + \frac{x^3}{3!} + \frac{x^4}{4!} + \cdots\right) - \left(\frac{x^2}{2!} + \frac{x^3}{2!} + \frac{x^4}{2!2!} + \frac{x^5}{2!3!} + \cdots\right)$$

$$+ \left(\frac{x^4}{4!} + \frac{x^5}{4!} + \frac{x^6}{2!4!} + \cdots\right) + \cdots$$

$$= 1 + x - \frac{x^3}{3} - \frac{x^4}{6} + \cdots$$

By Theorem 20, we can use the Taylor series of the function f to find the Taylor series of $f(u(x))$ where $u(x)$ is any continuous function. The Taylor series resulting from this substitution will converge for all x such that $u(x)$ lies within the interval of convergence of the Taylor series of f. For instance, we can find the Taylor series for $\cos 2x$ by substituting $2x$ for x in the Taylor series for $\cos x$:

$$\cos 2x = \sum_{k=0}^{\infty} \frac{(-1)^k (2x)^{2k}}{(2k)!} = 1 - \frac{(2x)^2}{2!} + \frac{(2x)^4}{4!} - \frac{(2x)^6}{6!} + \cdots \qquad \text{Eq. (5) with } 2x \text{ for } x$$

$$= 1 - \frac{2^2 x^2}{2!} + \frac{2^4 x^4}{4!} - \frac{2^6 x^6}{6!} + \cdots$$

$$= \sum_{k=0}^{\infty} (-1)^k \frac{2^{2k} x^{2k}}{(2k)!}.$$

EXAMPLE 5   For what values of x can we replace $\sin x$ by $x - (x^3/3!)$ and obtain an error whose magnitude is no greater than 3×10^{-4}?

Solution   Here we can take advantage of the fact that the Taylor series for $\sin x$ is an alternating series for every nonzero value of x. According to the Alternating Series Estimation Theorem (Section 10.6), the error in truncating

$$\sin x = x - \frac{x^3}{3!} + \frac{x^5}{5!} - \frac{x^7}{7!} + \cdots$$

after $(x^3/3!)$ is no greater than

$$\left| \frac{x^5}{5!} \right| = \frac{|x|^5}{120}.$$

Therefore the error will be less than or equal to 3×10^{-4} if

$$\frac{|x|^5}{120} < 3 \times 10^{-4} \qquad \text{or} \qquad |x| < \sqrt[5]{360 \times 10^{-4}} \approx 0.514. \qquad \text{Rounded down, to be safe}$$

The Alternating Series Estimation Theorem tells us something that the Remainder Estimation Theorem does not: namely, that the estimate $x - (x^3/3!)$ for $\sin x$ is an underestimate when x is positive, because then $x^5/120$ is positive.

Figure 10.25 shows the graph of $\sin x$, along with the graphs of a number of its approximating Taylor polynomials. The graph of $P_3(x) = x - (x^3/3!)$ is almost indistinguishable from the sine curve when $0 \le x \le 1$.   ■

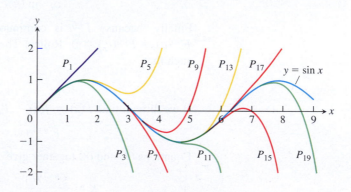

FIGURE 10.25 The polynomials

$$P_{2n+1}(x) = \sum_{k=0}^{n} \frac{(-1)^k x^{2k+1}}{(2k+1)!}$$

converge to $\sin x$ as $n \to \infty$. Notice how closely $P_3(x)$ approximates the sine curve for $x \le 1$ (Example 5).

A Proof of Taylor's Theorem

We prove Taylor's theorem assuming $a < b$. The proof for $a > b$ is nearly the same.

The Taylor polynomial

$$P_n(x) = f(a) + f'(a)(x - a) + \frac{f''(a)}{2!}(x - a)^2 + \cdots + \frac{f^{(n)}(a)}{n!}(x - a)^n$$

and its first n derivatives match the function f and its first n derivatives at $x = a$. We do not disturb that matching if we add another term of the form $K(x - a)^{n+1}$, where K is any constant, because such a term and its first n derivatives are all equal to zero at $x = a$. The new function

$$\phi_n(x) = P_n(x) + K(x - a)^{n+1}$$

and its first n derivatives still agree with f and its first n derivatives at $x = a$.

We now choose the particular value of K that makes the curve $y = \phi_n(x)$ agree with the original curve $y = f(x)$ at $x = b$. In symbols,

$$f(b) = P_n(b) + K(b - a)^{n+1}, \qquad \text{or} \qquad K = \frac{f(b) - P_n(b)}{(b - a)^{n+1}}. \tag{6}$$

With K defined by Equation (6), the function

$$F(x) = f(x) - \phi_n(x)$$

measures the difference between the original function f and the approximating function ϕ_n for each x in $[a, b]$.

We now use Rolle's Theorem (Section 4.2). First, because $F(a) = F(b) = 0$ and both F and F' are continuous on $[a, b]$, we know that

$$F'(c_1) = 0 \qquad \text{for some } c_1 \text{ in } (a, b).$$

Next, because $F'(a) = F'(c_1) = 0$ and both F' and F'' are continuous on $[a, c_1]$, we know that

$$F''(c_2) = 0 \qquad \text{for some } c_2 \text{ in } (a, c_1).$$

Rolle's Theorem, applied successively to $F'', F''', \ldots, F^{(n-1)}$, implies the existence of

$$c_3 \quad \text{in } (a, c_2) \qquad \text{such that } F'''(c_3) = 0,$$
$$c_4 \quad \text{in } (a, c_3) \qquad \text{such that } F^{(4)}(c_4) = 0,$$
$$\vdots$$
$$c_n \quad \text{in } (a, c_{n-1}) \qquad \text{such that } F^{(n)}(c_n) = 0.$$

Finally, because $F^{(n)}$ is continuous on $[a, c_n]$ and differentiable on (a, c_n), and $F^{(n)}(a) = F^{(n)}(c_n) = 0$, Rolle's Theorem implies that there is a number c_{n+1} in (a, c_n) such that

$$F^{(n+1)}(c_{n+1}) = 0. \tag{7}$$

If we differentiate $F(x) = f(x) - P_n(x) - K(x - a)^{n+1}$ a total of $n + 1$ times, we get

$$F^{(n+1)}(x) = f^{(n+1)}(x) - 0 - (n + 1)!K. \tag{8}$$

Equations (7) and (8) together give

$$K = \frac{f^{(n+1)}(c)}{(n + 1)!} \qquad \text{for some number } c = c_{n+1} \text{ in } (a, b). \tag{9}$$

Equations (6) and (9) give

$$f(b) = P_n(b) + \frac{f^{(n+1)}(c)}{(n + 1)!}(b - a)^{n+1}.$$

This concludes the proof.

EXERCISES 10.9

Finding Taylor Series

Use substitution (as in Example 4) to find the Taylor series at $x = 0$ of the functions in Exercises 1–12.

1. e^{-5x} **2.** $e^{-x/2}$ **3.** $5 \sin(-x)$

4. $\sin\left(\dfrac{\pi x}{2}\right)$ **5.** $\cos 5x^2$ **6.** $\cos\left(x^{2/3}/\sqrt{2}\right)$

7. $\ln(1 + x^2)$ **8.** $\tan^{-1}(3x^4)$ **9.** $\dfrac{1}{1 + \frac{3}{4}x^3}$

10. $\dfrac{1}{2 - x}$ **11.** $\ln(3 + 6x)$ **12.** $e^{-x^2 + \ln 5}$

Use power series operations to find the Taylor series at $x = 0$ for the functions in Exercises 13–30.

13. xe^x **14.** $x^2 \sin x$ **15.** $\dfrac{x^2}{2} - 1 + \cos x$

16. $\sin x - x + \dfrac{x^3}{3!}$ **17.** $x \cos \pi x$ **18.** $x^2 \cos(x^2)$

19. $\cos^2 x$ (*Hint:* $\cos^2 x = (1 + \cos 2x)/2$.)

20. $\sin^2 x$ **21.** $\dfrac{x^2}{1 - 2x}$ **22.** $x \ln(1 + 2x)$

23. $\dfrac{1}{(1 - x)^2}$ **24.** $\dfrac{2}{(1 - x)^3}$ **25.** $x \tan^{-1} x^2$

26. $\sin x \cdot \cos x$ **27.** $e^x + \dfrac{1}{1 + x}$ **28.** $\cos x - \sin x$

29. $\dfrac{x}{3} \ln(1 + x^2)$ **30.** $\ln(1 + x) - \ln(1 - x)$

Find the first four nonzero terms in the Maclaurin series for the functions in Exercises 31–38.

31. $e^x \sin x$ **32.** $\dfrac{\ln(1 + x)}{1 - x}$ **33.** $(\tan^{-1} x)^2$

34. $\cos^2 x \cdot \sin x$ **35.** $e^{\sin x}$ **36.** $\sin(\tan^{-1} x)$

37. $\cos(e^x - 1)$ **38.** $\cos\sqrt{x} + \ln(\cos x)$

Error Estimates

39. Estimate the error if $P_3(x) = x - (x^3/6)$ is used to estimate the value of $\sin x$ at $x = 0.1$.

40. Estimate the error if $P_4(x) = 1 + x + (x^2/2) + (x^3/6) + (x^4/24)$ is used to estimate the value of e^x at $x = 1/2$.

41. For approximately what values of x can you replace $\sin x$ by $x - (x^3/6)$ with an error of magnitude no greater than 5×10^{-4}? Give reasons for your answer.

42. If $\cos x$ is replaced by $1 - (x^2/2)$ and $|x| < 0.5$, what estimate can be made of the error? Does $1 - (x^2/2)$ tend to be too large, or too small? Give reasons for your answer.

43. How close is the approximation $\sin x = x$ when $|x| < 10^{-3}$? For which of these values of x is $x < \sin x$?

44. The estimate $\sqrt{1 + x} = 1 + (x/2)$ is used when x is small. Estimate the error when $|x| < 0.01$.

45. The approximation $e^x = 1 + x + (x^2/2)$ is used when x is small. Use the Remainder Estimation Theorem to estimate the error when $|x| < 0.1$.

46. (*Continuation of Exercise 45.*) When $x < 0$, the series for e^x is an alternating series. Use the Alternating Series Estimation Theorem to estimate the error that results from replacing e^x by $1 + x + (x^2/2)$ when $-0.1 < x < 0$. Compare your estimate with the one you obtained in Exercise 45.

Theory and Examples

47. Use the identity $\sin^2 x = (1 - \cos 2x)/2$ to obtain the Maclaurin series for $\sin^2 x$. Then differentiate this series to obtain the Maclaurin series for $2 \sin x \cos x$. Check that this is the series for $\sin 2x$.

48. (*Continuation of Exercise 47.*) Use the identity $\cos^2 x = \cos 2x + \sin^2 x$ to obtain a power series for $\cos^2 x$.

49. Taylor's Theorem and the Mean Value Theorem Explain how the Mean Value Theorem (Section 4.2, Theorem 4) is a special case of Taylor's Theorem.

50. Linearizations at inflection points Show that if the graph of a twice-differentiable function $f(x)$ has an inflection point at $x = a$, then the linearization of f at $x = a$ is also the quadratic approximation of f at $x = a$. This explains why tangent lines fit so well at inflection points.

51. The (second) second derivative test Use the equation

$$f(x) = f(a) + f'(a)(x - a) + \frac{f''(c_2)}{2}(x - a)^2$$

to establish the following test.

Let f have continuous first and second derivatives and suppose that $f'(a) = 0$. Then

a. f has a local maximum at a if $f'' \leq 0$ throughout an interval whose interior contains a;

b. f has a local minimum at a if $f'' \geq 0$ throughout an interval whose interior contains a.

52. A cubic approximation Use Taylor's formula with $a = 0$ and $n = 3$ to find the standard cubic approximation of $f(x) = 1/(1 - x)$ at $x = 0$. Give an upper bound for the magnitude of the error in the approximation when $|x| \leq 0.1$.

53. a. Use Taylor's formula with $n = 2$ to find the quadratic approximation of $f(x) = (1 + x)^k$ at $x = 0$ (k a constant).

b. If $k = 3$, for approximately what values of x in the interval $[0, 1]$ will the error in the quadratic approximation be less than $1/100$?

54. Improving approximations of π

a. Let P be an approximation of π accurate to n decimals. Show that $P + \sin P$ gives an approximation correct to $3n$ decimals. (*Hint:* Let $P = \pi + x$.)

[T] **b.** Try it with a calculator.

55. The Taylor series generated by $f(x) = \sum_{n=0}^{\infty} a_n x^n$ is $\sum_{n=0}^{\infty} a_n x^n$ A function defined by a power series $\sum_{n=0}^{\infty} a_n x^n$ with a radius of convergence $R > 0$ has a Taylor series that converges to the function at every point of $(-R, R)$. Show this by showing that the Taylor series generated by $f(x) = \sum_{n=0}^{\infty} a_n x^n$ is the series $\sum_{n=0}^{\infty} a_n x^n$ itself.

An immediate consequence of this is that series like

$$x \sin x = x^2 - \frac{x^4}{3!} + \frac{x^6}{5!} - \frac{x^8}{7!} + \cdots$$

and

$$x^2 e^x = x^2 + x^3 + \frac{x^4}{2!} + \frac{x^5}{3!} + \cdots,$$

obtained by multiplying Taylor series by powers of x, as well as series obtained by integration and differentiation of convergent power series, are themselves the Taylor series generated by the functions they represent.

56. Taylor series for even functions and odd functions (*Continuation of Section 10.7, Exercise 59.*) Suppose that $f(x) = \sum_{n=0}^{\infty} a_n x^n$ converges for all x in an open interval $(-R, R)$. Show that

a. If f is even, then $a_1 = a_3 = a_5 = \cdots = 0$, i.e., the Taylor series for f at $x = 0$ contains only even powers of x.

b. If f is odd, then $a_0 = a_2 = a_4 = \cdots = 0$, i.e., the Taylor series for f at $x = 0$ contains only odd powers of x.

COMPUTER EXPLORATIONS

Taylor's formula with $n = 1$ and $a = 0$ gives the linearization of a function at $x = 0$. With $n = 2$ and $n = 3$ we obtain the standard quadratic and cubic approximations. In these exercises we explore the errors associated with these approximations. We seek answers to two questions:

a. For what values of x can the function be replaced by each approximation with an error less than 10^{-2}?

b. What is the maximum error we could expect if we replace the function by each approximation over the specified interval?

Using a CAS, perform the following steps to aid in answering questions (a) and (b) for the functions and intervals in Exercises 57–62.

Step 1: Plot the function over the specified interval.

Step 2: Find the Taylor polynomials $P_1(x)$, $P_2(x)$, and $P_3(x)$ at $x = 0$.

Step 3: Calculate the $(n + 1)$st derivative $f^{(n+1)}(c)$ associated with the remainder term for each Taylor polynomial. Plot the derivative as a function of c over the specified interval and estimate its maximum absolute value, M.

Step 4: Calculate the remainder $R_n(x)$ for each polynomial. Using the estimate M from Step 3 in place of $f^{(n+1)}(c)$, plot $R_n(x)$ over the specified interval. Then estimate the values of x that answer question (a).

Step 5: Compare your estimated error with the actual error $E_n(x) = |f(x) - P_n(x)|$ by plotting $E_n(x)$ over the specified interval. This will help answer question (b).

Step 6: Graph the function and its three Taylor approximations together. Discuss the graphs in relation to the information discovered in Steps 4 and 5.

57. $f(x) = \dfrac{1}{\sqrt{1 + x}}, \quad |x| \le \dfrac{3}{4}$

58. $f(x) = (1 + x)^{3/2}, \quad -\dfrac{1}{2} \le x \le 2$

59. $f(x) = \dfrac{x}{x^2 + 1}, \quad |x| \le 2$

60. $f(x) = (\cos x)(\sin 2x), \quad |x| \le 2$

61. $f(x) = e^{-x} \cos 2x, \quad |x| \le 1$

62. $f(x) = e^{x/3} \sin 2x, \quad |x| \le 2$

10.10 Applications of Taylor Series

We can use Taylor series to solve problems that would otherwise be intractable. For example, many functions have antiderivatives that cannot be expressed using familiar functions. In this section we show how to evaluate integrals of such functions by giving them as Taylor series. We also show how to use Taylor series to evaluate limits that lead to indeterminate forms and how Taylor series can be used to extend the exponential function from real to complex numbers. We begin with a discussion of the binomial series, which comes from the Taylor series of the function $f(x) = (1 + x)^m$, and conclude the section with Table 10.1, which lists some commonly used Taylor series.

The Binomial Series for Powers and Roots

The Taylor series generated by $f(x) = (1 + x)^m$, when m is constant, is

$$1 + mx + \frac{m(m - 1)}{2!} x^2 + \frac{m(m - 1)(m - 2)}{3!} x^3 + \cdots$$

$$+ \frac{m(m - 1)(m - 2) \cdots (m - k + 1)}{k!} x^k + \cdots. \quad (1)$$

This series, called the **binomial series**, converges absolutely for $|x| < 1$. To derive the series, we first list the function and its derivatives:

$$f(x) = (1 + x)^m$$
$$f'(x) = m(1 + x)^{m-1}$$
$$f''(x) = m(m - 1)(1 + x)^{m-2}$$
$$f'''(x) = m(m - 1)(m - 2)(1 + x)^{m-3}$$
$$\vdots$$
$$f^{(k)}(x) = m(m - 1)(m - 2) \cdots (m - k + 1)(1 + x)^{m-k}.$$

We then evaluate these at $x = 0$ and substitute into the Taylor series formula to obtain Series (1).

If m is an integer greater than or equal to zero, the series stops after $(m + 1)$ terms because the coefficients from $k = m + 1$ on are zero.

If m is not a positive integer or zero, the series is infinite and converges for $|x| < 1$. To see why, let u_k be the term involving x^k. Then apply the Ratio Test for absolute convergence to see that

$$\left| \frac{u_{k+1}}{u_k} \right| = \left| \frac{m - k}{k + 1} x \right| \to |x| \qquad \text{as } k \to \infty.$$

Our derivation of the binomial series shows only that it is generated by $(1 + x)^m$ and converges for $|x| < 1$. The derivation does not show that the series converges to $(1 + x)^m$. It does, but we leave the proof to Exercise 58. The following formulation gives a succinct way to express the series.

The Binomial Series
For $-1 < x < 1$,

$$(1 + x)^m = 1 + \sum_{k=1}^{\infty} \binom{m}{k} x^k,$$

where we define

$$\binom{m}{1} = m, \qquad \binom{m}{2} = \frac{m(m - 1)}{2!},$$

and

$$\binom{m}{k} = \frac{m(m - 1)(m - 2) \cdots (m - k + 1)}{k!} \qquad \text{for } k \geq 3.$$

EXAMPLE 1   If $m = -1$,

$$\binom{-1}{1} = -1, \qquad \binom{-1}{2} = \frac{-1(-2)}{2!} = 1,$$

and

$$\binom{-1}{k} = \frac{-1(-2)(-3) \cdots (-1 - k + 1)}{k!} = (-1)^k \left(\frac{k!}{k!} \right) = (-1)^k.$$

With these coefficient values and with x replaced by $-x$, the binomial series formula gives the familiar geometric series

$$(1 + x)^{-1} = 1 + \sum_{k=1}^{\infty} (-1)^k x^k = 1 - x + x^2 - x^3 + \cdots + (-1)^k x^k + \cdots . \quad \blacksquare$$

EXAMPLE 2   We know from Section 3.11, Example 1, that $\sqrt{1 + x} \approx 1 + (x/2)$ for $|x|$ small. With $m = 1/2$, the binomial series gives quadratic and higher-order approximations as well, along with error estimates that come from the Alternating Series Estimation Theorem:

$$(1 + x)^{1/2} = 1 + \frac{x}{2} + \frac{\left(\frac{1}{2}\right)\left(-\frac{1}{2}\right)}{2!}x^2 + \frac{\left(\frac{1}{2}\right)\left(-\frac{1}{2}\right)\left(-\frac{3}{2}\right)}{3!}x^3$$

$$+ \frac{\left(\frac{1}{2}\right)\left(-\frac{1}{2}\right)\left(-\frac{3}{2}\right)\left(-\frac{5}{2}\right)}{4!}x^4 + \cdots$$

$$= 1 + \frac{x}{2} - \frac{x^2}{8} + \frac{x^3}{16} - \frac{5x^4}{128} + \cdots.$$

Substitution for x gives still other approximations. For example,

$$\sqrt{1 - x^2} \approx 1 - \frac{x^2}{2} - \frac{x^4}{8} \qquad \text{for } |x^2| \text{ small}$$

$$\sqrt{1 - \frac{1}{x}} \approx 1 - \frac{1}{2x} - \frac{1}{8x^2} \qquad \text{for } \left|\frac{1}{x}\right| \text{ small, that is, } |x| \text{ large.} \qquad \blacksquare$$

Evaluating Nonelementary Integrals

Sometimes we can use a familiar Taylor series to find the sum of a given power series in terms of a known function. For example,

$$x^2 - \frac{x^6}{3!} + \frac{x^{10}}{5!} - \frac{x^{14}}{7!} + \cdots = (x^2) - \frac{(x^2)^3}{3!} + \frac{(x^2)^5}{5!} - \frac{(x^2)^7}{7!} + \cdots = \sin x^2.$$

Additional examples are provided in Exercises 59–62.

Taylor series can be used to express nonelementary integrals in terms of series. Integrals like $\int \sin x^2 \, dx$ arise in the study of the diffraction of light.

EXAMPLE 3   Express $\int \sin x^2 \, dx$ as a power series.

Solution   From the series for $\sin x$ we substitute x^2 for x to obtain

$$\sin x^2 = x^2 - \frac{x^6}{3!} + \frac{x^{10}}{5!} - \frac{x^{14}}{7!} + \frac{x^{18}}{9!} - \cdots.$$

Therefore,

$$\int \sin x^2 \, dx = C + \frac{x^3}{3} - \frac{x^7}{7 \cdot 3!} + \frac{x^{11}}{11 \cdot 5!} - \frac{x^{15}}{15 \cdot 7!} + \frac{x^{19}}{19 \cdot 9!} - \cdots. \qquad \blacksquare$$

EXAMPLE 4   Estimate $\int_0^1 \sin x^2 \, dx$ with an error of less than 0.001.

Solution   From the indefinite integral in Example 3, we easily find that

$$\int_0^1 \sin x^2 \, dx = \frac{1}{3} - \frac{1}{7 \cdot 3!} + \frac{1}{11 \cdot 5!} - \frac{1}{15 \cdot 7!} + \frac{1}{19 \cdot 9!} - \cdots.$$

The series on the right-hand side alternates, and we find by numerical evaluations that

$$\frac{1}{11 \cdot 5!} \approx 0.00076$$

is the first term to be numerically less than 0.001. The sum of the preceding two terms gives

$$\int_0^1 \sin x^2 \, dx \approx \frac{1}{3} - \frac{1}{42} \approx 0.310.$$

With two more terms we could estimate

$$\int_0^1 \sin x^2 \, dx \approx 0.310268$$

with an error of less than 10^{-6}. With only one term beyond that we have

$$\int_0^1 \sin x^2 \, dx \approx \frac{1}{3} - \frac{1}{42} + \frac{1}{1320} - \frac{1}{75600} + \frac{1}{6894720} \approx 0.310268303,$$

with an error of about 1.08×10^{-9}. To guarantee this accuracy with the error formula for the Trapezoidal Rule would require using about 8000 subintervals. ∎

Arctangents

In Section 10.7, Example 5, we found a series for $\tan^{-1} x$ by differentiating to get

$$\frac{d}{dx}\tan^{-1} x = \frac{1}{1 + x^2} = 1 - x^2 + x^4 - x^6 + \cdots$$

and then integrating to get

$$\tan^{-1} x = x - \frac{x^3}{3} + \frac{x^5}{5} - \frac{x^7}{7} + \cdots.$$

However, we did not prove the term-by-term integration theorem on which this conclusion depended. We now derive the series again by integrating both sides of the finite formula

$$\frac{1}{1 + t^2} = 1 - t^2 + t^4 - t^6 + \cdots + (-1)^n t^{2n} + \frac{(-1)^{n+1} t^{2n+2}}{1 + t^2}, \tag{2}$$

in which the last term comes from adding the remaining terms as a geometric series with first term $a = (-1)^{n+1} t^{2n+2}$ and ratio $r = -t^2$. Integrating both sides of Equation (2) from $t = 0$ to $t = x$ gives

$$\tan^{-1} x = x - \frac{x^3}{3} + \frac{x^5}{5} - \frac{x^7}{7} + \cdots + (-1)^n \frac{x^{2n+1}}{2n + 1} + R_n(x),$$

where

$$R_n(x) = \int_0^x \frac{(-1)^{n+1} t^{2n+2}}{1 + t^2} \, dt.$$

The denominator of the integrand is greater than or equal to 1; hence

$$|R_n(x)| \leq \int_0^{|x|} t^{2n+2} \, dt = \frac{|x|^{2n+3}}{2n + 3}.$$

If $|x| \leq 1$, the right side of this inequality approaches zero as $n \to \infty$. Therefore $\lim_{n \to \infty} R_n(x) = 0$ if $|x| \leq 1$ and

$$\tan^{-1} x = \sum_{n=0}^{\infty} \frac{(-1)^n x^{2n+1}}{2n + 1}, \qquad |x| \leq 1.$$

$$\tan^{-1} x = x - \frac{x^3}{3} + \frac{x^5}{5} - \frac{x^7}{7} + \cdots, \qquad |x| \leq 1. \tag{3}$$

We take this route instead of finding the Taylor series directly because the formulas for the higher-order derivatives of $\tan^{-1} x$ are unmanageable. When we put $x = 1$ in Equation (3), we get **Leibniz's formula**:

$$\frac{\pi}{4} = 1 - \frac{1}{3} + \frac{1}{5} - \frac{1}{7} + \frac{1}{9} - \cdots + \frac{(-1)^n}{2n + 1} + \cdots .$$

Because this series converges very slowly, it is not used in approximating π to many decimal places. The series for $\tan^{-1} x$ converges most rapidly when x is near zero. For that reason, people who use the series for $\tan^{-1} x$ to compute π use various trigonometric identities.

For example, if

$$\alpha = \tan^{-1} \frac{1}{2} \quad \text{and} \quad \beta = \tan^{-1} \frac{1}{3},$$

then

$$\tan(\alpha + \beta) = \frac{\tan \alpha + \tan \beta}{1 - \tan \alpha \tan \beta} = \frac{\frac{1}{2} + \frac{1}{3}}{1 - \frac{1}{6}} = 1 = \tan \frac{\pi}{4}$$

and therefore

$$\frac{\pi}{4} = \alpha + \beta = \tan^{-1} \frac{1}{2} + \tan^{-1} \frac{1}{3}.$$

Now Equation (3) may be used with $x = 1/2$ to evaluate $\tan^{-1}(1/2)$ and with $x = 1/3$ to give $\tan^{-1}(1/3)$. The sum of these results, multiplied by 4, gives π.

Evaluating Indeterminate Forms

We can sometimes evaluate indeterminate forms by expressing the functions involved as Taylor series.

EXAMPLE 5   Evaluate

$$\lim_{x \to 1} \frac{\ln x}{x - 1}.$$

Solution   We represent $\ln x$ as a Taylor series in powers of $x - 1$. This can be accomplished by calculating the Taylor series generated by $\ln x$ at $x = 1$ directly or by replacing x by $x - 1$ in the series for $\ln(1 + x)$ in Section 10.7, Example 6. Either way, we obtain

$$\ln x = (x - 1) - \frac{1}{2}(x - 1)^2 + \cdots ,$$

from which we find that

$$\lim_{x \to 1} \frac{\ln x}{x - 1} = \lim_{x \to 1} \left(1 - \frac{1}{2}(x - 1) + \cdots \right) = 1.$$

Of course, this particular limit can be evaluated using l'Hôpital's Rule just as well.   ■

EXAMPLE 6   Evaluate

$$\lim_{x \to 0} \frac{\sin x - \tan x}{x^3}.$$

Solution   The Taylor series for $\sin x$ and $\tan x$, to terms in x^5, are

$$\sin x = x - \frac{x^3}{3!} + \frac{x^5}{5!} - \cdots, \qquad \tan x = x + \frac{x^3}{3} + \frac{2x^5}{15} + \cdots.$$

Subtracting the series term by term, it follows that

$$\sin x - \tan x = -\frac{x^3}{2} - \frac{x^5}{8} - \cdots = x^3 \left(-\frac{1}{2} - \frac{x^2}{8} - \cdots \right).$$

Division of both sides by x^3 and taking limits then gives

$$\lim_{x \to 0} \frac{\sin x - \tan x}{x^3} = \lim_{x \to 0} \left(-\frac{1}{2} - \frac{x^2}{8} - \cdots \right) = -\frac{1}{2}.$$ ■

If we apply series to calculate $\lim_{x \to 0}((1/\sin x) - (1/x))$, we not only find the limit successfully but also discover an approximation formula for $\csc x$.

EXAMPLE 7   Find $\lim_{x \to 0} \left(\dfrac{1}{\sin x} - \dfrac{1}{x} \right)$.

Solution   Using algebra and the Taylor series for $\sin x$, we have

$$\frac{1}{\sin x} - \frac{1}{x} = \frac{x - \sin x}{x \sin x} = \frac{x - \left(x - \dfrac{x^3}{3!} + \dfrac{x^5}{5!} - \cdots \right)}{x \cdot \left(x - \dfrac{x^3}{3!} + \dfrac{x^5}{5!} - \cdots \right)}$$

$$= \frac{x^3 \left(\dfrac{1}{3!} - \dfrac{x^2}{5!} + \cdots \right)}{x^2 \left(1 - \dfrac{x^2}{3!} + \cdots \right)} = x \cdot \frac{\dfrac{1}{3!} - \dfrac{x^2}{5!} + \cdots}{1 - \dfrac{x^2}{3!} + \cdots}.$$

Therefore,

$$\lim_{x \to 0} \left(\frac{1}{\sin x} - \frac{1}{x} \right) = \lim_{x \to 0} \left(x \cdot \frac{\dfrac{1}{3!} - \dfrac{x^2}{5!} + \cdots}{1 - \dfrac{x^2}{3!} + \cdots} \right) = 0.$$

From the quotient on the right, we can see that if $|x|$ is small, then

$$\frac{1}{\sin x} - \frac{1}{x} \approx x \cdot \frac{1}{3!} = \frac{x}{6} \qquad \text{or} \qquad \csc x \approx \frac{1}{x} + \frac{x}{6}.$$ ■

Euler's Identity

A complex number is a number of the form $a + bi$, where a and b are real numbers and $i = \sqrt{-1}$ (see Appendix 7). If we substitute $x = i\theta$ (θ real) in the Taylor series for e^x and use the relations

$$i^2 = -1, \qquad i^3 = i^2 i = -i, \qquad i^4 = i^2 i^2 = 1, \qquad i^5 = i^4 i = i,$$

and so on, to simplify the result, we obtain

$$e^{i\theta} = 1 + \frac{i\theta}{1!} + \frac{i^2\theta^2}{2!} + \frac{i^3\theta^3}{3!} + \frac{i^4\theta^4}{4!} + \frac{i^5\theta^5}{5!} + \frac{i^6\theta^6}{6!} + \cdots$$

$$= \left(1 - \frac{\theta^2}{2!} + \frac{\theta^4}{4!} - \frac{\theta^6}{6!} + \cdots\right) + i\left(\theta - \frac{\theta^3}{3!} + \frac{\theta^5}{5!} - \cdots\right) = \cos\theta + i\sin\theta.$$

This does not *prove* that $e^{i\theta} = \cos\theta + i\sin\theta$ because we have not yet defined what it means to raise e to an imaginary power. Rather, it tells us how to define $e^{i\theta}$ so that its properties are consistent with the properties of the exponential function for real numbers.

DEFINITION

For any real number θ, $e^{i\theta} = \cos\theta + i\sin\theta$.   (4)

Equation (4), called **Euler's identity**, enables us to define e^{a+bi} to be $e^a \cdot e^{bi}$ for any complex number $a + bi$. So

$$e^{a+ib} = e^a(\cos b + i\sin b).$$

One consequence of this identity is the equation

$$e^{i\pi} = -1.$$

When written in the form $e^{i\pi} + 1 = 0$, this equation combines five of the most important constants in mathematics.

TABLE 10.1 Frequently Used Taylor Series

$$\frac{1}{1-x} = 1 + x + x^2 + \cdots + x^n + \cdots = \sum_{n=0}^{\infty} x^n, \qquad |x| < 1$$

$$\frac{1}{1+x} = 1 - x + x^2 - \cdots + (-x)^n + \cdots = \sum_{n=0}^{\infty} (-1)^n x^n, \qquad |x| < 1$$

$$e^x = 1 + x + \frac{x^2}{2!} + \cdots + \frac{x^n}{n!} + \cdots = \sum_{n=0}^{\infty} \frac{x^n}{n!}, \qquad |x| < \infty$$

$$\sin x = x - \frac{x^3}{3!} + \frac{x^5}{5!} - \cdots + (-1)^n \frac{x^{2n+1}}{(2n+1)!} + \cdots = \sum_{n=0}^{\infty} \frac{(-1)^n x^{2n+1}}{(2n+1)!}, \qquad |x| < \infty$$

$$\cos x = 1 - \frac{x^2}{2!} + \frac{x^4}{4!} - \cdots + (-1)^n \frac{x^{2n}}{(2n)!} + \cdots = \sum_{n=0}^{\infty} \frac{(-1)^n x^{2n}}{(2n)!}, \qquad |x| < \infty$$

$$\ln(1+x) = x - \frac{x^2}{2} + \frac{x^3}{3} - \cdots + (-1)^{n-1} \frac{x^n}{n} + \cdots = \sum_{n=1}^{\infty} \frac{(-1)^{n-1} x^n}{n}, \qquad -1 < x \le 1$$

$$\tan^{-1} x = x - \frac{x^3}{3} + \frac{x^5}{5} - \cdots + (-1)^n \frac{x^{2n+1}}{2n+1} + \cdots = \sum_{n=0}^{\infty} \frac{(-1)^n x^{2n+1}}{2n+1}, \qquad |x| \le 1$$

EXERCISES 10.10

Binomial Series

Find the first four terms of the binomial series for the functions in Exercises 1–10.

1. $(1 + x)^{1/2}$

2. $(1 + x)^{1/3}$

3. $(1 - x)^{-3}$

4. $(1 - 2x)^{1/2}$

5. $\left(1 + \dfrac{x}{2}\right)^{-2}$

6. $\left(1 - \dfrac{x}{3}\right)^{4}$

7. $(1 + x^3)^{-1/2}$

8. $(1 + x^2)^{-1/3}$

9. $\left(1 + \dfrac{1}{x}\right)^{1/2}$

10. $\dfrac{x}{\sqrt[3]{1 + x}}$

Find the binomial series for the functions in Exercises 11–14.

11. $(1 + x)^4$

12. $(1 + x^2)^3$

13. $(1 - 2x)^3$

14. $\left(1 - \dfrac{x}{2}\right)^4$

Approximations and Nonelementary Integrals

T In Exercises 15–18, use series to estimate the integrals' values with an error of magnitude less than 10^{-5}. (The answer section gives the integrals' values rounded to seven decimal places.)

15. $\displaystyle\int_0^{0.6} \sin x^2 \, dx$

16. $\displaystyle\int_0^{0.4} \dfrac{e^{-x} - 1}{x} \, dx$

17. $\displaystyle\int_0^{0.5} \dfrac{1}{\sqrt{1 + x^4}} \, dx$

18. $\displaystyle\int_0^{0.35} \sqrt[3]{1 + x^2} \, dx$

T Use series to approximate the values of the integrals in Exercises 19–22 with an error of magnitude less than 10^{-8}.

19. $\displaystyle\int_0^{0.1} \dfrac{\sin x}{x} \, dx$

20. $\displaystyle\int_0^{0.1} e^{-x^2} \, dx$

21. $\displaystyle\int_0^{0.1} \sqrt{1 + x^4} \, dx$

22. $\displaystyle\int_0^{1} \dfrac{1 - \cos x}{x^2} \, dx$

23. Estimate the error if $\cos t^2$ is approximated by $1 - \dfrac{t^4}{2} + \dfrac{t^8}{4!}$ in the integral $\int_0^1 \cos t^2 \, dt$.

24. Estimate the error if $\cos \sqrt{t}$ is approximated by $1 - \dfrac{t}{2} + \dfrac{t^2}{4!} - \dfrac{t^3}{6!}$ in the integral $\int_0^1 \cos \sqrt{t} \, dt$.

In Exercises 25–28, find a polynomial that will approximate $F(x)$ throughout the given interval with an error of magnitude less than 10^{-3}.

25. $F(x) = \displaystyle\int_0^x \sin t^2 \, dt, \quad [0, 1]$

26. $F(x) = \displaystyle\int_0^x t^2 e^{-t^2} \, dt, \quad [0, 1]$

27. $F(x) = \displaystyle\int_0^x \tan^{-1} t \, dt, \quad$ **(a)** $[0, 0.5]$ **(b)** $[0, 1]$

28. $F(x) = \displaystyle\int_0^x \dfrac{\ln(1 + t)}{t} \, dt, \quad$ **(a)** $[0, 0.5]$ **(b)** $[0, 1]$

Indeterminate Forms

Use series to evaluate the limits in Exercises 29–40.

29. $\displaystyle\lim_{x \to 0} \dfrac{e^x - (1 + x)}{x^2}$

30. $\displaystyle\lim_{x \to 0} \dfrac{e^x - e^{-x}}{x}$

31. $\displaystyle\lim_{t \to 0} \dfrac{1 - \cos t - (t^2/2)}{t^4}$

32. $\displaystyle\lim_{\theta \to 0} \dfrac{\sin \theta - \theta + (\theta^3/6)}{\theta^5}$

33. $\displaystyle\lim_{y \to 0} \dfrac{y - \tan^{-1} y}{y^3}$

34. $\displaystyle\lim_{y \to 0} \dfrac{\tan^{-1} y - \sin y}{y^3 \cos y}$

35. $\displaystyle\lim_{x \to \infty} x^2 \left(e^{-1/x^2} - 1\right)$

36. $\displaystyle\lim_{x \to \infty} (x + 1) \sin \dfrac{1}{x + 1}$

37. $\displaystyle\lim_{x \to 0} \dfrac{\ln(1 + x^2)}{1 - \cos x}$

38. $\displaystyle\lim_{x \to 2} \dfrac{x^2 - 4}{\ln(x - 1)}$

39. $\displaystyle\lim_{x \to 0} \dfrac{\sin 3x^2}{1 - \cos 2x}$

40. $\displaystyle\lim_{x \to 0} \dfrac{\ln(1 + x^3)}{x \cdot \sin x^2}$

Using Table 10.1

In Exercises 41–52, use Table 10.1 to find the sum of each series.

41. $1 + 1 + \dfrac{1}{2!} + \dfrac{1}{3!} + \dfrac{1}{4!} + \cdots$

42. $\left(\dfrac{1}{4}\right)^3 + \left(\dfrac{1}{4}\right)^4 + \left(\dfrac{1}{4}\right)^5 + \left(\dfrac{1}{4}\right)^6 + \cdots$

43. $1 - \dfrac{3^2}{4^2 \cdot 2!} + \dfrac{3^4}{4^4 \cdot 4!} - \dfrac{3^6}{4^6 \cdot 6!} + \cdots$

44. $\dfrac{1}{2} - \dfrac{1}{2 \cdot 2^2} + \dfrac{1}{3 \cdot 2^3} - \dfrac{1}{4 \cdot 2^4} + \cdots$

45. $\dfrac{\pi}{3} - \dfrac{\pi^3}{3^3 \cdot 3!} + \dfrac{\pi^5}{3^5 \cdot 5!} - \dfrac{\pi^7}{3^7 \cdot 7!} + \cdots$

46. $\dfrac{2}{3} - \dfrac{2^3}{3^3 \cdot 3} + \dfrac{2^5}{3^5 \cdot 5} - \dfrac{2^7}{3^7 \cdot 7} + \cdots$

47. $x^3 + x^4 + x^5 + x^6 + \cdots$

48. $1 - \dfrac{3^2 x^2}{2!} + \dfrac{3^4 x^4}{4!} - \dfrac{3^6 x^6}{6!} + \cdots$

49. $x^3 - x^5 + x^7 - x^9 + x^{11} - \cdots$

50. $x^2 - 2x^3 + \dfrac{2^2 x^4}{2!} - \dfrac{2^3 x^5}{3!} + \dfrac{2^4 x^6}{4!} - \cdots$

51. $-1 + 2x - 3x^2 + 4x^3 - 5x^4 + \cdots$

52. $1 + \dfrac{x}{2} + \dfrac{x^2}{3} + \dfrac{x^3}{4} + \dfrac{x^4}{5} + \cdots$

Theory and Examples

53. Replace x by $-x$ in the Taylor series for $\ln(1 + x)$ to obtain a series for $\ln(1 - x)$. Then subtract this from the Taylor series for $\ln(1 + x)$ to show that for $|x| < 1$,

$$\ln\dfrac{1 + x}{1 - x} = 2\left(x + \dfrac{x^3}{3} + \dfrac{x^5}{5} + \cdots\right).$$

54. How many terms of the Taylor series for $\ln(1 + x)$ should you add to be sure of calculating $\ln(1.1)$ with an error of magnitude less than 10^{-8}? Give reasons for your answer.

55. According to the Alternating Series Estimation Theorem, how many terms of the Taylor series for $\tan^{-1} 1$ would you have to add to be sure of finding $\pi/4$ with an error of magnitude less than 10^{-3}? Give reasons for your answer.

56. Show that the Taylor series for $f(x) = \tan^{-1} x$ diverges for $|x| > 1$.

T **57. Estimating Pi** About how many terms of the Taylor series for $\tan^{-1} x$ would you have to use to evaluate each term on the right-hand side of the equation

$$\pi = 48 \tan^{-1}\frac{1}{18} + 32 \tan^{-1}\frac{1}{57} - 20 \tan^{-1}\frac{1}{239}$$

with an error of magnitude less than 10^{-6}? In contrast, the convergence of $\sum_{n=1}^{\infty}(1/n^2)$ to $\pi^2/6$ is so slow that even 50 terms will not yield two-place accuracy.

58. Use the following steps to prove that the binomial series in Equation (1) converges to $(1 + x)^m$.

a. Differentiate the series

$$f(x) = 1 + \sum_{k=1}^{\infty}\binom{m}{k}x^k$$

to show that

$$f'(x) = \frac{mf(x)}{1 + x}, \quad -1 < x < 1.$$

b. Define $g(x) = (1 + x)^{-m} f(x)$ and show that $g'(x) = 0$.

c. From part (b), show that

$$f(x) = (1 + x)^m.$$

59. a. Use the binomial series and the fact that

$$\frac{d}{dx}\sin^{-1} x = (1 - x^2)^{-1/2}$$

to generate the first four nonzero terms of the Taylor series for $\sin^{-1} x$. What is the radius of convergence?

b. **Series for $\cos^{-1} x$** Use your result in part (a) to find the first five nonzero terms of the Taylor series for $\cos^{-1} x$.

60. a. Series for $\sinh^{-1} x$ Find the first four nonzero terms of the Taylor series for

$$\sinh^{-1} x = \int_0^x \frac{dt}{\sqrt{1 + t^2}}.$$

T **b.** Use the first *three* terms of the series in part (a) to estimate $\sinh^{-1} 0.25$. Give an upper bound for the magnitude of the estimation error.

61. Obtain the Taylor series for $1/(1 + x)^2$ from the series for $-1/(1 + x)$.

62. Use the Taylor series for $1/(1 - x^2)$ to obtain a series for $2x/(1 - x^2)^2$.

T **63. Estimating Pi** The English mathematician Wallis discovered the formula

$$\frac{\pi}{4} = \frac{2 \cdot 4 \cdot 4 \cdot 6 \cdot 6 \cdot 8 \cdot \cdots}{3 \cdot 3 \cdot 5 \cdot 5 \cdot 7 \cdot 7 \cdot \cdots}.$$

Find π to two decimal places with this formula.

64. The complete elliptic integral of the first kind is the integral

$$K = \int_0^{\pi/2} \frac{d\theta}{\sqrt{1 - k^2 \sin^2 \theta}},$$

where $0 < k < 1$ is constant.

a. Show that the first four terms of the binomial series for $1/\sqrt{1 - x}$ are

$$(1 - x)^{-1/2} = 1 + \frac{1}{2}x + \frac{1 \cdot 3}{2 \cdot 4}x^2 + \frac{1 \cdot 3 \cdot 5}{2 \cdot 4 \cdot 6}x^3 + \cdots.$$

b. From part (a) and the reduction integral Formula 67 at the back of the book, show that

$$K = \frac{\pi}{2}\left[1 + \left(\frac{1}{2}\right)^2 k^2 + \left(\frac{1 \cdot 3}{2 \cdot 4}\right)^2 k^4 + \left(\frac{1 \cdot 3 \cdot 5}{2 \cdot 4 \cdot 6}\right)^2 k^6 + \cdots\right].$$

65. Series for $\sin^{-1} x$ Integrate the binomial series for $(1 - x^2)^{-1/2}$ to show that for $|x| < 1$,

$$\sin^{-1} x = x + \sum_{n=1}^{\infty} \frac{1 \cdot 3 \cdot 5 \cdot \cdots \cdot (2n - 1)}{2 \cdot 4 \cdot 6 \cdot \cdots \cdot (2n)} \frac{x^{2n+1}}{2n + 1}.$$

66. Series for $\tan^{-1} x$ for $|x| > 1$ Derive the series

$$\tan^{-1} x = \frac{\pi}{2} - \frac{1}{x} + \frac{1}{3x^3} - \frac{1}{5x^5} + \cdots, \quad x > 1$$

$$\tan^{-1} x = -\frac{\pi}{2} - \frac{1}{x} + \frac{1}{3x^3} - \frac{1}{5x^5} + \cdots, \quad x < -1,$$

by integrating the series

$$\frac{1}{1 + t^2} = \frac{1}{t^2} \cdot \frac{1}{1 + (1/t^2)} = \frac{1}{t^2} - \frac{1}{t^4} + \frac{1}{t^6} - \frac{1}{t^8} + \cdots$$

in the first case from x to ∞ and in the second case from $-\infty$ to x.

Euler's Identity

67. Use Equation (4) to write the following powers of e in the form $a + bi$.

a. $e^{-i\pi}$ b. $e^{i\pi/4}$ c. $e^{-i\pi/2}$

68. Use Equation (4) to show that

$$\cos \theta = \frac{e^{i\theta} + e^{-i\theta}}{2} \quad \text{and} \quad \sin \theta = \frac{e^{i\theta} - e^{-i\theta}}{2i}.$$

69. Establish the equations in Exercise 68 by combining the formal Taylor series for $e^{i\theta}$ and $e^{-i\theta}$.

70. Show that

a. $\cosh i\theta = \cos \theta$,   b. $\sinh i\theta = i \sin \theta$.

71. By multiplying the Taylor series for e^x and $\sin x$, find the terms through x^5 of the Taylor series for $e^x \sin x$. This series is the imaginary part of the series for

$$e^x \cdot e^{ix} = e^{(1+i)x}.$$

Use this fact to check your answer. For what values of x should the series for $e^x \sin x$ converge?

72. When a and b are real, we define $e^{(a+ib)x}$ with the equation

$$e^{(a+ib)x} = e^{ax} \cdot e^{ibx} = e^{ax}(\cos bx + i \sin bx).$$

Differentiate the right-hand side of this equation to show that

$$\frac{d}{dx}e^{(a+ib)x} = (a + ib)e^{(a+ib)x}.$$

Thus the familiar rule $(d/dx)e^{kx} = ke^{kx}$ holds for k complex as well as real.

73. Use the definition of $e^{i\theta}$ to show that for any real numbers θ, θ_1, and θ_2,

a. $e^{i\theta_1}e^{i\theta_2} = e^{i(\theta_1 + \theta_2)}$,

b. $e^{-i\theta} = 1/e^{i\theta}$.

74. Two complex numbers $a + ib$ and $c + id$ are equal if and only if $a = c$ and $b = d$. Use this fact to evaluate

$$\int e^{ax}\cos bx\, dx \quad \text{and} \quad \int e^{ax}\sin bx\, dx$$

from

$$\int e^{(a+ib)x}\, dx = \frac{a - ib}{a^2 + b^2}e^{(a+ib)x} + C,$$

where $C = C_1 + iC_2$ is a complex constant of integration.

CHAPTER 10   Questions to Guide Your Review

1. What is an infinite sequence? What does it mean for such a sequence to converge? To diverge? Give examples.

2. What is a monotonic sequence? Under what circumstances does such a sequence have a limit? Give examples.

3. What theorems are available for calculating limits of sequences? Give examples.

4. What theorem sometimes enables us to use l'Hôpital's Rule to calculate the limit of a sequence? Give an example.

5. What are the six commonly occurring limits in Theorem 5 that arise frequently when you work with sequences and series?

6. What is an infinite series? What does it mean for such a series to converge? To diverge? Give examples.

7. What is a geometric series? When does such a series converge? Diverge? When it does converge, what is its sum? Give examples.

8. Besides geometric series, what other convergent and divergent series do you know?

9. What is the nth-Term Test for Divergence? What is the idea behind the test?

10. What can be said about term-by-term sums and differences of convergent series? About constant multiples of convergent and divergent series?

11. What happens if you add a finite number of terms to a convergent series? A divergent series? What happens if you delete a finite number of terms from a convergent series? A divergent series?

12. How do you reindex a series? Why might you want to do this?

13. Under what circumstances will an infinite series of nonnegative terms converge? Diverge? Why study series of nonnegative terms?

14. What is the Integral Test? What is the reasoning behind it? Give an example of its use.

15. When do p-series converge? Diverge? How do you know? Give examples of convergent and divergent p-series.

16. What are the Direct Comparison Test and the Limit Comparison Test? What is the reasoning behind these tests? Give examples of their use.

17. What are the Ratio and Root Tests? Do they always give you the information you need to determine convergence or divergence? Give examples.

18. What is absolute convergence? Conditional convergence? How are the two related?

19. What is an alternating series? What theorem is available for determining the convergence of such a series?

20. How can you estimate the error involved in approximating the sum of an alternating series with one of the series' partial sums? What is the reasoning behind the estimate?

21. What do you know about rearranging the terms of an absolutely convergent series? Of a conditionally convergent series?

22. What is a power series? How do you test a power series for convergence? What are the possible outcomes?

23. What are the basic facts about

    a. sums, differences, and products of power series?

    b. substitution of a function for x in a power series?

    c. term-by-term differentiation of power series?

    d. term-by-term integration of power series?

    e. Give examples.

24. What is the Taylor series generated by a function $f(x)$ at a point $x = a$? What information do you need about f to construct the series? Give an example.

25. What is a Maclaurin series?

26. Does a Taylor series always converge to its generating function? Explain.

27. What are Taylor polynomials? Of what use are they?

28. What is Taylor's formula? What does it say about the errors involved in using Taylor polynomials to approximate functions? In particular, what does Taylor's formula say about the error in a linearization? A quadratic approximation?

29. What is the binomial series? On what interval does it converge? How is it used?

30. How can you sometimes use power series to estimate the values of nonelementary definite integrals? To find limits?

31. What are the Taylor series for $1/(1 - x)$, $1/(1 + x)$, e^x, $\sin x$, $\cos x$, $\ln(1 + x)$, and $\tan^{-1}x$? How do you estimate the errors involved in replacing these series with their partial sums?

CHAPTER 10 Practice Exercises

Determining Convergence of Sequences

Which of the sequences whose nth terms appear in Exercises 1–18 converge, and which diverge? Find the limit of each convergent sequence.

1. $a_n = 1 + \dfrac{(-1)^n}{n}$

2. $a_n = \dfrac{1 - (-1)^n}{\sqrt{n}}$

3. $a_n = \dfrac{1 - 2^n}{2^n}$

4. $a_n = 1 + (0.9)^n$

5. $a_n = \sin \dfrac{n\pi}{2}$

6. $a_n = \sin n\pi$

7. $a_n = \dfrac{\ln(n^2)}{n}$

8. $a_n = \dfrac{\ln(2n + 1)}{n}$

9. $a_n = \dfrac{n + \ln n}{n}$

10. $a_n = \dfrac{\ln(2n^3 + 1)}{n}$

11. $a_n = \left(\dfrac{n - 5}{n}\right)^n$

12. $a_n = \left(1 + \dfrac{1}{n}\right)^{-n}$

13. $a_n = \sqrt[n]{\dfrac{3^n}{n}}$

14. $a_n = \left(\dfrac{3}{n}\right)^{1/n}$

15. $a_n = n(2^{1/n} - 1)$

16. $a_n = \sqrt[n]{2n + 1}$

17. $a_n = \dfrac{(n + 1)!}{n!}$

18. $a_n = \dfrac{(-4)^n}{n!}$

Convergent Series

Find the sums of the series in Exercises 19–24.

19. $\displaystyle\sum_{n=3}^{\infty} \dfrac{1}{(2n - 3)(2n - 1)}$

20. $\displaystyle\sum_{n=2}^{\infty} \dfrac{-2}{n(n + 1)}$

21. $\displaystyle\sum_{n=1}^{\infty} \dfrac{9}{(3n - 1)(3n + 2)}$

22. $\displaystyle\sum_{n=3}^{\infty} \dfrac{-8}{(4n - 3)(4n + 1)}$

23. $\displaystyle\sum_{n=0}^{\infty} e^{-n}$

24. $\displaystyle\sum_{n=1}^{\infty} (-1)^n \dfrac{3}{4^n}$

Determining Convergence of Series

Which of the series in Exercises 25–44 converge absolutely, which converge conditionally, and which diverge? Give reasons for your answers.

25. $\displaystyle\sum_{n=1}^{\infty} \dfrac{1}{\sqrt{n}}$

26. $\displaystyle\sum_{n=1}^{\infty} \dfrac{-5}{n}$

27. $\displaystyle\sum_{n=1}^{\infty} \dfrac{(-1)^n}{\sqrt{n}}$

28. $\displaystyle\sum_{n=1}^{\infty} \dfrac{1}{2n^3}$

29. $\displaystyle\sum_{n=1}^{\infty} \dfrac{(-1)^n}{\ln(n + 1)}$

30. $\displaystyle\sum_{n=2}^{\infty} \dfrac{1}{n(\ln n)^2}$

31. $\displaystyle\sum_{n=1}^{\infty} \dfrac{\ln n}{n^3}$

32. $\displaystyle\sum_{n=3}^{\infty} \dfrac{\ln n}{\ln(\ln n)}$

33. $\displaystyle\sum_{n=1}^{\infty} \dfrac{(-1)^n}{n\sqrt{n^2 + 1}}$

34. $\displaystyle\sum_{n=1}^{\infty} \dfrac{(-1)^n 3n^2}{n^3 + 1}$

35. $\displaystyle\sum_{n=1}^{\infty} \dfrac{n + 1}{n!}$

36. $\displaystyle\sum_{n=1}^{\infty} \dfrac{(-1)^n(n^2 + 1)}{2n^2 + n - 1}$

37. $\displaystyle\sum_{n=1}^{\infty} \dfrac{(-3)^n}{n!}$

38. $\displaystyle\sum_{n=1}^{\infty} \dfrac{2^n 3^n}{n^n}$

39. $\displaystyle\sum_{n=1}^{\infty} \dfrac{1}{\sqrt{n(n + 1)(n + 2)}}$

40. $\displaystyle\sum_{n=2}^{\infty} \dfrac{1}{n\sqrt{n^2 - 1}}$

41. $1 - \left(\dfrac{1}{\sqrt{3}}\right)^2 + \left(\dfrac{1}{\sqrt{3}}\right)^4 - \left(\dfrac{1}{\sqrt{3}}\right)^6 + \left(\dfrac{1}{\sqrt{3}}\right)^8 - \cdots$

42. $\displaystyle\sum_{n=0}^{\infty} \dfrac{(-1)^n}{e^{-n} + 1}$

43. $\displaystyle\sum_{n=0}^{\infty} \dfrac{1}{1 + r + r^2 + \cdots + r^n}, \quad \text{for } -1 < r < 1$

44. $\displaystyle\sum_{n=1}^{\infty} \dfrac{(-1)^n}{\sqrt{n + 100} - \sqrt{n}}$

Power Series

In Exercises 45–54, **(a)** find the series' radius and interval of convergence. Then identify the values of x for which the series converges **(b)** absolutely and **(c)** conditionally.

45. $\displaystyle\sum_{n=1}^{\infty} \dfrac{(x + 4)^n}{n3^n}$

46. $\displaystyle\sum_{n=1}^{\infty} \dfrac{(x - 1)^{2n-2}}{(2n - 1)!}$

47. $\displaystyle\sum_{n=1}^{\infty} \dfrac{(-1)^{n-1}(3x - 1)^n}{n^2}$

48. $\displaystyle\sum_{n=0}^{\infty} \dfrac{(n + 1)(2x + 1)^n}{(2n + 1)2^n}$

49. $\displaystyle\sum_{n=1}^{\infty} \dfrac{x^n}{n^n}$

50. $\displaystyle\sum_{n=1}^{\infty} \dfrac{x^n}{\sqrt{n}}$

51. $\displaystyle\sum_{n=0}^{\infty} \dfrac{(n + 1)x^{2n-1}}{3^n}$

52. $\displaystyle\sum_{n=0}^{\infty} \dfrac{(-1)^n(x - 1)^{2n+1}}{2n + 1}$

53. $\displaystyle\sum_{n=1}^{\infty} (\operatorname{csch} n)x^n$

54. $\displaystyle\sum_{n=1}^{\infty} (\coth n)x^n$

Maclaurin Series

Each of the series in Exercises 55–60 is the value of the Taylor series at $x = 0$ of a function $f(x)$ at a particular point. What function and what point? What is the sum of the series?

55. $1 - \dfrac{1}{4} + \dfrac{1}{16} - \cdots + (-1)^n \dfrac{1}{4^n} + \cdots$

56. $\dfrac{2}{3} - \dfrac{4}{18} + \dfrac{8}{81} - \cdots + (-1)^{n-1} \dfrac{2^n}{n3^n} + \cdots$

57. $\pi - \dfrac{\pi^3}{3!} + \dfrac{\pi^5}{5!} - \cdots + (-1)^n \dfrac{\pi^{2n+1}}{(2n + 1)!} + \cdots$

58. $1 - \dfrac{\pi^2}{9 \cdot 2!} + \dfrac{\pi^4}{81 \cdot 4!} - \cdots + (-1)^n \dfrac{\pi^{2n}}{3^{2n}(2n)!} + \cdots$

59. $1 + \ln 2 + \dfrac{(\ln 2)^2}{2!} + \cdots + \dfrac{(\ln 2)^n}{n!} + \cdots$

60. $\dfrac{1}{\sqrt{3}} - \dfrac{1}{9\sqrt{3}} + \dfrac{1}{45\sqrt{3}} - \cdots$
$+ (-1)^{n-1} \dfrac{1}{(2n - 1)(\sqrt{3})^{2n-1}} + \cdots$

Find Taylor series at $x = 0$ for the functions in Exercises 61–68.

61. $\dfrac{1}{1 - 2x}$

62. $\dfrac{1}{1 + x^3}$

63. $\sin \pi x$

64. $\sin \dfrac{2x}{3}$

65. $\cos \left(x^{5/3}\right)$

66. $\cos \dfrac{x^3}{\sqrt{5}}$

67. $e^{(\pi x/2)}$

68. e^{-x^2}

Taylor Series

In Exercises 69–72, find the first four nonzero terms of the Taylor series generated by f at $x = a$.

69. $f(x) = \sqrt{3 + x^2}$ at $x = -1$

70. $f(x) = 1/(1 - x)$ at $x = 2$

71. $f(x) = 1/(x + 1)$ at $x = 3$

72. $f(x) = 1/x$ at $x = a > 0$

Nonelementary Integrals

Use series to approximate the values of the integrals in Exercises 73–76 with an error of magnitude less than 10^{-8}. (The answer section gives the integrals' values rounded to 10 decimal places.)

73. $\displaystyle\int_0^{1/2} e^{-x^3}\,dx$

74. $\displaystyle\int_0^1 x\sin\left(x^3\right)\,dx$

75. $\displaystyle\int_0^{1/2} \dfrac{\tan^{-1}x}{x}\,dx$

76. $\displaystyle\int_0^{1/64} \dfrac{\tan^{-1}x}{\sqrt{x}}\,dx$

Using Series to Find Limits

In Exercises 77–82:

a. Use power series to evaluate the limit.

T **b.** Then use a grapher to support your calculation.

77. $\displaystyle\lim_{x \to 0} \dfrac{7 \sin x}{e^{2x} - 1}$

78. $\displaystyle\lim_{\theta \to 0} \dfrac{e^\theta - e^{-\theta} - 2\theta}{\theta - \sin \theta}$

79. $\displaystyle\lim_{t \to 0}\left(\dfrac{1}{2 - 2\cos t} - \dfrac{1}{t^2}\right)$

80. $\displaystyle\lim_{h \to 0} \dfrac{(\sin h)/h - \cos h}{h^2}$

81. $\displaystyle\lim_{z \to 0} \dfrac{1 - \cos^2 z}{\ln(1 - z) + \sin z}$

82. $\displaystyle\lim_{y \to 0} \dfrac{y^2}{\cos y - \cosh y}$

Theory and Examples

83. Use a series representation of $\sin 3x$ to find values of r and s for which

$$\lim_{x \to 0}\left(\dfrac{\sin 3x}{x^3} + \dfrac{r}{x^2} + s\right) = 0.$$

T **84.** Compare the accuracies of the approximations $\sin x \approx x$ and $\sin x \approx 6x/(6 + x^2)$ by comparing the graphs of $f(x) = \sin x - x$ and $g(x) = \sin x - (6x/(6 + x^2))$. Describe what you find.

85. Find the radius of convergence of the series

$$\sum_{n=1}^{\infty} \dfrac{2 \cdot 5 \cdot 8 \cdot \cdots \cdot (3n - 1)}{2 \cdot 4 \cdot 6 \cdot \cdots \cdot (2n)}x^n.$$

86. Find the radius of convergence of the series

$$\sum_{n=1}^{\infty} \dfrac{3 \cdot 5 \cdot 7 \cdot \cdots \cdot (2n + 1)}{4 \cdot 9 \cdot 14 \cdot \cdots \cdot (5n - 1)}(x - 1)^n.$$

87. Find a closed-form formula for the nth partial sum of the series $\sum_{n=2}^{\infty}\ln\left(1 - (1/n^2)\right)$ and use it to determine the convergence or divergence of the series.

88. Evaluate $\sum_{k=2}^{\infty}\left(1/(k^2 - 1)\right)$ by finding the limits as $n \to \infty$ of the series' nth partial sum.

89. a. Find the interval of convergence of the series

$$y = 1 + \dfrac{1}{6}x^3 + \dfrac{1}{180}x^6 + \cdots$$

$$+ \dfrac{1 \cdot 4 \cdot 7 \cdot \cdots \cdot (3n - 2)}{(3n)!}x^{3n} + \cdots.$$

b. Show that the function defined by the series satisfies a differential equation of the form

$$\dfrac{d^2y}{dx^2} = x^a y + b$$

and find the values of the constants a and b.

90. a. Find the Maclaurin series for the function $x^2/(1 + x)$.

b. Does the series converge at $x = 1$? Explain.

91. If $\sum_{n=1}^{\infty} a_n$ and $\sum_{n=1}^{\infty} b_n$ are convergent series of nonnegative numbers, can anything be said about $\sum_{n=1}^{\infty} a_n b_n$? Give reasons for your answer.

92. If $\sum_{n=1}^{\infty} a_n$ and $\sum_{n=1}^{\infty} b_n$ are divergent series of nonnegative numbers, can anything be said about $\sum_{n=1}^{\infty} a_n b_n$? Give reasons for your answer.

93. Prove that the sequence $\{x_n\}$ and the series $\sum_{k=1}^{\infty}(x_{k+1} - x_k)$ both converge or both diverge.

94. Prove that $\sum_{n=1}^{\infty}(a_n/(1 + a_n))$ converges if $a_n > 0$ for all n and $\sum_{n=1}^{\infty} a_n$ converges.

95. Suppose that $a_1, a_2, a_3, \ldots, a_n$ are positive numbers satisfying the following conditions:

i) $a_1 \geq a_2 \geq a_3 \geq \cdots$;

ii) the series $a_2 + a_4 + a_8 + a_{16} + \cdots$ diverges.

Show that the series

$$\dfrac{a_1}{1} + \dfrac{a_2}{2} + \dfrac{a_3}{3} + \cdots$$

diverges.

96. Use the result in Exercise 95 to show that

$$1 + \sum_{n=2}^{\infty} \dfrac{1}{n \ln n}$$

diverges.

97. Show that if $a_n > 0$ and $\sum_{n=1}^{\infty} a_n$ converges, then $\sum_{n=1}^{\infty} \frac{\sqrt{a_n}}{n}$ converges.

98. Determine whether $\sum_{n=1}^{\infty} b_n$ converges or diverges.

    a. $b_1 = 1, \quad b_{n+1} = (-1)^n \frac{n+1}{3n+2} b_n$

    b. $b_1 = 3, \quad b_{n+1} = \frac{n}{\ln n} b_n$

99. Assume that $b_n > 0$ and $\sum_{n=1}^{\infty} b_n$ converges. What, if anything, can be said about the following series?

    a. $\sum_{n=1}^{\infty} \tan(b_n)$

    b. $\sum_{n=1}^{\infty} \ln(1 + b_n)$

    c. $\sum_{n=1}^{\infty} \ln(2 + b_n)$

100. Consider the convergent series $\sum_{n=1}^{\infty} \frac{(-1)^n}{e^n + e^{cn}}$, where c is a constant. What should c be so that the first 10 terms of the series estimate the sum of the entire series with an error of less than 0.00001?

101. Assume that the following sequence has a limit L. Find the value of L.

$$4^{1/3}, (4(4^{1/3}))^{1/3}, (4(4(4^{1/3}))^{1/3})^{1/3}, (4(4(4(4^{1/3}))^{1/3})^{1/3})^{1/3}, \dots$$

102. Consider the infinite sequence of shaded right triangles in the accompanying diagram. Compute the total area of the triangles.

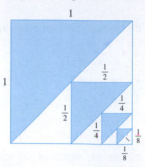

CHAPTER 10   Additional and Advanced Exercises

Determining Convergence of Series

Which of the series $\sum_{n=1}^{\infty} a_n$ defined by the formulas in Exercises 1–4 converge, and which diverge? Give reasons for your answers.

1. $\sum_{n=1}^{\infty} \frac{1}{(3n-2)^{n+(1/2)}}$ **2.** $\sum_{n=1}^{\infty} \frac{(\tan^{-1} n)^2}{n^2 + 1}$

3. $\sum_{n=1}^{\infty} (-1)^n \tanh n$ **4.** $\sum_{n=2}^{\infty} \frac{\log_n(n!)}{n^3}$

Which of the series $\sum_{n=1}^{\infty} a_n$ defined by the formulas in Exercises 5–8 converge, and which diverge? Give reasons for your answers.

5. $a_1 = 1, \quad a_{n+1} = \frac{n(n+1)}{(n+2)(n+3)} a_n$

    (*Hint:* Write out several terms, see which factors cancel, and then generalize.)

6. $a_1 = a_2 = 7, \quad a_{n+1} = \frac{n}{(n-1)(n+1)} a_n$ if $n \geq 2$

7. $a_1 = a_2 = 1, \quad a_{n+1} = \frac{1}{1 + a_n}$ if $n \geq 2$

8. $a_n = 1/3^n$ if n is odd, $a_n = n/3^n$ if n is even

Choosing Centers for Taylor Series

Taylor's formula

$$f(x) = f(a) + f'(a)(x - a) + \frac{f''(a)}{2!}(x - a)^2 + \cdots$$
$$+ \frac{f^{(n)}(a)}{n!}(x - a)^n + \frac{f^{(n+1)}(c)}{(n+1)!}(x - a)^{n+1}$$

expresses the value of f at x in terms of the values of f and its derivatives at $x = a$. In numerical computations, we therefore need a to be a

point where we know the values of f and its derivatives. We also need a to be close enough to the values of f we are interested in to make $(x - a)^{n+1}$ so small we can neglect the remainder.

In Exercises 9–14, what Taylor series would you choose to represent the function near the given value of x? (There may be more than one good answer.) Write out the first four nonzero terms of the series you choose.

9. $\cos x$ near $x = 1$ **10.** $\sin x$ near $x = 6.3$

11. e^x near $x = 0.4$ **12.** $\ln x$ near $x = 1.3$

13. $\cos x$ near $x = 69$ **14.** $\tan^{-1} x$ near $x = 2$

Theory and Examples

15. Let a and b be constants with $0 < a < b$. Does the sequence $\{(a^n + b^n)^{1/n}\}$ converge? If it does converge, what is the limit?

16. Find the sum of the infinite series

$$1 + \frac{2}{10} + \frac{3}{10^2} + \frac{7}{10^3} + \frac{2}{10^4} + \frac{3}{10^5} + \frac{7}{10^6} + \frac{2}{10^7}$$
$$+ \frac{3}{10^8} + \frac{7}{10^9} + \cdots.$$

17. Evaluate

$$\sum_{n=0}^{\infty} \int_n^{n+1} \frac{1}{1 + x^2} \, dx.$$

18. Find all values of x for which

$$\sum_{n=1}^{\infty} \frac{nx^n}{(n+1)(2x+1)^n}$$

converges absolutely.

19. a. Does the value of

$$\lim_{n \to \infty} \left(1 - \frac{\cos(a/n)}{n}\right)^n, \quad a \text{ constant,}$$

appear to depend on the value of a? If so, how?

b. Does the value of

$$\lim_{n \to \infty} \left(1 - \frac{\cos(a/n)}{bn}\right)^n, \quad a \text{ and } b \text{ constant, } b \neq 0,$$

appear to depend on the value of b? If so, how?

c. Use calculus to confirm your findings in parts (a) and (b).

20. Show that if $\sum_{n=1}^{\infty} a_n$ converges, then

$$\sum_{n=1}^{\infty} \left(\frac{1 + \sin(a_n)}{2}\right)^n$$

converges.

21. Find a value for the constant b that will make the radius of convergence of the power series

$$\sum_{n=2}^{\infty} \frac{b^n x^n}{\ln n}$$

equal to 5.

22. How do you know that the functions $\sin x$, $\ln x$, and e^x are not polynomials? Give reasons for your answer.

23. Find the value of a for which the limit

$$\lim_{x \to 0} \frac{\sin(ax) - \sin x - x}{x^3}$$

is finite and evaluate the limit.

24. Find values of a and b for which

$$\lim_{x \to 0} \frac{\cos(ax) - b}{2x^2} = -1.$$

25. Raabe's (or Gauss's) Test The following test, which we state without proof, is an extension of the Ratio Test.

Raabe's Test: If $\sum_{n=1}^{\infty} u_n$ is a series of positive constants and there exist constants C, K, and N such that

$$\frac{u_n}{u_{n+1}} = 1 + \frac{C}{n} + \frac{f(n)}{n^2},$$

where $|f(n)| < K$ for $n \geq N$, then $\sum_{n=1}^{\infty} u_n$ converges if $C > 1$ and diverges if $C \leq 1$.

Show that the results of Raabe's Test agree with what you know about the series $\sum_{n=1}^{\infty} (1/n^2)$ and $\sum_{n=1}^{\infty} (1/n)$.

26. (*Continuation of Exercise 25.*) Suppose that the terms of $\sum_{n=1}^{\infty} u_n$ are defined recursively by the formulas

$$u_1 = 1, \quad u_{n+1} = \frac{(2n-1)^2}{(2n)(2n+1)} u_n.$$

Apply Raabe's Test to determine whether the series converges.

27. If $\sum_{n=1}^{\infty} a_n$ converges, and if $a_n \neq 1$ and $a_n > 0$ for all n,

a. Show that $\sum_{n=1}^{\infty} a_n^2$ converges.

b. Does $\sum_{n=1}^{\infty} a_n/(1 - a_n)$ converge? Explain.

28. (*Continuation of Exercise 27.*) If $\sum_{n=1}^{\infty} a_n$ converges, and if $1 > a_n > 0$ for all n, show that $\sum_{n=1}^{\infty} \ln(1 - a_n)$ converges.

(*Hint:* First show that $|\ln(1 - a_n)| \leq a_n/(1 - a_n)$.)

29. Nicole Oresme's Theorem Prove Nicole Oresme's Theorem that

$$1 + \frac{1}{2} \cdot 2 + \frac{1}{4} \cdot 3 + \cdots + \frac{n}{2^{n-1}} + \cdots = 4.$$

(*Hint:* Differentiate both sides of the equation $1/(1 - x) = 1 + \sum_{n=1}^{\infty} x^n$.)

30. a. Show that

$$\sum_{n=1}^{\infty} \frac{n(n+1)}{x^n} = \frac{2x^2}{(x-1)^3}$$

for $|x| > 1$ by differentiating the identity

$$\sum_{n=1}^{\infty} x^{n+1} = \frac{x^2}{1 - x}$$

twice, multiplying the result by x, and then replacing x by $1/x$.

b. Use part (a) to find the real solution greater than 1 of the equation

$$x = \sum_{n=1}^{\infty} \frac{n(n+1)}{x^n}.$$

31. Quality control

a. Differentiate the series

$$\frac{1}{1 - x} = 1 + x + x^2 + \cdots + x^n + \cdots$$

to obtain a series for $1/(1 - x)^2$.

b. In one throw of two dice, the probability of getting a roll of 7 is $p = 1/6$. If you throw the dice repeatedly, the probability that a 7 will appear for the first time at the nth throw is $q^{n-1}p$, where $q = 1 - p = 5/6$. The expected number of throws until a 7 first appears is $\sum_{n=1}^{\infty} nq^{n-1}p$. Find the sum of this series.

c. As an engineer applying statistical control to an industrial operation, you inspect items taken at random from the assembly line. You classify each sampled item as either "good" or "bad." If the probability of an item's being good is p and of an item's being bad is $q = 1 - p$, the probability that the first bad item found is the nth one inspected is $p^{n-1}q$. The average number inspected up to and including the first bad item found is $\sum_{n=1}^{\infty} np^{n-1}q$. Evaluate this sum, assuming $0 < p < 1$.

32. Expected value Suppose that a random variable X may assume the values 1, 2, 3, . . . , with probabilities $p_1, p_2, p_3, \ldots$, where p_k is the probability that X equals k ($k = 1, 2, 3, \ldots$). Suppose also that $p_k \geq 0$ and that $\sum_{k=1}^{\infty} p_k = 1$. The **expected value** of X, denoted by $E(X)$, is the number $\sum_{k=1}^{\infty} kp_k$, provided the series converges. In each of the following cases, show that $\sum_{k=1}^{\infty} p_k = 1$ and find $E(X)$ if it exists. (*Hint:* See Exercise 31.)

a. $p_k = 2^{-k}$ **b.** $p_k = \dfrac{5^{k-1}}{6^k}$

c. $p_k = \dfrac{1}{k(k+1)} = \dfrac{1}{k} - \dfrac{1}{k+1}$

T **33. Safe and effective dosage** The concentration in the blood resulting from a single dose of a drug normally decreases with time as the drug is eliminated from the body. Doses may therefore need to be repeated periodically to keep the concentration from dropping below some particular level. One model for the effect of repeated doses gives the residual concentration just before the $(n + 1)$st dose as

$$R_n = C_0 e^{-kt_0} + C_0 e^{-2kt_0} + \cdots + C_0 e^{-nkt_0},$$

where C_0 = the change in concentration achievable by a single dose (mg/mL), k = the *elimination constant* (h^{-1}), and t_0 = time between doses (h). See the accompanying figure.

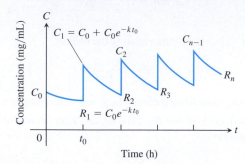

Time (h)

a. Write R_n in closed from as a single fraction, and find $R = \lim_{n \to \infty} R_n$.

b. Calculate R_1 and R_{10} for $C_0 = 1$ mg/mL, $k = 0.1$ h^{-1}, and $t_0 = 10$ h. How good an estimate of R is R_{10}?

c. If $k = 0.01$ h^{-1} and $t_0 = 10$ h, find the smallest n such that $R_n > (1/2)R$. Use $C_0 = 1$ mg/mL.
(*Source: Prescribing Safe and Effective Dosage*, B. Horelick and S. Koont, COMAP, Inc., Lexington, MA.)

34. Time between drug doses (*Continuation of Exercise 33.*) If a drug is known to be ineffective below a concentration C_L and harmful above some higher concentration C_H, one need to find values of C_0 and t_0 that will produce a concentration that is safe (not above C_H) but effective (not below C_L). See the accompanying figure. We therefore want to find values for C_0 and t_0 for which

$$R = C_L \quad \text{and} \quad C_0 + R = C_H.$$

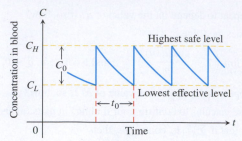

Thus $C_0 = C_H - C_L$. When these values are substituted in the equation for R obtained in part (a) of Exercise 33, the resulting equation simplifies to

$$t_0 = \frac{1}{k} \ln \frac{C_H}{C_L}.$$

To reach an effective level rapidly, one might administer a "loading" dose that would produce a concentration of C_H mg/mL. This could be followed every t_0 hours by a dose that raises the concentration by $C_0 = C_H - C_L$ mg/mL.

a. Verify the preceding equation for t_0.

b. If $k = 0.05$ h^{-1} and the highest safe concentration is e times the lowest effective concentration, find the length of time between doses that will ensure safe and effective concentrations.

c. Given $C_H = 2$ mg/mL, $C_L = 0.5$ mg/mL, and $k = 0.02$ h^{-1}, determine a scheme for administering the drug.

d. Suppose that $k = 0.2$ h^{-1} and that the smallest effective concentration is 0.03 mg/mL. A single dose that produces a concentration of 0.1 mg/mL is administered. About how long will the drug remain effective?

CHAPTER 10 Technology Application Projects

Mathematica/Maple Projects

Projects can be found within MyMathLab.

- **Bouncing Ball**
  The model predicts the height of a bouncing ball, and the time until it stops bouncing.

- **Taylor Polynomial Approximations of a Function**
  A graphical animation shows the convergence of the Taylor polynomials to functions having derivatives of all orders over an interval in their domains.

11

Parametric Equations and Polar Coordinates

OVERVIEW In this chapter we study new ways to define curves in the plane. Instead of thinking of a curve as the graph of a function or equation, we think of it as the path of a moving particle whose position is changing over time. Then each of the x- and y-coordinates of the particle's position becomes a function of a third variable t. We can also change the way in which points in the plane themselves are described by using *polar coordinates* rather than the rectangular or Cartesian system. Both of these new tools are useful for describing motion, like that of planets and satellites, or projectiles moving in the plane or space.

11.1 Parametrizations of Plane Curves

Parametric Equations

Figure 11.1 shows the path of a moving particle in the xy-plane. Notice that the path fails the vertical line test, so it cannot be described as the graph of a function of the variable x. However, we can sometimes describe the path by a pair of equations, $x = f(t)$ and $y = g(t)$, where f and g are continuous functions. When studying motion, t usually denotes time. Equations like these can describe more general curves than those described by a single function, and they provide not only the graph of the path traced out but also the location of the particle $(x, y) = (f(t), g(t))$ at any time t.

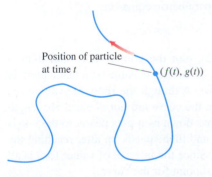

Position of particle at time t — $(f(t), g(t))$

FIGURE 11.1 The curve or path traced by a particle moving in the xy-plane is not always the graph of a function or single equation.

> **DEFINITION** If x and y are given as functions
>
> $$x = f(t), \qquad y = g(t)$$
>
> over an interval I of t-values, then the set of points $(x, y) = (f(t), g(t))$ defined by these equations is a **parametric curve**. The equations are **parametric equations** for the curve.

The variable t is a **parameter** for the curve, and its domain I is the **parameter interval**. If I is a closed interval, $a \le t \le b$, the point $(f(a), g(a))$ is the **initial point** of the curve and $(f(b), g(b))$ is the **terminal point**. When we give parametric equations and a parameter interval for a curve, we say that we have **parametrized** the curve. The equations and interval together constitute a **parametrization** of the curve. A given curve can be represented by different sets of parametric equations. (See Exercises 29 and 30.)

EXAMPLE 1 Sketch the curve defined by the parametric equations

$$x = \sin \pi t/2, \qquad y = t, \qquad 0 \le t \le 6.$$

Solution We make a table of values (Table 11.1), plot the points (x, y), and draw a smooth curve through them (Figure 11.2). If we think of the curve as the path of a moving particle, the particle starts at time $t = 0$ at the initial point $(0, 0)$ and then moves upward in a wavy path until at time $t = 6$ it reaches the terminal point $(0, 6)$. The direction of motion is shown by the arrows in Figure 11.2.

TABLE 11.1 Values of $x = \sin \pi t/2$ and $y = t$ for selected values of t.

t	x	y
0	0	0
1	1	1
2	0	2
3	−1	3
4	0	4
5	1	5
6	0	6

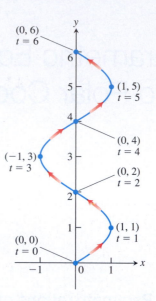

FIGURE 11.2 The curve given by the parametric equations $x = \sin \pi t/2$ and $y = t$ (Example 1).

EXAMPLE 2 Sketch the curve defined by the parametric equations

$$x = t^2, \qquad y = t + 1, \qquad -\infty < t < \infty.$$

Solution We make a table of values (Table 11.2), plot the points (x, y), and draw a smooth curve through them (Figure 11.3). We think of the curve as the path that a particle moves along the curve in the direction of the arrows. Although the time intervals in the table are equal, the consecutive points plotted along the curve are not at equal arc length distances. The reason for this is that the particle slows down as it gets nearer to the y-axis along the lower branch of the curve as t increases, and then speeds up after reaching the y-axis at $(0, 1)$ and moving along the upper branch. Since the interval of values for t is all real numbers, there is no initial point and no terminal point for the curve.

TABLE 11.2 Values of $x = t^2$ and $y = t + 1$ for selected values of t.

t	x	y
−3	9	−2
−2	4	−1
−1	1	0
0	0	1
1	1	2
2	4	3
3	9	4

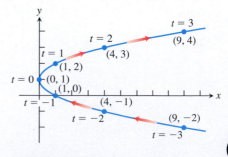

FIGURE 11.3 The curve given by the parametric equations $x = t^2$ and $y = t + 1$ (Example 2).

For this example we can use algebraic manipulation to eliminate the parameter t and obtain an algebraic equation for the curve in terms of x and y alone. We solve $y = t + 1$ for t and substitute the resulting equation $t = y - 1$ into the equation for x, which yields

$$x = t^2 = (y - 1)^2 = y^2 - 2y + 1.$$

The equation $x = y^2 - 2y + 1$ represents a parabola, as displayed in Figure 11.3. It is sometimes quite difficult, or even impossible, to eliminate the parameter from a pair of parametric equations, as we did here. ∎

EXAMPLE 3   Graph the parametric curves

(a) $x = \cos t,$ $y = \sin t,$ $0 \le t \le 2\pi.$

(b) $x = a \cos t,$ $y = a \sin t,$ $0 \le t \le 2\pi.$

Solution

(a) Since $x^2 + y^2 = \cos^2 t + \sin^2 t = 1$, the parametric curve lies along the unit circle $x^2 + y^2 = 1$. As t increases from 0 to 2π, the point $(x, y) = (\cos t, \sin t)$ starts at $(1, 0)$ and traces the entire circle once counterclockwise (Figure 11.4).

(b) For $x = a \cos t$, $y = a \sin t$, $0 \le t \le 2\pi$, we have $x^2 + y^2 = a^2 \cos^2 t + a^2 \sin^2 t = a^2$. The parametrization describes a motion that begins at the point $(a, 0)$ and traverses the circle $x^2 + y^2 = a^2$ once counterclockwise, returning to $(a, 0)$ at $t = 2\pi$. The graph is a circle centered at the origin with radius $r = |a|$ and coordinate points $(a \cos t, a \sin t)$. ∎

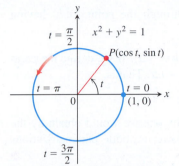

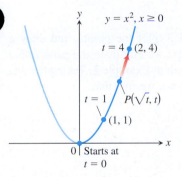

FIGURE 11.4   The equations $x = \cos t$ and $y = \sin t$ describe motion on the circle $x^2 + y^2 = 1$. The arrow shows the direction of increasing t (Example 3).

EXAMPLE 4   The position $P(x, y)$ of a particle moving in the xy-plane is given by the equations and parameter interval

$$x = \sqrt{t}, \qquad y = t, \qquad t \ge 0.$$

Identify the path traced by the particle and describe the motion.

Solution   We try to identify the path by eliminating t between the equations $x = \sqrt{t}$ and $y = t$, which might produce a re-cognizable algebraic relation between x and y. We find that

$$y = t = \left(\sqrt{t}\right)^2 = x^2.$$

Thus, the particle's position coordinates satisfy the equation $y = x^2$, so the particle moves along the parabola $y = x^2$.

It would be a mistake, however, to conclude that the particle's path is the entire parabola $y = x^2$; it is only half the parabola. The particle's x-coordinate is never negative. The particle starts at $(0, 0)$ when $t = 0$ and rises into the first quadrant as t increases (Figure 11.5). The parameter interval is $[0, \infty)$ and there is no terminal point. ∎

FIGURE 11.5   The equations $x = \sqrt{t}$ and $y = t$ and the interval $t \ge 0$ describe the path of a particle that traces the right-hand half of the parabola $y = x^2$ (Example 4).

The graph of any function $y = f(x)$ can always be given a **natural parametrization** $x = t$ and $y = f(t)$. The domain of the parameter in this case is the same as the domain of the function f.

EXAMPLE 5   A parametrization of the graph of the function $f(x) = x^2$ is given by

$$x = t, \qquad y = f(t) = t^2, \qquad -\infty < t < \infty.$$

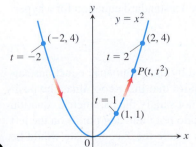

When $t \ge 0$, this parametrization gives the same path in the xy-plane as we had in Example 4. However, since the parameter t here can now also be negative, we obtain the left-hand part of the parabola as well; that is, we have the entire parabolic curve. For this parametrization, there is no starting point and no terminal point (Figure 11.6). ∎

FIGURE 11.6   The path defined by $x = t, y = t^2, -\infty < t < \infty$ is the entire parabola $y = x^2$ (Example 5).

Notice that a parametrization also specifies *when* a particle moving along the curve is *located* at a specific point along the curve. In Example 4, the point $(2, 4)$ is reached when $t = 4$; in Example 5, it is reached "earlier" when $t = 2$. You can see the implications of this aspect of parametrizations when considering the possibility of two objects coming into collision: they have to be at the exact same location point $P(x, y)$ for some (possibly different) values of their respective parameters. We will say more about this aspect of parametrizations when we study motion in Chapter 13.

EXAMPLE 6 Find a parametrization for the line through the point (a, b) having slope m.

Solution A Cartesian equation of the line is $y - b = m(x - a)$. If we define the parameter t by $t = x - a$, we find that $x = a + t$ and $y - b = mt$. That is,

$$x = a + t, \qquad y = b + mt, \qquad -\infty < t < \infty$$

parametrizes the line. This parametrization differs from the one we would obtain by the natural parametrization in Example 5 when $t = x$. However, both parametrizations describe the same line. ∎

TABLE 11.3 Values of $x = t + (1/t)$ and $y = t - (1/t)$ for selected values of t.

t	$1/t$	x	y
0.1	10.0	10.1	−9.9
0.2	5.0	5.2	−4.8
0.4	2.5	2.9	−2.1
1.0	1.0	2.0	0.0
2.0	0.5	2.5	1.5
5.0	0.2	5.2	4.8
10.0	0.1	10.1	9.9

EXAMPLE 7 Sketch and identify the path traced by the point $P(x, y)$ if

$$x = t + \frac{1}{t}, \qquad y = t - \frac{1}{t}, \qquad t > 0.$$

Solution We make a brief table of values in Table 11.3, plot the points, and draw a smooth curve through them, as we did in Example 1. Next we eliminate the parameter t from the equations. The procedure is more complicated than in Example 2. Taking the difference between x and y as given by the parametric equations, we find that

$$x - y = \left(t + \frac{1}{t}\right) - \left(t - \frac{1}{t}\right) = \frac{2}{t}.$$

If we add the two parametric equations, we get

$$x + y = \left(t + \frac{1}{t}\right) + \left(t - \frac{1}{t}\right) = 2t.$$

We can then eliminate the parameter t by multiplying these last equations together:

$$(x - y)(x + y) = \left(\frac{2}{t}\right)(2t) = 4.$$

Expanding the expression on the left-hand side, we obtain a standard equation for a hyperbola (reviewed in Section 11.6):

$$x^2 - y^2 = 4. \tag{1}$$

Thus the coordinates of all the points $P(x, y)$ described by the parametric equations satisfy Equation (1). However, Equation (1) does not require that the x-coordinate be positive. So there are points (x, y) on the hyperbola that do not satisfy the parametric equation $x = t + (1/t), t > 0$. In fact, the parametric equations do not yield any points on the left branch of the hyperbola given by Equation (1), points where the x-coordinate would be negative. For small positive values of t, the path lies in the fourth quadrant and rises into the first quadrant as t increases, crossing the x-axis when $t = 1$ (see Figure 11.7). The parameter domain is $(0, \infty)$ and there is no starting point and no terminal point for the path. ∎

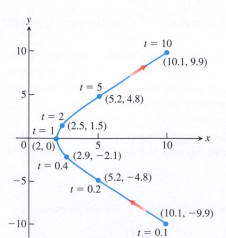

FIGURE 11.7 The curve for $x = t + (1/t)$, $y = t - (1/t)$, $t > 0$ in Example 7. (The part shown is for $0.1 \leq t \leq 10$.)

Examples 4, 5, and 6 illustrate that a given curve, or portion of it, can be represented by different parametrizations. In the case of Example 7, we can also represent the right-hand branch of the hyperbola by the parametrization

$$x = \sqrt{4 + t^2}, \qquad y = t, \qquad -\infty < t < \infty,$$

which is obtained by solving Equation (1) for $x \geq 0$ and letting y be the parameter. Still another parametrization for the right-hand branch of the hyperbola given by Equation (1) is

$$x = 2 \sec t, \qquad y = 2 \tan t, \qquad -\frac{\pi}{2} < t < \frac{\pi}{2}.$$

This parametrization follows from the trigonometric identity $\sec^2 t - \tan^2 t = 1$, because

$$x^2 - y^2 = 4 \sec^2 t - 4 \tan^2 t = 4(\sec^2 t - \tan^2 t) = 4.$$

As t runs between $-\pi/2$ and $\pi/2$, $x = \sec t$ remains positive and $y = \tan t$ runs between $-\infty$ and ∞, so P traverses the hyperbola's right-hand branch. It comes in along the branch's lower half as $t \to 0^-$, reaches $(2, 0)$ at $t = 0$, and moves out into the first quadrant as t increases steadily toward $\pi/2$. This is the same branch of the hyperbola shown in Figure 11.7.

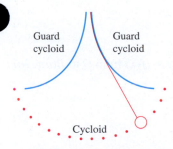

Guard cycloid   Guard cycloid

Cycloid

FIGURE 11.8 In Huygens' pendulum clock, the bob swings in a cycloid, so the frequency is independent of the amplitude.

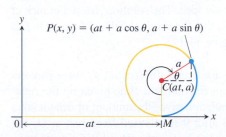

FIGURE 11.9 The position of $P(x, y)$ on the rolling wheel at angle t (Example 8).

Cycloids

The problem with a pendulum clock whose bob swings in a circular arc is that the frequency of the swing depends on the amplitude of the swing. The wider the swing, the longer it takes the bob to return to center (its lowest position).

This does not happen if the bob can be made to swing in a *cycloid*. In 1673, Christian Huygens designed a pendulum clock whose bob would swing in a cycloid, a curve we define in Example 8. He hung the bob from a fine wire constrained by guards that caused it to draw up as it swung away from center (Figure 11.8). We describe the path parametrically in the next example.

EXAMPLE 8 A wheel of radius a rolls along a horizontal straight line. Find parametric equations for the path traced by a point P on the wheel's circumference. The path is called a **cycloid**.

Solution We take the line to be the x-axis, mark a point P on the wheel, start the wheel with P at the origin, and roll the wheel to the right. As parameter, we use the angle t through which the wheel turns, measured in radians. Figure 11.9 shows the wheel a short while later when its base lies at units from the origin. The wheel's center C lies at (at, a) and the coordinates of P are

$$x = at + a \cos \theta, \qquad y = a + a \sin \theta.$$

To express θ in terms of t, we observe that $t + \theta = 3\pi/2$ in the figure, so that

$$\theta = \frac{3\pi}{2} - t.$$

This makes

$$\cos \theta = \cos \left(\frac{3\pi}{2} - t \right) = -\sin t, \qquad \sin \theta = \sin \left(\frac{3\pi}{2} - t \right) = -\cos t.$$

The equations we seek are

$$x = at - a \sin t, \qquad y = a - a \cos t.$$

These are usually written with the a factored out:

$$x = a(t - \sin t), \qquad y = a(1 - \cos t). \tag{2}$$

Figure 11.10 shows the first arch of the cycloid and part of the next. ∎

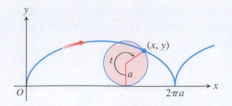

FIGURE 11.10 The cycloid curve $x = a(t - \sin t)$, $y = a(1 - \cos t)$, for $t \geq 0$.

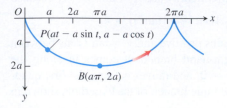

FIGURE 11.11 Turning Figure 11.10 upside down, the y-axis points downward, indicating the direction of the gravitational force. Equations (2) still describe the curve parametrically.

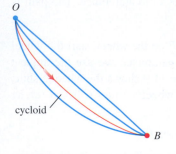

FIGURE 11.12 The cycloid is the unique curve which minimizes the time it takes for a frictionless bead to slide from point O to B.

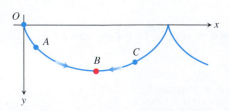

FIGURE 11.13 Beads released simultaneously on the upside-down cycloid at O, A, and C will reach B at the same time.

Brachistochrones and Tautochrones

If we turn Figure 11.10 upside down, Equations (2) still apply and the resulting curve (Figure 11.11) has two interesting physical properties. The first relates to the origin O and the point B at the bottom of the first arch. Among all smooth curves joining these points, the cycloid is the curve along which a frictionless bead, subject only to the force of gravity, will slide from O to B the fastest. This makes the cycloid a **brachistochrone** ("brah-*kiss*-toe-krone"), or shortest-time curve for these points. The second property is that even if you start the bead partway down the curve toward B, it will still take the bead the same amount of time to reach B. This makes the cycloid a **tautochrone** ("*taw*-toe-krone"), or same-time curve for O and B.

Are there any other brachistochrones joining O and B, or is the cycloid the only one? We can formulate this as a mathematical question in the following way. At the start, the kinetic energy of the bead is zero, since its velocity (speed) is zero. The work done by gravity in moving the bead from $(0, 0)$ to any other point (x, y) in the plane is mgy, and this must equal the change in kinetic energy. (See Exercise 25 in Section 6.5.) That is,

$$mgy = \frac{1}{2} mv^2 - \frac{1}{2} m(0)^2.$$

Thus, the speed of the bead when it reaches (x, y) has to be $v = \sqrt{2gy}$. That is,

$$\frac{ds}{dT} = \sqrt{2gy} \qquad \text{\footnotesize ds is the arc length differential along the bead's path and T represents time.}$$

or

$$dT = \frac{ds}{\sqrt{2gy}} = \frac{\sqrt{1 + (dy/dx)^2}\, dx}{\sqrt{2gy}}. \tag{3}$$

The time T_f it takes the bead to slide along a particular path $y = f(x)$ from O to $B(a\pi, 2a)$ is

$$T_f = \int_{x=0}^{x=a\pi} \sqrt{\frac{1 + (dy/dx)^2}{2gy}}\, dx. \tag{4}$$

What curves $y = f(x)$, if any, minimize the value of this integral?

At first sight, we might guess that the straight line joining O and B would give the shortest time, but perhaps not. There might be some advantage in having the bead fall vertically at first to build up its speed faster. With a higher speed, the bead could travel a longer path and still reach B first. Indeed, this is the right idea. The solution, from a branch of mathematics known as the *calculus of variations*, is that the original cycloid from O to B is the one and only brachistochrone for O and B (Figure 11.12).

In the next section we show how to find the arc length differential ds for a parametrized curve. Once we know how to find ds, we can calculate the time given by the right-hand side of Equation (4) for the cycloid. This calculation gives the amount of time it takes a frictionless bead to slide down the cycloid to B after it is released from rest at O. The time turns out to be equal to $\pi\sqrt{a/g}$, where a is the radius of the wheel defining the particular cycloid. Moreover, if we start the bead at some lower point on the cycloid, corresponding to a parameter value $t_0 > 0$, we can integrate the parametric form of $ds/\sqrt{2gy}$ in Equation (3) over the interval $[t_0, \pi]$ to find the time it takes the bead to reach the point B. That calculation results in the same time $T = \pi\sqrt{a/g}$. It takes the bead the same amount of time to reach B no matter where it starts, which makes the cycloid a tautochrone. Beads starting simultaneously from O, A, and C in Figure 11.13, for instance, will all reach B at exactly the same time. This is the reason why Huygens' pendulum clock in Figure 11.8 is independent of the amplitude of the swing.

EXERCISES 11.1

Finding Cartesian from Parametric Equations

Exercises 1–18 give parametric equations and parameter intervals for the motion of a particle in the xy-plane. Identify the particle's path by finding a Cartesian equation for it. Graph the Cartesian equation. (The graphs will vary with the equation used.) Indicate the portion of the graph traced by the particle and the direction of motion.

1. $x = 3t$, $y = 9t^2$, $-\infty < t < \infty$

2. $x = -\sqrt{t}$, $y = t$, $t \geq 0$

3. $x = 2t - 5$, $y = 4t - 7$, $-\infty < t < \infty$

4. $x = 3 - 3t$, $y = 2t$, $0 \leq t \leq 1$

5. $x = \cos 2t$, $y = \sin 2t$, $0 \leq t \leq \pi$

6. $x = \cos(\pi - t)$, $y = \sin(\pi - t)$, $0 \leq t \leq \pi$

7. $x = 4 \cos t$, $y = 2 \sin t$, $0 \leq t \leq 2\pi$

8. $x = 4 \sin t$, $y = 5 \cos t$, $0 \leq t \leq 2\pi$

9. $x = \sin t$, $y = \cos 2t$, $-\dfrac{\pi}{2} \leq t \leq \dfrac{\pi}{2}$

10. $x = 1 + \sin t$, $y = \cos t - 2$, $0 \leq t \leq \pi$

11. $x = t^2$, $y = t^6 - 2t^4$, $-\infty < t < \infty$

12. $x = \dfrac{t}{t - 1}$, $y = \dfrac{t - 2}{t + 1}$, $-1 < t < 1$

13. $x = t$, $y = \sqrt{1 - t^2}$, $-1 \leq t \leq 0$

14. $x = \sqrt{t + 1}$, $y = \sqrt{t}$, $t \geq 0$

15. $x = \sec^2 t - 1$, $y = \tan t$, $-\pi/2 < t < \pi/2$

16. $x = -\sec t$, $y = \tan t$, $-\pi/2 < t < \pi/2$

17. $x = -\cosh t$, $y = \sinh t$, $-\infty < t < \infty$

18. $x = 2 \sinh t$, $y = 2 \cosh t$, $-\infty < t < \infty$

In Exercises 19–24, match the parametric equations with the parametric curves labeled A through F.

19. $x = 1 - \sin t$, $y = 1 + \cos t$

20. $x = \cos t$, $y = 2 \sin t$

21. $x = \dfrac{1}{4} t \cos t$, $y = \dfrac{1}{4} t \sin t$

22. $x = \sqrt{t}$, $y = \sqrt{t} \cos t$

23. $x = \ln t$, $y = 3e^{-t/2}$

24. $x = \cos t$, $y = \sin 3t$

A.

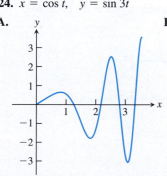

B.

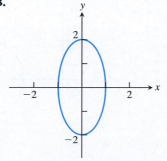

C.

D.

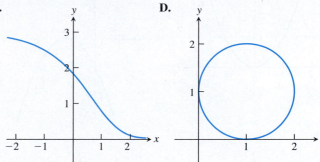

E.

F.

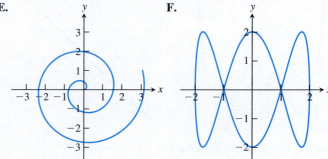

In Exercises 25–28, use the given graphs of $x = f(t)$ and $y = g(t)$ to sketch the corresponding parametric curve in the xy-plane.

25.

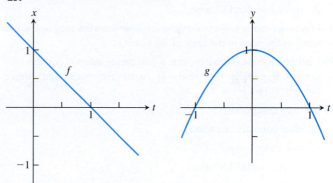

26.

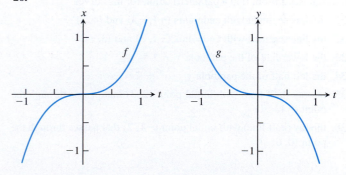

27.

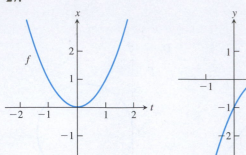

28.

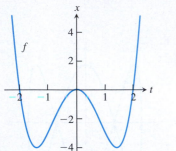

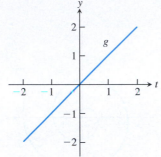

Finding Parametric Equations

29. Find parametric equations and a parameter interval for the motion of a particle that starts at $(a, 0)$ and traces the circle $x^2 + y^2 = a^2$

  a. once clockwise.

  b. once counterclockwise.

  c. twice clockwise.

  d. twice counterclockwise.

(There are many ways to do these, so your answers may not be the same as the ones in the back of the book.)

30. Find parametric equations and a parameter interval for the motion of a particle that starts at $(a, 0)$ and traces the ellipse $(x^2/a^2) + (y^2/b^2) = 1$

  a. once clockwise.

  b. once counterclockwise.

  c. twice clockwise.

  d. twice counterclockwise.

(As in Exercise 29, there are many correct answers.)

In Exercises 31–36, find a parametrization for the curve.

31. the line segment with endpoints $(-1, -3)$ and $(4, 1)$

32. the line segment with endpoints $(-1, 3)$ and $(3, -2)$

33. the lower half of the parabola $x - 1 = y^2$

34. the left half of the parabola $y = x^2 + 2x$

35. the ray (half line) with initial point $(2, 3)$ that passes through the point $(-1, -1)$

36. the ray (half line) with initial point $(-1, 2)$ that passes through the point $(0, 0)$

37. Find parametric equations and a parameter interval for the motion of a particle starting at the point $(2, 0)$ and tracing the top half of the circle $x^2 + y^2 = 4$ four times.

38. Find parametric equations and a parameter interval for the motion of a particle that moves along the graph of $y = x^2$ in the following way: Beginning at $(0, 0)$ it moves to $(3, 9)$, and then travels back and forth from $(3, 9)$ to $(-3, 9)$ infinitely many times.

39. Find parametric equations for the semicircle

$$x^2 + y^2 = a^2, \quad y > 0,$$

using as parameter the slope $t = dy/dx$ of the tangent to the curve at (x, y).

40. Find parametric equations for the circle

$$x^2 + y^2 = a^2,$$

using as parameter the arc length s measured counterclockwise from the point $(a, 0)$ to the point (x, y).

41. Find a parametrization for the line segment joining points $(0, 2)$ and $(4, 0)$ using the angle θ in the accompanying figure as the parameter.

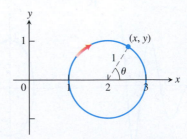

42. Find a parametrization for the curve $y = \sqrt{x}$ with terminal point $(0, 0)$ using the angle θ in the accompanying figure as the parameter.

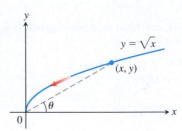

43. Find a parametrization for the circle $(x - 2)^2 + y^2 = 1$ starting at $(1, 0)$ and moving clockwise once around the circle, using the central angle θ in the accompanying figure as the parameter.

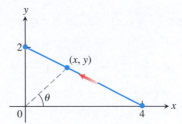

44. Find a parametrization for the circle $x^2 + y^2 = 1$ starting at $(1, 0)$ and moving counterclockwise to the terminal point $(0, 1)$, using the angle θ in the accompanying figure as the parameter.

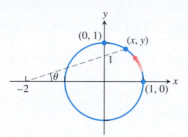

45. The witch of Maria Agnesi The bell-shaped witch of Maria Agnesi can be constructed in the following way. Start with a circle of radius 1, centered at the point $(0, 1)$, as shown in the accompanying figure. Choose a point A on the line $y = 2$ and connect it to the origin with a line segment. Call the point where the segment crosses the circle B. Let P be the point where the vertical line through A crosses the horizontal line through B. The witch is the curve traced by P as A moves along the line $y = 2$. Find parametric equations and a parameter interval for the witch by expressing the coordinates of P in terms of t, the radian measure of the angle that segment OA makes with the positive x-axis. The following equalities (which you may assume) will help.

a. $x = AQ$
b. $y = 2 - AB \sin t$
c. $AB \cdot OA = (AQ)^2$

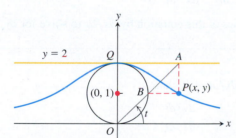

46. Hypocycloid When a circle rolls on the inside of a fixed circle, any point P on the circumference of the rolling circle describes a *hypocycloid*. Let the fixed circle be $x^2 + y^2 = a^2$, let the radius of the rolling circle be b, and let the initial position of the tracing point P be $A(a, 0)$. Find parametric equations for the hypocycloid, using as the parameter the angle θ from the positive x-axis to the line joining the circles' centers. In particular, if $b = a/4$, as in the accompanying figure, show that the hypocycloid is the astroid

$$x = a \cos^3 \theta, \quad y = a \sin^3 \theta.$$

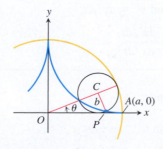

47. As the point N moves along the line $y = a$ in the accompanying figure, P moves in such a way that $OP = MN$. Find parametric equations for the coordinates of P as functions of the angle t that the line ON makes with the positive y-axis.

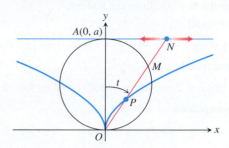

48. Trochoids A wheel of radius a rolls along a horizontal straight line without slipping. Find parametric equations for the curve traced out by a point P on a spoke of the wheel b units from its center. As parameter, use the angle θ through which the wheel turns. The curve is called a *trochoid*, which is a cycloid when $b = a$.

Distance Using Parametric Equations

49. Find the point on the parabola $x = t, y = t^2, -\infty < t < \infty$, closest to the point $(2, 1/2)$. (*Hint:* Minimize the square of the distance as a function of t.)

50. Find the point on the ellipse $x = 2 \cos t, y = \sin t, 0 \le t \le 2\pi$ closest to the point $(3/4, 0)$. (*Hint:* Minimize the square of the distance as a function of t.)

T GRAPHER EXPLORATIONS

If you have a parametric equation grapher, graph the equations over the given intervals in Exercises 51–58.

51. Ellipse $x = 4 \cos t, \quad y = 2 \sin t, \quad$ over
   a. $0 \le t \le 2\pi$
   b. $0 \le t \le \pi$
   c. $-\pi/2 \le t \le \pi/2$.

52. Hyperbola branch $x = \sec t$ (enter as $1/\cos(t)$), $y = \tan t$ (enter as $\sin(t)/\cos(t)$), over
   a. $-1.5 \le t \le 1.5$
   b. $-0.5 \le t \le 0.5$
   c. $-0.1 \le t \le 0.1$.

53. Parabola $x = 2t + 3, \quad y = t^2 - 1, \quad -2 \le t \le 2$

54. Cycloid $x = t - \sin t, \quad y = 1 - \cos t, \quad$ over
   a. $0 \le t \le 2\pi$
   b. $0 \le t \le 4\pi$
   c. $\pi \le t \le 3\pi$.

55. Deltoid

$$x = 2 \cos t + \cos 2t, \quad y = 2 \sin t - \sin 2t; \quad 0 \le t \le 2\pi$$

What happens if you replace 2 with -2 in the equations for x and y? Graph the new equations and find out.

56. A nice curve

$$x = 3 \cos t + \cos 3t, \quad y = 3 \sin t - \sin 3t; \quad 0 \le t \le 2\pi$$

What happens if you replace 3 with -3 in the equations for x and y? Graph the new equations and find out.

57. a. Epicycloid

$$x = 9 \cos t - \cos 9t, \quad y = 9 \sin t - \sin 9t; \quad 0 \le t \le 2\pi$$

b. Hypocycloid

$$x = 8 \cos t + 2 \cos 4t, \quad y = 8 \sin t - 2 \sin 4t; \quad 0 \le t \le 2\pi$$

c. Hypotrochoid

$$x = \cos t + 5 \cos 3t, \quad y = 6 \cos t - 5 \sin 3t; \quad 0 \le t \le 2\pi$$

58. a. $x = 6 \cos t + 5 \cos 3t, \quad y = 6 \sin t - 5 \sin 3t;$
$0 \le t \le 2\pi$

b. $x = 6 \cos 2t + 5 \cos 6t, \quad y = 6 \sin 2t - 5 \sin 6t;$
$0 \le t \le \pi$

c. $x = 6 \cos t + 5 \cos 3t, \quad y = 6 \sin 2t - 5 \sin 3t;$
$0 \le t \le 2\pi$

d. $x = 6 \cos 2t + 5 \cos 6t, \quad y = 6 \sin 4t - 5 \sin 6t;$
$0 \le t \le \pi$

11.2 Calculus with Parametric Curves

In this section we apply calculus to parametric curves. Specifically, we find slopes, lengths, and areas associated with parametrized curves.

Tangents and Areas

A parametrized curve $x = f(t)$ and $y = g(t)$ is **differentiable** at t if f and g are differentiable at t. At a point on a differentiable parametrized curve where y is also a differentiable function of x, the derivatives dy/dt, dx/dt, and dy/dx are related by the Chain Rule:

$$\frac{dy}{dt} = \frac{dy}{dx} \cdot \frac{dx}{dt}.$$

If $dx/dt \ne 0$, we may divide both sides of this equation by dx/dt to solve for dy/dx.

Parametric Formula for dy/dx

If all three derivatives exist and $dx/dt \ne 0$, then

$$\frac{dy}{dx} = \frac{dy/dt}{dx/dt}. \tag{1}$$

If parametric equations define y as a twice-differentiable function of x, we can apply Equation (1) to the function $dy/dx = y'$ to calculate d^2y/dx^2 as a function of t:

$$\frac{d^2y}{dx^2} = \frac{d}{dx}(y') = \frac{dy'/dt}{dx/dt}. \quad \text{Eq. (1) with } y' \text{ in place of } y$$

Parametric Formula for d^2y/dx^2

If the equations $x = f(t)$, $y = g(t)$ define y as a twice-differentiable function of x, then at any point where $dx/dt \ne 0$ and $y' = dy/dx$,

$$\frac{d^2y}{dx^2} = \frac{dy'/dt}{dx/dt}. \tag{2}$$

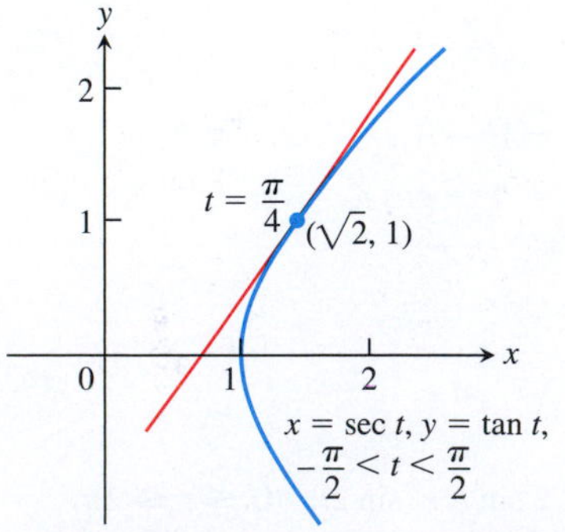

FIGURE 11.14 The curve in Example 1 is the right-hand branch of the hyperbola $x^2 - y^2 = 1$.

EXAMPLE 1 Find the tangent to the curve

$$x = \sec t, \quad y = \tan t, \quad -\frac{\pi}{2} < t < \frac{\pi}{2},$$

at the point $\left(\sqrt{2}, 1\right)$, where $t = \pi/4$ (Figure 11.14).

Solution The slope of the curve at t is

$$\frac{dy}{dx} = \frac{dy/dt}{dx/dt} = \frac{\sec^2 t}{\sec t \tan t} = \frac{\sec t}{\tan t}. \qquad \text{Eq. (1)}$$

Setting t equal to $\pi/4$ gives

$$\frac{dy}{dx}\Bigg|_{t=\pi/4} = \frac{\sec(\pi/4)}{\tan(\pi/4)} = \frac{\sqrt{2}}{1} = \sqrt{2}.$$

The tangent line is

$$y - 1 = \sqrt{2}\left(x - \sqrt{2}\right)$$
$$y = \sqrt{2}\,x - 2 + 1$$
$$y = \sqrt{2}\,x - 1.$$

■

Finding d^2y/dx^2 in Terms of t

1. Express $y' = dy/dx$ in terms of t.
2. Find dy'/dt.
3. Divide dy'/dt by dx/dt.

EXAMPLE 2 Find d^2y/dx^2 as a function of t if $x = t - t^2$ and $y = t - t^3$.

Solution

1. Express $y' = dy/dx$ in terms of t.

$$y' = \frac{dy}{dx} = \frac{dy/dt}{dx/dt} = \frac{1 - 3t^2}{1 - 2t}$$

2. Differentiate y' with respect to t.

$$\frac{dy'}{dt} = \frac{d}{dt}\left(\frac{1 - 3t^2}{1 - 2t}\right) = \frac{2 - 6t + 6t^2}{(1 - 2t)^2} \qquad \text{Derivative Quotient Rule}$$

3. Divide dy'/dt by dx/dt.

$$\frac{d^2y}{dx^2} = \frac{dy'/dt}{dx/dt} = \frac{(2 - 6t + 6t^2)/(1 - 2t)^2}{1 - 2t} = \frac{2 - 6t + 6t^2}{(1 - 2t)^3} \qquad \text{Eq. (2)}$$

■

EXAMPLE 3 Find the area enclosed by the astroid (Figure 11.15)

$$x = \cos^3 t, \qquad y = \sin^3 t, \qquad 0 \le t \le 2\pi.$$

Solution By symmetry, the enclosed area is 4 times the area beneath the curve in the first quadrant where $0 \le t \le \pi/2$. We can apply the definite integral formula for area studied in Chapter 5, using substitution to express the curve and differential dx in terms of the parameter t. Thus,

$$A = 4\int_0^1 y\,dx \qquad \text{4 times area under } y \text{ from } x = 0 \text{ to } x = 1$$

$$= 4\int_0^{\pi/2} (\sin^3 t)(3\cos^2 t \sin t)\,dt \qquad \text{Substitution for } y \text{ and } dx$$

$$= 12\int_0^{\pi/2} \left(\frac{1 - \cos 2t}{2}\right)^2 \left(\frac{1 + \cos 2t}{2}\right)dt \qquad \sin^4 t = \left(\frac{1 - \cos 2t}{2}\right)^2$$

$$= \frac{3}{2}\int_0^{\pi/2} (1 - 2\cos 2t + \cos^2 2t)(1 + \cos 2t)\,dt \qquad \text{Expand squared term.}$$

$$= \frac{3}{2}\int_0^{\pi/2} (1 - \cos 2t - \cos^2 2t + \cos^3 2t)\,dt \qquad \text{Multiply terms.}$$

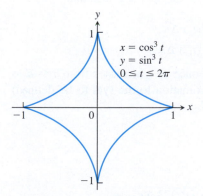

FIGURE 11.15 The astroid in Example 3.

$$= \frac{3}{2}\left[\int_0^{\pi/2}(1 - \cos 2t)\,dt - \int_0^{\pi/2}\cos^2 2t\,dt + \int_0^{\pi/2}\cos^3 2t\,dt\right]$$

Section 8.2, Example 3

$$= \frac{3}{2}\left[\left(t - \frac{1}{2}\sin 2t\right) - \frac{1}{2}\left(t + \frac{1}{4}\sin 2t\right) + \frac{1}{2}\left(\sin 2t - \frac{1}{3}\sin^3 2t\right)\right]_0^{\pi/2}$$

$$= \frac{3}{2}\left[\left(\frac{\pi}{2} - 0 - 0 - 0\right) - \frac{1}{2}\left(\frac{\pi}{2} + 0 - 0 - 0\right) + \frac{1}{2}(0 - 0 - 0 + 0)\right]$$ Evaluate.

$$= \frac{3\pi}{8}. \qquad \blacksquare$$

Length of a Parametrically Defined Curve

Let *C* be a curve given parametrically by the equations

$$x = f(t) \qquad \text{and} \qquad y = g(t), \qquad a \le t \le b.$$

We assume the functions *f* and *g* are **continuously differentiable** (meaning they have continuous first derivatives) on the interval $[a, b]$. We also assume that the derivatives $f'(t)$ and $g'(t)$ are not simultaneously zero, which prevents the curve *C* from having any corners or cusps. Such a curve is called a **smooth curve**. We subdivide the path (or arc) *AB* into *n* pieces at points $A = P_0, P_1, P_2, \ldots, P_n = B$ (Figure 11.16). These points correspond to a partition of the interval $[a, b]$ by $a = t_0 < t_1 < t_2 < \cdots < t_n = b$, where $P_k = (f(t_k), g(t_k))$. Join successive points of this subdivision by straight-line segments (Figure 11.16). A representative line segment has length

$$L_k = \sqrt{(\Delta x_k)^2 + (\Delta y_k)^2}$$
$$= \sqrt{[f(t_k) - f(t_{k-1})]^2 + [g(t_k) - g(t_{k-1})]^2}$$

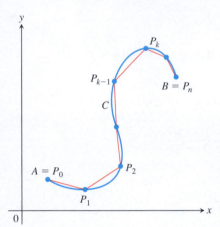

FIGURE 11.16 The length of the smooth curve C from *A* to *B* is approximated by the sum of the lengths of the polygonal path (straight-line segments) starting at $A = P_0$, then to P_1, and so on, ending at $B = P_n$.

(see Figure 11.17). If Δt_k is small, the length L_k is approximately the length of arc $P_{k-1}P_k$. By the Mean Value Theorem there are numbers t_k^* and t_k^{**} in $[t_{k-1}, t_k]$ such that

$$\Delta x_k = f(t_k) - f(t_{k-1}) = f'(t_k^*)\,\Delta t_k,$$
$$\Delta y_k = g(t_k) - g(t_{k-1}) = g'(t_k^{**})\,\Delta t_k.$$

Assuming the path from *A* to *B* is traversed exactly once as *t* increases from $t = a$ to $t = b$, with no doubling back or retracing, an approximation to the (yet to be defined) "length" of the curve *AB* is the sum of all the lengths L_k:

$$\sum_{k=1}^n L_k = \sum_{k=1}^n \sqrt{(\Delta x_k)^2 + (\Delta y_k)^2}$$

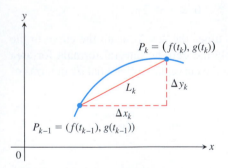

FIGURE 11.17 The arc $P_{k-1}P_k$ is approximated by the straight-line segment shown here, which has length $L_k = \sqrt{(\Delta x_k)^2 + (\Delta y_k)^2}$.

$$= \sum_{k=1}^n \sqrt{[f'(t_k^*)]^2 + [g'(t_k^{**})]^2}\,\Delta t_k.$$

Although this last sum on the right is not exactly a Riemann sum (because f' and g' are evaluated at different points), it can be shown that its limit, as the norm of the partition tends to zero and the number of segments $n \to \infty$, is the definite integral

$$\lim_{\|P\| \to 0} \sum_{k=1}^n \sqrt{[f'(t_k^*)]^2 + [g'(t_k^{**})]^2}\,\Delta t_k = \int_a^b \sqrt{[f'(t)]^2 + [g'(t)]^2}\,dt.$$

Therefore, it is reasonable to define the length of the curve from *A* to *B* to be this integral.

DEFINITION   If a curve C is defined parametrically by $x = f(t)$ and $y = g(t)$, $a \le t \le b$, where f' and g' are continuous and not simultaneously zero on $[a, b]$, and C is traversed exactly once as t increases from $t = a$ to $t = b$, then **the length of C** is the definite integral

$$L = \int_a^b \sqrt{[f'(t)]^2 + [g'(t)]^2}\, dt.$$

If $x = f(t)$ and $y = g(t)$, then using the Leibniz notation we can write the formula for arc length this way:

$$L = \int_a^b \sqrt{\left(\frac{dx}{dt}\right)^2 + \left(\frac{dy}{dt}\right)^2}\, dt. \tag{3}$$

A smooth curve C does not double back or reverse the direction of motion over the time interval $[a, b]$ since $(f')^2 + (g')^2 > 0$ throughout the interval. At a point where a curve does start to double back on itself, either the curve fails to be differentiable or both derivatives must simultaneously equal zero. We will examine this phenomenon in Chapter 13, where we study tangent vectors to curves.

If there are two different parametrizations for a curve C whose length we want to find, it does not matter which one we use. However, the parametrization we choose must meet the conditions stated in the definition of the length of C (see Exercise 41 for an example).

EXAMPLE 4   Using the definition, find the length of the circle of radius r defined parametrically by

$$x = r \cos t \quad \text{and} \quad y = r \sin t, \quad 0 \le t \le 2\pi.$$

Solution   As t varies from 0 to 2π, the circle is traversed exactly once, so the circumference is

$$L = \int_0^{2\pi} \sqrt{\left(\frac{dx}{dt}\right)^2 + \left(\frac{dy}{dt}\right)^2}\, dt.$$

We find

$$\frac{dx}{dt} = -r \sin t, \quad \frac{dy}{dt} = r \cos t$$

and

$$\left(\frac{dx}{dt}\right)^2 + \left(\frac{dy}{dt}\right)^2 = r^2(\sin^2 t + \cos^2 t) = r^2.$$

Therefore, the total arc length is

$$L = \int_0^{2\pi} \sqrt{r^2}\, dt = r\left[t\right]_0^{2\pi} = 2\pi r. \qquad \blacksquare$$

EXAMPLE 5   Find the length of the astroid (Figure 11.15)

$$x = \cos^3 t, \quad y = \sin^3 t, \quad 0 \le t \le 2\pi.$$

Solution Because of the curve's symmetry with respect to the coordinate axes, its length is four times the length of the first-quadrant portion. We have

$$x = \cos^3 t, \qquad\qquad y = \sin^3 t$$

$$\left(\frac{dx}{dt}\right)^2 = [3\cos^2 t(-\sin t)]^2 = 9\cos^4 t \sin^2 t$$

$$\left(\frac{dy}{dt}\right)^2 = [3\sin^2 t(\cos t)]^2 = 9\sin^4 t \cos^2 t$$

$$\sqrt{\left(\frac{dx}{dt}\right)^2 + \left(\frac{dy}{dt}\right)^2} = \sqrt{9\cos^2 t \sin^2 t \underbrace{(\cos^2 t + \sin^2 t)}_{1}}$$

$$= \sqrt{9\cos^2 t \sin^2 t}$$

$$= 3|\cos t \sin t| \qquad \cos t \sin t \ge 0 \text{ for } 0 \le t \le \pi/2$$

$$= 3\cos t \sin t.$$

Therefore,

$$\text{Length of first-quadrant portion} = \int_0^{\pi/2} 3\cos t \sin t \, dt$$

$$= \frac{3}{2}\int_0^{\pi/2} \sin 2t \, dt \qquad \cos t \sin t = (1/2)\sin 2t$$

$$= -\frac{3}{4}\cos 2t \Big]_0^{\pi/2} = \frac{3}{2}.$$

The length of the astroid is four times this: $4(3/2) = 6$. ∎

EXAMPLE 6 Find the perimeter of the ellipse $\dfrac{x^2}{a^2} + \dfrac{y^2}{b^2} = 1$.

Solution Parametrically, we represent the ellipse by the equations $x = a \sin t$ and $y = b \cos t$, $a > b$ and $0 \le t \le 2\pi$. Then,

$$\left(\frac{dx}{dt}\right)^2 + \left(\frac{dy}{dt}\right)^2 = a^2\cos^2 t + b^2\sin^2 t$$

$$= a^2 - (a^2 - b^2)\sin^2 t$$

$$= a^2[1 - e^2\sin^2 t] \qquad e = \sqrt{1 - \frac{b^2}{a^2}} \text{ (eccentricity,}$$
$$\text{not the number } 2.71828\ldots)$$

From Equation (3), the perimeter is given by

$$P = 4a\int_0^{\pi/2} \sqrt{1 - e^2\sin^2 t} \, dt.$$

(We investigate the meaning of the eccentricity e in Section 11.7.) The integral for P is nonelementary and is known as the *complete elliptic integral of the second kind*. We can compute its value to within any degree of accuracy using infinite series in the following way. From the binomial expansion for $\sqrt{1 - x^2}$ in Section 10.10, we have

$$\sqrt{1 - e^2\sin^2 t} = 1 - \frac{1}{2}e^2\sin^2 t - \frac{1}{2\cdot 4}e^4\sin^4 t - \cdots, \qquad |e\sin t| \le e < 1$$

Then to each term in this last expression we apply the integral Formula 157 (at the back of the book) for $\int_0^{\pi/2} \sin^n t \, dt$ when n is even, giving the perimeter

$$P = 4a \int_0^{\pi/2} \sqrt{1 - e^2 \sin^2 t} \, dt$$

$$= 4a \left[\frac{\pi}{2} - \left(\frac{1}{2} e^2\right)\left(\frac{1}{2} \cdot \frac{\pi}{2}\right) - \left(\frac{1}{2 \cdot 4} e^4\right)\left(\frac{1 \cdot 3}{2 \cdot 4} \cdot \frac{\pi}{2}\right) - \left(\frac{1 \cdot 3}{2 \cdot 4 \cdot 6} e^6\right)\left(\frac{1 \cdot 3 \cdot 5}{2 \cdot 4 \cdot 6} \cdot \frac{\pi}{2}\right) - \cdots \right]$$

$$= 2\pi a \left[1 - \left(\frac{1}{2}\right)^2 e^2 - \left(\frac{1 \cdot 3}{2 \cdot 4}\right)^2 \frac{e^4}{3} - \left(\frac{1 \cdot 3 \cdot 5}{2 \cdot 4 \cdot 6}\right)^2 \frac{e^6}{5} - \cdots \right].$$

Since $e < 1$, the series on the right-hand side converges by comparison with the geometric series $\sum_{n=1}^{\infty} (e^2)^n$. We do not have an explicit value for P, but we can estimate it as closely as we like by summing finitely many terms from the infinite series. ∎

Length of a Curve $y = f(x)$

We will show that the length formula in Section 6.3 is a special case of Equation (3). Given a continuously differentiable function $y = f(x)$, $a \leq x \leq b$, we can assign $x = t$ as a parameter. The graph of the function f is then the curve C defined parametrically by

$$x = t \quad \text{and} \quad y = f(t), \quad a \leq t \leq b,$$

which is a special case of what we have considered in this chapter. We have

$$\frac{dx}{dt} = 1 \quad \text{and} \quad \frac{dy}{dt} = f'(t).$$

From Equation (1),

$$\frac{dy}{dx} = \frac{dy/dt}{dx/dt} = f'(t),$$

giving

$$\left(\frac{dx}{dt}\right)^2 + \left(\frac{dy}{dt}\right)^2 = 1 + [f'(t)]^2$$

$$= 1 + [f'(x)]^2. \qquad t = x$$

Substitution into Equation (3) gives exactly the arc length formula for the graph of $y = f(x)$ that we found in Section 6.3.

The Arc Length Differential

As in Section 6.3, we define the arc length function for a parametrically defined curve $x = f(t)$ and $y = g(t)$, $a \leq t \leq b$, by

$$s(t) = \int_a^t \sqrt{[f'(z)]^2 + [g'(z)]^2} \, dz.$$

Then, by the Fundamental Theorem of Calculus,

$$\frac{ds}{dt} = \sqrt{[f'(t)]^2 + [g'(t)]^2} = \sqrt{\left(\frac{dx}{dt}\right)^2 + \left(\frac{dy}{dt}\right)^2}.$$

The differential of arc length is

$$ds = \sqrt{\left(\frac{dx}{dt}\right)^2 + \left(\frac{dy}{dt}\right)^2} \, dt. \qquad (4)$$

Equation (4) is often abbreviated as

$$ds = \sqrt{dx^2 + dy^2}.$$

Just as in Section 6.3, we can integrate the differential ds between appropriate limits to find the total length of a curve.

Here's an example where we use the arc length differential to find the centroid of an arc.

EXAMPLE 7   Find the centroid of the first-quadrant arc of the astroid in Example 5.

Solution   We take the curve's density to be $\delta = 1$ and calculate the curve's mass and moments about the coordinate axes as we did in Section 6.6.

The distribution of mass is symmetric about the line $y = x$, so $\bar{x} = \bar{y}$. A typical segment of the curve (Figure 11.18) has mass

$$dm = 1 \cdot ds = \sqrt{\left(\frac{dx}{dt}\right)^2 + \left(\frac{dy}{dt}\right)^2}\, dt = 3 \cos t \sin t\, dt. \qquad \text{From Example 5}$$

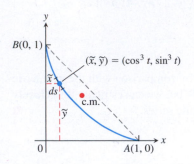

FIGURE 11.18 The centroid (c.m.) of the astroid arc in Example 7.

The curve's mass is

$$M = \int_0^{\pi/2} dm = \int_0^{\pi/2} 3 \cos t \sin t\, dt = \frac{3}{2}. \qquad \text{Again from Example 5}$$

The curve's moment about the x-axis is

$$M_x = \int \tilde{y}\, dm = \int_0^{\pi/2} \sin^3 t \cdot 3 \cos t \sin t\, dt$$

$$= 3 \int_0^{\pi/2} \sin^4 t \cos t\, dt = 3 \cdot \frac{\sin^5 t}{5}\bigg]_0^{\pi/2} = \frac{3}{5}.$$

It follows that

$$\bar{y} = \frac{M_x}{M} = \frac{3/5}{3/2} = \frac{2}{5}.$$

The centroid is the point $(2/5, 2/5)$.   ■

EXAMPLE 8   Find the time T_c it takes for a frictionless bead to slide along the cycloid $x = a(t - \sin t)$, $y = a(1 - \cos t)$ from $t = 0$ to $t = \pi$ (see Figure 11.13).

Solution   From Equation (3) in Section 11.1, we want to find the time

$$T_c = \int_{t=0}^{t=\pi} \frac{ds}{\sqrt{2gy}}.$$

We need to express ds parametrically in terms of the parameter t. For the cycloid, $dx/dt = a(1 - \cos t)$ and $dy/dt = a \sin t$, so

$$ds = \sqrt{\left(\frac{dx}{dt}\right)^2 + \left(\frac{dy}{dt}\right)^2}\, dt$$

$$= \sqrt{a^2\, (1 - 2\cos t + \cos^2 t + \sin^2 t)}\, dt$$

$$= \sqrt{a^2\, (2 - 2\cos t)}\, dt.$$

Substituting for ds and y in the integrand, it follows that

$$T_c = \int_0^{\pi} \sqrt{\frac{a^2(2 - 2\cos t)}{2ga\,(1 - \cos t)}}\, dt \qquad y = a(1 - \cos t)$$

$$= \int_0^{\pi} \sqrt{\frac{a}{g}}\, dt = \pi \sqrt{\frac{a}{g}}.$$

This is the amount of time it takes the frictionless bead to slide down the cycloid to B after it is released from rest at O (see Figure 11.13).   ■

Areas of Surfaces of Revolution

In Section 6.4 we found integral formulas for the area of a surface when a curve is revolved about a coordinate axis. Specifically, we found that the surface area is $S = \int 2\pi y \, ds$ for revolution about the x-axis, and $S = \int 2\pi x \, ds$ for revolution about the y-axis. If the curve is parametrized by the equations $x = f(t)$ and $y = g(t)$, $a \leq t \leq b$, where f and g are continuously differentiable and $(f')^2 + (g')^2 > 0$ on $[a, b]$, then the arc length differential ds is given by Equation (4). This observation leads to the following formulas for area of surfaces of revolution for smooth parametrized curves.

Area of Surface of Revolution for Parametrized Curves

If a smooth curve $x = f(t)$, $y = g(t)$, $a \leq t \leq b$, is traversed exactly once as t increases from a to b, then the areas of the surfaces generated by revolving the curve about the coordinate axes are as follows.

1. Revolution about the x-axis ($y \geq 0$):

$$S = \int_a^b 2\pi y \sqrt{\left(\frac{dx}{dt}\right)^2 + \left(\frac{dy}{dt}\right)^2} \, dt \tag{5}$$

2. Revolution about the y-axis ($x \geq 0$):

$$S = \int_a^b 2\pi x \sqrt{\left(\frac{dx}{dt}\right)^2 + \left(\frac{dy}{dt}\right)^2} \, dt \tag{6}$$

As with length, we can calculate surface area from any convenient parametrization that meets the stated criteria.

EXAMPLE 9   The standard parametrization of the circle of radius 1 centered at the point $(0, 1)$ in the xy-plane is

$$x = \cos t, \qquad y = 1 + \sin t, \qquad 0 \leq t \leq 2\pi.$$

Use this parametrization to find the area of the surface swept out by revolving the circle about the x-axis (Figure 11.19).

Solution   We evaluate the formula

$$S = \int_a^b 2\pi y \sqrt{\left(\frac{dx}{dt}\right)^2 + \left(\frac{dy}{dt}\right)^2} \, dt \qquad \text{Eq. (5) for revolution about the } x\text{-axis; } y = 1 + \sin t \geq 0$$

$$= \int_0^{2\pi} 2\pi (1 + \sin t) \underbrace{\sqrt{(-\sin t)^2 + (\cos t)^2}}_{1} \, dt$$

$$= 2\pi \int_0^{2\pi} (1 + \sin t) \, dt$$

$$= 2\pi \left[t - \cos t \right]_0^{2\pi} = 4\pi^2. \qquad \blacksquare$$

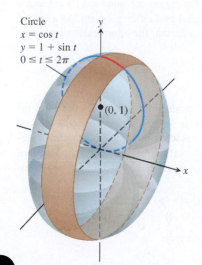

Circle
$x = \cos t$
$y = 1 + \sin t$
$0 \leq t \leq 2\pi$

$(0, 1)$

FIGURE 11.19   In Example 9 we calculate the area of the surface of revolution swept out by this parametrized curve.

EXERCISES 11.2

Tangents to Parametrized Curves

In Exercises 1–14, find an equation for the line tangent to the curve at the point defined by the given value of t. Also, find the value of d^2y/dx^2 at this point.

1. $x = 2 \cos t, \quad y = 2 \sin t, \quad t = \pi/4$

2. $x = \sin 2\pi t, \quad y = \cos 2\pi t, \quad t = -1/6$

3. $x = 4 \sin t, \quad y = 2 \cos t, \quad t = \pi/4$

4. $x = \cos t, \quad y = \sqrt{3} \cos t, \quad t = 2\pi/3$

5. $x = t, \quad y = \sqrt{t}, \quad t = 1/4$

6. $x = \sec^2 t - 1, \quad y = \tan t, \quad t = -\pi/4$

7. $x = \sec t, \quad y = \tan t, \quad t = \pi/6$

8. $x = -\sqrt{t+1}, \quad y = \sqrt{3t}, \quad t = 3$

9. $x = 2t^2 + 3, \quad y = t^4, \quad t = -1$

10. $x = 1/t, \quad y = -2 + \ln t, \quad t = 1$

11. $x = t - \sin t, \quad y = 1 - \cos t, \quad t = \pi/3$

12. $x = \cos t, \quad y = 1 + \sin t, \quad t = \pi/2$

13. $x = \dfrac{1}{t+1}, \quad y = \dfrac{t}{t-1}, \quad t = 2$

14. $x = t + e^t, \quad y = 1 - e^t, \quad t = 0$

Implicitly Defined Parametrizations

Assuming that the equations in Exercises 15–20 define x and y implicitly as differentiable functions $x = f(t), y = g(t)$, find the slope of the curve $x = f(t), y = g(t)$ at the given value of t.

15. $x^3 + 2t^2 = 9, \quad 2y^3 - 3t^2 = 4, \quad t = 2$

16. $x = \sqrt{5 - \sqrt{t}}, \quad y(t - 1) = \sqrt{t}, \quad t = 4$

17. $x + 2x^{3/2} = t^2 + t, \quad y\sqrt{t+1} + 2t\sqrt{y} = 4, \quad t = 0$

18. $x \sin t + 2x = t, \quad t \sin t - 2t = y, \quad t = \pi$

19. $x = t^3 + t, \quad y + 2t^3 = 2x + t^2, \quad t = 1$

20. $t = \ln(x - t), \quad y = te^t, \quad t = 0$

Area

21. Find the area under one arch of the cycloid

$$x = a(t - \sin t), \quad y = a(1 - \cos t).$$

22. Find the area enclosed by the y-axis and the curve

$$x = t - t^2, \quad y = 1 + e^{-t}.$$

23. Find the area enclosed by the ellipse

$$x = a \cos t, \quad y = b \sin t, \quad 0 \le t \le 2\pi.$$

24. Find the area under $y = x^3$ over $[0, 1]$ using the following parametrizations.

a. $x = t^2, \quad y = t^6$ **b.** $x = t^3, \quad y = t^9$

Lengths of Curves

Find the lengths of the curves in Exercises 25–30.

25. $x = \cos t, \quad y = t + \sin t, \quad 0 \le t \le \pi$

26. $x = t^3, \quad y = 3t^2/2, \quad 0 \le t \le \sqrt{3}$

27. $x = t^2/2, \quad y = (2t + 1)^{3/2}/3, \quad 0 \le t \le 4$

28. $x = (2t + 3)^{3/2}/3, \quad y = t + t^2/2, \quad 0 \le t \le 3$

29. $x = 8 \cos t + 8t \sin t$
      $y = 8 \sin t - 8t \cos t,$
      $0 \le t \le \pi/2$

30. $x = \ln(\sec t + \tan t) - \sin t$
      $y = \cos t, \quad 0 \le t \le \pi/3$

Surface Area

Find the areas of the surfaces generated by revolving the curves in Exercises 31–34 about the indicated axes.

31. $x = \cos t, \quad y = 2 + \sin t, \quad 0 \le t \le 2\pi; \quad x$-axis

32. $x = (2/3)t^{3/2}, \quad y = 2\sqrt{t}, \quad 0 \le t \le \sqrt{3}; \quad y$-axis

33. $x = t + \sqrt{2}, \quad y = (t^2/2) + \sqrt{2}t, \quad -\sqrt{2} \le t \le \sqrt{2}; \quad y$-axis

34. $x = \ln(\sec t + \tan t) - \sin t, y = \cos t, 0 \le t \le \pi/3; x$-axis

35. A cone frustum The line segment joining the points $(0, 1)$ and $(2, 2)$ is revolved about the x-axis to generate a frustum of a cone. Find the surface area of the frustum using the parametrization $x = 2t, y = t + 1, 0 \le t \le 1$. Check your result with the geometry formula: Area $= \pi(r_1 + r_2)$(slant height).

36. A cone The line segment joining the origin to the point (h, r) is revolved about the x-axis to generate a cone of height h and base radius r. Find the cone's surface area with the parametric equations $x = ht, y = rt, 0 \le t \le 1$. Check your result with the geometry formula: Area $= \pi r$(slant height).

Centroids

37. Find the coordinates of the centroid of the curve

$$x = \cos t + t \sin t, \quad y = \sin t - t \cos t, \quad 0 \le t \le \pi/2.$$

38. Find the coordinates of the centroid of the curve

$$x = e^t \cos t, \quad y = e^t \sin t, \quad 0 \le t \le \pi.$$

39. Find the coordinates of the centroid of the curve

$$x = \cos t, \quad y = t + \sin t, \quad 0 \le t \le \pi.$$

T **40.** Most centroid calculations for curves are done with a calculator or computer that has an integral evaluation program. As a case in point, find, to the nearest hundredth, the coordinates of the centroid of the curve

$$x = t^3, \quad y = 3t^2/2, \quad 0 \le t \le \sqrt{3}.$$

Theory and Examples

41. Length is independent of parametrization To illustrate the fact that the numbers we get for length do not depend on the way we parametrize our curves (except for the mild restrictions preventing doubling back mentioned earlier), calculate the length of the semicircle $y = \sqrt{1 - x^2}$ with these two different parametrizations:

a. $x = \cos 2t, \quad y = \sin 2t, \quad 0 \le t \le \pi/2.$

b. $x = \sin \pi t, \quad y = \cos \pi t, \quad -1/2 \le t \le 1/2.$

42. a. Show that the Cartesian formula

$$L = \int_c^d \sqrt{1 + \left(\frac{dx}{dy}\right)^2} \, dy$$

for the length of the curve $x = g(y)$, $c \le y \le d$ (Section 6.3, Equation 4), is a special case of the parametric length formula

$$L = \int_a^b \sqrt{\left(\frac{dx}{dt}\right)^2 + \left(\frac{dy}{dt}\right)^2} \, dt.$$

Use this result to find the length of each curve.

b. $x = y^{3/2}$, $0 \le y \le 4/3$

c. $x = \frac{3}{2} y^{2/3}$, $0 \le y \le 1$

43. The curve with parametric equations

$$x = (1 + 2 \sin \theta) \cos \theta, \quad y = (1 + 2 \sin \theta) \sin \theta$$

is called a *limaçon* and is shown in the accompanying figure. Find the points (x, y) and the slopes of the tangent lines at these points for

a. $\theta = 0$. **b.** $\theta = \pi/2$. **c.** $\theta = 4\pi/3$.

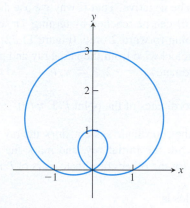

44. The curve with parametric equations

$$x = t, \quad y = 1 - \cos t, \quad 0 \le t \le 2\pi$$

is called a *sinusoid* and is shown in the accompanying figure. Find the point (x, y) where the slope of the tangent line is

a. largest. **b.** smallest.

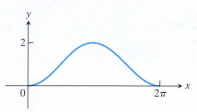

T The curves in Exercises 45 and 46 are called *Bowditch curves* or *Lissajous figures*. In each case, find the point in the interior of the first quadrant where the tangent to the curve is horizontal, and find the equations of the two tangents at the origin.

45.

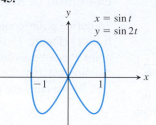
$$x = \sin t$$
$$y = \sin 2t$$

46.

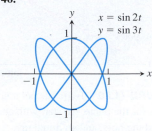
$$x = \sin 2t$$
$$y = \sin 3t$$

47. Cycloid

a. Find the length of one arch of the cycloid

$$x = a(t - \sin t), \quad y = a(1 - \cos t).$$

b. Find the area of the surface generated by revolving one arch of the cycloid in part (a) about the x-axis for $a = 1$.

48. Volume Find the volume swept out by revolving the region bounded by the x-axis and one arch of the cycloid

$$x = t - \sin t, \quad y = 1 - \cos t$$

about the x-axis.

49. Find the volume swept out by revolving the region bounded by the x-axis and the graph of

$$x = 2t, \quad y = t(2 - t)$$

about the x-axis.

50. Find the volume swept out by revolving the region bounded by the y-axis and the graph of

$$x = t(1 - t), \quad y = 1 + t^2$$

about the y-axis.

COMPUTER EXPLORATIONS

In Exercises 51–54, use a CAS to perform the following steps for the given curve over the closed interval.

a. Plot the curve together with the polygonal path approximations for $n = 2, 4, 8$ partition points over the interval. (See Figure 11.16.)

b. Find the corresponding approximation to the length of the curve by summing the lengths of the line segments.

c. Evaluate the length of the curve using an integral. Compare your approximations for $n = 2, 4, 8$ with the actual length given by the integral. How does the actual length compare with the approximations as n increases? Explain your answer.

51. $x = \frac{1}{3} t^3$, $y = \frac{1}{2} t^2$, $0 \le t \le 1$

52. $x = 2t^3 - 16t^2 + 25t + 5$, $y = t^2 + t - 3$, $0 \le t \le 6$

53. $x = t - \cos t$, $y = 1 + \sin t$, $-\pi \le t \le \pi$

54. $x = e^t \cos t$, $y = e^t \sin t$, $0 \le t \le \pi$

11.3 Polar Coordinates

In this section we study polar coordinates and their relation to Cartesian coordinates. You will see that polar coordinates are very useful for calculating many multiple integrals studied in Chapter 15. They are also useful in describing the paths of planets and satellites.

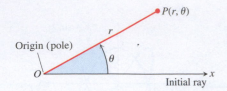

FIGURE 11.20 To define polar coordinates for the plane, we start with an origin, called the pole, and an initial ray.

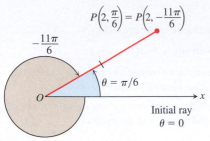

FIGURE 11.21 Polar coordinates are not unique.

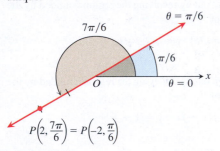

FIGURE 11.22 Polar coordinates can have negative r-values.

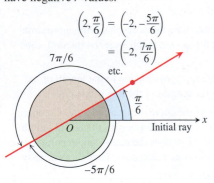

FIGURE 11.23 The point $P(2, \pi/6)$ has infinitely many polar coordinate pairs (Example 1).

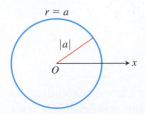

FIGURE 11.24 The polar equation for a circle is $r = a$.

Definition of Polar Coordinates

To define polar coordinates, we first fix an **origin** O (called the **pole**) and an **initial ray** from O (Figure 11.20). Usually the positive x-axis is chosen as the initial ray. Then each point P can be located by assigning to it a **polar coordinate pair** (r, θ) in which r gives the directed distance from O to P and θ gives the directed angle from the initial ray to ray OP. So we label the point P as

$$P(r, \theta)$$

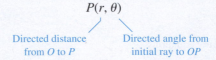

Directed distance
from O to P

Directed angle from
initial ray to OP

As in trigonometry, θ is positive when measured counterclockwise and negative when measured clockwise. The angle associated with a given point is not unique. While a point in the plane has just one pair of Cartesian coordinates, it has infinitely many pairs of polar coordinates. For instance, the point 2 units from the origin along the ray $\theta = \pi/6$ has polar coordinates $r = 2$, $\theta = \pi/6$. It also has coordinates $r = 2$, $\theta = -11\pi/6$ (Figure 11.21). In some situations we allow r to be negative. That is why we use directed distance in defining $P(r, \theta)$. The point $P(2, 7\pi/6)$ can be reached by turning $7\pi/6$ radians counterclockwise from the initial ray and going forward 2 units (Figure 11.22). It can also be reached by turning $\pi/6$ radians counterclockwise from the initial ray and going *backward* 2 units. So the point also has polar coordinates $r = -2$, $\theta = \pi/6$.

EXAMPLE 1 Find all the polar coordinates of the point $P(2, \pi/6)$.

Solution We sketch the initial ray of the coordinate system, draw the ray from the origin that makes an angle of $\pi/6$ radians with the initial ray, and mark the point $(2, \pi/6)$ (Figure 11.23). We then find the angles for the other coordinate pairs of P in which $r = 2$ and $r = -2$.

For $r = 2$, the complete list of angles is

$$\frac{\pi}{6}, \quad \frac{\pi}{6} \pm 2\pi, \quad \frac{\pi}{6} \pm 4\pi, \quad \frac{\pi}{6} \pm 6\pi, \ldots .$$

For $r = -2$, the angles are

$$-\frac{5\pi}{6}, \quad -\frac{5\pi}{6} \pm 2\pi, \quad -\frac{5\pi}{6} \pm 4\pi, \quad -\frac{5\pi}{6} \pm 6\pi, \ldots .$$

The corresponding coordinate pairs of P are

$$\left(2, \frac{\pi}{6} + 2n\pi\right), \qquad n = 0, \pm 1, \pm 2, \ldots$$

and

$$\left(-2, -\frac{5\pi}{6} + 2n\pi\right), \qquad n = 0, \pm 1, \pm 2, \ldots .$$

When $n = 0$, the formulas give $(2, \pi/6)$ and $(-2, -5\pi/6)$. When $n = 1$, they give $(2, 13\pi/6)$ and $(-2, 7\pi/6)$, and so on. ∎

Polar Equations and Graphs

If we hold r fixed at a constant value $r = a \neq 0$, the point $P(r, \theta)$ will lie $|a|$ units from the origin O. As θ varies over any interval of length 2π, P then traces a circle of radius $|a|$ centered at O (Figure 11.24).

If we hold θ fixed at a constant value $\theta = \theta_0$ and let r vary between $-\infty$ and ∞, the point $P(r, \theta)$ traces the line through O that makes an angle of measure θ_0 with the initial ray. (See Figure 11.22 for an example.)

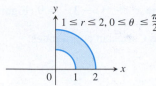

(b)

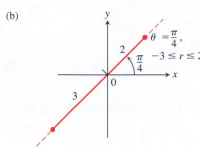

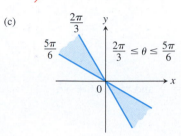

FIGURE 11.25 The graphs of typical inequalities in r and θ (Example 3).

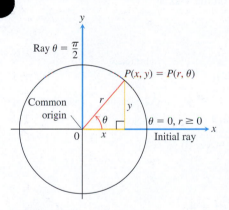

FIGURE 11.26 The usual way to relate polar and Cartesian coordinates.

EXAMPLE 2 A circle or line can have more than one polar equation.

(a) $r = 1$ and $r = -1$ are equations for the circle of radius 1 centered at O.

(b) $\theta = \pi/6, \theta = 7\pi/6$, and $\theta = -5\pi/6$ are equations for the line in Figure 11.23. ∎

Equations of the form $r = a$ and $\theta = \theta_0$ can be combined to define regions, segments, and rays.

EXAMPLE 3 Graph the sets of points whose polar coordinates satisfy the following conditions.

(a) $1 \le r \le 2$ and $0 \le \theta \le \dfrac{\pi}{2}$

(b) $-3 \le r \le 2$ and $\theta = \dfrac{\pi}{4}$

(c) $\dfrac{2\pi}{3} \le \theta \le \dfrac{5\pi}{6}$ (no restriction on r)

Solution The graphs are shown in Figure 11.25. ∎

Relating Polar and Cartesian Coordinates

When we use both polar and Cartesian coordinates in a plane, we place the two origins together and let the initial polar ray be the positive x-axis. The ray $\theta = \pi/2, r > 0$, becomes the positive y-axis (Figure 11.26). The two coordinate systems are then related by the following equations.

Equations Relating Polar and Cartesian Coordinates

$$x = r \cos \theta, \qquad y = r \sin \theta, \qquad r^2 = x^2 + y^2, \qquad \tan \theta = \frac{y}{x}$$

The first two of these equations uniquely determine the Cartesian coordinates x and y given the polar coordinates r and θ. On the other hand, if x and y are given, the third equation gives two possible choices for r (a positive and a negative value). For each $(x, y) \neq (0, 0)$, there is a unique $\theta \in [0, 2\pi)$ satisfying the first two equations, each then giving a polar coordinate representation of the Cartesian point (x, y). The other polar coordinate representations for the point can be determined from these two, as in Example 1.

EXAMPLE 4 Here are some plane curves expressed in terms of both polar coordinate and Cartesian coordinate equations.

Polar equation	Cartesian equivalent
$r \cos \theta = 2$	$x = 2$
$r^2 \cos \theta \sin \theta = 4$	$xy = 4$
$r^2 \cos^2\theta - r^2 \sin^2\theta = 1$	$x^2 - y^2 = 1$
$r = 1 + 2r \cos \theta$	$y^2 - 3x^2 - 4x - 1 = 0$
$r = 1 - \cos \theta$	$x^4 + y^4 + 2x^2y^2 + 2x^3 + 2xy^2 - y^2 = 0$

Some curves are more simply expressed with polar coordinates; others are not. ∎

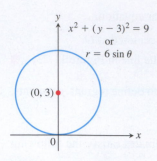

FIGURE 11.27 The circle in Example 5.

EXAMPLE 5 Find a polar equation for the circle $x^2 + (y - 3)^2 = 9$ (Figure 11.27)

Solution We apply the equations relating polar and Cartesian coordinates:

$$x^2 + (y - 3)^2 = 9$$
$$x^2 + y^2 - 6y + 9 = 9 \qquad \text{Expand } (y - 3)^2.$$
$$x^2 + y^2 - 6y = 0 \qquad \text{Cancelation}$$
$$r^2 - 6r \sin \theta = 0 \qquad x^2 + y^2 = r^2, \ y = r \sin \theta$$
$$r = 0 \quad \text{or} \quad r - 6 \sin \theta = 0$$
$$r = 6 \sin \theta \qquad \text{Includes both possibilities} \qquad ■$$

EXAMPLE 6 Replace the following polar equations by equivalent Cartesian equations and identify their graphs.

(a) $r \cos \theta = -4$

(b) $r^2 = 4r \cos \theta$

(c) $r = \dfrac{4}{2 \cos \theta - \sin \theta}$

Solution We use the substitutions $r \cos \theta = x$, $r \sin \theta = y$, and $r^2 = x^2 + y^2$.

(a) $r \cos \theta = -4$

The Cartesian equation: $r \cos \theta = -4$
$$x = -4 \qquad \text{Substitute.}$$

The graph: Vertical line through $x = -4$ on the x-axis

(b) $r^2 = 4r \cos \theta$

The Cartesian equation: $r^2 = 4r \cos \theta$
$$x^2 + y^2 = 4x \qquad \text{Substitute.}$$
$$x^2 - 4x + y^2 = 0$$
$$x^2 - 4x + 4 + y^2 = 4 \qquad \text{Complete the square.}$$
$$(x - 2)^2 + y^2 = 4 \qquad \text{Factor.}$$

The graph: Circle, radius 2, center $(h, k) = (2, 0)$

(c) $r = \dfrac{4}{2 \cos \theta - \sin \theta}$

The Cartesian equation: $r(2 \cos \theta - \sin \theta) = 4$
$$2r \cos \theta - r \sin \theta = 4 \qquad \text{Multiply by } r.$$
$$2x - y = 4 \qquad \text{Substitute.}$$
$$y = 2x - 4 \qquad \text{Solve for } y.$$

The graph: Line, slope $m = 2$, y-intercept $b = -4$ ■

EXERCISES 11.3

Polar Coordinates

1. Which polar coordinate pairs label the same point?

a. $(3, 0)$ **b.** $(-3, 0)$ **c.** $(2, 2\pi/3)$

d. $(2, 7\pi/3)$ **e.** $(-3, \pi)$ **f.** $(2, \pi/3)$

g. $(-3, 2\pi)$ **h.** $(-2, -\pi/3)$

2. Which polar coordinate pairs label the same point?

a. $(-2, \pi/3)$ **b.** $(2, -\pi/3)$ **c.** (r, θ)

d. $(r, \theta + \pi)$ **e.** $(-r, \theta)$ **f.** $(2, -2\pi/3)$

g. $(-r, \theta + \pi)$ **h.** $(-2, 2\pi/3)$

3. Plot the following points (given in polar coordinates). Then find all the polar coordinates of each point.

a. $(2, \pi/2)$ **b.** $(2, 0)$

c. $(-2, \pi/2)$ **d.** $(-2, 0)$

4. Plot the following points (given in polar coordinates). Then find all the polar coordinates of each point.

a. $(3, \pi/4)$ **b.** $(-3, \pi/4)$

c. $(3, -\pi/4)$ **d.** $(-3, -\pi/4)$

Polar to Cartesian Coordinates

5. Find the Cartesian coordinates of the points in Exercise 1.

6. Find the Cartesian coordinates of the following points (given in polar coordinates).

a. $\left(\sqrt{2}, \pi/4\right)$ **b.** $(1, 0)$

c. $(0, \pi/2)$ **d.** $\left(-\sqrt{2}, \pi/4\right)$

e. $(-3, 5\pi/6)$ **f.** $(5, \tan^{-1}(4/3))$

g. $(-1, 7\pi)$ **h.** $\left(2\sqrt{3}, 2\pi/3\right)$

Cartesian to Polar Coordinates

7. Find the polar coordinates, $0 \le \theta < 2\pi$ and $r \ge 0$, of the following points given in Cartesian coordinates.

a. $(1, 1)$ **b.** $(-3, 0)$

c. $\left(\sqrt{3}, -1\right)$ **d.** $(-3, 4)$

8. Find the polar coordinates, $-\pi \le \theta < \pi$ and $r \ge 0$, of the following points given in Cartesian coordinates.

a. $(-2, -2)$ **b.** $(0, 3)$

c. $\left(-\sqrt{3}, 1\right)$ **d.** $(5, -12)$

9. Find the polar coordinates, $0 \le \theta < 2\pi$ and $r \le 0$, of the following points given in Cartesian coordinates.

a. $(3, 3)$ **b.** $(-1, 0)$

c. $\left(-1, \sqrt{3}\right)$ **d.** $(4, -3)$

10. Find the polar coordinates, $-\pi \le \theta < 2\pi$ and $r \le 0$, of the following points given in Cartesian coordinates.

a. $(-2, 0)$ **b.** $(1, 0)$

c. $(0, -3)$ **d.** $\left(\frac{\sqrt{3}}{2}, \frac{1}{2}\right)$

Graphing Sets of Polar Coordinate Points

Graph the sets of points whose polar coordinates satisfy the equations and inequalities in Exercises 11–26.

11. $r = 2$ **12.** $0 \le r \le 2$

13. $r \ge 1$ **14.** $1 \le r \le 2$

15. $0 \le \theta \le \pi/6, \quad r \ge 0$ **16.** $\theta = 2\pi/3, \quad r \le -2$

17. $\theta = \pi/3, \quad -1 \le r \le 3$ **18.** $\theta = 11\pi/4, \quad r \ge -1$

19. $\theta = \pi/2, \quad r \ge 0$ **20.** $\theta = \pi/2, \quad r \le 0$

21. $0 \le \theta \le \pi, \quad r = 1$ **22.** $0 \le \theta \le \pi, \quad r = -1$

23. $\pi/4 \le \theta \le 3\pi/4, \quad 0 \le r \le 1$

24. $-\pi/4 \le \theta \le \pi/4, \quad -1 \le r \le 1$

25. $-\pi/2 \le \theta \le \pi/2, \quad 1 \le r \le 2$

26. $0 \le \theta \le \pi/2, \quad 1 \le |r| \le 2$

Polar to Cartesian Equations

Replace the polar equations in Exercises 27–52 with equivalent Cartesian equations. Then describe or identify the graph.

27. $r \cos \theta = 2$ **28.** $r \sin \theta = -1$

29. $r \sin \theta = 0$ **30.** $r \cos \theta = 0$

31. $r = 4 \csc \theta$ **32.** $r = -3 \sec \theta$

33. $r \cos \theta + r \sin \theta = 1$ **34.** $r \sin \theta = r \cos \theta$

35. $r^2 = 1$ **36.** $r^2 = 4r \sin \theta$

37. $r = \dfrac{5}{\sin \theta - 2 \cos \theta}$ **38.** $r^2 \sin 2\theta = 2$

39. $r = \cot \theta \csc \theta$ **40.** $r = 4 \tan \theta \sec \theta$

41. $r = \csc \theta \, e^{r \cos \theta}$ **42.** $r \sin \theta = \ln r + \ln \cos \theta$

43. $r^2 + 2r^2 \cos \theta \sin \theta = 1$ **44.** $\cos^2 \theta = \sin^2 \theta$

45. $r^2 = -4r \cos \theta$ **46.** $r^2 = -6r \sin \theta$

47. $r = 8 \sin \theta$ **48.** $r = 3 \cos \theta$

49. $r = 2 \cos \theta + 2 \sin \theta$ **50.** $r = 2 \cos \theta - \sin \theta$

51. $r \sin \left(\theta + \dfrac{\pi}{6}\right) = 2$

52. $r \sin \left(\dfrac{2\pi}{3} - \theta\right) = 5$

Cartesian to Polar Equations

Replace the Cartesian equations in Exercises 53–66 with equivalent polar equations.

53. $x = 7$ **54.** $y = 1$ **55.** $x = y$

56. $x - y = 3$ **57.** $x^2 + y^2 = 4$ **58.** $x^2 - y^2 = 1$

59. $\dfrac{x^2}{9} + \dfrac{y^2}{4} = 1$ **60.** $xy = 2$

61. $y^2 = 4x$ **62.** $x^2 + xy + y^2 = 1$

63. $x^2 + (y - 2)^2 = 4$ **64.** $(x - 5)^2 + y^2 = 25$

65. $(x - 3)^2 + (y + 1)^2 = 4$ **66.** $(x + 2)^2 + (y - 5)^2 = 16$

67. Find all polar coordinates of the origin.

68. Vertical and horizontal lines

a. Show that every vertical line in the xy-plane has a polar equation of the form $r = a \sec \theta$.

b. Find the analogous polar equation for horizontal lines in the xy-plane.

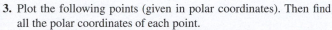

11.4 Graphing Polar Coordinate Equations

It is often helpful to graph an equation expressed in polar coordinates in the Cartesian xy-plane. This section describes some techniques for graphing these equations using symmetries and tangents to the graph.

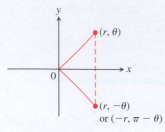

(a) About the x-axis

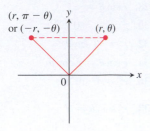

(b) About the y-axis

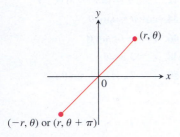

(c) About the origin

FIGURE 11.28 Three tests for symmetry in polar coordinates.

Symmetry

The following list shows how to test for three standard types of symmetries when using polar coordinates. These symmetries are illustrated in Figure 11.28.

Symmetry Tests for Polar Graphs in the Cartesian xy-Plane

1. *Symmetry about the x-axis:* If the point (r, θ) lies on the graph, then the point $(r, -\theta)$ or $(-r, \pi - \theta)$ lies on the graph (Figure 11.28a).

2. *Symmetry about the y-axis:* If the point (r, θ) lies on the graph, then the point $(r, \pi - \theta)$ or $(-r, -\theta)$ lies on the graph (Figure 11.28b).

3. *Symmetry about the origin:* If the point (r, θ) lies on the graph, then the point $(-r, \theta)$ or $(r, \theta + \pi)$ lies on the graph (Figure 11.28c).

Slope

The slope of a polar curve $r = f(\theta)$ in the xy-plane is dy/dx, but this is **not** given by the formula $r' = df/d\theta$. To see why, think of the graph of f as the graph of the parametric equations

$$x = r \cos \theta = f(\theta) \cos \theta, \qquad y = r \sin \theta = f(\theta) \sin \theta.$$

If f is a differentiable function of θ, then so are x and y and, when $dx/d\theta \neq 0$, we can calculate dy/dx from the parametric formula

$$\frac{dy}{dx} = \frac{dy/d\theta}{dx/d\theta} \qquad \text{Section 11.2, Eq. (1) with } t = \theta$$

$$= \frac{\dfrac{d}{d\theta}(f(\theta) \sin \theta)}{\dfrac{d}{d\theta}(f(\theta) \cos \theta)} \qquad \text{Substitute}$$

$$= \frac{\dfrac{df}{d\theta} \sin \theta + f(\theta) \cos \theta}{\dfrac{df}{d\theta} \cos \theta - f(\theta) \sin \theta} \qquad \text{Product Rule for derivatives}$$

Therefore we see that dy/dx is not the same as $df/d\theta$.

Slope of the Curve $r = f(\theta)$ in the Cartesian xy-Plane

$$\left.\frac{dy}{dx}\right|_{(r,\,\theta)} = \frac{f'(\theta) \sin \theta + f(\theta) \cos \theta}{f'(\theta) \cos \theta - f(\theta) \sin \theta} \qquad (1)$$

provided $dx/d\theta \neq 0$ at (r, θ).

If the curve $r = f(\theta)$ passes through the origin at $\theta = \theta_0$, then $f(\theta_0) = 0$, and the slope equation gives

$$\left.\frac{dy}{dx}\right|_{(0,\,\theta_0)} = \frac{f'(\theta_0) \sin \theta_0}{f'(\theta_0) \cos \theta_0} = \tan \theta_0.$$

That is, the slope at $(0, \theta_0)$ is $\tan \theta_0$. The reason we say "slope at $(0, \theta_0)$" and not just "slope at the origin" is that a polar curve may pass through the origin (or any point) more than once, with different slopes at different θ-values. This is not the case in our first example, however.

θ	$r = 1 - \cos\theta$
0	0
$\dfrac{\pi}{3}$	$\dfrac{1}{2}$
$\dfrac{\pi}{2}$	1
$\dfrac{2\pi}{3}$	$\dfrac{3}{2}$
π	2

(a)

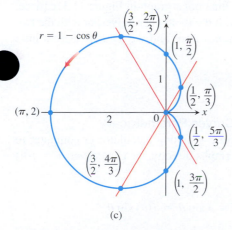

(b)

(c)

FIGURE 11.29 The steps in graphing the cardioid $r = 1 - \cos\theta$ (Example 1). The arrow shows the direction of increasing θ.

EXAMPLE 1   Graph the curve $r = 1 - \cos\theta$ in the Cartesian xy-plane.

Solution   The curve is symmetric about the x-axis because

$$(r, \theta) \text{ on the graph} \Rightarrow r = 1 - \cos\theta$$
$$\Rightarrow r = 1 - \cos(-\theta) \qquad \cos\theta = \cos(-\theta)$$
$$\Rightarrow (r, -\theta) \text{ on the graph.}$$

As θ increases from 0 to π, $\cos\theta$ decreases from 1 to -1, and $r = 1 - \cos\theta$ increases from a minimum value of 0 to a maximum value of 2. As θ continues on from π to 2π, $\cos\theta$ increases from -1 back to 1 and r decreases from 2 back to 0. The curve starts to repeat when $\theta = 2\pi$ because the cosine has period 2π.

The curve leaves the origin with slope $\tan(0) = 0$ and returns to the origin with slope $\tan(2\pi) = 0$.

We make a table of values from $\theta = 0$ to $\theta = \pi$, plot the points, draw a smooth curve through them with a horizontal tangent at the origin, and reflect the curve across the x-axis to complete the graph (Figure 11.29). The curve is called a *cardioid* because of its heart shape. ■

EXAMPLE 2   Graph the curve $r^2 = 4\cos\theta$ in the Cartesian xy-plane.

Solution   The equation $r^2 = 4\cos\theta$ requires $\cos\theta \geq 0$, so we get the entire graph by running θ from $-\pi/2$ to $\pi/2$. The curve is symmetric about the x-axis because

$$(r, \theta) \text{ on the graph} \Rightarrow r^2 = 4\cos\theta$$
$$\Rightarrow r^2 = 4\cos(-\theta) \qquad \cos\theta = \cos(-\theta)$$
$$\Rightarrow (r, -\theta) \text{ on the graph.}$$

The curve is also symmetric about the origin because

$$(r, \theta) \text{ on the graph} \Rightarrow r^2 = 4\cos\theta$$
$$\Rightarrow (-r)^2 = 4\cos\theta$$
$$\Rightarrow (-r, \theta) \text{ on the graph.}$$

Together, these two symmetries imply symmetry about the y-axis.

The curve passes through the origin when $\theta = -\pi/2$ and $\theta = \pi/2$. It has a vertical tangent both times because $\tan\theta$ is infinite.

For each value of θ in the interval between $-\pi/2$ and $\pi/2$, the formula $r^2 = 4\cos\theta$ gives two values of r:

$$r = \pm 2\sqrt{\cos\theta}.$$

We make a short table of values, plot the corresponding points, and use information about symmetry and tangents to guide us in connecting the points with a smooth curve (Figure 11.30).

θ	$\cos\theta$	$r = \pm 2\sqrt{\cos\theta}$
0	1	± 2
$\pm\dfrac{\pi}{6}$	$\dfrac{\sqrt{3}}{2}$	$\approx \pm 1.9$
$\pm\dfrac{\pi}{4}$	$\dfrac{1}{\sqrt{2}}$	$\approx \pm 1.7$
$\pm\dfrac{\pi}{3}$	$\dfrac{1}{2}$	$\approx \pm 1.4$
$\pm\dfrac{\pi}{2}$	0	0

(a)

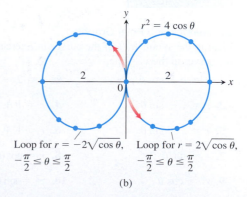

Loop for $r = -2\sqrt{\cos\theta}$, $-\dfrac{\pi}{2} \leq \theta \leq \dfrac{\pi}{2}$ Loop for $r = 2\sqrt{\cos\theta}$, $-\dfrac{\pi}{2} \leq \theta \leq \dfrac{\pi}{2}$

(b)

FIGURE 11.30 The graph of $r^2 = 4\cos\theta$. The arrows show the direction of increasing θ. The values of r in the table are rounded (Example 2). ■

(a) r^2

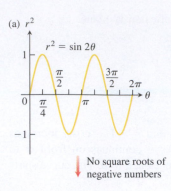

No square roots of
negative numbers

(b) r

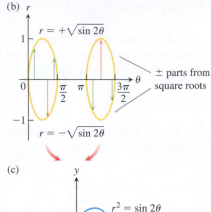

± parts from
square roots

(c)

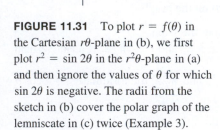

FIGURE 11.31 To plot $r = f(\theta)$ in the Cartesian $r\theta$-plane in (b), we first plot $r^2 = \sin 2\theta$ in the $r^2\theta$-plane in (a) and then ignore the values of θ for which $\sin 2\theta$ is negative. The radii from the sketch in (b) cover the polar graph of the lemniscate in (c) twice (Example 3).

Converting a Graph from the $r\theta$- to xy-Plane

One way to graph a polar equation $r = f(\theta)$ in the xy-plane is to make a table of (r, θ)-values, plot the corresponding points there, and connect them in order of increasing θ. This can work well if enough points have been plotted to reveal all the loops and dimples in the graph. Another method of graphing is to

1. first graph the function $r = f(\theta)$ in the *Cartesian $r\theta$-plane*,

2. then use that Cartesian graph as a "table" and guide to sketch the *polar* coordinate graph in the xy-plane.

This method is sometimes better than simple point plotting because the first Cartesian graph shows at a glance where r is positive, negative, and nonexistent, as well as where r is increasing and decreasing. Here is an example.

EXAMPLE 3 Graph the *lemniscate* curve $r^2 = \sin 2\theta$ in the Cartesian xy-plane.

Solution For this example it will be easier to first plot r^2, instead of r, as a function of θ in the Cartesian $r^2\theta$-plane (see Figure 11.31a). We pass from there to the graph of $r = \pm\sqrt{\sin 2\theta}$ in the $r\theta$-plane (Figure 11.31b), and then draw the polar graph (Figure 11.31c). The graph in Figure 11.31b "covers" the final polar graph in Figure 11.31c twice. We could have managed with either loop alone, with the two upper halves, or with the two lower halves. The double covering does no harm, however, and we actually learn a little more about the behavior of the function this way.

USING TECHNOLOGY   Graphing Polar Curves Parametrically

For complicated polar curves we may need to use a graphing calculator or computer to graph the curve. If the device does not plot polar graphs directly, we can convert $r = f(\theta)$ into parametric form using the equations

$$x = r\cos\theta = f(\theta)\cos\theta, \qquad y = r\sin\theta = f(\theta)\sin\theta.$$

Then we use the device to draw a parametrized curve in the Cartesian xy-plane.

EXERCISES 11.4

Symmetries and Polar Graphs

Identify the symmetries of the curves in Exercises 1–12. Then sketch the curves in the xy-plane.

1. $r = 1 + \cos\theta$
2. $r = 2 - 2\cos\theta$

3. $r = 1 - \sin\theta$
4. $r = 1 + \sin\theta$

5. $r = 2 + \sin\theta$
6. $r = 1 + 2\sin\theta$

7. $r = \sin(\theta/2)$
8. $r = \cos(\theta/2)$

9. $r^2 = \cos\theta$
10. $r^2 = \sin\theta$

11. $r^2 = -\sin\theta$
12. $r^2 = -\cos\theta$

Graph the lemniscates in Exercises 13–16. What symmetries do these curves have?

13. $r^2 = 4\cos 2\theta$
14. $r^2 = 4\sin 2\theta$

15. $r^2 = -\sin 2\theta$
16. $r^2 = -\cos 2\theta$

Slopes of Polar Curves in the xy-Plane

Find the slopes of the curves in Exercises 17–20 at the given points. Sketch the curves along with their tangents at these points.

17. **Cardioid** $r = -1 + \cos\theta$; $\theta = \pm\pi/2$

18. **Cardioid** $r = -1 + \sin\theta$; $\theta = 0, \pi$

19. **Four-leaved rose** $r = \sin 2\theta$; $\theta = \pm\pi/4, \pm3\pi/4$

20. **Four-leaved rose** $r = \cos 2\theta$; $\theta = 0, \pm\pi/2, \pi$

Concavity of Polar Curves in the xy-Plane

Equation (1) gives the formula for the derivative y' of a polar curve $r = f(\theta)$. The second derivative is $\dfrac{d^2y}{dx^2} = \dfrac{dy'/d\theta}{dx/d\theta}$ (see Equation (2) in Section 11.2). Find the slope and concavity of the curves in Exercises 21–24 at the given points.

21. $r = \sin\theta, \quad \theta = \pi/6, \pi/3$ **22.** $r = e^{\theta}, \quad \theta = 0, \pi$

23. $r = \theta, \quad \theta = 0, \pi/2$ **24.** $r = 1/\theta, \quad \theta = -\pi, 1$

Graphing Limaçons

Graph the limaçons in Exercises 25–28. Limaçon ("*lee*-ma-sahn") is Old French for "snail." You will understand the name when you graph the limaçons in Exercise 25. Equations for limaçons have the form $r = a \pm b\cos\theta$ or $r = a \pm b\sin\theta$. There are four basic shapes.

25. Limaçons with an inner loop

  a. $r = \dfrac{1}{2} + \cos\theta$ **b.** $r = \dfrac{1}{2} + \sin\theta$

26. Cardioids

  a. $r = 1 - \cos\theta$ **b.** $r = -1 + \sin\theta$

27. Dimpled limaçons

  a. $r = \dfrac{3}{2} + \cos\theta$ **b.** $r = \dfrac{3}{2} - \sin\theta$

28. Oval limaçons

  a. $r = 2 + \cos\theta$ **b.** $r = -2 + \sin\theta$

Graphing Polar Regions and Curves in the xy-Plane

29. Sketch the region defined by the inequalities $-1 \le r \le 2$ and $-\pi/2 \le \theta \le \pi/2$.

30. Sketch the region defined by the inequalities $0 \le r \le 2\sec\theta$ and $-\pi/4 \le \theta \le \pi/4$.

In Exercises 31 and 32, sketch the region defined by the inequality.

31. $0 \le r \le 2 - 2\cos\theta$ **32.** $0 \le r^2 \le \cos\theta$

T **33.** Which of the following has the same graph as $r = 1 - \cos\theta$?

  a. $r = -1 - \cos\theta$ **b.** $r = 1 + \cos\theta$

  Confirm your answer with algebra.

T **34.** Which of the following has the same graph as $r = \cos 2\theta$?

  a. $r = -\sin(2\theta + \pi/2)$ **b.** $r = -\cos(\theta/2)$

  Confirm your answer with algebra.

T **35. A rose within a rose** Graph the equation $r = 1 - 2\sin 3\theta$.

T **36. The nephroid of Freeth** Graph the nephroid of Freeth:

$$r = 1 + 2\sin\frac{\theta}{2}.$$

T **37. Roses** Graph the roses $r = \cos m\theta$ for $m = 1/3, 2, 3,$ and 7.

T **38. Spirals** Polar coordinates are just the thing for defining spirals. Graph the following spirals.

  a. $r = \theta$

  b. $r = -\theta$

  c. A *logarithmic spiral*: $r = e^{\theta/10}$

  d. A *hyperbolic spiral*: $r = 8/\theta$

  e. An *equilateral hyperbola*: $r = \pm 10/\sqrt{\theta}$

  (Use different colors for the two branches.)

T **39.** Graph the equation $r = \sin\left(\frac{8}{7}\theta\right)$ for $0 \le \theta \le 14\pi$.

T **40.** Graph the equation

$$r = \sin^2(2.3\theta) + \cos^4(2.3\theta)$$

for $0 \le \theta \le 10\pi$.

11.5 Areas and Lengths in Polar Coordinates

This section shows how to calculate areas of plane regions and lengths of curves in polar coordinates.

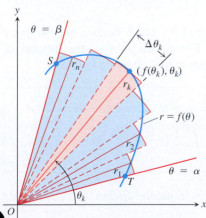

FIGURE 11.32 To derive a formula for the area of region OTS, we approximate the region with fan-shaped circular sectors.

Area in the Plane

The region OTS in Figure 11.32 is bounded by the rays $\theta = \alpha$ and $\theta = \beta$ and the curve $r = f(\theta)$. We approximate the region with n nonoverlapping fan-shaped circular sectors based on a partition P of angle TOS. The typical sector has radius $r_k = f(\theta_k)$ and central angle of radian measure $\Delta\theta_k$. Its area is $\Delta\theta_k/2\pi$ times the area of a circle of radius r_k, or

$$A_k = \frac{1}{2}r_k^2\,\Delta\theta_k = \frac{1}{2}\big(f(\theta_k)\big)^2\,\Delta\theta_k.$$

The area of region OTS is approximately

$$\sum_{k=1}^{n} A_k = \sum_{k=1}^{n} \frac{1}{2}\big(f(\theta_k)\big)^2\,\Delta\theta_k.$$

If f is continuous, we expect the approximations to improve as the norm of the partition P goes to zero, where the norm of P is the largest value of $\Delta\theta_k$. We are therefore led to the following formula for the region's area:

$$A = \lim_{\|P\| \to 0} \sum_{k=1}^{n} \frac{1}{2} \big(f(\theta_k)\big)^2 \, \Delta\theta_k = \int_{\alpha}^{\beta} \frac{1}{2} \big(f(\theta)\big)^2 \, d\theta.$$

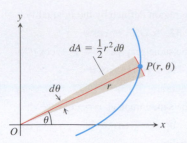

FIGURE 11.33 The area differential dA for the curve $r = f(\theta)$.

> **Area of the Fan-Shaped Region Between the Origin and the Curve**
> $r = f(\theta)$ when $\alpha \le \theta \le \beta, r \ge 0,$ and $\beta - \alpha \le 2\pi$.
>
> $$A = \int_{\alpha}^{\beta} \frac{1}{2} r^2 \, d\theta$$
>
> This is the integral of the **area differential** (Figure 11.33)
>
> $$dA = \frac{1}{2} r^2 \, d\theta = \frac{1}{2} \big(f(\theta)\big)^2 \, d\theta.$$

In the area formula above, we assumed that $r \ge 0$ and that the region does not sweep out an angle of more than 2π. This avoids issues with negatively signed areas or with regions that overlap themselves. More general regions can usually be handled by subdividing them into regions of this type if necessary.

EXAMPLE 1   Find the area of the region in the xy-plane enclosed by the cardioid $r = 2(1 + \cos\theta)$.

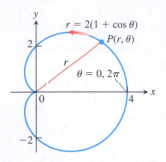

FIGURE 11.34 The cardioid in Example 1.

Solution   We graph the cardioid (Figure 11.34) and determine that the radius OP sweeps out the region exactly once as θ runs from 0 to 2π. The area is therefore

$$\int_{\theta=0}^{\theta=2\pi} \frac{1}{2} r^2 \, d\theta = \int_{0}^{2\pi} \frac{1}{2} \cdot 4(1 + \cos\theta)^2 \, d\theta$$

$$= \int_{0}^{2\pi} 2(1 + 2\cos\theta + \cos^2\theta) \, d\theta$$

$$= \int_{0}^{2\pi} \left(2 + 4\cos\theta + 2 \cdot \frac{1 + \cos 2\theta}{2} \right) d\theta$$

$$= \int_{0}^{2\pi} (3 + 4\cos\theta + \cos 2\theta) \, d\theta$$

$$= \left[3\theta + 4\sin\theta + \frac{\sin 2\theta}{2} \right]_{0}^{2\pi} = 6\pi - 0 = 6\pi. \quad \blacksquare$$

To find the area of a region like the one in Figure 11.35, which lies between two polar curves $r_1 = r_1(\theta)$ and $r_2 = r_2(\theta)$ from $\theta = \alpha$ to $\theta = \beta$, we subtract the integral of $(1/2)r_1^2 \, d\theta$ from the integral of $(1/2)r_2^2 \, d\theta$. This leads to the following formula.

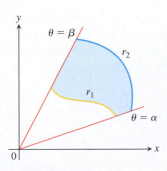

FIGURE 11.35 The area of the shaded region is calculated by subtracting the area of the region between r_1 and the origin from the area of the region between r_2 and the origin.

> **Area of the Region $0 \le r_1(\theta) \le r \le r_2(\theta), \alpha \le \theta \le \beta,$ and $\beta - \alpha \le 2\pi$.**
>
> $$A = \int_{\alpha}^{\beta} \frac{1}{2} r_2^2 \, d\theta - \int_{\alpha}^{\beta} \frac{1}{2} r_1^2 \, d\theta = \int_{\alpha}^{\beta} \frac{1}{2} \big(r_2^2 - r_1^2 \big) \, d\theta \qquad (1)$$

EXAMPLE 2   Find the area of the region that lies inside the circle $r = 1$ and outside the cardioid $r = 1 - \cos\theta$.

Solution   We sketch the region to determine its boundaries and find the limits of integration (Figure 11.36). The outer curve is $r_2 = 1$, the inner curve is $r_1 = 1 - \cos\theta$, and θ runs from $-\pi/2$ to $\pi/2$. The area, from Equation (1), is

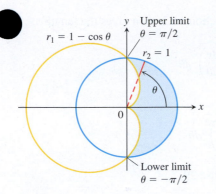

FIGURE 11.36 The region and limits of integration in Example 2.

$$A = \int_{-\pi/2}^{\pi/2} \frac{1}{2}\left(r_2{}^2 - r_1{}^2\right) d\theta \qquad \text{Eq. (1)}$$

$$= 2\int_{0}^{\pi/2} \frac{1}{2}\left(r_2{}^2 - r_1{}^2\right) d\theta \qquad \text{Symmetry}$$

$$= \int_{0}^{\pi/2} (1 - (1 - 2\cos\theta + \cos^2\theta)) \, d\theta \qquad r_2 = 1 \text{ and } r_1 = 1 - \cos\theta$$

$$= \int_{0}^{\pi/2} (2\cos\theta - \cos^2\theta) \, d\theta = \int_{0}^{\pi/2} \left(2\cos\theta - \frac{1 + \cos 2\theta}{2}\right) d\theta$$

$$= \left[2\sin\theta - \frac{\theta}{2} - \frac{\sin 2\theta}{4}\right]_{0}^{\pi/2} = 2 - \frac{\pi}{4}. \qquad \blacksquare$$

The fact that we can represent a point in different ways in polar coordinates requires extra care in deciding when a point lies on the graph of a polar equation and in determining the points in which polar graphs intersect. (We needed intersection points in Example 2.) In Cartesian coordinates, we can always find the points where two curves cross by solving their equations simultaneously. In polar coordinates, the story is different. Simultaneous solution may reveal some intersection points without revealing others, so it is sometimes difficult to find all points of intersection of two polar curves. One way to identify all the points of intersection is to graph the equations.

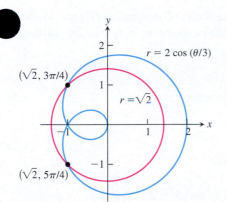

FIGURE 11.37 The curves $r = 2\cos(\theta/3)$ and $r = \sqrt{2}$ intersect at two points (Example 3).

EXAMPLE 3   Find all of the points where the curve $r = 2\cos(\theta/3)$ intersects the circle of radius $\sqrt{2}$ centered at the origin.

Solution   Note that the function $r = 2\cos(\theta/3)$ takes both positive and negative values. Therefore, when we look for the points where this curve intersects the circle, it is important to take into account that the circle is described both by the equation $r = \sqrt{2}$ *and* the equation $r = -\sqrt{2}$.

Solving $2\cos(\theta/3) = \sqrt{2}$ for θ yields:

$$2\cos(\theta/3) = \sqrt{2}, \quad \cos(\theta/3) = \sqrt{2}/2, \quad \theta/3 = \pi/4, \quad \theta = 3\pi/4.$$

This gives us one point, $\left(\sqrt{2}, 3\pi/4\right)$, where the two curves intersect. However, as we can see by looking at the graphs in Figure 11.37, there is a second intersection point. To find the second point, we solve $2\cos(\theta/3) = -\sqrt{2}$ for θ:

$$2\cos(\theta/3) = -\sqrt{2}, \quad \cos(\theta/3) = -\sqrt{2}/2, \quad \theta/3 = 3\pi/4, \quad \theta = 9\pi/4.$$

The second intersection point is located at $\left(-\sqrt{2}, 9\pi/4\right)$. We can specify this point in polar coordinates using a positive value of r and an angle between 0 and 2π. In polar coordinates, adding multiples of 2π to θ gives a second description of the same point in the plane. Similarly, changing the sign of r, while at the same time adding or subtracting π to θ, also gives a description of the same point. So in polar coordinates $\left(-\sqrt{2}, 9\pi/4\right)$ describes the same point in the plane as $\left(-\sqrt{2}, \pi/4\right)$ and also as $\left(\sqrt{2}, 5\pi/4\right)$. The second intersection point is located at $\left(\sqrt{2}, 5\pi/4\right)$. $\blacksquare$

Length of a Polar Curve

We can obtain a polar coordinate formula for the length of a curve $r = f(\theta)$, $\alpha \le \theta \le \beta$, by parametrizing the curve as

$$x = r\cos\theta = f(\theta)\cos\theta, \qquad y = r\sin\theta = f(\theta)\sin\theta, \qquad \alpha \le \theta \le \beta. \qquad (2)$$

The parametric length formula, Equation (3) from Section 11.2, then gives the length as

$$L = \int_\alpha^\beta \sqrt{\left(\frac{dx}{d\theta}\right)^2 + \left(\frac{dy}{d\theta}\right)^2} \, d\theta.$$

This equation becomes

$$L = \int_\alpha^\beta \sqrt{r^2 + \left(\frac{dr}{d\theta}\right)^2} \, d\theta$$

when Equations (2) are substituted for x and y (Exercise 29).

Length of a Polar Curve

If $r = f(\theta)$ has a continuous first derivative for $\alpha \leq \theta \leq \beta$ and if the point $P(r, \theta)$ traces the curve $r = f(\theta)$ exactly once as θ runs from α to β, then the length of the curve is

$$L = \int_\alpha^\beta \sqrt{r^2 + \left(\frac{dr}{d\theta}\right)^2} \, d\theta. \tag{3}$$

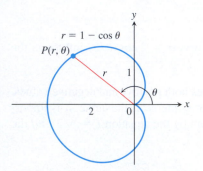

FIGURE 11.38 Calculating the length of a cardioid (Example 4).

EXAMPLE 4 Find the length of the cardioid $r = 1 - \cos\theta$.

Solution We sketch the cardioid to determine the limits of integration (Figure 11.38). The point $P(r, \theta)$ traces the curve once, counterclockwise as θ runs from 0 to 2π, so these are the values we take for α and β.

With

$$r = 1 - \cos\theta, \qquad \frac{dr}{d\theta} = \sin\theta,$$

we have

$$r^2 + \left(\frac{dr}{d\theta}\right)^2 = (1 - \cos\theta)^2 + (\sin\theta)^2$$

$$= 1 - 2\cos\theta + \underbrace{\cos^2\theta + \sin^2\theta}_{1} = 2 - 2\cos\theta$$

and

$$L = \int_\alpha^\beta \sqrt{r^2 + \left(\frac{dr}{d\theta}\right)^2} \, d\theta = \int_0^{2\pi} \sqrt{2 - 2\cos\theta} \, d\theta$$

$$= \int_0^{2\pi} \sqrt{4 \sin^2 \frac{\theta}{2}} \, d\theta \qquad 1 - \cos\theta = 2\sin^2(\theta/2)$$

$$= \int_0^{2\pi} 2 \left| \sin \frac{\theta}{2} \right| \, d\theta$$

$$= \int_0^{2\pi} 2 \sin \frac{\theta}{2} \, d\theta \qquad \sin(\theta/2) \geq 0 \quad \text{for} \quad 0 \leq \theta \leq 2\pi$$

$$= \left[-4 \cos \frac{\theta}{2} \right]_0^{2\pi} = 4 + 4 = 8. \qquad \blacksquare$$

EXERCISES 11.5

Finding Polar Areas

Find the areas of the regions in Exercises 1–8.

1. Bounded by the spiral $r = \theta$ for $0 \le \theta \le \pi$

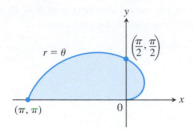

2. Bounded by the circle $r = 2 \sin \theta$ for $\pi/4 \le \theta \le \pi/2$

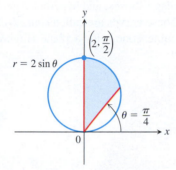

3. Inside the oval limaçon $r = 4 + 2 \cos \theta$

4. Inside the cardioid $r = a(1 + \cos \theta), \quad a > 0$

5. Inside one leaf of the four-leaved rose $r = \cos 2\theta$

6. Inside one leaf of the three-leaved rose $r = \cos 3\theta$

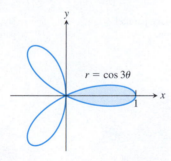

7. Inside one loop of the lemniscate $r^2 = 4 \sin 2\theta$

8. Inside the six-leaved rose $r^2 = 2 \sin 3\theta$

Find the areas of the regions in Exercises 9–18.

9. Shared by the circles $r = 2 \cos \theta$ and $r = 2 \sin \theta$

10. Shared by the circles $r = 1$ and $r = 2 \sin \theta$

11. Shared by the circle $r = 2$ and the cardioid $r = 2(1 - \cos \theta)$

12. Shared by the cardioids $r = 2(1 + \cos \theta)$ and $r = 2(1 - \cos \theta)$

13. Inside the lemniscate $r^2 = 6 \cos 2\theta$ and outside the circle $r = \sqrt{3}$

14. Inside the circle $r = 3a \cos \theta$ and outside the cardioid $r = a(1 + \cos \theta), a > 0$

15. Inside the circle $r = -2 \cos \theta$ and outside the circle $r = 1$

16. Inside the circle $r = 6$ above the line $r = 3 \csc \theta$

17. Inside the circle $r = 4 \cos \theta$ and to the right of the vertical line $r = \sec \theta$

18. Inside the circle $r = 4 \sin \theta$ and below the horizontal line $r = 3 \csc \theta$

19. a. Find the area of the shaded region in the accompanying figure.

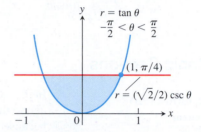

b. It looks as if the graph of $r = \tan \theta, -\pi/2 < \theta < \pi/2$, could be asymptotic to the lines $x = 1$ and $x = -1$. Is it? Give reasons for your answer.

20. The area of the region that lies inside the cardioid curve $r = \cos \theta + 1$ and outside the circle $r = \cos \theta$ is not

$$\frac{1}{2} \int_0^{2\pi} \left[(\cos \theta + 1)^2 - \cos^2 \theta \right] d\theta = \pi.$$

Why not? What *is* the area? Give reasons for your answers.

Finding Lengths of Polar Curves

Find the lengths of the curves in Exercises 21–28.

21. The spiral $r = \theta^2, \quad 0 \le \theta \le \sqrt{5}$

22. The spiral $r = e^\theta/\sqrt{2}, \quad 0 \le \theta \le \pi$

23. The cardioid $r = 1 + \cos \theta$

24. The curve $r = a \sin^2 (\theta/2), \quad 0 \le \theta \le \pi, \quad a > 0$

25. The parabolic segment $r = 6/(1 + \cos \theta), \quad 0 \le \theta \le \pi/2$

26. The parabolic segment $r = 2/(1 - \cos \theta), \quad \pi/2 \le \theta \le \pi$

27. The curve $r = \cos^3 (\theta/3), \quad 0 \le \theta \le \pi/4$

28. The curve $r = \sqrt{1 + \sin 2\theta}, \quad 0 \le \theta \le \pi\sqrt{2}$

29. The length of the curve $r = f(\theta), \alpha \le \theta \le \beta$ Assuming that the necessary derivatives are continuous, show how the substitutions

$$x = f(\theta) \cos \theta, \quad y = f(\theta) \sin \theta$$

(Equations 2 in the text) transform

$$L = \int_\alpha^\beta \sqrt{\left(\frac{dx}{d\theta}\right)^2 + \left(\frac{dy}{d\theta}\right)^2} \, d\theta$$

into

$$L = \int_\alpha^\beta \sqrt{r^2 + \left(\frac{dr}{d\theta}\right)^2} \, d\theta.$$

30. Circumferences of circles As usual, when faced with a new formula, it is a good idea to try it on familiar objects to be sure it gives results consistent with past experience. Use the length formula in Equation (3) to calculate the circumferences of the following circles ($a > 0$).

a. $r = a$ **b.** $r = a \cos \theta$ **c.** $r = a \sin \theta$

Theory and Examples

31. Average value If f is continuous, the average value of the polar coordinate r over the curve $r = f(\theta)$, $\alpha \le \theta \le \beta$, with respect to θ is given by the formula

$$r_{av} = \frac{1}{\beta - \alpha} \int_{\alpha}^{\beta} f(\theta) \, d\theta.$$

Use this formula to find the average value of r with respect to θ over the following curves ($a > 0$).

a. The cardioid $r = a(1 - \cos \theta)$

b. The circle $r = a$

c. The circle $r = a \cos \theta$, $-\pi/2 \le \theta \le \pi/2$

32. $r = f(\theta)$ vs. $r = 2f(\theta)$ Can anything be said about the relative lengths of the curves $r = f(\theta)$, $\alpha \le \theta \le \beta$, and $r = 2f(\theta)$, $\alpha \le \theta \le \beta$? Give reasons for your answer.

11.6 Conic Sections

HISTORICAL BIOGRAPHY

Gregory St. Vincent
(1584–1667)
www.goo.gl/WZD6Hz

In this section we define and review parabolas, ellipses, and hyperbolas geometrically and derive their standard Cartesian equations. These curves are called *conic sections* or *conics* because they are formed by cutting a double cone with a plane (Figure 11.39). This

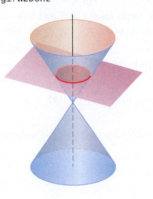

Circle: plane perpendicular to cone axis

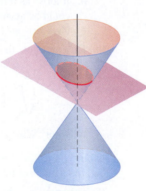

Ellipse: plane oblique to cone axis

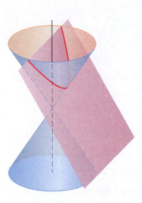

Parabola: plane parallel to side of cone

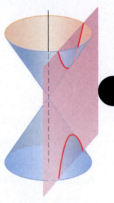

Hyperbola: plane parallel to cone axis

(a)

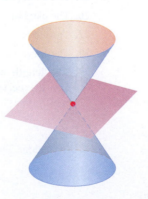

Point: plane through cone vertex only

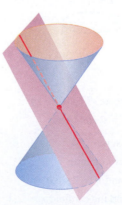

Single line: plane tangent to cone

Pair of intersecting lines

(b)

FIGURE 11.39 The standard conic sections (a) are the curves in which a plane cuts a *double* cone. Hyperbolas come in two parts, called *branches*. The point and lines obtained by passing the plane through the cone's vertex (b) are *degenerate* conic sections.

geometric method was the only way that conic sections could be described by Greek mathematicians, since they did not have our tools of Cartesian or polar coordinates. In the next section we express the conics in polar coordinates.

Parabolas

> **DEFINITIONS** A set that consists of all the points in a plane equidistant from a given fixed point and a given fixed line in the plane is a **parabola**. The fixed point is the **focus** of the parabola. The fixed line is the **directrix**.

If the focus F lies on the directrix L, the parabola is the line through F perpendicular to L. We consider this to be a degenerate case and assume henceforth that F does not lie on L.

A parabola has its simplest equation when its focus and directrix straddle one of the coordinate axes. For example, suppose that the focus lies at the point $F(0, p)$ on the positive y-axis and that the directrix is the line $y = -p$ (Figure 11.40). In the notation of the figure, a point $P(x, y)$ lies on the parabola if and only if $PF = PQ$. From the distance formula,

$$PF = \sqrt{(x - 0)^2 + (y - p)^2} = \sqrt{x^2 + (y - p)^2}$$
$$PQ = \sqrt{(x - x)^2 + (y - (-p))^2} = \sqrt{(y + p)^2}.$$

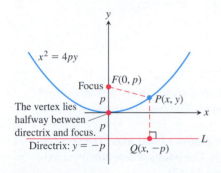

$x^2 = 4py$

Focus $F(0, p)$

The vertex lies halfway between directrix and focus.

Directrix: $y = -p$ $Q(x, -p)$

FIGURE 11.40 The standard form of the parabola $x^2 = 4py$, $p > 0$.

When we equate these expressions, square, and simplify, we get

$$y = \frac{x^2}{4p} \qquad \text{or} \qquad x^2 = 4py. \qquad \text{Standard form} \qquad (1)$$

These equations reveal the parabola's symmetry about the y-axis. We call the y-axis the axis of the parabola (short for "axis of symmetry").

The point where a parabola crosses its axis is the **vertex**. The vertex of the parabola $x^2 = 4py$ lies at the origin (Figure 11.40). The positive number p is the parabola's **focal length**.

If the parabola opens downward, with its focus at $(0, -p)$ and its directrix the line $y = p$, then Equations (1) become

$$y = -\frac{x^2}{4p} \qquad \text{and} \qquad x^2 = -4py.$$

By interchanging the variables x and y, we obtain similar equations for parabolas opening to the right or to the left (Figure 11.41).

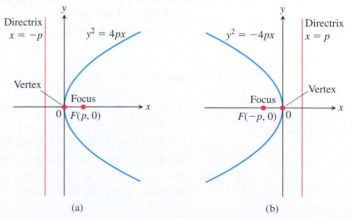

(a)   (b)

FIGURE 11.41 (a) The parabola $y^2 = 4px$. (b) The parabola $y^2 = -4px$.

EXAMPLE 1   Find the focus and directrix of the parabola $y^2 = 10x$.

Solution   We find the value of p in the standard equation $y^2 = 4px$:

$$4p = 10, \quad \text{so} \quad p = \frac{10}{4} = \frac{5}{2}.$$

Then we find the focus and directrix for this value of p:

$$\text{Focus:} \quad (p, 0) = \left(\frac{5}{2}, 0\right)$$

$$\text{Directrix:} \quad x = -p \quad \text{or} \quad x = -\frac{5}{2}. \quad \blacksquare$$

Ellipses

FIGURE 11.42   Points on the focal axis of an ellipse.

> **DEFINITIONS**   An **ellipse** is the set of points in a plane whose distances from two fixed points in the plane have a constant sum. The two fixed points are the foci of the ellipse.
>
> The line through the foci of an ellipse is the ellipse's **focal axis**. The point on the axis halfway be-tween the foci is the **center**. The points where the focal axis and ellipse cross are the ellipse's **vertices** (Figure 11.42).

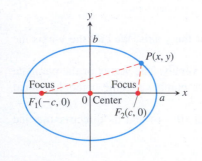

FIGURE 11.43   The ellipse defined by the equation $PF_1 + PF_2 = 2a$ is the graph of the equation $(x^2/a^2) + (y^2/b^2) = 1$, where $b^2 = a^2 - c^2$.

If the foci are $F_1(-c, 0)$ and $F_2(c, 0)$ (Figure 11.43), and $PF_1 + PF_2$ is denoted by $2a$, then the coordinates of a point P on the ellipse satisfy the equation

$$\sqrt{(x + c)^2 + y^2} + \sqrt{(x - c)^2 + y^2} = 2a.$$

To simplify this equation, we move the second radical to the right-hand side, square, isolate the remaining radical, and square again, obtaining

$$\frac{x^2}{a^2} + \frac{y^2}{a^2 - c^2} = 1. \tag{2}$$

Since $PF_1 + PF_2$ is greater than the length F_1F_2 (by the triangle inequality for triangle PF_1F_2), the number $2a$ is greater than $2c$. Accordingly, $a > c$ and the number $a^2 - c^2$ in Equation (2) is positive.

The algebraic steps leading to Equation (2) can be reversed to show that every point P whose coordinates satisfy an equation of this form with $0 < c < a$ also satisfies the equation $PF_1 + PF_2 = 2a$. A point therefore lies on the ellipse if and only if its coordinates satisfy Equation (2).

If we let b denote the positive square root of $a^2 - c^2$,

$$b = \sqrt{a^2 - c^2}, \tag{3}$$

then $a^2 - c^2 = b^2$ and Equation (2) takes the form

$$\frac{x^2}{a^2} + \frac{y^2}{b^2} = 1. \tag{4}$$

Equation (4) reveals that this ellipse is symmetric with respect to the origin and both coordinate axes. It lies inside the rectangle bounded by the lines $x = \pm a$ and $y = \pm b$. It crosses the axes at the points $(\pm a, 0)$ and $(0, \pm b)$. The tangents at these points are perpendicular to the axes because

$$\frac{dy}{dx} = -\frac{b^2 x}{a^2 y}, \qquad \text{\color{blue}Obtained from Eq. (4)}$$
$$\text{\color{blue}by implicit differentiation}$$

which is zero if $x = 0$ and infinite if $y = 0$.

The **major axis** of the ellipse in Equation (4) is the line segment of length $2a$ joining the points $(\pm a, 0)$. The **minor axis** is the line segment of length $2b$ joining the points $(0, \pm b)$. The number a itself is the **semimajor axis**, the number b the **semiminor axis**. The number c, found from Equation (3) as

$$c = \sqrt{a^2 - b^2},$$

is the **center-to-focus distance** of the ellipse. If $a = b$ then the ellipse is a circle.

EXAMPLE 2   The ellipse

$$\frac{x^2}{16} + \frac{y^2}{9} = 1 \tag{5}$$

shown in Figure 11.44 has

Semimajor axis: $\quad a = \sqrt{16} = 4, \qquad$ Semiminor axis: $\quad b = \sqrt{9} = 3,$

Center-to-focus distance: $\quad c = \sqrt{16 - 9} = \sqrt{7},$

Foci: $\quad (\pm c, 0) = \left(\pm\sqrt{7}, 0\right),$

Vertices: $\quad (\pm a, 0) = (\pm 4, 0),$

Center: $\quad (0, 0).$ ■

If we interchange x and y in Equation (5), we have the equation

$$\frac{x^2}{9} + \frac{y^2}{16} = 1. \tag{6}$$

The major axis of this ellipse is now vertical instead of horizontal, with the foci and vertices on the y-axis. We can determine which way the major axis runs simply by finding the intercepts of the ellipse with the coordinate axes. The longer of the two axes of the ellipse is the major axis.

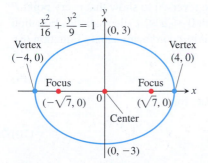

$$\frac{x^2}{16} + \frac{y^2}{9} = 1$$

FIGURE 11.44   An ellipse with its major axis horizontal (Example 2).

Standard-Form Equations for Ellipses Centered at the Origin

Foci on the x-axis: $\quad \dfrac{x^2}{a^2} + \dfrac{y^2}{b^2} = 1 \quad (a > b)$

Center-to-focus distance: $\quad c = \sqrt{a^2 - b^2}$

Foci: $\quad (\pm c, 0)$

Vertices: $\quad (\pm a, 0)$

Foci on the y-axis: $\quad \dfrac{x^2}{b^2} + \dfrac{y^2}{a^2} = 1 \quad (a > b)$

Center-to-focus distance: $\quad c = \sqrt{a^2 - b^2}$

Foci: $\quad (0, \pm c)$

Vertices: $\quad (0, \pm a)$

In each case, a is the semimajor axis and b is the semiminor axis.

Hyperbolas

DEFINITIONS   A **hyperbola** is the set of points in a plane whose distances from two fixed points in the plane have a constant difference. The two fixed points are the foci of the hyperbola.

The line through the foci of a hyperbola is the **focal axis**. The point on the axis halfway between the foci is the hyperbola's **center**. The points where the focal axis and hyperbola cross are the **vertices** (Figure 11.45).

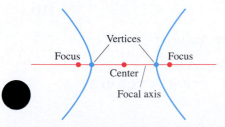

FIGURE 11.45   Points on the focal axis of a hyperbola.

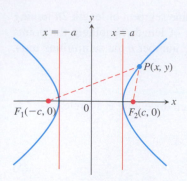

FIGURE 11.46 Hyperbolas have two branches. For points on the right-hand branch of the hyperbola shown here, $PF_1 - PF_2 = 2a$. For points on the left-hand branch, $PF_2 - PF_1 = 2a$. We then let $b = \sqrt{c^2 - a^2}$.

If the foci are $F_1(-c, 0)$ and $F_2(c, 0)$ (Figure 11.46) and the constant difference is $2a$, then a point (x, y) lies on the hyperbola if and only if

$$\sqrt{(x + c)^2 + y^2} - \sqrt{(x - c)^2 + y^2} = \pm 2a. \tag{7}$$

To simplify this equation, we move the second radical to the right-hand side, square, isolate the remaining radical, and square again, obtaining

$$\frac{x^2}{a^2} + \frac{y^2}{a^2 - c^2} = 1. \tag{8}$$

So far, this looks just like the equation for an ellipse. But now $a^2 - c^2$ is negative because $2a$, being the difference of two sides of triangle PF_1F_2, is less than $2c$, the third side.

The algebraic steps leading to Equation (8) can be reversed to show that every point P whose coordinates satisfy an equation of this form with $0 < a < c$ also satisfies Equation (7). A point therefore lies on the hyperbola if and only if its coordinates satisfy Equation (8).

If we let b denote the positive square root of $c^2 - a^2$,

$$b = \sqrt{c^2 - a^2}, \tag{9}$$

then $a^2 - c^2 = -b^2$ and Equation (8) takes the compact form

$$\frac{x^2}{a^2} - \frac{y^2}{b^2} = 1. \tag{10}$$

The differences between Equation (10) and the equation for an ellipse (Equation 4) are the minus sign and the new relation

$$c^2 = a^2 + b^2. \qquad \text{From Eq. (9)}$$

Like the ellipse, the hyperbola is symmetric with respect to the origin and coordinate axes. It crosses the x-axis at the points $(\pm a, 0)$. The tangents at these points are vertical because

$$\frac{dy}{dx} = \frac{b^2 x}{a^2 y} \qquad \begin{matrix}\text{Obtained from Eq. (10) by} \\ \text{implicit differentiation}\end{matrix}$$

and this is infinite when $y = 0$. The hyperbola has no y-intercepts; in fact, no part of the curve lies between the lines $x = -a$ and $x = a$.

The lines

$$y = \pm \frac{b}{a} x$$

are the two **asymptotes** of the hyperbola defined by Equation (10). The fastest way to find the equations of the asymptotes is to replace the 1 in Equation (10) by 0 and solve the new equation for y:

$$\underbrace{\frac{x^2}{a^2} - \frac{y^2}{b^2} = 1}_{\text{hyperbola}} \rightarrow \underbrace{\frac{x^2}{a^2} - \frac{y^2}{b^2} = 0}_{\text{0 for 1}} \rightarrow \underbrace{y = \pm \frac{b}{a} x.}_{\text{asymptotes}}$$

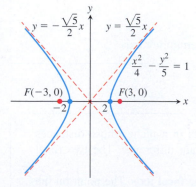

FIGURE 11.47 The hyperbola and its asymptotes in Example 3.

EXAMPLE 3   The equation

$$\frac{x^2}{4} - \frac{y^2}{5} = 1 \tag{11}$$

is Equation (10) with $a^2 = 4$ and $b^2 = 5$ (Figure 11.47). We have

Center-to-focus distance: $c = \sqrt{a^2 + b^2} = \sqrt{4 + 5} = 3$,

Foci: $(\pm c, 0) = (\pm 3, 0)$,   Vertices: $(\pm a, 0) = (\pm 2, 0)$,

Center: $(0, 0)$,

Asymptotes: $\dfrac{x^2}{4} - \dfrac{y^2}{5} = 0$ or $y = \pm \dfrac{\sqrt{5}}{2} x.$

If we interchange x and y in Equation (11), the foci and vertices of the resulting hyperbola will lie along the y-axis. We still find the asymptotes in the same way as before, but now their equations will be $y = \pm 2x/\sqrt{5}$.

Standard-Form Equations for Hyperbolas Centered at the Origin

Foci on the x-axis: $\dfrac{x^2}{a^2} - \dfrac{y^2}{b^2} = 1$

Center-to-focus distance: $c = \sqrt{a^2 + b^2}$

Foci: $(\pm c, 0)$

Vertices: $(\pm a, 0)$

Asymptotes: $\dfrac{x^2}{a^2} - \dfrac{y^2}{b^2} = 0$ or $y = \pm\dfrac{b}{a}x$

Foci on the y-axis: $\dfrac{y^2}{a^2} - \dfrac{x^2}{b^2} = 1$

Center-to-focus distance: $c = \sqrt{a^2 + b^2}$

Foci: $(0, \pm c)$

Vertices: $(0, \pm a)$

Asymptotes: $\dfrac{y^2}{a^2} - \dfrac{x^2}{b^2} = 0$ or $y = \pm\dfrac{a}{b}x$

Notice the difference in the asymptote equations (b/a in the first, a/b in the second).

We shift conics using the principles reviewed in Section 1.2, replacing x by $x + h$ and y by $y + k$.

EXAMPLE 4 Show that the equation $x^2 - 4y^2 + 2x + 8y - 7 = 0$ represents a hyperbola. Find its center, asymptotes, and foci.

Solution We reduce the equation to standard form by completing the square in x and y as follows:

$$(x^2 + 2x) - 4(y^2 - 2y) = 7$$
$$(x^2 + 2x + 1) - 4(y^2 - 2y + 1) = 7 + 1 - 4$$
$$\frac{(x + 1)^2}{4} - (y - 1)^2 = 1.$$

This is the standard form Equation (10) of a hyperbola with x replaced by $x + 1$ and y replaced by $y - 1$. The hyperbola is shifted one unit to the left and one unit upward, and it has center $x + 1 = 0$ and $y - 1 = 0$, or $x = -1$ and $y = 1$. Moreover,

$$a^2 = 4, \quad b^2 = 1, \quad c^2 = a^2 + b^2 = 5,$$

so the asymptotes are the two lines

$$\frac{x + 1}{2} - (y - 1) = 0 \quad \text{and} \quad \frac{x + 1}{2} + (y - 1) = 0,$$

or

$$y - 1 = \pm\frac{1}{2}(x + 1).$$

The shifted foci have coordinates $\left(-1 \pm \sqrt{5}, 1\right)$.

Identifying Graphs

Match the parabolas in Exercises 1–4 with the following equations:

$$x^2 = 2y, \quad x^2 = -6y, \quad y^2 = 8x, \quad y^2 = -4x.$$

Then find each parabola's focus and directrix.

1.

2.

3.

4.

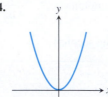

Match each conic section in Exercises 5–8 with one of these equations:

$$\frac{x^2}{4} + \frac{y^2}{9} = 1, \qquad \frac{x^2}{2} + y^2 = 1,$$

$$\frac{y^2}{4} - x^2 = 1, \qquad \frac{x^2}{4} - \frac{y^2}{9} = 1.$$

Then find the conic section's foci and vertices. If the conic section is a hyperbola, find its asymptotes as well.

5.

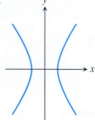

6.

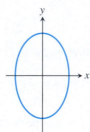

7.

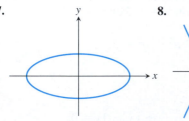

8.

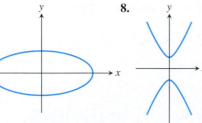

Parabolas

Exercises 9–16 give equations of parabolas. Find each parabola's focus and directrix. Then sketch the parabola. Include the focus and directrix in your sketch.

9. $y^2 = 12x$ **10.** $x^2 = 6y$ **11.** $x^2 = -8y$

12. $y^2 = -2x$ **13.** $y = 4x^2$ **14.** $y = -8x^2$

15. $x = -3y^2$ **16.** $x = 2y^2$

Ellipses

Exercises 17–24 give equations for ellipses. Put each equation in standard form. Then sketch the ellipse. Include the foci in your sketch.

17. $16x^2 + 25y^2 = 400$ **18.** $7x^2 + 16y^2 = 112$

19. $2x^2 + y^2 = 2$ **20.** $2x^2 + y^2 = 4$

21. $3x^2 + 2y^2 = 6$ **22.** $9x^2 + 10y^2 = 90$

23. $6x^2 + 9y^2 = 54$ **24.** $169x^2 + 25y^2 = 4225$

Exercises 25 and 26 give information about the foci and vertices of ellipses centered at the origin of the xy-plane. In each case, find the ellipse's standard-form equation from the given information.

25. Foci: $\left(\pm\sqrt{2}, 0\right)$ Vertices: $(\pm 2, 0)$

26. Foci: $(0, \pm 4)$ Vertices: $(0, \pm 5)$

Hyperbolas

Exercises 27–34 give equations for hyperbolas. Put each equation in standard form and find the hyperbola's asymptotes. Then sketch the hyperbola. Include the asymptotes and foci in your sketch.

27. $x^2 - y^2 = 1$ **28.** $9x^2 - 16y^2 = 144$

29. $y^2 - x^2 = 8$ **30.** $y^2 - x^2 = 4$

31. $8x^2 - 2y^2 = 16$ **32.** $y^2 - 3x^2 = 3$

33. $8y^2 - 2x^2 = 16$ **34.** $64x^2 - 36y^2 = 2304$

Exercises 35–38 give information about the foci, vertices, and asymptotes of hyperbolas centered at the origin of the xy-plane. In each case, find the hyperbola's standard-form equation from the information given.

35. Foci: $\left(0, \pm\sqrt{2}\right)$

Asymptotes: $y = \pm x$

36. Foci: $(\pm 2, 0)$

Asymptotes: $y = \pm\dfrac{1}{\sqrt{3}}x$

37. Vertices: $(\pm 3, 0)$

Asymptotes: $y = \pm\dfrac{4}{3}x$

38. Vertices: $(0, \pm 2)$

Asymptotes: $y = \pm\dfrac{1}{2}x$

Shifting Conic Sections

You may wish to review Section 1.2 before solving Exercises 39–56.

39. The parabola $y^2 = 8x$ is shifted down 2 units and right 1 unit to generate the parabola $(y + 2)^2 = 8(x - 1)$.

a. Find the new parabola's vertex, focus, and directrix.

b. Plot the new vertex, focus, and directrix, and sketch in the parabola.

40. The parabola $x^2 = -4y$ is shifted left 1 unit and up 3 units to generate the parabola $(x + 1)^2 = -4(y - 3)$.

a. Find the new parabola's vertex, focus, and directrix.

b. Plot the new vertex, focus, and directrix, and sketch in the parabola.

41. The ellipse $(x^2/16) + (y^2/9) = 1$ is shifted 4 units to the right and 3 units up to generate the ellipse

$$\frac{(x-4)^2}{16} + \frac{(y-3)^2}{9} = 1.$$

    a. Find the foci, vertices, and center of the new ellipse.

    b. Plot the new foci, vertices, and center, and sketch in the new ellipse.

42. The ellipse $(x^2/9) + (y^2/25) = 1$ is shifted 3 units to the left and 2 units down to generate the ellipse

$$\frac{(x+3)^2}{9} + \frac{(y+2)^2}{25} = 1.$$

    a. Find the foci, vertices, and center of the new ellipse.

    b. Plot the new foci, vertices, and center, and sketch in the new ellipse.

43. The hyperbola $(x^2/16) - (y^2/9) = 1$ is shifted 2 units to the right to generate the hyperbola

$$\frac{(x-2)^2}{16} - \frac{y^2}{9} = 1.$$

    a. Find the center, foci, vertices, and asymptotes of the new hyperbola.

    b. Plot the new center, foci, vertices, and asymptotes, and sketch in the hyperbola.

44. The hyperbola $(y^2/4) - (x^2/5) = 1$ is shifted 2 units down to generate the hyperbola

$$\frac{(y+2)^2}{4} - \frac{x^2}{5} = 1.$$

    a. Find the center, foci, vertices, and asymptotes of the new hyperbola.

    b. Plot the new center, foci, vertices, and asymptotes, and sketch in the hyperbola.

Exercises 45–48 give equations for parabolas and tell how many units up or down and to the right or left each parabola is to be shifted. Find an equation for the new parabola, and find the new vertex, focus, and directrix.

45. $y^2 = 4x$,   left 2, down 3   **46.** $y^2 = -12x$,   right 4, up 3

47. $x^2 = 8y$,   right 1, down 7   **48.** $x^2 = 6y$,   left 3, down 2

Exercises 49–52 give equations for ellipses and tell how many units up or down and to the right or left each ellipse is to be shifted. Find an equation for the new ellipse, and find the new foci, vertices, and center.

49. $\dfrac{x^2}{6} + \dfrac{y^2}{9} = 1$,   left 2, down 1

50. $\dfrac{x^2}{2} + y^2 = 1$,   right 3, up 4

51. $\dfrac{x^2}{3} + \dfrac{y^2}{2} = 1$,   right 2, up 3

52. $\dfrac{x^2}{16} + \dfrac{y^2}{25} = 1$,   left 4, down 5

Exercises 53–56 give equations for hyperbolas and tell how many units up or down and to the right or left each hyperbola is to be shifted. Find an equation for the new hyperbola, and find the new center, foci, vertices, and asymptotes.

53. $\dfrac{x^2}{4} - \dfrac{y^2}{5} = 1$, right 2, up 2

54. $\dfrac{x^2}{16} - \dfrac{y^2}{9} = 1$, left 2, down 1

55. $y^2 - x^2 = 1$, left 1, down 1

56. $\dfrac{y^2}{3} - x^2 = 1$, right 1, up 3

Find the center, foci, vertices, asymptotes, and radius, as appropriate, of the conic sections in Exercises 57–68.

57. $x^2 + 4x + y^2 = 12$

58. $2x^2 + 2y^2 - 28x + 12y + 114 = 0$

59. $x^2 + 2x + 4y - 3 = 0$ **60.** $y^2 - 4y - 8x - 12 = 0$

61. $x^2 + 5y^2 + 4x = 1$ **62.** $9x^2 + 6y^2 + 36y = 0$

63. $x^2 + 2y^2 - 2x - 4y = -1$

64. $4x^2 + y^2 + 8x - 2y = -1$

65. $x^2 - y^2 - 2x + 4y = 4$ **66.** $x^2 - y^2 + 4x - 6y = 6$

67. $2x^2 - y^2 + 6y = 3$ **68.** $y^2 - 4x^2 + 16x = 24$

Theory and Examples

69. If lines are drawn parallel to the coordinate axes through a point P on the parabola $y^2 = kx, k > 0$, the parabola partitions the rectangular region bounded by these lines and the coordinate axes into two smaller regions, A and B.

    a. If the two smaller regions are revolved about the y-axis, show that they generate solids whose volumes have the ratio 4:1.

    b. What is the ratio of the volumes generated by revolving the regions about the x-axis?

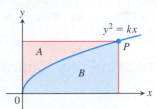

70. Suspension bridge cables hang in parabolas The suspension bridge cable shown in the accompanying figure supports a uniform load of w pounds per horizontal foot. It can be shown that if H is the horizontal tension of the cable at the origin, then the curve of the cable satisfies the equation

$$\frac{dy}{dx} = \frac{w}{H}x.$$

Show that the cable hangs in a parabola by solving this differential equation subject to the initial condition that $y = 0$ when $x = 0$.

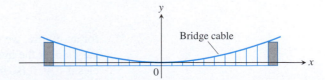

71. The width of a parabola at the focus Show that the number $4p$ is the *width* of the parabola $x^2 = 4py$ $(p > 0)$ at the focus by showing that the line $y = p$ cuts the parabola at points that are $4p$ units apart.

72. The asymptotes of $(x^2/a^2) - (y^2/b^2) = 1$ Show that the vertical distance between the line $y = (b/a)x$ and the upper half of the right-hand branch $y = (b/a)\sqrt{x^2 - a^2}$ of the hyperbola $(x^2/a^2) - (y^2/b^2) = 1$ approaches 0 by showing that

$$\lim_{x \to \infty} \left(\frac{b}{a}x - \frac{b}{a}\sqrt{x^2 - a^2} \right) = \frac{b}{a} \lim_{x \to \infty} \left(x - \sqrt{x^2 - a^2} \right) = 0.$$

Similar results hold for the remaining portions of the hyperbola and the lines $y = \pm(b/a)x$.

73. Area Find the dimensions of the rectangle of largest area that can be inscribed in the ellipse $x^2 + 4y^2 = 4$ with its sides parallel to the coordinate axes. What is the area of the rectangle?

74. Volume Find the volume of the solid generated by revolving the region enclosed by the ellipse $9x^2 + 4y^2 = 36$ about the (a) x-axis, (b) y-axis.

75. Volume The "triangular" region in the first quadrant bounded by the x-axis, the line $x = 4$, and the hyperbola $9x^2 - 4y^2 = 36$ is revolved about the x-axis to generate a solid. Find the volume of the solid.

76. Tangents Show that the tangents to the curve $y^2 = 4px$ from any point on the line $x = -p$ are perpendicular.

77. Tangents Find equations for the tangents to the circle $(x - 2)^2 + (y - 1)^2 = 5$ at the points where the circle crosses the coordinate axes.

78. Volume The region bounded on the left by the y-axis, on the right by the hyperbola $x^2 - y^2 = 1$, and above and below by the lines $y = \pm 3$ is revolved about the y-axis to generate a solid. Find the volume of the solid.

79. Centroid Find the centroid of the region that is bounded below by the x-axis and above by the ellipse $(x^2/9) + (y^2/16) = 1$.

80. Surface area The curve $y = \sqrt{x^2 + 1}, 0 \le x \le \sqrt{2}$, which is part of the upper branch of the hyperbola $y^2 - x^2 = 1$, is revolved about the x-axis to generate a surface. Find the area of the surface.

81. The reflective property of parabolas The accompanying figure shows a typical point $P(x_0, y_0)$ on the parabola $y^2 = 4px$. The line L is tangent to the parabola at P. The parabola's focus lies at $F(p, 0)$. The ray L' extending from P to the right is parallel to the x-axis. We show that light from F to P will be reflected out along L' by showing that β equals α. Establish this equality by taking the following steps.

a. Show that $\tan \beta = 2p/y_0$.

b. Show that $\tan \phi = y_0/(x_0 - p)$.

c. Use the identity

$$\tan \alpha = \frac{\tan \phi - \tan \beta}{1 + \tan \phi \tan \beta}$$

to show that $\tan \alpha = 2p/y_0$.

Since α and β are both acute, $\tan \beta = \tan \alpha$ implies $\beta = \alpha$.

This reflective property of parabolas is used in applications like car headlights, radio telescopes, and satellite TV dishes.

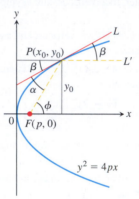

11.7 Conics in Polar Coordinates

Polar coordinates are especially important in astronomy and astronautical engineering because satellites, moons, planets, and comets all move approximately along ellipses, parabolas, and hyperbolas that can be described with a single relatively simple polar coordinate equation. We develop that equation here after first introducing the idea of a conic section's *eccentricity*. The eccentricity reveals the conic section's type (circle, ellipse, parabola, or hyperbola) and the degree to which it is "squashed" or flattened.

Eccentricity

Although the center-to-focus distance c does not appear in the standard Cartesian equation

$$\frac{x^2}{a^2} + \frac{y^2}{b^2} = 1, \quad (a > b)$$

for an ellipse, we can still determine c from the equation $c = \sqrt{a^2 - b^2}$. If we fix a and vary c over the interval $0 \le c \le a$, the resulting ellipses will vary in shape. They are circles if $c = 0$ (so that $a = b$) and flatten, becoming more oblong, as c increases. If $c = a$, the foci and vertices overlap and the ellipse degenerates into a line segment. Thus we are led to consider the ratio $e = c/a$. We use this ratio for hyperbolas as well, except in this case c equals $\sqrt{a^2 + b^2}$ instead of $\sqrt{a^2 - b^2}$. We refer to this ratio as the *eccentricity* of the ellipse or hyperbola.

DEFINITION

The **eccentricity** of the ellipse $(x^2/a^2) + (y^2/b^2) = 1$ $(a > b)$ is

$$e = \frac{c}{a} = \frac{\sqrt{a^2 - b^2}}{a}.$$

The **eccentricity** of the hyperbola $(x^2/a^2) - (y^2/b^2) = 1$ is

$$e = \frac{c}{a} = \frac{\sqrt{a^2 + b^2}}{a}.$$

The **eccentricity** of a parabola is $e = 1$.

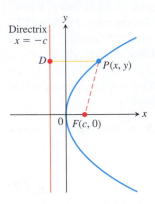

FIGURE 11.48 The distance from the focus F to any point P on a parabola equals the distance from P to the nearest point D on the directrix, so $PF = PD$.

Whereas a parabola has one focus and one directrix, each **ellipse** has two foci and two **directrices**. These are the lines perpendicular to the major axis at distances $\pm a/e$ from the center. From Figure 11.48 we see that a parabola has the property

$$PF = 1 \cdot PD \tag{1}$$

for any point P on it, where F is the focus and D is the point nearest P on the directrix. For an ellipse, it can be shown that the equations that replace Equation (1) are

$$PF_1 = e \cdot PD_1, \qquad PF_2 = e \cdot PD_2. \tag{2}$$

Here, e is the eccentricity, P is any point on the ellipse, F_1 and F_2 are the foci, and D_1 and D_2 are the points on the directrices nearest P (Figure 11.49).

In both Equations (2) the directrix and focus must correspond; that is, if we use the distance from P to F_1, we must also use the distance from P to the directrix at the same end of the ellipse. The directrix $x = -a/e$ corresponds to $F_1(-c, 0)$, and the directrix $x = a/e$ corresponds to $F_2(c, 0)$.

As with the ellipse, it can be shown that the lines $x = \pm a/e$ act as **directrices** for the **hyperbola** and that

$$PF_1 = e \cdot PD_1 \qquad \text{and} \qquad PF_2 = e \cdot PD_2. \tag{3}$$

Here P is any point on the hyperbola, F_1 and F_2 are the foci, and D_1 and D_2 are the points nearest P on the directrices (Figure 11.50).

In both the ellipse and the hyperbola, the eccentricity is the ratio of the distance between the foci to the distance between the vertices (because $c/a = 2c/2a$).

$$\text{Eccentricity} = \frac{\text{distance between foci}}{\text{distance between vertices}}$$

In an ellipse, the foci are closer together than the vertices and the ratio is less than 1. In a hyperbola, the foci are farther apart than the vertices and the ratio is greater than 1.

The "focus–directrix" equation $PF = e \cdot PD$ unites the parabola, ellipse, and hyperbola in the following way. Suppose that the distance PF of a point P from a fixed point F (the focus) is a constant multiple of its distance from a fixed line (the directrix). That is, suppose

$$PF = e \cdot PD, \tag{4}$$

where e is the constant of proportionality. Then the path traced by P is

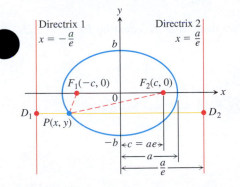

FIGURE 11.49 The foci and directrices of the ellipse $(x^2/a^2) + (y^2/b^2) = 1$. Directrix 1 corresponds to focus F_1 and directrix 2 to focus F_2.

(a) a *parabola* if $e = 1$,

(b) an *ellipse* of eccentricity e if $e < 1$, and

(c) a *hyperbola* of eccentricity e if $e > 1$.

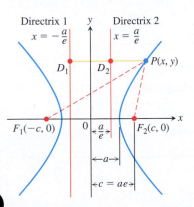

FIGURE 11.50 The foci and directrices of the hyperbola $(x^2/a^2) - (y^2/b^2) = 1$. No matter where P lies on the hyperbola, $PF_1 = e \cdot PD_1$ and $PF_2 = e \cdot PD_2$.

As e increases ($e \to 1^-$), ellipses become more oblong, and ($e \to \infty$) hyperbolas flatten toward two lines parallel to the directrix. There are no coordinates in Equation (4), and when we try to translate it into Cartesian coordinate form, it translates in different ways depending on the size of e. However, as we are about to see, in polar coordinates the equation $PF = e \cdot PD$ translates into a single equation regardless of the value of e.

Given the focus and corresponding directrix of a hyperbola centered at the origin and with foci on the x-axis, we can use the dimensions shown in Figure 11.50 to find e. Knowing e, we can derive a Cartesian equation for the hyperbola from the equation $PF = e \cdot PD$, as in the next example. We can find equations for ellipses centered at the origin and with foci on the x-axis in a similar way, using the dimensions shown in Figure 11.49.

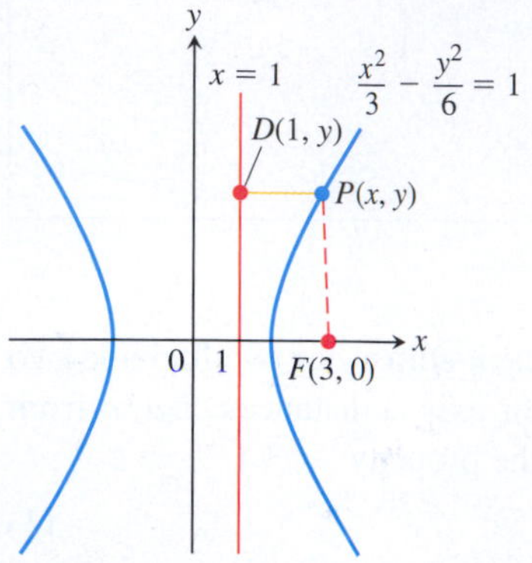

EXAMPLE 1 Find a Cartesian equation for the hyperbola centered at the origin that has a focus at $(3, 0)$ and the line $x = 1$ as the corresponding directrix.

Solution We first use the dimensions shown in Figure 11.50 to find the hyperbola's eccentricity. The focus is (see Figure 11.51)

$$(c, 0) = (3, 0), \quad \text{so} \quad c = 3.$$

Again from Figure 11.50, the directrix is the line

$$x = \frac{a}{e} = 1, \quad \text{so} \quad a = e.$$

When combined with the equation $e = c/a$ that defines eccentricity, these results give

$$e = \frac{c}{a} = \frac{3}{e}, \quad \text{so} \quad e^2 = 3 \quad \text{and} \quad e = \sqrt{3}.$$

Knowing e, we can now derive the equation we want from the equation $PF = e \cdot PD$. In the coordinates of Figure 11.51, we have

$$
\begin{aligned}
PF &= e \cdot PD &&\text{Eq. (4)} \\
\sqrt{(x - 3)^2 + (y - 0)^2} &= \sqrt{3}\,|x - 1| &&e = \sqrt{3} \\
x^2 - 6x + 9 + y^2 &= 3(x^2 - 2x + 1) &&\text{Square both sides.} \\
2x^2 - y^2 &= 6 &&\text{Simplify.} \\
\frac{x^2}{3} - \frac{y^2}{6} &= 1.
\end{aligned}
$$

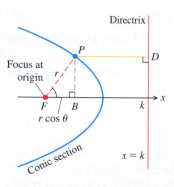

FIGURE 11.51 The hyperbola and directrix in Example 1.

Polar Equations

To find a polar equation for an ellipse, parabola, or hyperbola, we place one focus at the origin and the corresponding directrix to the right of the origin along the vertical line $x = k$ (Figure 11.52). In polar coordinates, this makes

$$PF = r$$

and

$$PD = k - FB = k - r\cos\theta.$$

The conic's focus–directrix equation $PF = e \cdot PD$ then becomes

$$r = e(k - r\cos\theta),$$

which can be solved for r to obtain the following expression.

FIGURE 11.52 If a conic section is put in the position with its focus placed at the origin and a directrix perpendicular to the initial ray and right of the origin, we can find its polar equation from the conic's focus–directrix equation.

Polar Equation for a Conic with Eccentricity e

$$r = \frac{ke}{1 + e\cos\theta}, \tag{5}$$

where $x = k > 0$ is the vertical directrix.

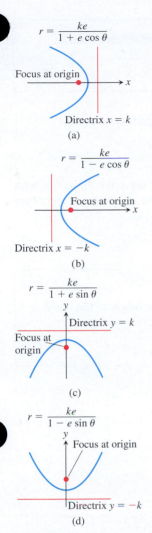

$$r = \frac{ke}{1 + e \cos \theta}$$

Focus at origin

Directrix $x = k$

(a)

$$r = \frac{ke}{1 - e \cos \theta}$$

Focus at origin

Directrix $x = -k$

(b)

$$r = \frac{ke}{1 + e \sin \theta}$$

Directrix $y = k$

Focus at origin

(c)

$$r = \frac{ke}{1 - e \sin \theta}$$

Focus at origin

Directrix $y = -k$

(d)

FIGURE 11.53 Equations for conic sections with eccentricity $e > 0$ but different locations of the directrix. The graphs here show a parabola, so $e = 1$.

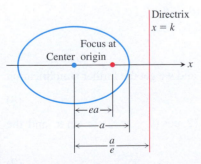

FIGURE 11.54 In an ellipse with semimajor axis a, the focus–directrix distance is $k = (a/e) - ea$, so $ke = a(1 - e^2)$.

EXAMPLE 2 Here are polar equations for three conics. The eccentricity values identifying the conic are the same for both polar and Cartesian coordinates.

$$e = \frac{1}{2}: \quad \text{ellipse} \quad r = \frac{k}{2 + \cos \theta}$$

$$e = 1: \quad \text{parabola} \quad r = \frac{k}{1 + \cos \theta}$$

$$e = 2: \quad \text{hyperbola} \quad r = \frac{2k}{1 + 2 \cos \theta}$$

You may see variations of Equation (5), depending on the location of the directrix. If the directrix is the line $x = -k$ to the left of the origin (the origin is still a focus), we replace Equation (5) with

$$r = \frac{ke}{1 - e \cos \theta}.$$

The denominator now has a $(-)$ instead of a $(+)$. If the directrix is either of the lines $y = k$ or $y = -k$, the equations have sines in them instead of cosines, as shown in Figure 11.53.

EXAMPLE 3 Find an equation for the hyperbola with eccentricity $3/2$ and directrix $x = 2$.

Solution We use Equation (5) with $k = 2$ and $e = 3/2$:

$$r = \frac{2(3/2)}{1 + (3/2) \cos \theta} \quad \text{or} \quad r = \frac{6}{2 + 3 \cos \theta}.$$

EXAMPLE 4 Find the directrix of the parabola $r = \frac{25}{10 + 10 \cos \theta}$.

Solution We divide the numerator and denominator by 10 to put the equation in standard polar form:

$$r = \frac{5/2}{1 + \cos \theta}.$$

This is the equation

$$r = \frac{ke}{1 + e \cos \theta}$$

with $k = 5/2$ and $e = 1$. The equation of the directrix is $x = 5/2$.

From the ellipse diagram in Figure 11.54, we see that k is related to the eccentricity e and the semimajor axis a by the equation

$$k = \frac{a}{e} - ea.$$

From this, we find that $ke = a(1 - e^2)$. Replacing ke in Equation (5) by $a(1 - e^2)$ gives the standard polar equation for an ellipse.

Polar Equation for the Ellipse with Eccentricity *e* and Semimajor Axis *a*

$$r = \frac{a(1 - e^2)}{1 + e \cos \theta} \qquad (6)$$

Notice that when $e = 0$, Equation (6) becomes $r = a$, which represents a circle.

Lines

Suppose the perpendicular from the origin to line L meets L at the point $P_0(r_0, \theta_0)$, with $r_0 \geq 0$ (Figure 11.55). Then, if $P(r, \theta)$ is any other point on L, the points P, P_0, and O are the vertices of a right triangle, from which we can read the relation

$$r_0 = r \cos (\theta - \theta_0).$$

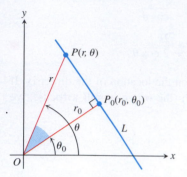

FIGURE 11.55 We can obtain a polar equation for line L by reading the relation $r_0 = r \cos (\theta - \theta_0)$ from the right triangle OP_0P.

The Standard Polar Equation for Lines

If the point $P_0(r_0, \theta_0)$ is the foot of the perpendicular from the origin to the line L, and $r_0 \geq 0$, then an equation for L is

$$r \cos (\theta - \theta_0) = r_0. \qquad (7)$$

For example, if $\theta_0 = \pi/3$ and $r_0 = 2$, we find that

$$r \cos \left(\theta - \frac{\pi}{3} \right) = 2$$

$$r \left(\cos \theta \cos \frac{\pi}{3} + \sin \theta \sin \frac{\pi}{3} \right) = 2$$

$$\frac{1}{2} r \cos \theta + \frac{\sqrt{3}}{2} r \sin \theta = 2, \qquad \text{or} \qquad x + \sqrt{3} y = 4.$$

Circles

To find a polar equation for the circle of radius a centered at $P_0(r_0, \theta_0)$, we let $P(r, \theta)$ be a point on the circle and apply the Law of Cosines to triangle OP_0P (Figure 11.56). This gives

$$a^2 = r_0{}^2 + r^2 - 2r_0 r \cos (\theta - \theta_0).$$

If the circle passes through the origin, then $r_0 = a$ and this equation simplifies to

$$a^2 = a^2 + r^2 - 2ar \cos (\theta - \theta_0)$$
$$r^2 = 2ar \cos (\theta - \theta_0)$$
$$r = 2a \cos (\theta - \theta_0).$$

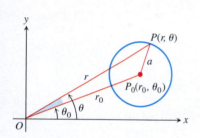

FIGURE 11.56 We can get a polar equation for this circle by applying the Law of Cosines to triangle OP_0P.

If the circle's center lies on the positive x-axis, $\theta_0 = 0$ and we get the further simplification

$$r = 2a \cos \theta. \qquad (8)$$

If the center lies on the positive y-axis, $\theta = \pi/2$, $\cos (\theta - \pi/2) = \sin \theta$, and the equation $r = 2a \cos (\theta - \theta_0)$ becomes

$$r = 2a \sin \theta. \qquad (9)$$

Equations for circles through the origin centered on the negative x- and y-axes can be obtained by replacing r with $-r$ in the above equations.

EXAMPLE 5   Here are several polar equations given by Equations (8) and (9) for circles through the origin and having centers that lie on the x- or y-axis.

Radius	Center (polar coordinates)	Polar equation
3	$(3, 0)$	$r = 6 \cos \theta$
2	$(2, \pi/2)$	$r = 4 \sin \theta$
1/2	$(-1/2, 0)$	$r = -\cos \theta$
1	$(-1, \pi/2)$	$r = -2 \sin \theta$

     ■

EXERCISES 11.7

Ellipses and Eccentricity

In Exercises 1–8, find the eccentricity of the ellipse. Then find and graph the ellipse's foci and directrices.

1. $16x^2 + 25y^2 = 400$ **2.** $7x^2 + 16y^2 = 112$

3. $2x^2 + y^2 = 2$ **4.** $2x^2 + y^2 = 4$

5. $3x^2 + 2y^2 = 6$ **6.** $9x^2 + 10y^2 = 90$

7. $6x^2 + 9y^2 = 54$ **8.** $169x^2 + 25y^2 = 4225$

Exercises 9–12 give the foci or vertices and the eccentricities of ellipses centered at the origin of the xy-plane. In each case, find the ellipse's standard-form equation in Cartesian coordinates.

9. Foci: $(0, \pm 3)$

     Eccentricity: 0.5

10. Foci: $(\pm 8, 0)$

     Eccentricity: 0.2

11. Vertices: $(0, \pm 70)$

     Eccentricity: 0.1

12. Vertices: $(\pm 10, 0)$

     Eccentricity: 0.24

Exercises 13–16 give foci and corresponding directrices of ellipses centered at the origin of the xy-plane. In each case, use the dimensions in Figure 11.49 to find the eccentricity of the ellipse. Then find the ellipse's standard-form equation in Cartesian coordinates.

13. Focus: $\left(\sqrt{5}, 0\right)$

     Directrix: $x = \dfrac{9}{\sqrt{5}}$

14. Focus: $(4, 0)$

     Directrix: $x = \dfrac{16}{3}$

15. Focus: $(-4, 0)$

     Directrix: $x = -16$

16. Focus: $\left(-\sqrt{2}, 0\right)$

     Directrix: $x = -2\sqrt{2}$

Hyperbolas and Eccentricity

In Exercises 17–24, find the eccentricity of the hyperbola. Then find and graph the hyperbola's foci and directrices.

17. $x^2 - y^2 = 1$ **18.** $9x^2 - 16y^2 = 144$

19. $y^2 - x^2 = 8$ **20.** $y^2 - x^2 = 4$

21. $8x^2 - 2y^2 = 16$ **22.** $y^2 - 3x^2 = 3$

23. $8y^2 - 2x^2 = 16$ **24.** $64x^2 - 36y^2 = 2304$

Exercises 25–28 give the eccentricities and the vertices or foci of hyperbolas centered at the origin of the xy-plane. In each case, find the hyperbola's standard-form equation in Cartesian coordinates.

25. Eccentricity: 3

     Vertices: $(0, \pm 1)$

26. Eccentricity: 2

     Vertices: $(\pm 2, 0)$

27. Eccentricity: 3

     Foci: $(\pm 3, 0)$

28. Eccentricity: 1.25

     Foci: $(0, \pm 5)$

Eccentricities and Directrices

Exercises 29–36 give the eccentricities of conic sections with one focus at the origin along with the directrix corresponding to that focus. Find a polar equation for each conic section.

29. $e = 1, \quad x = 2$ **30.** $e = 1, \quad y = 2$

31. $e = 5, \quad y = -6$ **32.** $e = 2, \quad x = 4$

33. $e = 1/2, \quad x = 1$ **34.** $e = 1/4, \quad x = -2$

35. $e = 1/5, \quad y = -10$ **36.** $e = 1/3, \quad y = 6$

Parabolas and Ellipses

Sketch the parabolas and ellipses in Exercises 37–44. Include the directrix that corresponds to the focus at the origin. Label the vertices with appropriate polar coordinates. Label the centers of the ellipses as well.

37. $r = \dfrac{1}{1 + \cos \theta}$ **38.** $r = \dfrac{6}{2 + \cos \theta}$

39. $r = \dfrac{25}{10 - 5 \cos \theta}$ **40.** $r = \dfrac{4}{2 - 2 \cos \theta}$

41. $r = \dfrac{400}{16 + 8 \sin \theta}$ **42.** $r = \dfrac{12}{3 + 3 \sin \theta}$

43. $r = \dfrac{8}{2 - 2 \sin \theta}$ **44.** $r = \dfrac{4}{2 - \sin \theta}$

Lines

Sketch the lines in Exercises 45–48 and find Cartesian equations for them.

45. $r \cos \left(\theta - \dfrac{\pi}{4} \right) = \sqrt{2}$ **46.** $r \cos \left(\theta + \dfrac{3\pi}{4} \right) = 1$

47. $r \cos \left(\theta - \dfrac{2\pi}{3} \right) = 3$ **48.** $r \cos \left(\theta + \dfrac{\pi}{3} \right) = 2$

Find a polar equation in the form $r \cos (\theta - \theta_0) = r_0$ for each of the lines in Exercises 49–52.

49. $\sqrt{2}x + \sqrt{2}y = 6$ **50.** $\sqrt{3}x - y = 1$

51. $y = -5$ **52.** $x = -4$

Circles

Sketch the circles in Exercises 53–56. Give polar coordinates for their centers and identify their radii.

53. $r = 4 \cos \theta$ **54.** $r = 6 \sin \theta$

55. $r = -2 \cos \theta$ **56.** $r = -8 \sin \theta$

Find polar equations for the circles in Exercises 57–64. Sketch each circle in the coordinate plane and label it with both its Cartesian and polar equations.

57. $(x - 6)^2 + y^2 = 36$ **58.** $(x + 2)^2 + y^2 = 4$
59. $x^2 + (y - 5)^2 = 25$ **60.** $x^2 + (y + 7)^2 = 49$
61. $x^2 + 2x + y^2 = 0$ **62.** $x^2 - 16x + y^2 = 0$

63. $x^2 + y^2 + y = 0$ **64.** $x^2 + y^2 - \dfrac{4}{3}y = 0$

Examples of Polar Equations

[T] Graph the lines and conic sections in Exercises 65–74.

65. $r = 3 \sec (\theta - \pi/3)$ **66.** $r = 4 \sec (\theta + \pi/6)$

67. $r = 4 \sin \theta$ **68.** $r = -2 \cos \theta$

69. $r = 8/(4 + \cos \theta)$ **70.** $r = 8/(4 + \sin \theta)$

71. $r = 1/(1 - \sin \theta)$ **72.** $r = 1/(1 + \cos \theta)$
73. $r = 1/(1 + 2 \sin \theta)$ **74.** $r = 1/(1 + 2 \cos \theta)$

75. Perihelion and aphelion A planet travels about its sun in an ellipse whose semimajor axis has length a. (See accompanying figure.)

  a. Show that $r = a(1 - e)$ when the planet is closest to the sun and that $r = a(1 + e)$ when the planet is farthest from the sun.

  b. Use the data in the table in Exercise 76 to find how close each planet in our solar system comes to the sun and how far away each planet gets from the sun.

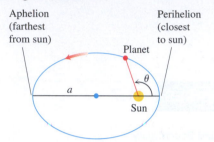

76. Planetary orbits Use the data in the table below and Equation (6) to find polar equations for the orbits of the planets.

Planet	Semimajor axis (astronomical units)	Eccentricity
Mercury	0.3871	0.2056
Venus	0.7233	0.0068
Earth	1.000	0.0167
Mars	1.524	0.0934
Jupiter	5.203	0.0484
Saturn	9.539	0.0543
Uranus	19.18	0.0460
Neptune	30.06	0.0082

CHAPTER 11   Questions to Guide Your Review

1. What is a parametrization of a curve in the xy-plane? Does a function $y = f(x)$ always have a parametrization? Are parametrizations of a curve unique? Give examples.

2. Give some typical parametrizations for lines, circles, parabolas, ellipses, and hyperbolas. How might the parametrized curve differ from the graph of its Cartesian equation?

3. What is a cycloid? What are typical parametric equations for cycloids? What physical properties account for the importance of cycloids?

4. What is the formula for the slope dy/dx of a parametrized curve $x = f(t), y = g(t)$? When does the formula apply? When can you expect to be able to find d^2y/dx^2 as well? Give examples.

5. How can you sometimes find the area bounded by a parametrized curve and one of the coordinate axes?

6. How do you find the length of a smooth parametrized curve $x = f(t), y = g(t), a \le t \le b$? What does smoothness have to do with length? What else do you need to know about the parametrization in order to find the curve's length? Give examples.

7. What is the arc length function for a smooth parametrized curve? What is its arc length differential?

8. Under what conditions can you find the area of the surface generated by revolving a curve $x = f(t), y = g(t), a \le t \le b$, about the x-axis? the y-axis? Give examples.

9. What are polar coordinates? What equations relate polar coordinates to Cartesian coordinates? Why might you want to change from one coordinate system to the other?

10. What consequence does the lack of uniqueness of polar coordinates have for graphing? Give an example.

11. How do you graph equations in polar coordinates? Include in your discussion symmetry, slope, behavior at the origin, and the use of Cartesian graphs. Give examples.

12. How do you find the area of a region $0 \le r_1(\theta) \le r \le r_2(\theta)$, $\alpha \le \theta \le \beta$, in the polar coordinate plane? Give examples.

13. Under what conditions can you find the length of a curve $r = f(\theta)$, $\alpha \le \theta \le \beta$, in the polar coordinate plane? Give an example of a typical calculation.

14. What is a parabola? What are the Cartesian equations for parabolas whose vertices lie at the origin and whose foci lie on the coordinate axes? How can you find the focus and directrix of such a parabola from its equation?

15. What is an ellipse? What are the Cartesian equations for ellipses centered at the origin with foci on one of the coordinate axes?

16. What is a hyperbola? What are the Cartesian equations for hyperbolas centered at the origin with foci on one of the coordinate axes? How can you find the foci, vertices, and directrices of such an ellipse from its equation?

17. What is the eccentricity of a conic section? How can you classify conic sections by eccentricity? How does eccentricity change the shape of ellipses and hyperbolas?

18. Explain the equation $PF = e \cdot PD$.

19. What are the standard equations for lines and conic sections in polar coordinates? Give examples.

CHAPTER 11   Practice Exercises

Identifying Parametric Equations in the Plane

Exercises 1–6 give parametric equations and parameter intervals for the motion of a particle in the xy-plane. Identify the particle's path by finding a Cartesian equation for it. Graph the Cartesian equation and indicate the direction of motion and the portion traced by the particle.

1. $x = t/2$, $\quad y = t + 1$; $\quad -\infty < t < \infty$

2. $x = \sqrt{t}$, $\quad y = 1 - \sqrt{t}$; $\quad t \ge 0$

3. $x = (1/2) \tan t$, $\quad y = (1/2) \sec t$; $\quad -\pi/2 < t < \pi/2$

4. $x = -2 \cos t$, $\quad y = 2 \sin t$; $\quad 0 \le t \le \pi$

5. $x = -\cos t$, $\quad y = \cos^2 t$; $\quad 0 \le t \le \pi$

6. $x = 4 \cos t$, $\quad y = 9 \sin t$; $\quad 0 \le t \le 2\pi$

Finding Parametric Equations and Tangent Lines

7. Find parametric equations and a parameter interval for the motion of a particle in the xy-plane that traces the ellipse $16x^2 + 9y^2 = 144$ once counterclockwise. (There are many ways to do this.)

8. Find parametric equations and a parameter interval for the motion of a particle that starts at the point $(-2, 0)$ in the xy-plane and traces the circle $x^2 + y^2 = 4$ three times clockwise. (There are many ways to do this.)

In Exercises 9 and 10, find an equation for the line in the xy-plane that is tangent to the curve at the point corresponding to the given value of t. Also, find the value of d^2y/dx^2 at this point.

9. $x = (1/2) \tan t$, $\quad y = (1/2) \sec t$; $\quad t = \pi/3$

10. $x = 1 + 1/t^2$, $\quad y = 1 - 3/t$; $\quad t = 2$

11. Eliminate the parameter to express the curve in the form $y = f(x)$.

  a. $x = 4t^2$, $\quad y = t^3 - 1$

  b. $x = \cos t$, $\quad y = \tan t$

12. Find parametric equations for the given curve.

  a. Line through $(1, -2)$ with slope 3

  b. $(x - 1)^2 + (y + 2)^2 = 9$

  c. $y = 4x^2 - x$

  d. $9x^2 + 4y^2 = 36$

Lengths of Curves

Find the lengths of the curves in Exercises 13–19.

13. $y = x^{1/2} - (1/3)x^{3/2}$, $\quad 1 \le x \le 4$

14. $x = y^{2/3}$, $\quad 1 \le y \le 8$

15. $y = (5/12)x^{6/5} - (5/8)x^{4/5}$, $\quad 1 \le x \le 32$

16. $x = (y^3/12) + (1/y)$, $\quad 1 \le y \le 2$

17. $x = 5 \cos t - \cos 5t$, $\quad y = 5 \sin t - \sin 5t$, $\quad 0 \le t \le \pi/2$

18. $x = t^3 - 6t^2$, $\quad y = t^3 + 6t^2$, $\quad 0 \le t \le 1$

19. $x = 3 \cos \theta$, $\quad y = 3 \sin \theta$, $\quad 0 \le \theta \le \dfrac{3\pi}{2}$

20. Find the length of the enclosed loop $x = t^2$, $y = (t^3/3) - t$ shown here. The loop starts at $t = -\sqrt{3}$ and ends at $t = \sqrt{3}$.

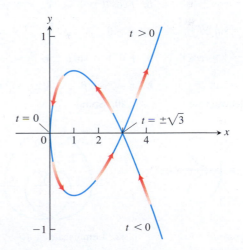

Surface Areas

Find the areas of the surfaces generated by revolving the curves in Exercises 21 and 22 about the indicated axes.

21. $x = t^2/2$, $\quad y = 2t$, $\quad 0 \le t \le \sqrt{5}$; $\quad x$-axis

22. $x = t^2 + 1/(2t)$, $\quad y = 4\sqrt{t}$, $\quad 1/\sqrt{2} \le t \le 1$; $\quad y$-axis

Polar to Cartesian Equations

Sketch the lines in Exercises 23–28. Also, find a Cartesian equation for each line.

23. $r \cos\left(\theta + \dfrac{\pi}{3}\right) = 2\sqrt{3}$ **24.** $r \cos\left(\theta - \dfrac{3\pi}{4}\right) = \dfrac{\sqrt{2}}{2}$

25. $r = 2 \sec \theta$ **26.** $r = -\sqrt{2} \sec \theta$

27. $r = -(3/2) \csc \theta$ **28.** $r = \left(3\sqrt{3}\right) \csc \theta$

Find Cartesian equations for the circles in Exercises 29–32. Sketch each circle in the coordinate plane and label it with both its Cartesian and polar equations.

29. $r = -4 \sin \theta$ **30.** $r = 3\sqrt{3} \sin \theta$

31. $r = 2\sqrt{2} \cos \theta$ **32.** $r = -6 \cos \theta$

Cartesian to Polar Equations

Find polar equations for the circles in Exercises 33–36. Sketch each circle in the coordinate plane and label it with both its Cartesian and polar equations.

33. $x^2 + y^2 + 5y = 0$ **34.** $x^2 + y^2 - 2y = 0$

35. $x^2 + y^2 - 3x = 0$ **36.** $x^2 + y^2 + 4x = 0$

Graphs in Polar Coordinates

Sketch the regions defined by the polar coordinate inequalities in Exercises 37 and 38.

37. $0 \le r \le 6 \cos \theta$ **38.** $-4 \sin \theta \le r \le 0$

Match each graph in Exercises 39–46 with the appropriate equation (a)–(l). There are more equations than graphs, so some equations will not be matched.

a. $r = \cos 2\theta$ **b.** $r \cos \theta = 1$ **c.** $r = \dfrac{6}{1 - 2 \cos \theta}$

d. $r = \sin 2\theta$ **e.** $r = \theta$ **f.** $r^2 = \cos 2\theta$

g. $r = 1 + \cos \theta$ **h.** $r = 1 - \sin \theta$ **i.** $r = \dfrac{2}{1 - \cos \theta}$

j. $r^2 = \sin 2\theta$ **k.** $r = -\sin \theta$ **l.** $r = 2 \cos \theta + 1$

39. Four-leaved rose   **40.** Spiral

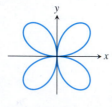   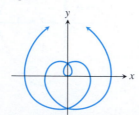

41. Limaçon   **42.** Lemniscate

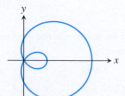   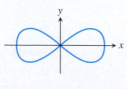

43. Circle   **44.** Cardioid

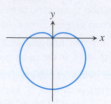   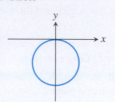

45. Parabola   **46.** Lemniscate

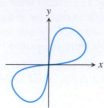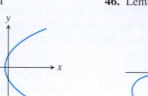

Area in Polar Coordinates

Find the areas of the regions in the polar coordinate plane described in Exercises 47–50.

47. Enclosed by the limaçon $r = 2 - \cos \theta$

48. Enclosed by one leaf of the three-leaved rose $r = \sin 3\theta$

49. Inside the "figure eight" $r = 1 + \cos 2\theta$ and outside the circle $r = 1$

50. Inside the cardioid $r = 2(1 + \sin \theta)$ and outside the circle $r = 2 \sin \theta$

Length in Polar Coordinates

Find the lengths of the curves given by the polar coordinate equations in Exercises 51–54.

51. $r = -1 + \cos \theta$

52. $r = 2 \sin \theta + 2 \cos \theta, \quad 0 \le \theta \le \pi/2$

53. $r = 8 \sin^3(\theta/3), \quad 0 \le \theta \le \pi/4$

54. $r = \sqrt{1 + \cos 2\theta}, \quad -\pi/2 \le \theta \le \pi/2$

Graphing Conic Sections

Sketch the parabolas in Exercises 55–58. Include the focus and directrix in each sketch.

55. $x^2 = -4y$ **56.** $x^2 = 2y$

57. $y^2 = 3x$ **58.** $y^2 = -(8/3)x$

Find the eccentricities of the ellipses and hyperbolas in Exercises 59–62. Sketch each conic section. Include the foci, vertices, and asymptotes (as appropriate) in your sketch.

59. $16x^2 + 7y^2 = 112$ **60.** $x^2 + 2y^2 = 4$

61. $3x^2 - y^2 = 3$ **62.** $5y^2 - 4x^2 = 20$

Exercises 63–68 give equations for conic sections and tell how many units up or down and to the right or left each curve is to be shifted. Find an equation for the new conic section, and find the new foci, vertices, centers, and asymptotes, as appropriate. If the curve is a parabola, find the new directrix as well.

63. $x^2 = -12y, \quad$ right 2, up 3   **64.** $y^2 = 10x, \quad$ left 1/2, down 1

65. $\dfrac{x^2}{9} + \dfrac{y^2}{25} = 1, \quad$ left 3, down 5

66. $\dfrac{x^2}{169} + \dfrac{y^2}{144} = 1$,   right 5, up 12

67. $\dfrac{y^2}{8} - \dfrac{x^2}{2} = 1$,   right 2, up $2\sqrt{2}$

68. $\dfrac{x^2}{36} - \dfrac{y^2}{64} = 1$,   left 10, down 3

Identifying Conic Sections

Complete the squares to identify the conic sections in Exercises 69–76. Find their foci, vertices, centers, and asymptotes (as appropriate). If the curve is a parabola, find its directrix as well.

69. $x^2 - 4x - 4y^2 = 0$ **70.** $4x^2 - y^2 + 4y = 8$

71. $y^2 - 2y + 16x = -49$ **72.** $x^2 - 2x + 8y = -17$

73. $9x^2 + 16y^2 + 54x - 64y = -1$

74. $25x^2 + 9y^2 - 100x + 54y = 44$

75. $x^2 + y^2 - 2x - 2y = 0$ **76.** $x^2 + y^2 + 4x + 2y = 1$

Conics in Polar Coordinates

Sketch the conic sections whose polar coordinate equations are given in Exercises 77–80. Give polar coordinates for the vertices and, in the case of ellipses, for the centers as well.

77. $r = \dfrac{2}{1 + \cos\theta}$ **78.** $r = \dfrac{8}{2 + \cos\theta}$

79. $r = \dfrac{6}{1 - 2\cos\theta}$ **80.** $r = \dfrac{12}{3 + \sin\theta}$

Exercises 81–84 give the eccentricities of conic sections with one focus at the origin of the polar coordinate plane, along with the directrix for that focus. Find a polar equation for each conic section.

81. $e = 2$, $r\cos\theta = 2$ **82.** $e = 1$, $r\cos\theta = -4$

83. $e = 1/2$, $r\sin\theta = 2$ **84.** $e = 1/3$, $r\sin\theta = -6$

Theory and Examples

85. Find the volume of the solid generated by revolving the region enclosed by the ellipse $9x^2 + 4y^2 = 36$ about **(a)** the x-axis, **(b)** the y-axis.

86. The "triangular" region in the first quadrant bounded by the x-axis, the line $x = 4$, and the hyperbola $9x^2 - 4y^2 = 36$ is revolved about the x-axis to generate a solid. Find the volume of the solid.

87. Show that the equations $x = r\cos\theta,\ y = r\sin\theta$ transform the polar equation

$$r = \frac{k}{1 + e\cos\theta}$$

into the Cartesian equation

$$(1 - e^2)x^2 + y^2 + 2kex - k^2 = 0.$$

88. Archimedes spirals The graph of an equation of the form $r = a\theta$, where a is a nonzero constant, is called an *Archimedes spiral*. Is there anything special about the widths between the successive turns of such a spiral?

CHAPTER 11   Additional and Advanced Exercises

Finding Conic Sections

1. Find an equation for the parabola with focus $(4, 0)$ and directrix $x = 3$. Sketch the parabola together with its vertex, focus, and directrix.

2. Find the vertex, focus, and directrix of the parabola

$$x^2 - 6x - 12y + 9 = 0.$$

3. Find an equation for the curve traced by the point $P(x, y)$ if the distance from P to the vertex of the parabola $x^2 = 4y$ is twice the distance from P to the focus. Identify the curve.

4. A line segment of length $a + b$ runs from the x-axis to the y-axis. The point P on the segment lies a units from one end and b units from the other end. Show that P traces an ellipse as the ends of the segment slide along the axes.

5. The vertices of an ellipse of eccentricity 0.5 lie at the points $(0, \pm2)$. Where do the foci lie?

6. Find an equation for the ellipse of eccentricity $2/3$ that has the line $x = 2$ as a directrix and the point $(4, 0)$ as the corresponding focus.

7. One focus of a hyperbola lies at the point $(0, -7)$ and the corresponding directrix is the line $y = -1$. Find an equation for the hyperbola if its eccentricity is **(a)** 2, **(b)** 5.

8. Find an equation for the hyperbola with foci $(0, -2)$ and $(0, 2)$ that passes through the point $(12, 7)$.

9. Show that the line

$$b^2xx_1 + a^2yy_1 - a^2b^2 = 0$$

is tangent to the ellipse $b^2x^2 + a^2y^2 - a^2b^2 = 0$ at the point (x_1, y_1) on the ellipse.

10. Show that the line

$$b^2xx_1 - a^2yy_1 - a^2b^2 = 0$$

is tangent to the hyperbola $b^2x^2 - a^2y^2 - a^2b^2 = 0$ at the point (x_1, y_1) on the hyperbola.

Equations and Inequalities

What points in the xy-plane satisfy the equations and inequalities in Exercises 11–16? Draw a figure for each exercise.

11. $(x^2 - y^2 - 1)(x^2 + y^2 - 25)(x^2 + 4y^2 - 4) = 0$

12. $(x + y)(x^2 + y^2 - 1) = 0$

13. $(x^2/9) + (y^2/16) \le 1$

14. $(x^2/9) - (y^2/16) \le 1$

15. $(9x^2 + 4y^2 - 36)(4x^2 + 9y^2 - 16) \le 0$

16. $(9x^2 + 4y^2 - 36)(4x^2 + 9y^2 - 16) > 0$

Polar Coordinates

17. a. Find an equation in polar coordinates for the curve

$$x = e^{2t} \cos t, \quad y = e^{2t} \sin t; \quad -\infty < t < \infty.$$

b. Find the length of the curve from $t = 0$ to $t = 2\pi$.

18. Find the length of the curve $r = 2 \sin^3(\theta/3)$, $0 \leq \theta \leq 3\pi$, in the polar coordinate plane.

Exercises 19–22 give the eccentricities of conic sections with one focus at the origin of the polar coordinate plane, along with the directrix for that focus. Find a polar equation for each conic section.

19. $e = 2$, $r \cos \theta = 2$ **20.** $e = 1$, $r \cos \theta = -4$

21. $e = 1/2$, $r \sin \theta = 2$ **22.** $e = 1/3$, $r \sin \theta = -6$

Theory and Examples

23. Epicycloids   When a circle rolls externally along the circumference of a second, fixed circle, any point P on the circumference of the rolling circle describes an *epicycloid*, as shown here. Let the fixed circle have its center at the origin O and have radius a.

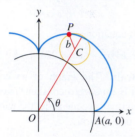

Let the radius of the rolling circle be b and let the initial position of the tracing point P be $A(a, 0)$. Find parametric equations for the epicycloid, using as the parameter the angle θ from the positive x-axis to the line through the circles' centers.

24. Find the centroid of the region enclosed by the x-axis and the cycloid arch

$$x = a(t - \sin t), \quad y = a(1 - \cos t); \quad 0 \leq t \leq 2\pi.$$

The Angle Between the Radius Vector and the Tangent Line to a Polar Coordinate Curve   In Cartesian coordinates, when we want to discuss the direction of a curve at a point, we use the angle ϕ measured counterclockwise from the positive x-axis to the tangent line. In polar coordinates, it is more convenient to calculate the angle ψ from the *radius vector* to the tangent line (see the accompanying figure). The angle ϕ can then be calculated from the relation

$$\phi = \theta + \psi, \tag{1}$$

which comes from applying the Exterior Angle Theorem to the triangle in the accompanying figure.

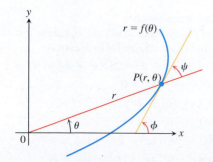

Suppose the equation of the curve is given in the form $r = f(\theta)$, where $f(\theta)$ is a differentiable function of θ. Then

$$x = r \cos \theta \quad \text{and} \quad y = r \sin \theta \tag{2}$$

are differentiable functions of θ with

$$\frac{dx}{d\theta} = -r \sin \theta + \cos \theta \frac{dr}{d\theta},$$

$$\frac{dy}{d\theta} = r \cos \theta + \sin \theta \frac{dr}{d\theta}. \tag{3}$$

Since $\psi = \phi - \theta$ from (1),

$$\tan \psi = \tan (\phi - \theta) = \frac{\tan \phi - \tan \theta}{1 + \tan \phi \tan \theta}.$$

Furthermore,

$$\tan \phi = \frac{dy}{dx} = \frac{dy/d\theta}{dx/d\theta}$$

because $\tan \phi$ is the slope of the curve at P. Also,

$$\tan \theta = \frac{y}{x}.$$

Hence

$$\tan \psi = \frac{\dfrac{dy/d\theta}{dx/d\theta} - \dfrac{y}{x}}{1 + \dfrac{y}{x}\dfrac{dy/d\theta}{dx/d\theta}} = \frac{x\dfrac{dy}{d\theta} - y\dfrac{dx}{d\theta}}{x\dfrac{dx}{d\theta} + y\dfrac{dy}{d\theta}}. \tag{4}$$

The numerator in the last expression in Equation (4) is found from Equations (2) and (3) to be

$$x\frac{dy}{d\theta} - y\frac{dx}{d\theta} = r^2.$$

Similarly, the denominator is

$$x\frac{dx}{d\theta} + y\frac{dy}{d\theta} = r\frac{dr}{d\theta}.$$

When we substitute these into Equation (4), we obtain

$$\tan \psi = \frac{r}{dr/d\theta}. \tag{5}$$

This is the equation we use for finding ψ as a function of θ.

25. Show, by reference to a figure, that the angle β between the tangents to two curves at a point of intersection may be found from the formula

$$\tan \beta = \frac{\tan \psi_2 - \tan \psi_1}{1 + \tan \psi_2 \tan \psi_1}. \tag{6}$$

When will the two curves intersect at right angles?

26. Find the value of $\tan \psi$ for the curve $r = \sin^4(\theta/4)$.

27. Find the angle between the radius vector to the curve $r = 2a \sin 3\theta$ and its tangent when $\theta = \pi/6$.

T **28. a.** Graph the hyperbolic spiral $r\theta = 1$. What appears to happen to ψ as the spiral winds in around the origin?

    b. Confirm your finding in part (a) analytically.

29. The circles $r = \sqrt{3} \cos \theta$ and $r = \sin \theta$ intersect at the point $\left(\sqrt{3}/2, \pi/3 \right)$. Show that their tangents are perpendicular there.

30. Find the angle at which the cardioid $r = a(1 - \cos \theta)$ crosses the ray $\theta = \pi/2$.

CHAPTER 11 **Technology Application Projects**

Mathematica/Maple Projects

Projects can be found within MyMathLab.

- *Radar Tracking of a Moving Object*
  Part I: Convert from polar to Cartesian coordinates.

- *Parametric and Polar Equations with a Figure Skater*
  Part I: Visualize position, velocity, and acceleration to analyze motion defined by parametric equations.
  Part II: Find and analyze the equations of motion for a figure skater tracing a polar plot.

Appendices

A.1 Real Numbers and the Real Line

This section reviews real numbers, inequalities, intervals, and absolute values.

Real Numbers

Much of calculus is based on properties of the real number system. **Real numbers** are numbers that can be expressed as decimals, such as

$$-\frac{3}{4} = -0.75000 \ldots$$

$$\frac{1}{3} = 0.33333 \ldots$$

$$\sqrt{2} = 1.4142 \ldots$$

The dots ... in each case indicate that the sequence of decimal digits goes on forever. Every conceivable decimal expansion represents a real number, although some numbers have two representations. For instance, the infinite decimals .999 ... and 1.000 ... represent the same real number 1. A similar statement holds for any number with an infinite tail of 9's.

The real numbers can be represented geometrically as points on a number line called the **real line**.

$$\begin{array}{c|c|c|c|c|c|c|c|c}
 & & & & & & & & \\
\hline
-2 & & -1\,\frac{3}{4} & 0 & \frac{1}{3} & 1 & \sqrt{2} & 2 & 3\,\pi & 4
\end{array}$$

The symbol $\mathbb{R}$ denotes either the real number system or, equivalently, the real line.

The properties of the real number system fall into three categories: algebraic properties, order properties, and completeness. The **algebraic properties** say that the real numbers can be added, subtracted, multiplied, and divided (except by 0) to produce more real numbers under the usual rules of arithmetic. *You can never divide by* 0.

The **order properties** of real numbers are given in Appendix 6. The useful rules at the left can be derived from them, where the symbol $\Rightarrow$ means "implies."

Notice the rules for multiplying an inequality by a number. Multiplying by a positive number preserves the inequality; multiplying by a negative number reverses the inequality. Also, reciprocation reverses the inequality for numbers of the same sign. For example, $2 < 5$ but $-2 > -5$ and $1/2 > 1/5$.

The **completeness property** of the real number system is deeper and harder to define precisely. However, the property is essential to the idea of a limit (Chapter 2). Roughly speaking, it says that there are enough real numbers to "complete" the real number line, in

RULES FOR INEQUALITIES

If a, b, and c are real numbers, then:

1. $a < b \Rightarrow a + c < b + c$
2. $a < b \Rightarrow a - c < b - c$
3. $a < b$ and $c > 0 \Rightarrow ac < bc$
4. $a < b$ and $c < 0 \Rightarrow bc < ac$
   Special case: $a < b \Rightarrow -b < -a$
5. $a > 0 \Rightarrow \frac{1}{a} > 0$
6. If a and b are both positive or both negative, then $a < b \Rightarrow \frac{1}{b} < \frac{1}{a}$.

the sense that there are no "holes" or "gaps" in it. Many theorems of calculus would fail if the real number system were not complete. Appendix 6 introduces the ideas involved and discusses how the real numbers are constructed.

We distinguish three special subsets of real numbers.

1. The **natural numbers**, namely 1, 2, 3, 4, . . .
2. The **integers**, namely 0, ± 1, ± 2, ± 3, . . .
3. The **rational numbers**, namely the numbers that can be expressed in the form of a fraction m/n, where m and n are integers and $n \neq 0$. Examples are

$$\frac{1}{3}, \quad -\frac{4}{9} = \frac{-4}{9} = \frac{4}{-9}, \quad \frac{200}{13}, \quad \text{and} \quad 57 = \frac{57}{1}.$$

The rational numbers are precisely the real numbers with decimal expansions that are either

    a. terminating (ending in an infinite string of zeros), for example,

$$\frac{3}{4} = 0.75000 \ldots = 0.75 \quad \text{or}$$

    b. eventually repeating (ending with a block of digits that repeats over and over), for example,

$$\frac{23}{11} = 2.090909 \ldots = 2.\overline{09} \qquad \text{The bar indicates the block of repeating digits.}$$

A terminating decimal expansion is a special type of repeating decimal, since the ending zeros repeat.

The set of rational numbers has all the algebraic and order properties of the real numbers but lacks the completeness property. For example, there is no rational number whose square is 2; there is a "hole" in the rational line where $\sqrt{2}$ should be.

Real numbers that are not rational are called **irrational numbers**. They are characterized by having nonterminating and nonrepeating decimal expansions. Examples are π, $\sqrt{2}$, $\sqrt[3]{5}$, and $\log_{10} 3$. Since every decimal expansion represents a real number, there are infinitely many irrational numbers. Both rational and irrational numbers are found arbitrarily close to any given point on the real line.

Set notation is very useful for specifying sets of real numbers. A **set** is a collection of objects, and these objects are the **elements** of the set. If S is a set, the notation $a \in S$ means that a is an element of S, and $a \notin S$ means that a is not an element of S. If S and T are sets, then $S \cup T$ is their **union** and consists of all elements belonging to either S or T (or to both S and T). The **intersection** $S \cap T$ consists of all elements belonging to both S and T. The **empty set** $\varnothing$ is the set that contains no elements. For example, the intersection of the rational numbers and the irrational numbers is the empty set.

Some sets can be described by *listing* their elements in braces. For instance, the set A consisting of the natural numbers (or positive integers) less than 6 can be expressed as

$$A = \{1, 2, 3, 4, 5\}.$$

The entire set of integers is written as

$$\{0, \pm 1, \pm 2, \pm 3, \ldots\}.$$

Another way to describe a set is to enclose in braces a rule that generates all the elements of the set. For instance, the set

$$A = \{x \mid x \text{ is an integer and } 0 < x < 6\}$$

is the set of positive integers less than 6.

Intervals

A subset of the real line is called an **interval** if it contains at least two numbers and contains all the real numbers lying between any two of its elements. For example, the set of all real numbers x such that $x > 6$ is an interval, as is the set of all x such that $-2 \le x \le 5$. The set of all nonzero real numbers is not an interval; since 0 is absent, the set fails to contain every real number between -1 and 1 (for example).

Geometrically, intervals correspond to rays and line segments on the real line, along with the real line itself. Intervals of numbers corresponding to line segments are **finite intervals**; intervals corresponding to rays and the real line are **infinite intervals**.

A finite interval is said to be **closed** if it contains both of its endpoints, **half-open** if it contains one endpoint but not the other, and **open** if it contains neither endpoint. The endpoints are also called **boundary points**; they make up the interval's **boundary**. The remaining points of the interval are **interior points** and together compose the interval's **interior**. Infinite intervals are closed if they contain a finite endpoint, and open otherwise. The entire real line $\mathbb{R}$ is an infinite interval that is both open and closed. Table A.1 summarizes the various types of intervals.

TABLE A.1 Types of intervals

Notation	Set description	Type	Picture
(a, b)	$\{x \mid a < x < b\}$	Open	
$[a, b]$	$\{x \mid a \le x \le b\}$	Closed	
$[a, b)$	$\{x \mid a \le x < b\}$	Half-open	
$(a, b]$	$\{x \mid a < x \le b\}$	Half-open	
(a, ∞)	$\{x \mid x > a\}$	Open	
$[a, \infty)$	$\{x \mid x \ge a\}$	Closed	
$(-\infty, b)$	$\{x \mid x < b\}$	Open	
$(-\infty, b]$	$\{x \mid x \le b\}$	Closed	
$(-\infty, \infty)$	$\mathbb{R}$ (set of all real numbers)	Both open and closed	

Solving Inequalities

The process of finding the interval or intervals of numbers that satisfy an inequality in x is called **solving** the inequality.

EXAMPLE 1   Solve the following inequalities and show their solution sets on the real line.

(a) $2x - 1 < x + 3$ **(b)** $\dfrac{6}{x - 1} \ge 5$

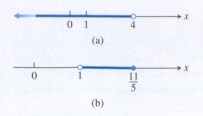

(a)

(b)

FIGURE A.1   Solution sets for the inequalities in Example 1. Hollow circles indicate endpoints that are not included in the interval, and solid dots indicate included endpoints.

Solution

(a)

$$2x - 1 < x + 3$$
$$2x < x + 4 \qquad \text{Add 1 to both sides.}$$
$$x < 4 \qquad \text{Subtract } x \text{ from both sides.}$$

The solution set is the open interval $(-\infty, 4)$ (Figure A.1a).

(b) The inequality $6/(x - 1) \geq 5$ can hold only if $x > 1$, because otherwise $6/(x - 1)$ is undefined or negative. Therefore, $(x - 1)$ is positive and the inequality will be preserved if we multiply both sides by $(x - 1)$:

$$\frac{6}{x - 1} \geq 5$$
$$6 \geq 5x - 5 \qquad \text{Multiply both sides by } (x - 1).$$
$$11 \geq 5x \qquad \text{Add 5 to both sides.}$$
$$\frac{11}{5} \geq x. \qquad \text{Or } x \leq \frac{11}{5}.$$

The solution set is the half-open interval $(1, 11/5]$ (Figure A.1b). ∎

Absolute Value

The **absolute value** of a number x, denoted by $|x|$, is defined by the formula

$$|x| = \begin{cases} x, & x \geq 0 \\ -x, & x < 0. \end{cases}$$

EXAMPLE 2 $|3| = 3, \quad |0| = 0, \quad |-5| = -(-5) = 5, \quad \big||-a|\big| = |a|$ ∎

Geometrically, the absolute value of x is the distance from x to 0 on the real number line. Since distances are always positive or 0, we see that $|x| \geq 0$ for every real number x, and $|x| = 0$ if and only if $x = 0$. Also,

$$|x - y| = \text{the distance between } x \text{ and } y$$

on the real line (Figure A.2).

Since the symbol $\sqrt{a}$ always denotes the *nonnegative* square root of a, an alternate definition of $|x|$ is

$$|x| = \sqrt{x^2}.$$

It is important to remember that $\sqrt{a^2} = |a|$. Do not write $\sqrt{a^2} = a$ unless you already know that $a \geq 0$.

The absolute value function has the following properties. (You are asked to prove these properties in the exercises.)

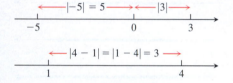

FIGURE A.2   Absolute values give distances between points on the number line.

Absolute Value Properties

1. $|-a| = |a|$ — A number and its negative have the same absolute value.

2. $|ab| = |a||b|$ — The absolute value of a product is the product of the absolute values.

3. $\left|\dfrac{a}{b}\right| = \dfrac{|a|}{|b|}$ — The absolute value of a quotient is the quotient of the absolute values.

4. $|a + b| \leq |a| + |b|$ — The **triangle inequality**. The absolute value of the sum of two numbers is less than or equal to the sum of their absolute values.

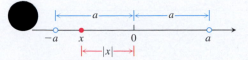

FIGURE A.3 $|x| < a$ means x lies between $-a$ and a.

Note that $|-a| \neq -|a|$. For example, $|-3| = 3$, whereas $-|3| = -3$. If a and b differ in sign, then $|a + b|$ is less than $|a| + |b|$. In all other cases, $|a + b|$ equals $|a| + |b|$. Absolute value bars in expressions like $|-3 + 5|$ work like parentheses: We do the arithmetic inside *before* taking the absolute value.

EXAMPLE 3

$$|-3 + 5| = |2| = 2 < |-3| + |5| = 8$$
$$|3 + 5| = |8| = |3| + |5|$$
$$|-3 - 5| = |-8| = 8 = |-3| + |-5|$$

∎

ABSOLUTE VALUES AND INTERVALS

If a is any positive number, then

5. $|x| = a \quad \Leftrightarrow \quad x = \pm a$
6. $|x| < a \quad \Leftrightarrow \quad -a < x < a$
7. $|x| > a \quad \Leftrightarrow \quad x > a$ or $x < -a$
8. $|x| \leq a \quad \Leftrightarrow \quad -a \leq x \leq a$
9. $|x| \geq a \quad \Leftrightarrow \quad x \geq a$ or $x \leq -a$

The inequality $|x| < a$ says that the distance from x to 0 is less than the positive number a. This means that x must lie between $-a$ and a, as we can see from Figure A.3.

Statements 5-9 in the table at left are all consequences of the definition of absolute value and are often helpful when solving equations or inequalities involving absolute values. The symbol $\Leftrightarrow$ that appears in the table is often used by mathematicians to denote the "if and only if" logical relationship. It also means "implies and is implied by."

EXAMPLE 4   Solve the equation $|2x - 3| = 7$.

Solution   By Property 5, $2x - 3 = \pm 7$, so there are two possibilities:

$2x - 3 = 7$	$2x - 3 = -7$	Equivalent equations without absolute values
$2x = 10$	$2x = -4$	Solve as usual.
$x = 5$	$x = -2$	

The solutions of $|2x - 3| = 7$ are $x = 5$ and $x = -2$.

∎

EXAMPLE 5   Solve the inequality $\left| 5 - \dfrac{2}{x} \right| < 1$.

Solution   We have

$$\left| 5 - \frac{2}{x} \right| < 1 \quad \Leftrightarrow \quad -1 < 5 - \frac{2}{x} < 1 \qquad \text{Property 6}$$

$$\Leftrightarrow \quad -6 < -\frac{2}{x} < -4 \qquad \text{Subtract 5.}$$

$$\Leftrightarrow \quad 3 > \frac{1}{x} > 2 \qquad \text{Multiply by } -\frac{1}{2}.$$

$$\Leftrightarrow \quad \frac{1}{3} < x < \frac{1}{2}. \qquad \text{Take reciprocals.}$$

Notice how the various rules for inequalities were used here. Multiplying by a negative number reverses the inequality. So does taking reciprocals in an inequality in which both sides are positive. The original inequality holds if and only if $(1/3) < x < (1/2)$. The solution set is the open interval $(1/3, 1/2)$.

∎

EXERCISES   A.1

1. Express $1/9$ as a repeating decimal, using a bar to indicate the repeating digits. What are the decimal representations of $2/9$? $3/9$? $8/9$? $9/9$?

2. If $2 < x < 6$, which of the following statements about x are necessarily true, and which are not necessarily true?

   a. $0 < x < 4$ b. $0 < x - 2 < 4$

   c. $1 < \dfrac{x}{2} < 3$ d. $\dfrac{1}{6} < \dfrac{1}{x} < \dfrac{1}{2}$

   e. $1 < \dfrac{6}{x} < 3$ f. $|x - 4| < 2$

   g. $-6 < -x < 2$ h. $-6 < -x < -2$

In Exercises 3–6, solve the inequalities and show the solution sets on the real line.

3. $-2x > 4$ 4. $5x - 3 \le 7 - 3x$

5. $2x - \dfrac{1}{2} \ge 7x + \dfrac{7}{6}$ 6. $\dfrac{4}{5}(x - 2) < \dfrac{1}{3}(x - 6)$

Solve the equations in Exercises 7–9.

7. $|y| = 3$ 8. $|2t + 5| = 4$ 9. $|8 - 3s| = \dfrac{9}{2}$

Solve the inequalities in Exercises 10–17, expressing the solution sets as intervals or unions of intervals. Also, show each solution set on the real line.

10. $|x| < 2$ 11. $|t - 1| \le 3$ 12. $|3y - 7| < 4$

13. $\left|\dfrac{z}{5} - 1\right| \le 1$ 14. $\left|3 - \dfrac{1}{x}\right| < \dfrac{1}{2}$ 15. $|2s| \ge 4$

16. $|1 - x| > 1$ 17. $\left|\dfrac{r + 1}{2}\right| \ge 1$

Solve the inequalities in Exercises 18–21. Express the solution sets as intervals or unions of intervals and show them on the real line. Use the result $\sqrt{a^2} = |a|$ as appropriate.

18. $x^2 < 2$ 19. $4 < x^2 < 9$

20. $(x - 1)^2 < 4$ 21. $x^2 - x < 0$

22. Do not fall into the trap of thinking $|-a| = a$. For what real numbers a is this equation true? For what real numbers is it false?

23. Solve the equation $|x - 1| = 1 - x$.

24. **A proof of the triangle inequality** Give the reason justifying each of the numbered steps in the following proof of the triangle inequality.

$$|a + b|^2 = (a + b)^2 \qquad (1)$$
$$= a^2 + 2ab + b^2$$
$$\le a^2 + 2|a||b| + b^2 \qquad (2)$$
$$= |a|^2 + 2|a||b| + |b|^2 \qquad (3)$$
$$= (|a| + |b|)^2$$
$$|a + b| \le |a| + |b| \qquad (4)$$

25. Prove that $|ab| = |a||b|$ for any numbers a and b.

26. If $|x| \le 3$ and $x > -1/2$, what can you say about x?

27. Graph the inequality $|x| + |y| \le 1$.

28. For any number a, prove that $|-a| = |a|$.

29. Let a be any positive number. Prove that $|x| > a$ if and only if $x > a$ or $x < -a$.

29. a. If b is any nonzero real number, prove that $|1/b| = 1/|b|$.

    b. Prove that $\left|\dfrac{a}{b}\right| = \dfrac{|a|}{|b|}$ for any numbers a and $b \neq 0$.

A.2 Mathematical Induction

Many formulas, like

$$1 + 2 + \cdots + n = \frac{n(n + 1)}{2},$$

can be shown to hold for every positive integer n by applying an axiom called the *mathematical induction principle*. A proof that uses this axiom is called a *proof by mathematical induction* or a *proof by induction*.

The steps in proving a formula by induction are the following:

1. Check that the formula holds for $n = 1$.

2. Prove that *if* the formula holds for any positive integer $n = k$, *then* it also holds for the next integer, $n = k + 1$.

The induction axiom says that once these steps are completed, the formula holds for all positive integers n. By Step 1 it holds for $n = 1$. By Step 2 it holds for $n = 2$, and therefore by Step 2 also for $n = 3$, and by Step 2 again for $n = 4$, and so on. If the first domino falls, and if the kth domino always knocks over the $(k + 1)$st when it falls, then all the dominoes fall.

From another point of view, suppose we have a sequence of statements $S_1, S_2, \ldots, S_n, \ldots$, one for each positive integer. Suppose we can show that assuming any one of the statements to be true implies that the next statement in line is true. Suppose that we can also show that S_1 is true. Then we may conclude that the statements are true from S_1 on.

EXAMPLE 1   Use mathematical induction to prove that for every positive integer n,

$$1 + 2 + \cdots + n = \frac{n(n + 1)}{2}.$$

Solution   We accomplish the proof by carrying out the two steps above.

1. The formula holds for $n = 1$ because

$$1 = \frac{1(1 + 1)}{2}.$$

2. If the formula holds for $n = k$, does it also hold for $n = k + 1$? The answer is yes, as we now show. If it is the case that

$$1 + 2 + \cdots + k = \frac{k(k + 1)}{2},$$

then it follows that

$$1 + 2 + \cdots + k + (k + 1) = \frac{k(k + 1)}{2} + (k + 1) = \frac{k^2 + k + 2k + 2}{2}$$

$$= \frac{(k + 1)(k + 2)}{2} = \frac{(k + 1)((k + 1) + 1)}{2}.$$

The last expression in this string of equalities is the expression $n(n + 1)/2$ for $n = (k + 1)$.

The mathematical induction principle now guarantees the original formula for all positive integers n.   ■

In Example 4 of Section 5.2 we gave another proof for the formula giving the sum of the first n integers. However, proofs by mathematical induction can also be used to find sums of the squares and cubes of the first n integers (Exercises 9 and 10). Here is another example of a proof by induction.

EXAMPLE 2   Show by mathematical induction that for all positive integers n,

$$\frac{1}{2^1} + \frac{1}{2^2} + \cdots + \frac{1}{2^n} = 1 - \frac{1}{2^n}.$$

Solution   We accomplish the proof by carrying out the two steps of mathematical induction.

1. The formula holds for $n = 1$ because

$$\frac{1}{2^1} = 1 - \frac{1}{2^1}.$$

2. If it is the case that

$$\frac{1}{2^1} + \frac{1}{2^2} + \cdots + \frac{1}{2^k} = 1 - \frac{1}{2^k},$$

then it follows that

$$\frac{1}{2^1} + \frac{1}{2^2} + \cdots + \frac{1}{2^k} + \frac{1}{2^{k+1}} = 1 - \frac{1}{2^k} + \frac{1}{2^{k+1}} = 1 - \frac{1 \cdot 2}{2^k \cdot 2} + \frac{1}{2^{k+1}}$$

$$= 1 - \frac{2}{2^{k+1}} + \frac{1}{2^{k+1}} = 1 - \frac{1}{2^{k+1}}.$$

Thus, the original formula holds for $n = (k + 1)$ whenever it holds for $n = k$.

With these steps verified, the mathematical induction principle now guarantees the formula for every positive integer n. ■

Other Starting Integers

Instead of starting at $n = 1$ some induction arguments start at another integer. The steps for such an argument are as follows.

1. Check that the formula holds for $n = n_1$ (the first appropriate integer).
2. Prove that if the formula holds for any integer $n = k \geq n_1$, then it also holds for $n = (k + 1)$.

Once these steps are completed, the mathematical induction principle guarantees the formula for all $n \geq n_1$.

EXAMPLE 3   Show that $n! > 3^n$ if n is large enough.

Solution   How large is large enough? We experiment:

n	1	2	3	4	5	6	7
$n!$	1	2	6	24	120	720	5040
3^n	3	9	27	81	243	729	2187

It looks as if $n! > 3^n$ for $n \geq 7$. To be sure, we apply mathematical induction. We take $n_1 = 7$ in Step 1 and complete Step 2.

Suppose $k! > 3^k$ for some $k \geq 7$. Then

$$(k + 1)! = (k + 1)(k!) > (k + 1)3^k > 7 \cdot 3^k > 3^{k+1}.$$

Thus, for $k \geq 7$,

$$k! > 3^k \quad \text{implies} \quad (k + 1)! > 3^{k+1}.$$

The mathematical induction principle now guarantees $n! \geq 3^n$ for all $n \geq 7$. ■

Proof of the Derivative Sum Rule for Sums of Finitely Many Functions

We prove the statement

$$\frac{d}{dx}(u_1 + u_2 + \cdots + u_n) = \frac{du_1}{dx} + \frac{du_2}{dx} + \cdots + \frac{du_n}{dx}$$

by mathematical induction. The statement is true for $n = 2$, as was proved in Section 3.3. This is Step 1 of the induction proof.

Step 2 is to show that if the statement is true for any positive integer $n = k$, where $k \geq n_0 = 2$, then it is also true for $n = k + 1$. So suppose that

$$\frac{d}{dx}(u_1 + u_2 + \cdots + u_k) = \frac{du_1}{dx} + \frac{du_2}{dx} + \cdots + \frac{du_k}{dx}. \tag{1}$$

Then

$$\frac{d}{dx}(\underbrace{u_1 + u_2 + \cdots + u_k}_{\substack{\text{Call the function} \\ \text{defined by this sum } u}} + \underbrace{u_{k+1}}_{\substack{\text{Call this} \\ \text{function } v.}})$$

$$= \frac{d}{dx}(u_1 + u_2 + \cdots + u_k) + \frac{du_{k+1}}{dx} \qquad \text{Sum Rule for } \frac{d}{dx}(u + v)$$

$$= \frac{du_1}{dx} + \frac{du_2}{dx} + \cdots + \frac{du_k}{dx} + \frac{du_{k+1}}{dx}. \qquad \text{Eq. (1)}$$

With these steps verified, the mathematical induction principle now guarantees the Sum Rule for every integer $n \geq 2$.

EXERCISES A.2

1. Assuming that the triangle inequality $|a + b| \leq |a| + |b|$ holds for any two numbers a and b, show that

$$|x_1 + x_2 + \cdots + x_n| \leq |x_1| + |x_2| + \cdots + |x_n|$$

for any n numbers.

2. Show that if $r \neq 1$, then

$$1 + r + r^2 + \cdots + r^n = \frac{1 - r^{n+1}}{1 - r}$$

for every positive integer n.

3. Use the Product Rule, $\frac{d}{dx}(uv) = u\frac{dv}{dx} + v\frac{du}{dx}$, and the fact that $\frac{d}{dx}(x) = 1$ to show that $\frac{d}{dx}(x^n) = nx^{n-1}$ for every positive integer n.

4. Suppose that a function $f(x)$ has the property that $f(x_1 x_2) = f(x_1) + f(x_2)$ for any two positive numbers x_1 and x_2. Show that

$$f(x_1 x_2 \cdots x_n) = f(x_1) + f(x_2) + \cdots + f(x_n)$$

for the product of any n positive numbers $x_1, x_2, \ldots, x_n$.

5. Show that

$$\frac{2}{3^1} + \frac{2}{3^2} + \cdots + \frac{2}{3^n} = 1 - \frac{1}{3^n}$$

for all positive integers n.

6. Show that $n! > n^3$ if n is large enough.

7. Show that $2^n > n^2$ if n is large enough.

8. Show that $2^n \geq 1/8$ for $n \geq -3$.

9. **Sums of squares** Show that the sum of the squares of the first n positive integers is

$$\frac{n\left(n + \frac{1}{2}\right)(n + 1)}{3}.$$

10. **Sums of cubes** Show that the sum of the cubes of the first n positive integers is $(n(n + 1)/2)^2$.

11. **Rules for finite sums** Show that the following finite sum rules hold for every positive integer n. (See Section 5.2.)

    a. $\displaystyle\sum_{k=1}^{n}(a_k + b_k) = \sum_{k=1}^{n} a_k + \sum_{k=1}^{n} b_k$

    b. $\displaystyle\sum_{k=1}^{n}(a_k - b_k) = \sum_{k=1}^{n} a_k - \sum_{k=1}^{n} b_k$

    c. $\displaystyle\sum_{k=1}^{n} ca_k = c \cdot \sum_{k=1}^{n} a_k \qquad$ (any number c)

    d. $\displaystyle\sum_{k=1}^{n} a_k = n \cdot c \qquad$ (if a_k has the constant value c)

12. Show that $|x^n| = |x|^n$ for every positive integer n and every real number x.

A.3 Lines, Circles, and Parabolas

This section reviews coordinates, lines, distance, circles, and parabolas in the plane. The notion of increment is also discussed.

Cartesian Coordinates in the Plane

In Appendix 1 we identified the points on the line with real numbers by assigning them coordinates. Points in the plane can be identified with ordered pairs of real numbers. To begin, we draw two perpendicular coordinate lines that intersect at the 0-point of each line.

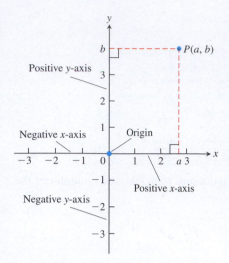

FIGURE A.4   Cartesian coordinates in the plane are based on two perpendicular axes intersecting at the origin.

HISTORICAL BIOGRAPHY

René Descartes
(1596–1650)
www.goo.gl/XSzlEA

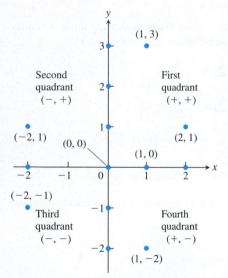

FIGURE A.5   Points labeled in the xy-coordinate or Cartesian plane. The points on the axes all have coordinate pairs but are usually labeled with single real numbers, (so (1, 0) on the x-axis is labeled as 1). Notice the coordinate sign patterns of the quadrants.

These lines are called **coordinate axes**. On the horizontal x-axis, numbers are denoted by x and increase to the right. On the vertical y-axis, numbers are denoted by y and increase upward (Figure A.4). Thus "upward" and "to the right" are positive directions, whereas "downward" and "to the left" are considered negative. The **origin** O, also labeled 0, of the coordinate system is the point in the plane where x and y are both zero.

If P is any point in the plane, it can be located by exactly one ordered pair of real numbers in the following way. Draw lines through P perpendicular to the two coordinate axes. These lines intersect the axes at points with coordinates a and b (Figure A.4). The ordered pair (a, b) is assigned to the point P and is called its **coordinate pair**. The first number a is the **x-coordinate** (or **abscissa**) of P; the second number b is the **y-coordinate** (or **ordinate**) of P. The x-coordinate of every point on the y-axis is 0. The y-coordinate of every point on the x-axis is 0. The origin is the point (0, 0).

Starting with an ordered pair (a, b), we can reverse the process and arrive at a corresponding point P in the plane. Often we identify P with the ordered pair and write P(a, b). We sometimes also refer to "the point (a, b)" and it will be clear from the context when (a, b) refers to a point in the plane and not to an open interval on the real line. Several points labeled by their coordinates are shown in Figure A.5.

This coordinate system is called the **rectangular coordinate system** or **Cartesian coordinate system** (after the sixteenth-century French mathematician René Descartes). The coordinate axes of this coordinate or Cartesian plane divide the plane into four regions called **quadrants**, numbered counterclockwise as shown in Figure A.5.

The **graph** of an equation or inequality in the variables x and y is the set of all points P(x, y) in the plane whose coordinates satisfy the equation or inequality. When we plot data in the coordinate plane or graph formulas whose variables have different units of measure, we do not need to use the same scale on the two axes. If we plot time vs. thrust for a rocket motor, for example, there is no reason to place the mark that shows 1 sec on the time axis the same distance from the origin as the mark that shows 1 lb on the thrust axis.

Usually when we graph functions whose variables do not represent physical measurements and when we draw figures in the coordinate plane to study their geometry and trigonometry, we make the scales on the axes identical. A vertical unit of distance then looks the same as a horizontal unit. As on a surveyor's map or a scale drawing, line segments that are supposed to have the same length will look as if they do and angles that are supposed to be congruent will look congruent.

Computer displays and calculator displays are another matter. The vertical and horizontal scales on machine-generated graphs usually differ, and there are corresponding distortions in distances, slopes, and angles. Circles may look like ellipses, rectangles may look like squares, right angles may appear to be acute or obtuse, and so on. We discuss these displays and distortions in greater detail in Section 1.4.

Increments and Straight Lines

When a particle moves from one point in the plane to another, the net changes in its coordinates are called *increments*. They are calculated by subtracting the coordinates of the starting point from the coordinates of the ending point. If x changes from x_1 to x_2, the **increment** in x is

$$\Delta x = x_2 - x_1.$$

EXAMPLE 1   As shown in Figure A.6, in going from the point A(4, −3) to the point B(2, 5) the increments in the x- and y-coordinates are

$$\Delta x = 2 - 4 = -2, \qquad \Delta y = 5 - (-3) = 8.$$

From C(5, 6) to D(5, 1) the coordinate increments are

$$\Delta x = 5 - 5 = 0, \qquad \Delta y = 1 - 6 = -5.$$

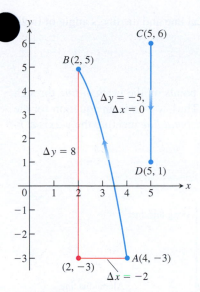

FIGURE A.6 Coordinate increments may be positive, negative, or zero (Example 1).

Given two points $P_1(x_1, y_1)$ and $P_2(x_2, y_2)$ in the plane, we call the increments $\Delta x = x_2 - x_1$ and $\Delta y = y_2 - y_1$ the **run** and the **rise**, respectively, between P_1 and P_2. Two such points always determine a unique straight line (usually called simply a line) passing through them both. We call the line P_1P_2.

Any nonvertical line in the plane has the property that the ratio

$$m = \frac{\text{rise}}{\text{run}} = \frac{\Delta y}{\Delta x} = \frac{y_2 - y_1}{x_2 - x_1}$$

has the same value for every choice of the two points $P_1(x_1, y_1)$ and $P_2(x_2, y_2)$ on the line (Figure A.7). This is because the ratios of corresponding sides for similar triangles are equal.

DEFINITION   The constant ratio

$$m = \frac{\text{rise}}{\text{run}} = \frac{\Delta y}{\Delta x} = \frac{y_2 - y_1}{x_2 - x_1}$$

is the **slope** of the nonvertical line P_1P_2.

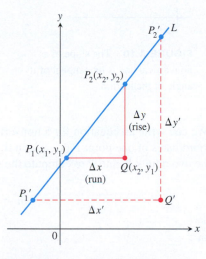

FIGURE A.7   Triangles P_1QP_2 and $P_1'Q'P_2'$ are similar, so the ratio of their sides has the same value for any two points on the line. This common value is the line's slope.

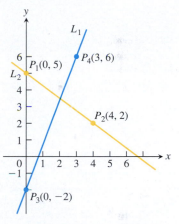

FIGURE A.8   The slope of L_1 is

$$m = \frac{\Delta y}{\Delta x} = \frac{6 - (-2)}{3 - 0} = \frac{8}{3}.$$

That is, y increases 8 units every time x increases 3 units. The slope of L_2 is

$$m = \frac{\Delta y}{\Delta x} = \frac{2 - 5}{4 - 0} = \frac{-3}{4}.$$

That is, y decreases 3 units every time x increases 4 units.

The slope tells us the direction (uphill, downhill) and steepness of a line. A line with positive slope rises uphill to the right; one with negative slope falls downhill to the right (Figure A.8). The greater the absolute value of the slope, the more rapid the rise or fall. The slope of a vertical line is *undefined*. Since the run Δx is zero for a vertical line, we cannot form the slope ratio m.

The direction and steepness of a line can also be measured with an angle. The **angle of inclination** of a line that crosses the x-axis is the smallest counterclockwise angle from the x-axis to the line (Figure A.9). The inclination of a horizontal line is 0°. The inclination of a vertical line is 90°. If ϕ (the Greek letter phi) is the inclination of a line, then $0 \le \phi < 180°$.

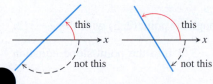

FIGURE A.9   Angles of inclination are measured counterclockwise from the x-axis.

The relationship between the slope m of a nonvertical line and the line's angle of inclination ϕ is shown in Figure A.10:

$$m = \tan\phi.$$

Straight lines have relatively simple equations. All points on the *vertical line* through the point a on the x-axis have x-coordinates equal to a. Thus, $x = a$ is an equation for the vertical line. Similarly, $y = b$ is an equation for the *horizontal line* meeting the y-axis at b. (See Figure A.11.)

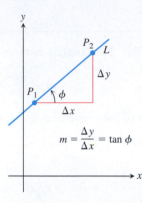

$$m = \frac{\Delta y}{\Delta x} = \tan\phi$$

FIGURE A.10 The slope of a nonvertical line is the tangent of its angle of inclination.

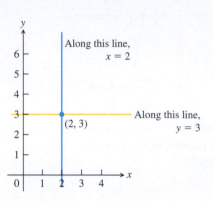

FIGURE A.11 The standard equations for the vertical and horizontal lines through $(2, 3)$ are $x = 2$ and $y = 3$.

We can write an equation for a nonvertical straight line L if we know its slope m and the coordinates of one point $P_1(x_1, y_1)$ on it. If $P(x, y)$ is *any* other point on L, then we can use the two points P_1 and P to compute the slope,

$$m = \frac{y - y_1}{x - x_1}$$

so that

$$y - y_1 = m(x - x_1), \qquad \text{or} \qquad y = y_1 + m(x - x_1).$$

The equation

$$y = y_1 + m(x - x_1)$$

is the **point-slope equation** of the line that passes through the point (x_1, y_1) and has slope m.

EXAMPLE 2 Write an equation for the line through the point $(2, 3)$ with slope $-3/2$.

Solution We substitute $x_1 = 2$, $y_1 = 3$, and $m = -3/2$ into the point-slope equation and obtain

$$y = 3 - \frac{3}{2}(x - 2), \qquad \text{or} \qquad y = -\frac{3}{2}x + 6.$$

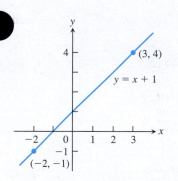

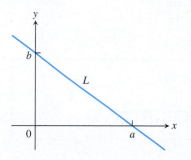

FIGURE A.12 The line in Example 3.

FIGURE A.13 Line L has x-intercept a and y-intercept b.

EXAMPLE 3 Write an equation for the line through $(-2, -1)$ and $(3, 4)$.

Solution The line's slope is

$$m = \frac{-1 - 4}{-2 - 3} = \frac{-5}{-5} = 1.$$

We can use this slope with either of the two given points in the point-slope equation:

With $(x_1, y_1) = (-2, -1)$

$y = -1 + 1 \cdot (x - (-2))$

$y = -1 + x + 2$

$y = x + 1$

With $(x_1, y_1) = (3, 4)$

$y = 4 + 1 \cdot (x - 3)$

$y = 4 + x - 3$

$y = x + 1$

Same result

Either way, we see that $y = x + 1$ is an equation for the line (Figure A.12).

The y-coordinate of the point where a nonvertical line intersects the y-axis is called the **y-intercept** of the line. Similarly, the **x-intercept** of a nonhorizontal line is the x-coordinate of the point where it crosses the x-axis (Figure A.13). A line with slope m and y-intercept b passes through the point $(0, b)$, so it has equation

$$y = b + m(x - 0), \qquad \text{or, more simply,} \qquad y = mx + b.$$

The equation

$$y = mx + b$$

is called the **slope-intercept equation** of the line with slope m and y-intercept b.

Lines with equations of the form $y = mx$ have y-intercept 0 and so pass through the origin. Equations of lines are called **linear** equations.

The equation

$$Ax + By = C \qquad (A \text{ and } B \text{ not both } 0)$$

is called the **general linear equation** in x and y because its graph always represents a line and every line has an equation in this form (including lines with undefined slope).

Parallel and Perpendicular Lines

Lines that are parallel have equal angles of inclination, so they have the same slope (if they are not vertical). Conversely, lines with equal slopes have equal angles of inclination and so are parallel.

If two nonvertical lines L_1 and L_2 are perpendicular, their slopes m_1 and m_2 satisfy $m_1 m_2 = -1$, so each slope is the *negative reciprocal* of the other:

$$m_1 = -\frac{1}{m_2}, \qquad m_2 = -\frac{1}{m_1}.$$

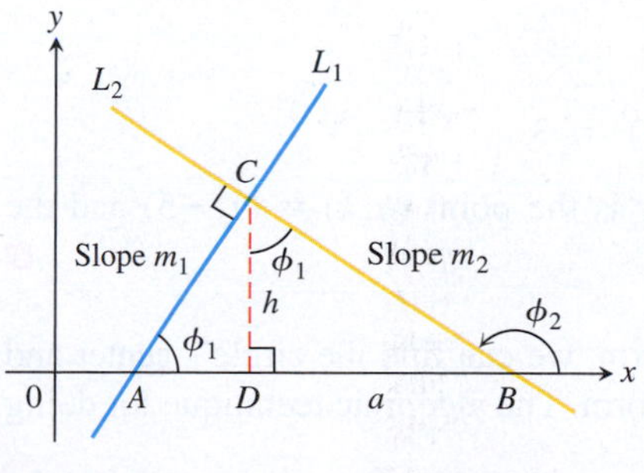

To see this, notice by inspecting similar triangles in Figure A.14 that $m_1 = a/h$, and $m_2 = -h/a$. Hence, $m_1 m_2 = (a/h)(-h/a) = -1$.

FIGURE A.14 $\triangle ADC$ is similar to $\triangle CDB$. Hence ϕ_1 is also the upper angle in $\triangle CDB$. From the sides of $\triangle CDB$, we read $\tan \phi_1 = a/h$.

Distance and Circles in the Plane

The distance between points in the plane is calculated with a formula that comes from the Pythagorean theorem (Figure A.15).

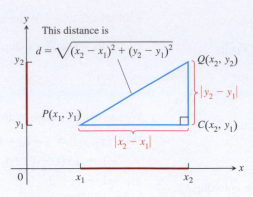

FIGURE A.15 To calculate the distance between $P(x_1, y_1)$ and $Q(x_2, y_2)$, apply the Pythagorean theorem to triangle PCQ.

Distance Formula for Points in the Plane

The distance between $P(x_1, y_1)$ and $Q(x_2, y_2)$ is

$$d = \sqrt{(\Delta x)^2 + (\Delta y)^2} = \sqrt{(x_2 - x_1)^2 + (y_2 - y_1)^2}.$$

By definition, a **circle** of radius a is the set of all points $P(x, y)$ whose distance from some center $C(h, k)$ equals a (Figure A.16). From the distance formula, P lies on the circle if and only if

$$\sqrt{(x - h)^2 + (y - k)^2} = a,$$

so

$$(x - h)^2 + (y - k)^2 = a^2. \qquad (1)$$

Equation (1) is the **standard equation** of a circle with center (h, k) and radius a. The circle of radius $a = 1$ and centered at the origin is the **unit circle** with equation

$$x^2 + y^2 = 1.$$

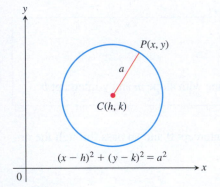

FIGURE A.16 A circle of radius a in the xy-plane, with center at (h, k).

EXAMPLE 4

(a) The standard equation for the circle of radius 2 centered at $(3, 4)$ is

$$(x - 3)^2 + (y - 4)^2 = 2^2 = 4.$$

(b) The circle

$$(x - 1)^2 + (y + 5)^2 = 3$$

has $h = 1, k = -5$, and $a = \sqrt{3}$. The center is the point $(h, k) = (1, -5)$ and the radius is $a = \sqrt{3}$. ∎

If an equation for a circle is not in standard form, we can find the circle's center and radius by first converting the equation to standard form. The algebraic technique for doing so is *completing the square*.

EXAMPLE 5   Find the center and radius of the circle

$$x^2 + y^2 + 4x - 6y - 3 = 0.$$

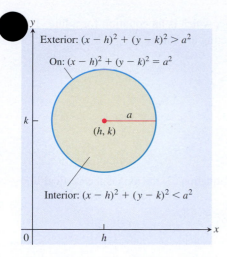

FIGURE A.17 The interior and exterior of the circle $(x - h)^2 + (y - k)^2 = a^2$.

Solution We convert the equation to standard form by completing the squares in x and y:

$$x^2 + y^2 + 4x - 6y - 3 = 0 \qquad \text{Start with the given equation.}$$

$$(x^2 + 4x) + (y^2 - 6y) = 3 \qquad \begin{array}{l}\text{Gather terms. Move the con-}\\ \text{stant to the right-hand side.}\end{array}$$

$$\left(x^2 + 4x + \left(\frac{4}{2}\right)^2\right) + \left(y^2 - 6y + \left(\frac{-6}{2}\right)^2\right) =$$

$$3 + \left(\frac{4}{2}\right)^2 + \left(\frac{-6}{2}\right)^2$$

Add the square of half the coefficient of x to each side of the equation. Do the same for y. The parenthetical expressions on the left-hand side are now perfect squares.

$$(x^2 + 4x + 4) + (y^2 - 6y + 9) = 3 + 4 + 9$$

$$(x + 2)^2 + (y - 3)^2 = 16$$

Write each quadratic as a squared linear expression.

The center is $(-2, 3)$ and the radius is $a = 4$. ∎

The points (x, y) satisfying the inequality

$$(x - h)^2 + (y - k)^2 < a^2$$

make up the **interior** region of the circle with center (h, k) and radius a (Figure A.17). The circle's **exterior** consists of the points (x, y) satisfying

$$(x - h)^2 + (y - k)^2 > a^2.$$

Parabolas

The geometric definition and properties of general parabolas are reviewed in Chapter 11. Here we look at parabolas arising as the graphs of equations of the form $y = ax^2 + bx + c$.

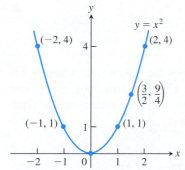

FIGURE A.18 The parabola $y = x^2$ (Example 6).

EXAMPLE 6 Consider the equation $y = x^2$. Some points whose coordinates satisfy this equation are $(0, 0)$, $(1, 1)$, $\left(\frac{3}{2}, \frac{9}{4}\right)$, $(-1, 1)$, $(2, 4)$, and $(-2, 4)$. These points (and all others satisfying the equation) make up a smooth curve called a parabola (Figure A.18). ∎

The graph of an equation of the form

$$y = ax^2$$

is a **parabola** whose **axis** (axis of symmetry) is the y-axis. The parabola's **vertex** (point where the parabola and axis cross) lies at the origin. The parabola opens upward if $a > 0$ and downward if $a < 0$. The larger the value of $|a|$, the narrower the parabola (Figure A.19).

Generally, the graph of $y = ax^2 + bx + c$ is a shifted and scaled version of the parabola $y = x^2$. We discuss shifting and scaling of graphs in more detail in Section 1.2.

The Graph of $y = ax^2 + bx + c$, $a \neq 0$

The graph of the equation $y = ax^2 + bx + c, a \neq 0$, is a parabola. The parabola opens upward if $a > 0$ and downward if $a < 0$. The **axis** is the line

$$x = -\frac{b}{2a}. \tag{2}$$

The **vertex** of the parabola is the point where the axis and parabola intersect. Its x-coordinate is $x = -b/2a$; its y-coordinate is found by substituting $x = -b/2a$ in the parabola's equation.

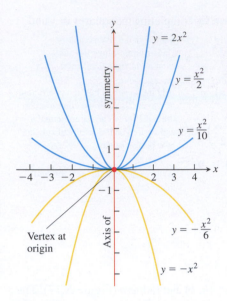

FIGURE A.19 Besides determining the direction in which the parabola $y = ax^2$ opens, the number a is a scaling factor. The parabola widens as a approaches zero and narrows as $|a|$ becomes large.

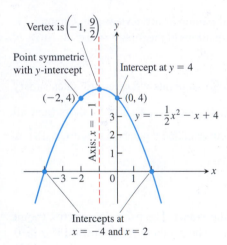

FIGURE A.20 The parabola in Example 7.

Notice that if $a = 0$, then we have $y = bx + c$, which is an equation for a line. Th[e] axis, given by Equation (2), can be found by completing the square.

EXAMPLE 7 Graph the equation $y = -\frac{1}{2}x^2 - x + 4$.

Solution Comparing the equation with $y = ax^2 + bx + c$ we see that

$$a = -\frac{1}{2}, \qquad b = -1, \qquad c = 4.$$

Since $a < 0$, the parabola opens downward. From Equation (2) the axis is the vertical line

$$x = -\frac{b}{2a} = -\frac{(-1)}{2(-1/2)} = -1.$$

When $x = -1$, we have

$$y = -\frac{1}{2}(-1)^2 - (-1) + 4 = \frac{9}{2}.$$

The vertex is $(-1, 9/2)$.

The x-intercepts are where $y = 0$:

$$-\frac{1}{2}x^2 - x + 4 = 0$$
$$x^2 + 2x - 8 = 0$$
$$(x - 2)(x + 4) = 0$$
$$x = 2, \qquad x = -4$$

We plot some points, sketch the axis, and use the direction of opening to complete the graph in Figure A.20.

Ellipses

The geometric definition and properties of general ellipses are reviewed in Chapter 11. Here we relate them to circles. Although they are not the graphs of functions, circles can be stretched horizontally or vertically in the same way as the graphs of functions. The standard equation for a circle of radius r centered at the origin is

$$x^2 + y^2 = r^2.$$

Substituting cx for x in the standard equation for a circle (Figure A.21) gives

$$c^2x^2 + y^2 = r^2. \tag{3}$$

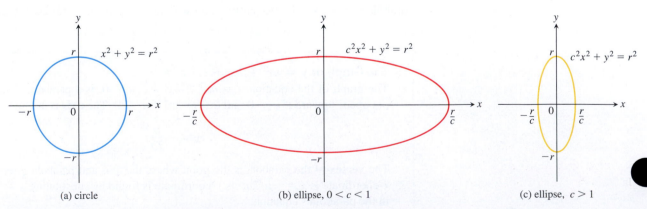

FIGURE A.21 Horizontal stretching or compression of a circle produces graphs of ellipses.

If $0 < c < 1$, the graph of Equation (3) horizontally stretches the circle; if $c > 1$ the circle is compressed horizontally. In either case, the graph of Equation (3) is an ellipse (Figure A.21). Notice in Figure A.21 that the y-intercepts of all three graphs are always $-r$ and r. In Figure A.21b, the line segment joining the points $(\pm r/c, 0)$ is called the **major axis** of the ellipse; the **minor axis** is the line segment joining $(0, \pm r)$. The axes of the ellipse are reversed in Figure A.21c: The major axis is the line segment joining the points $(0, \pm r)$, and the minor axis is the line segment joining the points $(\pm r/c, 0)$. In both cases, the major axis is the longer line segment.

If we divide both sides of Equation (3) by r^2, we obtain

$$\frac{x^2}{a^2} + \frac{y^2}{b^2} = 1 \tag{4}$$

where $a = r/c$ and $b = r$. If $a > b$, the major axis is horizontal; if $a < b$, the major axis is vertical. The **center** of the ellipse given by Equation (4) is the origin (Figure A.22). Substituting $x - h$ for x, and $y - k$ for y, in Equation (4) results in

$$\frac{(x - h)^2}{a^2} + \frac{(y - k)^2}{b^2} = 1. \tag{5}$$

Equation (5) is the **standard equation of an ellipse** with center at (h, k).

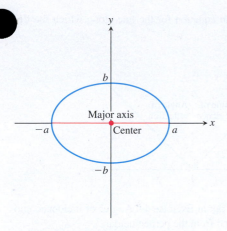

FIGURE A.22 Graph of the ellipse $\frac{x^2}{a^2} + \frac{y^2}{b^2} = 1$, $a > b$, where the major axis is horizontal.

EXERCISES A.3

Distance, Slopes, and Lines

In Exercises 1 and 2, a particle moves from A to B in the coordinate plane. Find the increments Δx and Δy in the particle's coordinates. Also find the distance from A to B.

1. $A(-3, 2),\quad B(-1, -2)$ **2.** $A(-3.2, -2),\quad B(-8.1, -2)$

Describe the graphs of the equations in Exercises 3 and 4.

3. $x^2 + y^2 = 1$ **4.** $x^2 + y^2 \le 3$

Plot the points in Exercises 5 and 6 and find the slope (if any) of the line they determine. Also find the common slope (if any) of the lines perpendicular to line AB.

5. $A(-1, 2),\quad B(-2, -1)$ **6.** $A(2, 3),\quad B(-1, 3)$

In Exercises 7 and 8, find an equation for **(a)** the vertical line and **(b)** the horizontal line through the given point.

7. $(-1, 4/3)$ **8.** $\left(0, -\sqrt{2}\right)$

In Exercises 9–15, write an equation for each line described.

9. Passes through $(-1, 1)$ with slope -1

10. Passes through $(3, 4)$ and $(-2, 5)$

11. Has slope $-5/4$ and y-intercept 6

12. Passes through $(-12, -9)$ and has slope 0

13. Has y-intercept 4 and x-intercept -1

14. Passes through $(5, -1)$ and is parallel to the line $2x + 5y = 15$

15. Passes through $(4, 10)$ and is perpendicular to the line $6x - 3y = 5$

In Exercises 16 and 17, find the line's x- and y-intercepts and use this information to graph the line.

16. $3x + 4y = 12$ **17.** $\sqrt{2}x - \sqrt{3}y = \sqrt{6}$

18. Is there anything special about the relationship between the lines $Ax + By = C_1$ and $Bx - Ay = C_2$ $(A \ne 0, B \ne 0)$? Give reasons for your answer.

19. A particle starts at $A(-2, 3)$ and its coordinates change by increments $\Delta x = 5$, $\Delta y = -6$. Find its new position.

20. The coordinates of a particle change by $\Delta x = 5$ and $\Delta y = 6$ as it moves from $A(x, y)$ to $B(3, -3)$. Find x and y.

Circles

In Exercises 21–23, find an equation for the circle with the given center $C(h, k)$ and radius a. Then sketch the circle in the xy-plane. Include the circle's center in your sketch. Also, label the circle's x- and y-intercepts, if any, with their coordinate pairs.

21. $C(0, 2),\quad a = 2$ **22.** $C(-1, 5),\quad a = \sqrt{10}$

23. $C\left(-\sqrt{3}, -2\right),\quad a = 2$

Graph the circles whose equations are given in Exercises 24–26. Label each circle's center and intercepts (if any) with their coordinate pairs.

24. $x^2 + y^2 + 4x - 4y + 4 = 0$

25. $x^2 + y^2 - 3y - 4 = 0$ **26.** $x^2 + y^2 - 4x + 4y = 0$

Parabolas

Graph the parabolas in Exercises 27–30. Label the vertex, axis, and intercepts in each case.

27. $y = x^2 - 2x - 3$ **28.** $y = -x^2 + 4x$

29. $y = -x^2 - 6x - 5$ **30.** $y = \frac{1}{2}x^2 + x + 4$

Inequalities

Describe the regions defined by the inequalities and pairs of inequalities in Exercises 31–34.

31. $x^2 + y^2 > 7$ **32.** $(x - 1)^2 + y^2 \leq 4$

33. $x^2 + y^2 > 1$, $x^2 + y^2 < 4$

34. $x^2 + y^2 + 6y < 0$, $y > -3$

35. Write an inequality that describes the points that lie inside the circle with center $(-2, 1)$ and radius $\sqrt{6}$.

36. Write a pair of inequalities that describe the points that lie inside or on the circle with center $(0, 0)$ and radius $\sqrt{2}$, and on or to the right of the vertical line through $(1, 0)$.

Theory and Examples

In Exercises 37–40, graph the two equations and find the points at which the graphs intersect.

37. $y = 2x$, $x^2 + y^2 = 1$ **38.** $y - x = 1$, $y = x^2$

39. $y = -x^2$, $y = 2x^2 - 1$

40. $x^2 + y^2 = 1$, $(x - 1)^2 + y^2 = 1$

41. Insulation By measuring slopes in the figure, estimate the temperature change in degrees per inch for **(a)** the gypsum wallboard; **(b)** the fiberglass insulation; **(c)** the wood sheathing.

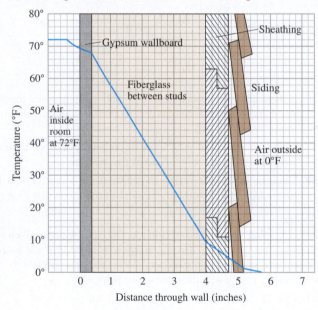

The temperature changes in the wall in Exercises 41 and 42.

42. Insulation According to the figure in Exercise 41, which of the materials is the best insulator? The poorest? Explain.

43. Pressure under water The pressure p experienced by a diver under water is related to the diver's depth d by an equation of the form $p = kd + 1$ (k a constant). At the surface, the pressure is 1 atmosphere. The pressure at 100 meters is about 10.94 atmospheres. Find the pressure at 50 meters.

44. Reflected light A ray of light comes in along the line $x + y = 1$ from the second quadrant and reflects off the x-axis (see the accompanying figure). The angle of incidence is equal to the angle

of reflection. Write an equation for the line along which the departing light travels.

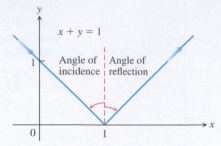

The path of the light ray in Exercise 44. Angles of incidence and reflection are measured from the perpendicular.

45. Fahrenheit vs. Celsius In the FC-plane, sketch the graph of the equation

$$C = \frac{5}{9}(F - 32)$$

linking Fahrenheit and Celsius temperatures. On the same graph sketch the line $C = F$. Is there a temperature at which a Celsius thermometer gives the same numerical reading as a Fahrenheit thermometer? If so, find it.

46. The Mt. Washington Cog Railway Civil engineers calculate the slope of roadbed as the ratio of the distance it rises or falls to the distance it runs horizontally. They call this ratio the **grade** of the roadbed, usually written as a percentage. Along the coast, commercial railroad grades are usually less than 2%. In the mountains, they may go as high as 4%. Highway grades are usually less than 5%.

The steepest part of the Mt. Washington Cog Railway in New Hampshire has an exceptional 37.1% grade. Along this part of the track, the seats in the front of the car are 14 ft above those in the rear. About how far apart are the front and rear rows of seats?

47. By calculating the lengths of its sides, show that the triangle with vertices at the points $A(1, 2)$, $B(5, 5)$, and $C(4, -2)$ is isosceles but not equilateral.

48. Show that the triangle with vertices $A(0, 0)$, $B\left(1, \sqrt{3}\right)$, and $C(2, 0)$ is equilateral.

49. Show that the points $A(2, -1)$, $B(1, 3)$, and $C(-3, 2)$ are vertices of a square, and find the fourth vertex.

50. Three different parallelograms have vertices at $(-1, 1)$, $(2, 0)$, and $(2, 3)$. Sketch them and find the coordinates of the fourth vertex of each.

51. For what value of k is the line $2x + ky = 3$ perpendicular to the line $4x + y = 1$? For what value of k are the lines parallel?

52. Midpoint of a line segment Show that the point with coordinates

$$\left(\frac{x_1 + x_2}{2}, \frac{y_1 + y_2}{2}\right)$$

is the midpoint of the line segment joining $P(x_1, y_1)$ to $Q(x_2, y_2)$.

A.4 Proofs of Limit Theorems

This appendix proves Theorem 1, Parts 2–5, and Theorem 4 from Section 2.2.

THEOREM 1—Limit Laws

If L, M, c, and k are real numbers and

$$\lim_{x \to c} f(x) = L \quad \text{and} \quad \lim_{x \to c} g(x) = M, \quad \text{then}$$

1. *Sum Rule:* $\quad\quad\quad\quad\quad\quad\quad \lim_{x \to c} (f(x) + g(x)) = L + M$

2. *Difference Rule:* $\quad\quad\quad\quad \lim_{x \to c} (f(x) - g(x)) = L - M$

3. *Constant Multiple Rule:* $\quad \lim_{x \to c} (k\, f(x)) = kL$

4. *Product Rule:* $\quad\quad\quad\quad\quad \lim_{x \to c} (f(x) g(x)) = LM$

5. *Quotient Rule:* $\quad\quad\quad\quad \lim_{x \to c} \dfrac{f(x)}{g(x)} = \dfrac{L}{M}, \quad M \neq 0$

6. *Power Rule:* $\quad\quad\quad\quad\quad \lim_{x \to c} [\,f(x)\,]^n = L^n$, n a positive integer

7. *Root Rule:* $\quad\quad\quad\quad\quad\quad \lim_{x \to c} \sqrt[n]{f(x)} = \sqrt[n]{L} = L^{1/n}$, n a positive integer

(If n is even, we assume that $\lim_{x \to c} f(x) = L > 0$.)

We proved the Sum Rule in Section 2.3, and the Power and Root Rules are proved in more advanced texts. We obtain the Difference Rule by replacing $g(x)$ by $-g(x)$ and M by $-M$ in the Sum Rule. The Constant Multiple Rule is the special case $g(x) = k$ of the Product Rule. This leaves only the Product and Quotient Rules.

Proof of the Limit Product Rule We show that for any $\varepsilon > 0$ there exists a $\delta > 0$ such that for all x in the intersection D of the domains of f and g,

$$|f(x) g(x) - LM| < \varepsilon \quad\quad \text{whenever} \quad\quad 0 < |x - c| < \delta.$$

Suppose then that ε is a positive number, and write $f(x)$ and $g(x)$ as

$$f(x) = L + (f(x) - L), \quad\quad g(x) = M + (g(x) - M).$$

Multiply these expressions together and subtract LM:

$$f(x) g(x) - LM = (L + (f(x) - L))(M + (g(x) - M)) - LM$$

$$= LM + L(g(x) - M) + M(f(x) - L)$$

$$\quad\quad + (f(x) - L)(g(x) - M) - LM$$

$$= L(g(x) - M) + M(f(x) - L) + (f(x) - L)(g(x) - M). \quad (1)$$

Since f and g have limits L and M as $x \to c$, there exist positive numbers δ_1, δ_2, δ_3, and δ_4 such that

$$\begin{array}{lll}
|f(x) - L| < \sqrt{\varepsilon/3} & \text{whenever} & 0 < |x - c| < \delta_1 \\
|g(x) - M| < \sqrt{\varepsilon/3} & \text{whenever} & 0 < |x - c| < \delta_2 \\
|f(x) - L| < \varepsilon/(3(1 + |M|)) & \text{whenever} & 0 < |x - c| < \delta_3 \\
|g(x) - M| < \varepsilon/(3(1 + |L|)) & \text{whenever} & 0 < |x - c| < \delta_4.
\end{array} \quad (2)$$

If we take δ to be the smallest of the numbers δ_1 through δ_4, the inequalities on the right-hand side of the Implications (2) will hold simultaneously for $0 < |x - c| < \delta$. Therefore, for all x in D, if $0 < |x - c| < \delta$ then

$$|f(x)\,g(x) - LM| \qquad \text{\color{blue}Triangle inequality applied to Eq. (1)}$$
$$\leq |L||g(x) - M| + |M||f(x) - L| + |f(x) - L||g(x) - M|$$
$$\leq (1 + |L|)|g(x) - M| + (1 + |M|)|f(x) - L| + |f(x) - L||g(x) - M|$$
$$< \frac{\varepsilon}{3} + \frac{\varepsilon}{3} + \sqrt{\frac{\varepsilon}{3}}\sqrt{\frac{\varepsilon}{3}} = \varepsilon. \qquad \text{\color{blue}Values from (2)}$$

This completes the proof of the Limit Product Rule.   ∎

Proof of the Limit Quotient Rule   We show that $\lim_{x \to c}(1/g(x)) = 1/M$. We can then conclude that

$$\lim_{x \to c} \frac{f(x)}{g(x)} = \lim_{x \to c}\left(f(x) \cdot \frac{1}{g(x)}\right) = \lim_{x \to c} f(x) \cdot \lim_{x \to c} \frac{1}{g(x)} = L \cdot \frac{1}{M} = \frac{L}{M}$$

by the Limit Product Rule.

Let $\varepsilon > 0$ be given. To show that $\lim_{x \to c}(1/g(x)) = 1/M$, we need to show that there exists a $\delta > 0$ such that

$$\left|\frac{1}{g(x)} - \frac{1}{M}\right| < \varepsilon \qquad \text{whenever} \qquad 0 < |x - c| < \delta.$$

Since g has the limit M as $x \to c$ and since $|M| > 0$, there exists a positive number δ_1 such that

$$|g(x) - M| < \frac{M}{2} \qquad \text{whenever} \qquad 0 < |x - c| < \delta_1. \qquad (3)$$

For any numbers A and B, the triangle inequality implies that $|A| - |B| \leq |A - B|$ and $|B| - |A| \leq |A - B|$, from which it follows that $\big||A| - |B|\big| \leq |A - B|$. With $A = g(x)$ and $B = M$, this becomes

$$\big||g(x)| - |M|\big| \leq |g(x) - M|,$$

which can be combined with the inequality on the right in Implication (3) to get, in turn,

$$\big||g(x)| - |M|\big| < \frac{|M|}{2}$$
$$-\frac{|M|}{2} < |g(x)| - |M| < \frac{|M|}{2}$$
$$\frac{|M|}{2} < |g(x)| < \frac{3|M|}{2}$$
$$|M| < 2|g(x)| < 3|M|$$
$$\frac{1}{|g(x)|} < \frac{2}{|M|} < \frac{3}{|g(x)|}. \qquad (4)$$

Therefore, $0 < |x - c| < \delta_1$ implies that

$$\left|\frac{1}{g(x)} - \frac{1}{M}\right| = \left|\frac{M - g(x)}{Mg(x)}\right| \leq \frac{1}{|M|} \cdot \frac{1}{|g(x)|} \cdot |M - g(x)|$$

$$< \frac{1}{|M|} \cdot \frac{2}{|M|} \cdot |M - g(x)|. \qquad \text{\color{blue}Inequality (4)} \qquad (5)$$

Since $(1/2)\,|M|^2\varepsilon > 0$, there exists a number $\delta_2 > 0$ such that

$$|M - g(x)| < \frac{\varepsilon}{2}|M|^2 \qquad \text{whenever} \qquad 0 < |x - c| < \delta_2. \qquad (6)$$

If we take δ to be the smaller of δ_1 and δ_2, the conclusions in (5) and (6) both hold whenever $0 < |x - c| < \delta$. Combining these conclusions gives

$$\left|\frac{1}{g(x)} - \frac{1}{M}\right| < \varepsilon \qquad \text{whenever} \qquad 0 < |x - c| < \delta.$$

This concludes the proof of the Limit Quotient Rule. ∎

THEOREM 4—The Sandwich Theorem

Suppose that $g(x) \le f(x) \le h(x)$ for all x in some open interval I containing c, except possibly at $x = c$ itself. Suppose also that $\lim_{x \to c} g(x) = \lim_{x \to c} h(x) = L$. Then $\lim_{x \to c} f(x) = L$.

Proof for Right-Hand Limits Suppose $\lim_{x \to c^+} g(x) = \lim_{x \to c^+} h(x) = L$. Then for any $\varepsilon > 0$ there exists a $\delta > 0$ such that the interval $(c, c + \delta)$ is contained in I and

$$L - \varepsilon < g(x) < L + \varepsilon \qquad \text{and} \qquad L - \varepsilon < h(x) < L + \varepsilon$$

whenever $c < x < c + \delta$. Since we always have $g(x) \le f(x) \le h(x)$ it follows that if $c < x < c + \delta$, then

$$L - \varepsilon < g(x) \le f(x) \le h(x) < L + \varepsilon,$$
$$L - \varepsilon < f(x) < L + \varepsilon,$$
$$-\varepsilon < f(x) - L < \varepsilon.$$

Therefore $|f(x) - L| < \varepsilon$ whenever $c < x < c + \delta$.

Proof for Left-Hand Limits Suppose $\lim_{x \to c^-} g(x) = \lim_{x \to c^-} h(x) = L$. Then for any $\varepsilon > 0$ there exists a $\delta > 0$ such that the interval $(c - \delta, c)$ is contained in I and

$$L - \varepsilon < g(x) < L + \varepsilon \qquad \text{and} \qquad L - \varepsilon < h(x) < L + \varepsilon$$

whenever $c - \delta < x < c$. We conclude as before that $|f(x) - L| < \varepsilon$ whenever $c - \delta < x < c$.

Proof for Two-Sided Limits If $\lim_{x \to c} g(x) = \lim_{x \to c} h(x) = L$, then $g(x)$ and $h(x)$ both approach L as $x \to c^+$ and as $x \to c^-$; so $\lim_{x \to c^+} f(x) = L$ and $\lim_{x \to c^-} f(x) = L$. Hence $\lim_{x \to c} f(x)$ exists and equals L. ∎

EXERCISES A.4

1. Suppose that functions $f_1(x)$, $f_2(x)$, and $f_3(x)$ have limits L_1, L_2, and L_3, respectively, as $x \to c$. Show that their sum has limit $L_1 + L_2 + L_3$. Use mathematical induction (Appendix 2) to generalize this result to the sum of any finite number of functions.

2. Use mathematical induction and the Limit Product Rule in Theorem 1 to show that if functions $f_1(x)$, $f_2(x)$, ..., $f_n(x)$ have limits $L_1, L_2, \ldots, L_n$ as $x \to c$, then

$$\lim_{x \to c} f_1(x) \cdot f_2(x) \cdot \cdots \cdot f_n(x) = L_1 \cdot L_2 \cdot \cdots \cdot L_n.$$

3. Use the fact that $\lim_{x \to c} x = c$ and the result of Exercise 2 to show that $\lim_{x \to c} x^n = c^n$ for any integer $n > 1$.

4. **Limits of polynomials** Use the fact that $\lim_{x \to c}(k) = k$ for any number k together with the results of Exercises 1 and 3 to show that $\lim_{x \to c} f(x) = f(c)$ for any polynomial function

$$f(x) = a_n x^n + a_{n-1} x^{n-1} + \cdots + a_1 x + a_0.$$

5. **Limits of rational functions**   Use Theorem 1 and the result of Exercise 4 to show that if $f(x)$ and $g(x)$ are polynomial functions and $g(c) \neq 0$, then

$$\lim_{x \to c} \frac{f(x)}{g(x)} = \frac{f(c)}{g(c)}.$$

6. **Composites of continuous functions**   Figure A.23 gives the diagram for a proof that the composite of two continuous func-

tions is continuous. Reconstruct the proof from the diagram. The statement to be proved is this: If f is continuous at $x = c$ and g is continuous at $f(c)$, then $g \circ f$ is continuous at c.

Assume that c is an interior point of the domain of f and that $f(c)$ is an interior point of the domain of g. This will make the limits involved two-sided. (The arguments for the cases that involve one-sided limits are similar.)

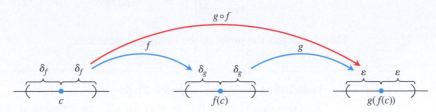

FIGURE A.23   The diagram for a proof that the composite of two continuous functions is continuous.

A.5   Commonly Occurring Limits

This appendix verifies limits (4)–(6) in Theorem 5 of Section 10.1.

Limit 4: If $|x| < 1$, $\lim\limits_{n \to \infty} x^n = 0$ We need to show that to each $\varepsilon > 0$ there corresponds an integer N so large that $|x^n| < \varepsilon$ for all n greater than N. Since $\varepsilon^{1/n} \to 1$, while $|x| < 1$, there exists an integer N for which $\varepsilon^{1/N} > |x|$. In other words,

$$|x^N| = |x|^N < \varepsilon. \tag{1}$$

This is the integer we seek because, if $|x| < 1$, then

$$|x^n| < |x^N| \quad \text{for all } n > N. \tag{2}$$

Combining (1) and (2) produces $|x^n| < \varepsilon$ for all $n > N$, concluding the proof.   ∎

Limit 5: For any number x, $\lim\limits_{n \to \infty} \left(1 + \dfrac{x}{n}\right)^n = e^x$ Let

$$a_n = \left(1 + \frac{x}{n}\right)^n.$$

Then

$$\ln a_n = \ln\left(1 + \frac{x}{n}\right)^n = n \ln\left(1 + \frac{x}{n}\right) \to x,$$

as we can see by the following application of L'Hôpital's Rule, in which we differentiate with respect to n:

$$\lim_{n \to \infty} n \ln\left(1 + \frac{x}{n}\right) = \lim_{n \to \infty} \frac{\ln(1 + x/n)}{1/n}$$

$$= \lim_{n \to \infty} \frac{\left(\dfrac{1}{1 + x/n}\right) \cdot \left(-\dfrac{x}{n^2}\right)}{-1/n^2} = \lim_{n \to \infty} \frac{x}{1 + x/n} = x.$$

Apply Theorem 3, Section 10.1, with $f(x) = e^x$ to conclude that

$$\left(1 + \frac{x}{n}\right)^n = a_n = e^{\ln a_n} \to e^x.$$

∎

Limit 6: For any number x, $\displaystyle \lim_{n\to\infty} \frac{x^n}{n!} = 0$ Since

$$-\frac{|x|^n}{n!} \le \frac{x^n}{n!} \le \frac{|x|^n}{n!},$$

all we need to show is that $|x|^n/n! \to 0$. We can then apply the Sandwich Theorem for Sequences (Section 10.1, Theorem 2) to conclude that $x^n/n! \to 0$.

The first step in showing that $|x|^n/n! \to 0$ is to choose an integer $M > |x|$, so that $(|x|/M) < 1$. By Limit 4, just proved, we then have $(|x|/M)^n \to 0$. We then restrict our attention to values of $n > M$. For these values of n, we can write

$$\frac{|x|^n}{n!} = \frac{|x|^n}{1 \cdot 2 \cdot \,\cdots\, \cdot \underbrace{M \cdot (M+1) \cdot (M+2) \cdot \,\cdots\, \cdot n}}$$

$$\underbrace{}_{(n-M)\ \text{factors}}$$

$$\le \frac{|x|^n}{M! M^{n-M}} = \frac{|x|^n M^M}{M! M^n} = \frac{M^M}{M!}\left(\frac{|x|}{M}\right)^n.$$

Thus,

$$0 \le \frac{|x|^n}{n!} \le \frac{M^M}{M!}\left(\frac{|x|}{M}\right)^n.$$

Now, the constant $M^M/M!$ does not change as n increases. Thus the Sandwich Theorem tells us that $|x|^n/n! \to 0$ because $(|x|/M)^n \to 0$. ∎

A.6 **Theory of the Real Numbers**

A rigorous development of calculus is based on properties of the real numbers. Many results about functions, derivatives, and integrals would be false if stated for functions defined only on the rational numbers. In this appendix we briefly examine some basic concepts of the theory of the reals that hint at what might be learned in a deeper, more theoretical study of calculus.

Three types of properties make the real numbers what they are. These are the **algebraic**, **order**, and **completeness** properties. The algebraic properties involve addition and multiplication, subtraction and division. They apply to rational or complex numbers (discussed in Appendix A.7) as well as to the reals.

The structure of numbers is built around a set with addition and multiplication operations. The following properties are required of addition and multiplication.

A1 $a + (b + c) = (a + b) + c$ for all a, b, c.

A2 $a + b = b + a$ for all a, b.

A3   There is a number called "0" such that $a + 0 = a$ for all a.

A4   For each number a, there is a number b such that $a + b = 0$.

M1 $a(bc) = (ab)c$ for all a, b, c.

M2 $ab = ba$ for all a, b.

M3   There is a number called "1" such that $a \cdot 1 = a$ for all a.

M4   For each nonzero number a, there is a number b such that $ab = 1$.

D $a(b + c) = ab + ac$ for all a, b, c.

A1 and M1 are *associative laws*, A2 and M2 are *commutativity laws*, A3 and M3 are *identity laws*, and D is the *distributive law*. Sets that have these algebraic properties are examples of **fields**, and are studied in depth in the area of theoretical mathematics called abstract algebra.

The **order** properties allow us to compare the size of any two numbers. The order properties are

O1 For any a and b, either $a \leq b$ or $b \leq a$ or both.

O2 If $a \leq b$ and $b \leq a$ then $a = b$.

O3 If $a \leq b$ and $b \leq c$ then $a \leq c$.

O4 If $a \leq b$ then $a + c \leq b + c$.

O5 If $a \leq b$ and $0 \leq c$ then $ac \leq bc$.

O3 is the *transitivity law*, and O4 and O5 relate ordering to addition and multiplication.

We can order the reals, the integers, and the rational numbers, but we cannot order the complex numbers (there is no reasonable way to decide whether a number like $i = \sqrt{-1}$ is bigger or smaller than zero). A field in which the size of any two elements can be compared as above is called an **ordered field**. Both the rational numbers and the real numbers are ordered fields, and there are many others.

We can think of real numbers geometrically, lining them up as points on a line. The **completeness property** says that the real numbers correspond to all points on the line, with no "holes" or "gaps." The rationals, in contrast, omit points such as $\sqrt{2}$ and π, and the integers even leave out fractions like $1/2$. The reals, having the completeness property, omit no points.

What exactly do we mean by this vague idea of missing holes? To answer this we must give a more precise description of completeness. A number M is an **upper bound** for a set of numbers if all numbers in the set are smaller than or equal to M. M is a **least upper bound** if it is the smallest upper bound. For example, $M = 2$ is an upper bound for the negative numbers. So is $M = 1$, showing that 2 is not a least upper bound. The least upper bound for the set of negative numbers is $M = 0$. We define a **complete** ordered field to be one in which every nonempty set bounded above has a least upper bound.

If we work with just the rational numbers, the set of numbers less than $\sqrt{2}$ is bounded, but it does not have a rational least upper bound, since any rational upper bound M can be replaced by a slightly smaller rational number that is still larger than $\sqrt{2}$. So the rationals are not complete. In the real numbers, a set that is bounded above always has a least upper bound. The reals are a complete ordered field.

The completeness property is at the heart of many results in calculus. One example occurs when searching for a maximum value for a function on a closed interval $[a, b]$, as in Section 4.1. The function $y = x - x^3$ has a maximum value on $[0, 1]$ at the point x satisfying $1 - 3x^2 = 0$, or $x = \sqrt{1/3}$. If we limited our consideration to functions defined only on rational numbers, we would have to conclude that the function has no maximum, since $\sqrt{1/3}$ is irrational (Figure A.24). The Extreme Value Theorem (Section 4.1), which implies that continuous functions on closed intervals $[a, b]$ have a maximum value, is not true for functions defined only on the rationals.

The Intermediate Value Theorem implies that a continuous function f on an interval $[a, b]$ with $f(a) < 0$ and $f(b) > 0$ must be zero somewhere in $[a, b]$. The function values cannot jump from negative to positive without there being some point x in $[a, b]$ where $f(x) = 0$. The Intermediate Value Theorem also relies on the completeness of the real numbers and is false for continuous functions defined only on the rationals. The function $f(x) = 3x^2 - 1$ has $f(0) = -1$ and $f(1) = 2$, but if we consider f only on the rational numbers, it never equals zero. The only value of x for which $f(x) = 0$ is $x = \sqrt{1/3}$, an irrational number.

We have captured the desired properties of the reals by saying that the real numbers are a complete ordered field. But we're not quite finished. Greek mathematicians in the

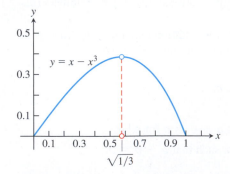

FIGURE A.24 The maximum value of $y = x - x^3$ on $[0, 1]$ occurs at the irrational number $x = \sqrt{1/3}$.

school of Pythagoras tried to impose another property on the numbers of the real line, the condition that all numbers are ratios of integers. They learned that their effort was doomed when they discovered irrational numbers such as $\sqrt{2}$. How do we know that our efforts to specify the real numbers are not also flawed, for some unseen reason? The artist Escher drew optical illusions of spiral staircases that went up and up until they rejoined themselves at the bottom. An engineer trying to build such a staircase would find that no structure realized the plans the architect had drawn. Could it be that our design for the reals contains some subtle contradiction, and that no construction of such a number system can be made?

We resolve this issue by giving a specific description of the real numbers and verifying that the algebraic, order, and completeness properties are satisfied in this model. This is called a **construction** of the reals, and just as stairs can be built with wood, stone, or steel, there are several approaches to constructing the reals. One construction treats the reals as all the infinite decimals,

$$a.d_1d_2d_3d_4 \ldots$$

In this approach a real number is an integer a followed by a sequence of decimal digits $d_1, d_2, d_3, \ldots$, each between 0 and 9. This sequence may stop, or repeat in a periodic pattern, or keep going forever with no pattern. In this form, $2.00, 0.3333333 \ldots$ and $3.1415926535898 \ldots$ represent three familiar real numbers. The real meaning of the dots "$\ldots$" following these digits requires development of the theory of sequences and series, as in Chapter 10. Each real number is constructed as the limit of a sequence of rational numbers given by its finite decimal approximations. An infinite decimal is then the same as a series

$$a + \frac{d_1}{10} + \frac{d_2}{100} + \cdots.$$

This decimal construction of the real numbers is not entirely straightforward. It's easy enough to check that it gives numbers that satisfy the completeness and order properties, but verifying the algebraic properties is rather involved. Even adding or multiplying two numbers requires an infinite number of operations. Making sense of division requires a careful argument involving limits of rational approximations to infinite decimals.

A different approach was taken by Richard Dedekind (1831–1916), a German mathematician, who gave the first rigorous construction of the real numbers in 1872. Given any real number x, we can divide the rational numbers into two sets: those less than or equal to x and those greater. Dedekind cleverly reversed this reasoning and defined a real number to be a division of the rational numbers into two such sets. This seems like a strange approach, but such indirect methods of constructing new structures from old are powerful tools in theoretical mathematics.

These and other approaches can be used to construct a system of numbers having the desired algebraic, order, and completeness properties. A final issue that arises is whether all the constructions give the same thing. Is it possible that different constructions result in different number systems satisfying all the required properties? If yes, which of these is the real numbers? Fortunately, the answer turns out to be no. The reals are the only number system satisfying the algebraic, order, and completeness properties.

Confusion about the nature of the numbers and about limits caused considerable controversy in the early development of calculus. Calculus pioneers such as Newton, Leibniz, and their successors, when looking at what happens to the difference quotient

$$\frac{\Delta y}{\Delta x} = \frac{f(x + \Delta x) - f(x)}{\Delta x}$$

as each of Δy and Δx approach zero, talked about the resulting derivative being a quotient of two infinitely small quantities. These "infinitesimals," written dx and dy, were thought to be some new kind of number, smaller than any fixed number but not zero. Similarly, a definite integral was thought of as a sum of an infinite number of infinitesimals

$$f(x) \cdot dx$$

as x varied over a closed interval. While the approximating difference quotients $\Delta y / \Delta x$ were understood much as today, it was the quotient of infinitesimal quantities, rather than a limit, that was thought to encapsulate the meaning of the derivative. This way of thinking led to logical difficulties, as attempted definitions and manipulations of infinitesimals ran into contradictions and inconsistencies. The more concrete and computable difference quotients did not cause such trouble, but they were thought of merely as useful calculation tools. Difference quotients were used to work out the numerical value of the derivative and to derive general formulas for calculation, but were not considered to be at the heart of the question of what the derivative actually was. Today we realize that the logical problems associated with infinitesimals can be avoided by *defining* the derivative to be the limit of its approximating difference quotients. The ambiguities of the old approach are no longer present, and in the standard theory of calculus, infinitesimals are neither needed nor used.

A.7   Complex Numbers

Complex numbers are expressed in the form $a + ib$, or $a + bi$, where a and b are real numbers and i is a symbol for $\sqrt{-1}$. Unfortunately, the words "real" and "imaginary" have connotations that somehow place $\sqrt{-1}$ in a less favorable position in our minds than $\sqrt{2}$. As a matter of fact, a good deal of imagination, in the sense of *inventiveness*, has been required to construct the *real* number system, which forms the basis of calculus (see Appendix 6). In this appendix we review the various stages of these inventions.

The Hierarchy of Numbers

The first stage of number development was the recognition of the counting numbers $1, 2, 3, \ldots$, which we now call the **natural numbers** or the **positive integers**. Certain arithmetical operations on the positive integers, such as addition and multiplication, keep us entirely within this system. That is, if m and n are any positive integers, then their sum $m + n$ and product mn are also positive integers.

Some equations can be solved entirely within the system of positive integers. For example, we can solve $3 + x = 7$ using only positive integers. But other simple equations, such as $7 + x = 3$, cannot be solved if positive integers are the only numbers at our disposal. The number zero and the negative numbers were invented to solve equations such as $7 + x = 3$. Using the **integers**

$$\ldots, -3, -2, -1, 0, 1, 2, 3, \ldots,$$

we can always find the missing integer x that solves the equation $m + x = p$ when we are given the other two integers m and p in the equation.

Addition and multiplication of integers always keep us within the system of integers. However, division does not, and so fractions m/n, where m and n are integers with n nonzero, were invented. This system, which is called the **rational numbers**, is rich enough to perform all of the **rational operations** of arithmetic, including addition, subtraction, multiplication, and division (although division by zero is excluded since it is meaningless).

Yet there are still simple polynomial equations that cannot be solved within the system of rational numbers. The ancient Greeks realized that there is no *rational* number that solves the equation $x^2 = 2$, even though the Pythagorean Theorem implies that the length x of the diagonal of the unit square satisfies this equation! (See Figure A.25.) To see why $x^2 = 2$ has no rational solution, consider the following argument.

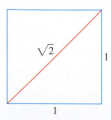

FIGURE A.25   The diagonal of the unit square has irrational length.

Suppose that there did exist some integers p and q with no common factor other than 1 such that the fraction $x = p/q$ satisfied $x^2 = 2$. Writing this out, we see that $p^2/q^2 = 2$, and therefore

$$p^2 = 2q^2.$$

Thus p^2 is an even integer. Since the square of an odd number is odd, we conclude that p itself must be an even number (for if p were odd, then p^2 would also be odd). Hence p is divisible by 2, and therefore $p = 2k$ for some integer k. Hence $p^2 = 4k^2$. Since we already saw that $p^2 = 2q^2$, it follows that $2q^2 = p^2 = 4k^2$, and therefore

$$q^2 = 2k^2.$$

Hence q^2 is an even number. This requires that q itself be even. Therefore, both p and q are divisible by 2, which is contrary to our assumption that they contain no common factors other than 1. Since we have arrived at a contradiction, there cannot exist any such integers p and q, and therefore there is no rational number that solves the equation $x^2 = 2$.

The invention of real numbers addressed this issue (and others). Using real numbers, we can represent every possible physical length. As we saw in Appendix A.6, each real number can be represented as an infinite decimal $a.d_1d_2d_3d_4\ldots$, where a is an integer followed by a sequence of decimal digits each between 0 and 9. If the sequence stops or repeats in a periodic pattern, then the decimal represents a rational number. An irrational number is represented by a nonterminating and nonrepeating decimal. The rational and irrational numbers together make up the real number system. Unlike the rational numbers, the real numbers have the **completeness property**, meaning that there are no "holes" or "gaps" in the real line. Yet for all of its utility, there are still simple equations that cannot be solved within the real number system alone. For example, the polynomial equation $x^2 + 1 = 0$ has no real solutions.

The Complex Numbers

We have discussed three invented systems of numbers that form a hierarchy in which each system contains the previous system. Each system is richer than its predecessor in that it permits additional operations to be performed without going outside the system.

1. Using the integer system we can solve all equations of the form

    $$x + a = 0, \tag{1}$$

    where a is an integer.

2. Using the rational numbers we can solve all equations of the form

    $$ax + b = 0, \tag{2}$$

    provided that a and b are rational numbers and $a \neq 0$.

3. Using the real numbers, we can solve all of Equations (1) and (2) and, in addition, all quadratic equations

    $$ax^2 + bx + c = 0 \quad \text{provided that} \quad a \neq 0 \quad \text{and} \quad b^2 - 4ac \geq 0. \tag{3}$$

    The **quadratic formula**

    $$x = \frac{-b \pm \sqrt{b^2 - 4ac}}{2a} \tag{4}$$

gives the solutions to Equation (3). When $b^2 - 4ac$ is negative there are no real number solutions to the equation $ax^2 + bx + c = 0$. In particular, the simple quadratic equation $x^2 + 1 = 0$ cannot be solved using any of the three invented systems of numbers that we have discussed.

Thus we come to the fourth invented system, which is the set of **complex numbers** $a + ib$. The symbol i represents a new number whose square equals -1. We call a the **real part** and b the **imaginary part** of the complex number $a + ib$. Sometimes it is convenient to write $a + bi$ instead of $a + ib$; both notations describe the same complex number.

We define equality and addition for complex numbers in the following way.

Equality $a + ib = c + id$
if and only if
$a = c$ and $b = d$

Two complex numbers $a + ib$ and $c + id$ are equal if and only if their real parts are equal and their imaginary parts are equal.

Addition $(a + ib) + (c + id)$
$= (a + c) + i(b + d)$

We sum the real parts and separately sum the imaginary parts.

To multiply two complex numbers, we multiply using the distributive rule and then simplify using $i^2 = -1$:

Multiplication $(a + ib)(c + id)$
$= ac + iad + ibc + i^2bd$
$= (ac - bd) + i(ad + bc)$ $i^2 = -1.$

The set of all complex numbers $a + i0$, where the second number b is zero, has all of the properties of the set of real numbers. For example, addition and multiplication as complex numbers give

$$(a + i0) + (c + i0) = (a + c) + i0, \qquad (a + i0)(c + i0) = ac + i0,$$

which are numbers of the same type with imaginary part zero. We usually just write a instead of $a + i0$, and in this sense the real number system is "embedded" into the complex number system.

If we multiply a "real number" $a = a + i0$ by a complex number $c + id$, we get $a(c + id) = (a + i0)(c + id) = ac + iad$. In particular, the number $0 = 0 + i0$ plays the role of zero in the complex number system, and the complex number $1 = 1 + i0$ plays the role of unity, or one, in the complex number system.

The complex number $i = 0 + i1$, which has real part zero and imaginary part one, has the property that its square is

$$i^2 = (0 + i1)^2 = (0 + i1)(0 + i1) = (-1) + i0 = -1.$$

Thus $x = i$ is a solution to the quadratic equation $x^2 + 1 = 0$. Using the complex number system, there are exactly two solutions to this equation, the other solution being $x = -i = 0 + i(-1)$.

We can divide any two complex numbers as long as we do not divide by the number $0 = 0 + i0$. As long as $a + ib \neq 0$ (meaning that *either $a \neq 0$ or $b \neq 0$ or both $a \neq 0$ and $b \neq 0$*), we carry out division as follows:

$$\frac{c + id}{a + ib} = \frac{(c + id)(a - ib)}{(a + ib)(a - ib)} = \frac{(ac + bd) + i(ad - bc)}{a^2 + b^2} = \frac{ac + bd}{a^2 + b^2} + i\frac{ad - bc}{a^2 + b^2}.$$

Note that $a^2 + b^2 \neq 0$ since we stipulated that a and b cannot both be zero.

The number $a - ib$ that is used as the multiplier to clear the i from the denominator is called the **complex conjugate** of $a + ib$. If we denote the original complex number by $z = a + ib$, then it is customary to write $\bar{z}$ (read "z bar") to denote its complex conjugate:

$$z = a + ib, \qquad \bar{z} = a - ib.$$

Multiplying the numerator and denominator of a fraction $(c + id)/(a + ib)$ by the complex conjugate of the denominator will always replace the denominator by a real number.

EXAMPLE 1   We give some illustrations of the arithmetic operations with complex numbers.

(a) $(2 + 3i) + (6 - 2i) = (2 + 6) + (3 - 2)i = 8 + i$

(b) $(2 + 3i) - (6 - 2i) = (2 - 6) + (3 - (-2))i = -4 + 5i$

(c) $(2 + 3i)(6 - 2i) = (2)(6) + (2)(-2i) + (3i)(6) + (3i)(-2i)$

$$= 12 - 4i + 18i - 6i^2 = 12 + 14i + 6 = 18 + 14i$$

(d) $\dfrac{2 + 3i}{6 - 2i} = \dfrac{2 + 3i}{6 - 2i}\dfrac{6 + 2i}{6 + 2i} = \dfrac{12 + 4i + 18i + 6i^2}{36 + 12i - 12i - 4i^2} = \dfrac{6 + 22i}{40} = \dfrac{3}{20} + \dfrac{11}{20}i$ ∎

Argand Diagrams

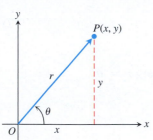

FIGURE A.26 This Argand diagram represents $z = x + iy$ both as a point $P(x, y)$ and as a vector $\overrightarrow{OP}$.

There are two geometric representations of the complex number $z = x + iy$:

1. as the point $P(x, y)$ in the xy-plane
2. as the vector $\overrightarrow{OP}$ from the origin to P.

In each representation, the x-axis is called the **real axis** and the y-axis is the **imaginary axis**. Both representations are **Argand diagrams** for $x + iy$ (Figure A.26).

In terms of the polar coordinates of x and y, we have

$$x = r \cos \theta, \qquad y = r \sin \theta,$$

and

$$z = x + iy = r(\cos \theta + i \sin \theta). \tag{5}$$

We define the **absolute value** of a complex number $x + iy$ to be the length r of a vector $\overrightarrow{OP}$ from the origin to $P(x, y)$. We denote the absolute value by vertical bars; thus,

$$|x + iy| = \sqrt{x^2 + y^2}.$$

If we always choose the polar coordinates r and θ so that r is nonnegative, then

$$r = |x + iy|.$$

The polar angle θ is called the **argument** of z and is written $\theta = \arg z$. Of course, any integer multiple of 2π may be added to θ to produce another appropriate angle.

The following equation gives a useful formula connecting a complex number z, its conjugate $\bar{z}$, and its absolute value $|z|$:

$$z \cdot \bar{z} = |z|^2.$$

Euler's Formula

The identity

$$e^{i\theta} = \cos \theta + i \sin \theta, \tag{6}$$

is called **Euler's formula**. We show an Argand diagram for $e^{i\theta}$ in Figure A.27. Using Equation (6), we can write Equation (5) as

$$z = re^{i\theta}.$$

The notation exp (A) is also used for e^A.

This formula, in turn, leads to the following rules for calculating products, quotients, powers, and roots of complex numbers.

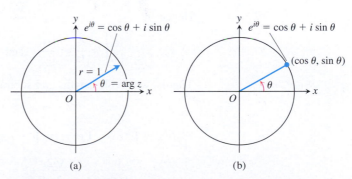

FIGURE A.27 Argand diagrams for $e^{i\theta} = \cos \theta + i \sin \theta$ (a) as a vector and (b) as a point.

Products

To multiply two complex numbers, we multiply their absolute values and add their angles. To see why, let

$$z_1 = r_1 e^{i\theta_1}, \qquad z_2 = r_2 e^{i\theta_2}, \tag{7}$$

so that

$$|z_1| = r_1, \qquad \arg z_1 = \theta_1; \qquad |z_2| = r_2, \qquad \arg z_2 = \theta_2.$$

Then

$$z_1 z_2 = r_1 e^{i\theta_1} \cdot r_2 e^{i\theta_2} = r_1 r_2 e^{i(\theta_1 + \theta_2)}$$

and hence

$$|z_1 z_2| = r_1 r_2 = |z_1| \cdot |z_2|$$
$$\arg (z_1 z_2) = \theta_1 + \theta_2 = \arg z_1 + \arg z_2. \tag{8}$$

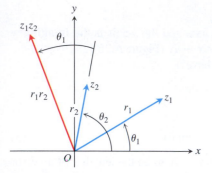

FIGURE A.28 When z_1 and z_2 are multiplied, $|z_1 z_2| = r_1 \cdot r_2$ and $\arg (z_1 z_2) = \theta_1 + \theta_2$.

Thus, the product of two complex numbers is represented by a vector whose length is the product of the lengths of the two factors and whose argument is the sum of their arguments (Figure A.28). In particular, from Equation (8) a vector may be rotated counterclockwise through an angle θ by multiplying it by $e^{i\theta}$. Multiplication by i rotates $90°$, by -1 rotates $180°$, by $-i$ rotates $270°$, and so on.

EXAMPLE 2   Let $z_1 = 1 + i, z_2 = \sqrt{3} - i$. We plot these complex numbers in an Argand diagram (Figure A.29) from which we read off the polar representations

$$z_1 = \sqrt{2}e^{i\pi/4}, \qquad z_2 = 2e^{-i\pi/6}.$$

Then

$$z_1 z_2 = 2\sqrt{2} \exp\left(\frac{i\pi}{4} - \frac{i\pi}{6}\right) = 2\sqrt{2} \exp\left(\frac{i\pi}{12}\right)$$

$$= 2\sqrt{2}\left(\cos \frac{\pi}{12} + i \sin \frac{\pi}{12}\right) \approx 2.73 + 0.73i. \qquad \blacksquare$$

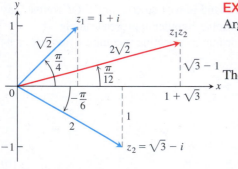

FIGURE A.29   To multiply two complex numbers, multiply their absolute values and add their arguments.

Quotients

Suppose $r_2 \neq 0$ in Equation (7). Then

$$\frac{z_1}{z_2} = \frac{r_1 e^{i\theta_1}}{r_2 e^{i\theta_2}} = \frac{r_1}{r_2} e^{i(\theta_1 - \theta_2)}.$$

Hence

$$\left|\frac{z_1}{z_2}\right| = \frac{r_1}{r_2} = \frac{|z_1|}{|z_2|} \quad \text{and} \quad \arg\left(\frac{z_1}{z_2}\right) = \theta_1 - \theta_2 = \arg z_1 - \arg z_2.$$

That is, we divide lengths and subtract angles for the quotient of complex numbers.

EXAMPLE 3   Let $z_1 = 1 + i$ and $z_2 = \sqrt{3} - i$, as in Example 2. Then

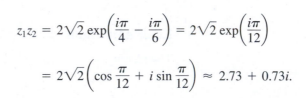

$$\frac{1 + i}{\sqrt{3} - i} = \frac{\sqrt{2}e^{i\pi/4}}{2e^{-i\pi/6}} = \frac{\sqrt{2}}{2} e^{5\pi i/12} \approx 0.707\left(\cos \frac{5\pi}{12} + i \sin \frac{5\pi}{12}\right)$$

$$\approx 0.183 + 0.683i. \qquad \blacksquare$$

Powers

If n is a positive integer, we may apply the product formulas in Equation (8) to find

$$z^n = z \cdot z \cdot \cdots \cdot z. \qquad \text{\small\color{blue}n factors}$$

With $z = re^{i\theta}$, we obtain

$$z^n = (re^{i\theta})^n = r^n e^{i(\theta + \theta + \cdots + \theta)} \qquad \text{\small\color{blue}n summands}$$

$$= r^n e^{in\theta}. \tag{9}$$

The length $r = |z|$ is raised to the nth power and the angle $\theta = \arg z$ is multiplied by n. If we take $r = 1$ in Equation (9), we obtain De Moivre's Theorem.

De Moivre's Theorem

$$(\cos\theta + i\sin\theta)^n = \cos n\theta + i\sin n\theta. \tag{10}$$

If we expand the left side of De Moivre's equation above by the Binomial Theorem and reduce it to the form $a + ib$, we obtain formulas for $\cos n\theta$ and $\sin n\theta$ as polynomials of degree n in $\cos\theta$ and $\sin\theta$.

EXAMPLE 4   If $n = 3$ in Equation (10), we have

$$(\cos\theta + i\sin\theta)^3 = \cos 3\theta + i\sin 3\theta.$$

The left side of this equation expands to

$$\cos^3\theta + 3i\cos^2\theta\sin\theta - 3\cos\theta\sin^2\theta - i\sin^3\theta.$$

The real part of this must equal $\cos 3\theta$ and the imaginary part must equal $\sin 3\theta$. Therefore,

$$\cos 3\theta = \cos^3\theta - 3\cos\theta\sin^2\theta,$$
$$\sin 3\theta = 3\cos^2\theta\sin\theta - \sin^3\theta. \qquad \blacksquare$$

Roots

If $z = re^{i\theta}$ is a complex number different from zero and n is a positive integer, then there are precisely n different complex numbers $w_0, w_1, \ldots, w_{n-1}$, that are nth roots of z. To see why, let $w = \rho e^{i\alpha}$ be an nth root of $z = re^{i\theta}$. Then

$$w^n = z$$

or

$$\rho^n e^{in\alpha} = re^{i\theta}.$$

Since both r and ρ^n are positive, this implies that $\rho^n = r$, and so

$$\rho = \sqrt[n]{r}$$

is the real, positive nth root of r. For the argument, although we cannot say that $n\alpha$ and θ must be equal, we can say that they may differ only by an integer multiple of 2π. That is,

$$n\alpha = \theta + 2k\pi, \qquad k = 0, \pm 1, \pm 2, \ldots.$$

Therefore,

$$\alpha = \frac{\theta}{n} + k\frac{2\pi}{n}.$$

Hence, all the nth roots of $z = re^{i\theta}$ are given by

$$\sqrt[n]{re^{i\theta}} = \sqrt[n]{r}\,\exp\,i\left(\frac{\theta}{n} + k\frac{2\pi}{n}\right), \qquad k = 0, \pm 1, \pm 2, \dots. \qquad (11)$$

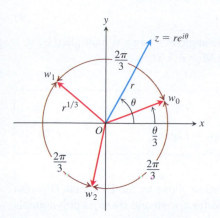

FIGURE A.30 The three cube roots of $z = re^{i\theta}$.

There might appear to be infinitely many different answers corresponding to the infinitely many possible values of k, but $k = n + m$ gives the same answer as $k = m$ in Equation (11). Thus, we need only take n consecutive values for k to obtain all the different nth roots of z. For convenience, we take

$$k = 0, 1, 2, \dots, n - 1.$$

All the nth roots of $re^{i\theta}$ lie on a circle centered at the origin and having radius equal to the real, positive nth root of r. One of them has argument $\alpha = \theta/n$. The others are uniformly spaced around the circle, each being separated from its neighbors by an angle equal to $2\pi/n$. Figure A.30 illustrates the placement of the three cube roots, w_0, w_1, w_2, of the complex number $z = re^{i\theta}$.

EXAMPLE 5 Find the four fourth roots of -16.

Solution As our first step, we plot the number -16 in an Argand diagram (Figure A.31) and determine its polar representation $re^{i\theta}$. Here, $z = -16$, $r = +16$, and $\theta = \pi$. One of the fourth roots of $16e^{i\pi}$ is $2e^{i\pi/4}$. We obtain others by successive additions of $2\pi/4 = \pi/2$ to the argument of this first one. Hence,

$$\sqrt[4]{16\,\exp\,i\pi} = 2\,\exp\,i\left(\frac{\pi}{4}, \frac{3\pi}{4}, \frac{5\pi}{4}, \frac{7\pi}{4}\right),$$

and the four roots are

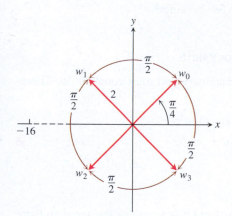

FIGURE A.31 The four fourth roots of -16.

$$w_0 = 2\left[\cos\frac{\pi}{4} + i\sin\frac{\pi}{4}\right] = \sqrt{2}(1 + i)$$

$$w_1 = 2\left[\cos\frac{3\pi}{4} + i\sin\frac{3\pi}{4}\right] = \sqrt{2}(-1 + i)$$

$$w_2 = 2\left[\cos\frac{5\pi}{4} + i\sin\frac{5\pi}{4}\right] = \sqrt{2}(-1 - i)$$

$$w_3 = 2\left[\cos\frac{7\pi}{4} + i\sin\frac{7\pi}{4}\right] = \sqrt{2}(1 - i). \qquad \blacksquare$$

The Fundamental Theorem of Algebra

One might say that the invention of $\sqrt{-1}$ is all well and good and leads to a number system that is richer than the real number system alone; but where will this process end? Are we also going to invent still more systems so as to obtain $\sqrt[4]{-1}$, $\sqrt[6]{-1}$, and so on? But it turns out this is not necessary. These numbers are already expressible in terms of the complex number system $a + ib$. In fact, the Fundamental Theorem of Algebra says that with the introduction of the complex numbers we now have enough numbers to factor every polynomial into a product of linear factors and so enough numbers to solve every possible polynomial equation.

The Fundamental Theorem of Algebra
Every polynomial equation of the form

$$a_n z^n + a_{n-1} z^{n-1} + \cdots + a_1 z + a_0 = 0,$$

in which the coefficients $a_0, a_1, \ldots, a_n$ are any complex numbers, whose degree n is greater than or equal to one, and whose leading coefficient a_n is not zero, has exactly n roots in the complex number system, provided each multiple root of multiplicity m is counted as m roots.

A proof of this theorem can be found in most texts on the theory of functions of a complex variable.

EXERCISES A.7

Operations with Complex Numbers

1. Find the following products of complex numbers
  a. $(2 + 3i)(4 - 2i)$ **b.** $(2 - i)(-2 - 3i)$
  c. $(-1 - 2i)(2 + i)$

2. Solve the following equations for the real numbers, x and y.
  a. $(3 + 4i)^2 - 2(x - iy) = x + iy$

  b. $\left(\dfrac{1+i}{1-i}\right)^2 + \dfrac{1}{x + iy} = 1 + i$

  c. $(3 - 2i)(x + iy) = 2(x - 2iy) + 2i - 1$

Graphing and Geometry

3. How may the following complex numbers be obtained from $z = x + iy$ geometrically? Sketch.
  a. $\bar{z}$ **b.** $\overline{(-z)}$
  c. $-z$ **d.** $1/z$

4. Show that the distance between the two points z_1 and z_2 in an Argand diagram is $|z_1 - z_2|$.

In Exercises 5–10, graph the points $z = x + iy$ that satisfy the given conditions.

5. a. $|z| = 2$ **b.** $|z| < 2$ **c.** $|z| > 2$
6. $|z - 1| = 2$ **7.** $|z + 1| = 1$
8. $|z + 1| = |z - 1|$ **9.** $|z + i| = |z - 1|$
10. $|z + 1| \geq |z|$

Express the complex numbers in Exercises 11–14 in the form $re^{i\theta}$, with $r \geq 0$ and $-\pi < \theta \leq \pi$. Draw an Argand diagram for each calculation.

11. $\left(1 + \sqrt{-3}\right)^2$ **12.** $\dfrac{1+i}{1-i}$

13. $\dfrac{1 + i\sqrt{3}}{1 - i\sqrt{3}}$ **14.** $(2 + 3i)(1 - 2i)$

Powers and Roots

Use De Moivre's Theorem to express the trigonometric functions in Exercises 15 and 16 in terms of $\cos\theta$ and $\sin\theta$.

15. $\cos 4\theta$ **16.** $\sin 4\theta$

17. Find the three cube roots of 1.

18. Find the two square roots of i.

19. Find the three cube roots of $-8i$.

20. Find the six sixth roots of 64.

21. Find the four solutions of the equation $z^4 - 2z^2 + 4 = 0$.

22. Find the six solutions of the equation $z^6 + 2z^3 + 2 = 0$.

23. Find all solutions of the equation $x^4 + 4x^2 + 16 = 0$.

24. Solve the equation $x^4 + 1 = 0$.

Theory and Examples

25. Complex numbers and vectors in the plane Show with an Argand diagram that the law for adding complex numbers is the same as the parallelogram law for adding vectors.

26. Complex arithmetic with conjugates Show that the conjugate of the sum (product, or quotient) of two complex numbers, z_1 and z_2, is the same as the sum (product, or quotient) of their conjugates.

27. Complex roots of polynomials with real coefficients come in complex-conjugate pairs
  a. Extend the results of Exercise 26 to show that $f(\bar{z}) = \overline{f(z)}$ when

$$f(z) = a_n z^n + a_{n-1} z^{n-1} + \cdots + a_1 z + a_0$$

  is a polynomial with real coefficients $a_0, \ldots, a_n$.

  b. If z is a root of the equation $f(z) = 0$, where $f(z)$ is a polynomial with real coefficients as in part (a), show that the conjugate $\bar{z}$ is also a root of the equation. (*Hint:* Let $f(z) = u + iv = 0$; then both u and v are zero. Use the fact that $f(\bar{z}) = \overline{f(z)} = u - iv$.)

28. **Absolute value of a conjugate**   Show that $|\bar{z}| = |z|$.

29. **When $\mathbf{z} = \bar{\mathbf{z}}$**   If z and $\bar{z}$ are equal, what can you say about the location of the point z in the complex plane?

30. **Real and imaginary parts**   Let Re(z) denote the real part of z and Im(z) the imaginary part. Show that the following relations hold for any complex numbers z, z_1, and z_2.

a. $z + \bar{z} = 2\text{Re}(z)$

b. $z - \bar{z} = 2i\text{Im}(z)$

c. $|\text{Re}(z)| \le |z|$

d. $|z_1 + z_2|^2 = |z_1|^2 + |z_2|^2 + 2\text{Re}(z_1\bar{z_2})$

e. $|z_1 + z_2| \le |z_1| + |z_2|$

A.8   The Distributive Law for Vector Cross Products

In this appendix we prove the Distributive Law

$$\mathbf{u} \times (\mathbf{v} + \mathbf{w}) = \mathbf{u} \times \mathbf{v} + \mathbf{u} \times \mathbf{w},$$

which is Property 2 in Section 12.4.

Proof   To derive the Distributive Law, we construct $\mathbf{u} \times \mathbf{v}$ a new way. We draw $\mathbf{u}$ and $\mathbf{v}$ from the common point O and construct a plane M perpendicular to $\mathbf{u}$ at O (Figure A.32). We then project $\mathbf{v}$ orthogonally onto M, yielding a vector $\mathbf{v}'$ with length $|\mathbf{v}|\sin\theta$. We rotate $\mathbf{v}'$ 90° about $\mathbf{u}$ in the positive sense to produce a vector $\mathbf{v}''$. Finally, we multiply $\mathbf{v}''$ by the length of $\mathbf{u}$. The resulting vector $|\mathbf{u}|\mathbf{v}''$ is equal to $\mathbf{u} \times \mathbf{v}$ since $\mathbf{v}''$ has the same direction as $\mathbf{u} \times \mathbf{v}$ by its construction (Figure A.32) and

$$|\mathbf{u}||\mathbf{v}''| = |\mathbf{u}||\mathbf{v}'| = |\mathbf{u}||\mathbf{v}|\sin\theta = |\mathbf{u} \times \mathbf{v}|.$$

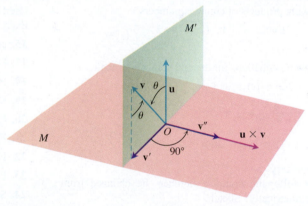

FIGURE A.32   As explained in the text, $\mathbf{u} \times \mathbf{v} = |\mathbf{u}|\mathbf{v}''$. (The primes used here are purely notational and do not denote derivatives.)

Now each of these three operations, namely,

1. projection onto M

2. rotation about $\mathbf{u}$ through 90°

3. multiplication by the scalar $|\mathbf{u}|$

when applied to a triangle whose plane is not parallel to $\mathbf{u}$, will produce another triangle. If we start with the triangle whose sides are $\mathbf{v}$, $\mathbf{w}$, and $\mathbf{v} + \mathbf{w}$ (Figure A.33) and apply these three steps, we successively obtain the following:

1. A triangle whose sides are $\mathbf{v}'$, $\mathbf{w}'$, and $(\mathbf{v} + \mathbf{w})'$ satisfying the vector equation

$$\mathbf{v}' + \mathbf{w}' = (\mathbf{v} + \mathbf{w})'$$

2. A triangle whose sides are $\mathbf{v}''$, $\mathbf{w}''$, and $(\mathbf{v} + \mathbf{w})''$ satisfying the vector equation

$$\mathbf{v}'' + \mathbf{w}'' = (\mathbf{v} + \mathbf{w})''$$

(The double prime on each vector has the same meaning as in Figure A.32.)

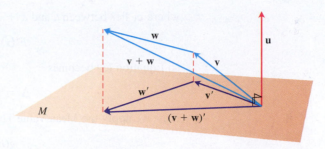

FIGURE A.33 The vectors, $\mathbf{v}$, $\mathbf{w}$, $\mathbf{v} + \mathbf{w}$, and their projections onto a plane perpendicular to $\mathbf{u}$.

3. A triangle whose sides are $|\mathbf{u}|\mathbf{v}''$, $|\mathbf{u}|\mathbf{w}''$, and $|\mathbf{u}|(\mathbf{v} + \mathbf{w})''$ satisfying the vector equation

$$|\mathbf{u}|\mathbf{v}'' + |\mathbf{u}|\mathbf{w}'' = |\mathbf{u}|(\mathbf{v} + \mathbf{w})''.$$

Substituting $|\mathbf{u}|\mathbf{v}'' = \mathbf{u} \times \mathbf{v}$, $|\mathbf{u}|\mathbf{w}'' = \mathbf{u} \times \mathbf{w}$, and $|\mathbf{u}|(\mathbf{v} + \mathbf{w})'' = \mathbf{u} \times (\mathbf{v} + \mathbf{w})$ from our discussion above into this last equation gives

$$\mathbf{u} \times \mathbf{v} + \mathbf{u} \times \mathbf{w} = \mathbf{u} \times (\mathbf{v} + \mathbf{w}),$$

which is the law we wanted to establish.   ∎

A.9   The Mixed Derivative Theorem and the Increment Theorem

This appendix derives the Mixed Derivative Theorem (Theorem 2, Section 14.3) and the Increment Theorem for Functions of Two Variables (Theorem 3, Section 14.3). Euler first published the Mixed Derivative Theorem in 1734, in a series of papers he wrote on hydrodynamics.

THEOREM 2—The Mixed Derivative Theorem

If $f(x, y)$ and its partial derivatives f_x, f_y, f_{xy}, and f_{yx} are defined throughout an open region containing a point (a, b) and are all continuous at (a, b), then

$$f_{xy}(a, b) = f_{yx}(a, b).$$

Proof   The equality of $f_{xy}(a, b)$ and $f_{yx}(a, b)$ can be established by four applications of the Mean Value Theorem (Theorem 4, Section 4.2). By hypothesis, the point (a, b) lies in the interior of a rectangle R in the xy-plane on which f, f_x, f_y, f_{xy}, and f_{yx} are all defined. We let h and k be the numbers such that the point $(a + h, b + k)$ also lies in R, and we consider the difference

$$\Delta = F(a + h) - F(a), \tag{1}$$

where

$$F(x) = f(x, b + k) - f(x, b). \tag{2}$$

We apply the Mean Value Theorem to F, which is continuous because it is differentiable. Then Equation (1) becomes

$$\Delta = hF'(c_1), \tag{3}$$

where c_1 lies between a and $a + h$. From Equation (2),

$$F'(x) = f_x(x, b + k) - f_x(x, b),$$

so Equation (3) becomes

$$\Delta = h[f_x(c_1, b + k) - f_x(c_1, b)]. \qquad (4)$$

Now we apply the Mean Value Theorem to the function $g(y) = f_x(c_1, y)$ and have

$$g(b + k) - g(b) = kg'(d_1),$$

or

$$f_x(c_1, b + k) - f_x(c_1, b) = kf_{xy}(c_1, d_1)$$

for some d_1 between b and $b + k$. By substituting this into Equation (4), we get

$$\Delta = hkf_{xy}(c_1, d_1) \qquad (5)$$

for some point (c_1, d_1) in the rectangle R' whose vertices are the four points (a, b), $(a + h, b)$, $(a + h, b + k)$, and $(a, b + k)$. (See Figure A.34.)

By substituting from Equation (2) into Equation (1), we may also write

$$
\begin{aligned}
\Delta &= f(a + h, b + k) - f(a + h, b) - f(a, b + k) + f(a, b) \\
&= [f(a + h, b + k) - f(a, b + k)] - [f(a + h, b) - f(a, b)] \\
&= \phi(b + k) - \phi(b),
\end{aligned}
\qquad (6)
$$

where

$$\phi(y) = f(a + h, y) - f(a, y). \qquad (7)$$

The Mean Value Theorem applied to Equation (6) now gives

$$\Delta = k\phi'(d_2) \qquad (8)$$

for some d_2 between b and $b + k$. By Equation (7),

$$\phi'(y) = f_y(a + h, y) - f_y(a, y). \qquad (9)$$

Substituting from Equation (9) into Equation (8) gives

$$\Delta = k[f_y(a + h, d_2) - f_y(a, d_2)].$$

Finally, we apply the Mean Value Theorem to the expression in brackets and get

$$\Delta = khf_{yx}(c_2, d_2) \qquad (10)$$

for some c_2 between a and $a + h$.

Together, Equations (5) and (10) show that

$$f_{xy}(c_1, d_1) = f_{yx}(c_2, d_2), \qquad (11)$$

where (c_1, d_1) and (c_2, d_2) both lie in the rectangle R' (Figure A.34). Equation (11) is not quite the result we want, since it says only that f_{xy} has the same value at (c_1, d_1) that f_{yx} has at (c_2, d_2). The numbers h and k in our discussion, however, may be made as small as we wish. The hypothesis that f_{xy} and f_{yx} are both continuous at (a, b) means that $f_{xy}(c_1, d_1) = f_{xy}(a, b) + \varepsilon_1$ and $f_{yx}(c_2, d_2) = f_{yx}(a, b) + \varepsilon_2$, where each of $\varepsilon_1, \varepsilon_2 \to 0$ as both $h, k \to 0$. Hence, if we let h and $k \to 0$, we have $f_{xy}(a, b) = f_{yx}(a, b)$. ∎

The equality of $f_{xy}(a, b)$ and $f_{yx}(a, b)$ can be proved with hypotheses weaker than the ones we assumed. For example, it is enough for f, f_x, and f_y to exist in R and for f_{xy} to be continuous at (a, b). Then f_{yx} will exist at (a, b) and equal f_{xy} at that point.

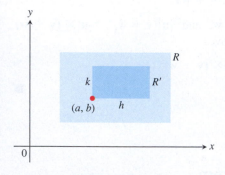

FIGURE A.34 The key to proving $f_{xy}(a, b) = f_{yx}(a, b)$ is that no matter how small R' is, f_{xy} and f_{yx} take on equal values somewhere inside R' (although not necessarily at the same point).

> **THEOREM 3—The Increment Theorem for Functions of Two Variables**
> Suppose that the first partial derivatives of $f(x, y)$ are defined throughout an open region R containing the point (x_0, y_0) and that f_x and f_y are continuous at (x_0, y_0). Then the change
>
> $$\Delta z = f(x_0 + \Delta x, y_0 + \Delta y) - f(x_0, y_0)$$
>
> in the value of f that results from moving from (x_0, y_0) to another point $(x_0 + \Delta x, y_0 + \Delta y)$ in R satisfies an equation of the form
>
> $$\Delta z = f_x(x_0, y_0)\,\Delta x + f_y(x_0, y_0)\,\Delta y + \varepsilon_1 \Delta x + \varepsilon_2 \Delta y$$
>
> in which each of $\varepsilon_1, \varepsilon_2 \to 0$ as both $\Delta x, \Delta y \to 0$.

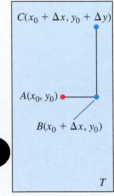

FIGURE A.35 The rectangular region T in the proof of the Increment Theorem. The figure is drawn for Δx and Δy positive, but either increment might be zero or negative.

Proof   We work within a rectangle T centered at $A(x_0, y_0)$ and lying within R, and we assume that Δx and Δy are already so small that the line segment joining A to $B(x_0 + \Delta x, y_0)$ and the line segment joining B to $C(x_0 + \Delta x, y_0 + \Delta y)$ lie in the interior of T (Figure A.35).

We may think of Δz as the sum $\Delta z = \Delta z_1 + \Delta z_2$ of two increments, where

$$\Delta z_1 = f(x_0 + \Delta x, y_0) - f(x_0, y_0)$$

is the change in the value of f from A to B and

$$\Delta z_2 = f(x_0 + \Delta x, y_0 + \Delta y) - f(x_0 + \Delta x, y_0)$$

is the change in the value of f from B to C (Figure A.36).

On the closed interval of x-values joining x_0 to $x_0 + \Delta x$, the function $F(x) = f(x, y_0)$ is a differentiable (and hence continuous) function of x, with derivative

$$F'(x) = f_x(x, y_0).$$

By the Mean Value Theorem (Theorem 4, Section 4.2), there is an x-value c between x_0 and $x_0 + \Delta x$ at which

$$F(x_0 + \Delta x) - F(x_0) = F'(c)\,\Delta x$$

or

$$f(x_0 + \Delta x, y_0) - f(x_0, y_0) = f_x(c, y_0)\,\Delta x$$

or

$$\Delta z_1 = f_x(c, y_0)\,\Delta x. \tag{12}$$

Similarly, $G(y) = f(x_0 + \Delta x, y)$ is a differentiable (and hence continuous) function of y on the closed y-interval joining y_0 and $y_0 + \Delta y$, with derivative

$$G'(y) = f_y(x_0 + \Delta x, y).$$

Hence, there is a y-value d between y_0 and $y_0 + \Delta y$ at which

$$G(y_0 + \Delta y) - G(y_0) = G'(d)\,\Delta y$$

or

$$f(x_0 + \Delta x, y_0 + \Delta y) - f(x_0 + \Delta x, y) = f_y(x_0 + \Delta x, d)\,\Delta y$$

or

$$\Delta z_2 = f_y(x_0 + \Delta x, d)\,\Delta y. \tag{13}$$

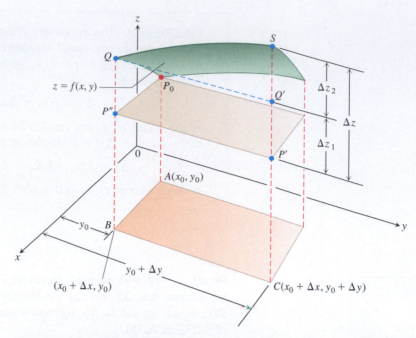

FIGURE A.36 Part of the surface $z = f(x, y)$ near $P_0(x_0, y_0, f(x_0, y_0))$. The points P_0, P', and P'' have the same height $z_0 = f(x_0, y_0)$ above the xy-plane. The change in z is $\Delta z = P'S$. The change

$$\Delta z_1 = f(x_0 + \Delta x, y_0) - f(x_0, y_0),$$

shown as $P''Q = P'Q'$, is caused by changing x from x_0 to $x_0 + \Delta x$ while holding y equal to y_0. Then, with x held equal to $x_0 + \Delta x$,

$$\Delta z_2 = f(x_0 + \Delta x, y_0 + \Delta y) - f(x_0 + \Delta x, y_0)$$

is the change in z caused by changing y_0 from $y_0 + \Delta y$, which is represented by $Q'S$. The total change in z is the sum of Δz_1 and Δz_2.

Now, as both Δx and $\Delta y \to 0$, we know that $c \to x_0$ and $d \to y_0$. Therefore, since f_x and f_y are continuous at (x_0, y_0), the quantities

$$\varepsilon_1 = f_x(c, y_0) - f_x(x_0, y_0),$$
$$\varepsilon_2 = f_y(x_0 + \Delta x, d) - f_y(x_0, y_0) \tag{14}$$

both approach zero as both Δx and $\Delta y \to 0$.

Finally,

$$\begin{aligned}
\Delta z &= \Delta z_1 + \Delta z_2 \\
&= f_x(c, y_0)\Delta x + f_y(x_0 + \Delta x, d)\Delta y && \text{From Eqs. (12) and (13)} \\
&= [f_x(x_0, y_0) + \varepsilon_1]\Delta x + [f_y(x_0, y_0) + \varepsilon_2]\Delta y && \text{From Eq. (14)} \\
&= f_x(x_0, y_0)\Delta x + f_y(x_0, y_0)\Delta y + \varepsilon_1\Delta x + \varepsilon_2\Delta y,
\end{aligned}$$

where both ε_1 and $\varepsilon_2 \to 0$ as both Δx and $\Delta y \to 0$, which is what we set out to prove. ∎

Analogous results hold for functions of any finite number of independent variables. Suppose that the first partial derivatives of $w = f(x, y, z)$ are defined throughout an open region containing the point (x_0, y_0, z_0) and that f_x, f_y, and f_z are continuous at (x_0, y_0, z_0). Then

$$\begin{aligned}
\Delta w &= f(x_0 + \Delta x, y_0 + \Delta y, z_0 + \Delta z) - f(x_0, y_0, z_0) \\
&= f_x\Delta x + f_y\Delta y + f_z\Delta z + \varepsilon_1\Delta x + \varepsilon_2\Delta y + \varepsilon_3\Delta z, \tag{15}
\end{aligned}$$

where $\varepsilon_1, \varepsilon_2, \varepsilon_3 \to 0$ as $\Delta x, \Delta y$, and $\Delta z \to 0$.

The partial derivatives f_x, f_y, f_z in Equation (15) are to be evaluated at the point (x_0, y_0, z_0).

Equation (15) can be proved by treating Δw as the sum of three increments,

$$\Delta w_1 = f(x_0 + \Delta x, y_0, z_0) - f(x_0, y_0, z_0) \tag{16}$$

$$\Delta w_2 = f(x_0 + \Delta x, y_0 + \Delta y, z_0) - f(x_0 + \Delta x, y_0, z_0) \tag{17}$$

$$\Delta w_3 = f(x_0 + \Delta x, y_0 + \Delta y, z_0 + \Delta z) - f(x_0 + \Delta x, y_0 + \Delta y, z_0), \tag{18}$$

and applying the Mean Value Theorem to each of these separately. Two coordinates remain constant and only one varies in each of these partial increments $\Delta w_1, \Delta w_2, \Delta w_3$. In Equation (17), for example, only y varies, since x is held equal to $x_0 + \Delta x$ and z is held equal to z_0. Since $f(x_0 + \Delta x, y, z_0)$ is a continuous function of y with a derivative f_y, it is subject to the Mean Value Theorem, and we have

$$\Delta w_2 = f_y(x_0 + \Delta x, y_1, z_0)\, \Delta y$$

for some y_1 between y_0 and $y_0 + \Delta y$.

Chapter 1

SECTION 1.1, pp. 11–13

1. $D: (-\infty, \infty)$, $R: [1, \infty)$ **3.** $D: [-2, \infty)$, $R: [0, \infty)$

5. $D: (-\infty, 3) \cup (3, \infty)$, $R: (-\infty, 0) \cup (0, \infty)$

7. **(a)** Not a function of x because some values of x have two values of y

(b) A function of x because for every x there is only one possible y

9. $A = \dfrac{\sqrt{3}}{4}x^2$, $p = 3x$

11. $x = \dfrac{d}{\sqrt{3}}$, $A = 2d^2$, $V = \dfrac{d^3}{3\sqrt{3}}$

13. $L = \dfrac{\sqrt{20x^2 - 20x + 25}}{4}$

15. $(-\infty, \infty)$ **17.** $(-\infty, \infty)$

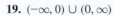

$f(x) = 5 - 2x$ $\qquad$ $g(x) = \sqrt{|x|}$

19. $(-\infty, 0) \cup (0, \infty)$

$F(t) = \dfrac{t}{|t|}$

21. $(-\infty, -5) \cup (-5, -3] \cup [3, 5) \cup (5, \infty)$

23. **(a)** For each positive value of x, there are two values of y. **(b)** For each value of $x \neq 0$, there are two values of y.

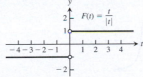

$|y| = x$ $\qquad$ $y^2 = x^2$

25.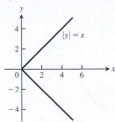

$f(x) = \begin{cases} x, & 0 \le x \le 1 \\ 2 - x, & 1 < x \le 2 \end{cases}$

27.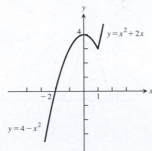

$y = x^2 + 2x$

$y = 4 - x^2$

29. **(a)** $f(x) = \begin{cases} x, & 0 \le x \le 1 \\ -x + 2, & 1 < x \le 2 \end{cases}$

(b) $f(x) = \begin{cases} 2, & 0 \le x < 1 \\ 0, & 1 \le x < 2 \\ 2, & 2 \le x < 3 \\ 0, & 3 \le x \le 4 \end{cases}$

31. **(a)** $f(x) = \begin{cases} -x, & -1 \le x < 0 \\ 1, & 0 < x \le 1 \\ -\frac{1}{2}x + \frac{3}{2}, & 1 < x < 3 \end{cases}$

(b) $f(x) = \begin{cases} \frac{1}{2}x, & -2 \le x \le 0 \\ -2x + 2, & 0 < x \le 1 \\ -1, & 1 < x \le 3 \end{cases}$

33. **(a)** $0 \le x < 1$ **(b)** $-1 < x \le 0$ **35.** Yes

37. Symmetric about the origin **39.** Symmetric about the origin

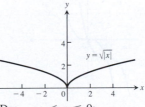

$y = -x^3$

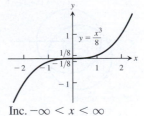

$y = -\dfrac{1}{x}$

Inc. $-\infty < x < 0$ and $0 < x < \infty$

Dec. $-\infty < x < \infty$

41. Symmetric about the y-axis **43.** Symmetric about the origin

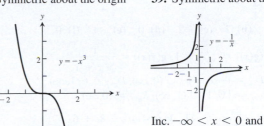

$y = \sqrt{|x|}$ $\qquad$ $y = \dfrac{x^3}{8}$

Dec. $-\infty < x \le 0$; Inc. $-\infty < x < \infty$
Inc. $0 \le x < \infty$

45. No symmetry

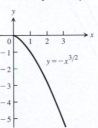

$y = -x^{3/2}$

Dec. $0 \le x < \infty$

47. Even **49.** Even **51.** Odd **53.** Even

55. Neither **57.** Neither **59.** Odd **61.** Even

63. $t = 180$ **65.** $s = 2.4$ **67.** $V = x(14 - 2x)(22 - 2x)$

69. **(a)** h **(b)** f **(c)** g **71.** **(a)** $(-2, 0) \cup (4, \infty)$

75. $C = 5(2 + \sqrt{2})h$

SECTION 1.2, pp. 18–21

1. $D_f : -\infty < x < \infty$, $D_g : x \geq 1$, $R_f : -\infty < y < \infty$,
$R_g : y \geq 0$, $D_{f+g} = D_{f\cdot g} = D_g$, $R_{f+g} : y \geq 1$, $R_{f\cdot g} : y \geq 0$

3. $D_f : -\infty < x < \infty$, $D_g : -\infty < x < \infty$, $R_f : y = 2$, $R_g : y \geq 1$,
$D_{f/g} : -\infty < x < \infty$, $R_{f/g} : 0 < y \leq 2$, $D_{g/f} : -\infty < x < \infty$,
$R_{g/f} : y \geq 1/2$

5. (a) 2 **(b)** 22 **(c)** $x^2 + 2$ **(d)** $x^2 + 10x + 22$ **(e)** 5
(f) -2 **(g)** $x + 10$ **(h)** $x^4 - 6x^2 + 6$

7. $13 - 3x$ **9.** $\sqrt{\dfrac{5x + 1}{4x + 1}}$

11. (a) $f(g(x))$ **(b)** $j(g(x))$ **(c)** $g(g(x))$ **(d)** $j(j(x))$
(e) $g(h(f(x)))$ **(f)** $h(j(f(x)))$

13.

	$g(x)$	$f(x)$	$(f \circ g)(x)$
(a)	$x - 7$	$\sqrt{x}$	$\sqrt{x - 7}$
(b)	$x + 2$	$3x$	$3x + 6$
(c)	x^2	$\sqrt{x - 5}$	$\sqrt{x^2 - 5}$
(d)	$\dfrac{x}{x - 1}$	$\dfrac{x}{x - 1}$	x
(e)	$\dfrac{1}{x - 1}$	$1 + \dfrac{1}{x}$	x
(f)	$\dfrac{1}{x}$	$\dfrac{1}{x}$	x

15. (a) 1 **(b)** 2 **(c)** -2 **(d)** 0 **(e)** -1 **(f)** 0

17. (a) $f(g(x)) = \sqrt{\dfrac{1}{x} + 1}$, $g(f(x)) = \dfrac{1}{\sqrt{x + 1}}$
(b) $D_{f\circ g} = (-\infty, -1] \cup (0, \infty)$, $D_{g\circ f} = (-1, \infty)$
(c) $R_{f\circ g} = [0, 1) \cup (1, \infty)$, $R_{g\circ f} = (0, \infty)$

19. $g(x) = \dfrac{2x}{x - 1}$ **21.** $V(t) = 4t^2 - 8t + 6$

23. (a) $y = -(x + 7)^2$ **(b)** $y = -(x - 4)^2$

25. (a) Position 4 **(b)** Position 1 **(c)** Position 2 **(d)** Position 3

27. $(x + 2)^2 + (y + 3)^2 = 49$ **29.** $y + 1 = (x + 1)^3$

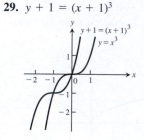

31. $y = \sqrt{x + 0.81}$ **33.** $y = 2x$

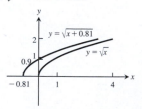

35. $y - 1 = \dfrac{1}{x - 1}$ **37.**

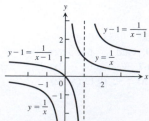

39.

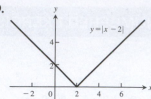

41.

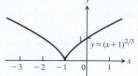

43.

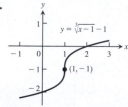

45.

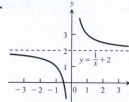

47.

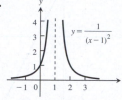

49.

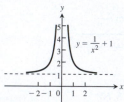

51.

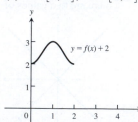

53.

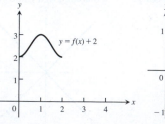

55.

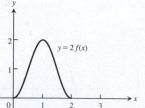

57. (a) $D : [0, 2]$, $R : [2, 3]$ **(b)** $D : [0, 2]$, $R : [-1, 0]$

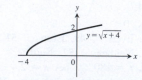

(c) $D : [0, 2]$, $R : [0, 2]$ **(d)** $D : [0, 2]$, $R : [-1, 0]$

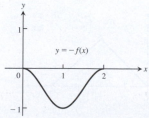

(e) $D: [-2, 0], \quad R: [0, 1]$ **(f)** $D: [1, 3], R: [0, 1]$

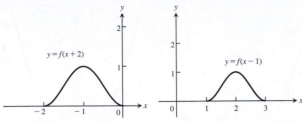

(g) $D: [-2, 0], \quad R: [0, 1]$ **(h)** $D: [-1, 1], \quad R: [0, 1]$

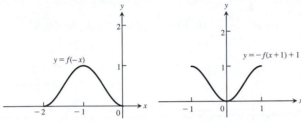

59. $y = 3x^2 - 3$ **61.** $y = \dfrac{1}{2} + \dfrac{1}{2x^2}$ **63.** $y = \sqrt{4x + 1}$

65. $y = \sqrt{4 - \dfrac{x^2}{4}}$ **67.** $y = 1 - 27x^3$

69.

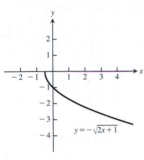

71.

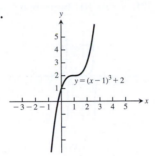

73.

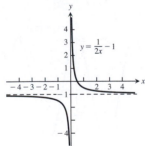

75.

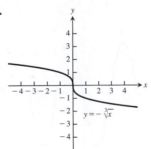

77.

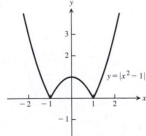

79. (a) Odd   **(b)** Odd   **(c)** Odd   **(d)** Even   **(e)** Even
(f) Even   **(g)** Even   **(h)** Even   **(i)** Odd

SECTION 1.3, pp. 27–29

1. (a) 8π m   **(b)** $\dfrac{55\pi}{9}$ m   **3.** 8.4 in.

5.

θ	$-\pi$	$-2\pi/3$	0	$\pi/2$	$3\pi/4$
$\sin\theta$	0	$-\dfrac{\sqrt{3}}{2}$	0	1	$\dfrac{1}{\sqrt{2}}$
$\cos\theta$	-1	$-\dfrac{1}{2}$	1	0	$-\dfrac{1}{\sqrt{2}}$
$\tan\theta$	0	$\sqrt{3}$	0	UND	-1
$\cot\theta$	UND	$\dfrac{1}{\sqrt{3}}$	UND	0	-1
$\sec\theta$	-1	-2	1	UND	$-\sqrt{2}$
$\csc\theta$	UND	$-\dfrac{2}{\sqrt{3}}$	UND	1	$\sqrt{2}$

7. $\cos x = -4/5, \tan x = -3/4$

9. $\sin x = -\dfrac{\sqrt{8}}{3}, \tan x = -\sqrt{8}$

11. $\sin x = -\dfrac{1}{\sqrt{5}}, \cos x = -\dfrac{2}{\sqrt{5}}$

13. Period π **15.** Period 2

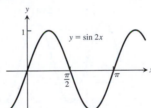

17. Period 6   **19.** Period 2π

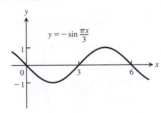

21. Period 2π **23.** Period $\pi/2$, symmetric about the origin

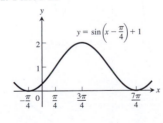

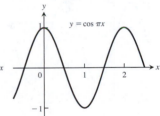

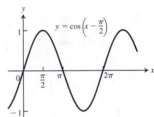

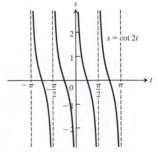

25. Period 4, symmetric about the y-axis

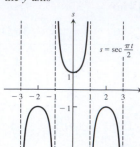

29. $D : (-\infty, \infty)$,
$R : y = -1, 0, 1$

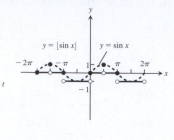

39. $-\cos x$ **41.** $-\cos x$ **43.** $\dfrac{\sqrt{6} + \sqrt{2}}{4}$ **45.** $\dfrac{\sqrt{2} + \sqrt{6}}{4}$

47. $\dfrac{2 + \sqrt{2}}{4}$ **49.** $\dfrac{2 - \sqrt{3}}{4}$ **51.** $\dfrac{\pi}{3}, \dfrac{2\pi}{3}, \dfrac{4\pi}{3}, \dfrac{5\pi}{3}$

53. $\dfrac{\pi}{6}, \dfrac{\pi}{2}, \dfrac{5\pi}{6}, \dfrac{3\pi}{2}$ **59.** $\sqrt{7} \approx 2.65$ **63.** $a = 1.464$

65. $r = \dfrac{\alpha \sin (\theta)}{1 - \sin (\theta)}$

67. $A = 2, B = 2\pi$,
$C = -\pi, D = -1$

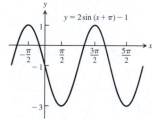

69. $A = -\dfrac{2}{\pi}, B = 4$,
$C = 0, D = \dfrac{1}{\pi}$

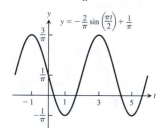

SECTION 1.4, P. 33
1. d   **3.** d
5. $[-3, 5]$ by $[-15, 40]$

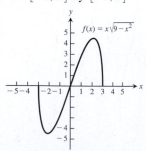

7. $[-3, 6]$ by $[-250, 50]$

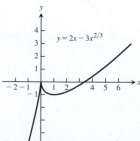

9. $[-5, 5]$ by $[-6, 6]$

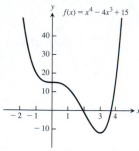

11. $[-2, 6]$ by $[-5, 4]$

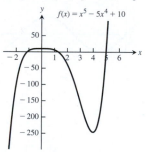

13. $[-2, 8]$ by $[-5, 10]$

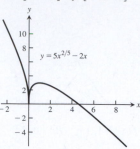

15. $[-3, 3]$ by $[0, 10]$

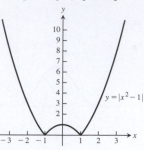

17. $[-10, 10]$ by $[-10, 10]$

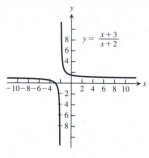

19. $[-4, 4]$ by $[0, 3]$

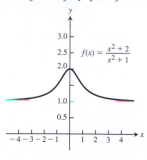

21. $[-10, 10]$ by $[-6, 6]$

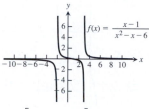

23. $[-6, 10]$ by $[-6, 6]$

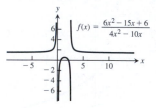

25. $\left[-\dfrac{\pi}{125}, \dfrac{\pi}{125} \right]$ by
$[-1.25, 1.25]$

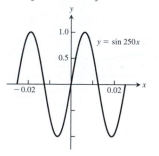

27. $[-100\pi, 100\pi]$ by
$[-1.25, 1.25]$

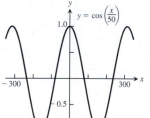

29. $\left[-\dfrac{\pi}{15}, \dfrac{\pi}{15} \right]$ by $[-0.25, 0.25]$

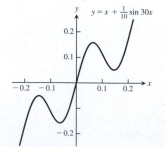

31.

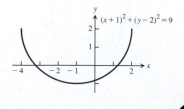

33.

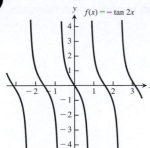

35.

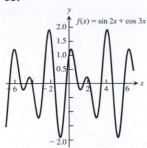

11. D: $(0, 1]$ R: $[0, \infty)$ **13.** D: $[-1, 1]$ R: $[-\pi/2, \pi/2]$

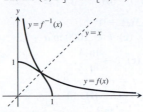

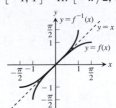

15. D: $[0, 6]$ R: $[0, 3]$ **17.** Symmetric about the line $y = x$

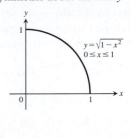

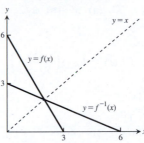

SECTION 1.5, pp. 37–38

1.

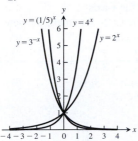

3.

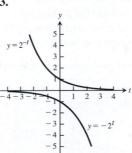

19. $f^{-1}(x) = \sqrt{x - 1}$ **21.** $f^{-1}(x) = \sqrt[3]{x + 1}$

23. $f^{-1}(x) = \sqrt{x} - 1$

25. $f^{-1}(x) = \sqrt[5]{x}$; D: $-\infty < x < \infty$; R: $-\infty < y < \infty$

27. $f^{-1}(x) = \sqrt[3]{x} - 1$; D: $-\infty < x < \infty$; R: $-\infty < y < \infty$

29. $f^{-1}(x) = \dfrac{1}{\sqrt{x}}$; D: $x > 0$; R: $y > 0$

31. $f^{-1}(x) = \dfrac{2x + 3}{x - 1}$; D: $-\infty < x < \infty, x \neq 1$;
R: $-\infty < y < \infty, y \neq 2$

33. $f^{-1}(x) = 1 - \sqrt{x + 1}$; D: $-1 \leq x < \infty$;
R: $-\infty < y \leq 1$

35. $f^{-1}(x) = \dfrac{2x + b}{x - 1}$;
D: $-\infty < x < \infty, x \neq 1$, R: $-\infty < y < \infty, y \neq 2$

5.

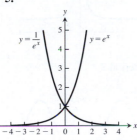

7.

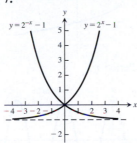

9.

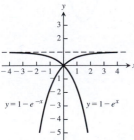

37. **(a)** $f^{-1}(x) = \dfrac{1}{m}x$
(b) The graph of f^{-1} is the line through the origin with slope $1/m$.

39. **(a)** $f^{-1}(x) = x - 1$

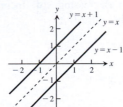

11. $16^{1/4} = 2$ **13.** $4^{1/2} = 2$ **15.** 5   **17.** $14^{\sqrt{3}}$ **19.** 4

21. D: $-\infty < x < \infty$; R: $0 < y < 1/2$

23. D: $-\infty < t < \infty$; R: $1 < y < \infty$

25. $x \approx 2.3219$ **27.** $x \approx -0.6309$ **29.** After 19 years

31. **(a)** $A(t) = 6.6\left(\dfrac{1}{2}\right)^{t/14}$ **(b)** About 38 days later

33. ≈ 11.433 years, or when interest is paid

35. $2^{48} \approx 2.815 \times 10^{14}$

(b) $f^{-1}(x) = x - b$. The graph of f^{-1} is a line parallel to the graph of f. The graphs of f and f^{-1} lie on opposite sides of the line $y = x$ and are equidistant from that line.

(c) Their graphs will be parallel to one another and lie on opposite sides of the line $y = x$ equidistant from that line.

41. **(a)** $\ln 3 - 2 \ln 2$ **(b)** $2(\ln 2 - \ln 3)$ **(c)** $-\ln 2$
(d) $\dfrac{2}{3}\ln 3$ **(e)** $\ln 3 + \dfrac{1}{2}\ln 2$ **(f)** $\dfrac{1}{2}(3 \ln 3 - \ln 2)$

43. **(a)** $\ln 5$ **(b)** $\ln(x - 3)$ **(c)** $\ln\left(\dfrac{2t^2}{b}\right)$

SECTION 1.6, pp. 48–50

1. One-to-one   **3.** Not one-to-one   **5.** One-to-one
7. Not one-to-one   **9.** One-to-one

45. (a) 7.2   (b) $\dfrac{1}{x^2}$ (c) $\dfrac{x}{y}$ **47.** (a) 1   (b) 1   (c) $-x^2 - y^2$

49. e^{2t+4} **51.** $e^{5t} + b$ **53.** $y = 2xe^x + 1$

55. (a) $k = \ln 2$ (b) $k = (1/10)\ln 2$ (c) $k = 1000 \ln a$

57. (a) $t = -10 \ln 3$ (b) $t = -\dfrac{\ln 2}{k}$ (c) $t = \dfrac{\ln .4}{\ln .2}$

59. $4(\ln |x|)^2$ **61.** $t = \ln 3$ **63.** $t = e^2/(e^2 - 1)$

65. (a) 7   (b) $\sqrt{2}$ (c) 75   (d) 2   (e) 0.5   (f) -1

67. (a) $\sqrt{x}$ (b) x^2 (c) $\sin x$ **69.** (a) $\dfrac{\ln 3}{\ln 2}$ (b) 3   (c) 2

71. (a) $-\pi/6$ (b) $\pi/4$ (c) $-\pi/3$ **73.** (a) π (b) $\pi/2$

75. Yes, $g(x)$ is also one-to-one.

77. Yes, $f \circ g$ is also one-to-one.

79. (a) $f^{-1}(x) = \log_2\left(\dfrac{x}{100 - x}\right)$ (b) $f^{-1}(x) = \log_{1.1}\left(\dfrac{x}{50 - x}\right)$

    (c) $f^{-1}(x) = \ln\left(\dfrac{x + 1}{1 - x}\right)$ (d) $f^{-1}(x) = e^{\frac{2x}{x+1}}$

81. (a) $y = \ln x - 3$ (b) $y = \ln(x - 1)$

    (c) $y = 3 + \ln(x + 1)$ (d) $y = \ln(x - 2) - 4$

    (e) $y = \ln(-x)$ (f) $y = e^x$

83. ≈ -0.7667

85. (a) Amount $= 8\left(\dfrac{1}{2}\right)^{t/12}$ (b) 36 hours

87. ≈ 44.081 years

PRACTICE EXERCISES, pp. 51–53

1. $A = \pi r^2, C = 2\pi r, A = \dfrac{C^2}{4\pi}$ **3.** $x = \tan \theta, y = \tan^2 \theta$

5. Origin   **7.** Neither   **9.** Even   **11.** Even

13. Odd   **15.** Neither

17. (a) Even   (b) Odd   (c) Odd   (d) Even   (e) Even

19. (a) Domain: all reals   (b) Range: $[-2, \infty)$

21. (a) Domain: $[-4, 4]$ (b) Range: $[0, 4]$

23. (a) Domain: all reals   (b) Range: $(-3, \infty)$

25. (a) Domain: all reals   (b) Range: $[-3, 1]$

27. (a) Domain: $(3, \infty)$ (b) Range: all reals

29. (a) Domain: $(-\infty, -1]$ and $[3, \infty)$ (b) Range: $(-\infty, 5]$

31. (a) Domain: $(-\infty, 0)$ and $(0, \infty)$ (b) Range: $[-4, 4]$

33. (a) Increasing   (b) Neither   (c) Decreasing   (d) Increasing

35. (a) Domain: $[-4, 4]$ (b) Range: $[0, 2]$

37. $f(x) = \begin{cases} 1 - x, & 0 \le x < 1 \\ 2 - x, & 1 \le x \le 2 \end{cases}$

39. (a) 1   (b) $\dfrac{1}{\sqrt{2.5}} = \sqrt{\dfrac{2}{5}}$ (c) $x, x \ne 0$

    (d) $\dfrac{1}{\sqrt{1/\sqrt{x + 2} + 2}}$

41. (a) $(f \circ g)(x) = -x, x \ge -2, (g \circ f)(x) = \sqrt{4 - x^2}$

    (b) Domain $(f \circ g)$: $[-2, \infty)$, domain $(g \circ f)$: $[-2, 2]$

    (c) Range $(f \circ g)$: $(-\infty, 2]$, range $(g \circ f)$: $[0, 2]$

43.

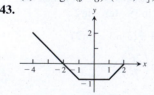
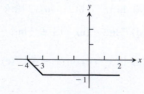

45. Replace the portion for $x < 0$ with the mirror image of the portion for $x > 0$ to make the new graph symmetric with respect to the y-axis.

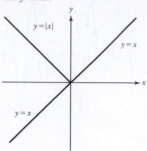

47. Reflects the portion for $y < 0$ across the x-axis

49. Reflects the portion for $y < 0$ across the x-axis

51. Adds the mirror image of the portion for $x > 0$ to make the new graph symmetric with respect to the y-axis

53. (a) $y = g(x - 3) + \dfrac{1}{2}$ (b) $y = g\left(x + \dfrac{2}{3}\right) - 2$

    (c) $y = g(-x)$ (d) $y = -g(x)$ (e) $y = 5g(x)$

    (f) $y = g(5x)$

55.   **57.**

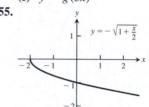
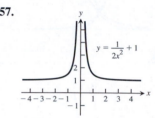

59. Period π **61.** Period 2

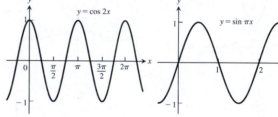

63.

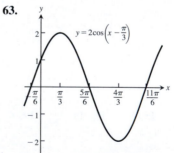

65. (a) $a = 1$ $b = \sqrt{3}$ (b) $a = 2\sqrt{3}/3$ $c = 4\sqrt{3}/3$

67. (a) $a = \dfrac{b}{\tan B}$ (b) $c = \dfrac{a}{\sin A}$

69. ≈ 16.98 m   **71.** (b) 4π

73. (a) Domain: $-\infty < x < \infty$ (b) Domain: $x > 0$

75. (a) Domain: $-3 \le x \le 3$ (b) Domain: $0 \le x \le 4$

77. $(f \circ g)(x) = \ln(4 - x^2)$ and domain: $-2 < x < 2$;
$(g \circ f)(x) = 4 - (\ln x)^2$ and domain: $x > 0$;
$(f \circ f)(x) = \ln(\ln x)$ and domain: $x > 1$;
$(g \circ g)(x) = -x^4 + 8x^2 - 12$ and domain: $-\infty < x < \infty$.

83. (a) D: $(-\infty, \infty)$ R: $\left[\dfrac{-\pi}{2}, \dfrac{\pi}{2}\right]$

(b) D: $[-1, 1]$ R: $[-1, 1]$

85. (a) No   **(b)** Yes

87. (a) $f(g(x)) = \left(\sqrt[3]{x}\right)^3 = x$, $g(f(x)) = \sqrt[3]{x^3} = x$

(b)

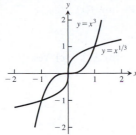

ADDITIONAL AND ADVANCED EXERCISES, pp. 53–55

1. Yes. For instance: $f(x) = 1/x$ and $g(x) = 1/x$, or $f(x) = 2x$ and $g(x) = x/2$, or $f(x) = e^x$ and $g(x) = \ln x$.

3. If $f(x)$ is odd, then $g(x) = f(x) - 2$ is not odd. Nor is $g(x)$ even, unless $f(x) = 0$ for all x. If f is even, then $g(x) = f(x) - 2$ is also even.

5.

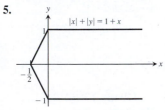

19. (a) Domain: all reals. Range: If $a > 0$, then (d, ∞); if $a < 0$, then $(-\infty, d)$.

(b) Domain: (c, ∞), range: all reals

21. (a) $y = 100{,}000 - 10{,}000x$, $0 \le x \le 10$ **(b)** After 4.5 years

23. After $\dfrac{\ln(10/3)}{\ln 1.08} \approx 15.6439$ years. (If the bank only pays interest at the end of the year, it will take 16 years.)

25. $x = 2, x = 1$ **27.** $1/2$

Chapter 2

SECTION 2.1, pp. 61–63

1. (a) 19   **(b)** 1

3. (a) $-\dfrac{4}{\pi}$ **(b)** $-\dfrac{3\sqrt{3}}{\pi}$ **5.** 1

7. (a) 4   **(b)** $y = 4x - 9$

9. (a) 2   **(b)** $y = 2x - 7$

11. (a) 12   **(b)** $y = 12x - 16$

13. (a) -9 **(b)** $y = -9x - 2$

15. (a) $-1/4$ **(b)** $y = -x/4 - 1$

17. (a) $1/4$ **(b)** $y = x/4 + 1$

19. Your estimates may not completely agree with these.

(a)

PQ_1	PQ_2	PQ_3	PQ_4
43	46	49	50

The appropriate units are m/sec.

(b) ≈ 50 m/sec or 180 km/h

21. (a)

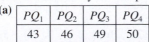

(b) $\approx \$56{,}000$/year

(c) $\approx \$42{,}000$/year

23. (a) $0.414213, 0.449489, \left(\sqrt{1 + h} - 1\right)/h$ **(b)** $g(x) = \sqrt{x}$

$1 + h$	1.1	1.01	1.001	1.0001
$\sqrt{1 + h}$	1.04880	1.004987	1.0004998	1.0000499
$\left(\sqrt{1 + h} - 1\right)/h$	0.4880	0.4987	0.4998	0.499

1.00001	1.000001
1.000005	1.0000005
0.5	0.5

(c) 0.5   **(d)** 0.5

25. (a) 15 mph, 3.3 mph, 10 mph   **(b)** 10 mph, 0 mph, 4 mph

(c) 20 mph when $t = 3.5$ hr

SECTION 2.2, pp. 71–74

1. (a) Does not exist. As x approaches 1 from the right, $g(x)$ approaches 0. As x approaches 1 from the left, $g(x)$ approaches 1. There is no single number L that all the values $g(x)$ get arbitrarily close to as $x \to 1$.

(b) 1   **(c)** 0   **(d)** 1/2

3. (a) True   **(b)** True   **(c)** False   **(d)** False   **(e)** False
(f) True   **(g)** True   **(h)** False   **(i)** True   **(j)** True   **(k)** False

5. As x approaches 0 from the left, $x/|x|$ approaches -1. As x approaches 0 from the right, $x/|x|$ approaches 1. There is no single number L that the function values all get arbitrarily close to as $x \to 0$.

7. Nothing can be said.   **9.** No; no; no   **11.** -4 **13.** -8

15. 3   **17.** $-25/2$ **19.** 16   **21.** 3/2   **23.** 1/10

25. -7 **27.** 3/2   **29.** $-1/2$ **31.** -1 **33.** 4/3

35. 1/6   **37.** 4   **39.** 1/2   **41.** 3/2   **43.** -1

45. 1   **47.** 1/3   **49.** $\sqrt{4 - \pi}$

51. (a) Quotient Rule   **(b)** Difference and Power Rules
(c) Sum and Constant Multiple Rules

53. (a) -10 **(b)** -20 **(c)** -1 **(d)** 5/7

55. (a) 4   **(b)** -21 **(c)** -12 **(d)** $-7/3$

57. 2   **59.** 3   **61.** $1/(2\sqrt{7})$ **63.** $\sqrt{5}$

65. (a) The limit is 1.

67. (a) $f(x) = (x^2 - 9)/(x + 3)$

x	-3.1	-3.01	-3.001	-3.0001	-3.00001	-3.000001
$f(x)$	-6.1	-6.01	-6.001	-6.0001	-6.00001	-6.000001

x	-2.9	-2.99	-2.999	-2.9999	-2.99999	-2.999999
$f(x)$	-5.9	-5.99	-5.999	-5.9999	-5.99999	-5.999999

(c) $\lim\limits_{x \to -3} f(x) = -6$

69. (a) $G(x) = (x + 6)/(x^2 + 4x - 12)$

x	-5.9	-5.99	-5.999	-5.9999
$G(x)$	$-.126582$	$-.1251564$	$-.1250156$	$-.1250015$

-5.99999	-5.999999
$-.1250001$	$-.1250000$

x	-6.1	-6.01	-6.001	-6.0001
$G(x)$	$-.123456$	$-.124843$	$-.124984$	$-.124998$

-6.00001	-6.000001
$-.124999$	$-.124999$

(c) $\lim\limits_{x \to -6} G(x) = -1/8 = -0.125$

71. (a) $f(x) = (x^2 - 1)/(|x| - 1)$

x	-1.1	-1.01	-1.001	-1.0001	-1.00001	-1.000001
$f(x)$	2.1	2.01	2.001	2.0001	2.00001	2.000001

x	$-.9$	$-.99$	$-.999$	$-.9999$	$-.99999$	$-.999999$
$f(x)$	1.9	1.99	1.999	1.9999	1.99999	1.999999

(c) $\lim\limits_{x \to -1} f(x) = 2$

73. (a) $g(\theta) = (\sin \theta)/\theta$

θ	$.1$	$.01$	$.001$	$.0001$	$.00001$	$.000001$
$g(\theta)$	$.998334$	$.999983$	$.999999$	$.999999$	$.999999$	$.999999$

θ	$-.1$	$-.01$	$-.001$	$-.0001$	$-.00001$	$-.000001$
$g(\theta)$	$.998334$	$.999983$	$.999999$	$.999999$	$.999999$	$.999999$

$\lim\limits_{\theta \to 0} g(\theta) = 1$

75. (a) $f(x) = x^{1/(1-x)}$

x	$.9$	$.99$	$.999$	$.9999$	$.99999$	$.999999$
$f(x)$	$.348678$	$.366032$	$.367695$	$.367861$	$.367877$	$.367879$

x	1.1	1.01	1.001	1.0001	1.00001	1.000001
$f(x)$	$.385543$	$.369711$	$.368063$	$.367897$	$.367881$	$.367878$

$\lim\limits_{x \to 1} f(x) \approx 0.36788$

77. $c = 0, 1, -1$; the limit is 0 at $c = 0$, and 1 at $c = 1, -1$.
79. 7   **81. (a)** 5 **(b)** 5   **83. (a)** 0 **(b)** 0

SECTION 2.3, pp. 79–82

1. $\delta = 2$

3. $\delta = 1/2$

5. $\delta = 1/18$

7. $\delta = 0.1$ **9.** $\delta = 7/16$ **11.** $\delta = \sqrt{5} - 2$
13. $\delta = 0.36$ **15.** $(3.99, 4.01)$, $\delta = 0.01$
17. $(-0.19, 0.21)$, $\delta = 0.19$ **19.** $(3, 15)$, $\delta = 5$
21. $(10/3, 5)$, $\delta = 2/3$
23. $(-\sqrt{4.5}, -\sqrt{3.5})$, $\delta = \sqrt{4.5} - 2 \approx 0.12$
25. $(\sqrt{15}, \sqrt{17})$, $\delta = \sqrt{17} - 4 \approx 0.12$
27. $\left(2 - \dfrac{0.03}{m}, 2 + \dfrac{0.03}{m}\right)$, $\delta = \dfrac{0.03}{m}$
29. $\left(\dfrac{1}{2} - \dfrac{c}{m}, \dfrac{c}{m} + \dfrac{1}{2}\right)$, $\delta = \dfrac{c}{m}$

31. $L = -3$, $\delta = 0.01$ **33.** $L = 4$, $\delta = 0.05$
35. $L = 4$, $\delta = 0.75$
55. $[3.384, 3.387]$. To be safe, the left endpoint was rounded up and the right endpoint rounded down.
59. The limit does not exist as x approaches 3.

SECTION 2.4, pp. 88–90

1. (a) True   **(b)** True   **(c)** False   **(d)** True
(e) True   **(f)** True   **(g)** False   **(h)** False
(i) False   **(j)** False   **(k)** True   **(l)** False
3. (a) 2, 1   **(b)** No, $\lim\limits_{x \to 2^+} f(x) \neq \lim\limits_{x \to 2^-} f(x)$
(c) 3, 3   **(d)** Yes, 3
5. (a) No   **(b)** Yes, 0   **(c)** No
7. (a)   **(b)** 1, 1   **(c)** Yes, 1

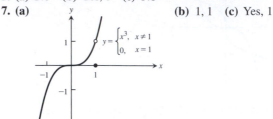

9. (a) $D : 0 \le x \le 2, R : 0 < y \le 1$ and $y = 2$
(b) $[0, 1) \cup (1, 2]$ **(c)** $x = 2$ **(d)** $x = 0$

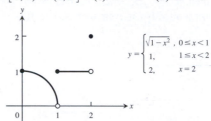

11. $\sqrt{3}$ **13.** 1   **15.** $2/\sqrt{5}$ **17. (a)** 1   **(b)** -1
19. (a) 1   **(b)** -1 **21. (a)** 1   **(b)** $2/3$ **23.** 1   **25.** $3/4$
27. 2   **29.** $1/2$ **31.** 2   **33.** 0   **35.** 1   **37.** $1/2$
39. 0   **41.** $3/8$ **43.** 3   **45.** 0
51. $\delta = \varepsilon^2$, $\lim\limits_{x \to 5^+} \sqrt{x - 5} = 0$
55. (a) 400   **(b)** 399   **(c)** The limit does not exist.

SECTION 2.5, pp. 100–102

1. No; not defined at $x = 2$
3. Continuous   **5. (a)** Yes   **(b)** Yes   **(c)** Yes   **(d)** Yes
7. (a) No   **(b)** No   **9.** 0
11. 1, nonremovable; 0, removable   **13.** All x except $x = 2$
15. All x except $x = 3, x = 1$ **17.** All x
19. All x except $x = 0$ **21.** All x except $n\pi/2$, n any integer
23. All x except $n\pi/2$, n an odd integer   **25.** All $x \ge -3/2$
27. All x **29.** All x **31.** All x except $x = 1$
33. 0; continuous at $x = \pi$ **35.** 1; continuous at $y = 1$
37. $\sqrt{2}/2$; continuous at $t = 0$ **39.** 1; continuous at $x = 0$
41. $g(3) = 6$ **43.** $f(1) = 3/2$ **45.** $a = 4/3$
47. $a = -2, 3$ **49.** $a = 5/2, b = -1/2$
73. $x \approx 1.8794, -1.5321, -0.3473$ **75.** $x \approx 1.7549$
77. $x \approx 3.5156$ **79.** $x \approx 0.7391$

SECTION 2.6, pp. 112–115

1. (a) 0   **(b)** -2 **(c)** 2   **(d)** Does not exist   **(e)** -1
(f) ∞ **(g)** Does not exist   **(h)** 1   **(i)** 0
3. (a) -3 **(b)** -3 **5. (a)** $1/2$ **(b)** $1/2$ **7. (a)** $-5/3$
(b) $-5/3$ **9.** 0   **11.** -1 **13. (a)** $2/5$ **(b)** $2/5$

15. (a) 0 (b) 0   **17.** (a) 7 (b) 7   **19.** (a) 0 (b) 0
21. (a) ∞ (b) ∞ **23.** 2   **25.** ∞ **27.** 0   **29.** 1
31. ∞ **33.** 1   **35.** 1/2   **37.** ∞ **39.** $-\infty$
41. $-\infty$ **43.** ∞ **45.** (a) ∞ (b) $-\infty$ **47.** ∞
49. ∞ **51.** $-\infty$ **53.** (a) ∞ (b) $-\infty$ (c) $-\infty$ (d) ∞
55. (a) $-\infty$ (b) ∞ (c) 0 (d) 3/2
57. (a) $-\infty$ (b) 1/4 (c) 1/4 (d) 1/4 (e) It will be $-\infty$.
59. (a) $-\infty$ (b) ∞ **61.** (a) ∞ (b) ∞ (c) ∞ (d) ∞
63.

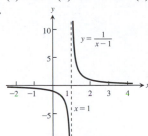

65.

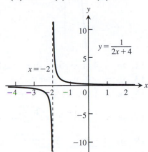

67.

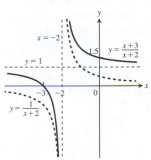

69. Domain: $(-\infty, \infty)$, Range: $[4, 7)$
71. Domain: $(-\infty, \infty)$, Range: $(-1, 4)$
73. Domain: $(-\infty, 0)$ and $(0, \infty)$, Range: $(-\infty, -1)$ and $(1, \infty)$
75. Here is one possibility.   **77.** Here is one possibility.

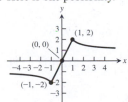

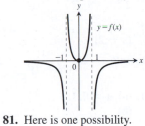

79. Here is one possibility.   **81.** Here is one possibility.

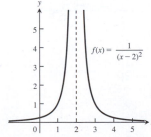

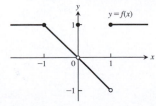

85. At most one   **87.** 0   **89.** $-3/4$ **91.** 5/2
99. (a) For every positive real number B there exists a corresponding number $\delta > 0$ such that for all x
$$c - \delta < x < c \implies f(x) > B.$$
(b) For every negative real number $-B$ there exists a corresponding number $\delta > 0$ such that for all x
$$c < x < c + \delta \implies f(x) < -B.$$

(c) For every negative real number $-B$ there exists a corresponding number $\delta > 0$ such that for all x
$$c - \delta < x < c \implies f(x) < -B.$$

105.

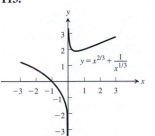

107.

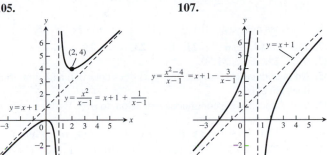

109.

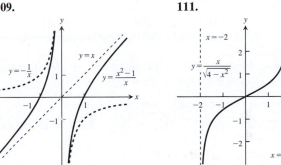

111.

113.

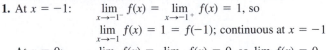

115. At ∞: ∞, at $-\infty$: 0

PRACTICE EXERCISES, pp. 116–117

1. At $x = -1$: $\lim\limits_{x \to -1^-} f(x) = \lim\limits_{x \to -1^+} f(x) = 1$, so
$\lim\limits_{x \to -1} f(x) = 1 = f(-1)$; continuous at $x = -1$

At $x = 0$: $\lim\limits_{x \to 0^-} f(x) = \lim\limits_{x \to 0^+} f(x) = 0$, so $\lim\limits_{x \to 0} f(x) = 0$.
However, $f(0) \neq 0$, so f is discontinuous at $x = 0$. The discontinuity can be removed by redefining $f(0)$ to be 0.

At $x = 1$: $\lim\limits_{x \to 1^-} f(x) = -1$ and $\lim\limits_{x \to 1^+} f(x) = 1$, so $\lim\limits_{x \to 1} f(x)$ does not exist. The function is discontinuous at $x = 1$, and the discontinuity is not removable.

3. (a) -21 **(b)** 49 **(c)** 0 **(d)** 1 **(e)** 1 **(f)** 7

(g) -7 **(h)** $-\dfrac{1}{7}$ **5.** 4

7. (a) $(-\infty, +\infty)$ **(b)** $[0, \infty)$ **(c)** $(-\infty, 0)$ and $(0, \infty)$
(d) $[0, \infty)$

9. (a) Does not exist **(b)** 0 **11.** $\dfrac{1}{2}$ **13.** $2x$ **15.** $-\dfrac{1}{4}$

17. $2/3$ **19.** $2/\pi$ **21.** 1 **23.** 4 **25.** $-\infty$

27. 0 **29.** 2 **31.** 0

35. No in both cases, because $\lim\limits_{x \to 1} f(x)$ does not exist, and $\lim\limits_{x \to -1} f(x)$ does not exist.

37. Yes, f does have a continuous extension, to $a = 1$ with $f(1) = 4/3$.

39. No   **41.** $2/5$ **43.** 0 **45.** $-\infty$ **47.** 0 **49.** 1

51. 1 **53.** $-\pi/2$ **55. (a)** $x = 3$ **(b)** $x = 1$ **(c)** $x = -4$

57. Domain: $[-4, 2)$ and $(2, 4]$, Range: $(-\infty, \infty)$

ADDITIONAL AND ADVANCED EXERCISES, pp. 118–120

3. 0; the left-hand limit was taken because the function is undefined for $v > c$.

5. $65 < t < 75$; within $5°F$ **13. (a)** B **(b)** A **(c)** A **(d)** A

21. (a) $\lim\limits_{a \to 0} r_+(a) = 0.5, \ \lim\limits_{a \to -1^+} r_+(a) = 1$

(b) $\lim\limits_{a \to 0} r_-(a)$ does not exist, $\lim\limits_{a \to -1^+} r_-(a) = 1$

25. 0 **27.** 1 **29.** 4 **31.** $y = 2x$ **33.** $y = x, y = -x$

37. $-4/3$

39. (a) Domain: $\{0, 1, 1/2, 1/3, 1/4, \dots\}$
(b) The domain intersects any interval (a, b) containing 0.
(c) 0

41. (a) Domain: $(-\infty, -1/\pi] \cup [-1/(2\pi),$
$-1/(3\pi)] \cup [-1/(4\pi), -1/(5\pi)] \cup \cdots \cup [1/(5\pi),$
$1/(4\pi)] \cup [1/(3\pi), 1/(2\pi)] \cup [1/\pi, \infty)$
(b) The domain intersects any interval (a, b) containing 0.
(c) 0

Chapter 3

SECTION 3.1, pp. 123–125

1. $P_1: m_1 = 1, P_2: m_2 = 5$ **3.** $P_1: m_1 = 5/2, P_2: m_2 = -1/2$

5. $y = 2x + 5$ **7.** $y = x + 1$

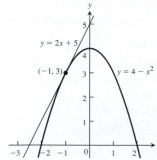

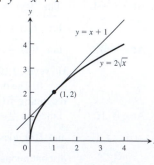

9. $y = 12x + 16$

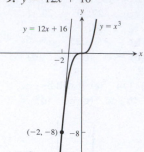

11. $m = 4, \ y - 5 = 4(x - 2)$

13. $m = -2, \ y - 3 = -2(x - 3)$

15. $m = 12, \ y - 8 = 12(t - 2)$

17. $m = \dfrac{1}{4}, \ y - 2 = \dfrac{1}{4}(x - 4)$

19. $m = -1$ **21.** $m = -1/4$

23. (a) It is the rate of change of the number of cells when $t = 5$. The units are the number of cells per hour.
(b) $P'(3)$ because the slope of the curve is greater there.
(c) $51.72 \approx 52$ cells/h

25. $(-2, -5)$ **27.** $y = -(x + 1), y = -(x - 3)$

29. 19.6 m/sec **31.** 6π **35.** Yes   **37.** No

39. (a) Nowhere   **41. (a)** At $x = 0$ **43. (a)** Nowhere

45. (a) At $x = 1$ **47. (a)** At $x = 0$

SECTION 3.2, pp. 130–134

1. $-2x, 6, 0, -2$ **3.** $-\dfrac{2}{t^3}, 2, -\dfrac{1}{4}, -\dfrac{2}{3\sqrt{3}}$

5. $\dfrac{3}{2\sqrt{3\theta}}, \dfrac{3}{2\sqrt{3}}, \dfrac{1}{2}, \dfrac{3}{2\sqrt{2}}$ **7.** $6x^2$ **9.** $\dfrac{1}{(2t + 1)^2}$

11. $\dfrac{3}{2}q^{1/2}$ **13.** $1 - \dfrac{9}{x^2}, 0$ **15.** $3t^2 - 2t, 5$

17. $\dfrac{-4}{(x - 2)\sqrt{x - 2}}, \ y - 4 = -\dfrac{1}{2}(x - 6)$ **19.** 6

21. $1/8$ **23.** $\dfrac{-1}{(x + 2)^2}$ **25.** $\dfrac{-1}{(x - 1)^2}$ **27. (b)**   **29. (d)**

31. (a) $x = 0, 1, 4$
(b)   **33.**

35. (a) i) $1.5 \ °F/hr$ **ii)** $2.9 \ °F/hr$
iii) $0 \ °F/hr$ **iv)** $-3.7 \ °F/hr$

(b) 7.3 °F/hr at 12 P.M., -11 °F/hr at 6 P.M.

(c)

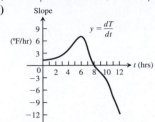

37. Since $\lim\limits_{h \to 0^+} \dfrac{f(0 + h) - f(0)}{h} = 1$

while $\lim\limits_{h \to 0^-} \dfrac{f(0 + h) - f(0)}{h} = 0,$

$f'(0) = \lim\limits_{h \to 0} \dfrac{f(0 + h) - f(0)}{h}$ does not exist and $f(x)$ is not

differentiable at $x = 0$.

39. Since $\lim\limits_{h \to 0^+} \dfrac{f(1 + h) - f(1)}{h} = 2$ while

$\lim\limits_{h \to 0^-} \dfrac{f(1 + h) - f(1)}{h} = \dfrac{1}{2},$ $f'(1) = \lim\limits_{h \to 0} \dfrac{f(1 + h) - f(1)}{h}$

does not exist and $f(x)$ is not differentiable at $x = 1$.

41. Since $f(x)$ is not continuous at $x = 0$, $f(x)$ is not differentiable at $x = 0$.

43. Since $\lim\limits_{h \to 0^+} \dfrac{f(0 + h) - f(0)}{h} = 3$ while

$\lim\limits_{h \to 0^-} \dfrac{f(0 + h) - f(0)}{h} = 0,$ f is not differentiable at $x = 0$.

45. (a) $-3 \le x \le 2$ **(b)** None   **(c)** None

47. (a) $-3 \le x < 0, 0 < x \le 3$ **(b)** None   **(c)** $x = 0$

49. (a) $-1 \le x < 0, 0 < x \le 2$ **(b)** $x = 0$ **(c)** None

SECTION 3.3, pp. 142–144

1. $\dfrac{dy}{dx} = -2x, \dfrac{d^2y}{dx^2} = -2$

3. $\dfrac{ds}{dt} = 15t^2 - 15t^4, \dfrac{d^2s}{dt^2} = 30t - 60t^3$

5. $\dfrac{dy}{dx} = 4x^2 - 1 + 2e^x, \dfrac{d^2y}{dx^2} = 8x + 2e^x$

7. $\dfrac{dw}{dz} = -\dfrac{6}{z^3} + \dfrac{1}{z^2}, \dfrac{d^2w}{dz^2} = \dfrac{18}{z^4} - \dfrac{2}{z^3}$

9. $\dfrac{dy}{dx} = 12x - 10 + 10x^{-3}, \dfrac{d^2y}{dx^2} = 12 - 30x^{-4}$

11. $\dfrac{dr}{ds} = \dfrac{-2}{3s^3} + \dfrac{5}{2s^2}, \dfrac{d^2r}{ds^2} = \dfrac{2}{s^4} - \dfrac{5}{s^3}$

13. $y' = -5x^4 + 12x^2 - 2x - 3$

15. $y' = 3x^2 + 10x + 2 - \dfrac{1}{x^2}$ **17.** $y' = \dfrac{-19}{(3x - 2)^2}$

19. $g'(x) = \dfrac{x^2 + x + 4}{(x + 0.5)^2}$ **21.** $\dfrac{dv}{dt} = \dfrac{t^2 - 2t - 1}{(1 + t^2)^2}$

23. $f'(s) = \dfrac{1}{\sqrt{s}(\sqrt{s} + 1)^2}$ **25.** $v' = -\dfrac{1}{x^2} + 2x^{-3/2}$

27. $y' = \dfrac{-4x^3 - 3x^2 + 1}{(x^2 - 1)^2(x^2 + x + 1)^2}$ **29.** $y' = -2e^{-x} + 3e^{3x}$

31. $y' = 3x^2e^x + x^3e^x$ **33.** $y' = \dfrac{9}{4}x^{5/4} - 2e^{-2x}$

35. $\dfrac{ds}{dt} = 3t^{1/2}$ **37.** $y' = \dfrac{2}{7x^{5/7}} - exe^{-1}$ **39.** $\dfrac{dr}{ds} = \dfrac{se^s - e^s}{s^2}$

41. $y' = 2x^3 - 3x - 1, y'' = 6x^2 - 3, y''' = 12x, y^{(4)} = 12,$
$y^{(n)} = 0$ for $n \ge 5$

43. $y' = 3x^2 + 8x + 1, y'' = 6x + 8, y''' = 6, y^{(n)} = 0$ for $n \ge 4$

45. $y' = 2x - 7x^{-2}, y'' = 2 + 14x^{-3}$

47. $\dfrac{dr}{d\theta} = 3\theta^{-4}, \dfrac{d^2r}{d\theta^2} = -12\theta^{-5}$ **49.** $\dfrac{dw}{dz} = -z^{-2} - 1, \dfrac{d^2w}{dz^2} = 2z^{-3}$

51. $\dfrac{dw}{dz} = 6ze^{2z}(1 + z), \dfrac{d^2w}{dz^2} = 6e^{2z}(1 + 4z + 2z^2)$

53. (a) 13   **(b)** -7 **(c)** 7/25   **(d)** 20

55. (a) $y = -\dfrac{x}{8} + \dfrac{5}{4}$ **(b)** $m = -4$ at $(0, 1)$

    (c) $y = 8x - 15, y = 8x + 17$

57. $y = 4x, y = 2$ **59.** $a = 1, b = 1, c = 0$

61. $(2, 4)$ **63.** $(0, 0), (4, 2)$ **65.** $y = -16x + 24$

67. (a) $y = 2x + 2$ **(c)** $(2, 6)$ **69.** 50   **71.** $a = -3$

73. $P'(x) = na_nx^{n-1} + (n - 1)a_{n-1}x^{n-2} + \cdots + 2a_2x + a_1$

75. The Product Rule is then the Constant Multiple Rule, so the latter is a special case of the Product Rule.

77. (a) $\dfrac{d}{dx}(uvw) = uvw' + uv'w + u'vw$

(b) $\dfrac{d}{dx}(u_1u_2u_3u_4) = u_1u_2u_3u_4' + u_1u_2u_3'u_4 + u_1u_2'u_3u_4 + $
    $u_1'u_2u_3u_4$

(c) $\dfrac{d}{dx}(u_1 \cdots u_n) = u_1u_2 \cdots u_{n-1}u_n' + u_1u_2 \cdots u_{n-2}u_{n-1}'u_n + $
    $\cdots + u_1'u_2 \cdots u_n$

79. $\dfrac{dP}{dV} = -\dfrac{nRT}{(V - nb)^2} + \dfrac{2an^2}{V^3}$

SECTION 3.4, pp. 150–154

1. (a) -2 m, -1 m/sec
   (b) 3 m/sec, 1 m/sec; 2 m/sec^2, 2 m/sec^2
   (c) Changes direction at $t = 3/2$ sec

3. (a) -9 m, -3 m/sec
   (b) 3 m/sec, 12 m/sec; 6 m/sec^2, -12 m/sec^2
   (c) No change in direction

5. (a) -20 m, -5 m/sec
   (b) 45 m/sec, $(1/5)$ m/sec; 140 m/sec^2, $(4/25)$ m/sec^2
   (c) No change in direction

7. (a) $a(1) = -6$ m/sec^2, $a(3) = 6$ m/sec^2
   (b) $v(2) = 3$ m/sec   **(c)** 6 m

9. Mars: ≈ 7.5 sec, Jupiter: ≈ 1.2 sec

11. $g_s = 0.75$ m/sec^2

13. (a) $v = -32t, |v| = 32t$ ft/sec, $a = -32$ ft/sec^2
   (b) $t \approx 3.3$ sec
   (c) $v \approx -107.0$ ft/sec

15. (a) $t = 2, t = 7$ **(b)** $3 \le t \le 6$
   (c)   **(d)**

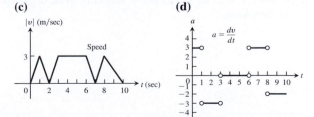

17. (a) 190 ft/sec   **(b)** 2 sec   **(c)** 8 sec, 0 ft/sec
    (d) 10.8 sec, 90 ft/sec   **(e)** 2.8 sec
    (f) Greatest acceleration happens 2 sec after launch
    (g) Constant acceleration between 2 and 10.8 sec, -32 ft/sec^2

19. (a) $\frac{4}{7}$ sec, 280 cm/sec   **(b)** 560 cm/sec, 980 cm/sec^2

    (c) 29.75 flashes/sec

21. C = position, A = velocity, B = acceleration

23. (a) \$110/machine   **(b)** \$80   **(c)** \$79.90

25. (a) $b'(0) = 10^4$ bacteria/h   **(b)** $b'(5) = 0$ bacteria/h
    (c) $b'(10) = -10^4$ bacteria/h

27. (a) $\dfrac{dy}{dt} = \dfrac{t}{12} - 1$

    (b) The largest value of $\dfrac{dy}{dt}$ is 0 m/h when $t = 12$ and the smallest

    value of $\dfrac{dy}{dt}$ is -1 m/h when $t = 0$.

    (c)
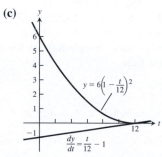

29. 4.88 ft, 8.66 ft, additional ft to stop car for 1 mph speed increase

31. $t = 25$ sec, $D = \dfrac{6250}{9}$ m

33.
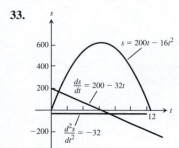

    (a) $v = 0$ when $t = 6.25$ sec
    (b) $v > 0$ when $0 \le t < 6.25 \Rightarrow$ the object moves up; $v < 0$
      when $6.25 < t \le 12.5 \Rightarrow$ the object moves down.
    (c) The object changes direction at $t = 6.25$ sec.
    (d) The object speeds up on $(6.25, 12.5]$ and slows down on
      $[0, 6.25)$.
    (e) The object is moving fastest at the endpoints $t = 0$ and
      $t = 12.5$ when it is traveling 200 ft/sec. It's moving slowest
      at $t = 6.25$ when the speed is 0.
    (f) When $t = 6.25$ the object is $s = 625$ m from the origin and
      farthest away.

35.
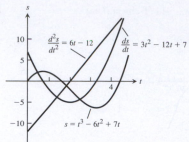

(a) $v = 0$ when $t = \dfrac{6 \pm \sqrt{15}}{3}$ sec

(b) $v < 0$ when $\dfrac{6 - \sqrt{15}}{3} < t < \dfrac{6 + \sqrt{15}}{3} \Rightarrow$

    the object moves left; $v > 0$ when $0 \le t < \dfrac{6 - \sqrt{15}}{3}$ or

    $\dfrac{6 + \sqrt{15}}{3} < t \le 4 \Rightarrow$ the object moves right.

(c) The object changes direction at $t = \dfrac{6 \pm \sqrt{15}}{3}$ sec.

(d) The object speeds up on $\left(\dfrac{6 - \sqrt{15}}{3}, 2\right) \cup \left(\dfrac{6 + \sqrt{15}}{3}, 4\right]$

    and slows down on $\left[0, \dfrac{6 - \sqrt{15}}{3}\right) \cup \left(2, \dfrac{6 + \sqrt{15}}{3}\right)$.

(e) The object is moving fastest at $t = 0$ and $t = 4$ when it is

    moving 7 units/sec and slowest at $t = \dfrac{6 \pm \sqrt{15}}{3}$ sec.

(f) When $t = \dfrac{6 + \sqrt{15}}{3}$ the object is at position $s \approx -6.303$

    units and farthest from the origin.

SECTION 3.5, pp. 158–160

1. $-10 - 3 \sin x$ **3.** $2x \cos x - x^2 \sin x$

5. $-\csc x \cot x - \dfrac{2}{\sqrt{x}} - \dfrac{7}{e^x}$ **7.** $\sin x \sec^2 x + \sin x$

9. $(e^{-x} \sec x)(1 - x + x \tan x)$ **11.** $\dfrac{-\csc^2 x}{(1 + \cot x)^2}$

13. $4 \tan x \sec x - \csc^2 x$ **15.** 0

17. $3x^2 \sin x \cos x + x^3 \cos^2 x - x^3 \sin^2 x$

19. $\sec^2 t + e^{-t}$ **21.** $\dfrac{-2 \csc t \cot t}{(1 - \csc t)^2}$ **23.** $-\theta(\theta \cos \theta + 2 \sin \theta)$

25. $\sec \theta \csc \theta (\tan \theta - \cot \theta) = \sec^2 \theta - \csc^2 \theta$ **27.** $\sec^2 q$

29. $\sec^2 q$ **31.** $\dfrac{q^3 \cos q - q^2 \sin q - q \cos q - \sin q}{(q^2 - 1)^2}$

33. (a) $2 \csc^3 x - \csc x$ **(b)** $2 \sec^3 x - \sec x$

35.
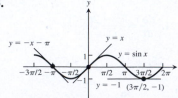

37.

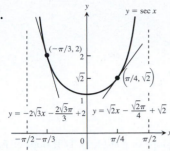

39. Yes, at $x = \pi$ **41.** No   **43.** Yes, at $x = 0, \pi,$ and 2π

45. $\left(-\dfrac{\pi}{4}, -1\right); \left(\dfrac{\pi}{4}, 1\right)$

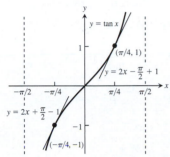

47. (a) $y = -x + \pi/2 + 2$ **(b)** $y = 4 - \sqrt{3}$

49. 0   **51.** $\sqrt{3}/2$ **53.** -1 **55.** 0

57. $-\sqrt{2}$ m/sec, $\sqrt{2}$ m/sec, $\sqrt{2}$ m/sec^2, $\sqrt{2}$ m/sec^3

59. $c = 9$ **61. (a)** $\sin x$ **(b)** $3 \cos x - \sin x$

   (c) $73 \sin x + x \cos x$

63. (a) i) 10 cm **ii)** 5 cm **iii)** $-5\sqrt{2} \approx -7.1$ cm

   (b) i) 0 cm/sec **ii)** $-5\sqrt{3} \approx -8.7$ cm/sec

   iii) $-5\sqrt{2} \approx -7.1$ cm/sec

SECTION 3.6, pp. 166–169

1. $12x^3$ **3.** $3 \cos(3x + 1)$ **5.** $\dfrac{\cos x}{2\sqrt{\sin x}}$

7. $2\pi x \sec^2(\pi x^2)$

9. With $u = (2x + 1), y = u^5 : \dfrac{dy}{dx} = \dfrac{dy}{du}\dfrac{du}{dx} = 5u^4 \cdot 2 =$

   $10(2x + 1)^4$

11. With $u = (1 - (x/7)), y = u^{-7} : \dfrac{dy}{dx} = \dfrac{dy}{du}\dfrac{du}{dx} =$

   $-7u^{-8} \cdot \left(-\dfrac{1}{7}\right) = \left(1 - \dfrac{x}{7}\right)^{-8}$

13. With $u = ((x^2/8) + x - (1/x)), y = u^4 : \dfrac{dy}{dx} = \dfrac{dy}{du}\dfrac{du}{dx} =$

   $4u^3 \cdot \left(\dfrac{x}{4} + 1 + \dfrac{1}{x^2}\right) = 4\left(\dfrac{x^2}{8} + x - \dfrac{1}{x}\right)^3 \left(\dfrac{x}{4} + 1 + \dfrac{1}{x^2}\right)$

15. With $u = \tan x, y = \sec u : \dfrac{dy}{dx} = \dfrac{dy}{du}\dfrac{du}{dx} =$

   $(\sec u \tan u)(\sec^2 x) = \sec(\tan x) \tan(\tan x) \sec^2 x$

17. With $u = \tan x, y = u^3 : \dfrac{dy}{dx} = \dfrac{dy}{du}\dfrac{du}{dx} = 3u^2 \sec^2 x =$

   $3 \tan^2 x (\sec^2 x)$

19. $y = e^u, u = -5x, \dfrac{dy}{dx} = -5e^{-5x}$

21. $y = e^u, u = 5 - 7x, \dfrac{dy}{dx} = -7e^{(5-7x)}$

23. $-\dfrac{1}{2\sqrt{3 - t}}$ **25.** $\dfrac{4}{\pi}(\cos 3t - \sin 5t)$ **27.** $\dfrac{\csc \theta}{\cot \theta + \csc \theta}$

29. $2x \sin^4 x + 4x^2 \sin^3 x \cos x + \cos^{-2} x + 2x \cos^{-3} x \sin x$

31. $(3x - 2)^5 - \dfrac{1}{x^3\left(4 - \dfrac{1}{2x^2}\right)^2}$ **33.** $\dfrac{(4x + 3)^3(4x + 7)}{(x + 1)^4}$

35. $(1 - x)e^{-x} + 3x^2 e^{x^3}$ **37.** $\left(\dfrac{5}{2}x^2 - 3x + 3\right)e^{5x/2}$

39. $\sqrt{x} \sec^2\left(2\sqrt{x}\right) + \tan\left(2\sqrt{x}\right)$ **41.** $\dfrac{x \sec x \tan x + \sec x}{2\sqrt{7 + x \sec x}}$

43. $\dfrac{2 \sin \theta}{(1 + \cos \theta)^2}$ **45.** $-2 \sin(\theta^2) \sin 2\theta + 2\theta \cos(2\theta) \cos(\theta^2)$

47. $\left(\dfrac{t + 2}{2(t + 1)^{3/2}}\right) \cos\left(\dfrac{t}{\sqrt{t + 1}}\right)$ **49.** $2\theta e^{-\theta^2} \sin\left(e^{-\theta^2}\right)$

51. $2\pi \sin(\pi t - 2) \cos(\pi t - 2)$ **53.** $\dfrac{8 \sin(2t)}{(1 + \cos 2t)^5}$

55. $10t^{10} \tan^9 t \sec^2 t + 10t^9 \tan^{10} t$

57. $\dfrac{dy}{dt} = -2\pi \sin(\pi t - 1) \cdot \cos(\pi t - 1) \cdot e^{\cos^2(\pi t-1)}$

59. $\dfrac{-3t^6(t^2 + 4)}{(t^3 - 4t)^4}$ **61.** $-2 \cos(\cos(2t - 5))(\sin(2t - 5))$

63. $\left(1 + \tan^4\left(\dfrac{t}{12}\right)\right)^2 \left(\tan^3\left(\dfrac{t}{12}\right)\sec^2\left(\dfrac{t}{12}\right)\right)$

65. $-\dfrac{t \sin(t^2)}{\sqrt{1 + \cos(t^2)}}$ **67.** $6 \tan(\sin^3 t) \sec^2(\sin^3 t) \sin^2 t \cos t$

69. $3(2t^2 - 5)^3 (18t^2 - 5)$ **71.** $\dfrac{6}{x^3}\left(1 + \dfrac{1}{x}\right)\left(1 + \dfrac{2}{x}\right)$

73. $2 \csc^2(3x - 1) \cot(3x - 1)$ **75.** $16(2x + 1)^2 (5x + 1)$

77. $2(2x^2 + 1)e^{x^2}$

79. $f'(x) = 0$ for $x = 1, 4; f''(x) = 0$ for $x = 2, 4$

81. $5/2$ **83.** $-\pi/4$ **85.** 0   **87.** -5

89. (a) $2/3$ **(b)** $2\pi + 5$ **(c)** $15 - 8\pi$ **(d)** $37/6$ **(e)** -1

   (f) $\sqrt{2}/24$ **(g)** $5/32$ **(h)** $-5/\left(3\sqrt{17}\right)$

91. 5   **93. (a)** 1 **(b)** 1   **95.** $y = 1 - 4x$

97. (a) $y = \pi x + 2 - \pi$ **(b)** $\pi/2$

99. It multiplies the velocity, acceleration, and jerk by 2, 4, and 8, respectively.

101. $v(6) = \dfrac{2}{5}$ m/sec, $a(6) = -\dfrac{4}{125}$ m/sec^2

SECTION 3.7, pp. 172–174

1. $\dfrac{-2xy - y^2}{x^2 + 2xy}$ **3.** $\dfrac{1 - 2y}{2x + 2y - 1}$

5. $\dfrac{-2x^3 + 3x^2 y - xy^2 + x}{x^2 y - x^3 + y}$ **7.** $\dfrac{1}{y(x + 1)^2}$ **9.** $\cos y \cot y$

11. $\dfrac{-\cos^2(xy) - y}{x}$ **13.** $\dfrac{-y^2}{y \sin\left(\dfrac{1}{y}\right) - \cos\left(\dfrac{1}{y}\right) + xy}$

15. $\dfrac{2e^{2x} - \cos(x + 3y)}{3\cos(x + 3y)}$ **17.** $-\dfrac{\sqrt{r}}{\sqrt{\theta}}$ **19.** $\dfrac{-r}{\theta}$

21. $y' = -\dfrac{x}{y},\ y'' = \dfrac{-y^2 - x^2}{y^3}$

23. $\dfrac{dy}{dx} = \dfrac{xe^{x^2} + 1}{y},\ \dfrac{d^2y}{dx^2} = \dfrac{(2x^2y^2 + y^2 - 2x)e^{x^2} - x^2e^{2x^2} - 1}{y^3}$

25. $y' = \dfrac{\sqrt{y}}{\sqrt{y} + 1},\ y'' = \dfrac{1}{2(\sqrt{y} + 1)^3}$

27. $y' = \dfrac{3x^2}{1 - \cos y},\ y'' = \dfrac{6x(1 - \cos y)^2 - 9x^4 \sin y}{(1 - \cos y)^3}$

29. -2 **31.** $(-2, 1): m = -1, (-2, -1): m = 1$

33. (a) $y = \dfrac{7}{4}x - \dfrac{1}{2}$ (b) $y = -\dfrac{4}{7}x + \dfrac{29}{7}$

35. (a) $y = 3x + 6$ (b) $y = -\dfrac{1}{3}x + \dfrac{8}{3}$

37. (a) $y = \dfrac{6}{7}x + \dfrac{6}{7}$ (b) $y = -\dfrac{7}{6}x - \dfrac{7}{6}$

39. (a) $y = -\dfrac{\pi}{2}x + \pi$ (b) $y = \dfrac{2}{\pi}x - \dfrac{2}{\pi} + \dfrac{\pi}{2}$

41. (a) $y = 2\pi x - 2\pi$ (b) $y = -\dfrac{x}{2\pi} + \dfrac{1}{2\pi}$

43. Points: $\left(-\sqrt{7}, 0\right)$ and $\left(\sqrt{7}, 0\right)$, Slope: -2

45. $m = -1$ at $\left(\dfrac{\sqrt{3}}{4}, \dfrac{\sqrt{3}}{2}\right),\quad m = \sqrt{3}$ at $\left(\dfrac{\sqrt{3}}{4}, \dfrac{1}{2}\right)$

47. $(-3, 2): m = -\dfrac{27}{8}; (-3, -2): m = \dfrac{27}{8}; (3, 2): m = \dfrac{27}{8};$

$(3, -2): m = -\dfrac{27}{8}$

49. $(3, -1)$

55. $\dfrac{dy}{dx} = -\dfrac{y^3 + 2xy}{x^2 + 3xy^2},\ \dfrac{dx}{dy} = -\dfrac{x^2 + 3xy^2}{y^3 + 2xy},\ \dfrac{dx}{dy} = \dfrac{1}{dy/dx}$

SECTION 3.8, pp. 183–184

1. (a) $f^{-1}(x) = \dfrac{x}{2} - \dfrac{3}{2}$

(b)

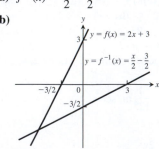

$y = f(x) = 2x + 3$

$y = f^{-1}(x) = \dfrac{x}{2} - \dfrac{3}{2}$

(c) $2, 1/2$

3. (a) $f^{-1}(x) = -\dfrac{x}{4} + \dfrac{5}{4}$

(b)

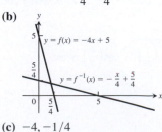

$y = f(x) = -4x + 5$

$y = f^{-1}(x) = -\dfrac{x}{4} + \dfrac{5}{4}$

(c) $-4, -1/4$

5. (b)

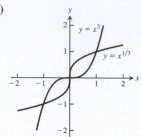

$y = x^3$

$y = x^{1/3}$

(c) Slope of f at $(1, 1)$: 3; slope of g at $(1, 1)$: $1/3$; slope of f at $(-1, -1)$: 3; slope of g at $(-1, -1)$: $1/3$

(d) $y = 0$ is tangent to $y = x^3$ at $x = 0$; $x = 0$ is tangent to $y = \sqrt[3]{x}$ at $x = 0$.

7. $1/9$ **9.** 3 **11.** $\dfrac{1}{x} + 1$ **13.** $2/t$ **15.** $-1/x$

17. $\dfrac{1}{\theta + 1} - e^{\theta}$ **19.** $3/x$ **21.** $2(\ln t) + (\ln t)^2$

23. $x^3 \ln x$ **25.** $\dfrac{1 - \ln t}{t^2}$ **27.** $\dfrac{1}{x(1 + \ln x)^2}$ **29.** $\dfrac{1}{x \ln x}$

31. $2\cos(\ln \theta)$ **33.** $-\dfrac{3x + 2}{2x(x + 1)}$ **35.** $\dfrac{2}{t(1 - \ln t)^2}$

37. $\dfrac{\tan(\ln \theta)}{\theta}$ **39.** $\dfrac{10x}{x^2 + 1} + \dfrac{1}{2(1 - x)}$

41. $\left(\dfrac{1}{2}\right)\sqrt{x(x + 1)}\left(\dfrac{1}{x} + \dfrac{1}{x + 1}\right) = \dfrac{2x + 1}{2\sqrt{x(x + 1)}}$

43. $\left(\dfrac{1}{2}\right)\sqrt{\dfrac{t}{t + 1}}\left(\dfrac{1}{t} - \dfrac{1}{t + 1}\right) = \dfrac{1}{2\sqrt{t}(t + 1)^{3/2}}$

45. $\sqrt{\theta + 3}(\sin \theta)\left(\dfrac{1}{2(\theta + 3)} + \cot \theta\right)$

47. $t(t + 1)(t + 2)\left[\dfrac{1}{t} + \dfrac{1}{t + 1} + \dfrac{1}{t + 2}\right] = 3t^2 + 6t + 2$

49. $\dfrac{\theta + 5}{\theta \cos \theta}\left[\dfrac{1}{\theta + 5} - \dfrac{1}{\theta} + \tan \theta\right]$

51. $\dfrac{x\sqrt{x^2 + 1}}{(x + 1)^{2/3}}\left[\dfrac{1}{x} + \dfrac{x}{x^2 + 1} - \dfrac{2}{3(x + 1)}\right]$

53. $\dfrac{1}{3}\sqrt[3]{\dfrac{x(x - 2)}{x^2 + 1}}\left(\dfrac{1}{x} + \dfrac{1}{x - 2} - \dfrac{2x}{x^2 + 1}\right)$ **55.** $-2\tan \theta$

57. $\dfrac{1 - t}{t}$ **59.** $1/(1 + e^{\theta})$ **61.** $e^{\cos t}(1 - t \sin t)$

63. $\dfrac{ye^y \cos x}{1 - ye^y \sin x}$ **65.** $\dfrac{dy}{dx} = \dfrac{y^2 - xy \ln y}{x^2 - xy \ln x}$ **67.** $2^x \ln x$

69. $\left(\dfrac{\ln 5}{2\sqrt{s}}\right)5^{\sqrt{s}}$ **71.** $\pi x^{(\pi - 1)}$ **73.** $\dfrac{1}{\theta \ln 2}$ **75.** $\dfrac{3}{x \ln 4}$

77. $\dfrac{2(\ln r)}{r(\ln 2)(\ln 4)}$ **79.** $\dfrac{-2}{(x + 1)(x - 1)}$

81. $\sin(\log_7 \theta) + \dfrac{1}{\ln 7}\cos(\log_7 \theta)$ **83.** $\dfrac{1}{\ln 5}$

85. $\dfrac{1}{t}(\log_2 3)3^{\log_2 t}$ **87.** $\dfrac{1}{t}$ **89.** $(x + 1)^x\left(\dfrac{x}{x + 1} + \ln(x + 1)\right)$

91. $\left(\sqrt{t}\right)^t\left(\dfrac{\ln t}{2} + \dfrac{1}{2}\right)$ **93.** $(\sin x)^x(\ln \sin x + x \cot x)$

95. $(x^{\ln x})\left(\dfrac{\ln x^2}{x}\right)$ **97.** $\dfrac{3y - xy \ln y}{x^2 - x}$ **99.** $\dfrac{1 - xy \ln y}{x^2(1 + \ln y)}$

SECTION 3.9, pp. 189–191

1. (a) $\pi/4$ (b) $-\pi/3$ (c) $\pi/6$

3. (a) $-\pi/6$ (b) $\pi/4$ (c) $-\pi/3$

5. (a) $\pi/3$ (b) $3\pi/4$ (c) $\pi/6$

7. (a) $3\pi/4$ (b) $\pi/6$ (c) $2\pi/3$

9. $1/\sqrt{2}$ **11.** $-1/\sqrt{3}$ **13.** $\pi/2$ **15.** $\pi/2$ **17.** $\pi/2$

19. 0 **21.** $\dfrac{-2x}{\sqrt{1-x^4}}$ **23.** $\dfrac{\sqrt{2}}{\sqrt{1-2t^2}}$

25. $\dfrac{1}{|2s+1|\sqrt{s^2+s}}$ **27.** $\dfrac{-2x}{(x^2+1)\sqrt{x^4+2x^2}}$

29. $\dfrac{-1}{\sqrt{1-t^2}}$ **31.** $\dfrac{-1}{2\sqrt{t}(1+t)}$ **33.** $\dfrac{1}{(\tan^{-1}x)(1+x^2)}$

35. $\dfrac{-e^t}{|e^t|\sqrt{(e^t)^2-1}} = \dfrac{-1}{\sqrt{e^{2t}-1}}$ **37.** $\dfrac{-2s^2}{\sqrt{1-s^2}}$ **39.** 0

41. $\sin^{-1}x$ **43.** 0 **45.** $\dfrac{8\sqrt{2}}{4+3\pi}$

51. (a) Defined; there is an angle whose tangent is 2.
   (b) Not defined; there is no angle whose cosine is 2.

53. (a) Not defined; no angle has secant 0.
   (b) Not defined; no angle has sine $\sqrt{2}$.

63. (a) Domain: all real numbers except those having the form $\dfrac{\pi}{2} + k\pi$ where k is an integer; range: $-\pi/2 < y < \pi/2$

   (b) Domain: $-\infty < x < \infty$; range: $-\infty < y < \infty$

65. (a) Domain: $-\infty < x < \infty$; range: $0 \le y \le \pi$
   (b) Domain: $-1 \le x \le 1$; range: $-1 \le y \le 1$

67. The graphs are identical.

SECTION 3.10, pp. 196–199

1. $\dfrac{dA}{dt} = 2\pi r \dfrac{dr}{dt}$ **3.** 10 **5.** -6 **7.** $-3/2$

9. $31/13$ **11.** (a) -180 m^2/min   (b) -135 m^3/min

13. (a) $\dfrac{dV}{dt} = \pi r^2 \dfrac{dh}{dt}$ (b) $\dfrac{dV}{dt} = 2\pi hr \dfrac{dr}{dt}$

   (c) $\dfrac{dV}{dt} = \pi r^2 \dfrac{dh}{dt} + 2\pi hr \dfrac{dr}{dt}$

15. (a) 1 volt/sec   (b) $-\dfrac{1}{3}$ amp/sec

   (c) $\dfrac{dR}{dt} = \dfrac{1}{I}\left(\dfrac{dV}{dt} - \dfrac{V}{I}\dfrac{dI}{dt}\right)$

   (d) $3/2$ ohms/sec, R is increasing.

17. (a) $\dfrac{ds}{dt} = \dfrac{x}{\sqrt{x^2+y^2}}\dfrac{dx}{dt}$

   (b) $\dfrac{ds}{dt} = \dfrac{x}{\sqrt{x^2+y^2}}\dfrac{dx}{dt} + \dfrac{y}{\sqrt{x^2+y^2}}\dfrac{dy}{dt}$

   (c) $\dfrac{dx}{dt} = -\dfrac{y}{x}\dfrac{dy}{dt}$

19. (a) $\dfrac{dA}{dt} = \dfrac{1}{2}ab\cos\theta \dfrac{d\theta}{dt}$

   (b) $\dfrac{dA}{dt} = \dfrac{1}{2}ab\cos\theta \dfrac{d\theta}{dt} + \dfrac{1}{2}b\sin\theta \dfrac{da}{dt}$

   (c) $\dfrac{dA}{dt} = \dfrac{1}{2}ab\cos\theta \dfrac{d\theta}{dt} + \dfrac{1}{2}b\sin\theta \dfrac{da}{dt} + \dfrac{1}{2}a\sin\theta \dfrac{db}{dt}$

21. (a) 14 cm^2/sec, increasing   (b) 0 cm/sec, constant
   (c) $-14/13$ cm/sec, decreasing

23. (a) -12 ft/sec   (b) -59.5 ft^2/sec   (c) -1 rad/sec

25. 20 ft/sec

27. (a) $\dfrac{dh}{dt} = 11.19$ cm/min   (b) $\dfrac{dr}{dt} = 14.92$ cm/min

29. (a) $\dfrac{-1}{24\pi}$ m/min   (b) $r = \sqrt{26y - y^2}$ m

   (c) $\dfrac{dr}{dt} = -\dfrac{5}{288\pi}$ m/min

31. 1 ft/min, 40π ft^2/min   **33.** 11 ft/sec

35. Increasing at $466/1681$ L/min^2

37. -5 m/sec   **39.** -1500 ft/sec

41. $\dfrac{5}{72\pi}$ in./min, $\dfrac{10}{3}$ in^2/min

43. (a) $-32/\sqrt{13} \approx -8.875$ ft/sec
   (b) $d\theta_1/dt = 8/65$ rad/sec, $d\theta_2/dt = -8/65$ rad/sec
   (c) $d\theta_1/dt = 1/6$ rad/sec, $d\theta_2/dt = -1/6$ rad/sec

45. -5.5 deg/min   **47.** 12π km/min

SECTION 3.11, pp. 209–211

1. $L(x) = 10x - 13$ **3.** $L(x) = 2$ **5.** $L(x) = x - \pi$

7. $2x$ **9.** $-x - 5$ **11.** $\dfrac{1}{12}x + \dfrac{4}{3}$ **13.** $1 - x$

15. $f(0) = 1$. Also, $f'(x) = k(1+x)^{k-1}$, so $f'(0) = k$. This means the linearization at $x = 0$ is $L(x) = 1 + kx$.

17. (a) 1.01 (b) 1.003

19. $\left(3x^2 - \dfrac{3}{2\sqrt{x}}\right)dx$ **21.** $\dfrac{2 - 2x^2}{(1+x^2)^2}dx$

23. $\dfrac{1-y}{3\sqrt{y}+x}dx$ **25.** $\dfrac{5}{2\sqrt{x}}\cos(5\sqrt{x})\,dx$

27. $(4x^2)\sec^2\!\left(\dfrac{x^3}{3}\right)dx$

29. $\dfrac{3}{\sqrt{x}}\left(\csc(1-2\sqrt{x})\cot(1-2\sqrt{x})\right)dx$

31. $\dfrac{1}{2\sqrt{x}}\cdot e^{\sqrt{x}}dx$ **33.** $\dfrac{2x}{1+x^2}dx$ **35.** $\dfrac{2xe^{x^2}}{1+e^{2x^2}}dx$

37. $\dfrac{-1}{\sqrt{e^{-2x}-1}}dx$

39. (a) 0.41 (b) 0.4 (c) 0.01

41. (a) 0.231 (b) 0.2 (c) 0.031

43. (a) $-1/3$ (b) $-2/5$ (c) $1/15$

45. $dV = 4\pi r_0^2\,dr$ **47.** $dS = 12x_0\,dx$ **49.** $dV = 2\pi r_0 h\,dr$

51. (a) 0.08π m^2 (b) 2% **53.** $dV \approx 565.5$ in^3

55. (a) 2% (b) 4% **57.** $\dfrac{1}{3}\%$ **59.** 3%

61. The ratio equals 37.87, so a change in the acceleration of gravity on the moon has about 38 times the effect that a change of the same magnitude has on Earth.

63. Increase $V \approx 40\%$

65. (a) i) $b_0 = f(a)$ ii) $b_1 = f'(a)$ iii) $b_2 = \dfrac{f''(a)}{2}$

   (b) $Q(x) = 1 + x + x^2$ (d) $Q(x) = 1 - (x-1) + (x-1)^2$

   (e) $Q(x) = 1 + \dfrac{x}{2} - \dfrac{x^2}{8}$

   (f) The linearization of any differentiable function $u(x)$ at $x = a$ is $L(x) = u(a) + u'(a)(x-a) = b_0 + b_1(x-a)$, where b_0 and b_1 are the coefficients of the constant and linear terms of the quadratic approximation. Thus, the linearization for $f(x)$ at $x = 0$ is $1 + x$; the linearization for $g(x)$ at $x = 1$ is $1 - (x-1)$ or $2 - x$; and the linearization for $h(x)$ at $x = 0$ is $1 + \dfrac{x}{2}$.

67. (a) $L(x) = x \ln 2 + 1 \approx 0.69x + 1$

(b)

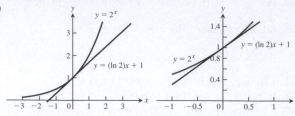

PRACTICE EXERCISES, pp. 212–217

1. $5x^4 - 0.25x + 0.25$ **3.** $3x(x - 2)$

5. $2(x + 1)(2x^2 + 4x + 1)$

7. $3(\theta^2 + \sec\theta + 1)^2(2\theta + \sec\theta\tan\theta)$

9. $\dfrac{1}{2\sqrt{t}\left(1 + \sqrt{t}\right)^2}$ **11.** $2\sec^2 x \tan x$

13. $8\cos^3(1 - 2t)\sin(1 - 2t)$ **15.** $5(\sec t)(\sec t + \tan t)^5$

17. $\dfrac{\theta\cos\theta + \sin\theta}{\sqrt{2\theta}\sin\theta}$ **19.** $\dfrac{\cos\sqrt{2\theta}}{\sqrt{2\theta}}$

21. $x\csc\left(\dfrac{2}{x}\right) + \csc\left(\dfrac{2}{x}\right)\cot\left(\dfrac{2}{x}\right)$

23. $\dfrac{1}{2}x^{1/2}\sec(2x)^2\left[16\tan(2x)^2 - x^{-2}\right]$

25. $-10x\csc^2(x^2)$ **27.** $8x^3\sin(2x^2)\cos(2x^2) + 2x\sin^2(2x^2)$

29. $\dfrac{-(t + 1)}{8t^3}$ **31.** $\dfrac{1 - x}{(x + 1)^3}$ **33.** $\dfrac{-1}{2x^2\left(1 + \dfrac{1}{x}\right)^{1/2}}$

35. $\dfrac{-2\sin\theta}{(\cos\theta - 1)^2}$ **37.** $3\sqrt{2x + 1}$ **39.** $-9\left[\dfrac{5x + \cos 2x}{(5x^2 + \sin 2x)^{5/2}}\right]$

41. $-2e^{-x/5}$ **43.** xe^{4x} **45.** $\dfrac{2\sin\theta\cos\theta}{\sin^2\theta} = 2\cot\theta$

47. $\dfrac{2}{(\ln 2)x}$ **49.** $-8^{-t}(\ln 8)$ **51.** $18x^{2.6}$

53. $(x + 2)^{x+2}(\ln(x + 2) + 1)$ **55.** $-\dfrac{1}{\sqrt{1 - u^2}}$

57. $\dfrac{-1}{\sqrt{1 - x^2}\cos^{-1}x}$ **59.** $\tan^{-1}(t) + \dfrac{t}{1 + t^2} - \dfrac{1}{2t}$

61. $\dfrac{1 - z}{\sqrt{z^2 - 1}} + \sec^{-1}z$ **63.** -1 **65.** $-\dfrac{y + 2}{x + 3}$

67. $\dfrac{-3x^2 - 4y + 2}{4x - 4y^{1/3}}$ **69.** $-\dfrac{y}{x}$ **71.** $\dfrac{1}{2y(x + 1)^2}$

73. $-1/2$ **75.** y/x **77.** $-\dfrac{2e^{-\tan^{-1}x}}{1 + x^2}$ **79.** $\dfrac{dp}{dq} = \dfrac{6q - 4p}{3p^2 + 4q}$

81. $\dfrac{dr}{ds} = (2r - 1)(\tan 2s)$

83. (a) $\dfrac{d^2y}{dx^2} = \dfrac{-2xy^3 - 2x^4}{y^5}$ **(b)** $\dfrac{d^2y}{dx^2} = \dfrac{-2xy^2 - 1}{x^4y^3}$

85. (a) 7   **(b)** -2 **(c)** $5/12$ **(d)** $1/4$ **(e)** 12   **(f)** $9/2$
   (g) $3/4$

87. 0   **89.** $\dfrac{3\sqrt{2}e^{\sqrt{3/2}}}{4}\cos\left(e^{\sqrt{3/2}}\right)$ **91.** $-\dfrac{1}{2}$ **93.** $\dfrac{-2}{(2t + 1)^2}$

95. (a)

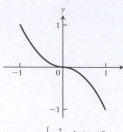

$f(x) = \begin{cases} x^2, & -1 \le x < 0 \\ -x^2, & 0 \le x < 1 \end{cases}$

(b) Yes   **(c)** Yes

97. (a)

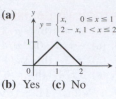

$y = \begin{cases} x, & 0 \le x \le 1 \\ 2 - x, & 1 < x \le 2 \end{cases}$

(b) Yes   **(c)** No

99. $\left(\dfrac{5}{2}, \dfrac{9}{4}\right)$ and $\left(\dfrac{3}{2}, -\dfrac{1}{4}\right)$

101. $(-1, 27)$ and $(2, 0)$

103. (a) $(-2, 16)$, $(3, 11)$ **(b)** $(0, 20)$, $(1, 7)$

105.

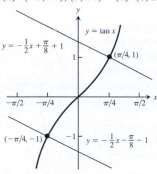

107. $\dfrac{1}{4}$ **109.** 4

111. Tangent: $y = -\dfrac{1}{4}x + \dfrac{9}{4}$, normal: $y = 4x - 2$

113. Tangent: $y = 2x - 4$, normal: $y = -\dfrac{1}{2}x + \dfrac{7}{2}$

115. Tangent: $y = -\dfrac{5}{4}x + 6$, normal: $y = \dfrac{4}{5}x - \dfrac{11}{5}$

117. $(1, 1)$: $m = -\dfrac{1}{2}$; $(1, -1)$: m not defined

119. $B =$ graph of f, $A =$ graph of f'

121.

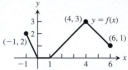

123. (a) 0, 0   **(b)** 1700 rabbits, ≈ 1400 rabbits

125. -1 **127.** $1/2$ **129.** 4   **131.** 1

133. To make g continuous at the origin, define $g(0) = 1$.

135. $\dfrac{2(x^2 + 1)}{\sqrt{\cos 2x}}\left[\dfrac{2x}{x^2 + 1} + \tan 2x\right]$

137. $5\left[\dfrac{(t + 1)(t - 1)}{(t - 2)(t + 3)}\right]^5\left[\dfrac{1}{t + 1} + \dfrac{1}{t - 1} - \dfrac{1}{t - 2} - \dfrac{1}{t + 3}\right]$

139. $\dfrac{1}{\sqrt{\theta}}(\sin\theta)^{\sqrt{\theta}}\left(\dfrac{\ln\sin\theta}{2} + \theta\cot\theta\right)$

141. (a) $\dfrac{dS}{dt} = (4\pi r + 2\pi h)\dfrac{dr}{dt}$

(b) $\dfrac{dS}{dt} = 2\pi r \dfrac{dh}{dt}$

(c) $\dfrac{dS}{dt} = (4\pi r + 2\pi h)\dfrac{dr}{dt} + 2\pi r \dfrac{dh}{dt}$

(d) $\dfrac{dr}{dt} = -\dfrac{r}{2r + h}\dfrac{dh}{dt}$

143. $-40 \text{ m}^2/\text{sec}$ **145.** $0.02 \text{ ohm}/\text{sec}$ **147.** $2 \text{ m}/\text{sec}$

149. (a) $r = \dfrac{2}{5}h$ **(b)** $-\dfrac{125}{144\pi} \text{ ft}/\text{min}$

151. (a) $\dfrac{3}{5} \text{ km}/\text{sec}$ or $600 \text{ m}/\text{sec}$ **(b)** $\dfrac{18}{\pi} \text{ rpm}$

153. (a) $L(x) = 2x + \dfrac{\pi - 2}{2}$

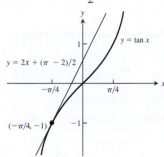

(b) $L(x) = -\sqrt{2}x + \dfrac{\sqrt{2}(4 - \pi)}{4}$

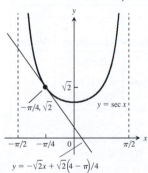

155. $L(x) = 1.5x + 0.5$ **157.** $dS = \dfrac{\pi r h_0}{\sqrt{r^2 + h_0^2}}\,dh$

159. (a) 4% **(b)** 8% **(c)** 12%

ADDITIONAL AND ADVANCED EXERCISES, pp. 217–219

1. (a) $\sin 2\theta = 2\sin\theta\cos\theta$; $2\cos 2\theta = 2\sin\theta(-\sin\theta) +$
$\cos\theta(2\cos\theta)$; $2\cos 2\theta = -2\sin^2\theta + 2\cos^2\theta$; $\cos 2\theta =$
$\cos^2\theta - \sin^2\theta$

(b) $\cos 2\theta = \cos^2\theta - \sin^2\theta$; $-2\sin 2\theta =$
$2\cos\theta(-\sin\theta) - 2\sin\theta(\cos\theta)$; $\sin 2\theta =$
$\cos\theta\sin\theta + \sin\theta\cos\theta$; $\sin 2\theta = 2\sin\theta\cos\theta$

3. (a) $a = 1, b = 0, c = -\dfrac{1}{2}$ **(b)** $b = \cos a, c = \sin a$

5. $h = -4, k = \dfrac{9}{2}, a = \dfrac{5\sqrt{5}}{2}$

7. (a) 0.09y **(b)** Increasing at 1% per year

9. Answers will vary. Here is one possibility.

11. (a) $2 \text{ sec}, 64 \text{ ft}/\text{sec}$ **(b)** $12.31 \text{ sec}, 393.85 \text{ ft}$

15. $y' = \dfrac{x^{y-1}y^2 - ye^x(x^y + 1)\ln y}{e^x(x^y + 1) - x^y y \ln x}$

17. (a) $m = -\dfrac{b}{\pi}$ **(b)** $m = -1, b = \pi$

19. (a) $a = \dfrac{3}{4}, b = \dfrac{9}{4}$ **21.** f odd $\Rightarrow f'$ is even

25. h' is defined but not continuous at $x = 0$; k' is defined *and* continuous at $x = 0$.

27. $\dfrac{-7}{75} \text{ rad}/\text{sec}$

31. (a) 0.8156 ft **(b)** 0.00613 sec
(c) It will lose about $8.83 \text{ min}/\text{day}$.

Chapter 4

SECTION 4.1, pp. 227–229

1. Absolute minimum at $x = c_2$; absolute maximum at $x = b$
3. Absolute maximum at $x = c$; no absolute minimum
5. Absolute minimum at $x = a$; absolute maximum at $x = c$
7. No absolute minimum; no absolute maximum
9. Absolute maximum at $(0, 5)$ **11.** (c)   **13.** (d)
15. Absolute minimum at   **17.** Absolute maximum at
    $x = 0$; no absolute $x = 2$; no absolute
    maximum   minimum

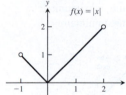   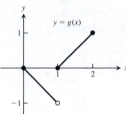

19. Absolute maximum at $x = \pi/2$; absolute minimum at $x = 3\pi/2$

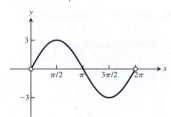

21. Absolute maximum: -3; absolute minimum: $-19/3$

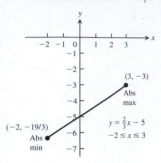

23. Absolute maximum: 3; absolute minimum: -1

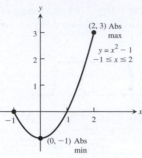

25. Absolute maximum: -0.25; absolute minimum: -4

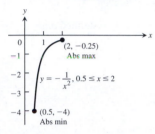

27. Absolute maximum: 2; absolute minimum: -1

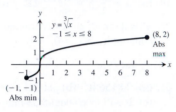

29. Absolute maximum: 2; absolute minimum: 0

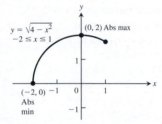

31. Absolute maximum: 1; absolute minimum: -1

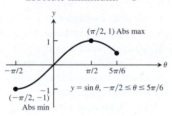

33. Absolute maximum: $2/\sqrt{3}$; absolute minimum: 1

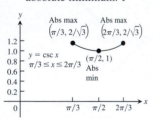

35. Absolute maximum: 2; absolute minimum: -1

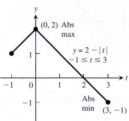

37. Absolute maximum is $1/e$ at $x = 1$; absolute minimum is $-e$ at $x = -1$.

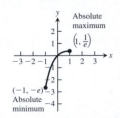

39. Absolute maximum value is $(1/4) + \ln 4$ at $x = 4$; absolute minimum value is 1 at $x = 1$; local maximum at $(1/2, 2 - \ln 2)$.

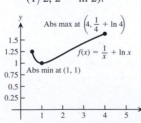

41. Increasing on $(0, 8)$, decreasing on $(-1, 0)$; absolute maximum: 16 at $x = 8$; absolute minimum: 0 at $x = 0$

43. Increasing on $(-32, 1)$; absolute maximum: 1 at $\theta = 1$; absolute minimum: -8 at $\theta = -32$

45. $x = 3$

47. $x = 1$, $x = 4$

49. $x = 1$

51. $x = 0$ and $x = 4$

53. $x = 0$ and $x = 1$

55. $x = 2$ and $x = -4$

57.

Critical point or endpoint	Derivative	Extremum	Value
$x = -\dfrac{4}{5}$	0	Local max	$\dfrac{12}{25}10^{1/3} \approx 1.034$
$x = 0$	Undefined	Local min	0

59.

Critical point or endpoint	Derivative	Extremum	Value
$x = -2$	Undefined	Local max	0
$x = -\sqrt{2}$	0	Minimum	-2
$x = \sqrt{2}$	0	Maximum	2
$x = 2$	Undefined	Local min	0

61.

Critical point or endpoint	Derivative	Extremum	Value
$x = 1$	Undefined	Minimum	2

63.

Critical point or endpoint	Derivative	Extremum	Value
$x = -1$	0	Maximum	5
$x = 1$	Undefined	Local min	1
$x = 3$	0	Maximum	5

65. (a) No

(b) The derivative is defined and nonzero for $x \neq 2$. Also, $f(2) = 0$ and $f(x) > 0$ for all $x \neq 2$.

(c) No, because $(-\infty, \infty)$ is not a closed interval.

(d) The answers are the same as parts (a) and (b), with 2 replaced by a.

67. y is increasing on $(-\infty, \infty)$ and so has no extrema.

69. y is decreasing on $(-\infty, \infty)$ and so has no extrema.

71. Yes

73. g assumes a local maximum at $-c$.

75. (a) Maximum value is 144 at $x = 2$.

(b) The largest volume of the box is 144 cubic units, and it occurs when $x = 2$.

77. $\dfrac{v_0{}^2}{2g} + s_0$

79. Maximum value is 11 at $x = 5$; minimum value is 5 on the interval $[-3, 2]$; local maximum at $(-5, 9)$.

81. Maximum value is 5 on the interval $[3, \infty)$; minimum value is -5 on the interval $(-\infty, -2]$.

SECTION 4.2, pp. 235–237

1. $1/2$ **3.** 1   **5.** $\pm\sqrt{1 - \dfrac{4}{\pi^2}} \approx \pm0.771$

7. $\frac{1}{3}\left(1 + \sqrt{7}\right) \approx 1.22, \frac{1}{3}\left(1 - \sqrt{7}\right) \approx -0.549$

9. Does not; f is not differentiable at the interior domain point $x = 0$.

11. Does   **13.** Does not; f is not differentiable at $x = -1$.

17. (a)

i)

ii)

iii)

iv)

29. Yes   **31. (a)** 4   **(b)** 3   **(c)** 3

33. (a) $\frac{x^2}{2} + C$ **(b)** $\frac{x^3}{3} + C$ **(c)** $\frac{x^4}{4} + C$

35. (a) $\frac{1}{x} + C$ **(b)** $x + \frac{1}{x} + C$ **(c)** $5x - \frac{1}{x} + C$

37. (a) $-\frac{1}{2}\cos 2t + C$ **(b)** $2\sin\frac{t}{2} + C$

(c) $-\frac{1}{2}\cos 2t + 2\sin\frac{t}{2} + C$

39. $f(x) = x^2 - x$ **41.** $f(x) = 1 + \frac{e^{2x}}{2}$

43. $s = 4.9t^2 + 5t + 10$ **45.** $s = \frac{1 - \cos(\pi t)}{\pi}$

47. $s = e^t + 19t + 4$ **49.** $s = \sin(2t) - 3$

51. If $T(t)$ is the temperature of the thermometer at time t, then $T(0) = -19\,°C$ and $T(14) = 100\,°C$. From the Mean Value Theorem, there exists a $0 < t_0 < 14$ such that $\frac{T(14) - T(0)}{14 - 0} = 8.5\,°C/\text{sec} = T'(t_0)$, the rate at which the temperature was changing at $t = t_0$ as measured by the rising mercury on the thermometer.

53. Because its average speed was approximately 7.667 knots, and by the Mean Value Theorem, it must have been going that speed at least once during the trip.

57. The conclusion of the Mean Value Theorem yields
$$\frac{\frac{1}{b} - \frac{1}{a}}{b - a} = -\frac{1}{c^2} \Rightarrow c^2\left(\frac{a - b}{ab}\right) = a - b \Rightarrow c = \sqrt{ab}.$$

61. $f(x)$ must be zero at least once between a and b by the Intermediate Value Theorem. Now suppose that $f(x)$ is zero twice between a and b. Then, by the Mean Value Theorem, $f'(x)$ would have to be zero at least once between the two zeros of $f(x)$, but this can't be true since we are given that $f'(x) \neq 0$ on this interval. Therefore, $f(x)$ is zero once and only once between a and b.

71. $1.09999 \leq f(0.1) \leq 1.1$

SECTION 4.3, pp. 241–242

1. (a) 0, 1

(b) Increasing on $(-\infty, 0)$ and $(1, \infty)$; decreasing on $(0, 1)$

(c) Local maximum at $x = 0$; local minimum at $x = 1$

3. (a) $-2, 1$

(b) Increasing on $(-2, 1)$ and $(1, \infty)$; decreasing on $(-\infty, -2)$

(c) No local maximum; local minimum at $x = -2$

5. (a) Critical point at $x = 1$

(b) Decreasing on $(-\infty, 1)$, increasing on $(1, \infty)$

(c) Local (and absolute) minimum at $x = 1$

7. (a) 0, 1

(b) Increasing on $(-\infty, -2)$ and $(1, \infty)$; decreasing on $(-2, 0)$ and $(0, 1)$

(c) Local minimum at $x = 1$

9. (a) $-2, 2$

(b) Increasing on $(-\infty, -2)$ and $(2, \infty)$; decreasing on $(-2, 0)$ and $(0, 2)$

(c) Local maximum at $x = -2$; local minimum at $x = 2$

11. (a) $-2, 0$

(b) Increasing on $(-\infty, -2)$ and $(0, \infty)$; decreasing on $(-2, 0)$

(c) Local maximum at $x = -2$; local minimum at $x = 0$

13. (a) $\frac{\pi}{2}, \frac{2\pi}{3}, \frac{4\pi}{3}$

(b) Increasing on $\left(\frac{2\pi}{3}, \frac{4\pi}{3}\right)$; decreasing on $\left(0, \frac{\pi}{2}\right), \left(\frac{\pi}{2}, \frac{2\pi}{3}\right)$, and $\left(\frac{4\pi}{3}, 2\pi\right)$

(c) Local maximum at $x = 0$ and $x = \frac{4\pi}{3}$; local minimum at $x = \frac{2\pi}{3}$ and $x = 2\pi$

15. (a) Increasing on $(-2, 0)$ and $(2, 4)$; decreasing on $(-4, -2)$ and $(0, 2)$

(b) Absolute maximum at $(-4, 2)$; local maximum at $(0, 1)$ and $(4, -1)$; absolute minimum at $(2, -3)$; local minimum at $(-2, 0)$

17. (a) Increasing on $(-4, -1)$, $(1/2, 2)$, and $(2, 4)$; decreasing on $(-1, 1/2)$

(b) Absolute maximum at $(4, 3)$; local maximum at $(-1, 2)$ and $(2, 1)$; no absolute minimum; local minimum at $(-4, -1)$ and $(1/2, -1)$

19. (a) Increasing on $(-\infty, -1.5)$; decreasing on $(-1.5, \infty)$

(b) Local maximum: 5.25 at $t = -1.5$; absolute maximum: 5.25 at $t = -1.5$

21. (a) Decreasing on $(-\infty, 0)$; increasing on $(0, 4/3)$; decreasing on $(4/3, \infty)$

(b) Local minimum at $x = 0$ $(0, 0)$; local maximum at $x = 4/3$ $(4/3, 32/27)$; no absolute extrema

23. (a) Decreasing on $(-\infty, 0)$; increasing on $(0, 1/2)$; decreasing on $(1/2, \infty)$

(b) Local minimum at $\theta = 0$ $(0, 0)$; local maximum at $\theta = 1/2$ $(1/2, 1/4)$; no absolute extrema

25. (a) Increasing on $(-\infty, \infty)$; never decreasing

(b) No local extrema; no absolute extrema

27. (a) Increasing on $(-2, 0)$ and $(2, \infty)$; decreasing on $(-\infty, -2)$ and $(0, 2)$

(b) Local maximum: 16 at $x = 0$; local minimum: 0 at $x = \pm 2$; no absolute maximum; absolute minimum: 0 at $x = \pm 2$

29. (a) Increasing on $(-\infty, -1)$; decreasing on $(-1, 0)$; increasing on $(0, 1)$; decreasing on $(1, \infty)$

(b) Local maximum: 0.5 at $x = \pm 1$; local minimum: 0 at $x = 0$; absolute maximum: 1/2 at $x = \pm 1$; no absolute minimum

31. (a) Increasing on $(10, \infty)$; decreasing on $(1, 10)$

(b) Local maximum: 1 at $x = 1$; local minimum: -8 at $x = 10$; absolute minimum: -8 at $x = 10$

33. (a) Decreasing on $\left(-2\sqrt{2}, -2\right)$; increasing on $(-2, 2)$; decreasing on $\left(2, 2\sqrt{2}\right)$

(b) Local minima: $g(-2) = -4$, $g\left(2\sqrt{2}\right) = 0$; local maxima: $g\left(-2\sqrt{2}\right) = 0$, $g(2) = 4$; absolute maximum: 4 at $x = 2$; absolute minimum: -4 at $x = -2$

35. (a) Increasing on $(-\infty, 1)$; decreasing when $1 < x < 2$, decreasing when $2 < x < 3$; discontinuous at $x = 2$; increasing on $(3, \infty)$
  (b) Local minimum at $x = 3$ $(3, 6)$; local maximum at $x = 1$ $(1, 2)$; no absolute extrema

37. (a) Increasing on $(-2, 0)$ and $(0, \infty)$; decreasing on $(-\infty, -2)$
  (b) Local minimum: $-6\sqrt[3]{2}$ at $x = -2$; no absolute maximum; absolute minimum: $-6\sqrt[3]{2}$ at $x = -2$

39. (a) Increasing on $\left(-\infty, -2/\sqrt{7}\right)$ and $\left(2/\sqrt{7}, \infty\right)$; decreasing on $\left(-2/\sqrt{7}, 0\right)$ and $\left(0, 2/\sqrt{7}\right)$
  (b) Local maximum: $24\sqrt[3]{2}/7^{7/6} \approx 3.12$ at $x = -2/\sqrt{7}$; local minimum: $-24\sqrt[3]{2}/7^{7/6} \approx -3.12$ at $x = 2/\sqrt{7}$; no absolute extrema

41. (a) Increasing on $((1/3)\ln(1/2), \infty)$, decreasing on $(-\infty, (1/3)\ln(1/2))$
  (b) Local minimum is $\dfrac{3}{2^{2/3}}$ at $x = (1/3)\ln(1/2)$; no local maximum; absolute minimum is $\dfrac{3}{2^{2/3}}$ at $x = (1/3)\ln(1/2)$; no absolute maximum

43. (a) Increasing on (e^{-1}, ∞), decreasing on $(0, e^{-1})$
  (b) A local minimum is $-e^{-1}$ at $x = e^{-1}$, no local maximum; absolute minimum is $-e^{-1}$ at $x = e^{-1}$, no absolute maximum

45. (a) Increasing on $(0, e^{-2})$ and $(1, \infty)$; decreasing on $(e^{-2}, 1)$
  (b) local maximum is $4/e^2$ at $x = e^{-2}$; absolute minimum is 0 at $x = 1$; no absolute maximum

47. (a) Local maximum: 1 at $x = 1$; local minimum: 0 at $x = 2$
  (b) Absolute maximum: 1 at $x = 1$; no absolute minimum

49. (a) Local maximum: 1 at $x = 1$; local minimum: 0 at $x = 2$
  (b) No absolute maximum; absolute minimum: 0 at $x = 2$

51. (a) Local maxima: -9 at $t = -3$ and 16 at $t = 2$; local minimum: -16 at $t = -2$
  (b) Absolute maximum: 16 at $t = 2$; no absolute minimum

53. (a) Local minimum: 0 at $x = 0$
  (b) No absolute maximum; absolute minimum: 0 at $x = 0$

55. (a) Local maximum: 5 at $x = 0$; local minimum: 0 at $x = -5$ and $x = 5$
  (b) Absolute maximum: 5 at $x = 0$; absolute minimum: 0 at $x = -5$ and $x = 5$

57. (a) Local maximum: 2 at $x = 0$;
  local minimum: $\dfrac{\sqrt{3}}{4\sqrt{3} - 6}$ at $x = 2 - \sqrt{3}$
  (b) No absolute maximum; an absolute minimum at $x = 2 - \sqrt{3}$

59. (a) Local maximum: 1 at $x = \pi/4$;
  local maximum: 0 at $x = \pi$;
  local minimum: 0 at $x = 0$;
  local minimum: -1 at $x = 3\pi/4$

61. Local maximum: 2 at $x = \pi/6$;
  local maximum: $\sqrt{3}$ at $x = 2\pi$;
  local minimum: -2 at $x = 7\pi/6$;
  local minimum: $\sqrt{3}$ at $x = 0$

63. (a) Local minimum: $(\pi/3) - \sqrt{3}$ at $x = 2\pi/3$;
  local maximum: 0 at $x = 0$;
  local maximum: π at $x = 2\pi$

65. (a) Local minimum: 0 at $x = \pi/4$

67. Local minimum at $x = 1$; no local maximum

69. Local maximum: 3 at $\theta = 0$;
  local minimum: -3 at $\theta = 2\pi$

71.

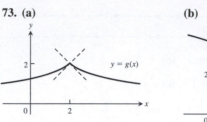

(a)   (b)   (c)   (d)

73. (a)   **(b)**

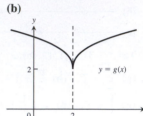

77. $a = -2, b = 4$

79. (a) Absolute minimum occurs at $x = \pi/3$ with $f(\pi/3) = -\ln 2$, and the absolute maximum occurs at $x = 0$ with $f(0) = 0$.
  (b) Absolute minimum occurs at $x = 1/2$ and $x = 2$ with $f(1/2) = f(2) = \cos(\ln 2)$, and the absolute maximum occurs at $x = 1$ with $f(1) = 1$.

81. Minimum of $2 - 2\ln 2 \approx 0.613706$ at $x = \ln 2$; maximum of 1 at $x = 0$

83. Absolute maximum value of $1/2e$ assumed at $x = 1/\sqrt{e}$

87. Increasing; $\dfrac{df^{-1}}{dx} = \dfrac{1}{9}x^{-2/3}$

89. Decreasing; $\dfrac{df^{-1}}{dx} = -\dfrac{1}{3}x^{-2/3}$

SECTION 4.4, pp. 251–255

1. Local maximum: $3/2$ at $x = -1$; local minimum: -3 at $x = 2$; point of inflection at $(1/2, -3/4)$; rising on $(-\infty, -1)$ and $(2, \infty)$; falling on $(-1, 2)$; concave up on $(1/2, \infty)$; concave down on $(-\infty, 1/2)$

3. Local maximum: $3/4$ at $x = 0$; local minimum: 0 at $x = \pm 1$; points of inflection at $\left(-\sqrt{3}, \dfrac{3\sqrt[3]{4}}{4}\right)$ and $\left(\sqrt{3}, \dfrac{3\sqrt[3]{4}}{4}\right)$; rising on $(-1, 0)$ and $(1, \infty)$; falling on $(-\infty, -1)$ and $(0, 1)$; concave up on $\left(-\infty, -\sqrt{3}\right)$ and $\left(\sqrt{3}, \infty\right)$; concave down on $\left(-\sqrt{3}, \sqrt{3}\right)$

5. Local maxima: $\dfrac{-2\pi}{3} + \dfrac{\sqrt{3}}{2}$ at $x = -2\pi/3$, $\dfrac{\pi}{3} + \dfrac{\sqrt{3}}{2}$ at $x = \pi/3$; local minima: $-\dfrac{\pi}{3} - \dfrac{\sqrt{3}}{2}$ at $x = -\pi/3$, $\dfrac{2\pi}{3} - \dfrac{\sqrt{3}}{2}$ at $x = 2\pi/3$; points of inflection at $(-\pi/2, -\pi/2)$, $(0, 0)$, and $(\pi/2, \pi/2)$; rising on $(-\pi/3, \pi/3)$; falling on $(-2\pi/3, -\pi/3)$ and $(\pi/3, 2\pi/3)$; concave up on $(-\pi/2, 0)$ and $(\pi/2, 2\pi/3)$; concave down on $(-2\pi/3, -\pi/2)$ and $(0, \pi/2)$

7. Local maxima: 1 at $x = -\pi/2$ and $x = \pi/2$, 0 at $x = -2\pi$ and $x = 2\pi$; local minima: -1 at $x = -3\pi/2$ and $x = 3\pi/2$, 0 at $x = 0$; points of inflection at $(-\pi, 0)$ and $(\pi, 0)$; rising on $(-3\pi/2, -\pi/2)$, $(0, \pi/2)$, and $(3\pi/2, 2\pi)$; falling on $(-2\pi, -3\pi/2)$, $(-\pi/2, 0)$, and $(\pi/2, 3\pi/2)$; concave up on $(-2\pi, -\pi)$ and $(\pi, 2\pi)$; concave down on $(-\pi, 0)$ and $(0, \pi)$

9.

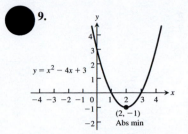

$y = x^2 - 4x + 3$

$(2, -1)$ Abs min

11.

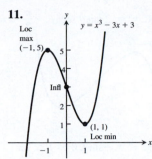

Loc max $(-1, 5)$

$y = x^3 - 3x + 3$

Infl

$(1, 1)$ Loc min

27.

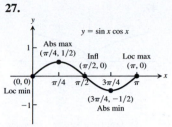

$y = \sin x \cos x$

Abs max $(\pi/4, 1/2)$

Infl $(\pi/2, 0)$

Loc max $(\pi, 0)$

$(0, 0)$ Loc min

$(3\pi/4, -1/2)$ Abs min

13.

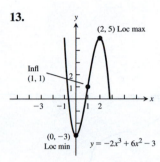

$(2, 5)$ Loc max

Infl $(1, 1)$

$(0, -3)$ Loc min

$y = -2x^3 + 6x^2 - 3$

15.

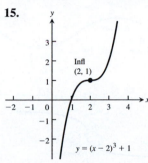

Infl $(2, 1)$

$y = (x - 2)^3 + 1$

29.

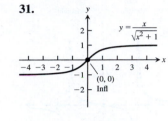

Vert tan at $x = 0$

$y = x^{1/5}$

$(0, 0)$ Infl

31.

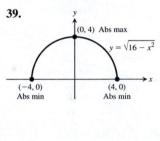

$y = \dfrac{x}{\sqrt{x^2 + 1}}$

$(0, 0)$ Infl

17.

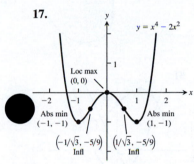

$y = x^4 - 2x^2$

Loc max $(0, 0)$

Abs min $(-1, -1)$

Abs min $(1, -1)$

$\left(-1/\sqrt{3}, -5/9\right)$ Infl

$\left(1/\sqrt{3}, -5/9\right)$ Infl

19.

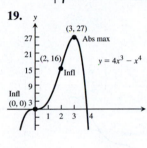

$(3, 27)$ Abs max

$(2, 16)$ Infl

$y = 4x^3 - x^4$

Infl $(0, 0)$

33.

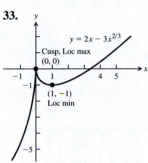

$y = 2x - 3x^{2/3}$

Cusp, Loc max $(0, 0)$

$(1, -1)$ Loc min

35.

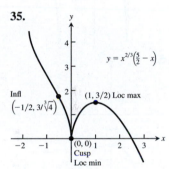

$y = x^{2/3}\left(\dfrac{5}{2} - x\right)$

Infl $\left(-1/2, 3/\sqrt[3]{4}\right)$

$(1, 3/2)$ Loc max

$(0, 0)$ Cusp Loc min

37.

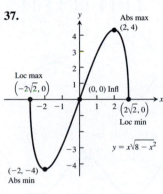

Abs max $(2, 4)$

Loc max $\left(-2\sqrt{2}, 0\right)$

$(0, 0)$ Infl

$\left(2\sqrt{2}, 0\right)$ Loc min

$(-2, -4)$ Abs min

$y = x\sqrt{8 - x^2}$

39.

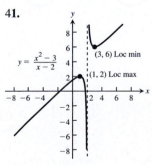

$(0, 4)$ Abs max

$y = \sqrt{16 - x^2}$

$(-4, 0)$ Abs min

$(4, 0)$ Abs min

21.

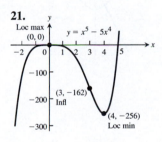

Loc max $(0, 0)$

$y = x^5 - 5x^4$

$(3, -162)$ Infl

$(4, -256)$ Loc min

23.

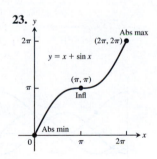

$(2\pi, 2\pi)$ Abs max

$y = x + \sin x$

(π, π) Infl

Abs min

41.

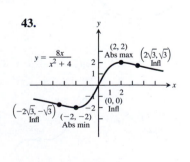

$(3, 6)$ Loc min

$y = \dfrac{x^2 - 3}{x - 2}$

$(1, 2)$ Loc max

43.

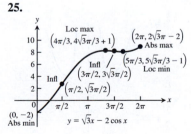

$y = \dfrac{8x}{x^2 + 4}$

$(2, 2)$ Abs max

$\left(2\sqrt{3}, \sqrt{3}\right)$ Infl

$\left(-2\sqrt{3}, -\sqrt{3}\right)$ Infl

$(-2, -2)$ Abs min

$(0, 0)$ Infl

25.

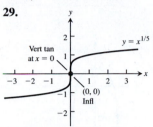

Loc max $\left(4\pi/3, 4\sqrt{3}\pi/3 + 1\right)$

$\left(2\pi, 2\sqrt{3}\pi - 2\right)$ Abs max

Infl $\left(3\pi/2, 3\sqrt{3}\pi/2\right)$

$\left(5\pi/3, 5\sqrt{3}\pi/3 - 1\right)$ Loc min

Infl $\left(\pi/2, \sqrt{3}\pi/2\right)$

$(0, -2)$ Abs min

$y = \sqrt{3}x - 2\cos x$

45.

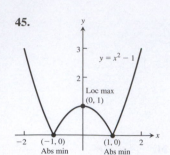

47.

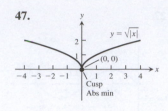

71. $y'' = 2\tan\theta\sec^2\theta, -\dfrac{\pi}{2} < \theta < \dfrac{\pi}{2}$

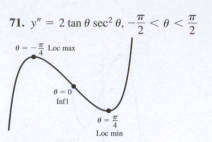

49.

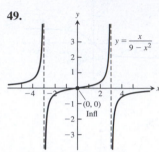

51.

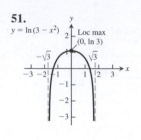

73. $y'' = -\sin t, 0 \le t \le 2\pi$

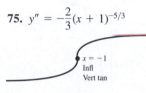

75. $y'' = -\dfrac{2}{3}(x+1)^{-5/3}$

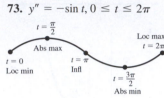

77. $y'' = \dfrac{1}{3}x^{-2/3} + \dfrac{2}{3}x^{-5/3}$

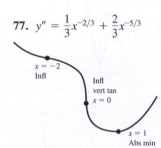

53.

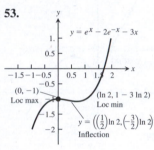

55.

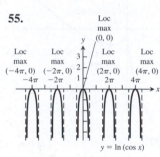

79. $y'' = \begin{cases} -2, & x < 0 \\ 2, & x > 0 \end{cases}$

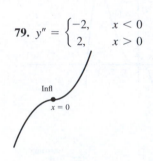

81.

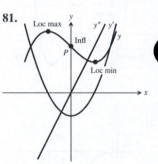

57.

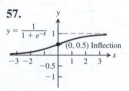

59. $y'' = 1 - 2x$

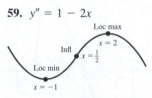

83.

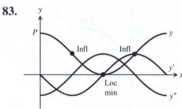

61. $y'' = 3(x-3)(x-1)$

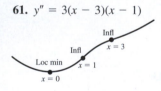

63. $y'' = 3(x-2)(x+2)$

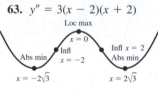

65. $y'' = 4(4-x)(5x^2 - 16x + 8)$

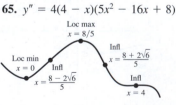

85.

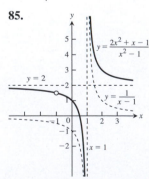

87.

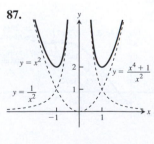

67. $y'' = 2\sec^2 x\tan x$

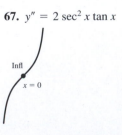

69. $y'' = -\dfrac{1}{2}\csc^2\dfrac{\theta}{2},$
$0 < \theta < 2\pi$

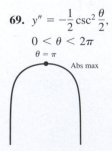

89.

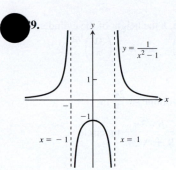

91.

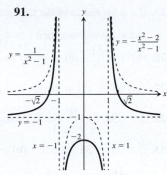

93.

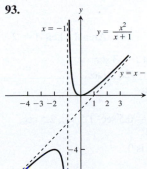

95.

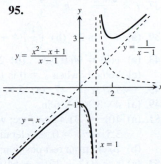

97.

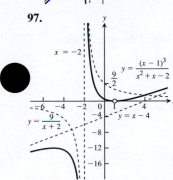

99.

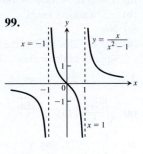

101.

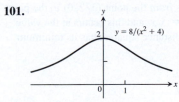

103.

Point	y'	y''
P	$-$	$+$
Q	$+$	0
R	$+$	$-$
S	0	$-$
T	$-$	$-$

105.

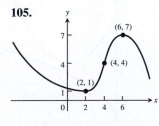

107.

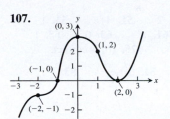

109.

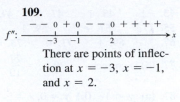

There are points of inflection at $x = -3$, $x = -1$, and $x = 2$.

111.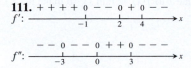

There are local maxima at $x = -1$ and $x = 4$. There is a local minimum at $x = 2$. There are points of inflection at $x = 0$ and $x = 3$.

113. (a) Towards origin: $0 \le t < 2$ and $6 \le t \le 10$; away from origin: $2 \le t \le 6$ and $10 \le t \le 15$

(b) $t = 2, t = 6, t = 10$

(c) $t = 5, t = 7, t = 13$

(d) Positive: $5 \le t \le 7$, $13 \le t \le 15$; negative: $0 \le t \le 5, 7 \le t \le 13$

115. ≈ 60 thousand units

117. Local minimum at $x = 2$; inflection points at $x = 1$ and $x = 5/3$

119. $-1, 2$ **121.** $b = -3$ **127.** $a = 1, b = 3, c = 9$

129. The zeros of $y' = 0$ and $y'' = 0$ are extrema and points of inflection, respectively. Inflection at $x = 3$, local maximum at $x = 0$, local minimum at $x = 4$.

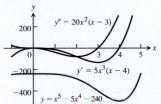

131. The zeros of $y' = 0$ and $y'' = 0$ are extrema and points of inflection, respectively. Inflection at $x = -\sqrt[3]{2}$; local maximum at $x = -2$; local minimum at $x = 0$.

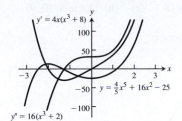

SECTION 4.5, pp. 262–263

1. $-1/4$ **3.** $5/7$ **5.** $1/2$ **7.** $1/4$ **9.** $-23/7$

11. $5/7$ **13.** 0 **15.** -16 **17.** -2 **19.** $1/4$

21. 2 **23.** 3 **25.** -1 **27.** $\ln 3$ **29.** $\dfrac{1}{\ln 2}$ **31.** $\ln 2$

33. 1 **35.** $1/2$ **37.** $\ln 2$ **39.** $-\infty$ **41.** $-1/2$

43. -1 **45.** 1 **47.** 0 **49.** 2 **51.** $1/e$ **53.** 1
55. $1/e$ **57.** $e^{1/2}$ **59.** 1 **61.** e^3 **63.** 0 **65.** 1
67. 3 **69.** 1 **71.** 0 **73.** ∞ **75.** (b) is correct.

77. (d) is correct.   **79.** $c = \dfrac{27}{10}$ **81.** (b) $\dfrac{-1}{2}$ **83.** -1

87. (a) $y = 1$ (b) $y = 0, y = \dfrac{3}{2}$

89. (a) We should assign the value 1 to $f(x) = (\sin x)^x$ to make it continuous at $x = 0$.

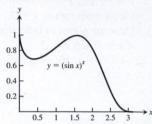

(c) The maximum value of $f(x)$ is close to 1 near the point $x \approx 1.55$ (see the graph in part (a)).

SECTION 4.6, pp. 269–276

1. 16 in., 4 in. by 4 in.

3. (a) $(x, 1 - x)$ (b) $A(x) = 2x(1 - x)$
(c) $\dfrac{1}{2}$ square units, 1 by $\dfrac{1}{2}$

5. $\dfrac{14}{3} \times \dfrac{35}{3} \times \dfrac{5}{3}$ in., $\dfrac{2450}{27}$ in^3

7. 80,000 m^2; 400 m by 200 m

9. (a) The optimum dimensions of the tank are 10 ft on the base edges and 5 ft deep.
(b) Minimizing the surface area of the tank minimizes its weight for a given wall thickness. The thickness of the steel walls would likely be determined by other considerations such as structural requirements.

11. 9×18 in.   **13.** $\dfrac{\pi}{2}$ **15.** $h : r = 8 : \pi$

17. (a) $V(x) = 2x(24 - 2x)(18 - 2x)$ (b) Domain: $(0, 9)$

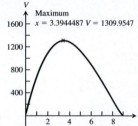

(c) Maximum volume ≈ 1309.95 in^3 when $x \approx 3.39$ in.
(d) $V'(x) = 24x^2 - 336x + 864$, so the critical point is at $x = 7 - \sqrt{13}$, which confirms the result in part (c).
(e) $x = 2$ in. or $x = 5$ in.

19. ≈ 2418.40 cm^3

21. (a) $h = 24, w = 18$
(b)

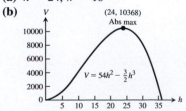

23. If r is the radius of the hemisphere, h the height of the cylinder, and V the volume, then $r = \left(\dfrac{3V}{8\pi}\right)^{1/3}$ and $h = \left(\dfrac{3V}{\pi}\right)^{1/3}$.

25. (b) $x = \dfrac{51}{8}$ (c) $L \approx 11$ in.

27. Radius $= \sqrt{2}$ m, height $= 1$ m, volume $= \dfrac{2\pi}{3}$ m^3

29. 1 **31.** $\dfrac{9b}{9 + \sqrt{3}\pi}$ m, triangle; $\dfrac{b\sqrt{3}\pi}{9 + \sqrt{3}\pi}$ m, circle

33. $\dfrac{3}{2} \times 2$ **35.** (a) 16 (b) -1

37. $r = \dfrac{2\sqrt{2}}{3}$ $h = \dfrac{4}{3}$

39. (a) $4 + \ln 2$ (b) $\left(\dfrac{1}{2}\right)(4 + \ln 2)$ **41.** Area 8 when $a = 2$

43. (a) $v(0) = 96$ ft/sec   (b) 256 ft at $t = 3$ sec
(c) Velocity when $s = 0$ is $v(7) = -128$ ft/sec.

45. ≈ 46.87 ft   **47.** (a) $6 \times 6\sqrt{3}$ in.

49. (a) $4\sqrt{3} \times 4\sqrt{6}$ in.

51. (a) $10\pi \approx 31.42$ cm/sec; when $t = 0.5$ sec, 1.5 sec, 2.5 sec, 3.5 sec; $s = 0$, acceleration is 0.
(b) 10 cm from rest position; speed is 0.

53. (a) $s = ((12 - 12t)^2 + 64t^2)^{1/2}$
(b) -12 knots, 8 knots
(c) No
(d) $4\sqrt{13}$. This limit is the square root of the sums of the squares of the individual speeds.

55. $x = \dfrac{a}{2}, v = \dfrac{ka^2}{4}$ **57.** $\dfrac{c}{2} + 50$

59. (a) $\sqrt{\dfrac{2km}{h}}$ (b) $\sqrt{\dfrac{2km}{h}}$ **63.** $4 \times 4 \times 3$ ft, \$288

65. $M = \dfrac{C}{2}$ **71.** (a) $y = -1$

73. (a) The minimum distance is $\dfrac{\sqrt{5}}{2}$.
(b) The minimum distance is from the point $(3/2, 0)$ to the point $(1, 1)$ on the graph of $y = \sqrt{x}$, and this occurs at the value $x = 1$, where $D(x)$, the distance squared, has its minimum value.

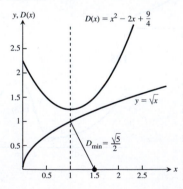

SECTION 4.7, pp. 279–281

1. $x_2 = -\dfrac{5}{3}, \dfrac{13}{21}$ **3.** $x_2 = -\dfrac{51}{31}, \dfrac{5763}{4945}$ **5.** $x_2 = \dfrac{2387}{2000}$

7. x_2 is approximately 2.20794   **9.** x_2 is approximately 0.68394

11. x_1, and all later approximations will equal x_0.

13.

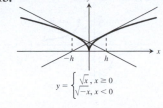

$$y = \begin{cases} \sqrt{x}, & x \ge 0 \\ \sqrt{-x}, & x < 0 \end{cases}$$

15. The points of intersection of $y = x^3$ and $y = 3x + 1$ or $y = x^3 - 3x$ and $y = 1$ have the same x-values as the roots of part (i) or the solutions of part (iv). **17.** 1.165561185

19. (a) Two (b) 0.35003501505249 and -1.0261731615301

21. ± 1.3065629648764, ± 0.5411961001462 **23.** $x \approx 0.45$

25. 0.8192 **27.** 0, 0.53485 **29.** The root is 1.17951.

31. (a) For $x_0 = -2$ or $x_0 = -0.8$, $x_i \to -1$ as i gets large.
  (b) For $x_0 = -0.5$ or $x_0 = 0.25$, $x_i \to 0$ as i gets large.
  (c) For $x_0 = 0.8$ or $x_0 = 2$, $x_i \to 1$ as i gets large.
  (d) For $x_0 = -\sqrt{21}/7$ or $x_0 = \sqrt{21}/7$, Newton's method does not converge. The values of x_i alternate between $-\sqrt{21}/7$ and $\sqrt{21}/7$ as i increases.

33. Answers will vary with machine speed.

SECTION 4.8, pp. 287–291

1. (a) x^2 (b) $\dfrac{x^3}{3}$ (c) $\dfrac{x^3}{3} - x^2 + x$

3. (a) x^{-3} (b) $-\dfrac{1}{3}x^{-3}$ (c) $-\dfrac{1}{3}x^{-3} + x^2 + 3x$

5. (a) $-\dfrac{1}{x}$ (b) $-\dfrac{5}{x}$ (c) $2x + \dfrac{5}{x}$

7. (a) $\sqrt{x^3}$ (b) $\sqrt{x}$ (c) $\dfrac{2\sqrt{x^3}}{3} + 2\sqrt{x}$

9. (a) $x^{2/3}$ (b) $x^{1/3}$ (c) $x^{-1/3}$

11. (a) $\ln x$ (b) $7 \ln x$ (c) $x - 5 \ln x$

13. (a) $\cos(\pi x)$ (b) $-3 \cos x$ (c) $-\dfrac{1}{\pi}\cos(\pi x) + \cos(3x)$

15. (a) $\tan x$ (b) $2\tan\left(\dfrac{x}{3}\right)$ (c) $-\dfrac{2}{3}\tan\left(\dfrac{3x}{2}\right)$

17. (a) $-\csc x$ (b) $\dfrac{1}{5}\csc(5x)$ (c) $2\csc\left(\dfrac{\pi x}{2}\right)$

19. (a) $\dfrac{1}{3}e^{3x}$ (b) $-e^{-x}$ (c) $2e^{x/2}$

21. (a) $\dfrac{1}{\ln 3}3^x$ (b) $\dfrac{-1}{\ln 2}2^{-x}$ (c) $\dfrac{1}{\ln(5/3)}\left(\dfrac{5}{3}\right)^x$

23. (a) $2\sin^{-1}x$ (b) $\dfrac{1}{2}\tan^{-1}x$ (c) $\dfrac{1}{2}\tan^{-1}2x$

25. $\dfrac{x^2}{2} + x + C$ **27.** $t^3 + \dfrac{t^2}{4} + C$ **29.** $\dfrac{x^4}{2} - \dfrac{5x^2}{2} + 7x + C$

31. $-\dfrac{1}{x} - \dfrac{x^3}{3} - \dfrac{x}{3} + C$ **33.** $\dfrac{3}{2}x^{2/3} + C$

35. $\dfrac{2}{3}x^{3/2} + \dfrac{3}{4}x^{4/3} + C$ **37.** $4y^2 - \dfrac{8}{3}y^{3/4} + C$

39. $x^2 + \dfrac{2}{x} + C$ **41.** $2\sqrt{t} - \dfrac{2}{\sqrt{t}} + C$ **43.** $-2\sin t + C$

45. $-21\cos\dfrac{\theta}{3} + C$ **47.** $3\cot x + C$ **49.** $-\dfrac{1}{2}\csc\theta + C$

51. $\dfrac{1}{3}e^{3x} - 5e^{-x} + C$ **53.** $-e^{-x} + \dfrac{4^x}{\ln 4} + C$

55. $4\sec x - 2\tan x + C$ **57.** $-\dfrac{1}{2}\cos 2x + \cot x + C$

59. $\dfrac{t}{2} + \dfrac{\sin 4t}{8} + C$ **61.** $\ln|x| - 5\tan^{-1}x + C$

63. $\dfrac{3x^{(\sqrt{3}+1)}}{\sqrt{3} + 1} + C$ **65.** $\tan\theta + C$ **67.** $-\cot x - x + C$

69. $-\cos\theta + \theta + C$

83. (a) Wrong: $\dfrac{d}{dx}\left(\dfrac{x^2}{2}\sin x + C\right) = \dfrac{2x}{2}\sin x + \dfrac{x^2}{2}\cos x = x\sin x + \dfrac{x^2}{2}\cos x$

  (b) Wrong: $\dfrac{d}{dx}(-x\cos x + C) = -\cos x + x\sin x$

  (c) Right: $\dfrac{d}{dx}(-x\cos x + \sin x + C) = -\cos x + x\sin x + \cos x = x\sin x$

85. (a) Wrong: $\dfrac{d}{dx}\left(\dfrac{(2x + 1)^3}{3} + C\right) = \dfrac{3(2x + 1)^2(2)}{3} = 2(2x + 1)^2$

  (b) Wrong: $\dfrac{d}{dx}((2x + 1)^3 + C) = 3(2x + 1)^2(2) = 6(2x + 1)^2$

  (c) Right: $\dfrac{d}{dx}((2x + 1)^3 + C) = 6(2x + 1)^2$

87. Right **89.** (b) **91.** $y = x^2 - 7x + 10$

93. $y = -\dfrac{1}{x} + \dfrac{x^2}{2} - \dfrac{1}{2}$ **95.** $y = 9x^{1/3} + 4$

97. $s = t + \sin t + 4$ **99.** $r = \cos(\pi\theta) - 1$

101. $v = \dfrac{1}{2}\sec t + \dfrac{1}{2}$ **103.** $v = 3\sec^{-1}t - \pi$

105. $y = x^2 - x^3 + 4x + 1$ **107.** $r = \dfrac{1}{t} + 2t - 2$

109. $y = x^3 - 4x^2 + 5$ **111.** $y = -\sin t + \cos t + t^3 - 1$

113. $y = 2x^{3/2} - 50$

115.

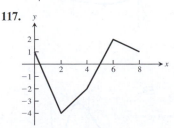

117.

119. $y = x - x^{4/3} + \dfrac{1}{2}$ **121.** $y = -\sin x - \cos x - 2$

123. (a) (i) 33.2 units, (ii) 33.2 units, (iii) 33.2 units
  (b) True

125. $t = 88/k$, $k = 16$

127. (a) $v = 10t^{3/2} - 6t^{1/2}$ (b) $s = 4t^{5/2} - 4t^{3/2}$

131. (a) $-\sqrt{x} + C$ **(b)** $x + C$ **(c)** $\sqrt{x} + C$
(d) $-x + C$ **(e)** $x - \sqrt{x} + C$ **(f)** $-x - \sqrt{x} + C$

PRACTICE EXERCISES, pp. 292–296

1. Minimum value is 1 at $x = 2$.

3. Local maximum at $(-2, 17)$; local minimum at $\left(\dfrac{4}{3}, -\dfrac{41}{27}\right)$

5. Minimum value is 0 at $x = -1$ and $x = 1$.

7. There is a local minimum at $(0, 1)$.

9. Maximum value is $\dfrac{1}{2}$ at $x = 1$; minimum value is $-\dfrac{1}{2}$ at $x = -1$.

11. The minimum value is 2 at $x = 0$.

13. The minimum value is $-\dfrac{1}{e}$ at $x = \dfrac{1}{e}$.

15. The maximum value is $\dfrac{\pi}{2}$ at $x = 0$; an absolute minimum value is 0 at $x = 1$ and $x = -1$.

17. No   **19.** No minimum; absolute maximum: $f(1) = 16$; critical points: $x = 1$ and $11/3$

21. Absolute minimum: $g(0) = 1$; no absolute maximum; critical point: $x = 0$

23. Absolute minimum: $2 - 2\ln 2$ at $x = 2$; absolute maximum: 1 at $x = 1$

25. Yes, except at $x = 0$ **27.** No   **31. (b)** one

33. (b) 0.8555 99677 2

39. Global minimum value of $\dfrac{1}{2}$ at $x = 2$

41. (a) $t = 0, 6, 12$ **(b)** $t = 3, 9$ **(c)** $6 < t < 12$
(d) $0 < t < 6, 12 < t < 14$

43.

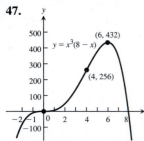

45.

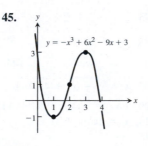

47.

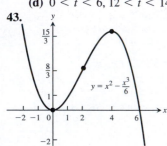

49.

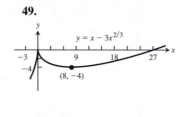

51.

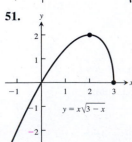

53.

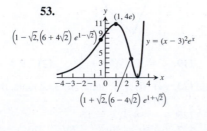

55.

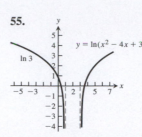

57.

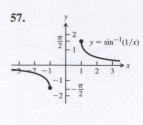

59. (a) Local maximum at $x = 4$, local minimum at $x = -4$, inflection point at $x = 0$

(b)

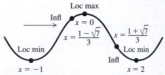

61. (a) Local maximum at $x = 0$, local minima at $x = -1$ and $x = 2$, inflection points at $x = (1 \pm \sqrt{7})/3$

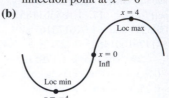

63. (a) Local maximum at $x = -\sqrt{2}$, local minimum at $x = \sqrt{2}$, inflection points at $x = \pm 1$ and 0

(b)

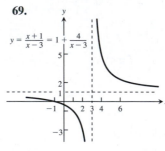

69.

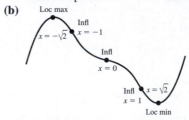

71.

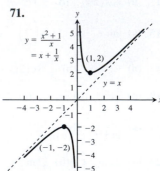

73.

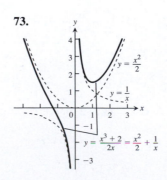

75.

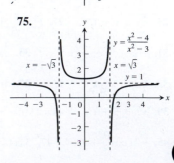

77. 5 **79.** 0 **81.** 1 **83.** 3/7 **85.** 0 **87.** 1

89. $\ln 10$ **91.** $\ln 2$ **93.** 5 **95.** $-\infty$ **97.** 1 **99.** e^{bk}

101. $-\infty$ **103. (a)** $0, 36$ **(b)** $18, 18$ **105.** 54 square units

107. height $= 2$, radius $= \sqrt{2}$

109. $x = 5 - \sqrt{5}$ hundred ≈ 276 tires,
$y = 2\left(5 - \sqrt{5}\right)$ hundred ≈ 553 tires

111. Dimensions: base is 6 in. by 12 in., height $= 2$ in.;
maximum volume $= 144$ in^3

113. $x_5 = 2.1958\ 23345$ **115.** $\dfrac{x^4}{4} + \dfrac{5}{2}x^2 - 7x + C$

117. $2t^{3/2} - \dfrac{4}{t} + C$ **119.** $-\dfrac{1}{r+5} + C$

121. $(\theta^2 + 1)^{3/2} + C$ **123.** $\dfrac{1}{3}(1 + x^4)^{3/4} + C$

125. $10\tan\dfrac{s}{10} + C$ **127.** $-\dfrac{1}{\sqrt{2}}\csc\sqrt{2}\theta + C$

129. $\dfrac{1}{2}x - \sin\dfrac{x}{2} + C$ **131.** $3\ln x - \dfrac{x^2}{2} + C$

133. $\dfrac{1}{2}e^t + e^{-t} + C$ **135.** $\dfrac{\theta^{2-\pi}}{2-\pi} + C$

137. $\dfrac{3}{2}\sec^{-1}|x| + C$ **139.** $y = x - \dfrac{1}{x} - 1$

141. $r = 4t^{5/2} + 4t^{3/2} - 8t$

143. Yes, $\sin^{-1}(x)$ and $-\cos^{-1}(x)$ differ by the constant $\pi/2$.

145. $1/\sqrt{2}$ units long by $1/\sqrt{e}$ units high, $A = 1/\sqrt{2e} \approx$
0.43 units^2

147. Absolute maximum $= 0$ at $x = e/2$, absolute minimum $=$
-0.5 at $x = 0.5$

149. $x = \pm 1$ are the critical points; $y = 1$ is a horizontal asymptote
in both directions; absolute minimum value of the function is
$e^{-\sqrt{2}/2}$ at $x = -1$, and absolute maximum value is $e^{\sqrt{2}/2}$ at
$x = 1$.

151. (a) Absolute maximum of $2/e$ at $x = e^2$, inflection point
$(e^{8/3}, (8/3)e^{-4/3})$, concave up on $(e^{8/3}, \infty)$, concave down
on $(0, e^{8/3})$

  (b) Absolute maximum of 1 at $x = 0$, inflection points
  $\left(\pm 1/\sqrt{2}, 1/\sqrt{e}\right)$, concave up on $\left(-\infty, -1/\sqrt{2}\right) \cup$
  $\left(1/\sqrt{2}, \infty\right)$, concave down on $\left(-1/\sqrt{2}, 1/\sqrt{2}\right)$

  (c) Absolute maximum of 1 at $x = 0$, inflection point $(1, 2/e)$,
  concave up on $(1, \infty)$, concave down on $(-\infty, 1)$

ADDITIONAL AND ADVANCED EXERCISES, pp. 296–299

1. The function is constant on the interval.

3. The extreme points will not be at the end of an open interval.

5. (a) A local minimum at $x = -1$, points of inflection at $x = 0$
and $x = 2$

  (b) A local maximum at $x = 0$ and local minima at $x = -1$ and
  $x = 2$, points of inflection at $x = \dfrac{1 \pm \sqrt{7}}{3}$

9. No **11.** $a = 1, b = 0, c = 1$ **13.** Yes

15. Drill the hole at $y = h/2$.

17. $r = \dfrac{RH}{2(H-R)}$ for $H > 2R, r = R$ if $H \le 2R$

19. $\dfrac{12}{5}$ and 5

21. (a) $\dfrac{10}{3}$ **(b)** $\dfrac{5}{3}$ **(c)** $\dfrac{1}{2}$ **(d)** 0 **(e)** $-\dfrac{1}{2}$ **(f)** 1

  (g) $\dfrac{1}{2}$ **(h)** 3

23. (a) $\dfrac{c-b}{2e}$ **(b)** $\dfrac{c+b}{2}$ **(c)** $\dfrac{b^2 - 2bc + c^2 + 4ae}{4e}$

  (d) $\dfrac{c+b+t}{2}$

25. $m_0 = 1 - \dfrac{1}{q}, m_1 = \dfrac{1}{q}$

27. $s = ce^{kt}$

29. (a) $k = -38.72$ **(b)** 25 ft

31. Yes, $y = x + C$ **33.** $v_0 = \dfrac{2\sqrt{2}}{3}b^{3/4}$ **39.** 3

Chapter 5

SECTION 5.1, pp. 308–310

1. (a) 0.125 **(b)** 0.21875 **(c)** 0.625 **(d)** 0.46875

3. (a) 1.066667 **(b)** 1.283333 **(c)** 2.666667 **(d)** 2.083333

5. $0.3125, 0.328125$ **7.** $1.5, 1.574603$

9. (a) 245 cm **(b)** 245 cm **11. (a)** 3490 ft **(b)** 3840 ft

13. (a) 74.65 ft/sec **(b)** 45.28 ft/sec **(c)** 146.59 ft

15. $\dfrac{31}{16}$ **17.** 1

19. (a) Upper $= 758$ gal, lower $= 543$ gal

  (b) Upper $= 2363$ gal, lower $= 1693$ gal

  (c) ≈ 31.4 h, ≈ 32.4 h

21. (a) 2 **(b)** $2\sqrt{2} \approx 2.828$ **(c)** $8\sin\left(\dfrac{\pi}{8}\right) \approx 3.061$

  (d) Each area is less than the area of the circle, π. As n
  increases, the polygon area approaches π.

SECTION 5.2, pp. 316–317

1. $\dfrac{6(1)}{1+1} + \dfrac{6(2)}{2+1} = 7$

3. $\cos(1)\pi + \cos(2)\pi + \cos(3)\pi + \cos(4)\pi = 0$

5. $\sin\pi - \sin\dfrac{\pi}{2} + \sin\dfrac{\pi}{3} = \dfrac{\sqrt{3} - 2}{2}$

7. All of them **9.** b

11. $\displaystyle\sum_{k=1}^{6} k$ **13.** $\displaystyle\sum_{k=1}^{4} \dfrac{1}{2^k}$ **15.** $\displaystyle\sum_{k=1}^{5} (-1)^{k+1}\dfrac{1}{k}$

17. (a) -15 **(b)** 1 **(c)** 1 **(d)** -11 **(e)** 16

19. (a) 55 **(b)** 385 **(c)** 3025 **21.** -56 **23.** -73

25. 240 **27.** 3376 **29. (a)** 21 **(b)** 3500 **(c)** 2620

31. (a) $4n$ **(b)** cn **(c)** $(n^2 - n)/2$ **33.** 2600 **35.** $-2\sqrt{3}$

37. (a)

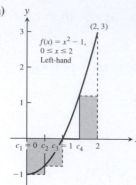

(b)

(c)

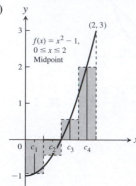

39. (a)

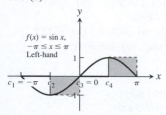

(b)

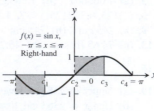

(c)

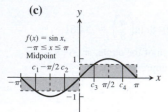

41. 1.2

43. $\dfrac{2}{3} - \dfrac{1}{2n} - \dfrac{1}{6n^2}$, $\dfrac{2}{3}$

45. $12 + \dfrac{27n + 9}{2n^2}$, 12

47. $\dfrac{5}{6} + \dfrac{6n + 1}{6n^2}$, $\dfrac{5}{6}$

49. $\dfrac{1}{2} + \dfrac{1}{n} + \dfrac{1}{2n^2}$, $\dfrac{1}{2}$

SECTION 5.3, pp. 326–329

1. $\displaystyle\int_0^2 x^2\,dx$ **3.** $\displaystyle\int_{-7}^5 (x^2 - 3x)\,dx$ **5.** $\displaystyle\int_2^3 \dfrac{1}{1 - x}\,dx$

7. $\displaystyle\int_{-\pi/4}^0 \sec x\,dx$

9. (a) 0   **(b)** -8 **(c)** -12 **(d)** 10   **(e)** -2 **(f)** 16
11. (a) 5   **(b)** $5\sqrt{3}$ **(c)** -5 **(d)** -5
13. (a) 4   **(b)** -4 **15.** Area $= 21$ square units
17. Area $= 9\pi/2$ square units   **19.** Area $= 2.5$ square units
21. Area $= 3$ square units   **23.** $b^2/4$ **25.** $b^2 - a^2$
27. (a) 2π **(b)** π **29.** $1/2$ **31.** $3\pi^2/2$ **33.** $7/3$
35. $1/24$ **37.** $3a^2/2$ **39.** $b/3$ **41.** -14
43. -2 **45.** $-7/4$ **47.** 7   **49.** 0

51. Using n subintervals of length $\Delta x = b/n$ and right-endpoint values:

$$\text{Area} = \int_0^b 3x^2\,dx = b^3$$

53. Using n subintervals of length $\Delta x = b/n$ and right-endpoint values:

$$\text{Area} = \int_0^b 2x\,dx = b^2$$

55. $\operatorname{av}(f) = 0$ **57.** $\operatorname{av}(f) = -2$ **59.** $\operatorname{av}(f) = 1$
61. (a) $\operatorname{av}(g) = -1/2$ **(b)** $\operatorname{av}(g) = 1$ **(c)** $\operatorname{av}(g) = 1/4$
63. $c(b - a)$ **65.** $b^3/3 - a^3/3$ **67.** 9
69. $b^4/4 - a^4/4$ **71.** $a = 0$ and $b = 1$ maximize the integral.
73. Upper bound $= 1$, lower bound $= 1/2$

75. For example, $\displaystyle\int_0^1 \sin(x^2)\,dx \le \int_0^1 dx = 1$

77. $\displaystyle\int_a^b f(x)\,dx \ge \int_a^b 0\,dx = 0$ **79.** Upper bound $= 1/2$

SECTION 5.4, pp. 339–341

1. $-10/3$ **3.** $124/125$ **5.** $753/16$ **7.** 1   **9.** $2\sqrt{3}$

11. 0   **13.** $-\pi/4$ **15.** $1 - \dfrac{\pi}{4}$ **17.** $\dfrac{2 - \sqrt{2}}{4}$ **19.** $-8/3$

21. $-3/4$ **23.** $\sqrt{2} - \sqrt[4]{8} + 1$ **25.** -1 **27.** 16

29. $7/3$ **31.** $2\pi/3$ **33.** $\dfrac{1}{\pi}(4^\pi - 2^\pi)$ **35.** $\dfrac{1}{2}(e - 1)$

37. $\sqrt{26} - \sqrt{5}$ **39.** $\left(\cos\sqrt{x}\right)\left(\dfrac{1}{2\sqrt{x}}\right)$ **41.** $4t^5$

43. $3x^2 e^{-x^3}$ **45.** $\sqrt{1 + x^2}$ **47.** $-\dfrac{1}{2}x^{-1/2}\sin x$ **49.** 0

51. 1   **53.** $2xe^{(1/2)x^2}$ **55.** 1   **57.** $28/3$ **59.** $1/2$

61. π **63.** $\dfrac{\sqrt{2}\pi}{2}$

65. d, since $y' = \dfrac{1}{x}$ and $y(\pi) = \displaystyle\int_\pi^\pi \dfrac{1}{t}\,dt - 3 = -3$

67. b, since $y' = \sec x$ and $y(0) = \displaystyle\int_0^0 \sec t\,dt + 4 = 4$

69. $y = \displaystyle\int_2^x \sec t\,dt + 3$ **71.** $(1 + \ln t)^2$

73. $\dfrac{2}{3}bh$ **75.** \$9.00

77. (a) $T(0) = 70°F$, $T(16) = 76°F$, $T(25) = 85°F$
(b) $\operatorname{av}(T) = 75°F$
79. $2x - 2$ **81.** $-3x + 5$
83. (a) True. Since f is continuous, g is differentiable by Part 1 of the Fundamental Theorem of Calculus.
(b) True: g is continuous because it is differentiable.
(c) True, since $g'(1) = f(1) = 0$.
(d) False, since $g''(1) = f'(1) > 0$.
(e) True, since $g'(1) = 0$ and $g''(1) = f'(1) > 0$.
(f) False: $g''(x) = f'(x) > 0$, so g'' never changes sign.
(g) True, since $g'(1) = f(1) = 0$ and $g'(x) = f(x)$ is an increasing function of x (because $f'(x) > 0$).

85. (a) $v = \dfrac{ds}{dt} = \dfrac{d}{dt}\displaystyle\int_0^t f(x)\,dx = f(t) \Rightarrow v(5) = f(5) = 2$ m/sec

(b) $a = df/dt$ is negative, since the slope of the tangent line at $t = 5$ is negative.

(c) $s = \int_0^3 f(x)\,dx = \frac{1}{2}(3)(3) = \frac{9}{2}$ m, since the integral is the area of the triangle formed by $y = f(x)$, the x-axis, and $x = 3$.

(d) $t = 6$, since after $t = 6$ to $t = 9$, the region lies below the x-axis.

(e) At $t = 4$ and $t = 7$, since there are horizontal tangents there.

(f) Toward the origin between $t = 6$ and $t = 9$, since the velocity is negative on this interval. Away from the origin between $t = 0$ and $t = 6$, since the velocity is positive there.

(g) Right or positive side, because the integral of f from 0 to 9 is positive, there being more area above the x-axis than below.

SECTION 5.5, pp. 348–349

1. $\frac{1}{6}(2x + 4)^6 + C$ 3. $-\frac{1}{3}(x^2 + 5)^{-3} + C$

5. $\frac{1}{10}(3x^2 + 4x)^5 + C$ 7. $-\frac{1}{3}\cos 3x + C$

9. $\frac{1}{2}\sec 2t + C$ 11. $-6(1 - r^3)^{1/2} + C$

13. $\frac{1}{3}(x^{3/2} - 1) - \frac{1}{6}\sin(2x^{3/2} - 2) + C$

15. (a) $-\frac{1}{4}(\cot^2 2\theta) + C$ (b) $-\frac{1}{4}(\csc^2 2\theta) + C$

17. $-\frac{1}{3}(3 - 2s)^{3/2} + C$ 19. $-\frac{2}{5}(1 - \theta^2)^{5/4} + C$

21. $\left(-2/(1 + \sqrt{x})\right) + C$ 23. $\frac{1}{3}\tan(3x + 2) + C$

25. $\frac{1}{2}\sin^6\left(\frac{x}{3}\right) + C$ 27. $\left(\frac{r^3}{18} - 1\right)^6 + C$

29. $-\frac{2}{3}\cos(x^{3/2} + 1) + C$ 31. $\frac{1}{2\cos(2t + 1)} + C$

33. $-\sin\left(\frac{1}{t} - 1\right) + C$ 35. $-\frac{\sin^2(1/\theta)}{2} + C$

37. $\frac{2}{3}(1 + x)^{3/2} - 2(1 + x)^{1/2} + C$ 39. $\frac{2}{3}\left(2 - \frac{1}{x}\right)^{3/2} + C$

41. $\frac{2}{27}\left(1 - \frac{3}{x^3}\right)^{3/2} + C$ 43. $\frac{1}{12}(x - 1)^{12} + \frac{1}{11}(x - 1)^{11} + C$

45. $-\frac{1}{8}(1 - x)^8 + \frac{4}{7}(1 - x)^7 - \frac{2}{3}(1 - x)^6 + C$

47. $\frac{1}{5}(x^2 + 1)^{5/2} - \frac{1}{3}(x^2 + 1)^{3/2} + C$ 49. $\frac{-1}{4(x^2 - 4)^2} + C$

51. $e^{\sin x} + C$ 53. $2\tan(e^{\sqrt{x}} + 1) + C$ 55. $\ln|\ln x| + C$

57. $z - \ln(1 + e^z) + C$ 59. $\frac{5}{6}\tan^{-1}\left(\frac{2r}{3}\right) + C$

61. $e^{\sin^{-1}x} + C$ 63. $\frac{1}{3}(\sin^{-1}x)^3 + C$ 65. $\ln|\tan^{-1}y| + C$

67. (a) $-\frac{6}{2 + \tan^3 x} + C$ (b) $-\frac{6}{2 + \tan^3 x} + C$

  (c) $-\frac{6}{2 + \tan^3 x} + C$

69. $\frac{1}{6}\sin\sqrt{3(2r - 1)^2 + 6} + C$ 73. $s = \frac{1}{2}(3t^2 - 1)^4 - 5$

75. $s = 4t - 2\sin\left(2t + \frac{\pi}{6}\right) + 9$

77. $s = \sin\left(2t - \frac{\pi}{2}\right) + 100t + 1$ 79. 6 m

SECTION 5.6, pp. 355–359

1. (a) $14/3$ (b) $2/3$ 3. (a) $1/2$ (b) $-1/2$

5. (a) $15/16$ (b) 0   7. (a) 0 (b) $1/8$ 9. (a) 4 (b) 0

11. (a) $506/375$ (b) $86{,}744/375$ 13. (a) 0 (b) 0

15. $2\sqrt{3}$ 17. $3/4$ 19. $3^{5/2} - 1$ 21. 3   23. $\pi/3$

25. e 27. $\ln 3$ 29. $(\ln 2)^2$ 31. $\frac{1}{\ln 4}$ 33. $\ln 2$

35. $\ln(2 + \sqrt{3}) - \frac{\sqrt{3}}{2}$ 37. π 39. $\pi/12$ 41. $2\pi/3$

43. $\sqrt{3} - 1$ 45. $-\pi/12$ 47. $\pi^2/32$ 49. $16/3$

51. $2^{5/2}$ 53. $\pi/2$ 55. $128/15$ 57. $4/3$ 59. $5/6$

61. $38/3$ 63. $49/6$ 65. $32/3$ 67. $48/5$ 69. $8/3$

71. 8   73. $5/3$ (There are three intersection points.)   75. 18

77. $243/8$ 79. $8/3$ 81. 2   83. $104/15$ 85. $56/15$

87. 4   89. $\frac{4}{3} - \frac{4}{\pi}$ 91. $\pi/2$ 93. 2   95. $1/2$

97. 1   99. $\ln 16$ 101. 2   103. $2\ln 5$

105. (a) $(\pm\sqrt{c}, c)$ (b) $c = 4^{2/3}$ (c) $c = 4^{2/3}$

107. $11/3$ 109. $3/4$ 111. Neither   113. $F(6) - F(2)$

115. (a) -3 (b) 3   117. $I = a/2$

PRACTICE EXERCISES, pp. 360–363

1. (a) About 680 ft (b) h (feet)

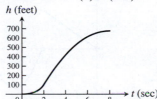
h (feet)

3. (a) $-1/2$ (b) 31 (c) 13 (d) 0

5. $\int_1^5 (2x - 1)^{-1/2}\,dx = 2$ 7. $\int_{-\pi}^0 \cos\frac{x}{2}\,dx = 2$

9. (a) 4 (b) 2 (c) -2 (d) -2π (e) $8/5$

11. $8/3$ 13. 62   15. 1   17. $1/6$ 19. 18

21. $9/8$ 23. $\frac{\pi^2}{32} + \frac{\sqrt{2}}{2} - 1$ 25. 4   27. $\frac{8\sqrt{2} - 7}{6}$

29. Min: -4, max: 0, area: $27/4$ 31. $6/5$ 33. 1

37. $y = \int_5^x \left(\frac{\sin t}{t}\right) dt - 3$ 39. $y = \sin^{-1}x$

41. $y = \sec^{-1}x + \frac{2\pi}{3},\, x > 1$ 43. $f(x) = e^{x^2/2 - 1/2}$

45. $-4(\cos x)^{1/2} + C$ 47. $\theta^2 + \theta + \sin(2\theta + 1) + C$

49. $\frac{t^3}{3} + \frac{4}{t} + C$ 51. $-\frac{1}{3}\cos(2t^{3/2}) + C$

53. $\tan(e^x - 7) + C$ 55. $e^{\tan x} + C$ 57. $\frac{-\ln 7}{3}$

59. $\ln(9/25)$ 61. $-\frac{1}{2}(\ln x)^{-2} + C$ 63. $\frac{1}{2\ln 3}(3^{x^2}) + C$

65. $\frac{3}{2}\sin^{-1}2(r - 1) + C$ 67. $\frac{\sqrt{2}}{2}\tan^{-1}\left(\frac{x - 1}{\sqrt{2}}\right) + C$

69. $\frac{1}{4}\sec^{-1}\left|\frac{2x - 1}{2}\right| + C$ 71. $e^{\sin^{-1}\sqrt{x}} + C$

73. $2\sqrt{\tan^{-1} y} + C$ **75.** $\dfrac{1}{4(\sin 2\theta + \cos 2\theta)^2} + C$ **77.** 16

79. 2   **81.** 1   **83.** 8   **85.** $27\sqrt{3}/160$ **87.** $\pi/2$

89. $\sqrt{3}$ **91.** $6\sqrt{3} - 2\pi$ **93.** -1 **95.** 2   **97.** 1

99. $15/16 + \ln 2$ **101.** $e - 1$ **103.** $1/6$ **105.** $9/14$

107. $\dfrac{9\ln 2}{4}$ **109.** π **111.** $\pi/\sqrt{3}$ **113.** $\pi/6$

115. $\pi/12$ **117.** (a) b (b) b

121. (a) $\dfrac{d}{dx}(x \ln x - x + C) = x \cdot \dfrac{1}{x} + \ln x - 1 + 0 = \ln x$

   (b) $\dfrac{1}{e - 1}$

123. 25°F   **125.** $\sqrt{2 + \cos^3 x}$ **127.** $\dfrac{-6}{3 + x^4}$

129. $\dfrac{dy}{dx} = \dfrac{-2}{x}e^{\cos(2 \ln x)}$ **131.** $\dfrac{dy}{dx} = \dfrac{1}{\sqrt{1 - x^2}\sqrt{1 - 2(\sin^{-1}x)^2}}$

133. Yes   **135.** $-\sqrt{1 + x^2}$

137. Cost $\approx \$10,899$ using a lower sum estimate

ADDITIONAL AND ADVANCED EXERCISES, pp. 364–367

1. (a) Yes   (b) No   **5.** (a) $1/4$ (b) $\sqrt[3]{12}$

7. $f(x) = \dfrac{x}{\sqrt{x^2 + 1}}$ **9.** $y = x^3 + 2x - 4$

11. $36/5$ **13.** $\dfrac{1}{2} - \dfrac{2}{\pi}$

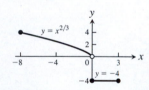

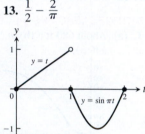

15. $13/3$ **21.** $\ln 2$ **23.** $1/6$

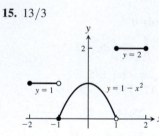

25. $\displaystyle\int_0^1 f(x)\,dx$ **27.** (b) πr^2

29. (a) 0   (b) -1
   (c) $-\pi$ (d) $x = 1$
   (e) $y = 2x + 2 - \pi$
   (f) $x = -1, x = 2$
   (g) $[-2\pi, 0]$

17. $1/2$ **19.** $\pi/2$

31. $2/x$ **33.** $\dfrac{\sin 4y}{\sqrt{y}} - \dfrac{\sin y}{2\sqrt{y}}$ **35.** $2x \ln |x| - x \ln \dfrac{|x|}{\sqrt{2}}$

37. $(\sin x)/x$ **39.** $x = 1$ **41.** $\dfrac{1}{\ln 2}, \dfrac{1}{2 \ln 2}, 2:1$

43. $2/17$

Chapter 6

SECTION 6.1, pp. 375–379

1. 16   **3.** $16/3$ **5.** (a) $2\sqrt{3}$ (b) 8   **7.** (a) 60   (b) 36

9. 8π **11.** 10   **13.** (a) $s^2 h$ (b) $s^2 h$ **15.** $8/3$

17. $\dfrac{2\pi}{3}$ **19.** $4 - \pi$ **21.** $\dfrac{32\pi}{5}$ **23.** 36π **25.** π

27. $\dfrac{\pi}{2}\left(1 - \dfrac{1}{e^2}\right)$ **29.** $\dfrac{\pi}{2}\ln 4$ **31.** $\pi\left(\dfrac{\pi}{2} + 2\sqrt{2} - \dfrac{11}{3}\right)$

33. 2π **35.** 2π **37.** $4\pi \ln 4$ **39.** $\pi^2 - 2\pi$ **41.** $\dfrac{2\pi}{3}$

43. $\dfrac{117\pi}{5}$ **45.** $\pi(\pi - 2)$ **47.** $\dfrac{4\pi}{3}$ **49.** 8π **51.** $\dfrac{7\pi}{6}$

53. (a) 8π (b) $\dfrac{32\pi}{5}$ (c) $\dfrac{8\pi}{3}$ (d) $\dfrac{224\pi}{15}$

55. (a) $\dfrac{16\pi}{15}$ (b) $\dfrac{56\pi}{15}$ (c) $\dfrac{64\pi}{15}$ **57.** $V = 2a^2 b\pi^2$

59. (a) $V = \dfrac{\pi h^2(3a - h)}{3}$ (b) $\dfrac{1}{120\pi}$ m/sec

63. $V = 3308$ cm^3 **65.** 2

SECTION 6.2, pp. 384–387

1. 6π **3.** 2π **5.** $14\pi/3$ **7.** 8π **9.** $5\pi/6$

11. $\dfrac{7\pi}{15}$ **13.** (b) 4π **15.** $\dfrac{16\pi}{15}(3\sqrt{2} + 5)$

17. $\dfrac{8\pi}{3}$ **19.** $\dfrac{4\pi}{3}$ **21.** $\dfrac{16\pi}{3}$

23. (a) 16π (b) 32π (c) 28π
   (d) 24π (e) 60π (f) 48π

25. (a) $\dfrac{27\pi}{2}$ (b) $\dfrac{27\pi}{2}$ (c) $\dfrac{72\pi}{5}$ (d) $\dfrac{108\pi}{5}$

27. (a) $\dfrac{6\pi}{5}$ (b) $\dfrac{4\pi}{5}$ (c) 2π (d) 2π

29. (a) About the x-axis: $V = \dfrac{2\pi}{15}$; about the y-axis: $V = \dfrac{\pi}{6}$

   (b) About the x-axis: $V = \dfrac{2\pi}{15}$; about the y-axis: $V = \dfrac{\pi}{6}$

31. (a) $\dfrac{5\pi}{3}$ (b) $\dfrac{4\pi}{3}$ (c) 2π (d) $\dfrac{2\pi}{3}$

33. (a) $\dfrac{4\pi}{15}$ (b) $\dfrac{7\pi}{30}$ **35.** (a) $\dfrac{24\pi}{5}$ (b) $\dfrac{48\pi}{5}$

37. (a) $\dfrac{9\pi}{16}$ (b) $\dfrac{9\pi}{16}$

39. Disk: 2 integrals; washer: 2 integrals; shell: 1 integral

41. (a) $\dfrac{256\pi}{3}$ (b) $\dfrac{244\pi}{3}$

47. $\pi\left(1 - \dfrac{1}{e}\right)$ **49.** 2

SECTION 6.3, pp. 391–393

1. 12   **3.** $\dfrac{53}{6}$ **5.** $\dfrac{123}{32}$ **7.** $\dfrac{99}{8}$ **9.** $\ln 2 + \dfrac{3}{8}$

11. $\dfrac{53}{6}$ **13.** $2^{3/2} - (5/4)^{3/2}$ **15.** 2

17. (a) $\displaystyle\int_{-1}^{2} \sqrt{1 + 4x^2}\,dx$ (c) ≈ 6.13

19. (a) $\displaystyle\int_{0}^{\pi} \sqrt{1 + \cos^2 y}\,dy$ (c) ≈ 3.82

21. (a) $\displaystyle\int_{-1}^{3} \sqrt{1 + (y + 1)^2}\,dy$ (c) ≈ 9.29

23. (a) $\displaystyle\int_{0}^{\pi/6} \sec x\,dx$ (c) ≈ 0.55

25. (a) $y = \sqrt{x}$ from $(1, 1)$ to $(4, 2)$

(b) Only one. We know the derivative of the function and the value of the function at one value of x.

27. 1   **29.** Yes, $f(x) = \pm x + C$ where C is any real number.

37. $\int_0^x \sqrt{1 + 9t}\, dt,\ \dfrac{2}{27}(10^{3/2} - 1)$

SECTION 6.4, pp. 396–398

1. (a) $2\pi \int_0^{\pi/4} (\tan x)\sqrt{1 + \sec^4 x}\, dx$ **(c)** $S \approx 3.84$

(b)
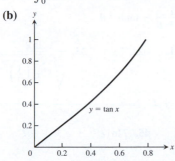

3. (a) $2\pi \int_1^2 \dfrac{1}{y}\sqrt{1 + y^{-4}}\, dy$ **(c)** $S \approx 5.02$

(b)
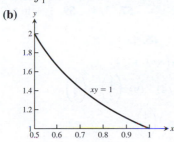

5. (a) $2\pi \int_1^4 (3 - x^{1/2})^2 \sqrt{1 + (1 - 3x^{-1/2})^2}\, dx$ **(c)** $S \approx 63.37$

(b)
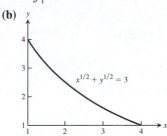

7. (a) $2\pi \int_0^{\pi/3}\left(\int_0^y \tan t\, dt\right)\sec y\, dy$ **(c)** $s \approx 2.08$

(b)
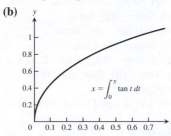

9. $4\pi\sqrt{5}$ **11.** $3\pi\sqrt{5}$ **13.** $98\pi/81$ **15.** 2π

17. $\pi(\sqrt{8} - 1)/9$ **19.** $35\pi\sqrt{5}/3$ **21.** $\pi\left(\dfrac{15}{16} + \ln 2\right)$

23. $253\pi/20$ **27.** Order 226.2 liters of each color.

SECTION 6.5, pp. 404–408

1. 116 J   **3.** 400 N/m   **5.** 4 cm, 0.08 J
7. (a) 7238 lb/in.   **(b)** 905 in.-lb, 2714 in.-lb
9. 780 J   **11.** 72,900 ft-lb   **13.** 490 J
15. (a) 1,497,600 ft-lb   **(b)** 1 hr, 40 min
  (d) At 62.26 lb/ft³: a) 1,494,240 ft-lb b) 1 hr, 40 min
     At 62.59 lb/ft³: a) 1,502,160 ft-lb b) 1 hr, 40.1 min
17. 37,306 ft-lb   **19.** 7,238,299.47 ft-lb   **21.** 2446.25 ft-lb
23. 15,073,099.75 J   **27.** 85.1 ft-lb   **29.** 151.3 J
31. 91.32 in.-oz   **33.** 5.144×10^{10} J   **35.** 1684.8 lb
37. (a) 6364.8 lb   **(b)** 5990.4 lb   **39.** 1164.8 lb   **41.** 1309 lb
43. (a) 12,480 lb   **(b)** 8580 lb   **(c)** 9722.3 lb

45. (a) 93.33 lb   **(b)** 3 ft   **47.** $\dfrac{wb}{2}$

49. No. The tank will overflow because the movable end will have moved only $3\frac{1}{3}$ ft by the time the tank is full.

SECTION 6.6, pp. 418–420

1. $M = 14/3,\ \bar{x} = 93/35$ **3.** $M = \ln 4,\ \bar{x} = (3 - \ln 4)/(\ln 4)$
5. $M = 13,\ \bar{x} = 41/26$ **7.** $\bar{x} = 0,\ \bar{y} = 12/5$
9. $\bar{x} = 1,\ \bar{y} = -3/5$ **11.** $\bar{x} = 16/105,\ \bar{y} = 8/15$
13. $\bar{x} = 0,\ \bar{y} = \pi/8$ **15.** $\bar{x} \approx 1.44,\ \bar{y} \approx 0.36$

17. $\bar{x} = \dfrac{\ln 4}{\pi},\ \bar{y} = 0$ **19.** $\bar{x} = 7,\ \bar{y} = \dfrac{\ln 16}{12}$

21. $\bar{x} = 5/7,\ \bar{y} = 10/33.\ (\bar{x})^4 < \bar{y}$, so the center of mass is outside the region.
23. $\bar{x} = 3/2,\ \bar{y} = 1/2$

25. (a) $\dfrac{224\pi}{3}$ **(b)** $\bar{x} = 2,\ \bar{y} = 0$

(c)
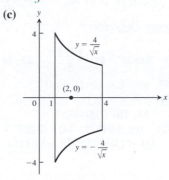

29. $\bar{x} = \bar{y} = 1/3$ **31.** $\bar{x} = a/3,\ \bar{y} = b/3$ **33.** $13\delta/6$

35. $\bar{x} = 0,\ \bar{y} = \dfrac{a\pi}{4}$ **37.** $\bar{x} = 1/2,\ \bar{y} = 4$

39. $\bar{x} = 6/5,\ \bar{y} = 8/7$ **41.** $V = 32\pi,\ S = 32\sqrt{2}\pi$ **45.** $4\pi^2$

47. $\bar{x} = 0,\ \bar{y} = \dfrac{2a}{\pi}$ **49.** $\bar{x} = 0,\ \bar{y} = \dfrac{4b}{3\pi}$

51. $\sqrt{2}\pi a^3(4 + 3\pi)/6$ **53.** $\bar{x} = \dfrac{a}{3},\ \bar{y} = \dfrac{b}{3}$

PRACTICE EXERCISES, pp. 421–423

1. $\dfrac{9\pi}{280}$ **3.** π^2 **5.** $\dfrac{72\pi}{35}$

7. (a) 2π **(b)** π **(c)** $12\pi/5$ **(d)** $26\pi/5$

9. (a) 8π **(b)** $1088\pi/15$ **(c)** $512\pi/15$

11. $\pi\left(3\sqrt{3} - \pi\right)/3$ **13.** π **15.** $\dfrac{28\pi}{3}$ ft^3

17. $(\pi/3)(a^2 + ab + b^2)h$ **19.** $\dfrac{10}{3}$ **21.** $3 + \dfrac{1}{8}\ln 2$

23. $\sqrt{2}$ **25.** $28\pi\sqrt{2}/3$ **27.** 4π **29.** 4640 J

31. $\dfrac{w}{2}\left(2ar - a^2\right)$ **33.** 418,208.81 ft-lb

35. $22{,}500\pi$ ft-lb, 257 sec **37. (a)** 128 ft-lb **(b)** 219.6 ft-lb

39. $\bar{x} = 0, \bar{y} = 8/5$ **41.** $\bar{x} = 3/2, \bar{y} = 12/5$

43. $\bar{x} = 9/5, \bar{y} = 11/10$ **45.** 332.8 lb **47.** 2196.48 lb

ADDITIONAL AND ADVANCED EXERCISES, pp. 423–424

1. $f(x) = \sqrt{\dfrac{2x - a}{\pi}}$ **3.** $f(x) = \sqrt{C^2 - 1}\, x + a$, where $C \geq 1$

5. $\dfrac{\pi}{30\sqrt{2}}$ **7.** 28/3 **9.** $\dfrac{4h\sqrt{3mh}}{3}$

11. $\bar{x} = 0, \bar{y} = \dfrac{n}{2n + 1}, (0, 1/2)$

15. (a) $\bar{x} = \bar{y} = 4(a^2 + ab + b^2)/(3\pi(a + b))$
(b) $(2a/\pi, 2a/\pi)$

17. ≈ 2329.6 lb

Chapter 7

SECTION 7.1, pp. 433–435

1. $\ln\left(\dfrac{2}{3}\right)$ **3.** $\ln\left|y^2 - 25\right| + C$ **5.** $\ln\left|6 + 3\tan t\right| + C$

7. $\ln\left(1 + \sqrt{x}\right) + C$ **9.** 1 **11.** $2(\ln 2)^4$ **13.** 2

15. $2e^{\sqrt{r}} + C$ **17.** $-e^{-t^2} + C$ **19.** $-e^{1/x} + C$

21. $\dfrac{1}{\pi}e^{\sec \pi t} + C$ **23.** 1 **25.** $\ln(1 + e^r) + C$ **27.** $\dfrac{1}{2\ln 2}$

29. $\dfrac{1}{\ln 2}$ **31.** $\dfrac{6}{\ln 7}$ **33.** 32760 **35.** $3^{\sqrt{2}+1}$

37. $\dfrac{1}{\ln 10}\left(\dfrac{(\ln x)^2}{2}\right) + C$ **39.** $2(\ln 2)^2$ **41.** $\dfrac{3\ln 2}{2}$ **43.** $\ln 10$

45. $(\ln 10)\ln\left|\ln x\right| + C$ **47.** $y = 1 - \cos(e^t - 2)$

49. $y = 2(e^{-x} + x) - 1$ **51.** $y = x + \ln|x| + 2$

53. $\pi\ln 16$ **55.** $6 + \ln 2$ **57. (b)** 0.00469

71. (a) 1.89279 **(b)** -0.35621 **(c)** 0.94575 **(d)** -2.80735
(e) 5.29595 **(f)** 0.97041 **(g)** -1.03972 **(h)** -1.61181

SECTION 7.2, pp. 442–444

9. $\dfrac{2}{3}y^{3/2} - x^{1/2} = C$ **11.** $e^y - e^x = C$

13. $-x + 2\tan\sqrt{y} = C$ **15.** $e^{-y} + 2e^{\sqrt{x}} = C$

17. $y = \sin(x^2 + C)$ **19.** $\dfrac{1}{3}\ln\left|y^3 - 2\right| = x^3 + C$

21. $4\ln\left(\sqrt{y} + 2\right) = e^{x^2} + C$

23. (a) -0.00001 **(b)** 10,536 years **(c)** 82%

25. 54.88 g **27.** 59.8 ft **29.** 2.8147498×10^{14}

31. (a) 8 years **(b)** 32.02 years **33.** Yes, $y(20) < 1$

35. 15.28 years **37.** 56,562 years

41. (a) 17.5 min **(b)** 13.26 min **43.** $-3°$C

45. About 6693 years **47.** 54.62% **49.** $\approx 15{,}683$ years

SECTION 7.3, pp. 450–453

1. $\cosh x = 5/4, \tanh x = -3/5, \coth x = -5/3,$
$\operatorname{sech} x = 4/5, \operatorname{csch} x = -4/3$

3. $\sinh x = 8/15, \tanh x = 8/17, \coth x = 17/8, \operatorname{sech} x = 15/17,$
$\operatorname{csch} x = 15/8$

5. $x + \dfrac{1}{x}$ **7.** e^{5x} **9.** e^{4x} **13.** $2\cosh\dfrac{x}{3}$

15. $\operatorname{sech}^2\sqrt{t} + \dfrac{\tanh\sqrt{t}}{\sqrt{t}}$ **17.** $\coth z$

19. $(\ln\operatorname{sech}\theta)(\operatorname{sech}\theta\tanh\theta)$ **21.** $\tanh^3 v$ **23.** 2

25. $\dfrac{1}{2\sqrt{x(1 + x)}}$ **27.** $\dfrac{1}{1 + \theta} - \tanh^{-1}\theta$

29. $\dfrac{1}{2\sqrt{t}} - \coth^{-1}\sqrt{t}$ **31.** $-\operatorname{sech}^{-1} x$ **33.** $\dfrac{\ln 2}{\sqrt{1 + \left(\dfrac{1}{2}\right)^{2\theta}}}$

35. $\left|\sec x\right|$ **41.** $\dfrac{\cosh 2x}{2} + C$

43. $12\sinh\left(\dfrac{x}{2} - \ln 3\right) + C$ **45.** $7\ln\left|e^{x/7} + e^{-x/7}\right| + C$

47. $\tanh\left(x - \dfrac{1}{2}\right) + C$ **49.** $-2\operatorname{sech}\sqrt{t} + C$ **51.** $\ln\dfrac{5}{2}$

53. $\dfrac{3}{32} + \ln 2$ **55.** $e - e^{-1}$ **57.** 3/4 **59.** $\dfrac{3}{8} + \ln\sqrt{2}$

61. $\ln(2/3)$ **63.** $\dfrac{-\ln 3}{2}$ **65.** $\ln 3$

67. (a) $\sinh^{-1}\left(\sqrt{3}\right)$ **(b)** $\ln\left(\sqrt{3} + 2\right)$

69. (a) $\coth^{-1}(2) - \coth^{-1}(5/4)$ **(b)** $\left(\dfrac{1}{2}\right)\ln\left(\dfrac{1}{3}\right)$

71. (a) $-\operatorname{sech}^{-1}\left(\dfrac{12}{13}\right) + \operatorname{sech}^{-1}\left(\dfrac{4}{5}\right)$

(b) $-\ln\left(\dfrac{1 + \sqrt{1 - (12/13)^2}}{(12/13)}\right) + \ln\left(\dfrac{1 + \sqrt{1 - (4/5)^2}}{(4/5)}\right)$

$= -\ln\left(\dfrac{3}{2}\right) + \ln(2) = \ln(4/3)$

73. (a) 0 **(b)** 0

77. (a) $\sqrt{\dfrac{mg}{k}}$ **(b)** $80\sqrt{5} \approx 178.89$ ft/sec

79. 2π **81.** $\dfrac{6}{5}$

SECTION 7.4, pp. 457–458

1. (a) Slower **(b)** Slower **(c)** Slower **(d)** Faster
(e) Slower **(f)** Slower **(g)** Same **(h)** Slower
3. (a) Same **(b)** Faster **(c)** Same **(d)** Same
(e) Slower **(f)** Faster **(g)** Slower **(h)** Same
5. (a) Same **(b)** Same **(c)** Same **(d)** Faster
(e) Faster **(f)** Same **(g)** Slower **(h)** Faster **7.** d, a, c, b
9. (a) False **(b)** False **(c)** True **(d)** True **(e)** True
(f) True **(g)** False **(h)** True
13. When the degree of f is less than or equal to the degree of g.
15. 1, 1
21. (b) $\ln(e^{17000000}) = 17{,}000{,}000 < (e^{17\times 10^6})^{1/10^6}$
$= e^{17} \approx 24{,}154{,}952.75$

(c) $x \approx 3.4306311 \times 10^{15}$

(d) They cross at $x \approx 3.4306311 \times 10^{15}$.

23. (a) The algorithm that takes $O(n \log_2 n)$ steps

(b)

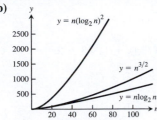

25. It could take one million for a sequential search; at most 20 steps for a binary search.

PRACTICE EXERCISES, pp. 459–460

1. $-\cos e^x + C$ **3.** $\ln 8$ **5.** $2 \ln 2$

7. $\frac{1}{2}(\ln(x-5))^2 + C$ **9.** $3 \ln 7$ **11.** $2(\sqrt{2} - 1)$

13. $y = \frac{\ln 2}{\ln (3/2)}$ **15.** $y = \ln x - \ln 3$ **17.** $y = \frac{1}{1 - e^x}$

19. (a) Same rate   **(b)** Same rate   **(c)** Faster   **(d)** Faster
(e) Same rate   **(f)** Same rate

21. (a) True   **(b)** False   **(c)** False   **(d)** True
(e) True   **(f)** True

23. $1/3$ **25.** $1/e$ m/sec   **27.** $\ln 5x - \ln 3x = \ln (5/3)$

29. $1/2$ **31.** $y = \left(\tan^{-1}\left(\frac{x + C}{2} \right) \right)^2$

33. $y^2 = \sin^{-1}(2 \tan x + C)$

35. $y = -2 + \ln (2 - e^{-x})$ **37.** $y = 4x - 4\sqrt{x} + 1$

39. 19,035 years   **41.** $\ln(16/9)$

ADDITIONAL AND ADVANCED EXERCISES, p. 460

1. (a) 1   **(b)** $\pi/2$ **(c)** π

3. $\tan^{-1}x + \tan^{-1}\left(\frac{1}{x} \right)$ is a constant and the constant is $\frac{\pi}{2}$ for $x > 0$; it is $-\frac{\pi}{2}$ for $x < 0$.

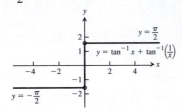

7. $\bar{x} = \frac{\ln 4}{\pi}, \bar{y} = 0$

Chapter 8

SECTION 8.1, pp. 465–466

1. $\ln 5$ **3.** $2 \tan x - 2 \sec x - x + C$

5. $\sin^{-1} x + \sqrt{1 - x^2} + C$ **7.** $e^{-\cot z} + C$

9. $\tan^{-1}(e^z) + C$ **11.** π **13.** $t + \cot t + \csc t + C$

15. $\sqrt{2}$ **17.** $\frac{1}{8}\ln(1 + 4 \ln^2 y) + C$

19. $\ln|1 + \sin \theta| + C$ **21.** $2t^2 - t + 2\tan^{-1}\left(\frac{t}{2} \right) + C$

23. $2(\sqrt{2} - 1) \approx 0.82843$ **25.** $\sec^{-1}(e^y) + C$

27. $\sin^{-1}(2 \ln x) + C$ **29.** $\ln|\sin x| + \ln|\cos x| + C$

31. $7 + \ln 8$ **33.** $\left(\sin^{-1} y - \sqrt{1 - y^2} \right)_{-1}^{0} = \frac{\pi}{2} - 1$

35. $\sec^{-1}\left| \frac{x - 1}{7} \right| + C$ **37.** $\frac{\theta^3}{3} - \frac{\theta^2}{2} + \theta + \frac{5}{2}\ln|2\theta - 5| + C$

39. $x - \ln(1 + e^x) + C$ **41.** $(1/2)e^{2x} - e^x + \ln(1 + e^x) + C$

43. $2 \arctan(\sqrt{x}) + C$ **45.** $2\sqrt{2} - \ln(3 + 2\sqrt{2})$

47. $\ln(2 + \sqrt{3})$ **49.** $\bar{x} = 0, \quad \bar{y} = \dfrac{1}{\ln(3 + 2\sqrt{2})}$

51. $xe^{x^3} + C$ **53.** $\frac{1}{30}(x^4 + 1)^{3/2}(3x^4 - 2) + C$

SECTION 8.2, pp. 471–474

1. $-2x \cos(x/2) + 4 \sin(x/2) + C$

3. $t^2 \sin t + 2t \cos t - 2 \sin t + C$

5. $\ln 4 - \frac{3}{4}$ **7.** $xe^x - e^x + C$

9. $-(x^2 + 2x + 2)e^{-x} + C$

11. $y \tan^{-1}(y) - \ln\sqrt{1 + y^2} + C$

13. $x \tan x + \ln|\cos x| + C$

15. $(x^3 - 3x^2 + 6x - 6)e^x + C$ **17.** $(x^2 - 7x + 7)e^x + C$

19. $(x^5 - 5x^4 + 20x^3 - 60x^2 + 120x - 120)e^x + C$

21. $\frac{1}{2}(-e^\theta \cos \theta + e^\theta \sin \theta) + C$

23. $\frac{e^{2x}}{13}(3 \sin 3x + 2 \cos 3x) + C$

25. $\frac{2}{3}\left(\sqrt{3s + 9}\, e^{\sqrt{3s+9}} - e^{\sqrt{3s+9}} \right) + C$

27. $\frac{\pi \sqrt{3}}{3} - \ln(2) - \frac{\pi^2}{18}$

29. $\frac{1}{2}\left[-x \cos(\ln x) + x \sin(\ln x) \right] + C$

31. $\frac{1}{2}\ln|\sec x^2 + \tan x^2| + C$

33. $\frac{1}{2}x^2 (\ln x)^2 - \frac{1}{2}x^2 \ln x + \frac{1}{4}x^2 + C$

35. $-\frac{1}{x}\ln x - \frac{1}{x} + C$ **37.** $\frac{1}{4}e^{x^4} + C$

39. $\frac{1}{3}x^2 (x^2 + 1)^{3/2} - \frac{2}{15}(x^2 + 1)^{5/2} + C$

41. $-\frac{2}{5}\sin 3x \sin 2x - \frac{3}{5}\cos 3x \cos 2x + C$

43. $\frac{2}{9}x^{3/2}(3 \ln x - 2) + C$

45. $2\sqrt{x} \sin\sqrt{x} + 2 \cos\sqrt{x} + C$

47. $\frac{\pi^2 - 4}{8}$ **49.** $\frac{5\pi - 3\sqrt{3}}{9}$

51. $\frac{1}{2}(x^2 + 1)\tan^{-1} x - \frac{x}{2} + C$ **53.** $xe^{x^2} + C$

55. $(2/3)x^{3/2} \arcsin(\sqrt{x}) + (2/9)x\sqrt{1 - x} + (4/9)\sqrt{1 - x} + C$

57. (a) π **(b)** 3π **(c)** 5π **(d)** $(2n + 1)\pi$

59. $2\pi(1 - \ln 2)$ **61. (a)** $\pi(\pi - 2)$ **(b)** 2π

63. (a) 1   **(b)** $(e - 2)\pi$ **(c)** $\frac{\pi}{2}(e^2 + 9)$

(d) $\bar{x} = \frac{1}{4}(e^2 + 1), \bar{y} = \frac{1}{2}(e - 2)$

65. $\frac{1}{2\pi}(1 - e^{-2\pi})$ **67.** $u = x^n, dv = \cos x\, dx$

69. $u = x^n, dv = e^{ax}\, dx$ **73.** $u = x^n, dv = (x + 1)^{-(1/2)}\, dx$

77. $x \sin^{-1}x + \cos(\sin^{-1}x) + C$

79. $x \sec^{-1}x - \ln|x + \sqrt{x^2 - 1}| + C$ **81.** Yes

83. (a) $x \sinh^{-1}x - \cosh(\sinh^{-1}x) + C$

     (b) $x \sinh^{-1}x - (1 + x^2)^{1/2} + C$

SECTION 8.3, pp. 479–480

1. $\frac{1}{2}\sin 2x + C$ **3.** $-\frac{1}{4}\cos^4 x + C$

5. $\frac{1}{3}\cos^3 x - \cos x + C$ **7.** $-\cos x + \frac{2}{3}\cos^3 x - \frac{1}{5}\cos^5 x + C$

9. $\sin x - \frac{1}{3}\sin^3 x + C$ **11.** $\frac{1}{4}\sin^4 x - \frac{1}{6}\sin^6 x + C$

13. $\frac{1}{2}x + \frac{1}{4}\sin 2x + C$ **15.** $16/35$ **17.** 3π

19. $-4\sin x \cos^3 x + 2\cos x \sin x + 2x + C$

21. $-\cos^4 2\theta + C$ **23.** 4 **25.** 2

27. $\sqrt{\frac{3}{2}} - \frac{2}{3}$ **29.** $\frac{4}{5}\left(\frac{3}{2}\right)^{5/2} - \frac{18}{35} - \frac{2}{7}\left(\frac{3}{2}\right)^{7/2}$ **31.** $\sqrt{2}$

33. $\frac{1}{2}\tan^2 x + C$ **35.** $\frac{1}{3}\sec^3 x + C$ **37.** $\frac{1}{3}\tan^3 x + C$

39. $2\sqrt{3} + \ln(2 + \sqrt{3})$ **41.** $\frac{2}{3}\tan\theta + \frac{1}{3}\sec^2\theta \tan\theta + C$

43. $4/3$ **45.** $2\tan^2 x - 2\ln(1 + \tan^2 x) + C$

47. $\frac{1}{4}\tan^4 x - \frac{1}{2}\tan^2 x + \ln|\sec x| + C$ **49.** $\frac{4}{3} - \ln\sqrt{3}$

51. $-\frac{1}{10}\cos 5x - \frac{1}{2}\cos x + C$ **53.** π

55. $\frac{1}{2}\sin x + \frac{1}{14}\sin 7x + C$

57. $\frac{1}{6}\sin 3\theta - \frac{1}{4}\sin\theta - \frac{1}{20}\sin 5\theta + C$

59. $-\frac{2}{5}\cos^5\theta + C$ **61.** $\frac{1}{4}\cos\theta - \frac{1}{20}\cos 5\theta + C$

63. $\sec x - \ln|\csc x + \cot x| + C$ **65.** $\cos x + \sec x + C$

67. $\frac{1}{4}x^2 - \frac{1}{4}x \sin 2x - \frac{1}{8}\cos 2x + C$ **69.** $\ln(2 + \sqrt{3})$

71. $\pi^2/2$ **73.** $\bar{x} = \frac{4\pi}{3}, \bar{y} = \frac{8\pi^2 + 3}{12\pi}$ **75.** $(\pi/4)(4 - \pi)$

SECTION 8.4, pp. 484–485

1. $\ln|\sqrt{9 + x^2} + x| + C$ **3.** $\pi/4$ **5.** $\pi/6$

7. $\frac{25}{2}\sin^{-1}\left(\frac{t}{5}\right) + \frac{t\sqrt{25 - t^2}}{2} + C$

9. $\frac{1}{2}\ln\left|\frac{2x}{7} + \frac{\sqrt{4x^2 - 49}}{7}\right| + C$

11. $7\left[\frac{\sqrt{y^2 - 49}}{7} - \sec^{-1}\left(\frac{y}{7}\right)\right] + C$ **13.** $\frac{\sqrt{x^2 - 1}}{x} + C$

15. $-\sqrt{9 - x^2} + C$ **17.** $\frac{1}{3}(x^2 + 4)^{3/2} - 4\sqrt{x^2 + 4} + C$

19. $\frac{-2\sqrt{4 - w^2}}{w} + C$ **21.** $\sin^{-1}x - \sqrt{1 - x^2} + C$

23. $4\sqrt{3} - \frac{4\pi}{3}$ **25.** $-\frac{x}{\sqrt{x^2 - 1}} + C$

27. $-\frac{1}{5}\left(\frac{\sqrt{1 - x^2}}{x}\right)^5 + C$ **29.** $2\tan^{-1}2x + \frac{4x}{(4x^2 + 1)} + C$

31. $\frac{1}{2}x^2 + \frac{1}{2}\ln|x^2 - 1| + C$ **33.** $\frac{1}{3}\left(\frac{v}{\sqrt{1 - v^2}}\right)^3 + C$

35. $\ln 9 - \ln(1 + \sqrt{10})$ **37.** $\pi/6$ **39.** $\sec^{-1}|x| + C$

41. $\sqrt{x^2 - 1} + C$ **43.** $\frac{1}{2}\ln|\sqrt{1 + x^4} + x^2| + C$

45. $4\sin^{-1}\frac{\sqrt{x}}{2} + \sqrt{x}\sqrt{4 - x} + C$

47. $\frac{1}{4}\sin^{-1}\sqrt{x} - \frac{1}{4}\sqrt{x}\sqrt{1 - x}(1 - 2x) + C$

49. $(9/2)\arcsin\left(\frac{x + 1}{3}\right) + (1/2)(x + 1)\sqrt{8 - 2x - x^2} + C$

51. $\sqrt{x^2 + 4x + 3} - \text{arcsec}(x + 2) + C$

53. $y = 2\left[\frac{\sqrt{x^2 - 4}}{2} - \sec^{-1}\left(\frac{x}{2}\right)\right]$

55. $y = \frac{3}{2}\tan^{-1}\left(\frac{x}{2}\right) - \frac{3\pi}{8}$ **57.** $3\pi/4$

59. (a) $\frac{1}{12}(\pi + 6\sqrt{3} - 12)$

     (b) $\bar{x} = \frac{3\sqrt{3} - \pi}{4(\pi + 6\sqrt{3} - 12)}, \bar{y} = \frac{\pi^2 + 12\sqrt{3}\pi - 72}{12(\pi + 6\sqrt{3} - 12)}$

61. (a) $-\frac{1}{3}x^2(1 - x^2)^{3/2} - \frac{2}{15}(1 - x^2)^{5/2} + C$

     (b) $-\frac{1}{3}(1 - x^2)^{3/2} + \frac{1}{5}(1 - x^2)^{5/2} + C$

     (c) $\frac{1}{5}(1 - x^2)^{5/2} - \frac{1}{3}(1 - x^2)^{3/2} + C$

63. $\sqrt{3} - \frac{\sqrt{2}}{2} + \frac{1}{2}\ln\left(\frac{2 + \sqrt{3}}{\sqrt{2} + 1}\right)$

SECTION 8.5, pp. 491–493

1. $\frac{2}{x - 3} + \frac{3}{x - 2}$ **3.** $\frac{1}{x + 1} + \frac{3}{(x + 1)^2}$

5. $\frac{-2}{z} + \frac{-1}{z^2} + \frac{2}{z - 1}$ **7.** $1 + \frac{17}{t - 3} + \frac{-12}{t - 2}$

9. $\frac{1}{2}[\ln|1 + x| - \ln|1 - x|] + C$

11. $\frac{1}{7}\ln|(x + 6)^2(x - 1)^5| + C$ **13.** $(\ln 15)/2$

15. $-\frac{1}{2}\ln|t| + \frac{1}{6}\ln|t + 2| + \frac{1}{3}\ln|t - 1| + C$ **17.** $3\ln 2 - 2$

19. $\frac{1}{4}\ln\left|\frac{x + 1}{x - 1}\right| - \frac{x}{2(x^2 - 1)} + C$ **21.** $(\pi + 2\ln 2)/8$

23. $\tan^{-1}y - \frac{1}{y^2 + 1} + C$

25. $-(s - 1)^{-2} + (s - 1)^{-1} + \tan^{-1}s + C$

27. $\frac{2}{3}\ln|x - 1| + \frac{1}{6}\ln|x^2 + x + 1| - \sqrt{3}\tan^{-1}\left(\frac{2x + 1}{\sqrt{3}}\right) + C$

29. $\frac{1}{4}\ln\left|\frac{x - 1}{x + 1}\right| + \frac{1}{2}\tan^{-1}x + C$

31. $\frac{-1}{\theta^2 + 2\theta + 2} + \ln(\theta^2 + 2\theta + 2) - \tan^{-1}(\theta + 1) + C$

33. $x^2 + \ln\left|\frac{x - 1}{x}\right| + C$

35. $9x + 2\ln|x| + \frac{1}{x} + 7\ln|x - 1| + C$

37. $\frac{y^2}{2} - \ln|y| + \frac{1}{2}\ln(1 + y^2) + C$ **39.** $\ln\left(\frac{e^t + 1}{e^t + 2}\right) + C$

41. $\frac{1}{5}\ln\left|\frac{\sin y - 2}{\sin y + 3}\right| + C$

43. $\dfrac{(\tan^{-1}2x)^2}{4} - 3\ln|x-2| + \dfrac{6}{x-2} + C$

45. $\ln\left|\dfrac{\sqrt{x}-1}{\sqrt{x}+1}\right| + C$

47. $2\sqrt{1+x} + \ln\left|\dfrac{\sqrt{x+1}-1}{\sqrt{x+1}+1}\right| + C$

49. $\dfrac{1}{4}\ln\left|\dfrac{x^4}{x^4+1}\right| + C$

51. $\dfrac{1}{\sqrt{2}}\ln\left|\dfrac{\sqrt{2}\cos\theta+1}{\sqrt{2}\cos\theta-1}\right| + \dfrac{1}{2}\ln\left|\dfrac{1-\cos\theta}{1+\cos\theta}\right| + C$

53. $4\sqrt{1+\sqrt{x}} + 2\ln\left|\dfrac{\sqrt{1+\sqrt{x}}-1}{\sqrt{1+\sqrt{x}}+1}\right| + C$

55. $\dfrac{1}{3}x^3 - 2x^2 + 5x - 10\ln|x+2| + C$

57. $\dfrac{1}{\ln 2}\ln(2^x+2^{-x}) + C$ **59.** $\dfrac{1}{4}\ln\left|\dfrac{x-1}{x+1}\right| - \dfrac{1}{2}\arctan x + C$

61. $\dfrac{1}{2}\ln|(\ln x+1)(\ln x+3)| + C$

63. $\ln|x+\sqrt{x^2-1}| + C$

65. $\dfrac{2}{9}x^3(x^3+1)^{3/2} - \dfrac{4}{45}(x^3+1)^{5/2} + C$

67. $x = \ln|t-2| - \ln|t-1| + \ln 2$

69. $x = \dfrac{6t}{t+2} - 1$ **71.** $3\pi\ln 25$

73. $\ln(3) - \dfrac{1}{2}$ **75.** 1.10

77. (a) $x = \dfrac{1000e^{4t}}{499+e^{4t}}$ (b) 1.55 days

SECTION 8.6, pp. 497–499

1. $\dfrac{2}{\sqrt{3}}\left(\tan^{-1}\sqrt{\dfrac{x-3}{3}}\right) + C$

3. $\sqrt{x-2}\left(\dfrac{2(x-2)}{3}+4\right) + C$ **5.** $\dfrac{(2x-3)^{3/2}(x+1)}{5} + C$

7. $\dfrac{-\sqrt{9-4x}}{x} - \dfrac{2}{3}\ln\left|\dfrac{\sqrt{9-4x}-3}{\sqrt{9-4x}+3}\right| + C$

9. $\dfrac{(x+2)(2x-6)\sqrt{4x-x^2}}{6} + 4\sin^{-1}\left(\dfrac{x-2}{2}\right) + C$

11. $-\dfrac{1}{\sqrt{7}}\ln\left|\dfrac{\sqrt{7}+\sqrt{7+x^2}}{x}\right| + C$

13. $\sqrt{4-x^2} - 2\ln\left|\dfrac{2+\sqrt{4-x^2}}{x}\right| + C$

15. $\dfrac{e^{2t}}{13}(2\cos 3t + 3\sin 3t) + C$

17. $\dfrac{x^2}{2}\cos^{-1}x + \dfrac{1}{4}\sin^{-1}x - \dfrac{1}{4}x\sqrt{1-x^2} + C$

19. $\dfrac{x^3}{3}\tan^{-1}x - \dfrac{x^2}{6} + \dfrac{1}{6}\ln(1+x^2) + C$

21. $-\dfrac{\cos 5x}{10} - \dfrac{\cos x}{2} + C$

23. $8\left[\dfrac{\sin(7t/2)}{7} - \dfrac{\sin(9t/2)}{9}\right] + C$

25. $6\sin(\theta/12) + \dfrac{6}{7}\sin(7\theta/12) + C$

27. $\dfrac{1}{2}\ln(x^2+1) + \dfrac{x}{2(1+x^2)} + \dfrac{1}{2}\tan^{-1}x + C$

29. $\left(x-\dfrac{1}{2}\right)\sin^{-1}\sqrt{x} + \dfrac{1}{2}\sqrt{x-x^2} + C$

31. $\sin^{-1}\sqrt{x} - \sqrt{x-x^2} + C$

33. $\sqrt{1-\sin^2 t} - \ln\left|\dfrac{1+\sqrt{1-\sin^2 t}}{\sin t}\right| + C$

35. $\ln\left|\ln y + \sqrt{3+(\ln y)^2}\right| + C$

37. $\ln|x+1+\sqrt{x^2+2x+5}| + C$

39. $\dfrac{x+2}{2}\sqrt{5-4x-x^2} + \dfrac{9}{2}\sin^{-1}\left(\dfrac{x+2}{3}\right) + C$

41. $-\dfrac{\sin^4 2x\cos 2x}{10} - \dfrac{2\sin^2 2x\cos 2x}{15} - \dfrac{4\cos 2x}{15} + C$

43. $\dfrac{\sin^3 2\theta\cos^2 2\theta}{10} + \dfrac{\sin^3 2\theta}{15} + C$

45. $\tan^2 2x - 2\ln|\sec 2x| + C$

47. $\dfrac{(\sec\pi x)(\tan\pi x)}{\pi} + \dfrac{1}{\pi}\ln|\sec\pi x + \tan\pi x| + C$

49. $\dfrac{-\csc^3 x\cot x}{4} - \dfrac{3\csc x\cot x}{8} - \dfrac{3}{8}\ln|\csc x + \cot x| + C$

51. $\dfrac{1}{2}\big[\sec(e^t-1)\tan(e^t-1) +$
$\ln|\sec(e^t-1)+\tan(e^t-1)|\big] + C$

53. $\sqrt{2}+\ln(\sqrt{2}+1)$ **55.** $\pi/3$

57. $2\pi\sqrt{3}+\pi\sqrt{2}\ln(\sqrt{2}+\sqrt{3})$ **59.** $\bar{x}=4/3, \bar{y}=\ln\sqrt{2}$

61. 7.62 **63.** $\pi/8$ **67.** $\pi/4$

SECTION 8.7, pp. 506–508

1. I: (a) 1.5, 0   (b) 1.5, 0   (c) 0%
II: (a) 1.5, 0   (b) 1.5, 0   (c) 0%
3. I: (a) 2.75, 0.08   (b) 2.67, 0.08   (c) 0.0312 ≈ 3%
II: (a) 2.67, 0   (b) 2.67, 0   (c) 0%
5. I: (a) 6.25, 0.5   (b) 6, 0.25   (c) 0.0417 ≈ 4%
II: (a) 6, 0   (b) 6, 0   (c) 0%
7. I: (a) 0.509, 0.03125   (b) 0.5, 0.009   (c) 0.018 ≈ 2%
II: (a) 0.5, 0.002604   (b) 0.5, 0.4794   (c) 0%
9. I: (a) 1.8961, 0.161   (b) 2, 0.1039   (c) 0.052 ≈ 5%
II: (a) 2.0045, 0.0066   (b) 2, 0.00454   (c) 0.2%
11. (a) 1   (b) 2   **13.** (a) 116   (b) 2
15. (a) 283   (b) 2   **17.** (a) 71   (b) 10
19. (a) 76   (b) 12   **21.** (a) 82   (b) 8
23. 15,990 ft^3 **25.** ≈10.63 ft
27. (a) ≈0.00021   (b) ≈1.37079   (c) ≈0.015%
31. (a) ≈5.870   (b) $|E_T| \le 0.0032$
33. 21.07 in.   **35.** 14.4   **39.** ≈28.7 mg

SECTION 8.8, pp. 517–519

1. $\pi/2$ **3.** 2   **5.** 6   **7.** $\pi/2$ **9.** $\ln 3$ **11.** $\ln 4$

13. 0   **15.** $\sqrt{3}$ **17.** π **19.** $\ln\left(1+\dfrac{\pi}{2}\right)$

21. -1 **23.** 1   **25.** $-1/4$ **27.** $\pi/2$ **29.** $\pi/3$

31. 6   **33.** $\ln 2$ **35.** Diverges   **37.** Diverges

39. Diverges   **41.** Diverges   **43.** Converges

45. Converges   **47.** Diverges   **49.** Converges

51. Converges   **53.** Diverges   **55.** Converges
57. Converges   **59.** Diverges   **61.** Converges
63. Diverges   **65.** Converges   **67.** Converges
69. (a) Converges when $p < 1$ **(b)** Converges when $p > 1$
71. 1   **73.** 2π **75.** $\ln 2$
77. (a) 1 **(b)** $\pi/3$ **(c)** Diverges
79. (a) $\pi/2$ **(b)** π **81. (b)** ≈ 0.88621
83. (a)

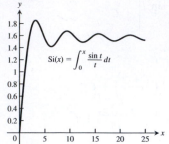
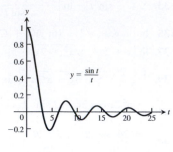

(b) $\pi/2$
85. (a)

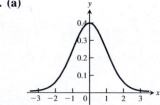

(b) $\approx 0.683,\ \approx 0.954,\ \approx 0.997$
91. ≈ 0.16462

SECTION 8.9, pp. 530–532

1. No   **3.** Yes   **5.** Yes   **7.** Yes   **11.** ≈ 0.537
13. ≈ 0.688 **15.** ≈ 0.0502 **17.** $\sqrt{21}$ **19.** $\dfrac{1}{2}\ln 2$
21. $\dfrac{1}{\pi},\ \dfrac{1}{\pi}\left(\tan^{-1}2 - \dfrac{\pi}{4}\right) \approx 0.10242$
25. mean $= \dfrac{8}{3} \approx 2.67$, median $= \sqrt{8} \approx 2.83$
27. mean $= 2$, median $= \sqrt{2} \approx 1.41$
29. $P\left(X < \dfrac{1}{2}\right) \approx 0.3935$
31. (a) ≈ 0.57, so about 57 in every 100 bulbs will fail.
   (b) ≈ 832 hr
33. ≈ 60 hydra   **35. (a)** ≈ 0.393 **(b)** ≈ 0.135 **(c)** 0
   (d) The probability that any customer waits longer than 3 minutes
   is $1 - (0.997521)^{200} \approx 0.391 < 1/2$. So the most likely
   outcome is that all 200 would be served within 3 minutes.
37. $10,256   **39.** $\approx 323,\ \approx 262$ **41.** ≈ 0.89435
43. (a) $\approx 16\%$ **(b)** ≈ 0.23832 **45.** ≈ 618 females
47. ≈ 61 adults   **49.** ≈ 289 shafts
51. (a) ≈ 0.977 **(b)** ≈ 0.159 **(c)** ≈ 0.838
55. (a) {LLL, LLD, LDL, DLL, LLU, LUL, ULL, LDD, LDU,
   LUD, LUU, DLD, DLU, ULD, ULU, DDL, DUL, UDL,
   UUL, DDD, DDU, DUD, UDD, DUU, UDU, UUD, UUU}
   (c) $7/27 \approx 0.26$ **(d)** $20/27 \approx 0.74$

PRACTICE EXERCISES, pp. 533–535

1. $(x + 1)(\ln(x + 1)) - (x + 1) + C$
3. $x\tan^{-1}(3x) - \dfrac{1}{6}\ln(1 + 9x^2) + C$
5. $(x + 1)^2 e^x - 2(x + 1)e^x + 2e^x + C$

7. $\dfrac{2e^x \sin 2x}{5} + \dfrac{e^x \cos 2x}{5} + C$
9. $2\ln|x - 2| - \ln|x - 1| + C$
11. $\ln|x| - \ln|x + 1| + \dfrac{1}{x + 1} + C$
13. $-\dfrac{1}{3}\ln\left|\dfrac{\cos\theta - 1}{\cos\theta + 2}\right| + C$
15. $4\ln|x| - \dfrac{1}{2}\ln(x^2 + 1) + 4\tan^{-1}x + C$
17. $\dfrac{1}{16}\ln\left|\dfrac{(v - 2)^5(v + 2)}{v^6}\right| + C$
19. $\dfrac{1}{2}\tan^{-1}t - \dfrac{\sqrt{3}}{6}\tan^{-1}\dfrac{t}{\sqrt{3}} + C$
21. $\dfrac{x^2}{2} + \dfrac{4}{3}\ln|x + 2| + \dfrac{2}{3}\ln|x - 1| + C$
23. $\dfrac{x^2}{2} - \dfrac{9}{2}\ln|x + 3| + \dfrac{3}{2}\ln|x + 1| + C$
25. $\dfrac{1}{3}\ln\left|\dfrac{\sqrt{x + 1} - 1}{\sqrt{x + 1} + 1}\right| + C$ **27.** $\ln|1 - e^{-s}| + C$
29. $-\sqrt{16 - y^2} + C$ **31.** $-\dfrac{1}{2}\ln|4 - x^2| + C$
33. $\ln\dfrac{1}{\sqrt{9 - x^2}} + C$ **35.** $\dfrac{1}{6}\ln\left|\dfrac{x + 3}{x - 3}\right| + C$
37. $-\dfrac{\cos^5 x}{5} + \dfrac{\cos^7 x}{7} + C$ **39.** $\dfrac{\tan^5 x}{5} + C$
41. $\dfrac{\cos\theta}{2} - \dfrac{\cos 11\theta}{22} + C$ **43.** $4\sqrt{1 - \cos(t/2)} + C$
45. At least 16   **47.** $T = \pi, S = \pi$ **49.** $25°F$
51. (a) ≈ 2.42 gal **(b)** ≈ 24.83 mi/gal
53. $\pi/2$ **55.** 6   **57.** $\ln 3$ **59.** 2   **61.** $\pi/6$
63. Diverges   **65.** Diverges   **67.** Converges
69. $\dfrac{1}{2}xe^{2x} - \dfrac{1}{4}e^{2x} + C$ **71.** $2\tan x - x + C$
73. $x\tan x - \ln|\sec x| + C$ **75.** $-\dfrac{1}{3}(\cos x)^3 + C$
77. $1 + \dfrac{1}{2}\ln\left(\dfrac{2}{1 + e^2}\right)$ **79.** $2\ln\left|1 - \dfrac{1}{x}\right| + \dfrac{4x + 1}{2x^2} + C$
81. $\dfrac{e^{2x} - 1}{e^x} + C$ **83.** $9/4$ **85.** $256/15$
87. $-\dfrac{1}{3}\csc^3 x + C$
89. $\dfrac{2x^{3/2}}{3} - x + 2\sqrt{x} - 2\ln(\sqrt{x} + 1) + C$
91. $\dfrac{1}{2}\sin^{-1}(x - 1) + \dfrac{1}{2}(x - 1)\sqrt{2x - x^2} + C$
93. $-2\cot x - \ln|\csc x + \cot x| + \csc x + C$
95. $\dfrac{1}{12}\ln\left|\dfrac{3 + v}{3 - v}\right| + \dfrac{1}{6}\tan^{-1}\dfrac{v}{3} + C$
97. $\dfrac{\theta\sin(2\theta + 1)}{2} + \dfrac{\cos(2\theta + 1)}{4} + C$
99. $\dfrac{1}{4}\sec^2\theta + C$ **101.** $2\left(\dfrac{(\sqrt{2 - x})^3}{3} - 2\sqrt{2 - x}\right) + C$
103. $\tan^{-1}(y - 1) + C$
105. $\dfrac{1}{4}\ln|z| - \dfrac{1}{4z} - \dfrac{1}{4}\left[\dfrac{1}{2}\ln(z^2 + 4) + \dfrac{1}{2}\tan^{-1}\left(\dfrac{z}{2}\right)\right] + C$

107. $-\dfrac{1}{4}\sqrt{9 - 4t^2} + C$ **109.** $\ln\left(\dfrac{e^t + 1}{e^t + 2}\right) + C$

111. $1/4$ **113.** $\dfrac{2}{3}x^{3/2} + C$ **115.** $-\dfrac{1}{5}\tan^{-1}(\cos 5t) + C$

117. $2\sqrt{r} - 2\ln\left(1 + \sqrt{r}\right) + C$

119. $\dfrac{1}{2}x^2 - \dfrac{1}{2}\ln(x^2 + 1) + C$

121. $\dfrac{2}{3}\ln|x + 1| + \dfrac{1}{6}\ln|x^2 - x + 1| + \dfrac{1}{\sqrt{3}}\tan^{-1}\left(\dfrac{2x - 1}{\sqrt{3}}\right) + C$

123. $\dfrac{4}{7}\left(1 + \sqrt{x}\right)^{7/2} - \dfrac{8}{5}\left(1 + \sqrt{x}\right)^{5/2} + \dfrac{4}{3}\left(1 + \sqrt{x}\right)^{3/2} + C$

125. $2\ln\left|\sqrt{x} + \sqrt{1 + x}\right| + C$

127. $\ln x - \ln|1 + \ln x| + C$

129. $\dfrac{1}{2}x^{\ln x} + C$ **131.** $\dfrac{1}{2}\ln\left|\dfrac{1 - \sqrt{1 - x^4}}{x^2}\right| + C$

133. (b) $\dfrac{\pi}{4}$ **135.** $x - \dfrac{1}{\sqrt{2}}\tan^{-1}\left(\sqrt{2}\tan x\right) + C$

ADDITIONAL AND ADVANCED EXERCISES, pp. 536–538

1. $x(\sin^{-1}x)^2 + 2(\sin^{-1}x)\sqrt{1 - x^2} - 2x + C$

3. $\dfrac{x^2\sin^{-1}x}{2} + \dfrac{x\sqrt{1 - x^2} - \sin^{-1}x}{4} + C$

5. $\dfrac{1}{2}\left(\ln\left(t - \sqrt{1 - t^2}\right) - \sin^{-1}t\right) + C$ **7.** 0

9. $\ln(4) - 1$ **11.** 1 **13.** $32\pi/35$ **15.** 2π

17. (a) π (b) $\pi(2e - 5)$

19. (b) $\pi\left(\dfrac{8(\ln 2)^2}{3} - \dfrac{16(\ln 2)}{9} + \dfrac{16}{27}\right)$

21. $\left(\dfrac{e^2 + 1}{4}, \dfrac{e - 2}{2}\right)$

23. $\sqrt{1 + e^2} - \ln\left(\dfrac{\sqrt{1 + e^2}}{e} + \dfrac{1}{e}\right) - \sqrt{2} + \ln\left(1 + \sqrt{2}\right)$

25. $\dfrac{12\pi}{5}$ **27.** $a = \dfrac{1}{2}, -\dfrac{\ln 2}{4}$ **29.** $\dfrac{1}{2} < p \le 1$

33. $\dfrac{2}{1 - \tan(x/2)} + C$ **35.** 1 **37.** $\dfrac{\sqrt{3}\pi}{9}$

39. $\dfrac{1}{\sqrt{2}}\ln\left|\dfrac{\tan(t/2) + 1 - \sqrt{2}}{\tan(t/2) + 1 + \sqrt{2}}\right| + C$

41. $\ln\left|\dfrac{1 + \tan(\theta/2)}{1 - \tan(\theta/2)}\right| + C$

Chapter 9

SECTION 9.1, pp. 546–548

1. (d)   **3.** (a)

5.

7. $y' = x - y; \ y(1) = -1$

9. $y' = -(1 + y)\sin x; \ y(0) = 2$

11. $y' = 1 + x\,e^y; \ y(-2) = 2$

13.

y	$f(y) = \dfrac{dy}{dx}$
-3	2
-2	1.5
-1	0
0	-1
1	-0.75
2	0
3	1

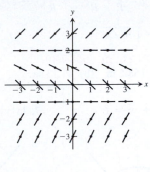

15. $y(\text{exact}) = -x^2, \ y_1 = -2, \ y_2 = -3.3333, \ y_3 = -5$

17. $y(\text{exact}) = 3e^{x(x+2)}, \ y_1 = 4.2, \ y_2 = 6.216, \ y_3 = 9.697$

19. $y(\text{exact}) = e^{x^2} + 1, \ y_1 = 2.0, \ y_2 = 2.0202, \ y_3 = 2.0618$

21. $y \approx 2.48832$, exact value is e.

23. $y \approx -0.2272$, exact value is $1/\left(1 - 2\sqrt{5}\right) \approx -0.2880$.

27.

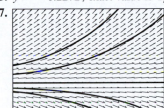

29.   **31.**

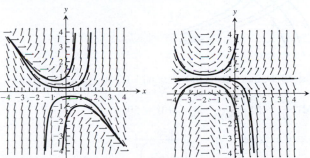

39. Euler's method gives $y \approx 3.45835$; the exact solution is $y = 1 + e \approx 3.71828$.

41. $y \approx 1.5000$; exact value is 1.5275.

SECTION 9.2, pp. 552–554

1. $y = \dfrac{e^x + C}{x}, \quad x > 0$ **3.** $y = \dfrac{C - \cos x}{x^3}, \quad x > 0$

5. $y = \dfrac{1}{2} - \dfrac{1}{x} + \dfrac{C}{x^2}, \quad x > 0$ **7.** $y = \dfrac{1}{2}xe^{x/2} + Ce^{x/2}$

9. $y = x(\ln x)^2 + Cx$

11. $s = \dfrac{t^3}{3(t - 1)^4} - \dfrac{t}{(t - 1)^4} + \dfrac{C}{(t - 1)^4}$

13. $r = (\csc\theta)(\ln|\sec\theta| + C), \quad 0 < \theta < \pi/2$

15. $y = \dfrac{3}{2} - \dfrac{1}{2}e^{-2t}$ **17.** $y = -\dfrac{1}{\theta}\cos\theta + \dfrac{\pi}{2\theta}$

19. $y = 6e^{x^2} - \dfrac{e^{x^2}}{x + 1}$ **21.** $y = y_0e^{kt}$

23. (b) is correct, but (a) is not.   **25.** $t = \dfrac{L}{R}\ln 2$ sec

27. (a) $i = \dfrac{V}{R} - \dfrac{V}{R}e^{-3} = \dfrac{V}{R}(1 - e^{-3}) \approx 0.95\dfrac{V}{R}$ amp **(b)** 86%

29. $y = \dfrac{1}{1 + Ce^{-x}}$ **31.** $y^3 = 1 + Cx^{-3}$

SECTION 9.3, pp. 559–560

1. (a) 168.5 m **(b)** 41.13 sec

3. $s(t) = 4.91\left(1 - e^{-(22.36/39.92)t}\right)$

5. $x^2 + y^2 = C$ **7.** $\ln|y| - \dfrac{1}{2}y^2 = \dfrac{1}{2}x^2 + C$

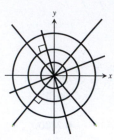 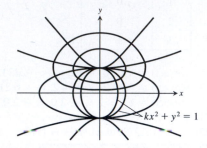

$kx^2 + y^2 = 1$

9. $y = \pm\sqrt{2x + C}$

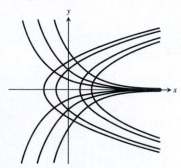

13. (a) 10 lb/min **(b)** $(100 + t)$ gal **(c)** $4\left(\dfrac{y}{100 + t}\right)$ lb/min

(d) $\dfrac{dy}{dt} = 10 - \dfrac{4y}{100 + t}, \quad y(0) = 50,$

$y = 2(100 + t) - \dfrac{150}{\left(1 + \dfrac{t}{100}\right)^4}$

(e) Concentration $= \dfrac{y(25)}{\text{amt. brine in tank}} = \dfrac{188.6}{125} \approx 1.5$ lb/gal

15. $y(27.8) \approx 14.8$ lb, $t \approx 27.8$ min

SECTION 9.4, pp. 566–567

1. $y' = (y + 2)(y - 3)$
  (a) $y = -2$ is a stable equilibrium value and $y = 3$ is an unstable equilibrium.
  (b) $y'' = 2(y + 2)\left(y - \dfrac{1}{2}\right)(y - 3)$

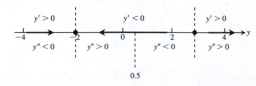

(c)

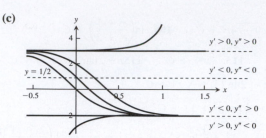

3. $y' = y^3 - y = (y + 1)y(y - 1)$
  (a) $y = -1$ and $y = 1$ are unstable equilibria and $y = 0$ is a stable equilibrium.
  (b) $y'' = (3y^2 - 1)y'$
  $= 3(y + 1)\left(y + 1/\sqrt{3}\right)y\left(y - 1/\sqrt{3}\right)(y - 1)$

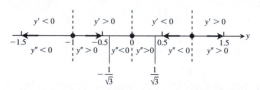

(c)

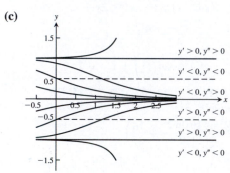

5. $y' = \sqrt{y}, y > 0$
  (a) There are no equilibrium values.
  (b) $y'' = \dfrac{1}{2}$

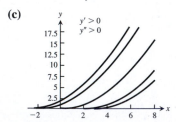

(c)

7. $y' = (y - 1)(y - 2)(y - 3)$
  (a) $y = 1$ and $y = 3$ are unstable equilibria and $y = 2$ is a stable equilibrium.
  (b) $y'' = (3y^2 - 12y + 11)(y - 1)(y - 2)(y - 3) =$
  $3(y - 1)\left(y - \dfrac{6 - \sqrt{3}}{3}\right)(y - 2)\left(y - \dfrac{6 + \sqrt{3}}{3}\right)(y - 3)$

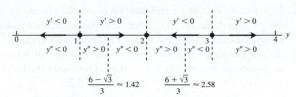

(c)

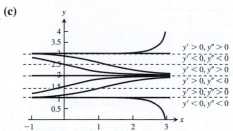

9. $\frac{dP}{dt} = 1 - 2P$ has a stable equilibrium at $P = \frac{1}{2}$;

$$\frac{d^2P}{dt^2} = -2\frac{dP}{dt} = -2(1 - 2P).$$

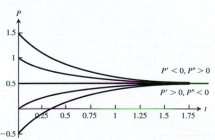

11. $\frac{dP}{dt} = 2P(P - 3)$ has a stable equilibrium at $P = 0$ and an unstable equilibrium at $P = 3$; $\frac{d^2P}{dt^2} = 2(2P - 3)\frac{dP}{dt} = 4P(2P - 3)(P - 3)$.

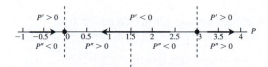

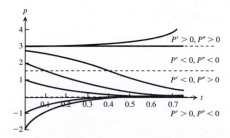

13. Before the catastrophe, the population exhibits logistic growth and $P(t)$ increases toward M_0, the stable equilibrium. After the catastrophe, the population declines logistically and $P(t)$ decreases toward M_1, the new stable equilibrium.

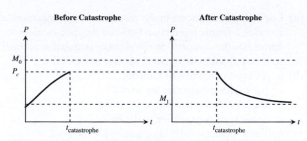

15. $\frac{dv}{dt} = g - \frac{k}{m}v^2$, $\quad g, k, m > 0$ and $v(t) \geq 0$

Equilibrium: $\frac{dv}{dt} = g - \frac{k}{m}v^2 = 0 \Rightarrow v = \sqrt{\frac{mg}{k}}$

Concavity: $\frac{d^2v}{dt^2} = -2\left(\frac{k}{m}v\right)\frac{dv}{dt} = -2\left(\frac{k}{m}v\right)\left(g - \frac{k}{m}v^2\right)$

(a)   **(b)**

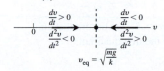

(c) $v_{\text{terminal}} = \sqrt{\frac{160}{0.005}} = 178.9 \text{ ft/sec} = 122 \text{ mph}$

17. $F = F_p - F_r; ma = 50 - 5|v|; \frac{dv}{dt} = \frac{1}{m}(50 - 5|v|)$. The maximum velocity occurs when $\frac{dv}{dt} = 0$ or $v = 10$ ft/sec.

19. Phase line:

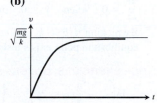

If the switch is closed at $t = 0$, then $i(0) = 0$, and the graph of the solution looks like this:

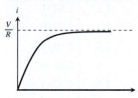

As $t \rightarrow \infty$, $i(t) \rightarrow i_{\text{steady state}} = \frac{V}{R}$.

SECTION 9.5, pp. 570–573

1. Seasonal variations, nonconformity of the environments, effects of other interactions, unexpected disasters, etc.

3. This model assumes that the number of interactions is proportional to the product of x and y:

$$\frac{dx}{dt} = (a - by)x, \quad a < 0,$$

$$\frac{dy}{dt} = m\left(1 - \frac{y}{M}\right)y - nxy = y\left(m - \frac{m}{M}y - nx\right).$$

Rest points are $(0, 0)$, unstable, and $(0, M)$, stable.

5. (a) Logistic growth occurs in the absence of the competitor, and involves a simple interaction between the species: Growth dominates the competition when either population is small, so it is difficult to drive either species to extinction.

(b) a: per capita growth rate for trout

m: per capita growth rate for bass

b: intensity of competition to the trout

n: intensity of competition to the bass

k_1: environmental carrying capacity for the trout

k_2: environmental carrying capacity for the bass

$\dfrac{a}{b}$: growth versus competition or net growth of trout

$\dfrac{m}{n}$: relative survival of bass

(c) $\dfrac{dx}{dt} = 0$ when $x = 0$ or $y = \dfrac{a}{b} - \dfrac{a}{bk_1}x$,

$\dfrac{dy}{dt} = 0$ when $y = 0$ or $y = k_2 - \dfrac{k_2 n}{m}x$.

By picking $a/b > k_2$ and $m/n > k_1$, we ensure that an equilibrium point exists inside the first quadrant.

PRACTICE EXERCISES, pp. 573–575

1. $y = -\ln\left(C - \dfrac{2}{5}(x - 2)^{5/2} - \dfrac{4}{3}(x - 2)^{3/2} \right)$

3. $\tan y = -x \sin x - \cos x + C$

5. $(y + 1)e^{-y} = -\ln|x| + C$

7. $y = C\dfrac{x - 1}{x}$ **9.** $y = \dfrac{x^2}{4}e^{x/2} + Ce^{x/2}$

11. $y = \dfrac{x^2 - 2x + C}{2x^2}$ **13.** $y = \dfrac{e^{-x} + C}{1 + e^x}$ **15.** $xy + y^3 = C$

17. $\tan y = -\cos x + (1/3)\cos^3 x + C$

19. $e^{x^2}y = (x - 1)e^x + C$

21. $\ln|\ln y| = (1/2)x^2 \ln x - (1/4)x^2 + C$

23. $y = \dfrac{2x^3 + 3x^2 + 6}{6(x + 1)^2}$ **25.** $y = \dfrac{1}{3}\left(1 - 4e^{-x^3}\right)$

27. $y = e^{-x}(3x^3 - 3x^2)$

29.

x	y	x	y
0	0	1.1	1.6241
0.1	0.1000	1.2	1.8319
0.2	0.2095	1.3	2.0513
0.3	0.3285	1.4	2.2832
0.4	0.4568	1.5	2.5285
0.5	0.5946	1.6	2.7884
0.6	0.7418	1.7	3.0643
0.7	0.8986	1.8	3.3579
0.8	1.0649	1.9	3.6709
0.9	1.2411	2.0	4.0057
1.0	1.4273		

31. $y(3) \approx 0.9131$

33. (a)

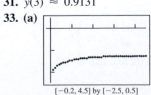

$[-0.2, 4.5]$ by $[-2.5, 0.5]$

(b) Note that we choose a small interval of x-values because the y-values decrease very rapidly and our calculator cannot handle the calculations for $x \le -1$. (This occurs because the analytic solution is $y = -2 + \ln(2 - e^{-x})$, which has an asymptote at $x = -\ln 2 \approx -0.69$. Obviously, the Euler approximations are misleading for $x \le -0.7$.)

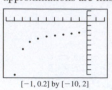

$[-1, 0.2]$ by $[-10, 2]$

35. $y(\text{exact}) = \dfrac{1}{2}x^2 - \dfrac{3}{2}$; $y(2) \approx 0.4$; exact value is $\dfrac{1}{2}$.

37. $y(\text{exact}) = -e^{(x^2-1)/2}$; $y(2) \approx -3.4192$; exact value is $-e^{3/2} \approx -4.4817$.

41. (a) $y = -1$ is stable and $y = 1$ is unstable.

(b) $\dfrac{d^2y}{dx^2} = 2y\dfrac{dy}{dx} = 2y(y^2 - 1)$

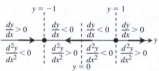

(c)

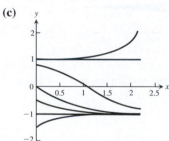

43. (a) 1 min: 102 gal; 10 min: 120 gal; 60 min: 220 gal

(b) $S(1) \approx 12.55$ lb, $S(10) \approx 32.22$ lb, $S(60) \approx 101.74$ lb

ADDITIONAL AND ADVANCED EXERCISES, p. 575

1. (a) $y = c + (y_0 - c)e^{-k(A/V)t}$

(b) Steady-state solution: $y_\infty = c$

5. $x^2(x^2 + 2y^2) = C$

7. $\ln|x| + e^{-y/x} = C$

9. $\ln|x| - \ln|\sec(y/x - 1) + \tan(y/x - 1)| = C$

Chapter 10

SECTION 10.1, pp. 586–590

1. $a_1 = 0, a_2 = -1/4, a_3 = -2/9, a_4 = -3/16$

3. $a_1 = 1, a_2 = -1/3, a_3 = 1/5, a_4 = -1/7$

5. $a_1 = 1/2, a_2 = 1/2, a_3 = 1/2, a_4 = 1/2$

7. $1, \dfrac{3}{2}, \dfrac{7}{4}, \dfrac{15}{8}, \dfrac{31}{16}, \dfrac{63}{32}, \dfrac{127}{64}, \dfrac{255}{128}, \dfrac{511}{256}, \dfrac{1023}{512}$

9. $2, 1, -\dfrac{1}{2}, -\dfrac{1}{4}, \dfrac{1}{8}, \dfrac{1}{16}, -\dfrac{1}{32}, -\dfrac{1}{64}, \dfrac{1}{128}, \dfrac{1}{256}$

11. $1, 1, 2, 3, 5, 8, 13, 21, 34, 55$

13. $a_n = (-1)^{n+1}, n \ge 1$

15. $a_n = (-1)^{n+1}(n)^2, n \geq 1$ **17.** $a_n = \dfrac{2^{n-1}}{3(n+2)}, n \geq 1$

19. $a_n = n^2 - 1, n \geq 1$ **21.** $a_n = 4n - 3, n \geq 1$

23. $a_n = \dfrac{3n+2}{n!}, n \geq 1$ **25.** $a_n = \dfrac{1 + (-1)^{n+1}}{2}, n \geq 1$

27. $a_n = \dfrac{1}{(n+1)(n+2)}$ **29.** $a_n = \sin\left(\dfrac{\sqrt{n+1}}{1 + (n+1)^2}\right)$

31. Converges, 2   **33.** Converges, -1 **35.** Converges, -5

37. Diverges   **39.** Diverges   **41.** Converges, $1/2$

43. Converges, 0   **45.** Converges, $\sqrt{2}$ **47.** Converges, 1

49. Converges, 0   **51.** Converges, 0   **53.** Converges, 0

55. Converges, 1   **57.** Converges, e^7 **59.** Converges, 1

61. Converges, 1   **63.** Diverges   **65.** Converges, 4

67. Converges, 0   **69.** Diverges   **71.** Converges, e^{-1}

73. Diverges   **75.** Converges, 0   **77.** Diverges

79. Converges, $e^{2/3}$ **81.** Converges, $x \, (x > 0)$

83. Converges, 0   **85.** Converges, 1   **87.** Converges, $1/2$

89. Converges, 1   **91.** Converges, $\pi/2$ **93.** Converges, 0

95. Converges, 0   **97.** Converges, $1/2$ **99.** Converges, 0

101. 8   **103.** 4   **105.** 5   **107.** $1 + \sqrt{2}$ **109.** $x_n = 2^{n-2}$

111. (a) $f(x) = x^2 - 2, 1.414213562 \approx \sqrt{2}$

     (b) $f(x) = \tan(x) - 1, 0.7853981635 \approx \pi/4$

     (c) $f(x) = e^x$, diverges

113. 1

121. Nondecreasing, bounded

123. Not nondecreasing, bounded

125. Converges, nondecreasing sequence theorem

127. Converges, nondecreasing sequence theorem

129. Diverges, definition of divergence

131. Converges

133. Converges

145. (b) $\sqrt{3}$

SECTION 10.2, pp. 597–599

1. $s_n = \dfrac{2(1 - (1/3)^n)}{1 - (1/3)}, 3$ **3.** $s_n = \dfrac{1 - (-1/2)^n}{1 - (-1/2)}, 2/3$

5. $s_n = \dfrac{1}{2} - \dfrac{1}{n+2}, \dfrac{1}{2}$ **7.** $1 - \dfrac{1}{4} + \dfrac{1}{16} - \dfrac{1}{64} + \cdots, \dfrac{4}{5}$

9. $-\dfrac{3}{4} + \dfrac{9}{16} + \dfrac{57}{64} + \dfrac{249}{256} + \cdots$, diverges.

11. $(5 + 1) + \left(\dfrac{5}{2} + \dfrac{1}{3}\right) + \left(\dfrac{5}{4} + \dfrac{1}{9}\right) + \left(\dfrac{5}{8} + \dfrac{1}{27}\right) + \cdots, \dfrac{23}{2}$

13. $(1 + 1) + \left(\dfrac{1}{2} - \dfrac{1}{5}\right) + \left(\dfrac{1}{4} + \dfrac{1}{25}\right) + \left(\dfrac{1}{8} - \dfrac{1}{125}\right) + \cdots, \dfrac{17}{6}$

15. Converges, $5/3$ **17.** Converges, $1/7$

19. Converges, $\dfrac{e}{e+2}$ **21.** Diverges   **23.** $23/99$

25. $7/9$ **27.** $1/15$ **29.** $41333/33300$ **31.** Diverges

33. Inconclusive   **35.** Diverges   **37.** Diverges

39. $s_n = 1 - \dfrac{1}{n+1}$; converges, 1

41. $s_n = \ln\sqrt{n+1}$; diverges

43. $s_n = \dfrac{\pi}{3} - \cos^{-1}\left(\dfrac{1}{n+2}\right)$; converges, $-\dfrac{\pi}{6}$ **45.** 1   **47.** 5

49. 1   **51.** $-\dfrac{1}{\ln 2}$ **53.** Converges, $2 + \sqrt{2}$

55. Converges, 1   **57.** Diverges

59. Converges, $\dfrac{e^2}{e^2 - 1}$

61. Converges, $2/9$ **63.** Converges, $3/2$ **65.** Diverges

67. Converges, 4   **69.** Diverges   **71.** Converges, $\dfrac{\pi}{\pi - e}$

73. Converges, $-5/6$ **75.** Diverges

77. $a = 1, r = -x$; converges to $1/(1 + x)$ for $|x| < 1$

79. $a = 3, r = (x - 1)/2$; converges to $6/(3 - x)$ for x in $(-1, 3)$

81. $|x| < \dfrac{1}{2}, \dfrac{1}{1 - 2x}$ **83.** $-2 < x < 0, \dfrac{1}{2 + x}$

85. $x \neq (2k + 1)\dfrac{\pi}{2}, k$ an integer; $\dfrac{1}{1 - \sin x}$

87. (a) $\displaystyle\sum_{n=-2}^{\infty} \dfrac{1}{(n+4)(n+5)}$ (b) $\displaystyle\sum_{n=0}^{\infty} \dfrac{1}{(n+2)(n+3)}$

     (c) $\displaystyle\sum_{n=5}^{\infty} \dfrac{1}{(n-3)(n-2)}$

97. (a) $r = 3/5$ (b) $r = -3/10$ **99.** $|r| < 1, \dfrac{1 + 2r}{1 - r^2}$

101. (a) 16.84 mg, 17.79 mg (b) 17.84 mg

103. (a) $0, \dfrac{1}{27}, \dfrac{2}{27}, \dfrac{1}{9}, \dfrac{2}{9}, \dfrac{7}{27}, \dfrac{8}{27}, \dfrac{1}{3}, \dfrac{2}{3}, \dfrac{7}{9}, \dfrac{8}{9}, 1$

     (b) $\displaystyle\sum_{n=1}^{\infty} \dfrac{1}{2}\left(\dfrac{2}{3}\right)^{n-1} = 1$ **105.** $(4/3)\pi$

SECTION 10.3, pp. 604–606

1. Converges   **3.** Converges   **5.** Converges   **7.** Diverges

9. Converges   **11.** Diverges

13. Converges; geometric series, $r = \dfrac{1}{10} < 1$

15. Diverges; $\displaystyle\lim_{n\to\infty} \dfrac{n}{n+1} = 1 \neq 0$

17. Diverges; p-series, $p < 1$

19. Converges; geometric series, $r = \dfrac{1}{8} < 1$

21. Diverges; Integral Test

23. Converges; geometric series, $r = 2/3 < 1$

25. Diverges; Integral Test

27. Diverges; $\displaystyle\lim_{n\to\infty} \dfrac{2^n}{n+1} \neq 0$

29. Diverges; $\lim_{n\to\infty}\left(\sqrt{n}/\ln n\right) \neq 0$

31. Diverges; geometric series, $r = \dfrac{1}{\ln 2} > 1$

33. Converges; Integral Test

35. Diverges; nth-Term Test

37. Converges; Integral Test

39. Diverges; nth-Term Test

41. Converges; by taking limit of partial sums

43. Converges; Integral Test

45. Converges; Integral Test   **47.** $a = 1$

49. (a)

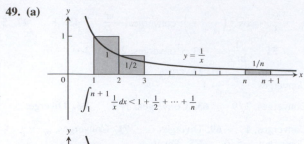

$$\int_1^{n+1} \frac{1}{x}\,dx < 1 + \frac{1}{2} + \cdots + \frac{1}{n}$$

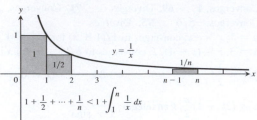

$$1 + \frac{1}{2} + \cdots + \frac{1}{n} < 1 + \int_1^n \frac{1}{x}\,dx$$

(b) ≈ 41.55

51. True **53.** $n \geq 251{,}415$

55. $s_8 = \sum_{n=1}^{8} \frac{1}{n^3} \approx 1.195$ **57.** 10^{60}

65. (a) $1.20166 \leq S \leq 1.20253$

(b) $S \approx 1.2021$, error < 0.0005

67. $\left(\frac{\pi^2}{6} - 1\right) \approx 0.64493$

SECTION 10.4, pp. 610–611

1. Converges; compare with $\sum(1/n^2)$
3. Diverges; compare with $\sum(1/\sqrt{n})$
5. Converges; compare with $\sum(1/n^{3/2})$
7. Converges; compare with $\sum\sqrt{\frac{n+4n}{n^4+0}} = \sqrt{5}\,\sum\frac{1}{n^{3/2}}$
9. Converges
11. Diverges; limit comparison with $\sum(1/n)$
13. Diverges; limit comparison with $\sum(1/\sqrt{n})$
15. Diverges
17. Diverges; limit comparison with $\sum(1/\sqrt{n})$
19. Converges; compare with $\sum(1/2^n)$
21. Diverges; nth-Term Test
23. Converges; compare with $\sum(1/n^2)$
25. Converges; $\left(\frac{n}{3n+1}\right)^n < \left(\frac{n}{3n}\right)^n = \left(\frac{1}{3}\right)^n$
27. Diverges; direct comparison with $\sum(1/n)$
29. Diverges; limit comparison with $\sum(1/n)$
31. Diverges; limit comparison with $\sum(1/n)$
33. Converges; compare with $\sum(1/n^{3/2})$
35. Converges; $\frac{1}{n2^n} \leq \frac{1}{2^n}$ **37.** Converges; $\frac{1}{3^{n-1}+1} < \frac{1}{3^{n-1}}$
39. Converges; comparison with $\sum(1/5n^2)$
41. Diverges; comparison with $\sum(1/n)$
43. Converges; comparison with $\sum\frac{1}{n(n-1)}$ or limit comparison with $\sum(1/n^2)$
45. Diverges; limit comparison with $\sum(1/n)$
47. Converges; $\frac{\tan^{-1}n}{n^{1.1}} < \frac{\pi/2}{n^{1.1}}$
49. Converges; compare with $\sum(1/n^2)$
51. Diverges; limit comparison with $\sum(1/n)$
53. Converges; limit comparison with $\sum(1/n^2)$
55. Diverges nth-Term Test
67. Converges **69.** Converges **71.** Converges

SECTION 10.5, pp. 616–617

1. Converges **3.** Diverges **5.** Converges
7. Converges **9.** Converges **11.** Diverges
13. Converges **15.** Converges
17. Converges; Ratio Test **19.** Diverges; Ratio Test
21. Converges; Ratio Test
23. Converges; compare with $\sum(3/(1.25)^n)$
25. Diverges; $\lim_{n\to\infty}\left(1 - \frac{3}{n}\right)^n = e^{-3} \neq 0$
27. Converges; compare with $\sum(1/n^2)$
29. Diverges; compare with $\sum(1/(2n))$ **31.** Diverges; $a_n \nrightarrow 0$
33. Converges; Ratio Test **35.** Converges; Ratio Test
37. Converges; Ratio Test **39.** Converges; Root Test
41. Converges; compare with $\sum(1/n^2)$
43. Converges; Ratio Test **45.** Diverges; Ratio Test
47. Converges; Ratio Test **49.** Diverges; Ratio Test
51. Converges; Ratio Test **53.** Converges; Ratio Test
55. Diverges; $a_n = \left(\frac{1}{3}\right)^{(1/n!)} \to 1$ **57.** Converges; Ratio Test
59. Diverges; Root Test **61.** Converges; Root Test
63. Converges; Ratio Test
65. (a) Diverges; nth-Term Test
(b) Diverges; Root Test
(c) Converges; Root Test
(d) Converges; Ratio Test
69. Yes

SECTION 10.6, pp. 622–624

1. Converges by Alternating Series Test
3. Converges; Alternating Series Test
5. Converges; Alternating Series Test
7. Diverges; $a_n \nrightarrow 0$
9. Diverges; $a_n \nrightarrow 0$
11. Converges; Alternating Series Test
13. Converges by Alternating Series Test
15. Converges absolutely. Series of absolute s is a convergent geometric series.
17. Converges conditionally; $1/\sqrt{n} \to 0$ but $\sum_{n=1}^{\infty}\frac{1}{\sqrt{n}}$ diverges.
19. Converges absolutely; compare with $\sum_{n=1}^{\infty}(1/n^2)$.
21. Converges conditionally; $1/(n+3) \to 0$ but $\sum_{n=1}^{\infty}\frac{1}{n+3}$ diverges (compare with $\sum_{n=1}^{\infty}(1/n)$).
23. Diverges; $\frac{3+n}{5+n} \to 1$
25. Converges conditionally; $\left(\frac{1}{n^2} + \frac{1}{n}\right) \to 0$ but $(1+n)/n^2 > 1/n$
27. Converges absolutely; Ratio Test
29. Converges absolutely by Integral Test
31. Diverges; $a_n \nrightarrow 0$
33. Converges absolutely by Ratio Test
35. Converges absolutely, since $\left|\frac{\cos n\pi}{n\sqrt{n}}\right| = \left|\frac{(-1)^{n+1}}{n^{3/2}}\right| = \frac{1}{n^{3/2}}$ (convergent p-series)
37. Converges absolutely by Root Test
39. Diverges; $a_n \to \infty$
41. Converges conditionally; $\sqrt{n+1} - \sqrt{n} = 1/(\sqrt{n} + \sqrt{n+1}) \to 0$, but series of absolute values diverges $\left(\text{compare with } \sum(1/\sqrt{n})\right)$.

43. Diverges, $a_n \to 1/2 \neq 0$

45. Converges absolutely; sech $n = \dfrac{2}{e^n + e^{-n}} = \dfrac{2e^n}{e^{2n} + 1} <$ $\dfrac{2e^n}{e^{2n}} = \dfrac{2}{e^n}$, a term from a convergent geometric series.

47. Converges conditionally; $\sum(-1)^{n+1}\dfrac{1}{2(n+1)}$ converges by Alternating Series Test; $\sum\dfrac{1}{2(n+1)}$ diverges by limit comparison with $\sum(1/n)$.

49. $|\text{Error}| < 0.2$ **51.** $|\text{Error}| < 2 \times 10^{-11}$
53. $n \geq 31$ **55.** $n \geq 4$ **57.** Converges; Root Test
59. Converges; Limit of Partial Sums
61. Converges; Ratio Test   **63.** Diverges; p-series Test
65. Converges; Root Test   **67.** Converges; Limit Comparison Test
69. Diverges; Limit of Partial Sums
71. Diverges; Limit Comparison Test
73. Diverges; nth-Term Test   **75.** Diverges; Limit of Partial Sums
77. Converges; Limit Comparison Test
79. Converges; Limit Comparison Test
81. Converges; Ratio Test
83. 0.54030   **85. (a)** $a_n \geq a_{n+1}$ **(b)** $-1/2$

SECTION 10.7, pp. 633–636

1. (a) $1, -1 < x < 1$ **(b)** $-1 < x < 1$ **(c)** none
3. (a) $1/4, -1/2 < x < 0$ **(b)** $-1/2 < x < 0$ **(c)** none
5. (a) $10, -8 < x < 12$ **(b)** $-8 < x < 12$ **(c)** none
7. (a) $1, -1 < x < 1$ **(b)** $-1 < x < 1$ **(c)** none
9. (a) $3, -3 \leq x \leq 3$ **(b)** $-3 \leq x \leq 3$ **(c)** none
11. (a) ∞, for all x **(b)** for all x **(c)** none
13. (a) $1/2, -1/2 < x < 1/2$ **(b)** $-1/2 < x < 1/2$ **(c)** none
15. (a) $1, -1 \leq x < 1$ **(b)** $-1 < x < 1$ **(c)** $x = -1$
17. (a) $5, -8 < x < 2$ **(b)** $-8 < x < 2$ **(c)** none
19. (a) $3, -3 < x < 3$ **(b)** $-3 < x < 3$ **(c)** none
21. (a) $1, -2 < x < 0$ **(b)** $-2 < x < 0$ **(c)** none
23. (a) $1, -1 < x < 1$ **(b)** $-1 < x < 1$ **(c)** none
25. (a) $0, x = 0$ **(b)** $x = 0$ **(c)** none
27. (a) $2, -4 < x \leq 0$ **(b)** $-4 < x < 0$ **(c)** $x = 0$
29. (a) $1, -1 \leq x \leq 1$ **(b)** $-1 \leq x \leq 1$ **(c)** none
31. (a) $1/4, 1 \leq x \leq 3/2$ **(b)** $1 \leq x \leq 3/2$ **(c)** none
33. (a) ∞, for all x **(b)** for all x **(c)** none
35. (a) $1, -1 \leq x < 1$ **(b)** $-1 < x < 1$ **(c)** -1
37. 3   **39.** 8   **41.** $-1/3 < x < 1/3, 1/(1 - 3x)$
43. $-1 < x < 3, 4/(3 + 2x - x^2)$
45. $0 < x < 16, 2/(4 - \sqrt{x})$
47. $-\sqrt{2} < x < \sqrt{2}, 3/(2 - x^2)$

49. $\dfrac{2}{x} = \displaystyle\sum_{n=0}^{\infty} 2(-1)^n(x - 1)^n, \; 0 < x < 2$

51. $\displaystyle\sum_{n=0}^{\infty}\left(-\tfrac{1}{3}\right)^n(x - 5)^n, \; 2 < x < 8$

53. $1 < x < 5, 2/(x - 1), \displaystyle\sum_{n=1}^{\infty}\left(-\tfrac{1}{2}\right)^n n(x - 3)^{n-1}$,

$1 < x < 5, -2/(x - 1)^2$

55. (a) $\cos x = 1 - \dfrac{x^2}{2!} + \dfrac{x^4}{4!} - \dfrac{x^6}{6!} + \dfrac{x^8}{8!} - \dfrac{x^{10}}{10!} + \cdots$; converges for all x
(b) Same answer as part (c)
(c) $2x - \dfrac{2^3 x^3}{3!} + \dfrac{2^5 x^5}{5!} - \dfrac{2^7 x^7}{7!} + \dfrac{2^9 x^9}{9!} - \dfrac{2^{11} x^{11}}{11!} + \cdots$

57. (a) $\dfrac{x^2}{2} + \dfrac{x^4}{12} + \dfrac{x^6}{45} + \dfrac{17x^8}{2520} + \dfrac{31x^{10}}{14175}, -\dfrac{\pi}{2} < x < \dfrac{\pi}{2}$

(b) $1 + x^2 + \dfrac{2x^4}{3} + \dfrac{17x^6}{45} + \dfrac{62x^8}{315} + \cdots, -\dfrac{\pi}{2} < x < \dfrac{\pi}{2}$

63. (a) T   **(b)** T   **(c)** F   **(d)** T   **(e)** N   **(f)** F   **(g)** N   **(h)** T

SECTION 10.8, pp. 640–641

1. $P_0(x) = 1, P_1(x) = 1 + 2x, P_2(x) = 1 + 2x + 2x^2,$
$P_3(x) = 1 + 2x + 2x^2 + \dfrac{4}{3}x^3$

3. $P_0(x) = 0, P_1(x) = x - 1, P_2(x) = (x - 1) - \dfrac{1}{2}(x - 1)^2,$
$P_3(x) = (x - 1) - \dfrac{1}{2}(x - 1)^2 + \dfrac{1}{3}(x - 1)^3$

5. $P_0(x) = \dfrac{1}{2}, P_1(x) = \dfrac{1}{2} - \dfrac{1}{4}(x - 2),$
$P_2(x) = \dfrac{1}{2} - \dfrac{1}{4}(x - 2) + \dfrac{1}{8}(x - 2)^2,$
$P_3(x) = \dfrac{1}{2} - \dfrac{1}{4}(x - 2) + \dfrac{1}{8}(x - 2)^2 - \dfrac{1}{16}(x - 2)^3$

7. $P_0(x) = \dfrac{\sqrt{2}}{2}, P_1(x) = \dfrac{\sqrt{2}}{2} + \dfrac{\sqrt{2}}{2}\left(x - \dfrac{\pi}{4}\right),$
$P_2(x) = \dfrac{\sqrt{2}}{2} + \dfrac{\sqrt{2}}{2}\left(x - \dfrac{\pi}{4}\right) - \dfrac{\sqrt{2}}{4}\left(x - \dfrac{\pi}{4}\right)^2,$
$P_3(x) = \dfrac{\sqrt{2}}{2} + \dfrac{\sqrt{2}}{2}\left(x - \dfrac{\pi}{4}\right) - \dfrac{\sqrt{2}}{4}\left(x - \dfrac{\pi}{4}\right)^2$
$\qquad - \dfrac{\sqrt{2}}{12}\left(x - \dfrac{\pi}{4}\right)^3$

9. $P_0(x) = 2, P_1(x) = 2 + \dfrac{1}{4}(x - 4),$
$P_2(x) = 2 + \dfrac{1}{4}(x - 4) - \dfrac{1}{64}(x - 4)^2,$
$P_3(x) = 2 + \dfrac{1}{4}(x - 4) - \dfrac{1}{64}(x - 4)^2 + \dfrac{1}{512}(x - 4)^3$

11. $\displaystyle\sum_{n=0}^{\infty}\dfrac{(-x)^n}{n!} = 1 - x + \dfrac{x^2}{2!} - \dfrac{x^3}{3!} + \dfrac{x^4}{4!} - \cdots$

13. $\displaystyle\sum_{n=0}^{\infty}(-1)^n x^n = 1 - x + x^2 - x^3 + \cdots$

15. $\displaystyle\sum_{n=0}^{\infty}\dfrac{(-1)^n 3^{2n+1} x^{2n+1}}{(2n + 1)!}$ **17.** $7\displaystyle\sum_{n=0}^{\infty}\dfrac{(-1)^n x^{2n}}{(2n)!}$ **19.** $\displaystyle\sum_{n=0}^{\infty}\dfrac{x^{2n}}{(2n)!}$

21. $x^4 - 2x^3 - 5x + 4$ **23.** $\displaystyle\sum_{n=1}^{\infty}(-1)^{n+1}\dfrac{x^{2n}}{(2n - 1)!}$

25. $8 + 10(x - 2) + 6(x - 2)^2 + (x - 2)^3$
27. $21 - 36(x + 2) + 25(x + 2)^2 - 8(x + 2)^3 + (x + 2)^4$

29. $\displaystyle\sum_{n=0}^{\infty}(-1)^n(n + 1)(x - 1)^n$ **31.** $\displaystyle\sum_{n=0}^{\infty}\dfrac{e^2}{n!}(x - 2)^n$

33. $\displaystyle\sum_{n=0}^{\infty}(-1)^{n+1}\dfrac{2^{2n}}{(2n)!}\left(x - \dfrac{\pi}{4}\right)^{2n}$

35. $-1 - 2x - \dfrac{5}{2}x^2 - \cdots, -1 < x < 1$

37. $x^2 - \dfrac{1}{2}x^3 + \dfrac{1}{6}x^4 + \cdots, -1 < x < 1$

39. $x^4 + x^6 + \dfrac{x^8}{2} + \cdots, (-\infty, \infty)$

45. $L(x) = 0, Q(x) = -x^2/2$ **47.** $L(x) = 1, Q(x) = 1 + x^2/2$
49. $L(x) = x, Q(x) = x$

SECTION 10.9, pp. 647–648

1. $\displaystyle\sum_{n=0}^{\infty} \frac{(-5x)^n}{n!} = 1 - 5x + \frac{5^2x^2}{2!} - \frac{5^3x^3}{3!} + \cdots$

3. $\displaystyle\sum_{n=0}^{\infty} \frac{5(-1)^n(-x)^{2n+1}}{(2n+1)!} = \sum_{n=0}^{\infty} \frac{5(-1)^{n+1}x^{2n+1}}{(2n+1)!}$

$= -5x + \frac{5x^3}{3!} - \frac{5x^5}{5!} + \frac{5x^7}{7!} + \cdots$

5. $\displaystyle\sum_{n=0}^{\infty} \frac{(-1)^n(5x^2)^{2n}}{(2n)!} = 1 - \frac{25x^4}{2!} + \frac{625x^8}{4!} - \cdots$

7. $\displaystyle\sum_{n=1}^{\infty} (-1)^{n+1} \frac{x^{2n}}{n} = x^2 - \frac{x^4}{2} + \frac{x^6}{3} - \frac{x^8}{4} + \cdots$

9. $\displaystyle\sum_{n=0}^{\infty} (-1)^n \left(\frac{3}{4}\right)^n x^{3n} = 1 - \frac{3}{4}x^3 + \frac{3^2}{4^2}x^6 - \frac{3^3}{4^3}x^9 + \cdots$

11. $\displaystyle\ln 3 + \sum_{n=1}^{\infty} (-1)^{n+1}\frac{2^n x^n}{n} = \ln 3 + 2x - 2x^2 + \frac{8}{3}x^3 - \cdots$

13. $\displaystyle\sum_{n=0}^{\infty} \frac{x^{n+1}}{n!} = x + x^2 + \frac{x^3}{2!} + \frac{x^4}{3!} + \frac{x^5}{4!} + \cdots$

15. $\displaystyle\sum_{n=2}^{\infty} \frac{(-1)^n x^{2n}}{(2n)!} = \frac{x^4}{4!} - \frac{x^6}{6!} + \frac{x^8}{8!} - \frac{x^{10}}{10!} + \cdots$

17. $\displaystyle x - \frac{\pi^2 x^3}{2!} + \frac{\pi^4 x^5}{4!} - \frac{\pi^6 x^7}{6!} + \cdots = \sum_{n=0}^{\infty} \frac{(-1)^n \pi^{2n} x^{2n+1}}{(2n)!}$

19. $\displaystyle 1 + \sum_{n=1}^{\infty} \frac{(-1)^n(2x)^{2n}}{2\cdot(2n)!} =$

$1 - \frac{(2x)^2}{2\cdot 2!} + \frac{(2x)^4}{2\cdot 4!} - \frac{(2x)^6}{2\cdot 6!} + \frac{(2x)^8}{2\cdot 8!} - \cdots$

21. $\displaystyle x^2 \sum_{n=0}^{\infty}(2x)^n = x^2 + 2x^3 + 4x^4 + \cdots$

23. $\displaystyle\sum_{n=1}^{\infty} nx^{n-1} = 1 + 2x + 3x^2 + 4x^3 + \cdots$

25. $\displaystyle\sum_{n=1}^{\infty} (-1)^{n+1} \frac{x^{4n-1}}{2n-1} = x^3 - \frac{x^7}{3} + \frac{x^{11}}{5} - \frac{x^{15}}{7} + \cdots$

27. $\displaystyle\sum_{n=0}^{\infty} \left(\frac{1}{n!} + (-1)^n\right)x^n = 2 + \frac{3}{2}x^2 - \frac{5}{6}x^3 + \frac{25}{24}x^4 - \cdots$

29. $\displaystyle\sum_{n=1}^{\infty} \frac{(-1)^{n-1}x^{2n+1}}{3n} = \frac{x^3}{3} - \frac{x^5}{6} + \frac{x^7}{9} - \cdots$

31. $\displaystyle x + x^2 + \frac{x^3}{3} - \frac{x^5}{30} + \cdots$

33. $\displaystyle x^2 - \frac{2}{3}x^4 + \frac{23}{45}x^6 - \frac{44}{105}x^8 + \cdots$

35. $\displaystyle 1 + x + \frac{1}{2}x^2 - \frac{1}{8}x^4 + \cdots$ **37.** $\displaystyle 1 - \frac{x^2}{2} - \frac{x^3}{2} - \frac{x^4}{4} - \cdots$

39. $\displaystyle |\text{Error}| \le \frac{1}{10^4\cdot 4!} < 4.2 \times 10^{-6}$

41. $|x| < (0.06)^{1/5} < 0.56968$

43. $|\text{Error}| < (10^{-3})^3/6 < 1.67 \times 10^{-10}, \; -10^{-3} < x < 0$

45. $|\text{Error}| < (3^{0.1})(0.1)^3/6 < 1.87 \times 10^{-4}$

53. (a) $\displaystyle Q(x) = 1 + kx + \frac{k(k-1)}{2}x^2$ (b) $0 \le x < 100^{-1/3}$

SECTION 10.10, pp. 655–657

1. $\displaystyle 1 + \frac{x}{2} - \frac{x^2}{8} + \frac{x^3}{16}$

3. $1 + 3x + 6x^2 + 10x^3$

5. $\displaystyle 1 - x + \frac{3x^2}{4} - \frac{x^3}{2}$

7. $\displaystyle 1 - \frac{x^3}{2} + \frac{3x^6}{8} - \frac{5x^9}{16}$

9. $\displaystyle 1 + \frac{1}{2x} - \frac{1}{8x^2} + \frac{1}{16x^3}$

11. $(1+x)^4 = 1 + 4x + 6x^2 + 4x^3 + x^4$

13. $(1 - 2x)^3 = 1 - 6x + 12x^2 - 8x^3$

15. 0.0713362 **17.** 0.4969536 **19.** 0.0999445 **21.** 0.10000

23. $\displaystyle\frac{1}{13\cdot 6!} \approx 0.00011$ **25.** $\displaystyle\frac{x^3}{3} - \frac{x^7}{7\cdot 3!} + \frac{x^{11}}{11\cdot 5!}$

27. (a) $\displaystyle\frac{x^2}{2} - \frac{x^4}{12}$

(b) $\displaystyle\frac{x^2}{2} - \frac{x^4}{3\cdot 4} + \frac{x^6}{5\cdot 6} - \frac{x^8}{7\cdot 8} + \cdots + (-1)^{15}\frac{x^{32}}{31\cdot 32}$

29. $1/2$ **31.** $-1/24$ **33.** $1/3$ **35.** -1 **37.** 2

39. $3/2$ **41.** e **43.** $\cos\frac{3}{4}$ **45.** $\frac{\sqrt{3}}{2}$ **47.** $\frac{x^3}{1-x}$

49. $\displaystyle\frac{x^3}{1+x^2}$ **51.** $\displaystyle\frac{-1}{(1+x)^2}$ **55.** 500 terms **57.** 4 terms

59. (a) $\displaystyle x + \frac{x^3}{6} + \frac{3x^5}{40} + \frac{5x^7}{112}$, radius of convergence = 1

(b) $\displaystyle\frac{\pi}{2} - x - \frac{x^3}{6} - \frac{3x^5}{40} - \frac{5x^7}{112}$

61. $1 - 2x + 3x^2 - 4x^3 + \cdots$

67. (a) -1 (b) $(1/\sqrt{2})(1+i)$ (c) $-i$

71. $\displaystyle x + x^2 + \frac{1}{3}x^3 - \frac{1}{30}x^5 + \cdots$, for all x

PRACTICE EXERCISES, pp. 658–660

1. Converges to 1 **3.** Converges to -1 **5.** Diverges
7. Converges to 0 **9.** Converges to 1 **11.** Converges o e^{-5}
13. Converges to 3 **15.** Converges to ln 2 **17.** Diverges
19. $1/6$ **21.** $3/2$ **23.** $e/(e-1)$ **25.** Diverges
27. Converges conditionally **29.** Converges conditionally
31. Converges absolutely **33.** Converges absolutely
35. Converges absolutely **37.** Converges absolutely
39. Converges absolutely **41.** Converges absolutely
43. Diverges
45. (a) $3, -7 \le x < -1$ (b) $-7 < x < -1$ (c) $x = -7$
47. (a) $1/3, 0 \le x \le 2/3$ (b) $0 \le x \le 2/3$ (c) None
49. (a) ∞, for all x (b) For all x (c) None
51. (a) $\sqrt{3}, -\sqrt{3} < x < \sqrt{3}$ (b) $-\sqrt{3} < x < \sqrt{3}$ (c) None
53. (a) $e, -e < x < e$ (b) $-e < x < e$ (c) Empty set

55. $\displaystyle\frac{1}{1+x}, \frac{1}{4}, \frac{4}{5}$ **57.** $\sin x, \pi, 0$ **59.** $e^x, \ln 2, 2$ **61.** $\displaystyle\sum_{n=0}^{\infty} 2^n x^n$

63. $\displaystyle\sum_{n=0}^{\infty} \frac{(-1)^n\pi^{2n+1}x^{2n+1}}{(2n+1)!}$ **65.** $\displaystyle\sum_{n=0}^{\infty} \frac{(-1)^n x^{10n/3}}{(2n)!}$ **67.** $\displaystyle\sum_{n=0}^{\infty} \frac{((\pi x)/2)^n}{n!}$

69. $\displaystyle 2 - \frac{(x+1)}{2\cdot 1!} + \frac{3(x+1)^2}{2^3\cdot 2!} - \frac{9(x+1)^3}{2^5\cdot 3!} + \cdots$

71. $\displaystyle\frac{1}{4} - \frac{1}{4^2}(x-3) + \frac{1}{4^3}(x-3)^2 - \frac{1}{4^4}(x-3)^3$

73. 0.4849171431 **75.** 0.4872223583 **77.** 7/2 **79.** 1/12

81. -2 **83.** $r = -3, s = 9/2$ **85.** $2/3$

87. $\displaystyle\ln\left(\frac{n+1}{2n}\right)$; the series converges to $\ln\left(\frac{1}{2}\right)$.

89. (a) ∞ (b) $a = 1, b = 0$ **91.** It converges.
99. (a) Converges; Limit Comparison Test
(b) Converges; Direct Comparison Test
(c) Diverges; nth-Term Test
101. 2

ADDITIONAL AND ADVANCED EXERCISES, pp. 660–662

1. Converges; Comparison Test
3. Diverges; nth-Term Test
5. Converges; Comparison Test
7. Diverges; nth-Term Test

9. With $a = \pi/3$, $\cos x = \dfrac{1}{2} - \dfrac{\sqrt{3}}{2}(x - \pi/3) - \dfrac{1}{4}(x - \pi/3)^2$

$\qquad + \dfrac{\sqrt{3}}{12}(x - \pi/3)^3 + \cdots$

11. With $a = 0$, $e^x = 1 + x + \dfrac{x^2}{2!} + \dfrac{x^3}{3!} + \cdots$

13. With $a = 22\pi$, $\cos x = 1 - \dfrac{1}{2}(x - 22\pi)^2 + \dfrac{1}{4!}(x - 22\pi)^4$

$\qquad - \dfrac{1}{6!}(x - 22\pi)^6 + \cdots$

15. Converges, limit $= b$ **17.** $\pi/2$ **21.** $b = \pm\dfrac{1}{5}$

23. $a = 2, L = -7/6$ **27. (b)** Yes

31. (a) $\displaystyle\sum_{n=1}^{\infty} nx^{n-1}$ **(b)** 6   **(c)** $\dfrac{1}{q}$

33. (a) $R_n = C_0 e^{-kt_0}(1 - e^{-nkt_0})/(1 - e^{-kt_0})$,

$\qquad R = C_0(e^{-kt_0})/(1 - e^{-kt_0}) = C_0/(e^{kt_0} - 1)$

$\quad$ **(b)** $R_1 = 1/e \approx 0.368$,

$\qquad R_{10} = R(1 - e^{-10}) \approx R(0.9999546) \approx 0.58195$;

$\qquad R \approx 0.58198; 0 < (R - R_{10})/R < 0.0001$

$\quad$ **(c)** 7

Chapter 11

SECTION 11.1, pp. 669–672

1.

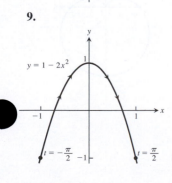

3.

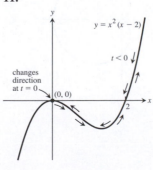

5.

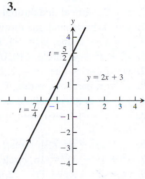

7.

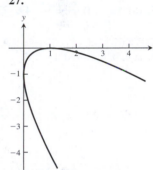

9.

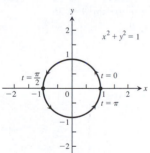

11.

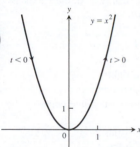

13.

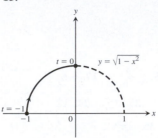

15.

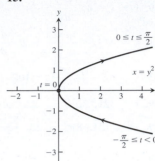

17.

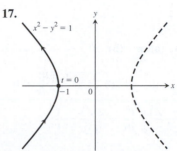

19. D   **21.** E   **23.** C

25.

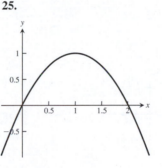

27.

29. (a) $x = a \cos t$, $\quad y = -a \sin t$, $\quad 0 \le t \le 2\pi$

$\quad$ **(b)** $x = a \cos t$, $\quad y = a \sin t$, $\quad 0 \le t \le 2\pi$

$\quad$ **(c)** $x = a \cos t$, $\quad y = -a \sin t$, $\quad 0 \le t \le 4\pi$

$\quad$ **(d)** $x = a \cos t$, $\quad y = a \sin t$, $\quad 0 \le t \le 4\pi$

31. Possible answer: $x = -1 + 5t$, $\quad y = -3 + 4t$, $\quad 0 \le t \le 1$

33. Possible answer: $x = t^2 + 1$, $\quad y = t$, $\quad t \le 0$

35. Possible answer: $x = 2 - 3t$, $\quad y = 3 - 4t$, $\quad t \ge 0$

37. Possible answer: $x = 2 \cos t$, $\quad y = 2 |\sin t|$, $\quad 0 \le t \le 4\pi$

39. Possible answer: $x = \dfrac{-at}{\sqrt{1 + t^2}}$, $\quad y = \dfrac{a}{\sqrt{1 + t^2}}$, $\quad -\infty < t < \infty$

41. Possible answer: $x = \dfrac{4}{1 + 2 \tan \theta}$, $\quad y = \dfrac{4 \tan \theta}{1 + 2 \tan \theta}$,

$\quad 0 \le \theta < \pi/2$ and $x = 0, y = 2$ if $\theta = \pi/2$

43. Possible answer: $x = 2 - \cos t$, $\quad y = \sin t$, $\quad 0 \le t \le 2\pi$

45. $x = 2 \cot t$, $\quad y = 2 \sin^2 t$, $\quad 0 < t < \pi$

47. $x = a \sin^2 t \tan t$, $\quad y = a \sin^2 t$, $\quad 0 \le t < \pi/2$ **49.** $(1, 1)$

SECTION 11.2, pp. 680–681

1. $y = -x + 2\sqrt{2}$, $\quad \dfrac{d^2 y}{dx^2} = -\sqrt{2}$

3. $y = -\dfrac{1}{2}x + 2\sqrt{2}$, $\quad \dfrac{d^2 y}{dx^2} = -\dfrac{\sqrt{2}}{4}$

5. $y = x + \dfrac{1}{4}$, $\quad \dfrac{d^2 y}{dx^2} = -2$ **7.** $y = 2x - \sqrt{3}$, $\quad \dfrac{d^2 y}{dx^2} = -3\sqrt{3}$

9. $y = x - 4$, $\dfrac{d^2y}{dx^2} = \dfrac{1}{2}$

11. $y = \sqrt{3}x - \dfrac{\pi\sqrt{3}}{3} + 2$, $\dfrac{d^2y}{dx^2} = -4$

13. $y = 9x - 1$, $\dfrac{d^2y}{dx^2} = 108$ **15.** $-\dfrac{3}{16}$ **17.** -6

19. 1   **21.** $3a^2\pi$ **23.** $|ab|\pi$ **25.** 4   **27.** 12

29. π^2 **31.** $8\pi^2$ **33.** $\dfrac{52\pi}{3}$ **35.** $3\pi\sqrt{5}$

37. $(\bar{x}, \bar{y}) = \left(\dfrac{12}{\pi} - \dfrac{24}{\pi^2}, \dfrac{24}{\pi^2} - 2\right)$

39. $(\bar{x}, \bar{y}) = \left(\dfrac{1}{3}, \pi - \dfrac{4}{3}\right)$ **41.** (a) π (b) π

43. (a) $x = 1$, $y = 0$, $\dfrac{dy}{dx} = \dfrac{1}{2}$ (b) $x = 0$, $y = 3$, $\dfrac{dy}{dx} = 0$

    (c) $x = \dfrac{\sqrt{3} - 1}{2}$, $y = \dfrac{3 - \sqrt{3}}{2}$, $\dfrac{dy}{dx} = \dfrac{2\sqrt{3} - 1}{\sqrt{3} - 2}$

45. $\left(\dfrac{\sqrt{2}}{2}, 1\right)$, $y = 2x$ at $t = 0$, $y = -2x$ at $t = \pi$

47. (a) $8a$ (b) $\dfrac{64\pi}{3}$ **49.** $32\pi/15$

SECTION 11.3, pp. 684–685

1. a, e; b, g; c, h; d, f   **3.**

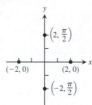

(a) $\left(2, \dfrac{\pi}{2} + 2n\pi\right)$ and $\left(-2, \dfrac{\pi}{2} + (2n + 1)\pi\right)$, n an integer

(b) $(2, 2n\pi)$ and $(-2, (2n + 1)\pi)$, n an integer

(c) $\left(2, \dfrac{3\pi}{2} + 2n\pi\right)$ and $\left(-2, \dfrac{3\pi}{2} + (2n + 1)\pi\right)$, n an integer

(d) $(2, (2n + 1)\pi)$ and $(-2, 2n\pi)$, n an integer

5. (a) $(3, 0)$ (b) $(-3, 0)$ (c) $\left(-1, \sqrt{3}\right)$ (d) $\left(1, \sqrt{3}\right)$
    (e) $(3, 0)$ (f) $\left(1, \sqrt{3}\right)$ (g) $(-3, 0)$ (h) $\left(-1, \sqrt{3}\right)$

7. (a) $\left(\sqrt{2}, \dfrac{\pi}{4}\right)$ (b) $(3, \pi)$ (c) $\left(2, \dfrac{11\pi}{6}\right)$

    (d) $\left(5, \pi - \tan^{-1}\dfrac{4}{3}\right)$

9. (a) $\left(-3\sqrt{2}, \dfrac{5\pi}{4}\right)$ (b) $(-1, 0)$ (c) $\left(-2, \dfrac{5\pi}{3}\right)$

    (d) $\left(-5, \pi - \tan^{-1}\dfrac{3}{4}\right)$

11.

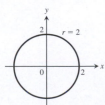

13.

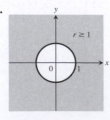

15.

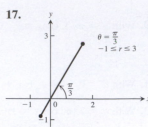

17.

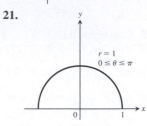

19.

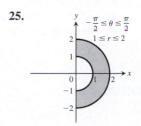

21.

23.   **25.**

27. $x = 2$, vertical line through $(2, 0)$ **29.** $y = 0$, the x-axis

31. $y = 4$, horizontal line through $(0, 4)$

33. $x + y = 1$, line, $m = -1$, $b = 1$

35. $x^2 + y^2 = 1$, circle, $C(0, 0)$, radius 1

37. $y - 2x = 5$, line, $m = 2$, $b = 5$

39. $y^2 = x$, parabola, vertex $(0, 0)$, opens right

41. $y = e^x$, graph of natural exponential function

43. $x + y = \pm1$, two straight lines of slope -1, y-intercepts
    $b = \pm1$

45. $(x + 2)^2 + y^2 = 4$, circle, $C(-2, 0)$, radius 2

47. $x^2 + (y - 4)^2 = 16$, circle, $C(0, 4)$, radius 4

49. $(x - 1)^2 + (y - 1)^2 = 2$, circle, $C(1, 1)$, radius $\sqrt{2}$

51. $\sqrt{3}y + x = 4$ **53.** $r\cos\theta = 7$ **55.** $\theta = \pi/4$

57. $r = 2$ or $r = -2$ **59.** $4r^2\cos^2\theta + 9r^2\sin^2\theta = 36$

61. $r\sin^2\theta = 4\cos\theta$ **63.** $r = 4\sin\theta$

65. $r^2 = 6r\cos\theta - 2r\sin\theta - 6$

67. $(0, \theta)$, where θ is any angle

SECTION 11.4, pp. 688–689

1. x-axis   **3.** y-axis

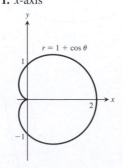

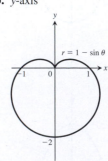

5. y-axis

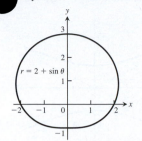

7. x-axis, y-axis, origin

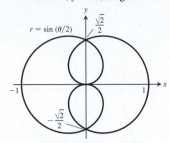

25. (a)

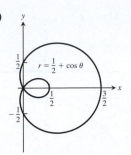

(b)

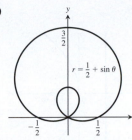

9. x-axis, y-axis, origin

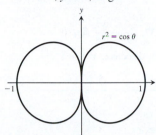

11. y-axis, x-axis, origin

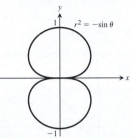

27. (a)

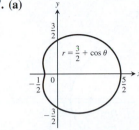

(b)

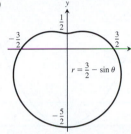

13. x-axis, y-axis, origin   **15.** Origin

17. The slope at $(-1, \pi/2)$ is -1, at $(-1, -\pi/2)$ is 1.

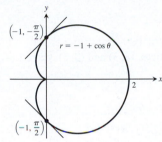

29.

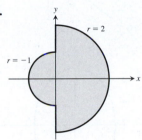

31.

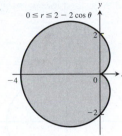

33. Equation (a)

SECTION 11.5, pp. 693–694

1. $\dfrac{1}{6}\pi^3$ **3.** 18π **5.** $\dfrac{\pi}{8}$ **7.** 2 **9.** $\dfrac{\pi}{2} - 1$

11. $5\pi - 8$ **13.** $3\sqrt{3} - \pi$ **15.** $\dfrac{\pi}{3} + \dfrac{\sqrt{3}}{2}$

17. $\dfrac{8\pi}{3} + \sqrt{3}$ **19. (a)** $\dfrac{3}{2} - \dfrac{\pi}{4}$ **21.** $19/3$ **23.** 8

25. $3\left(\sqrt{2} + \ln\left(1 + \sqrt{2}\right)\right)$ **27.** $\dfrac{\pi}{8} + \dfrac{3}{8}$

31. (a) a **(b)** a **(c)** $2a/\pi$

SECTION 11.6, pp. 700–702

1. $y^2 = 8x$, $F(2, 0)$, directrix: $x = -2$

3. $x^2 = -6y$, $F(0, -3/2)$, directrix: $y = 3/2$

5. $\dfrac{x^2}{4} - \dfrac{y^2}{9} = 1$, $F\left(\pm\sqrt{13}, 0\right)$, $V(\pm 2, 0)$,

asymptotes: $y = \pm\dfrac{3}{2}x$

7. $\dfrac{x^2}{2} + y^2 = 1$, $F(\pm 1, 0)$, $V\left(\pm\sqrt{2}, 0\right)$

19. The slope at $(1, \pi/4)$ is -1, at $(-1, -\pi/4)$ is 1, at $(-1, 3\pi/4)$ is 1, at $(1, -3\pi/4)$ is -1.

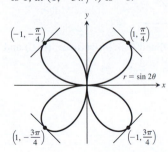

21. At $\pi/6$: slope $\sqrt{3}$, concavity 16 (concave up); at $\pi/3$: slope $-\sqrt{3}$, concavity -16 (concave down).

23. At 0: slope 0, concavity 2 (concave up); at $\pi/2$: slope $-2/\pi$, concavity $-2(8 + \pi^2)/\pi^3$ (concave down).

9.

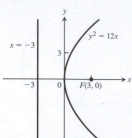

11.

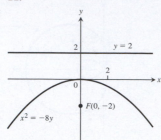

13.

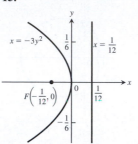

15.

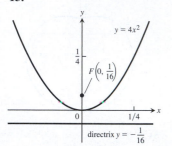

17.

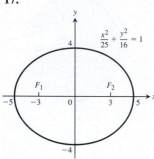

19.

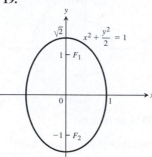

21.

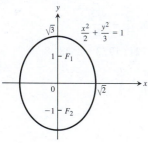

23.

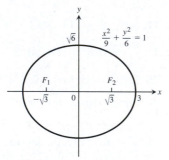

25. $\dfrac{x^2}{4} + \dfrac{y^2}{2} = 1$

27. Asymptotes: $y = \pm x$

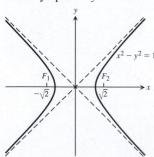

29. Asymptotes: $y = \pm x$

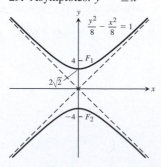

31. Asymptotes: $y = \pm 2x$

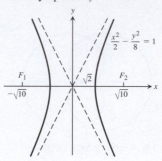

33. Asymptotes: $y = \pm x/2$

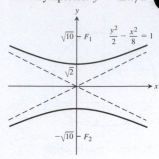

35. $y^2 - x^2 = 1$

37. $\dfrac{x^2}{9} - \dfrac{y^2}{16} = 1$

39. (a) Vertex: $(1, -2)$; focus: $(3, -2)$; directrix: $x = -1$

    (b)

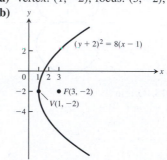

41. (a) Foci: $\left(4 \pm \sqrt{7}, 3\right)$; vertices: $(8, 3)$ and $(0, 3)$; center: $(4, 3)$

    (b)

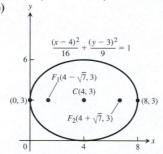

43. (a) Center: $(2, 0)$; foci: $(7, 0)$ and $(-3, 0)$; vertices: $(6, 0)$ and $(-2, 0)$; asymptotes: $y = \pm \dfrac{3}{4}(x - 2)$

    (b)

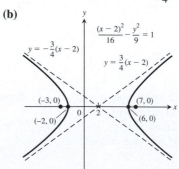

45. $(y + 3)^2 = 4(x + 2)$, $V(-2, -3)$, $F(-1, -3)$, directrix: $x = -3$

47. $(x - 1)^2 = 8(y + 7)$, $V(1, -7)$, $F(1, -5)$, directrix: $y = -9$

49. $\dfrac{(x + 2)^2}{6} + \dfrac{(y + 1)^2}{9} = 1$, $F\left(-2, \pm\sqrt{3} - 1\right)$, $V(-2, \pm 3 - 1)$, $C(-2, -1)$

51. $\dfrac{(x-2)^2}{3} + \dfrac{(y-3)^2}{2} = 1$, $F(3,3)$ and $F(1,3)$,

$V\left(\pm\sqrt{3} + 2, 3\right),\ C(2,3)$

53. $\dfrac{(x-2)^2}{4} - \dfrac{(y-2)^2}{5} = 1$, $C(2,2)$, $F(5,2)$ and $F(-1,2)$,

$V(4,2)$ and $V(0,2)$; asymptotes: $(y-2) = \pm\dfrac{\sqrt{5}}{2}(x-2)$

55. $(y+1)^2 - (x+1)^2 = 1$, $C(-1,-1)$, $F\left(-1, \sqrt{2}-1\right)$
and $F\left(-1, -\sqrt{2}-1\right)$, $V(-1,0)$ and $V(-1,-2)$; asymptotes
$(y+1) = \pm(x+1)$

57. $C(-2,0)$, $a = 4$ **59.** $V(-1,1)$, $F(-1,0)$

61. Ellipse: $\dfrac{(x+2)^2}{5} + y^2 = 1$, $C(-2,0)$, $F(0,0)$ and

$F(-4,0)$, $V\left(\sqrt{5}-2, 0\right)$ and $V\left(-\sqrt{5}-2, 0\right)$

63. Ellipse: $\dfrac{(x-1)^2}{2} + (y-1)^2 = 1$, $C(1,1)$, $F(2,1)$ and

$F(0,1)$, $V\left(\sqrt{2}+1, 1\right)$ and $V\left(-\sqrt{2}+1, 1\right)$

65. Hyperbola: $(x-1)^2 - (y-2)^2 = 1$, $C(1,2)$,
$F\left(1+\sqrt{2}, 2\right)$ and $F\left(1-\sqrt{2}, 2\right)$, $V(2,2)$ and
$V(0,2)$;   asymptotes: $(y-2) = \pm(x-1)$

67. Hyperbola: $\dfrac{(y-3)^2}{6} - \dfrac{x^2}{3} = 1$, $C(0,3)$, $F(0,6)$

and $F(0,0)$, $V\left(0, \sqrt{6}+3\right)$ and $V\left(0, -\sqrt{6}+3\right)$;
asymptotes: $y = \sqrt{2}x + 3$ or $y = -\sqrt{2}x + 3$

69. (b) 1:1   **73.** Length $= 2\sqrt{2}$, width $= \sqrt{2}$, area $= 4$

75. 24π

77. $x = 0, y = 0: y = -2x; x = 0, y = 2: y = 2x + 2;$
$x = 4, y = 0: y = 2x - 8;$

79. $\bar{x} = 0,\quad \bar{y} = \dfrac{16}{3\pi}$

SECTION 11.7, pp. 707–708

1. $e = \dfrac{3}{5}$, $F(\pm 3, 0)$;

directrices are $x = \pm\dfrac{25}{3}$.

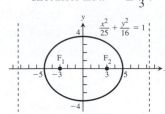

3. $e = \dfrac{1}{\sqrt{2}}$; $F(0, \pm 1)$;

directrices are $y = \pm 2$.

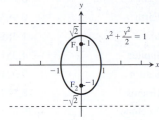

5. $e = \dfrac{1}{\sqrt{3}}$; $F(0, \pm 1)$;

directrices are $y = \pm 3$.

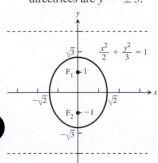

7. $e = \dfrac{\sqrt{3}}{3}$; $F\left(\pm\sqrt{3}, 0\right)$;

directrices are
$x = \pm 3\sqrt{3}$.

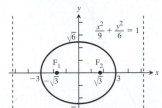

9. $\dfrac{x^2}{27} + \dfrac{y^2}{36} = 1$ **11.** $\dfrac{x^2}{4851} + \dfrac{y^2}{4900} = 1$

13. $\dfrac{x^2}{9} + \dfrac{y^2}{4} = 1$ **15.** $\dfrac{x^2}{64} + \dfrac{y^2}{48} = 1$

17. $e = \sqrt{2}$; $F\left(\pm\sqrt{2}, 0\right)$;

directrices are $x = \pm\dfrac{1}{\sqrt{2}}$.

19. $e = \sqrt{2}$; $F(0, \pm 4)$;
directrices are $y = \pm 2$.

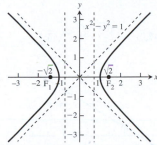

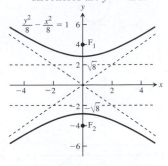

21. $e = \sqrt{5}$; $F\left(\pm\sqrt{10}, 0\right)$;

directrices are $x = \pm\dfrac{2}{\sqrt{10}}$.

23. $e = \sqrt{5}$; $F\left(0, \pm\sqrt{10}\right)$;

directrices are $y = \pm\dfrac{2}{\sqrt{10}}$.

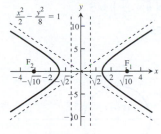

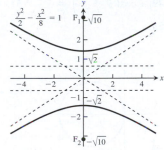

25. $y^2 - \dfrac{x^2}{8} = 1$ **27.** $x^2 - \dfrac{y^2}{8} = 1$ **29.** $r = \dfrac{2}{1 + \cos\theta}$

31. $r = \dfrac{30}{1 - 5\sin\theta}$ **33.** $r = \dfrac{1}{2 + \cos\theta}$ **35.** $r = \dfrac{10}{5 - \sin\theta}$

37.   **39.**

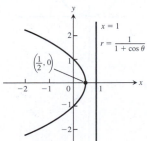

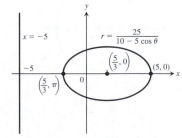

41.

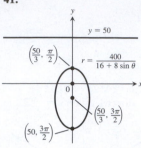

43.

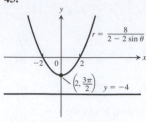

65.

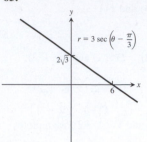

67.

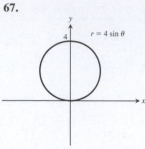

45. $y = 2 - x$

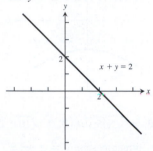

47. $y = \dfrac{\sqrt{3}}{3}x + 2\sqrt{3}$

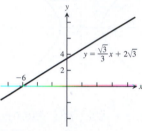

69.

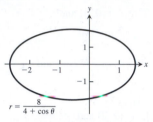

71.

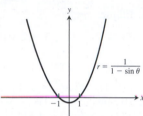

49. $r\cos\left(\theta - \dfrac{\pi}{4}\right) = 3$ **51.** $r\cos\left(\theta + \dfrac{\pi}{2}\right) = 5$

53.

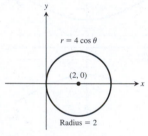

55.

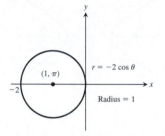

73.

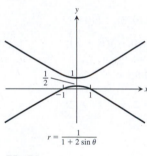

57. $r = 12\cos\theta$

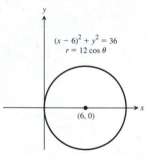

59. $r = 10\sin\theta$

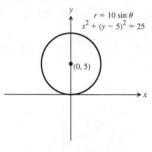

75. (b)

Planet	Perihelion	Aphelion
Mercury	0.3075 AU	0.4667 AU
Venus	0.7184 AU	0.7282 AU
Earth	0.9833 AU	1.0167 AU
Mars	1.3817 AU	1.6663 AU
Jupiter	4.9512 AU	5.4548 AU
Saturn	9.0210 AU	10.0570 AU
Uranus	18.2977 AU	20.0623 AU
Neptune	29.8135 AU	30.3065 AU

PRACTICE EXERCISES, pp. 709–711

1.

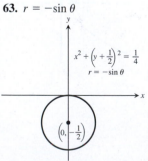

3.

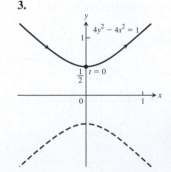

61. $r = -2\cos\theta$

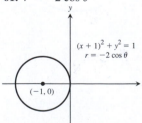

63. $r = -\sin\theta$

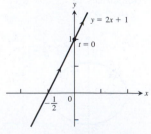

5.

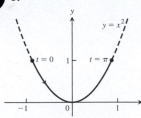

7. $x = 3 \cos t, \quad y = 4 \sin t, \quad 0 \le t \le 2\pi$

9. $y = \dfrac{\sqrt{3}}{2}x + \dfrac{1}{4}, \dfrac{1}{4}$

11. (a) $y = \dfrac{\pm |x|^{3/2}}{8} - 1$ **(b)** $y = \dfrac{\pm\sqrt{1 - x^2}}{x}$

13. $\dfrac{10}{3}$ **15.** $\dfrac{285}{8}$ **17.** 10   **19.** $\dfrac{9\pi}{2}$ **21.** $\dfrac{76\pi}{3}$

23. $y = \dfrac{\sqrt{3}}{3}x - 4$ **25.** $x = 2$

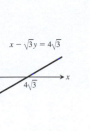

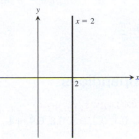

27. $y = -\dfrac{3}{2}$ **29.** $x^2 + (y + 2)^2 = 4$

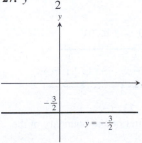

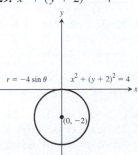

31. $\left(x - \sqrt{2}\right)^2 + y^2 = 2$ **33.** $r = -5 \sin \theta$

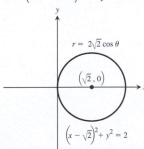

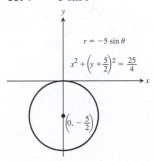

35. $r = 3 \cos \theta$

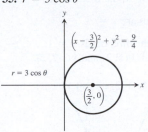

37.

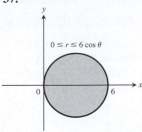

39. d   **41.** l   **43.** k   **45.** i   **47.** $\dfrac{9}{2}\pi$ **49.** $2 + \dfrac{\pi}{4}$

51. 8   **53.** $\pi - 3$

55. Focus is $(0, -1)$,
     directrix is $y = 1$.

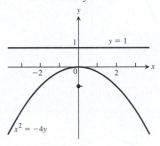

57. Focus is $\left(\dfrac{3}{4}, 0\right)$,
     directrix is $x = -\dfrac{3}{4}$.

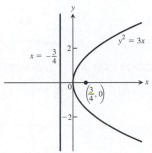

59. $e = \dfrac{3}{4}$

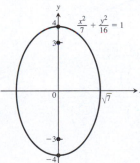

61. $e = 2$; the asymptotes are $y = \pm\sqrt{3}\,x$.

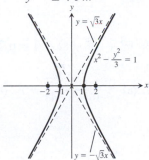

63. $(x - 2)^2 = -12(y - 3)$, $V(2, 3)$, $F(2, 0)$, directrix is $y = 6$.

65. $\dfrac{(x + 3)^2}{9} + \dfrac{(y + 5)^2}{25} = 1$, $C(-3, -5)$, $F(-3, -1)$ and
$F(-3, -9)$, $V(-3, -10)$ and $V(-3, 0)$.

67. $\dfrac{\left(y - 2\sqrt{2}\right)^2}{8} - \dfrac{(x - 2)^2}{2} = 1$, $C\left(2, 2\sqrt{2}\right)$,
$F\left(2, 2\sqrt{2} \pm \sqrt{10}\right)$, $V\left(2, 4\sqrt{2}\right)$ and $V(2, 0)$, the asymptotes
are $y = 2x - 4 + 2\sqrt{2}$ and $y = -2x + 4 + 2\sqrt{2}$.

69. Hyperbola: $C(2, 0)$, $V(0, 0)$ and $V(4, 0)$, the foci are

$F(2 \pm \sqrt{5}, 0)$, and the asymptotes are $y = \pm \dfrac{x - 2}{2}$.

71. Parabola: $V(-3, 1)$, $F(-7, 1)$, and the directrix is $x = 1$.
73. Ellipse: $C(-3, 2)$, $F(-3 \pm \sqrt{7}, 2)$, $V(1, 2)$ and $V(-7, 2)$
75. Circle: $C(1, 1)$ and radius $= \sqrt{2}$
77. $V(1, 0)$ **79.** $V(2, \pi)$ and $V(6, \pi)$

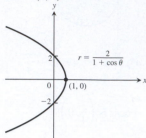

81. $r = \dfrac{4}{1 + 2 \cos \theta}$ **83.** $r = \dfrac{2}{2 + \sin \theta}$
85. (a) 24π (b) 16π

ADDITIONAL AND ADVANCED EXERCISES, pp. 711–713

1. $x - \dfrac{7}{2} = \dfrac{y^2}{2}$

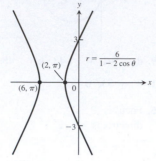

3. $3x^2 + 3y^2 - 8y + 4 = 0$ **5.** $F(0, \pm 1)$

7. (a) $\dfrac{(y - 1)^2}{16} - \dfrac{x^2}{48} = 1$ (b) $\dfrac{\left(y + \dfrac{3}{4}\right)^2}{\left(\dfrac{25}{16}\right)} - \dfrac{x^2}{\left(\dfrac{75}{2}\right)} = 1$

11.

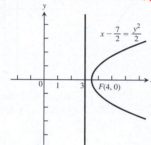

13.

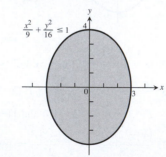

15.

17. (a) $r = e^{2\theta}$ (b) $\dfrac{\sqrt{5}}{2}\left(e^{4\pi} - 1\right)$

19. $r = \dfrac{4}{1 + 2 \cos \theta}$ **21.** $r = \dfrac{2}{2 + \sin \theta}$

23. $x = (a + b) \cos \theta - b \cos\left(\dfrac{a + b}{b}\theta\right),$

$y = (a + b) \sin \theta - b \sin\left(\dfrac{a + b}{b}\theta\right)$

27. $\dfrac{\pi}{2}$

Appendices

APPENDIX 1, p. AP-6

1. $0.\overline{1}$, $0.\overline{2}$, $0.\overline{3}$, $0.\overline{8}$, $0.\overline{9}$ or 1

3. $x < -2$ **5.** $x \le -\dfrac{1}{3}$

7. $3, -3$ **9.** $7/6, 25/6$
11. $-2 \le t \le 4$ **13.** $0 \le z \le 10$

15. $(-\infty, -2] \cup [2, \infty)$ **17.** $(-\infty, -3] \cup [1, \infty)$

19. $(-3, -2) \cup (2, 3)$ **21.** $(0, 1)$ **23.** $(-\infty, 1]$
27. The graph of $|x| + |y| \le 1$ is the interior and boundary of the "diamond-shaped" region.

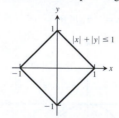

APPENDIX 3, pp. AP-17–AP-18

1. $2, -4; 2\sqrt{5}$ **3.** Unit circle

5. $m_\perp = -\dfrac{1}{3}$

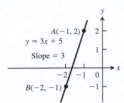

7. (a) $x = -1$ **(b)** $y = 4/3$ **9.** $y = -x$

11. $y = -\dfrac{5}{4}x + 6$ **13.** $y = 4x + 4$ **15.** $y = -\dfrac{x}{2} + 12$

17. x-intercept $= \sqrt{3}$, y-intercept $= -\sqrt{2}$

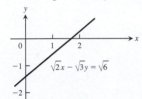

19. $(3, -3)$

21. $x^2 + (y - 2)^2 = 4$ **23.** $\left(x + \sqrt{3}\right)^2 + (y + 2)^2 = 4$

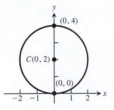
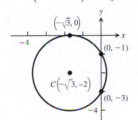

25. $x^2 + (y - 3/2)^2 = 25/4$ **27.**

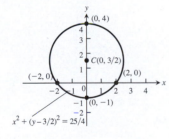
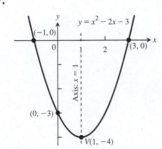

29.

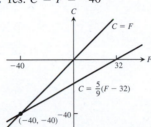

31. Exterior points of a circle of radius $\sqrt{7}$, centered at the origin

33. The washer between the circles $x^2 + y^2 = 1$ and $x^2 + y^2 = 4$ (points with distance from the origin between 1 and 2)

35. $(x + 2)^2 + (y - 1)^2 < 6$

37. $\left(\dfrac{1}{\sqrt{5}}, \dfrac{2}{\sqrt{5}}\right)$, $\left(-\dfrac{1}{\sqrt{5}}, -\dfrac{2}{\sqrt{5}}\right)$

39. $\left(-\dfrac{1}{\sqrt{3}}, -\dfrac{1}{3}\right)$, $\left(\dfrac{1}{\sqrt{3}}, -\dfrac{1}{3}\right)$

41. (a) ≈ -2.5 degrees/inch   **(b)** ≈ -16.1 degrees/inch
   (c) ≈ -8.3 degrees/inch

43. 5.97 atm

45. Yes: $C = F = -40°$

51. $k = -8$, $k = 1/2$

APPENDIX 7, pp. AP-33–AP-34

1. (a) $14 + 8i$ **(b)** $-7 - 4i$ **(c)** $-5i$

3. (a) By reflecting z across the real axis
   (b) By reflecting z across the imaginary axis
   (c) By reflecting z across the real axis and then multiplying the length of the vector by $1/|z|^2$

5. (a) Points on the circle $x^2 + y^2 = 4$
   (b) Points inside the circle $x^2 + y^2 = 4$
   (c) Points outside the circle $x^2 + y^2 = 4$

7. Points on a circle of radius 1, center $(-1, 0)$

9. Points on the line $y = -x$ **11.** $4e^{2\pi i/3}$ **13.** $1e^{2\pi i/3}$

15. $\cos^4\theta - 6\cos^2\theta\sin^2\theta + \sin^4\theta$

17. $1, -\dfrac{1}{2} \pm \dfrac{\sqrt{3}}{2}i$ **19.** $2i, -\sqrt{3} - i, \sqrt{3} - i$

21. $\dfrac{\sqrt{6}}{2} \pm \dfrac{\sqrt{2}}{2}i, -\dfrac{\sqrt{6}}{2} \pm \dfrac{\sqrt{2}}{2}i$ **23.** $1 \pm \sqrt{3}i, -1 \pm \sqrt{3}i$

Applications Index

Subject Index

Credits

A Brief Table of Integrals

Basic Forms

1. $\displaystyle\int k\,dx = kx + C, \quad k \text{ any number}$

2. $\displaystyle\int x^n\,dx = \frac{x^{n+1}}{n+1} + C, \quad n \neq -1$

3. $\displaystyle\int \frac{dx}{x} = \ln|x| + C$

4. $\displaystyle\int e^x\,dx = e^x + C$

5. $\displaystyle\int a^x\,dx = \frac{a^x}{\ln a} + C \quad (a > 0, a \neq 1)$

6. $\displaystyle\int \sin x\,dx = -\cos x + C$

7. $\displaystyle\int \cos x\,dx = \sin x + C$

8. $\displaystyle\int \sec^2 x\,dx = \tan x + C$

9. $\displaystyle\int \csc^2 x\,dx = -\cot x + C$

10. $\displaystyle\int \sec x \tan x\,dx = \sec x + C$

11. $\displaystyle\int \csc x \cot x\,dx = -\csc x + C$

12. $\displaystyle\int \tan x\,dx = \ln|\sec x| + C$

13. $\displaystyle\int \cot x\,dx = \ln|\sin x| + C$

14. $\displaystyle\int \sinh x\,dx = \cosh x + C$

15. $\displaystyle\int \cosh x\,dx = \sinh x + C$

16. $\displaystyle\int \frac{dx}{\sqrt{a^2 - x^2}} = \sin^{-1}\frac{x}{a} + C$

17. $\displaystyle\int \frac{dx}{a^2 + x^2} = \frac{1}{a}\tan^{-1}\frac{x}{a} + C$

18. $\displaystyle\int \frac{dx}{x\sqrt{x^2 - a^2}} = \frac{1}{a}\sec^{-1}\left|\frac{x}{a}\right| + C$

19. $\displaystyle\int \frac{dx}{\sqrt{a^2 + x^2}} = \sinh^{-1}\frac{x}{a} + C \quad (a > 0)$

20. $\displaystyle\int \frac{dx}{\sqrt{x^2 - a^2}} = \cosh^{-1}\frac{x}{a} + C \quad (x > a > 0)$

Forms Involving $ax + b$

21. $\displaystyle\int (ax + b)^n\,dx = \frac{(ax+b)^{n+1}}{a(n+1)} + C, \quad n \neq -1$

22. $\displaystyle\int x(ax + b)^n\,dx = \frac{(ax+b)^{n+1}}{a^2}\left[\frac{ax+b}{n+2} - \frac{b}{n+1}\right] + C, \quad n \neq -1, -2$

23. $\displaystyle\int (ax + b)^{-1}\,dx = \frac{1}{a}\ln|ax + b| + C$

24. $\displaystyle\int x(ax + b)^{-1}\,dx = \frac{x}{a} - \frac{b}{a^2}\ln|ax + b| + C$

25. $\displaystyle\int x(ax + b)^{-2}\,dx = \frac{1}{a^2}\left[\ln|ax + b| + \frac{b}{ax + b}\right] + C$

26. $\displaystyle\int \frac{dx}{x(ax + b)} = \frac{1}{b}\ln\left|\frac{x}{ax + b}\right| + C$

27. $\displaystyle\int \left(\sqrt{ax + b}\right)^n\,dx = \frac{2}{a}\frac{\left(\sqrt{ax+b}\right)^{n+2}}{n+2} + C, \quad n \neq -2$

28. $\displaystyle\int \frac{\sqrt{ax + b}}{x}\,dx = 2\sqrt{ax + b} + b\int \frac{dx}{x\sqrt{ax + b}}$

29. (a) $\displaystyle\int \frac{dx}{x\sqrt{ax+b}} = \frac{1}{\sqrt{b}}\ln\left|\frac{\sqrt{ax+b}-\sqrt{b}}{\sqrt{ax+b}+\sqrt{b}}\right| + C$ **(b)** $\displaystyle\int \frac{dx}{x\sqrt{ax-b}} = \frac{2}{\sqrt{b}}\tan^{-1}\sqrt{\frac{ax-b}{b}} + C$

30. $\displaystyle\int \frac{\sqrt{ax+b}}{x^2}dx = -\frac{\sqrt{ax+b}}{x} + \frac{a}{2}\int \frac{dx}{x\sqrt{ax+b}} + C$ **31.** $\displaystyle\int \frac{dx}{x^2\sqrt{ax+b}} = -\frac{\sqrt{ax+b}}{bx} - \frac{a}{2b}\int \frac{dx}{x\sqrt{ax+b}} + C$

Forms Involving $a^2 + x^2$

32. $\displaystyle\int \frac{dx}{a^2+x^2} = \frac{1}{a}\tan^{-1}\frac{x}{a} + C$ **33.** $\displaystyle\int \frac{dx}{(a^2+x^2)^2} = \frac{x}{2a^2(a^2+x^2)} + \frac{1}{2a^3}\tan^{-1}\frac{x}{a} + C$

34. $\displaystyle\int \frac{dx}{\sqrt{a^2+x^2}} = \sinh^{-1}\frac{x}{a} + C = \ln\left(x+\sqrt{a^2+x^2}\right) + C$

35. $\displaystyle\int \sqrt{a^2+x^2}\,dx = \frac{x}{2}\sqrt{a^2+x^2} + \frac{a^2}{2}\ln\left(x+\sqrt{a^2+x^2}\right) + C$

36. $\displaystyle\int x^2\sqrt{a^2+x^2}\,dx = \frac{x}{8}(a^2+2x^2)\sqrt{a^2+x^2} - \frac{a^4}{8}\ln\left(x+\sqrt{a^2+x^2}\right) + C$

37. $\displaystyle\int \frac{\sqrt{a^2+x^2}}{x}dx = \sqrt{a^2+x^2} - a\ln\left|\frac{a+\sqrt{a^2+x^2}}{x}\right| + C$

38. $\displaystyle\int \frac{\sqrt{a^2+x^2}}{x^2}dx = \ln\left(x+\sqrt{a^2+x^2}\right) - \frac{\sqrt{a^2+x^2}}{x} + C$

39. $\displaystyle\int \frac{x^2}{\sqrt{a^2+x^2}}dx = -\frac{a^2}{2}\ln\left(x+\sqrt{a^2+x^2}\right) + \frac{x\sqrt{a^2+x^2}}{2} + C$

40. $\displaystyle\int \frac{dx}{x\sqrt{a^2+x^2}} = -\frac{1}{a}\ln\left|\frac{a+\sqrt{a^2+x^2}}{x}\right| + C$ **41.** $\displaystyle\int \frac{dx}{x^2\sqrt{a^2+x^2}} = -\frac{\sqrt{a^2+x^2}}{a^2x} + C$

Forms Involving $a^2 - x^2$

42. $\displaystyle\int \frac{dx}{a^2-x^2} = \frac{1}{2a}\ln\left|\frac{x+a}{x-a}\right| + C$ **43.** $\displaystyle\int \frac{dx}{(a^2-x^2)^2} = \frac{x}{2a^2(a^2-x^2)} + \frac{1}{4a^3}\ln\left|\frac{x+a}{x-a}\right| + C$

44. $\displaystyle\int \frac{dx}{\sqrt{a^2-x^2}} = \sin^{-1}\frac{x}{a} + C$ **45.** $\displaystyle\int \sqrt{a^2-x^2}\,dx = \frac{x}{2}\sqrt{a^2-x^2} + \frac{a^2}{2}\sin^{-1}\frac{x}{a} + C$

46. $\displaystyle\int x^2\sqrt{a^2-x^2}\,dx = \frac{a^4}{8}\sin^{-1}\frac{x}{a} - \frac{1}{8}x\sqrt{a^2-x^2}\,(a^2-2x^2) + C$

47. $\displaystyle\int \frac{\sqrt{a^2-x^2}}{x}dx = \sqrt{a^2-x^2} - a\ln\left|\frac{a+\sqrt{a^2-x^2}}{x}\right| + C$ **48.** $\displaystyle\int \frac{\sqrt{a^2-x^2}}{x^2}dx = -\sin^{-1}\frac{x}{a} - \frac{\sqrt{a^2-x^2}}{x} + C$

49. $\displaystyle\int \frac{x^2}{\sqrt{a^2-x^2}}dx = \frac{a^2}{2}\sin^{-1}\frac{x}{a} - \frac{1}{2}x\sqrt{a^2-x^2} + C$ **50.** $\displaystyle\int \frac{dx}{x\sqrt{a^2-x^2}} = -\frac{1}{a}\ln\left|\frac{a+\sqrt{a^2-x^2}}{x}\right| + C$

51. $\displaystyle\int \frac{dx}{x^2\sqrt{a^2-x^2}} = -\frac{\sqrt{a^2-x^2}}{a^2x} + C$

Forms Involving $x^2 - a^2$

52. $\displaystyle\int \frac{dx}{\sqrt{x^2-a^2}} = \ln\left|x+\sqrt{x^2-a^2}\right| + C$

53. $\displaystyle\int \sqrt{x^2-a^2}\,dx = \frac{x}{2}\sqrt{x^2-a^2} - \frac{a^2}{2}\ln\left|x+\sqrt{x^2-a^2}\right| + C$

54. $\displaystyle\int \left(\sqrt{x^2 - a^2}\right)^n dx = \frac{x\left(\sqrt{x^2 - a^2}\right)^n}{n + 1} - \frac{na^2}{n + 1}\int \left(\sqrt{x^2 - a^2}\right)^{n-2} dx, \quad n \neq -1$

55. $\displaystyle\int \frac{dx}{\left(\sqrt{x^2 - a^2}\right)^n} = \frac{x\left(\sqrt{x^2 - a^2}\right)^{2-n}}{(2 - n)a^2} - \frac{n - 3}{(n - 2)a^2}\int \frac{dx}{\left(\sqrt{x^2 - a^2}\right)^{n-2}}, \quad n \neq 2$

56. $\displaystyle\int x\left(\sqrt{x^2 - a^2}\right)^n dx = \frac{\left(\sqrt{x^2 - a^2}\right)^{n+2}}{n + 2} + C, \quad n \neq -2$

57. $\displaystyle\int x^2\sqrt{x^2 - a^2}\, dx = \frac{x}{8}(2x^2 - a^2)\sqrt{x^2 - a^2} - \frac{a^4}{8}\ln\left|x + \sqrt{x^2 - a^2}\right| + C$

58. $\displaystyle\int \frac{\sqrt{x^2 - a^2}}{x}\, dx = \sqrt{x^2 - a^2} - a\sec^{-1}\left|\frac{x}{a}\right| + C$

59. $\displaystyle\int \frac{\sqrt{x^2 - a^2}}{x^2}\, dx = \ln\left|x + \sqrt{x^2 - a^2}\right| - \frac{\sqrt{x^2 - a^2}}{x} + C$

60. $\displaystyle\int \frac{x^2}{\sqrt{x^2 - a^2}}\, dx = \frac{a^2}{2}\ln\left|x + \sqrt{x^2 - a^2}\right| + \frac{x}{2}\sqrt{x^2 - a^2} + C$

61. $\displaystyle\int \frac{dx}{x\sqrt{x^2 - a^2}} = \frac{1}{a}\sec^{-1}\left|\frac{x}{a}\right| + C = \frac{1}{a}\cos^{-1}\left|\frac{a}{x}\right| + C$ **62.** $\displaystyle\int \frac{dx}{x^2\sqrt{x^2 - a^2}} = \frac{\sqrt{x^2 - a^2}}{a^2 x} + C$

Trigonometric Forms

63. $\displaystyle\int \sin ax\, dx = -\frac{1}{a}\cos ax + C$

64. $\displaystyle\int \cos ax\, dx = \frac{1}{a}\sin ax + C$

65. $\displaystyle\int \sin^2 ax\, dx = \frac{x}{2} - \frac{\sin 2ax}{4a} + C$

66. $\displaystyle\int \cos^2 ax\, dx = \frac{x}{2} + \frac{\sin 2ax}{4a} + C$

67. $\displaystyle\int \sin^n ax\, dx = -\frac{\sin^{n-1} ax \cos ax}{na} + \frac{n - 1}{n}\int \sin^{n-2} ax\, dx$

68. $\displaystyle\int \cos^n ax\, dx = \frac{\cos^{n-1} ax \sin ax}{na} + \frac{n - 1}{n}\int \cos^{n-2} ax\, dx$

69. (a) $\displaystyle\int \sin ax \cos bx\, dx = -\frac{\cos(a + b)x}{2(a + b)} - \frac{\cos(a - b)x}{2(a - b)} + C, \quad a^2 \neq b^2$

(b) $\displaystyle\int \sin ax \sin bx\, dx = \frac{\sin(a - b)x}{2(a - b)} - \frac{\sin(a + b)x}{2(a + b)} + C, \quad a^2 \neq b^2$

(c) $\displaystyle\int \cos ax \cos bx\, dx = \frac{\sin(a - b)x}{2(a - b)} + \frac{\sin(a + b)x}{2(a + b)} + C, \quad a^2 \neq b^2$

70. $\displaystyle\int \sin ax \cos ax\, dx = -\frac{\cos 2ax}{4a} + C$

71. $\displaystyle\int \sin^n ax \cos ax\, dx = \frac{\sin^{n+1} ax}{(n + 1)a} + C, \quad n \neq -1$

72. $\displaystyle\int \frac{\cos ax}{\sin ax}\, dx = \frac{1}{a}\ln|\sin ax| + C$

73. $\displaystyle\int \cos^n ax \sin ax\, dx = -\frac{\cos^{n+1} ax}{(n + 1)a} + C, \quad n \neq -1$

74. $\displaystyle\int \frac{\sin ax}{\cos ax}\, dx = -\frac{1}{a}\ln|\cos ax| + C$

75. $\displaystyle\int \sin^n ax \cos^m ax\, dx = -\frac{\sin^{n-1} ax \cos^{m+1} ax}{a(m + n)} + \frac{n - 1}{m + n}\int \sin^{n-2} ax \cos^m ax\, dx, \quad n \neq -m \quad \text{(reduces } \sin^n ax\text{)}$

76. $\displaystyle\int \sin^n ax \cos^m ax\, dx = \frac{\sin^{n+1} ax \cos^{m-1} ax}{a(m + n)} + \frac{m - 1}{m + n}\int \sin^n ax \cos^{m-2} ax\, dx, \quad m \neq -n \quad \text{(reduces } \cos^m ax\text{)}$

77. $\displaystyle\int \frac{dx}{b + c \sin ax} = \frac{-2}{a\sqrt{b^2 - c^2}} \tan^{-1}\left[\sqrt{\frac{b - c}{b + c}} \tan\left(\frac{\pi}{4} - \frac{ax}{2}\right)\right] + C, \quad b^2 > c^2$

78. $\displaystyle\int \frac{dx}{b + c \sin ax} = \frac{-1}{a\sqrt{c^2 - b^2}} \ln\left|\frac{c + b \sin ax + \sqrt{c^2 - b^2} \cos ax}{b + c \sin ax}\right| + C, \quad b^2 < c^2$

79. $\displaystyle\int \frac{dx}{1 + \sin ax} = -\frac{1}{a} \tan\left(\frac{\pi}{4} - \frac{ax}{2}\right) + C$

80. $\displaystyle\int \frac{dx}{1 - \sin ax} = \frac{1}{a} \tan\left(\frac{\pi}{4} + \frac{ax}{2}\right) + C$

81. $\displaystyle\int \frac{dx}{b + c \cos ax} = \frac{2}{a\sqrt{b^2 - c^2}} \tan^{-1}\left[\sqrt{\frac{b - c}{b + c}} \tan\frac{ax}{2}\right] + C, \quad b^2 > c^2$

82. $\displaystyle\int \frac{dx}{b + c \cos ax} = \frac{1}{a\sqrt{c^2 - b^2}} \ln\left|\frac{c + b \cos ax + \sqrt{c^2 - b^2} \sin ax}{b + c \cos ax}\right| + C, \quad b^2 < c^2$

83. $\displaystyle\int \frac{dx}{1 + \cos ax} = \frac{1}{a} \tan\frac{ax}{2} + C$

84. $\displaystyle\int \frac{dx}{1 - \cos ax} = -\frac{1}{a} \cot\frac{ax}{2} + C$

85. $\displaystyle\int x \sin ax\, dx = \frac{1}{a^2} \sin ax - \frac{x}{a} \cos ax + C$

86. $\displaystyle\int x \cos ax\, dx = \frac{1}{a^2} \cos ax + \frac{x}{a} \sin ax + C$

87. $\displaystyle\int x^n \sin ax\, dx = -\frac{x^n}{a} \cos ax + \frac{n}{a}\int x^{n-1} \cos ax\, dx$

88. $\displaystyle\int x^n \cos ax\, dx = \frac{x^n}{a} \sin ax - \frac{n}{a}\int x^{n-1} \sin ax\, dx$

89. $\displaystyle\int \tan ax\, dx = \frac{1}{a} \ln|\sec ax| + C$

90. $\displaystyle\int \cot ax\, dx = \frac{1}{a} \ln|\sin ax| + C$

91. $\displaystyle\int \tan^2 ax\, dx = \frac{1}{a} \tan ax - x + C$

92. $\displaystyle\int \cot^2 ax\, dx = -\frac{1}{a} \cot ax - x + C$

93. $\displaystyle\int \tan^n ax\, dx = \frac{\tan^{n-1} ax}{a(n - 1)} - \int \tan^{n-2} ax\, dx, \quad n \neq 1$

94. $\displaystyle\int \cot^n ax\, dx = -\frac{\cot^{n-1} ax}{a(n - 1)} - \int \cot^{n-2} ax\, dx, \quad n \neq 1$

95. $\displaystyle\int \sec ax\, dx = \frac{1}{a} \ln|\sec ax + \tan ax| + C$

96. $\displaystyle\int \csc ax\, dx = -\frac{1}{a} \ln|\csc ax + \cot ax| + C$

97. $\displaystyle\int \sec^2 ax\, dx = \frac{1}{a} \tan ax + C$

98. $\displaystyle\int \csc^2 ax\, dx = -\frac{1}{a} \cot ax + C$

99. $\displaystyle\int \sec^n ax\, dx = \frac{\sec^{n-2} ax \tan ax}{a(n - 1)} + \frac{n - 2}{n - 1}\int \sec^{n-2} ax\, dx, \quad n \neq 1$

100. $\displaystyle\int \csc^n ax\, dx = -\frac{\csc^{n-2} ax \cot ax}{a(n - 1)} + \frac{n - 2}{n - 1}\int \csc^{n-2} ax\, dx, \quad n \neq 1$

101. $\displaystyle\int \sec^n ax \tan ax\, dx = \frac{\sec^n ax}{na} + C, \quad n \neq 0$

102. $\displaystyle\int \csc^n ax \cot ax\, dx = -\frac{\csc^n ax}{na} + C, \quad n \neq 0$

Inverse Trigonometric Forms

103. $\displaystyle\int \sin^{-1} ax\, dx = x \sin^{-1} ax + \frac{1}{a}\sqrt{1 - a^2x^2} + C$

104. $\displaystyle\int \cos^{-1} ax\, dx = x \cos^{-1} ax - \frac{1}{a}\sqrt{1 - a^2x^2} + C$

105. $\displaystyle\int \tan^{-1} ax\, dx = x \tan^{-1} ax - \frac{1}{2a} \ln(1 + a^2x^2) + C$

106. $\displaystyle\int x^n \sin^{-1} ax\, dx = \frac{x^{n+1}}{n + 1} \sin^{-1} ax - \frac{a}{n + 1}\int \frac{x^{n+1}\, dx}{\sqrt{1 - a^2x^2}}, \quad n \neq -1$

107. $\int x^n \cos^{-1} ax \, dx = \dfrac{x^{n+1}}{n+1} \cos^{-1} ax + \dfrac{a}{n+1} \int \dfrac{x^{n+1} \, dx}{\sqrt{1 - a^2 x^2}}, \quad n \neq -1$

108. $\int x^n \tan^{-1} ax \, dx = \dfrac{x^{n+1}}{n+1} \tan^{-1} ax - \dfrac{a}{n+1} \int \dfrac{x^{n+1} \, dx}{1 + a^2 x^2}, \quad n \neq -1$

Exponential and Logarithmic Forms

109. $\int e^{ax} \, dx = \dfrac{1}{a} e^{ax} + C$

110. $\int b^{ax} \, dx = \dfrac{1}{a} \dfrac{b^{ax}}{\ln b} + C, \quad b > 0, b \neq 1$

111. $\int x e^{ax} \, dx = \dfrac{e^{ax}}{a^2} (ax - 1) + C$

112. $\int x^n e^{ax} \, dx = \dfrac{1}{a} x^n e^{ax} - \dfrac{n}{a} \int x^{n-1} e^{ax} \, dx$

113. $\int x^n b^{ax} \, dx = \dfrac{x^n b^{ax}}{a \ln b} - \dfrac{n}{a \ln b} \int x^{n-1} b^{ax} \, dx, \quad b > 0, b \neq 1$

114. $\int e^{ax} \sin bx \, dx = \dfrac{e^{ax}}{a^2 + b^2} (a \sin bx - b \cos bx) + C$

115. $\int e^{ax} \cos bx \, dx = \dfrac{e^{ax}}{a^2 + b^2} (a \cos bx + b \sin bx) + C$

116. $\int \ln ax \, dx = x \ln ax - x + C$

117. $\int x^n (\ln ax)^m \, dx = \dfrac{x^{n+1} (\ln ax)^m}{n+1} - \dfrac{m}{n+1} \int x^n (\ln ax)^{m-1} \, dx, \quad n \neq -1$

118. $\int x^{-1} (\ln ax)^m \, dx = \dfrac{(\ln ax)^{m+1}}{m+1} + C, \quad m \neq -1$

119. $\int \dfrac{dx}{x \ln ax} = \ln |\ln ax| + C$

Forms Involving $\sqrt{2ax - x^2}, \, a > 0$

120. $\int \dfrac{dx}{\sqrt{2ax - x^2}} = \sin^{-1} \left(\dfrac{x - a}{a} \right) + C$

121. $\int \sqrt{2ax - x^2} \, dx = \dfrac{x - a}{2} \sqrt{2ax - x^2} + \dfrac{a^2}{2} \sin^{-1} \left(\dfrac{x - a}{a} \right) + C$

122. $\int \left(\sqrt{2ax - x^2} \right)^n \, dx = \dfrac{(x - a) \left(\sqrt{2ax - x^2} \right)^n}{n+1} + \dfrac{na^2}{n+1} \int \left(\sqrt{2ax - x^2} \right)^{n-2} \, dx$

123. $\int \dfrac{dx}{\left(\sqrt{2ax - x^2} \right)^n} = \dfrac{(x - a) \left(\sqrt{2ax - x^2} \right)^{2-n}}{(n - 2) a^2} + \dfrac{n - 3}{(n - 2) a^2} \int \dfrac{dx}{\left(\sqrt{2ax - x^2} \right)^{n-2}}$

124. $\int x \sqrt{2ax - x^2} \, dx = \dfrac{(x + a)(2x - 3a) \sqrt{2ax - x^2}}{6} + \dfrac{a^3}{2} \sin^{-1} \left(\dfrac{x - a}{a} \right) + C$

125. $\int \dfrac{\sqrt{2ax - x^2}}{x} \, dx = \sqrt{2ax - x^2} + a \sin^{-1} \left(\dfrac{x - a}{a} \right) + C$

126. $\int \dfrac{\sqrt{2ax - x^2}}{x^2} \, dx = -2 \sqrt{\dfrac{2a - x}{x}} - \sin^{-1} \left(\dfrac{x - a}{a} \right) + C$

127. $\int \dfrac{x \, dx}{\sqrt{2ax - x^2}} = a \sin^{-1} \left(\dfrac{x - a}{a} \right) - \sqrt{2ax - x^2} + C$

128. $\int \dfrac{dx}{x \sqrt{2ax - x^2}} = -\dfrac{1}{a} \sqrt{\dfrac{2a - x}{x}} + C$

Hyperbolic Forms

129. $\int \sinh ax \, dx = \dfrac{1}{a} \cosh ax + C$

130. $\int \cosh ax \, dx = \dfrac{1}{a} \sinh ax + C$

131. $\int \sinh^2 ax \, dx = \dfrac{\sinh 2ax}{4a} - \dfrac{x}{2} + C$

132. $\int \cosh^2 ax \, dx = \dfrac{\sinh 2ax}{4a} + \dfrac{x}{2} + C$

133. $\displaystyle\int \sinh^n ax\, dx = \frac{\sinh^{n-1} ax \cosh ax}{na} - \frac{n-1}{n} \int \sinh^{n-2} ax\, dx, \quad n \neq 0$

134. $\displaystyle\int \cosh^n ax\, dx = \frac{\cosh^{n-1} ax \sinh ax}{na} + \frac{n-1}{n} \int \cosh^{n-2} ax\, dx, \quad n \neq 0$

135. $\displaystyle\int x \sinh ax\, dx = \frac{x}{a} \cosh ax - \frac{1}{a^2} \sinh ax + C$

136. $\displaystyle\int x \cosh ax\, dx = \frac{x}{a} \sinh ax - \frac{1}{a^2} \cosh ax + C$

137. $\displaystyle\int x^n \sinh ax\, dx = \frac{x^n}{a} \cosh ax - \frac{n}{a} \int x^{n-1} \cosh ax\, dx$

138. $\displaystyle\int x^n \cosh ax\, dx = \frac{x^n}{a} \sinh ax - \frac{n}{a} \int x^{n-1} \sinh ax\, dx$

139. $\displaystyle\int \tanh ax\, dx = \frac{1}{a} \ln (\cosh ax) + C$

140. $\displaystyle\int \coth ax\, dx = \frac{1}{a} \ln |\sinh ax| + C$

141. $\displaystyle\int \tanh^2 ax\, dx = x - \frac{1}{a} \tanh ax + C$

142. $\displaystyle\int \coth^2 ax\, dx = x - \frac{1}{a} \coth ax + C$

143. $\displaystyle\int \tanh^n ax\, dx = -\frac{\tanh^{n-1} ax}{(n-1)a} + \int \tanh^{n-2} ax\, dx, \quad n \neq 1$

144. $\displaystyle\int \coth^n ax\, dx = -\frac{\coth^{n-1} ax}{(n-1)a} + \int \coth^{n-2} ax\, dx, \quad n \neq 1$

145. $\displaystyle\int \operatorname{sech} ax\, dx = \frac{1}{a} \sin^{-1}(\tanh ax) + C$

146. $\displaystyle\int \operatorname{csch} ax\, dx = \frac{1}{a} \ln \left|\tanh \frac{ax}{2}\right| + C$

147. $\displaystyle\int \operatorname{sech}^2 ax\, dx = \frac{1}{a} \tanh ax + C$

148. $\displaystyle\int \operatorname{csch}^2 ax\, dx = -\frac{1}{a} \coth ax + C$

149. $\displaystyle\int \operatorname{sech}^n ax\, dx = \frac{\operatorname{sech}^{n-2} ax \tanh ax}{(n-1)a} + \frac{n-2}{n-1} \int \operatorname{sech}^{n-2} ax\, dx, \quad n \neq 1$

150. $\displaystyle\int \operatorname{csch}^n ax\, dx = -\frac{\operatorname{csch}^{n-2} ax \coth ax}{(n-1)a} - \frac{n-2}{n-1} \int \operatorname{csch}^{n-2} ax\, dx, \quad n \neq 1$

151. $\displaystyle\int \operatorname{sech}^n ax \tanh ax\, dx = -\frac{\operatorname{sech}^n ax}{na} + C, \quad n \neq 0$

152. $\displaystyle\int \operatorname{csch}^n ax \coth ax\, dx = -\frac{\operatorname{csch}^n ax}{na} + C, \quad n \neq 0$

153. $\displaystyle\int e^{ax} \sinh bx\, dx = \frac{e^{ax}}{2} \left[\frac{e^{bx}}{a+b} - \frac{e^{-bx}}{a-b} \right] + C, \quad a^2 \neq b^2$

154. $\displaystyle\int e^{ax} \cosh bx\, dx = \frac{e^{ax}}{2} \left[\frac{e^{bx}}{a+b} + \frac{e^{-bx}}{a-b} \right] + C, \quad a^2 \neq b^2$

Some Definite Integrals

155. $\displaystyle\int_0^\infty x^{n-1} e^{-x}\, dx = \Gamma(n) = (n-1)!, \quad n > 0$

156. $\displaystyle\int_0^\infty e^{-ax^2}\, dx = \frac{1}{2} \sqrt{\frac{\pi}{a}}, \quad a > 0$

157. $\displaystyle\int_0^{\pi/2} \sin^n x\, dx = \int_0^{\pi/2} \cos^n x\, dx = \begin{cases} \dfrac{1 \cdot 3 \cdot 5 \cdot \cdots \cdot (n-1)}{2 \cdot 4 \cdot 6 \cdot \cdots \cdot n} \cdot \dfrac{\pi}{2}, & \text{if } n \text{ is an even integer} \geq 2 \\[2mm] \dfrac{2 \cdot 4 \cdot 6 \cdot \cdots \cdot (n-1)}{3 \cdot 5 \cdot 7 \cdot \cdots \cdot n}, & \text{if } n \text{ is an odd integer} \geq 3 \end{cases}$

Basic Algebra Formulas

Arithmetic Operations

$$a(b + c) = ab + ac, \qquad \frac{a}{b} \cdot \frac{c}{d} = \frac{ac}{bd}$$

$$\frac{a}{b} + \frac{c}{d} = \frac{ad + bc}{bd}, \qquad \frac{a/b}{c/d} = \frac{a}{b} \cdot \frac{d}{c}$$

Laws of Signs

$$-(-a) = a, \qquad \frac{-a}{b} = -\frac{a}{b} = \frac{a}{-b}$$

Zero   Division by zero is not defined.

$$\text{If } a \neq 0: \quad \frac{0}{a} = 0, \quad a^0 = 1, \quad 0^a = 0$$

$$\text{For any number } a: a \cdot 0 = 0 \cdot a = 0$$

Laws of Exponents

$$a^m a^n = a^{m+n}, \qquad (ab)^m = a^m b^m, \qquad (a^m)^n = a^{mn}, \qquad a^{m/n} = \sqrt[n]{a^m} = \left(\sqrt[n]{a}\right)^m$$

If $a \neq 0$, then

$$\frac{a^m}{a^n} = a^{m-n}, \qquad a^0 = 1, \qquad a^{-m} = \frac{1}{a^m}.$$

The Binomial Theorem   For any positive integer n,

$$(a + b)^n = a^n + na^{n-1}b + \frac{n(n - 1)}{1 \cdot 2}a^{n-2}b^2$$

$$+ \frac{n(n - 1)(n - 2)}{1 \cdot 2 \cdot 3}a^{n-3}b^3 + \cdots + nab^{n-1} + b^n.$$

For instance,

$$(a + b)^2 = a^2 + 2ab + b^2, \qquad\qquad (a - b)^2 = a^2 - 2ab + b^2$$
$$(a + b)^3 = a^3 + 3a^2b + 3ab^2 + b^3, \qquad (a - b)^3 = a^3 - 3a^2b + 3ab^2 - b^3.$$

Factoring the Difference of Like Integer Powers, $n > 1$

$$a^n - b^n = (a - b)(a^{n-1} + a^{n-2}b + a^{n-3}b^2 + \cdots + ab^{n-2} + b^{n-1})$$

For instance,

$$a^2 - b^2 = (a - b)(a + b),$$
$$a^3 - b^3 = (a - b)(a^2 + ab + b^2),$$
$$a^4 - b^4 = (a - b)(a^3 + a^2b + ab^2 + b^3).$$

Completing the Square   If $a \neq 0$, then

$$ax^2 + bx + c = au^2 + C \qquad \left(u = x + (b/2a), C = c - \frac{b^2}{4a}\right)$$

The Quadratic Formula
If $a \neq 0$ and $ax^2 + bx + c = 0$, then

$$x = \frac{-b \pm \sqrt{b^2 - 4ac}}{2a}.$$

Geometry Formulas

A = area, B = area of base, C = circumference, S = surface area, V = volume

Triangle

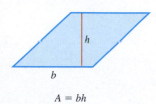

$$A = \frac{1}{2}bh$$

Similar Triangles

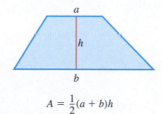

$$\frac{a'}{a} = \frac{b'}{b} = \frac{c'}{c}$$

Pythagorean Theorem

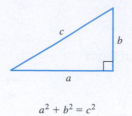

$$a^2 + b^2 = c^2$$

Parallelogram

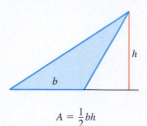

$$A = bh$$

Trapezoid

$$A = \frac{1}{2}(a + b)h$$

Circle

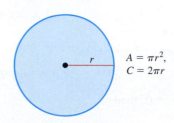

$$A = \pi r^2, \quad C = 2\pi r$$

Any Cylinder or Prism with Parallel Bases

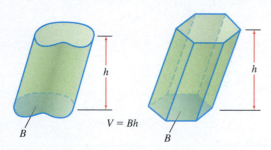

$$V = Bh$$

Right Circular Cylinder

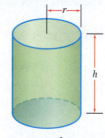

$$V = \pi r^2 h$$
$$S = 2\pi rh = \text{Area of side}$$

Any Cone or Pyramid

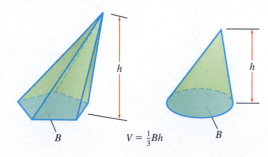

$$V = \frac{1}{3}Bh$$

Right Circular Cone

$$V = \frac{1}{3}\pi r^2 h$$
$$S = \pi rs = \text{Area of side}$$

Sphere

$$V = \frac{4}{3}\pi r^3, \quad S = 4\pi r^2$$

Trigonometry Formulas

Definitions and Fundamental Identities

Sine: $\quad \sin\theta = \dfrac{y}{r} = \dfrac{1}{\csc\theta}$

Cosine: $\quad \cos\theta = \dfrac{x}{r} = \dfrac{1}{\sec\theta}$

Tangent: $\quad \tan\theta = \dfrac{y}{x} = \dfrac{1}{\cot\theta}$

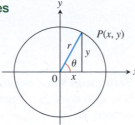

Identities

$\sin(-\theta) = -\sin\theta, \quad \cos(-\theta) = \cos\theta$

$\sin^2\theta + \cos^2\theta = 1, \quad \sec^2\theta = 1 + \tan^2\theta, \quad \csc^2\theta = 1 + \cot^2\theta$

$\sin 2\theta = 2\sin\theta\cos\theta, \quad \cos 2\theta = \cos^2\theta - \sin^2\theta$

$\cos^2\theta = \dfrac{1 + \cos 2\theta}{2}, \quad \sin^2\theta = \dfrac{1 - \cos 2\theta}{2}$

$\sin(A + B) = \sin A \cos B + \cos A \sin B$

$\sin(A - B) = \sin A \cos B - \cos A \sin B$

$\cos(A + B) = \cos A \cos B - \sin A \sin B$

$\cos(A - B) = \cos A \cos B + \sin A \sin B$

$\tan(A + B) = \dfrac{\tan A + \tan B}{1 - \tan A \tan B}$

$\tan(A - B) = \dfrac{\tan A - \tan B}{1 + \tan A \tan B}$

$\sin\left(A - \dfrac{\pi}{2}\right) = -\cos A, \qquad \cos\left(A - \dfrac{\pi}{2}\right) = \sin A$

$\sin\left(A + \dfrac{\pi}{2}\right) = \cos A, \qquad \cos\left(A + \dfrac{\pi}{2}\right) = -\sin A$

$\sin A \sin B = \dfrac{1}{2}\cos(A - B) - \dfrac{1}{2}\cos(A + B)$

$\cos A \cos B = \dfrac{1}{2}\cos(A - B) + \dfrac{1}{2}\cos(A + B)$

$\sin A \cos B = \dfrac{1}{2}\sin(A - B) + \dfrac{1}{2}\sin(A + B)$

$\sin A + \sin B = 2\sin\dfrac{1}{2}(A + B)\cos\dfrac{1}{2}(A - B)$

$\sin A - \sin B = 2\cos\dfrac{1}{2}(A + B)\sin\dfrac{1}{2}(A - B)$

$\cos A + \cos B = 2\cos\dfrac{1}{2}(A + B)\cos\dfrac{1}{2}(A - B)$

$\cos A - \cos B = -2\sin\dfrac{1}{2}(A + B)\sin\dfrac{1}{2}(A - B)$

Trigonometric Functions

Radian Measure

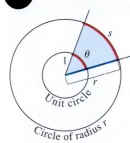

Unit circle

Circle of radius r

$\dfrac{s}{r} = \dfrac{\theta}{1} = \theta \quad$ or $\quad \theta = \dfrac{s}{r},$

$180° = \pi$ radians.

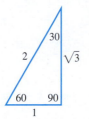

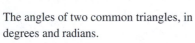

	Degrees	Radians

The angles of two common triangles, in degrees and radians.

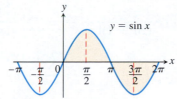

Domain: $(-\infty, \infty)$
Range: $[-1, 1]$

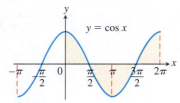

Domain: $(-\infty, \infty)$
Range: $[-1, 1]$

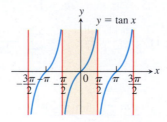

Domain: All real numbers except odd integer multiples of $\pi/2$
Range: $(-\infty, \infty)$

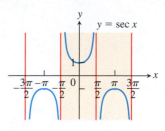

Domain: All real numbers except odd integer multiples of $\pi/2$
Range: $(-\infty, -1] \cup [1, \infty)$

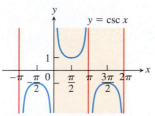

Domain: $x \neq 0, \pm\pi, \pm 2\pi, \dots$
Range: $(-\infty, -1] \cup [1, \infty)$

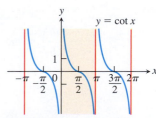

Domain: $x \neq 0, \pm\pi, \pm 2\pi, \dots$
Range: $(-\infty, \infty)$

Series

Tests for Convergence of Infinite Series

1. **The nth-Term Test:** Unless $a_n \to 0$, the series diverges.
2. **Geometric series:** $\sum ar^n$ converges if $|r| < 1$; otherwise it diverges.
3. **p-series:** $\sum 1/n^p$ converges if $p > 1$; otherwise it diverges.
4. **Series with nonnegative terms:** Try the Integral Test, Ratio Test, or Root Test. Try comparing to a known series with the Comparison Test or the Limit Comparison Test.

5. **Series with some negative terms:** Does $\sum |a_n|$ converge? If yes, so does $\sum a_n$ because absolute convergence implies convergence.
6. **Alternating series:** $\sum a_n$ converges if the series satisfies the conditions of the Alternating Series Test.

Taylor Series

$$\frac{1}{1-x} = 1 + x + x^2 + \cdots + x^n + \cdots = \sum_{n=0}^{\infty} x^n, \qquad |x| < 1$$

$$\frac{1}{1+x} = 1 - x + x^2 - \cdots + (-x)^n + \cdots = \sum_{n=0}^{\infty} (-1)^n x^n, \qquad |x| < 1$$

$$e^x = 1 + x + \frac{x^2}{2!} + \cdots + \frac{x^n}{n!} + \cdots = \sum_{n=0}^{\infty} \frac{x^n}{n!}, \qquad |x| < \infty$$

$$\sin x = x - \frac{x^3}{3!} + \frac{x^5}{5!} - \cdots + (-1)^n \frac{x^{2n+1}}{(2n+1)!} + \cdots = \sum_{n=0}^{\infty} \frac{(-1)^n x^{2n+1}}{(2n+1)!}, \qquad |x| < \infty$$

$$\cos x = 1 - \frac{x^2}{2!} + \frac{x^4}{4!} - \cdots + (-1)^n \frac{x^{2n}}{(2n)!} + \cdots = \sum_{n=0}^{\infty} \frac{(-1)^n x^{2n}}{(2n)!}, \qquad |x| < \infty$$

$$\ln(1+x) = x - \frac{x^2}{2} + \frac{x^3}{3} - \cdots + (-1)^{n-1} \frac{x^n}{n} + \cdots = \sum_{n=1}^{\infty} \frac{(-1)^{n-1} x^n}{n}, \qquad -1 < x \le 1$$

$$\ln \frac{1+x}{1-x} = 2 \tanh^{-1} x = 2\left(x + \frac{x^3}{3} + \frac{x^5}{5} + \cdots + \frac{x^{2n+1}}{2n+1} + \cdots \right) = 2 \sum_{n=0}^{\infty} \frac{x^{2n+1}}{2n+1}, \qquad |x| < 1$$

$$\tan^{-1} x = x - \frac{x^3}{3} + \frac{x^5}{5} - \cdots + (-1)^n \frac{x^{2n+1}}{2n+1} + \cdots = \sum_{n=0}^{\infty} \frac{(-1)^n x^{2n+1}}{2n+1}, \qquad |x| \le 1$$

Binomial Series

$$(1+x)^m = 1 + mx + \frac{m(m-1)x^2}{2!} + \frac{m(m-1)(m-2)x^3}{3!} + \cdots + \frac{m(m-1)(m-2)\cdots(m-k+1)x^k}{k!} + \cdots$$

$$= 1 + \sum_{k=1}^{\infty} \binom{m}{k} x^k, \qquad |x| < 1,$$

where

$$\binom{m}{1} = m, \qquad \binom{m}{2} = \frac{m(m-1)}{2!}, \qquad \binom{m}{k} = \frac{m(m-1)\cdots(m-k+1)}{k!} \qquad \text{for } k \ge 3.$$

Vector Operator Formulas (Cartesian Form)

Formulas for Grad, Div, Curl, and the Laplacian

Cartesian (x, y, z) $\mathbf{i}$, $\mathbf{j}$, and $\mathbf{k}$ are unit vectors in the directions of increasing x, y, and z. M, N, and P are the scalar components of $\mathbf{F}(x, y, z)$ in these directions.

Gradient $\quad \nabla f = \dfrac{\partial f}{\partial x}\mathbf{i} + \dfrac{\partial f}{\partial y}\mathbf{j} + \dfrac{\partial f}{\partial z}\mathbf{k}$

Divergence $\quad \nabla \cdot \mathbf{F} = \dfrac{\partial M}{\partial x} + \dfrac{\partial N}{\partial y} + \dfrac{\partial P}{\partial z}$

Curl $\quad \nabla \times \mathbf{F} = \begin{vmatrix} \mathbf{i} & \mathbf{j} & \mathbf{k} \\ \dfrac{\partial}{\partial x} & \dfrac{\partial}{\partial y} & \dfrac{\partial}{\partial z} \\ M & N & P \end{vmatrix}$

Laplacian $\quad \nabla^2 f = \dfrac{\partial^2 f}{\partial x^2} + \dfrac{\partial^2 f}{\partial y^2} + \dfrac{\partial^2 f}{\partial z^2}$

Vector Triple Products

$(\mathbf{u} \times \mathbf{v}) \cdot \mathbf{w} = (\mathbf{v} \times \mathbf{w}) \cdot \mathbf{u} = (\mathbf{w} \times \mathbf{u}) \cdot \mathbf{v}$

$\mathbf{u} \times (\mathbf{v} \times \mathbf{w}) = (\mathbf{u} \cdot \mathbf{w})\mathbf{v} - (\mathbf{u} \cdot \mathbf{v})\mathbf{w}$

The Fundamental Theorem of Line Integrals

Part 1 Let $\mathbf{F} = M\mathbf{i} + N\mathbf{j} + P\mathbf{k}$ be a vector field whose components are continuous throughout an open connected region D in space. Then there exists a differentiable function f such that

$$\mathbf{F} = \nabla f = \frac{\partial f}{\partial x}\mathbf{i} + \frac{\partial f}{\partial y}\mathbf{j} + \frac{\partial f}{\partial z}\mathbf{k}$$

if and only if for all points A and B in D, the value of $\int_A^B \mathbf{F} \cdot d\mathbf{r}$ is independent of the path joining A to B in D.

Part 2 If the integral is independent of the path from A to B, its value is

$$\int_A^B \mathbf{F} \cdot d\mathbf{r} = f(B) - f(A).$$

Green's Theorem and Its Generalization to Three Dimensions

Tangential form of Green's Theorem: $\quad \oint_C \mathbf{F} \cdot \mathbf{T}\, ds = \iint_R (\nabla \times \mathbf{F}) \cdot \mathbf{k}\, dA$

Stokes' Theorem: $\quad \oint_C \mathbf{F} \cdot \mathbf{T}\, ds = \iint_S (\nabla \times \mathbf{F}) \cdot \mathbf{n}\, d\sigma$

Normal form of Green's Theorem: $\quad \oint_C \mathbf{F} \cdot \mathbf{n}\, ds = \iint_R (\nabla \cdot \mathbf{F})\, dA$

Divergence Theorem: $\quad \iint_S \mathbf{F} \cdot \mathbf{n}\, d\sigma = \iiint_D \nabla \cdot \mathbf{F}\, dV$

Vector Identities

In the identities here, f and g are differentiable scalar functions; $\mathbf{F}$, $\mathbf{F}_1$, and $\mathbf{F}_2$ are differentiable vector fields; and a and b are real constants.

$\nabla \times (\nabla f) = \mathbf{0}$

$\nabla(fg) = f\nabla g + g\nabla f$

$\nabla \cdot (g\mathbf{F}) = g\nabla \cdot \mathbf{F} + \nabla g \cdot \mathbf{F}$

$\nabla \times (g\mathbf{F}) = g\nabla \times \mathbf{F} + \nabla g \times \mathbf{F}$

$\nabla \cdot (a\mathbf{F}_1 + b\mathbf{F}_2) = a\nabla \cdot \mathbf{F}_1 + b\nabla \cdot \mathbf{F}_2$

$\nabla \times (a\mathbf{F}_1 + b\mathbf{F}_2) = a\nabla \times \mathbf{F}_1 + b\nabla \times \mathbf{F}_2$

$\nabla(\mathbf{F}_1 \cdot \mathbf{F}_2) = (\mathbf{F}_1 \cdot \nabla)\mathbf{F}_2 + (\mathbf{F}_2 \cdot \nabla)\mathbf{F}_1 +$
$\mathbf{F}_1 \times (\nabla \times \mathbf{F}_2) + \mathbf{F}_2 \times (\nabla \times \mathbf{F}_1)$

$\nabla \cdot (\mathbf{F}_1 \times \mathbf{F}_2) = \mathbf{F}_2 \cdot (\nabla \times \mathbf{F}_1) - \mathbf{F}_1 \cdot (\nabla \times \mathbf{F}_2)$

$\nabla \times (\mathbf{F}_1 \times \mathbf{F}_2) = (\mathbf{F}_2 \cdot \nabla)\mathbf{F}_1 - (\mathbf{F}_1 \cdot \nabla)\mathbf{F}_2 +$
$\qquad (\nabla \cdot \mathbf{F}_2)\mathbf{F}_1 - (\nabla \cdot \mathbf{F}_1)\mathbf{F}_2$

$\nabla \times (\nabla \times \mathbf{F}) = \nabla(\nabla \cdot \mathbf{F}) - (\nabla \cdot \nabla)\mathbf{F} = \nabla(\nabla \cdot \mathbf{F}) - \nabla^2\mathbf{F}$

$(\nabla \times \mathbf{F}) \times \mathbf{F} = (\mathbf{F} \cdot \nabla)\mathbf{F} - \dfrac{1}{2}\nabla(\mathbf{F} \cdot \mathbf{F})$

Limits

General Laws

If L, M, c, and k are real numbers and

$$\lim_{x \to c} f(x) = L \quad \text{and} \quad \lim_{x \to c} g(x) = M, \quad \text{then}$$

Sum Rule: $\qquad\qquad\qquad\quad \lim_{x \to c} (f(x) + g(x)) = L + M$

Difference Rule: $\qquad\qquad\quad \lim_{x \to c} (f(x) - g(x)) = L - M$

Product Rule: $\qquad\qquad\quad\; \lim_{x \to c} (f(x) \cdot g(x)) = L \cdot M$

Constant Multiple Rule: $\qquad \lim_{x \to c} (k \cdot f(x)) = k \cdot L$

Quotient Rule: $\qquad\qquad\quad \lim_{x \to c} \dfrac{f(x)}{g(x)} = \dfrac{L}{M}, \quad M \neq 0$

The Sandwich Theorem

If $g(x) \leq f(x) \leq h(x)$ in an open interval containing c, except possibly at $x = c$, and if

$$\lim_{x \to c} g(x) = \lim_{x \to c} h(x) = L,$$

then $\lim_{x \to c} f(x) = L$.

Inequalities

If $f(x) \leq g(x)$ in an open interval containing c, except possibly at $x = c$, and both limits exist, then

$$\lim_{x \to c} f(x) \leq \lim_{x \to c} g(x).$$

Continuity

If g is continuous at L and $\lim_{x \to c} f(x) = L$, then

$$\lim_{x \to c} g(f(x)) = g(L).$$

Specific Formulas

If $P(x) = a_n x^n + a_{n-1} x^{n-1} + \cdots + a_0$, then

$$\lim_{x \to c} P(x) = P(c) = a_n c^n + a_{n-1} c^{n-1} + \cdots + a_0.$$

If $P(x)$ and $Q(x)$ are polynomials and $Q(c) \neq 0$, then

$$\lim_{x \to c} \frac{P(x)}{Q(x)} = \frac{P(c)}{Q(c)}.$$

If $f(x)$ is continuous at $x = c$, then

$$\lim_{x \to c} f(x) = f(c).$$

$$\lim_{x \to 0} \frac{\sin x}{x} = 1 \quad \text{and} \quad \lim_{x \to 0} \frac{1 - \cos x}{x} = 0$$

L'Hôpital's Rule

If $f(a) = g(a) = 0$, both f' and g' exist in an open interval I containing a, and $g'(x) \neq 0$ on I if $x \neq a$, then

$$\lim_{x \to a} \frac{f(x)}{g(x)} = \lim_{x \to a} \frac{f'(x)}{g'(x)},$$

assuming the limit on the right side exists.

Differentiation Rules

General Formulas

Assume u and v are differentiable functions of x.

Constant:
$$\frac{d}{dx}(c) = 0$$

Sum:
$$\frac{d}{dx}(u + v) = \frac{du}{dx} + \frac{dv}{dx}$$

Difference:
$$\frac{d}{dx}(u - v) = \frac{du}{dx} - \frac{dv}{dx}$$

Constant Multiple:
$$\frac{d}{dx}(cu) = c\frac{du}{dx}$$

Product:
$$\frac{d}{dx}(uv) = u\frac{dv}{dx} + \frac{du}{dx}v$$

Quotient:
$$\frac{d}{dx}\left(\frac{u}{v}\right) = \frac{v\frac{du}{dx} - u\frac{dv}{dx}}{v^2}$$

Power:
$$\frac{d}{dx}x^n = nx^{n-1}$$

Chain Rule:
$$\frac{d}{dx}(f(g(x)) = f'(g(x)) \cdot g'(x)$$

Trigonometric Functions

$$\frac{d}{dx}(\sin x) = \cos x \qquad \frac{d}{dx}(\cos x) = -\sin x$$

$$\frac{d}{dx}(\tan x) = \sec^2 x \qquad \frac{d}{dx}(\sec x) = \sec x \tan x$$

$$\frac{d}{dx}(\cot x) = -\csc^2 x \qquad \frac{d}{dx}(\csc x) = -\csc x \cot x$$

Exponential and Logarithmic Functions

$$\frac{d}{dx}e^x = e^x \qquad \frac{d}{dx}\ln x = \frac{1}{x}$$

$$\frac{d}{dx}a^x = a^x \ln a \qquad \frac{d}{dx}(\log_a x) = \frac{1}{x \ln a}$$

Inverse Trigonometric Functions

$$\frac{d}{dx}(\sin^{-1} x) = \frac{1}{\sqrt{1 - x^2}} \qquad \frac{d}{dx}(\cos^{-1} x) = -\frac{1}{\sqrt{1 - x^2}}$$

$$\frac{d}{dx}(\tan^{-1} x) = \frac{1}{1 + x^2} \qquad \frac{d}{dx}(\sec^{-1} x) = \frac{1}{|x|\sqrt{x^2 - 1}}$$

$$\frac{d}{dx}(\cot^{-1} x) = -\frac{1}{1 + x^2} \qquad \frac{d}{dx}(\csc^{-1} x) = -\frac{1}{|x|\sqrt{x^2 - 1}}$$

Hyperbolic Functions

$$\frac{d}{dx}(\sinh x) = \cosh x \qquad \frac{d}{dx}(\cosh x) = \sinh x$$

$$\frac{d}{dx}(\tanh x) = \operatorname{sech}^2 x \qquad \frac{d}{dx}(\operatorname{sech} x) = -\operatorname{sech} x \tanh x$$

$$\frac{d}{dx}(\coth x) = -\operatorname{csch}^2 x \qquad \frac{d}{dx}(\operatorname{csch} x) = -\operatorname{csch} x \coth x$$

Inverse Hyperbolic Functions

$$\frac{d}{dx}(\sinh^{-1} x) = \frac{1}{\sqrt{1 + x^2}} \qquad \frac{d}{dx}(\cosh^{-1} x) = \frac{1}{\sqrt{x^2 - 1}}$$

$$\frac{d}{dx}(\tanh^{-1} x) = \frac{1}{1 - x^2} \qquad \frac{d}{dx}(\operatorname{sech}^{-1} x) = -\frac{1}{x\sqrt{1 - x^2}}$$

$$\frac{d}{dx}(\coth^{-1} x) = \frac{1}{1 - x^2} \qquad \frac{d}{dx}(\operatorname{csch}^{-1} x) = -\frac{1}{|x|\sqrt{1 + x^2}}$$

Parametric Equations

If $x = f(t)$ and $y = g(t)$ are differentiable, then

$$y' = \frac{dy}{dx} = \frac{dy/dt}{dx/dt} \qquad \text{and} \qquad \frac{d^2y}{dx^2} = \frac{dy'/dt}{dx/dt}.$$

Integration Rules

General Formulas

Zero:
$$\int_a^a f(x)\,dx = 0$$

Order of Integration:
$$\int_b^a f(x)\,dx = -\int_a^b f(x)\,dx$$

Constant Multiples:
$$\int_a^b kf(x)\,dx = k\int_a^b f(x)\,dx, \qquad k \text{ any number}$$

$$\int_a^b -f(x)\,dx = -\int_a^b f(x)\,dx, \qquad k = -1$$

Sums and Differences:
$$\int_a^b (f(x) \pm g(x))\,dx = \int_a^b f(x)\,dx \pm \int_a^b g(x)\,dx$$

Additivity:
$$\int_a^b f(x)\,dx + \int_b^c f(x)\,dx = \int_a^c f(x)\,dx$$

Max-Min Inequality: If max f and min f are the maximum and minimum values of f on $[a, b]$, then
$$\min f \cdot (b - a) \le \int_a^b f(x)\,dx \le \max f \cdot (b - a).$$

Domination:
$$f(x) \ge g(x) \quad \text{on} \quad [a, b] \quad \text{implies} \quad \int_a^b f(x)\,dx \ge \int_a^b g(x)\,dx$$

$$f(x) \ge 0 \quad \text{on} \quad [a, b] \quad \text{implies} \quad \int_a^b f(x)\,dx \ge 0$$

The Fundamental Theorem of Calculus

Part 1 If f is continuous on $[a, b]$, then $F(x) = \int_a^x f(t)\,dt$ is continuous on $[a, b]$ and differentiable on (a, b) and its derivative is $f(x)$:

$$F'(x) = \frac{d}{dx}\int_a^x f(t)\,dt = f(x).$$

Part 2 If f is continuous at every point of $[a, b]$ and F is any antiderivative of f on $[a, b]$, then

$$\int_a^b f(x)\,dx = F(b) - F(a).$$

Substitution in Definite Integrals

$$\int_a^b f(g(x)) \cdot g'(x)\,dx = \int_{g(a)}^{g(b)} f(u)\,du$$

Integration by Parts

$$\int_a^b u(x)\,v'(x)\,dx = u(x)\,v(x)\Big]_a^b - \int_a^b v(x)\,u'(x)\,dx$$

A Brief Table of Integrals follows the Index at the back of the text.

Limits

General Laws

If L, M, c, and k are real numbers and

$$\lim_{x \to c} f(x) = L \quad \text{and} \quad \lim_{x \to c} g(x) = M, \quad \text{then}$$

Sum Rule: $\quad \lim_{x \to c} (f(x) + g(x)) = L + M$

Difference Rule: $\quad \lim_{x \to c} (f(x) - g(x)) = L - M$

Product Rule: $\quad \lim_{x \to c} (f(x) \cdot g(x)) = L \cdot M$

Constant Multiple Rule: $\quad \lim_{x \to c} (k \cdot f(x)) = k \cdot L$

Quotient Rule: $\quad \lim_{x \to c} \dfrac{f(x)}{g(x)} = \dfrac{L}{M}, \quad M \neq 0$

The Sandwich Theorem

If $g(x) \leq f(x) \leq h(x)$ in an open interval containing c, except possibly at $x = c$, and if

$$\lim_{x \to c} g(x) = \lim_{x \to c} h(x) = L,$$

then $\lim_{x \to c} f(x) = L$.

Inequalities

If $f(x) \leq g(x)$ in an open interval containing c, except possibly at $x = c$, and both limits exist, then

$$\lim_{x \to c} f(x) \leq \lim_{x \to c} g(x).$$

Continuity

If g is continuous at L and $\lim_{x \to c} f(x) = L$, then

$$\lim_{x \to c} g(f(x)) = g(L).$$

Specific Formulas

If $P(x) = a_n x^n + a_{n-1} x^{n-1} + \cdots + a_0$, then

$$\lim_{x \to c} P(x) = P(c) = a_n c^n + a_{n-1} c^{n-1} + \cdots + a_0.$$

If $P(x)$ and $Q(x)$ are polynomials and $Q(c) \neq 0$, then

$$\lim_{x \to c} \frac{P(x)}{Q(x)} = \frac{P(c)}{Q(c)}.$$

If $f(x)$ is continuous at $x = c$, then

$$\lim_{x \to c} f(x) = f(c).$$

$$\lim_{x \to 0} \frac{\sin x}{x} = 1 \quad \text{and} \quad \lim_{x \to 0} \frac{1 - \cos x}{x} = 0$$

L'Hôpital's Rule

If $f(a) = g(a) = 0$, both f' and g' exist in an open interval I containing a, and $g'(x) \neq 0$ on I if $x \neq a$, then

$$\lim_{x \to a} \frac{f(x)}{g(x)} = \lim_{x \to a} \frac{f'(x)}{g'(x)},$$

assuming the limit on the right side exists.

Differentiation Rules

General Formulas

Assume u and v are differentiable functions of x.

Constant:
$$\frac{d}{dx}(c) = 0$$

Sum:
$$\frac{d}{dx}(u + v) = \frac{du}{dx} + \frac{dv}{dx}$$

Difference:
$$\frac{d}{dx}(u - v) = \frac{du}{dx} - \frac{dv}{dx}$$

Constant Multiple:
$$\frac{d}{dx}(cu) = c\frac{du}{dx}$$

Product:
$$\frac{d}{dx}(uv) = u\frac{dv}{dx} + \frac{du}{dx}v$$

Quotient:
$$\frac{d}{dx}\left(\frac{u}{v}\right) = \frac{v\frac{du}{dx} - u\frac{dv}{dx}}{v^2}$$

Power:
$$\frac{d}{dx}x^n = nx^{n-1}$$

Chain Rule:
$$\frac{d}{dx}(f(g(x)) = f'(g(x)) \cdot g'(x)$$

Trigonometric Functions

$$\frac{d}{dx}(\sin x) = \cos x \qquad \frac{d}{dx}(\cos x) = -\sin x$$

$$\frac{d}{dx}(\tan x) = \sec^2 x \qquad \frac{d}{dx}(\sec x) = \sec x \tan x$$

$$\frac{d}{dx}(\cot x) = -\csc^2 x \qquad \frac{d}{dx}(\csc x) = -\csc x \cot x$$

Exponential and Logarithmic Functions

$$\frac{d}{dx}e^x = e^x \qquad\qquad \frac{d}{dx}\ln x = \frac{1}{x}$$

$$\frac{d}{dx}a^x = a^x \ln a \qquad \frac{d}{dx}(\log_a x) = \frac{1}{x \ln a}$$

Inverse Trigonometric Functions

$$\frac{d}{dx}(\sin^{-1} x) = \frac{1}{\sqrt{1 - x^2}} \qquad \frac{d}{dx}(\cos^{-1} x) = -\frac{1}{\sqrt{1 - x^2}}$$

$$\frac{d}{dx}(\tan^{-1} x) = \frac{1}{1 + x^2} \qquad \frac{d}{dx}(\sec^{-1} x) = \frac{1}{|x|\sqrt{x^2 - 1}}$$

$$\frac{d}{dx}(\cot^{-1} x) = -\frac{1}{1 + x^2} \qquad \frac{d}{dx}(\csc^{-1} x) = -\frac{1}{|x|\sqrt{x^2 - 1}}$$

Hyperbolic Functions

$$\frac{d}{dx}(\sinh x) = \cosh x \qquad \frac{d}{dx}(\cosh x) = \sinh x$$

$$\frac{d}{dx}(\tanh x) = \operatorname{sech}^2 x \qquad \frac{d}{dx}(\operatorname{sech} x) = -\operatorname{sech} x \tanh x$$

$$\frac{d}{dx}(\coth x) = -\operatorname{csch}^2 x \qquad \frac{d}{dx}(\operatorname{csch} x) = -\operatorname{csch} x \coth x$$

Inverse Hyperbolic Functions

$$\frac{d}{dx}(\sinh^{-1} x) = \frac{1}{\sqrt{1 + x^2}} \qquad \frac{d}{dx}(\cosh^{-1} x) = \frac{1}{\sqrt{x^2 - 1}}$$

$$\frac{d}{dx}(\tanh^{-1} x) = \frac{1}{1 - x^2} \qquad \frac{d}{dx}(\operatorname{sech}^{-1} x) = -\frac{1}{x\sqrt{1 - x^2}}$$

$$\frac{d}{dx}(\coth^{-1} x) = \frac{1}{1 - x^2} \qquad \frac{d}{dx}(\operatorname{csch}^{-1} x) = -\frac{1}{|x|\sqrt{1 + x^2}}$$

Parametric Equations

If $x = f(t)$ and $y = g(t)$ are differentiable, then

$$y' = \frac{dy}{dx} = \frac{dy/dt}{dx/dt} \qquad \text{and} \qquad \frac{d^2y}{dx^2} = \frac{dy'/dt}{dx/dt}.$$

Integration Rules

General Formulas

Zero:
$$\int_a^a f(x)\,dx = 0$$

Order of Integration:
$$\int_b^a f(x)\,dx = -\int_a^b f(x)\,dx$$

Constant Multiples:
$$\int_a^b kf(x)\,dx = k\int_a^b f(x)\,dx, \qquad k \text{ any number}$$

$$\int_a^b -f(x)\,dx = -\int_a^b f(x)\,dx, \qquad k = -1$$

Sums and Differences:
$$\int_a^b (f(x) \pm g(x))\,dx = \int_a^b f(x)\,dx \pm \int_a^b g(x)\,dx$$

Additivity:
$$\int_a^b f(x)\,dx + \int_b^c f(x)\,dx = \int_a^c f(x)\,dx$$

Max-Min Inequality: If max f and min f are the maximum and minimum values of f on $[a, b]$, then

$$\min f \cdot (b - a) \le \int_a^b f(x)\,dx \le \max f \cdot (b - a).$$

Domination: $\quad f(x) \ge g(x) \quad$ on $\quad [a, b] \quad$ implies $\quad \displaystyle\int_a^b f(x)\,dx \ge \int_a^b g(x)\,dx$

$$f(x) \ge 0 \quad \text{on} \quad [a, b] \quad \text{implies} \quad \int_a^b f(x)\,dx \ge 0$$

The Fundamental Theorem of Calculus

Part 1 If f is continuous on $[a, b]$, then $F(x) = \int_a^x f(t)\,dt$ is continuous on $[a, b]$ and differentiable on (a, b) and its derivative is $f(x)$:

$$F'(x) = \frac{d}{dx}\int_a^x f(t)\,dt = f(x).$$

Part 2 If f is continuous at every point of $[a, b]$ and F is any antiderivative of f on $[a, b]$, then

$$\int_a^b f(x)\,dx = F(b) - F(a).$$

Substitution in Definite Integrals

$$\int_a^b f(g(x)) \cdot g'(x)\,dx = \int_{g(a)}^{g(b)} f(u)\,du$$

Integration by Parts

$$\int_a^b u(x)v'(x)\,dx = u(x)v(x)\Big]_a^b - \int_a^b v(x)u'(x)\,dx$$

A Brief Table of Integrals follows the Index at the back of the text.